Tomato soup, 284
Towering antenna, 320

Environment

Air pollution, 552
Bear population, 372
Bitter cold, 24
Capture-recapture, 372
Corrugated waste, 372
Distribution of waste, 385
Diversity index, 672
Extracting metals from ore, 267
Falling temperatures, 24
Fast-food waste, 372
Foxes and rabbits, 371
Fuel efficiency, 380
Measuring ocean depth, 688
Neuse River pH, 664
Probability of rain, 218
Quality water, 679
Recycling metals, 267
Total waste, 386
Wildlife management, 685

Geometry

Acute angles, 434
Area, 237, 241, 246, 253
Badminton court, 274
Crop circles, 43
Circle formulas, 641
Circumference, 267
Diagonal of a box, 485
Diameter, 92, 267
Dimensions of a box, 315
Framed, 106
Glass prism, 314
Height, 93
Interior angles, 201
Perimeter, 58, 79, 92, 118
Racquetball, 320
Radius, 92, 485
Roundball court, 274
Square formulas, 641
Volume, 237, 290, 295

Investment

CD rollover, 259
Combined savings, 118
Comparing investments, 247
Compounded annually, 247
Compounded semiannually, 247
Diversification, 417
Emerging markets, 315
Financial planning, 222
Flat yield curve, 118
High-risk funds, 113
Increasing deposits, 237
Investing in bonds, 113

Investing in stocks, 268
Investing in treasury bills, 247
Living comfortably, 118
Loan shark, 113
Long-term investing, 258
Periodic deposits, 237
Present value, 267
Reading the curve, 118
Saving for retirement, 180
Simple interest, 100
Social Security benefits, 218
Venture capital, 315
Wise investments, 113

Science

Accident reconstruction, 619
Arecibo Observatory, 704
Celsius to Fahrenheit, 202
Celsius temperature, 78
Comparing wind chills, 476
Distance to the sun, 267
Enzyme concentration, 202
Expansion joint, 202
Factoring in the wind, 476
Fahrenheit temperature, 78
Falling objects, 627
Female femurs, 153
Forensics, 42
Height of a ball, 301
Height difference, 231
Kepler's third law, 508
Lens equation, 364
Maximum height, 552
Meters and kilometers, 372
Orbits of the planets, 486
Orbit of Venus, 508
Playing catch, 150
Probability of rain, 410
Resistance, 627
Seacoast artillery, 571
Skeletal remains, 153
Sonic boom, 723
Speed of light, 267
Stretching a spring, 202
Telephoto lens, 364
Time to swing, 534
Unidentified flying objects, 149
Using leverage, 627
Velocity of a bullet, 202
Velocity of a projectile, 202
Velocity and time, 212
World's largest telescope, 704

Sports

The 2.4-meter rule, 94
Adjusting the saddle, 101
America's Cup, 477
Another ace, 322

Basketball blowout, 371
Bicycle gear ratio, 627
Diving time, 477
Fly ball, 43
Football field, 43
Force on basketball shoes, 372
Force on running shoes, 372
Foul ball, 542
Games behind, 43
Hazards of depth, 168
Heel motion, 168
Maximum sail area, 485
Maximum sailing speed, 477
Ping pong, 578
Pole vaulting, 213, 534
Putting the shot, 571
Running backs, 380
Sailboat speed, 508
Sailboat stability, 507
Sky divers, 230, 289
Super Bowl contender, 410
Tennis, 106
Triathlon, 379
World records, 149

Statistics/Demographics

Accidental deaths, 72
Aiming for a C, 140
Bachelor's degrees, 131
C or better, 131
California growin', 486
Cigarette consumption, 72
Commuting students, 359
Comparing scores, 154
Farmland conversion, 51
Gross domestic product, 195
Higher education, 141
In-house training, 189
Increasing training, 180
Logistic growth, 664
Master's degrees, 131
Medicaid spending, 168
Monitoring AIDS, 113
Olympic swimming, 106
Population of Mexico, 36
Population of the United States, 36
Registered voters, 113
Senior citizens, 141
Stabilization ratio, 552
Super bowl score, 108
Tough on crime, 113
Two tests only, 140
Weighted average, 131
Women on the board, 168
Women and marriage, 189
World grain demand, 72

Elementary and Intermediate Algebra

Mark Dugopolski

Southeastern Louisiana University

Boston   Burr Ridge, IL   Dubuque, IA   Madison, WI
New York   San Francisco   St. Louis   Bangkok
Bogotá   Caracas   Kuala Lumpur   Lisbon   London   Madrid
Mexico City   Milan   Montreal   New Delhi   Santiago
Seoul   Singapore   Sydney   Taipei   Toronto

McGraw-Hill Higher Education ✕

*A Division of The **McGraw-Hill** Companies*

ELEMENTARY AND INTERMEDIATE ALGEBRA

Published by McGraw-Hill, a business unit of The McGraw-Hill Companies, Inc., 1221 Avenue of the Americas, New York, NY 10020. Copyright © 2002 by The McGraw-Hill Companies, Inc. All rights reserved. No part of this publication may be reproduced or distributed in any form or by any means, or stored in a database or retrieval system, without the prior written consent of The McGraw-Hill Companies, Inc., including, but not limited to, in any network or other electronic storage or transmission, or broadcast for distance learning.

Some ancillaries, including electronic and print components, may not be available to customers outside the United States.

This book is printed on acid-free paper.

1 2 3 4 5 6 7 8 9 0 VNH/VNH 0 9 8 7 6 5 4 3 2 1
1 2 3 4 5 6 7 8 9 0 VNH/VNH 0 9 8 7 6 5 4 3 2 1

ISBN 0–07–245028–2
ISBN 0–07–247602–8 (AIE)

Publisher: *William K. Barter*
Senior sponsoring editor: *David Dietz*
Developmental editor: *Beatrice Wikander*
Senior marketing manager: *Mary K. Kittell*
Project manager: *Vicki Krug*
Production supervisor: *Enboge Chong*
Designer: *K. Wayne Harms*
Cover/interior designer: *Sabrina Dupont*
Cover image: *Stone/David Madison*
Senior photo research coordinator: *Carrie K. Burger*
Photo research: *LouAnn K. Wilson*
Senior supplement producer: *Audrey A. Reiter*
Media technology lead producer: *Steve Metz*
Compositor: *Interactive Composition Corporation*
Typeface: *10.5/12 Times Roman*
Printer: *Von Hoffmann Press, Inc.*

Photo Credits
Pages: 16, 164, 228, 312, 400, 471, and Chapter Openers 7, 8, 14: © Susan Van Etten; Pages 88, 369, 538, 614 and 745: © McGraw-Hill Higher Education, Inc.; Chapter Opener 9: © Herb Snitzer/Stock Boston; p. 653: Courtesy Molly McCallister; Chapter Opener 13: © Reuters NewMedia, Inc./Corbis; p. 709: Courtesy Frederick von Huene, von Huene Workshop; All other photos: © PhotoDisc, Inc.

Library of Congress Cataloging-in-Publication Data

Dugopolski, Mark.
  Elementary and intermediate algebra / Mark Dugopolski.—1st ed.
    p. cm.
  Includes index.
  ISBN 0–07–245028–2 (acid-free paper)
  1. Algebra. I. Title.

QA152.3 .D84   2002
512.9—dc21   2001044173
                                              CIP

www.mhhe.com

CONTENTS

Chapter 9 · Radicals and Rational Exponents · 467

Chapter 10 · Quadratic Equations, Functions, and Inequalities · 525

Chapter 11 · Functions · 581

Chapter 12

Exponential and Logarithmic Functions 645

Chapter 13

Nonlinear Systems and the Conic Sections 689

Chapter 14

Sequences and Series 741

Appendix   A-1

Answers   A-7

Index   I-1

Elementary and Intermediate Algebra is designed to provide students with the algebra background needed for further college-level mathematics courses. This text combines the gentle pace of elementary algebra with the broad coverage of topics in intermediate algebra, eliminating topic overlap. The extent and sequence of the topics make this text suitable for a wide variety of courses of various lengths. The features, design, and pedagogy are the same as in the successful third editions of my *Elementary Algebra* and *Intermediate Algebra.*

Content

The unifying theme of this text is the development of algebraic skills and concepts, followed by the application of those skills and concepts to problem solving. Numerous real-world examples and exercises make the topics contemporary and interesting. Reading and understanding graphs is stressed in the examples and exercises of the text. Particular care has been given to achieving an appropriate balance of exercises that progressively increase in difficulty within each exercise set. Fractions and decimals are used in the exercises and throughout the text discussions to help reinforce the basic arithmetic skills needed for success in algebra. Geometric concepts and applications are also integrated throughout the book.

Features

Please see the walk-through preface, beginning on page xvii, for additional information about some of the features described below.

- Each chapter begins with introductory text that discusses a real application of algebra. The **Chapter Opener** is accompanied by a photograph and, in most cases, by a real-data application graph that helps students visualize algebra and more fully understand the concepts discussed in the chapter. In addition, each chapter contains a **Math at Work** feature (example on page 16), which profiles a real person and the mathematics that he or she uses on the job. These two features have corresponding real-data exercises.

- The text emphasizes real-data applications that involve graphs. Applications are distributed throughout the text to help demonstrate concepts, to motivate students, and to give students practice using new skills. Many of the real-data exercises involve data obtained from the Internet. (See Exercise 125 on page 36.) Internet addresses are provided in the textbook and the Online Learning Center (as hyperlinks) as a resource for both students and teachers. An **Index of Applications,** which lists applications by subject matter, is included at the front of the text.

- Every section begins with **In This Section,** a list of topics that shows the student what will be covered. Since the topics correspond to the headings within each section, students will find it easy to locate and study specific concepts.

- **Strategy Boxes** help students understand the step–by–step process of problem solving. These boxes provide a numbered list of concepts from a section or a set of steps to follow in problem solving.

- Important ideas, such as definitions, rules, summaries, and strategies, are set apart in boxes for quick reference. Color is used to highlight these boxes as well as other important points in the text.

- Three margin features appear throughout the text:

  1. **Calculator Close-Ups** give students an idea of how to use a graphing calculator and for what kinds of problems it should be used. The screens in this feature are done with a TI-83 and are not intended to replace the calculator manual (especially if students use a different make or model). Some **Calculator Close-Ups** simply introduce the features of a graphing calculator, but many are intended to enhance understanding of algebraic concepts. For this reason, many of the **Calculator Close-Ups** will benefit even those students who do not use a graphing calculator. *A graphing calculator is not required for studying from this text.*

  2. **Study Tips** are included in the margins throughout the text. These short tips are meant to continually reinforce good study habits and to remind students that it is never too late to make improvements in the manner in which they study.

  3. **Helpful Hints** are short comments that enhance the material in the text, provide another way of approaching a problem, or clear up misconceptions.

- At the end of every section are **Warm-up** exercises, a set of ten simple statements that are to be answered true or false. These exercises are designed to provide a smooth transition between the ideas and the exercise sets. They help students understand that statements in mathematics are either true or false. They are also good for discussion, review, or group work.

- The exercise sets in each section generally begin with six simple writing exercises. These exercises are designed to encourage students to review the definitions and rules of the section before doing more traditional exercises. For example, the student might simply be asked what properties of equality were discussed in this section.

- The end-of-section **Exercises** follow the same order as the textual material and contain exercises that are keyed to examples, as well as numerous exercises that are not keyed to examples. This organization allows the instructor to cover only part of a section if necessary and easily determine which exercises are appropriate to assign. The keyed exercises give the student a place to start practicing and building confidence, whereas the nonkeyed exercises are designed to wean the student from following examples in a "cookbook" manner. **Getting More Involved** exercises are designed to encourage writing, discussion, exploration, and cooperative learning. **Graphing Calculator Exercises** require a graphing calculator and are identified with a graphing calculator logo. Exercises for which a scientific calculator would be helpful are identified with a scientific calculator logo.

- At the end of each chapter are **Collaborative Activities** designed to encourage interaction and learning in groups. Instructions and suggestions for using these activities and answers to all problems can be found in the *Instructor's Solutions Manual.*

- Every chapter ends with a four-part **Wrap-up,** which includes the following:

  1. The **Chapter Summary** lists important concepts along with brief illustrative examples.

2. **Enriching Your Mathematical Word Power** consists of multiple-choice questions in which the important terms are to be matched with their meanings. This feature emphasizes the importance of proper terminology.

3. The **Review Exercises** contain problems that are keyed to the sections of the chapter as well as numerous miscellaneous exercises.

4. The **Chapter Test** is designed to help the student assess his or her readiness for a test. The **Chapter Test** has no keyed exercises, so the student can work independently of the sections and examples.

- The **Making Connections** cumulative review exercises at the end of each chapter are designed to help students review and synthesize the new material with ideas from previous chapters, and in some cases, review material necessary for success in the upcoming chapter. Every **Making Connections** exercise set includes at least one applied exercise that requires ideas from one or more of the previous chapters.

- The **Midtext Diagnostic Review** after Chapter 7 can be used to assess a student's readiness for the second half of the text. These review exercises are keyed to specific worked examples in the first seven chapters.

Coverage

For instructors who desire later coverage of linear equations in two variables, Chapter 4, "Linear Equations in Two Variables and Their Graphs," can be postponed. For those who want early coverage of linear equations in two variables, Chapter 8, "Systems of Linear Equations," could be covered after Chapter 4. For instructors who want less coverage of functions, Section 4.6 can be omitted or postponed.

Supplements for the Instructor

ANNOTATED INSTRUCTOR'S EDITION

This ancillary includes answers to all exercises and tests. These answers are printed in a second color for ease of use by instructors and are located adjacent to corresponding exercises.

COMPUTERIZED TEST BANK AND PDF TEST FILE

The Computerized Test Bank allows you to create well-formatted quizzes or tests using a large bank of algorithmically-generated and static questions through an intuitive Windows or Macintosh interface. When creating a test, you can manually choose individual questions or have the software randomly select questions based on section, question type, difficulty level, and other criteria. Instructors also have the ability to add or edit test bank questions to create their own customized test bank. Preformatted chapter tests (three forms per chapter) and final exams (four forms), are also provided, in printable PDF format.

INSTRUCTOR'S SOLUTIONS MANUAL

Prepared by Mark Dugopolski, this supplement contains detailed solutions to all the exercises in the text. The methods used to solve the problems in the manual are the same as those used to solve examples in the textbook. Instructions and suggestions for using the Collaborative Activities featured in the text are also included in the *Instructor's Solutions Manual*.

ONLINE LEARNING CENTER

Web-based, interactive learning is available for your students on the *Online Learning Center,* located at www.mhhe.com/dugopolski. Student resources are located in the OLC's **Student Center** and include interactive applications, algorithmically-generated practice exams, online quizzing, audio/visual tutorials, all chapters of the text in PDF format, and web links. Instructor resources are located in the **Instructor Center** and include PowerPoint® slides, an AMATYC standards correlation, and links to PageOut and author-recommended sites. The Course Integration Guide is also located on the OLC. The password for the **Instructor Center** is located in the preface to the *Instructor's Solutions Manual.*

COURSE INTEGRATION GUIDE

This supplement, located on the OLC, integrates the multimedia and print supplements that accompany the main text into a useful and well-organized guide. The *Course Integration Guide* includes helpful information about resources found in the text, video, CD-ROM, and Online Learning Center. The guide also contains a section-level correlation with the ELMC, CLAST, and TASP standards.

PAGEOUT

Need a course website? More than 50,000 professors have chosen PageOut to create a custom course website. Thier feedback is used to continually enhance PageOut.

New Features based on customer feedback:
- Timed tests and the ability to author original questions
- Ability to insert diacritical marks and html codes with the click of a button
- Ability to choose a pre-built page design, or create a custom design

Short on time? Let us do the work.
Our McGraw-Hill service team is ready to build your PageOut website and provide any necessary training.

   Learn more about PageOut and other McGraw-Hill digital solutions at www.mhhe.com/solutions.

Supplements for the Student

STUDENT'S SOLUTIONS MANUAL

Prepared by Mark Dugopolski, the *Student's Solutions Manual* contains complete worked-out solutions to almost all of the problems in the **Warm-ups,** the **Chapter Tests,** and the **Making Connections** reviews and to the odd-numbered problems in the end-of-section **Exercises** and the **Chapter Reviews.**

DUGOPOLSKI VIDEO SERIES (VIDEOTAPES OR VIDEO CDS)

The video series is composed of 14 video cassettes, one per chapter. On-screen instructors introduce topics and work through examples using the methods presented in the text. Students are encouraged to work examples on their own and check their results against those provided. The video series is also available on video CDs.

DUGOPOLSKI TUTORIAL CD-ROM

This interactive CD-ROM is a self-paced tutorial specifically linked to the text that reinforces topics through unlimited opportunities to review concepts and practice problem solving. The CD-ROM provides section-specific tutorials, questions with feedback, and algorithmically-generated questions. This product requires virtually no computer training on the part of students and supports Windows and Macintosh computers.

In addition, a number of other technology and web-based ancillaries are under development; they will support the ever-changing technology needs in developmental mathematics. For further information about these or other supplements, please contact your local McGraw-Hill sales representative.

ONLINE LEARNING CENTER

Student resources are located in the Student Center on the *Online Learning Center* and include interactive applications, algorithmically-generated practice exercises, online quizzing, audio/visual tutorials, all chapters of the text in PDF format, and links to useful and fun algebra websites. The OLC is located at www.mhhe.com/dugopolski, and a free password card is included with each new copy of this text.

NetTutor

NetTutor is a revolutionary system that enables students to interact with a live tutor over the World Wide Web. Students can receive instruction from live tutors using NetTutor's web-based, graphical chat capabilities. They can also submit questions and receive answers, browse previously answered questions, and view previous live chat sessions.

ALEKS®

ALEKS® (**A**ssessment and **LE**arning in **K**nowledge **S**paces) is an artificial intelligence-based system for individualized math learning, available over the World Wide Web. ALEKS® delivers precise, qualitative diagnostic assessments of students' math knowledge, guides them in the selection of appropriate new study material, and records their progress toward mastery of curricular goals in a robust classroom management system. It interacts with the student much as a skilled human tutor would, moving between explanation and practice as needed, correcting and analyzing errors, defining terms, and changing topics on request. By sophisticated modeling of a student's "knowledge state" for a given subject matter, ALEKS® can focus clearly on what the student is most ready to learn next, building a learning momentum that fuels success.

To learn more about ALEKS®, including purchasing information, visit the ALEKS® website at www.highed.aleks.com.

Acknowledgements

First I thank all of the students and professors who used *Elementary Algebra*, *Intermediate Algebra*, and *Algebra for College Students*. I sincerely appreciate the efforts of the reviewers of *Elementary and Intermediate Algebra* who made many helpful suggestions:

Susan Akers, *Northeast State Technical Community College*
Kathy Autrey, *Northwestern State University*
Walter Burlags, *Lake City Community College*
Marc Campbell, *Daytona Beach Community College*

Jo Cathey, *Columbia State Community College–Columbia Campus*
Dawn Dabney, *Northeast State Technical Community College*
Susan Davenport, *Blinn College*
Lenore DeSilets, *De Anza College*
Jacqueline Fesq, *Raritan Valley Community College*
Pat Foard, *South Plains College*
Judy Godwin, *Colin County Community College*
David Gorman, *Yavapai College*
Helen Harris, *Blinn College*
Margaret Hathaway, *Kansas City Community College*
Steven Howard
Matthew Hudock, *St. Philip's College*
Patricia Ann Hussy, *Triton College*
Patricia Jenkins, *South Carolina State University*
Bill Krant, *El Paso Community College–Via Verde Campus*
Anthony P. Malone, *Raymond Walters College*
Vince McGarry, *Austin Community College*
Leslie McGinnis, *Blinn College*
Barbara Napoli, *Our Lady of the Lake College*
Christopher O'Connor, *Shawnee State University*
Carol Olson, *Northwest Arkansas Community College*
Frank Pecchioni, *Jefferson Community College*
George Reed, *Angelina College*
Alan Sauter, *Collin County Community College*
James Smith, *Columbia State Community College–Lawrenceburg Campus*
Ann Thrower, *Kilgore Community College*
Danny Whited, *Virginia Intermont College*
Margorie Whitmore, *Northwest Arkansas Community College*
Jackie Wing, *Angelina College*
Mary Jane Wolfe, *University of Rio Grande*

I also thank Edgar Reyes of Southeastern Louisiana University for his work on the CD-ROM and Rebecca Muller of Southeastern Louisiana University for her work on the printed test bank. I thank the entire staff at McGraw-Hill for all of their help and encouragement during this project. Special thanks go to Bill Barter, David Dietz, and Bea Wikander. I also want to express my sincere appreciation to my wife, Cheryl, for her invaluable patience and support.

Hammond, Louisiana   M.D.

ALEKS®?

ALEKS is...

■ A comprehensive course management system. It tells you exactly what your students know and don't know.

■ Artificial intelligence. It totally individualized assessment and learning.

■ Customizable. Click on or off each course topic.

■ Web based. Use a standard browser for easy Internet access.

■ Inexpensive. There are no set up fees or site license fees.

ALEKS (Assessment and LEarning in Knowledge Spaces) is an artificial intelligence-based system for individualized math learning available via the World Wide Web.

http://www.highed.aleks.com

ALEKS delivers precise, qualitative diagnostic assessments of students' math knowledge, guides them in the selection of appropriate new study material, and records their progress toward mastery of curricular goals in a robust course management system.

ALEKS interacts with the student much as a skilled human tutor would, moving between explanation and practice as needed, correcting and analyzing errors, defining terms and changing topics on request. By sophisticated modeling of a student's "knowledge state" for a given subject matter, ALEKS can focus clearly on what the student is most ready to learn next, thereby building a learning momentum that fuels success.

The ALEKS system was developed with a multimillion-dollar grant from the National Science Foundation. It has little to do with what is commonly thought of as educational software. The theory behind ALEKS is a specialized field of mathematical cognitive science called "Knowledge Spaces."

Knowledge Space theory, which concerns itself with the mathematical dynamics of knowledge acquisition, has been under development by researchers in cognitive science since the early 1980s. The Chairman and founder of ALEKS Corporation, Jean-Claude Falmagne, is an internationally recognized leader in scientific work in this field.

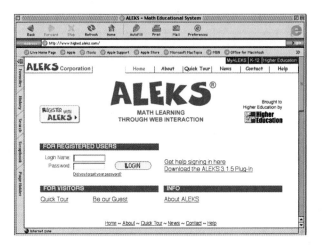

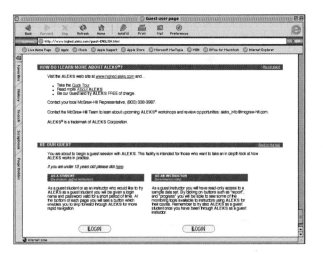

Please visit http://www.highed.aleks.com for a trial run.

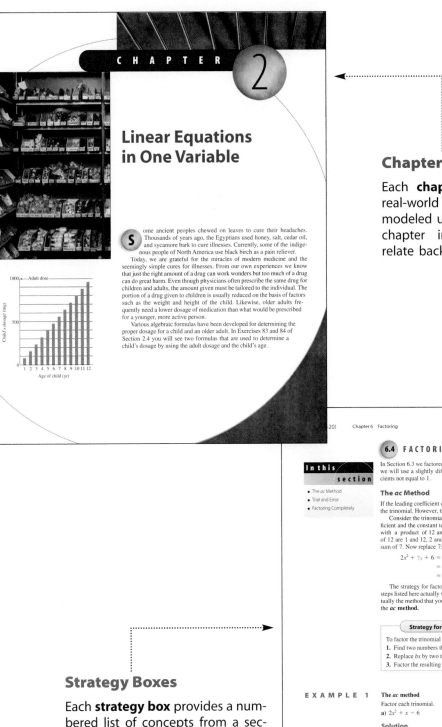

CHAPTER 2

Linear Equations in One Variable

Some ancient peoples chewed on leaves to cure their headaches. Thousands of years ago, the Egyptians used honey, salt, cedar oil, and sycamore bark to cure illnesses. Currently, some of the indigenous people of North America use black birch as a pain reliever.

Today, we are grateful for the miracles of modern medicine and the seemingly simple cures for illnesses. From our own experiences we know that just the right amount of a drug can work wonders but too much of a drug can do great harm. Even though physicians often prescribe the same drug for children and adults, the amount given must be tailored to the individual. The portion of a drug given to children is usually reduced on the basis of factors such as the weight and height of the child. Likewise, older adults frequently need a lower dosage of medication than what would be prescribed for a younger, more active person.

Various algebraic formulas have been developed for determining the proper dosage for a child and an older adult. In Exercises 83 and 84 of Section 2.4 you will see two formulas that are used to determine a child's dosage by using the adult dosage and the child's age.

Chapter Opener

Each **chapter opener** features a real-world situation that can be modeled using mathematics. Every chapter includes exercises that relate back to the chapter opener.

---

(20) Chapter 6 Factoring

6.4 FACTORING $ax^2 + bx + c$ WITH $a \neq 1$

In Section 6.3 we factored trinomials with a leading coefficient of 1. In this section we will use a slightly different technique to factor trinomials with leading coefficients not equal to 1.

In this section
- The ac Method
- Trial and Error
- Factoring Completely

The *ac* Method

If the leading coefficient of a trinomial is not 1, we can again use grouping to factor the trinomial. However, the procedure is slightly different.

Consider the trinomial $2x^2 + 7x + 6$. First find the product of the leading coefficient and the constant term. In this case it is $2 \cdot 6$, or 12. Now find two numbers with a product of 12 and a sum of 7. The pairs of numbers with a product of 12 are 1 and 12, 2 and 6, and 3 and 4. Only 3 and 4 have a product of 12 and a sum of 7. Now replace $7x$ by $3x + 4x$ and factor by grouping:

$$2x^2 + 7x + 6 = 2x^2 + 3x + 4x + 6 \qquad \text{Replace } 7x \text{ by } 3x + 4x.$$
$$= (2x + 3)x + (2x + 3)2 \qquad \text{Factor out the common factors.}$$
$$= (2x + 3)(x + 2) \qquad \text{Factor out } 2x + 3.$$

The strategy for factoring a trinomial is summarized in the following box. The steps listed here actually work whether or not the leading coefficient is 1. This is actually the method that you learned in Section 6.3 with $a = 1$. This method is called the *ac* **method.**

> **Strategy for Factoring $ax^2 + bx + c$ by the *ac* Method**
>
> To factor the trinomial $ax^2 + bx + c$:
> 1. Find two numbers that have a product equal to ac and a sum equal to b.
> 2. Replace bx by two terms using the two new numbers as coefficients.
> 3. Factor the resulting four-term polynomial by grouping.

EXAMPLE 1   **The *ac* method**
Factor each trinomial.
a) $2x^2 + x - 6$ **b)** $10x^2 + 13x - 3$

Solution
a) Because $2 \cdot (-6) = -12$, we need two integers with a product of -12 and a sum of 1. We can list the possible pairs of integers with a product of -12:

1 and -12	2 and -6	3 and -4
-1 and 12	-2 and 6	-3 and 4

Only -3 and 4 have a sum of 1. Replace x by $-3x + 4x$ and factor by grouping:

$$2x^2 + x - 6 = 2x^2 - 3x + 4x - 6 \qquad \text{Replace } x \text{ by } -3x + 4x.$$
$$= (2x - 3)x + (2x - 3)2 \qquad \text{Factor out the common factors.}$$
$$= (2x - 3)(x + 2) \qquad \text{Factor out } 2x - 3.$$

Check by FOIL.

Strategy Boxes

Each **strategy box** provides a numbered list of concepts from a section, or a set of steps to follow in problem solving. They can be used by students who prefer a more structured approach to problem solving or they can be used as a study tool to review important points within sections.

Margin Notes

Margin notes include: **Helpful Hints, Study Tips,** and **Calculator Close-ups.** The **Helpful Hints** point out common errors or reminders. The **Study Tips** provide practical suggestions for improving study habits. The optional **Calculator Close-ups** provide advice on using a graphing calculator to aid in understanding the mathematics. They also include insightful suggestions for increasing calculator proficiency.

---

helpful hint

Recall that the order of operations gives multiplication and division an equal ranking and says to do them in order from left to right. So without parentheses,

$$-6x^3 \div 2x^9$$

actually means

$$\frac{-6x^3}{2} \cdot x^9.$$

Solution

a) $\dfrac{y^9}{y^5} = y^{9-5} = y^4$

Use the definition of division to check that $y^4 \cdot y^5 = y^9$.

b) $\dfrac{12b^2}{3b^7} = \dfrac{12}{3} \cdot \dfrac{b^2}{b^7} = 4 \cdot \dfrac{1}{b^{7-2}} = \dfrac{4}{b^5}$

Use the definition of division to check that

$$\frac{4}{b^5} \cdot 3b^7 = \frac{12b^7}{b^5} = 12b^2.$$

c) $-6x^3 \div (2x^9) = \dfrac{-6x^3}{2x^9} = \dfrac{-3}{x^6}$

Use the definition of division to check that

$$\frac{-3}{x^6} \cdot 2x^9 = \frac{-6x^9}{x^6} = -6x^3.$$

d) $\dfrac{x^8y^2}{x^2y^2} = \dfrac{x^8}{x^2} \cdot \dfrac{y^2}{y^2} = x^6 \cdot y^0 = x^6$

Use the definition of division to check that $x^6 \cdot x^2y^2 = x^8y^2$.   ∎

We showed more steps in Example 2 than are necessary. For division problems like these you should try to write down only the quotient.

Dividing a Polynomial by a Monomial

We divided some simple polynomials by monomials in Chapter 1. For example,

$$\frac{6x + 8}{2} = \frac{1}{2}(6x + 8) = \frac{6x}{2} + \frac{8}{2} = 3x + 4.$$

We use the distributive property to take one-half of $6x$ and one-half of 8 to get $3x + 4$. So both $6x$ and 8 are divided by 2. To divide any polynomial by a monomial, we divide each term of the polynomial by the monomial.

EXAMPLE 3

study tip

Play offensive math, not defensive math. A student who says, "Give me a question and I'll see if I can answer it," is playing defensive math. The student is taking a passive approach to learning. A student who takes an active approach and knows the usual questions and answers for each topic is playing offensive math.

Dividing a polynomial by a monomial

Find the quotient for $(-8x^6 + 12x^4 - 4x^2) \div (4x^2)$.

Solution

$$\frac{-8x^6 + 12x^4 - 4x^2}{4x^2} = \frac{-8x^6}{4x^2} + \frac{12x^4}{4x^2} - \frac{4x^2}{4x^2}$$

$$= -2x^4 + 3x^2 - 1$$

The quotient is $-2x^4 + 3x^2 - 1$. We can check by multiplying.

$$4x^2(-2x^4 + 3x^2 - 1) = -8x^6 + 12x^4 - 4x^2.$$ ∎

Because division by zero is undefined, we will always assume that the divisor is nonzero in any quotient involving variables. For example, the division in Example 3 is valid only if $4x^2 \neq 0$, or $x \neq 0$.

Certain polynomials are given special names. A **monomial** is a polynomial that has one term, a **binomial** is a polynomial that has two terms, and a **trinomial** that has three terms. For example, $3x^5$ is a monomial, $2x - 1$ is a binomial, and $4x^6 - 3x + 2$ is a trinomial.

EXAMPLE 2

Types of polynomials

Identify each polynomial as a monomial, binomial, or trinomial and state its degree.

a) $5x^2 - 7x^3 + 2$ **b)** $x^{43} - x^2$ **c)** $5x$ **d)** -12

Solution

a) The polynomial $5x^2 - 7x^3 + 2$ is a third-degree trinomial.

b) The polynomial $x^{43} - x^2$ is a binomial with degree 43.

c) Because $5x = 5x^1$, this polynomial is a monomial with degree 1.

d) The polynomial -12 is a monomial with degree 0.   ∎

Value of a Polynomial

A polynomial is an algebraic expression. Like other algebraic expressions involving variables, a polynomial has no specific value unless the variables are replaced by numbers. A polynomial can be evaluated with or without the function notation discussed in Chapter 4.

EXAMPLE 3

Evaluating polynomials

a) Find the value of $-3x^4 - x^3 + 20x + 3$ when $x = 1$.

b) Find the value of $-3x^4 - x^3 + 20x + 3$ when $x = -2$.

c) If $P(x) = -3x^4 - x^3 + 20x + 3$, find $P(1)$.

calculator close-up

To evaluate the polynomial in Example 3 with a calculator, first use Y= to define the polynomial.

```
Plot1 Plot2 Plot3
\Y1◼-3X^4-X^3+20
X+3
\Y2=
\Y3=
\Y4=
\Y5=
\Y6=
```

Then find y₁(−2) and y₁(1).

```
Y1(-2)
            -77
Y1(1)
             19
```

Solution

a) Replace x by 1 in the polynomial:

$$-3x^4 - x^3 + 20x + 3 = -3(1)^4 - (1)^3 + 20(1) + 3$$
$$= -3 - 1 + 20 + 3$$
$$= 19$$

So the value of the polynomial is 19 when $x = 1$.

b) Replace x by -2 in the polynomial:

$$-3x^4 - x^3 + 20x + 3 = -3(-2)^4 - (-2)^3 + 20(-2) + 3$$
$$= -3(16) - (-8) - 40 + 3$$
$$= -48 + 8 - 40 + 3$$
$$= -77$$

So the value of the polynomial is -77 when $x = -2$.

c) This is a repeat of part (a) using the function notation from Chapter 4. $P(1)$, read "P of 1," is the value of the polynomial $P(x)$ when x is 1. To find $P(1)$, replace x by 1 in the formula for $P(x)$:

$$P(x) = -3x^4 - x^3 + 20x + 3$$
$$P(1) = -3(1)^4 - (1)^3 + 20(1) + 3$$
$$= 19$$

So $P(1) = 19$. The value of the polynomial when $x = 1$ is 19.   ∎

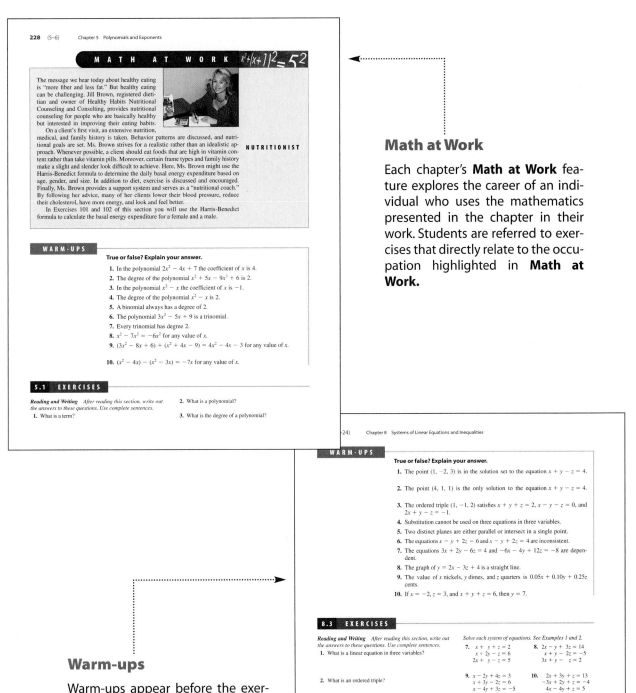

M A T H   A T   W O R K

$x^2 + (x+1)^2 = 5^2$

The message we hear today about healthy eating is "more fiber and less fat." But healthy eating can be challenging. Jill Brown, registered dietitian and owner of Healthy Habits Nutritional Counseling and Consulting, provides nutritional counseling for people who are basically healthy but interested in improving their eating habits.

On a client's first visit, an extensive nutrition, medical, and family history is taken. Behavior patterns are discussed, and nutritional goals are set. Ms. Brown strives for a realistic rather than an idealistic approach. Whenever possible, a client should eat foods that are high in vitamin content rather than take vitamin pills. Moreover, certain frame types and family history make a slight and slender look difficult to achieve. Here, Ms. Brown might use the Harris-Benedict formula to determine the daily basal energy expenditure based on age, gender, and size. In addition to diet, exercise is discussed and encouraged. Finally, Ms. Brown provides a support system and serves as a "nutritional coach." By following her advice, many of her clients lower their blood pressure, reduce their cholesterol, have more energy, and look and feel better.

In Exercises 101 and 102 of this section you will use the Harris-Benedict formula to calculate the basal energy expenditure for a female and a male.

NUTRITIONIST

WARM-UPS

True or false? Explain your answer.

1. In the polynomial $2x^2 - 4x + 7$ the coefficient of x is 4.
2. The degree of the polynomial $x^2 + 5x - 9x^3 + 6$ is 2.
3. In the polynomial $x^2 - x$ the coefficient of x is -1.
4. The degree of the polynomial $x^2 - x$ is 2.
5. A binomial always has a degree of 2.
6. The polynomial $3x^2 - 5x + 9$ is a trinomial.
7. Every trinomial has degree 2.
8. $x^2 - 7x^2 = -6x^2$ for any value of x.
9. $(3x^2 - 8x + 6) + (x^2 + 4x - 9) = 4x^2 - 4x - 3$ for any value of x.
10. $(x^2 - 4x) - (x^2 - 3x) = -7x$ for any value of x.

5.1   EXERCISES

Reading and Writing   *After reading this section, write out the answers to these questions. Use complete sentences.*

1. What is a term?

2. What is a polynomial?

3. What is the degree of a polynomial?

Math at Work

Each chapter's **Math at Work** feature explores the career of an individual who uses the mathematics presented in the chapter in their work. Students are referred to exercises that directly relate to the occupation highlighted in **Math at Work.**

WARM-UPS

True or false? Explain your answer.

1. The point $(1, -2, 3)$ is in the solution set to the equation $x + y - z = 4$.
2. The point $(4, 1, 1)$ is the only solution to the equation $x + y - z = 4$.
3. The ordered triple $(1, -1, 2)$ satisfies $x + y + z = 2$, $x - y - z = 0$, and $2x + y - z = -1$.
4. Substitution cannot be used on three equations in three variables.
5. Two distinct planes are either parallel or intersect in a single point.
6. The equations $x - y + 2z = 6$ and $x - y + 2z = 4$ are inconsistent.
7. The equations $3x + 2y - 6z = 4$ and $-6x - 4y + 12z = -8$ are dependent.
8. The graph of $y = 2x - 3z + 4$ is a straight line.
9. The value of x nickels, y dimes, and z quarters is $0.05x + 0.10y + 0.25z$ cents.
10. If $x = -2$, $z = 3$, and $x + y + z = 6$, then $y = 7$.

8.3   EXERCISES

Reading and Writing   *After reading this section, write out the answers to these questions. Use complete sentences.*

1. What is a linear equation in three variables?

2. What is an ordered triple?

3. What is a solution to a system of linear equations in three variables?

4. How do we solve systems of linear equations in three variables?

5. What does the graph of a linear equation in three variables look like?

6. How are the planes positioned when a system of linear equations in three variables is inconsistent?

Solve each system of equations. See Examples 1 and 2.

7. $x + y + z = 2$
   $x + 2y - z = 6$
   $2x + y - z = 5$

8. $2x - y + 3z = 14$
   $x + y - 2z = -5$
   $3x + y - z = 2$

9. $x - 2y + 4z = 3$
   $x + 3y - 2z = 6$
   $x - 4y + 3z = -5$

10. $2x + 3y + z = 13$
    $-3x + 2y + z = -4$
    $4x - 4y + z = 5$

11. $2x - y + z = 10$
    $3x - 2y - 2z = 7$
    $x - 3y - 2z = 10$

12. $x - 3y + 2z = -11$
    $2x - 4y + 3z = -15$
    $3x - 5y - 4z = 5$

13. $2x - 3y + z = -9$
    $-2x + y - 3z = 7$
    $x - y + 2z = -5$

14. $3x - 4y + z = 19$
    $2x + 4y + z = 0$
    $x - 2y + 5z = 17$

15. $2x - 5y + 2z = 16$
    $3x + 2y - 3z = -19$
    $4x - 3y + 4z = 18$

16. $-2x + 3y - 4z = 3$
    $3x - 5y + 2z = 4$
    $-4x + 2y - 3z = 0$

17. $x + y = 4$
    $y - z = -2$
    $x + y + z = 9$

18. $x + y - z = 0$
    $x - y = -2$
    $y + z = 10$

Warm-ups

Warm-ups appear before the exercises at the end of every section. They are true or false statements that can be used to check conceptual understanding of material within each section.

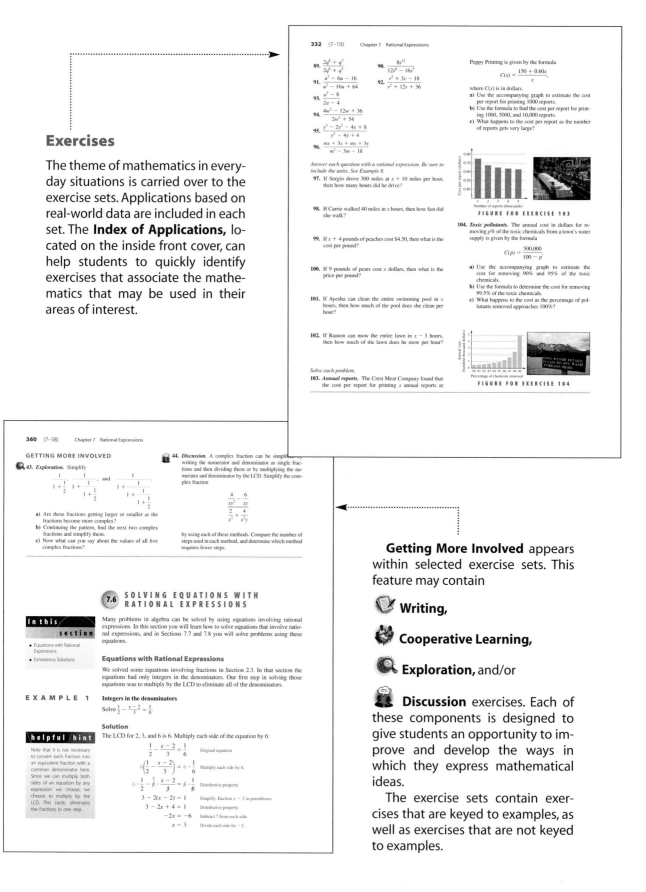

Exercises

The theme of mathematics in everyday situations is carried over to the exercise sets. Applications based on real-world data are included in each set. The **Index of Applications,** located on the inside front cover, can help students to quickly identify exercises that associate the mathematics that may be used in their areas of interest.

Getting More Involved appears within selected exercise sets. This feature may contain

Writing,

Cooperative Learning,

Exploration, and/or

Discussion exercises. Each of these components is designed to give students an opportunity to improve and develop the ways in which they express mathematical ideas.

The exercise sets contain exercises that are keyed to examples, as well as exercises that are not keyed to examples.

130 (3–10)   Chapter 3   Inequalities in One Variable

46. $4x \leq -8$

47. $-3x \leq 12$

48. $-2x > -6$

49. $2x - 3 > 7$

50. $3x - 2 < 6$

51. $3 - 5x \leq 18$

52. $5 - 4x \geq 19$

53. $\frac{x-3}{-5} < -2$

54. $\frac{2x-3}{4} > 6$

55. $\frac{5-3x}{4} \leq 2$

56. $\frac{7-5x}{-2} \geq -1$

57. $3 - \frac{1}{4}x \geq 2$

58. $5 - \frac{1}{3}x > 2$

59. $\frac{1}{4}x - \frac{1}{2} < \frac{1}{2}x - \frac{2}{3}$

60. $\frac{1}{3}x - \frac{1}{6} < \frac{1}{6}x - \frac{1}{2}$

61. $\frac{y-3}{2} > \frac{1}{2} - \frac{y-5}{4}$

62. $\frac{y-1}{3} - \frac{y+1}{5} > 1$

Solve each inequality and graph the solution set.

63. $2x + 3 > 2(x - 4)$

64. $-2(5x - 1) \leq -5(5 + 2x)$
65. $-4(2x - 5) \leq 2(6 - 4x)$
66. $-3(2x - 1) \geq 2(5 - 3x)$

67. $-\frac{1}{2}(x - 6) < \frac{1}{2}x + 2$

68. $-3\left(\frac{1}{2}x - \frac{1}{4}\right) > \frac{x}{2} - \frac{1}{4}$

69. $4.273 + 2.8x \leq 10.985$

70. $1.064 < 5.94 - 3.2x$

71. $3.25x - 27.39 > 4.06 + 5.1x$

72. $4.86(3.2x - 1.7) > 5.19 - x$

Identify the variable and write an inequality that describes each situation. See Example 6.
73. Tony is taller than 6 feet.

74. Glenda is under 60 years old.

75. Wilma makes less than $80,000 per year.

76. Bubba weighs over 80 pounds.

77. The maximum speed for the Concorde is 1450 miles per hour (mph).

78. The minimum speed on the freeway is 45 mph.

79. Julie can afford at most $400 per month.

Calculator Exercises

Calculator Exercises (and) are optional. They provide an opportunity for students to learn how a scientific or graphing calculator might be useful in solving various problems.

---

n hours of work. What is her revenue for 100 hours of work? What is her revenue for 101 hours of work? By how much did the one extra hour of work increase the revenue? (The increase in revenue is called the *marginal revenue* for the 101st hour.)

73. *In-house training.* The accompanying graph shows the percentage of U.S. workers receiving training by their employers. The percentage went from 5% in year 0 (1981) to 16% in year 14 (1995).
a) Find the slope of this line.
b) Write the equation of the line in slope-intercept form.
c) Use your equation to predict the percentage that will be receiving training in the year 2000.

FIGURE FOR EXERCISE 73

74. *Women and marriage.* The percentage of women in the 20 to 24 age group who have never married went from 64% in year 0 (1970) to 33% in year 26 (1996) (Census Bureau, www.census.gov).
a) Find the equation of the line through the two points (0, 0.64) and (26, 0.33) in slope-intercept form.
b) Use your equation to predict what the percentage will be in the year 2000.

75. *Pansies and snapdragons.* A nursery manager plans to spend $100 on 6-packs of pansies at 50 cents per pack and snapdragons at 25 cents per pack. The equation $0.50x + 0.25y = 100$ can be used to model this situation.
a) What do x and y represent?

b) Graph the equation.

c) Write the equation in slope-intercept form.

d) What is the slope of the line?
e) What does the slope tell you?

76. *Pens and pencils.* A bookstore manager plans to spend $60 on pens at 30 cents each and pencils at 10 cents each. The equation $0.10x + 0.30y = 60$ can be used to model this situation.
a) What do x and y represent?

b) Graph the equation.

c) Write the equation in slope-intercept form.

d) What is the slope of the line?

e) What does the slope tell you?

GRAPHING CALCULATOR EXERCISES

Graph each pair of straight lines on your graphing calculator using a viewing window that makes the lines look perpendicular. Answers may vary.

77. $y = 12x - 100,\ y = -\frac{1}{12}x + 50$

78. $2x - 3y = 300,\ 3x + 2y = -60$

Collaborative Activities

Collaborative Activities appear at the end of each chapter. The activities are designed to encourage interaction and learning in a group setting.

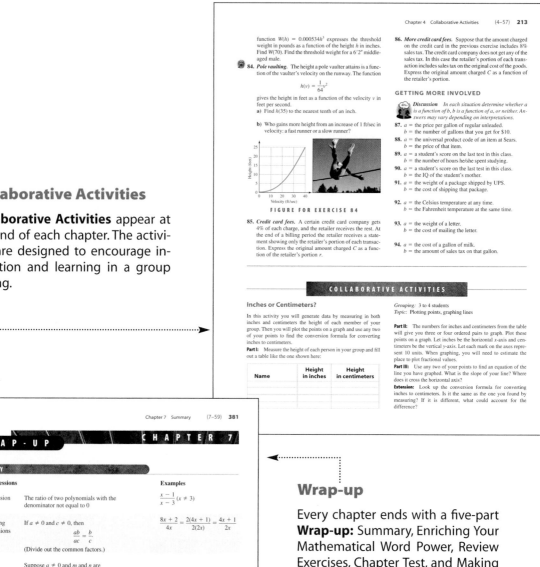

Chapter 4 Collaborative Activities   (4–57) **213**

function $W(h) = 0.000534h^3$ expresses the threshold weight in pounds as a function of the height h in inches. Find $W(70)$. Find the threshold weight for a 6′2″ middle-aged male.

84. *Pole vaulting.* The height a pole vaulter attains is a function of the vaulter's velocity on the runway. The function

$$h(v) = \frac{1}{64}v^2$$

gives the height in feet as a function of the velocity v in feet per second.

a) Find $h(35)$ to the nearest tenth of an inch.

b) Who gains more height from an increase of 1 ft/sec in velocity: a fast runner or a slow runner?

FIGURE FOR EXERCISE 84

85. *Credit card fees.* A certain credit card company gets 4% of each charge, and the retailer receives the rest. At the end of a billing period the retailer receives a statement showing only the retailer's portion of each transaction. Express the original amount charged C as a function of the retailer's portion r.

86. *More credit card fees.* Suppose that the amount charged on the credit card in the previous exercise includes 8% sales tax. The credit card company does not get any of the sales tax. In this case the retailer's portion of each transaction includes sales tax on the original cost of the goods. Express the original amount charged C as a function of the retailer's portion.

GETTING MORE INVOLVED

Discussion   In each situation determine whether a is a function of b, b is a function of a, or neither. Answers may vary depending on interpretations.

87. a = the price per gallon of regular unleaded.
b = the number of gallons that you get for $10.

88. a = the universal product code of an item at Sears.
b = the price of that item.

89. a = a student's score on the last test in this class.
b = the number of hours he/she spent studying.

90. a = a student's score on the last test in this class.
b = the IQ of the student's mother.

91. a = the weight of a package shipped by UPS.
b = the cost of shipping that package.

92. a = the Celsius temperature at any time.
b = the Fahrenheit temperature at the same time.

93. a = the weight of a letter.
b = the cost of mailing the letter.

94. a = the cost of a gallon of milk.
b = the amount of sales tax on that gallon.

COLLABORATIVE ACTIVITIES

Inches or Centimeters?

In this activity you will generate data by measuring in both inches and centimeters the height of each member of your group. Then you will plot the points on a graph and use any two of your points to find the conversion formula for converting inches to centimeters.

Part I: Measure the height of each person in your group and fill out a table like the one shown here:

Name	Height in inches	Height in centimeters

Grouping: 3 to 4 students
Topic: Plotting points, graphing lines

Part II: The numbers for inches and centimeters from the table will give you three or four ordered pairs to graph. Plot these points on a graph. Let inches be the horizontal x-axis and centimeters be the vertical y-axis. Let each mark on the axes represent 10 units. When graphing, you will need to estimate the place to plot fractional values.

Part III: Use any two of your points to find an equation of the line you have graphed. What is the slope of your line? Where does it cross the horizontal axis?

Extension: Look up the conversion formula for converting inches to centimeters. Is it the same as the one you found by measuring? If it is different, what could account for the difference?

Chapter 7 Summary   (7–59) **381**

W R A P - U P   **C H A P T E R 7**

SUMMARY

Rational Expressions

		Examples
Rational expression	The ratio of two polynomials with the denominator not equal to 0	$\frac{x-1}{x-3}\ (x \neq 3)$
Rule for reducing rational expressions	If $a \neq 0$ and $c \neq 0$, then $\frac{ab}{ac} = \frac{b}{c}.$ (Divide out the common factors.)	$\frac{8x+2}{4x} = \frac{2(4x+1)}{2(2x)} = \frac{4x+1}{2x}$
Quotient rule for exponents	Suppose $a \neq 0$ and m and n are positive integers. If $m \geq n$, then $\frac{a^m}{a^n} = a^{m-n}.$ If $m < n$, then $\frac{a^m}{a^n} = \frac{1}{a^{n-m}}.$	$\frac{x^7}{x^5} = x^2$ $\frac{x^2}{x^5} = \frac{1}{x^3}$

Multiplication and Division of Rational Expressions

		Examples
Multiplication	If $b \neq 0$ and $d \neq 0$, then $\frac{a}{b} \cdot \frac{c}{d} = \frac{ac}{bd}.$	$\frac{3}{x^3} \cdot \frac{6}{x^5} = \frac{18}{x^8}$
Division	If $b \neq 0$, $c \neq 0$, and $d \neq 0$, then $\frac{a}{b} \div \frac{c}{d} = \frac{a}{b} \cdot \frac{d}{c}.$ (Invert the divisor and multiply.)	$\frac{a}{x^3} \div \frac{5}{x^9} = \frac{a}{x^3} \cdot \frac{x^9}{5} = \frac{ax^6}{5}$

Addition and Subtraction of Rational Expressions

		Examples
Least common denominator	The LCD of a group of denominators is the smallest number that is a multiple of all of them.	8, 12 LCD = 24
Finding the least common denominator	1. Factor each denominator completely. Use exponent notation for repeated factors. 2. Write the product of all of the different factors that appear in the denominators. 3. On each factor, use the highest power that appears on that factor in any of the denominators.	$6a^2b, 4ab^3$ $4ab^3 = 2^2ab^3$ $6a^2b = 2 \cdot 3a^2b$ LCD $= 2^2 \cdot 3a^2b^3 = 12a^2b^3$

Wrap-up

Every chapter ends with a five-part **Wrap-up:** Summary, Enriching Your Mathematical Word Power, Review Exercises, Chapter Test, and Making Connections. These five items are illustrated here and on the following pages.

The **Summary** lists important concepts along with brief illustrative examples.

Chapter 8 Enriching Your Mathematical Word Power   (8–69) **461**

Linear Programming

Use the following steps to find the maximum or minimum value of a linear function subject to linear constraints.
1. Graph the region that satisfies all of the constraints.
2. Determine the coordinates of each vertex of the region.
3. Evaluate the function at each vertex of the region.
4. Identify which vertex gives the maximum or minimum value of the function.

ENRICHING YOUR MATHEMATICAL WORD POWER

For each mathematical term, choose the correct meaning.

1. system of equations
　　a. a systematic method for classifying equations
　　b. a method for solving an equation
　　c. two or more equations
　　d. the properties of equality

2. independent linear system
　　a. a system with exactly one solution
　　b. an equation that is satisfied by every real number
　　c. equations that are identical
　　d. a system of lines

3. inconsistent system
　　a. a system with no solution
　　b. a system of inconsistent equations
　　c. a system that is incorrect
　　d. a system that we are not sure how to solve

4. dependent system
　　a. a system that is independent
　　b. a system that depends on a variable
　　c. a system that has no solution
　　d. a system for which the graphs coincide

5. substitution method
　　a. replacing the variables by the correct answer
　　b. a method of eliminating a variable by substituting one equation into the other
　　c. the replacement method
　　d. any method of solving a system

6. addition method
　　a. adding the same number to each side of an equation
　　b. adding fractions
　　c. eliminating a variable by adding two equations
　　d. the sum of a number and its additive inverse is zero

7. linear equation in three variables
　　a. $Ax + By + Cz = D$ with A, B, and C not all zero
　　b. $Ax + By = C$ with A and B not both zero

　　c. the equation of a line
　　d. $A/x + B/y = C$ with A and B not both zero

8. linear inequality in two variables
　　a. when two lines are not equal
　　b. line segments that are unequal in length
　　c. an inequality of the form $Ax + By \geq C$ or with another symbol of inequality
　　d. an inequality of the form $Ax^2 + By^2 < C^2$

9. matrix
　　a. a television screen
　　b. a maze
　　c. a rectangular array of numbers
　　d. coordinates in four dimensions

10. augmented matrix
　　a. a matrix with a power booster
　　b. a matrix with no solution
　　c. a square matrix
　　d. a matrix containing the coefficients and constants of a system of equations

11. order
　　a. the length of a matrix
　　b. the number of rows and columns in a matrix
　　c. the highest power of a matrix
　　d. the lowest power of a matrix

12. determinant
　　a. a number corresponding to a square matrix
　　b. a number that is determined by any matrix
　　c. the first entry of a matrix
　　d. a number that determines whether a matrix has a solution

13. sign array
　　a. the signs of the entries of a matrix
　　b. the sign of the determinant
　　c. the signs of the answers
　　d. a matrix of $+$ and $-$ signs used in computing a determinant

---

Enriching Your Mathematical Word Power enables students to review terms introduced in each chapter. It is intended to help reinforce students' command of mathematical terminology.

Review Exercises contain problems that are keyed to each section of the chapter as well as **miscellaneous exercises,** which are not keyed to the sections. The **miscellaneous exercises** are designed to test the student's ability to synthesize various concepts.

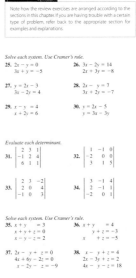

462 (8–70) Chapter 8 Systems of Linear Equations and Inequalities

REVIEW EXERCISES

8.1 *Solve by graphing. Indicate whether each system is independent, inconsistent, or dependent.*

1. $y = 2x - 1$
　　$x + y = 2$

2. $y = -x$
　　$y = -x + 3$

3. $y = 3x - 4$
　　$y = -2x + 1$

4. $x + y = 5$
　　$x - y = -1$

Solve each system by the substitution method. Indicate whether each system is independent, inconsistent, or dependent.

5. $y = 3x + 11$
　　$2x + 3y = 0$

6. $x - y = 3$
　　$3x - 2y = 3$

7. $x = y + 5$
　　$2x - 2y = 12$

8. $2x - y = 3$
　　$6x - 9 = 3y$

8.2 *Solve each system by the addition method. Indicate whether each system is independent, inconsistent, or dependent.*

9. $5x - 3y = -20$
　　$3x + 2y = 7$

10. $-3x + y = 3$
　　$2x - 3y = 5$

11. $2(y - 5) + 4 = 3(x - 6)$
　　$3x - 2y = 12$

12. $3x - 4(y - 5) = x + 2$
　　$2y - x = 7$

8.3 *Solve each system by elimination of variables.*

13. $2x - y - z = 3$
　　$3x + y + 2z = 4$
　　$4x + 2y - z = -4$

14. $2x + 3y - 2z = -11$
　　$3x - 2y + 3z = 7$
　　$x - 4y + 4z = 14$

15. $x - 3y + z = 5$
　　$2x - 4y - z = 7$
　　$2x - 6y + 2z = 6$

16. $x - y + z = 1$
　　$2x - 2y + 2z = 2$
　　$-3x + 3y - 3z = -3$

8.4 *Solve each system by using the Gaussian elimination method.*

17. $2x + y = 0$
　　$x - 3y = 14$

18. $2x - y = 8$
　　$3x + 2y = -2$

19. $x + y - z = 0$
　　$x - y + 2z = 4$
　　$2x + y - z = 1$

20. $2x - y + 2z = 9$
　　$x + 3y = 5$
　　$3x + z = 9$

8.5 *Evaluate each determinant.*

21. $\begin{vmatrix} 1 & 3 \\ 0 & 2 \end{vmatrix}$

22. $\begin{vmatrix} -1 & 2 \\ -3 & 5 \end{vmatrix}$

23. $\begin{vmatrix} 0.01 & 0.02 \\ 50 & 80 \end{vmatrix}$

24. $\begin{vmatrix} \frac{1}{2} & \frac{1}{3} \\ \frac{1}{4} & \frac{1}{5} \end{vmatrix}$

study tip

Note how the review exercises are arranged according to the sections in this chapter. If you are having trouble with a certain type of problem, refer back to the appropriate section for examples and explanations.

Solve each system. Use Cramer's rule.

25. $2x - y = 0$
　　$3x + y = -5$

26. $3x - 2y = 14$
　　$2x + 3y = -8$

27. $y = 2x - 3$
　　$3x - 2y = 4$

28. $2x - y = 7$
　　$3x + 2y = -7$

29. $x - y = 4$
　　$x + 2y = 6$

30. $y = 2x - 5$
　　$y = 3x - 3y$

Evaluate each determinant.

31. $\begin{vmatrix} 2 & 3 & 1 \\ -1 & 2 & 4 \\ 6 & 1 & 1 \end{vmatrix}$

32. $\begin{vmatrix} 1 & -1 & 0 \\ -2 & 0 & 0 \\ 3 & 1 & 5 \end{vmatrix}$

33. $\begin{vmatrix} 2 & 3 & -2 \\ 2 & 0 & 4 \\ -1 & 0 & 3 \end{vmatrix}$

34. $\begin{vmatrix} 3 & -1 & 4 \\ 2 & -1 & 1 \\ -2 & 0 & 1 \end{vmatrix}$

Solve each system. Use Cramer's rule.

35. $x + y = 3$
　　$x + y + z = 0$
　　$x - y - z = 2$

36. $x + y = 4$
　　$y + z = -3$
　　$x + z = -5$

37. $2x - y + z = 0$
　　$4x + 6y - 2z = 0$
　　$x - 2y - z = -9$

38. $x - y + z = 4$
　　$2x - 3y + z = 2$
　　$4x - y - z = 18$

The **Chapter Test** is designed to help the student assess his or her readiness for a test. The Chapter Test has no keyed exercises, which affords students an opportunity to synthesize concepts found within the chapter.

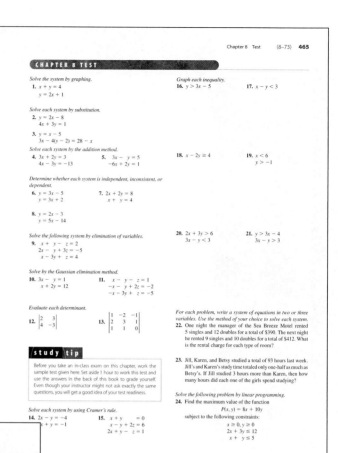

CHAPTER 8 TEST

Solve the system by graphing.

1. $x + y = 4$
$y = 2x + 1$

Solve each system by substitution.

2. $y = 2x - 8$
$4x + 3y = 1$

3. $y = x - 5$
$3x - 4(y - 2) = 28 - x$

Solve each system by the addition method.

4. $3x + 2y = 3$ **5.** $3x - y = 5$
$4x - 3y = -13$ $-6x + 2y = 1$

Determine whether each system is independent, inconsistent, or dependent.

6. $y = 3x - 5$ **7.** $2x + 2y = 8$
$y = 3x + 2$ $x + y = 4$

8. $y = 2x - 3$
$y = 5x - 14$

Solve the following system by elimination of variables.

9. $x + y - z = 2$
$2x - y + 3z = -5$
$x - 3y + z = 4$

Solve by the Gaussian elimination method.

10. $3x - y = 1$ **11.** $x - y - z = 1$
$x + 2y = 12$ $-x - y + 2z = -2$
$\qquad\qquad\qquad\qquad -x - 3y + z = -5$

Evaluate each determinant.

12. $\begin{vmatrix} 2 & 3 \\ 4 & -3 \end{vmatrix}$ **13.** $\begin{vmatrix} 1 & -2 & -1 \\ 2 & 3 & 1 \\ 1 & 1 & 0 \end{vmatrix}$

study tip

Before you take an in-class exam on this chapter, work the sample test given here. Set aside 1 hour to work this test and use the answers in the back of this book to grade yourself. Even though your instructor might not ask exactly the same questions, you will get a good idea of your test readiness.

Solve each system by using Cramer's rule.

14. $2x - y = -4$ **15.** $x + y = 0$
$x + y = -1$ $x - y + 2z = 6$
$\qquad\qquad\qquad\qquad 2x + y - z = 1$

Graph each inequality.

16. $y > 3x - 5$ **17.** $x - y < 3$

18. $x - 2y \geq 4$ **19.** $x < 6$
$\qquad\qquad\qquad\qquad y > -1$

20. $2x + 3y > 6$ **21.** $y > 3x - 4$
$3x - y < 3$ $3x - y > 3$

For each problem, write a system of equations in two or three variables. Use the method of your choice to solve each system.

22. One night the manager of the Sea Breeze Motel rented 5 singles and 12 doubles for a total of $390. The next night he rented 9 singles and 10 doubles for a total of $412. What is the rental charge for each type of room?

23. Jill, Karen, and Betsy studied a total of 93 hours last week. Jill's and Karen's study time totaled only one-half as much as Betsy's. If Jill studied 3 hours more than Karen, then how many hours did each one of the girls spend studying?

Solve the following problem by linear programming.

24. Find the maximum value of the function
$$P(x, y) = 8x + 10y$$
subject to the following constraints:
$$x \geq 0, y \geq 0$$
$$2x + 3y \leq 12$$
$$x + y \leq 5$$

MAKING CONNECTIONS   CHAPTERS 1–10

Solve each equation.

1. $2x - 15 = 0$ **2.** $2x^2 - 15 = 0$

3. $2x^2 + x - 15 = 0$

4. $2x^2 + 4x - 15 = 0$

5. $|4x + 11| = 3$

6. $|4x^2 + 11x| = 3$

7. $\sqrt{x} = x - 6$

8. $(2x - 5)^{2/3} = 4$

Solve each inequality.

9. $1 - 2x < 5 - x$

10. $(1 - 2x)(5 - x) \leq 0$

11. $\dfrac{1 - 2x}{5 - x} \leq 0$

12. $|5 - x| < 3$

13. $3x - 1 < 5$ and $-3 \leq x$

14. $x - 3 < 1$ or $2x \geq 8$

Solve each equation for y.

15. $2x - 3y = 9$

16. $\dfrac{y - 3}{x + 2} = \dfrac{1}{2}$

17. $3y^2 + cy + d = 0$

18. $my^2 - ny = w$

19. $\dfrac{1}{3}x - \dfrac{2}{5}y = \dfrac{5}{6}$

20. $y - 3 = -\dfrac{2}{3}(x - 4)$

Let $m = \dfrac{y_2 - y_1}{x_2 - x_1}$. Find the value of m for each of the following choices of $x_1, x_2, y_1,$ and y_2.

21. $x_1 = 2, x_2 = 5, y_1 = 3, y_2 = 7$

22. $x_1 = -3, x_2 = 4, y_1 = 5, y_2 = -6$

23. $x_1 = 0.3, x_2 = 0.5, y_1 = 0.8, y_2 = 0.4$

24. $x_1 = \dfrac{1}{2}, x_2 = \dfrac{1}{3}, y_1 = \dfrac{3}{5}, y_2 = -\dfrac{4}{3}$

Solve each problem.

25. *Ticket prices.* In the summer of 1994 the rock group Pearl Jam testified before a congressional committee that Ticketmaster was unfairly raising the prices of the group's concert tickets. One member of the group stated that fans should not have to pay more than $20 to see Pearl Jam. Of course, for any concert, as ticket prices rise, the number of tickets sold decreases, as shown in the figure. If you use the formula $n = 48,000 - 400p$ to predict the number sold depending on the price p, then how many will be sold at $20 per ticket? How many will be sold at $25 per ticket? Use the bar graph to estimate the price if 35,000 tickets were sold.

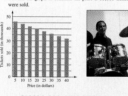

FIGURE FOR EXERCISE 25

26. *Increasing revenue.* Even though the number of tickets sold for a concert decreases with increasing price, the revenue generated does not necessarily decrease. Use the formula $R = p(48,000 - 400p)$ to determine the revenue when the price is $20 and when the price is $25. What price would produce a revenue of $1.28 million? Use the graph to find the price that determines the maximum revenue.

FIGURE FOR EXERCISE 26

Making Connections is a set of non-keyed exercises designed to help students synthesize new material with ideas from previous chapters and, in some cases, review material necessary for success in the upcoming chapter. They may serve as a cumulative review.

Real Numbers and Their Properties

I t has been said that baseball is the "great American pastime." All of us who have played the game or who have only been spectators believe we understand the game. But do we realize that a pitcher must aim for an invisible three-dimensional target that is about 20 inches wide by 23 inches high by 17 inches deep and that a pitcher must throw so that the batter has difficulty hitting the ball? A curve ball may deflect 14 inches to skim over the outside corner of the plate, or a knuckle ball can break 11 inches off center when it is 20 feet from the plate and then curve back over the center of the plate.

The batter is trying to hit a rotating ball that can travel up to 120 miles per hour and must make split-second decisions about shifting his weight, changing his stride, and swinging the bat. The size of the bat each batter uses depends on his strengths, and pitchers in turn try to capitalize on a batter's weaknesses.

Millions of baseball fans enjoy watching this game of strategy and numbers. Many watch their favorite teams at the local ball parks, while others cheer for the home team on television. Of course, baseball fans are always interested in which team is leading the division and the number of games that their favorite team is behind the leader. Finding the number of games behind for each team in the division involves both arithmetic and algebra. Algebra provides the formula for finding games behind, and arithmetic is used to do the computations. In Exercise 93 of Section 1.6 we will find the number of games behind for each team in the American League West.

1.1   THE REAL NUMBERS

In arithmetic we use only positive numbers and zero, but in algebra we use negative numbers also. The numbers that we use in algebra are called the real numbers. We start the discussion of the real numbers with some simpler sets of numbers.

The Integers

The most fundamental collection or **set** of numbers is the set of **counting numbers** or **natural numbers.** Of course, these are the numbers that we use for counting. The set of natural numbers is written in symbols as follows.

The Natural Numbers

$$\{1, 2, 3, \ldots\}$$

Braces, { }, are used to indicate a set of numbers. The three dots after 1, 2, and 3, which are read "and so on," mean that the pattern continues without end. There are infinitely many natural numbers.

The natural numbers, together with the number 0, are called the **whole numbers.** The set of whole numbers is written as follows.

The Whole Numbers

$$\{0, 1, 2, 3, \ldots\}$$

FIGURE 1.1

Although the whole numbers have many uses, they are not adequate for indicating losses or debts. A debt of $20 can be expressed by the negative number -20 (negative twenty). See Fig. 1.1. When a thermometer reads 10 degrees below zero on a Fahrenheit scale, we say that the temperature is $-10°$F. See Fig. 1.2. The whole numbers together with the negatives of the counting numbers form the set of **integers.**

The Integers

$$\{\ldots, -3, -2, -1, 0, 1, 2, 3, \ldots\}$$

The Rational Numbers

A **rational number** is any number that can be expressed as a ratio (or quotient) of two integers. The set of rational numbers includes both the positive and negative fractions. We cannot list the rational numbers as easily as we listed the numbers in the other sets we have been discussing. So we write the set of rational numbers in symbols using **set-builder notation** as follows.

Degrees
Fahrenheit

FIGURE 1.2

The Rational Numbers

$$\left\{ \frac{a}{b} \,\middle|\, a \text{ and } b \text{ are integers, with } b \neq 0 \right\}$$

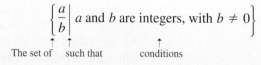

The set of   such that   conditions

helpful hint

Rational numbers are used for ratios. For example, if 2 out of 5 students surveyed attend summer school, then the ratio of students who attend summer school to the total number surveyed is 2/5. Note that the ratio 2/5 does not tell how many were surveyed or how many attend summer school.

We read this notation as "the set of numbers of the form $\frac{a}{b}$ such that a and b are integers, with $b \neq 0$." Note how we use the letters a and b to represent numbers here. A letter used to represent some numbers is called a **variable.**

Examples of rational numbers are

$$\frac{3}{1}, \quad \frac{5}{4}, \quad -\frac{7}{10}, \quad \frac{0}{6}, \quad \frac{5}{1}, \quad -\frac{77}{3}, \quad \text{and} \quad \frac{-3}{-6}.$$

Note that we usually use simpler forms for some of these rational numbers. For instance, $\frac{3}{1} = 3$ and $\frac{0}{6} = 0$. The integers are rational numbers because any integer can be written with a denominator of 1.

If you divide the denominator into the numerator, then you can convert a rational number to decimal form. As a decimal, every rational number either repeats indefinitely $\left(\text{for example, } \frac{1}{3} = 0.333 \ldots\right)$ or terminates $\left(\text{for example, } \frac{1}{8} = 0.125\right)$.

The Number Line

The number line is a diagram that helps us to visualize numbers and their relationships to each other. A number line is like the scale on the thermometer in Fig. 1.2. To construct a number line, we draw a straight line and label any convenient point with the number 0. Now we choose any convenient length and use it to locate other points. Points to the right of 0 correspond to the positive numbers, and points to the left of 0 correspond to the negative numbers. The number line is shown in Fig. 1.3.

FIGURE 1.3

The numbers corresponding to the points on the line are called the **coordinates** of the points. The distance between two consecutive integers is called a **unit** and is the same for any two consecutive integers. The point with coordinate 0 is called the **origin.** The numbers on the number line increase in size from left to right. *When we compare the size of any two numbers, the larger number lies to the right of the smaller on the number line.*

EXAMPLE 1

Comparing numbers on a number line

Determine which number is the larger in each given pair of numbers.

a) $-3, 2$ **b)** $0, -4$ **c)** $-2, -1$

Solution

a) The larger number is 2, because 2 lies to the right of -3 on the number line. In fact, any positive number is larger than any negative number.

b) The larger number is 0, because 0 lies to the right of -4 on the number line.

c) The larger number is -1, because -1 lies to the right of -2 on the number line.

The set of integers is illustrated or **graphed** in Fig. 1.4 by drawing a point for each integer. The three dots to the right and left below the number line and the blue arrows indicate that the numbers go on indefinitely in both directions.

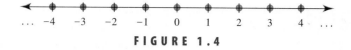

FIGURE 1.4

E X A M P L E   2

Graphing numbers on a number line

List the numbers described, and graph the numbers on a number line.

a) The whole numbers less than 4

b) The integers between 3 and 9

c) The integers greater than −3

Solution

a) The whole numbers less than 4 are 0, 1, 2, and 3. These numbers are shown in Fig. 1.5.

FIGURE 1.5

b) The integers between 3 and 9 are 4, 5, 6, 7, and 8. Note that 3 and 9 are not considered to be *between* 3 and 9. The graph is shown in Fig. 1.6.

FIGURE 1.6

c) The integers greater than −3 are −2, −1, 0, 1, and so on. To indicate the continuing pattern, we use three dots on the graph shown in Fig. 1.7.

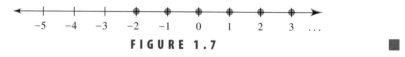

FIGURE 1.7

The Real Numbers

For every rational number there is a point on the number line. For example, the number $\frac{1}{2}$ corresponds to a point halfway between 0 and 1 on the number line, and $-\frac{5}{4}$ corresponds to a point one and one-quarter units to the left of 0, as shown in Fig. 1.8. Since there is a correspondence between numbers and points on the number line, the points are often referred to as numbers.

FIGURE 1.8

The set of numbers that corresponds to *all* points on a number line is called the set of **real numbers.** A graph of the real numbers is shown on a number line by shading all points as in Fig. 1.9. All rational numbers are real numbers, but there are

FIGURE 1.9

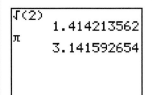

points on the number line that do not correspond to rational numbers. Those real numbers that are not rational are called **irrational.** An irrational number cannot be written as a ratio of integers. It can be shown that numbers such as $\sqrt{2}$ (the square root of 2) and π (Greek letter pi) are irrational. The number $\sqrt{2}$ is a number that can be multiplied by itself to obtain 2 ($\sqrt{2} \cdot \sqrt{2} = 2$). The number π is the ratio of the circumference and diameter of any circle. Irrational numbers are not as easy to represent as rational numbers. That is why we use symbols such as $\sqrt{2}$, $\sqrt{3}$, and π for irrational numbers. When we perform computations with irrational numbers, we use rational approximations for them. For example, $\sqrt{2} \approx 1.414$ and $\pi \approx 3.14$. The symbol $\approx$ means "is approximately equal to." Note that not all square roots are irrational. For example, $\sqrt{9} = 3$, because $3 \cdot 3 = 9$. We will deal with irrational numbers in greater depth when we discuss roots in Chapter 9.

Figure 1.10 summarizes the sets of numbers that make up the real numbers, and shows the relationships between them.

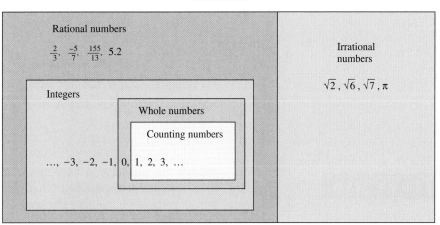

FIGURE 1.10

EXAMPLE 3

Types of numbers

Determine whether each statement is true or false.

a) Every rational number is an integer.

b) Every counting number is an integer.

c) Every irrational number is a real number.

Solution

a) False. For example, $\frac{1}{2}$ is a rational number that is not an integer.

b) True, because the integers consist of the counting numbers, the negatives of the counting numbers, and zero.

c) True, because the rational numbers together with the irrational numbers form the real numbers. ■

Absolute Value

The concept of absolute value will be used to define the basic operations with real numbers in Section 1.3. The **absolute value** of a number is the number's distance from 0 on the number line. For example, the numbers 5 and -5 are both five units away from 0 on the number line. So the absolute value of each of these numbers

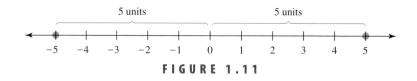

FIGURE 1.11

is 5. See Fig. 1.11. We write $|a|$ for "the absolute value of a." So

$$|5| = 5 \qquad \text{and} \qquad |-5| = 5.$$

The notation $|a|$ represents distance, and distance is never negative. So $|a|$ is greater than or equal to zero for any real number a.

E X A M P L E   4

Finding absolute value

Evaluate.

a) $|3|$ **b)** $|-3|$ **c)** $|0|$ **d)** $\left|\dfrac{2}{3}\right|$ **e)** $|-0.39|$

Solution

a) $|3| = 3$ because 3 is three units away from 0.

b) $|-3| = 3$ because -3 is three units away from 0.

c) $|0| = 0$ because 0 is zero units away from 0.

d) $\left|\dfrac{2}{3}\right| = \dfrac{2}{3}$ **e)** $|-0.39| = 0.39$ ■

    Two numbers that are located on opposite sides of zero and have the same absolute value are called **opposites** of each other. The numbers 5 and -5 are opposites of each other. We say that the opposite of 5 is -5 and the opposite of -5 is 5. The symbol "$-$" is used to indicate "opposite" as well as "negative." When the negative sign is used before a number, it should be read as "negative." When it is used in front of parentheses or a variable, it should be read as "opposite." For example, $-(5) = -5$ means "the opposite of 5 is negative 5," and $-(-5) = 5$ means "the opposite of negative 5 is 5." Zero does not have an opposite in the same sense as nonzero numbers. Zero is its own opposite. We read $-(0) = 0$ as the "the opposite of zero is zero."

    In general, $-a$ means "the opposite of a." If a is positive, $-a$ is negative. If a is negative, $-a$ is positive. Opposites have the following property.

Opposite of an Opposite

For any real number a,

$$-(-a) = a.$$

    Remember that we have defined $|a|$ to be the distance between 0 and a on the number line. Using opposites, we can give a symbolic definition of absolute value.

Absolute Value

$$|a| = \begin{cases} a & \text{if } a \text{ is positive or zero} \\ -a & \text{if } a \text{ is negative} \end{cases}$$

E X A M P L E   5   **Using the symbolic definition of absolute value**

Evaluate.

a) $|8|$

b) $|0|$

c) $|-8|$

Solution

a) If a is positive, then $|a| = a$. Since $8 > 0$, $|8| = 8$.

b) If a is 0, then $|a| = a$. So $|0| = 0$.

c) If a is negative, then $|a| = -a$. So $|-8| = -(-8) = 8$. ∎

WARM-UPS

True or false? Explain your answer.

1. The natural numbers and the counting numbers are the same.

2. The number 8,134,562,877,565 is a counting number.

3. Zero is a counting number.

4. Zero is not a rational number.

5. The opposite of negative 3 is positive 3.

6. The absolute value of 4 is -4.

7. $-(-9) = 9$

8. $-(-b) = b$ for any number b.

9. Negative six is greater than negative three.

10. Negative five is between four and six.

1.1   EXERCISES

Reading and Writing   *After reading this section write out the answers to these questions. Use complete sentences.*

1. What are the integers?

2. What are the rational numbers?

3. What is the difference between a rational and an irrational number?

4. What is a number line?

5. How do you know that one number is larger than another?

6. What is the ratio of the circumference and diameter of any circle?

Determine which number is the larger in each given pair of numbers. See Example 1.

7. $-3, 6$

8. $7, -10$

9. $0, -6$

10. $-8, 0$

11. $-3, -2$

12. $-5, -8$

13. $-12, -15$

14. $-13, -7$

List the numbers described and graph them on a number line. See Example 2.

15. The counting numbers smaller than 6

16. The natural numbers larger than 4

17. The whole numbers smaller than 5

18. The integers between -3 and 3

19. The whole numbers between -5 and 5

20. The integers smaller than -1

21. The counting numbers larger than -4

22. The natural numbers between -5 and 7

23. The integers larger than $\frac{1}{2}$

24. The whole numbers smaller than $\frac{7}{4}$

Determine whether each statement is true or false. Explain your answer. See Example 3.

25. Every integer is a rational number.

26. Every counting number is a whole number.

27. Zero is a counting number.

28. Every whole number is a counting number.

29. The ratio of the circumference and diameter of a circle is an irrational number.

30. Every rational number can be expressed as a ratio of integers.

31. Every whole number can be expressed as a ratio of integers.

32. Some of the rational numbers are integers.

33. Some of the integers are natural numbers.

34. There are infinitely many rational numbers.

35. Zero is an irrational number.

36. Every irrational number is a real number.

Determine the values of the following. See Examples 4 and 5.

37. $|-6|$ **38.** $|4|$ **39.** $|0|$

40. $|2|$ **41.** $|7|$ **42.** $|-7|$

43. $|-9|$ **44.** $|-2|$ **45.** $|-45|$

46. $|-30|$ **47.** $\left|\dfrac{3}{4}\right|$ **48.** $\left|-\dfrac{1}{2}\right|$

49. $|-5.09|$ **50.** $|0.00987|$

Select the smaller number in each given pair of numbers.

51. $-16, 9$ **52.** $-12, -7$

53. $-\dfrac{5}{2}, -\dfrac{9}{4}$ **54.** $\dfrac{5}{8}, \dfrac{6}{7}$

55. $|-3|, 2$ **56.** $|-6|, 0$

57. $|-4|, 3$ **58.** $|5|, -4$

Which number in each given pair has the larger absolute value?

59. $-5, -9$ **60.** $-12, -8$ **61.** $16, -9$ **62.** $-12, 7$

True or false? Explain your answer.

63. If we add the absolute values of -3 and -5, we get 8.

64. If we multiply the absolute values of -2 and 5, we get 10.

65. The absolute value of any negative number is greater than 0.

66. The absolute value of any positive number is less than 0.

67. The absolute value of -9 is larger than the absolute value of 6.

68. The absolute value of 12 is larger than the absolute value of -11.

GETTING MORE INVOLVED

69. *Writing.* Find a real-life question for which the answer is a rational number that is not an integer.

70. *Exploration.* a) Find a rational number between $\frac{1}{3}$ and $\frac{1}{4}$.

b) Find a rational number between -3.205 and -3.114.

c) Find a rational number between $\frac{2}{3}$ and 0.6667.

d) Explain how to find a rational number between any two given rational numbers.

71. *Discussion.* Suppose that a is a negative real number. Determine whether each of the following is positive or negative, and explain your answer.

a) $-a$ b) $|-a|$ c) $-|a|$ d) $-(-a)$ e) $-|-a|$

1.2 FRACTIONS

In this section and the next two, we will discuss operations performed with real numbers. We begin by reviewing the operations with fractions.

Equivalent Fractions

If a pizza is cut into 3 equal pieces and you eat 2, you have eaten $\frac{2}{3}$ of the pizza. If the pizza is cut into 6 equal pieces and you eat 4, you have still eaten 2 out of every 3 pieces. So the fraction $\frac{4}{6}$ is considered **equal** or **equivalent** to $\frac{2}{3}$. See Fig. 1.12. Every fraction can be written in infinitely many equivalent forms. Consider the following equivalent forms of $\frac{2}{3}$:

$$\frac{2}{3} = \frac{4}{6} = \frac{6}{9} = \frac{8}{12} = \frac{10}{15} = \cdots$$

The three dots mean "and so on."

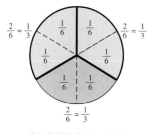

$$\frac{2}{6} = \frac{1}{3}$$

FIGURE 1.12

Notice that each equivalent form of $\frac{2}{3}$ can be obtained by multiplying the numerator (top number) and denominator (bottom number) of $\frac{2}{3}$ by a nonzero number. For example,

$$\frac{2}{3} = \frac{2 \cdot 5}{3 \cdot 5} = \frac{10}{15}.$$ The raised dot indicates multiplication.

Converting a fraction into an equivalent fraction with a larger denominator is called **building up** the fraction.

> **Building Up Fractions**
>
> If $b \neq 0$ and $c \neq 0$, then
>
> $$\frac{a}{b} = \frac{a \cdot c}{b \cdot c}.$$

Multiplying the numerator and denominator of a fraction by a nonzero number changes the fraction's appearance but not its value.

EXAMPLE 1

Building up fractions

Build up each fraction so that it is equivalent to the fraction with the indicated denominator.

a) $\dfrac{3}{4} = \dfrac{?}{28}$ **b)** $\dfrac{5}{3} = \dfrac{?}{30}$

helpful hint

In algebra it is best to build up fractions by multiplying both the numerator and denominator by the same number as shown in Example 1. So if you use an old method, be sure to learn this method.

Solution

a) Because $4 \cdot 7 = 28$, we multiply both the numerator and denominator by 7:

$$\frac{3}{4} = \frac{3 \cdot 7}{4 \cdot 7} = \frac{21}{28}$$

b) Because $3 \cdot 10 = 30$, we multiply both the numerator and denominator by 10:

$$\frac{5}{3} = \frac{5 \cdot 10}{3 \cdot 10} = \frac{50}{30}$$

Converting a fraction to an equivalent fraction with a smaller denominator is called **reducing** the fraction. For example, to reduce $\frac{10}{15}$, we *factor* 10 as $2 \cdot 5$ and 15 as $3 \cdot 5$, and then divide out the *common factor* 5:

$$\frac{10}{15} = \frac{2 \cdot 5}{3 \cdot 5} = \frac{2}{3}$$

The fraction $\frac{2}{3}$ cannot be reduced further because the numerator 2 and the denominator 3 have no factors (other than 1) in common. So we say that $\frac{2}{3}$ is in **lowest terms.**

Reducing Fractions

If $b \neq 0$ and $c \neq 0$, then

$$\frac{a \cdot c}{b \cdot c} = \frac{a}{b}.$$

Dividing the numerator and denominator of a fraction by a nonzero number changes the fraction's appearance but not its value.

EXAMPLE 2   **Reducing fractions**

Reduce each fraction to lowest terms.

a) $\dfrac{15}{24}$ 　　　　　　b) $\dfrac{42}{30}$

Solution

For each fraction, factor the numerator and denominator and then divide by the common factor:

a) $\dfrac{15}{24} = \dfrac{3 \cdot 5}{3 \cdot 8} = \dfrac{5}{8}$ 　　b) $\dfrac{42}{30} = \dfrac{7 \cdot 6}{5 \cdot 6} = \dfrac{7}{5}$ ■

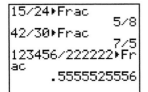

> **Strategy for Obtaining Equivalent Fractions**
>
> Equivalent fractions can be obtained by multiplying or dividing the numerator and denominator by the same nonzero number.

Multiplying Fractions

Suppose a pizza is cut into three equal pieces. If you eat $\frac{1}{2}$ of one piece, you have eaten $\frac{1}{6}$ of the pizza. See Fig. 1.13. You can obtain $\frac{1}{6}$ by multiplying $\frac{1}{2}$ and $\frac{1}{3}$:

$$\frac{1}{2} \cdot \frac{1}{3} = \frac{1 \cdot 1}{2 \cdot 3} = \frac{1}{6}$$

This example illustrates the definition of multiplication of fractions. To multiply two fractions, we multiply their numerators and multiply their denominators.

Multiplication of Fractions

If $b \neq 0$ and $d \neq 0$, then

$$\frac{a}{b} \cdot \frac{c}{d} = \frac{a \cdot c}{b \cdot d}.$$

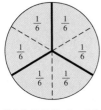

FIGURE 1.13

E X A M P L E   3

Multiplying fractions

Find the product, $\frac{2}{3} \cdot \frac{5}{8}$.

Solution

Multiply the numerators and the denominators:

$$\frac{2}{3} \cdot \frac{5}{8} = \frac{10}{24}$$

$$= \frac{2 \cdot 5}{2 \cdot 12} \qquad \text{Factor the numerator and denominator.}$$

$$= \frac{5}{12} \qquad \text{Divide out the common factor 2.} \qquad \blacksquare$$

It is usually easier to reduce before multiplying, as shown in the next example.

E X A M P L E   4

Reducing before multiplying

Find the indicated products.

a) $\frac{1}{3} \cdot \frac{3}{4}$ **b)** $\frac{4}{5} \cdot \frac{15}{22}$

calculator

close-up

A graphing calculator can multiply fractions and get fractional answers using the fraction feature. Note how a mixed number is written on a graphing calculator.

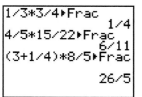

Solution

a) $\frac{1}{3} \cdot \frac{3}{4} = \frac{1}{3} \cdot \frac{3}{4} = \frac{1}{4}$

b) Factor the numerators and denominators, and then divide out the common factors before multiplying:

$$\frac{4}{5} \cdot \frac{15}{22} = \frac{2 \cdot 2}{5} \cdot \frac{3 \cdot 5}{2 \cdot 11} = \frac{6}{11} \qquad \blacksquare$$

Dividing Fractions

Again consider a pizza that is cut into three equal pieces. If one piece is divided among two people $\left(\frac{1}{3} \div 2\right)$, then each person gets $\frac{1}{6}$ of the pizza. Of course $\frac{1}{2}$ of $\frac{1}{3}$ is also $\frac{1}{6}$. So

$$\frac{1}{3} \div 2 = \frac{1}{3} \cdot \frac{1}{2} = \frac{1}{6}.$$

If $a \div b = c$, then b is called the **divisor** and c is called the **quotient** of a and b. We also refer to both $a \div b$ and $\frac{a}{b}$ as the quotient of a and b. To find the quotient for two fractions, we invert the divisor and multiply.

Division of Fractions

If $b \neq 0$ and $c \neq 0$ and $d \neq 0$, then

$$\frac{a}{b} \div \frac{c}{d} = \frac{a}{b} \cdot \frac{d}{c}.$$

EXAMPLE 5

Dividing fractions

Find the indicated quotients.

a) $\dfrac{1}{3} \div \dfrac{7}{6}$ **b)** $\dfrac{2}{3} \div 5$

Solution

In each case we invert the divisor (the number on the right) and multiply.

a) $\dfrac{1}{3} \div \dfrac{7}{6} = \dfrac{1}{3} \cdot \dfrac{6}{7}$ Invert the divisor.

$= \dfrac{1}{3} \cdot \dfrac{2 \cdot 3}{7}$ Reduce.

$= \dfrac{2}{7}$ Multiply.

b) $\dfrac{2}{3} \div 5 = \dfrac{2}{3} \div \dfrac{5}{1} = \dfrac{2}{3} \cdot \dfrac{1}{5} = \dfrac{2}{15}$ ■

calculator

close-up

When the divisor is a fraction on a graphing calculator, it must be in parentheses. A different result is obtained without using parentheses. Note that when the divisor is a whole number, parentheses are not necessary.

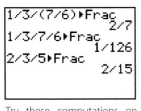

```
1/3/(7/6)▶Frac
                 2/7
1/3/7/6▶Frac
               1/126
2/3/5▶Frac
                2/15
```

Try these computations on your calculator.

Adding and Subtracting Fractions

To understand addition and subtraction of fractions, again consider the pizza that is cut into six equal pieces as shown in Fig. 1.14. If you eat $\frac{3}{6}$ and your friend eats $\frac{2}{6}$, together you have eaten $\frac{5}{6}$ of the pizza. Similarly, if you remove $\frac{1}{6}$ from $\frac{6}{6}$ you have $\frac{5}{6}$ left. To add or subtract fractions with identical denominators, we add or subtract their numerators and write the result over the common denominator.

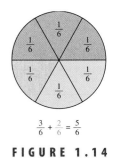

$\dfrac{3}{6} + \dfrac{2}{6} = \dfrac{5}{6}$

FIGURE 1.14

Addition and Subtraction of Fractions

If $b \ne 0$, then

$$\frac{a}{b} + \frac{c}{b} = \frac{a + c}{b} \quad \text{and} \quad \frac{a}{b} - \frac{c}{b} = \frac{a - c}{b}.$$

EXAMPLE 6

Adding and subtracting fractions

Perform the indicated operations.

a) $\dfrac{1}{7} + \dfrac{2}{7}$ **b)** $\dfrac{7}{10} - \dfrac{3}{10}$

helpful hint

A good way to remember that you need common denominators for addition is to think of a simple example. If you own 1/3 share of a car wash and your spouse owns 1/3, then together you own 2/3 of the business.

Solution

a) $\dfrac{1}{7} + \dfrac{2}{7} = \dfrac{3}{7}$ **b)** $\dfrac{7}{10} - \dfrac{3}{10} = \dfrac{4}{10} = \dfrac{2 \cdot 2}{2 \cdot 5} = \dfrac{2}{5}$ ■

If the fractions have different denominators, we must convert them to equivalent fractions with the same denominator and then add or subtract. For example, to add the fractions $\frac{1}{2}$ and $\frac{1}{3}$, we build up each fraction to get a denominator of 6. See

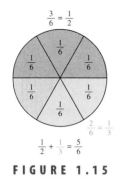

FIGURE 1.15

Fig. 1.15. The denominator 6 is the smallest number that is a multiple of both 2 and 3. For this reason, 6 is called the **least common denominator (LCD).** To find the LCD, use the following strategy.

> **Strategy for Finding the LCD**
>
> 1. Make a list of all multiples of one of the denominators.
> 2. The first number on the list that is evenly divisible by the other denominator is the LCD.

For example, for $\frac{1}{6}$ and $\frac{1}{8}$ consider all multiples of 6:

$$6, 12, 18, 24, 30, 36, \ldots$$

The first number in this list divisible by 8 is 24. So the LCD is 24. This method works well if the denominators are not too large. In Chapter 7 we will learn another method that is better suited for large numbers and algebra.

E X A M P L E   7   **Adding fractions**

Perform the indicated operations.

a) $\dfrac{1}{2} + \dfrac{1}{3}$ **b)** $\dfrac{1}{3} - \dfrac{1}{12}$

c) $\dfrac{3}{4} - \dfrac{1}{6}$ **d)** $2\dfrac{1}{3} + \dfrac{5}{9}$

Solution

a) In the multiples of 2 (2, 4, 6, 8, . . .), the first number divisible by 3 is 6. So 6 is the LCD.

$$\frac{1}{2} + \frac{1}{3} = \frac{1 \cdot 3}{2 \cdot 3} + \frac{1 \cdot 2}{3 \cdot 2} \quad \text{The LCD is 6.}$$

$$= \frac{3}{6} + \frac{2}{6} \qquad \text{Build each denominator to a denominator of 6.}$$

$$= \frac{5}{6} \qquad \text{Then add.}$$

b) In the multiples of 3 (3, 6, 9, 12, 15, . . .), the first number divisible by 12 is 12. So the LCD is 12.

$$\frac{1}{3} - \frac{1}{12} = \frac{1 \cdot 4}{3 \cdot 4} - \frac{1}{12} \quad \text{The LCD is 12.}$$

$$= \frac{4}{12} - \frac{1}{12} \qquad \text{Build up } \frac{1}{3} \text{ to get a denominator of 12.}$$

$$= \frac{3}{12} \qquad \text{Subtract.}$$

$$= \frac{1}{4} \qquad \text{Reduce to lowest terms.}$$

c) In the multiples of 4 (4, 8, 12, 16, . . .), the first number divisible by 6 is 12. So 12 is the LCD, the smallest multiple of 4 and 6.

$$\frac{3}{4} - \frac{1}{6} = \frac{3 \cdot 3}{4 \cdot 3} - \frac{1 \cdot 2}{6 \cdot 2} \qquad \text{The LCD is 12.}$$

$$= \frac{9}{12} - \frac{2}{12} \qquad \text{Build up each fraction to a denominator of 12.}$$

$$= \frac{7}{12}$$

d) To perform addition with the mixed number $2\frac{1}{3}$, first convert it into an improper fraction: $2\frac{1}{3} = 2 + \frac{1}{3} = \frac{6}{3} + \frac{1}{3} = \frac{7}{3}$.

$$2\frac{1}{3} + \frac{5}{9} = \frac{7}{3} + \frac{5}{9} \qquad \text{Write } 2\frac{1}{3} \text{ as an improper fraction.}$$

$$= \frac{7 \cdot 3}{3 \cdot 3} + \frac{5}{9} \qquad \text{The LCD is 9.}$$

$$= \frac{21}{9} + \frac{5}{9} = \frac{26}{9}$$

Fractions, Decimals, and Percents

In the decimal number system, fractions with a denominator of 10, 100, 1000, and so on are written as decimal numbers. For example,

$$\frac{3}{10} = 0.3, \qquad \frac{25}{100} = 0.25, \qquad \text{and} \qquad \frac{5}{1000} = 0.005.$$

Fractions with a denominator of 100 are often written as percents. Think of the percent symbol (%) as representing the denominator of 100. For example,

$$\frac{25}{100} = 25\%, \qquad \frac{5}{100} = 5\%, \qquad \text{and} \qquad \frac{300}{100} = 300\%.$$

The next example illustrates further how to convert from any one of the forms (fraction, decimal, percent) to the others.

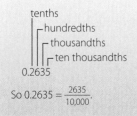
EXAMPLE 8

Changing forms

Convert each given fraction, decimal, or percent into its other two forms.

a) $\dfrac{1}{5}$ **b)** 6%   **c)** 0.1

Solution

a) $\dfrac{1}{5} = \dfrac{1 \cdot 20}{5 \cdot 20} = \dfrac{20}{100} = 20\%$ and $\dfrac{1}{5} = \dfrac{1 \cdot 2}{5 \cdot 2} = \dfrac{2}{10} = 0.2$

So $\dfrac{1}{5} = 0.2 = 20\%$.

b) $6\% = \dfrac{6}{100} = 0.06$ and $\dfrac{6}{100} = \dfrac{2 \cdot 3}{2 \cdot 50} = \dfrac{3}{50}$

So $6\% = 0.06 = \dfrac{3}{50}$.

c) $0.1 = \dfrac{1}{10} = \dfrac{1 \cdot 10}{10 \cdot 10} = \dfrac{10}{100} = 10\%$

So $0.1 = \dfrac{1}{10} = 10\%$.

calculator   close-up

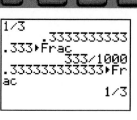

A calculator can convert fractions to decimals and decimals to fractions. The calculator shown here converts the terminating decimal 0.333333333333 into 1/3 even though 1/3 is a repeating decimal with infinitely many threes after the decimal point.

Applications

The dimensions for lumber used in construction are usually given in fractions. For example, a 2×4 stud used for framing a wall is actually $1\frac{1}{2}$ in. by $3\frac{1}{2}$ in. by $92\frac{5}{8}$ in. A 2×12 floor joist is actually $1\frac{1}{2}$ in. by $11\frac{1}{2}$ in.

E X A M P L E   9

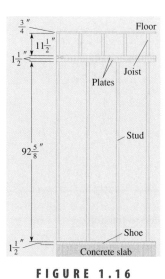

FIGURE 1.16

Framing a two-story house

In framing a two-story house, a carpenter uses a 2×4 shoe, a wall stud, two 2×4 plates, then 2×12 floor joists, and a $\frac{3}{4}$-in. plywood floor, before starting the second level. Use the dimensions in Fig. 1.16 to find the total height of the framing shown.

Solution

We can find the total height using multiplication and addition:

$$3 \cdot 1\frac{1}{2} + 92\frac{5}{8} + 11\frac{1}{2} + \frac{3}{4} = 4\frac{1}{2} + 92\frac{5}{8} + 11\frac{1}{2} + \frac{3}{4}$$

$$= 4\frac{4}{8} + 92\frac{5}{8} + 11\frac{4}{8} + \frac{6}{8}$$

$$= 107\frac{19}{8}$$

$$= 109\frac{3}{8}$$

The total height of the framing shown is $109\frac{3}{8}$ in.

M A T H   A T   W O R K $x^2 + (x+1)^2 = 5^2$

Building a new house can be a complicated and daunting task. Shirley Zaborowski, project manager for Court Construction, is responsible for estimating, pricing, negotiating, subcontracting, and scheduling all portions of new house construction.

BUILDING CONTRACTOR

Ms. Zaborowski works from drawings and first does a "take off" or estimate for the quantity of material needed. The quantity of concrete is measured in cubic yards and the amount of wood is measured in board feet. If masonry is being used, it is measured in bricks or blocks per square foot.

Scheduling is another important part of the project manager's responsibility and it is based on the take off. Certain industry standards help Ms. Zaborowski estimate how many carpenters are needed and how much time it takes to frame the house and how many electricians and plumbers are needed to wire the house, install the heating systems, and put in the bathrooms. Of course, common sense says that the foundation is done before the framing and the roof. However, some rough plumbing and electrical work can be done simultaneously with the framing. Ideally the estimates of time and cost are accurate and the homeowner can move in on schedule.

In Exercise 103 of this section you will use operations with fractions to find the volume of concrete needed to construct a rectangular patio.

WARM-UPS

True or false? Explain your answer.

1. Every fraction is equal to infinitely many equivalent fractions.
2. The fraction $\frac{8}{12}$ is equivalent to the fraction $\frac{4}{6}$.
3. The fraction $\frac{8}{12}$ reduced to lowest terms is $\frac{4}{6}$.

4. $\frac{1}{2} \cdot \frac{2}{3} = \frac{1}{3}$

5. $\frac{1}{2} \cdot \frac{3}{5} = \frac{3}{10}$

6. $\frac{1}{2} \cdot \frac{6}{5} = \frac{6}{10}$

7. $\frac{1}{2} \div 3 = \frac{1}{6}$

8. $5 \div \frac{1}{2} = 10$

9. $\frac{1}{2} + \frac{1}{4} = \frac{2}{6}$

10. $2 - \frac{1}{2} = \frac{3}{2}$

1.2   EXERCISES

Reading and Writing *After reading this section write out the answers to these questions. Use complete sentences.*

1. What are equivalent fractions?

2. How can you find all fractions that are equivalent to a given fraction?

3. What does it mean to reduce a fraction?

4. For which operations with fractions are you required to have common denominators? Why?

5. How do you convert a fraction to a decimal?

6. How do you convert a percent to a fraction?

Build up each fraction or whole number so that it is equivalent to the fraction with the indicated denominator. See Example 1.

7. $\dfrac{3}{4} = \dfrac{?}{8}$ **8.** $\dfrac{5}{7} = \dfrac{?}{21}$ **9.** $\dfrac{8}{3} = \dfrac{?}{12}$

10. $\dfrac{7}{2} = \dfrac{?}{8}$ **11.** $5 = \dfrac{?}{2}$ **12.** $9 = \dfrac{?}{3}$

13. $\dfrac{3}{4} = \dfrac{?}{100}$ **14.** $\dfrac{1}{2} = \dfrac{?}{100}$ **15.** $\dfrac{3}{10} = \dfrac{?}{100}$

16. $\dfrac{2}{5} = \dfrac{?}{100}$ **17.** $\dfrac{5}{3} = \dfrac{?}{42}$ **18.** $\dfrac{5}{7} = \dfrac{?}{98}$

Reduce each fraction to lowest terms. See Example 2.

19. $\dfrac{3}{6}$ **20.** $\dfrac{2}{10}$ **21.** $\dfrac{12}{18}$ **22.** $\dfrac{30}{40}$

23. $\dfrac{15}{5}$ **24.** $\dfrac{39}{13}$ **25.** $\dfrac{50}{100}$ **26.** $\dfrac{5}{1000}$

27. $\dfrac{200}{100}$ **28.** $\dfrac{125}{100}$ **29.** $\dfrac{18}{48}$ **30.** $\dfrac{34}{102}$

31. $\dfrac{26}{42}$ **32.** $\dfrac{70}{112}$ **33.** $\dfrac{84}{91}$ **34.** $\dfrac{121}{132}$

Find each product. See Examples 3 and 4.

35. $\dfrac{2}{3} \cdot \dfrac{5}{9}$ **36.** $\dfrac{1}{8} \cdot \dfrac{1}{8}$ **37.** $\dfrac{1}{3} \cdot 15$

38. $\dfrac{1}{4} \cdot 16$ **39.** $\dfrac{3}{4} \cdot \dfrac{14}{15}$ **40.** $\dfrac{5}{8} \cdot \dfrac{12}{35}$

41. $\dfrac{2}{5} \cdot \dfrac{35}{26}$ **42.** $\dfrac{3}{10} \cdot \dfrac{20}{21}$ **43.** $\dfrac{1}{2} \cdot \dfrac{6}{5}$

44. $\dfrac{1}{2} \cdot \dfrac{3}{5}$ **45.** $\dfrac{1}{2} \cdot \dfrac{1}{3}$ **46.** $\dfrac{3}{16} \cdot \dfrac{1}{7}$

Find each quotient. See Example 5.

47. $\dfrac{3}{4} \div \dfrac{1}{4}$ **48.** $\dfrac{2}{3} \div \dfrac{1}{2}$

49. $\dfrac{1}{3} \div 5$ **50.** $\dfrac{3}{5} \div 3$

51. $5 \div \dfrac{5}{4}$ **52.** $8 \div \dfrac{2}{3}$

53. $\dfrac{6}{10} \div \dfrac{3}{4}$ **54.** $\dfrac{2}{3} \div \dfrac{10}{21}$

55. $\dfrac{3}{16} \div \dfrac{5}{2}$ **56.** $\dfrac{1}{8} \div \dfrac{5}{16}$

Find each sum or difference. See Examples 6 and 7.

57. $\dfrac{1}{4} + \dfrac{1}{4}$ **58.** $\dfrac{1}{10} + \dfrac{1}{10}$

59. $\dfrac{5}{12} - \dfrac{1}{12}$ **60.** $\dfrac{17}{14} - \dfrac{5}{14}$

61. $\dfrac{1}{2} - \dfrac{1}{4}$ **62.** $\dfrac{1}{3} + \dfrac{1}{6}$

63. $\dfrac{1}{3} + \dfrac{1}{4}$ **64.** $\dfrac{1}{2} + \dfrac{3}{5}$

65. $\dfrac{3}{4} - \dfrac{2}{3}$ **66.** $\dfrac{4}{5} - \dfrac{3}{4}$

67. $\dfrac{1}{6} + \dfrac{5}{8}$ **68.** $\dfrac{3}{4} + \dfrac{1}{6}$

69. $\dfrac{5}{24} - \dfrac{1}{18}$ **70.** $\dfrac{3}{16} - \dfrac{1}{20}$

71. $3\dfrac{5}{6} + \dfrac{5}{16}$ **72.** $5\dfrac{3}{8} - \dfrac{15}{16}$

Convert each given fraction, decimal, or percent into its other two forms. See Example 8.

73. $\dfrac{3}{5}$ **74.** $\dfrac{19}{20}$

75. 9%   **76.** 60%

77. 0.08   **78.** 0.4

79. $\dfrac{3}{4}$ **80.** $\dfrac{5}{8}$

81. 2%   **82.** 120%

83. 0.01   **84.** 0.005

Perform the indicated operations.

85. $\dfrac{3}{8} \div \dfrac{1}{8}$ **86.** $\dfrac{7}{8} \div \dfrac{3}{14}$

87. $\dfrac{3}{4} \cdot \dfrac{28}{21}$ **88.** $\dfrac{5}{16} \cdot \dfrac{3}{10}$

89. $\dfrac{7}{12} + \dfrac{5}{32}$ **90.** $\dfrac{2}{15} + \dfrac{8}{21}$

91. $\dfrac{5}{24} - \dfrac{1}{15}$ **92.** $\dfrac{9}{16} - \dfrac{1}{12}$

93. $3\dfrac{1}{8} + \dfrac{15}{16}$ **94.** $5\dfrac{1}{4} - \dfrac{9}{16}$

95. $7\dfrac{2}{3} \cdot 2\dfrac{1}{4}$ **96.** $6\dfrac{1}{2} \div \dfrac{7}{2}$

97. $\dfrac{1}{2} + \dfrac{1}{3} + \dfrac{1}{4}$ **98.** $\dfrac{1}{2} + \dfrac{1}{3} - \dfrac{1}{6}$

99. $\dfrac{1}{2} \cdot \dfrac{1}{2} \cdot \dfrac{1}{2}$ **100.** $\dfrac{2}{3} \cdot \dfrac{2}{3} \cdot \dfrac{2}{3}$

Solve each problem. See Example 9.

101. *Stock prices.* On Monday, GM stock opened at $54\frac{3}{4}$ per share and closed up $\frac{3}{16}$. On Tuesday it closed down $\frac{1}{8}$. On Wednesday it gained $\frac{5}{16}$. On Thursday it fell $\frac{1}{4}$. On Friday there was no change. What was the closing price on Friday? What was the percent change for the week?

102. *Diversification.* Helen has $\frac{1}{5}$ of her portfolio in U.S. stocks, $\frac{1}{8}$ of her portfolio in European stocks, and $\frac{1}{10}$ of her portfolio in Japanese stocks. The remainder is invested in municipal bonds. What fraction of her portfolio is invested in municipal bonds? What percent is invested in municipal bonds?

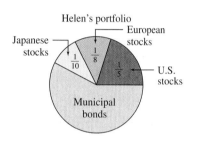

Helen's portfolio

FIGURE FOR EXERCISE 102

103. *Concrete patio.* A contractor plans to pour a concrete rectangular patio.
  a) Use the table to find the approximate volume of concrete in cubic yards for a 9 ft by 12 ft patio that is 4 inches thick.

Concrete required for 4 in. thick patio

L (ft)	W (ft)	V (yd³)
16	14	2.8
14	10	1.7
12	9	1.3
10	8	1.0

FIGURE FOR EXERCISE 103

  b) Find the exact volume of concrete in cubic feet and cubic yards for a patio that is $12\frac{1}{2}$ feet long, $8\frac{3}{4}$ feet wide, and 4 inches thick.

104. *Bundle of studs.* A lumber yard receives 2 × 4 studs in a bundle that contains 25 rows (or layers) of studs with 20 studs in each row. A 2 × 4 stud is actually $1\frac{1}{2}$ in. by $3\frac{1}{2}$ in. by $92\frac{5}{8}$ in. Find the cross-sectional area of a bundle in square inches. Find the volume of a bundle in cubic feet. (The formula $V = LWH$ gives the volume of a rectangular solid.)

GETTING MORE INVOLVED

105. *Writing.* Find an example of a real-life situation in which it is necessary to add two fractions.

106. *Cooperative learning.* Write a step-by-step procedure for adding two fractions with different denominators. Give your procedure to a classmate to try out on some addition problems. Refine your procedure as necessary.

107. *Fraction puzzle.* A wheat farmer in Manitoba left his L-shaped farm (shown in the diagram) to his four daughters. Divide the property into four pieces so that each piece is exactly the same size and shape.

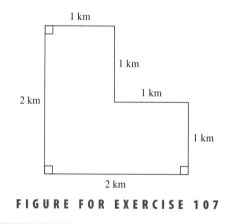

FIGURE FOR EXERCISE 107

1.3   ADDITION AND SUBTRACTION OF REAL NUMBERS

In arithmetic we add and subtract only positive numbers and zero. In Section 1.1 we introduced the concept of absolute value of a number. Now we will use absolute value to extend the operations of addition and subtraction to the real numbers. We will work only with rational numbers in this chapter. You will learn to perform operations with irrational numbers in Chapter 9.

Addition of Two Negative Numbers

A good way to understand positive and negative numbers is to *think of the positive numbers as assets and the negative numbers as debts.* For this illustration we can think of assets simply as cash. For example, if you have \$3 and \$5 in cash, then your total cash is \$8. You get the total by adding two positive numbers.

Think of debts as unpaid bills such as the electric bill or the phone bill. If you have debts of \$7 and \$8, then your total debt is \$15. You can get the total debt by adding negative numbers:

$$(-7) \quad + \quad (-8) \quad = \quad -15$$

$$\uparrow \qquad \uparrow \qquad \uparrow \qquad\qquad \uparrow$$

$$\text{\$7 debt} \quad \text{plus} \quad \text{\$8 debt} \qquad \text{\$15 debt}$$

We think of this addition as adding the absolute values of -7 and -8 ($7 + 8 = 15$), and then putting a negative sign on that result to get -15. These examples illustrate the following rule.

> ### Sum of Two Numbers with Like Signs
>
> To find the sum of two numbers with the same sign, add their absolute values. The sum has the same sign as the given numbers.

EXAMPLE 1

Adding numbers with like signs

Perform the indicated operations.

a) $23 + 56$ **b)** $(-12) + (-9)$ **c)** $(-3.5) + (-6.28)$ **d)** $\left(-\dfrac{1}{2}\right) + \left(-\dfrac{1}{4}\right)$

Solution

a) The sum of two positive numbers is a positive number: $23 + 56 = 79$.

b) The absolute values of -12 and -9 are 12 and 9, and $12 + 9 = 21$. So

$$(-12) + (-9) = -21.$$

c) Add the absolute values of -3.5 and -6.28, and put a negative sign on the sum. Remember to line up the decimal points when adding decimal numbers:

$$\begin{array}{r} 3.50 \\ \underline{6.28} \\ 9.78 \end{array}$$

So $(-3.5) + (-6.28) = -9.78$.

d) $\left(-\dfrac{1}{2}\right) + \left(-\dfrac{1}{4}\right) = \left(-\dfrac{2}{4}\right) + \left(-\dfrac{1}{4}\right) = -\dfrac{3}{4}$ ■

Addition of Numbers with Unlike Signs

If you have a debt of $5 and have only $5 in cash, then your debts equal your assets (in absolute value), and your net worth is $0. **Net worth** is the total of debts and assets. Symbolically,

$$-5 \quad + \quad 5 \quad = \quad 0.$$
$$\uparrow \qquad\qquad \uparrow \qquad\qquad \uparrow$$
$$\text{\$5 debt} \qquad \text{\$5 cash} \qquad \text{Net worth}$$

For any number a, a and its opposite, $-a$, have a sum of zero. For this reason, a and $-a$ are called **additive inverses** of each other. Note that the words "negative," "opposite," and "additive inverse" are often used interchangeably.

Additive Inverse Property

For any number a,

$$a + (-a) = 0 \qquad \text{and} \qquad (-a) + a = 0.$$

E X A M P L E   2   **Finding the sum of additive inverses**

Evaluate.

a) $34 + (-34)$ **b)** $-\dfrac{1}{4} + \dfrac{1}{4}$ **c)** $2.97 + (-2.97)$

Solution

a) $34 + (-34) = 0$

b) $-\dfrac{1}{4} + \dfrac{1}{4} = 0$

c) $2.97 + (-2.97) = 0$ ■

To understand the sum of a positive and a negative number that are not additive inverses of each other, consider the following situation. If you have a debt of $6 and $10 in cash, you may have $10 in hand, but your net worth is only $4. Your assets exceed your debts (in absolute value), and you have a positive net worth. In symbols,

$$-6 + 10 = 4.$$

Note that to get 4, we actually subtract 6 from 10.

If you have a debt of $7 but have only $5 in cash, then your debts exceed your assets (in absolute value). You have a negative net worth of $-\$2$. In symbols,

$$-7 + 5 = -2.$$

Note that to get the 2 in the answer, we subtract 5 from 7.

As you can see from these examples, the sum of a positive number and a negative number (with different absolute values) may be either positive or negative. These examples help us to understand the rule for adding numbers with unlike signs and different absolute values.

helpful   hint

We use the illustrations with debts and assets to make the rules for adding signed numbers understandable. However, in the end the carefully written rules tell us exactly how to perform operations with signed numbers, and we must obey the rules.

Sum of Two Numbers with Unlike Signs (and Different Absolute Values)

To find the sum of two numbers with unlike signs (and different absolute values), subtract their absolute values.

- The answer is positive if the number with the larger absolute value is positive.
- The answer is negative if the number with the larger absolute value is negative.

E X A M P L E   3

Adding numbers with unlike signs

Evaluate.

a) $-5 + 13$ **b)** $6 + (-7)$ **c)** $-6.4 + 2.1$

d) $-5 + 0.09$ **e)** $\left(-\dfrac{1}{3}\right) + \left(\dfrac{1}{2}\right)$ **f)** $\dfrac{3}{8} + \left(-\dfrac{5}{6}\right)$

Solution

a) The absolute values of -5 and 13 are 5 and 13. Subtract them to get 8. Since the number with the larger absolute value is 13 and it is positive, the result is positive:

$$-5 + 13 = 8$$

b) The absolute values of 6 and -7 are 6 and 7. Subtract them to get 1. Since -7 has the larger absolute value, the result is negative:

$$6 + (-7) = -1$$

c) Line up the decimal points and subtract 2.1 from 6.4.

$$\begin{array}{r} 6.4 \\ -2.1 \\ \hline 4.3 \end{array}$$

Since 6.4 is larger than 2.1, and 6.4 has a negative sign, the sign of the answer is negative. So $-6.4 + 2.1 = -4.3$.

d) Line up the decimal points and subtract 0.09 from 5.00.

$$\begin{array}{r} 5.00 \\ -0.09 \\ \hline 4.91 \end{array}$$

Since 5.00 is larger than 0.09, and 5.00 has the negative sign, the sign of the answer is negative. So $-5 + 0.09 = -4.91$.

e) $\left(-\dfrac{1}{3}\right) + \left(\dfrac{1}{2}\right) = \left(-\dfrac{2}{6}\right) + \left(\dfrac{3}{6}\right)$ **f)** $\dfrac{3}{8} + \left(-\dfrac{5}{6}\right) = \dfrac{9}{24} + \left(-\dfrac{20}{24}\right)$

$$= \dfrac{1}{6} \qquad\qquad = -\dfrac{11}{24}$$

calculator
close-up

Your calculator can add signed numbers. Most calculators have a key for subtraction and a different key for the negative sign.

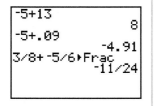

```
-5+13
              8
-5+.09
          -4.91
3/8+-5/6▶Frac
          -11/24
```

You should do the exercises in this section by hand and then check with a calculator.

Subtraction of Signed Numbers

Each subtraction problem with signed numbers is solved by doing an equivalent addition problem. So before attempting subtraction of signed numbers be sure that you understand addition of signed numbers.

Now think of subtraction as removing debts or assets, and think of addition as receiving debts or assets. If you have $10 in cash and $3 is taken from you, your resulting net worth is the same as if you have $10 cash and a phone bill for $3 arrives in the mail. In symbols,

$$10 \quad - \quad 3 \quad = \quad 10 \quad + \quad (-3).$$

↑	↑	↑	↑
Remove	Cash	Receive	Debt

Removing cash is equivalent to receiving a debt.

Suppose you have $15 but owe a friend $5. Your net worth is only $10. If the debt of $5 is canceled or forgiven, your net worth will go up to $15, the same as if you received $5 in cash. In symbols,

$$10 \quad - \quad (-5) \quad = \quad 10 \quad + \quad 5.$$

↑	↑	↑	↑
Remove	Debt	Receive	Cash

Removing a debt is equivalent to receiving cash.

Notice that each subtraction problem is equivalent to an addition problem in which we add the opposite of what we want to subtract. In other words, subtracting a number is the same as adding its opposite.

Subtraction of Real Numbers

For any real numbers a and b,

$$a - b = a + (-b).$$

E X A M P L E 4

Subtracting signed numbers

Perform each subtraction.

a) $-5 - 3$

b) $5 - (-3)$

c) $-5 - (-3)$

d) $\dfrac{1}{2} - \left(-\dfrac{1}{4}\right)$

e) $-3.6 - (-5)$

f) $0.02 - 8$

Solution

To do *any* subtraction, we can change it to addition of the opposite.

a) $-5 - 3 = -5 + (-3) = -8$

b) $5 - (-3) = 5 + (3) = 8$

c) $-5 - (-3) = -5 + 3 = -2$

d) $\dfrac{1}{2} - \left(-\dfrac{1}{4}\right) = \dfrac{2}{4} + \dfrac{1}{4} = \dfrac{3}{4}$

e) $-3.6 - (-5) = -3.6 + 5 = 1.4$

f) $0.02 - 8 = 0.02 + (-8) = -7.98$ ■

WARM-UPS

True or false? Explain your answer.

1. $-9 + 8 = -1$
2. $(-2) + (-4) = -6$
3. $0 - 7 = -7$
4. $5 - (-2) = 3$
5. $-5 - (-2) = -7$
6. The additive inverse of -3 is 0.
7. If b is a negative number, then $-b$ is a positive number.
8. The sum of a positive number and a negative number is a negative number.
9. The result of a subtracted from b is the same as b plus the opposite of a.
10. If a and b are negative numbers, then $a - b$ is a negative number.

1.3   EXERCISES

Reading and Writing   *After reading this section write out the answers to these questions. Use complete sentences.*

1. What operations did we study in this section?

2. How do you find the sum of two numbers with the same sign?

3. When can we say that two numbers are additive inverses of each other?

4. What is the sum of two numbers with opposite signs and the same absolute value?

5. How do we find the sum of two numbers with unlike signs?

6. What is the relationship between subtraction and addition?

Perform the indicated operation. See Example 1.

7. $3 + 10$
8. $81 + 19$
9. $(-3) + (-10)$
10. $(-81) + (-19)$
11. $-0.25 + (-0.9)$
12. $-0.8 + (-2.35)$

13. $\left(-\dfrac{1}{3}\right) + \left(-\dfrac{1}{6}\right)$
14. $\dfrac{2}{3} + \dfrac{1}{12}$

Evaluate. See Examples 2 and 3.

15. $-8 + 8$
16. $20 + (-20)$
17. $-\dfrac{17}{50} + \dfrac{17}{50}$
18. $\dfrac{12}{13} + \left(-\dfrac{12}{13}\right)$
19. $-7 + 9$
20. $10 + (-30)$
21. $7 + (-13)$
22. $-8 + 20$
23. $8.6 + (-3)$
24. $-9.5 + 12$
25. $3.9 + (-6.8)$
26. $-5.24 + 8.19$
27. $\dfrac{1}{4} + \left(-\dfrac{1}{2}\right)$
28. $-\dfrac{2}{3} + 2$

Fill in the parentheses to make each statement correct. See Example 4.

29. $8 - 2 = 8 + (?)$
30. $3.5 - 1.2 = 3.5 + (?)$
31. $4 - 12 = 4 + (?)$
32. $\dfrac{1}{2} - \dfrac{5}{6} = \dfrac{1}{2} + (?)$
33. $-3 - (-8) = -3 + (?)$
34. $-9 - (-2.3) = -9 + (?)$
35. $8.3 - (-1.5) = 8.3 + (?)$
36. $10 - (-6) = 10 + (?)$

Perform the indicated operation. See Example 4.

37. $6 - 10$
38. $3 - 19$
39. $-3 - 7$
40. $-3 - 12$
41. $5 - (-6)$
42. $5 - (-9)$

43. $-6 - 5$ **44.** $-3 - 6$ **45.** $\dfrac{1}{4} - \dfrac{1}{2}$

46. $\dfrac{2}{5} - \dfrac{2}{3}$ **47.** $\dfrac{1}{2} - \left(-\dfrac{1}{4}\right)$ **48.** $\dfrac{2}{3} - \left(-\dfrac{1}{6}\right)$

49. $10 - 3$ **50.** $13 - 3$ **51.** $1 - 0.07$

52. $0.03 - 1$ **53.** $7.3 - (-2)$ **54.** $-5.1 - 0.15$

55. $-0.03 - 5$ **56.** $0.7 - (-0.3)$

Perform the indicated operations. Do not use a calculator.

57. $-5 + 8$ **58.** $-6 + 10$

59. $-6 + (-3)$ **60.** $(-13) + (-12)$

61. $-80 - 40$ **62.** $44 - (-15)$

63. $61 - (-17)$ **64.** $-19 - 13$

65. $(-12) + (-15)$ **66.** $-12 + 12$

67. $13 + (-20)$ **68.** $15 + (-39)$

69. $-102 - 99$ **70.** $-94 - (-77)$

71. $-161 - 161$ **72.** $-19 - 88$

73. $-16 + 0.03$ **74.** $0.59 + (-3.4)$

75. $0.08 - 3$ **76.** $1.8 - 9$

77. $-3.7 + (-0.03)$ **78.** $0.9 + (-1)$

79. $-2.3 - (-6)$ **80.** $-7.08 - (-9)$

81. $\dfrac{3}{4} + \left(-\dfrac{3}{5}\right)$ **82.** $-\dfrac{1}{3} + \dfrac{3}{5}$

83. $-\dfrac{1}{12} - \left(-\dfrac{3}{8}\right)$ **84.** $-\dfrac{1}{17} - \left(-\dfrac{1}{17}\right)$

Use a calculator to perform the indicated operations.

85. $45.87 + (-49.36)$ **86.** $-0.357 + (-3.465)$

87. $0.6578 + (-1)$ **88.** $-2.347 + (-3.5)$

89. $-3.45 - 45.39$ **90.** $9.8 - 9.974$

91. $-5.79 - 3.06$ **92.** $0 - (-4.537)$

Solve each problem.

93. *Overdrawn.* Willard opened his checking account with a deposit of $97.86. He then wrote checks and had other charges as shown in his account register. Find his current balance.

Deposit		97.86
Wal-Mart	27.89	
Kmart	42.32	
ATM cash	25.00	
Service charge	3.50	
Check printing	8.00	

FIGURE FOR EXERCISE 93

94. *Net worth.* Melanie has a $125,000 house with a $78,422 mortgage. She has $21,236 in a savings account and has $9,477 in credit card debt. She owes $6,131 to the credit union and figures that her cars and other household items are worth a total of $15,000. What is Melanie's net worth?

95. *Falling temperatures.* At noon the temperature in Montreal was 5°C. By midnight the mercury had fallen 12°. What was the temperature at midnight?

96. *Bitter cold.* The overnight low temperature in Milwaukee was $-13°F$ for Monday night. The temperature went up 20° during the day on Tuesday and then fell 15° to reach Tuesday night's overnight low temperature.
a) What was the overnight low Tuesday night?
b) Judging from the accompanying graph, was the average low for the week above or below 0°F?

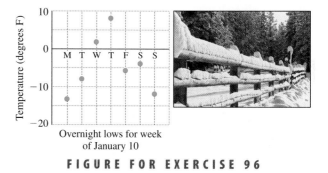

Overnight lows for week
of January 10

FIGURE FOR EXERCISE 96

GETTING MORE INVOLVED

97. *Writing.* What does absolute value have to do with adding signed numbers? Can you add signed numbers without using absolute value?

98. *Discussion.* Why do we learn addition of signed numbers before subtraction?

99. *Discussion.* Aimee and Joni are traveling south in separate cars on Interstate 5 near Stockton. While they are speaking to each other on cellular telephones, Aimee gives her location as mile marker x and Joni gives her location as mile marker y. Which of the following expressions gives the distance between them? Explain your answer.
a) $y - x$ b) $x - y$ c) $|x - y|$
d) $|y - x|$ e) $|x| + |y|$

1.4 MULTIPLICATION AND DIVISION OF REAL NUMBERS

In this section we will complete the study of the four basic operations with real numbers.

In this section
- Multiplication of Real Numbers
- Division of Real Numbers
- Division by Zero

Multiplication of Real Numbers

The result of multiplying two numbers is referred to as the **product** of the numbers. The numbers multiplied are called **factors.** In algebra we use a raised dot between the factors to indicate multiplication, or we place symbols next to one another to indicate multiplication. Thus $a \cdot b$ or ab are both referred to as the product of a and b. When multiplying numbers, we may enclose them in parentheses to make the meaning clear. To write 5 times 3, we may write it as $5 \cdot 3$, $5(3)$, $(5)3$, or $(5)(3)$. In multiplying a number and a variable, no sign is used between them. Thus $5x$ is used to represent the product of 5 and x.

Multiplication is just a short way to do repeated additions. Adding together five 3's gives

$$3 + 3 + 3 + 3 + 3 = 15.$$

So we have the multiplication fact $5 \cdot 3 = 15$. Adding together five -3's gives

$$(-3) + (-3) + (-3) + (-3) + (-3) = -15.$$

So we should have $5(-3) = -15$. We can think of $5(-3) = -15$ as saying that taking on five debts of \$3 each is equivalent to a debt of \$15. Losing five debts of \$3 each is equivalent to gaining \$15, so we should have $(-5)(-3) = 15$.

These examples illustrate the rule for multiplying signed numbers.

helpful hint

The product of two numbers with like signs is positive, but the product of three numbers with like signs could be negative. For example,

$$(-2)(-2)(-2) = 4(-2)$$
$$= -8.$$

Product of Signed Numbers

To find the product of two nonzero real numbers, multiply their absolute values.

- The product is *positive* if the numbers have *like* signs.
- The product is *negative* if the numbers have *unlike* signs.

EXAMPLE 1

Multiplying signed numbers

Evaluate each product.

a) $(-2)(-3)$ b) $3(-6)$ c) $-5 \cdot 10$

d) $\left(-\dfrac{1}{3}\right)\left(-\dfrac{1}{2}\right)$ e) $(-0.02)(0.08)$ f) $(-300)(-0.06)$

Solution

a) First find the product of the absolute values:

$$\left|-2\right| \cdot \left|-3\right| = 2 \cdot 3 = 6$$

Because -2 and -3 have the same sign, we get $(-2)(-3) = 6$.

b) First find the product of the absolute values:

$$\left|3\right| \cdot \left|-6\right| = 3 \cdot 6 = 18$$

Because 3 and -6 have unlike signs, we get $3(-6) = -18$.

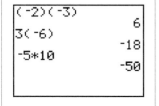

c) $-5 \cdot 10 = -50$ Unlike signs, negative result

d) $\left(-\dfrac{1}{3}\right)\left(-\dfrac{1}{2}\right) = \dfrac{1}{6}$ Like signs, positive result

e) When multiplying decimals, we total the number of decimal places in the factors to get the number of decimal places in the product. Thus

$$(-0.02)(0.08) = -0.0016.$$

f) $(-300)(-0.06) = 18$ ■

Division of Real Numbers

We say that $10 \div 5 = 2$ because $2 \cdot 5 = 10$. This example illustrates how division is defined in terms of multiplication.

Division of Real Numbers

If a, b, and c are any real numbers with $b \neq 0$, then

$$a \div b = c \qquad \text{provided that} \qquad c \cdot b = a.$$

Using the definition of division, we get

$$10 \div (-2) = -5$$

because $(-5)(-2) = 10$;

$$-10 \div 2 = -5$$

because $(-5)(2) = -10$; and

$$-10 \div (-2) = 5$$

because $(5)(-2) = -10$. From these examples we see that the rule for dividing signed numbers is similar to that for multiplying signed numbers.

Division of Signed Numbers

To find the quotient of nonzero real numbers, divide their absolute values.

- The quotient is *positive* if the numbers have *like* signs.
- The quotient is *negative* if the numbers have *unlike* signs.

Zero divided by any nonzero real number is zero.

E X A M P L E 2 **Dividing signed numbers**
Evaluate.

a) $(-8) \div (-4)$ **b)** $(-8) \div 8$ **c)** $8 \div (-4)$

d) $-4 \div \dfrac{1}{3}$ **e)** $-2.5 \div 0.05$ **f)** $0 \div (-6)$

Solution

a) $(-8) \div (-4) = 2$ Same sign, positive result

b) $(-8) \div 8 = -1$ Unlike signs, negative result

c) $8 \div (-4) = -2$

d) $-4 \div \dfrac{1}{3} = -4 \cdot \dfrac{3}{1}$ Invert and multiply.

$$= -4 \cdot 3$$

$$= -12$$

e) $-2.5 \div 0.05 = \dfrac{-2.5}{0.05}$ Write in fraction form.

$$= \dfrac{-2.5 \cdot 100}{0.05 \cdot 100}$$ Multiply by 100 to eliminate the decimals.

$$= \dfrac{-250}{5}$$ Simplify.

$$= -50$$ Divide.

f) $0 \div (-6) = 0$

Division can also be indicated by a fraction bar. For example,

$$24 \div 6 = \frac{24}{6} = 4.$$

If signed numbers occur in a fraction, we use the rules for dividing signed numbers. For example,

$$\frac{-9}{3} = -3, \qquad \frac{9}{-3} = -3, \qquad \frac{-1}{2} = \frac{1}{-2} = -\frac{1}{2}, \qquad \text{and} \qquad \frac{-4}{-2} = 2.$$

Note that if one negative sign appears in a fraction, the fraction has the same value whether the negative sign is in the numerator, in the denominator, or in front of the fraction. If the numerator and denominator of a fraction are both negative, then the fraction has a positive value.

Division by Zero

Why do we exclude division by zero from the definition of division? If we write $10 \div 0 = c$, we need to find a number c such that $c \cdot 0 = 10$. This is impossible. If we write $0 \div 0 = c$, we need to find a number c such that $c \cdot 0 = 0$. In fact, $c \cdot 0 = 0$ is true for any value of c. Having $0 \div 0$ equal to any number would be confusing in doing computations. Thus $a \div b$ is defined only for $b \neq 0$. Quotients such as

$$8 \div 0, \qquad 0 \div 0, \qquad \frac{8}{0}, \qquad \text{and} \qquad \frac{0}{0}$$

are said to be **undefined.**

WARM-UPS

True or false? Explain your answer.

1. The product of 7 and y is written as $7y$.

2. The product of -2 and 5 is 10.

3. The quotient of x and 3 can be written as $x \div 3$ or $\frac{x}{3}$.

4. $0 \div 6$ is undefined.

5. $(-9) \div (-3) = 3$ **6.** $6 \div (-2) = -3$

(continued)

7. $\left(-\dfrac{1}{2}\right)\left(-\dfrac{1}{2}\right) = \dfrac{1}{4}$

8. $(-0.2)(0.2) = -0.4$

9. $\left(-\dfrac{1}{2}\right) \div \left(-\dfrac{1}{2}\right) = 1$

10. $\dfrac{0}{0} = 0$

1.4 EXERCISES

Reading and Writing *After reading this section write out the answers to these questions. Use complete sentences.*

1. What operations did we study in this section?

2. What is a product?

3. How do you find the product of two signed numbers?

4. What is the relationship between division and multiplication?

5. How do you find the quotient of nonzero real numbers?

6. Why is division by zero undefined?

Evaluate. See Example 1.

7. $-3 \cdot 9$

8. $6(-4)$

9. $(-12)(-11)$

10. $(-9)(-15)$

11. $-\dfrac{3}{4} \cdot \dfrac{4}{9}$

12. $\left(-\dfrac{2}{3}\right)\left(-\dfrac{6}{7}\right)$

13. $0.5(-0.6)$

14. $(-0.3)(0.3)$

15. $(-12)(-12)$

16. $(-11)(-11)$

17. $-3 \cdot 0$

18. $0(-7)$

Evaluate. See Example 2.

19. $8 \div (-8)$

20. $-6 \div 2$

21. $(-90) \div (-30)$

22. $(-20) \div (-40)$

23. $\dfrac{44}{-66}$

24. $\dfrac{-33}{-36}$

25. $\left(-\dfrac{2}{3}\right) \div \left(-\dfrac{4}{5}\right)$

26. $-\dfrac{1}{3} \div \dfrac{4}{9}$

27. $\dfrac{-125}{0}$

28. $-37 \div 0$

29. $0 \div \left(-\dfrac{1}{3}\right)$

30. $0 \div 43.568$

31. $40 \div (-0.5)$

32. $3 \div (-0.1)$

33. $-0.5 \div (-2)$

34. $-0.75 \div (-0.5)$

Perform the indicated operations.

35. $(25)(-4)$

36. $(5)(-4)$

37. $(-3)(-9)$

38. $(-51) \div (-3)$

39. $-9 \div 3$

40. $86 \div (-2)$

41. $20 \div (-5)$

42. $(-8)(-6)$

43. $(-6)(5)$

44. $(-18) \div 3$

45. $(-57) \div (-3)$

46. $(-30)(4)$

47. $(0.6)(-0.3)$

48. $(-0.2)(-0.5)$

49. $(-0.03)(-10)$

50. $(0.05)(-1.5)$

51. $(-0.6) \div (0.1)$

52. $8 \div (-0.5)$

53. $(-0.6) \div (-0.4)$

54. $(-63) \div (-0.9)$

55. $-\dfrac{12}{5}\left(-\dfrac{55}{6}\right)$

56. $-\dfrac{9}{10} \cdot \dfrac{4}{3}$

57. $-2\dfrac{3}{4} \div 8\dfrac{1}{4}$

58. $-9\dfrac{1}{2} \div \left(-3\dfrac{1}{6}\right)$

59. $(0.45)(-365)$

60. $8.5 \div (-0.15)$

61. $(-52) \div (-0.034)$

62. $(-4.8)(5.6)$

Perform the indicated operations. Use a calculator to check.

63. $(-4)(-4)$

64. $-4 - 4$

65. $-4 + (-4)$

66. $-4 \div (-4)$

67. $-4 + 4$

68. $-4 \cdot 4$

69. $-4 - (-4)$

70. $0 \div (-4)$

71. $0.1 - 4$

72. $(0.1)(-4)$

73. $(-4) \div (0.1)$

74. $-0.1 - 4$

75. $(-0.1)(-4)$

76. $-0.1 + 4$

77. $|-0.4|$

78. $|0.4|$

79. $\dfrac{-0.06}{0.3}$

80. $\dfrac{2}{-0.04}$

81. $\dfrac{3}{-0.4}$

82. $\dfrac{-1.2}{-0.03}$

83. $-\dfrac{1}{5} + \dfrac{1}{6}$

84. $-\dfrac{3}{5} - \dfrac{1}{4}$

85. $\left(-\dfrac{3}{4}\right)\left(\dfrac{2}{15}\right)$

86. $-1 \div \left(-\dfrac{1}{4}\right)$

 *Use a calculator to perform the indicated operations. Round answers to three decimal places.*

87. $\dfrac{45.37}{6}$

88. $(-345) \div (28)$

89. $(-4.3)(-4.5)$

90. $\dfrac{-12.34}{-3}$

91. $\dfrac{0}{6.345}$

92. $0 \div (34.51)$

93. $199.4 \div 0$

94. $\dfrac{23.44}{0}$

GETTING MORE INVOLVED

95. *Discussion.* If you divide \$0 among five people, how much does each person get? If you divide \$5 among zero people, how much does each person get? What do these questions illustrate?

96. *Discussion.* What is the difference between the non-negative numbers and the positive numbers?

97. *Writing.* Why do we learn multiplication of signed numbers before division?

98. *Writing.* Try to rewrite the rules for multiplying and dividing signed numbers without using the idea of absolute value. Are your rewritten rules clearer than the original rules?

1.5 EXPONENTIAL EXPRESSIONS AND THE ORDER OF OPERATIONS

In Sections 1.3 and 1.4 you learned how to perform operations with a pair of real numbers to obtain a third real number. In this section you will learn to evaluate expressions involving several numbers and operations.

In this section

- Arithmetic Expressions
- Exponential Expressions
- The Order of Operations

Arithmetic Expressions

The result of writing numbers in a meaningful combination with the ordinary operations of arithmetic is called an **arithmetic expression** or simply an **expression.** Consider the expressions

$$(3 + 2) \cdot 5 \quad \text{and} \quad 3 + (2 \cdot 5).$$

The parentheses are used as **grouping symbols** and indicate which operation to perform first. Because of the parentheses, these expressions have different values:

$$(3 + 2) \cdot 5 = 5 \cdot 5 = 25$$
$$3 + (2 \cdot 5) = 3 + 10 = 13$$

Absolute value symbols and fraction bars are also used as grouping symbols. The numerator and denominator of a fraction are treated as if each is in parentheses.

EXAMPLE 1

Using grouping symbols

Evaluate each expression.

a) $(3 - 6)(3 + 6)$

b) $|\,3 - 4\,| - |\,5 - 9\,|$

c) $\dfrac{4 - (-8)}{5 - 9}$

Solution

a) $(3 - 6)(3 + 6) = (-3)(9)$ Evaluate within parentheses first.

$= -27$ Multiply.

b) $|3 - 4| - |5 - 9| = |-1| - |-4|$ Evaluate within absolute value symbols.

$= 1 - 4$ Find the absolute values.

$= -3$ Subtract.

c) $\dfrac{4 - (-8)}{5 - 9} = \dfrac{12}{-4}$ Evaluate the numerator and denominator.

$= -3$ Divide.

---

calculator close-up

One advantage of a graphing calculator is that you can enter an entire expression on its display and then evaluate it. If your calculator does not allow built-up form for fractions, then you must use parentheses around the numerator and denominator as shown here.

```
(3-6)(3+6)
              -27
abs(3-4)-abs(5-9
)
               -3
(4--8)/(5-9)
               -3
```

Exponential Expressions

An arithmetic expression with repeated multiplication can be written by using exponents. For example,

$$2 \cdot 2 \cdot 2 = 2^3 \qquad \text{and} \qquad 5 \cdot 5 = 5^2.$$

The 3 in 2^3 is the number of times that 2 occurs in the product $2 \cdot 2 \cdot 2$, while the 2 in 5^2 is the number of times that 5 occurs in $5 \cdot 5$. We read 2^3 as "2 cubed" or "2 to the third power." We read 5^2 as "5 squared" or "5 to the second power." In general, an expression of the form a^n is called an **exponential expression** and is defined as follows.

> **Exponential Expression**
>
> For any counting number n,
>
> $$a^n = \underbrace{a \cdot a \cdot a \cdot \ldots \cdot a.}_{n \text{ factors}}$$
>
> We call a the **base** and n the **exponent.**

The expression a^n is read "a to the nth power." If the exponent is 1, it is usually omitted. For example, $9^1 = 9$.

E X A M P L E 2

Using exponential notation

Write each product as an exponential expression.

a) $6 \cdot 6 \cdot 6 \cdot 6 \cdot 6$ **b)** $(-3)(-3)(-3)(-3)$ **c)** $\dfrac{3}{2} \cdot \dfrac{3}{2} \cdot \dfrac{3}{2}$

Solution

a) $6 \cdot 6 \cdot 6 \cdot 6 \cdot 6 = 6^5$

b) $(-3)(-3)(-3)(-3) = (-3)^4$

c) $\dfrac{3}{2} \cdot \dfrac{3}{2} \cdot \dfrac{3}{2} = \left(\dfrac{3}{2}\right)^3$

E X A M P L E   3   **Writing an exponential expression as a product**

Write each exponential expression as a product without exponents.

a) y^6 **b)** $(-2)^4$ **c)** $\left(\dfrac{5}{4}\right)^3$ **d)** $(-0.1)^2$

Solution

a) $y^6 = y \cdot y \cdot y \cdot y \cdot y \cdot y$

b) $(-2)^4 = (-2)(-2)(-2)(-2)$

c) $\left(\dfrac{5}{4}\right)^3 = \dfrac{5}{4} \cdot \dfrac{5}{4} \cdot \dfrac{5}{4}$

d) $(-0.1)^2 = (-0.1)(-0.1)$

To evaluate an exponential expression, write the base as many times as indicated by the exponent, then multiply the factors from left to right.

E X A M P L E   4   **Evaluating exponential expressions**

Evaluate.

a) 3^3 **b)** $(-2)^3$ **c)** $\left(\dfrac{2}{3}\right)^4$ **d)** $(0.4)^2$

calculator

close-up

You can use the power key for any power. Most calculators also have an x^2 key that gives the second power. Note that parentheses must be used when raising a fraction to a power.

```
(-2)^3
              -8
(2/3)^4▶Frac
           16/81
.4²
             .16
```

Solution

a) $3^3 = 3 \cdot 3 \cdot 3 = 9 \cdot 3 = 27$

b) $(-2)^3 = (-2)(-2)(-2)$
$\qquad\quad = 4(-2)$
$\qquad\quad = -8$

c) $\left(\dfrac{2}{3}\right)^4 = \dfrac{2}{3} \cdot \dfrac{2}{3} \cdot \dfrac{2}{3} \cdot \dfrac{2}{3}$

$\qquad\quad = \dfrac{4}{9} \cdot \dfrac{2}{3} \cdot \dfrac{2}{3}$

$\qquad\quad = \dfrac{8}{27} \cdot \dfrac{2}{3}$

$\qquad\quad = \dfrac{16}{81}$

d) $(0.4)^2 = (0.4)(0.4) = 0.16$

C A U T I O N   Note that $3^3 \neq 9$. We do not multiply the exponent and the base when evaluating an exponential expression.

Be especially careful with exponential expressions involving negative numbers. An exponential expression with a negative base is written with parentheses around the base as in $(-2)^4$:

$$(-2)^4 = (-2)(-2)(-2)(-2) = 16$$

To evaluate $-(2^4)$, use the base 2 as a factor four times, then find the opposite:

$$-(2^4) = -(2 \cdot 2 \cdot 2 \cdot 2) = -(16) = -16$$

We often omit the parentheses in $-(2^4)$ and simply write -2^4. So

$$-2^4 = -(2^4) = -16.$$

To evaluate $-(-2)^4$, use the base -2 as a factor four times, then find the opposite:

$$-(-2)^4 = -(16) = -16$$

E X A M P L E   5

Evaluating exponential expressions involving negative numbers

Evaluate.

a) $(-10)^4$ **b)** -10^4

c) $-(-0.5)^2$ **d)** $-(5-8)^2$

Solution

a) $(-10)^4 = (-10)(-10)(-10)(-10)$ Use -10 as a factor four times.

$$= 10,000$$

b) $-10^4 = -(10^4)$ Rewrite using parentheses.

$$= -(10,000) \quad \text{Find } 10^4.$$

$$= -10,000 \quad \text{Then find the opposite of 10,000.}$$

c) $-(-0.5)^2 = -(-0.5)(-0.5)$ Use -0.5 as a factor two times.

$$= -(0.25)$$

$$= -0.25$$

d) $-(5-8)^2 = -(-3)^2$ Evaluate within parentheses first.

$$= -(9) \quad \text{Square } -3 \text{ to get 9.}$$

$$= -9 \quad \text{Take the opposite of 9 to get } -9.$$

helpful hint

"Please Excuse My Dear Aunt Sally" is often used as a memory aid for the order of operations. Do Parentheses, Exponents, Multiplication, and Division, then Addition and Subtraction.   Multiplication and division have equal priority. The same goes for addition and subtraction.

The Order of Operations

When we evaluate expressions, operations within grouping symbols are always performed first. For example

$$(3 + 2) \cdot 5 = (5) \cdot 5 = 10 \quad \text{and} \quad (2 \cdot 3)^2 = 6^2 = 36.$$

To make expressions look simpler, we often omit some or all parentheses. In this case we must agree on the order in which to perform the operations. We agree to do multiplication before addition and exponential expressions before multiplication. So

$$3 + 2 \cdot 5 = 3 + 10 = 13 \quad \text{and} \quad 2 \cdot 3^2 = 2 \cdot 9 = 18.$$

We state the complete **order of operations** in the following box.

Order of Operations

If no grouping symbols are present, evaluate expressions in the following order:

1. Evaluate each exponential expression (in order from left to right).
2. Perform multiplication and division (in order from left to right).
3. Perform addition and subtraction (in order from left to right).

For operations within grouping symbols, use the above order within the grouping symbols.

Multiplication and division have equal priority in the order of operations. If both appear in an expression, they are performed in order from left to right. The same holds for addition and subtraction. For example,

$$8 \div 4 \cdot 3 = 2 \cdot 3 = 6 \quad \text{and} \quad 9 - 3 + 5 = 6 + 5 = 11.$$

E X A M P L E   6

Using the order of operations

Evaluate each expression.

a) $2^3 \cdot 3^2$ **b)** $2 \cdot 5 - 3 \cdot 4 + 4^2$ **c)** $2 \cdot 3 \cdot 4 - 3^3 + \dfrac{8}{2}$

calculator close-up

Most calculators follow the same order of operations shown here. Evaluate these expressions with your calculator.

```
2^3*3²
                72
2*5-3*4+4²
                14
2*3*4-3^3+8/2
                 1
```

Solution

a) $2^3 \cdot 3^2 = 8 \cdot 9$ Evaluate exponential expressions before multiplying.

$\qquad = 72$

b) $2 \cdot 5 - 3 \cdot 4 + 4^2 = 2 \cdot 5 - 3 \cdot 4 + 16$ Exponential expressions first

$\qquad\qquad\qquad\qquad = 10 - 12 + 16$ Multiplication second

$\qquad\qquad\qquad\qquad = 14$ Addition and subtraction from left to right

c) $2 \cdot 3 \cdot 4 - 3^3 + \dfrac{8}{2} = 2 \cdot 3 \cdot 4 - 27 + \dfrac{8}{2}$ Exponential expressions first

$\qquad\qquad\qquad\qquad = 24 - 27 + 4$ Multiplication and division second

$\qquad\qquad\qquad\qquad = 1$ Addition and subtraction from left to right

When grouping symbols are used, we perform operations within grouping symbols first. The order of operations is followed within the grouping symbols.

E X A M P L E   7

Grouping symbols and the order of operations

Evaluate.

a) $3 - 2(7 - 2^3)$ **b)** $3 - |7 - 3 \cdot 4|$ **c)** $\dfrac{9 - 5 + 8}{-5^2 - 3(-7)}$

Solution

a) $3 - 2(7 - 2^3) = 3 - 2(7 - 8)$ Evaluate within parentheses first.

$\qquad\qquad\qquad = 3 - 2(-1)$

$\qquad\qquad\qquad = 3 - (-2)$ Multiply.

$\qquad\qquad\qquad = 5$ Subtract.

b) $3 - |7 - 3 \cdot 4| = 3 - |7 - 12|$ Evaluate within the absolute value symbols first.

$\qquad\qquad\qquad = 3 - |-5|$

$\qquad\qquad\qquad = 3 - 5$ Evaluate the absolute value.

$\qquad\qquad\qquad = -2$ Subtract.

c) $\dfrac{9 - 5 + 8}{-5^2 - 3(-7)} = \dfrac{12}{-25 + 21} = \dfrac{12}{-4} = -3$ Numerator and denominator are treated as if in parentheses.

When grouping symbols occur within grouping symbols, we evaluate within the innermost grouping symbols first and then work outward. In this case, brackets [ ] can be used as grouping symbols along with parentheses to make the grouping clear.

E X A M P L E   8

Grouping within grouping

Evaluate each expression.

a) $6 - 4[5 - (7 - 9)]$ **b)** $-2|3 - (9 - 5)| - |-3|$

Solution

a) $6 - 4[5 - (7 - 9)] = 6 - 4[5 - (-2)]$ Innermost parentheses first

$\qquad\qquad\qquad\qquad\quad = 6 - 4[7]$ Next evaluate within the brackets.

$\qquad\qquad\qquad\qquad\quad = 6 - 28$ Multiply.

$\qquad\qquad\qquad\qquad\quad = -22$ Subtract.

b) $-2|3 - (9 - 5)| - |-3| = -2|3 - 4| - |-3|$ Innermost grouping first

$\qquad\qquad\qquad\qquad\qquad\quad = -2|-1| - |-3|$ Evaluate within the first absolute value.

$\qquad\qquad\qquad\qquad\qquad\quad = -2 \cdot 1 - 3$ Evaluate absolute values.

$\qquad\qquad\qquad\qquad\qquad\quad = -2 - 3$ Multiply.

$\qquad\qquad\qquad\qquad\qquad\quad = -5$ Subtract.

---

c a l c u l a t o r   c l o s e - u p

Graphing calculators can handle grouping symbols within grouping symbols. Since parentheses must occur in pairs, you should have the same number of left parentheses as right parentheses.

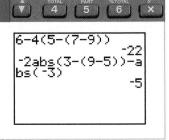

```
6-4(5-(7-9))
              -22
-2abs(3-(9-5))-a
bs(-3)
               -5
```

W A R M - U P S

True or false? Explain your answer.

1. $(-3)^2 = -6$ **2.** $5 - 3 \cdot 2 = 4$

3. $(5 - 3)2 = 4$ **4.** $|5 - 6| = |5| - |6|$

5. $5 + 6 \cdot 2 = (5 + 6) \cdot 2$ **6.** $(2 + 3)^2 = 2^2 + 3^2$

7. $5 - 3^3 = 8$ **8.** $(5 - 3)^3 = 8$

9. $6 - \dfrac{6}{2} = \dfrac{0}{2}$ **10.** $\dfrac{6 - 6}{2} = 0$

1.5 EXERCISES

Reading and Writing *After reading this section write out the answers to these questions. Use complete sentences.*

1. What is an arithmetic expression?

2. What is the purpose of grouping symbols?

3. What is an exponential expression?

4. What is the difference between -3^6 and $(-3)^6$?

5. What is the purpose of the order of operations?

6. What were the different types of grouping symbols used in this section?

Evaluate each expression. See Example 1.

7. $(4 - 3)(5 - 9)$ **8.** $(5 - 7)(-2 - 3)$

9. $|3 + 4| - |-2 - 4|$ **10.** $|-4 + 9| + |-3 - 5|$

11. $\dfrac{7 - (-9)}{3 - 5}$ **12.** $\dfrac{-8 + 2}{-1 - 1}$

13. $(-6 + 5)(7)$ **14.** $-6 + (5 \cdot 7)$

15. $(-3 - 7) - 6$ **16.** $-3 - (7 - 6)$

17. $-16 \div (8 \div 2)$ **18.** $(-16 \div 8) \div 2$

Write each product as an exponential expression. See Example 2.

19. $4 \cdot 4 \cdot 4 \cdot 4$ **20.** $1 \cdot 1 \cdot 1 \cdot 1 \cdot 1$

21. $(-5)(-5)(-5)(-5)$ **22.** $(-7)(-7)(-7)$

23. $(-y)(-y)(-y)$ **24.** $x \cdot x \cdot x \cdot x \cdot x$

25. $\dfrac{3}{7} \cdot \dfrac{3}{7} \cdot \dfrac{3}{7} \cdot \dfrac{3}{7} \cdot \dfrac{3}{7}$ **26.** $\dfrac{y}{2} \cdot \dfrac{y}{2} \cdot \dfrac{y}{2} \cdot \dfrac{y}{2}$

Write each exponential expression as a product without exponents. See Example 3.

27. 5^3 **28.** $(-8)^4$

29. b^2 **30.** $(-a)^5$

31. $\left(-\dfrac{1}{2}\right)^5$

32. $\left(-\dfrac{13}{12}\right)^3$

33. $(0.22)^4$

34. $(1.25)^6$

Evaluate each exponential expression. See Examples 4 and 5.

35. 3^4 **36.** 5^3 **37.** 0^9

38. 0^{12} **39.** $(-5)^4$ **40.** $(-2)^5$

41. $(-6)^3$ **42.** $(-12)^2$ **43.** $(10)^5$

44. $(-10)^6$ **45.** $(-0.1)^3$ **46.** $(-0.2)^2$

47. $\left(\dfrac{1}{2}\right)^3$ **48.** $\left(\dfrac{2}{3}\right)^3$ **49.** $\left(-\dfrac{1}{2}\right)^2$

50. $\left(-\dfrac{2}{3}\right)^2$ **51.** -8^2 **52.** -7^2

53. $-(-8)^4$ **54.** $-(-7)^3$

55. $-(7 - 10)^3$ **56.** $-(6 - 9)^4$

57. $(-2^2) - (3^2)$ **58.** $(-3^4) - (-5^2)$

Evaluate each expression. See Example 6.

59. $3^2 \cdot 2^2$ **60.** $5 \cdot 10^2$

61. $-3 \cdot 2 + 4 \cdot 6$ **62.** $-5 \cdot 4 - 8 \cdot 3$

63. $(-3)^3 + 2^3$ **64.** $3^2 - 5(-1)^3$

65. $-21 + 36 \div 3^2$ **66.** $-18 - 9^2 \div 3^3$

67. $-3 \cdot 2^3 - 5 \cdot 2^2$ **68.** $2 \cdot 5 - 3^2 + 4 \cdot 0$

69. $\dfrac{-8}{2} + 2 \cdot 3 \cdot 5 - 2^3$ **70.** $-4 \cdot 2 \cdot 6 - \dfrac{12}{3} + 3^3$

Evaluate each expression. See Example 7.

71. $(-3 + 4^2)(-6)$ **72.** $-3 \cdot (2^3 + 4) \cdot 5$

73. $(-3 \cdot 2 + 6)^3$ **74.** $5 - 2(-3 + 2)^3$

75. $2 - 5(3 - 4 \cdot 2)$ **76.** $(3 - 7)(4 - 6 \cdot 2)$

77. $3 - 2 \cdot |5 - 6|$ **78.** $3 - |6 - 7 \cdot 3|$

79. $(3^2 - 5) \cdot |3 \cdot 2 - 8|$

80. $|4 - 6 \cdot 3| + |6 - 9|$

81. $\dfrac{3 - 4 \cdot 6}{7 - 10}$ **82.** $\dfrac{6 - (-8)^2}{-3 - (-1)}$

83. $\dfrac{7 - 9 - 3^2}{9 - 7 - 3}$ **84.** $\dfrac{3^2 - 2 \cdot 4}{-30 + 2 \cdot 4^2}$

Evaluate each expression. See Example 8.

85. $3 + 4[9 - 6(2 - 5)]$

86. $9 + 3[5 - (3 - 6)^2]$

87. $6^2 - [(2 + 3)^2 - 10]$

88. $3[(2 - 3)^2 + (6 - 4)^2]$

89. $4 - 5 \cdot |3 - (3^2 - 7)|$

90. $2 + 3 \cdot |4 - (7^2 - 6^2)|$

91. $-2|3 - (7 - 3)| - |-9|$

92. $[3 - (2 - 4)][3 + |2 - 4|]$

Evaluate each expression. Use a calculator to check.

93. $1 + 2^3$ **94.** $(1 + 2)^3$

95. $(-2)^2 - 4(-1)(3)$ **96.** $(-2)^2 - 4(-2)(-3)$

97. $4^2 - 4(1)(-3)$ **98.** $3^2 - 4(-2)(3)$

99. $(-11)^2 - 4(5)(0)$ **100.** $(-12)^2 - 4(3)(0)$

101. $-5^2 - 3 \cdot 4^2$

102. $-6^2 - 5(-3)^2$

103. $[3 + 2(-4)]^2$

104. $[6 - 2(-3)]^2$

105. $|-1| - |-1|$

106. $4 - |1 - 7|$

107. $\dfrac{4 - (-4)}{-2 - 2}$

108. $\dfrac{3 - (-7)}{3 - 5}$

109. $3(-1)^2 - 5(-1) + 4$

110. $-2(1)^2 - 5(1) - 6$

111. $5 - 2^2 + 3^4$

112. $5 + (-2)^2 - 3^2$

113. $-2 \cdot |9 - 6^2|$

114. $8 - 3|5 - 4^2 + 1|$

115. $-3^2 - 5[4 - 2(4 - 9)]$

116. $-2[(3 - 4)^3 - 5] + 7$

117. $1 - 5|5 - (9 + 1)|$

118. $|6 - 3 \cdot 7| + |7 - (5 - 2)|$

Use a calculator to evaluate each expression.

119. $3.2^2 - 4(3.6)(-2.2)$

120. $(-4.5)^2 - 4(-2.8)(-4.6)$

121. $(5.63)^3 - [4.7 - (-3.3)^2]$

122. $9.8^3 - [1.2 - (4.4 - 9.6)^2]$

123. $\dfrac{3.44 - (-8.32)}{6.89 - 5.43}$

124. $\dfrac{-4.56 - 3.22}{3.44 - (-6.26)}$

Solve each problem.

125. *Population of the United States.* In 1998 the population of the United States was 270.1 million (U.S. Census Bureau, www.census.gov). If the population continues to grow at an annual rate of 0.86%, then the population in the year 2010 will be $270.1(1.0086)^{12}$ million. Find the predicted population in 2010 to the nearest tenth of a million people.

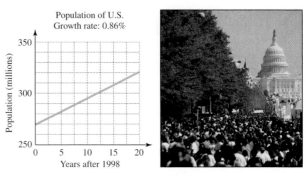

Population of U.S.
Growth rate: 0.86%

FIGURE FOR EXERCISE 125

126. *Population of Mexico.* In 1998 the population of Mexico was 97.2 million (World Resources 1997–1998, www.wri.org). If Mexico's population continues to grow at an annual rate of 2.0%, then the population in the year 2010 will be $97.2(1.02)^{12}$ million.

a) Find the predicted population in the year 2010 to the nearest tenth of a million people.

b) Will the U.S. or Mexico have the greater increase in population between the years 1998 and 2010? (See the previous exercise.)

GETTING MORE INVOLVED

127. *Discussion.* How do the expressions $(-5)^3$, $-(5^3)$, -5^3, $-(-5)^3$, and $-1 \cdot 5^3$ differ?

128. *Discussion.* How do the expressions $(-4)^4$, $-(4^4)$, -4^4, $-(-4)^4$, and $-1 \cdot 4^4$ differ?

In this

section

- Identifying Algebraic Expressions
- Translating Algebraic Expressions
- Evaluating Algebraic Expressions
- Equations
- Applications

1.6 ALGEBRAIC EXPRESSIONS

In Section 1.5 you studied arithmetic expressions. In this section you will study expressions that are more general—expressions that involve variables.

Identifying Algebraic Expressions

Since variables (or letters) are used to represent numbers, we can use variables in arithmetic expressions. The result of combining numbers and variables with the ordinary operations of arithmetic (in some meaningful way) is called an **algebraic**

expression or simply an **expression.** For example,

$$x + 2, \qquad \pi r^2, \qquad b^2 - 4ac, \qquad \text{and} \qquad \frac{a - b}{c - d}$$

are algebraic expressions.

Expressions are often named by the last operation to be performed in the expression. For example, the expression $x + 2$ is a **sum** because the only operation in the expression is addition. The expression $a - bc$ is referred to as a **difference** because subtraction is the last operation to be performed. The expression $3(x - 4)$ is a **product,** while $\frac{3}{x - 4}$ is a **quotient.** The expression $(a + b)^2$ is a **square** because the addition is performed before the square is found.

E X A M P L E   1

Naming expressions

Identify each expression as either a sum, difference, product, quotient, or square.

a) $3(x + 2)$ **b)** $b^2 - 4ac$ **c)** $\dfrac{a - b}{c - d}$ **d)** $(a - b)^2$

Solution

a) In $3(x + 2)$ we add before we multiply. So this expression is a product.

b) By the order of operations the last operation to perform in $b^2 - 4ac$ is subtraction. So this expression is a difference.

c) The last operation to perform in this expression is division. So this expression is a quotient.

d) In $(a - b)^2$ we subtract before we square. This expression is a square.   ∎

> **helpful hint**
>
> Sum, difference, product, and quotient are nouns. They are used as names for expressions. Add, subtract, multiply, and divide are verbs. They indicate an action to perform.

Translating Algebraic Expressions

Algebra is useful because it can be used to solve problems. Since problems are often communicated verbally, we must be able to translate verbal expressions into algebraic expressions and translate algebraic expressions into verbal expressions. Consider the following examples of verbal expressions and their corresponding algebraic expressions.

Verbal Expressions and Corresponding Algebraic Expressions	
Verbal Expression	**Algebraic Expression**
The sum of $5x$ and 3	$5x + 3$
The product of 5 and $x + 3$	$5(x + 3)$
The sum of 8 and $\dfrac{x}{3}$	$8 + \dfrac{x}{3}$
The quotient of $8 + x$ and 3	$\dfrac{8 + x}{3}$, $(8 + x)/3$, or $(8 + x) \div 3$
The difference of 3 and x^2	$3 - x^2$
The square of $3 - x$	$(3 - x)^2$

Because of the order of operations, reading from left to right does not always describe an expression accurately. For example, $5x + 3$ and $5(x + 3)$ can both be read as "5 times x plus 3." The next example shows how the terms sum, difference, product, quotient, and square are used to describe expressions. (You will study verbal and algebraic expressions further in Section 2.5.)

E X A M P L E 2 **Algebraic expressions to verbal expressions**

Translate each algebraic expression into a verbal expression. Use the word sum, difference, product, quotient, or square.

a) $\dfrac{3}{x}$ **b)** $2y + 1$ **c)** $3x - 2$ **d)** $(a - b)(a + b)$ **e)** $(a + b)^2$

Solution

a) The quotient of 3 and x **b)** The sum of $2y$ and 1

c) The difference of $3x$ and 2   **d)** The product of $a - b$ and $a + b$

e) The square of the sum $a + b$ ■

E X A M P L E 3 **Verbal expressions to algebraic expressions**

Translate each verbal expression into an algebraic expression.

a) The quotient of $a + b$ and 5   **b)** The difference of x^2 and y^2

c) The product of π and r^2 **d)** The square of the difference $x - y$

Solution

a) $\dfrac{a + b}{5}$, $(a + b) \div 5$, or $(a + b)/5$ **b)** $x^2 - y^2$

c) πr^2 **d)** $(x - y)^2$ ■

Evaluating Algebraic Expressions

The value of an algebraic expression depends on the values given to the variables. For example, the value of $x - 2y$ when $x = -2$ and $y = -3$ is found by replacing x and y by -2 and -3, respectively:

$$x - 2y = -2 - 2(-3) = -2 - (-6) = 4$$

If $x = 1$ and $y = 2$, the value of $x - 2y$ is found by replacing x by 1 and y by 2, respectively:

$$x - 2y = 1 - 2(2) = 1 - 4 = -3$$

Note that we use the order of operations when evaluating an algebraic expression.

E X A M P L E 4 **Evaluating algebraic expressions**

Evaluate each expression using $a = 3$, $b = -2$, and $c = -4$.

a) $a^2 + 2ab + b^2$ **b)** $(a - b)(a + b)$

c) $b^2 - 4ac$ **d)** $\dfrac{-a^2 - b^2}{c - b}$

Solution

a) $a^2 + 2ab + b^2 = 3^2 + 2(3)(-2) + (-2)^2$ Replace a by 3 and b by -2.

$\qquad\qquad\qquad = 9 + (-12) + 4$ Evaluate.

$\qquad\qquad\qquad = 1$ Add.

b) $(a - b)(a + b) = [3 - (-2)][3 + (-2)]$ Replace.

$\qquad\qquad\qquad = [5][1]$ Simplify within the brackets.

$\qquad\qquad\qquad = 5$ Multiply.

c) $b^2 - 4ac = (-2)^2 - 4(3)(-4)$ Replace.

$\qquad\qquad\quad = 4 - (-48)$ Square -2, and then multiply before subtracting.

$\qquad\qquad\quad = 52$ Subtract.

d) $\dfrac{-a^2 - b^2}{c - b} = \dfrac{-3^2 - (-2)^2}{-4 - (-2)} = \dfrac{-9 - 4}{-2} = \dfrac{13}{2}$ ■

Equations

An **equation** is a statement of equality of two expressions. For example,

$$11 - 5 = 6, \qquad x + 3 = 9, \qquad 2x + 5 = 13, \qquad \text{and} \qquad \frac{x}{2} - 4 = 1$$

are equations. In an equation involving a variable, any number that gives a true statement when we replace the variable by the number is said to **satisfy** the equation and is called a **solution** or **root** to the equation. For example, 6 is a solution to $x + 3 = 9$ because $6 + 3 = 9$ is true. Because $5 + 3 = 9$ is false, 5 is not a solution to the equation $x + 3 = 9$. We have **solved** an equation when we have found all solutions to the equation. You will learn how to solve certain equations in the next chapter.

E X A M P L E   5

Satisfying an equation

Determine whether the given number is a solution to the equation following it.

a) $6, 3x - 7 = 9$

b) $-3, \dfrac{2x - 4}{5} = -2$

c) $-5, -x - 2 = 3(x + 6)$

Solution

a) Replace x by 6 in the equation $3x - 7 = 9$:

$$3(6) - 7 = 9$$
$$18 - 7 = 9$$
$$11 = 9 \qquad \text{False.}$$

The number 6 is not a solution to the equation $3x - 7 = 9$.

b) Replace x by -3 in the equation $\dfrac{2x - 4}{5} = -2$:

$$\frac{2(-3) - 4}{5} = -2$$

$$\frac{-10}{5} = -2$$

$$-2 = -2 \qquad \text{True.}$$

The number -3 is a solution to the equation.

c) Replace x by -5 in $-x - 2 = 3(x + 6)$:

$$-(-5) - 2 = 3(-5 + 6)$$
$$5 - 2 = 3(1)$$
$$3 = 3 \qquad \text{True.}$$

The number -5 is a solution to the equation $-x - 2 = 3(x + 6)$. ∎

Just as we translated verbal expressions into algebraic expressions, we can translate verbal sentences into algebraic equations.

E X A M P L E   6

Writing equations

Translate each sentence into an equation.

a) The sum of x and 7 is 12.

b) The product of 4 and x is the same as the sum of y and 5.

c) The quotient of $x + 3$ and 5 is equal to -1.

Solution

a) $x + 7 = 12$ **b)** $4x = y + 5$ **c)** $\dfrac{x + 3}{5} = -1$ ∎

Applications

Algebraic expressions are used to describe or **model** real-life situations. We can evaluate an algebraic expression for many values of a variable to get a collection of data. A graph (picture) of this data can give us useful information. For example, a forensic scientist can use a graph to estimate the length of a person's femur from the person's height.

E X A M P L E   7

Reading a graph

A forensic scientist uses the expression $69.1 + 2.2F$ as an estimate of the height in centimeters of a male with a femur of length F centimeters (*American Journal of Physical Anthropology,* 1952).

a) If the femur of a male skeleton measures 50.6 cm, then what was the person's height?

b) Use the graph shown in Fig. 1.17 to estimate the length of a femur for a person who is 150 cm tall.

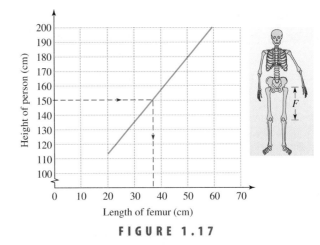

FIGURE 1.17

Solution

a) To find the height of the person, we use $F = 50.6$ in the expression $69.1 + 2.2F$:

$$69.1 + 2.2(50.6) \approx 180.4$$

So the person was approximately 180.4 cm tall.

b) To find the length of a femur for a person who is 150 cm tall, first locate 150 cm on the height scale of the graph in Fig. 1.17. Now draw a horizontal line to the graph and then a vertical line down to the length scale. So the length of femur for a person who is 150 cm tall is approximately 35 cm.   ■

WARM-UPS

True or false? Explain your answer.

1. The expression $2x + 3y$ is referred to as a sum.
2. The expression $5(y - 9)$ is a difference.
3. The expression $2(x + 3y)$ is a product.
4. The expression $\frac{x}{2} + \frac{y}{3}$ is a quotient.
5. The expression $(a - b)(a + b)$ is a product of a sum and a difference.

6. If x is -2, then the value of $2x + 4$ is 8.
7. If $a = -3$, then $a^3 - 5 = 22$.
8. The number 5 is a solution to the equation $2x - 3 = 13$.
9. The product of $x + 3$ and 5 is $(x + 3)5$.
10. The expression $2(x + 7)$ should be read as "the sum of 2 times x plus 7."

1.6   EXERCISES

Reading and Writing   *After reading this section write out the answers to these questions. Use complete sentences.*

1. What is an algebraic expression?

2. What is the difference between an algebraic expression and an arithmetic expression?

3. How can you tell whether an algebraic expression should be referred to as a sum, difference, product, quotient, or square?

4. How do you evaluate an algebraic expression?

5. What is an equation?

6. What is a solution to an equation?

Identify each expression as a sum, difference, product, quotient, square, or cube. See Example 1.

7. $a^3 - 1$

8. $b(b - 1)$

9. $(w - 1)^3$

10. $m^2 + n^2$

11. $3x + 5y$

12. $\dfrac{a - b}{b - a}$

13. $\dfrac{u}{v} - \dfrac{v}{u}$

14. $(s - t)^2$

15. $3(x + 5y)$

16. $a - \dfrac{a}{2}$

17. $\left(\dfrac{2}{z}\right)^2$

18. $(2q - p)^3$

Use the term sum, difference, product, quotient, square, or cube to translate each algebraic expression into a verbal expression. See Example 2.

19. $x^2 - a^2$

20. $a^3 + b^3$

21. $(x - a)^2$

22. $(a + b)^3$

23. $\dfrac{x - 4}{2}$

24. $2(x - 3)$

25. $\dfrac{x}{2} - 4$

26. $2x - 3$

27. $(ab)^3$

28. $a^3 b^3$

Translate each verbal expression into an algebraic expression. See Example 3.

29. The sum of $2x$ and $3y$

30. The product of $5x$ and z

31. The difference of 8 and $7x$

32. The quotient of 6 and $x + 4$

33. The square of $a + b$

34. The difference of a^3 and b^3

35. The product of $x + 9$ and $x + 12$

36. The cube of x

37. The quotient of $x - 7$ and $7 - x$

38. The product of -3 and $x - 1$

Evaluate each expression using $a = -1$, $b = 2$, and $c = -3$. See Example 4.

39. $-(a - b)$

40. $b - a$

41. $-b^2 + 7$

42. $-c^2 - b^2$

43. $c^2 - 2c + 1$

44. $b^2 - 2b + 4$

45. $a^3 - b^3$

46. $b^3 - c^3$

47. $(a - b)(a + b)$

48. $(a - c)(a + c)$

49. $b^2 - 4ac$

50. $a^2 - 4bc$

51. $\dfrac{a - c}{a - b}$

52. $\dfrac{b - c}{b + a}$

53. $\dfrac{2}{a} + \dfrac{6}{b} - \dfrac{9}{c}$

54. $\dfrac{c}{a} + \dfrac{6}{b} - \dfrac{b}{a}$

55. $a \div |-a|$

56. $|a| \div a$

57. $|b| - |a|$

58. $|c| + |b|$

59. $-|-a - c|$

60. $-|-a - b|$

61. $(3 - |a - b|)^2$

62. $(|b + c| - 2)^3$

Determine whether the given number is a solution to the equation following it. See Example 5.

63. 2, $3x + 7 = 13$

64. -1, $-3x + 7 = 10$

65. -2, $\dfrac{3x - 4}{2} = 5$

66. -3, $\dfrac{-2x + 9}{3} = 5$

67. -2, $-x + 4 = 6$

68. -9, $-x + 3 = 12$

69. 4, $3x - 7 = x + 1$

70. 5, $3x - 7 = 2x + 1$

71. 3, $-2(x - 1) = 2 - 2x$

72. -8, $x - 9 = -(9 - x)$

73. 1, $x^2 + 3x - 4 = 0$

74. -1, $x^2 + 5x + 4 = 0$

75. 8, $\dfrac{x}{x - 8} = 0$

76. 3, $\dfrac{x - 3}{x + 3} = 0$

77. -6, $\dfrac{x + 6}{x + 6} = 1$

78. 9, $\dfrac{9}{x - 9} = 0$

Translate each sentence into an equation. See Example 6.

79. The sum of $5x$ and $3x$ is $8x$.

80. The sum of $\dfrac{y}{2}$ and 3 is 7.

81. The product of 3 and $x + 2$ is equal to 12.

82. The product of -6 and $7y$ is equal to 13.

83. The quotient of x and 3 is the same as the product of x and 5.

84. The quotient of $x + 3$ and $5y$ is the same as the product of x and y.

85. The square of the sum of a and b is equal to 9.

86. The sum of the squares of a and b is equal to the square of c.

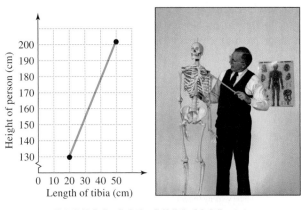 *Use a calculator to find the value of $b^2 - 4ac$ for each of the following choices of a, b, and c.*

87. $a = 4.2$, $b = 6.7$, $c = 1.8$

88. $a = -3.5$, $b = 9.1$, $c = 3.6$

89. $a = -1.2$, $b = 3.2$, $c = 5.6$

90. $a = 2.4$, $b = -8.5$, $c = -5.8$

Solve each problem. See Example 7.

91. *Forensics.* A forensic scientist uses the expression $81.7 + 2.4T$ to estimate the height in centimeters of a

FIGURE FOR EXERCISE 91

male with a tibia of length T centimeters. If a male skeleton has a tibia of length 36.5 cm, then what was the height of the person? Use the accompanying graph to estimate the length of a tibia for a male with a height of 180 cm.

92. Forensics. A forensic scientist uses the expression $72.6 + 2.5T$ to estimate the height in centimeters of a female with a tibia of length T centimeters. If a female skeleton has a tibia of length 32.4 cm, then what was the height of the person? Find the length of your tibia in centimeters, and use the expression from this exercise or the previous exercise to estimate your height.

93. Games behind. In baseball a team's standing is measured by its percentage of wins and by the number of games it is behind the leading team in its division. The expression

$$\frac{(X - x) + (y - Y)}{2}$$

gives the number of games behind for a team with x wins and y losses, where the division leader has X wins and Y losses. The table shown here gives the won-lost records for the American League East on July 9, 1998 (www.espnet.sportszone.com). On that date the Yankees led the division. Fill in the column for the games behind (GB).

	W	L	Pct	GB
NY Yankees	62	20	0.756	–
Boston	52	34	0.605	?
Toronto	46	43	0.517	?
Baltimore	39	50	0.438	?
Tampa Bay	34	53	0.391	?

TABLE FOR EXERCISE 93

94. Fly ball. The approximate distance in feet that a baseball travels when hit at an angle of 45° is given by the expression

$$\frac{(v_0)^2}{32}$$

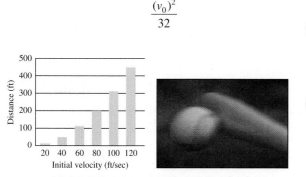

FIGURE FOR EXERCISE 94

where v_0 is the initial velocity in feet per second. If Barry Bonds of the Giants hits a ball at a 45° angle with an initial velocity of 120 feet per second, then how far will the ball travel? Use the accompanying graph to estimate the initial velocity for a ball that has traveled 370 feet.

95. Football field. The expression $2L + 2W$ gives the perimeter of a rectangle with length L and width W. What is the perimeter of a football field with length 100 yards and width 160 feet?

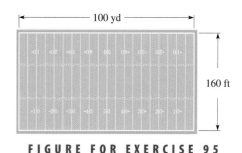

FIGURE FOR EXERCISE 95

96. Crop circles. The expression πr^2 gives the area of a circle with radius r. How many square meters of wheat were destroyed when an alien ship made a crop circle of diameter 25 meters in the wheat field at the Southwind Ranch?

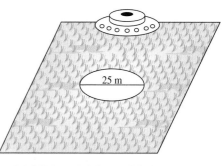

FIGURE FOR EXERCISE 96

GETTING MORE INVOLVED

97. Cooperative learning. Find some examples of algebraic expressions outside of this class, and explain to the class what they are used for.

98. Discussion. Why do we use letters to represent numbers? Wouldn't it be simpler to just use numbers?

99. Writing. Explain why the square of the sum of two numbers is different from the sum of the squares of two numbers.

1.7   PROPERTIES OF THE REAL NUMBERS

Everyone knows that the price of a hamburger plus the price of a Coke is the same as the price of a Coke plus the price of a hamburger. But do you know that this example illustrates the commutative property of addition? The properties of the real numbers are commonly used by anyone who performs the operations of arithmetic. In algebra we must have a thorough understanding of these properties.

The Commutative Properties

We get the same result whether we evaluate $3 + 5$ or $5 + 3$. This example illustrates the commutative property of addition. The fact that $4 \cdot 6$ and $6 \cdot 4$ are equal illustrates the commutative property of multiplication.

> **Commutative Properties**
>
> For any real numbers a and b,
>
> $$a + b = b + a \qquad \text{and} \qquad ab = ba.$$

EXAMPLE 1   **The commutative property of addition**
Use the commutative property of addition to rewrite each expression.
a) $2 + (-10)$ **b)** $8 + x^2$ **c)** $2y - 4x$

Solution
a) $2 + (-10) = -10 + 2$
b) $8 + x^2 = x^2 + 8$
c) $2y - 4x = 2y + (-4x) = -4x + 2y$ ■

EXAMPLE 2   **The commutative property of multiplication**
Use the commutative property of multiplication to rewrite each expression.
a) $n \cdot 3$ **b)** $(x + 2) \cdot 3$ **c)** $5 - yx$

Solution
a) $n \cdot 3 = 3 \cdot n = 3n$ **b)** $(x + 2) \cdot 3 = 3(x + 2)$
c) $5 - yx = 5 - xy$ ■

Addition and multiplication are commutative operations, but what about subtraction and division? Since $5 - 3 = 2$ and $3 - 5 = -2$, subtraction is not commutative. To see that division is not commutative, try dividing $8 among 4 people and $4 among 8 people.

The Associative Properties

Consider the computation of $2 + 3 + 6$. Using the order of operations, we add 2 and 3 to get 5 and then add 5 and 6 to get 11. If we add 3 and 6 first to get 9 and then add 2 and 9, we also get 11. So

$$(2 + 3) + 6 = 2 + (3 + 6).$$

We get the same result for either order of addition. This property is called the **associative property of addition.** The commutative and associative properties of addition are the reason that a hamburger, a Coke, and French fries cost the same as French fries, a hamburger, and a Coke.

We also have an **associative property of multiplication.** Consider the following two ways to find the product of 2, 3, and 4:

$$(2 \cdot 3)4 = 6 \cdot 4 = 24$$
$$2(3 \cdot 4) = 2 \cdot 12 = 24$$

We get the same result for either arrangement.

> ### Associative Properties
>
> For any real numbers a, b, and c,
> $$(a + b) + c = a + (b + c) \qquad \text{and} \qquad (ab)c = a(bc).$$

E X A M P L E   3

Using the properties of multiplication

Use the commutative and associative properties of multiplication and exponential notation to rewrite each product.

a) $(3x)(x)$ **b)** $(xy)(5yx)$

Solution

a) $(3x)(x) = 3(x \cdot x) = 3x^2$

b) The commutative and associative properties of multiplication allow us to rearrange the multiplication in any order. We generally write numbers before variables, and we usually write variables in alphabetical order:

$$(xy)(5yx) = 5xxyy = 5x^2y^2 \qquad \blacksquare$$

Consider the expression

$$3 - 9 + 7 - 5 - 8 + 4 - 13.$$

According to the accepted order of operations, we could evaluate this by computing from left to right. However, using the definition of subtraction, we can rewrite this expression as addition:

$$3 + (-9) + 7 + (-5) + (-8) + 4 + (-13)$$

The commutative and associative properties of addition allow us to add these numbers in any order we choose. It is usually faster to add the positive numbers, add the negative numbers, and then combine those two totals:

$$3 + 7 + 4 + (-9) + (-5) + (-8) + (-13) = 14 + (-35) = -21$$

Note that by performing the operations in this manner, we must subtract only once. There is no need to rewrite this expression as we have done here. We can sum the positive numbers and the negative numbers from the original expression and then combine their totals.

E X A M P L E   4

Using the properties of addition

Evaluate.

a) $3 - 7 + 9 - 5$ **b)** $4 - 5 - 9 + 6 - 2 + 4 - 8$

Solution

a) First add the positive numbers and the negative numbers:

$$3 - 7 + 9 - 5 = 12 + (-12)$$
$$= 0$$

b) $4 - 5 - 9 + 6 - 2 + 4 - 8 = 14 + (-24)$
$$= -10 \qquad \blacksquare$$

It is certainly not essential that we evaluate the expressions of Example 4 as shown. We get the same answer by adding and subtracting from left to right. However, in algebra, just getting the answer is not always the most important point. Learning new methods often increases understanding.

Even though addition is associative, subtraction is not an associative operation. For example, $(8 - 4) - 3 = 1$ and $8 - (4 - 3) = 7$. So

$$(8 - 4) - 3 \neq 8 - (4 - 3).$$

We can also use a numerical example to show that division is not associative. For instance, $(16 \div 4) \div 2 = 2$ and $16 \div (4 \div 2) = 8$. So

$$(16 \div 4) \div 2 \neq 16 \div (4 \div 2).$$

The Distributive Property

helpful hint

To visualize the distributive property, we can determine the number of circles shown here in two ways:

oooo ooooo
oooo ooooo
oooo ooooo

There are 3 · 9 or 27 circles, or there are 3 · 4 circles in the first group and 3 · 5 circles in the second group for a total of 27 circles.

If four men and five women pay $3 each for a movie, there are two ways to find the total amount spent:

$$3(4 + 5) = 3 \cdot 9 = 27$$
$$3 \cdot 4 + 3 \cdot 5 = 12 + 15 = 27$$

Since we get $27 either way, we can write

$$3(4 + 5) = 3 \cdot 4 + 3 \cdot 5.$$

We say that the multiplication by 3 is *distributed* over the addition. This example illustrates the **distributive property.**

Consider the following expressions involving multiplication and subtraction:

$$5(6 - 4) = 5 \cdot 2 = 10$$
$$5 \cdot 6 - 5 \cdot 4 = 30 - 20 = 10$$

Since both expressions have the same value, we can write

$$5(6 - 4) = 5 \cdot 6 - 5 \cdot 4.$$

Multiplication by 5 is distributed over each number in the parentheses. This example illustrates that multiplication distributes over subtraction.

> ### Distributive Property
>
> For any real numbers a, b, and c,
>
> $$a(b + c) = ab + ac \qquad \text{and} \qquad a(b - c) = ab - ac.$$

We can use the distributive property to remove parentheses. If we start with $4(x + 3)$ and write

$$4(x + 3) = 4x + 4 \cdot 3 = 4x + 12,$$

we are using it to multiply 4 and $x + 3$ or to remove the parentheses. We wrote the product $4(x + 3)$ as the sum $4x + 12$.

EXAMPLE 5

Writing a product as a sum or difference

Use the distributive property to remove the parentheses.

a) $a(3 - b)$ **b)** $-3(x - 2)$

Solution

a) $a(3 - b) = a3 - ab$ Distributive property
                $= 3a - ab$ $a3 = 3a$

b) $-3(x - 2) = -3x - (-3)(2)$ Distributive property
                 $= -3x - (-6)$ $(-3)(2) = -6$
                 $= -3x + 6$ Simplify.

    When we write a number or an expression as a product, we are **factoring.** If we start with $3x + 15$ and write

$$3x + 15 = 3x + 3 \cdot 5 = 3(x + 5),$$

we are using the distributive property to factor $3x + 15$. We factored out the common factor 3.

EXAMPLE 6

Writing a sum or difference as a product

Use the distributive property to factor each expression.

a) $7x - 21$ **b)** $5a + 5$

Solution

a) $7x - 21 = 7x - 7 \cdot 3$ Write 21 as $7 \cdot 3$.
              $= 7(x - 3)$ Distributive property

b) $5a + 5 = 5a + 5 \cdot 1$ Write 5 as $5 \cdot 1$.
            $= 5(a + 1)$ Factor out the common factor 5.

The Identity Properties

The numbers 0 and 1 have special properties. Multiplication of a number by 1 does not change the number, and addition of 0 to a number does not change the number. That is why 1 is called the **multiplicative identity** and 0 is called the **additive identity.**

> **Identity Properties**
>
> For any real number a,
> $$a \cdot 1 = 1 \cdot a = a \qquad \text{and} \qquad a + 0 = 0 + a = a.$$

The Inverse Properties

The idea of additive inverses was introduced in Section 1.3. Every real number a has an **additive inverse** or **opposite,** $-a$, such that $a + (-a) = 0$. Every nonzero real number a also has a **multiplicative inverse** or **reciprocal,** written $\frac{1}{a}$, such that $a \cdot \frac{1}{a} = 1$. Note that the sum of additive inverses is the additive identity and that the product of multiplicative inverses is the multiplicative identity.

Inverse Properties

For any real number a there is a number $-a$, such that

$$a + (-a) = 0.$$

For any nonzero real number a there is a number $\frac{1}{a}$ such that

$$a \cdot \frac{1}{a} = 1.$$

We are already familiar with multiplicative inverses for rational numbers. For example, the multiplicative inverse of $\frac{2}{3}$ is $\frac{3}{2}$ because

$$\frac{2}{3} \cdot \frac{3}{2} = \frac{6}{6} = 1.$$

E X A M P L E   7

Multiplicative inverses

Find the multiplicative inverse of each number.

a) 5   **b)** 0.3   **c)** $-\dfrac{3}{4}$ **d)** 1.7

Solution

a) The multiplicative inverse of 5 is $\frac{1}{5}$ because

$$5 \cdot \frac{1}{5} = 1.$$

b) To find the reciprocal of 0.3, we first write 0.3 as a ratio of integers:

$$0.3 = \frac{3}{10}$$

The multiplicative inverse of 0.3 is $\frac{10}{3}$ because

$$\frac{3}{10} \cdot \frac{10}{3} = 1.$$

c) The reciprocal of $-\frac{3}{4}$ is $-\frac{4}{3}$ because

$$\left(-\frac{3}{4}\right)\left(-\frac{4}{3}\right) = 1.$$

d) First convert 1.7 to a ratio of integers:

$$1.7 = 1\frac{7}{10} = \frac{17}{10}$$

The multiplicative inverse is $\frac{10}{17}$.

calculator close-up

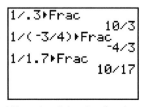

You can find multiplicative inverses with a calculator as shown here.

```
1/.3▶Frac
            10/3
1/(-3/4)▶Frac
            -4/3
1/1.7▶Frac
            10/17
```

When the divisor is a fraction, it must be in parentheses.

Multiplication Property of Zero

Zero has a property that no other number has. Multiplication involving zero always results in zero.

Multiplication Property of Zero

For any real number a,

$$0 \cdot a = 0 \qquad \text{and} \qquad a \cdot 0 = 0.$$

E X A M P L E   8

Identifying the properties

Name the property that justifies each equation.

a) $5 \cdot 7 = 7 \cdot 5$

b) $4 \cdot \dfrac{1}{4} = 1$

c) $1 \cdot 864 = 864$

d) $6 + (5 + x) = (6 + 5) + x$

e) $3x + 5x = (3 + 5)x$

f) $6 + (x + 5) = 6 + (5 + x)$

g) $\pi x^2 + \pi y^2 = \pi(x^2 + y^2)$

h) $325 + 0 = 325$

i) $-3 + 3 = 0$

j) $455 \cdot 0 = 0$

Solution

a) Commutative

b) Multiplicative inverse

c) Multiplicative identity

d) Associative

e) Distributive

f) Commutative

g) Distributive

h) Additive identity

i) Additive inverse

j) Multiplication property of 0

Applications

Reciprocals are important in problems involving work. For example, if you wax one car in 3 hours, then your rate is $\frac{1}{3}$ of a car per hour. If you can wash one car in 12 minutes $\left(\frac{1}{5}\text{ of an hour}\right)$, then you are washing cars at the rate of 5 cars per hour. In general, if you can complete a task in x hours, then your rate is $\frac{1}{x}$ tasks per hour.

E X A M P L E   9

Washing rates

A car wash has two machines. The old machine washes one car in 0.1 hour, while the new machine washes one car in 0.08 hour. If both machines are operating, then at what rate (in cars per hour) are the cars being washed?

helpful hint

When machines or people are working together, we can add their rates provided they do not interfere with each other's work. If operating both car wash machines causes a traffic jam, then the rate together might not be 22.5 cars per hour.

Solution

The old machine is working at the rate of $\frac{1}{0.1}$ cars per hour, and the new machine is working at the rate of $\frac{1}{0.08}$ cars per hour. Their rate working together is the sum of their individual rates:

$$\frac{1}{0.1} + \frac{1}{0.08} = 10 + 12.5 = 22.5$$

So working together, the machines are washing 22.5 cars per hour.

WARM-UPS

True or false? Explain your answer.

1. $24 \div (4 \div 2) = (24 \div 4) \div 2$

2. $1 \div 2 = 2 \div 1$

3. $6 - 5 = -5 + 6$

4. $9 - (4 - 3) = (9 - 4) - 3$

5. Multiplication is a commutative operation.

6. $5x + 5 = 5(x + 1)$ for any value of x.

7. The multiplicative inverse of 0.02 is 50.

8. $-3(x - 2) = -3x + 6$ for any value of x.

9. $3x + 2x = (3 + 2)x$ for any value of x.

10. The additive inverse of 0 is 0.

1.7   EXERCISES

Reading and Writing   *After reading this section write out the answers to these questions. Use complete sentences.*

1. What is the difference between the commutative property of addition and the associative property of addition?

2. Which property involves two different operations?

3. What is factoring?

4. Which two numbers play a prominent role in the properties studied here?

5. What is the purpose of studying the properties of real numbers?

6. What is the relationship between rate and time?

Use the commutative property of addition to rewrite each expression. See Example 1.

7. $9 + r$ 8. $t + 6$ 9. $3(2 + x)$

10. $P(1 + rt)$ 11. $4 - 5x$ 12. $b - 2a$

Use the commutative property of multiplication to rewrite each expression. See Example 2.

13. $x \cdot 6$ 14. $y \cdot (-9)$ 15. $(x - 4)(-2)$

16. $a(b + c)$ 17. $4 - y \cdot 8$ 18. $z \cdot 9 - 2$

Use the commutative and associative properties of multiplication and exponential notation to rewrite each product. See Example 3.

19. $(4w)(w)$ 20. $(y)(2y)$ 21. $3a(ba)$

22. $(x \cdot x)(7x)$ 23. $(x)(9x)(xz)$ 24. $y(y \cdot 5)(wy)$

Evaluate by finding first the sum of the positive numbers and then the sum of the negative numbers. See Example 4.

25. $8 - 4 + 3 - 10$

26. $-3 + 5 - 12 + 10$

27. $8 - 10 + 7 - 8 - 7$

28. $6 - 11 + 7 - 9 + 13 - 2$

29. $-4 - 11 + 7 - 8 + 15 - 20$

30. $-8 + 13 - 9 - 15 + 7 - 22 + 5$

31. $-3.2 + 2.4 - 2.8 + 5.8 - 1.6$

32. $5.4 - 5.1 + 6.6 - 2.3 + 9.1$

33. $3.26 - 13.41 + 5.1 - 12.35 - 5$

34. $5.89 - 6.1 + 8.58 - 6.06 - 2.34$

Use the distributive property to remove the parentheses. See Example 5.

35. $3(x - 5)$ 36. $4(b - 1)$

37. $a(2 + t)$ 38. $b(a + w)$

39. $-3(w - 6)$ 40. $-3(m - 5)$

41. $-4(5 - y)$ 42. $-3(6 - p)$

43. $-1(a - 7)$ 44. $-1(c - 8)$

45. $-1(t + 4)$ 46. $-1(x + 7)$

Use the distributive property to factor each expression. See Example 6.

47. $2m + 12$ 48. $3y + 6$

49. $4x - 4$ 50. $6y + 6$

51. $4y - 16$ 52. $5x + 15$

53. $4a + 8$ 54. $7a - 35$

Find the multiplicative inverse (reciprocal) of each number. See Example 7.

55. $\dfrac{1}{2}$ 56. $\dfrac{1}{3}$ 57. -5

58. -6 59. 7 60. 8

61. 1 62. -1 63. -0.25

64. 0.75 65. 2.5 66. 3.5

Name the property that justifies each equation. See Example 8.

67. $3 \cdot x = x \cdot 3$

68. $x + 5 = 5 + x$

69. $2(x - 3) = 2x - 6$

70. $a(bc) = (ab)c$

71. $-3(xy) = (-3x)y$

72. $3(x + 1) = 3x + 3$

73. $4 + (-4) = 0$

74. $1.3 + 9 = 9 + 1.3$

75. $x^2 \cdot 5 = 5x^2$

76. $0 \cdot \pi = 0$

77. $1 \cdot 3y = 3y$

78. $(0.1)(10) = 1$

79. $2a + 5a = (2 + 5)a$

80. $3 + 0 = 3$

81. $-7 + 7 = 0$

82. $1 \cdot b = b$

83. $(2346)0 = 0$

84. $4x + 4 = 4(x + 1)$

85. $ay + y = y(a + 1)$

86. $ab + bc = b(a + c)$

Complete each equation, using the property named.

87. $a + y =$ _____, commutative

88. $6x + 6 =$ _____, distributive

89. $5(aw) =$ _____, associative

90. $x + 3 =$ _____, commutative

91. $\dfrac{1}{2}x + \dfrac{1}{2} =$ _____, distributive

92. $-3(x - 7) =$ _____, distributive

93. $6x + 15 =$ _____, distributive

94. $(x + 6) + 1 =$ _____, associative

95. $4(0.25) =$ _____, inverse property

96. $-1(5 - y) =$ _____, distributive

97. $0 = 96(\underline{\quad})$, multiplication property of zero

98. $3 \cdot (\underline{\quad}) = 3$, identity property

99. $0.33(\underline{\quad}) = 1$, inverse property

100. $-8(1) =$ _____, identity property

Solve each problem. See Example 9.

101. *Laying bricks.* A bricklayer lays one brick in 0.04 hour, while his apprentice lays one brick in 0.05 hour.

    a) If both are working, then at what combined rate (in bricks per hour) are they laying bricks?

    b) Which person is working faster?

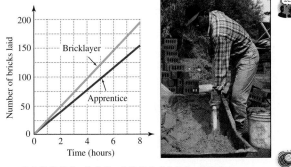

FIGURE FOR EXERCISE 101

102. *Recovering golf balls.* Susan and Joan are diving for golf balls in a large water trap. Susan recovers a golf ball every 0.016 hour while Joan recovers a ball every 0.025 hour. If both are working, then at what rate (in golf balls per hour) are they recovering golf balls?

103. *Population explosion.* In 1998, the population of the earth was increasing by one person every 0.3801 second (*World Population Data Sheet* 1998, www.prb.org).

    a) At what rate in people per second is the population of the earth increasing?

    b) At what rate in people per week is the population of the earth increasing?

104. *Farmland conversion.* The amount of farmland in the United States is decreasing by one acre every 0.00876 hours as farmland is being converted to nonfarm use (American Farmland Trust, www.farmland.org). At what rate in acres per day is the farmland decreasing?

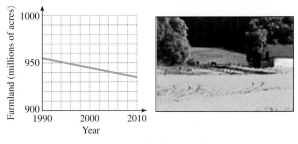

FIGURE FOR EXERCISE 104

GETTING MORE INVOLVED

105. *Writing.* The perimeter of a rectangle is the sum of twice the length and twice the width. Write in words another way to find the perimeter that illustrates the distributive property.

106. *Discussion.* Eldrid bought a loaf of bread for $1.69 and a gallon of milk for $2.29. Using a tax rate of 5%, he correctly figured that the tax on the bread would be 8 cents and the tax on the milk would be 11 cents, for a total of $4.17. However, at the cash register he was correctly charged $4.18. How could this happen? Which property of the real numbers is in question in this case?

107. *Exploration.* Determine whether each of the following pairs of tasks are "commutative." That is, does the order in which they are performed produce the same result?

    a) Put on your coat; put on your hat.

    b) Put on your shirt; put on your coat.

Find another pair of "commutative" tasks and another pair of "noncommutative" tasks.

1.8 USING THE PROPERTIES TO SIMPLIFY EXPRESSIONS

The properties of the real numbers can be helpful when we are doing computations. In this section we will see how the properties can be applied in arithmetic and algebra.

Using the Properties in Computation

The properties of the real numbers can often be used to simplify computations. For example, to find the product of 26 and 200, we can write

$$(26)(200) = (26)(2 \cdot 100)$$
$$= (26 \cdot 2)(100)$$
$$= 52 \cdot 100$$
$$= 5200$$

It is the associative property that allows us to multiply 26 by 2 to get 52, then multiply 52 by 100 to get 5200.

EXAMPLE 1

Using the properties

Use the appropriate property to aid you in evaluating each expression.

a) $347 + 35 + 65$ **b)** $3 \cdot 435 \cdot \dfrac{1}{3}$ **c)** $6 \cdot 28 + 4 \cdot 28$

Solution

a) Notice that the sum of 35 and 65 is 100. So apply the associative property as follows:

$$347 + (35 + 65) = 347 + 100$$
$$= 447$$

b) Use the commutative and associative properties to rearrange this product. We can then do the multiplication quickly:

$$3 \cdot 435 \cdot \frac{1}{3} = 435\left(3 \cdot \frac{1}{3}\right) \quad \text{Commutative and associative properties}$$
$$= 435 \cdot 1 \quad \text{Inverse property}$$
$$= 435 \quad \text{Identity property}$$

c) Use the distributive property to rewrite this expression.

$$6 \cdot 28 + 4 \cdot 28 = (6 + 4)28$$
$$= 10 \cdot 28$$
$$= 280$$ ∎

Like Terms

An expression containing a number or the product of a number and one or more variables raised to powers is called a **term.** For example,

$$-3, \qquad 5x, \qquad -3x^2y, \qquad a, \qquad \text{and} \qquad -abc$$

are terms. The number preceding the variables in a term is called the **coefficient.** In the term $5x$, the coefficient of x is 5. In the term $-3x^2y$ the coefficient of x^2y is -3. In the term a, the coefficient of a is 1 because $a = 1 \cdot a$. In the term $-abc$ the coefficient of abc is -1 because $-abc = -1 \cdot abc$. If two terms contain the same variables with the same exponents, they are called **like terms.** For example, $3x^2$ and $-5x^2$ are like terms, but $3x^2$ and $-5x^3$ are not like terms.

Combining Like Terms

Using the distributive property on an expression involving the sum of like terms allows us to combine the like terms as shown in the next example.

E X A M P L E 2

Combining like terms

Use the distributive property to perform the indicated operations.

a) $3x + 5x$ b) $-5xy - (-4xy)$

Solution

a) $3x + 5x = (3 + 5)x$ Distributive property

         $= 8x$ Add the coefficients.

Because the distributive property is valid for any real numbers, we have $3x + 5x = 8x$ no matter what number is used for x.

b) $-5xy - (-4xy) = [-5 - (-4)]xy$ Distributive property

              $= -1xy$ $-5 - (-4) = -5 + 4 = -1$

              $= -xy$ Multiplying by -1 is the same as taking the opposite.

Of course, we do not want to write out all of the steps shown in Example 2 every time we combine like terms. We can combine like terms as easily as we can add or subtract their coefficients.

E X A M P L E 3

Combining like terms

Perform the indicated operations.

a) $w + 2w$ b) $-3a + (-7a)$ c) $-9x + 5x$

d) $7xy - (-12xy)$ e) $2x^2 + 4x^2$

Solution

a) $w + 2w = 1w + 2w = 3w$ b) $-3a + (-7a) = -10a$

c) $-9x + 5x = -4x$ d) $7xy - (-12xy) = 19xy$

e) $2x^2 + 4x^2 = 6x^2$

C A U T I O N   There are no like terms in expressions such as

$$2 + 5x, \qquad 3xy + 5y, \qquad 3w + 5a, \qquad \text{and} \qquad 3z^2 + 5z$$

The terms in these expressions cannot be combined.

Products and Quotients

In the next example we use the associative property of multiplication to simplify the product of two expressions.

E X A M P L E   4

Finding products

Simplify.

a) $3(5x)$ b) $2\left(\dfrac{x}{2}\right)$ c) $(4x)(6x)$ d) $(-2a)(4b)$

Solution

a) $3(5x) = (3 \cdot 5)x$ Associative property

$\qquad\quad = (15)x$ Multiply

$\qquad\quad = 15x$ Remove unnecessary parentheses.

b) $2\left(\dfrac{x}{2}\right) = 2\left(\dfrac{1}{2} \cdot x\right)$ Multiplying by $\frac{1}{2}$ is the same as dividing by 2.

$\qquad\quad = \left(2 \cdot \dfrac{1}{2}\right)x$ Associative property

$\qquad\quad = 1 \cdot x$ Multiplicative inverse

$\qquad\quad = x$ Multiplicative identity is 1.

c) $(4x)(6x) = 4 \cdot 6 \cdot x \cdot x$ Commutative and associative properties

$\qquad\qquad = 24x^2$ Definition of exponent

d) $(-2a)(4b) = -2 \cdot 4 \cdot a \cdot b = -8ab$ ∎

C A U T I O N Be careful with expressions such as $3(5x)$ and $3(5 + x)$. In $3(5x)$ we multiply 5 by 3 to get $3(5x) = 15x$. In $3(5 + x)$, both 5 and x are multiplied by the 3 to get $3(5 + x) = 15 + 3x$.

In Example 4 we showed how the properties are used to simplify products. However, in practice we usually do not write out any steps for these problems—we can write just the answer.

E X A M P L E   5

Finding products quickly

Find each product.

a) $(-3)(4x)$ b) $(-4a)(-7a)$ c) $(-3a)\left(\dfrac{b}{3}\right)$ d) $6 \cdot \dfrac{x}{2}$

Solution

a) $-12x$ b) $28a^2$ c) $-ab$ d) $3x$ ∎

In Section 1.1 we found the quotient of two numbers by inverting the divisor and then multiplying. Since $a \div b = a \cdot \dfrac{1}{b}$, any quotient can be written as a product.

E X A M P L E   6

Simplifying quotients

Simplify.

a) $\dfrac{10x}{5}$ b) $\dfrac{4x + 8}{2}$

Solution

a) Since dividing by 5 is equivalent to multiplying by $\frac{1}{5}$, we have

$$\frac{10x}{5} = \frac{1}{5}(10x) = \left(\frac{1}{5} \cdot 10\right)x = (2)x = 2x.$$

Note that you can simply divide 10 by 5 to get 2.

b) Since dividing by 2 is equivalent to multiplying by $\frac{1}{2}$, we have

$$\frac{4x + 8}{2} = \frac{1}{2}(4x + 8) = 2x + 4.$$

Note that both 4 and 8 are divided by 2.   ■

C A U T I O N   It is not correct to divide only one term in the numerator by the denominator. For example,

$$\frac{4 + 7}{2} \neq 2 + 7$$

because $\frac{4 + 7}{2} = \frac{11}{2}$ and $2 + 7 = 9$.

Removing Parentheses

Multiplying a number by -1 merely changes the sign of the number. For example,

$$(-1)(7) = -7 \qquad \text{and} \qquad (-1)(-8) = 8.$$

So -1 times a number is the *opposite* of the number. Using variables, we write

$$(-1)x = -x \qquad \text{or} \qquad -1(y + 5) = -(y + 5).$$

When a minus sign appears in front of a sum, we can change the minus sign to -1 and use the distributive property. For example,

$$-(w + 4) = -1(w + 4)$$
$$= (-1)w + (-1)4 \quad \text{Distributive property}$$
$$= -w + (-4) \qquad \text{Note: } -1 \cdot w = -w, \; -1 \cdot 4 = -4$$
$$= -w - 4$$

Note how the minus sign in front of the parentheses caused all of the signs to change: $-(w + 4) = -w - 4$. As another example, consider the following:

$$-(x - 3) = -1(x - 3)$$
$$= (-1)x - (-1)3$$
$$= -x + 3$$

C A U T I O N   When removing parentheses preceded by a minus sign, you must change the sign of *every* term within the parentheses.

calculator

close-up

A negative sign in front of parentheses changes the sign of every term inside the parentheses.

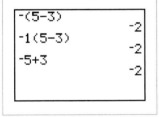

```
-(5-3)
                   -2
-1(5-3)
                   -2
-5+3
                   -2
```

E X A M P L E   7   **Removing parentheses**
Simplify each expression.

a) $5 - (x + 3)$ **b)** $3x - 6 - (2x - 4)$ **c)** $-6x - (-x + 2)$

Solution

a) $5 - (x + 3) = 5 - x - 3$ Change the sign of each term in parentheses.

$\qquad\qquad\qquad = 5 - 3 - x$ Commutative property

$\qquad\qquad\qquad = 2 - x$ Combine like terms.

b) $3x - 6 - (2x - 4) = 3x - 6 - 2x + 4$ — Remove parentheses and change signs.

$\qquad = 3x - 2x - 6 + 4$ — Commutative property

$\qquad = x - 2$ — Combine like terms.

c) $-6x - (-x + 2) = -6x + x - 2$ — Remove parentheses and change signs.

$\qquad = -5x - 2$ — Combine like terms. ■

The commutative and associative properties of addition allow us to rearrange the terms so that we may combine the like terms. However, it is not necessary to actually write down the rearrangement. We can identify the like terms and combine them without rearranging.

E X A M P L E 8 **Simplifying algebraic expressions**

Simplify.

a) $(-2x + 3) + (5x - 7)$ $\qquad$ **b)** $-3x + 6x + 5(4 - 2x)$

c) $-2x(3x - 7) - (x - 6)$ $\qquad$ **d)** $x - 0.02(x + 500)$

Solution

a) $(-2x + 3) + (5x - 7) = 3x - 4$ — Combine like terms.

b) $-3x + 6x + 5(4 - 2x) = -3x + 6x + 20 - 10x$ — Distributive property

$\qquad = -7x + 20$ — Combine like terms.

c) $-2x(3x - 7) - (x - 6) = -6x^2 + 14x - x + 6$ — Distributive property

$\qquad = -6x^2 + 13x + 6$ — Combine like terms.

d) $x - 0.02(x + 500) = 1x - 0.02x - 10$ — Distributive property

$\qquad = 0.98x - 10$ — Combine like terms. ■

WARM-UPS

True or false? Explain your answer.

A statement involving variables should be marked true only if it is true for all values of the variable.

1. $3(x + 6) = 3x + 18$ $\qquad$ **2.** $-3x + 9 = -3(x + 9)$

3. $-1(x - 4) = -x + 4$ $\qquad$ **4.** $3a + 4a = 7a$

5. $(3a)(4a) = 12a$ $\qquad$ **6.** $3(5 \cdot 2) = 15 \cdot 6$

7. $x + x = x^2$ $\qquad$ **8.** $x \cdot x = 2x$

9. $3 + 2x = 5x$ $\qquad$ **10.** $-(5x - 2) = -5x + 2$

1.8 EXERCISES

Reading and Writing *After reading this section write out the answers to these questions. Use complete sentences.*

1. What are like terms?

2. What is the coefficient of a term?

3. What can you do to like terms that you cannot do to unlike terms?

4. What operations can you perform with unlike terms?

5. What is the difference between a positive sign preceding a set of parentheses and a negative sign preceding a set of parentheses?

6. What happens when a number is multiplied by -1?

Use the appropriate properties to evaluate the expressions. See Example 1.

7. $35(200)$

8. $15(300)$

9. $\frac{4}{3}(0.75)$

10. $5(0.2)$

11. $256 + 78 + 22$

12. $12 + 88 + 376$

13. $35 \cdot 3 + 35 \cdot 7$

14. $98 \cdot 478 + 2 \cdot 478$

15. $18 \cdot 4 \cdot 2 \cdot \frac{1}{4}$

16. $19 \cdot 3 \cdot 2 \cdot \frac{1}{3}$

17. $(120)(300)$

18. $150 \cdot 200$

19. $12 \cdot 375(-6 + 6)$

20. $354^2(-2 \cdot 4 + 8)$

21. $78 + 6 + 8 + 4 + 2$

22. $-47 + 12 - 6 - 12 + 6$

Combine like terms where possible. See Examples 2 and 3.

23. $5w + 6w$

24. $4a + 10a$

25. $4x - x$

26. $a - 6a$

27. $2x - (-3x)$

28. $2b - (-5b)$

29. $-3a - (-2a)$

30. $-10m - (-6m)$

31. $-a - a$

32. $a - a$

33. $10 - 6t$

34. $9 - 4w$

35. $3x^2 + 5x^2$

36. $3r^2 + 4r^2$

37. $-4x + 2x^2$

38. $6w^2 - w$

39. $5mw^2 - 12mw^2$

40. $4ab^2 - 19ab^2$

Simplify the following products or quotients. See Examples 4–6.

41. $3(4h)$

42. $2(5h)$

43. $6b(-3)$

44. $-3m(-1)$

45. $(-3m)(3m)$

46. $(2x)(-2x)$

47. $(-3d)(-4d)$

48. $(-5t)(-2t)$

49. $(-y)(-y)$

50. $y(-y)$

51. $-3a(5b)$

52. $-7w(3r)$

53. $-3a(2 + b)$

54. $-2x(3 + y)$

55. $-k(1 - k)$

56. $-t(t - 1)$

57. $\frac{3y}{3}$

58. $\frac{-9t}{9}$

59. $\frac{-15y}{5}$

60. $\frac{-12b}{2}$

61. $2\left(\frac{y}{2}\right)$

62. $6\left(\frac{m}{3}\right)$

63. $8y\left(\frac{y}{4}\right)$

64. $10\left(\frac{2a}{5}\right)$

65. $\frac{6a - 3}{3}$

66. $\frac{-8x + 6}{2}$

67. $\frac{-9x + 6}{-3}$

68. $\frac{10 - 5x}{-5}$

Simplify each expression. See Example 7.

69. $x - (3x - 1)$

70. $4x - (2x - 5)$

71. $5 - (y - 3)$

72. $8 - (m - 6)$

73. $2m + 3 - (m + 9)$

74. $7 - 8t - (2t + 6)$

75. $-3 - (-w + 2)$

76. $-5x - (-2x + 9)$

Simplify the following expressions by combining like terms. See Example 8.

77. $3x + 5x + 6 + 9$

78. $2x + 6x + 7 + 15$

79. $-2x + 3 + 7x - 4$

80. $-3x + 12 + 5x - 9$

81. $3a - 7 - (5a - 6)$

82. $4m - 5 - (m - 2)$

83. $2(a - 4) - 3(-2 - a)$

84. $2(w + 6) - 3(-w - 5)$

85. $-5m + 6(m - 3) + 2m$

86. $-3a + 2(a - 5) + 7a$

87. $5 - 3(x + 2) - 6$

88. $7 + 2(k - 3) - k + 6$

89. $x - 0.05(x + 10)$

90. $x - 0.02(x + 300)$

91. $4.5 - 3.2(x - 5.3) - 8.75$

92. $0.03(4.5x - 3.9) + 0.06(9.8x - 45)$

Simplify each expression.

93. $3x - (4 - x)$

94. $2 + 8x - 11x$

95. $y - 5 - (-y - 9)$

96. $a - (b - c - a)$

97. $7 - (8 - 2y - m)$

98. $x - 8 - (-3 - x)$

99. $\frac{1}{2}(10 - 2x) + \frac{1}{3}(3x - 6)$

100. $\frac{1}{2}(x - 20) - \frac{1}{5}(x + 15)$

101. $0.2(x + 3) - 0.05(x + 20)$

102. $0.08x + 0.12(x + 100)$

103. $2k + 1 - 3(5k - 6) - k + 4$

104. $2w - 3 + 3(w - 4) - 5(w - 6)$

105. $-3m - 3[2m - 3(m + 5)]$

106. $6h + 4[2h - 3(h - 9) - (h - 1)]$

Solve each problem.

107. Married filing jointly. The expression

$$0.15(43,850) + 0.28(x - 43,850)$$

gives the 2000 federal income tax for a married couple filing jointly with a taxable income of x dollars, where x is over \$43,850 but not over \$105,950 (Internal Revenue Service, www.irs.gov).

a) Simplify the expression.

b) Use the expression to find the amount of tax for a couple with a taxable income of \$80,000.

c) Use the graph shown here to estimate the 2000 federal income tax for a couple with a taxable income of \$150,000.

d) Use the graph to find the approximate taxable income for a couple who paid \$70,000 in federal income tax.

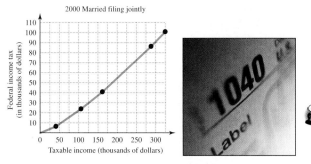

2000 Married filing jointly

FIGURE FOR EXERCISE 107

108. Marriage penalty. The expression

$$0.15(26,250) + 0.28(x - 26,250)$$

gives the 2000 federal income tax for a single taxpayer with taxable income of x dollars, where x is over \$26,250 but not over \$63,550.

a) Simplify the expression.

b) Find the amount of tax for a single taxpayer with taxable income of \$40,000.

c) Who pays more, two single taxpayers with taxable incomes of \$40,000 each or one married couple with taxable income of \$80,000 together? See Exercise 107.

109. Perimeter of a corral. The perimeter of a rectangular corral that has width x feet and length $x + 40$ feet is $2(x) + 2(x + 40)$. Simplify the expression for the perimeter. Find the perimeter if $x = 30$ feet.

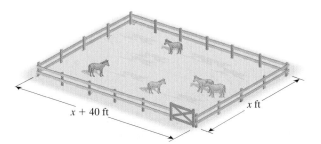

$x + 40$ ft $\qquad$ x ft

FIGURE FOR EXERCISE 109

GETTING MORE INVOLVED

110. Discussion. What is wrong with the way in which each of the following expressions is simplified?

a) $4(2 + x) = 8 + x$

b) $4(2x) = 8 \cdot 4x = 32x$

c) $\dfrac{4 + x}{2} = 2 + x$

d) $5 - (x - 3) = 5 - x - 3 = 2 - x$

111. Discussion. An instructor asked his class to evaluate the expression $1/2x$ for $x = 5$. Some students got 0.1; others got 2.5. Which answer is correct and why?

COLLABORATIVE ACTIVITIES

Remembering the Rules

This chapter reviews different types of numbers used in algebra. This activity will review the rules for the basic operations: addition, subtraction, multiplication, and division for fractions, decimals, and real numbers.

Part I: Remembering the rules. Have each member of your group choose an operation: addition, subtraction, multiplication, or division.

1. Fractions:
  a. Write the rules for working a fraction problem using the operation you have chosen. Use your book as a reference

Grouping: 4 students
Topic: Fractions, decimals, and signed numbers

and consider the following sample problems:

$$\frac{1}{2} + \frac{2}{5}, \quad \frac{1}{3} \cdot \frac{6}{7}, \quad \frac{3}{5} - \frac{1}{3}, \quad \frac{1}{3} \div \frac{2}{3}$$

  b. Starting with addition, each of you will share what he or she has written with the other members of the group. Make additions or corrections if needed.

Switch operations. Each member of the group now takes the operation of the person to his or her right.

2. Decimals: Repeat parts 1(a) and 1(b) for the following sample problems:

$$0.012 + 3, \ 2.1 - 0.25, \ 3.2 \cdot 0.23, \ 5.4 \div 1.2$$

Switch operations. Each member of the group now takes the operation of the person to his or her right.

3. Signed numbers: Repeat parts 1(a) and 1(b) for the following sample problems:

$$-3 + 5, \ 3 - (-2), \ -2 \cdot 3, \ -6 \div -2$$

Part II: Testing your rules. As a group, work through the following problems together, using the rules you have written from Part I. Add any new rules that come up while you work.

1. $\dfrac{1}{3} + \dfrac{2}{5}$

2. $0.076 + 7 + 2.005$

3. $-8 + 17$

4. $2\dfrac{9}{14} - 1\dfrac{5}{7}$

5. $8 - 3.024$

6. $-5 - (-19)$

7. $3\dfrac{1}{3} \cdot 2\dfrac{2}{5}$

8. $0.0723 \cdot 100$

9. $12(-3)(-2)$

10. $1\dfrac{2}{3} \div 5\dfrac{1}{2}$

11. $1.024 \div 3.2$

12. $-405 \div 15$

13. $-2\left(\dfrac{2}{3} - \dfrac{1}{4}\right) \div 2\dfrac{1}{2} + (3.052 - (-0.948))$

Extension: Before each exam, form a study group to review material. Write any new rules or definitions used in each chapter. Save these to study for your final exam.

WRAP-UP   CHAPTER 1

SUMMARY

The Real Numbers		**Examples**
Counting or natural numbers	$\{1, 2, 3, \ldots\}$	
Whole numbers	$\{0, 1, 2, 3, \ldots\}$	
Integers	$\{\ldots, -3, -2, -1, 0, 1, 2, 3, \ldots\}$	
Rational numbers	$\left\{\dfrac{a}{b} \,\middle|\, a \text{ and } b \text{ are integers with } b \neq 0\right\}$	$\dfrac{3}{2}, 5, -6, 0$
Irrational numbers	$\{x \mid x \text{ is a real number that is not rational}\}$	$\sqrt{2}, \sqrt{3}, \pi$
Real numbers	The set of real numbers consists of all rational numbers together with all irrational numbers.	

Fractions		**Examples**
Reducing fractions	$\dfrac{a \cdot c}{b \cdot c} = \dfrac{a}{b}$	$\dfrac{4}{6} = \dfrac{2 \cdot 2}{2 \cdot 3} = \dfrac{2}{3}$
Building up fractions	$\dfrac{a}{b} = \dfrac{a \cdot c}{b \cdot c}$	$\dfrac{3}{8} = \dfrac{3 \cdot 5}{8 \cdot 5} = \dfrac{15}{40}$
Multiplying fractions	$\dfrac{a}{b} \cdot \dfrac{c}{d} = \dfrac{ac}{bd}$	$\dfrac{2}{3} \cdot \dfrac{4}{5} = \dfrac{8}{15}$
Dividing fractions	$\dfrac{a}{b} \div \dfrac{c}{d} = \dfrac{a}{b} \cdot \dfrac{d}{c}$	$\dfrac{2}{3} \div \dfrac{4}{5} = \dfrac{2}{3} \cdot \dfrac{5}{4} = \dfrac{10}{12} = \dfrac{5}{6}$

Adding or subtracting fractions	$\dfrac{a}{b} + \dfrac{c}{b} = \dfrac{a+c}{b}$	$\dfrac{1}{5} + \dfrac{2}{5} = \dfrac{3}{5}$
	$\dfrac{a}{b} - \dfrac{c}{b} = \dfrac{a-c}{b}$	$\dfrac{3}{5} - \dfrac{2}{5} = \dfrac{1}{5}$
Least common denominator	The smallest number that is a multiple of all denominators.	$\dfrac{1}{4} + \dfrac{1}{6} = \dfrac{3}{12} + \dfrac{2}{12} = \dfrac{5}{12}$

Operations with Real Numbers

Examples

Absolute value	$	a	= \begin{cases} a & \text{if } a \text{ is positive or zero} \\ -a & \text{if } a \text{ is negative} \end{cases}$	$	3	= 3,	0	= 0$ $	-3	= 3$
Sum of two numbers with like signs	Add their absolute values. The sum has the same sign as the given numbers.	$-3 + (-4) = -7$
Sum of two numbers with unlike signs (and different absolute values)	Subtract the absolute values of the numbers. The answer is positive if the number with the larger absolute value is positive. The answer is negative if the number with the larger absolute value is negative.	$-4 + 7 = 3$ $-7 + 4 = -3$
Sum of opposites	The sum of any number and its opposite is 0.	$-6 + 6 = 0$
Subtraction of signed numbers	$a - b = a + (-b)$ Subtract any number by adding its opposite.	$3 - 5 = 3 + (-5) = -2$ $4 - (-3) = 4 + 3 = 7$
Product or quotient	Like signs $\leftrightarrow$ Positive result Unlike signs $\leftrightarrow$ Negative result	$(-3)(-2) = 6$ $(-8) \div 2 = -4$
Definition of exponents	For any counting number n, $a^n = \underbrace{a \cdot a \cdot a \cdot \ldots \cdot a}_{n \text{ factors}}.$	$2^3 = 2 \cdot 2 \cdot 2 = 8$
Order of operations	No parentheses or absolute value present: 1. Exponential expressions 2. Multiplication and division 3. Addition and subtraction With parentheses or absolute value: First evaluate within each set of parentheses or absolute value, using the order of operations.	 $5 + 2^3 = 13$ $2 + 3 \cdot 5 = 17$ $4 + 5 \cdot 3^2 = 49$ $(2 + 3)(5 - 7) = -10$ $2 + 3	2 - 5	= 11$

Properties of the Real Numbers

Examples

| Commutative properties | $a + b = b + a$
 $a \cdot b = b \cdot a$ | $5 + 7 = 7 + 5$
 $6 \cdot 3 = 3 \cdot 6$ |

Associative properties	$a + (b + c) = (a + b) + c$ $a \cdot (b \cdot c) = (a \cdot b) \cdot c$	$1 + (2 + 3) = (1 + 2) + 3$ $2(3 \cdot 4) = (2 \cdot 3)4$
Distributive properties	$a(b + c) = ab + ac$ $a(b - c) = ab - ac$	$2(3 + x) = 6 + 2x$ $-2(x - 5) = -2x + 10$
Identity properties	$a + 0 = a$ and $0 + a = a$ Zero is the additive identity. $1 \cdot a = a$ and $a \cdot 1 = a$ One is the multiplicative identity.	$5 + 0 = 0 + 5 = 5$ $7 \cdot 1 = 1 \cdot 7 = 7$
Inverse properties	For any real number a, there is a number $-a$ (additive inverse or opposite) such that $a + (-a) = 0$ and $-a + a = 0.$ For any nonzero real number a there is a number $\frac{1}{a}$ (multiplicative inverse or reciprocal) such that $\qquad a \cdot \dfrac{1}{a} = 1$ and $\dfrac{1}{a} \cdot a = 1.$	$3 + (-3) = 0$ $-3 + 3 = 0$ $3 \cdot \dfrac{1}{3} = 1$ $\dfrac{1}{3} \cdot 3 = 1$
Multiplication property of 0	$a \cdot 0 = 0$ and $0 \cdot a = 0$	$5 \cdot 0 = 0$ $0(-7) = 0$

ENRICHING YOUR MATHEMATICAL WORD POWER

For each mathematical term, choose the correct meaning.

1.   like terms
  a.   terms that are identical
  b.   the terms of a sum
  c.   terms that have the same variables with the same exponents
  d.   terms with the same variables

2.   equivalent fractions
  a.   identical fractions
  b.   fractions that represent the same number
  c.   fractions with the same denominator
  d.   fractions with the same numerator

3.   variable
  a.   a letter that is used to represent some numbers
  b.   the letter x
  c.   an equation with a letter in it
  d.   not the same

4.   reducing
  a.   less than
  b.   losing weight
  c.   making equivalent
  d.   dividing out common factors

5.   lowest terms
  a.   numerator is smaller than the denominator
  b.   no common factors
  c.   the best interest rate
  d.   when the numerator is 1

6.   additive inverse
  a.   the number -1
  b.   the number 0
  c.   the opposite of addition
  d.   opposite

7.   order of operations
  a.   the order in which operations are to be performed in the absence of grouping symbols
  b.   the order in which the operations were invented
  c.   the order in which operations are written
  d.   a list of operations in alphabetical order

8.   least common denominator
  a.   the smallest divisor of all denominators
  b.   the denominator that appears the least
  c.   the smallest identical denominator
  d.   the least common multiple of the denominators

9. absolute value
 a. definite value
 b. positive number
 c. distance from 0 on the number line
 d. the opposite of a number

10. natural numbers
 a. the counting numbers
 b. numbers that are not irrational
 c. the nonnegative numbers
 d. numbers that we find in nature

REVIEW EXERCISES

1.1 *Which of the numbers* $-\sqrt{5}, -2, 0, 1, 2, 3.14, \pi,$ *and* 10 *are*

1. whole numbers?

2. natural numbers?

3. integers?

4. rational numbers?

5. irrational numbers?

6. real numbers?

study tip

Note how the review exercises are arranged according to the sections in this chapter. If you are having trouble with a certain type of problem, refer back to the appropriate section for examples and explanations.

True or false? Explain your answer.

7. Every whole number is a rational number.

8. Zero is not a rational number.

9. The counting numbers between -4 and 4 are $-3, -2,$ $-1, 0, 1, 2,$ and 3.

10. There are infinitely many integers.

11. The set of counting numbers smaller than the national debt is infinite.

12. The decimal number 0.25 is a rational number.

13. Every integer greater than -1 is a whole number.

14. Zero is the only number that is neither rational nor irrational.

1.2 *Perform the indicated operations.*

15. $\dfrac{1}{3} + \dfrac{3}{8}$ **16.** $\dfrac{2}{3} - \dfrac{1}{4}$ **17.** $\dfrac{3}{5} \cdot 10$

18. $\dfrac{3}{5} \div 10$ **19.** $\dfrac{2}{5} \cdot \dfrac{15}{14}$ **20.** $7 \div \dfrac{1}{2}$

21. $4 + \dfrac{2}{3}$ **22.** $\dfrac{7}{12} - \dfrac{1}{4}$ **23.** $\dfrac{1}{2} + \dfrac{1}{3} + \dfrac{1}{4}$

24. $\dfrac{3}{4} \div 9$

1.3 *Evaluate.*

25. $-5 + 7$ **26.** $-9 + (-4)$

27. $35 - 48$ **28.** $-3 - 9$

29. $-12 + 5$ **30.** $-12 - 5$

31. $-12 - (-5)$ **32.** $-9 - (-9)$

33. $-0.05 + 12$ **34.** $-0.03 + (-2)$

35. $-0.1 - (-0.05)$ **36.** $-0.3 + 0.3$

37. $\dfrac{1}{3} - \dfrac{1}{2}$ **38.** $-\dfrac{2}{3} + \dfrac{1}{4}$

39. $-\dfrac{1}{3} + \left(-\dfrac{2}{5}\right)$ **40.** $\dfrac{1}{3} - \left(-\dfrac{1}{4}\right)$

1.4 *Evaluate.*

41. $(-3)(5)$ **42.** $(-9)(-4)$

43. $(-8) \div (-2)$ **44.** $50 \div (-5)$

45. $\dfrac{-20}{-4}$ **46.** $\dfrac{30}{-5}$

47. $\left(-\dfrac{1}{2}\right)\left(-\dfrac{1}{3}\right)$ **48.** $8 \div \left(-\dfrac{1}{3}\right)$

49. $-0.09 \div 0.3$ **50.** $4.2 \div (-0.3)$

51. $(0.3)(-0.8)$ **52.** $0 \div (-0.0538)$

53. $(-5)(-0.2)$ **54.** $\dfrac{1}{2}(-12)$

1.5 *Evaluate.*

55. $3 + 7(9)$ **56.** $(3 + 7)9$

57. $(3 + 4)^2$ **58.** $3 + 4^2$

59. $3 + 2 \cdot |5 - 6 \cdot 4|$ **60.** $3 - (8 - 9)$

61. $(3 - 7) - (4 - 9)$ **62.** $3 - 7 - 4 - 9$

63. $-2 - 4(2 - 3 \cdot 5)$ **64.** $3^2 - 7 + 5^2$

65. $3^2 - (7 + 5)^2$ **66.** $|4 - 6 \cdot 3| - |7 - 9|$

67. $\dfrac{-3 - 5}{2 - (-2)}$ **68.** $\dfrac{1 - 9}{4 - 6}$

69. $\dfrac{6 + 3}{3} - 5 \cdot 4 + 1$ **70.** $\dfrac{2 \cdot 4 + 4}{3} - 3(1 - 2)$

1.6 *Let* $a = -1$, $b = -2$, *and* $c = 3$. *Find the value of each algebraic expression.*

71. $b^2 - 4ac$ **72.** $a^2 - 4b$

73. $(c - b)(c + b)$ **74.** $(a + b)(a - b)$

75. $a^2 + 2ab + b^2$ **76.** $a^2 - 2ab + b^2$

77. $a^3 - b^3$ **78.** $a^3 + b^3$

79. $\dfrac{b + c}{a + b}$ **80.** $\dfrac{b - c}{2b - a}$

81. $|a - b|$ **82.** $|b - a|$

83. $(a + b)c$ **84.** $ac + bc$

Determine whether the given number is a solution to the equation following it.

85. $4, 3x - 2 = 10$ **86.** $1, 5(x + 3) = 20$

87. $-6, \dfrac{3x}{2} = 9$ **88.** $-30, \dfrac{x}{3} - 4 = 6$

89. $15, \dfrac{x + 3}{2} = 9$ **90.** $1, \dfrac{12}{2x + 1} = 4$

91. $4, -x - 3 = 1$ **92.** $7, -x + 1 = 6$

1.7 *Name the property that justifies each statement.*

93. $a(x + y) = ax + ay$

94. $3(4y) = (3 \cdot 4)y$

95. $(0.001)(1000) = 1$

96. $xy = yx$

97. $0 + y = y$

98. $325 \cdot 1 = 325$

99. $3 + (2 + x) = (3 + 2) + x$

100. $2x - 6 = 2(x - 3)$

101. $5 \cdot 200 = 200 \cdot 5$

102. $3 + (x + 2) = (x + 2) + 3$

103. $-50 + 50 = 0$

104. $43 \cdot 59 \cdot 82 \cdot 0 = 0$

105. $12 \cdot 1 = 12$

106. $3x + 1 = 1 + 3x$

1.8 *Simplify by combining like terms.*

107. $3a + 7 - (4a - 5)$

108. $2m + 6 - (m - 2)$

109. $2a(3a - 5) + 4a$

110. $3a(a - 5) + 5a(a + 2)$

111. $3(t - 2) - 5(3t - 9)$

112. $2(m + 3) - 3(3 - m)$

113. $0.1(a + 0.3) - (a + 0.6)$

114. $0.1(x + 0.3) - (x - 0.9)$

115. $0.05(x - 20) - 0.1(x + 30)$

116. $0.02(x - 100) + 0.2(x - 50)$

117. $5 - 3x(-5x - 2) + 12x^2$

118. $7 - 2x(3x - 7) - x^2$

119. $-(a - 2) - 2 - a$

120. $-(w - y) - 3(y - w)$

121. $x(x + 1) + 3(x - 1)$

122. $y(y - 2) + 3(y + 1)$

MISCELLANEOUS

Evaluate each expression. Use a calculator to check.

123. $752(-13) + 752(13)$ **124.** $75 - (-13)$

125. $|15 - 23|$ **126.** $4^2 - 6^2$

127. $-6^2 + 3(5)$ **128.** $(0.03)(-200)$

129. $\dfrac{2}{5} + \dfrac{1}{10}$ **130.** $\dfrac{2 + 1}{5 + 10}$

131. $(0.05) \div (-0.1)$

132. $(4 - 9)^2 + (2 \cdot 3 - 1)^2$

133. $2\left(-\dfrac{1}{2}\right)^2 + \left(-\dfrac{1}{2}\right) - 1$

134. $\left(-\dfrac{6}{7}\right)\left(\dfrac{21}{26}\right)$

Simplify each expression if possible.

135. $\dfrac{2x + 4}{2}$ **136.** $4(2x)$

137. $4 + 2x$ **138.** $4(2 + x)$

139. $4 \cdot \dfrac{x}{2}$ **140.** $4 - (x - 2)$

141. $-4(x - 2)$ **142.** $(4x)(2x)$

143. $4x + 2x$ **144.** $2 + (x + 4)$

145. $4 \cdot \dfrac{x}{4}$ **146.** $4 \cdot \dfrac{3x}{2}$

147. $2 \cdot x \cdot 4$ **148.** $4 - 2(2 - x)$

Solve each problem.

149. *Telemarketing.* Brenda and Nicki sell memberships in an automobile club over the telephone. Brenda sells one membership every 0.125 hour, and Nicki sells one membership every 0.1 hour. At what rate (in memberships per hour) are the memberships being sold when both are working?

150. *High-income bracket.* Greg Maddux of the Atlanta Braves had a taxable income of approximately \$11.5 million for 1997 (*The Nando Times,* www.nando.net). The

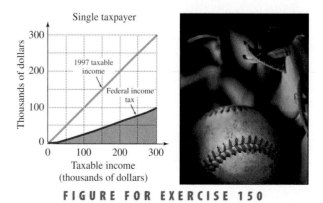

FIGURE FOR EXERCISE 150

expression

$$81,646.50 + 0.396(x - 271,050)$$

represents the 1997 federal income tax in dollars for a taxpayer with x dollars of taxable income, where x is over $271,050.

a) Simplify the expression.

b) Use the simplified expression to find the amount of federal income tax for Greg Maddux.

CHAPTER 1 TEST

Which of the numbers -3, $-\sqrt{3}$, $-\dfrac{1}{4}$, 0, $\sqrt{5}$, π, and 8 are

1. whole numbers?

2. integers?

3. rational numbers?

4. irrational numbers?

Evaluate each expression.

5. $6 + 3(-9)$

6. $(-2)^2 - 4(-2)(-1)$

7. $\dfrac{-3^2 - 9}{3 - 5}$

8. $-5 + 6 - 12 + 4$

9. $0.05 - 1$

10. $(5 - 9)(5 + 9)$

11. $(878 + 89) + 11$

12. $6 + |3 - 5(2)|$

13. $8 - 3|7 - 10|$

14. $(839 + 974)[3(-4) + 12]$

15. $974(7) + 974(3)$

16. $-\dfrac{2}{3} + \dfrac{3}{8}$

17. $(-0.05)(400)$

18. $\left(-\dfrac{3}{4}\right)\left(\dfrac{2}{9}\right)$

19. $13 \div \left(-\dfrac{1}{3}\right)$

Identify the property that justifies each equation.

20. $2(x + 7) = 2x + 14$

21. $48 \cdot 1000 = 1000 \cdot 48$

22. $2 + (6 + x) = (2 + 6) + x$

23. $-348 + 348 = 0$

24. $1 \cdot (-6) = -6$

25. $0 \cdot 388 = 0$

Use the distributive property to write each sum or difference as a product.

26. $3x + 30$

27. $7w - 7$

Simplify each expression.

28. $6 + 4x + 2x$

29. $6 + 4(x - 2)$

30. $5x - (3 - 2x)$

31. $x + 10 - 0.1(x + 25)$

32. $2a(4a - 5) - 3a(-2a - 5)$

33. $\dfrac{6x + 12}{6}$

34. $8 \cdot \dfrac{t}{2}$

35. $(-9xy)(-6xy)$

Evaluate each expression if $a = -2$, $b = 3$, and $c = 4$.

36. $b^2 - 4ac$

37. $\dfrac{a - b}{b - c}$

38. $(a - c)(a + c)$

Determine whether the given number is a solution to the equation following it.

39. -2, $3x - 4 = 2$

40. 13, $\dfrac{x + 3}{8} = 2$

41. -3, $-x + 5 = 8$

Solve each problem.

42. Burke and Nora deliver pizzas for Godmother's Pizza. Burke averages one delivery every 0.25 hour, and Nora averages one delivery every 0.2 hour. At what rate (in deliveries per hour) are the deliveries made when both are working?

43. A forensic scientist uses the expression $80.405 + 3.660R - 0.06(A - 30)$ to estimate the height in centimeters for a male with a radius (bone in the forearm) of length R centimeters and age A in years, where A is over 30. Simplify the expression. Use the expression to estimate the height of an 80-year-old male with a radius of length 25 cm.

Linear Equations in One Variable

 ome ancient peoples chewed on leaves to cure their headaches. Thousands of years ago, the Egyptians used honey, salt, cedar oil, and sycamore bark to cure illnesses. Currently, some of the indigenous people of North America use black birch as a pain reliever.

Today, we are grateful for the miracles of modern medicine and the seemingly simple cures for illnesses. From our own experiences we know that just the right amount of a drug can work wonders but too much of a drug can do great harm. Even though physicians often prescribe the same drug for children and adults, the amount given must be tailored to the individual. The portion of a drug given to children is usually reduced on the basis of factors such as the weight and height of the child. Likewise, older adults frequently need a lower dosage of medication than what would be prescribed for a younger, more active person.

Various algebraic formulas have been developed for determining the proper dosage for a child and an older adult. In Exercises 83 and 84 of Section 2.4 you will see two formulas that are used to determine a child's dosage by using the adult dosage and the child's age.

2.1 THE ADDITION AND MULTIPLICATION PROPERTIES OF EQUALITY

In Section 1.6, you learned that an equation is a statement that two expressions are equal. You also learned how to determine whether a number is a solution to an equation. In this section you will learn systematic procedures for finding solutions to equations.

The Addition Property of Equality

The equations that we work with in this section and the next two are called linear equations.

> **Linear Equation**
>
> A **linear equation in one variable** x is an equation that can be written in the form
>
> $$ax + b = 0,$$
>
> where a and b are real numbers and $a \neq 0$.

An equation such as $2x + 3 = 0$ is a linear equation. We also refer to equations such as

$$x + 8 = 0, \quad 3x = 7, \quad 2x + 5 = 9 - 5x, \quad \text{and} \quad 3 + 5(x - 1) = -7 + x$$

as linear equations, because these equations could be written in the form $ax + b = 0$ using the properties of equality, which we are about to discuss.

If two workers have equal salaries and each gets a $1000 raise, then they still have equal salaries after the raise. This example illustrates the addition property of equality.

> **The Addition Property of Equality**
>
> Adding the same number to both sides of an equation does not change the solution to the equation. In symbols, if $a = b$, then
>
> $$a + c = b + c.$$

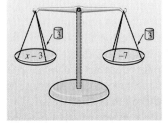
To **solve** an equation means to find all of the solutions to the equation. The set of all solutions to an equation is the **solution set** to the equation. Equations that have the same solution set are **equivalent equations.** In our first example, we will use the addition property of equality to solve an equation.

EXAMPLE 1

Adding the same number to both sides

Solve $x - 3 = -7$.

Solution

We can remove the 3 from the left side of the equation by adding 3 to each side of the equation:

$$x - 3 = -7$$
$$x - 3 + 3 = -7 + 3 \qquad \text{Add 3 to each side.}$$
$$x + 0 = -4 \qquad \text{Simplify each side.}$$
$$x = -4 \qquad \text{Zero is the additive identity.}$$

Since -4 satisfies the last equation, it should also satisfy the original equation because all of the previous equations are equivalent. Check that -4 satisfies the original equation by replacing x by -4:

$$x - 3 = -7 \quad \text{Original equation}$$
$$-4 - 3 = -7 \quad \text{Replace } x \text{ by } -4.$$
$$-7 = -7 \quad \text{Simplify.}$$

Since $-4 - 3 = -7$ is correct, $\{-4\}$ is the solution set to the equation.   ■

In Example 1, we used addition to isolate the variable on the left-hand side of the equation. Once the variable is isolated, we can determine the solution to the equation. Because subtraction is defined in terms of addition, we can also use subtraction to isolate the variable.

E X A M P L E   2

Subtracting the same number from both sides

Solve $9 + x = -2$.

Solution

We can remove the 9 from the left side by adding -9 to each side or by subtracting 9 from each side of the equation:

$$9 + x = -2$$
$$9 + x - 9 = -2 - 9 \quad \text{Subtract 9 from each side.}$$
$$x = -11 \quad \text{Simplify each side.}$$

> **study tip**
>
> Think! Thinking is the manipulation of facts and principles. Your thinking will be as clear as your understanding of the facts and principles.

Check that -11 satisfies the original equation by replacing x by -11:

$$9 + x = -2 \quad \text{Original equation}$$
$$9 + (-11) = -2 \quad \text{Replace } x \text{ by } -11.$$

Since $9 + (-11) = -2$ is correct, $\{-11\}$ is the solution set to the equation.   ■

Our goal in solving equations is to isolate the variable. In the first two examples, the variable was isolated on the left side of the equation. In the next example, we isolate the variable on the right side of the equation.

E X A M P L E   3

Isolating the variable on the right side

Solve $\frac{1}{2} = -\frac{1}{4} + y$.

Solution

We can remove $-\frac{1}{4}$ from the right side by adding $\frac{1}{4}$ to both sides of the equation:

$$\frac{1}{2} = -\frac{1}{4} + y$$
$$\frac{1}{2} + \frac{1}{4} = -\frac{1}{4} + y + \frac{1}{4} \quad \text{Add } \frac{1}{4} \text{ to each side.}$$
$$\frac{3}{4} = y \quad \text{Simplify each side.}$$

Check that $\frac{3}{4}$ satisfies the original equation by replacing y by $\frac{3}{4}$:

$$\frac{1}{2} = -\frac{1}{4} + y \qquad \text{Original equation}$$

$$\frac{1}{2} = -\frac{1}{4} + \frac{3}{4} \qquad \text{Replace } y \text{ by } \frac{3}{4}.$$

$$\frac{1}{2} = \frac{2}{4} \qquad \text{Simplify.}$$

Since $\frac{1}{2} = \frac{2}{4}$ is correct, $\left\{\frac{3}{4}\right\}$ is the solution set to the equation.

The Multiplication Property of Equality

To isolate a variable that is involved in a product or a quotient, we need the multiplication property of equality.

> **The Multiplication Property of Equality**
>
> Multiplying both sides of an equation by the same nonzero number does not change the solution to the equation. In symbols, if $a = b$ and $c \neq 0$, then
>
> $$ac = bc.$$

If the variable in an equation is divided by a number, we can isolate the variable by multiplying each side of the equation by the divisor as in the next example.

E X A M P L E   4   **Multiplying both sides by the same number**

Solve $\frac{z}{2} = 6$.

Solution

We isolate the variable z by multiplying each side of the equation by 2.

$$\frac{z}{2} = 6 \qquad \text{Original equation}$$

$$2 \cdot \frac{z}{2} = 2 \cdot 6 \qquad \text{Multiply each side by 2.}$$

$$1z = 12 \qquad \text{Because } 2 \cdot \frac{z}{2} = 2 \cdot \frac{1}{2}z = 1z$$

$$z = 12 \qquad \text{Multiplicative identity}$$

Because $\frac{12}{2} = 6$, $\{12\}$ is the solution set to the equation.

Because dividing by a number is the same as multiplying by its reciprocal, the multiplication property of equality allows us to divide each side of the equation by any nonzero number.

EXAMPLE 5

Dividing both sides by the same number

Solve $-5w = 30$.

Solution

Since w is multiplied by -5, we can isolate w by multiplying by $-\frac{1}{5}$ or by dividing each side by -5:

$$-5w = 30 \qquad \text{Original equation}$$

$$\frac{-5w}{-5} = \frac{30}{-5} \qquad \text{Divide each side by } -5.$$

$$1 \cdot w = -6 \qquad \text{Because } \frac{-5}{-5} = 1$$

$$w = -6 \qquad \text{Multiplicative identity}$$

Because $-5(-6) = 30$, $\{-6\}$ is the solution set to the equation. ■

In the next example, the coefficient of the variable is a fraction. We could divide each side by the coefficient as we did in Example 5, but it is easier to multiply each side by the reciprocal of the coefficient.

EXAMPLE 6

Multiplying by the reciprocal

Solve $\frac{2}{3}p = 40$.

Solution

Multiply each side by $\frac{3}{2}$, the reciprocal of $\frac{2}{3}$, to isolate p on the left side.

$$\frac{2}{3}p = 40$$

$$\frac{3}{2} \cdot \frac{2}{3}p = \frac{3}{2} \cdot 40 \qquad \text{Multiply each side by } \frac{3}{2}.$$

$$1 \cdot p = 60 \qquad \text{Multiplicative inverses}$$

$$p = 60 \qquad \text{Multiplicative identity}$$

Because $\frac{2}{3} \cdot 60 = 40$, we can be sure that the solution set is $\{60\}$. ■

> **helpful hint**
>
> You could solve this equation by multiplying each side by 3 to get $2p = 120$, and then dividing each side by 2 to get $p = 60$.

If the coefficient of the variable is an integer, we usually divide each side by that integer, as in Example 5. If the coefficient of the variable is a fraction, we usually multiply each side by the reciprocal of the fraction as in Example 6.

If $-x$ appears in an equation, we can multiply by -1 to get x, because $-1(-x) = -(-x) = x$.

EXAMPLE 7

Multiplying by -1

Solve $-h = 12$.

Solution

Multiply each side by -1 to get h on the left side.

$$-h = 12$$

$$-1(-h) = -1 \cdot 12$$

$$h = -12$$

Since $-(-12) = 12$, the solution set is $\{-12\}$. ■

Variables on Both Sides

In the next example, the variable occurs on both sides of the equation. Because the variable represents a real number, we can still isolate the variable by using the addition property of equality.

E X A M P L E   8

Subtracting an algebraic expression from both sides

Solve $-9 + 6y = 7y$.

Solution

The expression $6y$ can be removed from the left side of the equation by subtracting $6y$ from both sides.

$$-9 + 6y = 7y$$
$$-9 + 6y - 6y = 7y - 6y \qquad \text{Subtract } 6y \text{ from each side.}$$
$$-9 = y \qquad \text{Simplify each side.}$$

Check by replacing y by -9 in the original equation:

$$-9 + 6(-9) = 7(-9)$$
$$-63 = -63$$

The solution set to the equation is $\{-9\}$. ■

Applications

In the next example, we use the multiplication property of equality in an applied situation.

E X A M P L E   9

Population density

In 1990, San Francisco had $\frac{2}{3}$ as many people per hectare as New York (U.S. Bureau of Census, www.census.gov). The population density of San Francisco was 60 people per hectare. What was the population density of New York?

Solution

If p represents the population density of New York, then $\frac{2}{3}p = 60$. To find p, solve the equation:

$$\frac{2}{3}p = 60$$

$$\frac{3}{2} \cdot \frac{2}{3}p = \frac{3}{2} \cdot 60 \qquad \text{Multiply each side by } \frac{3}{2}.$$

$$p = 90 \qquad \text{Simplify.}$$

So the population density of New York was 90 people per hectare. ■

WARM-UPS

True or false? Explain your answer.

1. The solution to $x - 5 = 5$ is 10.
2. The equation $\frac{x}{2} = 4$ is equivalent to the equation $x = 8$.
3. To solve $\frac{3}{4}y = 12$, we should multiply each side by $\frac{3}{4}$.
4. The equation $\frac{x}{7} = 4$ is equivalent to $\frac{1}{7}x = 4$.
5. Multiplying each side of an equation by any real number will result in an equation that is equivalent to the original equation.
6. To isolate t in $2t = 7 + t$, subtract t from each side.
7. To solve $\frac{2r}{3} = 30$, we should multiply each side by $\frac{3}{2}$.
8. Adding any real number to both sides of an equation will result in an equation that is equivalent to the original equation.
9. The equation $5x = 0$ is equivalent to $x = 0$.
10. The solution to $2x - 3 = x + 1$ is 4.

2.1   EXERCISES

Reading and Writing   After reading this section, write out the answers to these questions. Use complete sentences.

1. What does the addition property of equality say?

2. What are equivalent equations?

3. What is the multiplication property of equality?

4. What is a linear equation in one variable?

5. How can you tell if your solution to an equation is correct?

6. To obtain an equivalent equation, what are you not allowed to do to both sides of the equation?

Solve each equation. Show your work and check your answer. See Example 1.

7. $x - 6 = -5$
8. $x - 7 = -2$
9. $-13 + x = -4$
10. $-8 + x = -12$

11. $y - \frac{1}{2} = \frac{1}{2}$
12. $y - \frac{1}{4} = \frac{1}{2}$
13. $w - \frac{1}{3} = \frac{1}{3}$
14. $w - \frac{1}{3} = \frac{1}{2}$

Solve each equation. Show your work and check your answer. See Example 2.

15. $x + 3 = -6$
16. $x + 4 = -3$
17. $12 + x = -7$
18. $19 + x = -11$
19. $t + \frac{1}{2} = \frac{3}{4}$
20. $t + \frac{1}{3} = 1$
21. $\frac{1}{19} + m = \frac{1}{19}$
22. $\frac{1}{3} + n = \frac{1}{2}$

Solve each equation. Show your work and check your answer. See Example 3.

23. $2 = x + 7$
24. $3 = x + 5$
25. $-13 = y - 9$
26. $-14 = z - 12$
27. $0.5 = -2.5 + x$
28. $0.6 = -1.2 + x$
29. $\frac{1}{8} = -\frac{1}{8} + r$
30. $\frac{1}{6} = -\frac{1}{6} + h$

Solve each equation. Show your work and check your answer. See Example 4.

31. $\frac{x}{2} = -4$
32. $\frac{x}{3} = -6$
33. $0.03 = \frac{y}{60}$
34. $0.05 = \frac{y}{80}$

35. $\dfrac{a}{2} = \dfrac{1}{3}$

36. $\dfrac{b}{2} = \dfrac{1}{5}$

37. $\dfrac{1}{6} = \dfrac{c}{3}$

38. $\dfrac{1}{12} = \dfrac{d}{3}$

Solve each equation. Show your work and check your answer. See Example 5.

39. $-3x = 15$

40. $-5x = -20$

41. $20 = 4y$

42. $18 = -3a$

43. $2w = 2.5$

44. $-2x = -5.6$

45. $5 = 20x$

46. $-3 = 27d$

Solve each equation. Show your work and check your answer. See Example 6.

47. $\dfrac{3}{2}x = -3$

48. $\dfrac{2}{3}x = -8$

49. $90 = \dfrac{3y}{4}$

50. $14 = \dfrac{7y}{8}$

51. $-\dfrac{3}{5}w = -\dfrac{1}{3}$

52. $-\dfrac{5}{2}t = -\dfrac{3}{5}$

53. $\dfrac{2}{3} = -\dfrac{4x}{3}$

54. $\dfrac{1}{14} = -\dfrac{6p}{7}$

Solve each equation. Show your work and check your answer. See Example 7.

55. $-x = 8$

56. $-x = 4$

57. $-y = -\dfrac{1}{3}$

58. $-y = -\dfrac{7}{8}$

59. $3.4 = -z$

60. $4.9 = -t$

61. $-k = -99$

62. $-m = -17$

Solve each equation. Show your work and check your answer. See Example 8.

63. $4x = 3x - 7$

64. $3x = 2x + 9$

65. $9 - 6y = -5y$

66. $12 - 18w = -17w$

67. $-6x = 8 - 7x$

68. $-3x = -6 - 4x$

69. $\dfrac{1}{2}c = 5 - \dfrac{1}{2}c$

70. $-\dfrac{1}{2}h = 13 - \dfrac{3}{2}h$

Use the appropriate property of equality to solve each equation.

71. $12 = x + 17$

72. $-3 = x + 6$

73. $\dfrac{3}{4}y = -6$

74. $\dfrac{5}{9}z = -10$

75. $-3.2 + x = -1.2$

76. $t - 3.8 = -2.9$

77. $2a = \dfrac{1}{3}$

78. $-3w = \dfrac{1}{2}$

79. $-9m = 3$

80. $-4h = -2$

81. $-b = -44$

82. $-r = 55$

83. $\dfrac{2}{3}x = \dfrac{1}{2}$

84. $\dfrac{3}{4}x = \dfrac{1}{3}$

85. $-5x = 7 - 6x$

86. $-\dfrac{1}{2} + 3y = 4y$

87. $\dfrac{5a}{7} = -10$

88. $\dfrac{7r}{12} = -14$

89. $\dfrac{1}{2}v = -\dfrac{1}{2}v + \dfrac{3}{8}$

90. $\dfrac{1}{3}s + \dfrac{7}{9} = \dfrac{4}{3}s$

Solve each problem by writing and solving an equation. See Example 9.

91. *Cigarette consumption.* In 1999, cigarette consumption in the U.S. was 125 packs per capita. This rate of consumption was $\dfrac{5}{8}$ of what it was in 1980. Find the rate of consumption in 1980.

92. *World grain demand.* Freeport McMoRan projects that in 2010 world grain supply will be 1.8 trillion metric tons and the supply will be only $\dfrac{3}{4}$ of world grain demand. What will world grain demand be in 2010?

FIGURE FOR EXERCISE 92

93. *Advancers and decliners.* On Thursday, $\dfrac{13}{25}$ of the stocks traded on the New York Stock Exchange advanced in price. If 1495 stocks advanced, then how many stocks were traded on that day?

94. *Accidental deaths.* In 1996, $\dfrac{23}{50}$ of all accidental deaths in the U.S. were the result of automobile accidents (National Center for Health Statistics, www.nchs.gov). If there were 43,194 deaths due to automobile accidents, then how many accidental deaths were there in 1996?

2.2 SOLVING GENERAL LINEAR EQUATIONS

All of the equations that we solved in Section 2.1 required only a single application of a property of equality. In this section you will solve equations that require more than one application of a property of equality.

Equations of the Form $ax + b = 0$

To solve an equation of the form $ax + b = 0$ we might need to apply both the addition property of equality and the multiplication property of equality.

E X A M P L E   1

Using the addition and multiplication properties of equality

Solve $3r - 5 = 0$.

Solution

To isolate r, first add 5 to each side, then divide each side by 3.

$$3r - 5 = 0 \qquad \text{Original equation}$$

$$3r - 5 + 5 = 0 + 5 \qquad \text{Add 5 to each side.}$$

$$3r = 5 \qquad \text{Combine like terms.}$$

$$\frac{3r}{3} = \frac{5}{3} \qquad \text{Divide each side by 3.}$$

$$r = \frac{5}{3} \qquad \text{Simplify.}$$

Checking $\frac{5}{3}$ in the original equation gives

$$3 \cdot \frac{5}{3} - 5 = 5 - 5 = 0.$$

So $\left\{ \frac{5}{3} \right\}$ is the solution set to the equation.

> **helpful hint**
>
> If we divide by 3 first, we would get $r - \frac{5}{3} = 0$. Then add $\frac{5}{3}$ to each side to get $r = \frac{5}{3}$. Although we get the correct answer, we usually save division to the last step so that fractions do not appear until necessary.

CAUTION   It is usually best to use the addition property of equality first and the multiplication property last.

E X A M P L E   2

Using the addition and multiplication properties of equality

Solve $-\frac{2}{3}x + 8 = 0$.

Solution

To isolate x, first subtract 8 from each side, then multiply each side by $-\frac{3}{2}$.

$$-\frac{2}{3}x + 8 = 0 \qquad \text{Original equation}$$

$$-\frac{2}{3}x + 8 - 8 = 0 - 8 \qquad \text{Subtract 8 from each side.}$$

$$-\frac{2}{3}x = -8 \qquad \text{Combine like terms.}$$

$$-\frac{3}{2}\left(-\frac{2}{3}x\right) = -\frac{3}{2}(-8) \qquad \text{Multiply each side by } -\frac{3}{2}.$$

$$x = 12 \qquad \text{Simplify.}$$

study tip

As you leave class, talk to a classmate about what happened in class. What was the class about? What new terms were mentioned and what do they mean? How does this lesson fit in with the last lesson?

Checking 12 in the original equation gives

$$-\frac{2}{3}(12) + 8 = -8 + 8 = 0.$$

So $\{12\}$ is the solution set to the equation.   ■

Equations of the Form $ax + b = cx + d$

In solving equations our goal is to isolate the variable. We use the addition property of equality to eliminate unwanted terms. Note that it does not matter whether the variable ends up on the right or left side. For some equations we will perform fewer steps if we isolate the variable on the right side.

E X A M P L E   3
Isolating the variable on the right side

Solve $3w - 8 = 7w$.

Solution

To eliminate the $3w$ from the left side, we can subtract $3w$ from both sides.

$$3w - 8 = 7w \qquad \text{Original equation}$$
$$3w - 8 - 3w = 7w - 3w \qquad \text{Subtract } 3w \text{ from each side.}$$
$$-8 = 4w \qquad \text{Simplify each side.}$$
$$-\frac{8}{4} = \frac{4w}{4} \qquad \text{Divide each side by 4.}$$
$$-2 = w \qquad \text{Simplify.}$$

To check, replace w with -2 in the original equation:

$$3w - 8 = 7w \qquad \text{Original equation}$$
$$3(-2) - 8 = 7(-2)$$
$$-14 = -14$$

Since -2 satisfies the original equation, the solution set is $\{-2\}$.   ■

You should solve the equation in Example 3 by isolating the variable on the left side to see that it takes more steps. In the next example, it is simplest to isolate the variable on the left side.

E X A M P L E   4
Isolating the variable on the left side

Solve $\frac{1}{2}b - 8 = 12$.

Solution

To eliminate the 8 from the left side, we add 8 to each side.

$$\frac{1}{2}b - 8 = 12 \qquad \text{Original equation}$$

$$\frac{1}{2}b - 8 + 8 = 12 + 8 \qquad \text{Add 8 to each side.}$$

$$\frac{1}{2}b = 20 \qquad \text{Simplify each side.}$$

$$2 \cdot \frac{1}{2}b = 2 \cdot 20 \qquad \text{Multiply each side by 2.}$$

$$b = 40 \qquad \text{Simplify.}$$

To check, replace b with 40 in the original equation:

$$\frac{1}{2}b - 8 = 12 \quad \text{Original equation}$$

$$\frac{1}{2}(40) - 8 = 12$$

$$12 = 12$$

Since 40 satisfies the original equation, the solution set is $\{40\}$.   ■

It does not matter whether the variable is isolated on the left side or the right side. However, you should decide where you want the variable isolated before you begin to solve the equation.

E X A M P L E   5

Solving $ax + b = cx + d$

Solve $2m - 4 = 4m - 10$.

Solution

First, we decide to isolate the variable on the left side. So we must eliminate the 4 from the left side and eliminate $4m$ from the right side:

$$2m - 4 = 4m - 10$$
$$2m - 4 + 4 = 4m - 10 + 4 \quad \text{Add 4 to each side.}$$
$$2m = 4m - 6 \quad \text{Simplify each side.}$$
$$2m - 4m = 4m - 6 - 4m \quad \text{Subtract } 4m \text{ from each side.}$$
$$-2m = -6 \quad \text{Simplify each side.}$$
$$\frac{-2m}{-2} = \frac{-6}{-2} \quad \text{Divide each side by } -2.$$
$$m = 3 \quad \text{Simplify.}$$

> **study tip**
>
> Take good notes. Good note taking is the key to mastering the material. It helps you to concentrate in class and provides a source for review. Learn to listen effectively.

To check, replace m by 3 in the original equation:

$$2m - 4 = 4m - 10 \quad \text{Original equation}$$
$$2 \cdot 3 - 4 = 4 \cdot 3 - 10$$
$$2 = 2$$

Since 3 satisfies the original equation, the solution set is $\{3\}$.   ■

Equations with Parentheses

Equations that contain parentheses or like terms on the same side should be simplified as much as possible before applying any properties of equality.

E X A M P L E   6

Simplifying before using properties of equality

Solve $2(q - 3) + 5q = 8(q - 1)$.

Solution

First remove parentheses and combine like terms on each side of the equation.

$$2(q - 3) + 5q = 8(q - 1) \quad \text{Original equation}$$
$$2q - 6 + 5q = 8q - 8 \quad \text{Distributive property}$$
$$7q - 6 = 8q - 8 \quad \text{Combine like terms.}$$
$$7q - 6 + 6 = 8q - 8 + 6 \quad \text{Add 6 to each side.}$$
$$7q = 8q - 2 \quad \text{Combine like terms.}$$
$$7q - 8q = 8q - 2 - 8q \quad \text{Subtract } 8q \text{ from each side.}$$

$$-q = -2$$
$$-1(-q) = -1(-2) \qquad \text{Multiply each side by } -1.$$
$$q = 2 \qquad \text{Simplify.}$$

To check, we replace q by 2 in the original equation and simplify:

$$2(q - 3) + 5q = 8(q - 1) \quad \text{Original equation}$$
$$2(2 - 3) + 5(2) = 8(2 - 1) \quad \text{Replace } q \text{ by 2.}$$
$$2(-1) + 10 = 8(1)$$
$$8 = 8$$

Because both sides have the same value, the solution set is {2}. ■

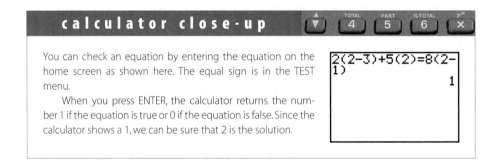

calculator close-up

You can check an equation by entering the equation on the home screen as shown here. The equal sign is in the TEST menu.

When you press ENTER, the calculator returns the number 1 if the equation is true or 0 if the equation is false. Since the calculator shows a 1, we can be sure that 2 is the solution.

```
2(2-3)+5(2)=8(2-
1)
                 1
```

Linear equations can vary greatly in appearance, but there is a strategy that you can use for solving any of them. The following strategy summarizes the techniques that we have been using in the examples. Keep it in mind when you are solving linear equations.

Strategy for Solving Equations

1. Remove parentheses and combine like terms to simplify each side as much as possible.
2. Use the addition property of equality to get like terms from opposite sides onto the same side so that they may be combined.
3. The multiplication property of equality is generally used last.
4. Multiply each side of $-x = a$ by -1 to get $x = -a$.
5. Check that the solution satisfies the original equation.

Applications

Linear equations occur in business situations where there is a fixed cost and a per item cost. A mail order company might charge $3 plus $2 per CD for shipping and handling. A lawyer might charge $300 plus $65 per hour for handling your lawsuit. AT&T might charge 10 cents per minute plus $4.95 for long distance calls. The next example illustrates the kind of problem that can be solved in this situation.

E X A M P L E 7

Long distance charges

With AT&T's One Rate plan you are charged 10 cents per minute plus $4.95 for long distance service for one month. If a long distance bill is $8.65, then what is the number of minutes used?

Solution

Let x represent the number of minutes of calls in the month. At \$0.10 per minute, the cost for x minutes is the product $0.10x$ dollars. Since there is a fixed cost of \$4.95, an expression for the total cost is $0.10x + 4.95$ dollars. Since the total cost is \$8.65, we have $0.10x + 4.95 = 8.65$. Solve this equation to find x.

$$0.10x + 4.95 = 8.65$$
$$0.10x + 4.95 - 4.95 = 8.65 - 4.95 \qquad \text{Subtract 4.95 from each side.}$$
$$0.10x = 3.70 \qquad \text{Simplify.}$$
$$\frac{0.10x}{0.10} = \frac{3.70}{0.10} \qquad \text{Divide each side by 0.10.}$$
$$x = 37 \qquad \text{Simplify.}$$

So the bill is for 37 minutes. ■

WARM-UPS

True or false? Explain your answer.

1. The solution to $4x - 3 = 3x$ is 3.
2. The equation $2x + 7 = 8$ is equivalent to $2x = 1$.
3. To solve $3x - 5 = 8x + 7$, you should add 5 to each side and subtract $8x$ from each side.
4. To solve $5 - 4x = 9 + 7x$, you should subtract 9 from each side and then subtract $7x$ from each side.
5. Multiplying each side of an equation by the same nonzero real number will result in an equation that is equivalent to the original equation.
6. To isolate y in $3y - 7 = 6$, divide each side by 3 and then add 7 to each side.
7. To solve $\frac{3w}{4} = 300$, we should multiply each side by $\frac{4}{3}$.
8. The equation $-n = 9$ is equivalent to $n = -9$.
9. The equation $-y = -7$ is equivalent to $y = 7$.
10. The solution to $7x = 5x$ is 0.

2.2 EXERCISES

Reading and Writing *After reading this section, write out the answers to these questions. Use complete sentences.*

1. What properties of equality do you apply to solve $ax + b = 0$?

2. Which property of equality is usually applied last?

3. What property of equality is used to solve $-x = 8$?

4. What is usually the first step in solving an equation involving parentheses?

Solve each equation. Show your work and check your answer. See Examples 1 and 2.

5. $5a - 10 = 0$
6. $8y + 24 = 0$
7. $-3y - 6 = 0$
8. $-9w - 54 = 0$
9. $3x - 2 = 0$
10. $5y + 1 = 0$
11. $2p + 5 = 0$
12. $9z - 8 = 0$
13. $\frac{1}{2}w - 3 = 0$
14. $\frac{3}{8}t + 6 = 0$
15. $-\frac{2}{3}x + 8 = 0$
16. $-\frac{1}{7}z - 5 = 0$
17. $-m + \frac{1}{2} = 0$
18. $-y - \frac{3}{4} = 0$

19. $3p + \dfrac{1}{2} = 0$

20. $9z - \dfrac{1}{4} = 0$

Solve each equation. See Examples 3 and 4.

21. $6x - 8 = 4x$

22. $9y + 14 = 2y$

23. $4z = 5 - 2z$

24. $3t = t - 3$

25. $4a - 9 = 7$

26. $7r + 5 = 47$

27. $9 = -6 - 3b$

28. $13 = 3 - 10s$

29. $\dfrac{1}{2}w - 4 = 13$

30. $\dfrac{1}{3}q + 13 = -5$

31. $6 - \dfrac{1}{3}d = \dfrac{1}{3}d$

32. $9 - \dfrac{1}{2}a = \dfrac{1}{4}a$

33. $2w - 0.4 = 2$

34. $10h - 1.3 = 6$

35. $x = 3.3 - 0.1x$

36. $y = 2.4 - 0.2y$

Solve each equation. See Example 5.

37. $3x - 3 = x + 5$

38. $9y - 1 = 6y + 5$

39. $4 - 7d = 13 - 4d$

40. $y - 9 = 12 - 6y$

41. $c + \dfrac{1}{2} = 3c - \dfrac{1}{2}$

42. $x - \dfrac{1}{4} = \dfrac{1}{2} - x$

43. $\dfrac{2}{3}a - 5 = \dfrac{1}{3}a + 5$

44. $\dfrac{1}{2}t - 3 = \dfrac{1}{4}t - 9$

Solve each equation. See Example 6.

45. $5(a - 1) + 3 = 28$

46. $2(w + 4) - 1 = 1$

47. $2 - 3(q - 1) = 10 - (q + 1)$

48. $-2(y - 6) = 3(7 - y) - 5$

49. $2(x - 1) + 3x = 6x - 20$

50. $3 - (r - 1) = 2(r + 1) - r$

51. $2\left(y - \dfrac{1}{2}\right) = 4\left(y - \dfrac{1}{4}\right) + y$

52. $\dfrac{1}{2}(4m - 6) = \dfrac{2}{3}(6m - 9) + 3$

Solve each linear equation. Show your work and check your answer.

53. $5t = -2 + 4t$

54. $8y = 6 + 7y$

55. $3x - 7 = 0$

56. $5x + 4 = 0$

57. $-x + 6 = 5$

58. $-x - 2 = 9$

59. $-9 - a = -3$

60. $4 - r = 6$

61. $2q + 5 = q - 7$

62. $3z - 6 = 2z - 7$

63. $-3x + 1 = 5 - 2x$

64. $5 - 2x = 6 - x$

65. $-12 - 5x = -4x + 1$

66. $-3x - 4 = -2x + 8$

67. $3x + 0.3 = 2 + 2x$

68. $2y - 0.05 = y + 1$

69. $k - 0.6 = 0.2k + 1$

70. $2.3h + 6 = 1.8h - 1$

71. $0.2x - 4 = 0.6 - 0.8x$

72. $0.3x = 1 - 0.7x$

73. $-3(k - 6) = 2 - k$

74. $-2(h - 5) = 3 - h$

75. $2(p + 1) - p = 36$

76. $3(q + 1) - q = 23$

77. $7 - 3(5 - u) = 5(u - 4)$

78. $v - 4(4 - v) = -2(2v - 1)$

79. $4(x + 3) = 12$

80. $5(x - 3) = -15$

81. $\dfrac{w}{5} - 4 = -6$

82. $\dfrac{q}{2} + 13 = -22$

83. $\dfrac{2}{3}y - 5 = 7$

84. $\dfrac{3}{4}u - 9 = -6$

85. $4 - \dfrac{2n}{5} = 12$

86. $9 - \dfrac{2m}{7} = 19$

87. $-\dfrac{1}{3}p - \dfrac{1}{2} = \dfrac{1}{2}$

88. $-\dfrac{3}{4}z - \dfrac{2}{3} = \dfrac{1}{3}$

89. $3.5x - 23.7 = -38.75$

90. $3(x - 0.87) - 2x = 4.98$

Solve each problem. See Example 7.

91. *The practice.* A lawyer charges $300 plus $65 per hour for a divorce. If the total charge for Bill's divorce was $1405, then for what number of hours did the lawyer work on the case?

92. *The plumber.* A plumber charges $45 plus $26 per hour to unclog drains. If the bill for unclogging Tamika's drain was $123, then for how many hours did the plumber work?

93. *Celsius temperature.* If the air temperature in Quebec is 68° Fahrenheit, then the solution to the equation $\dfrac{9}{5}C + 32 = 68$ gives the Celsius temperature of the air. Find the Celsius temperature.

94. *Fahrenheit temperature.* Water boils at 212°F.
   a) Use the accompanying graph to determine the Celsius temperature at which water boils.
   b) Find the Fahrenheit temperature of hot tap water at 70°C by solving the equation

$$70 = \dfrac{5}{9}(F - 32).$$

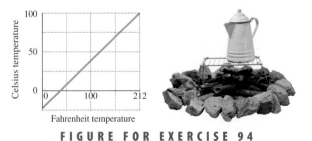

FIGURE FOR EXERCISE 94

95. *Rectangular patio.* If a rectangular patio has a length that is 3 feet longer than its width and a perimeter of 42 feet, then the width can be found by solving the equation $2x + 2(x + 3) = 42$. What is the width?

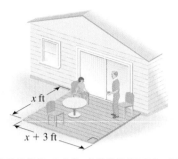

FIGURE FOR EXERCISE 95

96. *Perimeter of a triangle.* The perimeter of the triangle shown in the accompanying figure is 12 meters. Determine the values of x, $x + 1$, and $x + 2$ by solving the equation

$$x + (x + 1) + (x + 2) = 12.$$

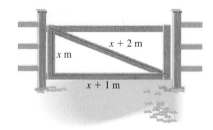

FIGURE FOR EXERCISE 96

97. *Cost of a car.* Jane paid 9% sales tax and a $150 title and license fee when she bought her new Saturn for a total of $16,009.50. If x represents the price of the car, then x satisfies $x + 0.09x + 150 = 16,009.50$. Find the price of the car by solving the equation.

98. *Cost of labor.* An electrician charged Eunice $29.96 for a service call plus $39.96 per hour for a total of $169.82 for installing her electric dryer. If n represents the number of hours for labor, then n satisfies

$$39.96n + 29.96 = 169.82.$$

Find n by solving this equation.

2.3   IDENTITIES, CONDITIONAL EQUATIONS, AND INCONSISTENT EQUATIONS

In this section, we will solve more equations of the type that we solved in Sections 2.1 and 2.2. However, some equations in this section have infinitely many solutions, and some have no solution.

In this section

- Identities
- Conditional Equations
- Inconsistent Equations
- Equations Involving Fractions
- Equations Involving Decimals
- Simplifying the Process

Identities

It is easy to find equations that are satisfied by any real number that we choose as a replacement for the variable. For example, the equations

$$x \div 2 = \frac{1}{2}x, \qquad x + x = 2x, \qquad \text{and} \qquad x + 1 = x + 1$$

are satisfied by all real numbers. The equation

$$\frac{5}{x} = \frac{5}{x}$$

is satisfied by any real number except 0 because division by 0 is undefined.

> **Identity**
>
> An equation that is satisfied by every real number for which both sides are defined is called an **identity.**

We cannot recognize that the equation in the next example is an identity until we have simplified each side.

E X A M P L E   1

Solving an identity

Solve $7 - 5(x - 6) + 4 = 3 - 2(x - 5) - 3x + 28$.

Solution

We first use the distributive property to remove the parentheses:

$$7 - 5(x - 6) + 4 = 3 - 2(x - 5) - 3x + 28$$
$$7 - 5x + 30 + 4 = 3 - 2x + 10 - 3x + 28$$
$$41 - 5x = 41 - 5x \qquad \text{Combine like terms.}$$

This last equation is true for any value of x because the two sides are identical. So the solution set to the original equation is the set of all real numbers. ■

> **CAUTION**   If you get an equation in which both sides are identical, as in Example 1, there is no need to continue to simplify the equation. If you do continue, you will eventually get $0 = 0$, from which you can still conclude that the equation is an identity.

Conditional Equations

The statement $2x + 4 = 10$ is true only on condition that we choose $x = 3$. The equation $x^2 = 4$ is satisfied only if we choose $x = 2$ or $x = -2$. These equations are called conditional equations.

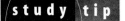
Conditional Equation

A **conditional equation** is an equation that is satisfied by at least one real number but is not an identity.

Every equation that we solved in Sections 2.1 and 2.2 is a conditional equation.

Inconsistent Equations

It is easy to find equations that are false no matter what number we use to replace the variable. Consider the equation

$$x = x + 1.$$

If we replace x by 3, we get $3 = 3 + 1$, which is false. If we replace x by 4, we get $4 = 4 + 1$, which is also false. Clearly, there is no number that will satisfy $x = x + 1$. Other examples of equations with no solutions include

$$x = x - 2, \qquad x - x = 5, \qquad \text{and} \qquad 0 \cdot x + 6 = 7.$$

Inconsistent Equation

An equation that has no solution is called an **inconsistent equation.**

The solution set to an inconsistent equation has no members. The set with no members is called the **empty set** and it is denoted by the symbol $\varnothing$.

E X A M P L E   2

Solving an inconsistent equation

Solve $2 - 3(x - 4) = 4(x - 7) - 7x$.

Solution

Use the distributive property to remove the parentheses:

$$2 - 3(x - 4) = 4(x - 7) - 7x \qquad \text{The original equation}$$
$$2 - 3x + 12 = 4x - 28 - 7x \qquad \text{Distributive property}$$
$$14 - 3x = -28 - 3x \qquad \text{Combine like terms on each side.}$$
$$14 - 3x + 3x = -28 - 3x + 3x \qquad \text{Add } 3x \text{ to each side.}$$
$$14 = -28 \qquad \text{Simplify.}$$

The last equation is not true for any x. So the solution set to the original equation is the empty set, $\varnothing$. The equation is inconsistent.   ■

Keep the following points in mind in solving equations.

Summary: Identities and Inconsistent Equations

1. An equation that is equivalent to an equation in which both sides are identical is an identity. The equation is satisfied by all real numbers for which both sides are defined.

2. An equation that is equivalent to an equation that is always false is inconsistent. The equation has no solution. The solution set is the empty set, $\varnothing$.

Equations Involving Fractions

We solved some equations involving fractions in Sections 2.1 and 2.2. Here, we will solve equations with fractions by eliminating all fractions in the first step. All of the fractions will be eliminated if we multiply each side by the least common denominator.

E X A M P L E   3

Multiplying by the least common denominator

Solve $\frac{y}{2} - 1 = \frac{y}{3} + 1$

Solution

The least common denominator (LCD) for the denominators 2 and 3 is 6. Since both 2 and 3 divide into 6 evenly, multiplying each side by 6 will eliminate the fractions:

$$6\left(\frac{y}{2} - 1\right) = 6\left(\frac{y}{3} + 1\right) \qquad \text{Multiply each side by 6.}$$

$$6 \cdot \frac{y}{2} - 6 \cdot 1 = 6 \cdot \frac{y}{3} + 6 \cdot 1 \qquad \text{Distributive property}$$

$$3y - 6 = 2y + 6 \qquad \text{Simplify: } 6 \cdot \frac{y}{2} = 3y$$

$$3y = 2y + 12 \qquad \text{Add 6 to each side.}$$

$$y = 12 \qquad \text{Subtract } 2y \text{ from each side.}$$

Check 12 in the original equation:

$$\frac{12}{2} - 1 = \frac{12}{3} + 1$$

$$5 = 5$$

Since 12 satisfies the original equation, the solution set is {12}.   ■

---

helpful hint

Note that the fractions in Example 3 will be eliminated if you multiply each side of the equation by any number divisible by both 2 and 3. For example, multiplying by 24 yields

$$12y - 24 = 8y + 24$$
$$4y = 48$$
$$y = 12.$$

Equations involving fractions are usually easier to solve if we first multiply each side by the LCD of the fractions.

Equations Involving Decimals

When an equation involves decimal numbers, we can work with the decimal numbers or we can eliminate all of the decimal numbers by multiplying both sides by 10, or 100, or 1000, and so on. Multiplying a decimal number by 10 moves the decimal point one place to the right. Multiplying by 100 moves the decimal point two places to the right, and so on.

E X A M P L E   4   **An equation involving decimals**

Solve $0.3p + 8.04 = 12.6$.

Solution

The largest number of decimal places appearing in the decimal numbers of the equation is two (in the number 8.04). Therefore we multiply each side of the equation by 100 because multiplying by 100 moves decimal points two places to the right:

$$0.3p + 8.04 = 12.6 \qquad \text{Original equation}$$

$$100(0.3p + 8.04) = 100(12.6) \qquad \text{Multiplication property of equality}$$

$$100(0.3p) + 100(8.04) = 100(12.6) \qquad \text{Distributive property}$$

$$30p + 804 = 1260$$

$$30p + 804 - 804 = 1260 - 804 \qquad \text{Subtract 804 from each side.}$$

$$30p = 456$$

$$\frac{30p}{30} = \frac{456}{30} \qquad \text{Divide each side by 30.}$$

$$p = 15.2$$

You can use a calculator to check that

$$0.3(15.2) + 8.04 = 12.6.$$

The solution set is $\{15.2\}$.

E X A M P L E   5   **Another equation with decimals**

Solve $0.5x + 0.4(x + 20) = 13.4$.

Solution

First use the distributive property to remove the parentheses:

$$0.5x + 0.4(x + 20) = 13.4 \qquad \text{Original equation}$$

$$0.5x + 0.4x + 8 = 13.4 \qquad \text{Distributive property}$$

$$10(0.5x + 0.4x + 8) = 10(13.4) \qquad \text{Multiply each side by 10.}$$

$$5x + 4x + 80 = 134 \qquad \text{Simplify.}$$

$$9x + 80 = 134 \qquad \text{Combine like terms.}$$

$$9x + 80 - 80 = 134 - 80 \qquad \text{Subtract 80 from each side.}$$

$$9x = 54 \qquad \text{Simplify.}$$

$$x = 6 \qquad \text{Divide each side by 9.}$$

Check 6 in the original equation:

$$0.5(6) + 0.4(6 + 20) = 13.4 \quad \text{Replace } x \text{ by 6.}$$
$$3 + 0.4(26) = 13.4$$
$$3 + 10.4 = 13.4$$

Since both sides of the equation have the same value, the solution set is $\{6\}$. ■

CAUTION If you multiply each side by 10 in Example 5 before using the distributive property, be careful how you handle the terms in parentheses:

$$10 \cdot 0.5x + 10 \cdot 0.4(x + 20) = 10 \cdot 13.4$$
$$5x + 4(x + 20) = 134$$

It is not correct to multiply 0.4 by 10 *and also* to multiply $x + 20$ by 10.

Simplifying the Process

It is very important to develop the skill of solving equations in a systematic way, writing down every step as we have been doing. As you become more skilled at solving equations, you will probably want to simplify the process a bit. One way to simplify the process is by writing only the result of performing an operation on each side. Another way is to isolate the variable on the side where the variable has the larger coefficient, when the variable occurs on both sides. We use these ideas in the next example and in future examples in this text.

E X A M P L E 6

Simplifying the process

Solve each equation.

a) $2a - 3 = 0$ **b)** $2k + 5 = 3k + 1$

Solution

a) Add 3 to each side, then divide each side by 2:

$$2a - 3 = 0$$
$$2a = 3 \quad \text{Add 3 to each side.}$$
$$a = \frac{3}{2} \quad \text{Divide each side by 2.}$$

Check that $\frac{3}{2}$ satisfies the original equation. The solution set is $\left\{\frac{3}{2}\right\}$.

b) For this equation we can get a single k on the right by subtracting $2k$ from each side. (If we subtract $3k$ from each side, we get $-k$, and then we need another step.)

$$2k + 5 = 3k + 1$$
$$5 = k + 1 \quad \text{Subtract } 2k \text{ from each side.}$$
$$4 = k \quad \text{Subtract 1 from each side.}$$

Check that 4 satisfies the original equation. The solution set is $\{4\}$. ■

> **study tip**
>
> I hear and I forget; I see and I remember; I do and I understand. There is no substitute for doing exercises, lots of exercises.

WARM-UPS

True or false? Explain your answer.

1. The equation $x - x = 99$ has no solution.
2. The equation $2n + 3n = 5n$ is an identity.
3. The equation $2y + 3y = 4y$ is inconsistent.

WARM-UPS

(continued)

4. All real numbers satisfy the equation $1 \div x = \frac{1}{x}$.
5. The equation $5a + 3 = 0$ is an inconsistent equation.
6. The equation $2t = t$ is a conditional equation.
7. The equation $w - 0.1w = 0.9w$ is an identity.
8. The equation $0.2x + 0.03x = 8$ is equivalent to $20x + 3x = 8$.
9. The equation $\frac{x}{x} = 1$ is an identity.
10. The solution to $3h - 8 = 0$ is $\frac{8}{3}$.

2.3   EXERCISES

Reading and Writing   After reading this section, write out the answers to these questions. Use complete sentences.

1. What is an identity?

2. What is a conditional equation?

3. What is an inconsistent equation?

4. What is the usual first step when solving an equation involving fractions?

5. What is a good first step for solving an equation involving decimals?

6. Where should the variable be when you are finished solving an equation?

Solve each equation. Identify each as a conditional equation, an inconsistent equation, or an identity. See Examples 1 and 2.

7. $x + x = 2x$

8. $2x - x = x$

9. $a - 1 = a + 1$

10. $r + 7 = r$

11. $3y + 4y = 12y$

12. $9t - 8t = 7$

13. $-4 + 3(w - 1) = w + 2(w - 2) - 1$

14. $4 - 5(w + 2) = 2(w - 1) - 7w - 4$

15. $3(m + 1) = 3(m + 3)$

16. $5(m - 1) - 6(m + 3) = 4 - m$

17. $x + x = 2$

18. $3x - 5 = 0$

19. $2 - 3(5 - x) = 3x$

20. $3 - 3(5 - x) = 0$

21. $(3 - 3)(5 - z) = 0$

22. $(2 \cdot 4 - 8)p = 0$

23. $\dfrac{0}{x} = 0$

24. $\dfrac{2x}{2} = x$

25. $x \cdot x = x^2$

26. $\dfrac{2x}{2x} = 1$

Solve each equation by first eliminating the fractions. See Example 3.

27. $\dfrac{x}{2} + 3 = x - \dfrac{1}{2}$

28. $13 - \dfrac{x}{2} = x - \dfrac{1}{2}$

29. $\dfrac{x}{2} + \dfrac{x}{3} = 20$

30. $\dfrac{x}{2} - \dfrac{x}{3} = 5$

31. $\dfrac{w}{2} + \dfrac{w}{4} = 12$

32. $\dfrac{a}{4} - \dfrac{a}{2} = -5$

33. $\dfrac{3z}{2} - \dfrac{2z}{3} = -10$

34. $\dfrac{3m}{4} + \dfrac{m}{2} = -5$

35. $\dfrac{1}{3}p - 5 = \dfrac{1}{4}p$

36. $\dfrac{1}{2}q - 6 = \dfrac{1}{5}q$

37. $\dfrac{1}{6}v + 1 = \dfrac{1}{4}v - 1$

38. $\dfrac{1}{15}k + 5 = \dfrac{1}{6}k - 10$

Solve each equation by first eliminating the decimal numbers. See Examples 4 and 5.

39. $x - 0.2x = 72$

40. $x - 0.1x = 63$

41. $0.3x + 1.2 = 0.5x$

42. $0.4x - 1.6 = 0.6x$

43. $0.02x - 1.56 = 0.8x$

44. $0.6x + 10.4 = 0.08x$

45. $0.1a - 0.3 = 0.2a - 8.3$

46. $0.5b + 3.4 = 0.2b + 12.4$

47. $0.05r + 0.4r = 27$

48. $0.08t + 28.3 = 0.5t - 9.5$

49. $0.05y + 0.03(y + 50) = 17.5$

50. $0.07y + 0.08(y - 100) = 44.5$

51. $0.1x + 0.05(x - 300) = 105$

52. $0.2x - 0.05(x - 100) = 35$

Solve each equation. If you feel proficient enough, try simplifying the process, as described in Example 6.

53. $2x - 9 = 0$

54. $3x + 7 = 0$

55. $-2x + 6 = 0$

56. $-3x - 12 = 0$

57. $\dfrac{z}{5} + 1 = 6$

58. $\dfrac{s}{2} + 2 = 5$

59. $\dfrac{c}{2} - 3 = -4$

60. $\dfrac{b}{3} - 4 = -7$

61. $3 = t + 6$

62. $-5 = y - 9$

63. $5 + 2q = 3q$

64. $-4 - 5p = -4p$

65. $8x - 1 = 9 + 9x$

66. $4x - 2 = -8 + 5x$

67. $-3x + 1 = -1 - 2x$

68. $-6x + 3 = -7 - 5x$

Solve each equation.

69. $3x - 5 = 2x - 9$

70. $5x - 9 = x - 4$

71. $x + 2(x + 4) = 3(x + 3) - 1$

72. $u + 3(u - 4) = 4(u - 5)$

73. $23 - 5(3 - n) = -4(n - 2) + 9n$

74. $-3 - 4(t - 5) = -2(t + 3) + 11$

75. $0.05x + 30 = 0.4x - 5$

76. $x - 0.08x = 460$

77. $-\dfrac{2}{3}a + 1 = 2$

78. $-\dfrac{3}{4}t = \dfrac{1}{2}$

79. $\dfrac{y}{2} + \dfrac{y}{6} = 20$

80. $\dfrac{3w}{5} - 1 = \dfrac{w}{2} + 1$

81. $0.09x - 0.2(x + 4) = -1.46$

82. $0.08x + 0.5(x + 100) = 73.2$

83. $436x - 789 = -571$

84. $0.08x + 4533 = 10x + 69$

85. $\dfrac{x}{344} + 235 = 292$

86. $34(x - 98) = \dfrac{x}{2} + 475$

Solve each problem.

87. *Sales commission.* Danielle sold her house through an agent who charged 8% of the selling price. After the commission was paid, Danielle received \$117,760. If x is the selling price, then x satisfies

$$x - 0.08x = 117,760.$$

Solve this equation to find the selling price.

88. *Raising rabbits.* Before Roland sold two female rabbits, half of his rabbits were female. After the sale, only one-third of his rabbits were female. If x represents his original number of rabbits, then

$$\frac{1}{2}x - 2 = \frac{1}{3}(x - 2).$$

Solve this equation to find the number of rabbits that he had before the sale.

89. *Eavesdropping.* Reginald overheard his boss complaining that his federal income tax for 2000 was \$34,276.

a) Use the accompanying graph to estimate his boss's taxable income for 2000.

b) Find his boss's exact taxable income for 2000 by solving the equation

$$23,965.5 + 0.31(x - 105,950) = 34,276.$$

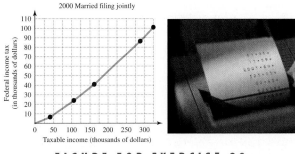

FIGURE FOR EXERCISE 89

90. *Federal taxes.* According to Bruce Harrell, CPA, the federal income tax for a class C corporation is found by solving a linear equation. The reason for the equation is that the amount x of federal tax is deducted before the state tax is figured, and the amount of state tax is deducted before the federal tax is figured. To find the amount of federal tax for a corporation with a taxable income of \$200,000, for which the federal tax rate is 25% and the state tax rate is 10%, Bruce must solve

$$x = 0.25[200,000 - 0.10(200,000 - x)].$$

Solve the equation for Bruce.

2.4 FORMULAS

In this section, you will learn to rewrite formulas using the same properties of equality that we used to solve equations. You will also learn how to find the value of one of the variables in a formula when we know the value of all of the others.

Solving for a Variable

Most drivers know the relationship between distance, rate, and time. For example, if you drive 70 mph for 3 hours, then you will travel 210 miles. At 60 mph a 300-mile trip will take 5 hours. If a 400-mile trip took 8 hours, then you averaged 50 mph. The relationship between distance D, rate R, and time T is expressed by the formula

$$D = R \cdot T.$$

A **formula** or **literal equation** is an equation involving two or more variables.

To find the time for a 300-mile trip at 60 mph, you are using the formula in the form $T = \frac{D}{R}$. The process of rewriting a formula for one variable in terms of the others is called **solving for a certain variable.** To solve for a certain variable, we use the same techniques that we use in solving equations.

EXAMPLE 1

Solving for a certain variable

Solve the formula $D = RT$ for T:

Solution

$$D = RT \qquad \text{Original formula}$$

$$\frac{D}{R} = \frac{R \cdot T}{R} \qquad \text{Divide each side by } R.$$

$$\frac{D}{R} = T \qquad \text{Divide out (or cancel) the common factor } R.$$

$$T = \frac{D}{R} \qquad \text{It is customary to write the single variable on the left.} \qquad ■$$

The formula $C = \frac{5}{9}(F - 32)$ is used to find the Celsius temperature for a given Fahrenheit temperature. If we solve this formula for F, then we have a formula for finding Fahrenheit temperature for a given Celsius temperature.

EXAMPLE 2

Solving for a certain variable

Solve the formula $C = \frac{5}{9}(F - 32)$ for F.

Solution

We could apply the distributive property to the right side of the equation, but it is simpler to proceed as follows:

$$C = \frac{5}{9}(F - 32)$$

$$\frac{9}{5}C = \frac{9}{5} \cdot \frac{5}{9}(F - 32) \qquad \text{Multiply each side by } \frac{9}{5}, \text{ the reciprocal of } \frac{5}{9}.$$

$$\frac{9}{5}C = F - 32 \qquad \text{Simplify.}$$

$$\frac{9}{5}C + 32 = F - 32 + 32 \quad \text{Add 32 to each side.}$$

$$\frac{9}{5}C + 32 = F \qquad \text{Simplify.}$$

The formula is usually written as $F = \frac{9}{5}C + 32$. ■

When solving for a variable that appears more than once in the equation, we must combine the terms to obtain a single occurrence of the variable. *When a formula has been solved for a certain variable, that variable will not occur on both sides of the equation.*

E X A M P L E   3

Solving for a variable that appears on both sides

Solve $5x - b = 3x + d$ for x.

Solution

First get all terms involving x onto one side and all other terms onto the other side:

$$
\begin{aligned}
5x - b &= 3x + d & &\text{Original formula} \\
5x - 3x - b &= d & &\text{Subtract } 3x \text{ from each side.} \\
5x - 3x &= b + d & &\text{Add } b \text{ to each side.} \\
2x &= b + d & &\text{Combine like terms.} \\
x &= \frac{b + d}{2} & &\text{Divide each side by 2.}
\end{aligned}
$$

The formula solved for x is $x = \frac{b + d}{2}$. ■

In Chapter 4, it will be necessary to solve an equation involving x and y for y.

E X A M P L E   4

Solving for y

Solve $x + 2y = 6$ for y. Write the answer in the form $y = mx + b$, where m and b are fixed real numbers.

helpful hint

If we simply wanted to solve $x + 2y = 6$ for y, we could have written

$$y = \frac{6 - x}{2} \text{ or } y = \frac{-x + 6}{2}.$$

However, in Example 4 we requested the form $y = mx + b$. This form is a popular form that we will study in detail in Chapter 4.

Solution

$$
\begin{aligned}
x + 2y &= 6 & &\text{Original equation} \\
2y &= 6 - x & &\text{Subtract } x \text{ from each side.} \\
\frac{1}{2} \cdot 2y &= \frac{1}{2}(6 - x) & &\text{Multiply each side by } \frac{1}{2}. \\
y &= 3 - \frac{1}{2}x & &\text{Distributive property} \\
y &= -\frac{1}{2}x + 3 & &\text{Rearrange to get } y = mx + b \text{ form.}
\end{aligned}
$$

■

Notice that in Example 4 we multiplied each side of the equation by $\frac{1}{2}$, and so we multiplied each term on the right-hand side by $\frac{1}{2}$. Instead of multiplying by $\frac{1}{2}$, we could have divided each side of the equation by 2. We would then divide each term on the right side by 2. This idea is illustrated in the next example.

M A T H   A T   W O R K $x^2 + (x+1)^2 = 5^2$

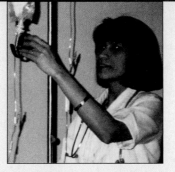

Even before the days of Florence Nightingale, nurses around the world were giving comfort and aid to the sick and injured. Continuing in this tradition, Asenet Craffey, staff nurse at the Massachusetts Eye and Ear Infirmary, works in the intensive care unit. During her 12-hour shifts, Ms. Craffey is responsible for the full nursing care of four to eight patients. In the intensive care unit, the nurse-to-patient ratio is usually one to one. When Ms. Craffey is assigned to this unit, she is responsible for over-all care of a patient as well as being prepared for crisis care. Staff scheduling is an additional duty that Ms. Craffey performs, making sure that there is adequate nursing coverage for the day's planned surgeries and quality patient care. Full care means being directly involved in all of the patient's care: monitoring vital signs, changing dressings, helping to feed, following the prescribed orders left by the physicians, and administering drugs.

NURSE

Many drugs come directly from the pharmacy in the exact dosage for a particular patient. Intravenous (IV) drugs, however, must be monitored so that the correct amount of drops per minute are administered. IV medications can be glucose solutions, antibiotics, or pain killers. Often the prescribed dosage is 1 gram per 100, 200, 500, or 1000 cubic centimeters of liquid. In Exercise 85 of this section you will calculate a drug dosage, just as Ms. Craffey would on the job.

E X A M P L E   5

Solving for y

Solve $2x - 3y = 9$ for y. Write the answer in the form $y = mx + b$, where m and b are real numbers. (When we study lines in Chapter 4 you will see that $y = mx + b$ is the slope-intercept form of the equation of a line.)

Solution

$2x - 3y = 9$	Original equation
$-3y = -2x + 9$	Subtract $2x$ from each side.
$\dfrac{-3y}{-3} = \dfrac{-2x + 9}{-3}$	Divide each side by -3.
$y = \dfrac{-2x}{-3} + \dfrac{9}{-3}$	By the distributive property, each term is divided by -3.
$y = \dfrac{2}{3}x - 3$	Simplify.

Even though we wrote $y = \frac{2}{3}x - 3$ in Example 5, the equation is still considered to be in the form $y = mx + b$ because we could have written $y = \frac{2}{3}x + (-3)$.

Finding the Value of a Variable

In many situations we know the values of all variables in a formula except one. We use the formula to determine the unknown value.

E X A M P L E   6

Finding the value of a variable in a formula

If $2x - 3y = 9$, find y when $x = 6$.

Solution

Method 1: First solve the equation for y. Because we have already solved this equation for y in Example 5 we will not repeat that process in this example. We have

$$y = \frac{2}{3}x - 3.$$

Now replace x by 6 in this equation:

$$y = \frac{2}{3}(6) - 3$$
$$= 4 - 3 = 1$$

So, when $x = 6$, we have $y = 1$.

Method 2: First replace x by 6 in the original equation, then solve for y:

$$2x - 3y = 9 \qquad \text{Original equation}$$
$$2 \cdot 6 - 3y = 9 \qquad \text{Replace } x \text{ by 6.}$$
$$12 - 3y = 9 \qquad \text{Simplify.}$$
$$-3y = -3 \qquad \text{Subtract 12 from each side.}$$
$$y = 1 \qquad \text{Divide each side by } -3.$$

So when $x = 6$, we have $y = 1$. ■

If we had to find the value of y for many different values of x, it would be best to solve the equation for y, then insert the various values of x. Method 1 of Example 6 would be the better method. If we must find only one value of y, it does not matter which method we use. When doing the exercises corresponding to this example, you should try both methods.

The next example involves the simple interest formula $I = Prt$, where I is the amount of interest, P is the principal or the amount invested, r is the annual interest rate, and t is the time in years. The interest rate is generally expressed as a percent. When using a rate in computations, you must convert it to a decimal.

E X A M P L E   7

Using the simple interest formula

If the simple interest is $120, the principal is $400, and the time is 2 years, find the rate.

Solution

First, solve the formula $I = Prt$ for r, then insert values of P, I, and t:

$$Prt = I \qquad \text{Simple interest formula}$$
$$\frac{Prt}{Pt} = \frac{I}{Pt} \qquad \text{Divide each side by } Pt.$$
$$r = \frac{I}{Pt} \qquad \text{Simplify.}$$
$$r = \frac{120}{400 \cdot 2} \qquad \text{Substitute the values of } I, P, \text{ and } t.$$
$$r = 0.15 \qquad \text{Simplify.}$$
$$r = 15\% \qquad \text{Move the decimal point two places to the right.}$$ ■

> **helpful hint**
>
> All interest computation is based on simple interest. However, depositors do not like to wait two years to get interest as in Example 7. More often the time is $\frac{1}{12}$ year or $\frac{1}{365}$ year. Simple interest computed every month is said to be compounded monthly. Simple interest computed every day is said to be compounded daily.

In solving a geometric problem, it is always helpful to draw a diagram, as we do in the next example.

E X A M P L E   8

Using a geometric formula

The perimeter of a rectangle is 36 feet. If the width is 6 feet, then what is the length?

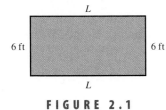

L

6 ft   6 ft

L

FIGURE 2.1

Solution

First, put the given information on a diagram as shown in Fig. 2.1. Substitute the given values into the formula for the perimeter of a rectangle found at the back of the book, and then solve for L. (We could solve for L first and then insert the given values.)

$$P = 2L + 2W \qquad \text{Perimeter of a rectangle}$$
$$36 = 2L + 2 \cdot 6 \qquad \text{Substitute 36 for } P \text{ and 6 for } W.$$
$$36 = 2L + 12 \qquad \text{Simplify.}$$
$$24 = 2L \qquad \text{Subtract 12 from each side.}$$
$$12 = L \qquad \text{Divide each side by 2.}$$

Check: If $L = 12$ and $W = 6$, then $P = 2(12) + 2(6) = 36$ feet. So we can be certain that the length is 12 feet. ■

The next example involves the sale-price formula $S = L - rL$, where S is the selling price, L is the list or original price, and r is the rate of discount. The rate of discount is generally expressed as a percent. In computations, rates must be written as decimals (or fractions).

E X A M P L E   9

Finding the original price

What was the original price of a stereo that sold for $560 after a 20% discount.

Solution

Express 20% as the decimal 0.20 or 0.2 and use the formula $S = L - rL$:

$$\text{Selling price} = \text{list price} - \text{amount of discount}$$
$$560 = L - 0.2L$$
$$10(560) = 10(L - 0.2L) \qquad \text{Multiply each side by 10.}$$
$$5600 = 10L - 2L \qquad \text{Remove the parentheses.}$$
$$5600 = 8L \qquad \text{Combine like terms.}$$
$$\frac{5600}{8} = \frac{8L}{8} \qquad \text{Divide each side by 8.}$$
$$700 = L$$

Check: We find that 20% of $700 is $140 and $700 − $140 = $560, the selling price. So we are certain that the original price was $700. ■

WARM-UPS

True or false? Explain your answer.

1. If we solve $D = R \cdot T$ for T, we get $T \cdot R = D$.
2. If we solve $a - b = 3a - m$ for a, we get $a = 3a - m + b$.
3. Solving $A = LW$ for L, we get $L = \frac{W}{A}$.
4. Solving $D = RT$ for R, we get $R = \frac{d}{t}$.
5. The perimeter of a rectangle is the product of its length and width.
6. The volume of a shoe box is the product of its length, width, and height.

WARM-UPS

(continued)

7. The sum of the length and width of a rectangle is one-half of its perimeter.

8. Solving $y - x = 5$ for y gives us $y = x + 5$.

9. If $x = -1$ and $y = -3x + 6$, then $y = 3$.

10. The circumference of a circle is the product of its diameter and the number π.

2.4   EXERCISES

Reading and Writing   *After reading this section, write out the answers to these questions. Use complete sentences.*

1. What is a formula?

2. What is a literal equation?

3. What does it mean to solve a formula for a certain variable?

4. How do you solve a formula for a variable that appears on both sides?

5. What are the two methods shown for finding the value of a variable in a formula?

6. What formula expresses the perimeter of a rectangle in terms of its length and width?

Solve each formula for the specified variable. See Examples 1 and 2.

7. $D = RT$ for R

8. $A = LW$ for W

9. $C = \pi D$ for D

10. $F = ma$ for a

11. $I = Prt$ for P

12. $I = Prt$ for t

13. $F = \dfrac{9}{5}C + 32$ for C

14. $y = \dfrac{3}{4}x - 7$ for x

15. $A = \dfrac{1}{2}bh$ for h

16. $A = \dfrac{1}{2}bh$ for b

17. $P = 2L + 2W$ for L

18. $P = 2L + 2W$ for W

19. $A = \dfrac{1}{2}(a + b)$ for a

20. $A = \dfrac{1}{2}(a + b)$ for b

21. $S = P + Prt$ for r

22. $S = P + Prt$ for t

23. $A = \dfrac{1}{2}h(a + b)$ for a

24. $A = \dfrac{1}{2}h(a + b)$ for b

Solve each equation for x. See Example 3.

25. $5x + a = 3x + b$

26. $2c - x = 4x + c - 5b$

27. $4(a + x) - 3(x - a) = 0$

28. $-2(x - b) - (5a - x) = a + b$

29. $3x - 2(a - 3) = 4x - 6 - a$

30. $2(x - 3w) = -3(x + w)$

31. $3x + 2ab = 4x - 5ab$

32. $x - a = -x + a + 4b$

Solve each equation for y. See Examples 4 and 5.

33. $x + y = -9$

34. $3x + y = -5$

35. $x + y - 6 = 0$

36. $4x + y - 2 = 0$

37. $2x - y = 2$

38. $x - y = -3$

39. $3x - y + 4 = 0$

40. $-2x - y + 5 = 0$

41. $x + 2y = 4$

42. $3x + 2y = 6$

43. $2x - 2y = 1$

44. $3x - 2y = -6$

45. $y + 2 = 3(x - 4)$
46. $y - 3 = -3(x - 1)$
47. $y - 1 = \dfrac{1}{2}(x - 2)$
48. $y - 4 = -\dfrac{2}{3}(x - 9)$
49. $\dfrac{1}{2}x - \dfrac{1}{3}y = -2$
50. $\dfrac{x}{2} + \dfrac{y}{4} = \dfrac{1}{2}$

For each equation that follows, find y given that x = 2. See Example 6.

51. $y = 3x - 4$ **52.** $y = -2x + 5$
53. $3x - 2y = -8$ **54.** $4x + 6y = 8$
55. $\dfrac{3x}{2} - \dfrac{5y}{3} = 6$ **56.** $\dfrac{2y}{5} - \dfrac{3x}{4} = \dfrac{1}{2}$
57. $y - 3 = \dfrac{1}{2}(x - 6)$ **58.** $y - 6 = -\dfrac{3}{4}(x - 2)$
59. $y - 4.3 = 0.45(x - 8.6)$
60. $y + 33.7 = 0.78(x - 45.6)$

Solve each of the following problems. Appendix A contains some geometric formulas that may be helpful. See Examples 7–9.

61. Finding the rate. If the simple interest on $5000 for 3 years is $600, then what is the rate?

62. Finding the rate. Wayne paid $420 in simple interest on a loan of $1000 for 7 years. What was the rate?

63. Finding the time. Kathy paid $500 in simple interest on a loan of $2500. If the annual interest rate was 5%, then what was the time?

64. Finding the time. Robert paid $240 in simple interest on a loan of $1000. If the annual interest rate was 8%, then what was the time?

65. Finding the length. The area of a rectangle is 28 square yards. The width is 4 yards. Find the length.

66. Finding the width. The area of a rectangle is 60 square feet. The length is 4 feet. Find the width.

67. Finding the length. If it takes 600 feet of wire fencing to fence a rectangular feed lot that has a width of 75 feet, then what is the length of the lot?

68. Finding the depth. If it takes 500 feet of fencing to enclose a rectangular lot that is 104 feet wide, then how deep is the lot?

69. Finding the original price. Find the original price if there is a 15% discount and the sale price is $255.

70. Finding the list price. Find the list price if there is a 12% discount and the sale price is $4400.

71. Rate of discount. Find the rate of discount if the discount is $40 and the original price is $200.

72. Rate of discount. Find the rate of discount if the discount is $20 and the original price is $250.

73. Width of a football field. The perimeter of a football field in the NFL, excluding the end zones, is 920 feet. How wide is the field?

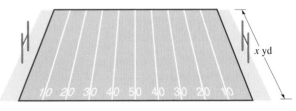

FIGURE FOR EXERCISE 73

74. Perimeter of a frame. If a picture frame is 16 inches by 20 inches, then what is its perimeter?

75. Volume of a box. A rectangular box measures 2 feet wide, 3 feet long, and 4 feet deep. What is its volume?

76. Volume of a refrigerator. The volume of a rectangular refrigerator is 20 cubic feet. If the top measures 2 feet by 2.5 feet, then what is the height?

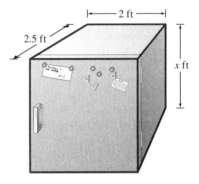

FIGURE FOR EXERCISE 76

77. Radius of a pizza. If the circumference of a pizza is 8π inches, then what is the radius?

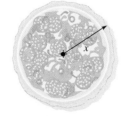

FIGURE FOR EXERCISE 77

78. Diameter of a circle. If the circumference of a circle is 4π meters, then what is the diameter?

79. *Height of a banner.* If a banner in the shape of a triangle has an area of 16 square feet with a base of 4 feet, then what is the height of the banner?

├────── 4 ft ──────┤

1996

Division Champs

│ x

FIGURE FOR EXERCISE 79

80. *Length of a leg.* If a right triangle has an area of 14 square meters and one leg is 4 meters in length, then what is the length of the other leg?

81. *Length of the base.* A trapezoid with height 20 inches and lower base 8 inches has an area of 200 square inches. What is the length of its upper base?

82. *Height of a trapezoid.* The end of a flower box forms the shape of a trapezoid. The area of the trapezoid is 300 square centimeters. The bases are 16 centimeters and 24 centimeters in length. Find the height.

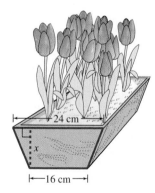

├── 24 cm ──┤

x

├──── 16 cm ────┤

FIGURE FOR EXERCISE 82

83. *Fried's rule.* Doctors often prescribe the same drugs for children as they do for adults. The formula $d = 0.08aD$ (Fried's rule) is used to calculate the child's dosage d, where a is the child's age and D is the adult dosage. If a doctor prescribes 1000 milligrams of acetaminophen for an adult, then how many milligrams would the doctor prescribe for an eight-year-old child? Use the bar graph to determine the age at which a child would get the same dosage as an adult.

84. *Cowling's rule.* Cowling's rule is another method for determining the dosage of a drug to prescribe to a child.

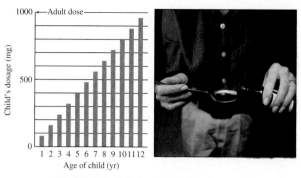

FIGURE FOR EXERCISE 83

For this rule, the formula

$$d = \frac{D(a + 1)}{24}$$

gives the child's dosage d, where D is the adult dosage and a is the age of the child in years. If the adult dosage of a drug is 600 milligrams and a doctor uses this formula to determine that a child's dosage is 200 milligrams, then how old is the child?

85. *Administering Vancomycin.* A patient is to receive 750 mg of the antibiotic Vancomycin. However, Vancomycin comes in a solution containing 1 gram (available dose) of Vancomycin per 5 milliliters (quantity) of solution. Use the formula

$$\text{Amount} = \frac{\text{desired dose}}{\text{available dose}} \times \text{quantity}$$

to find the amount of this solution that should be administered to the patient.

86. *International communications.* The global investment in telecom infrastructure since 1990 can be modeled by the formula

$$I = 7.5t + 115,$$

where I is in billions of dollars and t is the number of years since 1990 (*Fortune,* September 8, 1997). See the accompanying figure.

a) Use the formula to find the global investment in 1994.

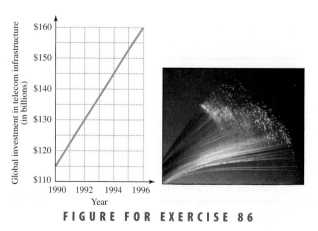

FIGURE FOR EXERCISE 86

b) Use the formula to predict the global investment in 2001.

c) Find the year in which the global investment will reach $250 billion.

87. *The 2.4-meter rule.* A 2.4-meter sailboat is a one-person boat that is about 13 feet in length, has a displacement of about 550 pounds, and a sail area of about 81 square feet. To compete in the 2.4-meter class, a boat must satisfy the formula

$$2.4 = \frac{L + 2D - F\sqrt{S}}{2.37},$$

where L = length, F = freeboard, D = girth, and S = sail area. Solve the formula for L.

PHOTO FOR EXERCISE 87

2.5 TRANSLATING VERBAL EXPRESSIONS INTO ALGEBRAIC EXPRESSIONS

You translated some verbal expressions into algebraic expressions in Section 1.6; in this section you will study translating in more detail.

In this section

- Writing Algebraic Expressions
- Pairs of Numbers
- Consecutive Integers
- Using Formulas
- Writing Equations

Writing Algebraic Expressions

The following box contains a list of some frequently occurring verbal expressions and their equivalent algebraic expressions.

Translating Words into Algebra		
	Verbal Phrase	**Algebraic Expression**
Addition:	The sum of a number and 8	$x + 8$
	Five is added to a number	$x + 5$
	Two more than a number	$x + 2$
	A number increased by 3	$x + 3$
Subtraction:	Four is subtracted from a number	$x - 4$
	Three less than a number	$x - 3$
	The difference between 7 and a number	$7 - x$
	A number decreased by 2	$x - 2$
Multiplication:	The product of 5 and a number	$5x$
	Twice a number	$2x$
	One-half of a number	$\frac{1}{2}x$
	Five percent of a number	$0.05x$
Division:	The ratio of a number to 6	$\frac{x}{6}$
	The quotient of 5 and a number	$\frac{5}{x}$
	Three divided by some number	$\frac{3}{x}$

E X A M P L E   1

Writing algebraic expressions

Translate each verbal expression into an algebraic expression.

a) The sum of a number and 9   **b)** Eighty percent of a number

c) A number divided by 4   **d)** The result of a number subtracted from 5

e) Three less than a number

Solution

a) If x is the number, then the sum of a number and 9 is $x + 9$.

b) If w is the number, then eighty percent of the number is $0.80w$.

c) If y is the number, then the number divided by 4 is $\frac{y}{4}$.

d) If z is the number, then the result of subtracting z from 5 is $5 - z$.

e) If a is the number, then 3 less than a is $a - 3$. ■

helpful   hint

We know that x and $10 - x$ have a sum of 10 for any value of x. We can easily check that fact by adding:

$$x + 10 - x = 10$$

In general it is not true that x and $x - 10$ have a sum of 10, because

$$x + x - 10 = 2x - 10.$$

For what value of x is the sum of x and $x - 10$ equal to 10?

Pairs of Numbers

There is often more than one unknown quantity in a problem, but a relationship between the unknown quantities is given. For example, if one unknown number is 5 more than another unknown number, we can use

$$x \quad \text{and} \quad x + 5,$$

to represent them. Note that x and $x + 5$ can also be used to represent two unknown numbers that differ by 5, for if two numbers differ by 5, one of the numbers is 5 more than the other.

How would you represent two numbers that have a sum of 10? If one of the numbers is 2, the other is certainly $10 - 2$, or 8. Thus if x is one of the numbers, then $10 - x$ is the other. The expressions

$$x \quad \text{and} \quad 10 - x$$

have a sum of 10 for any value of x.

E X A M P L E   2

Algebraic expressions for pairs of numbers

Write algebraic expressions for each pair of numbers.

a) Two numbers that differ by 12   **b)** Two numbers with a sum of -8

Solution

a) The expressions x and $x - 12$ represent two numbers that differ by 12. We can check by subtracting:

$$x - (x - 12) = x - x + 12 = 12$$

Of course, x and $x + 12$ also differ by 12 because $x + 12 - x = 12$.

b) The expressions x and $-8 - x$ have a sum of -8. We can check by addition:

$$x + (-8 - x) = x - 8 - x = -8$$ ■

Pairs of numbers occur in geometry in discussing measures of angles. You will need the following facts about degree measures of angles.

Degree Measures of Angles

Two angles are called **complementary** if the sum of their degree measures is $90°$.

Two angles are called **supplementary** if the sum of their degree measures is $180°$.

The sum of the degree measures of the three angles of any triangle is $180°$.

For complementary angles, we use x and $90 - x$ for their degree measures. For supplementary angles, we use x and $180 - x$. Complementary angles that share a common side form a right angle. Supplementary angles that share a common side form a straight angle or straight line.

E X A M P L E   3   **Degree measures**

Write algebraic expressions for each pair of angles shown.

a)

b)

c)

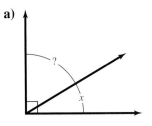
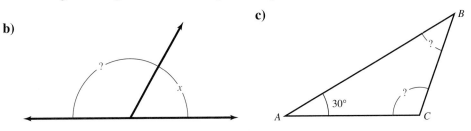

Solution

a) Since the angles shown are complementary, we can use x to represent the degree measure of the smaller angle and $90 - x$ to represent the degree measure of the larger angle.

b) Since the angles shown are supplementary, we can use x to represent the degree measure of the smaller angle and $180 - x$ to represent the degree measure of the larger angle.

c) If we let x represent the degree measure of angle B, then $180 - x - 30$, or $150 - x$, represents the degree measure of angle C. ■

Consecutive Integers

To gain practice in problem solving, we will solve problems about consecutive integers in Section 2.6. Note that each integer is 1 larger than the previous integer, while consecutive even integers as well as consecutive odd integers differ by 2.

E X A M P L E   4   **Expressions for integers**

Write algebraic expressions for the following unknown integers.

a) Three consecutive integers, the smallest of which is w

b) Three consecutive even integers, the smallest of which is z

helpful hint

If x is even, then x, $x + 2$, and $x + 4$ represent three consecutive even integers. If x is odd, then x, $x + 2$, and $x + 4$ represent three consecutive odd integers. Is it possible for x, $x + 1$, and $x + 3$ to represent three consecutive even integers or three consecutive odd integers?

Solution

a) Since each integer is 1 larger than the preceding integer, we can use w, $w + 1$, and $w + 2$ to represent them.

b) Since consecutive even integers differ by 2, these integers can be represented by z, $z + 2$, and $z + 4$. ■

Using Formulas

In writing expressions for unknown quantities, we often use standard formulas such as those given at the back of the book.

E X A M P L E   5   **Writing algebraic expressions using standard formulas**

Find an algebraic expression for

a) the distance if the rate is 30 miles per hour and the time is T hours.

b) the discount if the rate is 40% and the original price is p dollars.

Solution

a) Using the formula $D = RT$, we have $D = 30T$. So $30T$ is an expression that represents the distance in miles.

b) Since the discount is the rate times the original price, an algebraic expression for the discount is $0.40p$ dollars.   ■

Writing Equations

To solve a problem using algebra, we describe or **model** the problem with an equation. In this section we write the equations only, and in Section 2.6 we write and solve them. Sometimes we must write an equation from the information given in the problem and sometimes we use a standard model to get the equation. Some standard models are shown in the following box.

Uniform Motion Model

Distance = Rate · Time $D = R \cdot T$

Selling Price and Discount Model

Discount = Rate of discount · Original price
Selling Price = Original price − Discount

Real Estate Commission Model

Commission = Rate of commission · Selling price
Amount for owner = Selling price − Commission

Percentage Models

What number is 5% of 40? $x = 0.05 \cdot 40$
Ten is what percent of 80? $10 = x \cdot 80$
Twenty is 4% of what number? $20 = 0.04 \cdot x$

Geometric Models for Area

Rectangle: $A = LW$ Square: $A = s^2$
Parallelogram: $A = bh$ Triangle: $A = \frac{1}{2}bh$

Geometric Models for Perimeter

Perimeter of any figure = the sum of the lengths of the sides
Rectangle: $P = 2L + 2W$ Square: $P = 4s$

More geometric formulas can be found in Appendix A.

E X A M P L E   6

Writing equations

Identify the variable and write an equation that describes each situation.

a) Find two numbers that have a sum of 14 and a product of 45.

b) A coat is on sale for 25% off the list price. If the sale price is $87, then what is the list price?

c) What percent of 8 is 2?

d) The value of x dimes and $x - 3$ quarters is $2.05.

Solution

a) Let $x =$ one of the numbers and $14 - x =$ the other number. Since their product is 45, we have

$$x(14 - x) = 45.$$

b) Let $x =$ the list price and $0.25x =$ the amount of discount. We can write an equation expressing the fact that the selling price is the list price minus the discount:

$$\text{List price} - \text{discount} = \text{selling price}$$
$$x - 0.25x = 87$$

c) If we let x represent the percentage, then the equation is $x \cdot 8 = 2$, or $8x = 2$.

d) The value of x dimes at 10 cents each is $10x$ cents. The value of $x - 3$ quarters at 25 cents each is $25(x - 3)$ cents. We can write an equation expressing the fact that the total value of the coins is 205 cents:

$$\text{Value of dimes} + \text{value of quarters} = \text{total value}$$
$$10x + 25(x - 3) = 205$$

> **helpful** \ **hint**
>
> At this point we are simply learning to write equations that model certain situations. Don't worry about solving these equations now. In Section 2.6 we will solve problems by writing an equation and solving it.

C A U T I O N   The value of the coins in Example 6(d) is either 205 cents or 2.05 dollars. If the total value is expressed in dollars, then all of the values must be expressed in dollars. So we could also write the equation as

$$0.10x + 0.25(x - 3) = 2.05.$$

WARM-UPS

True or false? Explain your answer.

1. For any value of x, the numbers x and $x + 6$ differ by 6.

2. For any value of a, a and $10 - a$ have a sum of 10.

3. If Jack ran at x miles per hour for 3 hours, he ran $3x$ miles.

4. If Jill ran at x miles per hour for 10 miles, she ran for $10x$ hours.

5. If the realtor gets 6% of the selling price and the house sells for x dollars, the owner gets $x - 0.06x$ dollars.

6. If the owner got $50,000 and the realtor got 10% of the selling price, the house sold for $55,000.

7. Three consecutive odd integers can be represented by x, $x + 1$, and $x + 3$.

8. The value in cents of n nickels and d dimes is $0.05n + 0.10d$.

9. If the sales tax rate is 5% and x represents the price of the goods purchased, then the total bill is $1.05x$.

10. If the length of a rectangle is 4 feet more than the width w, then the perimeter is $w + (w + 4)$ feet.

2.5   EXERCISES

Reading and Writing   *After reading this section, write out the answers to these questions. Use complete sentences.*

1. What are the different ways of verbally expressing the operation of addition?

2. How can you algebraically express two numbers using only one variable?

3. What are complementary angles?

4. What are supplementary angles?

5. What is the relationship between distance, rate, and time?

6. What is the difference between expressing consecutive even integers and consecutive odd integers algebraically?

Translate each verbal expression into an algebraic expression. See Example 1.

7. The sum of a number and 3

8. Two more than a number

9. Three less than a number

10. Four subtracted from a number

11. The product of a number and 5

12. Five divided by some number

13. Ten percent of a number

14. Eight percent of a number

15. The ratio of a number and 3

16. The quotient of 12 and a number

17. One-third of a number

18. Three-fourths of a number

Write algebraic expressions for each pair of numbers. See Example 2.

19. Two numbers with a difference of 15

20. Two numbers that differ by 9

21. Two numbers with a sum of 6

22. Two numbers with a sum of 5

23. Two numbers with a sum of −4

24. Two numbers with a sum of −8

25. Two numbers such that one is 3 larger than the other

26. Two numbers such that one is 8 smaller than the other

27. Two numbers such that one is 5% of the other

28. Two numbers such that one is 40% of the other

29. Two numbers such that one is 30% more than the other

30. Two numbers such that one is 20% smaller than the other

Each of the following figures shows a pair of angles. Write algebraic expressions for the degree measures of each pair of angles. See Example 3.

31.

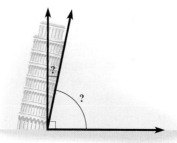

FIGURE FOR EXERCISE 31

32.

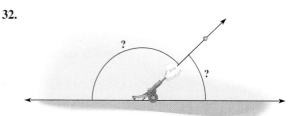

FIGURE FOR EXERCISE 32

33.

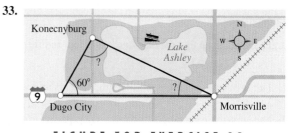

FIGURE FOR EXERCISE 33

34.

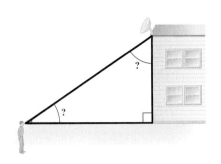

FIGURE FOR EXERCISE 34

Write algebraic expressions for the following unknown integers. See Example 4.

35. Two consecutive even integers, the smallest of which is n

36. Two consecutive odd integers, the smallest of which is x

37. Two consecutive integers

38. Three consecutive even integers

39. Three consecutive odd integers

40. Three consecutive integers

41. Four consecutive even integers

42. Four consecutive odd integers

Find an algebraic expression for the quantity in italics using the given information. See Example 5.

43. The *distance*, given that the rate is x miles per hour and the time is 3 hours

44. The *distance*, given that the rate is $x + 10$ miles per hour and the time is 5 hours

45. The *discount*, given that the rate is 25% and the original price is q dollars

46. The *discount*, given that the rate is 10% and the original price is t yen

47. The *time*, given that the distance is x miles and the rate is 20 miles per hour

48. The *time*, given that the distance is 300 kilometers and the rate is $x + 30$ kilometers per hour

49. The *rate*, given that the distance is $x - 100$ meters and the time is 12 seconds

50. The *rate*, given that the distance is 200 feet and the time is $x + 3$ seconds

51. The *area* of a rectangle with length x meters and width 5 meters

52. The *area* of a rectangle with sides b yards and $b - 6$ yards

53. The *perimeter* of a rectangle with length $w + 3$ inches and width w inches

54. The *perimeter* of a rectangle with length r centimeters and width $r - 1$ centimeters

55. The *width* of a rectangle with perimeter 300 feet and length x feet

56. The *length* of a rectangle with area 200 square feet and width w feet

57. The *length* of a rectangle, given that its width is x feet and its length is 1 foot longer than twice the width

58. The *length* of a rectangle, given that its width is w feet and its length is 3 feet shorter than twice the width

59. The *area* of a rectangle, given that the width is x meters and the length is 5 meters longer than the width

60. The *perimeter* of a rectangle, given that the length is x yards and the width is 10 yards shorter

61. The *simple interest*, given that the principal is $x + 1000$, the rate is 18%, and the time is 1 year

62. The *simple interest*, given that the principal is $3x$, the rate is 6%, and the time is 1 year

63. The *price per pound* of peaches, given that x pounds sold for $16.50

64. The *rate per hour* of a mechanic who gets $480 for working x hours

65. The *degree measure* of an angle, given that its complementary angle has measure x degrees

66. The *degree measure* of an angle, given that its supplementary angle has measure x degrees

Identify the variable and write an equation that describes each situation. Do not solve the equation. See Example 6.

67. Two numbers differ by 5 and have a product of 8

68. Two numbers differ by 6 and have a product of -9

69. Herman's house sold for x dollars. The real estate agent received 7% of the selling price, and Herman received $84,532.

70. Gwen sold her car on consignment for x dollars. The saleswoman's commission was 10% of the selling price, and Gwen received $6570.

71. What percent of 500 is 100?

72. What percent of 40 is 120?

73. The value of x nickels and $x + 2$ dimes is $3.80.

74. The value of d dimes and $d - 3$ quarters is $6.75.

75. The sum of a number and 5 is 13.

76. Twelve subtracted from a number is -6.

77. The sum of three consecutive integers is 42.

78. The sum of three consecutive odd integers is 27.

79. The product of two consecutive integers is 182.

80. The product of two consecutive even integers is 168.

81. Twelve percent of Harriet's income is $3000.

82. If 9% of the members buy tickets, then we will sell 252 tickets to this group.

83. Thirteen is 5% of what number?

84. Three hundred is 8% of what number?

85. The length of a rectangle is 5 feet longer than the width, and the area is 126 square feet.

86. The length of a rectangle is 1 yard shorter than twice the width, and the perimeter is 298 yards.

87. The value of n nickels and $n - 1$ dimes is 95 cents.

88. The value of q quarters, $q + 1$ dimes, and $2q$ nickels is 90 cents.

89. The measure of an angle is 38° smaller than the measure of its supplementary angle.

90. The measure of an angle is 16° larger than the measure of its complementary angle.

91. *Target heart rate.* For a cardiovascular work out, fitness experts recommend that you reach your target heart rate and stay at that rate for at least 20 minutes (*Cycling,* Burkett and Darst). To find your target heart rate, find the sum of your age and your resting heart rate, then subtract that sum from 220. Find 60% of that result and add it to your resting heart rate.
   a) Write an equation with variable r expressing the fact that the target heart rate for 30-year-old Bob is 144.

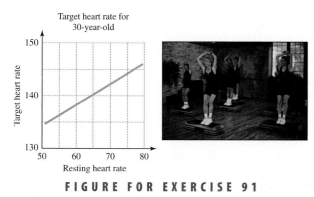

Target heart rate for 30-year-old

FIGURE FOR EXERCISE 91

b) Judging from the accompanying graph, does the target heart rate for a 30-year-old increase or decrease as the resting heart rate increases.

92. *Adjusting the saddle.* The saddle height on a bicycle should be 109% of the rider's inside leg measurement L (*Cycling,* Burkett and Darst). See the figure. Write an equation expressing the fact that the saddle height for Brenda is 36 in.

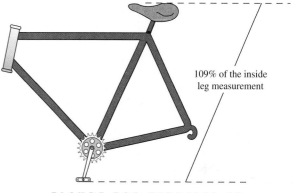

109% of the inside leg measurement

FIGURE FOR EXERCISE 92

Given that the area of each figure is 24 square feet, use the dimensions shown to write an equation expressing this fact. Do not solve the equation.

93.

x

$x + 3$

94.

$h + 2$

$h + 2$

95.

$w - 4$

w

96.

$y - 2$

y

2.6 NUMBER, GEOMETRIC, AND UNIFORM MOTION APPLICATIONS

In this section, we apply the ideas of Section 2.5 to solving problems. Many of the problems can be solved by using arithmetic only and not algebra. However, remember that we are not just trying to find the answer, we are trying to learn how to apply algebra. So even if the answer is obvious to you, set the problem up and solve it by using algebra as shown in the examples.

Number Problems

Algebra is often applied to problems involving time, rate, distance, interest, or discount. **Number problems** do not involve any physical situation. In number problems we simply find some numbers that satisfy some given conditions. Number problems can provide good practice for solving more complex problems.

EXAMPLE 1

A consecutive integer problem

The sum of three consecutive integers is 48. Find the integers.

Solution

We first represent the unknown quantities with variables. Let $x, x + 1,$ and $x + 2$ represent the three consecutive integers. We now write an equation that describes the problem and solve it. The equation expresses the fact that the sum of the integers is 48.

$$x + (x + 1) + (x + 2) = 48$$
$$3x + 3 = 48 \quad \text{Combine like terms.}$$
$$3x = 45 \quad \text{Subtract 3 from each side.}$$
$$x = 15 \quad \text{Divide each side by 3.}$$
$$x + 1 = 16 \quad \text{If } x \text{ is 15, then } x + 1 \text{ is 16 and } x + 2 \text{ is 17.}$$
$$x + 2 = 17$$

Because $15 + 16 + 17 = 48$, the three consecutive integers that have a sum of 48 are 15, 16, and 17. ∎

General Strategy for Solving Verbal Problems

You should use the following steps as a guide for solving problems.

Strategy for Solving Problems

1. Read the problem as many times as necessary. Guessing the answer and checking it will help you understand the problem.
2. If possible, draw a diagram to illustrate the problem.
3. Choose a variable and *write* what it represents.
4. Write algebraic expressions for any other unknowns in terms of that variable.
5. Write an equation that describes the situation.
6. Solve the equation.
7. Answer the original question.
8. Check your answer in the original problem (not the equation).

Geometric Problems

Geometric problems involve geometric figures. For these problems you should always draw the figure and label it.

E X A M P L E   2

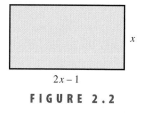

x

$2x - 1$

FIGURE 2.2

A perimeter problem

The length of a rectangular piece of property is 1 foot less than twice the width. If the perimeter is 748 feet, find the length and width.

Solution

Let $x =$ the width. Since the length is 1 foot less than twice the width, $2x - 1 =$ the length. Draw a diagram as in Fig. 2.2. We know that $2L + 2W = P$ is the formula for perimeter of a rectangle. Substituting $2x - 1$ for L and x for W in this formula yields an equation in x:

$$2L + 2W = P$$
$$2(2x - 1) + 2(x) = 748 \quad \text{Replace } L \text{ by } 2x - 1 \text{ and } W \text{ by } x.$$
$$4x - 2 + 2x = 748 \quad \text{Remove the parentheses.}$$
$$6x - 2 = 748 \quad \text{Combine like terms.}$$
$$6x = 750 \quad \text{Add 2 to each side.}$$
$$x = 125 \quad \text{Divide each side by 6.}$$
$$2x - 1 = 249 \quad \text{If } x = 125, \text{ then } 2x - 1 = 2(125) - 1 = 249.$$

Check these answers by computing $2L + 2W$:

$$2(249) + 2(125) = 748$$

So the width is 125 feet, and the length is 249 feet.   ■

The next geometric example involves the degree measures of angles. For this problem, the figure is given.

E X A M P L E   3

Complementary angles

In Fig. 2.3 the angle formed by the guy wire and the ground is 3.5 times as large as the angle formed by the guy wire and the antenna. Find the degree measure of each of these angles.

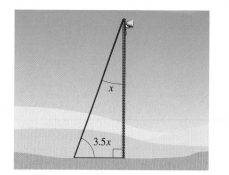

x

$3.5x$

FIGURE 2.3

Solution

Let $x =$ the degree measure of the smaller angle, and let $3.5x =$ the degree measure of the larger angle. Since the antenna meets the ground at a $90°$ angle, the sum of the degree measures of the other two angles of the right triangle is 90°. (They are complementary angles.) So we have the following equation:

$$x + 3.5x = 90$$
$$4.5x = 90 \quad \text{Combine like terms.}$$
$$x = 20 \quad \text{Divide each side by 4.5.}$$
$$3.5x = 70 \quad \text{Find the other angle.}$$

Check: 70° is 3.5 · 20° and 20° + 70° = 90°. So the smaller angle is 20°, and the larger angle is 70°. ■

Uniform Motion Problems

Problems involving motion at a constant rate are called **uniform motion problems.** In uniform motion problems we often use an average rate when the actual rate is not constant. For example, you can drive all day and average 50 miles per hour, but you are not driving at a constant 50 miles per hour.

E X A M P L E   4

Finding the rate

Bridgette drove her car for 2 hours on an icy road. When the road cleared up, she increased her speed by 35 miles per hour and drove 3 more hours, completing her 255-mile trip. How fast did she travel on the icy road?

Solution

It is helpful to draw a diagram and then make a table to classify the given information. Remember that $D = RT$.

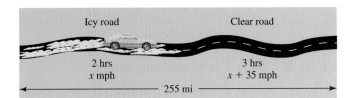

	Rate	Time	Distance
Icy road	$x \dfrac{\text{mi}}{\text{hr}}$	2 hr	$2x$ mi
Clear road	$x + 35 \dfrac{\text{mi}}{\text{hr}}$	3 hr	$3(x + 35)$ mi

The equation expresses the fact that her total distance traveled was 255 miles:

Icy road distance + clear road distance = total distance

$$2x + 3(x + 35) = 255$$
$$2x + 3x + 105 = 255$$
$$5x + 105 = 255$$
$$5x = 150$$
$$x = 30$$
$$x + 35 = 65$$

If she drove at 30 miles per hour for 2 hours on the icy road, she went 60 miles. If she drove at 65 miles per hour for 3 hours on the clear road, she went 195 miles. Since $60 + 195 = 255$, we can be sure that her speed on the icy road was 30 mph. ∎

In the next uniform motion problem we find the time.

E X A M P L E   5   **Finding the time**

Pierce drove from Allentown to Baker, averaging 55 miles per hour. His journey back to Allentown using the same route took 3 hours longer because he averaged only 40 miles per hour. How long did it take him to drive from Allentown to Baker? What is the distance between Allentown and Baker?

Solution

Draw a diagram and then make a table to classify the given information. Remember that $D = RT$.

	Rate	**Time**	**Distance**
Going	$55 \dfrac{\text{mi}}{\text{hr}}$	x hr	$55x$ mi
Returning	$40 \dfrac{\text{mi}}{\text{hr}}$	$x + 3$ hr	$40(x + 3)$ mi

We can write an equation expressing the fact that the distance either way is the same:

$$\text{Distance going} = \text{distance returning}$$
$$55x = 40(x + 3)$$
$$55x = 40x + 120$$
$$15x = 120$$
$$x = 8$$

The trip from Allentown to Baker took 8 hours. The distance between Allentown and Baker is $55 \cdot 8$, or 440 miles. ∎

W A R M - U P S

True or false? Explain your answer.

1. The first step in solving a word problem is to write the equation.

2. You should always write down what the variable represents.

3. Diagrams and tables are used as aids in solving problems.

4. To represent two consecutive odd integers, we use x and $x + 1$.

5. If $5x$ is 2 miles more than $3(x + 20)$, then $5x + 2 = 3(x + 20)$.

6. We can represent two numbers with a sum of 6 by x and $6 - x$.

7. Two numbers that differ by 7 can be represented by x and $x + 7$.

(*continued*)

8. The degree measures of two complementary angles can be represented by x and $90 - x$.

9. The degree measures of two supplementary angles can be represented by x and $x + 180$.

10. If x is half as large as $x + 50$, then $2x = x + 50$.

2.6 EXERCISES

Reading and Writing *After reading this section, write out the answers to these questions. Use complete sentences.*

1. What types of problems are discussed in this section?

2. Why do we solve number problems?

3. What is uniform motion?

4. What are supplementary angles?

5. What are complementary angles?

6. What should you always do when solving a geometric problem?

Show a complete solution to each problem. See Example 1.

7. *Consecutive integers.* Find three consecutive integers whose sum is 141.

8. *Consecutive even integers.* Find three consecutive even integers whose sum is 114.

9. *Consecutive odd integers.* Two consecutive odd integers have a sum of 152. What are the integers?

10. *Consecutive odd integers.* Four consecutive odd integers have a sum of 120. What are the integers?

11. *Consecutive integers.* Find four consecutive integers whose sum is 194.

12. *Consecutive even integers.* Find four consecutive even integers whose sum is 340.

Show a complete solution to each problem. See Examples 2 and 3.

13. *Olympic swimming.* If an Olympic swimming pool is twice as long as it is wide and the perimeter is 150 meters, then what are the length and width?

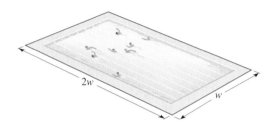

FIGURE FOR EXERCISE 13

14. *Wimbledon tennis.* If the perimeter of a tennis court is 228 feet and the length is 6 feet longer than twice the width, then what are the length and width?

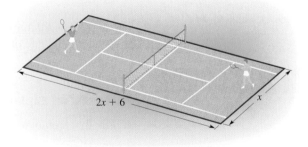

FIGURE FOR EXERCISE 14

15. *Framed.* Julia framed an oil painting that her uncle gave her. The painting was 4 inches longer than it was wide, and it took 176 inches of frame molding. What were the dimensions of the picture?

16. *Industrial triangle.* Geraldo drove his truck from Indianapolis to Chicago, then to St. Louis, and then back to Indianapolis. He observed that the second side of his triangular route was 81 miles short of being twice as long as the first side and that the third side was 61 miles longer than the first side. If he traveled a total of 720 miles, then how long is each side of this triangular route?

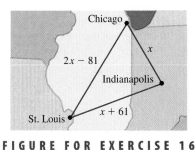

FIGURE FOR EXERCISE 16

17. *Triangular banner.* A banner in the shape of an isosceles triangle has a base that is 5 inches shorter than either of the equal sides. If the perimeter of the banner is 34 inches, then what is the length of the equal sides?

18. *Border paper.* Dr. Good's waiting room is 8 feet longer than it is wide. When Vincent wallpapered Dr. Good's waiting room, he used 88 feet of border paper. What are the dimensions of Dr. Good's waiting room?

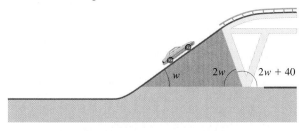

FIGURE FOR EXERCISE 18

19. *Ramping up.* A civil engineer is planning a highway overpass as shown in the figure. Find the degree measure of the angle marked w.

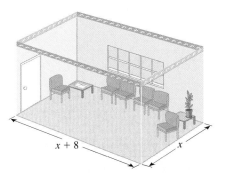

FIGURE FOR EXERCISE 19

20. *Ramping down.* For the other side of the overpass, the engineer has drawn the plans shown in the figure. Find the degree measure of the angle marked z.

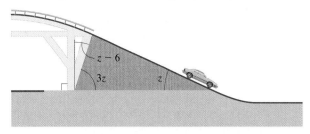

FIGURE FOR EXERCISE 20

Show a complete solution to each problem. See Examples 4 and 5.

21. *Highway miles.* Bret drove for 4 hours on the freeway, then decreased his speed by 20 miles per hour and drove for 5 more hours on a country road. If his total trip was 485 miles, then what was his speed on the freeway?

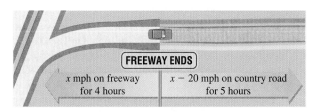

FIGURE FOR EXERCISE 21

22. *Walking and running.* On Saturday morning, Lynn walked for 2 hours and then ran for 30 minutes. If she ran twice as fast as she walked and she covered 12 miles altogether, then how fast did she walk?

23. *Driving all night.* Kathryn drove her rig 5 hours before dawn and 6 hours after dawn. If her average speed was 5 miles per hour more in the dark and she covered 630 miles altogether, then what was her speed after dawn?

24. *Commuting to work.* On Monday, Roger drove to work in 45 minutes. On Tuesday he averaged 12 miles per hour more, and it took him 9 minutes less to get to work. How far does he travel to work?

25. *Head winds.* A jet flew at an average speed of 640 mph from Los Angeles to Chicago. Because of head winds the jet averaged only 512 mph on the return trip, and the return trip took 48 minutes longer. How many hours was the flight from Chicago to Los Angeles? How far is it from Chicago to Los Angeles?

26. *Ride the Peaks.* Penny's bicycle trip from Colorado Springs to Pikes Peak took 1.5 hours longer than the return trip to Colorado Springs. If she averaged 6 mph on the way to Pikes Peak and 15 mph for the return trip, then how long was the ride from Colorado Springs to Pikes Peak?

Solve each problem.

27. ***Super Bowl score.*** The 1977 Super Bowl was played in the Rose Bowl in Pasadena. In that football game the Oakland Raiders scored 18 more points than the Minnesota Vikings. If the total number of points scored was 46, then what was the final score for the game?

28. ***Top three companies.*** Revenues for the top three companies in 1997, General Motors, Ford, and Exxon, totaled $453 billion (Fortune 500 List, www.fortune.com). If Ford's revenue was $31 billion greater that Exxon's, and General Motor's revenue was $25 billion greater than Ford's, then what was the 1997 revenue for each company?

29. ***Idabel to Lawton.*** Before lunch, Sally drove from Idabel to Ardmore, averaging 50 mph. After lunch she continued on to Lawton, averaging 53 mph. If her driving time after lunch was 1 hour less than her driving time before lunch and the total trip was 256 miles, then how many hours did she drive before lunch? How far is it from Ardmore to Lawton?

30. ***Norfolk to Chadron.*** On Monday, Chuck drove from Norfolk to Valentine, averaging 47 mph. On Tuesday, he continued on to Chadron, averaging 69 mph. His driving time on Monday was 2 hours longer than his driving time on Tuesday. If the total distance from Norfolk to Chadron is 326 miles, then how many hours did he drive on Monday? How far is it from Valentine to Chadron?

31. ***Golden oldies.*** Joan Crawford, John Wayne, and James Stewart were born in consecutive years (*Doubleday Almanac*). Joan Crawford was the oldest of the three, and James Stewart was the youngest. In 1950, after all three

had their birthdays, the sum of their ages was 129. In what years were they born?

32. ***Leading men.*** Bob Hope was born 2 years after Clark Gable and 2 years before Henry Fonda (*Doubleday Almanac*). In 1951, after all three of them had their birthdays, the sum of their ages was 144. In what years were they born?

33. ***Trimming a garage door.*** A carpenter used 30 ft of molding in three pieces to trim a garage door. If the long piece was 2 ft longer than twice the length of each shorter piece, then how long was each piece?

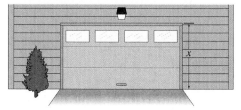

FIGURE FOR EXERCISE 33

34. ***Fencing dog pens.*** Clint is constructing two adjacent rectangular dog pens. Each pen will be three times as long as it is wide, and the pens will share a common long side. If Clint has 65 ft of fencing, what are the dimensions of each pen?

FIGURE FOR EXERCISE 34

2.7 DISCOUNT, INVESTMENT, AND MIXTURE APPLICATIONS

In this section, we continue our study of applications of algebra. The problems in this section involve percents.

Discount Problems

In this section

• Discount Problems
• Commission Problems
• Investment Problems
• Mixture Problems

When an item is sold at a discount, the amount of the discount is usually described as being a percentage of the original price. The percentage is called the **rate of discount.** Multiplying the rate of discount and the original price gives the amount of the discount.

EXAMPLE 1

Finding the original price

Ralph got a 12% discount when he bought his new 1999 Corvette Coupe. If the amount of his discount was $4584, then what was the original price of the Corvette?

Solution

Let x represent the original price. The discount is found by multiplying the 12% rate of discount and the original price:

$$\text{rate of discount} \cdot \text{original price} = \text{amount of discount}$$
$$0.12x = 4584$$
$$x = \frac{4584}{0.12} \qquad \text{Divide each side by 0.12.}$$
$$x = 38{,}200$$

To check, find 12% of $38,200. Since $0.12 \cdot 38{,}200 = 4584$, the original price of the Corvette was $38,200. ◼

E X A M P L E   2   **Finding the original price**

When Susan bought her new car, she also got a discount of 12%. She paid $17,600 for her car. What was the original price of Susan's car?

Solution

Let x represent the original price for Susan's car. The amount of discount is 12% of x, or $0.12x$. We can write an equation expressing the fact that the original price minus the discount is the price Susan paid.

$$\text{Original price} - \text{discount} = \text{sale price}$$
$$x - 0.12x = 17{,}600$$
$$0.88x = 17{,}600 \qquad 1.00x - 0.12x = 0.88x$$
$$x = \frac{17{,}600}{0.88} \qquad \text{Divide each side by 0.88.}$$
$$x = \$20{,}000$$

> **helpful hint**
>
> To get familiar with the problem, guess that the original price was $30,000. Then her discount is 0.12(30,000) or $3600. The price she paid would be 30,000 − 3600 or $26,400, which is incorrect.

Check: 12% of $20,000 is $2400, and $20,000 − $2400 = $17,600. The original price of Susan's car was $20,000. ◼

Commission Problems

A salesperson's commission for making a sale is often a percentage of the selling price. **Commission problems** are very similar to other problems involving percents. The commission is found by multiplying the rate of commission and the selling price.

E X A M P L E   3   **Real estate commission**

Sarah is selling her house through a real estate agent whose commission rate is 7%. What should the selling price be so that Sarah can get the $83,700 she needs to pay off the mortgage?

Solution

Let x be the selling price. The commission is 7% of x (not 7% of $83,700). Sarah receives the selling price less the sales commission:

$$\text{Selling price} - \text{commission} = \text{Sarah's share}$$
$$x - 0.07x = 83{,}700$$
$$0.93x = 83{,}700 \qquad 1.00x - 0.07x = 0.93x$$
$$x = \frac{83{,}700}{0.93}$$
$$x = 90{,}000$$

Check: 7% of $90,000 is $6300, and $90,000 − $6300 = $83,700. So the house should sell for $90,000. ◼

Investment Problems

The interest on an investment is a percentage of the investment, just as the sales commission is a percentage of the sale amount. However, in **investment problems** we must often account for more than one investment at different rates. So it is a good idea to make a table, as in the next example.

EXAMPLE 4

Diversified investing

Ruth Ann invested some money in a certificate of deposit with an annual yield of 9%. She invested twice as much in a mutual fund with an annual yield of 10%. Her interest from the two investments at the end of the year was $232. How much was invested at each rate?

Solution

When there are many unknown quantities, it is often helpful to identify them in a table. Since the time is 1 year, the amount of interest is the product of the interest rate and the amount invested.

helpful hint

To get familiar with the problem, guess that she invested $1000 at 9% and $2000 at 10%. Then her interest in one year would be

0.09(1000) + 0.10(2000)

or $290, which is close but incorrect.

	Interest rate	Amount invested	Interest for 1 year
CD	9%	x	$0.09x$
Mutual fund	10%	$2x$	$0.10(2x)$

Since the total interest from the investments was $232, we can write the following equation:

$$\text{CD interest} + \text{mutual fund interest} = \text{total interest}$$
$$0.09x + 0.10(2x) = 232$$
$$0.09x + 0.20x = 232$$
$$0.29x = 232$$
$$x = \frac{232}{0.29}$$
$$x = \$800$$
$$2x = \$1600$$

To check, we find the total interest:

$$0.09(800) + 0.10(1600) = 72 + 160$$
$$= 232$$

study tip

Finding out what happened in class and attending class are not the same. Attend every class and be attentive. Don't just take notes and let your mind wander. Use class time as a learning time.

So Ruth Ann invested $800 at 9% and $1600 at 10%. ◼

Mixture Problems

Mixture problems are concerned with the result of mixing two quantities, each of which contains another substance. Notice how similar the following mixture problem is to the last investment problem.

E X A M P L E   5

Mixing milk

How many gallons of milk containing 4% butterfat must be mixed with 80 gallons of 1% milk to obtain 2% milk?

Solution

It is helpful to draw a diagram and then make a table to classify the given information.

x gal milk 4% fat   +   80 gal milk 1% fat   = $x + 80$ gal milk 2% fat

	Percentage of fat	Amount of milk	Amount of fat
4% milk	4%	x	$0.04x$
1% milk	1%	80	$0.01(80)$
2% milk	2%	$x + 80$	$0.02(x + 80)$

The equation expresses the fact that the total fat from the first two types of milk is the same as the fat in the mixture:

$$\text{Fat in 4\% milk} + \text{fat in 1\% milk} = \text{fat in 2\% milk}$$

$0.04x + 0.01(80) = 0.02(x + 80)$	
$0.04x + 0.8 = 0.02x + 1.6$	Simplify.
$100(0.04x + 0.8) = 100(0.02x + 1.6)$	Multiply each side by 100.
$4x + 80 = 2x + 160$	Distributive property.
$2x + 80 = 160$	Subtract $2x$ from each side.
$2x = 80$	Subtract 80 from each side.
$x = 40$	Divide each side by 2.

To check, calculate the total fat:

$$2\% \text{ of } 120 \text{ gallons} = 0.02(120) = 2.4 \text{ gallons of fat}$$
$$0.04(40) + 0.01(80) = 1.6 + 0.8 = 2.4 \text{ gallons of fat}$$

So we mix 40 gallons of 4% milk with 80 gallons of 1% milk to get 120 gallons of 2% milk. ■

In mixture problems, the solutions might contain fat, alcohol, salt, or some other substance. We always assume that the substance neither appears nor disappears in the process. For example, if there are 3 grams of salt in one glass of water and 2 grams in another, then there are exactly 5 grams in a mixture of the two.

True or false? Explain your answer.

1. If the original price is w and the discount is 8%, then the selling price is $w - 0.08w$.

2. If x is the selling price and the commission is 8% of the selling price, then the commission is $0.08x$.

3. If you need $40,000 for your house and the agent gets 10% of the selling price, then the agent gets $4000, and the house sells for $44,000.

4. If you mix 10 liters of a 20% acid solution with x liters of a 30% acid solution, then the total amount of acid is $2 + 0.3x$ liters.

5. A 10% acid solution mixed with a 14% acid solution results in a 24% acid solution.

6. If a TV costs x dollars and sales tax is 5%, then the total bill is $1.05x$ dollars.

2.7 EXERCISES

Reading and Writing   *After reading this section, write out the answers to these questions. Use complete sentences.*

1. What types of problems are discussed in this section?

2. What is the difference between discount and rate of discount?

3. What is the relationship between discount, original price, rate of discount, and sale price?

4. What do mixture problems and investment problems have in common?

5. Why do we make a table when solving certain problems.

6. What is the relationship between amount of interest, amount invested, and interest rate?

Show a complete solution to each problem. See Examples 1 and 2.

7. *Close-out sale.* At a 25% off sale, Jose saved $80 on a 19-inch Panasonic TV. What was the original price of the television.

8. *Big bike.* A 12% discount on a Giant Perigee saved Melanie $46.68. What was the original price of the bike?

9. *Circuit city.* After getting a 20% discount, Robert paid $320 for a Pioneer CD player for his car. What was the original price of the CD player?

10. *Chrysler Sebring.* After getting a 15% discount on the price of a new Chrysler Sebring convertible, Helen paid $27,000. What was the original price of the convertible?

Show a complete solution to each problem. See Example 3.

11. *Selling price of a home.* Kirk wants to get $72,000 for his house. The real estate agent gets a commission equal to 10% of the selling price for selling the house. What should the selling price be?

FIGURE FOR EXERCISE 11

12. *Horse trading.* Gene is selling his palomino at an auction. The auctioneer's commission is 10% of the selling price. If Gene still owes $810 on the horse, then what must the horse sell for so that Gene can pay off his loan?

13. *Sales tax collection.* Merilee sells tomatoes at a roadside stand. Her total receipts including the 7% sales tax were $462.24. What amount of sales tax did she collect?

14. *Toyota Corolla.* Gwen bought a new Toyota Corolla. The selling price plus the 8% state sales tax was $15,714. What was the selling price?

Show a complete solution to each problem. See Example 4.

15. *Wise investments.* Wiley invested some money in the Berger 100 Fund and $3000 more than that amount in the Berger 101 Fund. For the year he was in the fund, the 100 Fund paid 18% simple interest and the 101 Fund paid 15% simple interest. If the income from the two investments totaled $3750 for one year, then how much did he invest in each fund?

16. *Loan shark.* Becky lent her brother some money at 8% simple interest, and she lent her sister twice as much at twice the interest rate. If she received a total of 20 cents interest, then how much did she lend to each of them?

17. *Investing in bonds.* David split his $25,000 inheritance between Fidelity Short-Term Bond Fund with an annual yield of 5% and T. Rowe Price Tax-Free Short-Intermediate Fund with an annual yield of 4%. If his total income for one year on the two investments was $1140, then how much did he invest in each fund?

18. *High-risk funds.* Of the $50,000 that Natasha pocketed on her last real estate deal, $20,000 went to charity. She invested part of the remainder in Dreyfus New Leaders Fund with an annual yield of 16% and the rest in Templeton Growth Fund with an annual yield of 25%. If she made $6060 on these investments in one year, then how much did she invest in each fund?

Show a complete solution to each problem. See Example 5.

19. *Mixing milk.* How many gallons of milk containing 1% butterfat must be mixed with 30 gallons of milk containing 3% butterfat to obtain a mixture containing 2% butterfat?

x gal $\quad$ 30 gal $\quad$ $x + 30$ gal
1% fat $\quad$ 3% fat $\quad$ 2% fat

FIGURE FOR EXERCISE 19

20. *Acid solutions.* How many gallons of a 5% acid solution should be mixed with 30 gallons of a 10% acid solution to obtain a mixture that is 8% acid?

21. *Alcohol solutions.* Gus has on hand a 5% alcohol solution and a 20% alcohol solution. He needs 30 liters of a 10% alcohol solution. How many liters of each solution should he mix together to obtain the 30 liters?

22. *Adjusting antifreeze.* Angela needs 20 quarts of 50% antifreeze solution in her radiator. She plans to obtain this by mixing some pure antifreeze with an appropriate amount of a 40% antifreeze solution. How many quarts of each should she use?

100% antifreeze ? qts   +   40% solution ? qts   =   50% solution 20 qts

FIGURE FOR EXERCISE 22

Solve each problem.

23. *Registered voters.* If 60% of the registered voters of Lancaster County voted in the November election and 33,420 votes were cast, then how many registered voters are there in Lancaster County?

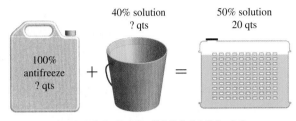

FIGURE FOR EXERCISE 23

24. *Tough on crime.* In a random sample of voters, 594 respondents said that they favored passage of a $33 billion crime bill. If the number in favor of the crime bill was 45% of the number of voters in the sample, then how many voters were in the sample?

25. *Ford Taurus.* At an 8% sales tax rate, the sales tax on Peter's new Ford Taurus was $1200. What was the price of the car?

26. *Taxpayer blues.* Last year, Faye paid 24% of her income to taxes. If she paid $9600 in taxes, then what was her income?

27. *Making a profit.* A retail store buys shirts for $8 and sells them for $14. What percent increase is this?

28. *Monitoring AIDS.* If 28 new AIDS cases were reported in Landon County this year and 35 new cases were reported last year, then what percent decrease in new cases is this?

29. *High school integration.* Wilson High School has 400 students, of whom 20% are African American. The school board plans to merge Wilson High with Jefferson High. This one school will then have a student population that is 44% African American. If Jefferson currently has a student population that is 60% African American, then how many students are at Jefferson?

30. *Junior high integration.* The school board plans to merge two junior high schools into one school of 800 students in which 40% of the students will be Caucasian. One of the schools currently has 58% Caucasian students; the other has only 10% Caucasian students. How many students are in each of the two schools?

31. *Hospital capacity.* When Memorial Hospital is filled to capacity, it has 18 more people in semiprivate rooms (two patients to a room) than in private rooms. The room rates are $200 per day for a private room and $150 per day for a semiprivate room. If the total receipts for rooms is $17,400 per day when all are full, then how many rooms of each type does the hospital have?

32. *Public relations.* Memorial Hospital is planning an advertising campaign. It costs the hospital $3000 each time a television ad is aired and $2000 each time a radio ad is aired. The administrator wants to air 60 more television ads than radio ads. If the total cost of airing the ads is $580,000, then how many ads of each type will be aired?

33. *Mixed nuts.* Cashews sell for $4.80 per pound, and pistachios sell for $6.40 per pound. How many pounds of pistachios should be mixed with 20 pounds of cashews to get a mixture that sells for $5.40 per pound?

34. *Premium blend.* Premium coffee sells for $6.00 per pound, and regular coffee sells for $4.00 per pound. How many pounds of each type of coffee should be blended to obtain 100 pounds of a blend that sells for $4.64 per pound?

35. *Nickels and dimes.* Candice paid her library fine with 10 coins consisting of nickels and dimes. If the fine was $0.80, then how many of each type of coin did she use?

36. *Dimes and quarters.* Jeremy paid for his breakfast with 36 coins consisting of dimes and quarters. If the bill was $4.50, then how many of each type of coin did he use?

37. *Cooking oil.* Crisco Canola Oil is 7% saturated fat. Crisco blends corn oil that is 14% saturated fat with Crisco Canola Oil to get Crisco Canola and Corn Oil, which is 11% saturated fat. How many gallons of corn oil must Crisco mix with 600 gallons of Crisco Canola Oil to get Crisco Canola and Corn Oil?

38. *Chocolate ripple.* The Delicious Chocolate Shop makes a dark chocolate that is 35% fat and a white chocolate that is 48% fat. How many kilograms of dark chocolate should be mixed with 50 kilograms of white chocolate to make a ripple blend that is 40% fat?

39. *Hawaiian Punch.* Hawaiian Punch is 10% fruit juice. How much water would you have to add to one gallon of Hawaiian Punch to get a drink that is 6% fruit juice?

40. *VCRs and CDs.* The manager of a stereo shop placed an order for $10,710 worth of VCRs at $120 each and CD players at $150 each. If the number of VCRs she ordered was three times the number of CD players, then how many of each did she order?

COLLABORATIVE ACTIVITIES

Finding the Better Deal?

For this activity, the students in your group should choose roles. Four standard roles are Moderator (keeps the group on task), Messenger (asks the group's questions to the instructor, tutor, or helper), Quality Manager (checks to see that the work is top quality), and Recorder (records the group's work). See the Instructor's Solution Manual for a description of these roles. After you have chosen roles, read through the activity completely, and answer the questions.

Scenario: You have decided to buy a new car and have asked some friends to help you choose the best deal and the best financing. You have already looked into your finances. You have $1700 from your summer job and $1500 that your parents will give you for a down payment on a car.

Grouping: 2 to 4 students per group
Topic: Percents

You found a car that you really liked that was 10% off the regular $9800 price. Your friends at the student Credit Union tell you it has a 48-month car loan at $7\frac{1}{2}$% annual simple interest.

At a second dealership you find a similar car on sale for $9000 if you finance it through the dealership. The dealer said that after the down payment you could pay it off in 5 years with monthly payments of $140. This second deal sounds good! (You have decided you could afford up to $160 a month in payments.) The idea of having an extra $20 a month is appealing. However, you wonder how much you will actually pay for the second car.

Which car should you buy?

Questions: The following questions will help you to work your way through the problem.

For the first car, if you finance it at the Credit Union:

1. How much will it cost after the discount?
2. How much will you need to borrow?
3. What will the total interest be for the 48 months?
4. How much will the monthly payments be?

For the second car:

5. Find the amount to be financed.
6. Find the interest and the interest rate.

For both cars:

7. Find the total cost for each car.

After reviewing the information, decide which car you would buy.

Extension:

1. What other costs would there be? Find these out for your city.
2. Do a comparison of car loans at your local banks.

WRAP-UP   CHAPTER 2

SUMMARY

Equations		**Examples**
Linear equation	An equation of the form $ax + b = 0$ with $a \neq 0$	$3x + 7 = 0$
Identity	An equation that is satisfied by every number for which both sides are defined	$x + x = 2x$
Conditional equation	An equation that has at least one solution but is not an identity	$5x - 10 = 0$
Inconsistent equation	An equation that has no solution	$x = x + 1$
Equivalent equations	Equations that have exactly the same solutions	$2x + 1 = 5$ $2x = 4$
Properties of equality	If the same number is added to or subtracted from each side of an equation, the resulting equation is equivalent to the original equation.	$x - 5 = -9$ $x = -4$
	If each side of an equation is multiplied or divided by the same nonzero number, the resulting equation is equivalent to the original equation.	$9x = 27$ $x = 3$
Solving equations	1. Remove parentheses and combine like terms to simplify each side as much as possible. 2. Get like terms from opposite sides onto the same side so that they may be combined. 3. Use the multiplication-division property of equality last to isolate the variable. 4. The equation $-x = a$ is equivalent to $x = -a$. 5. Check that the solution satisfies the original equation.	$2(x - 3) = -7 + 3(x - 1)$ $2x - 6 = -10 + 3x$ $-x - 6 = -10$ $-x = -4$ $x = 4$ *Check:* $2(4 - 3) = -7 + 3(4 - 1)$ $2 = 2$

Applications

Steps in solving applied problems

1. Read the problem.
2. If possible, draw a diagram to illustrate the problem.
3. Choose a variable and write down what it represents.
4. Represent any other unknowns in terms of that variable.
5. Write an equation that describes the situation.
6. Solve the equation.
7. Answer the original question.
8. Check your answer by using it to solve the original problem (not the equation).

ENRICHING YOUR MATHEMATICAL WORD POWER

For each mathematical term, choose the correct meaning.

1. **linear equation**
   a. an equation in which the terms are in line
   b. an equation of the form $ax + b = 0$ where $a \neq 0$
   c. the equation $a = b$
   d. an equation of the form $a^2 + b^2 = c^2$

2. **identity**
   a. an equation that is satisfied by all real numbers
   b. an equation that is satisfied by every real number
   c. an equation that is identical
   d. an equation that is satisfied by every real number for which both sides are defined

3. **conditional equation**
   a. an equation that has at least one real solution
   b. an equation that is correct
   c. an equation that is satisfied by at least one real number but is not an identity
   d. an equation that we are not sure how to solve

4. **inconsistent equation**
   a. an equation that is wrong
   b. an equation that is only sometimes consistent
   c. an equation that has no solution
   d. an equation with two variables

5. **equivalent equations**
   a. equations that are identical
   b. equations that are correct
   c. equations that are equal
   d. equations that have the same solution

6. **formula**
   a. an equation
   b. a type of race car
   c. a process
   d. an equation involving two or more variables

7. **literal equation**
   a. a formula
   b. an equation with words
   c. a false equation
   d. a fact

8. **complementary angles**
   a. angles that compliment each other
   b. angles whose degree measures total 90°
   c. angles whose degree measures total 180°
   d. angles with the same vertex

9. **supplementary angles**
   a. angles with soft flexible sides
   b. angles whose degree measures total 90°
   c. angles whose degree measures total 180°
   d. angles that form a square

10. **uniform motion**
    a. movement of an army
    b. movement in a straight line
    c. consistent motion
    d. motion at a constant rate

REVIEW EXERCISES

2.1 *Solve each equation and check your answer.*

1. $x - 23 = 12$

2. $14 = 18 + y$

3. $\dfrac{2}{3}u = -4$

4. $-\dfrac{3}{8}r = 15$

5. $-5y = 35$

6. $-12 = 6h$

7. $6m = 13 + 5m$

8. $19 - 3n = -2n$

2.2 *Solve each equation and check your answer.*

9. $2x - 5 = 9$

10. $5x - 8 = 38$

11. $3p - 14 = -4p$

12. $36 - 9y = 3y$

13. $2z + 12 = 5z - 9$

14. $15 - 4w = 7 - 2w$

15. $2(h - 7) = -14$

16. $2(t - 7) = 0$

17. $3(w - 5) = 6(w + 2) - 3$

18. $2(a - 4) + 4 = 5(9 - a)$

2.3 *Solve each equation. Identify each equation as a conditional equation, an inconsistent equation, or an identity.*

19. $2(x - 7) - 5 = 5 - (3 - 2x)$

20. $2(x - 7) + 5 = -(9 - 2x)$

21. $2(w - w) = 0$

22. $2y - y = 0$

23. $\dfrac{3r}{3r} = 1$

24. $\dfrac{3t}{3} = 1$

25. $\dfrac{1}{2}a - 5 = \dfrac{1}{3}a - 1$

26. $\dfrac{1}{2}b - \dfrac{1}{2} = \dfrac{1}{4}b$

27. $0.06q + 14 = 0.3q - 5.2$

28. $0.05(z + 20) = 0.1z - 0.5$

29. $0.05(x + 100) + 0.06x = 115$

30. $0.06x + 0.08(x + 1) = 0.41$

Solve each equation.

31. $2x + \dfrac{1}{2} = 3x + \dfrac{1}{4}$

32. $5x - \dfrac{1}{3} = 6x - \dfrac{1}{2}$

33. $\dfrac{x}{2} - \dfrac{3}{4} = \dfrac{x}{6} + \dfrac{1}{8}$

34. $\dfrac{1}{3} - \dfrac{x}{5} = \dfrac{1}{2} - \dfrac{x}{10}$

35. $\dfrac{5}{6}x = -\dfrac{2}{3}$

36. $-\dfrac{2}{3}x = \dfrac{3}{4}$

37. $-\dfrac{1}{2}(x - 10) = \dfrac{3}{4}x$

38. $-\dfrac{1}{3}(6x - 9) = 23$

39. $3 - 4(x - 1) + 6 = -3(x + 2) - 5$

40. $6 - 5(1 - 2x) + 3 = -3(1 - 2x) - 1$

41. $5 - 0.1(x - 30) = 18 + 0.05(x + 100)$

42. $0.6(x - 50) = 18 - 0.3(40 - 10x)$

2.4 *Solve each equation for x.*

43. $ax + b = 0$

44. $mx + e = t$

45. $ax - 2 = b$

46. $b = 5 - x$

47. $LWx = V$

48. $3xy = 6$

49. $2x - b = 5x$

50. $t - 5x = 4x$

Solve each equation for y. Write the answer in the form $y = mx + b$, where m and b are real numbers.

51. $5x + 2y = 6$

52. $5x - 3y + 9 = 0$

53. $y - 1 = -\dfrac{1}{2}(x - 6)$

54. $y + 6 = \dfrac{1}{2}(x + 8)$

55. $\dfrac{1}{2}x + \dfrac{1}{4}y = 4$

56. $-\dfrac{x}{3} + \dfrac{y}{2} = 1$

Find the value of y in each formula if $x = -3$.

57. $y = 3x - 4$

58. $2x - 3y = -7$

59. $5xy = 6$

60. $3xy - 2x = -12$

61. $y - 3 = -2(x - 4)$

62. $y + 1 = 2(x - 5)$

2.5 *Translate each verbal expression into an algebraic expression.*

63. The sum of a number and 9

64. The product of a number and 7

65. Two numbers that differ by 8

66. Two numbers with a sum of 12

67. Sixty-five percent of a number

68. One half of a number

Identify the variable, and write an equation that describes each situation. Do not solve the equation.

69. One side of a rectangle is 5 feet longer than the other, and the area is 98 square feet.

70. One side of a rectangle is 1 foot longer than twice the other side, and the perimeter is 56 feet.

71. By driving 10 miles per hour slower than Jim, Barbara travels the same distance in 3 hours as Jim does in 2 hours.

72. Gladys and Ned drove 840 miles altogether, with Gladys averaging 5 miles per hour more in her 6 hours at the wheel than Ned did in his 5 hours at the wheel.

73. The sum of three consecutive even integers is 88.

74. The sum of two consecutive odd integers is 40.

75. The three angles of a triangle have degree measures of t, $2t$, and $t - 10$.

76. Two complementary angles have degree measures p and $3p - 6$.

2.6–7 *Solve each problem.*

77. ***Odd integers.*** If the sum of three consecutive odd integers is 237, then what are the integers?

78. *Even integers.* Find two consecutive even integers that have a sum of 450.

79. *Driving to the shore.* Lawanda and Betty both drive the same distance to the shore. By driving 15 miles per hour faster than Betty, Lawanda can get there in 3 hours while Betty takes 4 hours. How fast does each of them drive?

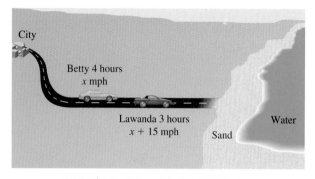

FIGURE FOR EXERCISE 79

80. *Rectangular lot.* The length of a rectangular lot is 50 feet more than the width. If the perimeter is 500 feet, then what are the length and width?

81. *Combined savings.* Wanda makes $6000 more per year than her husband does. Wanda saves 10% of her income for retirement, and her husband saves 6%. If together they save $5400 per year, then how much does each of them make per year?

82. *Layoffs looming.* American Products plans to lay off 10% of its employees in its aerospace division and 15% of its employees in its agricultural division. If altogether 12% of the 3000 employees in these two divisions will be laid off, then how many employees are in each division?

MISCELLANEOUS

Use an equation or formula to solve each problem.

83. *Flat yield curve.* The accompanying graph shows that the *yield curve* for U.S. Treasury bonds was relatively flat from 2 years out to 30 years on July 13, 1998 (Bloomberg, www.bloomberg.com). In this situation there is not much benefit to be obtained from long-term investing.

a) Use the interest rate in the graph to find the amount of interest earned in the first year on a 2-year bond of $10,000.

b) How much more interest would you earn in the first year on a 30-year bond of $10,000?

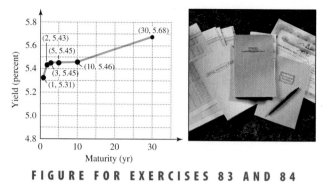

FIGURE FOR EXERCISES 83 AND 84

84. *Reading the curve.* Use the accompanying graph to find the maturity of a U.S. Treasury bond that had a yield of 5.46% on July 13, 1998.

85. *Combined videos.* The owners of ABC Video discovered that they had no movies in common with XYZ Video and bought XYZ's entire stock. Although XYZ had 200 titles, they had no children's movies, while 60% of ABC's titles were children's movies. If 40% of the movies in the combined stock are children's movies, then how many movies did ABC have before the merger?

86. *Living comfortably.* Gary has figured that he needs to take home $30,400 a year to live comfortably. If the government gets 24% of Gary's income, then what must his income be for him to live comfortably?

87. *Bracing a gate.* The diagonal brace on a rectangular gate forms an angle with the horizontal side with degree measure x and an angle with the vertical side with degree measure $2x - 3$. Find x.

88. *Digging up the street.* A contractor wants to install a pipeline connecting point A with point C on opposite sides of a road as shown in the figure for this exercise. To save money, the contractor has decided to lay the pipe to point B and then under the road to point C. Find the measure of the angle marked x in the figure.

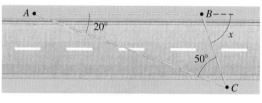

FIGURE FOR EXERCISE 88

CHAPTER 2 TEST

Solve each equation.

1. $6x - 7 = 0$

2. $-10x - 6 + 4x = -4x + 8$

3. $5(2x - 3) = x + 3$

4. $-\frac{2}{3}x + 1 = 7$

5. $2(x + 6) = 2x - 5$

6. $x + 7x = 8x$

7. $x + 0.06x = 742$

8. $\frac{1}{2}x - \frac{1}{3} = \frac{1}{4}x + \frac{1}{6}$

Determine whether each equation is a conditional equation, an inconsistent equation, or an identity.

9. $x + 2 = x$

10. $x + x = 2x$

11. $x + x = 2$

Solve for the indicated variable.

12. $2x - 3y = 9$ for y (in the form $y = mx + b$)

13. $m = aP - w$ for a

14. $2x - 3 = ax$ for x

Write a complete solution to each problem.

15. The perimeter of a rectangle is 72 meters. If the width is 8 meters less than the length, then what is the width of the rectangle?

16. If the area of a triangle is 54 square inches and the base is 12 inches, then what is the height?

17. How many liters of a 20% alcohol solution should Maria mix with 50 liters of a 60% alcohol solution to obtain a 30% solution?

18. If the degree measure of the smallest angle of a triangle is one-half of the degree measure of the second largest angle and one-third of the degree measure of the largest angle, then what is the degree measure of each angle?

Simplify each expression.

1. $3x + 5x$

2. $3x \cdot 5x$

3. $\dfrac{4x + 2}{2}$

4. $5 - 4(3 - x)$

5. $3x + 8 - 5(x - 1)$

6. $(-6)^2 - 4(-3)2$

7. $3^2 \cdot 2^3$

8. $4(-7) - (-6)(3)$

9. $-2x \cdot x \cdot x$

10. $(-1)(-1)(-1)(-1)(-1)$

Solve each equation.

11 $3x + 5x = 8$

12. $3x + 5x = 8x$

13. $3x + 5x = 7x$

14. $3x + 5 = 8$

15. $3x + 1 = 7$

16. $5 - 4(3 - x) = 1$

17. $3x + 8 = 5(x - 1)$

18. $x - 0.05x = 190$

Solve the problem.

19. *Linear depreciation.* In computing income taxes, a company is allowed to depreciate a $20,000 computer system over five years. Using *linear depreciation,* the value V of the computer system at any year t from 0 through 5 is given by

$$V = C - \frac{(C - S)}{5}t,$$

where C is the initial cost of the system and S is the scrap value of the system.

a) What is the value of the computer system after two years if its scrap value is $4000?

b) If the value of the system after three years is claimed to be $14,000, then what is the scrap value of the company's system?

c) If the accompanying graph models the depreciation of the system, then what is the scrap value of the system?

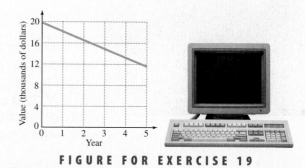

FIGURE FOR EXERCISE 19

Inequalities in One Variable

The practice of awarding degrees originated in the universities of medieval Europe. The first known degree, a degree in civil law, was awarded in Italy at the University of Bologna in the twelfth century. The University of Paris awarded its first bachelor's degree in the thirteenth century. By the time the first colleges were opened in the American colonies, the process of awarding degrees was firmly established. At first, American schools offered only a few types of degrees. The colleges established in the colonies were primarily to train young men for the ministry. Notable were Harvard (1636; Puritan), William and Mary (1693; Anglican), Yale (1701; Congregationalist), Princeton (1746; New Lights Presbyterian), Brown (1765; Baptist), and Rutgers (1766; Dutch Reformed).

The industrial revolution sparked a demand for training in many areas. Today, approximately 1500 types of degrees are granted by academic institutions in the United States. Over 1 million Bachelor of Arts (B.A.) or of Science (B.S.) degrees are granted annually. Over one-quarter of a million Master of Arts (M.A.) or Science (M.S.) degrees are awarded annually.

The growth of bachelor's and master's degrees is modeled with linear equations in Exercise 89 of Section 3.2. In that exercise we also use inequalities and compound inequalities to discuss the growth of these degrees.

3.1 INEQUALITIES

In this section

- Basic Ideas
- Interval Notation and Graphs
- Solving Linear Inequalities
- Applications

We worked with equations in the last chapter. Equations express the equality of two algebraic expressions. But we are often concerned with two algebraic expressions that are not equal, one expression being greater than or less than the other. In this section we will begin our study of inequalities.

Basic Ideas

Statements that express the inequality of algebraic expressions are called **inequalities.** The symbols that we use to express inequality are given below with their meanings.

Inequality Symbols

Symbol	Meaning
$<$	Is less than
$\le$	Is less than or equal to
$>$	Is greater than
$\ge$	Is greater than or equal to

helpful hint

A good way to learn inequality symbols is to notice that the inequality symbol always points at the smaller number. An inequality symbol can be read in either direction. For example, we can read $-4 < x$ as "-4 is less than x" or as "x is greater than -4." It is usually easier to understand an inequality if you read the variable first.

It is clear that 5 is less than 10, but how do we compare -5 and -10? If we think of negative numbers as debts, we would say that -10 is the larger debt. However, in algebra the size of a number is determined only by its position on the number line. For two numbers a and b we say that *a is less than b* if and only if a is to the *left* of b on the number line. To compare -5 and -10, we locate each point on the number line in Fig. 3.1. Because -10 is to the left of -5 on the number line, we say that -10 is less than -5. In symbols,

$$-10 < -5.$$

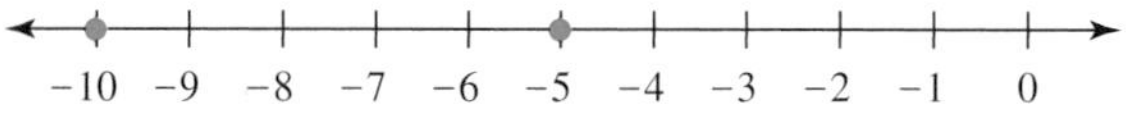

FIGURE 3.1

We say that a is greater than b if and only if a is to the *right* of b on the number line. Thus we can also write

$$-5 > -10.$$

The statement $a \le b$ is true if a is less than b or if a is equal to b. The statement $a \ge b$ is true if a is greater than b or if a equals b. For example, the statement $3 \le 5$ is true, and so is the statement $5 \le 5$.

EXAMPLE 1 **Inequalities**

Determine whether each statement is true or false.

a) $-5 < 3$ **b)** $-9 > -6$

c) $-3 \le 2$ **d)** $4 \ge 4$

Solution

a) The statement $-5 < 3$ is true because -5 is to the left of 3 on the number line. In fact, any negative number is less than any positive number.

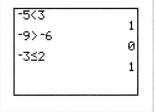
b) The statement $-9 > -6$ is false because -9 lies to the left of -6.

c) The statement $-3 \leq 2$ is true because -3 is less than 2.

d) The statement $4 \geq 4$ is true because $4 = 4$ is true.   ■

Interval Notation and Graphs

If an inequality involves a variable, then which real numbers can be used in place of the variable to obtain a correct statement? The set of all such numbers is the **solution set** to the inequality. For example, $x < 3$ is correct if x is replaced by any number that lies to the left of 3 on the number line:

$$1.5 < 3, \qquad 0 < 3, \qquad \text{and} \qquad -2 < 3$$

The set of real numbers to the left of 3 is written in set notation as $\{x \mid x < 3\}$, in **interval notation** as $(-\infty, 3)$, and graphed in Fig. 3.2:

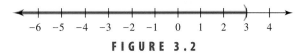

FIGURE 3.2

Note that $-\infty$ (negative infinity) is not a number, but it indicates that there is no end to the real numbers less than 3. The parenthesis used next to the 3 in the interval notation and on the graph means that 3 is not included in the solution set to $x < 3$.

An inequality such as $x \geq 1$ is satisfied by 1 and any real number that lies to the right of 1 on the number line. The solution set to $x \geq 1$ is written in set notation as $\{x \mid x \geq 1\}$, in interval notation as $[1, \infty)$, and graphed in Fig. 3.3:

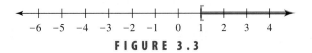

FIGURE 3.3

The bracket used next to the 1 in the interval notation and on the graph means that 1 is in the solution set to $x \geq 1$.

The solution set to an inequality can be stated symbolically with set notation and interval notation, or visually with a graph. Interval notation is popular because it is simpler to write than set notation. The interval notation and graph for each of the four basic inequalities is summarized as follows.

Basic Interval Notation (k any real number)

Inequality	Solution Set with Interval Notation	Graph
$x > k$	(k, ∞)	
$x \geq k$	$[k, \infty)$	
$x < k$	$(-\infty, k)$	
$x \leq k$	$(-\infty, k]$	

Note that a bracket is used next to k if k is in the interval and a parenthesis when k is not in the interval. A parenthesis is always used on the infinite end of the interval because ∞ is not a number that might or might not be in the interval.

E X A M P L E   2   **Interval notation and graphs**

Write the solution set to each inequality in interval notation and graph it.

a) $x > -5$

b) $x \leq 2$

Solution

a) The solution set to the inequality $x > -5$ is $\{x \mid x > -5\}$. The solution set is the interval of all numbers to the right of -5 on the number line. This set is written in interval notation as $(-5, \infty)$, and it is graphed in Fig. 3.4.

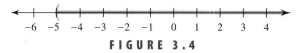

F I G U R E   3 . 4

b) The solution set to $x \leq 2$ is $\{x \mid x \leq 2\}$. This set includes 2 and all real numbers to the left of 2. Because 2 is included, we use a bracket at 2. The interval notation for this set is $(-\infty, 2]$. The graph is shown in Fig. 3.5.

F I G U R E   3 . 5   ■

Solving Linear Inequalities

In Section 2.1 we defined a linear equation as an equation of the form $ax + b = 0$. If we replace the equality symbol in a linear equation with an inequality symbol, we have a linear inequality.

> **Linear Inequality**
>
> A **linear inequality** in one variable x is any inequality of the form $ax + b < 0$, where a and b are real numbers, with $a \neq 0$. In place of $<$ we may also use $\leq$, $>$, or $\geq$.

Inequalities that can be rewritten as $ax + b > 0$ are also called linear inequalities.

   Equivalent inequalities have the same solution set. Any real number can be added to or subtracted from each side of an inequality to get an equivalent inequality. For example, if $x + 2 < 8$ then we can subtract 2 from each side to get $x < 6$. If $x - 3 > 4$, then we can add 3 to each side to get $x > 7$.

   We can also multiply or divide each side of an inequality by any nonzero real number to get an equivalent inequality. However, multiplying or dividing by a negative number reverses the inequality. To understand why, multiply each side of $5 < 6$ by -1. The correct result is $-5 > -6$.

These properties of inequality are summarized as follows.

Properties of Inequality

Addition Property of Inequality
If the same number is added to both sides of an inequality, then the solution set to the inequality is unchanged.

Multiplication Property of Inequality
If both sides of an inequality are multiplied by the same *positive number,* then the solution set to the inequality is unchanged.

If both sides of an inequality are multiplied by the same *negative number* and *the inequality symbol is reversed,* then the solution set to the inequality is unchanged.

Because subtraction is defined in terms of addition, the addition property of inequality also allows us to subtract the same number from both sides. Because division is defined in terms of multiplication, the multiplication property of inequality also allows us to divide both sides by the same nonzero real number *as long as we reverse the inequality symbol when dividing by a negative number.*

In the next example we use the properties of inequality to solve an inequality in the same manner that we solve equations.

E X A M P L E   3   **Solving inequalities**

Solve each inequality. State and graph the solution set.

a) $2x - 7 < -1$ **b)** $5 - 3x < 11$

Solution

a) We proceed exactly as we do when solving equations:

$$2x - 7 < -1 \quad \text{Original inequality}$$
$$2x < 6 \quad \text{Add 7 to each side.}$$
$$x < 3 \quad \text{Divide each side by 2.}$$

The solution set is written in set notation as $\{x \mid x < 3\}$ and in interval notation as $(-\infty, 3)$. The graph is shown in Fig. 3.6.

FIGURE 3.6   **FIGURE 3.7**

b) We divide by a negative number to solve this inequality.

$$5 - 3x < 11 \quad \text{Original equation}$$
$$-3x < 6 \quad \text{Subtract 5 from each side.}$$
$$x > -2 \quad \text{Divide each side by } -3 \text{ and reverse the inequality symbol.}$$

The solution set is written in set notation as $\{x \mid x > -2\}$ and in interval notation as $(-2, \infty)$. The graph is shown in Fig. 3.7. ■

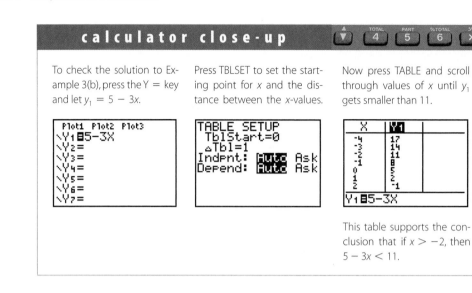

calculator close-up

To check the solution to Example 3(b), press the Y = key and let $y_1 = 5 - 3x$.

Press TBLSET to set the starting point for x and the distance between the x-values.

Now press TABLE and scroll through values of x until y_1 gets smaller than 11.

This table supports the conclusion that if $x > -2$, then $5 - 3x < 11$.

E X A M P L E   4

Solving inequalities

Solve $\frac{8 + 3x}{-5} \geq -4$. State and graph the solution set.

Solution

$$\frac{8 + 3x}{-5} \geq -4 \qquad \text{Original inequality}$$

$$-5\left(\frac{8 + 3x}{-5}\right) \leq -5(-4) \qquad \text{Multiply each side by } -5 \text{ and reverse the inequality symbol.}$$

$$8 + 3x \leq 20 \qquad \text{Simplify.}$$

$$3x \leq 12 \qquad \text{Subtract 8 from each side.}$$

$$x \leq 4 \qquad \text{Divide each side by 3.}$$

The solution set is $(-\infty, 4]$, and its graph is shown in Fig. 3.8.

F I G U R E   3 . 8

E X A M P L E   5

An inequality with fractions

Solve $\frac{1}{2}x - \frac{2}{3} \leq x + \frac{4}{3}$. State and graph the solution set.

helpful hint

For inequalities we usually try to isolate the variable on the left side. However, if you get $-4 \leq x$ you can still write the equivalent inequality $x \geq -4$.

Solution

First multiply each side of the inequality by 6, the LCD:

$$\frac{1}{2}x - \frac{2}{3} \leq x + \frac{4}{3} \qquad \text{Original inequality}$$

$$6\left(\frac{1}{2}x - \frac{2}{3}\right) \leq 6\left(x + \frac{4}{3}\right) \qquad \text{Multiplying by positive 6 does not reverse the inequality.}$$

$$3x - 4 \leq 6x + 8 \qquad \text{Distributive property}$$
$$3x \leq 6x + 12 \qquad \text{Add 4 to each side.}$$
$$-3x \leq 12 \qquad \text{Subtract } 6x \text{ from each side.}$$
$$x \geq -4 \qquad \text{Divide each side by } -3 \text{ and reverse the inequality.}$$

The solution set is the interval $[-4, \infty)$. Its graph is shown in Fig. 3.9.

FIGURE 3.9

Applications

Inequalities have applications just as equations do. To use inequalities, we must be able to translate a verbal problem into an algebraic inequality. Inequality can be expressed verbally in a variety of ways.

E X A M P L E   6

Writing inequalities

Identify the variable and write an inequality that describes the situation.

a) Chris paid more than $200 for a suit.

b) A candidate for president must be at least 35 years old.

c) The capacity of an elevator is at most 1500 pounds.

d) The company must hire no fewer than 10 programmers.

Solution

a) If c is the cost of the suit in dollars, then $c > 200$.

b) If a is the age of the candidate in years, then $a \geq 35$.

c) If x is the capacity of the elevator in pounds, then $x \leq 1500$.

d) If n represents the number of programmers and n is not less than 10, then $n \geq 10$.

In Example 6(d) we know that n is not less than 10. So there are exactly two other possibilities: n is greater than 10 or equal to 10. If $x \neq 4$, then either $x > 4$ or $x < 4$. If w is not greater than 5, then $w \leq 5$. The fact that there are only three possibilities when comparing two numbers is called the **trichotomy property.**

Trichotomy Property

For any two real numbers a and b, exactly one of the following is true:

$$a < b, \qquad a = b, \qquad \text{or} \qquad a > b$$

We follow the same steps to solve problems involving inequalities as we do to solve problems involving equations.

E X A M P L E   7

Price range

Lois plans to spend less than $500 on an electric dryer, including the 9% sales tax and a $64 setup charge. In what range is the selling price of the dryer that she can afford?

Solution

If we let x represent the selling price in dollars for the dryer, then the amount of sales tax is $0.09x$. Because her total cost must be less than $500, we can write the following inequality:

$$x + 0.09x + 64 < 500$$
$$1.09x < 436 \qquad \text{Subtract 64 from each side.}$$
$$x < \frac{436}{1.09} \qquad \text{Divide each side by 1.09.}$$
$$x < 400$$

The selling price of the dryer must be less than $400. ∎

Note that if we had written the equation $x + 0.09x + 64 = 500$ for the last example, we would have gotten $x = 400$. We could then have concluded that the selling price must be less than $400. This would certainly solve the problem, but it would not illustrate the use of inequalities. The original problem describes an inequality, and we should solve it as an inequality.

E X A M P L E   8

Paying off the mortgage

Tessie owns a piece of land on which she owes $12,760 to a bank. She wants to sell the land for enough money to at least pay off the mortgage. The real estate agent gets 6% of the selling price, and her city has a $400 real estate transfer tax paid by the seller. What should the range of the selling price be for Tessie to get at least enough money to pay off her mortgage?

Solution

If x is the selling price in dollars, then the commission is $0.06x$. We can write an inequality expressing the fact that the selling price minus the real estate commission minus the $400 tax must be at least $12,760:

$$x - 0.06x - 400 \geq 12,760$$
$$0.94x - 400 \geq 12,760 \qquad 1 - 0.06 = 0.94$$
$$0.94x \geq 13,160 \qquad \text{Add 400 to each side.}$$
$$x \geq \frac{13,160}{0.94} \qquad \text{Divide each side by 0.94.}$$
$$x \geq 14,000$$

The selling price must be at least $14,000 for Tessie to pay off the mortgage. ∎

WARM-UPS

True or false? Explain your answer.

1. $0 < 0$ 2. $-300 > -2$ 3. $-60 \leq -60$

4. The inequality $6 < x$ is equivalent to $x < 6$.

5. The inequality $-2x < 10$ is equivalent to $x < -5$.

6. The solution set to $3x \geq -12$ is $(-\infty, -4]$.

7. The solution set to $-x > 4$ is $(-\infty, -4)$.

8. If x is no larger than 8, then $x \leq 8$.

9. If m is any real number, then exactly one of the following is true: $m < 0$, $m = 0$, or $m > 0$.

10. The number -2 is a member of the solution set to the inequality $3 - 4x \leq 11$.

3.1   EXERCISES

Reading and Writing   *After reading this section, write out the answers to these questions. Use complete sentences.*

1. What is an inequality?

2. What symbols are used to express inequality?

3. What does it mean when we say that a is less than b?

4. What is a linear inequality?

5. How does solving linear inequalities differ from solving linear equations?

6. What verbal phrases are used to indicate an inequality?

Determine whether each inequality is true or false. See Example 1.

7. $-3 < -9$

8. $-8 > -7$

9. $0 \leq 8$

10. $-6 \geq -8$

11. $(-3)20 > (-3)40$

12. $(-1)(-3) < (-1)(5)$

13. $9 - (-3) \leq 12$

14. $(-4)(-5) + 2 \geq 21$

Determine whether each inequality is satisfied by the given number.

15. $2x - 4 < 8, -3$

16. $5 - 3x > -1, 6$

17. $2x - 3 \leq 3x - 9, 5$

18. $6 - 3x \geq 10 - 2x, -4$

19. $5 - x < 4 - 2x, -1$

20. $3x - 7 \geq 3x - 10, 9$

Write the solution set in interval notation and graph it. See Example 2.

21. $x \leq -1$

22. $x \geq -7$

23. $x > 20$

24. $x < 30$

25. $3 \leq x$

26. $-2 > x$

27. $x < 2.3$

28. $x \leq 4.5$

Rewrite each set in interval notation.

29. $\{x \mid x > 1\}$

30. $\{x \mid x < 3\}$

31. $\{x \mid x \leq -3\}$

32. $\{x \mid x \geq -2\}$

33. $\{x \mid x < 5\}$

34. $\{x \mid x > -7\}$

35. $\{x \mid x \geq -4\}$

36. $\{x \mid x \leq -9\}$

Fill in the blank with an inequality symbol so that the two statements are equivalent.

37. $x + 5 > 12$
    x ___ 7

38. $2x - 3 \leq -4$
    $2x$ ___ -1

39. $-x < 6$
    x ___ -6

40. $-5 \geq -x$
    5 ___ x

41. $-2x \geq 8$
    x ___ -4

42. $-5x > -10$
    x ___ 2

43. $4 < x$
    x ___ 4

44. $-9 \leq -x$
    x ___ 9

Solve each of the following inequalities. Express the solution set in interval notation and graph it. See Examples 3–5.

45. $7x > -14$

46. $4x \le -8$

47. $-3x \le 12$

48. $-2x > -6$

49. $2x - 3 > 7$

50. $3x - 2 < 6$

51. $3 - 5x \le 18$

52. $5 - 4x \ge 19$

53. $\dfrac{x - 3}{-5} < -2$

54. $\dfrac{2x - 3}{4} > 6$

55. $\dfrac{5 - 3x}{4} \le 2$

56. $\dfrac{7 - 5x}{-2} \ge -1$

57. $3 - \dfrac{1}{4}x \ge 2$

58. $5 - \dfrac{1}{3}x > 2$

59. $\dfrac{1}{4}x - \dfrac{1}{2} < \dfrac{1}{2}x - \dfrac{2}{3}$

60. $\dfrac{1}{3}x - \dfrac{1}{6} < \dfrac{1}{6}x - \dfrac{1}{2}$

61. $\dfrac{y - 3}{2} > \dfrac{1}{2} - \dfrac{y - 5}{4}$

62. $\dfrac{y - 1}{3} - \dfrac{y + 1}{5} > 1$

Solve each inequality and graph the solution set.

63. $2x + 3 > 2(x - 4)$

64. $-2(5x - 1) \le -5(5 + 2x)$

65. $-4(2x - 5) \le 2(6 - 4x)$

66. $-3(2x - 1) \le 2(5 - 3x)$

67. $-\dfrac{1}{2}(x - 6) < \dfrac{1}{2}x + 2$

68. $-3\left(\dfrac{1}{2}x - \dfrac{1}{4}\right) > \dfrac{x}{2} - \dfrac{1}{4}$

69. $4.273 + 2.8x \le 10.985$

70. $1.064 < 5.94 - 3.2x$

71. $3.25x - 27.39 > 4.06 + 5.1x$

72. $4.86(3.2x - 1.7) > 5.19 - x$

Identify the variable and write an inequality that describes each situation. See Example 6.

73. Tony is taller than 6 feet.

74. Glenda is under 60 years old.

75. Wilma makes less than $80,000 per year.

76. Bubba weighs over 80 pounds.

77. The maximum speed for the Concorde is 1450 miles per hour (mph).

78. The minimum speed on the freeway is 45 mph.

79. Julie can afford at most $400 per month.

80. Fred must have at least a 3.2 grade point average.

81. Burt is no taller than 5 feet.

82. Ernie cannot run faster than 10 mph.

83. Tina makes no more than $8.20 per hour.

84. Rita will not take less than $12,000 for the car.

Solve each problem by using an inequality. See Examples 7 and 8.

85. *Car shopping.* Jennifer is shopping for a new car. In addition to the price of the car, there is an 8% sales tax and a $172 title and license fee. If Jennifer decides that she will spend less than $10,000 total, then what is the price range for the car?

86. *Sewing machines.* Charles wants to buy a sewing machine in a city with a 10% sales tax. He has at most $700 to spend. In what price range should he look?

87. *Truck shopping.* Linda and Bob are shopping for a new truck in a city with a 9% sales tax. There is also an $80 title and license fee to pay. They want to get a good truck and plan to spend at least $10,000. What is the price range for the truck?

88. *Curly's contribution.* Larry, Curly, and Moe are going to buy their mother a color television set. Larry has a better job than Curly and agrees to contribute twice as much as Curly. Moe is unemployed and can spare only $50. If the kind of television Mama wants costs at least $600, then what is the price range for Curly's contribution?

89. *Renting a Mustang.* Hillary can rent a Ford Mustang from Alpha Car Rental for $45 per day with no charge for miles. From Beta Car Rental she can get the same car for $35 per day plus 25 cents per mile. For what number of daily miles is Beta cheaper?

90. *Renting a Cadillac.* George can rent a Cadillac from Gamma Car Rental for $50 per day plus 35 cents per mile. He can get the same car from Delta Car Rental for $35 per day plus 45 cents per mile. For what number of daily miles is Delta cheaper?

91. *Bachelor's degrees.* The graph shows the number of bachelor's degrees awarded in the United States each year since 1985 (National Center for Education Statistics, www.nces.ed.gov).
   a) Has the number of bachelor's degrees been increasing or decreasing since 1985?

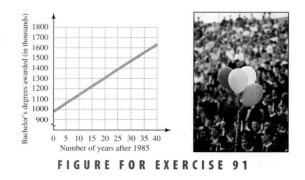

Bachelor's degrees awarded (in thousands)

Number of years after 1985

FIGURE FOR EXERCISE 91

b) The formula $B = 16.45n + 980.20$ can be used to approximate the number of degrees awarded in thousands in the year 1985 + n. What is the first year in which the number of bachelor's degrees will exceed 1.3 million?

92. *Master's degrees.* In 1985, 15.9% of all degrees awarded in U.S. higher education were master's degrees (National Center for Education Statistics). If the formulas $M = 7.79n + 287.87$ and $T = 30.95n + 1808.22$ give the number of master's degrees and the total number of higher education degrees awarded in thousands, respectively, in the year 1985 + n, then what is the first year in which more than 20% of all degrees awarded will be master's degrees?

93. *Weighted average.* Professor Jorgenson gives only a midterm exam and a final exam. The semester average is computed by taking $\frac{1}{3}$ of the midterm exam score plus $\frac{2}{3}$ of the final exam score. The grade is determined from the semester average by using the grading scale given in the table. If Stanley scored only 56 on the midterm, then for what range of scores on the final exam would he get a C or better in the course?

Grading	Scale
90–100	A
80–89	B
70–79	C
60–69	D

TABLE FOR EXERCISES 93 AND 94

94. *C or better.* Professor Brown counts her midterm as $\frac{2}{3}$ of the grade and her final as $\frac{1}{3}$ of the grade. Wilbert scored only 56 on the midterm. If Professor Brown also uses the grading scale given in the table, then what range of scores on the final exam would give Wilbert a C or better in the course?

95. *Designer jeans.* A pair of ordinary jeans at A-Mart costs $50 less than a pair of designer jeans at Enrico's. In fact, you can buy four pairs of A-Mart jeans for less than one pair of Enrico's jeans. What is the price range for a pair of A-Mart jeans?

96. *United Express.* Al and Rita both drive parcel delivery trucks for United Express. Al averages 20 mph less than Rita. In fact, Al is so slow that in 5 hours he covered fewer miles than Rita did in 3 hours. What are the possible values for Al's rate of speed?

GETTING MORE INVOLVED

97. *Discussion.* If 3 is added to every number in $(4, \infty)$, the resulting set is $(7, \infty)$. In each of the following cases, write the resulting set of numbers in interval notation. Explain your results.

a) The number -6 is subtracted from every number in $[2, \infty)$.

b) Every number in $(-\infty, -3)$ is multiplied by 2.

c) Every number in $(8, \infty)$ is divided by 4.

d) Every number in $(6, \infty)$ is multiplied by -2.

e) Every number in $(-\infty, -10)$ is divided by -5.

98. *Writing.* Explain why saying that x is *at least* 9 is equivalent to saying that x is *greater than or equal to* 9. Explain why saying that x is *at most* 5 is equivalent to saying that x is *less than or equal to* 5.

3.2   COMPOUND INEQUALITIES

In this section we will use our knowledge of inequalities from Section 3.1 to work with compound inequalities.

In this section

- Basics
- Graphing the Solution Set
- Applications

Basics

The inequalities that we studied in Section 3.1 are referred to as **simple inequalities.** If we join two simple inequalities with the connective "and" or the connective "or," we get a **compound inequality.** A compound inequality using the connective "and" is true if and only if *both* simple inequalities are true.

E X A M P L E   1   **Compound inequalities using the connective "and"**

Determine whether each compound inequality is true.

a) $3 > 2$ and $3 < 5$ **b)** $6 > 2$ and $6 < 5$

Solution

a) The compound inequality is true because $3 > 2$ is true and $3 < 5$ is true.

b) The compound inequality is false because $6 < 5$ is false.   ■

A compound inequality using the connective "or" is true if one or the other or both of the simple inequalities are true. It is false only if both simple inequalities are false.

E X A M P L E   2   **Compound inequalities using the connective "or"**

Determine whether each compound inequality is true.

a) $2 < 3$ or $2 > 7$ **b)** $4 < 3$ or $4 \geq 7$

Solution

a) The compound inequality is true because $2 < 3$ is true.

b) The compound inequality is false because both $4 < 3$ and $4 \geq 7$ are false.   ■

If a compound inequality involves a variable, then we are interested in the solution set to the inequality. The solution set to an "and" inequality consists of all numbers that satisfy both simple inequalities, whereas the solution set to an "or" inequality consists of all numbers that satisfy at least one of the simple inequalities.

E X A M P L E   3

Solutions of compound inequalities

Determine whether 5 satisfies each compound inequality.

a) $x < 6$ and $x < 9$

b) $2x - 9 \leq 5$ or $-4x \geq -12$

There is a big difference between "and" and "or." To get money from an automatic teller you must have a bank card *and* know a secret number (PIN). There would be a lot of problems if you could get money by having a bank card *or* knowing a PIN.

Solution

a) Because $5 < 6$ and $5 < 9$ are both true, 5 satisfies the compound inequality.

b) Because $2 \cdot 5 - 9 \leq 5$ is true, it does not matter that $-4 \cdot 5 \geq -12$ is false. So 5 satisfies the compound inequality. ∎

Graphing the Solution Set

If A and B are sets of numbers, then the **intersection** of A and B is the set of all numbers that are in both A and B. The intersection of A and B is denoted as $A \cap B$ (read "A intersect B"). For example, if $A = \{1, 2, 3\}$ and $B = \{2, 3, 4, 5\}$, then $A \cap B = \{2, 3\}$ because only 2 and 3 are in both A and B.

The solution set to a compound inequality using the connective "and" is the intersection of the solution sets to each of the simple inequalities. Using graphs, as shown in the next example, will help you understand compound inequalities.

E X A M P L E   4

Graphing compound inequalities

Graph the solution set to the compound inequality $x > 2$ and $x < 5$.

Solution

We first sketch the graph of $x > 2$ and then the graph of $x < 5$, as shown in the top two number lines in Fig. 3.10. The intersection of these two solution sets is the portion of the number line that is shaded on both graphs, just the part between 2 and 5, not including the endpoints. The graph of $\{x \mid x > 2$ and $x < 5\}$ is shown at the bottom of Fig. 3.10. We write this set in interval notation as $(2, 5)$.

study tip

Never leave an exam early. Most papers turned in early contain careless errors that could be found and corrected. Every point counts. Reread the questions to be sure that you don't have a nice solution to a question that wasn't asked. Check all arithmetic. You can check many problems by using a different method. Some problems can be checked with a calculator. Make sure that you did not forget to answer a question.

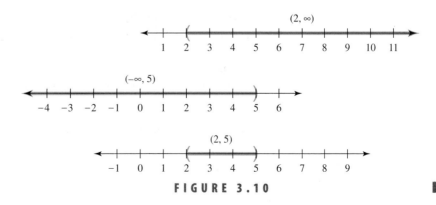

F I G U R E   3 . 1 0

∎

If A and B are sets of numbers, then the **union** of A and B is the set of all numbers that are in either A or B. The union of A and B is denoted as $A \cup B$ (read

"*A* union *B*"). For example, if $A = \{1, 2, 3\}$ and $B = \{2, 3, 4, 5\}$, then $A \cup B = \{1, 2, 3, 4, 5\}$ because all of these numbers are in *A* or *B*. Notice that the numbers in *A* and *B* are in $A \cup B$.

The solution set to a compound inequality using the connective "or" is the union of the solution sets to each of the simple inequalities.

EXAMPLE 5 **Graphing compound inequalities**

Graph the solution set to the compound inequality $x > 4$ or $x < -1$.

Solution

To find the union of the solution sets to the simple inequalities, we sketch their graphs as shown at the top of Fig. 3.11. We graph the union of these two sets by putting both shaded regions together on the same line as shown in the bottom graph in Fig. 3.11. This set is written in interval notation as $(-\infty, -1) \cup (4, \infty)$.

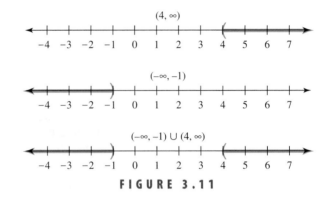

FIGURE 3.11

study tip

Make sure you know exactly how your grade in this course is determined. You should know how much weight is given to tests, quizzes, projects, homework, and the final exam. You should know the policy for making up work in case of illness. Record all scores and compute your final grade.

CAUTION When graphing the intersection of two simple inequalities, do not draw too much. For the intersection, graph only numbers that satisfy *both* inequalities. Omit numbers that satisfy one but not the other inequality. Graphing a union is usually easier because we can simply draw both solution sets on the same number line.

It is not always necessary to graph the solution set to each simple inequality before graphing the solution set to the compound inequality. We can save time and work if we learn to think of the two preliminary graphs but draw only the final one.

EXAMPLE 6 **Overlapping intervals**

Sketch the graph and write the solution set in interval notation to each compound inequality.

a) $x < 3$ and $x < 5$ **b)** $x > 4$ or $x > 0$

Solution

a) To graph $x < 3$ and $x < 5$, we shade only the numbers that are both less than 3 and less than 5. So numbers between 3 and 5 are not shaded in Fig. 3.12. The compound inequality $x < 3$ and $x < 5$ is equivalent to the simple inequality $x < 3$. The solution set can be written as $(-\infty, 3)$.

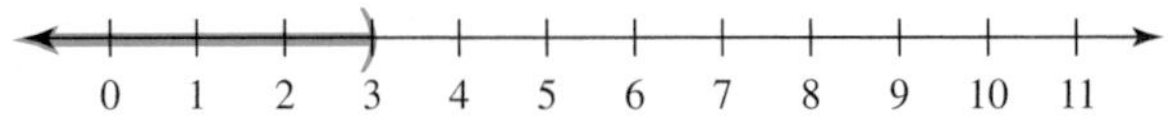

FIGURE 3.12

b) To graph $x > 4$ or $x > 0$, we shade both regions on the same number line as shown in Fig. 3.13. The compound inequality $x > 4$ or $x > 0$ is equivalent to the simple inequality $x > 0$. The solution set is $(0, \infty)$.

FIGURE 3.13

The next example shows a compound inequality that has no solution and one that is satisfied by every real number.

E X A M P L E   7   **All or nothing**

Sketch the graph and write the solution set in interval notation to each compound inequality.

a) $x < 2$ and $x > 6$ **b)** $x < 3$ or $x > 1$

Solution

a) A number satisfies $x < 2$ and $x > 6$ if it is both less than 2 *and* greater than 6. There are no such numbers. The solution set is the empty set, $\varnothing$.

b) To graph $x < 3$ or $x > 1$, we shade both regions on the same number line as shown in Fig. 3.14. Since the two regions cover the entire line, the solution set is the set of all real numbers $(-\infty, \infty)$.

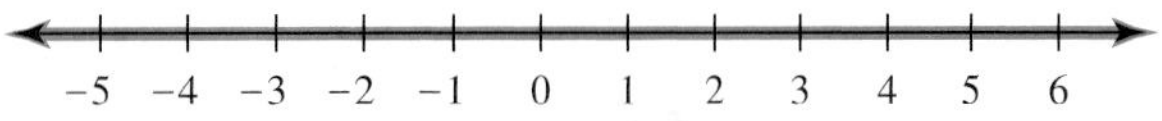

FIGURE 3.14

If we start with a more complicated compound inequality, we first simplify each part of the compound inequality and then find the union or intersection.

E X A M P L E   8   **Intersection**

Solve $x + 2 > 3$ and $x - 6 < 7$. Graph the solution set.

Solution

First simplify each simple inequality:

$$x + 2 - 2 > 3 - 2 \quad \text{and} \quad x - 6 + 6 < 7 + 6$$

$$x > 1 \quad \text{and} \quad x < 13$$

The intersection of these two solution sets is the set of numbers between (but not including) 1 and 13. Its graph is shown in Fig. 3.15. The solution set is written in interval notation as $(1, 13)$.

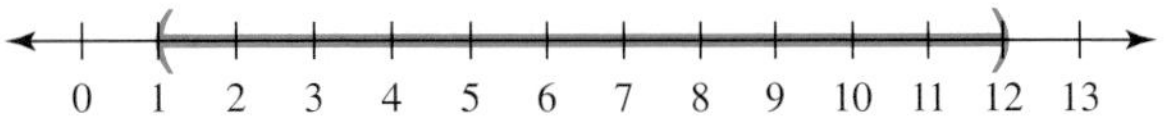

FIGURE 3.15

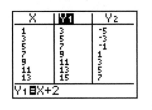

To check Example 8, press Y= and let $y_1 = x + 2$ and $y_2 = x - 6$. Now scroll through a table of values for y_1 and y_2. From the table you can see that y_1 is greater than 3 and y_2 is less than 7 precisely when x is between 1 and 13.

E X A M P L E   9

Union

Graph the solution set to the inequality

$$5 - 7x \geq 12 \quad \text{or} \quad 3x - 2 < 7.$$

Solution

First solve each of the simple inequalities:

$$5 - 7x - 5 \geq 12 - 5 \quad \text{or} \quad 3x - 2 + 2 < 7 + 2$$

$$-7x \geq 7 \quad \text{or} \quad 3x < 9$$

$$x \leq -1 \quad \text{or} \quad x < 3$$

The union of the two solution intervals is $(-\infty, 3)$. The graph is shown in Fig. 3.16.

FIGURE 3.16

To check Example 9, press Y= and let $y_1 = 5 - 7x$ and $y_2 = 3x - 2$. Now scroll through a table of values for y_1 and y_2. From the table you can see that either $y_1 \geq 12$ or $y_2 < 7$ is true for $x < 3$. Note also that for $x \geq 3$ both $y_1 \geq 12$ and $y_2 < 7$ are incorrect. The table supports the conclusion of Example 9.

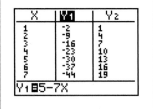

An inequality may be read from left to right or from right to left. Consider the inequality $1 < x$. If we read it in the usual way, we say, "1 is less than x." The meaning is clearer if we read the variable first. Reading from right to left, we say, "x is greater than 1."

Another notation is commonly used for the compound inequality

$$x > 1 \quad \text{and} \quad x < 13.$$

This compound inequality can also be written as

$$1 < x < 13.$$

Reading from left to right, we read $1 < x < 13$ as "1 is less than x is less than 13." The meaning of this inequality is clearer if we read the variable first and read the first inequality symbol from right to left. Reading the variable first, $1 < x < 13$ is read as "x is greater than 1 and less than 13." So x is between 1 and 13, and reading x first makes it clear.

C A U T I O N   We write $a < x < b$ only if $a < b$, and we write $a > x > b$ only if $a > b$. Similar rules hold for $\leq$ and $\geq$. So $4 < x < 9$ and $-6 \geq x \geq -8$ are correct uses of this notation, but $5 < x < 2$ is not correct. Also, the inequalities should *not* point in opposite directions as in $5 < x > 7$.

E X A M P L E   10   **Another notation**

Solve the inequality and graph the solution set:

$$-2 \leq 2x - 3 < 7$$

Solution

This inequality could be written as the compound inequality

$$2x - 3 \geq -2 \qquad \text{and} \qquad 2x - 3 < 7.$$

However, there is no need to rewrite the inequality because we can solve it in its original form.

$$-2 + 3 \leq 2x - 3 + 3 < 7 + 3 \quad \text{Add 3 to each part.}$$

$$1 \leq 2x \leq 10$$

$$\frac{1}{2} \leq \frac{2x}{2} < \frac{10}{2} \qquad\qquad\qquad \text{Divide each part by 2.}$$

$$\frac{1}{2} \leq x < 5$$

The solution set is $\left[\frac{1}{2}, 5\right)$, and its graph is shown in Fig. 3.17.

FIGURE 3.17

E X A M P L E   11   **Solving a compound inequality**

Solve the inequality $-1 < 3 - 2x < 9$ and graph the solution set.

Solution

$$-1 - 3 < 3 - 2x - 3 < 9 - 3 \quad \text{Subtract 3 from each part of the inequality.}$$

$$-4 < -2x < 6$$

$$2 > x > -3 \qquad\qquad\qquad \text{Divide each part by } -2 \text{ and reverse both inequality symbols.}$$

$$-3 < x < 2 \qquad\qquad\qquad \text{Rewrite the inequality with the smallest number on the left.}$$

The solution set is $(-3, 2)$, and its graph is shown in Fig. 3.18.

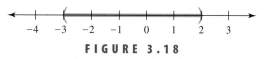

FIGURE 3.18

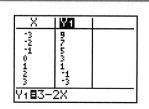

calculator close-up

Let $y_1 = 3 - 2x$ and make a table. Scroll through the table to see that y_1 is between -1 and 9 when x is between -3 and 2. The table supports the conclusion of Example 11.

Applications

When final exams are approaching, students are often interested in finding the final exam score that would give them a certain grade for a course.

E X A M P L E   12

Final exam scores

Fiana made a score of 76 on her midterm exam. For her to get a B in the course, the average of her midterm exam and final exam must be between 80 and 89 inclusive. What possible scores on the final exam would give Fiana a B in the course?

Solution

> **helpful hint**
>
> When you use two inequality symbols as in Example 12, they must both point in the same direction. In fact, we usually have them both point to the left so that the numbers increase in size from left to right.

Let x represent her final exam score. Between 80 and 89 inclusive means that an average between 80 and 89 as well as an average of exactly 80 or 89 will get a B. So the average of the two scores must be greater than or equal to 80 and less than or equal to 89.

$$80 \leq \frac{x + 76}{2} \leq 89$$

$$160 \leq x + 76 \leq 178 \quad \text{Multiply by 2.}$$

$$160 - 76 \leq x \leq 178 - 76 \quad \text{Subtract 76.}$$

$$84 \leq x \leq 102$$

If Fiana scores between 84 and 102 inclusive, she will get a B in the course. ■

WARM-UPS

True or false? Explain your answer.

1. $3 < 5$ and $3 \leq 10$

2. $3 < 5$ or $3 < 10$

3. $3 > 5$ and $3 < 10$

4. $3 \geq 5$ or $3 \leq 10$

5. $4 < 8$ and $4 > 2$

6. $4 < 8$ or $4 > 2$

7. $-3 < 0 < -2$

8. $(3, \infty) \cap (8, \infty) = (8, \infty)$

9. $(3, \infty) \cup [8, \infty) = [8, \infty)$

10. $(-2, \infty) \cap (-\infty, 9) = (-2, 9)$

3.2   EXERCISES

Reading and Writing   *After reading this section, write out the answers to these questions. Use complete sentences.*

1. What is a compound inequality?

2. When is a compound inequality using "and" true?

3. When is a compound inequality using "or" true?

4. How do we solve compound inequalities?

5. What is the meaning of $a < b < c$?

6. What is the meaning of $5 < x > 7$?

Determine whether each compound inequality is true. See Examples 1 and 2.

7. $-6 < 5$ and $-6 > -3$

8. $3 < 5$ or $0 < -3$

9. $4 \leq 4$ and $-4 \leq 0$

10. $1 < 5$ and $1 > -3$

11. $6 < 5$ or $-4 > -3$

12. $4 \leq -4$ or $0 \leq 0$

Determine whether -4 satisfies each compound inequality. See Example 3.

13. $x < 5$ and $x > -3$

14. $x < 5$ or $x > -3$

15. $x - 3 \geq -7$ or $x + 1 > 1$

16. $2x \leq -8$ and $5x \leq 0$

17. $2x - 1 < -7$ or $-2x > 18$

18. $-3x > 0$ and $3x - 4 < 11$

Graph the solution set to each compound inequality. See Examples 4–7.

19. $x > -1$ and $x < 4$

20. $x \leq 3$ and $x \leq 0$

21. $x \geq 2$ or $x \geq 5$

22. $x < -1$ or $x < 3$

23. $x \leq 6$ or $x > -2$

24. $x > -2$ and $x \leq 4$

25. $x \leq 6$ and $x > 9$

26. $x < 7$ or $x > 0$

27. $x \leq 6$ or $x > 9$

28. $x \geq 4$ and $x \leq -4$

29. $x \geq 6$ and $x \leq 1$

30. $x > 3$ or $x < -3$

Solve each compound inequality. Write the solution set using interval notation and graph it. See Examples 8 and 9.

31. $x - 3 > 7$ or $3 - x > 2$

32. $x - 5 > 6$ or $2 - x > 4$

33. $3 < x$ and $1 + x > 10$

34. $-0.3x < 9$ and $0.2x > 2$

35. $\frac{1}{2}x > 5$ or $-\frac{1}{3}x < 2$

36. $5 < x$ or $3 - \frac{1}{2}x < 7$

37. $2x - 3 \leq 5$ and $x - 1 > 0$

38. $\frac{3}{4}x < 9$ and $-\frac{1}{3}x \leq -15$

39. $\frac{1}{2}x - \frac{1}{3} \geq -\frac{1}{6}$ or $\frac{2}{7}x \leq \frac{1}{10}$

40. $\frac{1}{4}x - \frac{1}{3} > -\frac{1}{5}$ and $\frac{1}{2}x < 2$

41. $0.5x < 2$ and $-0.6x < -3$
42. $0.3x < 0.6$ or $0.05x > -4$

Solve each compound inequality. Write the solution set in interval notation and graph it. See Examples 10 and 11.

43. $5 < 2x - 3 < 11$

44. $-2 < 3x + 1 < 10$

45. $-1 < 5 - 3x \leq 14$

46. $-1 \leq 3 - 2x < 11$

47. $-3 < \dfrac{3m + 1}{2} \le 5$

48. $0 \le \dfrac{3 - 2x}{2} < 5$

49. $-2 < \dfrac{1 - 3x}{-2} < 7$

50. $-3 < \dfrac{2x - 1}{3} < 7$

51. $3 \le 3 - 5(x - 3) \le 8$

52. $2 \le 4 - \dfrac{1}{2}(x - 8) \le 10$

Write each union or intersection of intervals as a single interval if possible.

53. $(2, \infty) \cup (4, \infty)$ **54.** $(-3, \infty) \cup (-6, \infty)$

55. $(-\infty, 5) \cap (-\infty, 9)$ **56.** $(-\infty, -2) \cap (-\infty, 1)$

57. $(-\infty, 4] \cap [2, \infty)$ **58.** $(-\infty, 8) \cap [3, \infty)$

59. $(-\infty, 5) \cup [-3, \infty)$ **60.** $(-\infty, -2] \cup (2, \infty)$

61. $(3, \infty) \cap (-\infty, 3]$ **62.** $[-4, \infty) \cap (-\infty, -6]$

63. $(3, 5) \cap [4, 8)$ **64.** $[-2, 4] \cap (0, 9]$

65. $[1, 4) \cup (2, 6]$ **66.** $[1, 3) \cup (0, 5)$

Write either a simple or a compound inequality that has the given graph as its solution set.

67.

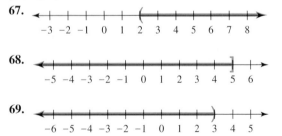

68.

69.

70.

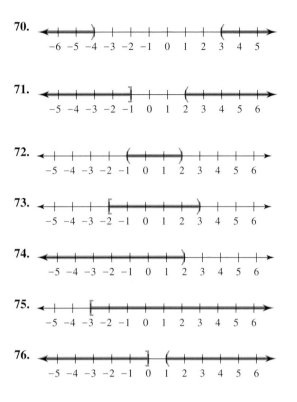

71.

72.

73.

74.

75.

76.

Solve each compound inequality and write the solution set using interval notation.

77. $2 < x < 7$ and $2x > 10$

78. $3 < 5 - x < 8$ or $-3x < 0$

79. $-1 < 3x + 2 \le 5$ or $\dfrac{3}{2}x - 6 > 9$

80. $0 < 5 - 2x \le 10$ and $-6 < 4 - x < 0$

81. $-3 < \dfrac{x - 1}{2} < 5$ and $-1 < \dfrac{1 - x}{2} < 2$

82. $-3 < \dfrac{3x - 1}{5} < \dfrac{1}{2}$ and $\dfrac{1}{3} < \dfrac{3 - 2x}{6} < \dfrac{9}{2}$

Solve each problem by using a compound inequality. See Example 12.

83. ***Aiming for a C.*** Professor Johnson gives only a midterm exam and a final exam. The semester average is computed by taking $\frac{1}{3}$ of the midterm exam score plus $\frac{2}{3}$ of the final exam score. To get a C, Beth must have a semester average between 70 and 79 inclusive. If Beth scored only 64 on the midterm, then for what range of scores on the final exam would Beth get a C?

84. ***Two tests only.*** Professor Davis counts his midterm as $\frac{2}{3}$ of the grade, and his final as $\frac{1}{3}$ of the grade. Jason scored only 64 on the midterm. What range of scores on the final exam would put Jason's average between 70 and 79 inclusive?

85. ***Keep on truckin'.*** Abdul is shopping for a new truck in a city with an 8% sales tax. There is also an $84 title and license fee to pay. He wants to get a good truck and plans to spend at least $12,000 but no more than $15,000. What is the price range for the truck?

86. ***Selling-price range.*** Renee wants to sell her car through a broker who charges a commission of 10% of the selling price. The book value of the car is $14,900, but Renee still owes $13,104 on it. Although the car is in only fair condition and will not sell for more than the book value, Renee must get enough to at least pay off the loan. What is the range of the selling price?

87. ***Hazardous to her health.*** Trying to break her smoking habit, Jane calculates that she smokes only three full cigarettes a day, one after each meal. The rest of the time she smokes on the run and smokes only half of the cigarette. She estimates that she smokes the equivalent of 5 to 12 cigarettes per day. How many times a day does she light up on the run?

88. ***Possible width.*** The length of a rectangle is 20 meters longer than the width. The perimeter must be between 80 and 100 meters. What are the possible values for the width of the rectangle?

89. ***Higher education.*** The annual numbers of bachelor's and master's degrees awarded can be approximated using the formulas

$$B = 16.45n + 980.20$$
and $$M = 7.79n + 287.87,$$

where n is the number of years after 1985 (National Center for Education Statistics, www.nces.ed.gov). For example, $n = 2$ gives the numbers in 1987.
a) How many bachelor's degrees were awarded in 1995?
b) In what year will the number of bachelor's degrees that are awarded reach 1.26 million?
c) What is the first year in which both B is greater than 1.3 million and M is greater than 0.5 million?
d) What is the first year in which either B is greater than 1.3 million or M is greater than 0.5 million?

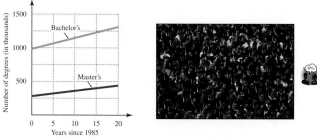

FIGURE FOR EXERCISE 89

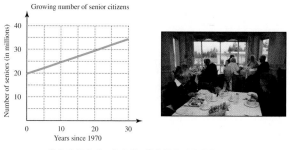

FIGURE FOR EXERCISE 90

90. ***Senior citizens.*** The number of senior citizens (65 years old and over) in the United States in millions in the year $1970 + n$ can be estimated by using the formula

$$S = 0.48n + 19.71$$

(U.S. Bureau of the Census, www.census.gov). The percentage of senior citizens living below the poverty level in the year $1970 + n$ can be estimated by using the formula

$$p = -0.72n + 24.2.$$

The variable n is the number of years after 1970.
a) How many senior citizens were there in 1998?
b) In what year did the percentage of seniors living below the poverty level reach 2.6%?
c) What is the first year in which we can expect both the number of seniors to be greater than 36 million and fewer than 2.6% living below the poverty level?

GETTING MORE INVOLVED

91. ***Discussion.*** If $-x$ is between a and b, then what can you say about x?

92. ***Discussion.*** For which of the inequalities is the notation used correctly?
a) $-2 \leq x < 3$
b) $-4 \geq x < 7$
c) $-1 \leq x > 0$
d) $6 < x \leq -8$
e) $5 \geq x \geq -9$

93. ***Discussion.*** In each case, write the resulting set of numbers in interval notation. Explain your answers.
a) Every number in $(3, 8)$ is multiplied by 4.
b) Every number in $[-2, 4)$ is multiplied by -5.

c) Three is added to every number in $(-3, 6)$.
d) Every number in $[3, 9]$ is divided by -3.

94. ***Discussion.*** Write the solution set using interval notation for each of the following inequalities in terms of s and t. State any restrictions on s and t. For what values of s and t is the solution set empty?
a) $x > s$ and $x < t$
b) $x > s$ and $x > t$

3.3 ABSOLUTE VALUE EQUATIONS AND INEQUALITIES

In Chapter 1 we learned that absolute value measures the distance of a number from 0 on the number line. In this section we will learn to solve equations and inequalities involving absolute value.

Absolute Value Equations

Solving equations involving absolute value requires some techniques that are different from those studied in previous sections. For example, the solution set to the equation

$$|x| = 5$$

is $\{-5, 5\}$ because both 5 and -5 are five units from 0 on the number line, as shown in Fig. 3.19. So $|x| = 5$ is equivalent to the compound equation

$$x = 5 \text{ or } x = -5.$$

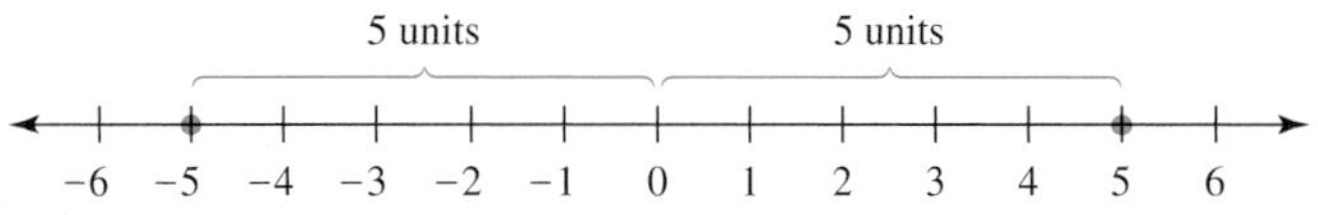

FIGURE 3.19

The equation $|x| = 0$ is equivalent to the equation $x = 0$ because 0 is the only number whose distance from 0 is zero. The solution set to $|x| = 0$ is $\{0\}$.

The equation $|x| = -7$ is inconsistent because absolute value measures distance, and distance is never negative. So the solution set is empty. These ideas are summarized as follows.

calculator

close-up

Use Y= to set y_1 = abs(x − 7). Make a table to see that y_1 has value 2 when x = 5 or x = 9. The table supports the conclusion of Example 1(a).

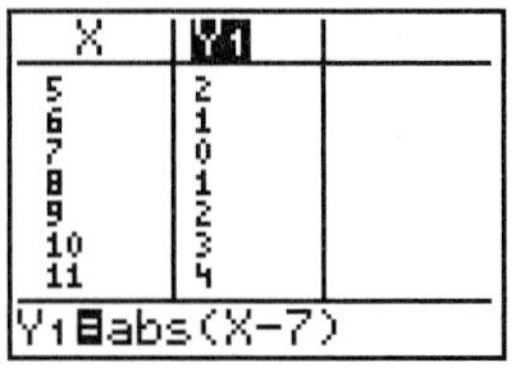

X	Y1
5	2
6	1
7	0
8	1
9	2
10	3
11	4

Y1∎abs(X−7)

Basic Absolute Value Equations		
Absolute Value Equation	**Equivalent Equation**	**Solution Set**
$\lvert x \rvert = k \ (k > 0)$	$x = k$ or $x = -k$	$\{k, -k\}$
$\lvert x \rvert = 0$	$x = 0$	$\{0\}$
$\lvert x \rvert = k \ (k < 0)$		$\varnothing$

We can use these ideas to solve more complicated absolute value equations.

EXAMPLE 1

Absolute value equal to a positive number

Solve each equation.

a) $|x - 7| = 2$　　　　**b)** $|3x - 5| = 7$

Solution

a) First rewrite $|x - 7| = 2$ without absolute value:

$$x - 7 = 2 \quad \text{or} \quad x - 7 = -2 \quad \text{Equivalent equation}$$
$$x = 9 \quad \text{or} \quad x = 5$$

The solution set is $\{5, 9\}$. The distance from 5 to 7 or from 9 to 7 is 2 units.

b) First rewrite $|3x - 5| = 7$ without absolute value:

$$3x - 5 = 7 \qquad \text{or} \qquad 3x - 5 = -7 \quad \text{Equivalent equation}$$

$$3x = 12 \qquad \text{or} \qquad 3x = -2$$

$$x = 4 \qquad \text{or} \qquad x = -\frac{2}{3}$$

The solution set is $\left\{-\frac{2}{3}, 4\right\}$.

E X A M P L E   2   Absolute value equal to zero

Solve $|2(x - 6) + 7| = 0$.

Solution

Since 0 is the only number whose absolute value is 0, the expression within the absolute value bars must be 0.

$$2(x - 6) + 7 = 0 \quad \text{Equivalent equation}$$

$$2x - 12 + 7 = 0$$

$$2x - 5 = 0$$

$$2x = 5$$

$$x = \frac{5}{2}$$

The solution set is $\left\{\frac{5}{2}\right\}$.

Solving absolute value equations is based on the form $|x| = k$, in which the absolute value expression is isolated on one side. In Examples 1 and 2, the equations were given in that form. In the next example, we must first isolate the absolute value expression to solve the equation.

E X A M P L E   3   Absolute value equal to a negative number

Solve $-5|3x - 7| + 4 = 14$.

Solution

First subtract 4 from each side to isolate the absolute value expression:

$$-5|3x - 7| + 4 = 14 \quad \text{Original equation}$$

$$-5|3x - 7| = 10 \quad \text{Subtract 4 from each side.}$$

$$|3x - 7| = -2 \quad \text{Divide each side by } -5.$$

There is no solution because no number has a negative absolute value.

The equation in the next example has an absolute value on both sides.

E X A M P L E   4

Absolute value on both sides

Solve $|2x - 1| = |x + 3|$.

Solution

Two quantities have the same absolute value only if they are equal or opposites. So we can write an equivalent compound equation:

$$2x - 1 = x + 3 \quad \text{or} \quad 2x - 1 = -(x + 3)$$
$$x - 1 = 3 \quad \text{or} \quad 2x - 1 = -x - 3$$
$$x = 4 \quad \text{or} \quad 3x = -2$$
$$x = 4 \quad \text{or} \quad x = -\frac{2}{3}$$

Check that both 4 and $-\frac{2}{3}$ satisfy the original equation. The solution set is $\left\{-\frac{2}{3}, 4\right\}$.

Absolute Value Inequalities

Since absolute value measures distance from 0 on the number line, $|x| > 5$ indicates that x is more than five units from 0. Any number on the number line to the right of 5 or to the left of -5 is more than five units from 0. So $|x| > 5$ is equivalent to

$$x > 5 \quad \text{or} \quad x < -5.$$

The solution set to this inequality is the union of the solution sets to the two simple inequalities. The solution set is $(-\infty, -5) \cup (5, \infty)$. The graph of $|x| > 5$ is shown in Fig. 3.20.

FIGURE 3.20

The inequality $|x| \leq 3$ indicates that x is less than or equal to three units from 0. Any number between -3 and 3 inclusive satisfies that condition. So $|x| \leq 3$ is equivalent to

$$-3 \leq x \leq 3.$$

The graph of $|x| \leq 3$ is shown in Fig. 3.21. These examples illustrate the basic types of absolute value inequalities.

FIGURE 3.21

We can solve more complicated inequalities in the same manner as simple ones.

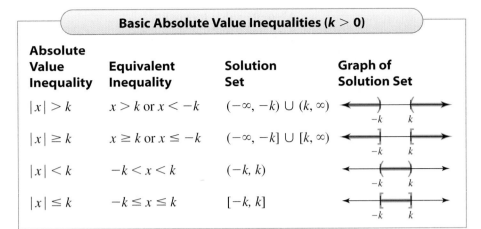

Basic Absolute Value Inequalities ($k > 0$)			
Absolute Value Inequality	**Equivalent Inequality**	**Solution Set**	**Graph of Solution Set**
$\lvert x \rvert > k$	$x > k$ or $x < -k$	$(-\infty, -k) \cup (k, \infty)$	
$\lvert x \rvert \geq k$	$x \geq k$ or $x \leq -k$	$(-\infty, -k] \cup [k, \infty)$	
$\lvert x \rvert < k$	$-k < x < k$	$(-k, k)$	
$\lvert x \rvert \leq k$	$-k \leq x \leq k$	$[-k, k]$	

E X A M P L E   5

Absolute value inequality

Solve $\lvert x - 9 \rvert < 2$ and graph the solution set.

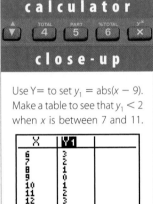

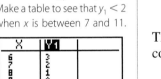

calculator

close-up

Use Y= to set $y_1 = \text{abs}(x - 9)$.
Make a table to see that $y_1 < 2$
when x is between 7 and 11.

Solution

Because $\lvert x \rvert < k$ is equivalent to $-k < x < k$, we can rewrite $\lvert x - 9 \rvert < 2$ as follows:

$$-2 < x - 9 < 2$$
$$-2 + 9 < x - 9 + 9 < 2 + 9 \quad \text{Add 9 to each part of the inequality.}$$
$$7 < x < 11$$

The graph of the solution set $(7, 11)$ is shown in Fig. 3.22. Note that the graph consists of all real numbers that are within two units of 9.

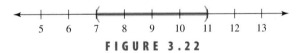

FIGURE 3.22

E X A M P L E   6

Absolute value inequality

Solve $\lvert 3x + 5 \rvert > 2$ and graph the solution set.

Solution

$$3x + 5 > 2 \qquad \text{or} \qquad 3x + 5 < -2 \quad \text{Equivalent compound inequality}$$
$$3x > -3 \qquad \text{or} \qquad 3x < -7$$
$$x > -1 \qquad \text{or} \qquad x < -\frac{7}{3}$$

The solution set is $\left(-\infty, -\frac{7}{3}\right) \cup (-1, \infty)$, and its graph is shown in Fig. 3.23.

FIGURE 3.23

E X A M P L E   7

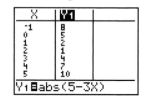

Use Y= to set $y_1 = abs(5 - 3x)$. The table supports the conclusion that $y \leq 6$ when x is between $-\frac{1}{3}$ and $\frac{11}{3}$ even though $-\frac{1}{3}$ and $\frac{11}{3}$ do not appear in the table. For more accuracy, make a table in which the change in x is $\frac{1}{3}$.

Absolute value inequality

Solve $|5 - 3x| \leq 6$ and graph the solution set.

Solution

$$-6 \leq 5 - 3x \leq 6 \quad \text{Equivalent inequality}$$
$$-11 \leq -3x \leq 1 \quad \text{Subtract 5 from each part.}$$
$$\frac{11}{3} \geq x \geq -\frac{1}{3} \quad \text{Divide by } -3 \text{ and reverse each inequality symbol.}$$
$$-\frac{1}{3} \leq x \leq \frac{11}{3} \quad \text{Write } -\frac{1}{3} \text{ on the left because it is smaller than } \frac{11}{3}.$$

The solution set is $\left[-\frac{1}{3}, \frac{11}{3}\right]$ and its graph is shown in Fig. 3.24.

FIGURE 3.24

There are a few absolute value inequalities that do not fit the preceding categories. They are easy to solve using the definition of absolute value.

E X A M P L E   8

Special case

Solve $3 + |7 - 2x| \geq 3$.

Solution

Subtract 3 from each side to isolate the absolute value expression.

$$|7 - 2x| \geq 0$$

Because the absolute value of any real number is greater than or equal to 0, the solution set is R, the set of all real numbers.

E X A M P L E   9

An impossible case

Solve $|5x - 12| < -2$.

Solution

We write an equivalent inequality only when the value of k is positive. With -2 on the right-hand side, we do not write an equivalent inequality. Since the absolute value of any quantity is greater than or equal to 0, no value for x can make this absolute value less than -2. The solution set is $\varnothing$, the empty set.

Applications

A simple example will show how absolute value equations and inequalities can be used in applications.

E X A M P L E   1 0

Controlling water temperature

The water temperature in a certain production process should be 143°F.

a) If the actual temperature differs by 13° from what it should be, then what is the actual temperature?

b) The product is defective if the actual temperature is more than 7° away from what it should be. For what temperatures is the product defective?

Solution

a) If x represents the actual water temperature, then $x - 143$ represents the difference between the actual temperature and the desired temperature. Since $x - 143$ could be positive or negative, we use absolute value:

$$|x - 143| = 13$$
$$x - 143 = 13 \quad \text{or} \quad x - 143 = -13$$
$$x = 156 \quad \text{or} \quad x = 130$$

The actual temperature is either 130°F or 156°F.

b) The product is defective if the absolute value of $x - 143$ is greater than 7:

$$|x - 143| > 7$$
$$x - 143 > 7 \quad \text{or} \quad x - 143 < -7$$
$$x > 150 \quad \text{or} \quad x < 136$$

The product is defective for temperatures greater than 150°F or less than 136°F.

study tip

When taking a test, try not to spend too much time on a single problem. If a problem is taking a long time, then you might be approaching it incorrectly. Move on to another problem and make sure that you get finished with the test.

WARM-UPS

True or false? Explain your answer.

1. The equation $|x| = 2$ is equivalent to $x = 2$ or $x = -2$.
2. All absolute value equations have two solutions.
3. The equation $|2x - 3| = 7$ is equivalent to $2x - 3 = 7$ or $2x + 3 = 7$.
4. The inequality $|x| > 5$ is equivalent to $x > 5$ or $x < -5$.
5. The equation $|x| = -5$ is equivalent to $x = 5$ or $x = -5$.
6. There is only one solution to the equation $|3 - x| = 0$.
7. We should write the inequality $x > 3$ or $x < -3$ as $3 < x < -3$.
8. The inequality $|x| < 7$ is equivalent to $-7 \leq x \leq 7$.
9. The equation $|x| + 2 = 5$ is equivalent to $|x| = 3$.
10. The inequality $|x| < -2$ is equivalent to $x < 2$ and $x > -2$.

3.3   EXERCISES

Reading and Writing   *After reading this section, write out the answers to these questions. Use complete sentences.*

1. What does absolute value measure?

2. Why does $|x| = 0$ have only one solution?

3. Why does $|x| = 4$ have two solutions?

4. Why is $|x| = -3$ inconsistent?

5. Why do all real numbers satisfy $|x| \geq 0$?

6. Why do no real numbers satisfy $|x| < -3$?

Solve each absolute value equation. See Examples 1–3.

7. $|a| = 5$ **8.** $|x| = 2$ **9.** $|x - 3| = 1$

10. $|x - 5| = 2$ **11.** $|3 - x| = 6$ **12.** $|7 - x| = 6$

13. $|3x - 4| = 12$ **14.** $|5x + 2| = -3$

15. $\left|\dfrac{2}{3}x - 8\right| = 0$ **16.** $\left|3 - \dfrac{3}{4}x\right| = \dfrac{1}{4}$

17. $|6 - 0.2x| = 10$
18. $|5 - 0.1x| = 0$
19. $|7(x - 6)| = -3$
20. $|2(a + 3)| = 15$
21. $|2(x - 4) + 3| = 5$

22. $|3(x - 2) + 7| = 6$

23. $|7.3x - 5.26| = 4.215$

24. $|5.74 - 2.17x| = 10.28$

Solve each absolute value equation. See Examples 3 and 4.
25. $3 + |x| = 5$
26. $|x| - 10 = -3$
27. $2 - |x + 3| = -6$
28. $4 - 3|x - 2| = -8$

29. $5 - \dfrac{|3 - 2x|}{3} = 4$

30. $3 - \dfrac{1}{2}\left|\dfrac{1}{2}x - 4\right| = 2$

31. $|x - 5| = |2x + 1|$

32. $|w - 6| = |3 - 2w|$

33. $\left|\dfrac{5}{2} - x\right| = \left|2 - \dfrac{x}{2}\right|$

34. $\left|x - \dfrac{1}{4}\right| = \left|\dfrac{1}{2}x - \dfrac{3}{4}\right|$

35. $|x - 3| = |3 - x|$
36. $|a - 6| = |6 - a|$

Write an absolute value inequality whose solution set is shown by the graph. See Examples 5–7.

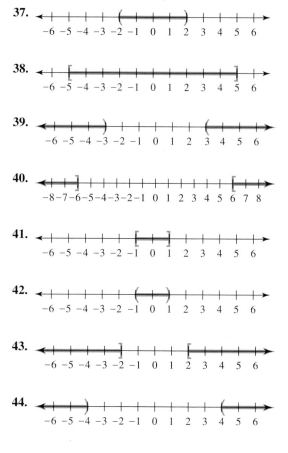

Determine whether each absolute value inequality is equivalent to the inequality following it. See Examples 5–7.

45. $|x| < 3,\, x < 3$
46. $|x| > 3,\, x > 3$
47. $|x - 3| > 1,\, x - 3 > 1$ or $x - 3 < -1$
48. $|x - 3| \leq 1,\, -1 \leq x - 3 \leq 1$
49. $|x - 3| \geq 1,\, x - 3 \geq 1$ or $x - 3 \leq 1$
50. $|x - 3| > 0,\, x - 3 > 0$
51. $|4 - x| < 1,\, 4 - x < 1$ and $-(4 - x) < 1$
52. $|4 - x| > 1,\, 4 - x > 1$ or $-(4 - x) > 1$

Solve each absolute value inequality and graph the solution set. See Examples 5–7.

53. $|x| > 6$

54. $|w| > 3$

55. $|2a| < 6$

56. $|3x| < 21$

57. $|x - 2| \geq 3$

58. $|x - 5| \geq 1$

59. $\dfrac{1}{5}|2x - 4| < 1$

60. $\dfrac{1}{3}|2x - 1| < 1$

61. $-2|5 - x| \geq -14$

62. $-3|6 - x| \geq -3$

63. $2|3 - 2x| - 6 \geq 18$

64. $2|5 - 2x| - 15 \geq 5$

Solve each absolute value inequality and graph the solution set. See Examples 8 and 9.

65. $|x - 2| > 0$

66. $|6 - x| \geq 0$
67. $|x - 5| \geq 0$
68. $|3x - 7| \geq -3$
69. $-2|3x - 7| > 6$
70. $-3|7x - 42| > 18$
71. $|2x + 3| + 6 > 0$

72. $|5 - x| + 5 > 5$

Solve each inequality. Write the solution set using interval notation.

73. $1 < |x + 2|$
74. $5 \geq |x - 4|$
75. $5 > |x| + 1$
76. $4 \leq |x| - 6$
77. $3 - 5|x| > -2$
78. $1 - 2|x| < -7$
79. $|5.67x - 3.124| < 1.68$

80. $|4.67 - 3.2x| \geq 1.43$
81. $|2x - 1| < 3$ and $2x - 3 > 2$

82. $|5 - 3x| \geq 3$ and $5 - 2x > 3$

83. $|x - 2| < 3$ and $|x - 7| < 3$
84. $|x - 5| < 4$ and $|x - 6| > 2$

Solve each problem by using an absolute value equation or inequality. See Example 10.

85. *Famous battles.* In the Hundred Years' War, Henry V defeated a French army in the battle of Agincourt and Joan of Arc defeated an English army in the battle of Orleans (*The Doubleday Almanac*). Suppose you know only that these two famous battles were 14 years apart and that the battle of Agincourt occurred in 1415. Use an absolute value equation to find the possibilities for the year in which the battle of Orleans occurred.

86. *World records.* In July 1985 Steve Cram of Great Britain set a world record of 3 minutes 29.67 seconds for the 1500-meter race and a world record of 3 minutes 46.31 seconds for the 1-mile race (*The Doubleday Almanac*). Suppose you know only that these two events occurred 11 days apart and that the 1500-meter record was set on July 16. Use an absolute value equation to find the possible dates for the 1-mile record run.

87. *Weight difference.* Research at a major university has shown that identical twins generally differ by less than 6 pounds in body weight. If Kim weighs 127 pounds, then in what range is the weight of her identical twin sister Kathy?

88. *Intelligence quotient.* Jude's IQ score is more than 15 points away from Sherry's. If Sherry scored 110, then in what range is Jude's score?

89. *Unidentified flying objects.* The formula

$$S = -16t^2 + v_0 t + s_0$$

gives height in feet above the earth at time t seconds for an object projected into the air with an initial velocity of v_0 feet per second (ft/sec) from an initial height of s_0 feet. Two balls are tossed into the air simultaneously, one from the ground at 50 ft/sec and one from a height of 10 feet at 40 ft/sec. See the accompanying graph.

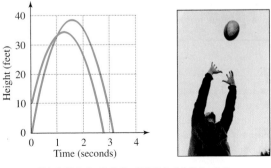

FIGURE FOR EXERCISE 89

a) Use the graph to estimate the time at which the balls are at the same height.

b) Find the time from part (a) algebraically.

c) For what values of t will their heights above the ground differ by less than 5 feet (while they are both in the air)?

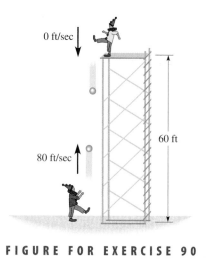

FIGURE FOR EXERCISE 90

90. *Playing catch.* A circus clown at the top of a 60-foot platform is playing catch with another clown on the ground. The clown on the platform drops a ball at the same time as the one on the ground tosses a ball upward at 80 ft/sec. For what length of time is the distance between the balls less than or equal to 10 feet? (*Hint:* Use the formula given in Exercise 89. The initial velocity of a ball that is dropped is 0 ft/sec.)

GETTING MORE INVOLVED

91. *Discussion.* For which real numbers m and n is each equation satisfied?

a) $|m - n| = |n - m|$

b) $|mn| = |m| \cdot |n|$

c) $\left|\dfrac{m}{n}\right| = \dfrac{|m|}{|n|}$

92. *Exploration.* **a)** Evaluate $|m + n|$ and $|m| + |n|$ for

i) $m = 3$ and $n = 5$

ii) $m = -3$ and $n = 5$

iii) $m = 3$ and $n = -5$

iv) $m = -3$ and $n = -5$

b) What can you conclude about the relationship between $|m + n|$ and $|m| + |n|$?

COLLABORATIVE ACTIVITIES

Everyday Algebra

Every day, people use algebra without even knowing it. Any time you solve for an unknown quantity, you are using algebra. Here is an example of a simple problem that you could solve without even thinking of algebra.

• While shopping Joe notices a store brand that is available at a lower price than the name brand. If the name brand product costs $3.79 and Joe has a coupon for 50 cents, what would the store brand price need to be for Joe to save money with the coupon?

This is a problem you could solve mentally by subtracting the two quantities and finding that if the store brand costs $3.29 or more, Joe would save money with the coupon. If the store brand is less than $3.29, then Joe would save money by buying it instead. The beauty of algebra is not apparent in cases like this because you already know what operation to perform to find the unknown quantity. Algebra becomes useful when it is not clear what to do.

Grouping: 2 to 4 students per group
Topic: Use of algebra in common occurrences

We will consider another situation in which we want to find the best price of an item. For this situation we will use the following formulas.

The formula for markup of an item is

$$P = C + rC,$$

where P is the price of the item, C is the wholesale cost of the item, and r is the percent of markup.

The formula for discounting an item is

$$S = P - dP,$$

where S is the discounted price and d is the percent discount.

• Lane belongs to a wholesale buying club. She can order items through the club with a markup of 8% above the wholesale cost. She also can buy the same items in a store where she can get a 10% discount off the shelf price. Lane wants to know what the store markup on any particular item must be for it to be cheaper to order through the club.

Form groups of two to four people. Assign a role to each person: **Recorder, Moderator, Messenger,** or **Quality Manager** (roles may be combined if there are fewer than four people in a group). In your groups:

1. Decide how to rewrite Lane's problem using algebra. Decide what the unknown quantities are and assign them variable names.

2. Write the equations or inequalities on your paper using the variables that you just defined.

3. Solve the problem and state the group's decision on what Lane should do.

Extension: Pick a similar problem from your own lives to use algebra to solve.

WRAP-UP   CHAPTER 3

SUMMARY

Inequalities   **Examples**

Linear inequality in one variable	Any inequality of the form $ax + b < 0$ with $a \neq 0$ In place of $<$ we can use $\leq$, $>$ or $\geq$.	$2x + 9 < 0$ $x - 2 \geq 7$ $-3x - 1 \geq 2x + 5$
Properties of inequality	Addition, subtraction, multiplication, and division may be performed on each side of an inequality, just as we do in solving equations, with one exception. When multiplying or dividing by a negative number, the inequality symbol is reversed.	$-3x + 1 > 7$ $-3x > 6$ $x < -2$

Compound inequality	Two simple inequalities connected with the word "and" or "or"
	    And corresponds to *intersection*.
	    Or corresponds to *union*.
		$x > 1$ and $x < 5$
		 $x > 3$ or $x < 1$

Absolute value

	Absolute Value Equation	**Equivalent Equation**	**Solution Set**
Basic absolute value equations	$\lvert x \rvert = k \ (k > 0)$ $\lvert x \rvert = 0$ $\lvert x \rvert = k \ (k < 0)$	$x = k$ or $x = -k$ $x = 0$	$\{k, -k\}$ $\{0\}$ $\varnothing$

	Absolute Value Inequality	**Equivalent Inequality**	**Solution Set**	**Graph of Solution Set**
Basic absolute value inequalities $(k > 0)$	$\lvert x \rvert > k$	$x > k$ or $x < -k$	$(-\infty, -k) \cup (k, \infty)$	
	$\lvert x \rvert \geq k$	$x \geq k$ or $x \leq -k$	$(-\infty, -k] \cup [k, \infty)$	
	$\lvert x \rvert < k$	$-k < x < k$	$(-k, k)$	
	$\lvert x \rvert \leq k$	$-k \leq x \leq k$	$[-k, k]$	

ENRICHING YOUR MATHEMATICAL WORD POWER

For each mathematical term, choose the correct meaning.

1. inequality
   a. an equation that is not correct
   b. two different numbers
   c. a statement that expresses the inequality of two algebraic expressions
   d. a larger number

2. equivalent inequalities
   a. the inequality reverses when dividing by a negative number

   b. $a < b$ and $b < c$
   c. $a < b$ and $a \le b$
   d. inequalities that have the same solution set

3. compound inequality
   a. an inequality that is complicated
   b. an inequality that reverses when divided by a negative number
   c. an inequality of negative numbers
   d. two simple inequalities joined with "and" or "or"

REVIEW EXERCISES

3.1 *Solve each inequality. State the solution set using interval notation and graph it.*

1. $3 - 4x < 15$

2. $5 - 6x > 35$

3. $2(x - 3) > -6$

4. $4(5 - x) < 20$

5. $-\dfrac{3}{4}x \ge 6$

6. $-\dfrac{2}{3}x \le 4$

7. $3(x + 2) > 5(x - 1)$

8. $4 - 2(x - 3) < 0$

9. $\dfrac{1}{2}x + 7 \le \dfrac{3}{4}x - 5$

10. $\dfrac{5}{6}x - 3 \ge \dfrac{2}{3}x + 7$

3.2 *Solve each compound inequality. State the solution set using interval notation and graph it.*

11. $x + 2 > 3$ or $x - 6 < -10$

12. $x - 2 > 5$ or $x - 2 < -1$

13. $x > 0$ and $x - 6 < 3$

14. $x \le 0$ and $x + 6 > 3$

15. $6 - x < 3$ or $-x < 0$

16. $-x > 0$ or $x + 2 < 7$

17. $2x < 8$ and $2(x - 3) < 6$

18. $\dfrac{1}{3}x > 2$ and $\dfrac{1}{4}x > 2$

19. $x - 6 > 2$ and $6 - x > 0$

20. $-\dfrac{1}{2}x < 6$ or $\dfrac{2}{3}x < 4$

21. $0.5x > 10$ or $0.1x < 3$

22. $0.02x > 4$ and $0.2x < 3$

23. $-2 \le \dfrac{2x - 3}{10} \le 1$

24. $-3 < \dfrac{4 - 3x}{5} < 2$

Write each union or intersection of intervals as a single interval.

25. $[1, 4) \cup (2, \infty)$

26. $(2, 5) \cup (-1, \infty)$

27. $(3, 6) \cap [2, 8]$

28. $[-1, 3] \cap [0, 8]$

29. $(-\infty, 5) \cup [5, \infty)$

30. $(-\infty, 1) \cup (0, \infty)$

31. $(-3, -1] \cap [-2, 5]$

32. $[-2, 4] \cap (4, 7]$

3.3 *Solve each absolute value equation or inequality and graph the solution set.*

33. $|2x| \ge 8$

34. $|5x - 1| \le 14$

35. $|x| = -5$

36. $|x| + 2 = 16$

37. $\left|1 - \dfrac{x}{5}\right| > \dfrac{9}{5}$

38. $\left|1 - \dfrac{1}{6}x\right| < \dfrac{1}{2}$

39. $|x - 3| < -3$ **40.** $|x - 7| \le -4$

41. $\left|\dfrac{x}{2}\right| - 5 = -1$

42. $\left|\dfrac{x}{2} - 5\right| = -1$

43. $|x + 4| \ge -1$ **44.** $|6x - 1| \ge 0$

45. $1 - \dfrac{3}{2}|x - 2| < -\dfrac{1}{2}$

46. $1 > \dfrac{1}{2}|6 - x| - \dfrac{3}{4}$

MISCELLANEOUS

Solve each problem.

47. *Rockbuster video.* Stephen plans to open a video rental store in Edmonton. Industry statistics show that 45% of the rental price goes for overhead. If the maximum that anyone will pay to rent a tape is $5 and Stephen wants a profit of at least $1.65 per tape, then in what range should the rental price be?

48. *Working girl.* Regina makes $6.80 per hour working in the snack bar. To keep her grant, she may not earn more than $51 per week. What is the range of the number of hours per week that she may work?

49. *Skeletal remains.* Forensic scientists use the formula $h = 60.089 + 2.238F$ to predict the height h (in centimeters) for a male whose femur measures F centimeters. (See the accompanying figure.) In what range is the length of the femur for males between 150 centimeters and 180 centimeters in height? Round to the nearest tenth of a centimeter.

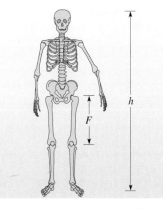

FIGURE FOR EXERCISE 49

50. *Female femurs.* Forensic scientists use the formula $h = 61.412 + 2.317F$ to predict the height h in centimeters for a female whose femur measures F centimeters.

a) Use the accompanying graph to estimate the femur length for a female with height of 160 centimeters.

b) In what range is the length of the femur for females who are over 170 centimeters tall?

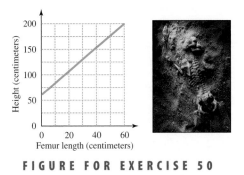

FIGURE FOR EXERCISE 50

51. *Car trouble.* Dane's car was found abandoned at mile marker 86 on the interstate. If Dane was picked up by the police on the interstate exactly 5 miles away, then at what mile marker was he picked up?

52. *Comparing scores.* Scott scored 72 points on the midterm, and Katie's score was more than 16 points away from Scott's. What was Katie's score?

For each graph in Exercises 53–70, write an equation or inequality that has the solution set shown by the graph. Use absolute value when possible.

53.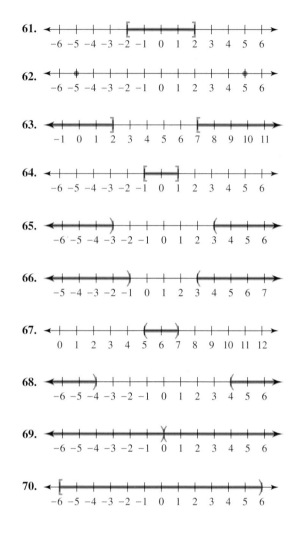

54.

55.

56.

57.

58.

59.

60.

61.

62.

63.

64.

65.

66.

67.

68.

69.

70.

Write an inequality that describes the graph.

1.

2.

Write the solution set to each inequality using interval notation.

3. $x \geq 3$

4. $x > 1$ and $x \leq 6$

5. $x < 5$ or $x > 9$

6. $|x| < 3$

7. $|x| \geq 2$

Solve each inequality. State the solution set using interval notation and graph the solution set.

8. $2x + 3 > 1$

9. $|m - 6| \leq 2$

10. $2|x - 3| - 5 > 15$

11. $2 - 3(w - 1) < -2w$

12. $2 < \dfrac{5 - 2x}{3} < 7$

13. $3x - 2 < 7$ and $-3x \le 15$

14. $\dfrac{2}{3}y < 4$ or $y - 3 < 12$

Solve each equation or inequality.

15. $|2x - 7| = 3$

16. $x - 4 > 1$ or $x < 12$

17. $3x < 0$ and $x - 5 > 2$

18. $|2x - 5| \le 0$

19. $|x - 3| < 0$

20. $|x - 6| > -6$

Write a complete solution to each problem.

21. Brandon gets a 40% discount on loose diamonds where he works. The cost of the setting is $250. If he plans to spend at most $1450, then what is the price range (list price) of the diamonds that he can afford?

22. *Perimeter of a triangle.* One side of a triangle is 1 foot longer than the shortest side, and the third side is twice as long as the shortest side. If the perimeter is less than 25 feet, then what is the range of the length of the shortest side?

23. Al and Brenda do the same job, but their annual salaries differ by more than $3000. Assume Al makes $28,000 per year and write an absolute value inequality to describe this situation. What are the possibilities for Brenda's salary?

Simplify each expression.

1. $5x + 6x$

2. $5x \cdot 6x$

3. $\dfrac{6x + 2}{2}$

4. $5 - 4(2 - x)$

5. $(30 - 1)(30 + 1)$

6. $(30 + 1)^2$

7. $(30 - 1)^2$

8. $(2 + 3)^2$

9. $2^2 + 3^2$

10. $(8 - 3)(3 - 8)$

11. $(-1)(3 - 8)$

12. -2^2

13. $3x + 8 - 5(x - 1)$

14. $(-6)^2 - 4(-3)2$

15. $3^2 \cdot 2^3$

16. $4(-6) - (-5)(3)$

17. $-3x \cdot x \cdot x$

18. $(-1)(-1)(-1)(-1)(-1)(-1)$

Solve each equation.

19. $5x + 6x = 8x$

20. $5x + 6x = 11x$

21. $5x + 6x = 0$

22. $5x + 6 = 11x$

23. $3x + 1 = 0$

24. $5 - 4(2 - x) = 1$

25. $3x + 6 = 3(x + 2)$

26. $x - 0.01x = 990$

27. $|5x + 6| = 11$

Solve the problem.

28. *Cost analysis.* Diller Electronics can rent a copy machine for 5 years from American Business Supply for $75 per month plus 6 cents per copy. The same copier can be purchased for $8000, but then it costs only 2 cents per copy for supplies and maintenance. The purchased copier has no value after 5 years.

a) Use the accompanying graph to estimate the number of copies for 5 years for which the cost of renting would equal the cost of buying.

b) Write a formula for the 5-year cost under each plan.

c) Algebraically find the number of copies for which the 5-year costs would be equal.

d) If Diller makes 120,000 copies in 5 years, which plan is cheaper and by how much?

e) For what range of copies do the two plans differ by less than $500?

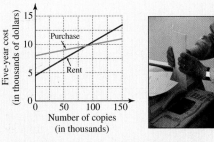

FIGURE FOR EXERCISE 28

Linear Equations in Two Variables and Their Graphs

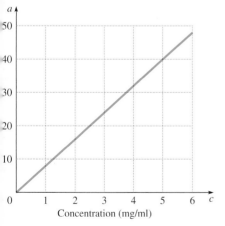

Concentration (mg/ml)

I f you pick up any package of food and read the label, you will find a long list that usually ends with some mysterious looking names. Many of these strange elements are food additives. A food additive is a substance or a mixture of substances other than basic foodstuffs that is present in food as a result of production, processing, storage, or packaging. They can be natural or synthetic and are categorized in many ways: preservatives, coloring agents, processing aids, and nutritional supplements, to name a few.

Food additives have been around since prehistoric humans discovered that salt would help to preserve meat. Today, food additives can include simple ingredients such as red color from Concord grape skins, calcium, or an enzyme. Throughout the centuries there have been lively discussions on what is healthy to eat. At the present time the food industry is working to develop foods that have less cholesterol, fats, and other unhealthy ingredients.

Although they frequently have different viewpoints, the food industry and the Food and Drug Administration (FDA) are working to provide consumers with information on a healthier diet. Recent developments such as the synthetically engineered tomato stirred great controversy, even though the FDA declared the tomato safe to eat.

In Exercise 35 of Section 4.5 you will see how a food chemist uses a linear equation in testing the concentration of an enzyme in a fruit juice.

4.1 GRAPHING LINES IN THE COORDINATE PLANE

In Chapter 1 you learned to graph numbers on a number line. We also used number lines to illustrate the solution to inequalities in Chapter 3. In this section you will learn to graph pairs of numbers in a coordinate system made up of a pair of number lines. We will use this coordinate system to illustrate the solution to equations and inequalities in two variables.

Ordered Pairs

The equation $y = 2x - 1$ is an equation in two variables. This equation is satisfied if we choose a value for x and a value for y that make it true. If we choose $x = 2$ and $y = 3$, then $y = 2x - 1$ becomes

$$\underset{\downarrow}{y} \qquad \underset{\downarrow}{x}$$
$$3 = 2(2) - 1.$$
$$3 = 3$$

helpful hint

In this chapter you will be doing a lot of graphing. Using graph paper will help you understand the concepts and help you recognize errors. For your convenience, a page of graph paper can be found at the end of this chapter. Make as many copies of it as you wish.

Because the last statement is true, we say that the pair of numbers 2 and 3 **satisfies the equation** or is a **solution to the equation.** We use the **ordered pair** $(2, 3)$ to represent $x = 2$ and $y = 3$. The format is always to write the value for x first and the value for y second. The first number of the ordered pair is called the ***x*-coordinate,** and the second number is called the ***y*-coordinate.** Note that the ordered pair $(3, 2)$ does not satisfy the equation $y = 2x - 1$, because for $x = 3$ and $y = 2$ we have

$$2 \neq 2(3) - 1.$$

The ordered pair $(2, 3)$ is a solution to $y = 2x - 1$. We can find as many solutions as we please by simply choosing any value for x or y and then using the equation to find the other coordinate of the ordered pair.

EXAMPLE 1

Finding solutions to an equation

Each of the ordered pairs below is missing one coordinate. Complete each ordered pair so that it satisfies the equation $y = -3x + 4$.

a) $(2, \quad)$ **b)** $(\quad, -5)$ **c)** $(0, \quad)$

Solution

a) The x-coordinate of $(2, \quad)$ is 2. Let $x = 2$ in the equation $y = -3x + 4$:

$$y = -3 \cdot 2 + 4$$
$$= -6 + 4$$
$$= -2$$

The ordered pair $(2, -2)$ satisfies the equation.

b) The y-coordinate of $(\quad, -5)$ is -5. Let $y = -5$ in the equation $y = -3x + 4$:

$$-5 = -3x + 4$$
$$-9 = -3x$$
$$3 = x$$

The ordered pair $(3, -5)$ satisfies the equation.

c) Replace x by 0 in the equation $y = -3x + 4$:

$$y = -3 \cdot 0 + 4$$
$$= 4$$

So $(0, 4)$ satisfies the equation. ◼

The Rectangular Coordinate System

We use the **rectangular** (or **Cartesian**) **coordinate system** to get a visual image of ordered pairs of real numbers. The rectangular coordinate system consists of two number lines drawn at a right angle to one another, intersecting at zero on each number line, as shown in Fig. 4.1. The plane containing these number lines is called the **coordinate plane.** On the horizontal number line the positive numbers are to the right of zero, and on the vertical number line the positive numbers are above zero.

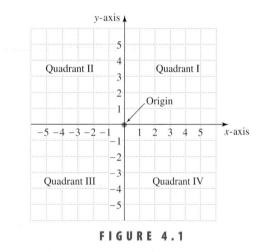

FIGURE 4.1

The horizontal number line is called the **x-axis,** and the vertical number line is called the **y-axis.** The point at which they intersect is called the **origin.** The two number lines divide the plane into four regions called **quadrants.** They are numbered as shown in Fig. 4.1. The quadrants do not include any points on the axes.

Plotting Points

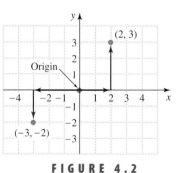

FIGURE 4.2

Just as every real number corresponds to a point on the number line, *every pair of real numbers corresponds to a point in the rectangular coordinate system.* For example, the point corresponding to the pair $(2, 3)$ is found by starting at the origin and moving two units to the right and then three units up. The point corresponding to the pair $(-3, -2)$ is found by starting at the origin and moving three units to the left and then two units down. Both of these points are shown in Fig. 4.2.

When we locate a point in the rectangular coordinate system, we are **plotting** or **graphing** the point. Because ordered pairs of numbers correspond to points in the coordinate plane, we frequently refer to an ordered pair as a point.

E X A M P L E   2

Plotting points

Plot the points $(2, 5)$, $(-1, 4)$, $(-3, -4)$, and $(3, -2)$.

Solution

To locate $(2, 5)$, start at the origin, move two units to the right, and then move up five units. To locate $(-1, 4)$, start at the origin, move one unit to the left, and then move up four units. All four points are shown in Fig. 4.3. ∎

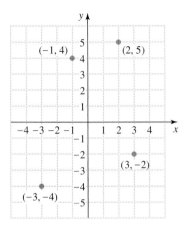

FIGURE 4.3

Graphing a Linear Equation

The **graph** of an equation is an illustration in the coordinate plane that shows all of the ordered pairs that satisfy the equation. When we draw the graph, we are **graphing the equation.**

Consider again the equation $y = 2x - 1$. The following table shows some ordered pairs that satisfy this equation.

x	-3	-2	-1	0	1	2	3
$y = 2x - 1$	-7	-5	-3	-1	1	3	5

The ordered pairs in this table are graphed in Fig. 4.4. Notice that the points lie in a straight line. If we choose any real number for x and find the point (x, y) that satisfies $y = 2x - 1$, we get another point along this line. Likewise, any point along this line satisfies the equation. So the graph of $y = 2x - 1$ is the straight line shown in Fig. 4.5. Because it is not possible to actually show all of the line, the arrows on the ends of the line indicate that it goes on indefinitely in both directions. The equation $y = 2x - 1$ is an example of a linear equation in two variables.

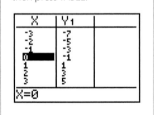
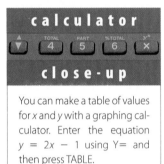

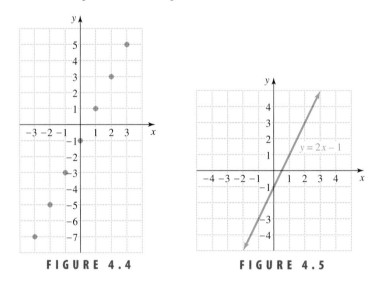

FIGURE 4.4   **FIGURE 4.5**

Linear Equation in Two Variables

A **linear equation in two variables** is an equation that can be written in the form

$$Ax + By = C,$$

where A, B, and C are real numbers, with A and B not both equal to zero.

Equations such as

$$x - y = 5, \quad y = 2x + 3, \quad 2x - 5y - 9 = 0, \quad \text{and} \quad x = 8$$

are linear equations because they could all be rewritten in the form $Ax + By = C$. The graph of any linear equation is a straight line.

EXAMPLE 3

Graphing an equation

Graph the equation $3x + y = 2$. Plot at least five points.

Solution

It is easier to make a table of ordered pairs if the equation is solved for y:

$$y = -3x + 2$$

Next, arbitrarily select values for x and then calculate the corresponding value for y. The following table of values shows five ordered pairs that satisfy the equation:

x	-2	-1	0	1	2
$y = -3x + 2$	8	5	2	-1	-4

The graph of the line through these points is shown in Fig. 4.6.

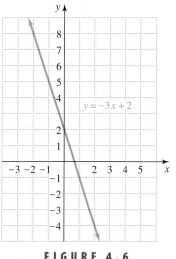

FIGURE 4.6

---

calculator close-up

To graph $y = -3x + 2$, enter the equation using the Y= key:

Next, set the viewing window (WINDOW) to get the desired view of the graph. Xmin and Xmax indicate the minimum

and maximum x-values used for the graph; likewise for Ymin and Ymax. Xscl and Yscl (scale) give the distance between tick marks on the respective axes.

Press GRAPH to get the graph:

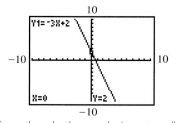

Even though the graph is not really "straight," it is consistent with the graph of $y = -3x + 2$ in Fig. 4.6.

E X A M P L E   4

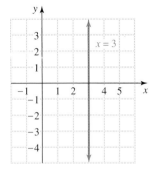

FIGURE 4.7

A vertical line

Graph the equation $x + 0 \cdot y = 3$. Plot at least five points.

Solution

If we choose a value of 3 for x, then we can choose any number for y, since y is multiplied by 0. The equation $x + 0 \cdot y = 3$ is usually written simply as $x = 3$. The following table shows five ordered pairs that satisfy the equation:

$x = 3$	3	3	3	3	3
y	-2	-1	0	1	2

Figure 4.7 shows the line through these points. ■

close-up

You cannot graph the vertical line $x = 3$ on most graphing calculators. The only equations that can be graphed are ones in which y is written in terms of x.

CAUTION   If an equation such as $x = 3$ is discussed in the context of equations in two variables, then we assume that it is a simplified form of $x + 0 \cdot y = 3$, and there are infinitely many ordered pairs that satisfy the equation. If the equation $x = 3$ is discussed in the context of equations in a single variable, then $x = 3$ has only one solution, 3.

All of the equations we have considered so far have involved single-digit numbers. If an equation involves large numbers, then we must change the scale on the x-axis, the y-axis, or both to accommodate the numbers involved. The change of scale is arbitrary, and the graph will look different for different scales.

E X A M P L E   5

Adjusting the scale

Graph the equation $y = 20x + 500$. Plot at least five points.

Solution

The following table shows five ordered pairs that satisfy the equation.

x	-20	-10	0	10	20
$y = 20x + 500$	100	300	500	700	900

To fit these points onto a graph, we change the scale on the x-axis to let each division represent 10 units and change the scale on the y-axis to let each division represent 200 units. The graph is shown in Fig. 4.8.

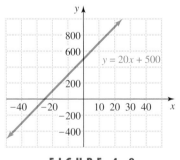

FIGURE 4.8 ■

Graphing a Line Using Intercepts

We know that the graph of a linear equation is a straight line. Because it takes only two points to determine a line, we can graph a linear equation using only two points. The two points that are the easiest to locate are usually the points where the line crosses the axes. The point where the graph crosses the x-axis is the **x-intercept,** and the point where the graph crosses the y-axis is the **y-intercept.**

E X A M P L E   6

Graphing a line using intercepts

Graph the equation $2x - 3y = 6$ by using the x- and y-intercepts.

Solution

To find the x-intercept, let $y = 0$ in the equation $2x - 3y = 6$:

$$2x - 3 \cdot 0 = 6$$
$$2x = 6$$
$$x = 3$$

The x-intercept is $(3, 0)$. To find the y-intercept, let $x = 0$ in $2x - 3y = 6$:

$$2 \cdot 0 - 3y = 6$$
$$-3y = 6$$
$$y = -2$$

The y-intercept is $(0, -2)$. Locate the intercepts and draw a line through them as shown in Fig. 4.9. To check, find one additional point that satisfies the equation, say $(6, 2)$, and see whether the line goes through that point.

helpful hint

You can find the intercepts for $2x - 3y = 6$ using the *cover-up method.* Cover up $-3y$ with your pencil, then solve $2x = 6$ mentally to get $x = 3$ and an x-intercept of $(3, 0)$. Now cover up $2x$ and solve $-3y = 6$ to get $y = -2$ and a y-intercept of $(0, -2)$.

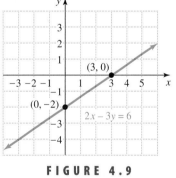

F I G U R E   4 . 9

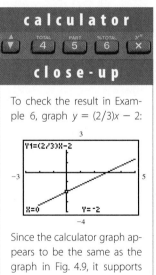

calculator

close-up

To check the result in Example 6, graph $y = (2/3)x - 2$:

Since the calculator graph appears to be the same as the graph in Fig. 4.9, it supports the conclusion that Fig. 4.9 is correct.

Applications

Linear equations occur in many real-life situations. If the cost of plans for a house is $475 for one copy plus $30 for each additional copy, then $C = 475 + 30x$, where x is the number of additional copies. If you have $1000 budgeted for landscaping with trees at $50 each and bushes at $20 each, then $50t + 20b = 1000$, where t is the number of trees and b is the number of bushes. In the next example we see a linear equation that models ticket demand.

E X A M P L E   7

Ticket demand

The demand for tickets to see the Ice Gators play hockey can be modeled by the equation $d = 8000 - 100p$, where d is the number of tickets sold and p is the price per ticket in dollars.

a) How many tickets will be sold at $20 per ticket?

b) Find the intercepts and interpret them.

c) Graph the linear equation.

d) What happens to the demand as the price increases?

Solution

a) If tickets are $20 each, then $d = 8000 - 100 \cdot 20 = 6000$. So at $20 per ticket, the demand will be 6000 tickets.

b) Replace d with 0 in the equation $d = 8000 - 100p$ and solve for p:

$$0 = 8000 - 100p$$
$$100p = 8000 \qquad \text{Add } 100p \text{ to each side.}$$
$$p = 80 \qquad \text{Divide each side by 100.}$$

If $p = 0$, then $d = 8000 - 100 \cdot 0 = 8000$. So the intercepts are (0, 8000) and (80, 0). If the tickets are free, the demand will be 8000 tickets. At $80 per ticket, no tickets will be sold.

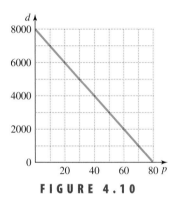

FIGURE 4.10

c) Graph the line using the intercepts (0, 8000) and (80, 0) as shown in Fig. 4.10.

d) When the tickets are free, the demand is high. As the price increases, the demand goes down. At $80 per ticket, there will be no demand. ■

M A T H   A T   W O R K $x^2 + (x+1)^2 = 5^2$

Christopher J. Edington, manager of the Biomechanics Laboratory at Converse, Inc., is a specialist in studying human movements from the hip down. In the past he has worked with diabetics, helping to educate them about the role of shoes and stress points in the shoes and their relationship to preventing foot injuries. More recently, he has helped to design and run tests for Converse's new athletic and leisure shoes. The latest development is a new basketball shoe that combines the lightweight characteristic of a running shoe with the support and durability of a standard basketball sneaker.

BIOMECHANIST

   Information on how the foot strikes the ground, the length of contact time, and movements of the foot, knee, and hip can be recorded by using high-speed video equipment. This information is then used to evaluate the performance and design requirements of a lightweight, flexible, and well-fitting shoe. To meet the requirements of a good basketball shoe, Mr. Edington helped design and test the "React" shock-absorbing technology that is in Converse's latest sneakers.

   In Exercise 84 of this section you will see the motion of a runner's heel as Mr. Edington does.

WARM-UPS

True or false? Explain your answer.

1. The point $(2, 4)$ satisfies the equation $2y - 3x = -8$.
2. If $(1, 5)$ satisfies an equation, then $(5, 1)$ also satisfies the equation.
3. The origin is in quadrant I.
4. The point $(4, 0)$ is on the y-axis.
5. The graph of $x + 0 \cdot y = 9$ is the same as the graph of $x = 9$.
6. The graph of $x = -5$ is a vertical line.
7. The graph of $0 \cdot x + y = 6$ is a horizontal line.
8. The y-intercept for the line $x + 2y = 5$ is $(5, 0)$.
9. The point $(5, -3)$ is in quadrant II.
10. The point $(-349, 0)$ is on the x-axis.

4.1   EXERCISES

Reading and Writing   After reading this section, write out the answers to these questions. Use complete sentences.

1. What is an ordered pair?

2. What is the rectangular coordinate system?

3. What name is given to the point of intersection of the x-axis and the y-axis?

4. What is the graph of an equation?

5. What is a linear equation in two variables?

6. What are intercepts?

Complete each ordered pair so that it satisfies the given equation. See Example 1.

7. $y = 3x + 9$: $(0, \ )$, $(\ , 24)$, $(2, \ )$

8. $y = 2x + 5$: $(8, \ )$, $(-1, \ )$, $(\ , -1)$

9. $y = -3x - 7$: $(0, \ )$, $(-4, \ )$, $(\ , -1)$

10. $y = -5x - 3$: $(\ , 2)$, $(-3, \ )$, $(0, \ )$

11. $y = -12x + 5$: $(0, \ )$, $(10, \ )$, $(\ , 17)$

12. $y = 18x + 200$: $(1, \ )$, $(-10, \ )$, $(\ , 200)$

13. $2x - 3y = 6$: $(3, \ )$, $(\ , -2)$, $(12, \ )$

14. $3x + 5y = 0$: $(-5, \ )$, $(\ , -3)$, $(10, \ )$

15. $0 \cdot y + x = 5$: $(\ , -3)$, $(\ , 5)$, $(\ , 0)$

16. $0 \cdot x + y = -6$: $(3, \ )$, $(-1, \ )$, $(4, \ )$

Plot the points on a rectangular coordinate system. See Example 2.

17. $(1, 5)$ 18. $(4, 3)$ 19. $(-2, 1)$

20. $(-3, 5)$ 21. $\left(3, -\dfrac{1}{2}\right)$ 22. $\left(2, -\dfrac{1}{3}\right)$

23. $(-2, -4)$ 24. $(-3, -5)$ 25. $(0, 3)$
26. $(0, -2)$ 27. $(-3, 0)$ 28. $(5, 0)$
29. $(\pi, 1)$ 30. $(-2, \pi)$ 31. $(1.4, 4)$
32. $(-3, 0.4)$

Graph each equation. Plot at least five points for each equation. Use graph paper. See Examples 3 and 4. If you have a graphing calculator, use it to check your graphs.

33. $y = x + 1$

34. $y = x - 1$

35. $y = 2x + 1$

36. $y = 3x - 1$

37. $y = 3x - 2$

38. $y = 2x + 3$

39. $y = x$

40. $y = -x$

41. $y = 1 - x$

42. $y = 2 - x$

43. $y = -2x + 3$

44. $y = -3x + 2$

45. $y = -3$

46. $y = 2$

47. $x = 2$

48. $x = -4$

49. $2x + y = 5$

50. $3x + y = 5$

51. $x + 2y = 4$

52. $x - 2y = 6$

53. $x - 3y = 6$ **54.** $x + 4y = 5$ **73.** $y = -400x + 2000$ **74.** $y = 500x + 3$

55. $y = 0.36x + 0.4$ **56.** $y = 0.27x - 0.42$

Graph each equation using the x- and y-intercepts. See Example 6. Use a third point to check.

75. $3x + 2y = 6$ **76.** $2x + y = 6$

For each point, name the quadrant in which it lies or the axis on which it lies.

57. $(-3, 45)$ **58.** $(-33, 47)$ **59.** $(-3, 0)$

60. $(0, -9)$ **61.** $(-2.36, -5)$ **62.** $(89.6, 0)$

77. $x - 4y = 4$ **78.** $-2x + y = 4$

63. $(3.4, 8.8)$ **64.** $(4.1, 44)$ **65.** $\left(-\dfrac{1}{2}, 50\right)$

66. $\left(-6, -\dfrac{1}{2}\right)$ **67.** $(0, -99)$ **68.** $(\pi, 0)$

Graph each equation. Plot at least five points for each equation. Use graph paper. See Example 5. If you have a graphing calculator, use it to check your graphs.

69. $y = x + 1200$ **70.** $y = 2x - 3000$

79. $y = \dfrac{3}{4}x - 9$ **80.** $y = -\dfrac{1}{2}x + 5$

71. $y = 50x - 2000$ **72.** $y = -300x + 4500$

81. $\dfrac{1}{2}x + \dfrac{1}{4}y = 1$ **82.** $\dfrac{1}{3}x - \dfrac{1}{2}y = 3$

Solve each problem. See Example 7.

83. *Percentage of full benefit.* The age at which you retire affects your Social Security benefits. The accompanying graph gives the percentage of full benefit for each age from 62 through 70, based on current legislation and retirement after the year 2005 (Source: Social Security Administration). What percentage of full benefit does a person receive if that person retires at age 63? At what age will a retiree receive the full benefit? For what ages do you receive more than the full benefit?

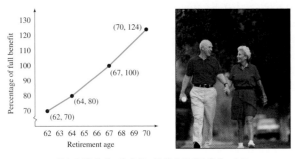

FIGURE FOR EXERCISE 83

84. *Heel motion.* When designing running shoes, Chris Edington studies the motion of a runner's foot. The following data gives the coordinates of the heel (in centimeters) at intervals of 0.05 millisecond during one cycle of level treadmill running at 3.8 meters per second (*Sagittal Plane Kinematics, Milliron and Cavanagh*):

$$(31.7, 5.7), (48.0, 5.7), (68.3, 5.8), (88.9, 6.9),$$
$$(107.2, 13.3), (119.4, 24.7), (127.2, 37.8),$$
$$(125.7, 52.0), (116.1, 60.2), (102.2, 59.5)$$
$$(88.7, 50.2), (73.9, 35.8), (52.6, 20.6),$$
$$(29.6, 10.7), (22.4, 5.9).$$

Graph these ordered pairs to see the heel motion.

85. *Medicaid spending.* The payment in billions by Medicaid (health care for the poor) can be modeled by the equation

$$P = 12.6n + 81.3,$$

where n is the number of years since 1990 (Health Care Financing Administration, www.hcfa.gov).
a) What amount was paid out by Medicaid in 1995?
b) In what year will the payment reach $220 billion?
c) Graph the equation for n ranging from 0 through 20.

86. *Women on the board.* The percentage of companies with at least one woman on the board is growing steadily (*Forbes,* February 10, 1997). The percentage can be approximated with the linear equation

$$p = 2n + 36,$$

where n is the number of years since 1980.
a) Find and interpret the p-intercept for the line.
b) Find and interpret the n-intercept.
c) Graph the line for n ranging from 0 through 20.
d) If this trend continues, then in what year would you expect to find nearly all companies having at least one woman on the board?

87. *Hazards of depth.* The table on page 169 shows the depth below sea level and atmospheric pressure (*Encyclopedia of Sports Science,* 1997). The equation

$$A = 0.03d + 1$$

expresses the atmospheric pressure in terms of the depth d.
a) Find the atmospheric pressure at the depth where nitrogen narcosis begins.
b) Find the maximum depth for intermediate divers.
c) Graph the equation for d ranging from 0 to 250 feet.

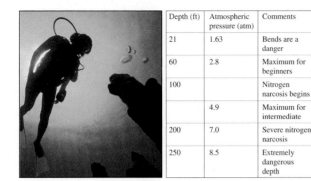

Depth (ft)	Atmospheric pressure (atm)	Comments
21	1.63	Bends are a danger
60	2.8	Maximum for beginners
100		Nitrogen narcosis begins
	4.9	Maximum for intermediate
200	7.0	Severe nitrogen narcosis
250	8.5	Extremely dangerous depth

FIGURE FOR EXERCISE 87

88. *Demand equation.* Helen's Health Foods usually sells 400 cans of ProPac Muscle Punch per week when the price is $5 per can. After experimenting with prices for some time, Helen has determined that the weekly demand can be found by using the equation

$$d = 600 - 40p,$$

where d is the number of cans and p is the price per can.
a) Will Helen sell more or less Muscle Punch if she raises her price from $5?
b) What happens to her sales every time she raises her price by $1?
c) Graph the equation.

d) What is the maximum price that she can charge and still sell at least one can?
89. *Advertising blitz.* Furniture City in Toronto had $24,000 to spend on advertising a year-end clearance sale. A 30-second radio ad costs $300, and a 30-second local television ad costs $400. To model this situation, the advertising manager wrote the equation $300x + 400y = 24,000$. What do x and y represent? Graph the equation. How many solutions are there to the equation, given that the number of ads of each type must be a whole number?

90. *Material allocation.* A tent maker had 4500 square yards of nylon tent material available. It takes 45 square yards of nylon to make an 8 × 10 tent and 50 square yards to make a 9 × 12 tent. To model this situation, the manager wrote the equation $45x + 50y = 4500$. What do x and y represent? Graph the equation. How many solutions are there to the equation, given that the number of tents of each type must be a whole number?

GRAPHING CALCULATOR EXERCISES

Graph each straight line on your graphing calculator using a viewing window that shows both intercepts. Answers may vary.

91. $2x + 3y = 1200$

92. $3x - 700y = 2100$

93. $200x - 300y = 6$

95. $y = 300x - 1$

94. $300x + 5y = 20$

96. $y = 300x - 6000$

4.2 SLOPE

In Section 4.1 you learned that the graph of a linear equation is a straight line. In this section, we will continue our study of lines in the coordinate plane.

In this section

- Slope Concepts
- Slope Using Coordinates
- Graphing a Line Given a Point and Its Slope
- Parallel Lines
- Perpendicular Lines
- Interpreting Slope

Slope Concepts

If a highway rises 6 feet in a horizontal run of 100 feet, then the grade is $\frac{6}{100}$ or 6%. See Fig. 4.11. The grade of a road is a measurement of the steepness of the road. It is the rate at which the road is going upward.

The steepness of a line is called the **slope** of the line and it is measured like the grade of a road. As you move from $(1, 1)$ to $(4, 3)$ in Fig. 4.12 on page 171 the x-coordinate increases by 3 and the y-coordinate increases by 2. The line rises 2 units in a horizontal run of 3 units. So the slope of the line is $\frac{2}{3}$. The slope is the rate at which the y-coordinate is increasing. It increases 2 units for every 3-unit increase in x or it increases $\frac{2}{3}$ of a unit for every 1-unit increase in x. In general, we have the following definition of slope.

Slope

$$\text{Slope} = \frac{\text{change in } y\text{-coordinate}}{\text{change in } x\text{-coordinate}}$$

FIGURE 4.11

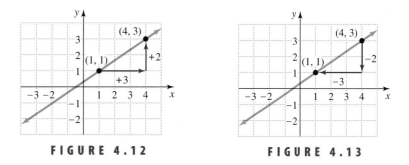

FIGURE 4.12   **FIGURE 4.13**

If we move from the point (4, 3) to the point (1, 1), there is a change of -2 in the y-coordinate and a change of -3 in the x-coordinate. See Fig. 4.13. In this case we get

$$\text{Slope} = \frac{-2}{-3} = \frac{2}{3}.$$

Note that going from (4, 3) to (1, 1) gives the same slope as going from (1, 1) to (4, 3).

We call the change in y-coordinate the **rise** and the change in x-coordinate the **run.** Moving up is a positive rise, and moving down is a negative rise. Moving to the right is a positive run, and moving to the left is a negative run. We usually use the letter m to stand for slope. So we have

$$m = \frac{\text{change in } y}{\text{change in } x} = \frac{\text{rise}}{\text{run}}.$$

E X A M P L E   1   **Finding the slope of a line**

Find the slopes of the given lines by going from point A to point B.

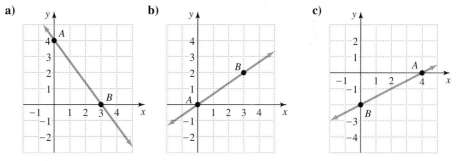

a)   b)   c)

Solution

a) The coordinates of point A are (0, 4), and the coordinates of point B are (3, 0). Going from A to B, the change in y is -4, and the change in x is $+3$. So

$$m = \frac{-4}{3} = -\frac{4}{3}.$$

b) Going from A to B, the rise is 2, and the run is 3. So

$$m = \frac{2}{3}.$$

c) Going from A to B, the rise is -2, and the run is -4. So

$$m = \frac{-2}{-4} = \frac{1}{2}.$$

CAUTION The change in y is always in the numerator, and the change in x is always in the denominator.

The ratio of rise to run is the ratio of the lengths of the two legs of any right triangle whose hypotenuse is on the line. As long as one leg is vertical and the other is horizontal, all such triangles for a certain line have the same shape. These triangles are similar triangles. The ratio of the length of the vertical side to the length of the horizontal side for any two such triangles is the same number. So we get the same value for the slope no matter which two points of the line are used to calculate it or in which order the points are used.

EXAMPLE 2

Finding slope

Find the slope of the line shown here using points A and B, points A and C, and points B and C.

Solution

Using A and B, we get

$$m = \frac{\text{rise}}{\text{run}} = \frac{1}{4}.$$

Using A and C, we get

$$m = \frac{\text{rise}}{\text{run}} = \frac{2}{8} = \frac{1}{4}.$$

Using B and C, we get

$$m = \frac{\text{rise}}{\text{run}} = \frac{1}{4}.$$

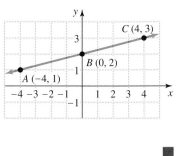

> **helpful hint**
>
> It is good to think of what the slope represents when x and y are measured quantities rather than just numbers. For example, if the change in y is 50 miles and the change in x is 2 hours, then the slope is 25 mph (or 25 miles per 1 hour). So the slope is the amount of change in y for a unit change in x.

Slope Using Coordinates

One way to obtain the rise and run is from a graph. The rise and run can also be found by using the coordinates of two points on the line as shown in Fig. 4.14.

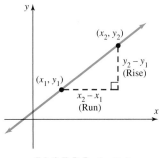

FIGURE 4.14

Coordinate Formula for Slope

The slope of the line containing the points (x_1, y_1) and (x_2, y_2) is given by

$$m = \frac{y_2 - y_1}{x_2 - x_1},$$

provided that $x_2 - x_1 \neq 0$.

EXAMPLE 3

Using coordinates to find slope

Find the slope of each of the following lines.

a) The line through $(0, 5)$ and $(6, 3)$

b) The line through $(-3, 4)$ and $(-5, -2)$

c) The line through $(-4, 2)$ and the origin

Solution

a) If $(x_1, y_1) = (0, 5)$ and $(x_2, y_2) = (6, 3)$ then

$$m = \frac{y_2 - y_1}{x_2 - x_1} = \frac{3 - 5}{6 - 0} = \frac{-2}{6} = -\frac{1}{3}.$$

/study \tip

Students who have difficulty with algebra often schedule it in a class that meets one day per week so they do not have to see it as often. However, many students do better in classes that meet more often for shorter time periods. So schedule your classes to maximize your chances of success.

If $(x_1, y_1) = (6, 3)$ and $(x_2, y_2) = (0, 5)$ then

$$m = \frac{y_2 - y_1}{x_2 - x_1} = \frac{5 - 3}{0 - 6} = \frac{2}{-6} = -\frac{1}{3}.$$

Note that it does not matter which point is called (x_1, y_1) and which is called (x_2, y_2). In either case the slope is $-\frac{1}{3}$.

b) Let $(x_1, y_1) = (-3, 4)$ and $(x_2, y_2) = (-5, -2)$:

$$m = \frac{y_2 - y_1}{x_2 - x_1} = \frac{-2 - 4}{-5 - (-3)}$$

$$= \frac{-6}{-2} = 3$$

c) Let $(x_1, y_1) = (0, 0)$ and $(x_2, y_2) = (-4, 2)$:

$$m = \frac{2 - 0}{-4 - 0} = \frac{2}{-4} = -\frac{1}{2}$$

CAUTION It does not matter which point is called (x_1, y_1) and which is called (x_2, y_2), but if you divide $y_2 - y_1$ by $x_1 - x_2$, the slope will have the wrong sign.

Note that slope is not defined if $x_2 - x_1 = 0$. So slope is not defined if the x-coordinates of the two points are equal. The x-coordinates for two points are equal only for points on a vertical line. So *slope is undefined for vertical lines.*

If $y_2 - y_1 = 0$, then the points have equal y-coordinates and lie on a horizontal line. *The slope for any horizontal line is zero.*

EXAMPLE 4

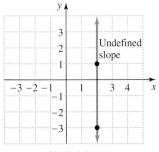

Vertical line

FIGURE 4.15

Slope for vertical and horizontal lines

Find the slope of the line through each pair of points.

a) $(2, 1)$ and $(2, -3)$

b) $(-2, 2)$ and $(4, 2)$

Solution

a) The points $(2, 1)$ and $(2, -3)$ are on the vertical line shown in Fig. 4.15. Since slope is undefined for vertical lines, this line does not have a slope. Using the slope formula we get

$$m = \frac{-3 - 1}{2 - 2} = \frac{-4}{0}.$$

Since division by zero is undefined, we can again conclude that slope is undefined for the vertical line through the given points.

b) The points $(-2, 2)$ and $(4, 2)$ are on the horizontal line shown in Fig. 4.16. Using the slope formula we get

$$m = \frac{2 - 2}{-2 - 4} = \frac{0}{-6} = 0.$$

So the slope of the horizontal line through these points is 0.

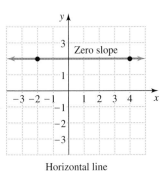

Horizontal line

FIGURE 4.16

Note that for a line with *positive slope,* the y-values increase as the x-values increase. For a line with *negative slope,* the y-values decrease as the x-values increase. See Fig. 4.17 on page 174.

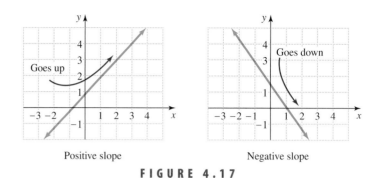

Positive slope — Negative slope

FIGURE 4.17

Graphing a Line Given a Point and Its Slope

We can find the slope of a line by examining its graph. We can also draw the graph of a line if we know its slope and a point on the line.

E X A M P L E 5

Graphing a line given a point and its slope
Graph each line.
a) The line through $(2, 1)$ with slope $\frac{3}{4}$
b) The line through $(-2, 4)$ with slope -3

Solution

a) First locate the point $(2, 1)$. Because the slope is $\frac{3}{4}$, we can find another point on the line by going up three units and to the right four units to get the point $(6, 4)$, as shown in Fig. 4.18. Now draw the line through $(2, 1)$ and $(6, 4)$.

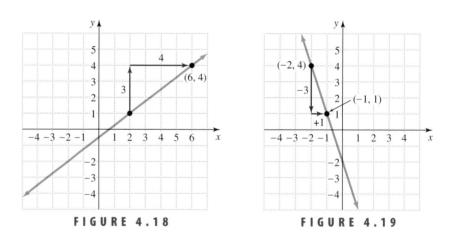

FIGURE 4.18 — **FIGURE 4.19**

b) First locate the point $(-2, 4)$. Because the slope is -3, or $\frac{-3}{1}$, we can locate another point on the line by starting at $(-2, 4)$ and moving down three units and then one unit to the right to get the point $(-1, 1)$. Now draw a line through $(-2, 4)$ and $(-1, 1)$, as shown in Fig. 4.19. ∎

Parallel Lines

Every nonvertical line has a unique slope, but there are infinitely many lines with a given slope. All lines that have a given slope are parallel.

Parallel Lines

Nonvertical lines are parallel if and only if they have equal slopes. Any two vertical lines are parallel to each other.

E X A M P L E   6   **Graphing parallel lines**

Draw a line through the point $(-2, 1)$ with slope $\frac{1}{2}$ and a line through $(3, 0)$ with slope $\frac{1}{2}$.

Solution

Because slope is the ratio of rise to run, a slope of $\frac{1}{2}$ means that we can locate a second point of the line by starting at $(-2, 1)$ and going up one unit and to the right two units. For the line through $(3, 0)$ we start at $(3, 0)$ and go up one unit and to the right two units. See Fig. 4.20.

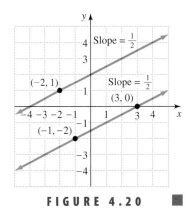

F I G U R E   4 . 2 0

Perpendicular Lines

Slope can also be used to determine whether lines are perpendicular. If the slope of one line is the opposite of the reciprocal of the slope of another line, then the lines are perpendicular. For example, lines with slopes $\frac{3}{4}$ and $-\frac{4}{3}$ are perpendicular.

helpful hint

The relationship between the slopes of perpendicular lines can also be remembered as

$$m_1 \cdot m_2 = -1.$$

For example, lines with slopes -3 and $\frac{1}{3}$ are perpendicular because $-3 \cdot \frac{1}{3} = -1$.

Perpendicular Lines

Two lines with slopes m_1 and m_2 are perpendicular if and only if

$$m_1 = -\frac{1}{m_2}.$$

Any vertical line is perpendicular to any horizontal line.

E X A M P L E   7   **Graphing perpendicular lines**

Draw two lines through the point $(-1, 2)$, one with slope $-\frac{1}{3}$ and the other with slope 3.

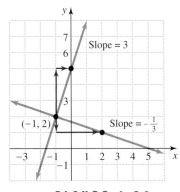

F I G U R E   4 . 2 1

Solution

Because slope is the ratio of rise to run, a slope of $-\frac{1}{3}$ means that we can locate a second point on the line by starting at $(-1, 2)$ and going down one unit and to the right three units. For the line with slope 3, we start at $(-1, 2)$ and go up three units and to the right one unit. See Fig. 4.21.

Interpreting Slope

Slope of a line is the ratio of the rise and the run. If the rise is measured in dollars and the run in days, then the slope is measured in dollars per day or dollars/day.

The slope of a line is the rate at which the dependent variable is increasing or decreasing.

E X A M P L E   8   **Interpreting slope**

A car goes from 60 mph to 0 mph in 120 feet after applying the brakes.

a) Find and interpret the slope of the line shown here.

b) What is the velocity at a distance of 80 feet?

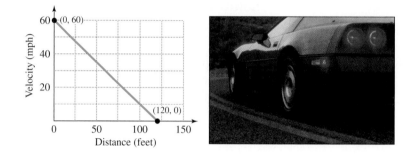

Solution

a) Find the slope of the line through (0, 60) and (120, 0):

$$m = \frac{60 - 0}{0 - 120} = -0.5$$

Because the vertical axis is miles per hour and the horizontal axis is feet, the slope is -0.5 mph/ft, which means the car is losing 0.5 mph of velocity for every foot it travels after the brakes are applied.

b) If the velocity is decreasing 0.5 mph for every foot the car travels, then in 80 feet the velocity goes down 0.5(80) or 40 mph. So the velocity at 80 feet is $60 - 40$ or 20 mph.   ■

WARM-UPS

True or false? Explain your answer.

1. Slope is a measurement of the steepness of a line.

2. Slope is rise divided by run.

3. Every line in the coordinate plane has a slope.

4. The line through the point (1, 1) and the origin has slope 1.

5. Slope can never be negative.

6. A line with slope 2 is perpendicular to any line with slope -2.

7. The slope of the line through (0, 3) and (4, 0) is $\frac{3}{4}$.

8. Two different lines cannot have the same slope.

9. The line through (1, 3) and $(-5, 3)$ has zero slope.

10. Slope can have units such as feet per second.

4.2   **EXERCISES**

Reading and Writing   *After reading this section, write out the answers to these questions. Use complete sentences.*

1. What is the slope of a line?

2. What is the difference between rise and run?

3. For which lines is slope undefined?

4. Which lines have zero slope?

5. What is the difference between lines with positive slope and lines with negative slope?

6. What is the relationship between the slopes of perpendicular lines?

In Exercises 7–18, find the slope of each line. See Examples 1 and 2.

7.
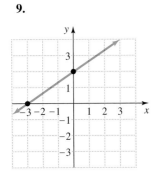

8.
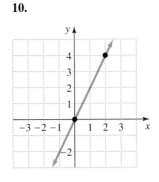

9.

10.
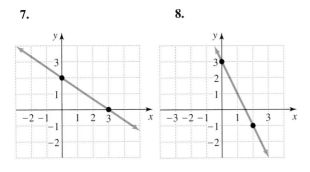

11.
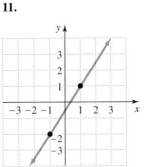

12.
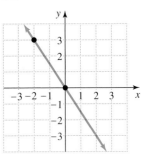

13.
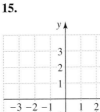

14.
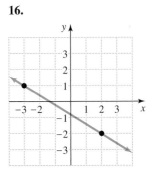

15.

16.

17.
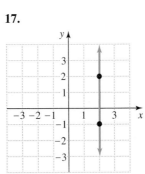

18.
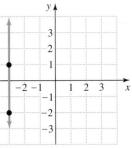

Find the slope of the line that goes through each pair of points. See Examples 3 and 4.

19. (1, 2), (3, 6)

20. (2, 5), (6, 10)

21. (2, 4), (5, −1)

22. (3, 1), (6, −2)

23. (−2, 4), (5, 9)

24. (−1, 3), (3, 5)

25. (−2, −3), (−5, 1)

26. (−6, −3), (−1, 1)

27. (−3, 4), (3, −2)

28. (−1, 3), (5, −2)

29. $\left(\frac{1}{2}, 2\right), \left(-1, \frac{1}{2}\right)$

30. $\left(\frac{1}{3}, 2\right), \left(-\frac{1}{3}, 1\right)$

31. (2, 3), (2, −9)

32. (−3, 6), (8, 6)

33. (−2, −5), (9, −5)

34. (4, −9), (4, 6)

35. (0.3, 0.9), (−0.1, −0.3)

36. (−0.1, 0.2), (0.5, 0.8)

Graph the line with the given point and slope. See Example 5.

37. The line through (1, 1) with slope $\frac{2}{3}$

38. The line through (2, 3) with slope $\frac{1}{2}$

39. The line through (−2, 3) with slope −2

40. The line through (−2, 5) with slope −1

41. The line through (0, 0) with slope $-\frac{2}{5}$

42. The line through (−1, 4) with slope $-\frac{2}{3}$

Solve each problem. See Examples 6 and 7.

43. Draw line l_1 through (1, −2) with slope $\frac{1}{2}$ and line l_2 through (−1, 1) with slope $\frac{1}{2}$.

44. Draw line l_1 through (0, 3) with slope 1 and line l_2 through (0, 0) with slope 1.

45. Draw l_1 through $(1, 2)$ with slope $\frac{1}{2}$, and draw l_2 through $(1, 2)$ with slope -2.

46. Draw l_1 through $(-2, 1)$ with slope $\frac{2}{3}$, and draw l_2 through $(-2, 1)$ with slope $-\frac{3}{2}$.

47. Draw any line l_1 with slope $\frac{3}{4}$. What is the slope of any line perpendicular to l_1? Draw any line l_2 perpendicular to l_1.

48. Draw any line l_1 with slope -1. What is the slope of any line perpendicular to l_1? Draw any line l_2 perpendicular to l_1.

49. Draw l_1 through $(-2, -3)$ and $(4, 0)$. What is the slope of any line parallel to l_1? Draw l_2 through $(1, 2)$ so that it is parallel to l_1.

50. Draw l_1 through $(-4, 0)$ and $(0, 6)$. What is the slope of any line parallel to l_1? Draw l_2 through the origin and parallel to l_1.

51. Draw l_1 through $(-2, 4)$ and $(3, -1)$. What is the slope of any line perpendicular to l_1? Draw l_2 through $(1, 3)$ so that it is perpendicular to l_1.

52. Draw l_1 through $(0, -3)$ and $(3, 0)$. What is the slope of any line perpendicular to l_1? Draw l_2 through the origin so that it is perpendicular to l_1.

Solve each problem. See Example 8.

53. Super cost. The average cost of a 30-second ad during the 1995 super bowl was $1 million, and in 1998 it was $1.3 million (*Detroit Free Press,* January 6, 1998, www.freep.com).

a) Find the slope of the line through (95, 1,000,000) and (98, 1,300,000) and interpret your result.

b) Use the accompanying graph to estimate the average cost of an ad in 1997.

c) What do you think the average cost will be in 2005?

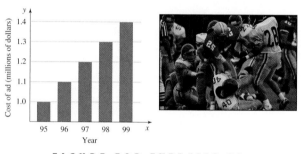

FIGURE FOR EXERCISE 53

54. Retirement pay. The annual Social Security benefit of a retiree depends on the age at the time of retirement. The accompanying graph gives the annual benefit for persons retiring at ages 62 through 70 in the year 2005 or later (Source: Social Security Administration). What is the annual benefit for a person who retires at age 64? At what retirement age does a person receive an annual benefit of $11,600? Find the slope of each line segment on the graph, and interpret your results. Why do people who postpone retirement until 70 years of age get the highest benefit?

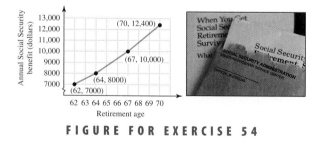

FIGURE FOR EXERCISE 54

55. Increasing training. The accompanying graph shows the percentage of U.S. workers receiving training by their employers. The percentage went from 5% in 1981 to 16% in 1995. Find the slope of this line. Interpret your result.

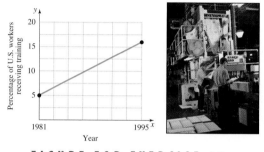

FIGURE FOR EXERCISE 55

56. Saving for retirement. Financial advisors at Fidelity Investments, Boston, use the accompanying table as a measure of whether a client is on the road to a comfortable retirement.

Age (a)	Years of Salary saved (y)	
35	0.5	
40	1.0	
45	1.5	
50	2.0	

a) Graph these points and draw a line through them.

b) What is the slope of the line?

c) By what percentage of your salary should you be increasing your savings every year?

4.3 # EQUATIONS OF LINES IN SLOPE-INTERCEPT AND STANDARD FORM

In Section 4.1 you learned that the graph of all solutions to a linear equation in two variables is a straight line. In this section we start with a line or a description of a line and write an equation for the line. The equation of a line in any form is called a **linear equation in two variables.**

Slope-Intercept Form

Consider the line through $(0, 1)$ with slope $\frac{2}{3}$ shown in Fig. 4.22. If we use the points (x, y) and $(0, 1)$ in the slope formula, we get an equation that is satisfied by every point on the line:

$$\frac{y_2 - y_1}{x_2 - x_1} = m \quad \text{Slope formula}$$

$$\frac{y - 1}{x - 0} = \frac{2}{3} \quad \text{Let } (x_1, y_1) = (0, 1) \text{ and } (x_2, y_2) = (x, y).$$

$$\frac{y - 1}{x} = \frac{2}{3}$$

Now solve the equation for y:

$$x \cdot \frac{y - 1}{x} = \frac{2}{3} \cdot x \quad \text{Multiply each side by } x.$$

$$y - 1 = \frac{2}{3}x$$

$$y = \frac{2}{3}x + 1 \quad \text{Add 1 to each side.}$$

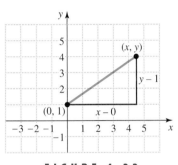

FIGURE 4.22

Because $(0, 1)$ is on the y-axis, it is called the **y-intercept** of the line. Note how the slope $\frac{2}{3}$ and the y-coordinate of the y-intercept $(0, 1)$ appear in $y = \frac{2}{3}x + 1$. For this reason it is called the **slope-intercept form** of the equation of the line.

> **Slope-Intercept Form**
>
> The equation of the line with y-intercept $(0, b)$ and slope m is
> $$y = mx + b.$$

EXAMPLE 1   **Using slope-intercept form**

Write the equation of each line in slope-intercept form.

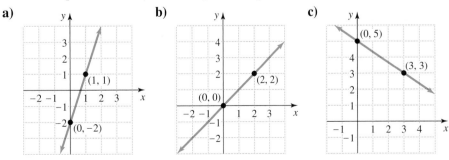

a)   b)   c)

close-up

Since a graphing calculator screen has a finite number of pixels, a graphing calculator plots only a finite number of ordered pairs that satisfy an equation. A graph is supposed to be a picture of all of the ordered pairs that satisfy an equation. Because a calculator plots many ordered pairs accurately and quickly, it can be a great help in drawing the graph of an equation.

Solution

a) The y-intercept is $(0, -2)$, and the slope is 3. Use the form $y = mx + b$ with $b = -2$ and $m = 3$. The equation in slope-intercept form is

$$y = 3x - 2.$$

b) The y-intercept is $(0, 0)$, and the slope is 1. So the equation is

$$y = x.$$

c) The y-intercept is $(0, 5)$, and the slope is $-\frac{2}{3}$. So the equation is

$$y = -\frac{2}{3}x + 5.$$

The equation of a line may take many different forms. The easiest way to find the slope and y-intercept for a line is to rewrite the equation in slope-intercept form.

EXAMPLE 2

Finding slope and y-intercept

Determine the slope and y-intercept of the line $3x - 2y = 6$.

Solution

Solve for y to get slope-intercept form:

$$3x - 2y = 6$$
$$-2y = -3x + 6$$
$$y = \frac{3}{2}x - 3$$

The slope is $\frac{3}{2}$, and the y-intercept is $(0, -3)$.

Standard Form

The graph of the equation $x = 3$ is a vertical line. Because slope is not defined for vertical lines, this line does not have an equation in slope-intercept form. Only non-vertical lines have equations in slope-intercept form. However, there is a form that includes all lines. It is called **standard form.**

helpful hint

In geometry we learn that two points determine a line. However, if you locate two points that are close together and draw a line through them, your line can have a lot of error in it at locations far from the two chosen points. Locating five points on the graph will improve your accuracy. If you draw a straight line with two points, they should be chosen as far as possible from each other.

> **Standard Form**
>
> Every line has an equation in the form
>
> $$Ax + By = C$$
>
> where A, B, and C are real numbers with A and B not both zero.

To write the equation $x = 3$ in this form, let $A = 1$, $B = 0$, and $C = 3$. We get

$$1 \cdot x + 0 \cdot y = 3,$$

which is equivalent to

$$x = 3.$$

In Example 2 we converted an equation in standard form to slope-intercept form. Any linear equation in standard form with $B \neq 0$ can be written in slope-intercept form by solving for y. In the next example we convert an equation in slope-intercept form to standard form.

EXAMPLE 3

Converting to standard form

Write the equation of the line $y = \frac{2}{5}x + 3$ in standard form using only integers.

Solution

To get standard form, first subtract $\frac{2}{5}x$ from each side:

$$y = \frac{2}{5}x + 3$$

$$-\frac{2}{5}x + y = 3$$

$$-5\left(-\frac{2}{5}x + y\right) = -5 \cdot 3 \quad \text{Multiply each side by } -5 \text{ to eliminate the fraction}$$
$$\text{and get positive } 2x.$$

$$2x - 5y = -15$$ ■

The answer $2x - 5y = -15$ in Example 3 is not the only answer using only integers. Equations such as $-2x + 5y = 15$ and $4x - 10y = -30$ are equivalent equations in standard form. We prefer to write $2x - 5y = -15$ because the greatest common factor of 2, 5, and 15 is 1 and the coefficient of x is positive.

Using Slope-Intercept Form for Graphing

One way to graph a linear equation is to find several points that satisfy the equation and then draw a straight line through them. We can also graph a linear equation by using the y-intercept and the slope.

> ### Strategy for Graphing a Line Using Slope and *y*-Intercept
>
> 1. Write the equation in slope-intercept form if necessary.
> 2. Plot the y-intercept.
> 3. Starting from the y-intercept, use the rise and run to locate a second point.
> 4. Draw a line through the two points.

EXAMPLE 4

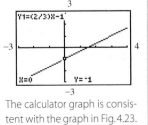
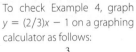

calculator

close-up

To check Example 4, graph $y = (2/3)x - 1$ on a graphing calculator as follows:

The calculator graph is consistent with the graph in Fig. 4.23.

Graphing a line using *y*-intercept and slope

Graph the line $2x - 3y = 3$.

Solution

First write the equation in slope-intercept form:

$$2x - 3y = 3$$

$$-3y = -2x + 3 \quad \text{Subtract } 2x \text{ from each side.}$$

$$y = \frac{2}{3}x - 1 \quad \text{Divide each side by } -3.$$

The slope is $\frac{2}{3}$, and the y-intercept is $(0, -1)$. A slope of $\frac{2}{3}$ means a rise of 2 and a run of 3. Start at $(0, -1)$ and go up two units and to the right three units to locate a second point on the line. Now draw a line through the two points. See Fig. 4.23 for the graph of $2x - 3y = 3$.

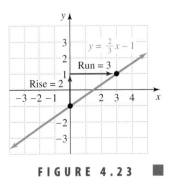

FIGURE 4.23 ■

CAUTION When using the slope to find a second point on the line, be sure to start at the *y*-intercept, not at the origin.

EXAMPLE 5

Graphing a line using *y*-intercept and slope

Graph the line $y = -3x + 4$.

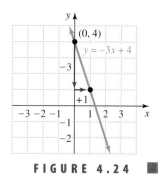

Solution

The slope is -3, and the *y*-intercept is $(0, 4)$. Because $-3 = \frac{-3}{1}$, we use a rise of -3 and a run of 1. To locate a second point on the line, start at $(0, 4)$ and go down three units and to the right one unit. Draw a line through the two points. See Fig. 4.24.

FIGURE 4.24 ■

Writing the Equation for a Line

In Example 1 we wrote the equation of a line by finding its slope and *y*-intercept from a graph. In the next example we write the equation of a line from a description of the line.

EXAMPLE 6

Writing an equation

Write the equation in slope-intercept form for the line through $(0, 4)$ that is perpendicular to the line $2x - 4y = 1$.

Solution

First find the slope of $2x - 4y = 1$:

$$2x - 4y = 1$$
$$-4y = -2x + 1$$
$$y = \frac{1}{2}x - \frac{1}{4} \qquad \text{The slope of this line is } \tfrac{1}{2}.$$

The slope of the line that we are interested in is the opposite of the reciprocal of $\frac{1}{2}$. So the line has slope -2 and *y*-intercept $(0, 4)$. Its equation is $y = -2x + 4$. ■

---

calculator close-up

If you use the same minimum and maximum window values for *x* and *y*, then the length of one unit on the *x*-axis is larger than on the *y*-axis because the screen is longer in the *x*-direction. In this case, perpendicular lines will not look perpendicular. The viewing window chosen here for the lines in Example 6 makes them look perpendicular.

Any viewing window proportional to this one will also produce approximately the same unit length on each axis. Some

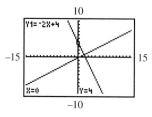

calculators have a square feature that automatically makes the unit length the same on both axes.

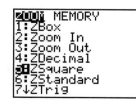

Applications

The slope-intercept and standard forms are both important in applications.

E X A M P L E   7   **Changing forms**

A landscaper has a total of $800 to spend on bushes at $20 each and trees at $50 each. So if x is the number of bushes and y is the number of trees he can buy, then $20x + 50y = 800$. Write this equation in slope-intercept form. Find and interpret the y-intercept and the slope.

Solution

Write in slope-intercept form:

$$20x + 50y = 800$$
$$50y = -20x + 800$$
$$y = -\frac{2}{5}x + 16$$

The slope is $-\frac{2}{5}$ and the intercept is $(0, 16)$. So he can get 16 trees if he buys no bushes and he loses $\frac{2}{5}$ of a tree for each additional bush that he purchases.   ■

WARM-UPS

True or false? Explain your answer.

1. There is only one line with y-intercept $(0, 3)$ and slope $-\frac{4}{3}$.
2. The equation of the line through $(1, 2)$ with slope 3 is $y = 3x + 2$.
3. The vertical line $x = -2$ has no y-intercept.
4. The equation $x = 5$ has a graph that is a vertical line.
5. The line $y = x - 3$ is perpendicular to the line $y = 5 - x$.
6. The line $y = 2x - 3$ is parallel to the line $y = 4x - 3$.
7. The line $2y = 3x - 8$ has a slope of 3.
8. Every straight line in the coordinate plane has an equation in standard form.
9. The line $x = 2$ is perpendicular to the line $y = 5$.
10. The line $y = x$ has no y-intercept.

4.3   EXERCISES

Reading and Writing   *After reading this section, write out the answers to these questions. Use complete sentences.*

1. What is the slope-intercept form for the equation of a line?

2. How can you determine the slope and y-intercept from the slope-intercept form.

3. What is the standard form for the equation of a line?

4. How can you graph a line when the equation is in slope-intercept form?

5. What form is used in this section to write an equation of a line from a description of the line?

6. What makes lines look perpendicular on a graph?

Write an equation for each line. Use slope-intercept form if possible. See Example 1.

7.

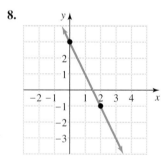

8.

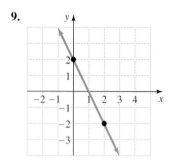

9.

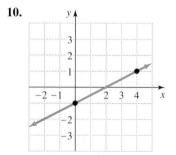

10.

11.

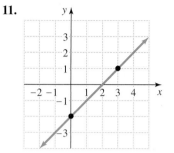

12.

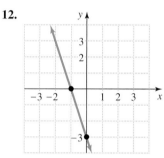

13.

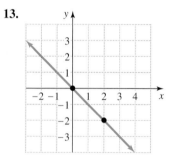

14.

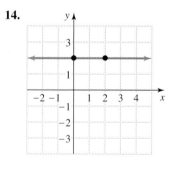

15.

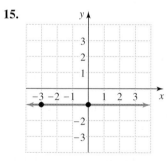

16.

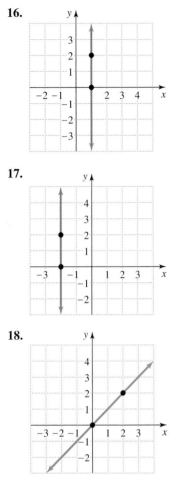

17.

18.

Find the slope and y-intercept for each line that has a slope and y-intercept. See Example 2.

19. $y = 3x - 9$ **20.** $y = -5x + 4$ **21.** $y = 4$

22. $y = -5$ **23.** $y = -3x$ **24.** $y = 2x$

25. $x + y = 5$ **26.** $x - y = 4$ **27.** $x - 2y = 4$

28. $x + 2y = 3$ **29.** $2x - 5y = 10$ **30.** $2x + 3y = 9$

31. $2x - y + 3 = 0$ **32.** $3x - 4y - 8 = 0$

33. $x = -3$

34. $\dfrac{2}{3}x = 4$

Write each equation in standard form using only integers. See Example 3.

35. $y = -x + 2$ **36.** $y = 3x - 5$

37. $y = \dfrac{1}{2}x + 3$ **38.** $y = \dfrac{2}{3}x - 4$

39. $y = \dfrac{3}{2}x - \dfrac{1}{3}$ **40.** $y = \dfrac{4}{5}x + \dfrac{2}{3}$

41. $y = -\dfrac{3}{5}x + \dfrac{7}{10}$ **42.** $y = -\dfrac{2}{3}x - \dfrac{5}{6}$

43. $\dfrac{3}{5}x + 6 = 0$ **44.** $\dfrac{1}{2}x - 9 = 0$

45. $\dfrac{3}{4}y = \dfrac{5}{2}$ **46.** $\dfrac{2}{3}y = \dfrac{1}{9}$

47. $\dfrac{x}{2} = \dfrac{3y}{5}$ **48.** $\dfrac{x}{8} = -\dfrac{4y}{5}$

49. $y = 0.02x + 0.5$ **50.** $0.2x = 0.03y - 0.1$

Draw the graph of each line using its y-intercept and its slope. See Examples 4 and 5.

51. $y = 2x - 1$ **52.** $y = 3x - 2$

53. $y = -3x + 5$ **54.** $y = -4x + 1$

55. $y = \dfrac{3}{4}x - 2$ **56.** $y = \dfrac{3}{2}x - 4$

57. $2y + x = 0$

58. $2x + y = 0$

59. $3x - 2y = 10$

60. $4x + 3y = 9$

61. $y - 2 = 0$

62. $y + 5 = 0$

Write an equation in slope-intercept form, if possible, for each line. See Example 6. In each case, make a sketch.

63. The line through $(0, 6)$ that is perpendicular to the line $y = 3x - 5$

64. The line through $(0, -1)$ that is perpendicular to the line $y = x$

65. The line with y-intercept $(0, 3)$ that is parallel to the line $2x + y = 5$

66. The line through the origin that is parallel to the line $2x - 5y = 8$

67. The line through $(2, 3)$ that runs parallel to the x-axis

68. The line through $(-3, 5)$ that runs parallel to the y-axis

69. The line through $(0, 4)$ and $(5, 0)$

70. The line through $(0, -3)$ and $(4, 0)$

Solve each problem. See Example 7.

71. *Marginal cost.* A manufacturer plans to spend $150,000 on research and development for a new lawn mower and then $200 to manufacture each mower. The formula $C = 200n + 150,000$ gives the cost in dollars of n mowers. What is the cost of 5000 mowers? What is the cost of 5001 mowers? By how much did the one extra lawn mower increase the cost? (The increase in cost is called the *marginal cost* of the 5001st lawn mower.)

72. *Marginal revenue.* A defense attorney charges her client $4000 plus $120 per hour. The formula $R = 120n + 4000$ gives her revenue in dollars for

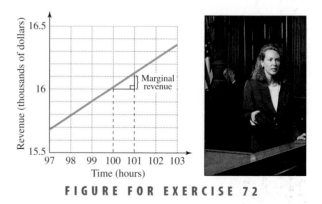

FIGURE FOR EXERCISE 72

n hours of work. What is her revenue for 100 hours of work? What is her revenue for 101 hours of work? By how much did the one extra hour of work increase the revenue? (The increase in revenue is called the *marginal revenue* for the 101st hour.)

73. *In-house training.* The accompanying graph shows the percentage of U.S. workers receiving training by their employers. The percentage went from 5% in year 0 (1981) to 16% in year 14 (1995).

a) Find the slope of this line.

b) Write the equation of the line in slope-intercept form.

c) Use your equation to predict the percentage that will be receiving training in the year 2000.

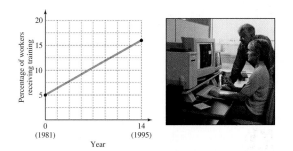

FIGURE FOR EXERCISE 73

74. *Women and marriage.* The percentage of women in the 20 to 24 age group who have never married went from 64% in year 0 (1970) to 33% in year 26 (1996) (Census Bureau, www.census.gov).

a) Find the equation of the line through the two points (0, 0.64) and (26, 0.33) in slope-intercept form.

b) Use your equation to predict what the percentage will be in the year 2000.

75. *Pansies and snapdragons.* A nursery manager plans to spend $100 on 6-packs of pansies at 50 cents per pack and snapdragons at 25 cents per pack. The equation $0.50x + 0.25y = 100$ can be used to model this situation.

a) What do x and y represent?

b) Graph the equation.

c) Write the equation in slope-intercept form.

d) What is the slope of the line?

e) What does the slope tell you?

76. *Pens and pencils.* A bookstore manager plans to spend $60 on pens at 30 cents each and pencils at 10 cents each. The equation $0.10x + 0.30y = 60$ can be used to model this situation.

a) What do x and y represent?

b) Graph the equation.

c) Write the equation in slope-intercept form.

d) What is the slope of the line?

e) What does the slope tell you?

GRAPHING CALCULATOR EXERCISES

Graph each pair of straight lines on your graphing calculator using a viewing window that makes the lines look perpendicular. Answers may vary.

77. $y = 12x - 100, \ y = -\frac{1}{12}x + 50$

78. $2x - 3y = 300, \ 3x + 2y = -60$

4.4 THE POINT-SLOPE FORM

In Section 4.3 we wrote the equation of a line given its slope and *y*-intercept. In this section you will learn to write the equation of a line given the slope and *any* other point on the line.

Point-Slope Form

Consider a line through the point (4, 1) with slope $\frac{2}{3}$ as shown in Fig. 4.25. Because the slope can be found by using any two points on the line, we use (4, 1) and an arbitrary point (*x*, *y*) in the formula for slope:

$$\frac{y_2 - y_1}{x_2 - x_1} = m \qquad \text{Slope formula}$$

$$\frac{y - 1}{x - 4} = \frac{2}{3} \qquad \text{Let } m = \tfrac{2}{3}, (x_1, y_1) = (4, 1), \text{ and } (x_2, y_2) = (x, y).$$

$$y - 1 = \frac{2}{3}(x - 4) \qquad \text{Multiply each side by } x - 4.$$

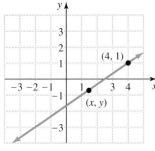

FIGURE 4.25

Note how the coordinates of the point (4, 1) and the slope $\frac{2}{3}$ appear in the above equation. We can use the same procedure to get the equation of any line given one point on the line and the slope. The resulting equation is called the **point-slope form** of the equation of the line.

helpful hint

If a point (*x*, *y*) is on a line with slope *m* through (x_1, y_1), then

$$\frac{y - y_1}{x - x_1} = m.$$

Multiplying each side of this equation by $x - x_1$ gives us the point-slope form.

Point-Slope Form

The equation of the line through the point (x_1, y_1) with slope *m* is

$$y - y_1 = m(x - x_1).$$

EXAMPLE 1

Writing an equation given a point and a slope

Find the equation of the line through (−2, 3) with slope $\frac{1}{2}$, and write it in slope-intercept form.

Solution

Because we know a point and the slope, we can use the point-slope form:

$$y - y_1 = m(x - x_1) \qquad \text{Point-slope form}$$

$$y - 3 = \frac{1}{2}[x - (-2)] \qquad \text{Substitute } m = \tfrac{1}{2} \text{ and } (x_1, y_1) = (-2, 3).$$

$$y - 3 = \frac{1}{2}(x + 2) \qquad \text{Simplify.}$$

$$y - 3 = \frac{1}{2}x + 1 \qquad \text{Distributive property}$$

$$y = \frac{1}{2}x + 4 \qquad \text{Slope-intercept form}$$

Alternate Solution

Replace m by $\frac{1}{2}$, x by -2, and y by 3 in the slope-intercept form:

$$y = mx + b \qquad \text{Slope-intercept form}$$

$$3 = \frac{1}{2}(-2) + b \qquad \text{Substitute } m = \text{ and } (x, y) = (-2, 3).$$

$$3 = -1 + b \qquad \text{Simplify.}$$

$$4 = b$$

Since $b = 4$, we can write $y = \frac{1}{2}x + 4$. ◼

C A U T I O N   The point-slope form can be used to find the equation of a line for *any* given point and slope. However, if the given point is the y-intercept, then it is simpler to use the slope-intercept form.

E X A M P L E   2

Writing an equation given two points

Find the equation of the line that contains the points $(-3, -2)$ and $(4, -1)$, and write it in standard form.

Solution

First find the slope using the two given points:

$$m = \frac{-2 - (-1)}{-3 - 4} = \frac{-1}{-7} = \frac{1}{7}$$

Now use one of the points, say $(-3, -2)$, and slope $\frac{1}{7}$ in the point-slope form:

$$y - y_1 = m(x - x_1) \qquad \text{Point-slope form}$$

$$y - (-2) = \frac{1}{7}[x - (-3)] \qquad \text{Substitute.}$$

$$y + 2 = \frac{1}{7}(x + 3) \qquad \text{Simplify.}$$

$$7(y + 2) = 7 \cdot \frac{1}{7}(x + 3) \qquad \text{Multiply each side by 7.}$$

$$7y + 14 = x + 3$$

$$7y = x - 11 \qquad \text{Subtract 14 from each side.}$$

$$-x + 7y = -11 \qquad \text{Subtract } x \text{ from each side.}$$

$$x - 7y = 11 \qquad \text{Multiply each side by } -1.$$

The equation in standard form is $x - 7y = 11$. Using the other given point, $(4, -1)$, would give the same final equation in standard form. Try it. ◼

calculator

close-up

Graph $y = (x + 3)/7 - 2$ to see that the line goes through $(-3, -2)$ and $(4, -1)$.

Note that the form of the equation does not matter on the calculator as long as it is solved for y.

Parallel Lines

In Section 4.2 you learned that parallel lines have the same slope. For example, the lines $y = 6x - 4$ and $y = 6x + 7$ are parallel because each has slope 6. In the next example we write the equation of a line that is parallel to a given line and contains a given point.

E X A M P L E   3

Writing an equation given a point and a parallel line

Write the equation of the line that is parallel to the line $3x + y = 9$ and contains the point $(2, -1)$. Give the answer in slope-intercept form.

Solution

We want the equation of the line through $(2, -1)$ that is parallel to $3x + y = 9$, as shown in Fig. 4.26. First write $3x + y = 9$ in slope-intercept form to determine its slope:

$$3x + y = 9$$
$$y = -3x + 9$$

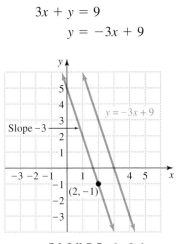

F I G U R E   4.26

The slope of $3x + y = 9$ and any line parallel to it is -3. So we want the equation of the line that has slope -3 and contains $(2, -1)$. Use the point-slope form:

$$y - y_1 = m(x - x_1) \qquad \text{Point-slope form}$$
$$y - (-1) = -3(x - 2) \qquad \text{Substitute.}$$
$$y + 1 = -3x + 6 \qquad \text{Simplify.}$$
$$y = -3x + 5 \qquad \text{Slope-intercept form}$$

The line $y = -3x + 5$ has slope -3 and contains the point $(2, -1)$. Check that $(2, -1)$ satisfies $y = -3x + 5$. ■

Perpendicular Lines

In Section 4.2 you learned that lines with slopes m and $-\frac{1}{m}$ (for $m \neq 0$) are perpendicular to each other. For example, the lines

$$y = -2x + 7 \qquad \text{and} \qquad y = \frac{1}{2}x - 8$$

are perpendicular to each other. In the next example we will write the equation of a line that is perpendicular to a given line and contains a given point.

E X A M P L E   4

Writing an equation given a point and a perpendicular line

Write the equation of the line that is perpendicular to $3x + 2y = 8$ and contains the point $(1, -3)$. Write the answer in slope-intercept form.

---

calculator

close-up

Graph $y = -3x + 9$ and $y = -3x + 5$ as follows:

Because the lines look parallel and $y = -3x + 5$ goes through $(2, -1)$, this graph supports the answer to Example 3.

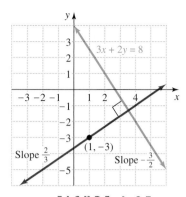

FIGURE 4.27

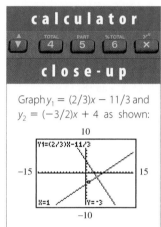

calculator

close-up

Graph $y_1 = (2/3)x - 11/3$ and $y_2 = (-3/2)x + 4$ as shown:

Because the lines look perpendicular and y_1 goes through $(1, -3)$, the graph supports the answer to Example 4.

Solution

First graph $3x + 2y = 8$ and a line through $(1, -3)$ that is perpendicular to $3x + 2y = 8$ as shown in Fig. 4.27. The right angle symbol is used in the figure to indicate that the lines are perpendicular. Now write $3x + 2y = 8$ in slope-intercept form to determine its slope:

$$3x + 2y = 8$$
$$2y = -3x + 8$$
$$y = -\frac{3}{2}x + 4 \quad \text{Slope-intercept form}$$

The slope of the given line is $-\frac{3}{2}$. The slope of any line perpendicular to it is $\frac{2}{3}$. Now we use the point-slope form with the point $(1, -3)$ and the slope $\frac{2}{3}$:

$$y - y_1 = m(x - x_1) \quad \text{Point-slope form}$$

$$y - (-3) = \frac{2}{3}(x - 1)$$

$$y + 3 = \frac{2}{3}x - \frac{2}{3}$$

$$y = \frac{2}{3}x - \frac{2}{3} - 3 \quad \text{Subtract 3 from each side.}$$

$$y = \frac{2}{3}x - \frac{11}{3} \quad \text{Slope-intercept form}$$

So $y = \frac{2}{3}x - \frac{11}{3}$ is the equation of the line that contains $(1, -3)$ and is perpendicular to $3x + 2y = 8$. Check that $(1, -3)$ satisfies $y = \frac{2}{3}x - \frac{11}{3}$.   ■

WARM-UPS

True or false? Explain your answer.

1. The formula $y = m(x - x_1)$ is the point-slope form for a line.
2. It is impossible to find the equation of a line through $(2, 5)$ and $(-3, 1)$.
3. The point-slope form will not work for the line through $(3, 4)$ and $(3, 6)$.
4. The equation of the line through the origin with slope 1 is $y = x$.
5. The slope of the line $5x + y = 4$ is 5.
6. The slope of any line perpendicular to the line $y = 4x - 3$ is $-\frac{1}{4}$.
7. The slope of any line parallel to the line $x + y = 1$ is -1.
8. The line $2x - y = -1$ goes through the point $(-2, -3)$.
9. The line is $2x + y = 4$ and $y = -2x + 7$ are parallel.
10. The equation of the line through $(0, 0)$ perpendicular to $y = x$ is $y = -x$.

4.4 EXERCISES

Reading and Writing *After reading this section, write out the answers to these questions. Use complete sentences.*

1. What is the point-slope form for the equation of a line?

2. For what is the point-slope form used?

3. What is the procedure for finding the equation of a line when given two points on the line?

4. How can you find the slope of a line when given the equation of the line?

5. What is the relationship between the slopes of parallel lines?

6. What is the relationship between the slopes of perpendicular lines?

Write each equation in slope-intercept form. See Example 1.

7. $y - 1 = 5(x + 2)$

8. $y + 3 = -3(x - 6)$

9. $3x - 4y = 80$

10. $2x + 3y = 90$

11. $y - \dfrac{1}{2} = \dfrac{2}{3}\left(x - \dfrac{1}{4}\right)$

12. $y + \dfrac{2}{3} = -\dfrac{1}{2}\left(x - \dfrac{2}{5}\right)$

Find the equation of each line. Write each answer in slope-intercept form. See Example 1.

13. The line through $(2, 3)$ with slope $\dfrac{1}{3}$

14. The line through $(1, 4)$ with slope $\dfrac{1}{4}$

15. The line through $(-2, 5)$ with slope $-\dfrac{1}{2}$

16. The line through $(-3, 1)$ with slope $-\dfrac{1}{3}$

17. The line with slope -6 that goes through $(-1, -7)$

18. The line with slope -8 that goes through $(-1, -5)$

Write each equation in standard form using only integers. See Example 2.

19. $y - 3 = 2(x - 5)$

20. $y + 2 = -3(x - 1)$

21. $y = \dfrac{1}{2}x - 3$

22. $y = \dfrac{1}{3}x + 5$

23. $y - 2 = \dfrac{2}{3}(x - 4)$

24. $y + 1 = \dfrac{3}{2}(x + 4)$

Find the equation of each line. Write each answer in standard form using only integers. See Example 2.

25. The line through the points $(1, 2)$ and $(5, 8)$

26. The line through the points $(3, 5)$ and $(8, 15)$

27. The line through the points $(-2, -1)$ and $(3, -4)$

28. The line through the points $(-1, -3)$ and $(2, -1)$

29. The line through the points $(-2, 0)$ and $(0, 2)$

30. The line through the points $(0, 3)$ and $(5, 0)$

The lines in each figure are perpendicular. Find the equation (in slope-intercept form) for the solid line.

31.

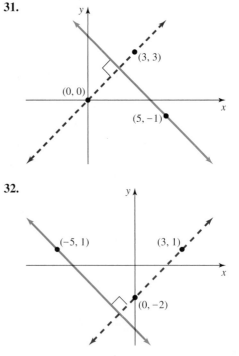

32.

33.

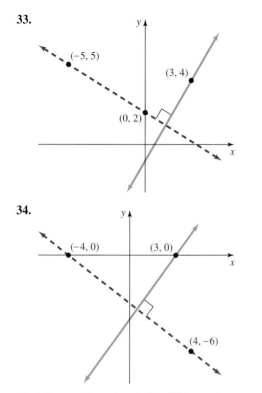

34.

Find the equation of each line. Write each answer in slope-intercept form. See Examples 3 and 4.

35. The line contains the point $(3, 4)$ and is perpendicular to $y = 3x - 1$.

36. The line contains the point $(-2, 3)$ and is perpendicular to $y = 2x + 7$.

37. The line is parallel to $y = x - 9$ and goes through the point $(7, 10)$.

38. The line is parallel to $y = -x + 5$ and goes through the point $(-3, 6)$.

39. The line is perpendicular to $3x - 2y = 10$ and passes through the point $(1, 1)$.

40. The line is perpendicular to $x - 5y = 4$ and passes through the point $(-1, 1)$.

41. The line is parallel to $2x + y = 8$ and contains the point $(-1, -3)$.

42. The line is parallel to $-3x + 2y = 9$ and contains the point $(-2, 1)$.

43. The line goes through $(-1, 2)$ and is perpendicular to $3x + y = 5$.

44. The line goes through $(1, 2)$ and is perpendicular to $y = \frac{1}{2}x - 3$.

45. The line goes through $(2, 3)$ and is parallel to $-2x + y = 6$.

46. The line goes through $(1, 4)$ and is parallel to $x - 2y = 6$.

Solve each problem.

47. *Automated tellers.* ATM volume reached 10.6 billion transactions in 1996 (Electronic Commerce Data Base). The accompanying graph shows the steady growth of automatic tellers.
   a) Write the equation of the line through $(92, 7.0)$ and $(96, 10.6)$.
   b) Use the equation to predict the number of transactions at automated teller machines in the year 2005?

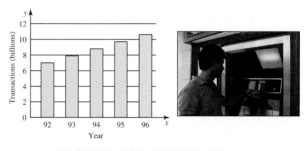

FIGURE FOR EXERCISE 47

48. *Direct deposit.* In 1994, one-third of all workers participated in direct deposit of their paychecks and this number is expected to reach three-fourths by the year 2000. (New York Automated Clearing House, www.nyach.org).
   a) Write the equation of the line through $(1993, 1/3)$ and $(2000, 3/4)$.
   b) Use the accompanying graph to predict the year in which 100% of all workers will participate in direct deposit of their paychecks.
   c) Use the equation from part (a) to predict the year in which 100% of all workers will participate in direct deposit of their paychecks.

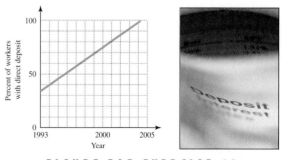

FIGURE FOR EXERCISE 48

49. *Gross domestic product.* The U.S. per capita gross domestic product went from \$14,000 in 1970 to \$18,000 in 1992 (*World Resources,* 1997).
   a) Write the equation of the line through the points $(1970, 14{,}000)$ and $(1992, 18{,}000)$.

   b) What do x and y represent in your equation?

c) Graph the equation

d) By how much is the gross domestic product increasing per year?

e) Use your equation to predict the per capita gross domestic product in the year 2000.

50. Body-mass index. The body mass index BMI is used to assess the level of fat in a person's body. When Tim weighed 147 pounds his BMI was 23.4. When his weight went to 185, his BMI was 29.5.

a) Find the equation of the line through (147, 23.4) and (185, 29.5).

b) What do x and y represent in your equation?

c) Graph the equation.

d) Interpret the slope of this line.

e) What is his BMI when his weight is 160?

GETTING MORE INVOLVED

51. Exploration. What is the slope of the line $2x + 3y = 9$? What is the slope of $4x - 5y = 6$? Write a formula for the slope of $Ax + By = C$, where $B \neq 0$.

GRAPHING CALCULATOR EXERCISES

52. Graph each equation on a graphing calculator. Choose a viewing window that includes both the x- and y-intercepts. Use the calculator output to help you draw the graph on paper.
a) $y = 20x - 300$
b) $y = -30x + 500$
c) $2x - 3y = 6000$

53. Graph $y = 2x + 1$ and $y = 1.99x - 1$ on a graphing calculator. Are these lines parallel? Explain your answer.

54. Graph $y = 0.5x + 0.8$ and $y = 0.5x + 0.7$ on a graphing calculator. Find a viewing window in which the two lines are separate.

55. Graph $y = 3x + 1$ and $y = -\frac{1}{3}x + 2$ on a graphing calculator. Do the lines look perpendicular? Explain.

4.5   APPLICATIONS OF LINEAR EQUATIONS

The linear equation $y = mx + b$ is a formula that determines a value of y for each given value of x. In this section you will study linear equations used as formulas.

Applied Examples

The daily rental charge for renting a 1988 Buick at Wrenta-Wreck is $30 plus 25 cents per mile. The rental charge depends on the number of miles you drive. If x represents the number of miles driven in one day, then $0.25x + 30$ will give the rental charge in dollars for that day. If we let y represent the rental charge, then we can write the equation

$$y = 0.25x + 30.$$

Because the value of y depends on the value of x, we call x the **independent variable** and y the **dependent variable.** Since $y = 0.25x + 30$ is in slope-intercept form, 0.25 is the slope of the line and (0, 30) is the y-intercept. The rental charge starts at $30 and is increasing at a rate of 25 cents per mile.

For variables in applications, we generally use letters that help us to remember what the variables represent. In the rental example above, we could let m represent the number of miles and R the rental charge. Then R is determined from m by the formula

$$R = 0.25m + 30.$$

There are many examples in which the value of one variable is determined from the value of another variable by means of a linear equation. When this is the case, we say that the dependent variable is a *linear function* of the independent variable. For example, the formula

$$F = \frac{9}{5}C + 32$$

is a linear equation that expresses Fahrenheit temperature in terms of Celsius temperature. In other words, the Fahrenheit temperature F is a linear function of the Celsius temperature C.

E X A M P L E   1

Distance as a function of time

A car is averaging 50 miles per hour. Write an equation that expresses the distance it travels as a linear function of the time it travels.

Solution

If the speed is 50 miles per hour, then from the formula $D = RT$, we can write

$$D = 50T.$$

This linear equation expresses D as a linear function of T.   ■

Graphing

The graph of a linear equation is a straight line. A linear equation used as a formula is graphed the same way that any linear equation is graphed. If a formula is in slope-intercept form, then we can graph it using the slope and intercept as in

Section 4.3. If we use letters other than x and y for the variables, then we label the axes with these letters. For example, to graph $F = \frac{9}{5}C + 32$ we label the horizontal axis with C and the vertical axis with F. When a formula is solved for a variable, that variable generally goes on the vertical axis of the graph.

E X A M P L E   2   Graphing a formula

Graph the linear equation $R = 0.25m + 30$ for $0 \leq m \leq 500$. R represents the rental charge in dollars, and m represents the number of miles. Find the rental charge for driving 200 miles.

Solution

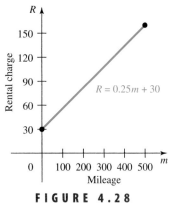

FIGURE 4.28

We label the x-axis with the letter m and the y-axis with the letter R. See Fig. 4.28. The ordered pairs are of the form (m, R). We adjust the scale on the m-axis to graph the values from 0 to 500. The slope of this line is 0.25, or $\frac{25}{100}$. To sketch the graph, start at the R-intercept $(0, 30)$. Move 100 units to the right and up 25 units to locate a second point on the line. Draw the line as in Fig. 4.28. To find the rental charge for 200 miles, let $m = 200$ in $R = 0.25m + 30$:

$$R = 0.25(200) + 30 = 80$$

So the rental charge for 200 miles is $80. ◼

Finding a Formula

If one variable is a linear function of another, then there is a linear equation expressing one variable in terms of the other. In Section 4.4 we used the point-slope form to find the equation of a line given two points on the line. We can use that same procedure to find a linear function for two variables.

E X A M P L E   3   Writing a linear function given two points

A contractor found that his labor cost for installing 100 feet of pipe was $30. He also found that his labor cost for installing 500 feet of pipe was $120. If the cost C in dollars is a linear function of the length L in feet, then what is the formula for this function? What would his labor cost be for installing 240 feet of pipe?

Solution

Because C is determined from L, we let C take the place of the dependent variable y and let L take the place of the independent variable x. So the ordered pairs are in the form (L, C). We can use the slope formula to find the slope of the line through the two points $(100, 30)$ and $(500, 120)$ shown in Fig. 4.29.

$$m = \frac{120 - 30}{500 - 100}$$
$$= \frac{90}{400}$$
$$= \frac{9}{40}$$

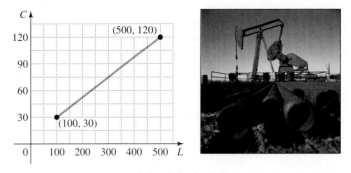

Now we use the point-slope form with the point (100, 30) and slope $\frac{9}{40}$:

$$y - y_1 = m(x - x_1)$$

$$C - 30 = \frac{9}{40}(L - 100)$$

$$C - 30 = \frac{9}{40}L - \frac{45}{2}$$

$$C = \frac{9}{40}L - \frac{45}{2} + 30$$

$$C = \frac{9}{40}L + \frac{15}{2} \quad C \text{ is a linear function of } L.$$

s t u d y   t i p

When working a test, scan the problems and pick out the ones that are the easiest for you. Do them first. Save the harder problems till last.

Now that we have a formula for C in terms of L, we can find C for any value of L. If $L = 240$ feet, then

$$C = \frac{9}{40} \cdot 240 + \frac{15}{2}$$

$$C = 54 + 7.5$$

$$C = 61.5$$

The labor cost to install 240 feet of pipe would be $61.50.

WARM-UPS

True or false? Explain your answer.

1. If $z = 3r - 9$, then z is a linear function of r.
2. The circumference of a circle is a linear function of its radius.
3. The area of a circle is a linear function of its radius.
4. The distance driven in 8 hours is a linear function of your average speed.

5. Celsius temperature is a linear function of Fahrenheit temperature.
6. The slope of the line through (1980, 3000) and (1990, 2000) is 100.
7. If your lawyer charges $90 per hour, then your bill is a linear function of the time spent on the case.
8. The area of a square is a linear function of the length of a side.
9. The perimeter of a square is a linear function of the length of a side.
10. The perimeter of a rectangle with a length of 5 meters is a linear function of its width.

4.5 EXERCISES

Reading and Writing   *After reading this section, write out the answers to these questions. Use complete sentences.*

1. What is the difference between the independent variable and the dependent variable?

2. What does it mean to say that one variable is a linear function of another?

3. Why should we use letters other than x and y in applications?

4. Which axis is usually used for the independent variable?

5. What is the procedure for writing a linear function when given two points?

6. Why can we say that Fahrenheit temperature is a linear function of Celsius temperature?

Write an equation that expresses one variable as a linear function of the other. See Example 1.

7. Express length in feet as a linear function of length in yards.

8. Express length in yards as a linear function of length in feet.

9. For a car averaging 65 miles per hour, express the distance it travels as a linear function of the time spent traveling.

10. For a car traveling 6 hours, express the distance it travels as a linear function of its average speed.

11. Express the circumference of a circle as a linear function of its diameter.

12. For a rectangle with a fixed width of 12 feet, express the perimeter as a linear function of its length.

13. Rodney makes $7.80 per hour. Express his weekly pay as a linear function of the number of hours he works.

14. A triangle has a base of 5 feet. Express its area as a linear function of its height.

Graph each formula for the given values of the independent variable. See Example 2.

15. $P = 40n + 300, 0 \le n \le 200$

16. $C = -50r + 500, 0 \le r \le 10$

17. $R = 30t + 1000, 100 \le t \le 900$

18. $W = 3m - 4000, 1000 \le m \le 5000$

19. $C = 2\pi r, 1 \le r \le 10$

20. $P = 4s$, $100 \leq s \leq 500$

21. $h = -7.5d + 350$, $0 \leq d \leq 40$

22. $a = -50g + 2500$, $0 \leq g \leq 50$

Solve each problem. See Example 2.

23. Profit per share. In the 1980s, People's Gas had a profit per share, P, that was determined by the equation $P = 0.35x + 4.60$, where x ranges from 0 to 9 corresponding to the years 1980 to 1989. What was the profit per share in 1987? Sketch the graph of this formula for x ranging from 0 to 9.

24. Loan value of a car. For the first 6 years the loan value of a $30,000 Corvette is determined by the formula $V = -4000a + 30,000$, where a is the age in years of the Corvette. What is the loan value of this automobile when it is 5 years old? Sketch the graph of this formula for a between 0 and 6 inclusive.

In Exercises 25–36, solve each problem. See Example 3.

25. Plumbing problems. When Millie called Pete's Plumbing, Pete worked 2 hours and charged her $70. When her neighbor Rosalee called Pete, he worked 4 hours and charged her $110. Pete's charge is a linear function of the number of hours he works. Find a formula for this function. How much will Pete charge for working 7 hours at Fred's house?

26. Interior angles. The sum of the measures of the interior angles of a triangle is 180°. The sum of the measures of the interior angles of a square is 360°. The sum S of the measures of the interior angles of any n-sided polygon is a linear function of the number of sides n. Express S as a linear function of n. What is the sum of the measures of the interior angles of an octagon?

FIGURE FOR EXERCISE 26

27. If the shoe fits. If a child's foot is 7.75 inches long, then the child wears a size 13 shoe. If the child has a foot that is 5.75 inches long, then the child wears a size 7 shoe. The shoe size S is a linear function of the length of the foot L.

a) Write a linear equation expressing S as a function of L.

b) What size shoe fits a child with a 6.25-inch foot?

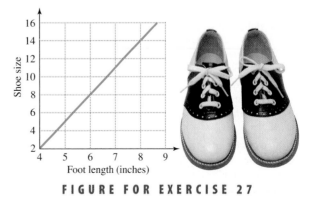

FIGURE FOR EXERCISE 27

28. Celsius to Fahrenheit. Fahrenheit temperature F is a linear function of Celsius temperature C. When $C = 0$, $F = 32$. When $C = 100$, $F = 212$. Use the point-slope form to write F as a linear function of C. What is the Fahrenheit temperature when $C = 45$?

29. Velocity of a projectile. The velocity v of a projectile is a linear function of the time t that it is in the air. A ball is thrown downward from the top of a tall building. Its velocity is 42 feet per second after 1 second and 74 feet per second after 2 seconds. Write v as a linear function of t. What is the velocity when $t = 3.5$ seconds?

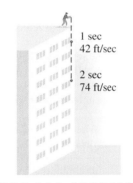

1 sec
42 ft/sec

2 sec
74 ft/sec

FIGURE FOR EXERCISE 29

30. Natural gas. The cost C of natural gas is a linear function of the number n of cubic feet of gas used. The cost of 1000 cubic feet of gas is $39, and the cost of 3000 cubic feet of gas is $99. Express C as a linear function of n. What is the cost of 2400 cubic feet of gas?

31. Expansion joint. The width of an expansion joint on the Carl T. Hull bridge is a linear function of the temperature of the roadway. When the temperature is 90°F, the width is 0.75 inch. When the temperature is 30°F, the width is 1.25 inches. Express w as a linear function of t. What is the width of the joint when the temperature is 80°F?

32. Perimeter of a rectangle. The perimeter P of a rectangle with a fixed width is a linear function of its length. The perimeter is 28 inches when the length is 6.5 inches, and the perimeter is 36 inches when the length is 10.5 inches. Write P as a linear function of L. What is the perimeter when $L = 40$ feet? What is the fixed width of the rectangle?

33. Stretching a spring. The amount A that a spring stretches beyond its natural length is a linear function of the weight w placed on the spring. A weight of 3 pounds stretches a certain spring 1.8 inches and a weight of

5 pounds stretches the same spring 3 inches. Express A as a linear function of w. How much will the spring stretch with a weight of 6 pounds?

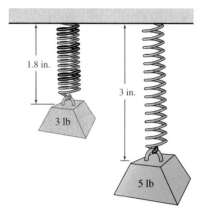

1.8 in.

3 in.

3 lb

5 lb

FIGURE FOR EXERCISE 33

34. Velocity of a bullet. If a gun is fired straight upward, then the velocity v of the bullet is a linear function of the time t that has elapsed since the gun was fired. Suppose that the bullet leaves the gun at 100 feet per second (time $t = 0$) and that after 2 seconds its velocity is 36 feet per second. Express v as a linear function of t. What is the velocity after 3 seconds?

35. Enzyme concentration. The amount of light absorbed by a certain liquid is a linear function of the concentration of an enzyme in the liquid. A concentration of 2 mg/ml (milligrams per milliliter) produces an absorption of 0.16 and a concentration of 5 mg/ml produces an absorption of 0.40. Express the absorption a as a linear function of the concentration c. What should the absorption be if the concentration is 3 mg/ml? Use the accompanying graph to estimate the concentration when the absorption is 0.50.

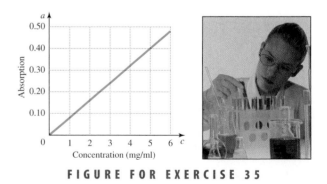

FIGURE FOR EXERCISE 35

36. Basal energy requirement. The basal energy requirement B is the number of calories that a person needs to maintain the life processes. B depends on the height, weight, and age of the person. For a 28-year-old female with a height of 160 cm, B is a linear function of the person's weight w (in kilograms). For a weight of 45 kg, B is 1300 calories. For a weight of 50 kg, B is 1365 calories. Express B as a linear function of w. What is B for a 28-year-old 160-cm female who weighs 53.2 kg?

GRAPHING CALCULATOR EXERCISES

Most calculators use the variables x and y for graphing. So to graph an equation such as $W = 9R - 21$, you must graph $y = 9x - 21$.

37. *Energy decreasing with age.* The basal energy requirement (in calories) for a 55-kg 160-cm male at age A is given by $B = 1481 - 4.7A$; the basal energy requirement for a female with the same weight and height is given by $B = 1623 - 6.9A$. Graph these functions on your calculator, and use the graphs to answer the

following questions.

a) Which person has a higher basal energy requirement at age 25?

b) Which person has a higher basal energy requirement at age 72?

c) Use the graph to estimate the age at which the basal energy requirements are equal.

d) At what age does the female require no calories?

38. *Equality of energy.* The basal energy requirement B for a 70-kg 160-cm male at age A is given by $B = 1620 - 4.7A$. The basal energy requirement B for a 65-cm female at age A is given by $B = 1786 - 6.8A$. Graph these linear functions on your calculator, and use the graphs to estimate the age at which these two people have the same basal energy requirement.

4.6   INTRODUCTION TO FUNCTIONS

In Section 4.5 you learned that if y is determined from the value of x by an equation of the form $y = mx + b$, then y is a linear function of x. In this section we will discuss other types of functions, but the idea is the same.

Functions Expressed by Formulas

If you get a speeding ticket, then your speed determines the cost of the ticket. You may not know exactly how the judge determines the cost, but the judge is using some rule to determine a cost from knowing your speed. The cost of the ticket is a function of your speed.

> **Function (as a Rule)**
>
> A function is a rule by which any allowable value of one variable (the **independent variable**) determines a *unique* value of a second variable (the **dependent variable**).

helpful hint

According to the dictionary, "determine" means to settle conclusively. If the value of the dependent variable is inconclusive or there is more than one, then the rule is not a function.

One way to express a function is to use a formula. For example, the formula

$$A = \pi r^2$$

gives the area of a circle as a function of its radius. The formula gives us a rule for finding a *unique* area for any given radius. A is the dependent variable, and r is the independent variable. The formula

$$S = -16t^2 + v_0 t + s_0$$

expresses altitude S of a projectile as a function of time t, where v_0 is the initial velocity and s_0 is the initial altitude. S is the dependent variable, and t is the independent variable.

In many areas of study, formulas are used to describe relationships between variables. In the next example we write a formula that describes or **models** a real situation.

E X A M P L E   1

Writing a formula for a function

A carpet layer charges $25 plus $4 per square yard for installing carpet. Write the total charge C as a function of the number n of square yards of carpet installed.

Solution

At $4 per square yard, n square yards installed cost $4n$ dollars. If we include the $25 charge, then the total cost is $4n + 25$ dollars. Thus the equation

$$C = 4n + 25$$

expresses C as a function of n. ■

In the next example, we modify a well-known geometric formula.

E X A M P L E   2

A function in geometry

Express the area of a circle as a function of its diameter.

Solution

The area of a circle is given by $A = \pi r^2$. Because the radius of a circle is one-half of the diameter, we have $r = \frac{d}{2}$. Now replace r by $\frac{d}{2}$ in the formula $A = \pi r^2$:

$$A = \pi \left(\frac{d}{2}\right)^2$$

$$= \frac{\pi d^2}{4}$$

So $A = \frac{\pi d^2}{4}$ expresses the area of a circle as a function of its diameter. ■

Functions Expressed by Tables

Another way to express a function is with a table. For example, Table 4.1 can be used to determine the cost at United Freight Service for shipping a package that weighs under 100 pounds. For any *allowable* weight, the table gives us a rule for finding the unique shipping cost. The weight is the independent variable, and the cost is the dependent variable.

Weight in Pounds	Cost
0 to 10	$4.60
11 to 30	$12.75
31 to 79	$32.90
80 to 99	$55.82

T A B L E   4 . 1

Weight in Pounds	Cost
0 to 15	$4.60
10 to 30	$12.75
31 to 79	$32.90
80 to 99	$55.82

TABLE 4.2

Now consider Table 4.2. It does not look much different from Table 4.1, but there is an important difference. The cost for shipping a 12-pound package according to Table 4.2 is either $4.60 or $12.75. Either the table has an error or perhaps $4.60 and $12.75 are costs for shipping to different destinations. In any case the weight does not determine a unique cost. So Table 4.2 does not express the cost as a function of the weight.

E X A M P L E   3

Functions defined by tables

Which of the following tables expresses y as a function of x?

a)
x	y
1	3
2	6
3	9
4	12
5	15

b)
x	y
1	1
-1	1
2	2
-2	2
3	3
-3	3

c)
x	y
1988	27,000
1989	27,000
1990	28,500
1991	29,000
1992	30,000
1993	30,750

d)
x	y
23	48
35	27
19	28
23	37
41	56
22	34

helpful hint

In a function, every value for the independent variable determines conclusively a corresponding value for the dependent variable. If there is more than one possible value for the dependent variable, then the set of ordered pairs is not a function.

Solution

In Tables a), b), and c), every value of x corresponds to only one value of y. Tables a), b), and c) each express y as a function of x. Notice that different values of x may correspond to the same value of y. In Table d) we have the value of 23 for x corresponding to two different values of y, 48 and 37. So Table d) does not express y as a function of x. ∎

helpful hint

A computer at your grocery store determines the price of each item from the universal product code. This function consists of a long list of ordered pairs in the computer memory. The function that pairs your total with the amount of tax is also handled by computer, but in this case the computer determines the second coordinate by using a formula.

Functions Expressed by Ordered Pairs

If the value of the independent variable is written as the first coordinate of an ordered pair and the value of the dependent variable is written as the second coordinate of an ordered pair, then a function can be expressed by a set of ordered pairs. In a function, each value of the independent variable corresponds to a unique value of the dependent variable. So no two ordered pairs can have the same first coordinate and different second coordinates.

Function (as a Set of Ordered Pairs)

A function is a set of ordered pairs of real numbers such that no two ordered pairs have the same first coordinates and different second coordinates.

E X A M P L E   4

Functions expressed by a set of ordered pairs

Determine whether each set of ordered pairs is a function.

a) $\{(1, 2), (1, 5), (-4, 6)\}$ **b)** $\{(-1, 3), (0, 3), (6, 3), (-3, 2)\}$

Solution

a) This set of ordered pairs is not a function because $(1, 2)$ and $(1, 5)$ have the same first coordinates but different second coordinates.

b) This set of ordered pairs is a function. Note that the same second coordinate with different first coordinates is permitted in a function.   ■

If there are infinitely many ordered pairs in a function, then we can use set-builder notation from Chapter 1 along with an equation to express the function. For example,

$$\{(x, y) \mid y = x^2\}$$

is the set of ordered pairs in which the y-coordinate is the square of the x-coordinate. Ordered pairs such as $(0, 0)$, $(2, 4)$, and $(-2, 4)$ belong to this set. This set is a function because every value of x determines only one value of y.

E X A M P L E   5

Functions expressed by set-builder notation

Determine whether each set of ordered pairs is a function.

a) $\{(x, y) \mid y = 3x^2 - 2x + 1\}$ **b)** $\{(x, y) \mid y^2 = x\}$ **c)** $\{(x, y) \mid x + y = 6\}$

Solution

a) This set is a function because each value we select for x determines only one value for y.

b) If $x = 9$, then we have $y^2 = 9$. Because both 3 and -3 satisfy $y^2 = 9$, both $(9, 3)$ and $(9, -3)$ belong to this set. So the set is not a function.

c) If we solve $x + y = 6$ for y, we get $y = -x + 6$. Because each value of x determines only one value for y, this set is a function. In fact, this set is a linear function.   ■

We often omit the set notation when discussing functions. For example, the equation

$$y = 3x^2 - 2x + 1$$

expresses y as a function of x because the set of ordered pairs determined by the equation is a function. However, the equation

$$y^2 = x$$

does not express y as a function of x because ordered pairs such as $(9, 3)$ and $(9, -3)$ satisfy the equation.

E X A M P L E   6

Functions expressed by equations

Determine whether each equation expresses y as a function of x.

a) $y = |x|$ **b)** $y = x^3$ **c)** $x = |y|$

Solution

a) Because every number has a unique absolute value, $y = |x|$ is a function.

b) Because every number has a unique cube, $y = x^3$ is a function.

c) The equation $x = |y|$ does not express y as a function of x because both $(4, -4)$ and $(4, 4)$ satisfy this equation. These ordered pairs have the same first coordinate but different second coordinates.   ■

Graphs of Functions

Every function determines a set of ordered pairs, and any set of ordered pairs has a graph in the rectangular coordinate system. For example, the set of ordered pairs determined by the linear function $y = 2x - 1$ is shown in Fig. 4.30.

Every graph illustrates a set of ordered pairs, but not every graph is a graph of a function. For example, the circle in Fig. 4.31 is not a graph of a function because the ordered pairs $(0, 4)$ and $(0, -4)$ are both on the graph, and these two ordered pairs have the same first coordinate and different second coordinates. Whether a graph has such ordered pairs can be determined by a simple visual test called the **vertical-line test.**

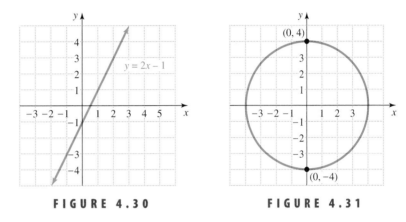

FIGURE 4.30   **FIGURE 4.31**

Vertical-Line Test

If it is possible to draw a vertical line that crosses a graph two or more times, then the graph is not the graph of a function.

If there is a vertical line that crosses a graph twice (or more), then we have two points (or more) with the same x-coordinate and different y-coordinates, and so the graph is not the graph of a function. If you mentally consider every possible vertical line and none of them cross the graph more than once, then you can conclude that the graph is the graph of a function.

E X A M P L E   7

Using the vertical-line test

Which of the following graphs are graphs of functions?

a)   **b)**   **c)**

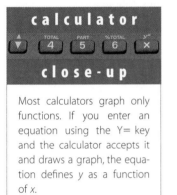

Solution

Neither a) nor c) is the graph of a function, since we can draw vertical lines that cross these graphs twice. Graph b) is the graph of a function, since no vertical line crosses it twice. ■

Domain and Range

The set of all possible numbers that can be used for the independent variable is called the **domain** of the function. For example, the domain of the function

$$y = \frac{1}{x}$$

is the set of all nonzero real numbers because $\frac{1}{x}$ is undefined for $x = 0$. For some functions the domain is clearly stated when the function is given. The set of all values of the dependent variable is called the **range** of the function.

E X A M P L E 8

Domain and range

State the domain and range of each function.

a) $\{(3, -1), (2, 5), (1, 5)\}$

b) $y = |x|$

c) $A = \pi r^2$ for $r > 0$

Solution

a) The domain is the set of numbers used as first coordinates, $\{1, 2, 3\}$. The range is the set of second coordinates, $\{-1, 5\}$.

b) Because $|x|$ is a real number for any real number x, the domain is the set of all real numbers. The range is the set of numbers that result from taking the absolute value of every real number. Thus the range is the set of nonnegative real numbers, $[0, \infty)$.

c) The condition $r > 0$ specifies the domain of the function. The domain is $(0, \infty)$, the positive real numbers. Because $A = \pi r^2$, the value of A is also greater than zero. So the range is also the set of positive real numbers. ■

Function Notation

When the variable y is a function of x, we may use the notation $f(x)$ to represent y. The symbol $f(x)$ is read as "f of x." So if x is the independent variable, we may use y or $f(x)$ to represent the dependent variable. For example, the function

$$y = 2x + 3$$

can also be written as

$$f(x) = 2x + 3.$$

We use y and $f(x)$ interchangeably. We think of f as the name of the function. We may use letters other than f. For example, the function $g(x) = 2x + 3$ is the same function as $f(x) = 2x + 3$.

The expression $f(x)$ represents the second coordinate when the first coordinate is x; it does not mean f times x. For example, if we replace x by 4 in $f(x) = 2x + 3$, we get

$$f(4) = 2 \cdot 4 + 3 = 11.$$

So if the first coordinate is 4, then the second coordinate is $f(4)$, or 11. The ordered pair (4, 11) belongs to the function f. This statement means that the function f pairs 4 with 11. We can use the diagram in Fig. 4.32 to picture this situation.

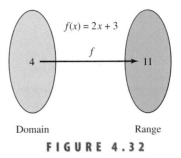

Domain   Range

FIGURE 4.32

E X A M P L E   9

Using function notation

Suppose $f(x) = x^2 - 1$ and $g(x) = -3x + 2$. Find the following:

a) $f(-2)$ **b)** $f(-1)$ **c)** $g(0)$ **d)** $g(6)$

Solution

a) Replace x by -2 in the formula $f(x) = x^2 - 1$:

$$f(-2) = (-2)^2 - 1$$
$$= 4 - 1$$
$$= 3$$

So $f(-2) = 3$.

b) Replace x by -1 in the formula $f(x) = x^2 - 1$:

$$f(-1) = (-1)^2 - 1$$
$$= 1 - 1$$
$$= 0$$

So $f(-1) = 0$.

c) Replace x by 0 in the formula $g(x) = -3x + 2$:

$$g(0) = -3 \cdot 0 + 2 = 2$$

So $g(0) = 2$.

d) Replace x by 6 in $g(x) = -3x + 2$ to get $g(6) = -16$.   ■

E X A M P L E   1 0

Using function notation in an application

The formula $C(n) = 0.10n + 4.95$ gives the monthly cost in dollars for n minutes of long-distance calls. Find $C(40)$ and $C(100)$.

Solution

Replace n with 40 in the formula:

$$C(n) = 0.10n + 4.95$$
$$C(40) = 0.10(40) + 4.95$$
$$= 8.95$$

So $C(40) = 8.95$. The cost for 40 minutes of calls is $8.95. Now

$$C(100) = 0.10(100) + 4.95 = 14.95.$$

So $C(100) = 14.95$. The cost of 100 minutes of calls is $14.95.

CAUTION $C(h)$ is not C times h. In the context of formulas, $C(h)$ represents the value of C corresponding to a value of h.

---

calculator close-up

A graphing calculator can be used to evaluate a formula in the same manner as in Example 10. To evaluate

$$C = 0.10n + 4.95$$

enter the formula into your calculator as $y_1 = 0.10x + 4.95$ using the Y= key:

```
Plot1 Plot2 Plot3
\Y1◼.10X+4.95
\Y2=
\Y3=
\Y4=
\Y5=
\Y6=
\Y7=
```

To find the cost of 40 minutes of calls, enter $y_1(40)$ on the home screen and press ENTER:

```
Y1(40)
              8.95
Y1(100)
             14.95
```

WARM-UPS

True or false? Explain your answer.

1. Any set of ordered pairs is a function.
2. The area of a square is a function of the length of a side.
3. The set $\{(-1, 3), (-3, 1), (-1, -3)\}$ is a function.
4. The set $\{(1, 5), (3, 5), (7, 5)\}$ is a function.
5. The domain of $f(x) = x^3$ is the set of all real numbers.
6. The domain of $y = |x|$ is the set of nonnegative real numbers.
7. The range of $y = |x|$ is the set of all real numbers.
8. The set $\{(x, y) \,|\, x = 2y\}$ is a function.
9. The set $\{(x, y) \,|\, x = y^2\}$ is a function.
10. If $f(x) = x^2 - 5$, then $f(-2) = -1$.

4.6 EXERCISES

Reading and Writing *After reading this section, write out the answers to these questions. Use complete sentences.*

1. What is a function?

2. What are the different ways to express functions?

3. What do all descriptions of functions have in common?

4. How can you tell at a glance if a graph is a graph of a function?

5. What is the domain of a function?

6. What is function notation?

Write a formula that describes the function for each of the following. See Examples 1 and 2.

7. A small pizza costs $5.00 plus 50 cents for each topping. Express the total cost C as a function of the number of toppings t.

8. A developer prices condominiums in Florida at $20,000 plus $40 per square foot of living area. Express the cost C as a function of the number of square feet of living area s.

9. The sales tax rate on groceries in Mayberry is 9%. Express the total cost T (including tax) as a function of the total price of the groceries S.

10. With a GM MasterCard, 5% of the amount charged is credited toward a rebate on the purchase of a new car. Express the rebate R as a function of the amount charged A.

11. Express the circumference of a circle as a function of its radius.

12. Express the circumference of a circle as a function of its diameter.

13. Express the perimeter P of a square as a function of the length s of a side.

14. Express the perimeter P of a rectangle with width 10 ft as a function of its length L.

15. Express the area A of a triangle with a base of 10 m as a function of its height h.

16. Express the area A of a trapezoid with bases 12 cm and 10 cm as a function of its height h.

Determine whether each table expresses the second variable as a function of the first variable. See Example 3.

17.

x	y
1	1
4	2
9	3
16	4
25	5
36	6
49	8

18.

x	y
2	4
3	9
4	16
5	25
8	36
9	49
10	100

19.

t	v
2	2
−2	2
3	3
−3	3
4	4
−4	4
5	5

20.

s	W
5	17
6	17
−1	17
−2	17
−3	17
7	17
8	17

21.

a	P
2	2
2	−2
3	3
3	−3
4	4
4	−4
5	5

22.

n	r
17	5
17	6
17	−1
17	−2
17	−3
17	−4
17	−5

23.

b	q
1970	0.14
1972	0.18
1974	0.18
1976	0.22
1978	0.25
1980	0.28

24.

c	h
345	0.3
350	0.4
355	0.5
360	0.6
365	0.7
370	0.8
380	0.9

Determine whether each set of ordered pairs is a function. See Example 4.

25. $\{(1, 2), (2, 3), (3, 4)\}$

26. $\{(1, -3), (1, 3), (2, 12)\}$

27. $\{(-1, 4), (2, 4), (3, 4)\}$

28. $\{(1, 7), (7, 1)\}$ **29.** $\{(0, -1), (0, 1)\}$

30. $\{(1, 7), (-2, 7), (3, 7), (4, 7)\}$

31. $\{(50, 50)\}$ **32.** $\{(0, 0)\}$

Determine whether each set is a function. See Example 5.

33. $\{(x, y) \mid y = x - 3\}$ **34.** $\{(x, y) \mid y = x^2 - 2x - 1\}$

35. $\{(x, y) \mid x = |y|\}$ **36.** $\{(x, y) \mid x = y^2 + 1\}$

37. $\{(x, y) \mid x = y + 1\}$ **38.** $\left\{(x, y) \,\middle|\, y = \dfrac{1}{x}\right\}$

39. $\{(x, y) \mid x = y^2 - 1\}$ **40.** $\{(x, y) \mid x = 3y\}$

Determine whether each equation expresses y as a function of x. See Example 6.

41. $x = 4y$ **42.** $x = -3y$

43. $y = \dfrac{2}{x}$ **44.** $y = \dfrac{x}{2}$

45. $y = x^3 - 1$ **46.** $y = |x - 1|$

47. $x^2 + y^2 = 25$ **48.** $x^2 - y^2 = 9$

Which of the following graphs are graphs of functions? See Example 7.

49.

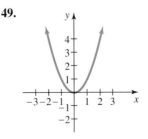

50.

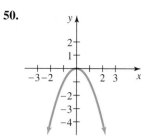

51.

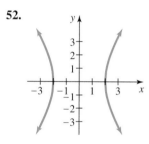

52.

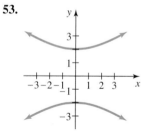

53.

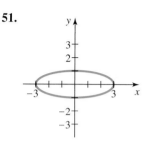

54.

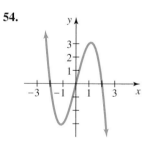

Determine the domain and range of each function. See Example 8.

55. $\{(3, 3), (2, 5), (1, 7)\}$

56. $\{(0, 1), (2, 1), (4, 1)\}$

57. $y = |x + 3|$

58. $y = |x - 1|$

59. $y = x$

60. $y = 2x + 1$

61. $y = x^2$

62. $y = x^3$

63. $A = s^2$ for $s > 0$

64. $S = -16t^2$ for $t \geq 0$

Let $f(x) = 2x - 1$, $g(x) = x^2 - 3$, *and* $h(x) = |x - 1|$. *Find the following. See Example 9.*

65. $f(0)$

66. $f(-1)$

67. $f\left(\dfrac{1}{2}\right)$

68. $f\left(\dfrac{3}{4}\right)$

69. $g(4)$

70. $g(-4)$

71. $g(0.5)$

72. $g(-1.5)$

73. $h(3)$

74. $h(-1)$

75. $h(0)$

76. $h(1)$

Let $f(x) = x^3 - x^2$ *and* $g(x) = x^2 - 4.2x + 2.76$. *Find the following. Round each answer to three decimal places.*

77. $f(5.68)$

78. $g(-2.7)$

79. $g(3.5)$

80. $f(67.2)$

Solve each problem.

81. *Velocity and time.* If a ball is thrown straight upward into the air with a velocity of 100 ft/sec, then its velocity t seconds later is given by

$$v(t) = -32t + 100.$$

a) Find $v(0)$, $v(1)$, and $v(2)$.

b) Is the velocity increasing or decreasing as the time increases?

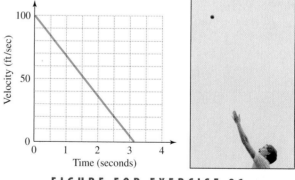

FIGURE FOR EXERCISE 81

82. *Cost and toppings.* The cost c in dollars for a pizza with n toppings is given by

$$c(n) = 0.75n + 6.99.$$

a) Find $c(2)$, $c(4)$, and $c(5)$.

b) Is the cost increasing or decreasing as the number of toppings increases?

83. *Threshold weight.* The threshold weight for an individual is the weight beyond which the risk of death increases significantly. For middle-aged males the function $W(h) = 0.000534h^3$ expresses the threshold weight in pounds as a function of the height h in inches. Find $W(70)$. Find the threshold weight for a 6'2" middle-aged male.

84. Pole vaulting. The height a pole vaulter attains is a function of the vaulter's velocity on the runway. The function

$$h(v) = \frac{1}{64}v^2$$

gives the height in feet as a function of the velocity v in feet per second.

a) Find $h(35)$ to the nearest tenth of an inch.

b) Who gains more height from an increase of 1 ft/sec in velocity: a fast runner or a slow runner?

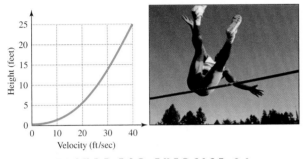

FIGURE FOR EXERCISE 84

85. Credit card fees. A certain credit card company gets 4% of each charge, and the retailer receives the rest. At the end of a billing period the retailer receives a statement showing only the retailer's portion of each transaction. Express the original amount charged C as a function of the retailer's portion r.

86. More credit card fees. Suppose that the amount charged on the credit card in the previous exercise includes 8% sales tax. The credit card company does not get any of the sales tax. In this case the retailer's portion of each transaction includes sales tax on the original cost of the goods. Express the original amount charged C as a function of the retailer's portion.

GETTING MORE INVOLVED

Discussion *In each situation determine whether a is a function of b, b is a function of a, or neither. Answers may vary depending on interpretations.*

87. a = the price per gallon of regular unleaded.
 b = the number of gallons that you get for $10.

88. a = the universal product code of an item at Sears.
 b = the price of that item.

89. a = a student's score on the last test in this class.
 b = the number of hours he/she spent studying.

90. a = a student's score on the last test in this class.
 b = the IQ of the student's mother.

91. a = the weight of a package shipped by UPS.
 b = the cost of shipping that package.

92. a = the Celsius temperature at any time.
 b = the Fahrenheit temperature at the same time.

93. a = the weight of a letter.
 b = the cost of mailing the letter.

94. a = the cost of a gallon of milk.
 b = the amount of sales tax on that gallon.

COLLABORATIVE ACTIVITIES

Inches or Centimeters?

In this activity you will generate data by measuring in both inches and centimeters the height of each member of your group. Then you will plot the points on a graph and use any two of your points to find the conversion formula for converting inches to centimeters.

Part I: Measure the height of each person in your group and fill out a table like the one shown here:

Name	Height in inches	Height in centimeters

Grouping: 3 to 4 students
Topic: Plotting points, graphing lines

Part II: The numbers for inches and centimeters from the table will give you three or four ordered pairs to graph. Plot these points on a graph. Let inches be the horizontal x-axis and centimeters be the vertical y-axis. Let each mark on the axes represent 10 units. When graphing, you will need to estimate the place to plot fractional values.

Part III: Use any two of your points to find an equation of the line you have graphed. What is the slope of your line? Where does it cross the horizontal axis?

Extension: Look up the conversion formula for converting inches to centimeters. Is it the same as the one you found by measuring? If it is different, what could account for the difference?

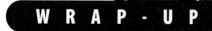

WRAP-UP   **CHAPTER 4**

SUMMARY

| **Slope of a Line** | | **Examples** |

Slope — The slope of the line through (x_1, y_1) and (x_2, y_2) is given by

$$m = \frac{y_2 - y_1}{x_2 - x_1}, \text{ provided that } x_2 - x_1 \neq 0.$$

Slope is the ratio of the rise to the run for any two points on the line:

$$m = \frac{\text{change in } y}{\text{change in } x} = \frac{\text{rise}}{\text{run}}$$

$(0, 1), (3, 5)$
$$m = \frac{5 - 1}{3 - 0} = \frac{4}{3}$$

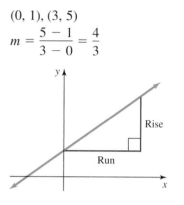

Types of slope

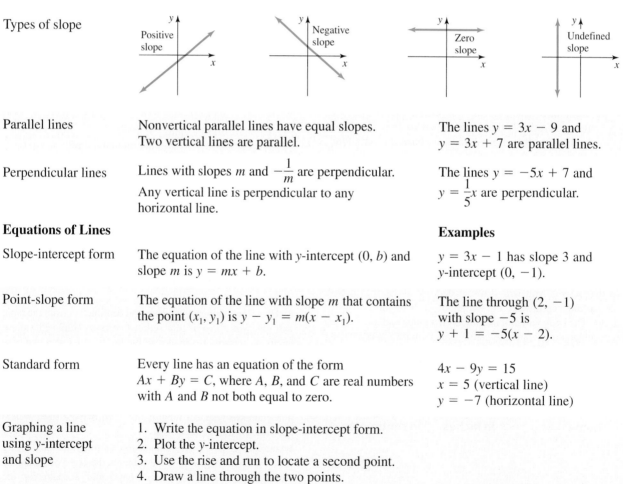

Parallel lines — Nonvertical parallel lines have equal slopes. Two vertical lines are parallel.

The lines $y = 3x - 9$ and $y = 3x + 7$ are parallel lines.

Perpendicular lines — Lines with slopes m and $-\dfrac{1}{m}$ are perpendicular. Any vertical line is perpendicular to any horizontal line.

The lines $y = -5x + 7$ and $y = \dfrac{1}{5}x$ are perpendicular.

Equations of Lines   **Examples**

Slope-intercept form — The equation of the line with y-intercept $(0, b)$ and slope m is $y = mx + b$.

$y = 3x - 1$ has slope 3 and y-intercept $(0, -1)$.

Point-slope form — The equation of the line with slope m that contains the point (x_1, y_1) is $y - y_1 = m(x - x_1)$.

The line through $(2, -1)$ with slope -5 is $y + 1 = -5(x - 2)$.

Standard form — Every line has an equation of the form $Ax + By = C$, where A, B, and C are real numbers with A and B not both equal to zero.

$4x - 9y = 15$
$x = 5$ (vertical line)
$y = -7$ (horizontal line)

Graphing a line using y-intercept and slope

1. Write the equation in slope-intercept form.
2. Plot the y-intercept.
3. Use the rise and run to locate a second point.
4. Draw a line through the two points.

Functions		**Examples**
Definition of a function	A function is a rule by which any allowable value of one variable (the independent variable) determines a unique value of a second variable (the dependent variable).	$A = \pi r^2$
Equivalent definition of a function	A function is a set of ordered pairs such that no two ordered pairs have the same first coordinates and different second coordinates. To say that y is a function of x means that y is determined uniquely by x.	$\{(1, 0), (3, 8)\}$ $\{(x, y) \mid y = x^2\}$
Domain	The set of values of the independent variable, x	$y = x^2$ Domain: all real numbers
Range	The set of values of the dependent variable, y	$y = x^2$ Range: nonnegative real numbers
Linear functions	If $y = mx + b$, we say that y is a linear function of x.	$F = \dfrac{9}{5}C + 32$
Function notation	If x is the independent variable, then we use the notation $f(x)$ to represent the dependent variable.	$y = 2x + 3$ $f(x) = 2x + 3$

ENRICHING YOUR MATHEMATICAL WORD POWER

For each mathematical term, choose the correct meaning.

1. **graph of an equation**
   a. the Cartesian coordinate system
   b. two number lines that intersect at a right angle
   c. the x-axis and y-axis
   d. an illustration in the coordinate plane that shows all ordered pairs that satisfy an equation

2. **x-coordinate**
   a. the first number in an ordered pair
   b. the second number in an ordered pair
   c. a point on the x-axis
   d. a point where a graph crosses the x-axis

3. **y-intercept**
   a. the second number in an ordered pair
   b. a point at which a graph intersects the y-axis
   c. any point on the y-axis
   d. the point where the y-axis intersects the x-axis

4. **coordinate plane**
   a. a matching plane
   b. when the x-axis is coordinated with the y-axis
   c. a plane with a rectangular coordinate system
   d. a coordinated system for graphs

5. **slope**
   a. the change in x divided by the change in y
   b. a measure of the steepness of a line
   c. the run divided by the rise
   d. the slope of a line

6. **slope-intercept form**
   a. $y = mx + b$
   b. rise over run
   c. the point at which a line crosses the y-axis
   d. $y - y_1 = m(x - x_1)$

7. **point-slope form**
   a. $Ax + By = C$
   b. rise over run
   c. $y - y_1 = m(x - x_1)$
   d. the slope of a line at a single point

8. **function**
   a. domain and range
   b. a set of ordered pairs
   c. a rule by which any allowable value of one variable determines a unique value of a second variable
   d. a graph

9. **domain**
   a. the set of first coordinates of a function
   b. the set of second coordinates of a function
   c. the set of real numbers
   d. the integers

10. **range**
    a. all of the possibilities
    b. the coordinates of a function
    c. the entire set of numbers
    d. the set of second coordinates of a function

REVIEW EXERCISES

4.1 *For each point, name the quadrant in which it lies or the axis on which it lies.*

1. $(-2, 5)$ **2.** $(-3, -5)$

3. $(3, 0)$ **4.** $(9, 10)$

5. $(0, -6)$ **6.** $(0, \pi)$

7. $(1.414, -3)$ **8.** $(-4, 1.732)$

14. $y = 2x - 6$

/study \tip

Note how the review exercises are arranged according to the sections in this chapter. If you are having trouble with a certain type of problem, refer back to the appropriate section for examples and explanations.

15. $x + y = 7$

Complete the given ordered pairs so that each ordered pair satisfies the given equation.

9. $y = 3x - 5$: $(0, \quad), (-3, \quad), (4, \quad)$

10. $y = -2x + 1$: $(9, \quad), (3, \quad), (-1, \quad)$

11. $2x - 3y = 8$: $(0, \quad), (3, \quad), (-6, \quad)$

16. $x - y = 4$

12. $x + 2y = 1$: $(0, \quad), (-2, \quad), (2, \quad)$

Sketch the graph of each equation by finding three ordered pairs that satisfy each equation.

13. $y = -3x + 4$

4.2 *Determine the slope of the line that goes through each pair of points.*

17. $(0, 0)$ and $(1, 1)$ **18.** $(-1, 1)$ and $(2, -2)$

19. $(-2, -3)$ and $(0, 0)$ **20.** $(-1, -2)$ and $(4, -1)$

21. $(-4, -2)$ and $(3, 1)$ **22.** $(0, 4)$ and $(5, 0)$

4.3 *Find the slope and y-intercept for each line.*

23. $y = 3x - 18$ **24.** $y = -x + 5$

25. $2x - y = 3$

26. $x - 2y = 1$

27. $4x - 2y - 8 = 0$

28. $3x + 5y + 10 = 0$

Sketch the graph of each equation.

29. $y = \dfrac{2}{3}x - 5$

30. $y = \dfrac{3}{2}x + 1$

31. $2x + y = -6$

32. $-3x - y = 2$

33. $y = -4$

34. $x = 9$

Determine the equation of each line. Write the answer in standard form using only integers as the coefficients.

35. The line through $(0, 4)$ with slope $\dfrac{1}{3}$

36. The line through $(-2, 0)$ with slope $-\dfrac{3}{4}$

37. The line through the origin that is perpendicular to the line $y = 2x - 1$

38. The line through $(0, 9)$ that is parallel to the line $3x + 5y = 15$

39. The line through $(3, 5)$ that is parallel to the x-axis

40. The line through $(-2, 4)$ that is perpendicular to the x-axis

4.4 *Write each equation in slope-intercept form.*

41. $y - 3 = \dfrac{2}{3}(x + 6)$

42. $y + 2 = -6(x - 1)$

43. $3x - 7y - 14 = 0$

44. $1 - x - y = 0$

45. $y - 5 = -\dfrac{3}{4}(x + 1)$

46. $y + 8 = -\dfrac{2}{5}(x - 2)$

Determine the equation of each line. Write the answer in slope-intercept form.

47. The line through $(-4, 7)$ with slope -2

48. The line through $(9, 0)$ with slope $\dfrac{1}{2}$

49. The line through the two points $(-2, 1)$ and $(3, 7)$

50. The line through the two points $(4, 0)$ and $(-3, -5)$

51. The line through $(3, -5)$ that is parallel to the line $y = 3x - 1$

52. The line through $(4, 0)$ that is perpendicular to the line $x + y = 3$

4.5 *Graph each linear equation for the indicated values of the independent variable.*

53. $P = -3t + 400, 10 \le t \le 90$

54. $R = 40w - 300, 20 \le w \le 80$

55. $v = 50n + 30, 0.1 \le n \le 0.9$

56. $w = -40q + 8000, 0 \le q \le 1000$

Solve each problem.

57. *Rental charge.* The charge C for renting an air hammer from Taylor and Son Equipment Rental is a linear function of the number n of days in the rental period. The charge is $113 for two days and $209 for five days. Write C as a linear function of n. What would the charge be for four days?

58. *Time on a treadmill.* After 2 minutes on a treadmill, Jenny has a heart rate of 82. After 3 minutes she has a heart rate of 86. Assuming that Jenny's heart rate h is a linear function of the time t on the treadmill, write h as a linear function of t. What heart rate could be expected for Jenny after 10 minutes on the treadmill?

59. *Probability of rain.* When the probability p of rain is 90%, the probability q that it does not rain is 10%. When the probability of rain is 80%, then the probability that it does not rain is 20%.

 a) Assuming that the probability q that it does not rain is a linear function of the probability p of rain, write q as a function of p.

 b) Use the accompanying graph to determine the probability of rain when the probability that it does not rain is 0.

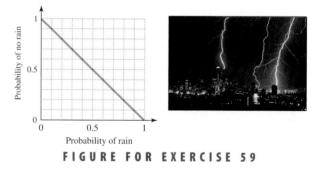

FIGURE FOR EXERCISE 59

60. *Social Security benefits.* On the basis of current legislation, if you earned an average salary of $25,000 over your working life and you retire after the year 2005 at age 62, 63, or 64, then your annual Social Security benefit will be $7000, $7500, or $8000, respectively (Social Security Administration, www.ssa.gov). Write a formula that gives annual benefit as a linear function of age for these three ages.

4.6 *Determine whether each set of ordered pairs is a function.*

61. $\{(4, 3), (5, 3)\}$ **62.** $\{(0, 0), (0, 1), (0, 2)\}$

63. $\{(3, 4), (3, 5)\}$ **64.** $\{(1, 2), (2, 3), (3, 4)\}$

65. $\{(x, y) \mid y = 45x\}$ **66.** $\{(x, y) \mid x = y^3\}$

Determine whether each equation expresses y as a function of x.

67. $y = x^2 + 10$ **68.** $y = 2x - 7$

69. $x^2 + y^2 = 1$ **70.** $x^2 = y^2$

Determine the domain and range of each function.

71. $f(x) = 2x - 3$ **72.** $g(x) = -|x|$

73. $\{(1, 2), (2, 0), (3, 0)\}$

74. $\{(2, 3), (4, 3), (6, 3)\}$

75. $\{(x, y) \mid y = x^2\}$

76. $\{(x, y) \mid y = x^4\}$

CHAPTER 4 TEST

For each point, name the quadrant in which it lies or the axis on which it lies.

1. $(-2, 7)$ **2.** $(-\pi, 0)$

3. $(3, -6)$ **4.** $(0, 1785)$

Find the slope of the line through each pair of points.

5. $(3, 3)$ and $(4, 4)$ **6.** $(-2, -3)$ and $(4, -8)$

13. $y = 4$

14. $x = -2$

/**study** **tip**

Before you take an in-class exam on this chapter, work the sample test given here. Set aside 1 hour to work this test and use the answers in the back of this book to grade yourself. Even though your instructor might not ask exactly the same questions, you will get a good idea of your test readiness

Write the equation of each line. Give the answer in slope-intercept form.

7. The line through $(0, 3)$ with slope $-\dfrac{1}{2}$

8. The line through $(-1, -2)$ with slope $\dfrac{3}{7}$

Write the equation of each line. Give the answer in standard form using only integers as the coefficients.

9. The line through $(2, -3)$ that is perpendicular to the line $y = -3x + 12$

10. The line through $(3, 4)$ that is parallel to the line $5x + 3y = 9$

Sketch the graph of each equation.

11. $y = \dfrac{1}{2}x - 3$ **12.** $2x - 3y = 6$

Determine whether each set is a function.

15. $\{(1, 3), (2, 3), (5, 7)\}$

16. $\{(x, y) \mid x = y^4\}$

Determine the domain and range of each function.

17. $f(x) = |x|$

18. $y = x^2$

Let $f(x) = 2x + 5$ and $g(x) = x^2 - 4$. Find the following.

19. $f(-2)$

20. $g(3)$

Solve each problem.

21. Julie's mail-order CD club charges a shipping and handling fee of $2.50 plus 75 cents per CD for each order shipped. Write the shipping and handling fee S as a function of the number n of CDs in the order.

22. The price P of a soft drink is a linear function of the volume v of the cup. A 10-ounce drink sells for 50 cents, and a 16-ounce drink sells for 68 cents. Write P as a linear function of v. What should the price be for a 20-ounce drink?

Graph Paper

Use these grids for graphing. Make as many copies of this page as you need.

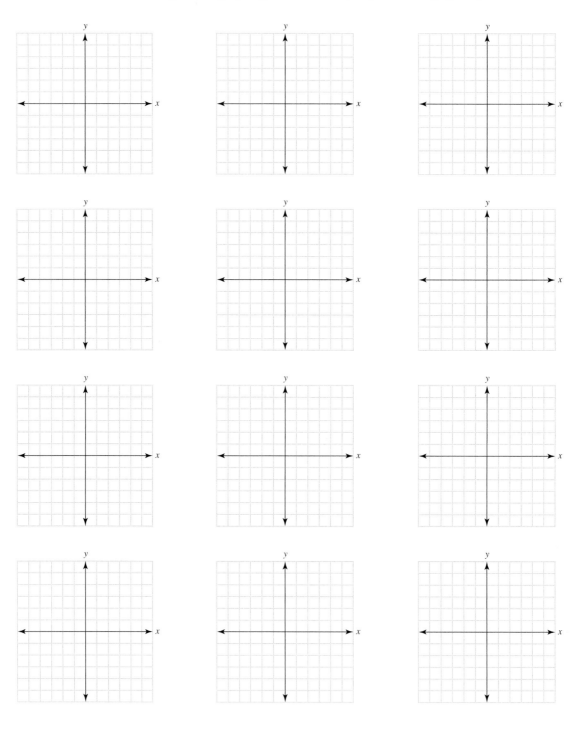

Simplify each arithmetic expression.

1. $9 - 5 \cdot 2$

2. $-4 \cdot 5 - 7 \cdot 2$

3. $3^2 - 2^3$

4. $3^2 \cdot 2^3$

5. $(-4)^2 - 4(1)(5)$

6. $-4^2 - 4 \cdot 3$

7. $\dfrac{-5 - 9}{2 - (-2)}$

8. $\dfrac{6 - 3.6}{6}$

9. $\dfrac{1 - \dfrac{1}{2}}{4 - (-1)}$

10. $\dfrac{4 - (-6)}{1 - \dfrac{1}{3}}$

Simplify each algebraic expression.

11. $4x - (-9x)$

12. $4(x - 9) - x$

13. $4x - x(x - 9)$

14. $-4(x - 9) - 4$

15. $4(9 - x) - 9(x - 4)$

16. $\dfrac{1}{2}(x - 3) + \dfrac{1}{4}(2x - 6)$

17. $\dfrac{4x - 8}{2}$

18. $\dfrac{-5x - 10}{-5}$

19. $\dfrac{2(x - 3) + 6}{2}$

20. $\dfrac{20 + 3(x - 5)}{5}$

Sketch the graph of each equation in the rectangular coordinate system.

21. $y = \dfrac{1}{3}x$

22. $y = 3x$

23. $y = -3x$

24. $y = -\dfrac{1}{3}x$

25. $y = 3x + 1$

26. $y = 3x - 2$

27. $y = 3$

28. $x = 3$

Solve each equation for y.

29. $3\pi y + 2 = t$

30. $x = \dfrac{y - b}{m}$

31. $3x - 3y - 12 = 0$

32. $2y - 3 = 9$

33. $\dfrac{y}{2} - \dfrac{y}{4} = \dfrac{1}{5}$

34. $0.6y - 0.06y = 108$

Solve each equation.

35. $5 = 4w - 7$

36. $3a - 6 = 5a + 4$

37. $4(x - 9) - x = 0$

38. $-4(x - 9) - 4 = -4x$

39. $4(9 - x) + 9(x - 4) = 0$

40. $4(x - 6) = -4(6 - x)$

Solve.

41. *Financial planning.* Financial advisors at Fidelity Investments use the information in the accompanying graph as a guide for retirement investing.

a) What is the slope of the line segment for ages 35 through 50?

b) What is the slope of the line segment for ages 50 through 65?

c) If a 38-year-old man is making $40,000 per year, then what percent of his income should he be saving?

d) If a 58-year-old woman has an annual salary of $60,000, then how much should she have saved and how much should she be saving per year?

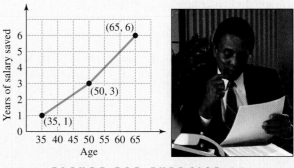

FIGURE FOR EXERCISE 41

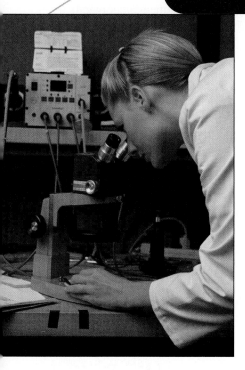

Polynomials and Exponents

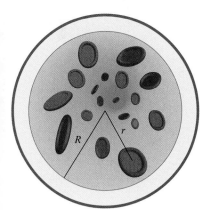

The nineteenth-century physician and physicist, Jean Louis Marie Poiseuille (1799–1869) is given credit for discovering a formula associated with the circulation of blood through arteries. Poiseuille's law, as it is known, can be used to determine the velocity of blood in an artery at a given distance from the center of the artery. The formula states that the flow of blood in an artery is faster toward the center of the blood vessel and is slower toward the outside. Blood flow can also be affected by a person's blood pressure, the length of the blood vessel, and the viscosity of the blood itself.

In later years, Poiseuille's continued interest in blood circulation led him to experiments to show that blood pressure rises and falls when a person exhales and inhales. In modern medicine, physicians can use Poiseuille's law to determine how much the radius of a blocked blood vessel must be widened to create a healthy flow of blood.

In this chapter you will study polynomials, the fundamental expressions of algebra. Polynomials are to algebra what integers are to arithmetic. We use polynomials to represent quantities in general, such as perimeter, area, revenue, and the volume of blood flowing through an artery. In Exercise 85 of Section 5.4, you will see Poiseuille's law represented by a polynomial.

5.1 ADDITION AND SUBTRACTION OF POLYNOMIALS

We first used polynomials in Chapter 1 but did not identify them as polynomials. Polynomials also occurred in the equations of Chapter 2. In this section we will define polynomials and begin a thorough study of polynomials.

Polynomials

In Chapter 1 we defined a **term** as an expression containing a number or the product of a number and one or more variables raised to powers. Some examples of terms are

$$4x^3, \quad -x^2y^3, \quad 6ab, \quad \text{and} \quad -2.$$

A **polynomial** is a single term or a finite sum of terms. The powers of the variables in a polynomial must be positive integers. For example,

$$4x^3 + (-15x^2) + x + (-2)$$

is a polynomial. Because it is simpler to write addition of a negative as subtraction, this polynomial is usually written as

$$4x^3 - 15x^2 + x - 2.$$

The **degree of a polynomial** in one variable is the highest power of the variable in the polynomial. So $4x^3 - 15x^2 + x - 2$ has degree 3 and $7w - w^2$ has degree 2. The **degree of a term** is the power of the variable in the term. Because the last term has no variable, its degree is 0.

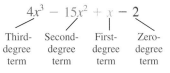

A single number is called a **constant** and so the last term is the **constant term.** The degree of a polynomial consisting of a single number such as 8 is 0.

The number preceding the variable in each term is called the **coefficient** of that variable or the coefficient of that term. In $4x^3 - 15x^2 + x - 2$ the coefficient of x^3 is 4, the coefficient of x^2 is -15, and the coefficient of x is 1 because $x = 1 \cdot x$.

EXAMPLE 1

Identifying coefficients

Determine the coefficients of x^3 and x^2 in each polynomial:

a) $x^3 + 5x^2 - 6$ **b)** $4x^6 - x^3 + x$

Solution

a) Write the polynomial as $1 \cdot x^3 + 5x^2 - 6$ to see that the coefficient of x^3 is 1 and the coefficient of x^2 is 5.

b) The x^2-term is missing in $4x^6 - x^3 + x$. Because $4x^6 - x^3 + x$ can be written as

$$4x^6 - 1 \cdot x^3 + 0 \cdot x^2 + x,$$

the coefficient of x^3 is -1 and the coefficient of x^2 is 0. ■

For simplicity we generally write polynomials with the exponents decreasing from left to right and the constant term last. So we write

$$x^3 - 4x^2 + 5x + 1 \qquad \text{rather than} \qquad -4x^2 + 1 + 5x + x^3.$$

When a polynomial is written with decreasing exponents, the coefficient of the first term is called the **leading coefficient.**

Certain polynomials are given special names. A **monomial** is a polynomial that has one term, a **binomial** is a polynomial that has two terms, and a **trinomial** is a polynomial that has three terms. For example, $3x^5$ is a monomial, $2x - 1$ is a binomial, and $4x^6 - 3x + 2$ is a trinomial.

E X A M P L E   2

Types of polynomials

Identify each polynomial as a monomial, binomial, or trinomial and state its degree.

a) $5x^2 - 7x^3 + 2$ **b)** $x^{43} - x^2$ **c)** $5x$ **d)** -12

Solution

a) The polynomial $5x^2 - 7x^3 + 2$ is a third-degree trinomial.

b) The polynomial $x^{43} - x^2$ is a binomial with degree 43.

c) Because $5x = 5x^1$, this polynomial is a monomial with degree 1.

d) The polynomial -12 is a monomial with degree 0.   ■

Value of a Polynomial

A polynomial is an algebraic expression. Like other algebraic expressions involving variables, a polynomial has no specific value unless the variables are replaced by numbers. A polynomial can be evaluated with or without the function notation discussed in Chapter 4.

E X A M P L E   3

Evaluating polynomials

a) Find the value of $-3x^4 - x^3 + 20x + 3$ when $x = 1$.

b) Find the value of $-3x^4 - x^3 + 20x + 3$ when $x = -2$.

c) If $P(x) = -3x^4 - x^3 + 20x + 3$, find $P(1)$.

Solution

a) Replace x by 1 in the polynomial:

$$-3x^4 - x^3 + 20x + 3 = -3(1)^4 - (1)^3 + 20(1) + 3$$
$$= -3 - 1 + 20 + 3$$
$$= 19$$

So the value of the polynomial is 19 when $x = 1$.

b) Replace x by -2 in the polynomial:

$$-3x^4 - x^3 + 20x + 3 = -3(-2)^4 - (-2)^3 + 20(-2) + 3$$
$$= -3(16) - (-8) - 40 + 3$$
$$= -48 + 8 - 40 + 3$$
$$= -77$$

So the value of the polynomial is -77 when $x = -2$.

c) This is a repeat of part (a) using the function notation from Chapter 4. $P(1)$, read "P of 1," is the value of the polynomial $P(x)$ when x is 1. To find $P(1)$, replace x by 1 in the formula for $P(x)$:

$$P(x) = -3x^4 - x^3 + 20x + 3$$
$$P(1) = -3(1)^4 - (1)^3 + 20(1) + 3$$
$$= 19$$

So $P(1) = 19$. The value of the polynomial when $x = 1$ is 19.   ■

calculator

close-up

To evaluate the polynomial in Example 3 with a calculator, first use Y= to define the polynomial.

```
Plot1  Plot2  Plot3
\Y1◻-3X^4-X^3+20
X+3
\Y2=
\Y3=
\Y4=
\Y5=
\Y6=
```

Then find $y_1(-2)$ and $y_1(1)$.

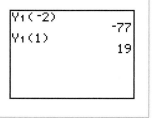

```
Y1(-2)
              -77
Y1(1)
               19
```

Addition of Polynomials

You learned how to combine like terms in Chapter 1. Also, you combined like terms when solving equations in Chapter 2. Addition of polynomials is done simply by adding the like terms.

> ### Addition of Polynomials
>
> To add two polynomials, add the like terms.

Polynomials can be added horizontally or vertically, as shown in the next example.

E X A M P L E   4

Adding polynomials

Perform the indicated operation.

a) $(x^2 - 6x + 5) + (-3x^2 + 5x - 9)$

b) $(-5a^3 + 3a - 7) + (4a^2 - 3a + 7)$

Solution

helpful / **hint**

When we perform operations with polynomials and write the results as equations, those equations are identities. For example,

$(x + 1) + (3x + 5) = 4x + 6$

is an identity. This equation is satisfied by every real number.

a) We can use the commutative and associative properties to get the like terms next to each other and then combine them:

$$(x^2 - 6x + 5) + (-3x^2 + 5x - 9) = x^2 - 3x^2 - 6x + 5x + 5 - 9$$
$$= -2x^2 - x - 4$$

b) When adding vertically, we line up the like terms:

$$
\begin{array}{r}
-5a^3 \qquad\quad + 3a - 7 \\
4a^2 - 3a + 7 \\
\hline
-5a^3 + 4a^2 \qquad\qquad
\end{array}
\quad \text{Add.}
$$

Subtraction of Polynomials

When we subtract polynomials, we subtract the like terms. Because $a - b = a + (-b)$, we can subtract by adding the opposite of the second polynomial to the first polynomial. Remember that a negative sign in front of parentheses changes the sign of each term in the parentheses. For example,

$$-(x^2 - 2x + 8) = -x^2 + 2x - 8.$$

Polynomials can be subtracted horizontally or vertically, as shown in the next example.

E X A M P L E   5

Subtracting polynomials

Perform the indicated operation.

a) $(x^2 - 5x - 3) - (4x^2 + 8x - 9)$ b) $(4y^3 - 3y + 2) - (5y^2 - 7y - 6)$

Solution

a) $(x^2 - 5x - 3) - (4x^2 + 8x - 9) = x^2 - 5x - 3 - 4x^2 - 8x + 9$ Change signs.
$$= -3x^2 - 13x + 6 \qquad\qquad \text{Add.}$$

b) To subtract $5y^2 - 7y - 6$ from $4y^3 - 3y + 2$ vertically, we line up the like terms as we do for addition:

$$\begin{array}{r} 4y^3 \quad\;\; - 3y + 2 \\ -\quad (5y^2 - 7y - 6) \\ \hline \end{array}$$

Now change the signs of $5y^2 - 7y - 6$ and add the like terms:

$$\begin{array}{r} 4y^3 \quad\;\; - 3y + 2 \\ -5y^2 + 7y + 6 \\ \hline 4y^3 - 5y^2 + 4y + 8 \end{array}$$

CAUTION   When adding or subtracting polynomials vertically, be sure to line up the like terms.

In the next example we combine addition and subtraction of polynomials.

EXAMPLE 6   **Adding and subtracting**
Perform the indicated operations:

$$(2x^2 - 3x) + (x^3 + 6) - (x^4 - 6x^2 - 9)$$

Solution
Remove the parentheses and combine the like terms:

$$(2x^2 - 3x) + (x^3 + 6) - (x^4 - 6x^2 - 9) = 2x^2 - 3x + x^3 + 6 - x^4 + 6x^2 + 9$$
$$= -x^4 + x^3 + 8x^2 - 3x + 15$$

Applications

Polynomials are often used to represent unknown quantities. In certain situations it is necessary to add or subtract such polynomials.

EXAMPLE 7   **Profit from prints**
Trey pays $60 per day for a permit to sell famous art prints in the Student Union Mall. Each print costs him $4, so the polynomial $4x + 60$ represents his daily cost in dollars for x prints sold. He sells the prints for $10 each. So the polynomial $10x$ represents his daily revenue for x prints sold. Find a polynomial that represents his daily profit from selling x prints. Evaluate the profit polynomial for $x = 30$.

Solution
Because profit is revenue minus cost, we can subtract the corresponding polynomials to get a polynomial that represents the daily profit:

$$\begin{aligned} \text{Profit} &= \text{Revenue} - \text{Cost} \\ &= 10x - (4x + 60) \\ &= 10x - 4x - 60 \\ &= 6x - 60 \end{aligned}$$

His daily profit from selling x prints is $6x - 60$ dollars. Evaluate this profit polynomial for $x = 30$:

$$6x - 60 = 6(30) - 60 = 120$$

So if Trey sells 30 prints, his profit is $120.

M A T H A T W O R K $x^2 + (x+1)^2 = 5^2$

The message we hear today about healthy eating is "more fiber and less fat." But healthy eating can be challenging. Jill Brown, registered dietitian and owner of Healthy Habits Nutritional Counseling and Consulting, provides nutritional counseling for people who are basically healthy but interested in improving their eating habits.

On a client's first visit, an extensive nutrition, medical, and family history is taken. Behavior patterns are discussed, and nutritional goals are set. Ms. Brown strives for a realistic rather than an idealistic approach. Whenever possible, a client should eat foods that are high in vitamin content rather than take vitamin pills. Moreover, certain frame types and family history make a slight and slender look difficult to achieve. Here, Ms. Brown might use the Harris-Benedict formula to determine the daily basal energy expenditure based on age, gender, and size. In addition to diet, exercise is discussed and encouraged. Finally, Ms. Brown provides a support system and serves as a "nutritional coach." By following her advice, many of her clients lower their blood pressure, reduce their cholesterol, have more energy, and look and feel better.

In Exercises 101 and 102 of this section you will use the Harris-Benedict formula to calculate the basal energy expenditure for a female and a male.

NUTRITIONIST

WARM-UPS

True or false? Explain your answer.

1. In the polynomial $2x^2 - 4x + 7$ the coefficient of x is 4.
2. The degree of the polynomial $x^2 + 5x - 9x^3 + 6$ is 2.
3. In the polynomial $x^2 - x$ the coefficient of x is -1.
4. The degree of the polynomial $x^2 - x$ is 2.
5. A binomial always has a degree of 2.
6. The polynomial $3x^2 - 5x + 9$ is a trinomial.
7. Every trinomial has degree 2.
8. $x^2 - 7x^2 = -6x^2$ for any value of x.
9. $(3x^2 - 8x + 6) + (x^2 + 4x - 9) = 4x^2 - 4x - 3$ for any value of x.
10. $(x^2 - 4x) - (x^2 - 3x) = -7x$ for any value of x.

5.1 EXERCISES

Reading and Writing *After reading this section, write out the answers to these questions. Use complete sentences.*

1. What is a term?

2. What is a polynomial?

3. What is the degree of a polynomial?

4. What is the value of a polynomial?

5. How do we add polynomials?

6. How do we subtract polynomials?

Determine the coefficients of x^3 and x^2 in each polynomial. See Example 1.

7. $-3x^3 + 7x^2$ **8.** $10x^3 - x^2$

9. $x^4 + 6x^2 - 9$ **10.** $x^5 - x^3 + 3$

11. $\dfrac{x^3}{3} + \dfrac{7x^2}{2} - 4$ **12.** $\dfrac{x^3}{2} - \dfrac{x^2}{4} + 2x + 1$

Identify each polynomial as a monomial, binomial, or trinomial and state its degree. See Example 2.

13. -1 **14.** 5 **15.** m^3

16. $3a^8$ **17.** $4x + 7$ **18.** $a + 6$

19. $x^{10} - 3x^2 + 2$ **20.** $y^6 - 6y^3 + 9$

21. $x^6 + 1$ **22.** $b^2 - 4$

23. $a^3 - a^2 + 5$ **24.** $-x^2 + 4x - 9$

Evaluate each polynomial as indicated. See Example 3.

25. Evaluate $2x^2 - 3x + 1$ for $x = -1$.

26. Evaluate $3x^2 - x + 2$ for $x = -2$.

27. Evaluate $-3x^3 - x^2 + 3x - 4$ for $x = 3$.

28. Evaluate $-2x^4 - 3x^2 + 5x - 9$ for $x = 2$.

29. If $P(x) = 3x^4 - 2x^3 + 7$, find $P(-2)$.

30. If $P(x) = -2x^3 + 5x^2 - 12$, find $P(5)$.

31. If $P(x) = 1.2x^3 - 4.3x - 2.4$, find $P(1.45)$.

32. If $P(x) = -3.5x^4 - 4.6x^3 + 5.5$, find $P(-2.36)$.

Perform the indicated operation. See Example 4.

33. $(x - 3) + (3x - 5)$ **34.** $(x - 2) + (x + 3)$

35. $(q - 3) + (q + 3)$ **36.** $(q + 4) + (q + 6)$

37. $(3x + 2) + (x^2 - 4)$ **38.** $(5x^2 - 2) + (-3x^2 - 1)$

39. $(4x - 1) + (x^3 + 5x - 6)$

40. $(3x - 7) + (x^2 - 4x + 6)$

41. $(a^2 - 3a + 1) + (2a^2 - 4a - 5)$

42. $(w^2 - 2w + 1) + (2w - 5 + w^2)$

43. $(w^2 - 9w - 3) + (w - 4w^2 + 8)$

44. $(a^3 - a^2 - 5a) + (6 - a - 3a^2)$

45. $(5.76x^2 - 3.14x - 7.09) + (3.9x^2 + 1.21x + 5.6)$

46. $(8.5x^2 + 3.27x - 9.33) + (x^2 - 4.39x - 2.32)$

Perform the indicated operation. See Example 5.

47. $(x - 2) - (5x - 8)$

48. $(x - 7) - (3x - 1)$

49. $(m - 2) - (m + 3)$

50. $(m + 5) - (m + 9)$

51. $(2z^2 - 3z) - (3z^2 - 5z)$

52. $(z^2 - 4z) - (5z^2 - 3z)$

53. $(w^5 - w^3) - (-w^4 + w^2)$

54. $(w^6 - w^3) - (-w^2 + w)$

55. $(t^2 - 3t + 4) - (t^2 - 5t - 9)$

56. $(t^2 - 6t + 7) - (5t^2 - 3t - 2)$

57. $(9 - 3y - y^2) - (2 + 5y - y^2)$

58. $(4 - 5y + y^3) - (2 - 3y + y^2)$

59. $(3.55x - 879) - (26.4x - 455.8)$

60. $(345.56x - 347.4) - (56.6x + 433)$

Add or subtract the polynomials as indicated. See Examples 4 and 5.

61. Add:
$$3a - 4$$
$$\underline{a + 6}$$

62. Add:
$$2w - 8$$
$$\underline{w + 3}$$

63. Subtract:
$$3x + 11$$
$$\underline{5x + 7}$$

64. Subtract:
$$4x + 3$$
$$\underline{2x + 9}$$

65. Add:
$$a - b$$
$$\underline{a + b}$$

66. Add:
$$s - 6$$
$$\underline{s - 1}$$

67. Subtract:
$$-3m + 1$$
$$\underline{2m - 6}$$

68. Subtract:
$$-5n + 2$$
$$\underline{3n - 4}$$

69. Add:
$$2x^2 - x - 3$$
$$\underline{2x^2 + x + 4}$$

70. Add:
$$-x^2 + 4x - 6$$
$$\underline{3x^2 - x - 5}$$

71. Subtract:
$$3a^3 - 5a^2 + 7$$
$$\underline{2a^3 + 4a^2 - 2a}$$

72. Subtract:
$$-2b^3 + 7b^2 \qquad - 9$$
$$\underline{b^3 \qquad - 4b - 2}$$

73. Subtract:
$$x^2 - 3x + 6$$
$$\underline{x^2 \qquad - 3}$$

74. Subtract:
$$x^4 - 3x^2 + 2$$
$$\underline{3x^4 - 2x^2}$$

75. Add:
$$y^3 + 4y^2 - 6y - 5$$
$$\underline{y^3 + 3y^2 + 2y - 9}$$

76. Add:
$$q^2 - 4q + 9$$
$$\underline{-3q^2 - 7q + 5}$$

Perform the operations indicated.

77. Find the sum of $2m - 9$ and $3m + 4$.

78. Find the sum of $-3n - 2$ and $6m - 3$.

79. Find the difference when $7y - 3$ is subtracted from $9y - 2$.

80. Find the difference when $-2y - 1$ is subtracted from $3y - 4$.

81. Subtract $x^2 - 3x - 1$ from the sum of $2x^2 - x + 3$ and $x^2 + 5x - 9$.

82. Subtract $-2y^2 + 3y - 8$ from the sum of $-3y^2 - 2y + 6$ and $7y^2 + 8y - 3$.

Perform the indicated operations. See Example 6.

83. $(4m - 2) + (2m + 4) - (9m - 1)$

84. $(-5m - 6) + (8m - 3) - (-5m + 3)$

85. $(6y - 2) - (8y + 3) - (9y - 2)$

86. $(-5y - 1) - (8y - 4) - (y + 3)$

87. $(-x^2 - 5x + 4) + (6x^2 - 8x + 9) - (3x^2 - 7x + 1)$

88. $(-8x^2 + 5x - 12) + (-3x^2 - 9x + 18) - (-3x^2 + 9x - 4)$

89. $(-6z^4 - 3z^3 + 7z^2) - (5z^3 + 3z^2 - 2) + (z^4 - z^2 + 5)$

90. $(-v^3 - v^2 - 1) - (v^4 - v^2 - v - 1) + (v^3 - 3v^2 + 6)$

Solve each problem. See Example 7.

91. *Profitable pumps.* Walter Waterman, of Walter's Water Pumps in Winnipeg has found that when he produces x water pumps per month, his revenue is $x^2 + 400x + 300$ dollars. His cost for producing x water pumps per month is $x^2 + 300x - 200$ dollars. Write a polynomial that represents his monthly profit for x water pumps. Evaluate this profit polynomial for $x = 50$.

92. *Manufacturing costs.* Ace manufacturing has determined that the cost of labor for producing x transmissions is $0.3x^2 + 400x + 550$ dollars, while the cost of materials is $0.1x^2 + 50x + 800$ dollars.
  a) Write a polynomial that represents the total cost of materials and labor for producing x transmissions.
  b) Evaluate the total cost polynomial for $x = 500$.

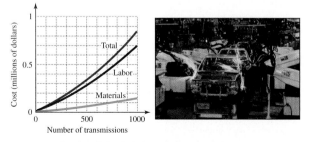

FIGURE FOR EXERCISE 92

c) Find the cost of labor for 500 transmissions and the cost of materials for 500 transmissions.

93. *Perimeter of a triangle.* The shortest side of a triangle is x meters, and the other two sides are $3x - 1$ and $2x + 4$ meters. Write a polynomial that represents the perimeter and then evaluate the perimeter polynomial if x is 4 meters.

94. *Perimeter of a rectangle.* The width of a rectangular playground is $2x - 5$ feet, and the length is $3x + 9$ feet. Write a polynomial that represents the perimeter and then evaluate this perimeter polynomial if x is 4 feet.

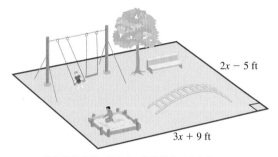

$2x - 5$ ft

$3x + 9$ ft

FIGURE FOR EXERCISE 94

95. *Before and after.* Jessica traveled $2x + 50$ miles in the morning and $3x - 10$ miles in the afternoon. Write a polynomial that represents the total distance that she traveled. Find the total distance if $x = 20$.

96. *Total distance.* Hanson drove his rig at x mph for 3 hours, then increased his speed to $x + 15$ mph and drove for 2 more hours. Write a polynomial that represents the total distance that he traveled. Find the total distance if $x = 45$ mph.

97. *Sky divers.* Bob and Betty simultaneously jump from two airplanes at different altitudes. Bob's altitude t seconds after leaving the plane is $-16t^2 + 6600$ feet. Betty's altitude t seconds after leaving the plane is $-16t^2 + 7400$ feet. Write a polynomial that represents the difference between their altitudes t seconds after leaving the planes. What is the difference between their altitudes 3 seconds after leaving the planes?

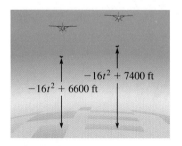

$-16t^2 + 7400$ ft

$-16t^2 + 6600$ ft

FIGURE FOR EXERCISE 97

98. *Height difference.* A red ball and a green ball are simultaneously tossed into the air. The red ball is given an initial velocity of 96 feet per second, and its height t seconds after it is tossed is $-16t^2 + 96t$ feet. The green ball is given an initial velocity of 30 feet per second, and its height t seconds after it is tossed is $-16t^2 + 80t$ feet.

a) Find a polynomial that represents the difference in the heights of the two balls.

b) How much higher is the red ball 2 seconds after the balls are tossed?

c) In reality, when does the difference in the heights stop increasing?

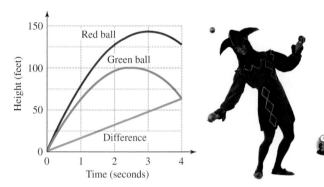

FIGURE FOR EXERCISE 98

99. *Total interest.* Donald received $0.08(x + 554)$ dollars interest on one investment and $0.09(x + 335)$ interest on another investment. Write a polynomial that represents the total interest he received. What is the total interest if $x = 1000$?

100. *Total acid.* Deborah figured that the amount of acid in one bottle of solution is $0.12x$ milliliters and the amount of acid in another bottle of solution is $0.22(75 - x)$ milliliters. Find a polynomial that represents the total amount of acid. What is the total amount of acid if $x = 50$?

101. *Harris-Benedict for females.* The Harris-Benedict polynomial

$$655.1 + 9.56w + 1.85h - 4.68a$$

represents the number of calories needed to maintain a female at rest for 24 hours, where w is her weight in kilograms, h is her height in centimeters, and a is her age. Find the number of calories needed by a 30-year-old 54-kilogram female who is 157 centimeters tall.

102. *Harris-Benedict for males.* The Harris-Benedict polynomial

$$66.5 + 13.75w + 5.0h - 6.78a$$

represents the number of calories needed to maintain a male at rest for 24 hours, where w is his weight in kilograms, h is his height in centimeters, and a is his age. Find the number of calories needed by a 40-year-old 90-kilogram male who is 185 centimeters tall.

GETTING MORE INVOLVED

103. *Discussion.* Is the sum of two natural numbers always a natural number? Is the sum of two integers always an integer? Is the sum of two polynomials always a polynomial? Explain.

104. *Discussion.* Is the difference of two natural numbers always a natural number? Is the difference of two rational numbers always a rational number? Is the difference of two polynomials always a polynomial? Explain.

105. *Writing.* Explain why the polynomial $2^4 - 7x^3 + 5x^2 - x$ has degree 3 and not degree 4.

106. *Discussion.* Which of the following polynomials does not have degree 2? Explain.
a) πr^2 **b)** $\pi^2 - 4$ **c)** $y^2 - 4$ **d)** $x^2 - x^4$
e) $a^2 - 3a + 9$

5.2 MULTIPLICATION OF POLYNOMIALS

You learned to multiply some polynomials in Chapter 1. In this section you will learn how to multiply any two polynomials.

In this section

- Multiplying Monomials with the Product Rule
- Multiplying Polynomials
- The Opposite of a Polynomial
- Applications

Multiplying Monomials with the Product Rule

To multiply two monomials, such as x^3 and x^5, recall that

$$x^3 = x \cdot x \cdot x \qquad \text{and} \qquad x^5 = x \cdot x \cdot x \cdot x \cdot x.$$

So

$$\underbrace{x^3 \cdot x^5 = \overbrace{(x \cdot x \cdot x)}^{3 \text{ factors}}\overbrace{(x \cdot x \cdot x \cdot x \cdot x)}^{5 \text{ factors}}}_{8 \text{ factors}} = x^8.$$

The exponent of the product of x^3 and x^5 is the sum of the exponents 3 and 5. This example illustrates the **product rule** for multiplying exponential expressions.

> **Product Rule**
>
> If a is any real number and m and n are any positive integers, then
>
> $$a^m \cdot a^n = a^{m+n}.$$

E X A M P L E   1   **Multiplying monomials**

Find the indicated products.

a) $x^2 \cdot x^4 \cdot x$ b) $(-2ab)(-3ab)$ c) $-4x^2y^2 \cdot 3xy^5$ d) $(3a)^2$

Solution

a) $x^2 \cdot x^4 \cdot x = x^2 \cdot x^4 \cdot x^1$
$\qquad\qquad = x^7$ Product rule

b) $(-2ab)(-3ab) = (-2)(-3) \cdot a \cdot a \cdot b \cdot b$
$\qquad\qquad\qquad = 6a^2b^2$ Product rule

c) $(-4x^2y^2)(3xy^5) = (-4)(3)x^2 \cdot x \cdot y^2 \cdot y^5$
$\qquad\qquad\qquad\quad = -12x^3y^7$ Product rule

d) $(3a)^2 = 3a \cdot 3a$
$\qquad\quad = 9a^2$

C A U T I O N   Be sure to distinguish between adding and multiplying monomials. You can add like terms to get $3x^4 + 2x^4 = 5x^4$, but you cannot combine the terms in $3w^5 + 6w^2$. However, you can multiply any two monomials: $3x^4 \cdot 2x^4 = 6x^8$ and $3w^5 \cdot 6w^2 = 18w^7$.

Multiplying Polynomials

To multiply a monomial and a polynomial, we use the distributive property.

E X A M P L E   2   **Multiplying monomials and polynomials**

Find each product.

a) $3x^2(x^3 - 4x)$ b) $(y^2 - 3y + 4)(-2y)$ c) $-a(b - c)$

Solution

a) $3x^2(x^3 - 4x) = 3x^2 \cdot x^3 - 3x^2 \cdot 4x$ Distributive property
$\qquad\qquad\quad = 3x^5 - 12x^3$

b) $(y^2 - 3y + 4)(-2y) = y^2(-2y) - 3y(-2y) + 4(-2y)$ Distributive property
$\qquad\qquad\qquad\qquad = -2y^3 - (-6y^2) + (-8y)$
$\qquad\qquad\qquad\qquad = -2y^3 + 6y^2 - 8y$

c) $-a(b - c) = (-a)b - (-a)c$ Distributive property

$$= -ab + ac$$

$$= ac - ab$$

Note in part (c) that either of the last two binomials is the correct answer. The last one is just a little simpler to read.

Just as we use the distributive property to find the product of a monomial and a polynomial, we can use the distributive property to find the product of two binomials and the product of a binomial and a trinomial. Before simplifying, the product of two binomials has four terms and the product of a binomial and a trinomial has six terms.

E X A M P L E   3   **Multiplying polynomials**

Use the distributive property to find each product.

a) $(x + 2)(x + 5)$ **b)** $(x + 3)(x^2 + 2x - 7)$

Solution

a) First multiply each term of $x + 5$ by $x + 2$:

$$(x + 2)(x + 5) = (x + 2)x + (x + 2)5$$ Distributive property

$$= x^2 + 2x + 5x + 10$$ Distributive property

$$= x^2 + 7x + 10$$ Combine like terms.

b) First multiply each term of the trinomial by $x + 3$:

$$(x + 3)(x^2 + 2x - 7) = (x + 3)x^2 + (x + 3)2x + (x + 3)(-7)$$ Distributive property

$$= x^3 + 3x^2 + 2x^2 + 6x - 7x - 21$$ Distributive property

$$= x^3 + 5x^2 - x - 21$$ Combine like terms.

Products of polynomials can also be found by arranging the multiplication vertically like multiplication of whole numbers.

E X A M P L E   4   **Multiplying vertically**

Find each product.

a) $(x - 2)(3x + 7)$ **b)** $(x + 2)(x^2 - x + 1)$

helpful hint

Many students find vertical multiplication easier than applying the distributive property twice horizontally. However, you should learn both methods because horizontal multiplication will help you with factoring by grouping in Section 6.2.

Solution

a)
$$
\begin{array}{r}
3x + 7 \\
x - 2 \\
\hline
-6x - 14 \\
3x^2 + 7x \\
\hline
3x^2 + x - 14
\end{array}
$$
$\leftarrow -2$ times $3x + 7$
$\leftarrow x$ times $3x + 7$
Add.

b)
$$
\begin{array}{r}
x^2 - x + 1 \\
x + 2 \\
\hline
2x^2 - 2x + 2 \\
x^3 - x^2 + x \\
\hline
x^3 + x^2 - x + 2
\end{array}
$$

These examples illustrate the following rule.

Multiplication of Polynomials

To multiply polynomials, multiply each term of one polynomial by every term of the other polynomial, then combine like terms.

The Opposite of a Polynomial

The opposite or additive inverse of y is $-y$ because $y + (-y) = 0$. *To find the opposite of a polynomial we change the sign of every term of the polynomial.* For example, the opposite of $x^2 - 3x + 1$ is $-x^2 + 3x - 1$ because

$$(x^2 - 3x + 1) + (-x^2 + 3x - 1) = 0.$$

We also write $-(x^2 - 3x + 1) = -x^2 + 3x - 1$. As another example, we find the opposite of $a - b$:

$$-(a - b) = -a + b = b - a$$

So $-(a - b) = b - a$. Since $(a - b) + (b - a) = 0$, we can be sure that $a - b$ and $b - a$ are opposites or additive inverses of each other.

CAUTION The opposite $a + b$ is not $a - b$ because $(a + b) + (a - b) = 2a$. The opposite of $a + b$ is $-a - b$ because $(a + b) + (-a - b) = 0$.

E X A M P L E   5

Opposite of a polynomial

Find the opposite of each polynomial.

a) $x - 2$ **b)** $9 - y^2$ **c)** $a + 4$ **d)** $-x^2 + 6x - 3$

Solution

a) $-(x - 2) = 2 - x$
b) $-(9 - y^2) = y^2 - 9$
c) $-(a + 4) = -a - 4$
d) $-(-x^2 + 6x - 3) = x^2 - 6x + 3$ ∎

Applications

E X A M P L E   6

Multiplying polynomials

A parking lot is 20 yards wide and 30 yards long. If the college increases the length and width by the same amount to handle an increasing number of cars, then what polynomial represents the area of the new lot? What is the new area if the increase is 15 yards?

Solution

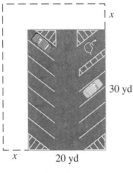

If x is the amount of increase, then the new lot will be $x + 20$ yards wide and $x + 30$ yards long as shown in Fig. 5.1. Multiply the length and width to get the area:

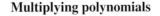

$$(x + 20)(x + 30) = (x + 20)x + (x + 20)30$$
$$= x^2 + 20x + 30x + 600$$
$$= x^2 + 50x + 600$$

F I G U R E   5 . 1

The polynomial $x^2 + 50x + 600$ represents the area of the new lot. If $x = 15$, then

$$x^2 + 50x + 600 = (15)^2 + 50(15) + 600 = 1575.$$

If the increase is 15 yards, then the area of the lot will be 1575 square yards. ■

WARM-UPS

True or false? Explain your answer.

1. $3x^3 \cdot 5x^4 = 15x^{12}$ for any value of x.
2. $3x^2 \cdot 2x^7 = 5x^9$ for any value of x.
3. $(3y^3)^2 = 9y^6$ for any value of y.
4. $-3x(5x - 7x^2) = -15x^3 + 21x^2$ for any value of x.
5. $2x(x^2 - 3x + 4) = 2x^3 - 6x^2 + 8x$ for any number x.
6. $-2(3 - x) = 2x - 6$ for any number x.
7. $(a + b)(c + d) = ac + ad + bc + bd$ for any values of $a, b, c,$ and d.
8. $-(x - 7) = 7 - x$ for any value of x.
9. $83 - 37 = -(37 - 83)$
10. The opposite of $x + 3$ is $x - 3$ for any number x.

5.2   EXERCISES

Reading and Writing   *After reading this section, write out the answers to these questions. Use complete sentences.*

1. What is the product rule for exponents?

2. Why is the sum of two monomials not necessarily a monomial?

3. What property of the real numbers is used when multiplying a monomial and a polynomial?

4. What property of the real numbers is used when multiplying two binomials?

5. How do we multiply any two polynomials?

6. How do we find the opposite of a polynomial?

Find each product. See Example 1.

7. $3x^2 \cdot 9x^3$ 8. $5x^7 \cdot 3x^5$ 9. $2a^3 \cdot 7a^8$

10. $3y^{12} \cdot 5y^{15}$ 11. $-6x^2 \cdot 5x^2$ 12. $-2x^2 \cdot 8x^5$

13. $(-9x^{10})(-3x^7)$ 14. $(-2x^2)(-8x^9)$ 15. $-6st \cdot 9st$

16. $-12sq \cdot 3s$ 17. $3wt \cdot 8w^7t^6$ 18. $h^8k^3 \cdot 5h$

19. $(5y)^2$ 20. $(6x)^2$

21. $(2x^3)^2$ 22. $(3y^5)^2$

Find each product. See Example 2.
23. $4y^2(y^5 - 2y)$
24. $6t^3(t^5 + 3t^2)$
25. $-3y(6y - 4)$
26. $-9y(y^2 - 1)$
27. $(y^2 - 5y + 6)(-3y)$
28. $(x^3 - 5x^2 - 1)7x^2$
29. $-x(y^2 - x^2)$

30. $-ab(a^2 - b^2)$

31. $(3ab^3 - a^2b^2 - 2a^3b)5a^3$

32. $(3c^2d - d^3 + 1)8cd^2$

33. $-\dfrac{1}{2}t^2v(4t^3v^2 - 6tv - 4v)$

34. $-\dfrac{1}{3}m^2n^3(-6mn^2 + 3mn - 12)$

Use the distributive property to find each product. See Example 3.

35. $(x + 1)(x + 2)$

36. $(x + 6)(x + 3)$

37. $(x - 3)(x + 5)$

38. $(y - 2)(y + 4)$

39. $(t - 4)(t - 9)$

40. $(w - 3)(w - 5)$

41. $(x + 1)(x^2 + 2x + 2)$

42. $(x - 1)(x^2 + x + 1)$

43. $(3y + 2)(2y^2 - y + 3)$

44. $(4y + 3)(y^2 + 3y + 1)$

45. $(y^2z - 2y^4)(y^2z + 3z^2 - y^4)$

46. $(m^3 - 4mn^2)(6m^4n^2 - 3m^6 + m^2n^4)$

Find each product vertically. See Example 4.

47.
$$\begin{array}{r} 2a - 3 \\ a + 5 \\ \hline \end{array}$$

48.
$$\begin{array}{r} 2w - 6 \\ w + 5 \\ \hline \end{array}$$

49.
$$\begin{array}{r} 7x + 30 \\ 2x + 5 \\ \hline \end{array}$$

50.
$$\begin{array}{r} 5x + 7 \\ 3x + 6 \\ \hline \end{array}$$

51.
$$\begin{array}{r} 5x + 2 \\ 4x - 3 \\ \hline \end{array}$$

52.
$$\begin{array}{r} 4x + 3 \\ 2x - 6 \\ \hline \end{array}$$

53.
$$\begin{array}{r} m - 3n \\ 2a + b \\ \hline \end{array}$$

54.
$$\begin{array}{r} 3x + 7 \\ a - 2b \\ \hline \end{array}$$

55.
$$\begin{array}{r} x^2 + 3x - 2 \\ x + 6 \\ \hline \end{array}$$

56.
$$\begin{array}{r} -x^2 + 3x - 5 \\ x - 7 \\ \hline \end{array}$$

57.
$$\begin{array}{r} 2a^3 - 3a^2 + 4 \\ -2a - 3 \\ \hline \end{array}$$

58.
$$\begin{array}{r} -3x^2 + 5x - 2 \\ -5x - 6 \\ \hline \end{array}$$

59.
$$\begin{array}{r} x - y \\ x + y \\ \hline \end{array}$$

60.
$$\begin{array}{r} a^2 + b^2 \\ a^2 - b^2 \\ \hline \end{array}$$

61.
$$\begin{array}{r} x^2 - xy + y^2 \\ x + y \\ \hline \end{array}$$

62.
$$\begin{array}{r} 4w^2 + 2wv + v^2 \\ 2w - v \\ \hline \end{array}$$

Find the opposite of each polynomial. See Example 5.

63. $3t - u$

64. $-3t - u$

65. $3x + y$

66. $x - 3y$

67. $-3a^2 - a + 6$

68. $3b^2 - b - 6$

69. $3v^2 + v - 6$

70. $-3t^2 + t - 6$

Perform the indicated operation.

71. $-3x(2x - 9)$

72. $-1(2 - 3x)$

73. $2 - 3x(2x - 9)$

74. $6 - 3(4x - 8)$

75. $(2 - 3x) + (2x - 9)$

76. $(2 - 3x) - (2x - 9)$

77. $(6x^6)^2$

78. $(-3a^3b)^2$

79. $3ab^3(-2a^2b^7)$

80. $-4xst \cdot 8xs$

81. $(5x + 6)(5x + 6)$

82. $(5x - 6)(5x - 6)$

83. $(5x - 6)(5x + 6)$

84. $(2x - 9)(2x + 9)$

85. $2x^2(3x^5 - 4x^2)$

86. $4a^3(3ab^3 - 2ab^3)$

87. $(m - 1)(m^2 + m + 1)$

88. $(a + b)(a^2 - ab + b^2)$

89. $(3x - 2)(x^2 - x - 9)$

90. $(5 - 6y)(3y^2 - y - 7)$

Solve each problem. See Example 6.

91. *Office space.* The length of a professor's office is x feet, and the width is $x + 4$ feet. Write a polynomial that represents the area. Find the area if $x = 10$ ft.

92. *Swimming space.* The length of a rectangular swimming pool is $2x - 1$ meters, and the width is $x + 2$ meters. Write a polynomial that represents the area. Find the area if x is 5 meters.

93. *Area of a truss.* A roof truss is in the shape of a triangle with a height of x feet and a base of $2x + 1$ feet. Write a polynomial that represents the area of the triangle. What is the area if x is 5 feet?

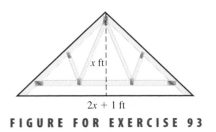

x ft

$2x + 1$ ft

FIGURE FOR EXERCISE 93

94. *Volume of a box.* The length, width, and height of a box are x, $2x$, and $3x - 5$ inches, respectively. Write a polynomial that represents its volume.

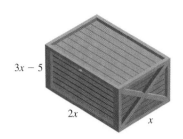

FIGURE FOR EXERCISE 94

95. *Number pairs.* If two numbers differ by 5, then what polynomial represents their product?

96. *Number pairs.* If two numbers have a sum of 9, then what polynomial represents their product?

97. *Area of a rectangle.* The length of a rectangle is $2.3x + 1.2$ meters, and its width is $3.5x + 5.1$ meters. What polynomial represents its area?

98. *Patchwork.* A quilt patch cut in the shape of a triangle has a base of $5x$ inches and a height of $1.732x$ inches. What polynomial represents its area?

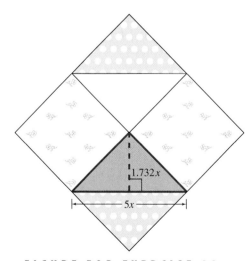

FIGURE FOR EXERCISE 98

99. *Total revenue.* If a promoter charges p dollars per ticket for a concert in Tulsa, then she expects to sell $40,000 - 1000p$ tickets to the concert. How many tickets will she sell if the tickets are $10 each? Find the total revenue when the tickets are $10 each. What polynomial represents the total revenue expected for the concert when the tickets are p dollars each?

100. *Manufacturing shirts.* If a manufacturer charges p dollars each for rugby shirts, then he expects to sell $2000 - 100p$ shirts per week. What polynomial represents the total revenue expected for a week? How many shirts will be sold if the manufacturer charges $20 each for the shirts? Find the total revenue when the shirts are sold for $20 each. Use the bar graph to determine the price that will give the maximum total revenue.

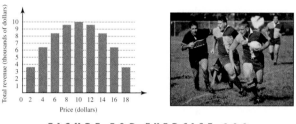

FIGURE FOR EXERCISE 100

101. *Periodic deposits.* At the beginning of each year for 5 years, an investor invests $10 in a mutual fund with an average annual return of r. If we let $x = 1 + r$, then at the end of the first year (just before the next investment) the value is $10x$ dollars. Because $10 is then added to the $10x$ dollars, the amount at the end of the second year is $(10x + 10)x$ dollars. Find a polynomial that represents the value of the investment at the end of the fifth year. Evaluate this polynomial if $r = 10\%$.

102. *Increasing deposits.* At the beginning of each year for 5 years, an investor invests in a mutual fund with an average annual return of r. The first year, she invests $10; the second year, she invests $20; the third year, she invests $30; the fourth year, she invests $40; the fifth year, she invests $50. Let $x = 1 + r$ as in Exercise 101 and write a polynomial in x that represents the value of the investment at the end of the fifth year. Evaluate this polynomial for $r = 8\%$.

GETTING MORE INVOLVED

103. *Discussion.* Name all properties of the real numbers that are used in finding the following products:
a) $-2ab^3c^2 \cdot 5a^2bc$ **b)** $(x^2 + 3)(x^2 - 8x - 6)$

104. *Discussion.* Find the product of 27 and 436 without using a calculator. Then use the distributive property to find the product $(20 + 7)(400 + 30 + 6)$ as you would find the product of a binomial and a trinomial. Explain how the two methods are related.

5.3 MULTIPLICATION OF BINOMIALS

In this

section

• The FOIL Method

• Multiplying Binomials Quickly

In Section 5.2 you learned to multiply polynomials. In this section you will learn a rule that makes multiplication of binomials simpler.

The FOIL Method

We can use the distributive property to find the product of two binomials. For example,

$$(x + 2)(x + 3) = (x + 2)x + (x + 2)3 \quad \text{Distributive property}$$
$$= x^2 + 2x + 3x + 6 \quad \text{Distributive property}$$
$$= x^2 + 5x + 6 \quad \text{Combine like terms.}$$

There are four terms in $x^2 + 2x + 3x + 6$. The term x^2 is the product of the *first* term of each binomial, x and x. The term $3x$ is the product of the two *outer* terms, 3 and x. The term $2x$ is the product of the two *inner* terms, 2 and x. The term 6 is the product of the last term of each binomial, 2 and 3. We can connect the terms multiplied by lines as follows:

$$(x + 2)(x + 3)$$

F = First terms
O = Outer terms
I = Inner terms
L = Last terms

If you remember the word FOIL, you can get the product of the two binomials much faster than writing out all of the steps above. This method is called the **FOIL method.** The name should make it easier to remember.

EXAMPLE 1

Using the FOIL method

Find each product.

a) $(x + 2)(x - 4)$

b) $(2x + 5)(3x - 4)$

c) $(a - b)(2a - b)$

d) $(x + 3)(y + 5)$

Solution

helpful hint

You may have to practice FOIL a while to get good at it. However, the better you are at FOIL, the easier you will find factoring in Chapter 6.

a) 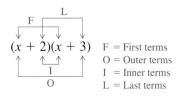 $(x + 2)(x - 4) = x^2 - 4x + 2x - 8$

$$= x^2 - 2x - 8 \quad \text{Combine the like terms.}$$

b) $(2x + 5)(3x - 4) = 6x^2 - 8x + 15x - 20$

$$= 6x^2 + 7x - 20 \quad \text{Combine the like terms.}$$

c) $(a - b)(2a - b) = 2a^2 - ab - 2ab + b^2$

$$= 2a^2 - 3ab + b^2$$

d) $(x + 3)(y + 5) = xy + 5x + 3y + 15 \quad \text{There are no like terms to combine.}$

FOIL can be used to multiply any two binomials. The binomials in the next example have higher powers than those of Example 1.

E X A M P L E   2

Using the FOIL method

Find each product.

a) $(x^3 - 3)(x^3 + 6)$ b) $(2a^2 + 1)(a^2 + 5)$

Solution

a) $(x^3 - 3)(x^3 + 6) = x^6 + 6x^3 - 3x^3 - 18$

$\qquad\qquad\qquad\quad = x^6 + 3x^3 - 18$

b) $(2a^2 + 1)(a^2 + 5) = 2a^4 + 10a^2 + a^2 + 5$

$\qquad\qquad\qquad\qquad = 2a^4 + 11a^2 + 5$

∎

Multiplying Binomials Quickly

The outer and inner products in the FOIL method are often like terms, and we can combine them without writing them down. Once you become proficient at using FOIL, you can find the product of two binomials without writing anything except the answer.

E X A M P L E   3

Using FOIL to find a product quickly

Find each product. Write down only the answer.

a) $(x + 3)(x + 4)$ b) $(2x - 1)(x + 5)$ c) $(a - 6)(a + 6)$

Solution

a) $(x + 3)(x + 4) = x^2 + 7x + 12$ Combine like terms: $3x + 4x = 7x$.

b) $(2x - 1)(x + 5) = 2x^2 + 9x - 5$ Combine like terms: $10x - x = 9x$.

c) $(a - 6)(a + 6) = a^2 - 36$ Combine like terms: $6a - 6a = 0$.

∎

E X A M P L E   4

More products

Find each product.

a) $\left(\frac{1}{2}x - 2\right)\left(\frac{1}{3}x + 1\right)$ b) $(x - 1)(x + 3)(x - 4)$

Solution

a) $\left(\frac{1}{2}x - 2\right)\left(\frac{1}{3}x + 1\right) = \frac{1}{6}x^2 - \frac{2}{3}x + \frac{1}{2}x - 2$ $\frac{1}{2} \cdot \frac{1}{3} = \frac{1}{6}$

$\qquad\qquad\qquad\qquad\quad = \frac{1}{6}x^2 - \frac{1}{6}x - 2$ $-\frac{2}{3} + \frac{1}{2} = -\frac{1}{6}$

b) Multiply the first two binomials and then multiply that result by $x - 4$:

$$(x - 1)(x + 3)(x - 4) = (x^2 + 2x - 3)(x - 4)$$
$$= x(x^2 + 2x - 3) - 4(x^2 + 2x - 3)$$
$$= x^3 + 2x^2 - 3x - 4x^2 - 8x + 12$$
$$= x^3 - 2x^2 - 11x + 12$$

∎

EXAMPLE 5

Area of a garden

Sheila has a square garden with sides of length x feet. If she increases the length by 7 feet and decreases the width by 2 feet, then what trinomial represents the area of the new rectangular garden?

Solution

The length of the new garden is $x + 7$ and the width is $x - 2$ as shown in Fig. 5.2. The area is $(x + 7)(x - 2)$ or $x^2 + 5x - 14$ square feet.

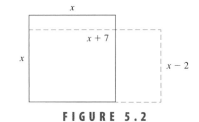

FIGURE 5.2

WARM-UPS

True or false? Answer true only if the equation is true for all values of the variable or variables. Explain your answer.

1. $(x + 3)(x + 2) = x^2 + 6$

2. $(x + 2)(y + 1) = xy + x + 2y + 2$

3. $(3a - 5)(2a + 1) = 6a^2 + 3a - 10a - 5$

4. $(y + 3)(y - 2) = y^2 + y - 6$

5. $(x^2 + 2)(x^2 + 3) = x^4 + 5x^2 + 6$

6. $(3a^2 - 2)(3a^2 + 2) = 9a^2 - 4$

7. $(t + 3)(t + 5) = t^2 + 8t + 15$

8. $(y - 9)(y - 2) = y^2 - 11y - 18$

9. $(x + 4)(x - 7) = x^2 + 4x - 28$

10. It is not necessary to learn FOIL as long as you can get the answer.

5.3 EXERCISES

Reading and Writing *After reading this section, write out the answers to these questions. Use complete sentences.*

1. What property of the real numbers do we usually use to find the product of two binomials?

2. What does FOIL stand for?

3. What is the purpose of FOIL?

4. What is the maximum number of terms that can be obtained when two binomials are multiplied?

Use FOIL to find each product. See Example 1.

5. $(x + 2)(x + 4)$

6. $(x + 3)(x + 5)$

7. $(a - 3)(a + 2)$

8. $(b - 1)(b + 2)$

9. $(2x - 1)(x - 2)$

10. $(2y - 5)(y - 2)$

11. $(2a - 3)(a + 1)$

12. $(3x - 5)(x + 4)$

13. $(w - 50)(w - 10)$

14. $(w - 30)(w - 20)$

15. $(y - a)(y + 5)$

16. $(a + t)(3 - y)$

17. $(5 - w)(w + m)$

18. $(a - h)(b + t)$

19. $(2m - 3t)(5m + 3t)$

20. $(2x - 5y)(x + y)$

21. $(5a + 2b)(9a + 7b)$

22. $(11x + 3y)(x + 4y)$

Use FOIL to find each product. See Example 2.

23. $(x^2 - 5)(x^2 + 2)$

24. $(y^2 + 1)(y^2 - 2)$

25. $(h^3 + 5)(h^3 + 5)$

26. $(y^6 + 1)(y^6 - 4)$

27. $(3b^3 + 2)(b^3 + 4)$

28. $(5n^4 - 1)(n^4 + 3)$

29. $(y^2 - 3)(y - 2)$

30. $(x - 1)(x^2 - 1)$

31. $(3m^3 - n^2)(2m^3 + 3n^2)$

32. $(6y^4 - 2z^2)(6y^4 - 3z^2)$

33. $(3u^2v - 2)(4u^2v + 6)$

34. $(5y^3w^2 + z)(2y^3w^2 + 3z)$

Find each product. Try to write only the answer. See Example 3.

35. $(b + 4)(b + 5)$

36. $(y + 8)(y + 4)$

37. $(x - 3)(x + 9)$

38. $(m + 7)(m - 8)$

39. $(a + 5)(a + 5)$

40. $(t - 4)(t - 4)$

41. $(2x - 1)(2x - 1)$

42. $(3y + 4)(3y + 4)$

43. $(z - 10)(z + 10)$

44. $(3h - 5)(3h + 5)$

45. $(a + b)(a + b)$

46. $(x - y)(x - y)$

47. $(a - b)(a - 2b)$

48. $(b - 8c)(b - c)$

49. $(2x - y)(x + 3y)$

50. $(3y + 5z)(y - 3z)$

51. $(5t - 2)(t - 1)$

52. $(2t - 3)(2t - 1)$

53. $(h - 7)(h - 9)$

54. $(h - 7w)(h - 7w)$

55. $(h + 7w)(h + 7w)$

56. $(h - 7q)(h + 7q)$

57. $(2h^2 - 1)(2h^2 - 1)$

58. $(3h^2 + 1)(3h^2 + 1)$

Perform the indicated operations. See Example 4.

59. $\left(2a + \dfrac{1}{2}\right)\left(4a - \dfrac{1}{2}\right)$

60. $\left(3b + \dfrac{2}{3}\right)\left(6b - \dfrac{1}{3}\right)$

61. $\left(\dfrac{1}{2}x - \dfrac{1}{3}\right)\left(\dfrac{1}{4}x + \dfrac{1}{2}\right)$

62. $\left(\dfrac{2}{3}t - \dfrac{1}{4}\right)\left(\dfrac{1}{2}t - \dfrac{1}{2}\right)$

63. $-2x^4(3x - 1)(2x + 5)$

64. $4xy^3(2x - y)(3x + y)$

65. $(x - 1)(x + 1)(x + 3)$

66. $(a - 3)(a + 4)(a - 5)$

67. $(3x - 2)(3x + 2)(x + 5)$

68. $(x - 6)(9x + 4)(9x - 4)$

69. $(x - 1)(x + 2) - (x + 3)(x - 4)$

70. $(k - 4)(k + 9) - (k - 3)(k + 7)$

Solve each problem. See Example 5.

71. *Area of a rug.* Find a trinomial that represents the area of a rectangular rug whose sides are $x + 3$ feet and $2x - 1$ feet.

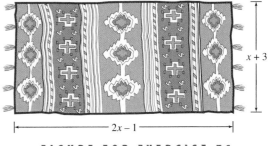

$x + 3$

$2x - 1$

FIGURE FOR EXERCISE 71

72. *Area of a parallelogram.* Find a trinomial that represents the area of a parallelogram whose base is $3x + 2$ meters and whose height is $2x + 3$ meters.

73. *Area of a sail.* The sail of a tall ship is triangular in shape with a base of $4.57x + 3$ meters and a height of $2.3x - 1.33$ meters. Find a polynomial that represents the area of the triangle.

74. *Area of a square.* A square has a side of length $1.732x + 1.414$ meters. Find a polynomial that represents its area.

GETTING MORE INVOLVED

 75. *Exploration.* Find the area of each of the four regions shown in the figure. What is the total area of the four regions? What does this exercise illustrate?

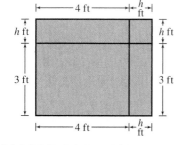

FIGURE FOR EXERCISE 75

 76. *Exploration.* Find the area of each of the four regions shown in the figure. What is the total area of the four regions? What does this exercise illustrate?

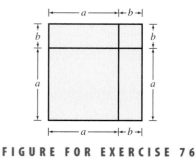

FIGURE FOR EXERCISE 76

5.4   SPECIAL PRODUCTS

In Section 5.3 you learned the FOIL method to make multiplying binomials simpler. In this section you will learn rules for squaring binomials and for finding the product of a sum and a difference. These products are called **special products.**

The Square of a Binomial

To compute $(a + b)^2$, the square of a binomial, we can write it as $(a + b)(a + b)$ and use FOIL:

$$(a + b)^2 = (a + b)(a + b)$$
$$= a^2 + ab + ab + b^2$$
$$= a^2 + 2ab + b^2$$

So to square $a + b$, *we square the first term* (a^2), *add twice the product of the two terms* $(2ab)$, *then add the square of the last term* (b^2). The square of a binomial occurs so frequently that it is helpful to learn this new rule to find it. The rule for squaring a sum is given symbolically as follows.

> **The Square of a Sum**
> $$(a + b)^2 = a^2 + 2ab + b^2$$

E X A M P L E   1   **Using the rule for squaring a sum**
Find the square of each sum.
a) $(x + 3)^2$ **b)** $(2a + 5)^2$

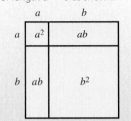
Solution

a) $(x + 3)^2 = x^2 + \underbrace{2(x)(3)} + 3^2 = x^2 + 6x + 9$

                ↑   ↑

          Square   ↑   Square

          of   Twice   of

        first   the   last

              product

b) $(2a + 5)^2 = (2a)^2 + 2(2a)(5) + 5^2$

              $= 4a^2 + 20a + 25$ ∎

CAUTION   Do not forget the middle term when squaring a sum. The equation $(x + 3)^2 = x^2 + 6x + 9$ is an identity, but $(x + 3)^2 = x^2 + 9$ is not an identity. For example, if $x = 1$ in $(x + 3)^2 = x^2 + 9$, then we get $4^2 = 1^2 + 9$, which is false.

When we use FOIL to find $(a - b)^2$, we see that

$$(a - b)^2 = (a - b)(a - b)$$
$$= a^2 - ab - ab + b^2$$
$$= a^2 - 2ab + b^2.$$

So to square $a - b$, *we square the first term* (a^2), *subtract twice the product of the two terms* $(-2ab)$, *and add the square of the last term* (b^2). The rule for squaring a difference is given symbolically as follows.

The Square of a Difference

$$(a - b)^2 = a^2 - 2ab + b^2$$

E X A M P L E   2

Using the rule for squaring a difference
Find the square of each difference.
a) $(x - 4)^2$
b) $(4b - 5y)^2$

Solution

a) $(x - 4)^2 = x^2 - 2(x)(4) + 4^2$
           $= x^2 - 8x + 16$

b) $(4b - 5y)^2 = (4b)^2 - 2(4b)(5y) + (5y)^2$
               $= 16b^2 - 40by + 25y^2$ ∎

Product of a Sum and a Difference

If we multiply the sum $a + b$ and the difference $a - b$ by using FOIL, we get

$$(a + b)(a - b) = a^2 - ab + ab - b^2$$
$$= a^2 - b^2.$$

The inner and outer products have a sum of 0. So *the product of a sum and a difference of the same two terms is equal to the difference of two squares.*

> **The Product of a Sum and a Difference**
> $$(a + b)(a - b) = a^2 - b^2$$

E X A M P L E   3   **Product of a sum and a difference**
Find each product.

a) $(x + 2)(x - 2)$ **b)** $(b + 7)(b - 7)$ **c)** $(3x - 5)(3x + 5)$

Solution
a) $(x + 2)(x - 2) = x^2 - 4$
b) $(b + 7)(b - 7) = b^2 - 49$
c) $(3x - 5)(3x + 5) = 9x^2 - 25$ ■

Higher Powers of Binomials

To find a power of a binomial that is higher than 2, we can use the rule for squaring a binomial along with the method of multiplying binomials using the distributive property. Finding the second or higher power of a binomial is called **expanding the binomial** because the result has more terms than the original.

E X A M P L E   4   **Higher powers of a binomial**
Expand each binomial.

a) $(x + 4)^3$ **b)** $(y - 2)^4$

Solution

> **study tip**
>
> Correct answers often have more than one form. If your answer to an exercise doesn't agree with the one in the back of this text, try to determine if it is simply a different form of the answer. For example, $\frac{1}{2}x$ and $\frac{x}{2}$ look different but they are equivalent expressions.

a)
$$\begin{aligned}
(x + 4)^3 &= (x + 4)^2(x + 4) \\
&= (x^2 + 8x + 16)(x + 4) \\
&= (x^2 + 8x + 16)x + (x^2 + 8x + 16)4 \\
&= x^3 + 8x^2 + 16x + 4x^2 + 32x + 64 \\
&= x^3 + 12x^2 + 48x + 64
\end{aligned}$$

b)
$$\begin{aligned}
(y - 2)^4 &= (y - 2)^2(y - 2)^2 \\
&= (y^2 - 4y + 4)(y^2 - 4y + 4) \\
&= (y^2 - 4y + 4)(y^2) + (y^2 - 4y + 4)(-4y) + (y^2 - 4y + 4)(4) \\
&= y^4 - 4y^3 + 4y^2 - 4y^3 + 16y^2 - 16y + 4y^2 - 16y + 16 \\
&= y^4 - 8y^3 + 24y^2 - 32y + 16
\end{aligned}$$
■

Applications to Area

E X A M P L E   5   **Area of a pizza**
A pizza parlor saves money by making all of its round pizzas one inch smaller in radius than advertised. Write a trinomial for the actual area of a pizza with an advertised radius of r inches.

Solution
A pizza advertised as r inches has an actual radius of $r - 1$ inches. The actual area is $\pi(r - 1)^2$:

$$\pi(r - 1)^2 = \pi(r^2 - 2r + 1) = \pi r^2 - 2\pi r + \pi.$$

So $\pi r^2 - 2\pi r + \pi$ is a trinomial representing the actual area. ■

True or false? Explain your answer.

1. $(2 + 3)^2 = 2^2 + 3^2$
2. $(x + 3)^2 = x^2 + 6x + 9$ for any value of x.
3. $(3 + 5)^2 = 9 + 25 + 30$
4. $(2x + 7)^2 = 4x^2 + 28x + 49$ for any value of x.
5. $(y + 8)^2 = y^2 + 64$ for any value of y.
6. The product of a sum and a difference of the same two terms is equal to the difference of two squares.
7. $(40 - 1)(40 + 1) = 1599$
8. $49 \cdot 51 = 2499$
9. $(x - 3)^2 = x^2 - 3x + 9$ for any value of x.
10. The square of a sum is equal to a sum of two squares.

5.4   EXERCISES

Reading and Writing *After reading this section, write out the answers to these questions. Use complete sentences.*

1. What are the special products?

2. What is the rule for squaring a sum?

3. Why do we need a new rule to find the square of a sum when we already have FOIL?

4. What happens to the inner and outer products in the product of a sum and a difference?

5. What is the rule for finding the product of a sum and a difference?

6. How can you find higher powers of binomials?

Square each binomial. See Example 1.

7. $(x + 1)^2$ 8. $(y + 2)^2$

9. $(y + 4)^2$ 10. $(z + 3)^2$

11. $(3x + 8)^2$ 12. $(2m + 7)^2$

13. $(s + t)^2$ 14. $(x + z)^2$

15. $(2x + y)^2$ 16. $(3t + v)^2$

17. $(2t + 3h)^2$ 18. $(3z + 5k)^2$

Square each binomial. See Example 2.

19. $(a - 3)^2$ 20. $(w - 4)^2$

21. $(t - 1)^2$ 22. $(t - 6)^2$

23. $(3t - 2)^2$ 24. $(5a - 6)^2$

25. $(s - t)^2$ 26. $(r - w)^2$

27. $(3a - b)^2$ 28. $(4w - 7)^2$

29. $(3z - 5y)^2$ 30. $(2z - 3w)^2$

Find each product. See Example 3.

31. $(a - 5)(a + 5)$ 32. $(x - 6)(x + 6)$

33. $(y - 1)(y + 1)$ 34. $(p + 2)(p - 2)$

35. $(3x - 8)(3x + 8)$ 36. $(6x + 1)(6x - 1)$

37. $(r + s)(r - s)$ 38. $(b - y)(b + y)$

39. $(8y - 3a)(8y + 3a)$ 40. $(4u - 9v)(4u + 9v)$

41. $(5x^2 - 2)(5x^2 + 2)$ 42. $(3y^2 + 1)(3y^2 - 1)$

Expand each binomial. See Example 4.

43. $(x + 1)^3$
44. $(y - 1)^3$
45. $(2a - 3)^3$
46. $(3w - 1)^3$

47. $(a - 3)^4$

48. $(2b + 1)^4$

49. $(a + b)^4$

50. $(2a - 3b)^4$

Find each product.

51. $(a - 20)(a + 20)$ **52.** $(1 - x)(1 + x)$

53. $(x + 8)(x + 7)$ **54.** $(x - 9)(x + 5)$

55. $(4x - 1)(4x + 1)$ **56.** $(9y - 1)(9y + 1)$

57. $(9y - 1)^2$ **58.** $(4x - 1)^2$

59. $(2t - 5)(3t + 4)$ **60.** $(2t + 5)(3t - 4)$

61. $(2t - 5)^2$ **62.** $(2t + 5)^2$

63. $(2t + 5)(2t - 5)$ **64.** $(3t - 4)(3t + 4)$

65. $(x^2 - 1)(x^2 + 1)$ **66.** $(y^3 - 1)(y^3 + 1)$

67. $(2y^3 - 9)^2$ **68.** $(3z^4 - 8)^2$

69. $(2x^3 + 3y^2)^2$ **70.** $(4y^5 + 2w^3)^2$

71. $\left(\dfrac{1}{2}x + \dfrac{1}{3}\right)^2$ **72.** $\left(\dfrac{2}{3}y - \dfrac{1}{2}\right)^2$

73. $(0.2x - 0.1)^2$

74. $(0.1y + 0.5)^2$

75. $(a + b)^3$

76. $(2a - 3b)^3$

77. $(1.5x + 3.8)^2$

78. $(3.45a - 2.3)^2$

79. $(3.5t - 2.5)(3.5t + 2.5)$

80. $(4.5h + 5.7)(4.5h - 5.7)$

In Exercises 81–90, solve each problem.

81. *Shrinking garden.* Rose's garden is a square with sides of length x feet. Next spring she plans to make it rectangular by lengthening one side 5 feet and shortening the other side by 5 feet. What polynomial represents the new area? By how much will the area of the new garden differ from that of the old garden?

82. *Square lot.* Sam lives on a lot that he thought was a square, 157 feet by 157 feet. When he had it surveyed, he discovered that one side was actually 2 feet longer than he thought and the other was actually 2 feet shorter than he thought. How much less area does he have than he thought he had?

83. *Area of a circle.* Find a polynomial that represents the area of a circle whose radius is $b + 1$ meters. Use the value 3.14 for π.

84. *Comparing dartboards.* A toy store sells two sizes of circular dartboards. The larger of the two has a radius that is 3 inches greater than that of the other. The radius of the smaller dartboard is t inches. Find a polynomial that represents the difference in area between the two dartboards.

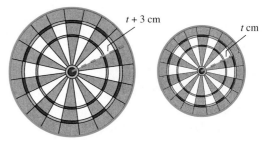

FIGURE FOR EXERCISE 84

85. *Poiseuille's law.* According to the nineteenth-century physician Poiseuille, the velocity (in centimeters per second) of blood r centimeters from the center of an artery of radius R centimeters is given by

$$v = k(R - r)(R + r),$$

where k is a constant. Rewrite the formula using a special product rule.

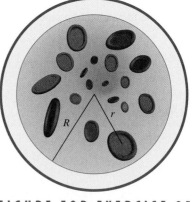

FIGURE FOR EXERCISE 85

86. *Going in circles.* A promoter is planning a circular race track with an inside radius of r feet and a width of w feet. The cost in dollars for paving the track is given by the formula

$$C = 1.2\pi[(r + w)^2 - r^2].$$

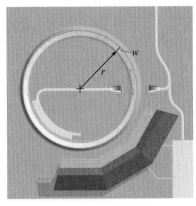

FIGURE FOR EXERCISE 86

Use a special product rule to simplify this formula. What is the cost of paving the track if the inside radius is 1000 feet and the width of the track is 40 feet?

87. Compounded annually. P dollars is invested at annual interest rate r for 2 years. If the interest is compounded annually, then the polynomial $P(1 + r)^2$ represents the value of the investment after 2 years. Rewrite this expression without parentheses. Evaluate the polynomial if $P = \$200$ and $r = 10\%$.

88. Compounded semiannually. P dollars is invested at annual interest rate r for 1 year. If the interest is compounded semiannually, then the polynomial $P\left(1 + \frac{r}{2}\right)^2$ represents the value of the investment after 1 year. Rewrite this expression without parentheses. Evaluate the polynomial if $P = \$200$ and $r = 10\%$.

 **89. Investing in treasury bills.** An investment advisor uses the polynomial $P(1 + r)^{10}$ to predict the value in

10 years of a client's investment of P dollars with an average annual return r. The accompanying graph shows historic average annual returns for the last 20 years for various asset classes (T. Rowe Price, www.troweprice.com). Use the historical average return to predict the value in 10 years of an investment of $10,000 in U.S. treasury bills?

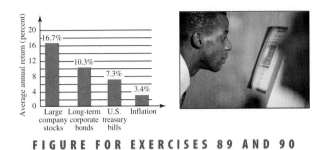

FIGURE FOR EXERCISES 89 AND 90

90. Comparing investments. How much more would the investment in Exercise 89 be worth in 10 years if the client invests in large company stocks rather than U.S. treasury bills?

GETTING MORE INVOLVED

91. Writing. What is the difference between the equations $(x + 5)^2 = x^2 + 10x + 25$ and $(x + 5)^2 = x^2 + 25$?

92. Writing. Is it possible to square a sum or a difference without using the rules presented in this section? Why should you learn the rules given in this section?

5.5   DIVISION OF POLYNOMIALS

You multiplied polynomials in Section 5.2. In this section you will learn to divide polynomials.

Dividing Monomials Using the Quotient Rule

In Chapter 1 we used the definition of division to divide signed numbers. Because the definition of division applies to any division, we restate it here.

In this section

- Dividing Monomials Using the Quotient Rule
- Dividing a Polynomial by a Monomial
- Dividing a Polynomial by a Binomial

Division of Real Numbers

If a, b, and c are any numbers with $b \neq 0$, then

$$a \div b = c \qquad \text{provided that} \qquad c \cdot b = a.$$

If $a \div b = c$, we call a the **dividend,** b the **divisor,** and c (or $a \div b$) the **quotient.**

You can find the quotient of two monomials by writing the quotient as a fraction and then reducing the fraction. For example,

$$x^5 \div x^2 = \frac{x^5}{x^2} = \frac{x \cdot x \cdot x \cdot \not{x} \cdot \not{x}}{\not{x} \cdot \not{x}} = x^3.$$

You can be sure that x^3 is correct by checking that $x^3 \cdot x^2 = x^5$. You can also divide x^2 by x^5, but the result is not a monomial:

$$x^2 \div x^5 = \frac{x^2}{x^5} = \frac{1 \cdot x \cdot x}{x \cdot x \cdot x \cdot x \cdot x} = \frac{1}{x^3}$$

Note that the exponent 3 can be obtained in either case by subtracting 5 and 2. These examples illustrate the quotient rule for exponents.

> **Quotient Rule**
>
> Suppose $a \neq 0$, and m and n are positive integers.
>
> $$\text{If } m \geq n, \text{ then } \frac{a^m}{a^n} = a^{m-n}.$$
>
> $$\text{If } n > m, \text{ then } \frac{a^m}{a^n} = \frac{1}{a^{n-m}}.$$

Note that if you use the quotient rule to subtract the exponents in $x^4 \div x^4$, you get the expression x^{4-4}, or x^0, which has not been defined yet. Because we must have $x^4 \div x^4 = 1$ if $x \neq 0$, we define the zero power of a nonzero real number to be 1. We do not define the expression 0^0.

> **Zero Exponent**
>
> For any nonzero real number a,
>
> $$a^0 = 1.$$

E X A M P L E   1

Using the definition of zero exponent

Simplify each expression. Assume that all variables are nonzero real numbers.

a) 5^0 **b)** $(3xy)^0$ **c)** $a^0 + b^0$

Solution

a) $5^0 = 1$ **b)** $(3xy)^0 = 1$ **c)** $a^0 + b^0 = 1 + 1 = 2$ ∎

With the definition of zero exponent the quotient rule is valid for all positive integers as stated.

E X A M P L E   2

Using the quotient rule in dividing monomials

Find each quotient.

a) $\dfrac{y^9}{y^5}$ **b)** $\dfrac{12b^2}{3b^7}$ **c)** $-6x^3 \div (2x^9)$ **d)** $\dfrac{x^8y^2}{x^2y^2}$

Solution

a) $\dfrac{y^9}{y^5} = y^{9-5} = y^4$

Use the definition of division to check that $y^4 \cdot y^5 = y^9$.

b) $\dfrac{12b^2}{3b^7} = \dfrac{12}{3} \cdot \dfrac{b^2}{b^7} = 4 \cdot \dfrac{1}{b^{7-2}} = \dfrac{4}{b^5}$

Use the definition of division to check that

$$\frac{4}{b^5} \cdot 3b^7 = \frac{12b^7}{b^5} = 12b^2.$$

c) $-6x^3 \div (2x^9) = \dfrac{-6x^3}{2x^9} = \dfrac{-3}{x^6}$

Use the definition of division to check that

$$\frac{-3}{x^6} \cdot 2x^9 = \frac{-6x^9}{x^6} = -6x^3.$$

d) $\dfrac{x^8y^2}{x^2y^2} = \dfrac{x^8}{x^2} \cdot \dfrac{y^2}{y^2} = x^6 \cdot y^0 = x^6$

Use the definition of division to check that $x^6 \cdot x^2y^2 = x^8y^2$.

We showed more steps in Example 2 than are necessary. For division problems like these you should try to write down only the quotient.

Dividing a Polynomial by a Monomial

We divided some simple polynomials by monomials in Chapter 1. For example,

$$\frac{6x + 8}{2} = \frac{1}{2}(6x + 8) = \frac{6x}{2} + \frac{8}{2} = 3x + 4.$$

We use the distributive property to take one-half of $6x$ and one-half of 8 to get $3x + 4$. So both $6x$ and 8 are divided by 2. To divide any polynomial by a monomial, we divide each term of the polynomial by the monomial.

E X A M P L E   3

Dividing a polynomial by a monomial

Find the quotient for $(-8x^6 + 12x^4 - 4x^2) \div (4x^2)$.

Solution

$$\frac{-8x^6 + 12x^4 - 4x^2}{4x^2} = \frac{-8x^6}{4x^2} + \frac{12x^4}{4x^2} - \frac{4x^2}{4x^2}$$

$$= -2x^4 + 3x^2 - 1$$

The quotient is $-2x^4 + 3x^2 - 1$. We can check by multiplying.

$$4x^2(-2x^4 + 3x^2 - 1) = -8x^6 + 12x^4 - 4x^2.$$

Because division by zero is undefined, we will always assume that the divisor is nonzero in any quotient involving variables. For example, the division in Example 3 is valid only if $4x^2 \neq 0$, or $x \neq 0$.

Dividing a Polynomial by a Binomial

Division of whole numbers is often done with a procedure called **long division.** For example, 253 is divided by 7 as follows:

$$
\begin{array}{r}
36 \quad \leftarrow \text{Quotient} \\
\text{Divisor} \rightarrow \quad 7\overline{)253} \quad \leftarrow \text{Dividend} \\
\underline{21} \\
43 \\
\underline{42} \\
1 \quad \leftarrow \text{Remainder}
\end{array}
$$

Note that $36 \cdot 7 + 1 = 253$. It is always true that

$$(\text{quotient})(\text{divisor}) + (\text{remainder}) = \text{dividend}.$$

To divide a polynomial by a binomial, we perform the division like long division of whole numbers. For example, to divide $x^2 - 3x - 10$ by $x + 2$, we get the first term of the quotient by dividing the first term of $x + 2$ into the first term of $x^2 - 3x - 10$. So divide x^2 by x to get x, then multiply and subtract as follows:

1 Divide:
2 Multiply:
3 Subtract:

$$
\begin{array}{r}
x \\
x + 2\overline{)x^2 - 3x - 10} \\
\underline{x^2 + 2x} \\
-5x
\end{array}
$$

$x^2 \div x = x$

$x \cdot (x + 2) = x^2 + 2x$

$-3x - 2x = -5x$

Now bring down -10 and continue the process. We get the second term of the quotient (below) by dividing the first term of $x + 2$ into the first term of $-5x - 10$. So divide $-5x$ by x to get -5:

1 Divide:
2 Multiply:
3 Subtract:

$$
\begin{array}{r}
x - 5 \\
x + 2\overline{)x^2 - 3x - 10} \\
\underline{x^2 + 2x} \quad \downarrow \\
-5x - 10 \\
\underline{-5x - 10} \\
0
\end{array}
$$

$-5x \div x = -5$

Bring down -10.

$-5(x + 2) = -5x - 10$

$-5x - (-5x) = 0, -10 - (-10) = 0$

So the quotient is $x - 5$, and the remainder is 0.

In the next example we must rearrange the dividend before dividing.

E X A M P L E   4

Dividing a polynomial by a binomial

Divide $2x^3 - 4 - 7x^2$ by $2x - 3$, and identify the quotient and the remainder.

Solution

Rearrange the dividend as $2x^3 - 7x^2 - 4$. Because the x-term in the dividend is missing, we write $0 \cdot x$ for it:

$$
\begin{array}{r}
x^2 - 2x - 3 \\
2x - 3\overline{)2x^3 - 7x^2 + 0 \cdot x - 4} \\
\underline{2x^3 - 3x^2} \\
-4x^2 + 0 \cdot x \\
\underline{-4x^2 + 6x} \\
-6x - 4 \\
\underline{-6x + 9} \\
-13
\end{array}
$$

$2x^3 \div (2x) = x^2$

$-7x^2 - (-3x^2) = -4x^2$

$0 \cdot x - 6x = -6x$

$-4 - (9) = -13$

The quotient is $x^2 - 2x - 3$, and the remainder is -13. Note that the degree of the remainder is 0 and the degree of the divisor is 1. To check, we must verify that

$$(2x - 3)(x^2 - 2x - 3) - 13 = 2x^3 - 7x^2 - 4.$$

CAUTION   To avoid errors, always write the terms of the divisor and the dividend in descending order of the exponents and insert a zero for any term that is missing.

If we divide both sides of the equation

$$\text{dividend} = (\text{quotient})(\text{divisor}) + (\text{remainder})$$

by the divisor, we get the equation

$$\frac{\text{dividend}}{\text{divisor}} = \text{quotient} + \frac{\text{remainder}}{\text{divisor}}.$$

This fact is used in expressing improper fractions as mixed numbers. For example, if 19 is divided by 5, the quotient is 3 and the remainder is 4. So

$$\frac{19}{5} = 3 + \frac{4}{5} = 3\frac{4}{5}.$$

We can also use this form to rewrite algebraic fractions.

EXAMPLE 5   **Rewriting algebraic fractions**

Express $\frac{-3x}{x - 2}$ in the form

$$\text{quotient} + \frac{\text{remainder}}{\text{divisor}}.$$

Solution

Use long division to get the quotient and remainder:

$$\begin{array}{r} -3 \\ x - 2\overline{)-3x + 0} \\ \underline{-3x + 6} \\ -6 \end{array}$$

Because the quotient is -3 and the remainder is -6, we can write

$$\frac{-3x}{x - 2} = -3 + \frac{-6}{x - 2}.$$

To check, we must verify that $-3(x - 2) - 6 = -3x$.

CAUTION   When dividing polynomials by long division, we do not stop until the remainder is 0 or the degree of the remainder is smaller than the degree of the divisor. For example, we stop dividing in Example 5 because the degree of the remainder -6 is 0 and the degree of the divisor $x - 2$ is 1.

WARM-UPS

True or false? Explain your answer.

1. $y^{10} \div y^2 = y^5$ for any nonzero value of y.

2. $\frac{7x + 2}{7} = x + 2$ for any value of x.

WARM-UPS

(*continued*)

3. $\frac{7x^2}{7} = x^2$ for any value of x.

4. If $3x^2 + 6$ is divided by 3, the quotient is $x^2 + 6$.

5. If $4y^2 - 6y$ is divided by $2y$, the quotient is $2y - 3$.

6. The quotient times the remainder plus the dividend equals the divisor.

7. $(x + 2)(x + 1) + 3 = x^2 + 3x + 5$ for any value of x.

8. If $x^2 + 3x + 5$ is divided by $x + 2$, then the quotient is $x + 1$.

9. If $x^2 + 3x + 5$ is divided by $x + 2$, the remainder is 3.

10. If the remainder is zero, then (divisor)(quotient) = dividend.

5.5 EXERCISES

Reading and Writing *After reading this section, write out the answers to these questions. Use complete sentences.*

1. What rule is important for dividing monomials?

2. What is the meaning of a zero exponent?

3. How many terms should you get when dividing a polynomial by a monomial?

4. How should the terms of the polynomials be written when dividing with long division?

5. How do you know when to stop the process in long division of polynomials?

6. How do you handle missing terms in the dividend polynomial when doing long division?

Simplify each expression. See Example 1.

7. 9^0

8. m^0

9. $(-2x^3)^0$

10. $(5a^3b)^0$

11. $2 \cdot 5^0 - 3^0$

12. $-4^0 - 8^0$

13. $(2x - y)^0$

14. $(a^2 + b^2)^0$

Find each quotient. Try to write only the answer. See Example 2.

15. $\dfrac{x^8}{x^2}$

16. $\dfrac{y^9}{y^3}$

17. $\dfrac{6a^7}{2a^{12}}$

18. $\dfrac{30b^2}{3b^6}$

19. $-12x^5 \div (3x^9)$

20. $-6y^5 \div (-3y^{10})$

21. $-6y^2 \div (6y)$

22. $-3a^2b \div (3ab)$

23. $\dfrac{-6x^3y^2}{2x^2y^2}$

24. $\dfrac{-4h^2k^4}{-2hk^3}$

25. $\dfrac{-9x^2y^2}{3x^5y^2}$

26. $\dfrac{-12z^4y^2}{-2z^{10}y^2}$

Find the quotients. See Example 3.

27. $\dfrac{3x - 6}{3}$

28. $\dfrac{5y - 10}{-5}$

29. $\dfrac{x^5 + 3x^4 - x^3}{x^2}$

30. $\dfrac{6y^6 - 9y^4 + 12y^2}{3y^2}$

31. $\dfrac{-8x^2y^2 + 4x^2y - 2xy^2}{-2xy}$

32. $\dfrac{-9ab^2 - 6a^3b^3}{-3ab^2}$

33. $(x^2y^3 - 3x^3y^2) \div (x^2y)$

34. $(4h^5k - 6h^2k^2) \div (-2h^2k)$

Find the quotient and remainder for each division. Check by using the fact that dividend = (divisor)(quotient) + remainder. See Example 4.

35. $(x^2 + 5x + 13) \div (x + 3)$

36. $(x^2 + 3x + 6) \div (x + 3)$

37. $(2x) \div (x + 5)$

38. $(5x) \div (x - 1)$

39. $(a^3 + 4a - 3) \div (a - 2)$

40. $(w^3 + 2w^2 - 3) \div (w - 2)$

41. $(x^2 - 3x) \div (x + 1)$

42. $(3x^2) \div (x + 1)$

43. $(h^3 - 27) \div (h - 3)$

44. $(w^3 + 1) \div (w + 1)$

45. $(6x^2 - 13x + 7) \div (3x - 2)$

46. $(4b^2 + 25b - 3) \div (4b + 1)$

47. $(x^3 - x^2 + x - 2) \div (x - 1)$

48. $(a^3 - 3a^2 + 4a - 4) \div (a - 2)$

Write each expression in the form

$$quotient + \frac{remainder}{divisor}.$$

See Example 5.

49. $\dfrac{3x}{x - 5}$ **50.** $\dfrac{2x}{x - 1}$

51. $\dfrac{-x}{x + 3}$ **52.** $\dfrac{-3x}{x + 1}$

53. $\dfrac{x - 1}{x}$ **54.** $\dfrac{a - 5}{a}$

55. $\dfrac{3x + 1}{x}$ **56.** $\dfrac{2y + 1}{y}$

57. $\dfrac{x^2}{x + 1}$

58. $\dfrac{x^2}{x - 1}$

59. $\dfrac{x^2 + 4}{x + 2}$

60. $\dfrac{x^2 + 1}{x - 1}$

61. $\dfrac{x^3}{x - 2}$

62. $\dfrac{x^3 - 1}{x + 1}$

63. $\dfrac{x^3 + 3}{x}$ **64.** $\dfrac{2x^2 + 4}{2x}$

Find each quotient.

65. $-6a^3b \div (2a^2b)$ **66.** $-14x^7 \div (-7x^2)$

67. $-8w^4t^7 \div (-2w^9t^3)$

68. $-9y^7z^4 \div (3y^3z^{11})$

69. $(3a - 12) \div (-3)$

70. $(-6z + 3z^2) \div (-3z)$

71. $(3x^2 - 9x) \div (3x)$

72. $(5x^3 + 15x^2 - 25x) \div (5x)$

73. $(12x^4 - 4x^3 + 6x^2) \div (-2x^2)$

74. $(-9x^3 + 3x^2 - 15x) \div (-3x)$

75. $(t^2 - 5t - 36) \div (t - 9)$

76. $(b^2 + 2b - 35) \div (b - 5)$

77. $(6w^2 - 7w - 5) \div (3w - 5)$

78. $(4z^2 + 23z - 6) \div (4z - 1)$

79. $(8x^3 + 27) \div (2x + 3)$

80. $(8y^3 - 1) \div (2y - 1)$

81. $(t^3 - 3t^2 + 5t - 6) \div (t - 2)$

82. $(2u^3 - 13u^2 - 8u + 7) \div (u - 7)$

83. $(-6v^2 - 4 + 9v + v^3) \div (v - 4)$

84. $(14y + 8y^2 + y^3 + 12) \div (6 + y)$

Solve each problem.

85. *Area of a rectangle.* The area of a rectangular billboard is $x^2 + x - 30$ square meters. If the length is $x + 6$ meters, find a binomial that represents the width.

$x + 6$ meters

FIGURE FOR EXERCISE 85

86. *Perimeter of a rectangle.* The perimeter of a rectangular backyard is $6x + 6$ yards. If the width is x yards, find a binomial that represents the length.

x yards

FIGURE FOR EXERCISE 86

GETTING MORE INVOLVED

87. *Exploration.* Divide $x^3 - 1$ by $x - 1$, $x^4 - 1$ by $x - 1$, and $x^5 - 1$ by $x - 1$. What is the quotient when $x^9 - 1$ is divided by $x - 1$?

88. *Exploration.* Divide $a^3 - b^3$ by $a - b$ and $a^4 - b^4$ by $a - b$. What is the quotient when $a^8 - b^8$ is divided by $a - b$?

89. *Discussion.* Are the expressions $\frac{10x}{5x}$, $10x \div 5x$, and $(10x) \div (5x)$ equivalent? Before you answer, review the order of operations in Section 1.5 and evaluate each expression for $x = 3$.

5.6 POSITIVE INTEGRAL EXPONENTS

The product rule for positive integral exponents was presented in Section 5.2, and the quotient rule was presented in Section 5.5. In this section we review those rules and then further investigate the properties of exponents.

The Product and Quotient Rules

The rules that we have already discussed are summarized below.

The following rules hold for nonnegative integers m and n and $a \neq 0$.

$$a^m \cdot a^n = a^{m+n} \qquad \text{Product rule}$$

$$\frac{a^m}{a^n} = a^{m-n} \quad \text{if } m \geq n \quad \text{Quotient rule}$$

$$\frac{a^m}{a^n} = \frac{1}{a^{n-m}} \quad \text{if } n > m$$

$$a^0 = 1 \qquad \text{Zero exponent}$$

CAUTION The product and quotient rules apply only if the bases of the expressions are identical. For example, $3^2 \cdot 3^4 = 3^6$, but the product rule cannot be applied to $5^2 \cdot 3^4$. Note also that the bases are not multiplied: $3^2 \cdot 3^4 \neq 9^6$.

Note that in the quotient rule the exponents are always subtracted, as in

$$\frac{x^7}{x^3} = x^4 \qquad \text{and} \qquad \frac{y^5}{y^8} = \frac{1}{y^3}.$$

If the larger exponent is in the denominator, then the result is placed in the denominator.

EXAMPLE 1 **Using the product and quotient rules**

Use the rules of exponents to simplify each expression. Assume that all variables represent nonzero real numbers.

a) $2^3 \cdot 2^2$ **b)** $(3x)^0(5x^2)(4x)$

c) $\dfrac{8x^2}{-2x^5}$ **d)** $\dfrac{(3a^2b)b^9}{(6a^5)a^3b^2}$

Solution

a) Because the bases are both 2, we can use the product rule:

$$2^3 \cdot 2^2 = 2^5 \qquad \text{Product rule}$$

$$= 32 \qquad \text{Simplify.}$$

b) $(3x)^0(5x^2)(4x) = 1 \cdot 5x^2 \cdot 4x \quad \text{Definition of zero exponent}$

$$= 20x^3 \qquad \text{Product rule}$$

c) $\dfrac{8x^2}{-2x^5} = -\dfrac{4}{x^3} \qquad \text{Quotient rule}$

d) First use the product rule to simplify the numerator and denominator:

$$\frac{(3a^2b)b^9}{(6a^5)a^3b^2} = \frac{3a^2b^{10}}{6a^8b^2} \quad \text{Product rule}$$

$$= \frac{b^8}{2a^6} \quad \text{Quotient rule}$$

Raising an Exponential Expression to a Power

When we raise an exponential expression to a power, we can use the product rule to find the result, as shown in the following example:

$$(w^4)^3 = w^4 \cdot w^4 \cdot w^4 \quad \text{Three factors of } w^4 \text{ because of the exponent 3}$$

$$= w^{12} \quad \text{Product rule}$$

By the product rule we add the three 4's to get 12, but 12 is also the product of 4 and 3. This example illustrates the **power rule** for exponents.

> **Power Rule**
>
> If m and n are nonnegative integers and $a \neq 0$, then
>
> $$(a^m)^n = a^{mn}.$$

In the next example we use the new rule along with the other rules.

E X A M P L E   2

Using the power rule

Use the rules of exponents to simplify each expression. Assume that all variables represent nonzero real numbers.

a) $3x^2(x^3)^5$

b) $\dfrac{(2^3)^4 \cdot 2^7}{2^5 \cdot 2^9}$

c) $\dfrac{3(x^5)^4}{15x^{22}}$

Solution

a) $3x^2(x^3)^5 = 3x^2x^{15} \quad$ Power rule

$\qquad\qquad\quad = 3x^{17} \quad$ Product rule

b) $\dfrac{(2^3)^4 \cdot 2^7}{2^5 \cdot 2^9} = \dfrac{2^{12} \cdot 2^7}{2^{14}} \quad$ Power rule and product rule

$\qquad\qquad\quad = \dfrac{2^{19}}{2^{14}} \quad$ Product rule

$\qquad\qquad\quad = 2^5 \quad$ Quotient rule

$\qquad\qquad\quad = 32 \quad$ Evaluate 2^5.

c) $\dfrac{3(x^5)^4}{15x^{22}} = \dfrac{3x^{20}}{15x^{22}} = \dfrac{1}{5x^2}$

Power of a Product

Consider an example of raising a monomial to a power. We will use known rules to rewrite the expression.

$$(2x)^3 = 2x \cdot 2x \cdot 2x \qquad \text{Definition of exponent 3}$$

$$= 2 \cdot 2 \cdot 2 \cdot x \cdot x \cdot x \quad \text{Commutative and associative properties}$$

$$= 2^3x^3 \qquad\qquad\qquad \text{Definition of exponents}$$

Note that the power 3 is applied to each factor of the product. This example illustrates the **power of a product rule.**

Power of a Product Rule

If a and b are real numbers and n is a positive integer, then

$$(ab)^n = a^n b^n.$$

E X A M P L E   3

Using the power of a product rule

Simplify. Assume that the variables are nonzero.

a) $(xy^3)^5$ b) $(-3m)^3$ c) $(2x^3y^2z^7)^3$

Solution

a) $(xy^3)^5 = x^5(y^3)^5$ Power of a product rule

         $= x^5 y^{15}$ Power rule

b) $(-3m)^3 = (-3)^3 m^3$ Power of a product rule

           $= -27m^3$ $(-3)(-3)(-3) = -27$

c) $(2x^3y^2z^7)^3 = 2^3(x^3)^3(y^2)^3(z^7)^3 = 8x^9y^6z^{21}$

Power of a Quotient

Raising a quotient to a power is similar to raising a product to a power:

$$\left(\frac{x}{5}\right)^3 = \frac{x}{5} \cdot \frac{x}{5} \cdot \frac{x}{5}$$ Definition of exponent 3

$$= \frac{x \cdot x \cdot x}{5 \cdot 5 \cdot 5}$$ Definition of multiplication of fractions

$$= \frac{x^3}{5^3}$$ Definition of exponents

The power is applied to both the numerator and denominator. This example illustrates the **power of a quotient rule.**

Power of a Quotient Rule

If a and b are real numbers, $b \neq 0$, and n is a positive integer, then

$$\left(\frac{a}{b}\right)^n = \frac{a^n}{b^n}.$$

E X A M P L E   4

Using the power of a quotient rule

Simplify. Assume that the variables are nonzero.

a) $\left(\dfrac{2}{5x^3}\right)^2$ b) $\left(\dfrac{3x^4}{2y^3}\right)^3$ c) $\left(\dfrac{-12a^5b}{4a^2b^7}\right)^3$

Solution

a) $\left(\dfrac{2}{5x^3}\right)^2 = \dfrac{2^2}{(5x^3)^2}$ Power of a quotient rule

           $= \dfrac{4}{25x^6}$ $(5x^3)^2 = 5^2(x^3)^2 = 25x^6$

b) $\left(\dfrac{3x^4}{2y^3}\right)^3 = \dfrac{3^3 x^{12}}{2^3 y^9}$ Power of a quotient and power of a product rule

$\qquad\qquad = \dfrac{27x^{12}}{8y^9}$ Simplify.

c) Use the quotient rule to simplify the expression inside the parentheses before using the power of a quotient rule.

$$\left(\dfrac{-12a^5 b}{4a^2 b^7}\right)^3 = \left(\dfrac{-3a^3}{b^6}\right)^3 \quad \text{Use the quotient rule first.}$$

$$= \dfrac{-27a^9}{b^{18}} \quad \text{Power of a quotient rule}$$

Summary of Rules

The rules for exponents are summarized in the following box.

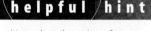

Note that the rules of exponents show how exponents behave with respect to multiplication and division only. We studied the more complicated problem of using exponents with addition and subtraction in Section 5.4 when we learned rules for $(a + b)^2$ and $(a - b)^2$.

Rules for Nonnegative Integral Exponents

The following rules hold for nonzero real numbers a and b and nonnegative integers m and n.

1. $a^0 = 1$ Definition of zero exponent

2. $a^m \cdot a^n = a^{m+n}$ Product rule

3. $\dfrac{a^m}{a^n} = a^{m-n}$ for $m \geq n$,

$\qquad \dfrac{a^m}{a^n} = \dfrac{1}{a^{n-m}}$ for $n > m$ Quotient rule

4. $(a^m)^n = a^{mn}$ Power rule

5. $(ab)^n = a^n \cdot b^n$ Power of a product rule

6. $\left(\dfrac{a}{b}\right)^n = \dfrac{a^n}{b^n}$ Power of a quotient rule

---

WARM-UPS

True or false? Assume that all variables represent nonzero real numbers. A statement involving variables is to be marked true only if it is an identity. Explain your answer.

1. $-3^0 = 1$

2. $2^5 \cdot 2^8 = 4^{13}$

3. $2^3 \cdot 3^3 = 6^5$

4. $(2x)^4 = 2x^4$

5. $(q^3)^5 = q^8$

6. $(-3x^2)^3 = 27x^6$

7. $(ab^3)^4 = a^4 b^{12}$

8. $\dfrac{a^{12}}{a^4} = a^3$

9. $\dfrac{6w^4}{3w^9} = 2w^5$

10. $\left(\dfrac{2y^3}{9}\right)^2 = \dfrac{4y^6}{81}$

5.6 EXERCISES

Reading and Writing *After reading this section, write out the answers to these questions. Use complete sentences.*

1. What is the product rule for exponents?

2. What is the quotient rule for exponents?

3. Why must the bases be the same in these rules?

4. What is the power rule for exponents?

5. What is the power of a product rule?

6. What is the power of a quotient rule?

For all exercises in this section, assume that the variables represent nonzero real numbers.

Simplify the exponential expressions. See Example 1.

7. $2^2 \cdot 2^5$

8. $x^6 \cdot x^7$

9. $(-3u^8)(-2u^2)$

10. $(3r^4)(-6r^2)$

11. $a^3b^4 \cdot ab^6(ab)^0$

12. $x^2y \cdot x^3y^6(x + y)^0$

13. $\dfrac{-2a^3}{4a^7}$

14. $\dfrac{-3t^9}{6t^{18}}$

15. $\dfrac{2a^5b \cdot 3a^7b^3}{15a^6b^8}$

16. $\dfrac{3xy^8 \cdot 5xy^9}{20x^3y^{14}}$

17. $2^3 \cdot 5^2$

18. $2^2 \cdot 10^3$

Simplify. See Example 2.

19. $(x^2)^3$

20. $(y^2)^4$

21. $2x^2 \cdot (x^2)^5$

22. $(y^2)^6 \cdot 3y^5$

23. $\dfrac{(t^2)^5}{(t^3)^4}$

24. $\dfrac{(r^4)^2}{(r^5)^3}$

25. $\dfrac{3x(x^5)^2}{6x^3(x^2)^4}$

26. $\dfrac{5y^3(y^5)^2}{10y^5(y^2)^6}$

Simplify. See Example 3.

27. $(xy^2)^3$

28. $(wy^2)^6$

29. $(-2t^5)^3$

30. $(-3r^3)^3$

31. $(-2x^2y^5)^3$

32. $(-3y^2z^3)^3$

33. $\dfrac{(a^4b^2c^5)^3}{a^3b^4c}$

34. $\dfrac{(2ab^2c^3)^5}{(2a^3bc)^4}$

Simplify. See Example 4.

35. $\left(\dfrac{x^4}{4}\right)^3$

36. $\left(\dfrac{y^2}{2}\right)^3$

37. $\left(\dfrac{-2a^2}{b^3}\right)^4$

38. $\left(\dfrac{-9r^3}{t^5}\right)^2$

39. $\left(\dfrac{2x^2y}{-4y^2}\right)^3$

40. $\left(\dfrac{3y^8}{2zy^2}\right)^4$

41. $\left(\dfrac{-6x^2y^4z^9}{3x^6y^4z^3}\right)^2$

42. $\left(\dfrac{-10rs^9t^4}{2rs^2t^7}\right)^3$

Simplify each expression. Your answer should be an integer or a fraction. Do not use a calculator.

43. $3^2 + 6^2$

44. $(5 - 3)^2$

45. $(3 + 6)^2$

46. $5^2 - 3^2$

47. $2^3 - 3^3$

48. $3^3 + 4^3$

49. $(2 - 3)^3$

50. $(3 + 4)^3$

51. $\left(\dfrac{2}{5}\right)^3$

52. $\left(\dfrac{3}{4}\right)^3$

53. $5^2 \cdot 2^3$

54. $10^3 \cdot 3^3$

55. $2^3 \cdot 2^4$

56. $10^2 \cdot 10^4$

57. $\left(\dfrac{2^3}{2^5}\right)^2$

58. $\left(\dfrac{3}{3^3}\right)^2$

Simplify each expression.

59. $3x^4 \cdot 5x^7$

60. $-2y^3(3y)$

61. $(-5x^4)^3$

62. $(4z^3)^3$

63. $-3y^5z^{12} \cdot 9yz^7$

64. $2a^4b^5 \cdot 2a^9b^2$

65. $\dfrac{-9u^4v^9}{-3u^5v^8}$

66. $\dfrac{-20a^5b^{13}}{5a^4b^{13}}$

67. $(-xt^2)(-2x^2t)^4$

68. $(-ab)^3(-3ba^2)^4$

69. $\left(\dfrac{2x^2}{x^4}\right)^3$

70. $\left(\dfrac{3y^8}{y^5}\right)^2$

71. $\left(\dfrac{-8a^3b^4}{4c^5}\right)^5$

72. $\left(\dfrac{-10a^5c}{5a^5b^4}\right)^5$

73. $\left(\dfrac{-8x^4y^7}{-16x^5y^6}\right)^5$

74. $\left(\dfrac{-5x^2yz^3}{-5x^2yz}\right)^5$

Solve each problem.

75. ***Long-term investing.*** Sheila invested P dollars at annual rate r for 10 years. At the end of 10 years her investment was worth $P(1 + r)^{10}$ dollars. She then reinvested this money for another 5 years at annual rate r. At the end of the second time period her investment

was worth $P(1 + r)^{10}(1 + r)^5$ dollars. Which law of exponents can be used to simplify the last expression? Simplify it.

76. **CD rollover.** Ronnie invested P dollars in a 2-year CD with an annual rate of return of r. After the CD rolled over two times, its value was $P((1 + r)^2)^3$. Which law of exponents can be used to simplify the expression? Simplify it.

GETTING MORE INVOLVED

77. **Writing.** When we square a product, we square each factor in the product. For example, $(3b)^2 = 9b^2$. Explain why we cannot square a sum by simply squaring each term of the sum.

78. **Writing.** Explain why we define 2^0 to be 1. Explain why $-2^0 \neq 1$.

5.7

NEGATIVE EXPONENTS AND SCIENTIFIC NOTATION

In this section

- Negative Integral Exponents
- Rules for Integral Exponents
- Converting from Scientific Notation
- Converting to Scientific Notation
- Computations with Scientific Notation

We defined exponential expressions with positive integral exponents in Chapter 1 and learned the rules for positive integral exponents in Section 5.6. In this section you will first study negative exponents and then see how positive and negative integral exponents are used in scientific notation.

Negative Integral Exponents

If x is nonzero, the reciprocal of x is written as $\frac{1}{x}$. For example, the reciprocal of 2^3 is written as $\frac{1}{2^3}$. To write the reciprocal of an exponential expression in a simpler way, we use a negative exponent. So $2^{-3} = \frac{1}{2^3}$. In general we have the following definition.

Negative Integral Exponents

If a is a nonzero real number and n is a positive integer, then

$$a^{-n} = \frac{1}{a^n}. \quad \text{(If } n \text{ is positive, } -n \text{ is negative.)}$$

Since a^{-n} and a^n are reciprocals, their product is 1. Using a negative exponent for the reciprocal allows us to get this result with the product rule for exponents:

$$a^{-n} \cdot a^n = a^{-n+n} = a^0 = 1$$

EXAMPLE 1

Simplifying expressions with negative exponents

Simplify.

a) 2^{-5} b) $(-2)^{-5}$ c) $\dfrac{2^{-3}}{3^{-2}}$

Solution

a) $2^{-5} = \dfrac{1}{2^5} = \dfrac{1}{32}$

b) $(-2)^{-5} = \dfrac{1}{(-2)^5}$ Definition of negative exponent

$= \dfrac{1}{-32} = -\dfrac{1}{32}$

c) $\dfrac{2^{-3}}{3^{-2}} = 2^{-3} \div 3^{-2}$

$= \dfrac{1}{2^3} \div \dfrac{1}{3^2}$

$= \dfrac{1}{8} \div \dfrac{1}{9} = \dfrac{1}{8} \cdot \dfrac{9}{1} = \dfrac{9}{8}$

calculator

close-up

You can evaluate expressions with negative exponents on a calculator as shown here.

```
2^-5▸Frac
            1/32
(-2)^-5▸Frac
           -1/32
2^-3/3^-2▸Frac
            9/8
```

CAUTION In simplifying -5^{-2}, the negative sign preceding the 5 is used after 5 is squared and the reciprocal is found. So $-5^{-2} = -(5^{-2}) = -\frac{1}{25}$.

To evaluate a^{-n}, you can first find the nth power of a and then find the reciprocal. However, the result is the same if you first find the reciprocal of a and then find the nth power of the reciprocal. For example,

$$3^{-2} = \frac{1}{3^2} = \frac{1}{9} \quad \text{or} \quad 3^{-2} = \left(\frac{1}{3}\right)^2 = \frac{1}{3} \cdot \frac{1}{3} = \frac{1}{9}.$$

So the power and the reciprocal can be found in either order. If the exponent is -1, we simply find the reciprocal. For example,

$$5^{-1} = \frac{1}{5}, \quad \left(\frac{1}{4}\right)^{-1} = 4, \quad \text{and} \quad \left(-\frac{3}{5}\right)^{-1} = -\frac{5}{3}.$$

Because $3^{-2} \cdot 3^2 = 1$, the reciprocal of 3^{-2} is 3^2, and we have

$$\frac{1}{3^{-2}} = 3^2.$$

These examples illustrate the following rules.

helpful hint

Just because the exponent is negative, it doesn't mean the expression is negative. Note that $(-2)^{-3} = -\frac{1}{8}$ while $(-2)^{-4} = \frac{1}{16}$.

Rules for Negative Exponents

If a is a nonzero real number and n is a positive integer, then

$$a^{-n} = \left(\frac{1}{a}\right)^n, \quad a^{-1} = \frac{1}{a}, \quad \frac{1}{a^{-n}} = a^n, \quad \text{and} \quad \left(\frac{a}{b}\right)^{-n} = \left(\frac{b}{a}\right)^n.$$

EXAMPLE 2

Using the rules for negative exponents
Simplify.

a) $\left(\frac{3}{4}\right)^{-3}$ **b)** $10^{-1} + 10^{-1}$ **c)** $\dfrac{2}{10^{-3}}$

Solution

a) $\left(\dfrac{3}{4}\right)^{-3} = \left(\dfrac{4}{3}\right)^3 = \dfrac{64}{27}$

b) $10^{-1} + 10^{-1} = \dfrac{1}{10} + \dfrac{1}{10} = \dfrac{2}{10} = \dfrac{1}{5}$

c) $\dfrac{2}{10^{-3}} = 2 \cdot \dfrac{1}{10^{-3}} = 2 \cdot 10^3 = 2 \cdot 1000 = 2000$

calculator

close-up

You can use a calculator to demonstrate that the product rule for exponents holds when the exponents are negative numbers.

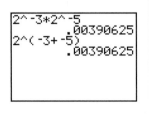

Rules for Integral Exponents

Negative exponents are used to make expressions involving reciprocals simpler looking and easier to write. Negative exponents have the added benefit of working in conjunction with all of the rules of exponents that you learned in Section 5.6. For example, we can use the product rule to get

$$x^{-2} \cdot x^{-3} = x^{-2+(-3)} = x^{-5}$$

and the quotient rule to get

$$\frac{y^3}{y^5} = y^{3-5} = y^{-2}.$$

With negative exponents there is no need to state the quotient rule in two parts as we did in Section 5.6. It can be stated simply as

$$\frac{a^m}{a^n} = a^{m-n}$$

for any integers m and n. We list the rules of exponents here for easy reference.

Rules for Integral Exponents

The following rules hold for nonzero real numbers a and b and any integers m and n.

1. $a^0 = 1$ Definition of zero exponent
2. $a^m \cdot a^n = a^{m+n}$ Product rule
3. $\dfrac{a^m}{a^n} = a^{m-n}$ Quotient rule
4. $(a^m)^n = a^{mn}$ Power rule
5. $(ab)^n = a^n \cdot b^n$ Power of a product rule
6. $\left(\dfrac{a}{b}\right)^n = \dfrac{a^n}{b^n}$ Power of a quotient rule

E X A M P L E 3

The product and quotient rules for integral exponents

Simplify. Write your answers without negative exponents. Assume that the variables represent nonzero real numbers.

a) $b^{-3}b^5$

b) $-3x^{-3} \cdot 5x^2$

c) $\dfrac{m^{-6}}{m^{-2}}$

d) $\dfrac{4y^5}{-12y^{-3}}$

Solution

a) $b^{-3}b^5 = b^{-3+5}$ Product rule

       $= b^2$ Simplify.

b) $-3x^{-3} \cdot 5x^2 = -15x^{-1}$ Product rule

              $= -\dfrac{15}{x}$ Definition of negative exponent

c) $\dfrac{m^{-6}}{m^{-2}} = m^{-6-(-2)}$ Quotient rule

        $= m^{-4}$ Simplify.

        $= \dfrac{1}{m^4}$ Definition of negative exponent

d) $\dfrac{4y^5}{-12y^{-3}} = \dfrac{y^{5-(-3)}}{-3} = \dfrac{-y^8}{3}$

> **helpful hint**
>
> Example 3(c) could be done using the rules for negative exponents and the old quotient rule:
>
> $$\frac{m^{-6}}{m^{-2}} = \frac{m^2}{m^6} = \frac{1}{m^4}$$
>
> It is always good to look at alternative methods. The more tools in your toolbox the better.

In the next example we use the power rules with negative exponents.

EXAMPLE 4

The power rules for integral exponents

Simplify each expression. Write your answers with positive exponents only. Assume that all variables represent nonzero real numbers.

a) $(a^{-3})^2$ **b)** $(10x^{-3})^{-2}$ **c)** $\left(\dfrac{4x^{-5}}{y^2}\right)^{-2}$

Solution

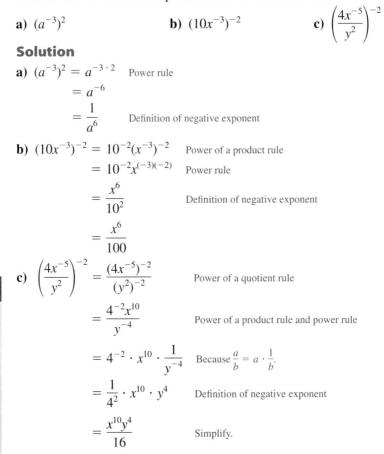

a) $(a^{-3})^2 = a^{-3 \cdot 2}$ Power rule

$\qquad\qquad = a^{-6}$

$\qquad\qquad = \dfrac{1}{a^6}$ Definition of negative exponent

b) $(10x^{-3})^{-2} = 10^{-2}(x^{-3})^{-2}$ Power of a product rule

$\qquad\qquad\qquad = 10^{-2}x^{(-3)(-2)}$ Power rule

$\qquad\qquad\qquad = \dfrac{x^6}{10^2}$ Definition of negative exponent

$\qquad\qquad\qquad = \dfrac{x^6}{100}$

c) $\left(\dfrac{4x^{-5}}{y^2}\right)^{-2} = \dfrac{(4x^{-5})^{-2}}{(y^2)^{-2}}$ Power of a quotient rule

$\qquad\qquad\qquad = \dfrac{4^{-2}x^{10}}{y^{-4}}$ Power of a product rule and power rule

$\qquad\qquad\qquad = 4^{-2} \cdot x^{10} \cdot \dfrac{1}{y^{-4}}$ Because $\dfrac{a}{b} = a \cdot \dfrac{1}{b}$.

$\qquad\qquad\qquad = \dfrac{1}{4^2} \cdot x^{10} \cdot y^4$ Definition of negative exponent

$\qquad\qquad\qquad = \dfrac{x^{10}y^4}{16}$ Simplify. ■

> **helpful hint**
>
> The exponent rules in this section apply to expressions that involve only multiplication and division. This is not too surprising since exponents, multiplication, and division are closely related. Recall that $a^3 = a \cdot a \cdot a$ and $a \div b = a \cdot b^{-1}$.

Converting from Scientific Notation

Many of the numbers occurring in science are either very large or very small. The speed of light is 983,569,000 feet per second. One millimeter is equal to 0.000001 kilometer. In scientific notation, numbers larger than 10 or smaller than 1 are written by using positive or negative exponents.

Scientific notation is based on multiplication by integral powers of 10. Multiplying a number by a positive power of 10 moves the decimal point to the right:

$$10(5.32) = 53.2$$
$$10^2(5.32) = 100(5.32) = 532$$
$$10^3(5.32) = 1000(5.32) = 5320$$

Multiplying by a negative power of 10 moves the decimal point to the left:

$$10^{-1}(5.32) = \frac{1}{10}(5.32) = 0.532$$

$$10^{-2}(5.32) = \frac{1}{100}(5.32) = 0.0532$$

$$10^{-3}(5.32) = \frac{1}{1000}(5.32) = 0.00532$$

close-up

On a graphing calculator you can write scientific notation by actually using the power of 10 or press EE to get the letter E, which indicates that the following number is the power of 10.

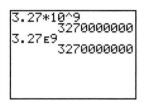

Note that if the exponent is not too large, scientific notation is converted to standard notation when you press ENTER.

So if n is a positive integer, multiplying by 10^n moves the decimal point n places to the right and multiplying by 10^{-n} moves it n places to the left.

A number in scientific notation is written as a product of a number between 1 and 10 and a power of 10. The times symbol $\times$ indicates multiplication. For example, 3.27×10^9 and 2.5×10^{-4} are numbers in scientific notation. In scientific notation there is one digit to the left of the decimal point.

To convert 3.27×10^9 to standard notation, move the decimal point nine places to the right:

$$3.27 \times 10^9 = 3,270,000,000$$

9 places to the right

Of course, it is not necessary to put the decimal point in when writing a whole number.

To convert 2.5×10^{-4} to standard notation, the decimal point is moved four places to the left:

$$2.5 \times 10^{-4} = 0.00025$$

4 places to the left

In general, we use the following strategy to convert from scientific notation to standard notation.

> ### Strategy for Converting from Scientific Notation to Standard Notation
>
> 1. Determine the number of places to move the decimal point by examining the exponent on the 10.
> 2. Move to the right for a positive exponent and to the left for a negative exponent.

E X A M P L E   5

Converting scientific notation to standard notation

Write in standard notation.

a) 7.02×10^6 **b)** 8.13×10^{-5}

Solution

a) Because the exponent is positive, move the decimal point six places to the right:

$$7.02 \times 10^6 = 7020000. = 7,020,000$$

b) Because the exponent is negative, move the decimal point five places to the left.

$$8.13 \times 10^{-5} = 0.0000813$$ ■

Converting to Scientific Notation

To convert a positive number to scientific notation, we just reverse the strategy for converting from scientific notation.

> ### Strategy for Converting to Scientific Notation
>
> 1. Count the number of places (n) that the decimal must be moved so that it will follow the first nonzero digit of the number.
> 2. If the original number was larger than 10, multiply by 10^n.
> 3. If the original number was smaller than 1, multiply by 10^{-n}.

Remember that the scientific notation for a number larger than 10 will have a positive power of 10 and the scientific notation for a number between 0 and 1 will have a negative power of 10.

EXAMPLE 6

close-up

To convert to scientific notation, set the mode to scientific. In scientific mode all results are given in scientific notation.

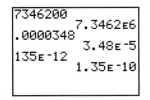

Converting numbers to scientific notation

Write in scientific notation.

a) 7,346,200   b) 0.0000348   c) 135×10^{-12}

Solution

a) Because 7,346,200 is larger than 10, the exponent on the 10 will be positive:

$$7{,}346{,}200 = 7.3462 \times 10^6$$

b) Because 0.0000348 is smaller than 1, the exponent on the 10 will be negative:

$$0.0000348 = 3.48 \times 10^{-5}$$

c) There should be only one nonzero digit to the left of the decimal point:

$$135 \times 10^{-12} = 1.35 \times 10^2 \times 10^{-12} \quad \text{Convert 135 to scientific notation.}$$
$$= 1.35 \times 10^{-10} \quad \text{Product rule}$$

Computations with Scientific Notation

An important feature of scientific notation is its use in computations. Numbers in scientific notation are nothing more than exponential expressions, and you have already studied operations with exponential expressions in this section. We use the same rules of exponents on numbers in scientific notation that we use on any other exponential expressions.

EXAMPLE 7

close-up

With a calculator's built-in scientific notation, some parentheses can be omitted as shown below. Writing out the powers of 10 can lead to errors.

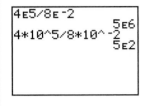

Try these computations with your calculator.

Using the rules of exponents with scientific notation

Perform the indicated computations. Write the answers in scientific notation.

a) $(3 \times 10^6)(2 \times 10^8)$ b) $\dfrac{4 \times 10^5}{8 \times 10^{-2}}$ c) $(5 \times 10^{-7})^3$

Solution

a) $(3 \times 10^6)(2 \times 10^8) = 3 \cdot 2 \cdot 10^6 \cdot 10^8 = 6 \times 10^{14}$

b) $\dfrac{4 \times 10^5}{8 \times 10^{-2}} = \dfrac{4}{8} \cdot \dfrac{10^5}{10^{-2}} = \dfrac{1}{2} \cdot 10^{5-(-2)}$ Quotient rule

$$= (0.5)10^7 \qquad \frac{1}{2} = 0.5$$
$$= 5 \times 10^{-1} \cdot 10^7 \quad \text{Write 0.5 in scientific notation.}$$
$$= 5 \times 10^6 \qquad \text{Product rule}$$

c) $(5 \times 10^{-7})^3 = 5^3(10^{-7})^3$ Power of a product rule

$$= 125 \cdot 10^{-21} \qquad \text{Power rule}$$
$$= 1.25 \times 10^2 \times 10^{-21} \quad 125 = 1.25 \times 10^2$$
$$= 1.25 \times 10^{-19} \qquad \text{Product rule}$$

E X A M P L E   8   **Converting to scientific notation for computations**

Perform these computations by first converting each number into scientific notation. Give your answer in scientific notation.

a) $(3,000,000)(0.0002)$ b) $(20,000,000)^3(0.0000003)$

Solution

a) $(3,000,000)(0.0002) = 3 \times 10^6 \cdot 2 \times 10^{-4}$ Scientific notation

$= 6 \times 10^2$ Product rule

b) $(20,000,000)^3(0.0000003) = (2 \times 10^7)^3(3 \times 10^{-7})$ Scientific notation

$= 8 \times 10^{21} \cdot 3 \times 10^{-7}$ Power of a product rule

$= 24 \times 10^{14}$

$= 2.4 \times 10^1 \times 10^{14}$ $24 = 2.4 \times 10^1$

$= 2.4 \times 10^{15}$ Product rule   ■

WARM-UPS

True or false? Explain your answer.

1. $10^{-2} = \dfrac{1}{100}$

2. $\left(-\dfrac{1}{5}\right)^{-1} = 5$

3. $3^{-2} \cdot 2^{-1} = 6^{-3}$

4. $\dfrac{3^{-2}}{3^{-1}} = \dfrac{1}{3}$

5. $23.7 = 2.37 \times 10^{-1}$

6. $0.000036 = 3.6 \times 10^{-5}$

7. $25 \cdot 10^7 = 2.5 \times 10^8$

8. $0.442 \times 10^{-3} = 4.42 \times 10^{-4}$

9. $(3 \times 10^{-9})^2 = 9 \times 10^{-18}$

10. $(2 \times 10^{-5})(4 \times 10^4) = 8 \times 10^{-20}$

5.7   EXERCISES

Reading and Writing   *After reading this section, write out the answers to these questions. Use complete sentences.*

1. What does a negative exponent mean?

2. What is the correct order for evaluating the operations indicated by a negative exponent?

3. What is the new quotient rule for exponents?

4. How do you convert a number from scientific notation to standard notation?

5. How do you convert a number from standard notation to scientific notation?

6. Which numbers are not usually written in scientific notation?

Variables in all exercises represent positive real numbers. Evaluate each expression. See Example 1.

7. 3^{-1}

8. 3^{-3}

9. $(-2)^{-4}$

10. $(-3)^{-4}$

11. -4^{-2}

12. -2^{-4}

13. $\dfrac{5^{-2}}{10^{-2}}$

14. $\dfrac{3^{-4}}{6^{-2}}$

Simplify. See Example 2.

15. $\left(\dfrac{5}{2}\right)^{-3}$

16. $\left(\dfrac{4}{3}\right)^{-2}$

17. $6^{-1} + 6^{-1}$

18. $2^{-1} + 4^{-1}$

19. $\dfrac{10}{5^{-3}}$

20. $\dfrac{1}{25 \cdot 10^{-4}}$

21. $\dfrac{1}{4^{-3}} + \dfrac{3^2}{2^{-1}}$

22. $\dfrac{2^3}{10^{-2}} - \dfrac{2}{7^{-2}}$

Simplify. Write answers without negative exponents. See Example 3.

23. $x^{-1}x^2$

24. $y^{-3}y^5$

25. $-2x^2 \cdot 8x^{-6}$

26. $5y^5(-6y^{-7})$

27. $-3a^{-2}(-2a^{-3})$

28. $(-b^{-3})(-b^{-5})$

29. $\dfrac{u^{-5}}{u^3}$

30. $\dfrac{w^{-4}}{w^6}$

31. $\dfrac{8t^{-3}}{-2t^{-5}}$

32. $\dfrac{-22w^{-4}}{-11w^{-3}}$

33. $\dfrac{-6x^5}{-3x^{-6}}$

34. $\dfrac{-51y^6}{17y^{-9}}$

Simplify each expression. Write answers without negative exponents. See Example 4.

35. $(x^2)^{-5}$

36. $(y^{-2})^4$

37. $(a^{-3})^{-3}$

38. $(b^{-5})^{-2}$

39. $(2x^{-3})^{-4}$

40. $(3y^{-1})^{-2}$

41. $(4x^2y^{-3})^{-2}$

42. $(6s^{-2}t^4)^{-1}$

43. $\left(\dfrac{2x^{-1}}{y^{-3}}\right)^{-2}$

44. $\left(\dfrac{a^{-2}}{3b^3}\right)^{-3}$

45. $\left(\dfrac{2a^{-3}}{ac^{-2}}\right)^{-4}$

46. $\left(\dfrac{3w^2}{w^4x^3}\right)^{-2}$

Simplify. Write answers without negative exponents.

47. $2^{-1} \cdot 3^{-1}$

48. $2^{-1} + 3^{-1}$

49. $(2 \cdot 3^{-1})^{-1}$

50. $(2^{-1} + 3)^{-1}$

51. $(x^{-2})^{-3} + 3x^7(-5x^{-1})$

52. $(ab^{-1})^2 - ab(-ab^{-3})$

53. $\dfrac{a^3b^{-2}}{a^{-1}} + \left(\dfrac{b^6a^{-2}}{b^5}\right)^{-2}$

54. $\left(\dfrac{x^{-3}y^{-1}}{2x}\right)^{-3} + \dfrac{6x^9y^3}{-3x^{-3}}$

Write each number in standard notation. See Example 5.

55. 9.86×10^9

56. 4.007×10^4

57. 1.37×10^{-3}

58. 9.3×10^{-5}

59. 1×10^{-6}

60. 3×10^{-1}

61. 6×10^5

62. 8×10^6

Write each number in scientific notation. See Example 6.

63. 9000

64. 5,298,000

65. 0.00078

66. 0.000214

67. 0.0000085

68. 5,670,000,000

69. 525×10^9

70. 0.0034×10^{-8}

Perform the computations. Write answers in scientific notation. See Example 7.

71. $(3 \times 10^5)(2 \times 10^{-15})$

72. $(2 \times 10^{-9})(4 \times 10^{23})$

73. $\dfrac{4 \times 10^{-8}}{2 \times 10^{30}}$

74. $\dfrac{9 \times 10^{-4}}{3 \times 10^{-6}}$

75. $\dfrac{3 \times 10^{20}}{6 \times 10^{-8}}$

76. $\dfrac{1 \times 10^{-8}}{4 \times 10^7}$

77. $(3 \times 10^{12})^2$

78. $(2 \times 10^{-5})^3$

79. $(5 \times 10^4)^3$

80. $(5 \times 10^{14})^{-1}$

81. $(4 \times 10^{32})^{-1}$

82. $(6 \times 10^{11})^2$

Perform the following computations by first converting each number into scientific notation. Write answers in scientific notation. See Example 8.

83. $(4300)(2,000,000)$

84. $(40,000)(4,000,000,000)$

85. $(4,200,000)(0.00005)$

86. $(0.00075)(4,000,000)$

87. $(300)^3(0.000001)^5$

88. $(200)^4(0.0005)^3$

89. $\dfrac{(4000)(90,000)}{0.00000012}$

90. $\dfrac{(30,000)(80,000)}{(0.000006)(0.002)}$

Perform the following computations with the aid of a calculator. Write answers in scientific notation. Round to three decimal places.

91. $(6.3 \times 10^6)(1.45 \times 10^{-4})$

92. $(8.35 \times 10^9)(4.5 \times 10^3)$

93. $(5.36 \times 10^{-4}) + (3.55 \times 10^{-5})$

94. $(8.79 \times 10^8) + (6.48 \times 10^9)$

95. $\dfrac{(3.5 \times 10^5)(4.3 \times 10^{-6})}{3.4 \times 10^{-8}}$

96. $\dfrac{(3.5 \times 10^{-8})(4.4 \times 10^{-4})}{2.43 \times 10^{45}}$

97. $(3.56 \times 10^{85})(4.43 \times 10^{96})$

98. $(8 \times 10^{99}) + (3 \times 10^{99})$

Solve each problem.

99. *Distance to the sun.* The distance from the earth to the sun is 93 million miles. Express this distance in feet. (1 mile = 5280 feet.)

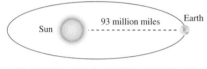

FIGURE FOR EXERCISE 99

100. *Speed of light.* The speed of light is 9.83569×10^8 feet per second. How long does it take light to travel from the sun to the earth? See Exercise 99.

101. *Warp drive, Scotty.* How long does it take a spacecraft traveling at 2×10^{35} miles per hour (warp factor 4) to travel 93 million miles.

102. *Area of a dot.* If the radius of a very small circle is 2.35×10^{-8} centimeters, then what is the circle's area?

103. *Circumference of a circle.* If the circumference of a circle is 5.68×10^9 feet, then what is its radius?

104. *Diameter of a circle.* If the diameter of a circle is 1.3×10^{-12} meters, then what is its radius?

105. *Extracting metals from ore.* Thomas Sherwood studied the relationship between the concentration of a metal in commercial ore and the price of the metal. The accompanying graph shows the Sherwood plot with the locations of several metals marked. Even though the scales on this graph are not typical, the graph can be read in the same manner as other graphs. Note also that a concentration of 100 is 100%.

a) Use the figure to estimate the price of copper (Cu) and its concentration in commercial ore.

b) Use the figure to estimate the price of a metal that has a concentration of 10^{-6} percent in commercial ore.

c) Would the four points shown in the graph lie along a straight line if they were plotted in our usual co-ordinate system?

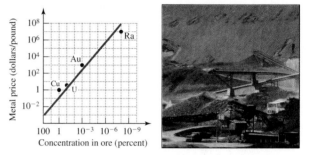

FIGURE FOR EXERCISE 105

106. *Recycling metals.* The accompanying graph shows the prices of various metals that are being recycled and the minimum concentration in waste required for recycling. The straight line is the line from the figure for Exercise 105. Points above the line correspond to metals for which it is economically feasible to increase recycling efforts.

a) Use the figure to estimate the price of mercury (Hg) and the minimum concentration in waste required for recycling mercury.

b) Use the figure to estimate the price of silver (Ag) and the minimum concentration in waste required for recycling silver.

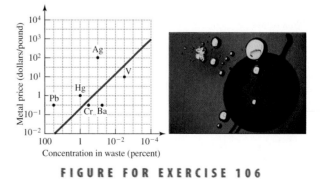

FIGURE FOR EXERCISE 106

107. *Present value.* The present value P that will amount to A dollars in n years with interest compounded annually at annual interest rate r, is given by

$$P = A(1 + r)^{-n}.$$

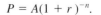

Find the present value that will amount to $50,000 in 20 years at 8% compounded annually.

108. *Investing in stocks.* U.S. small company stocks have returned an average of 14.9% annually for the last 50 years (T. Rowe Price, www.troweprice.com). Use the present value formula from the previous exercise to find the amount invested today in small company stocks that would be worth $1 million in 50 years,

assuming that small company stocks continue to return 14.9% annually for the next 50 years.

GETTING MORE INVOLVED

109. *Exploration.* **a)** If $w^{-3} < 0$, then what can you say about w? **b)** If $(-5)^m < 0$, then what can you say about m? **c)** What restriction must be placed on w and m so that $w^m < 0$?

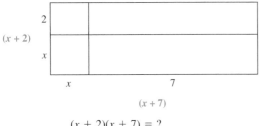

FIGURE FOR EXERCISE 108

110. *Discussion.* Which of the following expressions is not equal to -1? Explain your answer.

   a) -1^{-1} **b)** -1^{-2}

   c) $(-1^{-1})^{-1}$ **d)** $(-1)^{-1}$

   e) $(-1)^{-2}$

COLLABORATIVE ACTIVITIES

Area as a Model of FOIL

Sometimes we can use drawings to represent mathematical operations. The area of a rectangle can represent the process we use when multiplying binomials. The rectangle below represents the multiplication of the binomials $(x + 3)$ and $(x + 5)$:

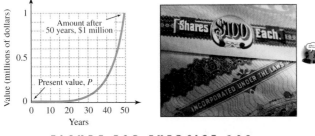

The areas of the inner rectangles are x^2, $3x$, $5x$, and 15.

The area of the red rectangle equals the sum of the areas of the four inner rectangles.

Area of red rectangle:

$$(x + 3)(x + 5) = x^2 + 3x + 5x + 15$$
$$= x^2 + 8x + 15$$

1. a. With your partner, find the areas of the inner rectangles to find the product $(x + 2)(x + 7)$ below:

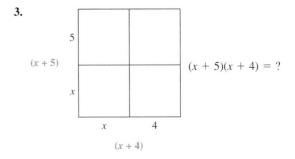

$$(x + 2)(x + 7) = ?$$

   b. Find the same product $(x + 2)(x + 7)$ using FOIL.

Grouping: Pairs

Topic: Multiplying polynomials

For problem 2, student A uses FOIL to find the given product while student B finds the area with the diagram.

2.

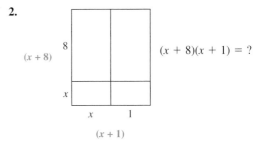

$$(x + 8)(x + 1) = ?$$

For problem 3, student B uses FOIL and A uses the diagram.

3.

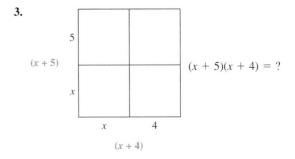

$$(x + 5)(x + 4) = ?$$

4. Student A draws a diagram to find the product $(x + 3)(x + 7)$. Student B finds $(x + 3)(x + 7)$ using FOIL.

5. Student B draws a diagram to find the product $(x + 2)(x + 1)$.
Student A finds $(x + 2)(x + 1)$ using FOIL.

Thinking in reverse: Work together to complete the product that is represented by the given diagram.

6.

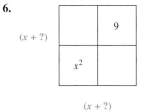

$(x + ?)$

7.

	10
x^2	

$(x + ?)$

$(x + ?)$

Extension: Make up a FOIL problem, then have your partner draw a diagram of it.

W R A P - U P

C H A P T E R   5

S U M M A R Y

Polynomials		**Examples**
Term	A number or the product of a number and one or more variables raised to powers	$5x^3, -4x, 7$
Polynomial	A single term or a finite sum of terms	$2x^5 - 9x^2 + 11$
Degree of a polynomial	The highest degree of any of the terms	Degree of $2x - 9$ is 1. Degree of $5x^3 - x^2$ is 3.

Adding, Subtracting, and Multiplying Polynomials		**Examples**
Add or subtract polynomials	Add or subtract the like terms.	$(x + 1) + (x - 4) = 2x - 3$ $(x^2 - 3x) - (4x^2 - x)$ $= -3x^2 - 2x$
Multiply monomials	Use the product rule for exponents.	$-2x^5 \cdot 6x^8 = -12x^{13}$
Multiply polynomials	Multiply each term of one polynomial by every term of the other polynomial, then combine like terms.	$$\begin{array}{r} x^2 + 2x + 5 \\ x - 1 \\ \hline -x^2 - 2x - 5 \\ x^3 + 2x^2 + 5x \\ \hline x^3 + x^2 + 3x - 5 \end{array}$$

Binomials		**Examples**
FOIL	A method for multiplying two binomials quickly	$(x - 2)(x + 3) = x^2 + x - 6$
Square of a sum	$(a + b)^2 = a^2 + 2ab + b^2$	$(x + 3)^2 = x^2 + 6x + 9$
Square of a difference	$(a - b)^2 = a^2 - 2ab + b^2$	$(m - 5)^2 = m^2 - 10m + 25$

| Product of a sum and a difference | $(a - b)(a + b) = a^2 - b^2$ | $(x + 2)(x - 2) = x^2 - 4$ |

Dividing Polynomials

Examples

| Dividing monomials | Use the quotient rule for exponents | $8x^5 \div (2x^2) = 4x^3$ |
| Divide a polynomial by a monomial | Divide each term of the polynomial by the monomial. | $\dfrac{3x^5 + 9x}{3x} = x^4 + 3$ |

Divide a polynomial by a binomial — If the divisor is a binomial, use long division.

(divisor)(quotient) + (remainder) = dividend

$$
\begin{array}{r}
x - 7 \leftarrow \text{Quotient} \\
\text{Divisor} \rightarrow x + 2\overline{)x^2 - 5x - 4} \leftarrow \text{Dividend} \\
\underline{x^2 + 2x} \\
-7x - 4 \\
\underline{-7x - 14} \\
10 \leftarrow \text{Remainder}
\end{array}
$$

Rules of Exponents

Examples

The following rules hold for any integers m and n, and nonzero real numbers a and b.

Zero exponent	$a^0 = 1$	$2^0 = 1, (-34)^0 = 1$
Product rule	$a^m \cdot a^n = a^{m+n}$	$a^2 \cdot a^3 = a^5$ $3x^6 \cdot 4x^9 = 12x^{15}$
Quotient rule	$\dfrac{a^m}{a^n} = a^{m-n}$	$x^8 \div x^2 = x^6, \dfrac{y^3}{y^3} = y^0 = 1$ $\dfrac{c^7}{c^9} = c^{-2} = \dfrac{1}{c^2}$
Power rule	$(a^m)^n = a^{mn}$	$(2^2)^3 = 2^6, (w^5)^3 = w^{15}$
Power of a product rule	$(ab)^n = a^n b^n$	$(2t)^3 = 8t^3$
Power of a quotient rule	$\left(\dfrac{a}{b}\right)^n = \dfrac{a^n}{b^n}$	$\left(\dfrac{x}{3}\right)^3 = \dfrac{x^3}{27}$

Negative Exponents

Examples

$$3^{-2} = \dfrac{1}{3^2}, x^{-5} = \dfrac{1}{x^5}$$

| Negative integral exponents | If n is a positive integer and a is a nonzero real number, then $a^{-n} = \dfrac{1}{a^n}$ | |
| Rules for negative exponents | If a is a nonzero real number and n is a positive integer, then $a^{-n} = \left(\dfrac{1}{a}\right)^n$, $a^{-1} = \dfrac{1}{a}$, and $\dfrac{1}{a^{-n}} = a^n$. | $\left(\dfrac{2}{3}\right)^{-3} = \left(\dfrac{3}{2}\right)^3, 5^{-1} = \dfrac{1}{5}$
 $\dfrac{1}{w^{-8}} = w^8$ |

Scientific Notation		**Examples**

Converting from scientific notation

1. Find the number of places to move the decimal point by examining the exponent on the 10.

   $5.6 \times 10^3 = 5600$

2. Move to the right for a positive exponent and to the left for a negative exponent.

   $9 \times 10^{-4} = 0.0009$

Converting into scientific notation (positive numbers)

1. Count the number of places (n) that the decimal point must be moved so that it will follow the first nonzero digit of the number.

2. If the original number was larger than 10, multiply by 10^n.

   $304.6 = 3.046 \times 10^2$

3. If the original number was smaller than 1, multiply by 10^{-n}.

   $0.0035 = 3.5 \times 10^{-3}$

ENRICHING YOUR MATHEMATICAL WORD POWER

For each mathematical term, choose the correct meaning.

1. **term**
   a. an expression containing a number or the product of a number and one or more variables
   b. the amount of time spent in this course
   c. a word that describes a number
   d. a variable

2. **polynomial**
   a. four or more terms
   b. many numbers
   c. a sum of four or more numbers
   d. a single term or a finite sum of terms

3. **degree of a polynomial**
   a. the number of terms in a polynomial
   b. the highest degree of any of the terms of a polynomial
   c. the value of a polynomial when $x = 0$
   d. the largest coefficient of any of the terms of a polynomial

4. **leading coefficient**
   a. the first coefficient
   b. the largest coefficient
   c. the coefficient of the first term when a polynomial is written with decreasing exponents
   d. the most important coefficient

5. **monomial**
   a. a single polynomial
   b. one number
   c. an equation that has only one solution
   d. a polynomial that has one term

6. **FOIL**
   a. a method for adding polynomials
   b. first, outer, inner, last

   c. an equation with no solution
   d. a polynomial with five terms

7. **dividend**
   a. a in a/b
   b. b in a/b
   c. the result of a/b
   d. what a bank pays on deposits

8. **divisor**
   a. a in a/b
   b. b in a/b
   c. the result of a/b
   d. two visors

9. **quotient**
   a. a in a/b
   b. b in a/b
   c. a/b
   d. the divisor plus the remainder

10. **binomial**
    a. a polynomial with two terms
    b. any two numbers
    c. the two coordinates in an ordered pair
    d. an equation with two variables

11. **integral exponent**
    a. an exponent that is an integer
    b. a positive exponent
    c. a rational exponent
    d. a fractional exponent

12. **scientific notation**
    a. the notation of rational exponents
    b. the notation of algebra
    c. a notation for expressing large or small numbers with powers of 10
    d. radical notation

REVIEW EXERCISES

5.1 *Perform the indicated operations.*

1. $(2w - 6) + (3w + 4)$

2. $(1 - 3y) + (4y - 6)$

3. $(x^2 - 2x - 5) - (x^2 + 4x - 9)$

4. $(3 - 5x - x^2) - (x^2 - 7x + 8)$

5. $(5 - 3w + w^2) + (w^2 - 4w - 9)$

6. $(-2t^2 + 3t - 4) + (t^2 - 7t + 2)$

7. $(4 - 3m - m^2) - (m^2 - 6m + 5)$

8. $(n^3 - n^2 + 9) - (n^4 - n^3 + 5)$

5.2 *Perform the indicated operations.*

9. $5x^2 \cdot (-10x^9)$ **10.** $3h^3t^2 \cdot 2h^2t^5$

11. $(-11a^7)^2$ **12.** $(12b^3)^2$

13. $x - 5(x - 3)$

14. $x - 4(x - 9)$

15. $5x + 3(x^2 - 5x + 4)$

16. $5 + 4x^2(x - 5)$

17. $3m^2(5m^3 - m + 2)$

18. $-4a^4(a^2 + 2a + 4)$

19. $(x - 5)(x^2 - 2x + 10)$

20. $(x + 2)(x^2 - 2x + 4)$

21. $(x^2 - 2x + 4)(3x - 2)$

22. $(5x + 3)(x^2 - 5x + 4)$

study tip

Note how the review exercises are arranged according to the sections in this chapter. If you are having trouble with a certain type of problem, refer back to the appropriate section for examples and explanations.

5.3 *Perform the indicated operations.*

23. $(q - 6)(q + 8)$

24. $(w + 5)(w + 12)$

25. $(2t - 3)(t - 9)$

26. $(5r + 1)(5r + 2)$

27. $(4y - 3)(5y + 2)$

28. $(11y + 1)(y + 2)$

29. $(3x^2 + 5)(2x^2 + 1)$

30. $(x^3 - 7)(2x^3 + 7)$

5.4 *Perform the indicated operations. Try to write only the answers.*

31. $(z - 7)(z + 7)$

32. $(a - 4)(a + 4)$

33. $(y + 7)^2$

34. $(a + 5)^2$

35. $(w - 3)^2$

36. $(a - 6)^2$

37. $(x^2 - 3)(x^2 + 3)$

38. $(2b^2 - 1)(2b^2 + 1)$

39. $(3a + 1)^2$

40. $(1 - 3c)^2$

41. $(4 - y)^2$

42. $(9 - t)^2$

5.5 *In Exercises 43–54, find each quotient.*

43. $-10x^5 \div (2x^3)$

44. $-6x^4y^2 \div (-2x^2y^2)$

45. $\dfrac{6a^5b^7c^6}{-3a^3b^9c^6}$

46. $\dfrac{-9h^5t^9r^2}{3h^7t^6r^2}$

47. $\dfrac{3x - 9}{-3}$

48. $\dfrac{7 - y}{-1}$

49. $\dfrac{9x^3 - 6x^2 + 3x}{-3x}$

50. $\dfrac{-8x^3y^5 + 4x^2y^4 - 2xy^3}{2xy^2}$

51. $(a - 1) \div (1 - a)$

52. $(t - 3) \div (3 - t)$

53. $(m^4 - 16) \div (m - 2)$

54. $(x^4 - 1) \div (x - 1)$

Find the quotient and remainder.

55. $(3m^3 - 9m^2 + 18m) \div (3m)$

56. $(8x^3 - 4x^2 - 18x) \div (2x)$

57. $(b^2 - 3b + 5) \div (b + 2)$

58. $(r^2 - 5r + 9) \div (r - 3)$

59. $(4x^2 - 9) \div (2x + 1)$

60. $(9y^3 + 2y) \div (3y + 2)$

61. $(x^3 + x^2 - 11x + 10) \div (x - 1)$

62. $(y^3 - 9y^2 + 3y - 6) \div (y + 1)$

Write each expression in the form

$$quotient + \frac{remainder}{divisor}.$$

63. $\dfrac{2x}{x - 3}$

64. $\dfrac{3x}{x - 4}$

65. $\dfrac{2x}{1 - x}$

66. $\dfrac{3x}{5 - x}$

67. $\dfrac{x^2 - 3}{x + 1}$

68. $\dfrac{x^2 + 3x + 1}{x - 3}$

69. $\dfrac{x^2}{x + 1}$

70. $\dfrac{-2x^2}{x - 3}$

5.6 *Simplify each expression.*

71. $2y^{10} \cdot 3y^{20}$

72. $(-3a^5)(5a^3)$

73. $\dfrac{-10b^5c^3}{2b^5c^9}$

74. $\dfrac{-30k^3y^9}{15k^3y^2}$

75. $(b^5)^6$

76. $(y^5)^8$

77. $(-2x^3y^2)^3$

78. $(-3a^4b^6)^4$

79. $\left(\dfrac{2a}{b}\right)^3$

80. $\left(\dfrac{3y}{2}\right)^3$

81. $\left(\dfrac{-6x^2y^5}{-3z^6}\right)^3$

82. $\left(\dfrac{-3a^4b^8}{6a^3b^{12}}\right)^4$

For the following exercises, assume that all of the variables represent positive real numbers.

5.7 *Simplify each expression. Use only positive exponents in answers.*

83. 2^{-5}

84. -2^{-4}

85. 10^{-3}

86. $5^{-1} \cdot 5^0$

87. $x^5 x^{-8}$

88. $a^{-3}a^{-9}$

89. $\dfrac{a^{-8}}{a^{-12}}$

90. $\dfrac{a^{10}}{a^{-4}}$

91. $\dfrac{a^3}{a^{-7}}$

92. $\dfrac{b^{-2}}{b^{-6}}$

93. $(x^{-3})^4$

94. $(x^5)^{-10}$

95. $(2x^{-3})^{-3}$

96. $(3y^{-5})^2$

97. $\left(\dfrac{a}{3b^{-3}}\right)^{-2}$

98. $\left(\dfrac{a^{-2}}{5b}\right)^{-3}$

Convert each number in scientific notation to a number in standard notation, and convert each number in standard notation to a number in scientific notation.

  99. 5000

100. 0.00009

101. 3.4×10^5

102. 5.7×10^{-8}

103. 0.0000461

104. $44{,}000$

105. 5.69×10^{-6}

106. 5.5×10^9

Perform each computation without using a calculator. Write answers in scientific notation.

107. $(3.5 \times 10^8)(2.0 \times 10^{-12})$

108. $(9 \times 10^{12})(2 \times 10^{17})$

109. $(2 \times 10^{-4})^4$

110. $(-3 \times 10^5)^3$

111. $(0.00000004)(2{,}000{,}000{,}000)$

112. $(3{,}000{,}000{,}000) \div (0.000002)$

113. $(0.0000002)^5$

114. $(50{,}000{,}000{,}000)^3$

MISCELLANEOUS
Perform the indicated operations.

115. $(x + 3)(x + 7)$

116. $(k + 5)(k + 4)$

117. $(t - 3y)(t - 4y)$

118. $(t + 7z)(t + 6z)$

119. $(2x^3)^0 + (2y)^0$

120. $(4y^2 - 9)^0$

121. $(-3ht^6)^3$

122. $(-9y^3c^4)^2$

123. $(2w + 3)(w - 6)$

124. $(3x + 5)(2x - 6)$

125. $(3u - 5v)(3u + 5v)$

126. $(9x^2 - 2)(9x^2 + 2)$

127. $(3h + 5)^2$

128. $(4v - 3)^2$

129. $(x + 3)^3$

130. $(k - 10)^3$

131. $(-7s^2t)(-2s^3t^5)$

132. $-5w^3r^2 \cdot 2w^4r^8$

133. $\left(\dfrac{k^4m^2}{2k^2m^2}\right)^4$

134. $\left(\dfrac{-6h^3y^5}{2h^7y^2}\right)^4$

135. $(5x^2 - 8x - 8) - (4x^2 + x - 3)$

136. $(4x^2 - 6x - 8) - (9x^2 - 5x + 7)$

137. $(2x^2 - 2x - 3) + (3x^2 + x - 9)$

138. $(x^2 - 3x - 1) + (x^2 - 2x + 1)$

139. $(x + 4)(x^2 - 5x + 1)$

140. $(2x^2 - 7x + 4)(x + 3)$

141. $(x^2 + 4x - 12) \div (x - 2)$

142. $(a^2 - 3a - 10) \div (a - 5)$

Solve each problem.

143. *Roundball court.* The length of a basketball court is 44 feet larger than its width. Find polynomials that represent its perimeter and area. The actual width of a basketball court is 50 feet. Evaluate these polynomials to find the actual perimeter and area of the court.

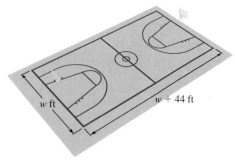

FIGURE FOR EXERCISE 143

144. *Badminton court.* The width of a badminton court is 24 feet less than its length. Find polynomials that represent its perimeter and area. The actual length of a badminton court is 44 feet. Evaluate these polynomials to find the perimeter and area of the court.

145. *Smoke alert.* A retailer of smoke alarms knows that at a price of p dollars each, she can sell $600 - 15p$ smoke alarms per week. Find a polynomial that represents the weekly revenue for the smoke alarms. Find the revenue for a week in which the price is $12 per smoke alarm. Use the bar graph to find the price per smoke alarm that gives the maximum weekly revenue.

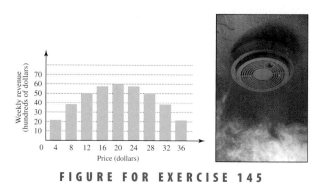

FIGURE FOR EXERCISE 145

146. *Boom box sales.* A retailer of boom boxes knows that at a price of q dollars each, he can sell $900 - 3q$ boom boxes per month. Find a polynomial that represents the monthly revenue for the boom boxes? How many boom boxes will he sell if the price is $300 each?

CHAPTER 5 TEST

Perform the indicated operations.

1. $(7x^3 - x^2 - 6) + (5x^2 + 2x - 5)$

2. $(x^2 - 3x - 5) - (2x^2 + 6x - 7)$

3. $\dfrac{6y^3 - 9y^2}{-3y}$

4. $(x - 2) \div (2 - x)$

5. $(x^3 - 2x^2 - 4x + 3) \div (x - 3)$

6. $3x^2(5x^3 - 7x^2 + 4x - 1)$

Find the products.

7. $(x + 5)(x - 2)$

8. $(3a - 7)(2a + 5)$

9. $(a - 7)^2$

10. $(4x + 3y)^2$

11. $(b - 3)(b + 3)$

12. $(3t^2 - 7)(3t^2 + 7)$

13. $(4x^2 - 3)(x^2 + 2)$

14. $(x - 2)(x + 3)(x - 4)$

Write each expression in the form

$$quotient + \frac{remainder}{divisor}.$$

15. $\dfrac{2x}{x - 3}$

16. $\dfrac{x^2 - 3x + 5}{x + 2}$

Use the rules of exponents to simplify each expression. Write answers without negative exponents.

17. $-5x^3 \cdot 7x^5$

18. $3x^3y \cdot (2xy^4)^2$

19. $-4a^6b^5 \div (2a^5b)$

20. $3x^{-2} \cdot 5x^7$

21. $\left(\dfrac{-2a}{b^2}\right)^5$

22. $\dfrac{-6a^7b^6c^2}{-2a^3b^8c^2}$

23. $\dfrac{6t^{-7}}{2t^9}$

24. $\dfrac{w^{-6}}{w^{-4}}$

25. $(-3s^{-3}t^2)^{-2}$

26. $(-2x^{-6}y)^3$

Convert to scientific notation.

27. 5,433,000

28. 0.0000065

Perform each computation by converting to scientific notation. Give answers in scientific notation.

29. $(80{,}000)(0.000006)$

30. $(0.0000003)^4$

Solve each problem.

31. Find the quotient and remainder when $x^2 - 5x + 9$ is divided by $x - 3$.

32. Subtract $3x^2 - 4x - 9$ from $x^2 - 3x + 6$.

33. The width of a pool table is x feet, and the length is 4 feet longer than the width. Find polynomials that represent the area and the perimeter of the pool table. Evaluate these polynomials for a width of 4 feet.

34. If a manufacturer charges q dollars each for footballs, then he can sell $3000 - 150q$ footballs per week. Find a polynomial that represents the revenue for 1 week. Find the weekly revenue if the price is $8 for each football.

Simplify each expression.

1. $-16 \div (-2)$
2. $(-2)^3 - 1$
3. $(-5)^2 - 3(-5) + 1$
4. $2^{10} \cdot 2^{15}$
5. $2^{15} \div 2^{10}$
6. $2^{10} - 2^5$
7. $3^2 \cdot 4^2$
8. $(172 - 85) \div (85 - 172)$
9. $(5 + 3)^2$
10. $5^2 + 3^2$
11. $(30 - 1)(30 + 1)$
12. $(30 + 1)^2$

Perform the indicated operations.

13. $(x + 3)(x + 5)$
14. $(x^2 + 8x + 15) \div (x + 5)$
15. $x + 3(x + 5)$
16. $(x^2 + 8x + 15)(x + 5)$
17. $-5t^3v \cdot 3t^2v^6$
18. $(-10t^3v^2) \div (-2t^2v)$
19. $(-6y^3 + 8y^2) \div (-2y^2)$
20. $(y^2 - 3y - 9) - (-3y^2 + 2y - 6)$

Solve each equation.

21. $2x + 1 = 0$
22. $x - 7 = 0$
23. $2x - 3 = 0$
24. $3x - 7 = 5$
25. $8 - 3x = x + 20$
26. $4 - 3(x + 2) = 0$

Solve the problem.

27. **Average cost.** Pineapple Recording plans to spend $100,000 to record a new CD by the Woozies and $2.25 per CD to manufacture the disks. The polynomial $2.25n + 100,000$ represents the total cost in dollars for recording and manufacturing n disks. Find an expression that represents the average cost per disk by dividing the total cost by n. Find the average cost per disk for $n = 1000, 100,000$, and $1,000,000$. What happens to the large initial investment of $100,000 if the company sells one million CDs?

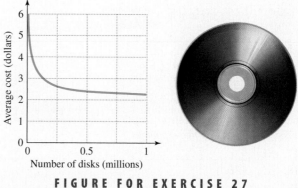

FIGURE FOR EXERCISE 27

Factoring

he sport of skydiving was born in the 1930s soon after the military began using parachutes as a means of deploying troops. Today, skydiving is a popular sport around the world.

With as little as 8 hours of ground instruction, first-time jumpers can be ready to make a solo jump. Without the assistance of oxygen, sky divers can jump from as high as 14,000 feet and reach speeds of more than 100 miles per hour as they fall toward the earth. Jumpers usually open their parachutes between 2000 and 3000 feet and then gradually glide down to their landing area. If the jump and the parachute are handled correctly, the landing can be as gentle as jumping off two steps.

Making a jump and floating to earth are only part of the sport of sky-diving. For example, in an activity called "relative work skydiving," a team of as many as 920 free-falling sky divers join together to make geometrically shaped formations. In a related exercise called "canopy relative work," the team members form geometric patterns after their parachutes or canopies have opened. This kind of skydiving takes skill and practice, and teams are not always successful in their attempts.

The amount of time a sky diver has for a free fall depends on the height of the jump and how much the sky diver uses the air to slow the fall. In Exercises 67 and 68 of Section 6.6 we find the amount of time that it takes a sky diver to fall from a given height.

6.1 FACTORING OUT COMMON FACTORS

In Chapter 5 you learned how to multiply a monomial and a polynomial. In this section you will learn how to reverse that multiplication by finding the greatest common factor for the terms of a polynomial and then factoring the polynomial.

Prime Factorization of Integers

To **factor** an expression means to write the expression as a product. For example, if we start with 12 and write $12 = 4 \cdot 3$, we have factored 12. Both 4 and 3 are **factors** or **divisors** of 12. There are other factorizations of 12:

$$12 = 2 \cdot 6 \qquad 12 = 1 \cdot 12 \qquad 12 = 2 \cdot 2 \cdot 3 = 2^2 \cdot 3$$

The one that is most useful to us is $12 = 2^2 \cdot 3$, because it expresses 12 as a product of *prime numbers*.

> **Prime Number**
>
> A positive integer larger than 1 that has no integral factors other than itself and 1 is called a **prime number.**

The numbers 2, 3, 5, 7, 11, 13, 17, 19, and 23 are the first nine prime numbers. A positive integer larger than 1 that is not a prime is a **composite number.** The numbers 4, 6, 8, 9, 10, and 12 are the first six composite numbers. Every composite number is a product of prime numbers. The **prime factorization** for 12 is $2^2 \cdot 3$.

EXAMPLE 1

helpful hint

The prime factorization of 36 can be found also with a *factoring tree:*

```
        36
       /  \
      2    18
          /  \
         2    9
             / \
            3   3
```

So $36 = 2 \cdot 2 \cdot 3 \cdot 3$.

Prime factorization

Find the prime factorization for 36.

Solution

We start by writing 36 as a product of two integers:

$$
\begin{aligned}
36 &= 2 \cdot 18 && \text{Write 36 as } 2 \cdot 18. \\
&= 2 \cdot 2 \cdot 9 && \text{Replace 18 by } 2 \cdot 9. \\
&= 2 \cdot 2 \cdot 3 \cdot 3 && \text{Replace 9 by } 3 \cdot 3. \\
&= 2^2 \cdot 3^2 && \text{Use exponential notation.}
\end{aligned}
$$

The prime factorization of 36 is $2^2 \cdot 3^2$. ∎

For larger numbers it is helpful to use the method shown in the next example.

EXAMPLE 2

Factoring a large number

Find the prime factorization for 420.

Solution

Start by dividing 420 by the smallest prime number that will divide into it evenly (without remainder). The smallest prime divisor of 420 is 2.

$$2\overline{)420} \atop 210$$

If a number is even, then it is divisible by 2. If the sum of the digits of a number is divisible by 3, then the number is divisible by 3. A number that ends in 0 or 5 is divisible by 5.

Now find the smallest prime that will divide evenly into the quotient, 210. The smallest prime divisor of 210 is 2. Continue this procedure, as follows, until the quotient is a prime number:

$$
\begin{array}{r}
7 \\
5\overline{)35} \\
3\overline{)105} \\
2\overline{)210} \\
\text{Start here} \rightarrow \ 2\overline{)420}
\end{array}
\qquad
\begin{array}{l}
35 \div 5 = 7 \\
105 \div 3 = 35 \\
210 \div 2 = 105
\end{array}
$$

The prime factorization of 420 is $2 \cdot 2 \cdot 3 \cdot 5 \cdot 7$, or $2^2 \cdot 3 \cdot 5 \cdot 7$. Note that it is really not necessary to divide by the smallest prime divisor at each step. We obtain the same factorization if we divide by any prime divisor at each step. ∎

Greatest Common Factor

The largest integer that is a factor of two or more integers is called the **greatest common factor (GCF)** of the integers. For example, 1, 2, 3, and 6 are common factors of 18 and 24. Because 6 is the largest, 6 is the GCF of 18 and 24. We can use prime factorizations to find the GCF. For example, to find the GCF of 8 and 12, we first factor 8 and 12:

$$8 = 2 \cdot 2 \cdot 2 = 2^3 \qquad 12 = 2 \cdot 2 \cdot 3 = 2^2 \cdot 3$$

We see that the factor 2 appears twice in both 8 and 12. So 2^2, or 4, is the GCF of 8 and 12. Notice that 2 is a factor in both 2^3 and $2^2 \cdot 3$ and that 2^2 is the smallest power of 2 in these factorizations. In general, we can use the following strategy to find the GCF.

Strategy for Finding the GCF for Positive Integers

1. Find the prime factorization of each integer.
2. Determine which primes appear in all of the factorizations and the smallest exponent that appears on each of the common prime factors.
3. The GCF is the product of the common prime factors using the exponents from part (2).

If two integers have no common prime factors, then their greatest common factor is 1, because 1 is a factor of every integer. For example, 6 and 35 have no common prime factors ($6 = 2 \cdot 3$ and $35 = 5 \cdot 7$). So the GCF for 6 and 35 is 1.

E X A M P L E   3

Greatest common factor

Find the GCF for each group of numbers.

a) 150, 225
b) 216, 360, 504
c) 55, 168

Solution

a) First find the prime factorization for each number:

$$
\begin{array}{r}
5 \\
5\overline{)25} \\
3\overline{)75} \\
2\overline{)150}
\end{array}
\qquad
\begin{array}{r}
5 \\
5\overline{)25} \\
3\overline{)75} \\
3\overline{)225}
\end{array}
$$

$$150 = 2 \cdot 3 \cdot 5^2 \qquad 225 = 3^2 \cdot 5^2$$

Because 2 is not a factor of 225, it is not a common factor of 150 and 225. Only 3 and 5 appear in both factorizations. Looking at both $2 \cdot 3 \cdot 5^2$ and $3^2 \cdot 5^2$, we see that the smallest power of 5 is 2 and the smallest power of 3 is 1. So the GCF of 150 and 225 is $3 \cdot 5^2$, or 75.

b) First find the prime factorization for each number:

$$216 = 2^3 \cdot 3^3 \qquad 360 = 2^3 \cdot 3^2 \cdot 5 \qquad 504 = 2^3 \cdot 3^2 \cdot 7$$

The only common prime factors are 2 and 3. The smallest power of 2 in the factorizations is 3, and the smallest power of 3 is 2. So the GCF is $2^3 \cdot 3^2$, or 72.

c) First find the prime factorization for each number:

$$55 = 5 \cdot 11 \qquad 168 = 2^3 \cdot 3 \cdot 7$$

Because there are no common factors other than 1, the GCF is 1.

helpful hint

The fact that every composite number has a unique prime factorization is known as the fundamental theorem of arithmetic.

Finding the Greatest Common Factor for Monomials

To find the GCF for a group of monomials, we use the same procedure as that used for integers.

Strategy for Finding the GCF for Monomials

1. Find the GCF for the coefficients of the monomials.
2. Form the product of the GCF of the coefficients and each variable that is common to all of the monomials, where the exponent on each variable is the smallest power of that variable in any of the monomials.

E X A M P L E   4

Greatest common factor of monomials

Find the greatest common factor for each group of monomials.

a) $15x^2, 9x^3$ **b)** $12x^2y^2, 30x^2yz, 42x^3y$

Solution

a) The GCF for 15 and 9 is 3, and the smallest power of x is 2. So the GCF for the monomials is $3x^2$. If we write these monomials as

$$15x^2 = 5 \cdot 3 \cdot x \cdot x \qquad \text{and} \qquad 9x^3 = 3 \cdot 3 \cdot x \cdot x \cdot x,$$

we can see that $3x^2$ is the GCF.

b) The GCF for 12, 30, and 42 is 6. For the common variables x and y, 2 is the smallest power of x and 1 is the smallest power of y. So the GCF for the monomials is $6x^2y$.

Factoring Out the Greatest Common Factor

In Chapter 5 we used the distributive property to multiply monomials and polynomials. For example,

$$6(5x - 3) = 30x - 18.$$

If we start with $30x - 18$ and write

$$30x - 18 = 6(5x - 3),$$

we have factored $30x - 18$. Because multiplication is the last operation to be performed in $6(5x - 3)$, the expression $6(5x - 3)$ is a product. Because 6 is the GCF of 30 and 18, we have **factored out** the GCF.

EXAMPLE 5

Factoring out the greatest common factor

Factor the following polynomials by factoring out the GCF.

a) $25a^2 + 40a$

b) $6x^4 - 12x^3 + 3x^2$

c) $x^2y^5 + x^6y^3$

d) $(a + b)w + (a + b)6$

Solution

a) The GCF of the coefficients 25 and 40 is 5. Because the smallest power of the common factor a is 1, we can factor $5a$ out of each term:

$$25a^2 + 40a = 5a \cdot 5a + 5a \cdot 8$$
$$= 5a(5a + 8)$$

b) The GCF of 6, 12, and 3 is 3. We can factor x^2 out of each term, since the smallest power of x in the three terms is 2. So factor $3x^2$ out of each term as follows:

$$6x^4 - 12x^3 + 3x^2 = 3x^2 \cdot 2x^2 - 3x^2 \cdot 4x + 3x^2 \cdot 1$$
$$= 3x^2(2x^2 - 4x + 1)$$

Check by multiplying: $3x^2(2x^2 - 4x + 1) = 6x^4 - 12x^3 + 3x^2$.

c) The GCF of the numerical coefficients is 1. Both x and y are common to each term. Using the lowest powers of x and y, we get

$$x^2y^5 + x^6y^3 = x^2y^3 \cdot y^2 + x^2y^3 \cdot x^4$$
$$= x^2y^3(y^2 + x^4).$$

Check by multiplying.

d) Even though this expression looks different from the rest, we can factor it in the same way. The binomial $a + b$ is a common factor, and we can factor it out just as we factor out a monomial:

$$(a + b)w + (a + b)6 = (a + b)(w + 6)$$

■

> **CAUTION**   If the GCF is one of the terms of the polynomial, then you must remember to leave a 1 in place of that term when the GCF is factored out. For example,
>
> $$ab + b = a \cdot b + 1 \cdot b = b(a + 1).$$

You should always check your answer by multiplying the factors.

Factoring Out the Opposite of the GCF

Because the greatest common factor for $-4x + 2xy$ is $2x$, we write

$$-4x + 2xy = 2x(-2 + y).$$

We could factor out $-2x$, the opposite of the greatest common factor:

$$-4x + 2xy = -2x(2 - y).$$

It will be necessary to factor out the opposite of the greatest common factor when you learn factoring by grouping in Section 6.2. Remember that you can check all factoring by multiplying the factors to see whether you get the original polynomial.

EXAMPLE 6

Factoring out the opposite of the GCF

Factor each polynomial twice. First factor out the greatest common factor, and then factor out the opposite of the GCF.

a) $3x - 3y$

b) $a - b$

c) $-x^3 + 2x^2 - 8x$

Solution

a) $3x - 3y = 3(x - y)$ Factor out 3.

           $= -3(-x + y)$ Factor out -3.

Note that the signs of the terms in parentheses change when -3 is factored out. Check the answers by multiplying.

b) $a - b = 1(a - b)$ Factor out 1, the GCF of a and b.

           $= -1(-a + b)$ Factor out -1.

We can also write $a - b = -1(b - a)$.

c) $-x^3 + 2x^2 - 8x = x(-x^2 + 2x - 8)$ Factor out x.

                  $= -x(x^2 - 2x + 8)$ Factor out $-x$. ■

> **CAUTION**   Be sure to change the sign of each term in parentheses when you factor out the opposite of the greatest common factor.

In the next example we factor to find the length of a rectangle.

EXAMPLE 7   **An application of factoring**

The width of a rectangle is w meters and its area is $w^2 + 30w$ square meters. Find an expression for the length of the rectangle.

Solution

The area of a rectangle is the product of its length and width. Since

$$A = w^2 + 30w = w(w + 30)$$

and w is the width, the length is $w + 30$ meters. ■

WARM-UPS

True or false? Explain your answer.

1. There are only nine prime numbers.
2. The prime factorization of 32 is $2^3 \cdot 3$.
3. The integer 51 is a prime number.
4. The GCF of the integers 12 and 16 is 4.
5. The GCF of the integers 10 and 21 is 1.
6. The GCF of the polynomial $x^5y^3 - x^4y^7$ is x^4y^3.
7. For the polynomial $2x^2y - 6xy^2$ we can factor out either $2xy$ or $-2xy$.
8. The greatest common factor of the polynomial $8a^3b - 12a^2b$ is $4ab$.
9. $x - 7 = 7 - x$ for any real number x.
10. $-3x^2 + 6x = -3x(x - 2)$ for any real number x.

6.1   EXERCISES

Reading and Writing   *After reading this section, write out the answers to these questions. Use complete sentences.*

1. What does it mean to factor an expression?

2. What is a prime number?

3. How do you find the prime factorization of a number?

4. What is the greatest common factor for two numbers?

5. What is the greatest common factor for two monomials?

6. How can you check if you have factored an expression correctly?

Find the prime factorization of each integer. See Examples 1 and 2.

7. 18   **8.** 20
9. 52   **10.** 76
11. 98   **12.** 100
13. 460   **14.** 345
15. 924   **16.** 585

Find the greatest common factor (GCF) for each group of integers. See Example 3.

17. 8, 20   **18.** 18, 42
19. 36, 60   **20.** 42, 70
21. 40, 48, 88   **22.** 15, 35, 45
23. 76, 84, 100   **24.** 66, 72, 120
25. 39, 68, 77   **26.** 81, 200, 539

Find the greatest common factor (GCF) for each group of monomials. See Example 4.

27. $6x, 8x^3$ **28.** $12x^2, 4x^3$
29. $12x^3, 4x^2, 6x$ **30.** $3y^5, 9y^4, 15y^3$
31. $3x^2y, 2xy^2$ **32.** $7a^2x^3, 5a^3x$
33. $24a^2bc, 60ab^2$ **34.** $30x^2yz^3, 75x^3yz^6$
35. $12u^3v^2, 25s^2t^4$ **36.** $45m^2n^5, 56a^4b^8$
37. $18a^3b, 30a^2b^2, 54ab^3$ **38.** $16x^2z, 40xz^2, 72z^3$

Complete the factoring of each monomial.

39. $27x = 9(\quad)$ **40.** $51y = 3y(\quad)$
41. $24t^2 = 8t(\quad)$ **42.** $18u^2 = 3u(\quad)$
43. $36y^5 = 4y^2(\quad)$
44. $42z^4 = 3z^2(\quad)$
45. $u^4v^3 = uv(\quad)$
46. $x^5y^3 = x^2y(\quad)$
47. $-14m^4n^3 = 2m^4(\quad)$
48. $-8y^3z^4 = 4z^3(\quad)$
49. $-33x^4y^3z^2 = -3x^3yz(\quad)$
50. $-96a^3b^4c^5 = -12ab^3c^3(\quad)$

Factor out the GCF in each expression. See Example 5.

51. $x^3 - 6x$ **52.** $10y^4 - 30y^2$
53. $5ax + 5ay$ **54.** $6wz + 15wa$
55. $h^5 - h^3$ **56.** $y^6 + y^5$
57. $-2k^7m^4 + 4k^3m^6$
58. $-6h^5t^2 + 3h^3t^6$
59. $2x^3 - 6x^2 + 8x$
60. $6x^3 + 18x^2 - 24x$
61. $12x^4t + 30x^3t - 24x^2t^2$
62. $15x^2y^2 - 9xy^2 + 6x^2y$

63. $(x - 3)a + (x - 3)b$
64. $(y + 4)3 + (y + 4)z$
65. $a(y + 1)^2 + b(y + 1)^2$
66. $w(w + 2)^2 + 8(w + 2)^2$
67. $36a^3b^5 - 27a^2b^4 + 18a^2b^9$
68. $56x^3y^5 - 40x^2y^6 + 8x^2y^3$

First factor out the GCF, and then factor out the opposite of the GCF. See Example 6.

69. $8x - 8y$
70. $2a - 6b$
71. $-4x + 8x^2$
72. $-5x^2 + 10x$
73. $x - 5$
74. $a - 6$
75. $4 - 7a$
76. $7 - 5b$
77. $-24a^3 + 16a^2$
78. $-30b^4 + 75b^3$
79. $-12x^2 - 18x$
80. $-20b^2 - 8b$
81. $-2x^3 - 6x^2 + 14x$

82. $-8x^4 + 6x^3 - 2x^2$

83. $4a^3b - 6a^2b^2 - 4ab^3$

84. $12u^5v^6 + 18u^2v^3 - 15u^4v^5$

Solve each problem by factoring. See Example 7.

85. ***Uniform motion.*** Helen traveled a distance of $20x + 40$ miles at 20 miles per hour on the Yellowhead Highway. Find a binomial that represents the time that she traveled.

86. ***Area of a painting.*** A rectangular painting with a width of x centimeters has an area of $x^2 + 50x$ square centimeters. Find a binomial that represents the length.

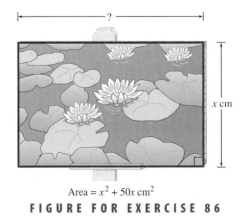

Area $= x^2 + 50x$ cm^2

FIGURE FOR EXERCISE 86

87. *Tomato soup.* The amount of metal S (in square inches) that it takes to make a can for tomato soup is a function of the radius r and height h:

$$S = 2\pi r^2 + 2\pi rh$$

a) Rewrite this formula by factoring out the greatest common factor on the right-hand side.

b) If $h = 5$ in., then S is a function of r. Write a formula for that function.

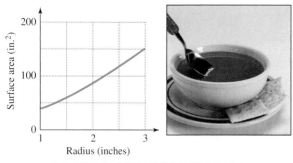

FIGURE FOR EXERCISE 87

c) The accompanying graph shows S for r between 1 in. and 3 in. (with $h = 5$ in.). Which of these r-values gives the maximum surface area?

88. *Amount of an investment.* The amount of an investment of P dollars for t years at simple interest rate r is given by $A = P + Prt$.

a) Rewrite this formula by factoring out the greatest common factor on the right-hand side.

b) Find A if $8300 is invested for 3 years at a simple interest rate of 15%.

GETTING MORE INVOLVED

89. *Discussion.* Is the greatest common factor of $-6x^2 + 3x$ positive or negative? Explain.

90. *Writing.* Explain in your own words why you use the smallest power of each common prime factor when finding the GCF of two or more integers.

6.2

FACTORING THE SPECIAL PRODUCTS AND FACTORING BY GROUPING

In Section 5.4 you learned how to find the special products: the square of a sum, the square of a difference, and the product of a sum and a difference. In this section you will learn how to reverse those operations.

Factoring a Difference of Two Squares

In Section 5.4 you learned that the product of a sum and a difference is a difference of two squares:

$$(a + b)(a - b) = a^2 - ab + ab - b^2 = a^2 - b^2$$

So a difference of two squares can be factored as a product of a sum and a difference, using the following rule.

Factoring a Difference of Two Squares

For any real numbers a and b,

$$a^2 - b^2 = (a + b)(a - b).$$

Note that the square of an integer is a perfect square. For example, 64 is a perfect square because $64 = 8^2$. The square of a monomial in which the coefficient is an integer is also called a **perfect square** or simply a **square**. For example, $9m^2$ is a perfect square because $9m^2 = (3m)^2$.

EXAMPLE 1

Factoring a difference of two squares

Factor each polynomial.

a) $y^2 - 81$ **b)** $9m^2 - 16$ **c)** $4x^2 - 9y^2$

Solution

a) Because $81 = 9^2$, the binomial $y^2 - 81$ is a difference of two squares:

$$y^2 - 81 = y^2 - 9^2 \qquad \text{Rewrite as a difference of two squares.}$$
$$= (y + 9)(y - 9) \quad \text{Factor.}$$

Check by multiplying.

b) Because $9m^2 = (3m)^2$ and $16 = 4^2$, the binomial $9m^2 - 16$ is a difference of two squares:

$$9m^2 - 16 = (3m)^2 - 4^2 \qquad \text{Rewrite as a difference of two squares.}$$
$$= (3m + 4)(3m - 4) \quad \text{Factor.}$$

Check by multiplying.

c) Because $4x^2 = (2x)^2$ and $9y^2 = (3y)^2$, the binomial $4x^2 - 9y^2$ is a difference of two squares:

$$4x^2 - 9y^2 = (2x + 3y)(2x - 3y) \qquad \blacksquare$$

Factoring a Perfect Square Trinomial

In Section 5.4 you learned how to square a binomial using the rule

$$(a + b)^2 = a^2 + 2ab + b^2.$$

You can reverse this rule to factor a trinomial such as $x^2 + 6x + 9$. Notice that

$$x^2 + 6x + 9 = x^2 + \underline{2 \cdot x \cdot 3} + 3^2.$$
$$\uparrow \qquad\qquad \uparrow$$
$$a^2 \qquad 2ab \qquad b^2$$

So if $a = x$ and $b = 3$, then $x^2 + 6x + 9$ fits the form $a^2 + 2ab + b^2$, and

$$x^2 + 6x + 9 = (x + 3)^2.$$

A trinomial that is of the form $a^2 + 2ab + b^2$ or $a^2 - 2ab + b^2$ is called a **perfect square trinomial.** A perfect square trinomial is the square of a binomial. Perfect square trinomials can be identified by using the following strategy.

Strategy for Identifying a Perfect Square Trinomial

A trinomial is a perfect square trinomial if:
1. the first and last terms are of the form a^2 and b^2 (perfect squares), and
2. the middle term is $2ab$ or $-2ab$.

E X A M P L E 2

Identifying the special products

Determine whether each binomial is a difference of two squares and whether each trinomial is a perfect square trinomial.

a) $x^2 - 14x + 49$ **b)** $4x^2 - 81$

c) $4a^2 + 24a + 25$ **d)** $9y^2 - 24y - 16$

Solution

a) The first term is x^2, and the last term is 7^2. The middle term, $-14x$, is $-2 \cdot x \cdot 7$. So this trinomial is a perfect square trinomial.

b) Both terms of $4x^2 - 81$ are perfect squares, $(2x)^2$ and 9^2. So $4x^2 - 81$ is a difference of two squares.

c) The first term of $4a^2 + 24a + 25$ is $(2a)^2$ and the last term is 5^2. However, $2 \cdot 2a \cdot 5$ is $20a$. Because the middle term is $24a$, this trinomial is not a perfect square trinomial.

d) The first and last terms in a perfect square trinomial are both positive. Because the last term in $9y^2 - 24y - 16$ is negative, the trinomial is not a perfect square trinomial. ■

Note that the middle term in a perfect square trinomial may have a positive or a negative coefficient, while the first and last terms must be positive. Any perfect square trinomial can be factored as the square of a binomial by using the following rule.

Factoring Perfect Square Trinomials

For any real numbers a and b,

$$a^2 + 2ab + b^2 = (a + b)^2$$
$$a^2 - 2ab + b^2 = (a - b)^2.$$

E X A M P L E   3

Factoring perfect square trinomials

Factor.

a) $x^2 - 4x + 4$ 　　　　**b)** $a^2 + 16a + 64$ 　　　　**c)** $4x^2 - 12x + 9$

Solution

a) The first term is x^2, and the last term is 2^2. Because the middle term is $-2 \cdot 2 \cdot x$, or $-4x$, this polynomial is a perfect square trinomial:

$$x^2 - 4x + 4 = (x - 2)^2$$

Check by expanding $(x - 2)^2$.

b) $a^2 + 16a + 64 = (a + 8)^2$

Check by expanding $(a + 8)^2$.

c) The first term is $(2x)^2$, and the last term is 3^2. Because $-2 \cdot 2x \cdot 3 = -12x$, the polynomial is a perfect square trinomial. So

$$4x^2 - 12x + 9 = (2x - 3)^2.$$

Check by expanding $(2x - 3)^2$. ■

Factoring Completely

To factor a polynomial means to write it as a product of simpler polynomials. A polynomial that cannot be factored is called a **prime** or **irreducible polynomial.** The polynomials $3x$, $w + 1$, and $4m - 5$ are prime polynomials. A polynomial is **factored completely** when it is written as a product of prime polynomials. So $(y - 8)(y + 1)$ is a complete factorization. When factoring polynomials, we usually do not factor integers that occur as common factors. So $6x(x - 7)$ is considered to be factored completely even though 6 could be factored.

Some polynomials have a factor common to all terms. To factor such polynomials completely, it is simpler to factor out the greatest common factor (GCF) and then factor the remaining polynomial. The following example illustrates factoring completely.

E X A M P L E   4

Factoring completely

Factor each polynomial completely.

a) $2x^3 - 50x$ **b)** $8x^2y - 32xy + 32y$

Solution

a) The greatest common factor of $2x^3$ and $50x$ is $2x$:

$$2x^3 - 50x = 2x(x^2 - 25) \qquad \text{Check this step by multiplying.}$$
$$= 2x(x + 5)(x - 5) \qquad \text{Difference of two squares}$$

b) $8x^2y - 32xy + 32y = 8y(x^2 - 4x + 4)$ Check this step by multiplying.
$$= 8y(x - 2)^2 \qquad \text{Perfect square trinomial}$$ ■

Remember that factoring reverses multiplication and *every step of factoring can be checked by multiplication.*

Factoring by Grouping

The product of two binomials may be a polynomial with four terms. For example,

$$(x + a)(x + 3) = (x + a)x + (x + a)3$$
$$= x^2 + ax + 3x + 3a.$$

We can factor a four-term polynomial of this type by simply reversing the steps we used to find the product. To reverse these steps, we factor out common factors from the first two terms and from the last two terms. For example,

$$w^2 - bw + 3w - 3b = w(w - b) + 3(w - b)$$
$$= (w + 3)(w - b)$$

This procedure is called **factoring by grouping.** Sometimes we must factor out a negative common factor or rearrange the terms as shown in the next example.

E X A M P L E   5

Factoring by grouping

Use grouping to factor each polynomial completely.

a) $xy + 2y + 3x + 6$ **b)** $2x^3 - 3x^2 - 2x + 3$ **c)** $ax + 3y - 3x - ay$

Solution

a) Notice that the first two terms have a common factor of y and the last two terms have a common factor of 3:

$$xy + 2y + 3x + 6 = (xy + 2y) + (3x + 6) \qquad \begin{array}{l}\text{Use the associative property}\\\text{to group the terms.}\end{array}$$
$$= y(x + 2) + 3(x + 2) \qquad \begin{array}{l}\text{Factor out the common}\\\text{factors in each group.}\end{array}$$
$$= (y + 3)(x + 2) \qquad \text{Factor out } x + 2.$$

b) We can factor x^2 out of the first two terms and 1 out of the last two terms:

$$2x^3 - 3x^2 - 2x + 3 = (2x^3 - 3x^2) + (-2x + 3) \qquad \text{Group the terms.}$$
$$= x^2(2x - 3) + 1(-2x + 3)$$

However, we cannot proceed any further because $2x - 3$ and $-2x + 3$ are not the same. To get $2x - 3$ as a common factor, we must factor out -1 from the last two terms:

$$2x^3 - 3x^2 - 2x + 3 = x^2(2x - 3) - 1(2x - 3) \qquad \text{Factor out the common factors.}$$
$$= (x^2 - 1)(2x - 3) \qquad \text{Factor out } 2x - 3.$$
$$= (x - 1)(x + 1)(2x - 3) \qquad \text{Difference of two squares}$$

c) In $ax + 3y - 3x - ay$ there are no common factors in the first two or the last two terms. However, if we use the commutative property to rewrite the polynomial as $ax - 3x - ay + 3y$, then we can factor by grouping:

$$ax + 3y - 3x - ay = ax - 3x - ay + 3y \qquad \text{Rearrange the terms.}$$
$$= x(a - 3) - y(a - 3) \qquad \text{Factor out } x \text{ and } -y.$$
$$= (x - y)(a - 3) \qquad \text{Factor out } a - 3.$$

WARM-UPS

True or false? Explain your answer.

1. The polynomial $x^2 + 16$ is a difference of two squares.
2. The polynomial $x^2 - 8x + 16$ is a perfect square trinomial.
3. The polynomial $9x^2 + 21x + 49$ is a perfect square trinomial.
4. $4x^2 + 4 = (2x + 2)^2$ for any real number x.
5. A difference of two squares is equal to a product of a sum and a difference.

6. The polynomial $16y + 1$ is a prime polynomial.
7. The polynomial $x^2 + 9$ can be factored as $(x + 3)(x + 3)$.
8. The polynomial $4x^2 - 4$ is factored completely as $4(x^2 - 1)$.
9. $y^2 - 2y + 1 = (y - 1)^2$ for any real number y.
10. $2x^2 - 18 = 2(x - 3)(x + 3)$ for any real number x.

6.2 EXERCISES

Reading and Writing *After reading this section, write out the answers to these questions. Use complete sentences.*

1. What is a perfect square?

2. How do we factor a difference of two squares?

3. How can you recognize if a trinomial is a perfect square?

4. What is a prime polynomial.

5. When is a polynomial factored completely?

6. What should you always look for first when attempting to factor a polynomial completely?

Factor each polynomial. See Example 1.

7. $a^2 - 4$
8. $h^2 - 9$
9. $x^2 - 49$
10. $y^2 - 36$
11. $y^2 - 9x^2$
12. $16x^2 - y^2$
13. $25a^2 - 49b^2$
14. $9a^2 - 64b^2$
15. $121m^2 - 1$
16. $144n^2 - 1$
17. $9w^2 - 25c^2$
18. $144w^2 - 121a^2$

Determine whether each binomial is a difference of two squares and whether each trinomial is a perfect square trinomial. See Example 2.

19. $x^2 - 20x + 100$ 20. $x^2 - 10x - 25$

21. $y^2 - 40$ 22. $a^2 - 49$

23. $4y^2 + 12y + 9$ 24. $9a^2 - 30a - 25$

25. $x^2 - 8x + 64$ 26. $x^2 + 4x + 4$

27. $9y^2 - 25c^2$

28. $9x^2 + 4$

29. $9a^2 + 6ab + b^2$

30. $4x^2 - 4xy + y^2$

Factor each perfect square trinomial. See Example 3.

31. $x^2 + 12x + 36$ **32.** $y^2 + 14y + 49$

33. $a^2 - 4a + 4$ **34.** $b^2 - 6b + 9$

35. $4w^2 + 4w + 1$ **36.** $9m^2 + 6m + 1$

37. $16x^2 - 8x + 1$ **38.** $25y^2 - 10y + 1$

39. $4t^2 + 20t + 25$ **40.** $9y^2 - 12y + 4$

41. $9w^2 + 42w + 49$ **42.** $144x^2 + 24x + 1$

43. $n^2 + 2nt + t^2$ **44.** $x^2 - 2xy + y^2$

Factor each polynomial completely. See Example 4.

45. $5x^2 - 125$ **46.** $3y^2 - 27$

47. $-2x^2 + 18$ **48.** $-5y^2 + 20$

49. $a^3 - ab^2$ **50.** $x^2y - y$

51. $3x^2 + 6x + 3$ **52.** $12a^2 + 36a + 27$

53. $-5y^2 + 50y - 125$ **54.** $-2a^2 - 16a - 32$

55. $x^3 - 2x^2y + xy^2$ **56.** $x^3y + 2x^2y^2 + xy^3$

57. $-3x^2 + 3y^2$
58. $-8a^2 + 8b^2$
59. $2ax^2 - 98a$
60. $32x^2y - 2y^3$
61. $3ab^2 - 18ab + 27a$
62. $-2a^2b + 8ab - 8b$
63. $-4m^3 + 24m^2n - 36mn^2$
64. $10a^3 - 20a^2b + 10ab^2$

Use grouping to factor each polynomial completely. See Example 5.

65. $bx + by + cx + cy$
66. $3x + 3z + ax + az$
67. $x^3 + x^2 - 4x - 4$
68. $x^3 + x^2 - x - 1$
69. $3a - 3b - xa + xb$
70. $ax - bx - 4a + 4b$
71. $a^3 + 3a^2 + a + 3$
72. $y^3 - 5y^2 + 8y - 40$
73. $xa + ay + 3y + 3x$
74. $x^3 + ax + 3a + 3x^2$

75. $abc - 3 + c - 3ab$

76. $xa + tb + ba + tx$

77. $x^2a - b + bx^2 - a$

78. $a^2m - b^2n + a^2n - b^2m$

79. $y^2 + y + by + b$

80. $ac + mc + aw^2 + mw^2$

Factor each polynomial completely.

81. $6a^3y + 24a^2y^2 + 24ay^3$

82. $8b^5c - 8b^4c^2 + 2b^3c^3$

83. $24a^3y - 6ay^3$

84. $27b^3c - 12bc^3$

85. $2a^3y^2 - 6a^2y$

86. $9x^3y - 18x^2y^2$

87. $ab + 2bw - 4aw - 8w^2$

88. $3am - 6n - an + 18m$

Use factoring to solve each problem.

89. *Skydiving.* The height (in feet) above the earth for a sky diver t seconds after jumping from an airplane at 6400 ft is approximated by the formula $h = -16t^2 + 6400$, provided that $t < 5$. Rewrite the formula with the right-hand side factored completely. Use your revised formula to find h when $t = 2$.

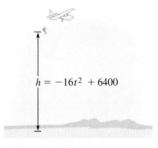

$h = -16t^2 + 6400$

FIGURE FOR EXERCISE 89

90. *Demand for pools.* Tropical Pools sells an above-ground model for p dollars each. The monthly revenue from the sale of this model is a function of the price, given by

$$R = -0.08p^2 + 300p.$$

Revenue is the product of the price p and the demand (quantity sold).

a) Factor out the price on the right-hand side of the formula.

b) What is an expression for the monthly demand?

c) What is the monthly demand for this pool when the price is $3000?

d) Use the graph on page 290 to estimate the price at which the revenue is maximized. Approximately how many pools will be sold monthly at this price?

e) What is the approximate maximum revenue?

f) Use the accompanying graph to estimate the price at which the revenue is zero.

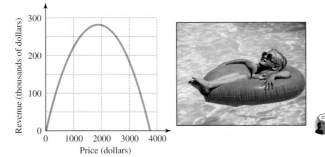

FIGURE FOR EXERCISE 91

FIGURE FOR EXERCISE 90

91. *Volume of a tank.* The volume of a fish tank with a square base and height y is $y^3 - 6y^2 + 9y$ cubic inches. Find the length of a side of the square base.

GETTING MORE INVOLVED

92. *Discussion.* For what real number k, does $3x^2 - k$ factor as $3(x - 2)(x + 2)$?

93. *Writing.* Explain in your own words how to factor a four-term polynomial by grouping.

94. *Writing.* Explain how you know that $x^2 + 1$ is a prime polynomial.

6.3 FACTORING $ax^2 + bx + c$ WITH $a = 1$

In this section we will factor the type of trinomials that result from multiplying two different binomials. We will do this only for trinomials in which the coefficient of x^2, the leading coefficient, is 1. Factoring trinomials with leading coefficient not equal to 1 will be done in Section 6.4.

Factoring $ax^2 + bx + c$ with $a = 1$

Let's look closely at an example of finding the product of two binomials using the distributive property:

$$(x + 2)(x + 3) = (x + 2)x + (x + 2)3 \quad \text{Distributive property}$$
$$= x^2 + 2x + 3x + 6 \quad \text{Distributive property}$$
$$= x^2 + 5x + 6 \quad \text{Combine like terms.}$$

To factor $x^2 + 5x + 6$, we reverse these steps as shown in our first example.

EXAMPLE 1

Factoring a trinomial

Factor.

a) $x^2 + 5x + 6$ **b)** $x^2 + 8x + 12$ **c)** $a^2 - 9a + 20$

Solution

a) The coefficient 5 is the sum of two numbers that have a product of 6. The only integers that have a product of 6 and a sum of 5 are 2 and 3. So write $5x$ as $2x + 3x$, then factor by grouping:

$$x^2 + 5x + 6 = x^2 + 2x + 3x + 6 \quad \text{Replace } 5x \text{ by } 2x + 3x.$$
$$= (x^2 + 2x) + (3x + 6) \quad \text{Group terms together.}$$
$$= x(x + 2) + 3(x + 2) \quad \text{Factor out common factors.}$$
$$= (x + 2)(x + 3) \quad \text{Factor out } x + 2.$$

b) To factor $x^2 + 8x + 12$, we must find two integers that have a product of 12 and a sum of 8. The pairs of integers with a product of 12 are 1 and 12, 2 and 6, and 3 and 4. Only 2 and 6 have a sum of 8. So write $8x$ as $2x + 6x$ and factor by grouping:

$$x^2 + 8x + 12 = x^2 + 2x + 6x + 12$$
$$= (x + 2)x + (x + 2)6 \quad \text{Factor out the common factors.}$$
$$= (x + 2)(x + 6) \quad \text{Factor out } x + 2.$$

Check by using FOIL: $(x + 2)(x + 6) = x^2 + 8x + 12$.

c) To factor $a^2 - 9a + 20$, we need two integers that have a product of 20 and a sum of -9. The integers are -4 and -5. Now replace $-9a$ by $-4a - 5a$ and factor by grouping:

$$a^2 - 9a + 20 = a^2 - 4a - 5a + 20 \quad \text{Replace } -9a \text{ by } -4a - 5a.$$
$$= a(a - 4) - 5(a - 4) \quad \text{Factor by grouping.}$$
$$= (a - 5)(a - 4) \quad \text{Factor out } a - 4. \qquad ■$$

After sufficient practice factoring trinomials, you may be able to skip most of the steps shown in these examples. For example, to factor $x^2 + x - 6$, simply find a pair of integers with a product of -6 and a sum of 1. The integers are 3 and -2, so we can write

$$x^2 + x - 6 = (x + 3)(x - 2)$$

and check by using FOIL.

E X A M P L E   2   **Factoring trinomials**

Factor.

a) $x^2 + 5x + 4$ **b)** $y^2 + 6y - 16$ **c)** $w^2 - 5w - 24$

Solution

a) To get a product of 4 and a sum of 5, use 1 and 4:

$$x^2 + 5x + 4 = (x + 1)(x + 4)$$

Check by using FOIL on $(x + 1)(x + 4)$.

b) To get a product of -16 we need a positive number and a negative number. To also get a sum of 6, use 8 and -2:

$$y^2 + 6y - 16 = (y + 8)(y - 2)$$

Check by using FOIL on $(y + 8)(y - 2)$.

c) To get a product of -24 and a sum of -5, use -8 and 3:

$$w^2 - 5w - 24 = (w - 8)(w + 3)$$

Check by using FOIL.   ■

Polynomials are easiest to factor when they are in the form $ax^2 + bx + c$. So if a polynomial can be rewritten into that form, rewrite it before attempting to factor it. In the next example we factor polynomials that need to be rewritten.

E X A M P L E   3   **Factoring trinomials**

Factor.

a) $2x - 8 + x^2$ **b)** $-36 + t^2 - 9t$

Solution

a) Before factoring, write the trinomial as $x^2 + 2x - 8$. Now, to get a product of -8 and a sum of 2, use -2 and 4:

$$2x - 8 + x^2 = x^2 + 2x - 8 \qquad \text{Write in } ax^2 + bx + c \text{ form.}$$
$$= (x + 4)(x - 2) \qquad \text{Factor and check by multiplying.}$$

b) Before factoring, write the trinomial as $t^2 - 9t - 36$. Now, to get a product of -36 and a sum of -9, use -12 and 3:

$$-36 + t^2 - 9t = t^2 - 9t - 36 \qquad \text{Write in } ax^2 + bx + c \text{ form.}$$
$$= (t - 12)(t + 3) \qquad \text{Factor and check by multiplying.} \quad ■$$

Prime Polynomials

To factor $x^2 + bx + c$, we try pairs of integers that have a product of c until we find a pair that has a sum of b. If there is no such pair of integers, then the polynomial cannot be factored and it is a prime polynomial. Before you can conclude that a polynomial is prime, you must try *all* possibilities.

EXAMPLE 4 **Prime polynomials**

Factor.

a) $x^2 + 7x - 6$ **b)** $x^2 + 9$

Solution

a) Because the last term is -6, we want a positive integer and a negative integer that have a product of -6 and a sum of 7. Check all possible pairs of integers:

Product	Sum
$-6 = (-1)(6)$	$-1 + 6 = 5$
$-6 = (1)(-6)$	$1 + (-6) = -5$
$-6 = (2)(-3)$	$2 + (-3) = -1$
$-6 = (-2)(3)$	$-2 + 3 = 1$

None of these possible factors of -6 have a sum of 7, so we can be certain that $x^2 + 7x - 6$ cannot be factored. It is a prime polynomial.

b) Because the x-term is missing in $x^2 + 9$, its coefficient is 0. That is, $x^2 + 9 = x^2 + 0x + 9$. So we seek two positive integers or two negative integers that have a product of 9 and a sum of 0. Check all possibilities:

Product	Sum
$9 = (3)(3)$	$3 + 3 = 6$
$9 = (-3)(-3)$	$-3 + (-3) = -6$
$9 = (9)(1)$	$9 + 1 = 10$
$9 = (-9)(-1)$	$-9 + (-1) = -10$

None of these pairs of integers has a sum of 0, so we can conclude that $x^2 + 9$ is a prime polynomial. Note that $x^2 + 9$ does not factor as $(x + 3)^2$ because $(x + 3)^2$ has a middle term: $(x + 3)^2 = x^2 + 6x + 9$. ■

The prime polynomial $x^2 + 9$ in Example 4(b) is a sum of two squares. It can be shown that any sum of two squares (in which there are no common factors) is a prime polynomial.

Sum of Two Squares

If a sum of two squares, $a^2 + b^2$, has no common factor other than 1, then it is a prime polynomial.

Factoring with Two Variables

In the next example we factor polynomials that have two variables using the same technique that we used for one variable.

E X A M P L E   5

Polynomials with two variables

Factor.

a) $x^2 + 2xy - 8y^2$ **b)** $a^2 - 7ab + 10b^2$

Solution

a) To get a product of -8 and a sum of 2, use 4 and -2. To get a product of $-8y^2$ use $4y$ and $-2y$:

$$x^2 + 2xy - 8y^2 = (x + 4y)(x - 2y)$$

Check by multiplying $(x + 4y)(x - 2y)$.

b) To get a product of 10 and a sum of -7, use -5 and -2. To get a product of $10b^2$, we use $-5b$ and $-2b$:

$$a^2 - 7ab + 10b^2 = (a - 5b)(a - 2b)$$

Check by multiplying.   ■

Factoring Completely

In Section 6.2 you learned that binomials such as $3x - 5$ (with no common factor) are prime polynomials. In Example 4 of this section we saw a trinomial that is a prime polynomial. There are infinitely many prime trinomials. When factoring a polynomial completely, we could have a factor that is a prime trinomial.

E X A M P L E   6

Factoring completely

Factor each polynomial completely.

a) $x^3 - 6x^2 - 16x$ **b)** $4x^3 + 4x^2 + 4x$ **c)** $3wy^2 + 18wy + 27w$

Solution

a) $x^3 - 6x^2 - 16x = x(x^2 - 6x - 16)$ Factor out the GCF.

                      $= x(x - 8)(x + 2)$ Factor $x^2 - 6x - 16$.

b) First factor out $4x$, the greatest common factor:

$$4x^3 + 4x^2 + 4x = 4x(x^2 + x + 1)$$

To factor $x^2 + x + 1$, we would need two integers with a product of 1 and a sum of 1. Because there are no such integers, $x^2 + x + 1$ is prime, and the factorization is complete.

c) $3wy^2 + 18wy + 27w = 3w(y^2 + 6y + 9)$ Factor out the GCF.

                            $= 3w(y + 3)^2$ Perfect square trinomial   ■

True or false? Answer true if the correct factorization is given and false if the factorization is incorrect. Explain your answer.

1. $x^2 - 6x + 9 = (x - 3)^2$
2. $x^2 + 6x + 9 = (x + 3)^2$
3. $x^2 + 10x + 9 = (x - 9)(x - 1)$
4. $x^2 - 8x - 9 = (x - 8)(x - 9)$
5. $x^2 + 8x - 9 = (x + 9)(x - 1)$
6. $x^2 + 8x + 9 = (x + 3)^2$
7. $x^2 - 10xy + 9y^2 = (x - y)(x - 9y)$
8. $x^2 + x + 1 = (x + 1)(x + 1)$
9. $x^2 + xy + 20y^2 = (x + 5y)(x - 4y)$
10. $x^2 + 1 = (x + 1)(x + 1)$

6.3 EXERCISES

Reading and Writing *After reading this section, write out the answers to these questions. Use complete sentences.*

1. What types of polynomials did we factor in this section?

2. How can you check if you have factored a trinomial correctly?

3. How can you determine if $x^2 + bx + c$ is prime?

4. How do you factor a sum of two squares?

5. When is a polynomial factored completely?

6. What should you always look for first when attempting to factor a polynomial completely?

Factor each trinomial. Write out all of the steps as shown in Example 1.

7. $x^2 + 4x + 3$ 8. $y^2 + 6y + 5$

9. $x^2 + 9x + 18$ 10. $w^2 + 6w + 8$

11. $a^2 - 7a + 12$ 12. $m^2 - 9m + 14$

13. $b^2 - 5b - 6$ 14. $a^2 + 5a - 6$

Factor each polynomial. See Examples 2–4. If the polynomial is prime, say so.

15. $y^2 + 7y + 10$
16. $x^2 + 8x + 15$
17. $a^2 - 6a + 8$
18. $b^2 - 8b + 15$
19. $m^2 - 10m + 16$
20. $m^2 - 17m + 16$
21. $w^2 + 9w - 10$
22. $m^2 + 6m - 16$
23. $w^2 - 8 - 2w$
24. $-16 + m^2 - 6m$
25. $a^2 - 2a - 12$
26. $x^2 + 3x + 3$
27. $15m - 16 + m^2$
28. $3y + y^2 - 10$
29. $a^2 - 4a + 12$
30. $y^2 - 6y - 8$
31. $z^2 - 25$
32. $p^2 - 1$
33. $h^2 + 49$
34. $q^2 + 4$
35. $m^2 + 12m + 20$
36. $m^2 + 21m + 20$
37. $t^2 - 3t + 10$
38. $x^2 - 5x - 3$
39. $m^2 - 18 - 17m$

40. $h^2 - 36 + 5h$

41. $m^2 - 23m + 24$

42. $m^2 + 23m + 24$

43. $5t - 24 + t^2$

44. $t^2 - 24 - 10t$

45. $t^2 - 2t - 24$

46. $t^2 + 14t + 24$

47. $t^2 - 10t - 200$

48. $t^2 + 30t + 200$

49. $x^2 - 5x - 150$

50. $x^2 - 25x + 150$

51. $13y + 30 + y^2$

52. $18z + 45 + z^2$

Factor each polynomial. See Example 5.

53. $x^2 + 5ax + 6a^2$

54. $a^2 + 7ab + 10b^2$

55. $x^2 - 4xy - 12y^2$

56. $y^2 + yt - 12t^2$

57. $x^2 - 13xy + 12y^2$

58. $h^2 - 9hs + 9s^2$

59. $x^2 + 4xz - 33z^2$

60. $x^2 - 5xs - 24s^2$

Factor each polynomial completely. Use the methods discussed in Sections 6.1 through 6.3. See Example 6.

61. $w^2 - 8w$

62. $x^4 - x^3$

63. $2w^2 - 162$

64. $6w^4 - 54w^2$

65. $x^2w^2 + 9x^2$

66. $a^4b + a^2b^3$

67. $w^2 - 18w + 81$

68. $w^2 + 30w + 81$

69. $6w^2 - 12w - 18$

70. $9w - w^3$

71. $32x^2 - 2x^4$

72. $20w^2 + 100w + 40$

73. $3w^2 + 27w + 54$

74. $w^3 - 3w^2 - 18w$

75. $18w^2 + w^3 + 36w$

76. $18a^2 + 3a^3 + 36a$

77. $8vw^2 + 32vw + 32v$

78. $3h^2t + 6ht + 3t$

79. $6x^3y + 30x^2y^2 + 36xy^3$

80. $3x^3y^2 - 3x^2y^2 + 3xy^2$

Use factoring to solve each problem.

81. **Area of a deck.** A rectangular deck has an area of $x^2 + 6x + 8$ square feet and a width of $x + 2$ feet. Find the length of the deck.

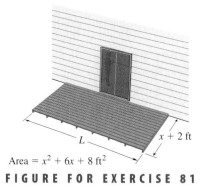

FIGURE FOR EXERCISE 81

82. **Area of a sail.** A triangular sail has an area of $x^2 + 5x + 6$ square meters and a height of $x + 3$ meters. Find the length of the sail's base.

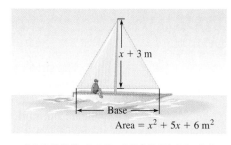

FIGURE FOR EXERCISE 82

83. **Volume of a cube.** Hector designed a cubic box with volume x^3 cubic feet. After increasing the dimensions of the bottom, the box has a volume of $x^3 + 8x^2 + 15x$ cubic feet. If each of the dimensions of the bottom was increased by a whole number of feet, then how much was each increase?

84. **Volume of a container.** A cubic shipping container had a volume of a^3 cubic meters. The height was decreased by a whole number of meters and the width was increased by a whole number of meters so that the volume of the container is now $a^3 + 2a^2 - 3a$ cubic meters. By how many meters were the height and width changed?

GETTING MORE INVOLVED

85. *Discussion.* Which of the following products is not equivalent to the others. Explain your answer.
a) $(2x - 4)(x + 3)$ b) $(x - 2)(2x + 6)$
c) $2(x - 2)(x + 3)$ d) $(2x - 4)(2x + 6)$

86. *Discussion.* When asked to factor completely a certain polynomial, four students gave the following answers. Only one student gave the correct answer. Which one must it be? Explain your answer.
a) $3(x^2 - 2x - 15)$ b) $(3x - 5)(5x - 15)$
c) $3(x - 5)(x - 3)$ d) $(3x - 15)(x - 3)$

6.4 FACTORING $ax^2 + bx + c$ WITH $a \neq 1$

In Section 6.3 we factored trinomials with a leading coefficient of 1. In this section we will use a slightly different technique to factor trinomials with leading coefficients not equal to 1.

The *ac* Method

If the leading coefficient of a trinomial is not 1, we can again use grouping to factor the trinomial. However, the procedure is slightly different.

Consider the trinomial $2x^2 + 7x + 6$. First find the product of the leading coefficient and the constant term. In this case it is $2 \cdot 6$, or 12. Now find two numbers with a product of 12 and a sum of 7. The pairs of numbers with a product of 12 are 1 and 12, 2 and 6, and 3 and 4. Only 3 and 4 have a product of 12 and a sum of 7. Now replace $7x$ by $3x + 4x$ and factor by grouping:

$$2x^2 + 7x + 6 = 2x^2 + 3x + 4x + 6 \qquad \text{Replace } 7x \text{ by } 3x + 4x.$$
$$= (2x + 3)x + (2x + 3)2 \qquad \text{Factor out the common factors.}$$
$$= (2x + 3)(x + 2) \qquad \text{Factor out } 2x + 3.$$

The strategy for factoring a trinomial is summarized in the following box. The steps listed here actually work whether or not the leading coefficient is 1. This is actually the method that you learned in Section 6.3 with $a = 1$. This method is called the *ac* **method.**

> **Strategy for Factoring $ax^2 + bx + c$ by the *ac* Method**
>
> To factor the trinomial $ax^2 + bx + c$:
> 1. Find two numbers that have a product equal to ac and a sum equal to b.
> 2. Replace bx by two terms using the two new numbers as coefficients.
> 3. Factor the resulting four-term polynomial by grouping.

EXAMPLE 1

The *ac* method

Factor each trinomial.

a) $2x^2 + x - 6$ **b)** $10x^2 + 13x - 3$

Solution

a) Because $2 \cdot (-6) = -12$, we need two integers with a product of -12 and a sum of 1. We can list the possible pairs of integers with a product of -12:

$$
\begin{array}{ccc}
1 \text{ and } -12 & 2 \text{ and } -6 & 3 \text{ and } -4 \\
-1 \text{ and } 12 & -2 \text{ and } 6 & -3 \text{ and } 4
\end{array}
$$

Only -3 and 4 have a sum of 1. Replace x by $-3x + 4x$ and factor by grouping:

$$2x^2 + x - 6 = 2x^2 - 3x + 4x - 6 \qquad \text{Replace } x \text{ by } -3x + 4x.$$
$$= (2x - 3)x + (2x - 3)2 \qquad \text{Factor out the common factors.}$$
$$= (2x - 3)(x + 2) \qquad \text{Factor out } 2x - 3.$$

Check by FOIL.

b) Because $10 \cdot (-3) = -30$, we need two integers with a product of -30 and a sum of 13. The product is negative, so the integers must have opposite signs. We can list all pairs of factors of -30 as follows:

$$
\begin{array}{llll}
1 \text{ and } -30 & 2 \text{ and } -15 & 3 \text{ and } -10 & 5 \text{ and } -6 \\
-1 \text{ and } 30 & -2 \text{ and } 15 & -3 \text{ and } 10 & -5 \text{ and } 6
\end{array}
$$

The only pair that has a sum of 13 is -2 and 15:

$$
\begin{aligned}
10x^2 + 13x - 3 &= 10x^2 - 2x + 15x - 3 && \text{Replace } 13x \text{ by } -2x + 15x. \\
&= (5x - 1)2x + (5x - 1)3 && \text{Factor out the common factors.} \\
&= (5x - 1)(2x + 3) && \text{Factor out } 5x - 1.
\end{aligned}
$$

Check by FOIL.   ■

Trial and Error

After you have gained some experience at factoring by the *ac* method, you can often find the factors without going through the steps of grouping. For example, consider the polynomial

$$3x^2 + 7x - 6.$$

The factors of $3x^2$ can only be $3x$ and x. The factors of 6 could be 2 and 3 or 1 and 6. We can list all of the possibilities that give the correct first and last terms, without regard to the signs:

$$(3x \quad 3)(x \quad 2) \qquad (3x \quad 2)(x \quad 3) \qquad (3x \quad 6)(x \quad 1) \qquad (3x \quad 1)(x \quad 6)$$

Because the factors of -6 have unlike signs, one binomial factor is a sum and the other binomial is a difference. Now we try some products to see if we get a middle term of $7x$:

$$
\begin{aligned}
(3x + 3)(x - 2) &= 3x^2 - 3x - 6 && \text{Incorrect.} \\
(3x - 3)(x + 2) &= 3x^2 + 3x - 6 && \text{Incorrect.}
\end{aligned}
$$

Actually, there is no need to try $(3x \quad 3)(x \quad 2)$ or $(3x \quad 6)(x \quad 1)$ because each contains a binomial with a common factor. As you can see from the above products, a common factor in the binomial causes a common factor in the product. But $3x^2 + 7x - 6$ has no common factor. So the factors must come from either $(3x \quad 2)(x \quad 3)$ or $(3x \quad 1)(x \quad 6)$. So we try again:

$$
\begin{aligned}
(3x + 2)(x - 3) &= 3x^2 - 7x - 6 && \text{Incorrect.} \\
(3x - 2)(x + 3) &= 3x^2 + 7x - 6 && \text{Correct.}
\end{aligned}
$$

Even though there may be many possibilities in some factoring problems, it is often possible to find the correct factors without writing down every possibility. We can use a bit of guesswork in factoring trinomials. *Try* whichever possibility you think might work. *Check* it by multiplying. If it is not right, then *try again*. That is why this method is called **trial and error.**

E X A M P L E   2

Trial and error

Factor each trinomial using trial and error.

a) $2x^2 + 5x - 3$ **b)** $3x^2 - 11x + 6$

Solution

a) Because $2x^2$ factors only as $2x \cdot x$ and 3 factors only as $1 \cdot 3$, there are only two possible ways to get the correct first and last terms, without regard to the signs:

$$(2x \quad 1)(x \quad 3) \quad \text{and} \quad (2x \quad 3)(x \quad 1)$$

Because the last term of the trinomial is negative, one of the missing signs must be $+$, and the other must be $-$. The trinomial is factored correctly as

$$2x^2 + 5x - 3 = (2x - 1)(x + 3).$$

Check by using FOIL.

b) There are four possible ways to factor $3x^2 - 11x + 6$:

$$(3x \quad 1)(x \quad 6) \qquad (3x \quad 2)(x \quad 3)$$
$$(3x \quad 6)(x \quad 1) \qquad (3x \quad 3)(x \quad 2)$$

Because the last term in $3x^2 - 11x + 6$ is positive and the middle term is negative, both signs in the factors must be negative. Because $3x^2 - 11x + 6$ has no common factor, we can rule out $(3x \quad 6)(x \quad 1)$ and $(3x \quad 3)(x \quad 2)$. So the only possibilities left are $(3x - 1)(x - 6)$ and $(3x - 2)(x - 3)$. The trinomial is factored correctly as

$$3x^2 - 11x + 6 = (3x - 2)(x - 3).$$

Check by using FOIL.

Factoring by trial and error is not just guessing. In fact, if the trinomial has a positive leading coefficient, we can determine in advance whether its factors are sums or differences.

Using Signs in Trial and Error

1. If the signs of the terms of a trinomial are $+\ +\ +$, then both factors are sums: $x^2 + 5x + 6 = (x + 2)(x + 3)$.
2. If the signs are $+\ -\ +$, then both factors are differences: $x^2 - 5x + 6 = (x - 2)(x - 3)$.
3. If the signs are $+\ +\ -$ or $+\ -\ -$, then one factor is a sum and the other is a difference: $x^2 + x - 6 = (x + 3)(x - 2)$ and $x^2 - x - 6 = (x - 3)(x + 2)$.

In the next example we factor a trinomial that has two variables.

E X A M P L E   3

Factoring a trinomial with two variables by trial and error

Factor $6x^2 - 7xy + 2y^2$.

Solution

We list the possible ways to factor the trinomial:

$$(3x \quad 2y)(2x \quad y) \qquad (3x \quad y)(2x \quad 2y) \qquad (6x \quad 2y)(x \quad y) \qquad (6x \quad y)(x \quad 2y)$$

Because the last term of the trinomial is positive and the middle term is negative, both factors must contain subtraction symbols. To get the middle term of $-7xy$, we use the first possibility listed:

$$6x^2 - 7xy + 2y^2 = (3x - 2y)(2x - y)$$

Factoring Completely

You can use the latest factoring technique along with the techniques that you learned earlier to factor polynomials completely. Remember always to first factor out the greatest common factor (if it is not 1).

E X A M P L E   4

Factoring completely

Factor each polynomial completely.

a) $4x^3 + 14x^2 + 6x$ b) $12x^2y + 6xy + 6y$

Solution

a) $4x^3 + 14x^2 + 6x = 2x(2x^2 + 7x + 3)$ Factor out the GCF, $2x$.

$\qquad\qquad\qquad\quad = 2x(2x + 1)(x + 3)$ Factor $2x^2 + 7x + 3$.

Check by multiplying.

b) $12x^2y + 6xy + 6y = 6y(2x^2 + x + 1)$ Factor out the GCF, $6y$.

To factor $2x^2 + x + 1$ by the ac method, we need two numbers with a product of 2 and a sum of 1. Because there are no such numbers, $2x^2 + x + 1$ is prime and the factorization is complete.   ■

Usually, our first step in factoring is to factor out the greatest common factor (if it is not 1). If the first term of a polynomial has a negative coefficient, then it is better to factor out the opposite of the GCF so that the resulting polynomial will have a positive leading coefficient.

E X A M P L E   5

Factoring out the opposite of the GCF

Factor each polynomial completely.

a) $-18x^3 + 51x^2 - 15x$ b) $-3a^2 + 2a + 21$

Solution

a) The GCF is $3x$. Because the first term has a negative coefficient, we factor out $-3x$:

$$-18x^3 + 51x^2 - 15x = -3x(6x^2 - 17x + 5)\quad \text{Factor out } -3x.$$

$$= -3x(3x - 1)(2x - 5)\quad \text{Factor } 6x^2 - 17x + 5.$$

b) The GCF for $-3a^2 + 2a + 21$ is 1. Because the first term has a negative coefficient, factor out -1:

$$-3a^2 + 2a + 21 = -1(3a^2 - 2a - 21)\quad \text{Factor out } -1.$$

$$= -1(3a + 7)(a - 3)\quad \text{Factor } 3a^2 - 2a - 21.\quad ■$$

WARM-UPS

True or false? Answer true if the correct factorization is given and false if the factorization is incorrect. Explain your answer.

1. $2x^2 + 3x + 1 = (2x + 1)(x + 1)$
2. $2x^2 + 5x + 3 = (2x + 1)(x + 3)$
3. $3x^2 + 10x + 3 = (3x + 1)(x + 3)$
4. $15x^2 + 31x + 14 = (3x + 7)(5x + 2)$
5. $2x^2 - 7x - 9 = (2x - 9)(x + 1)$
6. $2x^2 + 3x - 9 = (2x + 3)(x - 3)$
7. $2x^2 - 16x - 9 = (2x - 9)(2x + 1)$
8. $8x^2 - 22x - 5 = (4x - 1)(2x + 5)$
9. $9x^2 + x - 1 = (5x - 1)(4x + 1)$
10. $12x^2 - 13x + 3 = (3x - 1)(4x - 3)$

6.4   EXERCISES

Reading and Writing   *After reading this section, write out the answers to these questions. Use complete sentences.*

1. What types of polynomials did we factor in this section?

2. What is the *ac* method of factoring?

3. How can you determine if $ax^2 + bx + c$ is prime?

4. What is the trial-and-error method of factoring?

Find the following. See Example 1.

5. Two integers that have a product of 20 and a sum of 12

6. Two integers that have a product of 36 and a sum of -20

7. Two integers that have a product of -12 and a sum of -4

8. Two integers that have a product of -8 and a sum of 7

Each of the following trinomials is in the form $ax^2 + bx + c$. For each trinomial, find two integers that have a product of ac and a sum of b. Do not factor the trinomials. See Example 1.

9. $6x^2 + 7x + 2$ 10. $5x^2 + 17x + 6$

11. $6y^2 - 11y + 3$ 12. $6z^2 - 19z + 10$

13. $12w^2 + w - 1$ 14. $15t^2 - 17t - 4$

Factor each trinomial using the ac method. See Example 1.

15. $2x^2 + 3x + 1$ 16. $2x^2 + 11x + 5$

17. $2x^2 + 9x + 4$ 18. $2h^2 + 7h + 3$

19. $3t^2 + 7t + 2$ 20. $3t^2 + 8t + 5$

21. $2x^2 + 5x - 3$ 22. $3x^2 - x - 2$

23. $6x^2 + 7x - 3$ 24. $21x^2 + 2x - 3$

25. $2x^2 - 7x + 6$ 26. $3a^2 - 14a + 15$

27. $5b^2 - 13b + 6$ 28. $7y^2 + 16y - 15$

29. $4y^2 - 11y - 3$ 30. $35x^2 - 2x - 1$

31. $3x^2 + 2x + 1$ 32. $6x^2 - 4x - 5$

33. $8x^2 - 2x - 1$ 34. $8x^2 - 10x - 3$

35. $9t^2 - 9t + 2$ 36. $9t^2 + 5t - 4$

37. $15x^2 + 13x + 2$ 38. $15x^2 - 7x - 2$

39. $15x^2 - 13x + 2$ 40. $15x^2 + x - 2$

Complete the factoring.

41. $3x^2 + 7x + 2 = (x + 2)(\quad)$

42. $2x^2 - x - 15 = (x - 3)(\quad)$

43. $5x^2 + 11x + 2 = (5x + 1)(\quad)$

44. $4x^2 - 19x - 5 = (4x + 1)(\quad)$

45. $6a^2 - 17a + 5 = (3a - 1)(\quad)$

46. $4b^2 - 16b + 15 = (2b - 5)(\quad)$

Factor each trinomial using trial and error. See Examples 2 and 3.

47. $5a^2 + 11a + 2$ 48. $3y^2 + 10y + 7$

49. $4w^2 + 8w + 3$ 50. $6z^2 + 13z + 5$

51. $15x^2 - x - 2$ 52. $15x^2 + 13x - 2$

53. $8x^2 - 6x + 1$ 54. $8x^2 - 22x + 5$

55. $15x^2 - 31x + 2$ 56. $15x^2 + 31x + 2$

57. $2x^2 + 18x - 90$ 58. $3x^2 + 11x + 10$

59. $3x^2 + x - 10$ 60. $3x^2 - 17x + 10$

61. $10x^2 - 3xy - y^2$ 62. $8x^2 - 2xy - y^2$

63. $42a^2 - 13ab + b^2$ 64. $10a^2 - 27ab + 5b^2$

Factor each polynomial completely. See Examples 4 and 5.

65. $81w^3 - w$ 66. $81w^3 - w^2$

67. $4w^2 + 2w - 30$ 68. $2x^2 - 28x + 98$

69. $27 + 12x^2 + 36x$ 70. $24y + 12y^2 + 12$

71. $6w^2 - 11w - 35$ 72. $18x^2 - 6x + 6$

73. $3x^2z - 3zx - 18z$ 74. $a^2b + 2ab - 15b$

75. $10x^2y^2 + xy^2 - 9y^2$

76. $2x^2y^2 + xy^2 + 3y^2$

77. $a^2 + 2ab - 15b^2$

78. $a^2b^2 - 2a^2b - 15a^2$

79. $-6t^3 - t^2 + 2t$

80. $-36t^2 - 6t + 12$

81. $12t^4 - 2t^3 - 4t^2$

82. $12t^3 + 14t^2 + 4t$

83. $4x^2y - 8xy^2 + 3y^3$

84. $9x^2 + 24xy - 9y^2$

85. $-4w^2 + 7w - 3$

86. $-30w^2 + w + 1$

87. $-12a^3 + 22a^2b - 6ab^2$

88. $-36a^2b + 21ab^2 - 3b^3$

Solve each problem.

89. *Height of a ball.* If a ball is thrown upward at 40 feet per second from a rooftop 24 feet above the ground, then its height above the ground t seconds after it is thrown is given by $h = -16t^2 + 40t + 24$. Rewrite this formula with the polynomial on the right-hand side factored completely. Use the factored version of the formula to find h when $t = 3$.

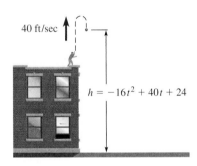

FIGURE FOR EXERCISE 89

90. *Worker efficiency.* In a study of worker efficiency at Wong Laboratories it was found that the number of components assembled per hour by the average worker t hours after starting work could be modeled by the function

$$N(t) = -3t^3 + 23t^2 + 8t.$$

a) Rewrite the formula by factoring the right-hand side completely.

b) Use the factored version of the formula to find $N(3)$.

c) Use the accompanying graph to estimate the time at which the workers are most efficient.

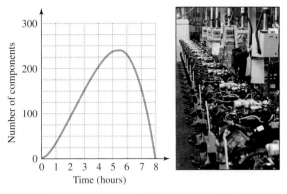

FIGURE FOR EXERCISE 90

d) Use the accompanying graph to estimate the maximum number of components assembled per hour during an 8-hour shift.

GETTING MORE INVOLVED

91. *Exploration.* Find all positive and negative integers b for which each polynomial can be factored.

a) $x^2 + bx + 3$

b) $3x^2 + bx + 5$

c) $2x^2 + bx - 15$

92. *Exploration.* Find two integers c (positive or negative) for which each polynomial can be factored. Many answers are possible.

a) $x^2 + x + c$

b) $x^2 - 2x + c$

c) $2x^2 - 3x + c$

93. *Cooperative learning.* Working in groups, cut two large squares, three rectangles, and one small square out of paper that are exactly the same size as shown in the accompanying figure. Then try to place the six figures next to one another so that they form a large rectangle. Do not overlap the pieces or leave any gaps. Explain how factoring $2x^2 + 3x + 1$ can help you solve this puzzle.

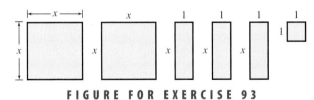

FIGURE FOR EXERCISE 93

94. *Cooperative learning.* Working in groups, cut four squares and eight rectangles out of paper as in the previous exercise to illustrate the trinomial $4x^2 + 7x + 3$. Select one group to demonstrate how to arrange the 12 pieces to form a large rectangle. Have another group explain how factoring the trinomial can help you solve this puzzle.

6.5   THE FACTORING STRATEGY

In previous sections we established the general idea of factoring and some special cases. In this section we will see how division relates to factoring and see two more special cases. We will then summarize all of the factoring that we have done with a factoring strategy.

Using Division in Factoring

To find the prime factorization for a large integer such as 1001, you could divide possible factors (prime numbers) into 1001 until you find one that leaves no remainder. If you are told that 13 is a factor (or make a lucky guess), then you could divide 1001 by 13 to get the quotient 77. With this information you can factor 1001:

$$1001 = 77 \cdot 13$$

Now you can factor 77 to get the prime factorization of 1001:

$$1001 = 7 \cdot 11 \cdot 13$$

We can use this same idea with polynomials that are of higher degree than the ones we have been factoring. If we can guess a factor or if we are given a factor, we can use division to find the other factor and then proceed to factor the polynomial completely. Of course, it is harder to guess a factor of a polynomial than it is to guess a factor of an integer. In the next example we will factor a third-degree polynomial completely, given one factor.

E X A M P L E   1

Using division in factoring

Factor the polynomial $x^3 + 2x^2 - 5x - 6$ completely, given that the binomial $x + 1$ is a factor of the polynomial.

Solution

Divide the polynomial by the binomial:

$$
\begin{array}{r}
x^2 + x - 6 \\
x + 1 \overline{)x^3 + 2x^2 - 5x - 6} \\
\underline{x^3 + x^2} \\
x^2 - 5x \\
\underline{x^2 + x} \\
-6x - 6 \qquad {\scriptstyle -5x - x = -6x} \\
\underline{-6x - 6} \\
0 \qquad {\scriptstyle -6 - (-6) = 0}
\end{array}
$$

Because the remainder is 0, the dividend is the divisor times the quotient:

$$x^3 + 2x^2 - 5x - 6 = (x + 1)(x^2 + x - 6)$$

Now we factor the remaining trinomial to get the complete factorization:

$$x^3 + 2x^2 - 5x - 6 = (x + 1)(x + 3)(x - 2)$$

Factoring a Difference or Sum of Two Cubes

We can use division to discover that $a - b$ is a factor of $a^3 - b^3$ (a difference of two cubes) and $a + b$ is a factor of $a^3 + b^3$ (a sum of two cubes):

$$
a - b \overline{)\begin{array}{l} a^2 + ab + b^2 \\ a^3 + 0a^2b + 0ab^2 - b^3 \end{array}}
\qquad
a + b \overline{)\begin{array}{l} a^2 - ab + b^2 \\ a^3 + 0a^2b + 0ab^2 + b^3 \end{array}}
$$

$$
\begin{array}{r}
\underline{a^3 -\ a^2b} \\
a^2b + 0ab^2 \\
\underline{a^2b -\ ab^2} \\
ab^2 - b^3 \\
\underline{ab^2 - b^3} \\
0
\end{array}
\qquad
\begin{array}{r}
\underline{a^3 +\ a^2b} \\
-a^2b + 0ab^2 \\
\underline{-a^2b -\ ab^2} \\
ab^2 + b^3 \\
\underline{ab^2 + b^3} \\
0
\end{array}
$$

So $a - b$ is a factor of $a^3 - b^3$, and $a + b$ is a factor of $a^3 + b^3$. These results give us two more factoring rules.

Factoring a Difference or Sum of Two Cubes

$$a^3 - b^3 = (a - b)(a^2 + ab + b^2)$$
$$a^3 + b^3 = (a + b)(a^2 - ab + b^2)$$

Note that $a^2 + ab + b^2$ and $a^2 - ab + b^2$ are prime. Do not confuse them with $a^2 + 2ab + b^2$ and $a^2 - 2ab + b^2$, which are not prime because

$$a^2 + 2ab + b^2 = (a + b)^2 \qquad \text{and} \qquad a^2 - 2ab + b^2 = (a - b)^2.$$

These similarities can help you remember the rules for factoring $a^3 - b^3$ and $a^3 + b^3$. Note also how $a^3 - b^3$ compares with $a^2 - b^2$:

$$a^2 - b^2 = (a - b)(a + b)$$
$$a^3 - b^3 = (a - b)(a^2 + ab + b^2)$$

E X A M P L E   2

Factoring a difference or sum of two cubes

Factor each polynomial.

a) $w^3 - 8$ **b)** $x^3 + 1$ **c)** $8y^3 - 27$

Solution

a) Because $8 = 2^3$, $w^3 - 8$ is a difference of two cubes. To factor $w^3 - 8$, let $a = w$ and $b = 2$ in the formula $a^3 - b^3 = (a - b)(a^2 + ab + b^2)$:

$$w^3 - 8 = (w - 2)(w^2 + 2w + 4)$$

b) Because $1 = 1^3$, the binomial $x^3 + 1$ is a sum of two cubes. Let $a = x$ and $b = 1$ in the formula $a^3 + b^3 = (a + b)(a^2 - ab + b^2)$:

$$x^3 + 1 = (x + 1)(x^2 - x + 1)$$

c) $8y^3 - 27 = (2y)^3 - 3^3$ This is a difference of two cubes.

$\qquad\qquad\ = (2y - 3)(4y^2 + 6y + 9)$ Let $a = 2y$ and $b = 3$ in the formula.

CAUTION The polynomial $(a - b)^3$ is not equivalent to $a^3 - b^3$ because if $a = 2$ and $b = 1$, then

$$(a - b)^3 = (2 - 1)^3 = 1^3 = 1$$

and

$$a^3 - b^3 = 2^3 - 1^3 = 8 - 1 = 7.$$

Likewise, $(a + b)^3$ is not equivalent to $a^3 + b^3$.

The Factoring Strategy

The following is a summary of the ideas that we use to factor a polynomial completely.

Strategy for Factoring Polynomials Completely

1. If there are any common factors, factor them out first.
2. When factoring a binomial, check to see whether it is a difference of two squares, a difference of two cubes, or a sum of two cubes. *A sum of two squares does not factor.*
3. When factoring a trinomial, check to see whether it is a perfect square trinomial.
4. When factoring a trinomial that is not a perfect square, use the *ac* method or the trial-and-error method.
5. If the polynomial has four terms, try factoring by grouping.
6. Check to see whether any of the factors can be factored again.

We will use the factoring strategy in the next two examples.

E X A M P L E   3 **Factoring polynomials**

Factor each polynomial completely.

a) $2a^2b - 24ab + 72b$

b) $3x^3 + 6x^2 - 75x - 150$

Solution

a) $\begin{aligned} 2a^2b - 24ab + 72b &= 2b(a^2 - 12a + 36) \qquad &&\text{First factor out the GCF, } 2b. \\ &= 2b(a - 6)^2 \qquad &&\text{Factor the perfect square trinomial.} \end{aligned}$

b) $\begin{aligned} 3x^3 + 6x^2 - 75x - 150 &= 3[x^3 + 2x^2 - 25x - 50] \qquad &&\text{Factor out the GCF, 3.} \\ &= 3[x^2(x + 2) - 25(x + 2)] \qquad &&\text{Factor out common factors.} \\ &= 3(x^2 - 25)(x + 2) \qquad &&\text{Factor by grouping.} \\ &= 3(x + 5)(x - 5)(x + 2) \qquad &&\text{Factor the difference of two squares.} \end{aligned}$

E X A M P L E   4   **Factoring polynomials**

Factor each polynomial completely.

a) $ax^3 + ax$ **b)** $by^3 + b$ **c)** $8m^3 + 22m^2 - 6m$

Solution

a) $ax^3 + ax = ax(x^2 + 1)$ The sum of two squares is prime.

b) $by^3 + b = b(y^3 + 1)$ Factor out the GCF.

$= b(y + 1)(y^2 - y + 1)$ Factor the sum of two cubes.

c) $8m^3 + 22m^2 - 6m = 2m(4m^2 + 11m - 3)$ Factor out the GCF.

$= 2m(4m - 1)(m + 3)$ Trial and error   ■

WARM-UPS

True or false? Explain your answer.

1. $x^2 - 4 = (x - 2)^2$ for any real number x.

2. The trinomial $4x^2 + 6x + 9$ is a perfect square trinomial.

3. The polynomial $4y^2 + 25$ is a prime polynomial.

4. $3y + ay + 3x + ax = (x + y)(3 + a)$ for any values of the variables.

5. The polynomial $3x^2 + 51$ cannot be factored.

6. If the GCF is not 1, then you should factor it out first.

7. $x^2 + 9 = (x + 3)^2$ for any real number x.

8. The polynomial $x^2 - 3x - 5$ is a prime polynomial.

9. The polynomial $y^2 - 5y - my + 5m$ can be factored by grouping.

10. The polynomial $x^2 + ax - 3x + 3a$ can be factored by grouping.

6.5   EXERCISES

Reading and Writing *After reading this section, write out the answers to these questions. Use complete sentences.*

1. What is the relationship between division and factoring?

2. How do we know that $a - b$ is a factor of $a^3 - b^3$?

3. How do we know that $a + b$ is a factor of $a^3 + b^3$?

4. How do you recognize if a polynomial is a sum of two cubes?

5. How do you factor a sum of two cubes?

6. How do you factor a difference of two cubes?

Factor each polynomial completely, given that the binomial following it is a factor of the polynomial. See Example 1.

7. $x^3 + 3x^2 - 10x - 24, x + 4$

8. $x^3 - 7x + 6, x - 1$

9. $x^3 + 4x^2 + x - 6, x - 1$

10. $x^3 - 5x^2 - 2x + 24, x + 2$

11. $x^3 - 8, x - 2$

12. $x^3 + 27, x + 3$

13. $x^3 + 4x^2 - 3x + 10, x + 5$

14. $2x^3 - 5x^2 - x - 6, x - 3$

15. $x^3 + 2x^2 + 2x + 1, x + 1$

16. $x^3 + 2x^2 - 5x - 6, x + 3$

Factor each difference or sum of cubes. See Example 2.

17. $m^3 - 1$

18. $z^3 - 27$

19. $x^3 + 8$

20. $y^3 + 27$

21. $8w^3 + 1$

22. $125m^3 + 1$

23. $8t^3 - 27$

24. $125n^3 - 8$

25. $x^3 - y^3$

26. $m^3 + n^3$

27. $8t^3 + y^3$

28. $u^3 - 125v^3$

Factor each polynomial completely. If a polynomial is prime, say so. See Examples 3 and 4.

29. $2x^2 - 18$

30. $3x^3 - 12x$

31. $4x^2 + 8x - 60$

32. $3x^2 + 18x + 27$

33. $x^3 + 4x^2 + 4x$

34. $a^3 - 5a^2 + 6a$

35. $5max^2 + 20ma$

36. $3bmw^2 - 12bm$

37. $9x^2 + 6x + 1$

38. $9x^2 + 6x + 3$

39. $6x^2y + xy - 2y$

40. $5x^2y^2 - xy^2 - 6y^2$

41. $y^2 + 10y - 25$

42. $8b^2 + 24b + 18$

43. $16m^2 - 4m - 2$

44. $32a^2 + 4a - 6$

45. $9a^2 + 24a + 16$

46. $3x^2 - 18x - 48$

47. $24x^2 - 26x + 6$

48. $4x^2 - 6x - 12$

49. $3a^2 - 27a$

50. $a^2 - 25a$

51. $8 - 2x^2$

52. $x^3 + 6x^2 + 9x$

53. $6x^3 - 5x^2 + 12x$

54. $x^3 + 2x^2 - x - 2$

55. $a^3b - 4ab$

56. $2m^2 - 1800$

57. $x^3 + 2x^2 - 4x - 8$

58. $m^2a + 2ma^2 + a^3$

59. $2w^4 - 16w$

60. $m^4n + mn^4$

61. $3a^2w - 18aw + 27w$

62. $8a^3 + 4a$

63. $5x^2 - 500$

64. $25x^2 - 16y^2$

65. $2m + 2n - wm - wn$

66. $aw - 5b - bw + 5a$

67. $3x^4 + 3x$

68. $3a^5 - 81a^2$

69. $4w^2 + 4w - 4$

70. $4w^2 + 8w - 5$

71. $a^4 + 7a^3 - 30a^2$

72. $2y^5 + 3y^4 - 20y^3$

73. $4aw^3 - 12aw^2 + 9aw$

74. $9bn^3 + 15bn^2 - 14bn$

75. $t^2 + 6t + 9$

76. $t^3 + 12t^2 + 36t$

Solve each problem.

77. ***Increasing cube.*** Each of the three dimensions of a cube with a volume of x^3 cubic centimeters is increased by a whole number of centimeters. If the new volume is $x^3 + 10x^2 + 31x + 30$ cubic centimeters and the new height is $x + 2$ centimeters, then what are the new length and width?

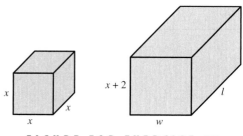

FIGURE FOR EXERCISE 77

78. ***Decreasing cube.*** Each of the three dimensions of a cube with a volume of y^3 cubic centimeters is decreased by a whole number of centimeters. If the new volume is $y^3 - 13y^2 + 54y - 72$ cubic centimeters and the new width is $y - 6$ centimeters, then what are the new length and height?

GETTING MORE INVOLVED

79. ***Discussion.*** Are there any values for a and b for which $(a + b)^3 = a^3 + b^3$? Find a pair of values for a and b for which $(a + b)^3 \neq a^3 + b^3$. Is $(a + b)^3$ equivalent to $a^3 + b^3$? Explain your answers.

80. ***Writing.*** Explain why $a^2 + ab + b^2$ and $a^2 - ab + b^2$ are prime polynomials.

6.6   SOLVING QUADRATIC EQUATIONS BY FACTORING

The techniques of factoring can be used to solve equations involving polynomials. These equations cannot be solved by the other methods that you have learned. After you learn to solve equations by factoring, you will use this technique to solve some new types of problems.

In this

section

- The Zero Factor Property
- Applications

The Zero Factor Property

In this chapter you learned to factor polynomials such as $x^2 + x - 6$. The equation $x^2 + x - 6 = 0$ is called a *quadratic equation*.

helpful   hint

The prefix "quad" means four. So why is a polynomial with three terms called quadratic? Perhaps it is because a quadratic polynomial can be factored into a product of two binomials, which contain a total of four terms.

> **Quadratic Equation**
>
> If a, b, and c are real numbers with $a \neq 0$, then
> $$ax^2 + bx + c = 0$$
> is called a **quadratic equation.**

The main idea used to solve quadratic equations is the **zero factor property,** which says that if a product is zero, then one or the other of the factors is zero.

> **The Zero Factor Property**
>
> The equation $a \cdot b = 0$ is equivalent to
> $$a = 0 \quad \text{or} \quad b = 0.$$

In our first example, we use the zero factor property and factoring to reduce a quadratic equation into two linear equations. Solving the linear equations gives us the solutions to the quadratic equation.

EXAMPLE 1   **Using the zero factor property**

Solve $x^2 + x - 6 = 0$.

Solution

First factor the polynomial on the left side:

$$x^2 + x - 6 = 0$$
$$(x + 3)(x - 2) = 0 \qquad \text{Factor the left side.}$$
$$x + 3 = 0 \quad \text{or} \quad x - 2 = 0 \qquad \text{Zero factor property}$$
$$x = -3 \quad \text{or} \quad x = 2 \qquad \text{Solve each equation.}$$

We now check that -3 and 2 satisfy the original equation.

$$\text{For } x = -3:$$

$$x^2 + x - 6 = (-3)^2 + (-3) - 6$$
$$= 9 - 3 - 6$$
$$= 0$$

$$\text{For } x = 2:$$

$$x^2 + x - 6 = (2)^2 + (2) - 6$$
$$= 4 + 2 - 6$$
$$= 0$$

The solutions to $x^2 + x - 6 = 0$ are -3 and 2. ∎

A sentence such as $x = -3$ or $x = 2$, which is made up of two or more equations connected with the word "or" is called a **compound equation.** In the next example we again solve a quadratic equation by using the zero factor property to write a compound equation.

E X A M P L E   2

Using the zero factor property

Solve the equation $3x^2 = -3x$.

Solution

First rewrite the equation with 0 on the right-hand side:

$$3x^2 = -3x$$
$$3x^2 + 3x = 0 \qquad \text{Add } 3x \text{ to each side.}$$
$$3x(x + 1) = 0 \qquad \text{Factor the left-hand side.}$$
$$3x = 0 \quad \text{or} \quad x + 1 = 0 \qquad \text{Zero factor property}$$
$$x = 0 \quad \text{or} \quad x = -1 \qquad \text{Solve each equation.}$$

Check 0 and -1 in the original equation $3x^2 = -3x$.

$$\text{For } x = 0: \qquad \text{For } x = -1:$$
$$3(0)^2 = -3(0) \qquad 3(-1)^2 = -3(-1)$$
$$0 = 0 \qquad 3 = 3$$

There are two solutions to the original equation, 0 and -1. ∎

C A U T I O N   If in Example 2 you divide each side of $3x^2 = -3x$ by $3x$, you would get $x = -1$ but not the solution $x = 0$. For this reason we usually do not divide each side of an equation by a variable.

The basic strategy for solving an equation by factoring follows.

Strategy for Solving an Equation by Factoring

1. Rewrite the equation with 0 on the right-hand side.
2. Factor the left-hand side completely.
3. Use the zero factor property to get simple linear equations.
4. Solve the linear equations.
5. Check the answers in the original equation.
6. State the solution(s) to the original equation.

E X A M P L E   3   **Using the zero factor property**

Solve $(2x + 1)(x - 1) = 14$.

Solution

To write the equation with 0 on the right-hand side, multiply the binomials on the left and then subtract 14 from each side:

$$
\begin{aligned}
(2x + 1)(x - 1) &= 14 && \text{Original equation} \\
2x^2 - x - 1 &= 14 && \text{Multiply the binomials.} \\
2x^2 - x - 15 &= 0 && \text{Subtract 14 from each side.} \\
(2x + 5)(x - 3) &= 0 && \text{Factor.} \\
2x + 5 = 0 \quad \text{or} \quad x - 3 &= 0 && \text{Zero factor property} \\
2x = -5 \quad \text{or} \quad x &= 3 \\
x = -\frac{5}{2} \quad \text{or} \quad x &= 3
\end{aligned}
$$

Check $-\frac{5}{2}$ and 3 in the original equation:

$$
\left(2 \cdot -\frac{5}{2} + 1\right)\left(-\frac{5}{2} - 1\right) = (-5 + 1)\left(-\frac{5}{2} - \frac{2}{2}\right)
$$

$$
= (-4)\left(-\frac{7}{2}\right)
$$

$$
= 14
$$

$$
(2 \cdot 3 + 1)(3 - 1) = (7)(2)
$$

$$
= 14
$$

So the solutions are $-\frac{5}{2}$ and 3.   ■

> **C A U T I O N**   In Example 3 we started with a product of two factors equal to 14. Because there are many pairs of factors that have a product of 14, we *cannot make any conclusion about the factors.* If the product of two factors is 0, then we can conclude that one or the other factor is 0.

If a perfect square trinomial occurs in a quadratic equation, then there are two identical factors of the trinomial. In this case it is not necessary to set both factors equal to zero. The solution can be found from one factor.

E X A M P L E   4   **An equation with a repeated factor**

Solve $5x^2 - 30x + 45 = 0$.

Solution

Notice that the trinomial on the left-hand side has a common factor:

$$
\begin{aligned}
5x^2 - 30x + 45 &= 0 \\
5(x^2 - 6x + 9) &= 0 && \text{Factor out the GCF.} \\
5(x - 3)^2 &= 0 && \text{Factor the perfect square trinomial.} \\
(x - 3)^2 &= 0 && \text{Divide each side by 5.} \\
x - 3 &= 0 && \text{Zero factor property} \\
x &= 3
\end{aligned}
$$

> **study tip**
>
> Set short-term goals and reward yourself for accomplishing them. When you have solved 10 problems, take a short break and listen to your favorite music.

You should check that 3 satisfies the original equation. Even though $x - 3$ occurs twice as a factor, it is not necessary to write $x - 3 = 0$ or $x - 3 = 0$. The only solution to the equation is 3. ∎

> **CAUTION**   To simplify $5(x - 3)^2 = 0$ in Example 4, we divided each side by 5. If we had used the zero factor property, we would have gotten $5 = 0$ or $(x - 3)^2 = 0$. Since $5 = 0$ has no solution, we can ignore it and continue to solve $(x - 3)^2 = 0$.

If the left-hand side of the equation has more than two factors, we can write an equivalent equation by setting each factor equal to zero.

EXAMPLE 5

An equation with three solutions

Solve $2x^3 - x^2 - 8x + 4 = 0$.

Solution

We can factor the four-term polynomial by grouping:

$$2x^3 - x^2 - 8x + 4 = 0$$
$$x^2(2x - 1) - 4(2x - 1) = 0 \quad \text{Factor out the common factors.}$$
$$(x^2 - 4)(2x - 1) = 0 \quad \text{Factor out } 2x - 1.$$
$$(x - 2)(x + 2)(2x - 1) = 0 \quad \text{Difference of two squares}$$
$$x - 2 = 0 \quad \text{or} \quad x + 2 = 0 \quad \text{or} \quad 2x - 1 = 0 \quad \text{Zero factor property}$$
$$x = 2 \quad \text{or} \quad x = -2 \quad \text{or} \quad x = \frac{1}{2} \quad \text{Solve each equation.}$$

> **helpful hint**
>
> Compare the number of solutions in Examples 1 through 5 to the degree of the polynomial. The number of real solutions to any polynomial equation is less than or equal to the degree of the polynomial. This fact is known as the fundamental theorem of algebra.

You should check that all three numbers satisfy the original equation. The solutions to this equation are -2, $\frac{1}{2}$, and 2. ∎

Applications

There are many problems that can be solved by equations like those we have just discussed.

EXAMPLE 6

Area of a garden

Merida's garden has a rectangular shape with a length that is 1 foot longer than twice the width. If the area of the garden is 55 square feet, then what are the dimensions of the garden?

Solution

If x represents the width of the garden, then $2x + 1$ represents the length. See Fig. 6.1. Because the area of a rectangle is the length times the width, we can write the equation

$$x(2x + 1) = 55.$$

FIGURE 6.1

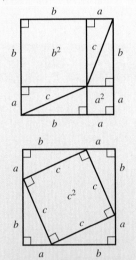

helpful hint

To prove the Pythagorean theorem, draw two squares with sides of length $a + b$, and partition them as shown.

Erasing the four identical triangles from each picture will subtract the same amount of area from each original square. Since we started with equal areas, we will have equal areas after erasing the triangles:

$$a^2 + b^2 = c^2$$

We must have zero on the right-hand side of the equation to use the zero factor property. So we rewrite the equation and then factor:

$$2x^2 + x - 55 = 0$$
$$(2x + 11)(x - 5) = 0 \quad \text{Factor.}$$
$$2x + 11 = 0 \quad \text{or} \quad x - 5 = 0 \quad \text{Zero factor property}$$
$$x = -\frac{11}{2} \quad \text{or} \quad x = 5$$

The width is certainly not $-\frac{11}{2}$. So we use $x = 5$ to get the length:

$$2x + 1 = 2(5) + 1 = 11$$

We check by multiplying 11 feet and 5 feet to get the area of 55 square feet. So the width is 5 ft, and the length is 11 ft. ∎

The next application involves a theorem from geometry called the **Pythagorean theorem.** This theorem says that in any right triangle the sum of the squares of the lengths of the legs is equal to the square of the length of the hypotenuse.

The Pythagorean Theorem

The triangle shown in Fig. 6.2 is a right triangle if and only if

$$a^2 + b^2 = c^2.$$

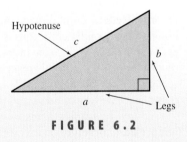

FIGURE 6.2

EXAMPLE 7

Using the Pythagorean theorem

The length of a rectangle is 1 meter longer than the width, and the diagonal measures 5 meters. What are the length and width?

Solution

If x represents the width of the rectangle, then $x + 1$ represents the length. Because the two sides are the legs of a right triangle, we can use the Pythagorean theorem to get a relationship between the length, width, and diagonal. See Fig. 6.3.

$$x^2 + (x + 1)^2 = 5^2 \quad \text{Pythagorean theorem}$$
$$x^2 + x^2 + 2x + 1 = 25 \quad \text{Simplify.}$$
$$2x^2 + 2x - 24 = 0$$
$$x^2 + x - 12 = 0 \quad \text{Divide each side by 2.}$$
$$(x - 3)(x + 4) = 0$$
$$x - 3 = 0 \quad \text{or} \quad x + 4 = 0 \quad \text{Zero factor property}$$
$$x = 3 \quad \text{or} \quad x = -4 \quad \text{The length cannot be negative.}$$
$$x + 1 = 4$$

FIGURE 6.3

To check this answer, we compute $3^2 + 4^2 = 5^2$, or $9 + 16 = 25$. So the rectangle is 3 meters by 4 meters. ■

CAUTION The hypotenuse is the longest side of a right triangle. So if the lengths of the sides of a right triangle are 5 meters, 12 meters, and 13 meters, then the length of the hypotenuse is 13 meters, and $5^2 + 12^2 = 13^2$.

M A T H   A T   W O R K $x^2 + (x + 1)^2 = 5^2$

Can you successfully invest money and at the same time be socially responsible? Geeta Bhide, president and founder of Walden Capital Management, answers with an emphatic "yes." Ms. Bhide helps clients integrate their social values with a portfolio of stocks, bonds, cash, and cash equivalents.

With the client's consent, Ms. Bhide might invest in bonds backed by the Department of Housing and Urban Development for housing in specific inner-city areas. Other choices might be environmentally conscious companies or companies that have a proven record of equal employment practices. In many cases, categorizing a particular company as socially conscious is a judgment call. For example, many oil companies provide a good return on investment, but they might not have an unblemished record on oil spills. In this instance, picking the best of the worst might be the correct choice. Because such trade-offs are necessary, clients are encouraged to define both their investment goals and the social ideals to which they subscribe.

INVESTMENT ADVISOR

As any investment advisor would, Ms. Bhide tries to minimize risk and maximize reward, or return on investment. In Exercises 81 and 82 of this section you will see how solving a quadratic equation by factoring can be used to find the average annual return on an investment.

WARM-UPS

True or false? Explain your answer.

1. The equation $x(x + 2) = 3$ is equivalent to $x = 3$ or $x + 2 = 3$.

2. Equations solved by factoring always have two different solutions.

3. The equation $a \cdot d = 0$ is equivalent to $a = 0$ or $d = 0$.

4. If x is the width in feet of a rectangular room and the length is 5 feet longer than the width, then the area is $x^2 + 5x$ square feet.

5. Both 1 and -4 are solutions to the equation $(x - 1)(x + 4) = 0$.

6. If a, b, and c are the sides of any triangle, then $a^2 + b^2 = c^2$.

7. If the perimeter of a rectangular room is 50 feet, then the sum of the length and width is 25 feet.

8. Equations solved by factoring may have more than two solutions.

9. Both 0 and 2 are solutions to the equation $x(x - 2) = 0$.

10. The solutions to $3(x - 2)(x + 5) = 0$ are 3, 2, and -5.

6.6 EXERCISES

Reading and Writing   *After reading this section, write out the answers to these questions. Use complete sentences.*

1. What is a quadratic equation?

2. What is a compound equation?

3. What is the zero factor property?

4. What method is used to solve quadratic equations in this section?

5. Why don't we usually divide each side of an equation by a variable?

6. What is the Pythagorean theorem?

Solve each equation. See Example 1.

7. $(x + 5)(x + 4) = 0$

8. $(a + 6)(a + 5) = 0$

9. $(2x + 5)(3x - 4) = 0$

10. $(3k - 8)(4k + 3) = 0$

11. $w^2 - 9w + 14 = 0$

12. $t^2 + 6t - 27 = 0$

13. $y^2 - 2y - 24 = 0$

14. $q^2 + 3q - 18 = 0$

15. $2m^2 + m - 1 = 0$

16. $2h^2 - h - 3 = 0$

Solve each equation. See Examples 2 and 3.

17. $m^2 = -7m$

18. $h^2 = -5h$

19. $a^2 + a = 20$

20. $p^2 + p = 42$

21. $2x^2 + 5x = 3$

22. $3x^2 - 10x = -7$

23. $(x + 2)(x + 6) = 12$

24. $(x + 2)(x - 6) = 20$

25. $(a + 3)(2a - 1) = 15$

26. $(b - 3)(3b + 4) = 10$

27. $2(4 - 5h) = 3h^2$

28. $2w(4w + 1) = 1$

Solve each equation. See Examples 4 and 5.

29. $2x^2 + 50 = 20x$

30. $3x^2 + 48 = 24x$

31. $4m^2 - 12m + 9 = 0$

32. $25y^2 + 20y + 4 = 0$

33. $x^3 - 9x = 0$

34. $25x - x^3 = 0$

35. $w^3 + 4w^2 - 4w = 16$

36. $a^3 + 2a^2 - a = 2$

37. $n^3 - 3n^2 + 3 = n$

38. $w^3 + w^2 - 25w = 25$

39. $y^3 - 9y^2 + 20y = 0$

40. $m^3 + 2m^2 - 3m = 0$

Solve each equation.

41. $x^2 - 16 = 0$

42. $x^2 - 36 = 0$

43. $x^2 = 9$

44. $x^2 = 25$

45. $a^3 = a$

46. $x^3 = 4x$

47. $3x^2 + 15x + 18 = 0$

48. $-2x^2 - 2x + 24 = 0$

49. $z^2 + \dfrac{11}{2}z = -6$

50. $m^2 + \dfrac{8}{3}m = 1$

51. $(t - 3)(t + 5) = 9$

52. $3x(2x + 1) = 18$

53. $(x - 2)^2 + x^2 = 10$

54. $(x - 3)^2 + (x + 2)^2 = 17$

55. $\dfrac{1}{16}x^2 + \dfrac{1}{8}x = \dfrac{1}{2}$

56. $\dfrac{1}{18}h^2 - \dfrac{1}{2}h + 1 = 0$

57. $a^3 + 3a^2 - 25a = 75$

58. $m^4 + m^3 = 100m^2 + 100m$

Solve each problem. See Examples 6 and 7.

59. *Dimensions of a rectangle.* The perimeter of a rectangle is 34 feet, and the diagonal is 13 feet long. What are the length and width of the rectangle?

60. *Address book.* The perimeter of the cover of an address book is 14 inches, and the diagonal measures 5 inches. What are the length and width of the cover?

FIGURE FOR EXERCISE 60

61. *Violla's bathroom.* The length of Violla's bathroom is 2 feet longer than twice the width. If the diagonal measures 13 feet, then what are the length and width?

62. *Rectangular stage.* One side of a rectangular stage is 2 meters longer than the other. If the diagonal is 10 meters, then what are the lengths of the sides?

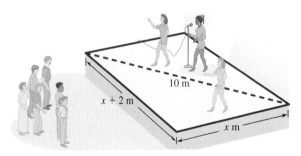

FIGURE FOR EXERCISE 62

63. *Consecutive integers.* The sum of the squares of two consecutive integers is 13. Find the integers.

64. *Consecutive integers.* The sum of the squares of two consecutive even integers is 52. Find the integers.

65. *Two numbers.* The sum of two numbers is 11, and their product is 30. Find the numbers.

66. *Missing ages.* Molly's age is twice Anita's. If the sum of the squares of their ages is 80, then what are their ages?

67. *Skydiving.* If there were no air resistance, then the height (in feet) above the earth for a sky diver t seconds after jumping from an airplane at 10,000 feet would be given by

$$h(t) = -16t^2 + 10,000.$$

a) Find the time that it would take to fall to earth with no air resistance, that is, find t for which $h(t) = 0$. A sky diver actually gets about twice as much free fall time due to air resistance.

b) Use the accompanying graph to determine whether the sky diver (with no air resistance) falls farther in the first 5 seconds or the last 5 seconds of the fall.

c) Is the sky diver's velocity increasing or decreasing as she falls?

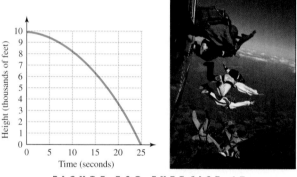

FIGURE FOR EXERCISE 67

68. *Skydiving.* If a sky diver jumps from an airplane at a height of 8256 feet, then for the first 5 seconds, her height above the earth is approximated by the formula $h = -16t^2 + 8256$. How many seconds does it take her to reach 8000 feet.

69. *Throwing a sandbag.* If a balloonist throws a sandbag downward at 24 feet per second from an altitude of 720 feet, then its height (in feet) above the ground after t seconds is given by $S = -16t^2 - 24t + 720$. How long does it take for the sandbag to reach the earth? (On the ground, $S = 0$.)

70. *Throwing a sandbag.* If the balloonist of the previous exercise throws his sandbag downward from an altitude of 128 feet with an initial velocity of 32 feet per second, then its altitude after t seconds is given by the formula $S = -16t^2 - 32t + 128$. How long does it take for the sandbag to reach the earth?

71. *Glass prism.* One end of a glass prism is in the shape of a triangle with a height that is 1 inch longer than twice the base. If the area of the triangle is 39 square inches, then how long are the base and height?

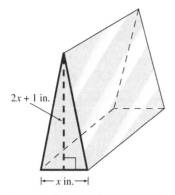

FIGURE FOR EXERCISE 71

72. Areas of two circles. The radius of a circle is 1 meter longer than the radius of another circle. If their areas differ by 5π square meters, then what is the radius of each?

73. Changing area. Last year Otto's garden was square. This year he plans to make it smaller by shortening one side 5 feet and the other 8 feet. If the area of the smaller garden will be 180 square feet, then what was the size of Otto's garden last year?

74. Dimensions of a box. Rosita's Christmas present from Carlos is in a box that has a width that is 3 inches shorter than the height. The length of the base is 5 inches longer than the height. If the area of the base is 84 square inches, then what is the height of the package?

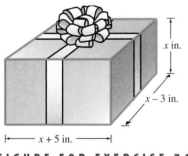

FIGURE FOR EXERCISE 74

75. Flying a kite. Imelda and Gordon have designed a new kite. While Imelda is flying the kite, Gordon is standing directly below it. The kite is designed so that its altitude is always 20 feet larger than the distance between Imelda and Gordon. What is the altitude of the kite when it is 100 feet from Imelda?

76. Avoiding a collision. A car is traveling on a road that is perpendicular to a railroad track. When the car is 30 meters from the crossing, the car's new collision detector warns the driver that there is a train 50 meters from the car and heading toward the same crossing. How far is the train from the crossing?

77. Carpeting two rooms. Virginia is buying carpet for two square rooms. One room is 3 yards wider than the other. If she needs 45 square yards of carpet, then what are the dimensions of each room?

78. Winter wheat. While finding the amount of seed needed to plant his three square wheat fields, Hank observed that the side of one field was 1 kilometer longer than the side of the smallest field and that the side of the largest field was 3 kilometers longer than the side of the smallest field. If the total area of the three fields is 38 square kilometers, then what is the area of each field?

79. Sailing to Miami. At point A the captain of a ship determined that the distance to Miami was 13 miles. If she sailed north to point B and then west to Miami, the distance would be 17 miles. If the distance from point A to point B is greater than the distance from point B to Miami, then how far is it from point A to point B?

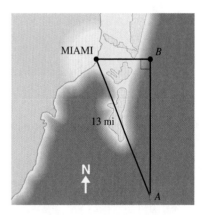

FIGURE FOR EXERCISE 79

80. Buried treasure. Ahmed has half of a treasure map, which indicates that the treasure is buried in the desert $2x + 6$ paces from Castle Rock. Vanessa has the other half of the map. Her half indicates that to find the treasure, one must get to Castle Rock, walk x paces to the north, and then walk $2x + 4$ paces to the east. If they share their information, then they can find x and save a lot of digging. What is x?

81. Emerging markets. Catarina's investment of $16,000 in an emerging market fund grew to $25,000 in 2 years. Find the average annual rate of return by solving the equation $16,000(1 + r)^2 = 25,000$.

82. Venture capital. Henry invested $12,000 in a new restaurant. When the restaurant was sold 2 years later, he received $27,000. Find his average annual return by solving the equation $12,000(1 + r)^2 = 27,000$.

The Puzzle Box

After graduating from college you get a job working for a small business that makes jigsaw puzzles. You are on the team to design the cover of a puzzle box. Your design will use 525 square centimeters. The production manager has found a great deal on the price of cardboard. The cardboard is precut to 29 cm by 33 cm. The production manager tells your team that the depth of the box can vary to fit the needs of your design. He asks your team not only to come up with a design, but also to determine what the depth of the box should be so that production can fold it correctly to allow for your design.

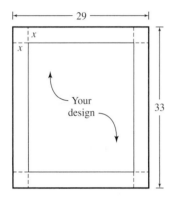

Getting started: Choose roles (consult your instructor). Decide on a design and a name for your puzzle.

Grouping: 4 students
Topic: Multiplying and factoring polynomials

Part I: Finding the depth and dimensions

1. Write an equation using the given information and the diagram.

2. Solve the equation to find the depth of the box.

3. Use the depth of the box to find the dimensions for your design.

Part II: Presenting your final design. For shipping purposes, management has decided to make the boxes only 3 cm deep. This will change the dimensions of the box top. There isn't time for you to come up with a new design. Your team must find a way to incorporate your present design with its original dimensions onto the new box top. You will need to make an exact drawing in centimeters of your solution, using the new dimensions for the box top.

1. Using the same precut cardboard with the new depth, what will be the dimensions of the top of the box?

2. Make an *exact* drawing of your puzzle box top using the new dimensions.

3. Include the puzzle name in your drawing.

Report: Write all your work and solutions as a report. List the names of the members of your group. Include answers and work for all questions.

WRAP-UP CHAPTER 6

SUMMARY

Factoring		Examples
Prime number	A positive integer larger than 1 that has no integral factors other than 1 and itself	2, 3, 5, 7, 11
Prime polynomial	A polynomial that cannot be factored is prime.	$x^2 + 3$ and $x^2 - x + 5$ are prime.
Strategy for finding the GCF for monomials	1. Find the GCF for the coefficients of the monomials. 2. Form the product of the GCF of the coefficients and each variable that is common to all of the monomials, where the exponent on each variable equals the smallest power of that variable in any of the monomials.	$12x^3yz, 8x^2y^3$ GCF $= 4x^2y$

Factoring out the GCF	Use the distributive property to factor out the GCF from all terms of a polynomial.	$2x^3 - 4x = 2x(x^2 - 2)$

Special Cases **Examples**

Difference of two squares	$a^2 - b^2 = (a + b)(a - b)$	$m^2 - 9 = (m - 3)(m + 3)$
Perfect square trinomial	$a^2 + 2ab + b^2 = (a + b)^2$ $a^2 - 2ab + b^2 = (a - b)^2$	$x^2 + 6x + 9 = (x + 3)^2$ $4h^2 - 12h + 9 = (2h - 3)^2$
Difference or sum of two cubes	$a^3 - b^3 = (a - b)(a^2 + ab + b^2)$ $a^3 + b^3 = (a + b)(a^2 - ab + b^2)$	$t^3 - 8 = (t - 2)(t^2 + 2t + 4)$ $p^3 + 1 = (p + 1)(p^2 - p + 1)$

Factoring Polynomials **Examples**

Factoring by grouping	Factor out common factors from groups of terms.	$6x + 6w + ax + aw$ $= 6(x + w) + a(x + w)$ $= (6 + a)(x + w)$
Strategy for factoring $ax^2 + bx + c$ by the ac method	1. Find two numbers that have a product equal to ac and a sum equal to b. 2. Replace bx by two terms using the two new numbers as coefficients. 3. Factor the resulting four-term polynomial by grouping.	$6x^2 + 17x + 12$ $= 6x^2 + 9x + 8x + 12$ $= (2x + 3)3x + (2x + 3)4$ $= (2x + 3)(3x + 4)$
Factoring by trial and error	Try possible factors of the trinomial and check by using FOIL. If incorrect, try again.	$2x^2 + 5x - 12$ $= (2x - 3)(x + 4)$
Strategy for factoring polynomials completely	1. First factor out any common factors. 2. When factoring a binomial, check to see whether it is a difference of two squares, a difference of two cubes, or a sum of two cubes. Remember that a sum of two squares (with no common factor) is prime. 3. When factoring a trinomial, check to see whether it is a perfect square trinomial. 4. When factoring a trinomial that is not a perfect square, use the ac method or trial and error. 5. If the polynomial has four terms, try factoring by grouping. 6. Check to see whether any factors can be factored again.	

Solving Equations

		Examples
Zero factor property	The equation $a \cdot b = 0$ is equivalent to $$a = 0 \quad \text{or} \quad b = 0.$$	$x(x - 1) = 0$ $x = 0$ or $x - 1 = 0$

Strategy for solving an equation by factoring

1. Rewrite the equation with 0 on the right-hand side.
2. Factor the left-hand side completely.
3. Set each factor equal to zero to get linear equations.
4. Solve the linear equations.
5. Check the answers in the original equation.
6. State the solution(s) to the original equation.

$x^2 + 3x = 18$
$x^2 + 3x - 18 = 0$
$(x + 6)(x - 3) = 0$
$x + 6 = 0 \quad \text{or} \quad x - 3 = 0$
$x = -6 \quad \text{or} \quad x = 3$

ENRICHING YOUR MATHEMATICAL WORD POWER

For each mathematical term, choose the correct meaning.

1. **factor**
   a. to write an expression as a product
   b. to multiply
   c. what two numbers have in common
   d. to FOIL

2. **prime number**
   a. a polynomial that cannot be factored
   b. a number with no divisors
   c. an integer between 1 and 10
   d. an integer larger than 1 that has no integral factors other than itself and 1

3. **greatest common factor**
   a. the least common multiple
   b. the least common denominator
   c. the largest integer that is a factor of two or more integers
   d. the largest number in a product

4. **prime polynomial**
   a. a polynomial that has no factors
   b. a product of prime numbers
   c. a first-degree polynomial
   d. a monomial

5. **factor completely**
   a. to factor by grouping
   b. to factor out a prime number
   c. to write as a product of primes
   d. to factor by trial and error

6. **sum of two cubes**
   a. $(a + b)^3$
   b. $a^3 + b^3$
   c. $a^3 - b^3$
   d. $a^3 b^3$

7. **quadratic equation**
   a. $ax + b = 0$ where $a \neq 0$
   b. $ax + b = cx + d$
   c. $ax^2 + bx + c = 0$ where $a \neq 0$
   d. any equation with four terms

8. **zero factor property**
   a. If $ab = 0$ then $a = 0$ or $b = 0$
   b. $a \cdot 0 = 0$ for any a
   c. $a = a + 0$ for any real number a
   d. $a + (-a) = 0$ for any real number a

9. **Pythagorean theorem**
   a. $a^2 + b^2 = (a + b)^2$
   b. a triangle is a right triangle if and only if it has one right angle
   c. the legs of a right triangle meet at a 90° angle
   d. a theorem that gives a relationship between the two legs and the hypotenuse of a right triangle

10. **difference of two squares**
    a. $a^3 - b^3$
    b. $2a - 2b$
    c. $a^2 - b^2$
    d. $(a - b)^2$

REVIEW EXERCISES

6.1 *Find the prime factorization for each integer.*

1. 144
2. 121
3. 58
4. 76
5. 150
6. 200

Find the greatest common factor for each group.

7. 36, 90
8. 30, 42, 78
9. $8x$, $12x^2$
10. $6a^2b$, $9ab^2$, $15a^2b^2$

Complete the factorization of each binomial.

11. $3x + 6 = 3(\quad)$
12. $7x^2 + x = x(\quad)$
13. $2a - 20 = -2(\quad)$
14. $a^2 - a = -a(\quad)$

Factor each polynomial by factoring out the GCF.

15. $2a - a^2$
16. $9 - 3b$
17. $6x^2y^2 - 9x^5y$
18. $a^3b^5 + a^3b^2$
19. $3x^2y - 12xy - 9y^2$
20. $2a^2 - 4ab^2 - ab$

6.2 *Factor each polynomial completely.*

21. $y^2 - 400$
22. $4m^2 - 9$
23. $w^2 - 8w + 16$
24. $t^2 + 20t + 100$
25. $4y^2 + 20y + 25$
26. $2a^2 - 4a - 2$
27. $r^2 - 4r + 4$
28. $3m^2 - 75$
29. $8t^3 - 24t^2 + 18t$
30. $t^2 - 9w^2$
31. $x^2 + 12xy + 36y^2$
32. $9y^2 - 12xy + 4x^2$
33. $x^2 + 5x - xy - 5y$
34. $x^2 + xy + ax + ay$

6.3 *Factor each polynomial.*

35. $b^2 + 5b - 24$
36. $a^2 - 2a - 35$

37. $r^2 - 4r - 60$
38. $x^2 + 13x + 40$
39. $y^2 - 6y - 55$
40. $a^2 + 6a - 40$
41. $u^2 + 26u + 120$
42. $v^2 - 22v - 75$

Factor completely.

43. $3t^3 + 12t^2$
44. $-4m^4 - 36m^2$
45. $5w^3 + 25w^2 + 25w$
46. $-3t^3 + 3t^2 - 6t$
47. $2a^3b + 3a^2b^2 + ab^3$
48. $6x^2y^2 - xy^3 - y^4$
49. $9x^3 - xy^2$
50. $h^4 - 100h^2$

6.4 *Factor each polynomial completely.*

51. $14t^2 + t - 3$
52. $15x^2 - 22x - 5$
53. $6x^2 - 19x - 7$
54. $2x^2 - x - 10$
55. $6p^2 + 5p - 4$
56. $3p^2 + 2p - 5$
57. $-30p^3 + 8p^2 + 8p$
58. $-6q^2 - 40q - 50$
59. $6x^2 - 29xy - 5y^2$
60. $10a^2 + ab - 2b^2$
61. $32x^2 + 16xy + 2y^2$
62. $8a^2 + 40ab + 50b^2$

6.5 *Factor completely.*

63. $5x^3 + 40x$
64. $w^2 + 6w + 9$
65. $9x^2 + 3x - 2$
66. $ax^3 + ax$
67. $x^3 + 2x^2 - x - 2$
68. $16x^2 - 2x - 3$
69. $x^2y - 16xy^2$
70. $-3x^2 + 27$
71. $a^2 + 2a + 1$
72. $-2w^2 - 12w - 18$
73. $x^3 - x^2 + x - 1$
74. $9x^2y^2 - 9y^2$
75. $a^2 + ab + 2a + 2b$
76. $4m^2 + 20m + 25$

77. $-2x^2 + 16x - 24$

78. $6x^2 + 21x - 45$

79. $m^3 - 1000$

80. $8p^3 + 1$

Factor each polynomial completely, given that the binomial following it is a factor of the polynomial.

81. $x^3 + x + 10, x + 2$

82. $x^3 - 5x - 12, x - 3$

83. $x^3 + 6x^2 - 7x - 60, x + 4$

84. $x^3 - 4x^2 - 3x - 10, x - 5$

6.6 *Solve each equation.*

85. $x^3 = 5x^2$

86. $2m^2 + 10m = -12$

87. $(a - 2)(a - 3) = 6$

88. $(w - 2)(w + 3) = 50$

89. $2m^2 - 9m - 5 = 0$

90. $12x^2 + 5x - 3 = 0$

91. $m^3 + 4m^2 - 9m = 36$

92. $w^3 + 5w^2 - w = 5$

93. $(x + 3)^2 + x^2 = 5$

94. $(h - 2)^2 + (h + 1)^2 = 9$

95. $p^2 + \frac{1}{4}p - \frac{1}{8} = 0$

96. $t^2 + 1 = \frac{13}{6}t$

Solve each problem.

97. *Positive numbers.* Two positive numbers differ by 6, and their squares differ by 96. Find the numbers.

98. *Consecutive integers.* Find three consecutive integers such that the sum of their squares is 77.

99. *Dimensions of a notebook.* The perimeter of a notebook is 28 inches, and the diagonal measures 10 inches. What are the length and width of the notebook?

100. *Two numbers.* The sum of two numbers is 8.5, and their product is 18. Find the numbers.

101. *Poiseuille's law.* According to the nineteenth-century physician Poiseuille, the velocity (in centimeters per second) of blood r centimeters from the center of an artery of radius R centimeters is given by $v = kR^2 - kr^2$, where k is a constant. Rewrite the formula by factoring the right-hand side completely.

102. *Racquetball.* The volume of rubber (in cubic centimeters) in a hollow rubber ball used in racquetball is given by

$$V = \frac{4}{3}\pi R^3 - \frac{4}{3}\pi r^3,$$

where the inside radius is r centimeters and the outside radius is R centimeters.

a) Rewrite the formula by factoring the right-hand side completely.

b) The accompanying graph shows the relationship between r and V when $R = 3$. Use the graph to estimate the value of r for which $V = 100\text{ cm}^3$.

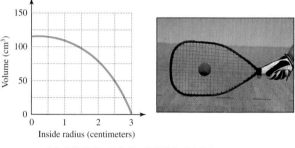

FIGURE FOR EXERCISE 102

103. *Leaning ladder.* A 10-foot ladder is placed against a building so that the distance from the bottom of the ladder to the building is 2 feet less than the distance from the top of the ladder to the ground. What is the distance from the bottom of the ladder to the building?

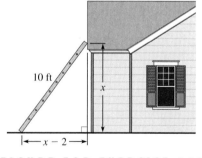

FIGURE FOR EXERCISE 103

104. *Towering antenna.* A guy wire of length 50 feet is attached to the ground and to the top of an antenna. The height of the antenna is 10 feet larger than the distance from the base of the antenna to the point where the guy wire is attached to the ground. What is the height of the antenna?

CHAPTER 6 TEST

Give the prime factorization for each integer.

1. 66

2. 336

Find the greatest common factor (GCF) for each group.

3. 48, 80

4. 42, 66, 78

5. $6y^2, 15y^3$

6. $12a^2b, 18ab^2, 24a^3b^3$

Factor each polynomial completely.

7. $5x^2 - 10x$

8. $6x^2y^2 + 12xy^2 + 12y^2$

9. $3a^3b - 3ab^3$

10. $a^2 + 2a - 24$

11. $4b^2 - 28b + 49$

12. $3m^3 + 27m$

13. $ax - ay + bx - by$

14. $ax - 2a - 5x + 10$

15. $6b^2 - 7b - 5$

16. $m^2 + 4mn + 4n^2$

17. $2a^2 - 13a + 15$

18. $z^3 + 9z^2 + 18z$

Factor the polynomial completely, given that $x - 1$ is a factor.

19. $x^3 - 6x^2 + 11x - 6$

Solve each equation.

20. $2x^2 + 5x - 12 = 0$

21. $3x^3 = 12x$

22. $(2x - 1)(3x + 5) = 5$

Write a complete solution to each problem.

23. If the length of a rectangle is 3 feet longer than the width and the diagonal is 15 feet, then what are the length and width?

24. The sum of two numbers is 4, and their product is -32. Find the numbers.

Simplify each expression.

1. $\dfrac{91 - 17}{17 - 91}$

2. $\dfrac{4 - 18}{-6 - 1}$

3. $5 - 2(7 - 3)$

4. $3^2 - 4(6)(-2)$

5. $2^5 - 2^4$

6. $0.07(37) + 0.07(63)$

Perform the indicated operations.

7. $x \cdot 2x$

8. $x + 2x$

9. $\dfrac{6 + 2x}{2}$

10. $\dfrac{6 \cdot 2x}{2}$

11. $2 \cdot 3y \cdot 4z$

12. $2(3y + 4z)$

13. $2 - (3 - 4z)$

14. $t^8 \div t^2$

15. $t^8 \cdot t^2$

16. $\dfrac{8t^8}{2t^2}$

Solve and graph each inequality.

17. $2x - 5 > 3x + 4$

18. $4 - 5x \le -11$

19. $-\dfrac{2}{3}x + 3 < -5$

20. $0.05(x - 120) - 24 < 0$

Solve each equation.

21. $2x - 3 = 0$

22. $2x + 1 = 0$

23. $(x - 3)(x + 5) = 0$

24. $(2x - 3)(2x + 1) = 0$

25. $3x(x - 3) = 0$

26. $x^2 = x$

27. $3x - 3x = 0$

28. $3x - 3x = 1$

29. $0.01x - x + 14.9 = 0.5x$

30. $0.05x + 0.04(x - 40) = 2$

31. $2x^2 = 18$

32. $2x^2 + 7x - 15 = 0$

Solve the problem.

33. ***Another ace.*** Professional tennis players can serve a tennis ball at speeds over 120 mph into a rectangular region that has a perimeter of 69 feet and an area of 283.5 square feet. Find the length and width of the service region.

Rational Expressions

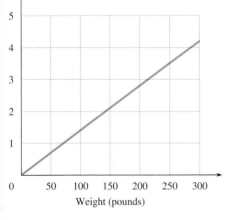

Weight (pounds)

Advanced technical developments have made sports equipment faster, lighter, and more responsive to the human body. Behind the more flexible skis, lighter bats, and comfortable athletic shoes lies the science of biomechanics, which is the study of human movement and the factors that influence it.

Designing and testing an athletic shoe go hand in hand. While a shoe is being designed, it is tested in a multitude of ways, including long-term wear, rear foot stability, and strength of materials. Testing basketball shoes usually includes an evaluation of the force applied to the ground by the foot during running, jumping, and landing. Many biomechanics laboratories have a special platform that can measure the force exerted when a player cuts from side to side as well as the force against the bottom of the shoe. Force exerted in landing from a lay-up shot can be as high as 14 times the weight of the body. Side-to-side force is usually about 1 to 2 body weights in a cutting movement.

In Exercises 57 and 58 of Section 7.7 you will see how designers of athletic shoes use proportions to find the amount of force on the foot and soles of shoes for activities such as running and jumping.

7.1 REDUCING RATIONAL EXPRESSIONS

Rational expressions in algebra are similar to the rational numbers in arithmetic. In this section you will learn the basic ideas of rational expressions.

Rational Expressions

A rational number is the ratio of two integers with the denominator not equal to 0. For example,

$$\frac{3}{4}, \quad \frac{-9}{-6}, \quad 7, \quad \text{and} \quad 0$$

are rational numbers. A **rational expression** is the ratio of two polynomials with the denominator not equal to 0. Because an integer is a monomial, a rational number is a rational expression. The following expressions are rational expressions:

$$\frac{x^2 - 1}{x + 8}, \quad \frac{3a^2 + 5a - 3}{a - 9}, \quad \frac{3}{7}, \quad w$$

We say that w is a rational expression because w can be written as $\frac{w}{1}$.

A rational expression involving a variable has no value unless we assign a value to the variable. Once the variable is given a value we can evaluate the expression. We can discuss the value of a rational expression with or without the function notation introduced in Chapter 4.

EXAMPLE 1

calculator

close-up

To evaluate the rational expression in Example 1(a) with a calculator, first use Y= to define the rational expression. Be sure to enclose both numerator and denominator in parentheses.

```
Plot1 Plot2 Plot3
\Y1◘(4X-1)/(X+2)
\Y2=
\Y3=
\Y4=
\Y5=
\Y6=
```

Then find $y_1(-3)$.

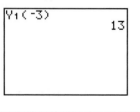

```
Y1(-3)
            13
```

Evaluating a rational expression

a) Find the value of $\frac{4x - 1}{x + 2}$ for $x = -3$.

b) If $R(x) = \frac{3x + 2}{2x - 1}$, find $R(4)$.

Solution

a) To find the value of $\frac{4x - 1}{x + 2}$ for $x = -3$, replace x by -3 in the rational expression:

$$\frac{4(-3) - 1}{-3 + 2} = \frac{-13}{-1} = 13$$

So the value of the rational expression is 13.

b) $R(4)$ is the function notation for the value of the expression when $x = 4$. To find $R(4)$ replace x by 4 in $R(x) = \frac{3x + 2}{2x - 1}$:

$$R(4) = \frac{3(4) + 2}{2(4) - 1}$$

$$R(4) = \frac{14}{7} = 2$$

So the value of the rational expression is 2 when $x = 4$, or $R(4) = 2$ (read "R of 4 is 2"). ■

Because the denominator cannot be zero, any number can be used in place of the variable *except* numbers that cause the denominator to be zero.

E X A M P L E   2

Ruling out values for x

Which numbers cannot be used in place of x in each rational expression?

a) $\dfrac{x^2 - 1}{x + 8}$

b) $\dfrac{x + 2}{2x + 1}$

c) $\dfrac{x + 5}{x^2 - 4}$

Solution

a) The denominator is 0 if $x + 8 = 0$, or $x = -8$. So -8 cannot be used in place of x.

b) The denominator is zero if $2x + 1 = 0$, or $x = -\frac{1}{2}$. So we cannot use $-\frac{1}{2}$ in place of x.

c) The denominator is zero if $x^2 - 4 = 0$. Solve this equation:

$$x^2 - 4 = 0$$
$$(x - 2)(x + 2) = 0 \qquad \text{Factor.}$$
$$x - 2 = 0 \quad \text{or} \quad x + 2 = 0 \qquad \text{Zero factor property}$$
$$x = 2 \quad \text{or} \quad x = -2$$

So 2 and -2 cannot be used in place of x.   ■

The **domain** of an expression is the set of all real numbers that can be used in place of the variable. So the domain of $\frac{x^2 - 1}{x + 8}$ is the set of all real numbers except -8. The domain of $\frac{x + 2}{2x + 1}$ is the set of all real numbers except $-\frac{1}{2}$. When dealing with rational expressions we generally assume that the variables represent numbers in the domain of the expression.

Reducing to Lowest Terms

Rational expressions are a generalization of rational numbers. The operations that we perform on rational numbers can be performed on rational expressions in exactly the same manner.

Each rational number can be written in infinitely many equivalent forms. For example,

$$\frac{3}{5} = \frac{6}{10} = \frac{9}{15} = \frac{12}{20} = \frac{15}{25} = \cdots.$$

Each equivalent form of $\frac{3}{5}$ is obtained from $\frac{3}{5}$ by multiplying both numerator and denominator by the same nonzero number. This is equivalent to multiplying the fraction by 1, which does not change its value. For example,

$$\frac{3}{5} = \frac{3}{5} \cdot 1 = \frac{3}{5} \cdot \frac{2}{2} = \frac{6}{10} \qquad \text{and} \qquad \frac{3}{5} = \frac{3 \cdot 3}{5 \cdot 3} = \frac{9}{15}.$$

If we start with $\frac{6}{10}$ and convert it into $\frac{3}{5}$, we say that we are **reducing $\frac{6}{10}$ to lowest terms.** We reduce by dividing the numerator and denominator by the common factor 2:

$$\frac{6}{10} = \frac{2 \cdot 3}{2 \cdot 5} = \frac{3}{5}$$

A rational number is expressed in lowest terms when the numerator and the denominator have no common factors other than 1.

> **CAUTION** We can reduce fractions only by dividing the numerator and the denominator by a common factor. Although it is true that
>
> $$\frac{6}{10} = \frac{2 + 4}{2 + 8},$$
>
> we cannot eliminate the 2's, because they are not factors. Removing them from the sums in the numerator and denominator would not result in $\frac{3}{5}$. Likewise, you cannot eliminate the 2's in an expression such as $\frac{y + 2}{x + 2}$.

Reducing Fractions

If $a \neq 0$ and $c \neq 0$, then

$$\frac{ab}{ac} = \frac{b}{c}.$$

To reduce rational expressions to lowest terms, we use exactly the same procedure as with fractions:

Reducing Rational Expressions

1. Factor the numerator and denominator completely.
2. Divide the numerator and denominator by the greatest common factor.

Dividing the numerator and denominator by the GCF is often referred to as **dividing out the GCF.**

E X A M P L E   3

Reducing

Reduce to lowest terms.

a) $\dfrac{30}{42}$

b) $\dfrac{x^2 - 9}{6x + 18}$

Solution

a) $\dfrac{30}{42} = \dfrac{2 \cdot 3 \cdot 5}{2 \cdot 3 \cdot 7}$ Factor.

$\phantom{\dfrac{30}{42}} = \dfrac{5}{7}$ Divide out the GCF: $2 \cdot 3$ or 6.

b) $\dfrac{x^2 - 9}{6x + 18} = \dfrac{(x - 3)(x + 3)}{6(x + 3)}$ Factor.

$\phantom{\dfrac{x^2 - 9}{6x + 18}} = \dfrac{x - 3}{6}$ Divide out the GCF: $x + 3$.

■

If two rational expressions are equivalent, then they have the same numerical value for any replacement of the variables. Of course, the replacement must not give us an undefined expression (0 in the denominator). So in Example 2(b) the equation

$$\frac{x^2 - 9}{6x + 18} = \frac{x - 3}{6}$$

is satisfied by all real numbers except $x = -3$. The equation is an identity.

Reducing with the Quotient Rule

To reduce rational expressions involving exponential expressions, we use the quotient rule for exponents from Chapter 5. We restate it here for reference.

Quotient Rule

Suppose $a \neq 0$, and m and n are positive integers.

$$\text{If } m \geq n, \text{ then } \frac{a^m}{a^n} = a^{m-n}.$$

$$\text{If } m < n, \text{ then } \frac{a^m}{a^n} = \frac{1}{a^{n-m}}.$$

E X A M P L E   4

Using the quotient rule in reducing

Reduce to lowest terms.

a) $\dfrac{3a^{15}}{6a^7}$

b) $\dfrac{6x^4y^2}{4xy^5}$

Solution

a) $\dfrac{3a^{15}}{6a^7} = \dfrac{\cancel{3}a^{15}}{\cancel{3} \cdot 2a^7}$ Factor.

  $= \dfrac{a^{15-7}}{2}$ Quotient rule

  $= \dfrac{a^8}{2}$

b) $\dfrac{6x^4y^2}{4xy^5} = \dfrac{\cancel{2} \cdot 3x^4y^2}{\cancel{2} \cdot 2xy^5}$ Factor.

  $= \dfrac{3x^{4-1}}{2y^{5-2}}$ Quotient rule

  $= \dfrac{3x^3}{2y^3}$

The essential part of reducing is getting a complete factorization for the numerator and denominator. To get a complete factorization, you must use the techniques for factoring from Chapter 6. If there are large integers in the numerator and denominator, you can use the technique shown in Section 6.1 to get a prime factorization of each integer.

E X A M P L E   5

Reducing expressions involving large integers

Reduce $\frac{420}{616}$ to lowest terms.

Solution

Use the method of Section 6.1 to get a prime factorization of 420 and 616:

$$
\begin{array}{cc}
7 & 11 \\
5\overline{)35} & 7\overline{)77} \\
3\overline{)105} & 2\overline{)154} \\
2\overline{)210} & 2\overline{)308} \\
\text{Start here} \rightarrow 2\overline{)420} & 2\overline{)616}
\end{array}
$$

The complete factorization for 420 is $2^2 \cdot 3 \cdot 5 \cdot 7$, and the complete factorization for 616 is $2^3 \cdot 7 \cdot 11$. To reduce the fraction, we divide out the common factors:

$$\frac{420}{616} = \frac{2^2 \cdot 3 \cdot 5 \cdot 7}{2^3 \cdot 7 \cdot 11}$$

$$= \frac{3 \cdot 5}{2 \cdot 11}$$

$$= \frac{15}{22}$$

Dividing $a - b$ by $b - a$

In Section 5.2 you learned that $a - b = -(b - a) = -1(b - a)$. So if $a - b$ is divided by $b - a$, the quotient is -1:

$$\frac{a - b}{b - a} = \frac{-1(b - a)}{b - a}$$

$$= -1$$

We will use this fact in the next example.

E X A M P L E   6   **Expressions with $a - b$ and $b - a$**

Reduce to lowest terms.

a) $\dfrac{5x - 5y}{4y - 4x}$
 b) $\dfrac{m^2 - n^2}{n - m}$

Solution

a) $\dfrac{5x - 5y}{4y - 4x} = \dfrac{5(x - y)}{4(y - x)}$ Factor.

$$= \frac{5}{4} \cdot (-1) \qquad \frac{x - y}{y - x} = -1$$

$$= -\frac{5}{4}$$

b) $\dfrac{m^2 - n^2}{n - m} = \dfrac{\overset{-1}{\cancel{(m - n)}}(m + n)}{\cancel{n - m}}$ Factor.

$$= -1(m + n) \qquad \frac{m - n}{n - m} = -1$$

$$= -m - n$$

> **CAUTION**   We can reduce $\frac{a - b}{b - a}$ to -1, but we cannot reduce $\frac{a - b}{a + b}$. There is no factor that is common to the numerator and denominator of $\frac{a - b}{a + b}$.

Factoring Out the Opposite of a Common Factor

If we can factor out a common factor, we can also factor out the opposite of that common factor. For example, from $-3x - 6y$ we can factor out the common factor 3 or the common factor -3:

$$-3x - 6y = 3(-x - 2y) \qquad \text{or} \qquad -3x - 6y = -3(x + 2y)$$

To reduce an expression, it is sometimes necessary to factor out the opposite of a common factor.

EXAMPLE 7

Factoring out the opposite of a common factor

Reduce $\dfrac{-3w - 3w^2}{w^2 - 1}$ to lowest terms.

Solution

We can factor $3w$ or $-3w$ from the numerator. If we factor out $-3w$, we get a common factor in the numerator and denominator:

$$\frac{-3w - 3w^2}{w^2 - 1} = \frac{-3w(1 + w)}{(w - 1)(w + 1)} \qquad \text{Factor.}$$

$$= \frac{-3w}{w - 1} \qquad \text{Since } 1 + w = w + 1, \text{ we divide out } w + 1.$$

$$= \frac{3w}{1 - w} \qquad \text{Multiply numerator and denominator by } -1.$$

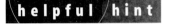

Note that $w + 1$ and $1 + w$ are equivalent, but $w - 1$ and $1 - w$ are not equivalent.

The last step in this reduction is not absolutely necessary, but we usually perform it to make the answer look a little simpler. ■

The main points to remember for reducing rational expressions are summarized in the following reducing strategy.

Strategy for Reducing Rational Expressions

1. Reducing is done by dividing out all common factors.
2. Factor the numerator and denominator completely to see the common factors.
3. Use the quotient rule to reduce a ratio of two monomials.
4. You may have to factor out a common factor with a negative sign to get identical factors in the numerator and denominator.
5. The quotient of $a - b$ and $b - a$ is -1.

Writing Rational Expressions

Rational expressions occur naturally in applications involving rates.

EXAMPLE 8

Writing rational expressions

Answer each question with a rational expression.

a) If a trucker drives 500 miles in $x + 1$ hours, then what is his average speed?

b) If a wholesaler buys 100 pounds of shrimp for x dollars, then what is the price per pound?

c) If a painter completes an entire house in $2x$ hours, then at what rate is she painting?

Solution

a) Because $R = \frac{D}{T}$, he is averaging $\frac{500}{x+1}$ mph.

b) At x dollars for 100 pounds, the wholesaler is paying $\frac{x}{100}$ dollars per pound or $\frac{x}{100}$ dollars/pound.

c) By completing 1 house in $2x$ hours, her rate is $\frac{1}{2x}$ house/hour. ■

WARM-UPS

True or false? Explain your answer.

1. A complete factorization of 3003 is $2 \cdot 3 \cdot 7 \cdot 11 \cdot 13$.

2. A complete factorization of 120 is $2^3 \cdot 3 \cdot 5$.

3. Any number can be used in place of x in the expression $\frac{x-2}{5}$.

4. We cannot replace x by -1 or 3 in the expression $\frac{x+1}{x-3}$.

5. The rational expression $\frac{x+2}{2}$ reduces to x.

6. $\frac{2x}{2} = x$ for any real number x.

7. $\frac{x^{13}}{x^{20}} = \frac{1}{x^7}$ for any nonzero value of x.

8. $\frac{a^2 + b^2}{a + b}$ reduced to lowest terms is $a + b$.

9. If $a \neq b$, then $\frac{a-b}{b-a} = 1$.

10. The expression $\frac{-3x-6}{x+2}$ reduces to -3.

7.1 EXERCISES

Reading and Writing *After reading this section, write out the answers to these questions. Use complete sentences.*

1. What is a rational number?

2. What is a rational expression?

3. How do you reduce a rational number to lowest terms?

4. How do you reduce a rational expression to lowest terms?

5. How is the quotient rule used in reducing rational expressions?

6. What is the relationship between $a - b$ and $b - a$?

Evaluate each rational expression. See Example 1.

7. Evaluate $\dfrac{3x-3}{x+5}$ for $x = -2$.

8. Evaluate $\dfrac{3x+1}{4x-4}$ for $x = 5$.

9. If $R(x) = \dfrac{2x+9}{x}$, find $R(3)$.

10. If $R(x) = \dfrac{-20x-2}{x-8}$, find $R(-1)$

11. If $R(x) = \dfrac{x-5}{x+3}$, find $R(2)$, $R(-4)$, $R(-3.02)$, and $R(-2.96)$.

12. If $R(x) = \dfrac{x^2 - 2x - 3}{x-2}$, find $R(3)$, $R(5)$, $R(2.05)$, and $R(1.999)$.

Which numbers cannot be used in place of the variable in each rational expression? See Example 2.

13. $\dfrac{x}{x + 1}$

14. $\dfrac{3x}{x - 7}$

15. $\dfrac{7a}{3a - 5}$

16. $\dfrac{84}{3 - 2a}$

17. $\dfrac{2x + 3}{x^2 - 16}$

18. $\dfrac{2y + 1}{y^2 - y - 6}$

19. $\dfrac{p - 1}{2}$

20. $\dfrac{m + 31}{5}$

Reduce each rational expression to lowest terms. Assume that the variables represent only numbers for which the denominators are nonzero. See Example 3.

21. $\dfrac{6}{27}$

22. $\dfrac{14}{21}$

23. $\dfrac{42}{90}$

24. $\dfrac{42}{54}$

25. $\dfrac{36a}{90}$

26. $\dfrac{56y}{40}$

27. $\dfrac{78}{30w}$

28. $\dfrac{68}{44y}$

29. $\dfrac{6x + 2}{6}$

30. $\dfrac{2w + 2}{2w}$

31. $\dfrac{2x + 4y}{6y + 3x}$

32. $\dfrac{3m + 9w}{3m - 6w}$

33. $\dfrac{w^2 - 49}{w + 7}$

34. $\dfrac{a^2 - b^2}{a - b}$

35. $\dfrac{a^2 - 1}{a^2 + 2a + 1}$

36. $\dfrac{x^2 - y^2}{x^2 + 2xy + y^2}$

37. $\dfrac{2x^2 + 4x + 2}{4x^2 - 4}$

38. $\dfrac{2x^2 + 10x + 12}{3x^2 - 27}$

39. $\dfrac{3x^2 + 18x + 27}{21x + 63}$

40. $\dfrac{x^3 - 3x^2 - 4x}{x^2 - 4x}$

Reduce each expression to lowest terms. Assume that all denominators are nonzero. See Example 4.

41. $\dfrac{x^{10}}{x^7}$

42. $\dfrac{y^8}{y^5}$

43. $\dfrac{z^3}{z^8}$

44. $\dfrac{w^9}{w^{12}}$

45. $\dfrac{4x^7}{-2x^5}$

46. $\dfrac{-6y^3}{3y^9}$

47. $\dfrac{-12m^9n^{18}}{8m^6n^{16}}$

48. $\dfrac{-9u^9v^{19}}{6u^9v^{14}}$

49. $\dfrac{6b^{10}c^4}{-8b^{10}c^7}$

50. $\dfrac{9x^{20}y}{-6x^{25}y^3}$

51. $\dfrac{30a^3bc}{18a^7b^{17}}$

52. $\dfrac{15m^{10}n^3}{24m^{12}np}$

Reduce each expression to lowest terms. See Example 5.

53. $\dfrac{210}{264}$

54. $\dfrac{616}{660}$

55. $\dfrac{231}{168}$

56. $\dfrac{936}{624}$

57. $\dfrac{630x^5}{300x^9}$

58. $\dfrac{96y^2}{108y^5}$

59. $\dfrac{924a^{23}}{448a^{19}}$

60. $\dfrac{270b^{75}}{165b^{12}}$

Reduce each expression to lowest terms. See Example 6.

61. $\dfrac{3a - 2b}{2b - 3a}$

62. $\dfrac{5m - 6n}{6n - 5m}$

63. $\dfrac{h^2 - t^2}{t - h}$

64. $\dfrac{r^2 - s^2}{s - r}$

65. $\dfrac{2g - 6h}{9h^2 - g^2}$

66. $\dfrac{5a - 10b}{4b^2 - a^2}$

67. $\dfrac{x^2 - x - 6}{9 - x^2}$

68. $\dfrac{1 - a^2}{a^2 + a - 2}$

Reduce each expression to lowest terms. See Example 7.

69. $\dfrac{-x - 6}{x + 6}$

70. $\dfrac{-5x - 20}{3x + 12}$

71. $\dfrac{-2y - 6y^2}{3 + 9y}$

72. $\dfrac{y^2 - 16}{-8 - 2y}$

73. $\dfrac{-3x - 6}{3x - 6}$

74. $\dfrac{8 - 4x}{-8x - 16}$

75. $\dfrac{-12a - 6}{2a^2 + 7a + 3}$

76. $\dfrac{-2b^2 - 6b - 4}{b^2 - 1}$

Reduce each expression to lowest terms.

77. $\dfrac{2x^{12}}{4x^8}$

78. $\dfrac{4x^2}{2x^9}$

79. $\dfrac{2x + 4}{4x}$

80. $\dfrac{2x + 4x^2}{4x}$

81. $\dfrac{a - 4}{4 - a}$

82. $\dfrac{2b - 4}{2b + 4}$

83. $\dfrac{2c - 4}{4 - c^2}$

84. $\dfrac{-2t - 4}{4 - t^2}$

85. $\dfrac{x^2 + 4x + 4}{x^2 - 4}$

86. $\dfrac{3x - 6}{x^2 - 4x + 4}$

87. $\dfrac{-2x - 4}{x^2 + 5x + 6}$

88. $\dfrac{-2x - 8}{x^2 + 2x - 8}$

89. $\dfrac{2q^8 + q^7}{2q^6 + q^5}$

90. $\dfrac{8s^{12}}{12s^6 - 16s^5}$

91. $\dfrac{u^2 - 6u - 16}{u^2 - 16u + 64}$

92. $\dfrac{v^2 + 3v - 18}{v^2 + 12v + 36}$

93. $\dfrac{a^3 - 8}{2a - 4}$

94. $\dfrac{4w^2 - 12w + 36}{2w^3 + 54}$

95. $\dfrac{y^3 - 2y^2 - 4y + 8}{y^2 - 4y + 4}$

96. $\dfrac{mx + 3x + my + 3y}{m^2 - 3m - 18}$

Answer each question with a rational expression. Be sure to include the units. See Example 8.

97. If Sergio drove 300 miles at $x + 10$ miles per hour, then how many hours did he drive?

98. If Carrie walked 40 miles in x hours, then how fast did she walk?

99. If $x + 4$ pounds of peaches cost $4.50, then what is the cost per pound?

100. If 9 pounds of pears cost x dollars, then what is the price per pound?

101. If Ayesha can clean the entire swimming pool in x hours, then how much of the pool does she clean per hour?

102. If Ramon can mow the entire lawn in $x - 3$ hours, then how much of the lawn does he mow per hour?

Solve each problem.

103. *Annual reports.* The Crest Meat Company found that the cost per report for printing x annual reports at Peppy Printing is given by the formula

$$C(x) = \frac{150 + 0.60x}{x},$$

where $C(x)$ is in dollars.

a) Use the accompanying graph to estimate the cost per report for printing 1000 reports.

b) Use the formula to find the cost per report for printing 1000, 5000, and 10,000 reports.

c) What happens to the cost per report as the number of reports gets very large?

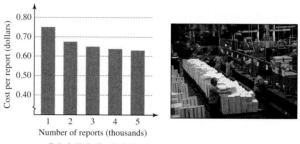

FIGURE FOR EXERCISE 103

104. *Toxic pollutants.* The annual cost in dollars for removing $p\%$ of the toxic chemicals from a town's water supply is given by the formula

$$C(p) = \frac{500,000}{100 - p}.$$

a) Use the accompanying graph to estimate the cost for removing 90% and 95% of the toxic chemicals.

b) Use the formula to determine the cost for removing 99.5% of the toxic chemicals.

c) What happens to the cost as the percentage of pollutants removed approaches 100%?

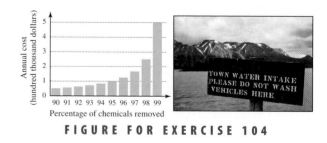

FIGURE FOR EXERCISE 104

7.2 **MULTIPLICATION AND DIVISION**

In Section 7.1 you learned to reduce rational expressions in the same way that we reduce rational numbers. In this section we will multiply and divide rational expressions using the same procedures that we use for rational numbers.

Multiplication of Rational Numbers

Two rational numbers are multiplied by multiplying their numerators and multiplying their denominators.

Multiplication of Rational Numbers

If $b \neq 0$ and $d \neq 0$, then

$$\frac{a}{b} \cdot \frac{c}{d} = \frac{ac}{bd}.$$

EXAMPLE 1 **Multiplying rational numbers**

Find the product $\frac{6}{7} \cdot \frac{14}{15}$.

Solution

The product is found by multiplying the numerators and multiplying the denominators:

$$\frac{6}{7} \cdot \frac{14}{15} = \frac{84}{105}$$

$$= \frac{21 \cdot 4}{21 \cdot 5} \quad \text{Factor the numerator and denominator.}$$

$$= \frac{4}{5} \quad \text{Divide out the GCF 21.}$$

helpful hint

Did you know that the line separating the numerator and denominator in a fraction is called the *vinculum*?

The reducing that we did after multiplying is easier to do before multiplying. First factor all terms, reduce, and then multiply:

$$\frac{6}{7} \cdot \frac{14}{15} = \frac{2 \cdot \cancel{3}}{\cancel{7}} \cdot \frac{2 \cdot \cancel{7}}{\cancel{3} \cdot 5}$$

$$= \frac{4}{5}$$

Multiplication of Rational Expressions

We multiply rational expressions in the same way we multiply rational numbers. As with rational numbers, we can factor, reduce, and then multiply.

E X A M P L E   2   **Multiplying rational expressions**

Find the indicated products.

a) $\dfrac{9x}{5y} \cdot \dfrac{10y}{3xy}$

b) $\dfrac{-8xy^4}{3z^3} \cdot \dfrac{15z}{2x^5y^3}$

Solution

a) $\dfrac{9x}{5y} \cdot \dfrac{10y}{3xy} = \dfrac{3 \cdot 3x}{5y} \cdot \dfrac{2 \cdot 5y}{3xy}$ Factor.

$= \dfrac{6}{y}$

b) $\dfrac{-8xy^4}{3z^3} \cdot \dfrac{15z}{2x^5y^3} = \dfrac{-2 \cdot 2 \cdot 2xy^4}{3z^3} \cdot \dfrac{3 \cdot 5z}{2x^5y^3}$ Factor.

$= \dfrac{-20xy^4z}{z^3x^5y^3}$ Reduce.

$= \dfrac{-20y}{z^2x^4}$ Quotient rule

E X A M P L E   3   **Multiplying rational expressions**

Find the indicated products.

a) $\dfrac{2x - 2y}{4} \cdot \dfrac{2x}{x^2 - y^2}$

b) $\dfrac{x^2 + 7x + 12}{2x + 6} \cdot \dfrac{x}{x^2 - 16}$

c) $\dfrac{a + b}{6a} \cdot \dfrac{8a^2}{a^2 + 2ab + b^2}$

Solution

a) $\dfrac{2x - 2y}{4} \cdot \dfrac{2x}{x^2 - y^2} = \dfrac{2(x - y)}{2 \cdot 2} \cdot \dfrac{2 \cdot x}{(x - y)(x + y)}$ Factor.

$= \dfrac{x}{x + y}$ Reduce.

b) $\dfrac{x^2 + 7x + 12}{2x + 6} \cdot \dfrac{x}{x^2 - 16} = \dfrac{(x + 3)(x + 4)}{2(x + 3)} \cdot \dfrac{x}{(x - 4)(x + 4)}$ Factor.

$= \dfrac{x}{2(x - 4)}$ Reduce.

$= \dfrac{x}{2x - 8}$

c) $\dfrac{a + b}{6a} \cdot \dfrac{8a^2}{a^2 + 2ab + b^2} = \dfrac{a + b}{2 \cdot 3a} \cdot \dfrac{2 \cdot 4a^2}{(a + b)^2}$ Factor.

$= \dfrac{4a}{3(a + b)}$ Reduce.

$= \dfrac{4a}{3a + 3b}$

Division of Rational Numbers

Division of rational numbers can be accomplished by multiplying by the reciprocal of the divisor.

Division of Rational Numbers

If $b \neq 0$, $c \neq 0$, and $d \neq 0$, then

$$\frac{a}{b} \div \frac{c}{d} = \frac{a}{b} \cdot \frac{d}{c}.$$

E X A M P L E   4   **Dividing rational numbers**
Find each quotient.

a) $5 \div \dfrac{1}{2}$ **b)** $\dfrac{6}{7} \div \dfrac{3}{14}$

Solution

a) $5 \div \dfrac{1}{2} = 5 \cdot 2 = 10$

b) $\dfrac{6}{7} \div \dfrac{3}{14} = \dfrac{6}{7} \cdot \dfrac{14}{3} = \dfrac{2 \cdot \cancel{3}}{\cancel{7}} \cdot \dfrac{2 \cdot \cancel{7}}{\cancel{3}} = 4$ ■

Division of Rational Expressions

We divide rational expressions in the same way we divide rational numbers: Invert the divisor and multiply.

E X A M P L E   5   **Dividing rational expressions**
Find each quotient.

a) $\dfrac{5}{3x} \div \dfrac{5}{6x}$ **b)** $\dfrac{x^7}{2} \div (2x^2)$

c) $\dfrac{4 - x^2}{x^2 + x} \div \dfrac{x - 2}{x^2 - 1}$

helpful hint

A doctor told a nurse to give a patient half of the usual dose of a certain medicine. The nurse figured, "dividing in half means dividing by 1/2 which means multiply by 2." So the patient got four times the prescribed amount and died (true story). There is a big difference between dividing a quantity in half and dividing by one-half.

Solution

a) $\dfrac{5}{3x} \div \dfrac{5}{6x} = \dfrac{5}{3x} \cdot \dfrac{6x}{5}$ Invert the divisor and multiply.

$= \dfrac{\cancel{5}}{\cancel{3x}} \cdot \dfrac{2 \cdot \cancel{3x}}{\cancel{5}}$ Factor.

$= 2$ Divide out the common factors.

b) $\dfrac{x^7}{2} \div (2x^2) = \dfrac{x^7}{2} \cdot \dfrac{1}{2x^2}$ Invert and multiply.

$= \dfrac{x^5}{4}$ Quotient rule

c) $\dfrac{4 - x^2}{x^2 + x} \div \dfrac{x - 2}{x^2 - 1} = \dfrac{4 - x^2}{x^2 + x} \cdot \dfrac{x^2 - 1}{x - 2}$ Invert and multiply.

$$= \dfrac{\overset{-1}{\cancel{(2 - x)}}\,(2 + x)}{x\cancel{(x + 1)}} \cdot \dfrac{\cancel{(x + 1)}(x - 1)}{\cancel{x - 2}}$$ Factor.

$$= \dfrac{-1(2 + x)(x - 1)}{x}$$ $\dfrac{2 - x}{x - 2} = -1$

$$= \dfrac{-1(x^2 + x - 2)}{x}$$ Simplify.

$$= \dfrac{-x^2 - x + 2}{x}$$ ∎

We sometimes write division of rational expressions using the fraction bar. For example, we can write

$$\dfrac{a + b}{3} \div \dfrac{1}{6} \quad \text{as} \quad \dfrac{\dfrac{a + b}{3}}{\dfrac{1}{6}}.$$

No matter how division is expressed, we invert the divisor and multiply.

E X A M P L E   6   **Division expressed with a fraction bar**
Find each quotient.

a) $\dfrac{\dfrac{a + b}{3}}{\dfrac{1}{6}}$ **b)** $\dfrac{\dfrac{x^2 - 1}{2}}{\dfrac{x - 1}{3}}$

c) $\dfrac{\dfrac{a^2 + 5}{3}}{2}$

Solution

a) $\dfrac{\dfrac{a + b}{3}}{\dfrac{1}{6}} = \dfrac{a + b}{3} \div \dfrac{1}{6}$ Rewrite as division.

$$= \dfrac{a + b}{3} \cdot \dfrac{6}{1}$$ Invert and multiply.

$$= \dfrac{a + b}{\cancel{3}} \cdot \dfrac{2 \cdot \cancel{3}}{1}$$ Factor.

$$= (a + b)2$$ Reduce.

$$= 2a + 2b$$

b) $\dfrac{\dfrac{x^2 - 1}{2}}{\dfrac{x - 1}{3}} = \dfrac{x^2 - 1}{2} \div \dfrac{x - 1}{3}$ Rewrite as division.

$= \dfrac{x^2 - 1}{2} \cdot \dfrac{3}{x - 1}$ Invert and multiply.

$= \dfrac{(x - 1)(x + 1)}{2} \cdot \dfrac{3}{x - 1}$ Factor.

$= \dfrac{3x + 3}{2}$ Reduce.

c) $\dfrac{\dfrac{a^2 + 5}{3}}{2} = \dfrac{a^2 + 5}{3} \div 2$ Rewrite as division.

$= \dfrac{a^2 + 5}{3} \cdot \dfrac{1}{2}$

$= \dfrac{a^2 + 5}{6}$

Applications

We saw in the last section that rational expressions can be used to represent rates. Note that there are several ways to write rates. For example, miles per hour is written mph, mi/hr, or $\frac{\text{mi}}{\text{hr}}$. The last way is best when doing operations with rates because it helps us reconcile our answers. Notice how hours "cancels" when we multiply miles per hour and hours in the next example, giving an answer in miles, as it should be.

E X A M P L E   7   **Writing rational expressions**

Answer each question with a rational expression.

a) Shasta averaged $\frac{200}{x}$ mph for x hours before she had lunch. How many miles did she drive in the first 3 hours after lunch assuming that she continued to average $\frac{200}{x}$ mph?

b) If a bathtub can be filled in x minutes, then the rate at which it is filling is $\frac{1}{x}$ tub/min. How much of the tub is filled in 10 minutes?

Solution

a) Because $R \cdot T = D$, the distance she traveled after lunch is the product of the rate and time:

$$\frac{200}{x} \, \frac{\text{mi}}{\text{hr}} \cdot 3 \text{ hr} = \frac{600}{x} \text{ mi}$$

b) Because $R \cdot T = W$, the work completed is the product of the rate and time:

$$\frac{1}{x} \, \frac{\text{tub}}{\text{min}} \cdot 10 \text{ min} = \frac{10}{x} \text{ tub}$$

WARM-UPS

True or false? Explain your answer.

1. $\frac{2}{3} \cdot \frac{5}{3} = \frac{10}{9}$.

2. The product of $\frac{x-7}{3}$ and $\frac{6}{7-x}$ is -2.

3. Dividing by 2 is equivalent to multiplying by $\frac{1}{2}$.

4. $3 \div x = \frac{1}{3} \cdot x$ for any nonzero number x.

5. Factoring polynomials is essential in multiplying rational expressions.

6. One-half of one-fourth is one-sixth.

7. One-half divided by three is three-halves.

8. The quotient of $(839 - 487)$ and $(487 - 839)$ is -1.

9. $\frac{a}{3} \div 3 = \frac{a}{9}$ for any value of a.

10. $\frac{a}{b} \cdot \frac{b}{a} = 1$ for any nonzero values of a and b.

7.2 EXERCISES

Reading and Writing *After reading this section, write out the answers to these questions. Use complete sentences.*

1. How do you multiply rational numbers?

2. How do you multiply rational expressions?

3. What can be done to simplify the process of multiplying rational numbers or rational expressions?

4. How do you divide rational numbers or rational expressions?

Perform the indicated operation. See Example 1.

5. $\frac{8}{15} \cdot \frac{35}{24}$

6. $\frac{3}{4} \cdot \frac{8}{21}$

7. $\frac{12}{17} \cdot \frac{51}{10}$

8. $\frac{25}{48} \cdot \frac{56}{35}$

9. $24 \cdot \frac{7}{20}$

10. $\frac{3}{10} \cdot 35$

Perform the indicated operation. See Example 2.

11. $\frac{5a}{12b} \cdot \frac{3ab}{55a}$

12. $\frac{3m}{7p} \cdot \frac{35p}{6mp}$

13. $\frac{-2x^6}{7a^5} \cdot \frac{21a^2}{6x}$

14. $\frac{5z^3w}{-9y^3} \cdot \frac{-6y^5}{20z^9}$

15. $\frac{15t^3y^5}{20w^7} \cdot 24t^5w^3y^2$

16. $22x^2y^3z \cdot \frac{6x^5}{33y^3z^4}$

Perform the indicated operation. See Example 3.

17. $\frac{3a + 3b}{15} \cdot \frac{10a}{a^2 - b^2}$

18. $\frac{b^3 + b}{5} \cdot \frac{10}{b^2 + b}$

19. $(x^2 - 6x + 9) \cdot \frac{3}{x - 3}$

20. $\frac{12}{4x + 10} \cdot (4x^2 + 20x + 25)$

21. $\frac{16a + 8}{5a^2 + 5} \cdot \frac{2a^2 + a - 1}{4a^2 - 1}$

22. $\frac{6x - 18}{2x^2 - 5x - 3} \cdot \frac{4x^2 + 4x + 1}{6x + 3}$

Perform the indicated operation. See Example 4.

23. $12 \div \frac{2}{5}$

24. $32 \div \frac{1}{4}$

25. $\frac{5}{7} \div \frac{15}{14}$

26. $\frac{3}{4} \div \frac{15}{2}$

27. $\frac{40}{3} \div 12$

28. $\frac{22}{9} \div 9$

Perform the indicated operation. See Example 5.

29. $\frac{5x^2}{3} \div \frac{10x}{21}$

30. $\frac{4u^2}{3v} \div \frac{14u}{15v^6}$

31. $\frac{8m^3}{n^4} \div (12mn^2)$

32. $\frac{2p^4}{3q^3} \div (4pq^5)$

33. $\frac{y - 6}{2} \div \frac{6 - y}{6}$

34. $\frac{4 - a}{5} \div \frac{a^2 - 16}{3}$

35. $\dfrac{x^2 + 4x + 4}{8} \div \dfrac{(x + 2)^3}{16}$

36. $\dfrac{a^2 + 2a + 1}{3} \div \dfrac{a^2 - 1}{a}$

37. $\dfrac{t^2 + 3t - 10}{t^2 - 25} \div (4t - 8)$

38. $\dfrac{w^2 - 7w + 12}{w^2 - 4w} \div (w^2 - 9)$

39. $(2x^2 - 3x - 5) \div \dfrac{2x - 5}{x - 1}$

40. $(6y^2 - y - 2) \div \dfrac{2y + 1}{3y - 2}$

Perform the indicated operation. See Example 6.

41. $\dfrac{\dfrac{x - 2y}{5}}{\dfrac{1}{10}}$

42. $\dfrac{\dfrac{3m + 6n}{8}}{\dfrac{3}{4}}$

43. $\dfrac{\dfrac{x^2 - 4}{12}}{\dfrac{x - 2}{6}}$

44. $\dfrac{\dfrac{6a^2 + 6}{5}}{\dfrac{6a + 6}{5}}$

45. $\dfrac{\dfrac{x^2 + 9}{3}}{\dfrac{5}{}}$

46. $\dfrac{\dfrac{1}{a - 3}}{4}$

47. $\dfrac{\dfrac{x^2 - y^2}{x - y}}{9}$

48. $\dfrac{\dfrac{x^2 + 6x + 8}{x + 2}}{x + 1}$

Perform the indicated operation.

49. $\dfrac{x - 1}{3} \cdot \dfrac{9}{1 - x}$

50. $\dfrac{2x - 2y}{3} \cdot \dfrac{1}{y - x}$

51. $\dfrac{3a + 3b}{a} \cdot \dfrac{1}{3}$

52. $\dfrac{a - b}{2b - 2a} \cdot \dfrac{2}{5}$

53. $\dfrac{\dfrac{b}{a}}{\dfrac{1}{2}}$

54. $\dfrac{\dfrac{2g}{3h}}{\dfrac{1}{h}}$

55. $\dfrac{6y}{3} \div (2x)$

56. $\dfrac{8x}{9} \div (18x)$

57. $\dfrac{a^3 b^4}{-2ab^2} \cdot \dfrac{a^5 b^7}{ab}$

58. $\dfrac{-2a^2}{3a^2} \cdot \dfrac{20a}{15a^3}$

59. $\dfrac{2mn^4}{6mn^2} \div \dfrac{3m^5 n^7}{m^2 n^4}$

60. $\dfrac{rt^2}{rt^2} \div \dfrac{rt^2}{r^3 t^2}$

61. $\dfrac{3x^2 + 16x + 5}{x} \cdot \dfrac{x^2}{9x^2 - 1}$

62. $\dfrac{x^2 + 6x + 5}{x} \cdot \dfrac{x^4}{3x + 3}$

63. $\dfrac{a^2 - 2a + 4}{a^2 - 4} \cdot \dfrac{(a + 2)^3}{2a + 4}$

64. $\dfrac{w^2 - 1}{(w - 1)^2} \cdot \dfrac{w - 1}{w^2 + 2w + 1}$

65. $\dfrac{2x^2 + 19x - 10}{x^2 - 100} \div \dfrac{4x^2 - 1}{2x^2 - 19x - 10}$

66. $\dfrac{x^3 - 1}{x^2 + 1} \div \dfrac{9x^2 + 9x + 9}{x^2 - x}$

67. $\dfrac{9 + 6m + m^2}{9 - 6m + m^2} \cdot \dfrac{m^2 - 9}{m^2 + mk + 3m + 3k}$

68. $\dfrac{3x + 3w + bx + bw}{x^2 - w^2} \cdot \dfrac{6 - 2b}{9 - b^2}$

Solve each problem. Answers could be rational expressions. Be sure to give your answer with appropriate units. See Example 7.

69. Distance. Florence averaged $\dfrac{26.2}{x}$ mph for the x hours in which she ran the Boston Marathon. If she ran at that same rate for $\frac{1}{2}$ hour in the Manchac Fun Run, then how many miles did she run at Manchac?

70. Work. Henry sold 120 magazine subscriptions in $x + 2$ days. If he sold at the same rate for another week, then how many magazines did he sell in the extra week?

71. Area of a rectangle. If the length of a rectangular flag is x meters and its width is $\frac{5}{x}$ meters, then what is the area of the rectangle?

$\frac{5}{x}$ m

x m

FIGURE FOR EXERCISE 71

72. Area of a triangle. If the base of a triangle is $8x + 16$ yards and its height is $\dfrac{1}{x + 2}$ yards, then what is the area of the triangle?

$\frac{1}{x + 2}$ yd

$8x + 16$ yd

FIGURE FOR EXERCISE 72

GETTING MORE INVOLVED

73. *Discussion.* Evaluate each expression.

a) One-half of $\frac{1}{4}$ b) One-third of 4

c) One-half of $\frac{4x}{3}$ d) One-half of $\frac{3x}{2}$

74. *Exploration.* Let $R = \frac{6x^2 + 23x + 20}{24x^2 + 29x - 4}$ and $H = \frac{2x + 5}{8x - 1}$.

a) Find R when $x = 2$ and $x = 3$. Find H when $x = 2$ and $x = 3$.

b) How are these values of R and H related and why?

<table>
<tr><td>**7.3**</td><td># FINDING THE LEAST COMMON DENOMINATOR</td></tr>
</table>

In this section

- Building Up the Denominator
- Finding the Least Common Denominator
- Converting to the LCD

Every rational expression can be written in infinitely many equivalent forms. Because we can add or subtract only fractions with identical denominators, we must be able to change the denominator of a fraction. You have already learned how to change the denominator of a fraction by reducing. In this section you will learn the opposite of reducing, which is called **building up the denominator.**

Building Up the Denominator

To convert the fraction $\frac{2}{3}$ into an equivalent fraction with a denominator of 21, we factor 21 as $21 = 3 \cdot 7$. Because $\frac{2}{3}$ already has a 3 in the denominator, multiply the numerator and denominator of $\frac{2}{3}$ by the missing factor 7 to get a denominator of 21:

$$\frac{2}{3} = \frac{2}{3} \cdot \frac{7}{7} = \frac{14}{21}$$

For rational expressions the process is the same. To convert the rational expression

$$\frac{5}{x + 3}$$

into an equivalent rational expression with a denominator of $x^2 - x - 12$, first factor $x^2 - x - 12$:

$$x^2 - x - 12 = (x + 3)(x - 4)$$

From the factorization we can see that the denominator $x + 3$ needs only a factor of $x - 4$ to have the required denominator. So multiply the numerator and denominator by the missing factor $x - 4$:

$$\frac{5}{x + 3} = \frac{5}{x + 3} \cdot \frac{x - 4}{x - 4} = \frac{5x - 20}{x^2 - x - 12}$$

EXAMPLE 1

Building up the denominator

Build each rational expression into an equivalent rational expression with the indicated denominator.

a) $3 = \dfrac{?}{12}$ b) $\dfrac{3}{w} = \dfrac{?}{wx}$ c) $\dfrac{2}{3y^3} = \dfrac{?}{12y^8}$

Solution

a) Because $3 = \frac{3}{1}$, we get a denominator of 12 by multiplying the numerator and denominator by 12:

$$3 = \frac{3}{1} = \frac{3 \cdot 12}{1 \cdot 12} = \frac{36}{12}$$

b) Multiply the numerator and denominator by x:

$$\frac{3}{w} = \frac{3 \cdot x}{w \cdot x} = \frac{3x}{wx}$$

c) To build the denominator $3y^3$ up to $12y^8$, multiply by $4y^5$:

$$\frac{2}{3y^3} = \frac{2 \cdot 4y^5}{3y^3 \cdot 4y^5} = \frac{8y^5}{12y^8}$$

In the next example we must factor the original denominator before building up the denominator.

E X A M P L E   2

Building up the denominator

Build each rational expression into an equivalent rational expression with the indicated denominator.

a) $\dfrac{7}{3x - 3y} = \dfrac{?}{6y - 6x}$ **b)** $\dfrac{x - 2}{x + 2} = \dfrac{?}{x^2 + 8x + 12}$

Solution

a) Because $3x - 3y = 3(x - y)$, we factor -6 out of $6y - 6x$. This will give a factor of $x - y$ in each denominator:

$$3x - 3y = 3(x - y)$$
$$6y - 6x = -6(x - y) = -2 \cdot 3(x - y)$$

To get the required denominator, we multiply the numerator and denominator by -2 only:

$$\frac{7}{3x - 3y} = \frac{7(-2)}{(3x - 3y)(-2)}$$

$$= \frac{-14}{6y - 6x}$$

b) Because $x^2 + 8x + 12 = (x + 2)(x + 6)$, we multiply the numerator and denominator by $x + 6$, the missing factor:

$$\frac{x - 2}{x + 2} = \frac{(x - 2)(x + 6)}{(x + 2)(x + 6)}$$

$$= \frac{x^2 + 4x - 12}{x^2 + 8x + 12}$$

> **helpful / hint**
>
> Notice that reducing and building up are exactly the opposite of each other. In reducing you remove a factor that is common to the numerator and denominator, and in building up you put a common factor into the numerator and denominator.

C A U T I O N   When building up a denominator, *both* the numerator and the denominator must be multiplied by the appropriate expression, because that is how we build up fractions.

Finding the Least Common Denominator

We can use the idea of building up the denominator to convert two fractions with different denominators into fractions with identical denominators. For example,

$$\frac{5}{6} \quad \text{and} \quad \frac{1}{4}$$

can both be converted into fractions with a denominator of 12, since $12 = 2 \cdot 6$ and $12 = 3 \cdot 4$:

$$\frac{5}{6} = \frac{5 \cdot 2}{6 \cdot 2} = \frac{10}{12} \qquad \frac{1}{4} = \frac{1 \cdot 3}{4 \cdot 3} = \frac{3}{12}$$

The smallest number that is a multiple of all of the denominators is called the **least common denominator (LCD).** The LCD for the denominators 6 and 4 is 12.

To find the LCD in a systematic way, we look at a complete factorization of each denominator. Consider the denominators 24 and 30:

$$24 = 2 \cdot 2 \cdot 2 \cdot 3 = 2^3 \cdot 3$$
$$30 = 2 \cdot 3 \cdot 5$$

Any multiple of 24 must have three 2's in its factorization, and any multiple of 30 must have one 2 as a factor. So a number with three 2's in its factorization will have enough to be a multiple of both 24 and 30. The LCD must also have one 3 and one 5 in its factorization. *We use each factor the maximum number of times it appears in either factorization.* So the LCD is $2^3 \cdot 3 \cdot 5$:

$$2^3 \cdot 3 \cdot 5 = \overbrace{2 \cdot 2 \cdot \underbrace{2 \cdot 3 \cdot 5}_{30}}^{24} = 120$$

If we omitted any one of the factors in $2 \cdot 2 \cdot 2 \cdot 3 \cdot 5$, we would not have a multiple of both 24 and 30. That is what makes 120 the *least* common denominator. To find the LCD for two polynomials, we use the same strategy.

> ### Strategy for Finding the LCD for Polynomials
>
> **1.** Factor each denominator completely. Use exponent notation for repeated factors.
> **2.** Write the product of all of the different factors that appear in the denominators.
> **3.** On each factor, use the highest power that appears on that factor in any of the denominators.

E X A M P L E   3   **Finding the LCD**

If the given expressions were used as denominators of rational expressions, then what would be the LCD for each group of denominators?

a) $20, 50$ **b)** x^3yz^2, x^5y^2z, xyz^5

c) $a^2 + 5a + 6, a^2 + 4a + 4$

Solution

a) First factor each number completely:

$$20 = 2^2 \cdot 5 \qquad 50 = 2 \cdot 5^2$$

The highest power of 2 is 2, and the highest power of 5 is 2. So the LCD of 20 and 50 is $2^2 \cdot 5^2$, or 100.

b) The expressions x^3yz^2, x^5y^2z, and xyz^5 are already factored. For the LCD, use the highest power of each variable. So the LCD is $x^5y^2z^5$.

c) First factor each polynomial.

$$a^2 + 5a + 6 = (a + 2)(a + 3) \qquad a^2 + 4a + 4 = (a + 2)^2$$

The highest power of $(a + 3)$ is 1, and the highest power of $(a + 2)$ is 2. So the LCD is $(a + 3)(a + 2)^2$. ∎

Converting to the LCD

When adding or subtracting rational expressions, we must convert the expressions into expressions with identical denominators. To keep the computations as simple as possible, we use the least common denominator.

EXAMPLE 4

Converting to the LCD

Find the LCD for the rational expressions, and convert each expression into an equivalent rational expression with the LCD as the denominator.

a) $\dfrac{4}{9xy}, \dfrac{2}{15xz}$

b) $\dfrac{5}{6x^2}, \dfrac{1}{8x^3y}, \dfrac{3}{4y^2}$

Solution

a) Factor each denominator completely:

$$9xy = 3^2xy \qquad 15xz = 3 \cdot 5xz$$

The LCD is $3^2 \cdot 5xyz$. Now convert each expression into an expression with this denominator. We must multiply the numerator and denominator of the first rational expression by $5z$ and the second by $3y$:

$$\left.\begin{array}{l}\dfrac{4}{9xy} = \dfrac{4 \cdot 5z}{9xy \cdot 5z} = \dfrac{20z}{45xyz}\\[2mm]\dfrac{2}{15xz} = \dfrac{2 \cdot 3y}{15xz \cdot 3y} = \dfrac{6y}{45xyz}\end{array}\right\} \text{Same denominator}$$

b) Factor each denominator completely:

$$6x^2 = 2 \cdot 3x^2 \qquad 8x^3y = 2^3x^3y \qquad 4y^2 = 2^2y^2$$

The LCD is $2^3 \cdot 3 \cdot x^3y^2$ or $24x^3y^2$. Now convert each expression into an expression with this denominator:

$$\dfrac{5}{6x^2} = \dfrac{5 \cdot 4xy^2}{6x^2 \cdot 4xy^2} = \dfrac{20xy^2}{24x^3y^2}$$

$$\dfrac{1}{8x^3y} = \dfrac{1 \cdot 3y}{8x^3y \cdot 3y} = \dfrac{3y}{24x^3y^2}$$

$$\dfrac{3}{4y^2} = \dfrac{3 \cdot 6x^3}{4y^2 \cdot 6x^3} = \dfrac{18x^3}{24x^3y^2}$$

∎

helpful hint

What is the difference between LCD, GCF, CBS, and NBC? The LCD for the denominators 4 and 6 is 12. The *least* common denominator is *greater than* or equal to both numbers. The GCF for 4 and 6 is 2. The *greatest* common factor is *less than* or equal to both numbers. CBS and NBC are TV networks.

E X A M P L E   5   **Converting to the LCD**

Find the LCD for the rational expressions

$$\frac{5x}{x^2 - 4} \quad \text{and} \quad \frac{3}{x^2 + x - 6}$$

and convert each into an equivalent rational expression with that denominator.

Solution

First factor the denominators:

$$x^2 - 4 = (x - 2)(x + 2)$$
$$x^2 + x - 6 = (x - 2)(x + 3)$$

The LCD is $(x - 2)(x + 2)(x + 3)$. Now we multiply the numerator and denominator of the first rational expression by $(x + 3)$ and those of the second rational expression by $(x + 2)$. Because each denominator already has one factor of $(x - 2)$, there is no reason to multiply by $(x - 2)$. We multiply each denominator by the factors in the LCD that are missing from that denominator:

$$\frac{5x}{x^2 - 4} = \frac{5x(x + 3)}{(x - 2)(x + 2)(x + 3)} = \frac{5x^2 + 15x}{(x - 2)(x + 2)(x + 3)} \left.\right\} \begin{array}{l}\text{Same} \\ \text{denominator}\end{array}$$

$$\frac{3}{x^2 + x - 6} = \frac{3(x + 2)}{(x - 2)(x + 3)(x + 2)} = \frac{3x + 6}{(x - 2)(x + 2)(x + 3)}$$ ∎

Note that in Example 5 we multiplied the expressions in the numerators but left the denominators in factored form. The numerators are simplified because it is the numerators that must be added when we add rational expressions in Section 7.4. Because we can add rational expressions with identical denominators, there is no need to multiply the denominators.

WARM-UPS

True or false? Explain your answer.

1. To convert $\frac{2}{3}$ into an equivalent fraction with a denominator of 18, we would multiply only the denominator of $\frac{2}{3}$ by 6.

2. Factoring has nothing to do with finding the least common denominator.

3. $\frac{3}{2ab^2} = \frac{15a^2b^2}{10a^3b^4}$ for any nonzero values of a and b.

4. The LCD for the denominators $2^5 \cdot 3$ and $2^4 \cdot 3^2$ is $2^5 \cdot 3^2$.

5. The LCD for the fractions $\frac{1}{6}$ and $\frac{1}{10}$ is 60.

6. The LCD for the denominators $6a^2b$ and $4ab^3$ is $2ab$.

7. The LCD for the denominators $a^2 + 1$ and $a + 1$ is $a^2 + 1$.

8. $\frac{x}{2} = \frac{x + 7}{2 + 7}$ for any real number x.

9. The LCD for the rational expressions $\frac{1}{x - 2}$ and $\frac{3}{x + 2}$ is $x^2 - 4$.

10. $x = \frac{3x}{3}$ for any real number x.

7.3   EXERCISES

Reading and Writing   *After reading this section, write out the answers to these questions. Use complete sentences.*

1. What is building up the denominator?

2. How do we build up the denominator of a rational expression?

3. What is the least common denominator for fractions?

4. How do you find the LCD for two polynomial denominators?

Build each rational expression into an equivalent rational expression with the indicated denominator. See Example 1.

5. $\dfrac{1}{3} = \dfrac{?}{27}$

6. $\dfrac{2}{5} = \dfrac{?}{35}$

7. $7 = \dfrac{?}{2x}$

8. $6 = \dfrac{?}{4y}$

9. $\dfrac{5}{b} = \dfrac{?}{3bt}$

10. $\dfrac{7}{2ay} = \dfrac{?}{2ayz}$

11. $\dfrac{-9z}{2aw} = \dfrac{?}{8awz}$

12. $\dfrac{-7yt}{3x} = \dfrac{?}{18xyt}$

13. $\dfrac{2}{3a} = \dfrac{?}{15a^3}$

14. $\dfrac{7b}{12c^5} = \dfrac{?}{36c^8}$

15. $\dfrac{4}{5xy^2} = \dfrac{?}{10x^2y^5}$

16. $\dfrac{5y^2}{8x^3z} = \dfrac{?}{24x^5z^3}$

Build each rational expression into an equivalent rational expression with the indicated denominator. See Example 2.

17. $\dfrac{5}{2x + 2} = \dfrac{?}{-8x - 8}$

18. $\dfrac{3}{m - n} = \dfrac{?}{2n - 2m}$

19. $\dfrac{8a}{5b^2 - 5b} = \dfrac{?}{20b^2 - 20b^3}$

20. $\dfrac{5x}{-6x - 9} = \dfrac{?}{18x^2 + 27x}$

21. $\dfrac{3}{x + 2} = \dfrac{?}{x^2 - 4}$

22. $\dfrac{a}{a + 3} = \dfrac{?}{a^2 - 9}$

23. $\dfrac{3x}{x + 1} = \dfrac{?}{x^2 + 2x + 1}$

24. $\dfrac{-7x}{2x - 3} = \dfrac{?}{4x^2 - 12x + 9}$

25. $\dfrac{y - 6}{y - 4} = \dfrac{?}{y^2 + y - 20}$

26. $\dfrac{z - 6}{z + 3} = \dfrac{?}{z^2 - 2z - 15}$

If the given expressions were used as denominators of rational expressions, then what would be the LCD for each group of denominators? See Example 3.

27. 12, 16

28. 28, 42

29. 12, 18, 20

30. 24, 40, 48

31. $6a^2$, $15a$

32. $18x^2$, $20xy$

33. $2a^4b$, $3ab^6$, $4a^3b^2$

34. $4m^3nw$, $6mn^5w^8$, $9m^6nw$

35. $x^2 - 16$, $x^2 + 8x + 16$

36. $x^2 - 9$, $x^2 + 6x + 9$

37. x, $x + 2$, $x - 2$

38. y, $y - 5$, $y + 2$

39. $x^2 - 4x$, $x^2 - 16$, $2x$

40. y, $y^2 - 3y$, $3y$

Find the LCD for the given rational expressions, and convert each rational expression into an equivalent rational expression with the LCD as the denominator. See Example 4.

41. $\dfrac{1}{6}, \dfrac{3}{8}$

42. $\dfrac{5}{12}, \dfrac{3}{20}$

43. $\dfrac{3}{84a}, \dfrac{5}{63b}$

44. $\dfrac{4b}{75a}, \dfrac{6}{105ab}$

45. $\dfrac{1}{3x^2}, \dfrac{3}{2x^5}$

46. $\dfrac{3}{8a^3b^9}, \dfrac{5}{6a^2c}$

47. $\dfrac{x}{9y^5z}, \dfrac{y}{12x^3}, \dfrac{1}{6x^2y}$

48. $\dfrac{5}{12a^6b}, \dfrac{3b}{14a^3}, \dfrac{1}{2ab^3}$

In Exercises 49–60, find the LCD for the given rational expressions, and convert each rational expression into an equivalent rational expression with the LCD as the denominator. See Example 5.

49. $\dfrac{2x}{x - 3}, \dfrac{5x}{x + 2}$

50. $\dfrac{2a}{a - 5}, \dfrac{3a}{a + 2}$

51. $\dfrac{4}{a-6}, \dfrac{5}{6-a}$

52. $\dfrac{4}{x-y}, \dfrac{5x}{2y-2x}$

53. $\dfrac{x}{x^2-9}, \dfrac{5x}{x^2-6x+9}$

54. $\dfrac{5x}{x^2-1}, \dfrac{-4}{x^2-2x+1}$

55. $\dfrac{w+2}{w^2-2w-15}, \dfrac{-2w}{w^2-4w-5}$

56. $\dfrac{z-1}{z^2+6z+8}, \dfrac{z+1}{z^2+5z+6}$

57. $\dfrac{-5}{6x-12}, \dfrac{x}{x^2-4}, \dfrac{3}{2x+4}$

58. $\dfrac{3}{4b^2-9}, \dfrac{2b}{2b+3}, \dfrac{-5}{2b^2-3b}$

59. $\dfrac{2}{2q^2-5q-3}, \dfrac{3}{2q^2+9q+4}, \dfrac{4}{q^2+q-12}$

60. $\dfrac{-3}{2p^2+7p-15}, \dfrac{p}{2p^2-11p+12}, \dfrac{2}{p^2+p-20}$

GETTING MORE INVOLVED

61. *Discussion.* Why do we learn how to convert two rational expressions into equivalent rational expressions with the same denominator?

62. *Discussion.* Which expression is the LCD for

$$\frac{3x-1}{2^2 \cdot 3 \cdot x^2(x+2)} \quad \text{and} \quad \frac{2x+7}{2 \cdot 3^2 \cdot x(x+2)^2} ?$$

a) $2 \cdot 3 \cdot x(x+2)$
b) $36x(x+2)$
c) $36x^2(x+2)^2$
d) $2^3 \cdot 3^3 x^3(x+2)^2$

7.4 **ADDITION AND SUBTRACTION**

In Section 7.3 you learned how to find the LCD and build up the denominators of rational expressions. In this section we will use that knowledge to add and subtract rational expressions with different denominators.

Addition and Subtraction of Rational Numbers

We can add or subtract rational numbers (or fractions) only with identical denominators according to the following definition.

In this
section

- Addition and Subtraction of Rational Numbers
- Addition and Subtraction of Rational Expressions
- Applications

Addition and Subtraction of Rational Numbers

If $b \neq 0$, then

$$\frac{a}{b} + \frac{c}{b} = \frac{a+c}{b} \quad \text{and} \quad \frac{a}{b} - \frac{c}{b} = \frac{a-c}{b}.$$

E X A M P L E   1

Adding or subtracting fractions with the same denominator

Perform the indicated operations. Reduce answers to lowest terms.

a) $\dfrac{1}{12} + \dfrac{7}{12}$

b) $\dfrac{1}{4} - \dfrac{3}{4}$

Solution

a) $\dfrac{1}{12} + \dfrac{7}{12} = \dfrac{8}{12} = \dfrac{4 \cdot 2}{4 \cdot 3} = \dfrac{2}{3}$

b) $\dfrac{1}{4} - \dfrac{3}{4} = \dfrac{-2}{4} = -\dfrac{1}{2}$ ∎

If the rational numbers have different denominators, we must convert them to equivalent rational numbers that have identical denominators and then add or subtract. Of course, it is most efficient to use the least common denominator (LCD), as in the following example.

E X A M P L E   2

Adding or subtracting fractions with different denominators

Find each sum or difference.

a) $\dfrac{3}{20} + \dfrac{7}{12}$

b) $\dfrac{1}{6} - \dfrac{4}{15}$

Solution

a) Because $20 = 2^2 \cdot 5$ and $12 = 2^2 \cdot 3$, the LCD is $2^2 \cdot 3 \cdot 5$, or 60. Convert each fraction to an equivalent fraction with a denominator of 60:

$$\dfrac{3}{20} + \dfrac{7}{12} = \dfrac{3 \cdot 3}{20 \cdot 3} + \dfrac{7 \cdot 5}{12 \cdot 5} \qquad \text{Build up the denominators.}$$

$$= \dfrac{9}{60} + \dfrac{35}{60} \qquad \text{Simplify numerators and denominators.}$$

$$= \dfrac{44}{60} \qquad \text{Add the fractions.}$$

$$= \dfrac{4 \cdot 11}{4 \cdot 15} \qquad \text{Factor.}$$

$$= \dfrac{11}{15} \qquad \text{Reduce.}$$

b) Because $6 = 2 \cdot 3$ and $15 = 3 \cdot 5$, the LCD is $2 \cdot 3 \cdot 5$ or 30:

$$\dfrac{1}{6} - \dfrac{4}{15} = \dfrac{1}{2 \cdot 3} - \dfrac{4}{3 \cdot 5} \qquad \text{Factor the denominators.}$$

$$= \dfrac{1 \cdot 5}{2 \cdot 3 \cdot 5} - \dfrac{4 \cdot 2}{3 \cdot 5 \cdot 2} \qquad \text{Build up the denominators.}$$

$$= \dfrac{5}{30} - \dfrac{8}{30} \qquad \text{Simplify the numerators and denominators.}$$

$$= \dfrac{-3}{30} \qquad \text{Subtract.}$$

$$= \dfrac{-1 \cdot 3}{10 \cdot 3} \qquad \text{Factor.}$$

$$= -\dfrac{1}{10} \qquad \text{Factor.}$$ ∎

helpful hint

Note how all of the operations with rational expressions are performed according to the rules for fractions that we studied in Chapter 1. So keep thinking of how you perform operations with fractions and you will improve your skills with fractions and with rational expressions.

Addition and Subtraction of Rational Expressions

Rational expressions are added or subtracted just like rational numbers. We can add or subtract only rational expressions that have identical denominators.

EXAMPLE 3 **Rational expressions with the same denominator**

Perform the indicated operations and reduce answers to lowest terms.

a) $\dfrac{2}{3y} + \dfrac{4}{3y}$ **b)** $\dfrac{2x}{x+2} + \dfrac{4}{x+2}$

c) $\dfrac{x^2 + 2x}{(x-1)(x+3)} - \dfrac{2x+1}{(x-1)(x+3)}$

Solution

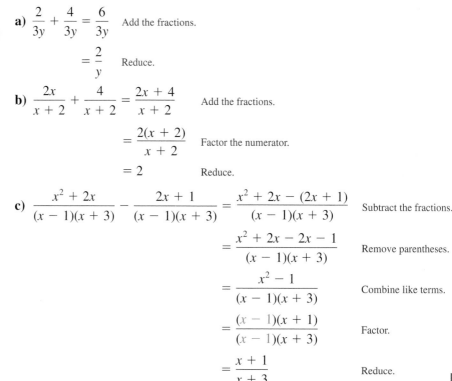

a) $\dfrac{2}{3y} + \dfrac{4}{3y} = \dfrac{6}{3y}$ Add the fractions.

$= \dfrac{2}{y}$ Reduce.

b) $\dfrac{2x}{x+2} + \dfrac{4}{x+2} = \dfrac{2x+4}{x+2}$ Add the fractions.

$= \dfrac{2(x+2)}{x+2}$ Factor the numerator.

$= 2$ Reduce.

c) $\dfrac{x^2+2x}{(x-1)(x+3)} - \dfrac{2x+1}{(x-1)(x+3)} = \dfrac{x^2+2x-(2x+1)}{(x-1)(x+3)}$ Subtract the fractions.

$= \dfrac{x^2+2x-2x-1}{(x-1)(x+3)}$ Remove parentheses.

$= \dfrac{x^2-1}{(x-1)(x+3)}$ Combine like terms.

$= \dfrac{(x-1)(x+1)}{(x-1)(x+3)}$ Factor.

$= \dfrac{x+1}{x+3}$ Reduce.   ■

CAUTION   When subtracting a numerator containing more than one term, be sure to enclose it in parentheses, as in Example 3(c). Because that numerator is a binomial, the sign of each of its terms must be changed for the subtraction.

In the next example the rational expressions have different denominators.

EXAMPLE 4 **Rational expressions with different denominators**

Perform the indicated operations.

a) $\dfrac{5}{2x} + \dfrac{2}{3}$ **b)** $\dfrac{4}{x^3 y} + \dfrac{2}{xy^3}$ **c)** $\dfrac{a+1}{6} - \dfrac{a-2}{8}$

Solution

a) The LCD for $2x$ and 3 is $6x$:

$$\frac{5}{2x} + \frac{2}{3} = \frac{5 \cdot 3}{2x \cdot 3} + \frac{2 \cdot 2x}{3 \cdot 2x} \qquad \text{Build up both denominators to } 6x.$$

$$= \frac{15}{6x} + \frac{4x}{6x} \qquad \text{Simplify numerators and denominators.}$$

$$= \frac{15 + 4x}{6x} \qquad \text{Add the rational expressions.}$$

b) The LCD is $x^3 y^3$.

$$\frac{4}{x^3 y} + \frac{2}{xy^3} = \frac{4 \cdot y^2}{x^3 y \cdot y^2} + \frac{2 \cdot x^2}{xy^3 \cdot x^2} \qquad \text{Build up both denominators to the LCD.}$$

$$= \frac{4y^2}{x^3 y^3} + \frac{2x^2}{x^3 y^3} \qquad \text{Simplify numerators and denominators.}$$

$$= \frac{4y^2 + 2x^2}{x^3 y^3} \qquad \text{Add the rational expressions.}$$

c) Because $6 = 2 \cdot 3$ and $8 = 2^3$, the LCD is $2^3 \cdot 3$, or 24:

$$\frac{a + 1}{6} - \frac{a - 2}{8} = \frac{(a + 1)4}{6 \cdot 4} - \frac{(a - 2)3}{8 \cdot 3} \qquad \begin{array}{l}\text{Build up both denominators to}\\ \text{the LCD 24.}\end{array}$$

$$= \frac{4a + 4}{24} - \frac{3a - 6}{24} \qquad \begin{array}{l}\text{Simplify numerators and}\\ \text{denominators.}\end{array}$$

$$= \frac{4a + 4 - (3a - 6)}{24} \qquad \text{Subtract the rational expressions.}$$

$$= \frac{4a + 4 - 3a + 6}{24} \qquad \text{Remove the parentheses.}$$

$$= \frac{a + 10}{24} \qquad \text{Combine like terms.}$$

E X A M P L E   5 **Rational expressions with different denominators**

Perform the indicated operations:

a) $\dfrac{1}{x^2 - 9} + \dfrac{2}{x^2 + 3x}$ **b)** $\dfrac{4}{5 - a} - \dfrac{2}{a - 5}$

Solution

a)

$$\frac{1}{x^2 - 9} + \frac{2}{x^2 + 3x} = \underbrace{\frac{1}{(x - 3)(x + 3)}}_{\text{Needs } x} + \underbrace{\frac{2}{x(x + 3)}}_{\text{Needs } x - 3} \qquad \begin{array}{l}\text{The LCD is}\\ x(x - 3)(x + 3).\end{array}$$

$$= \frac{1 \cdot x}{(x - 3)(x + 3)x} + \frac{2(x - 3)}{x(x + 3)(x - 3)}$$

$$= \frac{x}{x(x - 3)(x + 3)} + \frac{2x - 6}{x(x - 3)(x + 3)}$$

$$= \frac{3x - 6}{x(x - 3)(x + 3)} \qquad \begin{array}{l}\text{We usually leave the}\\ \text{denominator in}\\ \text{factored form.}\end{array}$$

b) Because $-1(5 - a) = a - 5$, we can get identical denominators by multiplying only the first expression by -1 in the numerator and denominator:

$$\frac{4}{5 - a} - \frac{2}{a - 5} = \frac{4(-1)}{(5 - a)(-1)} - \frac{2}{a - 5}$$

$$= \frac{-4}{a - 5} - \frac{2}{a - 5}$$

$$= \frac{-6}{a - 5} \qquad -4 - 2 = -6$$

$$= -\frac{6}{a - 5} \qquad\blacksquare$$

In the next example we combine three rational expressions by addition and subtraction.

EXAMPLE 6 **Rational expressions with different denominators**

Perform the indicated operations.

$$\frac{x + 1}{x^2 + 2x} + \frac{2x + 1}{6x + 12} - \frac{1}{6}$$

Solution

The LCD for $x(x + 2)$, $6(x + 2)$, and 6 is $6x(x + 2)$.

$$\frac{x + 1}{x^2 + 2x} + \frac{2x + 1}{6x + 12} - \frac{1}{6} = \frac{x + 1}{x(x + 2)} + \frac{2x + 1}{6(x + 2)} - \frac{1}{6} \qquad \text{Factor denominators.}$$

$$= \frac{6(x + 1)}{6x(x + 2)} + \frac{x(2x + 1)}{6x(x + 2)} - \frac{1x(x + 2)}{6x(x + 2)} \qquad \text{Build up to the LCD.}$$

$$= \frac{6x + 6}{6x(x + 2)} + \frac{2x^2 + x}{6x(x + 2)} - \frac{x^2 + 2x}{6x(x + 2)} \qquad \text{Simplify numerators.}$$

$$= \frac{6x + 6 + 2x^2 + x - x^2 - 2x}{6x(x + 2)} \qquad \text{Combine the numerators.}$$

$$= \frac{x^2 + 5x + 6}{6x(x + 2)} \qquad \text{Combine like terms.}$$

$$= \frac{(x + 3)(x + 2)}{6x(x + 2)} \qquad \text{Factor.}$$

$$= \frac{x + 3}{6x} \qquad \text{Reduce.} \qquad\blacksquare$$

Applications

Rational expressions occur in problems involving rates. In the next two examples we have situations where we must add rational expressions.

EXAMPLE 7 **Adding time**

Alice drove at a rate of x miles per hour for the first 20 miles and then increased her speed to $x + 10$ miles per hour for the next 40 miles. Write a rational expression for the total time of the trip.

Solution

Time is distance divided by rate. So her time for the first part of the trip was $20/x$ hours and her time for the last part of the trip was $40/(x + 10)$ hours. Find her total time as follows.

$$\frac{20}{x} + \frac{40}{x + 10} = \frac{20\,(x + 10)}{x(x + 10)} + \frac{40 \cdot x}{(x + 10)x} = \frac{60x + 200}{x(x + 10)}$$

So the total time is $\frac{60x + 200}{x(x + 10)}$ hours.   ■

E X A M P L E   8   ### Adding work

Harry takes twice as long as Lucy to proofread a manuscript. Write a rational expression for the amount of work they do in 3 hours working together on a manuscript.

Solution

Let $x =$ the number of hours it would take Lucy to complete the manuscript alone and $2x =$ the number of hours it would take Harry to complete the manuscript alone. Make a table showing rate, time, and work completed:

	Rate	Time	Work
Lucy	$\frac{1}{x}\frac{\text{msp}}{\text{hr}}$	3 hr	$\frac{3}{x}$ msp
Harry	$\frac{1}{2x}\frac{\text{msp}}{\text{hr}}$	3 hr	$\frac{3}{2x}$ msp

Now find the sum of each person's work.

$$\frac{3}{x} + \frac{3}{2x} = \frac{2 \cdot 3}{2 \cdot x} + \frac{3}{2x}$$

$$= \frac{6}{2x} + \frac{3}{2x}$$

$$= \frac{9}{2x}$$

So in 3 hours working together they will complete $\frac{9}{2x}$ of the manuscript.   ■

True or false? Explain your answer.

1. $\dfrac{1}{2} + \dfrac{1}{3} = \dfrac{2}{5}$ 2. $\dfrac{7}{12} - \dfrac{1}{12} = \dfrac{1}{2}$

3. $\dfrac{3}{5} + \dfrac{4}{3} = \dfrac{29}{15}$ 4. $\dfrac{4}{5} - \dfrac{5}{7} = \dfrac{3}{35}$

5. $\dfrac{5}{20} + \dfrac{3}{4} = 1$

6. $\dfrac{2}{x} + 1 = \dfrac{3}{x}$ for any nonzero value of x.

WARM-UPS

(continued)

7. $1 + \dfrac{1}{a} = \dfrac{a+1}{a}$ for any nonzero value of a.

8. $a - \dfrac{1}{4} = \dfrac{3}{4}a$ for any value of a.

9. $\dfrac{a}{2} + \dfrac{b}{3} = \dfrac{3a+2b}{6}$ for any values of a and b.

10. The LCD for the rational expressions $\dfrac{1}{x}$ and $\dfrac{3x}{x-1}$ is $x^2 - 1$.

7.4 EXERCISES

Reading and Writing *After reading this section, write out the answers to these questions. Use complete sentences.*

1. How do you add or subtract rational numbers?

2. How do you add or subtract rational expressions?

3. What is the least common denominator?

4. Why do we use the *least* common denominator when adding rational expressions?

Perform the indicated operation. Reduce each answer to lowest terms. See Example 1.

5. $\dfrac{1}{10} + \dfrac{1}{10}$

6. $\dfrac{1}{8} + \dfrac{3}{8}$

7. $\dfrac{7}{8} - \dfrac{1}{8}$

8. $\dfrac{4}{9} - \dfrac{1}{9}$

9. $\dfrac{1}{6} - \dfrac{5}{6}$

10. $-\dfrac{3}{8} - \dfrac{7}{8}$

11. $-\dfrac{7}{8} + \dfrac{1}{8}$

12. $-\dfrac{9}{20} + \left(-\dfrac{3}{20}\right)$

Perform the indicated operation. Reduce each answer to lowest terms. See Example 2.

13. $\dfrac{1}{3} + \dfrac{2}{9}$

14. $\dfrac{1}{4} + \dfrac{5}{6}$

15. $\dfrac{7}{16} + \dfrac{5}{18}$

16. $\dfrac{7}{6} + \dfrac{4}{15}$

17. $\dfrac{1}{8} - \dfrac{9}{10}$

18. $\dfrac{2}{15} - \dfrac{5}{12}$

19. $-\dfrac{1}{6} - \left(-\dfrac{3}{8}\right)$

20. $-\dfrac{1}{5} - \left(-\dfrac{1}{7}\right)$

Perform the indicated operation. Reduce each answer to lowest terms. See Example 3.

21. $\dfrac{3}{2w} + \dfrac{7}{2w}$

22. $\dfrac{5x}{3y} + \dfrac{7x}{3y}$

23. $\dfrac{3a}{a+5} + \dfrac{15}{a+5}$

24. $\dfrac{a+7}{a-4} + \dfrac{9-5a}{a-4}$

25. $\dfrac{q-1}{q-4} - \dfrac{3q-9}{q-4}$

26. $\dfrac{3-a}{3} - \dfrac{a-5}{3}$

27. $\dfrac{4h-3}{h(h+1)} - \dfrac{h-6}{h(h+1)}$

28. $\dfrac{2t-9}{t(t-3)} - \dfrac{t-9}{t(t-3)}$

29. $\dfrac{x^2-x-5}{(x+1)(x+2)} + \dfrac{1-2x}{(x+1)(x+2)}$

30. $\dfrac{2x-5}{(x-2)(x+6)} + \dfrac{x^2-2x+1}{(x-2)(x+6)}$

Perform the indicated operation. Reduce each answer to lowest terms. See Example 4.

31. $\dfrac{3}{2a} + \dfrac{1}{5a}$

32. $\dfrac{5}{6y} - \dfrac{3}{8y}$

33. $\dfrac{w-3}{9} - \dfrac{w-4}{12}$

34. $\dfrac{y+4}{10} - \dfrac{y-2}{14}$

35. $\dfrac{b^2}{4a} - c$

36. $y + \dfrac{3}{7b}$

37. $\dfrac{2}{wz^2} + \dfrac{3}{w^2z}$

38. $\dfrac{1}{a^5b} - \dfrac{5}{ab^3}$

Perform the indicated operation. Reduce each answer to lowest terms. See Examples 5 and 6.

39. $\dfrac{2}{x+1} - \dfrac{3}{x}$

40. $\dfrac{1}{a-1} - \dfrac{2}{a}$

41. $\dfrac{2}{a-b} + \dfrac{1}{a+b}$

42. $\dfrac{3}{x+1} + \dfrac{2}{x-1}$

43. $\dfrac{3}{x^2+x} - \dfrac{4}{5x+5}$

44. $\dfrac{3}{a^2+3a} - \dfrac{2}{5a+15}$

45. $\dfrac{2a}{a^2-9} + \dfrac{a}{a-3}$

46. $\dfrac{x}{x^2-1} + \dfrac{3}{x-1}$

47. $\dfrac{4}{a-b} + \dfrac{4}{b-a}$

48. $\dfrac{2}{x-3} + \dfrac{3}{3-x}$

49. $\dfrac{3}{2a-2} - \dfrac{2}{1-a}$

50. $\dfrac{5}{2x-4} - \dfrac{3}{2-x}$

51. $\dfrac{1}{x^2-4} - \dfrac{3}{x^2-3x-10}$

52. $\dfrac{2x}{x^2-9} + \dfrac{3x}{x^2+4x+3}$

53. $\dfrac{3}{x^2+x-2} + \dfrac{4}{x^2+2x-3}$

54. $\dfrac{x-1}{x^2-x-12} + \dfrac{x+4}{x^2+5x+6}$

55. $\dfrac{2}{x} - \dfrac{1}{x-1} + \dfrac{1}{x+2}$

56. $\dfrac{1}{a} - \dfrac{2}{a+1} + \dfrac{3}{a-1}$

57. $\dfrac{5}{3a-9} - \dfrac{3}{2a} + \dfrac{4}{a^2-3a}$

58. $\dfrac{3}{4c+2} - \dfrac{c-4}{2c^2+c} - \dfrac{5}{6c}$

Solve each problem. See Examples 7 and 8.

59. *Perimeter of a rectangle.* Suppose that the length of a rectangle is $\frac{3}{x}$ feet and its width is $\frac{5}{2x}$ feet. Find a rational expression for the perimeter of the rectangle.

60. *Perimeter of a triangle.* The lengths of the sides of a triangle are $\frac{1}{x}$, $\frac{1}{2x}$, and $\frac{2}{3x}$ meters. Find a rational expression for the perimeter of the triangle.

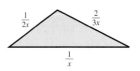

FIGURE FOR EXERCISE 60

61. *Traveling time.* Janet drove 120 miles at x mph before 6:00 A.M. After 6:00 A.M., she increased her speed by 5 mph and drove 195 additional miles. Use the fact that $T = \dfrac{D}{R}$ to complete the following table.

	Rate	Time	Distance
Before	$x \dfrac{\text{mi}}{\text{hr}}$		120 mi
After	$x+5 \dfrac{\text{mi}}{\text{hr}}$		195 mi

Write a rational expression for her total traveling time. Evaluate the expression for $x = 60$.

62. *Traveling time.* After leaving Moose Jaw, Hanson drove 200 kilometers at x km/hr and then decreased his speed by 20 km/hr and drove 240 additional kilometers. Make a table like the one in Exercise 61. Write a rational expression for his total traveling time. Evaluate the expression for $x = 100$.

63. *House painting.* Kent can paint a certain house by himself in x days. His helper Keith can paint the same house by himself in $x + 3$ days. Suppose that they work together on the job for 2 days. To complete the table, use the fact that the work completed is the product of the rate and the time. Write a rational expression for the

	Rate	Time	Work
Kent	$\dfrac{1}{x} \dfrac{\text{job}}{\text{day}}$	2 days	
Keith	$\dfrac{1}{x+3} \dfrac{\text{job}}{\text{day}}$	2 days	

fraction of the house that they complete by working together for 2 days. Evaluate the expression for $x = 6$.

64. *Barn painting.* Melanie can paint a certain barn by her-self in x days. Her helper Melissa can paint the same barn by herself in $2x$ days. Write a rational expression for the fraction of the barn that they complete in one day by working together. Evaluate the expression for $x = 5$.

GETTING MORE INVOLVED

65. *Writing.* Write a step-by-step procedure for adding rational expressions.

66. *Writing.* Explain why fractions must have the same denominator to be added. Use real-life examples.

FIGURE FOR EXERCISE 64

7.5 COMPLEX FRACTIONS

In this section we will use the idea of least common denominator to simplify complex fractions. Also we will see how complex fractions can arise in applications.

Complex Fractions

A **complex fraction** is a fraction having rational expressions in the numerator, denominator, or both. Consider the following complex fraction:

$$\frac{\dfrac{1}{2} + \dfrac{2}{3}}{\dfrac{1}{4} - \dfrac{5}{8}}$$

$\leftarrow$ Numerator of complex fraction

$\leftarrow$ Denominator of complex fraction

To simplify it, we can combine the fractions in the numerator as follows:

$$\frac{1}{2} + \frac{2}{3} = \frac{1 \cdot 3}{2 \cdot 3} + \frac{2 \cdot 2}{3 \cdot 2} = \frac{3}{6} + \frac{4}{6} = \frac{7}{6}$$

We can combine the fractions in the denominator as follows:

$$\frac{1}{4} - \frac{5}{8} = \frac{1 \cdot 2}{4 \cdot 2} - \frac{5}{8} = \frac{2}{8} - \frac{5}{8} = -\frac{3}{8}$$

Now divide the numerator by the denominator:

$$\frac{\dfrac{1}{2} + \dfrac{2}{3}}{\dfrac{1}{4} - \dfrac{5}{8}} = \frac{\dfrac{7}{6}}{-\dfrac{3}{8}} = \frac{7}{6} \div \left(-\frac{3}{8}\right)$$

$$= \frac{7}{6} \cdot \left(-\frac{8}{3}\right)$$

$$= -\frac{56}{18}$$

$$= -\frac{28}{9}$$

Using the LCD to Simplify Complex Fractions

A complex fraction can be simplified by writing the numerator and denominator as single fractions and then dividing, as we just did. However, there is a better method. The next example shows how to simplify a complex fraction by using the LCD of all of the single fractions in the complex fraction.

E X A M P L E   1   **Using the LCD to simplify a complex fraction**

Use the LCD to simplify

$$\frac{\dfrac{1}{2} + \dfrac{2}{3}}{\dfrac{1}{4} - \dfrac{5}{8}}.$$

Solution

The LCD of 2, 3, 4, and 8 is 24. Now multiply the numerator and denominator of the complex fraction by the LCD:

$$\frac{\dfrac{1}{2} + \dfrac{2}{3}}{\dfrac{1}{4} - \dfrac{5}{8}} = \frac{\left(\dfrac{1}{2} + \dfrac{2}{3}\right)24}{\left(\dfrac{1}{4} - \dfrac{5}{8}\right)24} \qquad \text{Multiply the numerator and denominator by the LCD.}$$

$$= \frac{\dfrac{1}{2} \cdot 24 + \dfrac{2}{3} \cdot 24}{\dfrac{1}{4} \cdot 24 - \dfrac{5}{8} \cdot 24} \qquad \text{Distributive property}$$

$$= \frac{12 + 16}{6 - 15} \qquad \text{Simplify.}$$

$$= \frac{28}{-9}$$

$$= -\frac{28}{9}$$

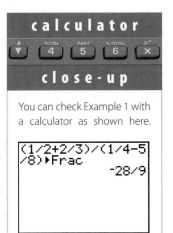

calculator

close-up

You can check Example 1 with a calculator as shown here.

(1/2+2/3)/(1/4-5
/8)▶Frac
                -28/9

CAUTION   We simplify a complex fraction by multiplying the numerator and denominator of the *complex fraction* by the LCD. Do not multiply the numerator and denominator of each fraction in the complex fraction by the LCD.

In the next example we simplify a complex fraction involving variables.

E X A M P L E   2   **A complex fraction with variables**

Simplify

$$\frac{2 - \dfrac{1}{x}}{\dfrac{1}{x^2} - \dfrac{1}{2}}.$$

Solution

The LCD of the denominators x, x^2, and 2 is $2x^2$:

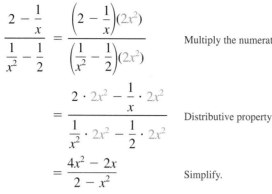

$$\frac{2 - \dfrac{1}{x}}{\dfrac{1}{x^2} - \dfrac{1}{2}} = \frac{\left(2 - \dfrac{1}{x}\right)(2x^2)}{\left(\dfrac{1}{x^2} - \dfrac{1}{2}\right)(2x^2)}$$ Multiply the numerator and denominator by $2x^2$.

$$= \frac{2 \cdot 2x^2 - \dfrac{1}{x} \cdot 2x^2}{\dfrac{1}{x^2} \cdot 2x^2 - \dfrac{1}{2} \cdot 2x^2}$$ Distributive property

$$= \frac{4x^2 - 2x}{2 - x^2}$$ Simplify.

The numerator of this answer can be factored, but the rational expression cannot be reduced. ◼

The general strategy for simplifying a complex fraction is stated as follows.

Strategy for Simplifying a Complex Fraction

1. Find the LCD for all the denominators in the complex fraction.
2. Multiply both the numerator and the denominator of the complex fraction by the LCD. Use the distributive property if necessary.
3. Combine like terms if possible.
4. Reduce to lowest terms when possible.

E X A M P L E   3

Simplifying a complex fraction

Simplify

$$\frac{\dfrac{1}{x - 2} - \dfrac{2}{x + 2}}{\dfrac{3}{2 - x} + \dfrac{4}{x + 2}}.$$

Solution

Because $x - 2$ and $2 - x$ are opposites, we can use $(x - 2)(x + 2)$ as the LCD. Multiply the numerator and denominator by $(x - 2)(x + 2)$:

$$\frac{\dfrac{1}{x - 2} - \dfrac{2}{x + 2}}{\dfrac{3}{2 - x} + \dfrac{4}{x + 2}} = \frac{\dfrac{1}{x - 2}(x - 2)(x + 2) - \dfrac{2}{x + 2}(x - 2)(x + 2)}{\dfrac{3}{2 - x}(x - 2)(x + 2) + \dfrac{4}{x + 2}(x - 2)(x + 2)}$$

$$= \frac{x + 2 - 2(x - 2)}{3(-1)(x + 2) + 4(x - 2)} \qquad \frac{x - 2}{2 - x} = -1$$

$$= \frac{x + 2 - 2x + 4}{-3x - 6 + 4x - 8}$$ Distributive property

$$= \frac{-x + 6}{x - 14}$$ Combine like terms. ◼

Applications

As their name suggests, complex fractions arise in some fairly complex situations.

EXAMPLE 4

Fast-food workers

A survey of college students found that $\frac{1}{2}$ of the female students had jobs and $\frac{2}{3}$ of the male students had jobs. It was also found that $\frac{1}{4}$ of the female students worked in fast-food restaurants and $\frac{1}{6}$ of the male students worked in fast-food restaurants. If equal numbers of male and female students were surveyed, then what fraction of the working students worked in fast-food restaurants?

Solution

Let x represent the number of males surveyed. The number of females surveyed is also x. The total number of students working in fast-food restaurants is

$$\frac{1}{4}x + \frac{1}{6}x.$$

The total number of working students in the survey is

$$\frac{1}{2}x + \frac{2}{3}x.$$

So the fraction of working students who work in fast-food restaurants is

$$\frac{\dfrac{1}{4}x + \dfrac{1}{6}x}{\dfrac{1}{2}x + \dfrac{2}{3}x}.$$

The LCD of the denominators 2, 3, 4, and 6 is 12. Multiply the numerator and denominator by 12 to eliminate the fractions as follows:

$$\frac{\dfrac{1}{4}x + \dfrac{1}{6}x}{\dfrac{1}{2}x + \dfrac{2}{3}x} = \frac{\left(\dfrac{1}{4}x + \dfrac{1}{6}x\right)12}{\left(\dfrac{1}{2}x + \dfrac{2}{3}x\right)12} \qquad \text{Multiply numerator and denominator by 12.}$$

$$= \frac{3x + 2x}{6x + 8x} \qquad \text{Distributive property}$$

$$= \frac{5x}{14x} \qquad \text{Combine like terms.}$$

$$= \frac{5}{14} \qquad \text{Reduce.}$$

So $\frac{5}{14}$ (or about 36%) of the working students work in fast-food restaurants. ■

True or false? Explain your answer.

1. The LCD for the denominators 4, x, 6, and x^2 is $12x^3$.
2. The LCD for the denominators $a - b$, $2b - 2a$, and 6 is $6a - 6b$.
3. The fraction $\frac{4117}{7983}$ is a complex fraction.
4. The LCD for the denominators $a - 3$ and $3 - a$ is $a^2 - 9$.
5. The largest common denominator for the fractions $\frac{1}{2}$, $\frac{1}{3}$, and $\frac{1}{4}$ is 24.

WARM-UPS

(continued)

Questions 6–10 refer to the following complex fractions:

a) $\dfrac{\dfrac{1}{2} + \dfrac{x}{3}}{\dfrac{1}{4} + \dfrac{1}{5}}$
 b) $\dfrac{1 + \dfrac{2}{b}}{\dfrac{2}{a} + 5}$

c) $\dfrac{x - \dfrac{1}{2}}{x + \dfrac{3}{2}}$
 d) $\dfrac{\dfrac{1}{2} + \dfrac{1}{3}}{1 + \dfrac{1}{2}}$

6. To simplify (a), we multiply the numerator and denominator by $60x$.

7. To simplify (b), we multiply the numerator and denominator by $\frac{ab}{ab}$.

8. The complex fraction (c) is equivalent to $\frac{2x - 1}{2x + 3}$.

9. If $x \neq -\frac{3}{2}$, then (c) represents a real number.

10. The complex fraction (d) can be written as $\frac{5}{6} \div \frac{3}{2}$.

7.5 EXERCISES

Reading and Writing *After reading this section, write out the answers to these questions. Use complete sentences.*

1. What is a complex fraction?

2. What are the two ways to simplify a complex fraction?

Simplify each complex fraction. See Example 1.

3. $\dfrac{\dfrac{1}{2} + \dfrac{1}{3}}{\dfrac{1}{4} - \dfrac{1}{2}}$
 4. $\dfrac{\dfrac{1}{3} - \dfrac{1}{4}}{\dfrac{1}{3} + \dfrac{1}{6}}$

5. $\dfrac{\dfrac{2}{5} + \dfrac{5}{6} - \dfrac{1}{2}}{\dfrac{1}{2} - \dfrac{1}{3} + \dfrac{1}{15}}$
 6. $\dfrac{\dfrac{2}{5} - \dfrac{2}{9} - \dfrac{1}{3}}{\dfrac{1}{3} + \dfrac{1}{5} + \dfrac{2}{15}}$

7. $\dfrac{3 + \dfrac{1}{2}}{5 - \dfrac{3}{4}}$
 8. $\dfrac{1 + \dfrac{1}{12}}{1 - \dfrac{1}{12}}$

9. $\dfrac{1 - \dfrac{1}{6} + \dfrac{2}{3}}{1 + \dfrac{1}{15} - \dfrac{3}{10}}$
 10. $\dfrac{3 - \dfrac{2}{9} - \dfrac{1}{6}}{\dfrac{5}{18} - \dfrac{1}{3} - 2}$

Simplify each complex fraction. See Example 2.

11. $\dfrac{\dfrac{1}{a} + \dfrac{3}{b}}{\dfrac{1}{b} - \dfrac{3}{a}}$
 12. $\dfrac{\dfrac{1}{x} - \dfrac{3}{2}}{\dfrac{3}{4} + \dfrac{1}{x}}$

13. $\dfrac{5 - \dfrac{3}{a}}{3 + \dfrac{1}{a}}$
 14. $\dfrac{4 + \dfrac{3}{y}}{1 - \dfrac{2}{y}}$

15. $\dfrac{\dfrac{1}{2} - \dfrac{2}{x}}{3 - \dfrac{1}{x^2}}$
 16. $\dfrac{\dfrac{2}{a} + \dfrac{5}{3}}{\dfrac{3}{a} - \dfrac{3}{a^2}}$

17. $\dfrac{\dfrac{3}{2b} + \dfrac{1}{b}}{\dfrac{3}{4} - \dfrac{1}{b^2}}$
 18. $\dfrac{\dfrac{3}{2w} + \dfrac{4}{3w}}{\dfrac{1}{4w} - \dfrac{5}{9w}}$

Simplify each complex fraction. See Example 3.

19. $\dfrac{1 - \dfrac{3}{y + 1}}{3 + \dfrac{1}{y + 1}}$
 20. $\dfrac{2 - \dfrac{1}{a - 3}}{3 - \dfrac{1}{a - 3}}$

21. $\dfrac{x + \dfrac{4}{x-2}}{x - \dfrac{x+1}{x-2}}$

22. $\dfrac{x - \dfrac{x-6}{x-1}}{x - \dfrac{x+15}{x-1}}$

23. $\dfrac{\dfrac{1}{3-x} - 5}{\dfrac{1}{x-3} - 2}$

24. $\dfrac{\dfrac{2}{x-5} - x}{\dfrac{3x}{5-x} - 1}$

25. $\dfrac{1 - \dfrac{5}{a-1}}{3 - \dfrac{2}{1-a}}$

26. $\dfrac{\dfrac{1}{3} - \dfrac{2}{9-x}}{\dfrac{1}{6} - \dfrac{1}{x-9}}$

27. $\dfrac{\dfrac{1}{m-3} - \dfrac{4}{m}}{\dfrac{3}{m-3} + \dfrac{1}{m}}$

28. $\dfrac{\dfrac{1}{y+3} - \dfrac{4}{y}}{\dfrac{1}{y} - \dfrac{2}{y+3}}$

29. $\dfrac{\dfrac{2}{w-1} - \dfrac{3}{w+1}}{\dfrac{4}{w+1} + \dfrac{5}{w-1}}$

30. $\dfrac{\dfrac{1}{x+2} - \dfrac{3}{x+3}}{\dfrac{2}{x+3} + \dfrac{3}{x+2}}$

31. $\dfrac{\dfrac{1}{a-b} - \dfrac{1}{a+b}}{\dfrac{1}{b-a} + \dfrac{1}{b+a}}$

32. $\dfrac{\dfrac{1}{2+x} - \dfrac{1}{2-x}}{\dfrac{1}{x+2} - \dfrac{1}{x-2}}$

Simplify each complex fraction.

33. $\dfrac{\dfrac{2x-9}{6}}{\dfrac{2x-3}{9}}$

34. $\dfrac{\dfrac{a-5}{12}}{\dfrac{a+2}{15}}$

35. $\dfrac{\dfrac{2x-4y}{xy^2}}{\dfrac{3x-6y}{x^3y}}$

36. $\dfrac{\dfrac{ab+b^2}{4ab^5}}{\dfrac{a+b}{6a^2b^4}}$

37. $\dfrac{\dfrac{a^2+2a-24}{a+1}}{\dfrac{a^2-a-12}{(a+1)^2}}$

38. $\dfrac{\dfrac{y^2-3y-18}{y^2-4}}{\dfrac{y^2+5y+6}{y-2}}$

39. $\dfrac{\dfrac{x}{x+1}}{\dfrac{1}{x^2-1} - \dfrac{1}{x-1}}$

40. $\dfrac{\dfrac{a}{a^2-b^2}}{\dfrac{1}{a+b} + \dfrac{1}{a-b}}$

Solve each problem. See Example 4.

41. *Sophomore math.* A survey of college sophomores showed that $\frac{5}{6}$ of the males were taking a mathematics class and $\frac{3}{4}$ of the females were taking a mathematics class. One-third of the males were enrolled in calculus, and $\frac{1}{5}$ of the females were enrolled in calculus. If just as many males as females were surveyed, then what fraction of the surveyed students taking mathematics were enrolled in calculus? Rework this problem assuming that the number of females in the survey was twice the number of males.

42. *Commuting students.* At a well-known university, $\frac{1}{4}$ of the undergraduate students commute, and $\frac{1}{3}$ of the graduate students commute. One-tenth of the undergraduate students drive more than 40 miles daily, and $\frac{1}{6}$ of the graduate students drive more than 40 miles daily. If there are twice as many undergraduate students as there are graduate students, then what fraction of the commuters drive more than 40 miles daily?

FIGURE FOR EXERCISE 42

GETTING MORE INVOLVED

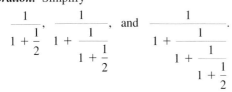 **43. *Exploration.*** Simplify

$$\frac{1}{1 + \dfrac{1}{2}}, \quad \frac{1}{1 + \dfrac{1}{1 + \dfrac{1}{2}}}, \quad \text{and} \quad \frac{1}{1 + \dfrac{1}{1 + \dfrac{1}{1 + \dfrac{1}{2}}}}.$$

a) Are these fractions getting larger or smaller as the fractions become more complex?

b) Continuing the pattern, find the next two complex fractions and simplify them.

c) Now what can you say about the values of all five complex fractions?

44. *Discussion.* A complex fraction can be simplified by writing the numerator and denominator as single fractions and then dividing them or by multiplying the numerator and denominator by the LCD. Simplify the complex fraction

$$\frac{\dfrac{4}{xy^2} - \dfrac{6}{xy}}{\dfrac{2}{x^2} + \dfrac{4}{x^2 y}}$$

by using each of these methods. Compare the number of steps used in each method, and determine which method requires fewer steps.

7.6 SOLVING EQUATIONS WITH RATIONAL EXPRESSIONS

Many problems in algebra can be solved by using equations involving rational expressions. In this section you will learn how to solve equations that involve rational expressions, and in Sections 7.7 and 7.8 you will solve problems using these equations.

Equations with Rational Expressions

We solved some equations involving fractions in Section 2.3. In that section the equations had only integers in the denominators. Our first step in solving those equations was to multiply by the LCD to eliminate all of the denominators.

EXAMPLE 1

Integers in the denominators

Solve $\frac{1}{2} - \frac{x-2}{3} = \frac{1}{6}$.

Solution

The LCD for 2, 3, and 6 is 6. Multiply each side of the equation by 6:

$$\frac{1}{2} - \frac{x-2}{3} = \frac{1}{6} \qquad \text{Original equation}$$

$$6\left(\frac{1}{2} - \frac{x-2}{3}\right) = 6 \cdot \frac{1}{6} \qquad \text{Multiply each side by 6.}$$

$$6 \cdot \frac{1}{2} - \overset{2}{\cancel{6}} \cdot \frac{x-2}{\cancel{3}} = \cancel{6} \cdot \frac{1}{\cancel{6}} \qquad \text{Distributive property}$$

$$3 - 2(x - 2) = 1 \qquad \text{Simplify. Enclose } x - 2 \text{ in parentheses.}$$

$$3 - 2x + 4 = 1 \qquad \text{Distributive property}$$

$$-2x = -6 \qquad \text{Subtract 7 from each side.}$$

$$x = 3 \qquad \text{Divide each side by } -2.$$

helpful hint

Note that it is not necessary to convert each fraction into an equivalent fraction with a common denominator here. Since we can multiply both sides of an equation by any expression we choose, we choose to multiply by the LCD. This tactic eliminates the fractions in one step.

Check $x = 3$ in the original equation:

$$\frac{1}{2} - \frac{3 - 2}{3} = \frac{1}{2} - \frac{1}{3} = \frac{3}{6} - \frac{2}{6} = \frac{1}{6}$$

The solution to the equation is 3.

C A U T I O N   When a numerator contains a binomial, as in Example 1, the numerator must be enclosed in parentheses when the denominator is eliminated.

To solve an equation involving rational expressions, we usually multiply each side of the equation by the LCD for all the denominators involved, just as we do for an equation with fractions.

E X A M P L E   2   **Variables in the denominators**

Solve $\frac{1}{x} + \frac{1}{6} = \frac{1}{4}$.

Solution

We multiply each side of the equation by $12x$, the LCD for 4, 6, and x:

$$\frac{1}{x} + \frac{1}{6} = \frac{1}{4} \qquad \text{Original equation}$$

$$12x\left(\frac{1}{x} + \frac{1}{6}\right) = 12x\left(\frac{1}{4}\right) \qquad \text{Multiply each side by } 12x.$$

$$12x \cdot \frac{1}{x} + \overset{2}{12x} \cdot \frac{1}{6} = \overset{3}{12x} \cdot \frac{1}{4} \qquad \text{Distributive property}$$

$$12 + 2x = 3x \qquad \text{Simplify.}$$

$$12 = x \qquad \text{Subtract } 2x \text{ from each side.}$$

Check that 12 satisfies the original equation:

$$\frac{1}{12} + \frac{1}{6} = \frac{1}{12} + \frac{2}{12} = \frac{3}{12} = \frac{1}{4}$$

E X A M P L E   3   **An equation with two solutions**

Solve the equation $\frac{100}{x} + \frac{100}{x + 5} = 9$.

Solution

The LCD for the denominators x and $x + 5$ is $x(x + 5)$:

$$\frac{100}{x} + \frac{100}{x + 5} = 9 \qquad \text{Original equation}$$

$$x(x + 5)\frac{100}{x} + x(x + 5)\frac{100}{x + 5} = x(x + 5)9 \qquad \begin{array}{l}\text{Multiply each side by}\\ x(x + 5).\end{array}$$

$$(x + 5)100 + x(100) = (x^2 + 5x)9 \qquad \begin{array}{l}\text{All denominators are}\\ \text{eliminated.}\end{array}$$

$$100x + 500 + 100x = 9x^2 + 45x \qquad \text{Simplify.}$$

$$500 + 200x = 9x^2 + 45x$$

$$0 = 9x^2 - 155x - 500 \qquad \text{Get 0 on one side.}$$

$$0 = (9x + 25)(x - 20) \qquad \text{Factor.}$$

$$9x + 25 = 0 \qquad \text{or} \qquad x - 20 = 0 \qquad \text{Zero factor property}$$

$$x = -\frac{25}{9} \qquad \text{or} \qquad x = 20$$

A check will show that both $-\frac{25}{9}$ and 20 satisfy the original equation.

Extraneous Solutions

In a rational expression we can replace the variable only by real numbers that do not cause the denominator to be 0. When solving equations involving rational expressions, we must check every solution to see whether it causes 0 to appear in a denominator. If a number causes the denominator to be 0, then it cannot be a solution to the equation. A number that appears to be a solution but causes 0 in a denominator is called an **extraneous solution.**

E X A M P L E   4

An equation with an extraneous solution

Solve the equation $\frac{1}{x-2} = \frac{x}{2x-4} + 1$.

Solution

Because the denominator $2x - 4$ factors as $2(x - 2)$, the LCD is $2(x - 2)$.

$$2(x - 2)\frac{1}{x - 2} = 2(x - 2)\frac{x}{2(x - 2)} + 2(x - 2) \cdot 1 \qquad \text{Multiply each side of the original equation by } 2(x - 2).$$

$$2 = x + 2x - 4 \qquad\qquad\qquad \text{Simplify.}$$

$$2 = 3x - 4$$

$$6 = 3x$$

$$2 = x$$

Check 2 in the original equation:

$$\frac{1}{2-2} = \frac{2}{2 \cdot 2 - 4} + 1$$

The denominator $2 - 2$ is 0. So 2 does not satisfy the equation, and it is an extraneous solution. The equation has no solutions. ■

E X A M P L E   5

Another extraneous solution

Solve the equation $\frac{1}{x} + \frac{1}{x-3} = \frac{x-2}{x-3}$.

Solution

The LCD for the denominators x and $x - 3$ is $x(x - 3)$:

$$\frac{1}{x} + \frac{1}{x - 3} = \frac{x - 2}{x - 3} \qquad\qquad \text{Original equation}$$

$$x(x - 3) \cdot \frac{1}{x} + x(x - 3) \cdot \frac{1}{x - 3} = x(x - 3) \cdot \frac{x - 2}{x - 3} \qquad \text{Multiply each side by } x(x - 3).$$

$$x - 3 + x = x(x - 2)$$

$$2x - 3 = x^2 - 2x$$

$$0 = x^2 - 4x + 3$$

$$0 = (x - 3)(x - 1)$$

$$x - 3 = 0 \qquad \text{or} \qquad x - 1 = 0$$

$$x = 3 \qquad \text{or} \qquad x = 1$$

If $x = 3$, then the denominator $x - 3$ has a value of 0. If $x = 1$, the original equation is satisfied. The only solution to the equation is 1. ■

CAUTION   Always be sure to check your answers in the original equation to determine whether they are extraneous solutions.

WARM-UPS

True or false? Explain your answers.

1. The LCD is not used in solving equations with rational expressions.
2. To solve the equation $x^2 = 8x$, we divide each side by x.
3. An extraneous solution is an irrational number.

Use the following equations for Questions 4–10.

a) $\dfrac{3}{x} + \dfrac{5}{x-2} = \dfrac{2}{3}$ b) $\dfrac{1}{x} + \dfrac{1}{2} = \dfrac{3}{4}$ c) $\dfrac{1}{x-1} + 2 = \dfrac{1}{x+1}$

4. To solve Eq. (a), we must add the expressions on the left-hand side.
5. Both 0 and 2 satisfy Eq. (a).
6. To solve Eq. (a), we multiply each side by $3x^2 - 6x$.
7. The only solution to Eq. (b) is 4.
8. Equation (b) is equivalent to $4 + 2x = 3x$.
9. To solve Eq. (c), we multiply each side by $x^2 - 1$.
10. The numbers 1 and -1 do not satisfy Eq. (c).

7.6   EXERCISES

Reading and Writing   *After reading this section, write out the answers to these questions. Use complete sentences.*

1. What is the typical first step for solving an equation involving rational expressions?

2. What is the difference in procedure for solving an equation involving rational expressions and adding rational expressions?

3. What is an extraneous solution?

4. Why do extraneous solutions sometimes occur for equations with rational expressions?

Solve each equation. See Example 1.

5. $\dfrac{x}{3} - 5 = \dfrac{x}{2} - 7$

6. $\dfrac{x}{3} - \dfrac{x}{2} = \dfrac{x}{5} - 11$

7. $\dfrac{y}{5} - \dfrac{2}{3} = \dfrac{y}{6} + \dfrac{1}{3}$

8. $\dfrac{z}{6} + \dfrac{5}{4} = \dfrac{z}{2} - \dfrac{3}{4}$

9. $\dfrac{3}{4} - \dfrac{t-4}{3} = \dfrac{t}{12}$

10. $\dfrac{4}{5} - \dfrac{v-1}{10} = \dfrac{v-5}{30}$

11. $\dfrac{1}{5} - \dfrac{w+10}{15} = \dfrac{1}{10} - \dfrac{w+1}{6}$

12. $\dfrac{q}{5} - \dfrac{q-1}{2} = \dfrac{13}{20} - \dfrac{q+1}{4}$

Solve each equation. See Example 2.

13. $\dfrac{1}{x} + \dfrac{1}{2} = \dfrac{3}{4}$

14. $\dfrac{3}{x} + \dfrac{1}{4} = \dfrac{5}{8}$

15. $\dfrac{2}{3x} + \dfrac{1}{2x} = \dfrac{7}{24}$

16. $\dfrac{1}{6x} - \dfrac{1}{8x} = \dfrac{1}{72}$

17. $\dfrac{1}{2} + \dfrac{a-2}{a} = \dfrac{a+2}{2a}$

18. $\dfrac{1}{b} + \dfrac{1}{5} = \dfrac{b-1}{5b} + \dfrac{3}{10}$

19. $\dfrac{1}{3} - \dfrac{k+3}{6k} = \dfrac{1}{3k} - \dfrac{k-1}{2k}$

20. $\dfrac{3}{p} - \dfrac{p+3}{3p} = \dfrac{2p-1}{2p} - \dfrac{5}{6}$

Solve each equation. See Example 3.

21. $\dfrac{x}{2} = \dfrac{5}{x + 3}$

22. $\dfrac{x}{3} = \dfrac{4}{x + 1}$

23. $\dfrac{2}{x + 1} = \dfrac{1}{x} + \dfrac{1}{6}$

24. $\dfrac{1}{w + 1} - \dfrac{1}{2w} = \dfrac{3}{40}$

25. $\dfrac{a - 1}{a^2 - 4} + \dfrac{1}{a - 2} = \dfrac{a + 4}{a + 2}$

26. $\dfrac{b + 17}{b^2 - 1} - \dfrac{1}{b + 1} = \dfrac{b - 2}{b - 1}$

Solve each equation. Watch for extraneous solutions. See Examples 4 and 5.

27. $\dfrac{1}{x - 1} + \dfrac{2}{x} = \dfrac{x}{x - 1}$

28. $\dfrac{4}{x} + \dfrac{3}{x - 3} = \dfrac{x}{x - 3} - \dfrac{1}{3}$

29. $\dfrac{5}{x + 2} + \dfrac{2}{x - 3} = \dfrac{x - 1}{x - 3}$

30. $\dfrac{6}{y - 2} + \dfrac{7}{y - 8} = \dfrac{y - 1}{y - 8}$

31. $1 + \dfrac{3y}{y - 2} = \dfrac{6}{y - 2}$

32. $\dfrac{5}{y - 3} = \dfrac{y + 7}{2y - 6} + 1$

33. $\dfrac{z}{z + 1} - \dfrac{1}{z + 2} = \dfrac{2z + 5}{z^2 + 3z + 2}$

34. $\dfrac{z}{z - 2} - \dfrac{1}{z + 5} = \dfrac{7}{z^2 + 3z - 10}$

In Exercises 35–56, solve each equation.

35. $\dfrac{a}{4} = \dfrac{5}{2}$

36. $\dfrac{y}{3} = \dfrac{6}{5}$

37. $\dfrac{w}{6} = \dfrac{3w}{11}$

38. $\dfrac{2m}{3} = \dfrac{3m}{2}$

39. $\dfrac{5}{x} = \dfrac{x}{5}$

40. $\dfrac{-3}{x} = \dfrac{x}{-3}$

41. $\dfrac{x - 3}{5} = \dfrac{x - 3}{x}$

42. $\dfrac{a + 4}{2} = \dfrac{a + 4}{a}$

43. $\dfrac{1}{x + 2} = \dfrac{x}{x + 2}$

44. $\dfrac{-3}{w + 2} = \dfrac{w}{w + 2}$

45. $\dfrac{1}{2x - 4} + \dfrac{1}{x - 2} = \dfrac{3}{2}$

46. $\dfrac{7}{3x - 9} - \dfrac{1}{x - 3} = \dfrac{4}{3}$

47. $\dfrac{3}{a^2 - a - 6} = \dfrac{2}{a^2 - 4}$

48. $\dfrac{8}{a^2 + a - 6} = \dfrac{6}{a^2 - 9}$

49. $\dfrac{4}{c - 2} - \dfrac{1}{2 - c} = \dfrac{25}{c + 6}$

50. $\dfrac{3}{x + 1} - \dfrac{1}{1 - x} = \dfrac{10}{x^2 - 1}$

51. $\dfrac{1}{x^2 - 9} + \dfrac{3}{x + 3} = \dfrac{4}{x - 3}$

52. $\dfrac{3}{x - 2} - \dfrac{5}{x + 3} = \dfrac{1}{x^2 + x - 6}$

53. $\dfrac{3}{2x + 4} - \dfrac{1}{x + 2} = \dfrac{1}{3x + 1}$

54. $\dfrac{5}{2m + 6} - \dfrac{1}{m + 1} = \dfrac{1}{m + 3}$

55. $\dfrac{2t - 1}{3t + 3} + \dfrac{3t - 1}{6t + 6} = \dfrac{t}{t + 1}$

56. $\dfrac{4w - 1}{3w + 6} - \dfrac{w - 1}{3} = \dfrac{w - 1}{w + 2}$

Solve each problem.

57. **Lens equation.** The focal length f for a camera lens is related to the object distance o and the image distance i by the formula

$$\frac{1}{f} = \frac{1}{o} + \frac{1}{i}.$$

See the accompanying figure. The image is in focus at distance i from the lens. For an object that is 600 mm from a 50-mm lens, use $f = 50$ mm and $o = 600$ mm to find i.

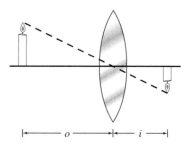

FIGURE FOR EXERCISE 57

58. **Telephoto lens.** Use the formula from Exercise 57 to find the image distance i for an object that is 2,000,000 mm from a 250-mm telephoto lens.

FIGURE FOR EXERCISE 58

7.7   APPLICATIONS OF RATIOS AND PROPORTIONS

In this section we will use the ideas of rational expressions in ratio and proportion problems. We will solve proportions in the same way we solved equations in Section 7.6.

Ratios

In Chapter 1 we defined a rational number as the *ratio of two integers*. We will now give a more general definition of ratio. If a and b are any real numbers (not just integers), with $b \neq 0$, then the expression $\frac{a}{b}$ is called the **ratio of a and b** or the **ratio of a to b.** The ratio of a to b is also written as $a:b$. A ratio is a comparison of two numbers. Some examples of ratios are

$$\frac{3}{4}, \quad \frac{4.2}{2.1}, \quad \frac{\frac{1}{4}}{\frac{1}{2}}, \quad \frac{3.6}{5}, \quad \text{and} \quad \frac{100}{1}.$$

Ratios are treated just like fractions. We can reduce ratios, and we can build them up. We generally express ratios as ratios of integers. When possible, we will convert a ratio into an equivalent ratio of integers in lowest terms.

E X A M P L E   1   **Finding equivalent ratios**

Find an equivalent ratio of integers in lowest terms for each ratio.

a) $\dfrac{4.2}{2.1}$ **b)** $\dfrac{\frac{1}{4}}{\frac{1}{2}}$ **c)** $\dfrac{3.6}{5}$

Solution

a) Because both the numerator and the denominator have one decimal place, we will multiply the numerator and denominator by 10 to eliminate the decimals:

$$\frac{4.2}{2.1} = \frac{4.2(10)}{2.1(10)} = \frac{42}{21} = \frac{21 \cdot 2}{21 \cdot 1} = \frac{2}{1} \quad \text{Do not omit the 1 in a ratio.}$$

So the ratio of 4.2 to 2.1 is equivalent to the ratio 2 to 1.

b) This ratio is a complex fraction. We can simplify this expression using the LCD method as shown in Section 7.5. Multiply the numerator and denominator of this ratio by 4:

$$\frac{\frac{1}{4}}{\frac{1}{2}} = \frac{\frac{1}{4} \cdot 4}{\frac{1}{2} \cdot 4} = \frac{1}{2}$$

c) We can get a ratio of integers if we multiply the numerator and denominator by 10.

$$\frac{3.6}{5} = \frac{3.6(10)}{5(10)} = \frac{36}{50}$$

$$= \frac{18}{25} \qquad \text{Reduce to lowest terms.}$$

In the next example a ratio is used to compare quantities.

E X A M P L E   2   **Nitrogen to potash**

In a 50-pound bag of lawn fertilizer there are 8 pounds of nitrogen and 12 pounds of potash. What is the ratio of nitrogen to potash?

Solution

The nitrogen and potash occur in this fertilizer in the ratio of 8 pounds to 12 pounds:

$$\frac{8}{12} = \frac{2 \cdot 4}{3 \cdot 4} = \frac{2}{3}$$

So the ratio of nitrogen to potash is 2 to 3.   ■

E X A M P L E   3   **Males to females**

In a class of 50 students, there were exactly 20 male students. What was the ratio of males to females in this class?

Solution

Because there were 20 males in the class of 50, there were 30 females. The ratio of males to females was 20 to 30, or 2 to 3.   ■

Ratios give us a means of comparing the size of two quantities. For this reason *the numbers compared in a ratio should be expressed in the same units.* For example, if one dog is 24 inches high and another is 1 foot high, then the ratio of their heights is 2 to 1, not 24 to 1.

E X A M P L E   4   **Quantities with different units**

What is the ratio of length to width for a poster with a length of 30 inches and a width of 2 feet?

Solution

Because the width is 2 feet, or 24 inches, the ratio of length to width is 30 to 24. Reduce as follows:

$$\frac{30}{24} = \frac{5 \cdot 6}{4 \cdot 6} = \frac{5}{4}$$

So the ratio of length to width is 5 to 4.   ■

Proportions

A **proportion** is any statement expressing the equality of two ratios. The statement

$$\frac{a}{b} = \frac{c}{d} \qquad \text{or} \qquad a:b = c:d$$

is a proportion. In any proportion the numbers in the positions of a and d above are called the **extremes.** The numbers in the positions of b and c above are called the **means.** In the proportion

$$\frac{30}{24} = \frac{5}{4},$$

the means are 24 and 5, and the extremes are 30 and 4. Note that $30 \cdot 4 = 5 \cdot 24$.

If we multiply each side of the proportion

$$\frac{a}{b} = \frac{c}{d}$$

by the LCD, bd, we get

$$\frac{a}{b} \cdot bd = \frac{c}{d} \cdot bd$$

or

$$a \cdot d = b \cdot c.$$

We can express this result by saying that *the product of the extremes is equal to the product of the means.* We call this fact the **extremes-means property** or **cross-multiplying.**

Extremes-Means Property (Cross-Multiplying)

Suppose a, b, c, and d are real numbers with $b \neq 0$ and $d \neq 0$. If

$$\frac{a}{b} = \frac{c}{d}, \text{ then } ad = bc.$$

We use the extremes-means property to solve proportions.

E X A M P L E   5

Using the extremes-means property

Solve the proportion $\frac{3}{x} = \frac{5}{x+5}$ for x.

Solution

Instead of multiplying each side by the LCD, we use the extremes-means property:

$$\frac{3}{x} = \frac{5}{x+5} \qquad \text{Original proportion}$$

$$3(x + 5) = 5x \qquad \text{Extremes-means property}$$

$$3x + 15 = 5x \qquad \text{Distributive property}$$

$$15 = 2x$$

$$\frac{15}{2} = x$$

> **helpful hint**
>
> The extremes-means property or cross-multiplying is nothing new. You can accomplish the same thing by multiplying each side of the equation by the LCD.

Check:

$$\frac{3}{\dfrac{15}{2}} = 3 \cdot \frac{2}{15} = \frac{2}{5}$$

$$\frac{5}{\dfrac{15}{2} + 5} = \frac{5}{\dfrac{25}{2}} = 5 \cdot \frac{2}{25} = \frac{2}{5}$$

So $\frac{15}{2}$ is the solution to the equation or the solution to the proportion. ■

E X A M P L E   6   **The capture-recapture proportion**

To estimate the number of catfish in her pond, a catfish farmer caught, tagged, and released 30 of them. Later, only 3 tagged catfish were found in a sample of 500. Estimate the number of catfish in the pond.

Solution

Let x be the number of catfish in the pond. The ratio $\frac{30}{x}$ is the ratio of tagged catfish to the total population. The ratio $\frac{3}{500}$ is the ratio of tagged catfish in the sample to the sample size. If the tagged catfish are well-mixed and the sample is truly random, then these ratios should be equal:

$$\frac{30}{x} = \frac{3}{500}$$

$$3x = 15{,}000 \quad \text{\small Extremes-means property}$$

$$x = 5000$$

So there are approximately 5000 catfish in the pond. ◾

Note that any proportion can be solved by multiplying each side by the LCD as we did when we solved other equations involving rational expressions. The extremes-means property gives us a shortcut for solving proportions.

E X A M P L E   7   **Solving a proportion**

In a conservative portfolio the ratio of the amount invested in bonds to the amount invested in stocks should be 3 to 1. A conservative investor invested $2850 more in bonds than she did in stocks. How much did she invest in each category?

Solution

Because the ratio of the amount invested in bonds to the amount invested in stocks is 3 to 1, we have

$$\frac{\text{Amount invested in bonds}}{\text{Amount invested in stocks}} = \frac{3}{1}.$$

If x represents the amount invested in stocks and $x + 2850$ represents the amount invested in bonds, then we can write and solve the following proportion:

$$\frac{x + 2850}{x} = \frac{3}{1}$$

$$3x = x + 2850 \quad \text{\small Extremes-means property}$$

$$2x = 2850$$

$$x = 1425$$

$$x + 2850 = 4275$$

So she invested $4275 in bonds and $1425 in stocks. Note that these amounts are in the ratio of 3 to 1. ◾

The next example shows how conversions from one unit of measurement to another can be done by using proportions.

E X A M P L E   8   **Converting measurements**

There are 3 feet in 1 yard. How many feet are there in 12 yards?

Solution

Let x represent the number of feet in 12 yards. There are two proportions that we can write to solve the problem:

$$\frac{3 \text{ feet}}{x \text{ feet}} = \frac{1 \text{ yard}}{12 \text{ yards}} \qquad \frac{3 \text{ feet}}{1 \text{ yard}} = \frac{x \text{ feet}}{12 \text{ yards}}$$

The ratios in the second proportion violate the rule of comparing only measurements that are expressed in the same units. Note that each side of the second proportion is actually the ratio 1 to 1, since 3 feet = 1 yard and x feet = 12 yards. For doing conversions we can use ratios like this to compare measurements in different units. Applying the extremes-means property to either proportion gives

$$3 \cdot 12 = x \cdot 1,$$

or

$$x = 36.$$

So there are 36 feet in 12 yards.   ■

M A T H   A T   W O R K $x^2 + (x+1)^2 = 5^2$

Did you ever wonder how your local store calculates how much of your favorite cosmetic to stock on the shelf? Mike Pittman, National Account Manager for a major cosmetic company, is responsible for providing more than 2000 stores across the United States with personal care products such as skin lotions, fragrances, and cosmetics.

Data on what has been sold is transmitted from the point of sale across a number of satellite dishes and computers to Mr. Pittman. The data

**S A L E S
A N A L Y S T**

usually includes size, color, and other pertinent facts. The information is then combined with demographics for certain geographic areas and movement data, as well as advertising and promotional information to answer questions such as: What color is selling best? Is it time to stock sunscreen? Is this a trend-setting area of the country? On the basis of his analysis of these and many other questions, Mr. Pittman recommends changes in packaging, promotional programs, and the quantities of products to be shipped.

Mr. Pittman's job requires a unique blend of sales, marketing, and quantitative skills. Of course, knowledge of computers and an understanding of people help. So the next time you see a whole aisle of personal care and cosmetic products, think of all the information that has been analyzed to put it there.

In Exercise 63 of this section you will see how Mr. Pittman uses a proportion to determine the quantity of mascara needed in a warehouse.

WARM-UPS

True or false? Explain your answer.

1. The ratio of 40 men to 30 women can be expressed as the ratio 4 to 3.
2. The ratio of 3 feet to 2 yards can be expressed as the ratio 3 to 2.
3. If the ratio of men to women in the Chamber of Commerce is 3 to 2 and there are 20 men, then there must be 30 women.
4. The ratio of 1.5 to 2 is equivalent to the ratio of 3 to 4.
5. A statement that two ratios are equal is called a proportion.
6. The product of the extremes is equal to the product of the means.
7. If $\frac{2}{x} = \frac{3}{5}$, then $5x = 6$.
8. The ratio of the height of a 12-inch cactus to the height of a 3-foot cactus is 4 to 1.
9. If 30 out of 100 lawyers preferred aspirin and the rest did not, then the ratio of lawyers that preferred aspirin to those who did not is 30 to 100.
10. If $\frac{x + 5}{x} = \frac{2}{3}$, then $3x + 15 = 2x$.

7.7   EXERCISES

Reading and Writing   *After reading this section, write out the answers to these questions. Use complete sentences.*

1. What is a ratio?

2. What are the different ways of expressing a ratio?

3. What are equivalent ratios?

4. What is a proportion?

5. What are the means and what are the extremes?

6. What is the extremes-means property?

For each ratio, find an equivalent ratio of integers in lowest terms. See Example 1.

7. $\dfrac{2.5}{3.5}$ 8. $\dfrac{4.8}{1.2}$ 9. $\dfrac{0.32}{0.6}$

10. $\dfrac{0.05}{0.8}$ 11. $\dfrac{35}{10}$ 12. $\dfrac{88}{33}$

13. $\dfrac{4.5}{7}$ 14. $\dfrac{3}{2.5}$ 15. $\dfrac{\frac{1}{2}}{\frac{1}{5}}$

16. $\dfrac{\frac{2}{3}}{\frac{3}{4}}$ 17. $\dfrac{\frac{5}{1}}{\frac{1}{3}}$ 18. $\dfrac{\frac{4}{1}}{\frac{1}{4}}$

Find a ratio for each of the following, and write it as a ratio of integers in lowest terms. See Examples 2–4.

19. **Men and women.** Find the ratio of men to women in a bowling league containing 12 men and 8 women.

20. **Coffee drinkers.** Among 100 coffee drinkers, 36 said that they preferred their coffee black and the rest did not prefer their coffee black. Find the ratio of those who prefer black coffee to those who prefer nonblack coffee.

FIGURE FOR EXERCISE 20

21. Smokers. A life insurance company found that among its last 200 claims, there were six dozen smokers. What is the ratio of smokers to nonsmokers in this group of claimants?

22. Hits and misses. A woman threw 60 darts and hit the target a dozen times. What is her ratio of hits to misses?

23. Violence and kindness. While watching television for one week, a consumer group counted 1240 acts of violence and 40 acts of kindness. What is the violence to kindness ratio for television, according to this group?

24. Length to width. What is the ratio of length to width for the rectangle shown below?

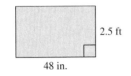

2.5 ft

48 in.

FIGURE FOR EXERCISE 24

25. Rise to run. What is the ratio of rise to run for the stairway shown in the accompanying figure?

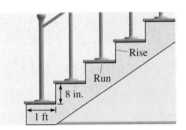

Rise

Run

8 in.

1 ft

FIGURE FOR EXERCISE 25

26. Rise and run. If the rise is $\frac{3}{2}$ and the run is 5, then what is the ratio of the rise to the run?

Solve each proportion. See Example 5.

27. $\dfrac{4}{x} = \dfrac{2}{3}$

28. $\dfrac{9}{x} = \dfrac{3}{2}$

29. $\dfrac{a}{2} = \dfrac{-1}{5}$

30. $\dfrac{b}{3} = \dfrac{-3}{4}$

31. $-\dfrac{5}{9} = \dfrac{3}{x}$

32. $-\dfrac{3}{4} = \dfrac{5}{x}$

33. $\dfrac{x+2}{x} = \dfrac{6}{5}$

34. $\dfrac{x}{x-3} = \dfrac{2}{3}$

35. $\dfrac{7}{5} = \dfrac{2x+1}{x+2}$

36. $\dfrac{5}{8} = \dfrac{3x+1}{2x+10}$

37. $\dfrac{10}{x} = \dfrac{34}{x+12}$

38. $\dfrac{x}{3} = \dfrac{x+1}{2}$

39. $\dfrac{a}{a+1} = \dfrac{a+3}{a}$

40. $\dfrac{c+3}{c-1} = \dfrac{c+2}{c-3}$

41. $\dfrac{m-1}{m-2} = \dfrac{m-3}{m+4}$

42. $\dfrac{h}{h-3} = \dfrac{h}{h-9}$

Use a proportion to solve each problem. See Examples 6–8.

43. New shows and reruns. The ratio of new shows to reruns on cable TV is 2 to 27. If Frank counted only eight new shows one evening, then how many reruns were there?

44. Fast food. If four out of five doctors prefer fast food, then at a convention of 445 doctors, how many prefer fast food?

45. Voting. If 220 out of 500 voters surveyed said that they would vote for the incumbent, then how many votes could the incumbent expect out of the 400,000 voters in the state?

FIGURE FOR EXERCISE 45

46. New product. A taste test with 200 randomly selected people found that only three of them said that they would buy a box of new Sweet Wheats cereal. How many boxes could the manufacturer expect to sell in a country of 280 million people?

47. Basketball blowout. As the final buzzer signaled the end of the basketball game, the Lions were 34 points ahead of the Tigers. If the Lions scored 5 points for every 3 scored by the Tigers, then what was the final score?

48. The golden ratio. The ancient Greeks thought that the most pleasing shape for a rectangle was one for which the ratio of the length to the width was 8 to 5, the golden ratio. If the length of a rectangular painting is 2 ft longer than its width, then for what dimensions would the length and width have the golden ratio?

49. Automobile sales. The ratio of sports cars to luxury cars sold in Wentworth one month was 3 to 2. If 20 more sports cars were sold than luxury cars, then how many of each were sold that month?

50. Foxes and rabbits. The ratio of foxes to rabbits in the Deerfield Forest Preserve is 2 to 9. If there are 35 fewer foxes than rabbits, then how many of each are there?

51. *Inches and feet.* If there are 12 inches in 1 foot, then how many inches are there in 7 feet?

52. *Feet and yards.* If there are 3 feet in 1 yard, then how many yards are there in 28 feet?

53. *Minutes and hours.* If there are 60 minutes in 1 hour, then how many minutes are there in 0.25 hour?

54. *Meters and kilometers.* If there are 1000 meters in 1 kilometer, then how many meters are there in 2.33 kilometers.

55. *Miles and hours.* If Alonzo travels 230 miles in 3 hours, then how many miles does he travel in 7 hours?

56. *Hiking time.* If Evangelica can hike 19 miles in 2 days on the Appalachian Trail, then how many days will it take her to hike 63 miles?

57. *Force on basketball shoes.* The designers of Converse shoes know that the force exerted on shoe soles in a jump shot is proportional to the weight of the person jumping. If a 70-pound boy exerts a force of 980 pounds on his shoe soles when he returns to the court after a jump, then what force does a 6 ft 8 in. professional ball player weighing 280 pounds exert on the soles of his shoes when he returns to the court after a jump? Use the accompanying graph to estimate the force for a 150-pound player.

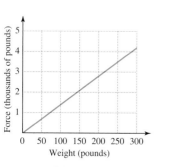

FIGURE FOR EXERCISE 57

58. *Force on running shoes.* The designers of Converse shoes know that the ratio of the force on the shoe soles to the weight of a runner is 3 to 1. What force does a 130-pound jogger exert on the soles of her shoes.

59. *Capture-recapture.* To estimate the number of trout in Trout Lake, rangers used the capture-recapture method. They caught, tagged, and released 200 trout. One week later, they caught a sample of 150 trout and found that 5 of them were tagged. Assuming that the ratio of tagged trout to the total number of trout in the lake is the same as the ratio of tagged trout in the sample to the number

of trout in the sample, find the number of trout in the lake.

60. *Bear population.* To estimate the size of the bear population on the Keweenaw Peninsula, conservationists captured, tagged, and released 50 bears. One year later, a random sample of 100 bears included only 2 tagged bears. What is the conservationist's estimate of the size of the bear population?

61. *Fast-food waste.* The accompanying figure shows the typical distribution of waste at a fast-food restaurant (U.S. Environmental Protection Agency, www.epa.gov).
  a) What is the ratio of customer waste to food waste?

  b) If a typical McDonald's generates 67 more pounds of food waste than customer waste per day, then how many pounds of customer waste does it generate?

WASTE GENERATION AT A FAST-FOOD RESTAURANT

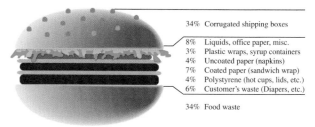

34% Corrugated shipping boxes

8% Liquids, office paper, misc.
3% Plastic wraps, syrup containers
4% Uncoated paper (napkins)
7% Coated paper (sandwich wrap)
4% Polystyrene (hot cups, lids, etc.)
6% Customer's waste (Diapers, etc.)

34% Food waste

FIGURE FOR EXERCISES 61 AND 62

62. *Corrugated waste.* Use the accompanying figure to find the ratio of waste from corrugated shipping boxes to waste not from corrugated shipping boxes. If a typical McDonald's generates 81 pounds of waste per day from corrugated shipping boxes, then how many pounds of waste per day does it generate that is not from corrugated shipping boxes?

63. *Mascara needs.* In determining warehouse needs for a particular mascara for a chain of 2000 stores, Mike Pittman first determines a need B based on sales figures for the past 52 weeks. He then determines the actual need A from the equation $\frac{A}{B} = k$, where

$$k = 1 + V + C + X - D.$$

He uses $V = 0.22$ if there is a national TV ad and $V = 0$ if not, $C = 0.26$ if there is a national coupon and $C = 0$ if not, $X = 0.36$ if there is a chain-specific ad and $X = 0$ if not, and $D = 0.29$ if there is a special display in the chain and $D = 0$ if not. (D is subtracted because less product is needed in the warehouse when more is on display in the store.) If $B = 4200$ units and there is a special display and a national coupon but no national TV ad and no chain-specific ad, then what is the value of A?

GETTING MORE INVOLVED

64. *Discussion.* Which of the following equations is not a proportion? Explain.

a) $\dfrac{1}{2} = \dfrac{1}{2}$

b) $\dfrac{x}{x + 2} = \dfrac{4}{5}$

c) $\dfrac{x}{4} = \dfrac{9}{x}$

d) $\dfrac{8}{x + 2} - 1 = \dfrac{5}{x + 2}$

65. *Discussion.* Find all of the errors in the following solution to an equation.

$$\frac{7}{x} = \frac{8}{x + 3} + 1$$

$$7(x + 3) = 8x + 1$$

$$7x + 3 = 8x$$

$$-x = -3$$

$$x = 3$$

7.8   APPLICATIONS OF RATIONAL EXPRESSIONS

In this section we will study additional applications of rational expressions.

In this section

- Formulas
- Uniform Motion Problems
- Work Problems
- Purchasing Problems

Formulas

Many formulas involve rational expressions. When solving a formula of this type for a certain variable, we usually multiply each side by the LCD to eliminate the denominators.

EXAMPLE 1

An equation of a line

The equation for the line through $(-2, 4)$ with slope $3/2$ can be written as

$$\frac{y - 4}{x + 2} = \frac{3}{2}.$$

We studied equations of this type in Chapter 4. Solve this equation for y.

Solution

To isolate y on the left-hand side of the equation, we multiply each side by $x + 2$:

$$\frac{y - 4}{x + 2} = \frac{3}{2} \qquad \text{Original equation}$$

$$(x + 2) \cdot \frac{y - 4}{x + 2} = (x + 2) \cdot \frac{3}{2} \qquad \text{Multiply by } x + 2.$$

$$y - 4 = \frac{3}{2}x + 3 \qquad \text{Simplify.}$$

$$y = \frac{3}{2}x + 7 \qquad \text{Add 4 to each side.}$$

Because the original equation is a proportion, we could have used the extremes-means property to solve it for y.   ■

helpful hint

When this equation was written in the form
$$y - y_1 = m(x - x_1)$$
in Chapter 4, we called it the point-slope formula for the equation of a line.

EXAMPLE 2

Distance, rate, and time

Solve the formula $\dfrac{D}{T} = R$ for T.

Solution

Because the only denominator is T, we multiply each side by T:

$$\frac{D}{T} = R \qquad \text{Original formula}$$

$$T \cdot \frac{D}{T} = T \cdot R \quad \text{Multiply each side by } T.$$

$$D = TR$$

$$\frac{D}{R} = \frac{TR}{R} \qquad \text{Divide each side by } R.$$

$$\frac{D}{R} = T \qquad \text{Simplify.}$$

The formula solved for T is $T = \frac{D}{R}$. ∎

E X A M P L E   3   **Focal length of a lens**

The formula $\frac{1}{f} = \frac{1}{o} + \frac{1}{i}$ gives the relationship between the focal length f, the object distance o, and the image distance i for a lens. Solve the formula for i.

Solution

The LCD for f, o, and i is foi.

$$\frac{1}{f} = \frac{1}{o} + \frac{1}{i} \qquad \text{Original formula}$$

$$foi \cdot \frac{1}{f} = foi \cdot \frac{1}{o} + foi \cdot \frac{1}{i} \quad \text{Multiply each side by the LCD, } foi.$$

$$oi = fi + fo \qquad \text{All denominators are eliminated.}$$

$$oi - fi = fo \qquad \text{Get all terms involving } i \text{ onto the left side.}$$

$$i(o - f) = fo \qquad \text{Factor out } i.$$

$$i = \frac{fo}{o - f} \qquad \text{Divide each side by } o - f.$$

∎

E X A M P L E   4   **Finding the value of a variable**

In the formula of Example 1, find x if $y = -3$.

Solution

Substitute $y = -3$ into the formula, then solve for x:

$$\frac{y - 4}{x + 2} = \frac{3}{2} \qquad \text{Original formula}$$

$$\frac{-3 - 4}{x + 2} = \frac{3}{2} \qquad \text{Replace } y \text{ by } -3.$$

$$\frac{-7}{x + 2} = \frac{3}{2} \qquad \text{Simplify.}$$

$$3x + 6 = -14 \qquad \text{Extremes-means property}$$

$$3x = -20$$

$$x = -\frac{20}{3}$$

Uniform Motion Problems

In uniform motion problems we use the formula $D = RT$. In some problems in which the time is unknown, we can use the formula $T = \frac{D}{R}$ to get an equation involving rational expressions.

E X A M P L E   5   **Driving to Florida**

Susan drove 1500 miles to Daytona Beach for spring break. On the way back she averaged 10 miles per hour less, and the drive back took her 5 hours longer. Find Susan's average speed on the way to Daytona Beach.

Solution

If x represents her average speed going there, then $x - 10$ is her average speed for the return trip. See Fig. 7.1. We use the formula $T = \frac{D}{R}$ to make the following table.

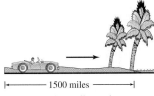

1500 miles
Speed = x miles per hour

	D	R	T	
Going	1500	x	$\frac{1500}{x}$	← Shorter time
Returning	1500	$x - 10$	$\frac{1500}{x - 10}$	← Longer time

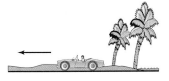

Speed = $x - 10$ miles per hour

FIGURE 7.1

Because the difference between the two times is 5 hours, we have

$$\text{longer time} - \text{shorter time} = 5.$$

Using the time expressions from the table, we get the following equation:

$$\frac{1500}{x - 10} - \frac{1500}{x} = 5$$

$$x(x - 10)\frac{1500}{x - 10} - x(x - 10)\frac{1500}{x} = x(x - 10)5 \quad \text{Multiply by } x(x - 10).$$

$$1500x - 1500(x - 10) = 5x^2 - 50x$$

$$15{,}000 = 5x^2 - 50x \quad \text{Simplify.}$$

$$3000 = x^2 - 10x \quad \text{Divide each side by 5.}$$

$$0 = x^2 - 10x - 3000$$

$$(x + 50)(x - 60) = 0 \quad \text{Factor.}$$

$$x + 50 = 0 \quad \text{or} \quad x - 60 = 0$$

$$x = -50 \quad \text{or} \quad x = 60$$

The answer $x = -50$ is a solution to the equation, but it cannot indicate the average speed of the car. Her average speed going to Daytona Beach was 60 mph. ∎

helpful hint

Notice that a work rate is the same as a slope from Chapter 4. The only difference is that the work rates here can contain a variable.

Work Problems

If you can complete a job in 3 hours, then you are working at the rate of $\frac{1}{3}$ of the job per hour. If you work for 2 hours at the rate of $\frac{1}{3}$ of the job per hour, then you will complete $\frac{2}{3}$ of the job. The product of the rate and time is the amount of work completed. For problems involving work, we will always assume that the work is done at a constant rate. So if a job takes x hours to complete, then the rate is $\frac{1}{x}$ of the job per hour.

E X A M P L E   6

Shoveling snow

After a heavy snowfall, Brian can shovel all of the driveway in 30 minutes. If his younger brother Allen helps, the job takes only 20 minutes. How long would it take Allen to do the job by himself?

Solution

Let x represent the number of minutes it would take Allen to do the job by himself. Brian's rate for shoveling is $\frac{1}{30}$ of the driveway per minute, and Allen's rate for shoveling is $\frac{1}{x}$ of the driveway per minute. We organize all of the information in a table like the table in Example 5.

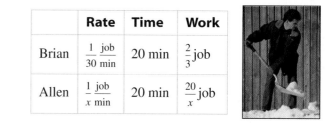

	Rate	Time	Work
Brian	$\frac{1 \text{ job}}{30 \text{ min}}$	20 min	$\frac{2}{3}$job
Allen	$\frac{1 \text{ job}}{x \text{ min}}$	20 min	$\frac{20}{x}$job

If Brian works for 20 min at the rate $\frac{1}{30}$ of the job per minute, then he does $\frac{20}{30}$ or $\frac{2}{3}$ of the job, as shown in Fig. 7.2. The amount of work that each boy does is a fraction of the whole job. So the expressions for work in the last column of the table have a sum of 1:

$$\frac{2}{3} + \frac{20}{x} = 1$$

$$3x \cdot \frac{2}{3} + 3x \cdot \frac{20}{x} = 3x \cdot 1 \quad \text{Multiply each side by } 3x.$$

$$2x + 60 = 3x$$

$$60 = x$$

If it takes Allen 60 minutes to do the job by himself, then he works at the rate of $\frac{1}{60}$ of the job per minute. In 20 minutes he does $\frac{1}{3}$ of the job while Brian does $\frac{2}{3}$. So it would take Allen 60 minutes to shovel the driveway by himself. ∎

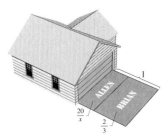

FIGURE 7.2

Notice the similarities between the uniform motion problem in Example 5 and the work problem in Example 6. In both cases it is beneficial to make a table. We use $D = R \cdot T$ in uniform motion problems and $W = R \cdot T$ in work problems. The main points to remember when solving work problems are summarized in the following strategy.

Strategy for Solving Work Problems

1. If a job is completed in x hours, then the rate is $\frac{1}{x}$ job/hr.
2. Make a table showing rate, time, and work completed ($W = R \cdot T$) for each person or machine.
3. The total work completed is the sum of the individual amounts of work completed.
4. If the job is completed, then the total work done is 1 job.

Purchasing Problems

Rates are used in uniform motion and work problems. But rates also occur in purchasing problems. If gasoline is 99.9 cents/gallon, then that is the rate at which your bill is increasing as you pump the gallons into your tank. In purchasing problems the product of the rate and the quantity purchased is the total cost.

E X A M P L E 7

Oranges and grapefruit

Tamara bought 50 pounds of fruit consisting of Florida oranges and Texas grapefruit. She paid twice as much per pound for the grapefruit as she did for the oranges. If Tamara bought $12 worth of oranges and $16 worth of grapefruit, then how many pounds of each did she buy?

x lb

Oranges

$50 - x$ lb

Grapefruit

FIGURE 7.3

Solution

Let x represent the number of pounds of oranges and $50 - x$ represent the number of pounds of grapefruit. See Fig. 7.3. Make a table.

	Rate	**Quantity**	**Total cost**
Oranges	$\dfrac{12}{x}$ dollars/pound	x pounds	12 dollars
Grapefruit	$\dfrac{16}{50 - x}$ dollars/pound	$50 - x$ pounds	16 dollars

Since the price per pound for the grapefruit is twice that for the oranges, we have:

$$2(\text{price per pound for oranges}) = \text{price per pound for grapefruit}$$

$$2\left(\frac{12}{x}\right) = \frac{16}{50 - x}$$

$$\frac{24}{x} = \frac{16}{50 - x}$$

$$16x = 1200 - 24x \quad \text{Extremes-means property}$$

$$40x = 1200$$

$$x = 30$$

$$50 - x = 20$$

If Tamara purchased 20 pounds of grapefruit for $16, then she paid $0.80 per pound. If she purchased 30 pounds of oranges for $12, then she paid $0.40 per pound. Because $0.80 is twice $0.40, we can be sure that she purchased 20 pounds of grapefruit and 30 pounds of oranges. ■

WARM-UPS

True or false? Explain your answer.

1. The formula $t = \dfrac{1-t}{m}$, solved for m, is $m = \dfrac{1-t}{t}$.
2. To solve $\dfrac{1}{m} + \dfrac{1}{n} = \dfrac{1}{2}$ for m, we multiply each side by $2mn$.
3. If Fiona drives 300 miles in x hours, then her average speed is $\dfrac{x}{300}$ mph.

(continued)

4. If Miguel drives 20 hard bargains in x hours, then he is driving $\frac{20}{x}$ hard bargains per hour.

5. If Fred can paint a house in y days, then he paints $\frac{1}{y}$ of the house per day.

6. If $\frac{1}{x}$ is 1 less than $\frac{2}{x+3}$, then $\frac{1}{x} - 1 = \frac{2}{x+3}$.

7. If a and b are nonzero and $a = \frac{m}{b}$, then $b = am$.

8. If $D = RT$, then $T = \frac{D}{R}$.

9. Solving $P + Prt = I$ for P gives $P = I - Prt$.

10. To solve $3R + yR = m$ for R, we must first factor the left-hand side.

7.8   EXERCISES

Solve each equation for y. See Example 1.

1. $\dfrac{y-1}{x-3} = 2$

2. $\dfrac{y-2}{x-4} = -2$

3. $\dfrac{y-1}{x+6} = -\dfrac{1}{2}$

4. $\dfrac{y+5}{x-2} = -\dfrac{1}{2}$

5. $\dfrac{y+a}{x-b} = m$

6. $\dfrac{y-h}{x+k} = a$

7. $\dfrac{y-1}{x+4} = -\dfrac{1}{3}$

8. $\dfrac{y-1}{x+3} = -\dfrac{3}{4}$

Solve each formula for the indicated variable. See Examples 2 and 3.

9. $A = \dfrac{B}{C}$ for C

10. $P = \dfrac{A}{C+D}$ for A

11. $\dfrac{1}{a} + m = \dfrac{1}{p}$ for p

12. $\dfrac{2}{f} + t = \dfrac{3}{m}$ for m

13. $F = k\dfrac{m_1 m_2}{r^2}$ for m_1

14. $F = \dfrac{mv^2}{r}$ for r

15. $\dfrac{1}{a} + \dfrac{1}{b} = \dfrac{1}{f}$ for a

16. $\dfrac{1}{R} = \dfrac{1}{R_1} + \dfrac{1}{R_2}$ for R

17. $S = \dfrac{a}{1-r}$ for r

18. $I = \dfrac{E}{R+r}$ for R

19. $\dfrac{P_1 V_1}{T_1} = \dfrac{P_2 V_2}{T_2}$ for P_2

20. $\dfrac{P_1 V_1}{T_1} = \dfrac{P_2 V_2}{T_2}$ for T_1

21. $V = \dfrac{4}{3}\pi r^2 h$ for h

22. $h = \dfrac{S - 2\pi r^2}{2\pi r}$ for S

Find the value of the indicated variable. See Example 4.

23. In the formula of Exercise 9, if $A = 12$ and $B = 5$, find C.

24. In the formula of Exercise 10, if $A = 500$, $P = 100$, and $C = 2$, find D.

25. In the formula of Exercise 11, if $p = 6$ and $m = 4$, find a.

26. In the formula of Exercise 12, if $m = 4$ and $t = 3$, find f.

27. In the formula of Exercise 13, if $F = 32$, $r = 4$, $m_1 = 2$, and $m_2 = 6$, find k.

28. In the formula of Exercise 14, if $F = 10$, $v = 8$, and $r = 6$, find m.

29. In the formula of Exercise 15, if $f = 3$ and $a = 2$, find b.

30. In the formula of Exercise 16, if $R = 3$ and $R_1 = 5$, find R_2.

31. In the formula of Exercise 17, if $S = \dfrac{3}{2}$ and $r = \dfrac{1}{5}$, find a.

32. In the formula of Exercise 18, if $I = 15$, $E = 3$, and $R = 2$, find r.

Show a complete solution to each problem. See Example 5.

33. *Fast walking.* Marcie can walk 8 miles in the same time as Frank walks 6 miles. If Marcie walks 1 mile per hour faster than Frank, then how fast does each person walk?

34. *Upstream, downstream.* Junior's boat will go 15 miles per hour in still water. If he can go 12 miles downstream in the same amount of time as it takes to go 9 miles upstream, then what is the speed of the current?

35. *Delivery routes.* Pat travels 70 miles on her milk route, and Bob travels 75 miles on his route. Pat travels 5 miles per hour slower than Bob, and her route takes her one-half hour longer than Bob's. How fast is each one traveling?

36. *Ride the peaks.* Smith bicycled 45 miles going east from Durango, and Jones bicycled 70 miles. Jones averaged 5 miles per hour more than Smith, and his trip took one-half hour longer than Smith's. How fast was each one traveling?

37. *Walking and running.* Raffaele ran 8 miles and then walked 6 miles. If he ran 5 miles per hour faster than he walked and the total time was 2 hours, then how fast did he walk?

38. *Triathlon.* Luisa participated in a triathlon in which she swam 3 miles, ran 5 miles, and then bicycled 10 miles. Luisa ran twice as fast as she swam, and she cycled three times as fast as she swam. If her total time for the triathlon was 1 hour and 46 minutes, then how fast did she swim?

FIGURE FOR EXERCISE 36

Show a complete solution to each problem. See Example 6.

39. *Fence painting.* Kiyoshi can paint a certain fence in 3 hours by himself. If Red helps, the job takes only 2 hours. How long would it take Red to paint the fence by himself?

40. *Envelope stuffing.* Every week, Linda must stuff 1000 envelopes. She can do the job by herself in 6 hours. If Laura helps, they get the job done in $5\frac{1}{2}$ hours. How long would it take Laura to do the job by herself?

41. *Garden destroying.* Mr. McGregor has discovered that a large dog can destroy his entire garden in 2 hours and that a small boy can do the same job in 1 hour. How long would it take the large dog and the small boy working together to destroy Mr. McGregor's garden?

42. *Draining the vat.* With only the small valve open, all of the liquid can be drained from a large vat in 4 hours. With only the large valve open, all of the liquid can be drained from the same vat in 2 hours. How long would it take to drain the vat with both valves open?

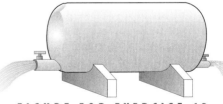

FIGURE FOR EXERCISE 42

43. *Cleaning sidewalks.* Edgar can blow the leaves off the sidewalks around the capitol building in 2 hours using a gasoline-powered blower. Ellen can do the same job in 8 hours using a broom. How long would it take them working together?

44. *Computer time.* It takes a computer 8 days to print all of the personalized letters for a national sweepstakes. A new computer is purchased that can do the same job in 5 days. How long would it take to do the job with both computers working on it?

Show a complete solution to each problem. See Example 7.

45. *Apples and bananas.* Bertha bought 18 pounds of fruit consisting of apples and bananas. She paid $9 for the apples and $2.40 for the bananas. If the price per pound of the apples was 3 times that of the bananas, then how many pounds of each type of fruit did she buy?

46. *Running backs.* In the playoff game the ball was carried by either Anderson or Brown on 21 plays. Anderson gained 36 yards, and Brown gained 54 yards. If Brown averaged twice as many yards per carry as Anderson, then on how many plays did Anderson carry the ball?

47. *Fuel efficiency.* Last week, Joe's Electric Service used 110 gallons of gasoline in its two trucks. The large truck was driven 800 miles, and the small truck was driven 600 miles. If the small truck gets twice as many miles per gallon as the large truck, then how many gallons of gasoline did the large truck use?

48. *Repair work.* Sally received a bill for a total of 8 hours labor on the repair of her bulldozer. She paid $50 to the master mechanic and $90 to his apprentice. If the master mechanic gets $10 more per hour than his apprentice, then how many hours did each work on the bulldozer?

FIGURE FOR EXERCISE 46

COLLABORATIVE ACTIVITIES

How Do I Get There from Here?

You want to decide whether to ride your bicycle, drive your car, or take the bus to school this year. The best thing to do is to analyze each of your options. Read all the information given below. Then have each person in your group pick one mode of transportation. Working individually, answer the questions using the given information, the map, and the distance formula $d = rt$. Present a case to your group for your type of transportation.

Information available (see the map for distances):

Bicycle: You would need to buy a new bike lock for $15 and two new tubes at $2.50 apiece. Determine how fast you would have to bike to beat the car.

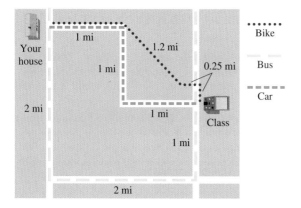

Grouping: 3 students
Topic: Distance formula

Car: You would need to pay for a parking permit, which costs $40. Traffic has increased, so it takes 12 minutes to get to school. What is your average speed?

Bus: The bus stops at the end of your block and has a new student rate of $1 a week. It leaves at 8:30 A.M. and will get to the college at 8:55 A.M. On Mondays and Wednesdays you have a 9:00 A.M. class, four blocks from the bus stop. Find the average speed of the bus and figure the cost for the 16-week semester.

When presenting your case: Include the time needed to get there, speed, cost, and convenience. Have at least three reasons why this would be the best way to travel. Consider unique features of your area such as traffic, weather, and terrain.

After each of you has presented your case, decide as a group which type of transportation you would choose.

3. **building up the denominator**
   a. the opposite of reducing a fraction
   b. finding the least common denominator
   c. adding the same number to the numerator and denominator
   d. writing a fraction larger

4. **least common denominator**
   a. the largest number that is a multiple of all denominators
   b. the sum of the denominators
   c. the product of the denominators
   d. the smallest number that is a multiple of all denominators

5. **extraneous solution**
   a. a number that appears to be a solution to an equation but does not satisfy the equation
   b. an extra solution to an equation
   c. the second solution
   d. a nonreal solution

6. **ratio of *a* to *b***
   a. b/a
   b. a/b
   c. $a/(a + b)$
   d. ab

7. **proportion**
   a. a ratio
   b. two ratios
   c. the product of the means equals the product of the extremes
   d. a statement expressing the equality of two ratios

8. **extremes**
   a. a and d in $a/b = c/d$
   b. b and c in $a/b = c/d$
   c. the extremes-means property
   d. if $a/b = c/d$ then $ad = bc$

9. **means**
   a. the average of a, b, c, and d
   b. a and d in $a/b = c/d$
   c. b and c in $a/b = c/d$
   d. if $a/b = c/d$, then $(a + b)/2 = (c + d)/2$

10. **cross-multiplying**
    a. $ab = ba$ for any real numbers a and b
    b. $(a - b)^2 = (b - a)^2$ for any real numbers a and b
    c. if $a/b = c/d$, then $ab = cd$
    d. if $a/b = c/d$, then $ad = bc$

REVIEW EXERCISES

7.1 *Reduce each rational expression to lowest terms.*

1. $\dfrac{24}{28}$

2. $\dfrac{42}{18}$

3. $\dfrac{2a^3c^3}{8a^5c}$

4. $\dfrac{39x^6}{15x}$

5. $\dfrac{6w - 9}{9w - 12}$

6. $\dfrac{3t - 6}{8 - 4t}$

7. $\dfrac{x^2 - 1}{3 - 3x}$

8. $\dfrac{3x^2 - 9x + 6}{10 - 5x}$

7.2 *Perform the indicated operation.*

9. $\dfrac{1}{6k} \cdot 3k^2$

10. $\dfrac{1}{15abc} \cdot 5a^3b^5c^2$

11. $\dfrac{2xy}{3} \div y^2$

12. $4ab \div \dfrac{1}{2a^4}$

13. $\dfrac{a^2 - 9}{a - 2} \cdot \dfrac{a^2 - 4}{a + 3}$

14. $\dfrac{x^2 - 1}{3x} \cdot \dfrac{6x}{2x - 2}$

15. $\dfrac{w - 2}{3w} \div \dfrac{4w - 8}{6w}$

16. $\dfrac{2y + 2x}{x - xy} \div \dfrac{x^2 + 2xy + y^2}{y^2 - y}$

7.3 *Find the least common denominator for each group of denominators.*

17. 36, 54

18. 10, 15, 35

19. $6ab^3$, $8a^7b^2$

20. $20u^4v$, $18uv^5$, $12u^2v^3$

21. $4x$, $6x - 6$

22. $8a$, $6a$, $2a^2 + 2a$

23. $x^2 - 4$, $x^2 - x - 2$

24. $x^2 - 9$, $x^2 + 6x + 9$

Convert each rational expression into an equivalent rational expression with the indicated denominator.

25. $\dfrac{5}{12} = \dfrac{?}{36}$

26. $\dfrac{2a}{15} = \dfrac{?}{45}$

27. $\dfrac{2}{3xy} = \dfrac{?}{15x^2y}$

28. $\dfrac{3z}{7x^2y} = \dfrac{?}{42x^3y^8}$

29. $\dfrac{5}{y-6} = \dfrac{?}{12-2y}$

30. $\dfrac{-3}{2-t} = \dfrac{?}{2t-4}$

31. $\dfrac{x}{x-1} = \dfrac{?}{x^2-1}$

32. $\dfrac{t}{t-3} = \dfrac{?}{t^2+2t-15}$

7.4 *Perform the indicated operation.*

33. $\dfrac{5}{36} + \dfrac{9}{28}$

34. $\dfrac{7}{30} - \dfrac{11}{42}$

35. $3 - \dfrac{4}{x}$

36. $1 + \dfrac{3a}{2b}$

37. $\dfrac{2}{ab^2} - \dfrac{1}{a^2 b}$

38. $\dfrac{3}{4x^3} + \dfrac{5}{6x^2}$

39. $\dfrac{9a}{2a-3} + \dfrac{5}{3a-2}$

40. $\dfrac{3}{x-2} - \dfrac{5}{x+3}$

41. $\dfrac{1}{a-8} - \dfrac{2}{8-a}$

42. $\dfrac{5}{x-14} + \dfrac{4}{14-x}$

43. $\dfrac{3}{2x-4} + \dfrac{1}{x^2-4}$

44. $\dfrac{x}{x^2-2x-3} - \dfrac{3x}{x^2-9}$

7.5 *Simplify each complex fraction.*

45. $\dfrac{\dfrac{1}{2} - \dfrac{3}{4}}{\dfrac{2}{3} + \dfrac{1}{2}}$

46. $\dfrac{\dfrac{2}{3} + \dfrac{5}{8}}{\dfrac{1}{2} - \dfrac{3}{8}}$

47. $\dfrac{\dfrac{1}{a} + \dfrac{2}{3b}}{\dfrac{1}{2b} - \dfrac{3}{a}}$

48. $\dfrac{\dfrac{3}{xy} - \dfrac{1}{3y}}{\dfrac{1}{6x} - \dfrac{3}{5y}}$

49. $\dfrac{\dfrac{1}{x-2} - \dfrac{3}{x+3}}{\dfrac{2}{x+3} + \dfrac{1}{x-2}}$

50. $\dfrac{\dfrac{4}{a+1} + \dfrac{5}{a^2-1}}{\dfrac{1}{a^2-1} - \dfrac{3}{a-1}}$

51. $\dfrac{\dfrac{x-1}{x-3}}{\dfrac{1}{x^2-x-6} - \dfrac{4}{x+2}}$

52. $\dfrac{\dfrac{6}{a^2+5a+6} - \dfrac{8}{a+2}}{\dfrac{2}{a+3} - \dfrac{4}{a+2}}$

7.6 *Solve each equation.*

53. $\dfrac{-2}{5} = \dfrac{3}{x}$

54. $\dfrac{3}{x} + \dfrac{5}{3x} = 1$

55. $\dfrac{14}{a^2-1} + \dfrac{1}{a-1} = \dfrac{3}{a+1}$

56. $2 + \dfrac{3}{y-5} = \dfrac{2y}{y-5}$

57. $z - \dfrac{3z}{2-z} = \dfrac{6}{z-2}$

58. $\dfrac{1}{x} + \dfrac{1}{3} = \dfrac{1}{2}$

7.7 *Solve each proportion.*

59. $\dfrac{3}{x} = \dfrac{2}{7}$

60. $\dfrac{4}{x} = \dfrac{x}{4}$

61. $\dfrac{2}{w-3} = \dfrac{5}{w}$

62. $\dfrac{3}{t-3} = \dfrac{5}{t+4}$

Solve each problem by using a proportion.

63. ***Taxis in Times Square.*** The ratio of taxis to private automobiles in Times Square at 6:00 P.M. on New Year's Eve was estimated to be 15 to 2. If there were 60 taxis, then how many private automobiles were there?

FIGURE FOR EXERCISE 63

64. ***Student-teacher ratio.*** The student-teacher ratio for Washington High was reported to be 27.5 to 1. If there are 42 teachers, then how many students are there?

65. ***Water and rice.*** At Wong's Chinese Restaurant the secret recipe for white rice calls for a 2 to 1 ratio of water to rice. In one batch the chef used 28 more cups of water than rice. How many cups of each did he use?

FIGURE FOR EXERCISE 65

66. *Oil and gas.* An outboard motor calls for a fuel mixture that has a gasoline-to-oil ratio of 50 to 1. How many pints of oil should be added to 6 gallons of gasoline?

7.8 *Solve each formula for the indicated variable.*

67. $\dfrac{y - b}{m} = x$ for y

68. $\dfrac{A}{h} = \dfrac{a + b}{2}$ for a

69. $F = \dfrac{mv + 1}{m}$ for m

70. $m = \dfrac{r}{1 + rt}$ for r

71. $\dfrac{y + 1}{x - 3} = 4$ for y

72. $\dfrac{y - 3}{x + 2} = \dfrac{-1}{3}$ for y

Solve each problem.

73. *Making a puzzle.* Tracy, Stacy, and Fred assembled a very large puzzle together in 40 hours. If Stacy worked twice as fast as Fred and Tracy worked just as fast as Stacy, then how long would it have taken Fred to assemble the puzzle alone?

74. *Going skiing.* Leon drove 270 miles to the lodge in the same time as Pat drove 330 miles to the lodge. If Pat drove 10 miles per hour faster than Leon, then how fast did each of them drive?

FIGURE FOR EXERCISE 74

75. *Merging automobiles.* When Bert and Ernie merged their automobile dealerships, Bert had 10 more cars than Ernie. While 36% of Ernie's stock consisted of new cars, only 25% of Bert's stock consisted of new cars. If they had 33 new cars on the lot after the merger, then how many cars did each one have before the merger?

76. *Magazine sales.* A company specializing in magazine sales over the telephone found that in 2500 phone calls, 360 resulted in sales and were made by male callers, and 480 resulted in sales and were made by female callers. If the company gets twice as many sales per call with a woman's voice than with a man's voice, then how many of the 2500 calls were made by females?

77. *Distribution of waste.* The accompanying figure shows the distribution of the total municipal solid waste into various categories in 1995 (U.S. Environmental Protection Agency, www.epa.gov). If the paper waste was 51.7 million tons greater than the yard waste, then what was the amount of yard waste generated?

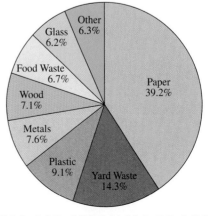

1995 Total Waste Generation (before recycling)

FIGURE FOR EXERCISES 77 AND 78

78. Total waste. Use the information given in the previous exercise to find the total waste generated in 1995 and the amount of food waste.

MISCELLANEOUS

In place of each question mark, put an expression that makes each equation an identity.

79. $\dfrac{5}{x} = \dfrac{?}{2x}$

80. $\dfrac{?}{a} = \dfrac{6}{3a}$

81. $\dfrac{2}{a - 5} = \dfrac{?}{5 - a}$

82. $\dfrac{-1}{a - 7} = \dfrac{1}{?}$

83. $3 = \dfrac{?}{x}$

84. $2a = \dfrac{?}{b}$

85. $m \div \dfrac{1}{2} = ?$

86. $5x \div \dfrac{1}{x} = ?$

87. $2a \div ? = 12a$

88. $10x \div ? = 20x^2$

89. $\dfrac{a - 1}{a^2 - 1} = \dfrac{1}{?}$

90. $\dfrac{?}{x^2 - 9} = \dfrac{1}{x - 3}$

91. $\dfrac{1}{a} - \dfrac{1}{5} = ?$

92. $\dfrac{3}{7} - \dfrac{2}{b} = ?$

93. $\dfrac{a}{2} - 1 = \dfrac{?}{2}$

94. $\dfrac{1}{a} - 1 = \dfrac{?}{a}$

95. $(a - b) \div (-1) = ?$

96. $(a - 7) \div (7 - a) = ?$

97. $\dfrac{\frac{1}{5a}}{2} = ?$

98. $\dfrac{3a}{\frac{1}{2}} = ?$

For each expression in Exercises 99–118, either perform the indicated operation or solve the equation, whichever is appropriate.

99. $\dfrac{1}{x} + \dfrac{1}{2x}$

100. $\dfrac{1}{y} + \dfrac{1}{3y} = 2$

101. $\dfrac{2}{3xy} + \dfrac{1}{6x}$

102. $\dfrac{3}{x - 1} - \dfrac{3}{x}$

103. $\dfrac{5}{a - 5} - \dfrac{3}{5 - a}$

104. $\dfrac{2}{x - 2} - \dfrac{3}{x} = \dfrac{-1}{x}$

105. $\dfrac{2}{x - 1} - \dfrac{2}{x} = 1$

106. $\dfrac{2}{x - 2} \cdot \dfrac{6x - 12}{14}$

107. $\dfrac{-3}{x + 2} \cdot \dfrac{5x + 10}{9}$

108. $\dfrac{3}{10} = \dfrac{5}{x}$

109. $\dfrac{1}{-3} = \dfrac{-2}{x}$

110. $\dfrac{x^2 - 4}{x} \div \dfrac{4x - 8}{x}$

111. $\dfrac{ax + am + 3x + 3m}{a^2 - 9} \div \dfrac{2x + 2m}{a - 3}$

112. $\dfrac{-2}{x} = \dfrac{3}{x + 2}$

113. $\dfrac{2}{x^2 - 25} + \dfrac{1}{x^2 - 4x - 5}$

114. $\dfrac{4}{a^2 - 1} + \dfrac{1}{2a + 2}$

115. $\dfrac{-3}{a^2 - 9} - \dfrac{2}{a^2 + 5a + 6}$

116. $\dfrac{-5}{a^2 - 4} - \dfrac{2}{a^2 - 3a + 2}$

117. $\dfrac{1}{a^2 - 1} + \dfrac{2}{1 - a} = \dfrac{3}{a + 1}$

118. $3 + \dfrac{1}{x - 2} = \dfrac{2x - 3}{x - 2}$

CHAPTER 7 TEST

What numbers cannot be used for x in each rational expression?

1. $\dfrac{2x - 1}{x^2 - 1}$ **2.** $\dfrac{5}{2 - 3x}$ **3.** $\dfrac{1}{x}$

Perform the indicated operation. Write each answer in lowest terms.

4. $\dfrac{2}{15} - \dfrac{4}{9}$

5. $\dfrac{1}{y} + 3$

6. $\dfrac{3}{a - 2} - \dfrac{1}{2 - a}$

7. $\dfrac{2}{x^2 - 4} - \dfrac{3}{x^2 + x - 2}$

8. $\dfrac{m^2 - 1}{(m - 1)^2} \cdot \dfrac{2m - 2}{3m + 3}$

9. $\dfrac{a - b}{3} \div \dfrac{b^2 - a^2}{6}$

10. $\dfrac{5a^2 b}{12a} \cdot \dfrac{2a^3 b}{15ab^6}$

Simplify each complex fraction.

11. $\dfrac{\dfrac{2}{3} + \dfrac{4}{5}}{\dfrac{2}{5} - \dfrac{3}{2}}$ **12.** $\dfrac{\dfrac{2}{x} + \dfrac{1}{x - 2}}{\dfrac{1}{x - 2} - \dfrac{3}{x}}$

Solve each equation.

13. $\dfrac{3}{x} = \dfrac{7}{5}$

14. $\dfrac{x}{x - 1} - \dfrac{3}{x} = \dfrac{1}{2}$

15. $\dfrac{1}{x} + \dfrac{1}{6} = \dfrac{1}{4}$

Solve each formula for the indicated variable.

16. $\dfrac{y - 3}{x + 2} = \dfrac{-1}{5}$ for y

17. $M = \dfrac{1}{3}b(c + d)$ for c

Solve each problem.

18. When all of the grocery carts escape from the supermarket, it takes Reginald 12 minutes to round them up and bring them back. Because Norman doesn't make as much per hour as Reginald, it takes Norman 18 minutes to do the same job. How long would it take them working together to complete the roundup?

19. Brenda and her husband Randy bicycled cross-country together. One morning, Brenda rode 30 miles. By traveling only 5 miles per hour faster and putting in one more hour, Randy covered twice the distance Brenda covered. What was the speed of each cyclist?

20. For a certain time period the ratio of the dollar value of exports to the dollar value of imports for the United States was 2 to 3. If the value of exports during that time period was 48 billion dollars, then what was the value of imports?

Solve each equation.

1. $3x - 2 = 5$

2. $\dfrac{3}{5}x = -2$

3. $2(x - 2) = 4x$

4. $2(x - 2) = 2x$

5. $2(x + 3) = 6x + 6$

6. $2(3x + 4) + x^2 = 0$

7. $4x - 4x^3 = 0$

8. $\dfrac{3}{x} = \dfrac{-2}{5}$

9. $\dfrac{3}{x} = \dfrac{x}{12}$

10. $\dfrac{x}{2} = \dfrac{4}{x - 2}$

11. $\dfrac{w}{18} - \dfrac{w - 1}{9} = \dfrac{4 - w}{6}$

12. $\dfrac{x}{x + 1} + \dfrac{1}{2x + 2} = \dfrac{7}{8}$

Solve each equation for y.

13. $2x + 3y = c$

14. $\dfrac{y - 3}{x - 5} = \dfrac{1}{2}$

15. $2y = ay + c$

16. $\dfrac{A}{y} = \dfrac{C}{B}$

17. $\dfrac{A}{y} + \dfrac{1}{3} = \dfrac{B}{y}$

18. $\dfrac{A}{y} - \dfrac{1}{2} = \dfrac{1}{3}$

19. $3y - 5ay = 8$

20. $y^2 - By = 0$

21. $A = \dfrac{1}{2}h(b + y)$

22. $2(b + y) = b$

Calculate the value of $b^2 - 4ac$ for each choice of a, b, and c.

23. $a = 1, b = 2, c = -15$

24. $a = 1, b = 8, c = 12$

25. $a = 2, b = 5, c = -3$

26. $a = 6, b = 7, c = -3$

Perform each indicated operation.

27. $(3x - 5) - (5x - 3)$

28. $(2a - 5)(a - 3)$

29. $x^7 \div x^3$

30. $\dfrac{x - 3}{5} + \dfrac{x + 4}{5}$

31. $\dfrac{1}{2} \cdot \dfrac{1}{x}$

32. $\dfrac{1}{2} + \dfrac{1}{x}$

33. $\dfrac{1}{2} \div \dfrac{1}{x}$

34. $\dfrac{1}{2} - \dfrac{1}{x}$

35. $\dfrac{x - 3}{5} - \dfrac{x + 4}{5}$

36. $\dfrac{3a}{2} \div 2$

37. $(x - 8)(x + 8)$

38. $3x(x^2 - 7)$

39. $2a^5 \cdot 5a^9$

40. $x^2 \cdot x^8$

41. $(k - 6)^2$

42. $(j + 5)^2$

43. $(g - 3) \div (3 - g)$

44. $(6x^3 - 8x^2) \div (2x)$

Solve.

45. *Present value.* An investor is interested in the amount or present value that she would have to invest today to receive periodic payments in the future. The present value of \$1 in 1 year and \$1 in 2 years with interest rate r compounded annually is given by the formula

$$P = \dfrac{1}{1 + r} + \dfrac{1}{(1 + r)^2}.$$

a) Rewrite the formula so that the right-hand side is a single rational expression.

b) Find P if $r = 7\%$.

c) The present value of \$1 per year for the next 10 years is given by the formula

$$P = \dfrac{1}{1 + r} + \dfrac{1}{(1 + r)^2} + \dfrac{1}{(1 + r)^3} + \cdots + \dfrac{1}{(1 + r)^{10}}.$$

Use this formula to find P if $r = 5\%$.

Use this review to check your understanding of Chapters 1–7. Immediately following this review you will find all answers along with references to examples from this text. If you answer a problem incorrectly, then you can refer to the cited example for a similar problem and solution.

Chapter 1

Evaluate each expression.

1. $-3^2 + 4 \cdot 2$

2. $-3 - 2 \,|\, 3 - 5 \,|$

3. $-\dfrac{2}{5} + \dfrac{3}{4}$

4. $-\dfrac{3}{5} \cdot \dfrac{20}{21}$

Identify the property that justifies each equation.

5. $3(x + 4) = 3x + 12$

6. $x \cdot 7 = 7x$

7. $4 + (9 + y) = (4 + 9) + y$

8. $0 + 3 = 3$

Simplify each expression.

9. $5x - (3 - 8x)$

10. $x + 3 - 0.2(5x - 30)$

11. $(-3x)(-5x)$

12. $\dfrac{3x + 12}{3}$

Chapter 2

Solve each equation.

13. $11x - 2 = 3$

14. $4x - 5 = 12x + 11$

15. $3(x - 6) = 3x - 6$

16. $x - 0.1x = 0.9x$

Solve each equation for y.

17. $5x - 3y = 9$

18. $ay + b = 0$

19. $a = t - by$

20. $\dfrac{a}{2} + \dfrac{y}{3} = \dfrac{3a}{4}$

Write a complete solution to each problem.

21. The sum of three consecutive integers is 102. What are the integers?

22. The perimeter of a rectangular painting is 100 inches. If the width is 4 inches less than the length, then what is the width?

23. The area of a triangular piece of property is 44,000 square feet. If the base of the triangle is 400 feet, then what is the height?

24. Ivan has 400 pounds of mixed nuts that contain no peanuts. How many pounds of peanuts should he put into the mixed nuts so that 20% of the mixture is peanuts?

Chapter 3

Solve each inequality. State the solution set using interval notation and graph the solution set.

25. $3x - 4 \le 11$

26. $5 - 7w > 26$

27. $|\, y - 6 \,| > 8$

28. $|\, 5 - 2q \,| \le 7$

Solve each compound inequality. State the solution set using interval notation.

29. $x + 2 > 5$ and $-3x \le 6$

30. $3x - 1 > 5$ or $5(x + 1) \le 5$

31. $x - 6 \ge 8$ and $x + 5 < 9$

32. $2x \le 7$ or $5x - 1 > 9$

Write a complete solution to each problem.

33. Charlie gets a 20% employees discount on clothes at the Toggery Shoppe. If he can afford at most $300 for a new suit and the store does not sell suits for less than $200, then what is the price range for the list price of suits that he can afford? Use interval notation to express your answer.

34. Kim's and Kurt's ages differ by 5 years and Kim is 23 years old. Write an absolute value equation that describes this situation. What are the possibilities for Kurt's age?

35. A rectangular flower bed is twice as long as it is wide. If the perimeter of the flower bed is less than 120 feet, then what are the possibilities for the width? Use interval notation to express your answer.

36. Alan's and Linda's grades on a test differed by less than 6 points and Linda's grade was 86. Write an absolute value inequality that describes this situation. What are the possibilities for Alan's grade? Use interval notation to express your answer.

Chapter 4

Find the slope of each line.

37. The line passing through the points $(1, 2)$ and $(3, 6)$

38. The line whose equation is $y = \frac{1}{2}x - 4$

39. The line parallel to $2x + 3y = 9$

40. The line perpendicular to $y = -3x + 5$

Find the equation of each line.

41. The line passing through the points $(0, 3)$ and $(2, 11)$

42. The line passing through the points $(-2, 4)$ and $(1, -2)$

43. The line through $(3, 5)$ that is parallel to $x = 4$

44. The line through $(0, 8)$ that is perpendicular to $y = \frac{1}{2}x$

Sketch each graph.

45. The graph of the equation $y = \frac{2}{3}x - 2$

46. The graph of the equation $3x - 5y = 150$

47. The graph of $y = 2$ in the coordinate plane

48. The graph of $x = 2$ in the coordinate plane

Chapter 5

Perform the indicated operations.

49. $(x^2 - 3x + 2) - (3x^2 + 9x - 4)$

50. $(x + 7)(x - 9)$

51. $(4w^2 - 3)^2$

52. $(x^3 - 2x^2 - x - 6) \div (x - 3)$

Use the rules of exponents to simplify each expression. Write the answers without negative exponents.

53. $-8x^4 \cdot 4x^3$

54. $3x(5x^2)^3$

55. $\dfrac{-6x^2y^3}{-2x^{-3}y^4}$

56. $\left(\dfrac{2a^2}{a^{-3}}\right)^3$

Perform each operation without a calculator. Write the answer in scientific notation.

57. $400{,}000 \cdot 600$

58. $(9 \times 10^3)(2 \times 10^6)$

59. $(2 \times 10^{-3})^4$

60. $\dfrac{2 \times 10^{-9}}{2000}$

Chapter 6

Factor each polynomial completely.

61. $24x^2y^3 + 18xy^5$

62. $x^2 - 3x - 54$

63. $4w^2 - 36w + 81$

64. $2a^3 - 6a^2 - 108a$

Solve each equation.

65. $x^2 = x$

66. $2x^3 - 8x = 0$

67. $a^2 + a - 6 = 0$

68. $(b - 2)(b + 3) = 24$

Write a complete solution to each problem.

69. The sum of two numbers is 10 and their product is 21. Find the numbers.

70. The length of a new television screen is 14 inches larger than the width and the diagonal is 26 inches. What are the length and width?

Chapter 7

Perform the indicated operation. Write each answer in lowest terms.

71. $\dfrac{5x}{2} + \dfrac{3x}{4}$

72. $\dfrac{5}{x - 2} - \dfrac{3}{2 - x}$

73. $\dfrac{9}{x^2 - 9} + \dfrac{2x}{x - 3}$

74. $\dfrac{2}{a - 5} + \dfrac{3}{a + 4}$

75. $\dfrac{w^3}{2w - 4} \cdot \dfrac{w^2 - 4}{w}$

76. $\dfrac{5ab^2}{6a^2b^3} \div \dfrac{10a}{21b^6}$

Solve each equation.

77. $\dfrac{2}{x} = \dfrac{3}{4}$

78. $\dfrac{1}{w - 3} = \dfrac{2}{w + 5}$

79. $\dfrac{1}{x} + \dfrac{3}{7} = \dfrac{1}{3x}$

80. $\dfrac{3}{a - 1} + \dfrac{1}{a + 2} = \dfrac{17}{10}$

Solve each formula for y.

81. $\dfrac{3}{y} = \dfrac{5}{x}$

82. $a = \dfrac{1}{2}y(w - c)$

83. $\dfrac{y - 3}{x + 5} = -3$

84. $\dfrac{3}{y} + \dfrac{1}{2} = \dfrac{1}{t}$

ANSWERS TO MIDTEXT DIAGNOSTIC REVIEW

If your answer is wrong, you can refer to the given example for a similar exercise and explanation.

1. -1; Section 1.5 Example 6
2. -7; Section 1.5 Example 1
3. $\frac{7}{20}$; Section 1.2 Example 7
4. $-\frac{4}{7}$; Section 1.2 Example 4
5. Distributive property; Section 1.7 Examples 5 and 8
6. Commutative property of multiplication; Section 1.7 Examples 2 and 8
7. Associative property of addition; Section 1.7 Example 8
8. Identity property; Section 1.7 Example 8
9. $13x - 3$; Section 1.8 Examples 7 and 8
10. 9; Section 1.8 Examples 7 and 8
11. $15x^2$; Section 1.8 Example 5
12. $x + 4$; Section 1.8 Example 6
13. $\left\{\frac{5}{11}\right\}$; Section 2.2 Examples 1 and 2
14. $\{-2\}$; Section 2.2 Example 5
15. No solution, $\varnothing$; Section 2.3 Example 2
16. All real numbers; Section 2.3 Example 1
17. $y = \frac{5}{3}x - 3$; Section 2.4 Examples 1–5
18. $y = -\frac{b}{a}$; Section 2.4 Examples 1–5
19. $y = \frac{t - a}{b}$; Section 2.4 Examples 1–5
20. $y = \frac{3}{4}a$; Section 2.4 Examples 1–5
21. 33, 34, 35; Section 2.6 Example 1
22. 23 in.; Section 2.6 Example 2
23. 220 ft; Section 2.6 Example 2
24. 100 pounds; Section 2.7 Example 5
25. $(-\infty, 5]$; Section 3.1 Examples 3 and 4

26. $(-\infty, -3)$; Section 3.1 Examples 3 and 4

27. $(-\infty, -2) \cup (14, \infty)$; Section 3.3 Examples 5–7

28. $[-1, 6]$; Section 3.3 Examples 5–7

29. $(3, \infty)$; Section 3.2 Example 8
30. $(-\infty, 0] \cup (2, \infty)$; Section 3.2 Example 9
31. No solution; Section 3.2 Example 7
32. $(-\infty, \infty)$; Section 3.2 Example 7
33. $[200, 375]$; Section 3.1 Examples 7 and 8
34. $|x - 23| = 5$, 18 or 28; Section 3.3 Example 10
35. $(0, 20)$; Section 3.1 Examples 7 and 8
36. $|x - 86| < 6$, $(80, 92)$; Section 3.3 Example 10
37. 2; Section 4.2 Examples 2 and 3
38. $\frac{1}{2}$; Section 4.3 Examples 1 and 2
39. $-\frac{2}{3}$; Section 4.4 Example 3
40. $\frac{1}{3}$; Section 4.3 Example 6
41. $y = 4x + 3$; Section 4.4 Example 2
42. $y = -2x$; Section 4.4 Example 2
43. $x = 3$; Section 4.4 Example 3
44. $y = -2x + 8$; Section 4.4 Example 4
45. Section 4.3 Example 4

46. Section 4.1 Example 6

47. Section 4.1 Example 4

48. Section 4.1 Example 5

49. $-2x^2 - 12x + 6$; Section 5.1 Example 5

50. $x^2 - 2x - 63$; Section 5.2 Example 3

51. $16w^4 - 24w^2 + 9$; Section 5.4 Example 1

52. $x^2 + x + 2$; Section 5.5 Example 4

53. $-32x^7$; Section 5.2 Example 1

54. $375x^7$; Section 5.6 Example 3

55. $\frac{3x^5}{y}$; Section 5.6 Example 1

56. $8a^{15}$; Section 5.6 Example 4

57. 2.4×10^8; Section 5.7 Example 8

58. 1.8×10^{10}; Section 5.7 Example 7

59. 1.6×10^{-11}; Section 5.7 Example 7

60. 1×10^{-12}; Section 5.7 Example 7

61. $6xy^3(4x + 3y^2)$; Section 6.1 Example 5

62. $(x - 9)(x + 6)$; Section 6.3 Examples 1–3

63. $(2w - 9)^2$; Section 6.2 Examples 2 and 3

64. $2a(a - 9)(a + 6)$; Section 6.3 Example 6

65. 0, 1; Section 6.6 Example 2

66. $-2, 0, 2$; Section 6.6 Example 5

67. $-3, 2$; Section 6.6 Example 1

68. $-6, 5$; Section 6.6 Example 3

69. 3 and 7; Section 6.6 Example 6

70. Length 24 in., width 10 in.; Section 6.6 Example 7

71. $\frac{13x}{4}$; Section 7.4 Example 4

72. $\frac{8}{x - 2}$; Section 7.4 Example 5

73. $\frac{2x^2 + 6x + 9}{x^2 - 9}$; Section 7.4 Example 5

74. $\frac{5a - 7}{(a - 5)(a + 4)}$; Section 7.4 Examples 4–6

75. $\frac{w^3 + 2w^2}{2}$; Section 7.2 Example 3

76. $\frac{7b^5}{4a^2}$; Section 7.2 Example 5

77. $\frac{8}{3}$; Section 7.6 Examples 1 and 2

78. 11; Section 7.7 Example 5

79. $-\frac{14}{9}$; Section 7.6 Examples 1 and 2

80. $-\frac{23}{13}, 2$; Section 7.6 Example 3

81. $y = \frac{3}{5}x$; Section 7.8 Example 2

82. $y = \frac{2a}{w - c}$; Section 7.8 Examples 1–3

83. $y = -2x - 12$; Section 7.8 Example 1

84. $y = \frac{6t}{2 - t}$; Section 7.8 Example 3

Systems of Linear Equations and Inequalities

I n his letter to M. Leroy in 1789 Benjamin Franklin said, "in this world nothing is certain but death and taxes." Since that time taxes have become not only inevitable, but also intricate and complex.

Each year the U.S. Congress revises parts of the Federal Income Tax Code. To help clarify these revisions, the Internal Revenue Service issues frequent revenue rulings. In addition, there are seven tax courts that further interpret changes and revisions, sometimes in entirely different ways. Is it any wonder that tax preparation has become complicated and few individuals actually prepare their own taxes? Both corporate and individual tax preparation is a growing business, and there are over 500,000 tax counselors helping more than 60 million taxpayers to file their returns correctly.

Everyone knows that doing taxes involves a lot of arithmetic, but not everyone knows that computing taxes can also involve algebra. In fact, to find state and federal taxes for certain corporations, you must solve a system of equations. You will see an example of using algebra to find amounts of income taxes in Exercises 53 and 54 of Section 8.1.

8.1 SOLVING SYSTEMS BY GRAPHING AND SUBSTITUTION

In Chapter 4 we studied linear equations in two variables, but we have usually considered only one equation at a time. In this chapter we will see problems that involve more than one equation. Any collection of two or more equations is called a **system** of equations. If the equations of a system involve two variables, then the set of ordered pairs that satisfy all of the equations is the **solution set of the system.** In this section we solve systems of linear equations in two variables and use systems to solve problems.

Solving a System by Graphing

Because the graph of each linear equation is a line, points that satisfy both equations lie on both lines. For some systems these points can be found by graphing.

EXAMPLE 1

A system with only one solution

Solve the system by graphing:

$$y = x + 2$$
$$x + y = 4$$

Solution

First write the equations in slope-intercept form:

$$y = x + 2$$
$$y = -x + 4$$

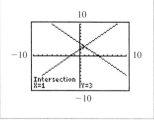

calculator

close-up

To check Example 1, graph

$$y_1 = x + 2$$

and

$$y_2 = -x + 4.$$

From the CALC menu, choose intersect to have the calculator locate the point of intersection of the two lines. After choosing intersect, you must indicate which two lines you want to intersect and then guess the point of intersection.

Use the y-intercept and the slope to graph each line. The graph of the system is shown in Fig. 8.1. From the graph it appears that these lines intersect at $(1, 3)$. To be certain, we can check that $(1, 3)$ satisfies both equations. Let $x = 1$ and $y = 3$ in $y = x + 2$ to get

$$3 = 1 + 2.$$

Let $x = 1$ and $y = 3$ in $x + y = 4$ to get

$$1 + 3 = 4.$$

Because $(1, 3)$ satisfies both equations, the solution set to the system is $\{(1, 3)\}$.

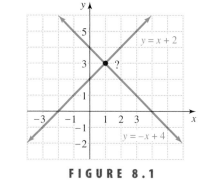

FIGURE 8.1

The graphs of the equations in the next example are parallel lines, and there is no point of intersection.

E X A M P L E   2

A system with no solution

Solve the system by graphing:

$$2x - 3y = 6$$
$$3y - 2x = 3$$

Solution

First write each equation in slope-intercept form:

$2x - 3y = 6$	$3y - 2x = 3$
$-3y = -2x + 6$	$3y = 2x + 3$
$y = \dfrac{2}{3}x - 2$	$y = \dfrac{2}{3}x + 1$

The graph of the system is shown in Fig. 8.2. Because the two lines in Fig. 8.2 are parallel, there is no ordered pair that satisfies both equations. The solution set to the system is the empty set, $\varnothing$.

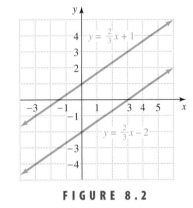

FIGURE   8.2

The equations in the next example look different, but their graphs are the same straight line.

E X A M P L E   3

A system with infinitely many solutions

Solve the system by graphing:

$$2(y + 2) = x$$
$$x - 2y = 4$$

Solution

Write each equation in slope-intercept form:

$2(y + 2) = x$	$x - 2y = 4$
$2y + 4 = x$	$-2y = -x + 4$
$y = \dfrac{1}{2}x - 2$	$y = \dfrac{1}{2}x - 2$

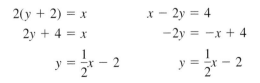

FIGURE   8.3

Because the equations have the same slope-intercept form, the original equations are equivalent. Their graphs are the same straight line as shown in Fig. 8.3. Every point on the line satisfies both equations of the system. There are infinitely many points in the solution set. The solution set is $\{(x, y) \mid x - 2y = 4\}$.

Independent, Inconsistent, and Dependent Equations

Our first three examples illustrate the three possible ways in which two lines can be positioned in a plane. In Example 1 the lines intersect in a single point. In this case we say that the equations are **independent** or the system is independent. If the two lines are parallel, as in Example 2, then there is no solution to the system, and the equations are **inconsistent** or the system is inconsistent. If the two equations of a system are equivalent, as in Example 3, the equations are **dependent** or the system is dependent. Figure 8.4 shows the types of graphs that correspond to independent, inconsistent, and dependent systems.

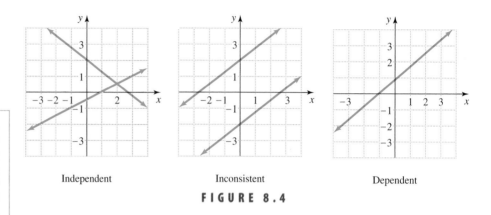

Independent   Inconsistent   Dependent

FIGURE 8.4

Solving by Substitution

Solving a system by graphing is certainly limited by the accuracy of the graph. If the lines intersect at a point whose coordinates are not integers, then it is difficult to determine those coordinates from the graph. The method of solving a system by **substitution** does not depend on a graph and is totally accurate. For substitution we replace a variable in one equation with an equivalent expression obtained from the other equation. Our intention in this substitution step is to eliminate a variable and to give us an equation involving only one variable.

EXAMPLE 4

An independent system solved by substitution

Solve the system by substitution:

$$2x + 3y = 8$$
$$y = -2x + 6$$

Solution

Since $y = -2x + 6$ we can replace y in $2x + 3y = 8$ by $-2x + 6$:

$$2x + 3y = 8$$
$$2x + 3(-2x + 6) = 8 \quad \text{Substitute } -2x + 6 \text{ for } y.$$
$$2x - 6x + 18 = 8$$
$$-4x = -10$$
$$x = \frac{5}{2}$$

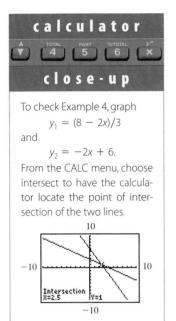

calculator

close-up

To check Example 4, graph

$$y_1 = (8 - 2x)/3$$

and

$$y_2 = -2x + 6.$$

From the CALC menu, choose intersect to have the calculator locate the point of intersection of the two lines.

To find y, we let $x = \frac{5}{2}$ in the equation $y = -2x + 6$:

$$y = -2\left(\frac{5}{2}\right) + 6 = -5 + 6 = 1$$

The next step is to check $x = \frac{5}{2}$ and $y = 1$ in each equation. If $x = \frac{5}{2}$ and $y = 1$ in $2x + 3y = 8$, we get

$$2\left(\frac{5}{2}\right) + 3(1) = 8.$$

If $x = \frac{5}{2}$ and $y = 1$ in $y + 2x = 6$, we get

$$1 + 2\left(\frac{5}{2}\right) = 6.$$

Because both of these equations are true, the solution set to the system is $\left\{\left(\frac{5}{2}, 1\right)\right\}$. The equations of this system are independent. ■

E X A M P L E   5   **An inconsistent system solved by substitution**

Solve by substitution:

$$x - 2y = 3$$
$$2x - 4y = 7$$

Solution

Solve the first equation for x to get $x = 2y + 3$. Substitute $2y + 3$ for x in the second equation:

$$2x - 4y = 7$$
$$2(2y + 3) - 4y = 7$$
$$4y + 6 - 4y = 7$$
$$6 = 7$$

helpful hint

The purpose of Example 5 is to show what happens when you try to solve an inconsistent system by substitution. If we had first written the equations in slope-intercept form, we would have known that the lines are parallel and the solution set is the empty set.

Because $6 = 7$ is incorrect no matter what values are chosen for x and y, there is no solution to this system of equations. The equations are inconsistent. To check, we write each equation in slope-intercept form:

$$x - 2y = 3 \qquad\qquad 2x - 4y = 7$$
$$-2y = -x + 3 \qquad\qquad -4y = -2x + 7$$
$$y = \frac{1}{2}x - \frac{3}{2} \qquad\qquad y = \frac{1}{2}x - \frac{7}{4}$$

The graphs of these equations are parallel lines with different y-intercepts. The solution set to the system is the empty set, $\varnothing$. ■

E X A M P L E   6   **A dependent system solved by substitution**

Solve by substitution:

$$2x + 3y = 5 + x + 4y$$
$$y = x - 5$$

Solution

Substitute $y = x - 5$ into the first equation:

$$2x + 3(x - 5) = 5 + x + 4(x - 5)$$
$$2x + 3x - 15 = 5 + x + 4x - 20$$
$$5x - 15 = 5x - 15$$

Because the last equation is an identity, any ordered pair that satisfies $y = x - 5$ will also satisfy $2x + 3y = 5 + x + 4y$. The equations of this system are dependent. The solution set to the system is the set of all points that satisfy $y = x - 5$. We write the solution set in set notation as

$$\{(x, y) \mid y = x - 5\}.$$

We can verify this result by writing $2x + 3y = 5 + x + 4y$ in slope-intercept form:

$$2x + 3y = 5 + x + 4y$$
$$3y = -x + 5 + 4y$$
$$-y = -x + 5$$
$$y = x - 5$$

Because this slope-intercept form is identical to the slope-intercept form of the other equation, they are two equations that look different for the same straight line. ∎

If a system is dependent, then an identity will result after the substitution. If the system is inconsistent, then an inconsistent equation will result after the substitution. The strategy for solving an independent system by substitution can be summarized as follows.

The Substitution Method

1. Solve one of the equations for one variable in terms of the other.
2. Substitute into the other equation to get an equation in one variable.
3. Solve for the remaining variable (if possible).
4. Insert the value just found into one of the original equations to find the value of the other variable.
5. Check the two values in both equations.

Applications

Many of the problems that we solved in previous chapters involved more than one unknown quantity. To solve them, we wrote expressions for all of the unknowns in terms of one variable. Now we can solve problems involving two unknowns by using two variables and writing a system of equations.

E X A M P L E   7 **Perimeter of a rectangle**

The length of a rectangular swimming pool is twice the width. If the perimeter is 120 feet, then what are the length and width?

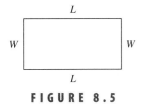

FIGURE 8.5

Solution

Draw a diagram as shown in Fig. 8.5. If L represents the length and W represents the width, then we can write the following system.

$$L = 2W$$
$$2L + 2W = 120$$

Since $L = 2W$, we can replace L in $2L + 2W = 120$ with $2W$:

$$2(2W) + 2W = 120$$
$$4W + 2W = 120$$
$$6W = 120$$
$$W = 20$$

So the width is 20 feet and the length is 2(20) or 40 feet. ■

E X A M P L E   8

helpful hint

In Chapter 2 we would have done Example 8 with one variable by letting x represent the amount invested at 10% and $20{,}000 - x$ represent the amount invested at 12%.

Tale of two investments

Belinda had $20,000 to invest. She invested part of it at 10% and the remainder at 12%. If her income from the two investments was $2160, then how much did she invest at each rate?

Solution

Let x be the amount invested at 10% and y be the amount invested at 12%. We can summarize all of the given information in a table:

	Amount	Rate	Interest
First investment	x	10%	$0.10x$
Second investment	y	12%	$0.12y$

We can write one equation about the amounts invested and another about the interest from the investments:

$$x + y = 20{,}000 \quad \text{Total amount invested}$$
$$0.10x + 0.12y = 2160 \quad \text{Total interest}$$

Solve the first equation for x to get $x = 20{,}000 - y$. Substitute $20{,}000 - y$ for x in the second equation:

$$0.10x + 0.12y = 2160$$
$$0.10(20{,}000 - y) + 0.12y = 2160 \quad \text{Replace } x \text{ by } 20{,}000 - y.$$
$$2000 - 0.10y + 0.12y = 2160 \quad \text{Solve for } y.$$
$$0.02y = 160$$
$$y = 8000$$
$$x = 12{,}000 \quad \text{Because } x = 20{,}000 - y$$

calculator

close-up

To check Example 8, graph
$$y_1 = 20{,}000 - x$$
and
$$y_2 = (2160 - 0.1x)/0.12.$$
The viewing window needs to be large enough to contain the point of intersection. Use the intersection feature to find the point of intersection.

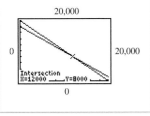

To check this answer, find 10% of $12,000 and 12% of $8000:

$$0.10(12{,}000) = 1{,}200$$
$$0.12(8000) = 960$$

Because $1200 + $960 = $2160 and $8000 + $12,000 = $20,000, we can be certain that Belinda invested $12,000 at 10% and $8000 at 12%. ■

M A T H   A T   W O R K

$x^2 + (x + 1)^2 = 5^2$

Accountants work with both people and numbers. When Maria L. Manning, an auditor at Deloitte & Touche LLP, is preparing for an audit, she studies the numbers on the balance sheet and income statement, comparing the current fiscal year to the prior one. The purpose of an independent audit is to give investors a realistic view of a company's finances. To determine that a company's financial statements are fairly stated in accordance with the General Accepted Accounting Procedures (GAAP), she first

ACCOUNTANT

interviews the comptroller to get the story behind the numbers. Typical questions she could ask are: How productive was your year? Are there any new products? What was behind the big stories in the newspapers? Then Ms. Manning and members of the audit team test the financial statement in detail, closely examining accounts relating to unusual losses or profits.

Ms. Manning is responsible for both manufacturing and mutual fund companies. At a manufacturing company, accounts receivable and inventory are two key components of an audit. For example, to test inventory, Ms. Manning visits a company's warehouse and physically counts all the items for sale to verify a company's assets. For a mutual fund company the audit team pays close attention to current events, for they indirectly affect the financial industry.

In Exercises 55 and 56 of this section you will work problems that involve one aspect of cost accounting: calculating the amount of taxes and bonuses paid by a company.

WARM-UPS

True or false? Explain your answer.

1. The ordered pair (1, 2) is in the solution set to the equation $2x + y = 4$.

2. The ordered pair (1, 2) satisfies $2x + y = 4$ and $3x - y = 6$.

3. The ordered pair (2, 3) satisfies $4x - y = 5$ and $4x - y = -5$.

4. If two distinct straight lines in the coordinate plane are not parallel, then they intersect in exactly one point.

5. The substitution method is used to eliminate a variable.

6. No ordered pair satisfies $y = 3x - 5$ and $y = 3x + 1$.

7. The equations $y = 3x - 6$ and $y = 2x + 4$ are independent.

8. The equations $y = 2x + 7$ and $y = 2x + 8$ are inconsistent.

9. The graphs of dependent equations are the same.

10. The graphs of independent linear equations intersect at exactly one point.

8.1   EXERCISES

Reading and Writing   *After reading this section, write out the answers to these questions. Use complete sentences.*

1. How do we solve a system of linear equations by graphing?

2. How can you determine whether a system has no solution by graphing?

3. What is the major disadvantage to solving a system by graphing?

4. How do we solve systems by substitution?

5. How can you identify an inconsistent system when solving by substitution?

6. How can you identify a dependent system when solving by substitution?

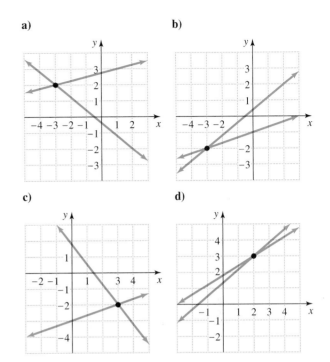

Solve each system by graphing. See Examples 1–3.

7. $y = 2x$
    $y = -x + 3$

8. $y = x - 3$
    $y = -x + 1$

9. $y = 2x - 1$
    $2y = x - 2$

10. $y = 2x + 1$
    $x + y = -2$

11. $y = x - 3$
    $x - 2y = 4$

12. $y = -3x$
    $x + y = 2$

13. $2y - 2x = 2$
    $2y - 2x = 6$

14. $3y - 3x = 9$
    $x - y = 1$

15. $y = -\dfrac{1}{2}x + 4$
    $x + 2y = 8$

16. $2x - 3y = 6$
    $y = \dfrac{2}{3}x - 2$

The graphs of the following systems are given in (a) through (d). Match each system with the correct graph.

17. $5x + 4y = 7$
    $x - 3y = 9$

18. $3x - 5y = -9$
    $5x - 6y = -8$

19. $4x - 5y = -2$
    $3y - x = -3$

20. $4x + 5y = -2$
    $4y - x = 11$

Solve each system by the substitution method. Determine whether the equations are independent, dependent, or inconsistent. See Examples 4–6.

21. $y = x - 5$
    $2x - 5y = 1$

22. $y = x + 4$
    $3y - 5x = 6$

23. $x = 2y - 7$
    $3x + 2y = -5$

24. $x = y + 3$
    $3x - 2y = 4$

25. $x - y = 5$
    $2x = 2y + 14$

26. $2x - y = 3$
    $2y = 4x - 6$

27. $y = 2x - 5$
    $y + 1 = 2(x - 2)$

28. $3x - 6y = 5$
    $2y = 4x - 6$

29. $2x + y = 9$
    $2x - 5y = 15$

30. $3y - x = 0$
    $x - 4y = -2$

31. $x - y = 0$
    $2x + 3y = 35$

32. $2y = x + 6$
    $-3x + 2y = -2$

33. $x + y = 40$
    $0.1x + 0.08y = 3.5$

34. $x - y = 10$
    $0.2x + 0.05y = 7$

35. $y = 2x - 30$
    $\dfrac{1}{5}x - \dfrac{1}{2}y = -1$

36. $3x - 5y = 4$
    $y = \dfrac{3}{4}x - 2$

37. $x + y = 4$
$x - y = 5$

38. $y = 2x - 3$
$y = 3x - 3$

39. $2x - y = 4$
$2x - y = 3$

40. $y = 3(x - 4)$
$3x - y = 12$

41. $3(y - 1) = 2(x - 3)$
$3y - 2x = -3$

42. $y = 3x$
$y = 3x + 1$

43. $x - y = -0.375$
$1.5x - 3y = -2.25$

44. $y - 2x = 1.875$
$2.5y - 3.5x = 11.8125$

In Exercises 45–58, write a system of two equations in two unknowns for each problem. Solve each system by substitution. See Examples 7 and 8.

45. Perimeter of a rectangle. The length of a rectangular swimming pool is 15 feet longer than the width. If the perimeter is 82 feet, then what are the length and width?

46. Household income. Alkena and Hsu together earn $84,326 per year. If Alkena earns $12,468 more per year than Hsu, then how much does each of them earn per year?

47. Different interest rates. Mrs. Brighton invested $30,000 and received a total of $2,300 in interest. If she invested part of the money at 10% and the remainder at 5%, then how much did she invest at each rate?

48. Different growth rates. The combined population of Marysville and Springfield was 25,000 in 1990. By 1995 the population of Marysville had increased by 10%, while Springfield had increased by 9%. If the total population increased by 2380 people, then what was the population of each city in 1990?

49. Finding numbers. The sum of two numbers is 2, and their difference is 26. Find the numbers.

50. Finding more numbers. The sum of two numbers is −16, and their difference is 8. Find the numbers.

51. Toasters and vacations. During one week a land developer gave away Florida vacation coupons or toasters to 100 potential customers who listened to a sales presentation. It costs the developer $6 for a toaster and $24 for a Florida vacation coupon. If his bill for prizes that week was $708, then how many of each prize did he give away?

52. Ticket sales. Tickets for a concert were sold to adults for $3 and to students for $2. If the total receipts were

$824 and twice as many adult tickets as student tickets were sold, then how many of each were sold?

53. Corporate taxes. According to Bruce Harrell, CPA, the amount of federal income tax for a class C corporation is deductible on the Louisiana state tax return, and the amount of state income tax for a class C corporation is deductible on the federal tax return. So for a state tax rate of 5% and a federal tax rate of 30%, we have

state tax = 0.05(taxable income − federal tax)

and

federal tax = 0.30(taxable income − state tax).

Find the amounts of state and federal income tax for a class C corporation that has a taxable income of $100,000.

54. More taxes. Use information given in Exercise 53 to find the amounts of state and federal income tax for a class C corporation that has a taxable income of $300,000. Use a state tax rate of 6% and a federal tax rate of 40%.

55. Cost accounting. The problems presented in this exercise and the next are encountered in cost accounting. A company has agreed to distribute 20% of its net income N to its employees as a bonus; $B = 0.20N$. If the company has income of $120,000 before the bonus, the bonus B is deducted from the $120,000 as an expense to determine net income; $N = 120,000 - B$. Solve the system of two equations in N and B to find the amount of the bonus.

56. Bonus and taxes. A company has an income of $100,000 before paying taxes and a bonus. The bonus B is to be 20% of the income after deducting income taxes T but before deducting the bonus. So

$$B = 0.20(100,000 - T).$$

Because the bonus is a deductible expense, the amount of income tax T at a 40% rate is 40% of the income after deducting the bonus. So

$$T = 0.40(100,000 - B).$$

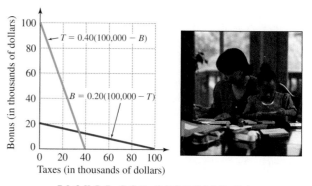

FIGURE FOR EXERCISE 56

a) Use the accompanying graph to estimate the values of T and B that satisfy both equations.

b) Solve the system algebraically to find the bonus and the amount of tax.

57. *Textbook case.* The accompanying graph shows the cost of producing textbooks and the revenue from the sale of those textbooks.

a) What is the cost of producing 10,000 textbooks?

b) What is the revenue when 10,000 textbooks are sold?

c) For what number of textbooks is the cost equal to the revenue?

d) The cost of producing zero textbooks is called the *fixed cost*. Find the fixed cost.

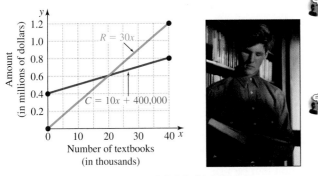

FIGURE FOR EXERCISE 57

58. *Free market.* The function $S = 5000 + 200x$ and $D = 9500 - 100x$ express the supply S and the demand D, respectively, for a popular compact disk brand as a function of its price x (in dollars).

a) Graph the functions on the same coordinate system.

b) What happens to the supply as the price increases?

c) What happens to the demand as the price increases?

d) The price at which supply and demand are equal is called the *equilibrium price*. What is the equilibrium price?

GETTING MORE INVOLVED

59. *Discussion.* Which of the following equations is not equivalent to $2x - 3y = 6$?

a) $3y - 2x = 6$ **b)** $y = \frac{2}{3}x - 2$

c) $x = \frac{3}{2}y + 3$ **d)** $2(x - 5) = 3y - 4$

60. *Discussion.* Which of the following equations is inconsistent with the equation $3x + 4y = 8$?

a) $y = \frac{3}{4}x + 2$ **b)** $6x + 8y = 16$

c) $y = -\frac{3}{4}x + 8$ **d)** $3x - 4y = 8$

GRAPHING CALCULATOR EXERCISES

61. Solve each system by graphing each pair of equations on a graphing calculator and using the trace feature or intersect feature to estimate the point of intersection. Find the coordinates of the intersection to the nearest tenth.

a) $y = 3.5x - 7.2$ **b)** $2.3x - 4.1y = 3.3$
$y = -2.3x + 9.1$ $3.4x + 9.2y = 1.3$

In this

section

- The Addition Method
- Equations Involving Fractions or Decimals
- Applications

8.2 **THE ADDITION METHOD**

In Section 8.1 you used substitution to eliminate a variable in a system of equations. In this section we see another method for eliminating a variable in a system of equations.

The Addition Method

In the **addition method** we eliminate a variable by adding the equations.

EXAMPLE 1

An independent system solved by addition

Solve the system by the addition method:

$$3x - 5y = -9$$
$$4x + 5y = 23$$

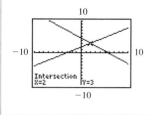

To check Example 1, graph
$$y_1 = (-9 - 3x)/-5$$
and
$$y_2 = (23 - 4x)/5.$$
Use the intersect feature to find the point of intersection of the two lines.

Solution

The addition property of equality allows us to add the same number to each side of an equation. We can also use the addition property of equality to add the two left sides and add the two right sides:

$$
\begin{aligned}
3x - 5y &= -9 \\
4x + 5y &= 23 \\
\hline
7x \phantom{{}- 5y} &= 14 \quad \text{Add.} \\
x &= 2
\end{aligned}
$$

The y-term was eliminated when we added the equations because the coefficients of the y-terms were opposites. Now use $x = 2$ in one of the original equations to find y. It does not matter which original equation we use. In this example we will use both equations to see that we get the same y in either case.

$$
\begin{aligned}
3x - 5y &= -9 & 4x + 5y &= 23 \\
3(2) - 5y &= -9 \quad \text{Replace } x \text{ by 2.} & 4(2) + 5y &= 23 \\
6 - 5y &= -9 \quad \text{Solve for } y. & 8 + 5y &= 23 \\
-5y &= -15 & 5y &= 15 \\
y &= 3 & y &= 3
\end{aligned}
$$

Because $3(2) - 5(3) = -9$ and $4(2) + 5(3) = 23$ are both true, $(2, 3)$ satisfies both equations. The solution set is $\{(2, 3)\}$. ∎

Actually the addition method can be used to eliminate any variable whose coefficients are opposites. If neither variable has coefficients that are opposites, then we use the multiplication property of equality to change the coefficients of the variables, as shown in Examples 2 and 3.

E X A M P L E   2

Using multiplication and addition

Solve the system by the addition method:

$$
\begin{aligned}
2x - 3y &= -13 \\
5x - 12y &= -46
\end{aligned}
$$

Solution

study tip

Keep on reviewing. After you have done your current assignment, go back a section or two and try a few problems. You will be amazed at how much your knowledge will improve with a regular review.

If we multiply both sides of the first equation by -4, the coefficients of y will be 12 and -12, and y will be eliminated by addition.

$$
\begin{aligned}
(-4)(2x - 3y) &= (-4)(-13) \quad \text{Multiply each side by } -4. \\
5x - 12y &= -46
\end{aligned}
$$

$$
\begin{aligned}
-8x + 12y &= 52 \\
5x - 12y &= -46 \quad \text{Add.} \\
\hline
-3x \phantom{{}+ 12y} &= 6 \\
x &= -2
\end{aligned}
$$

Replace x by -2 in one of the original equations to find y:

$$
\begin{aligned}
2x - 3y &= -13 \\
2(-2) - 3y &= -13 \\
-4 - 3y &= -13 \\
-3y &= -9 \\
y &= 3
\end{aligned}
$$

Because $2(-2) - 3(3) = -13$ and $5(-2) - 12(3) = -46$ are both true, the solution set is $\{(-2, 3)\}$. ■

EXAMPLE 3

Multiplying both equations before adding

Solve the system by the addition method:

$$-2x + 3y = 6$$
$$3x - 5y = -11$$

Solution

To eliminate x, we multiply the first equation by 3 and the second by 2:

$$3(-2x + 3y) = 3(6) \qquad \text{Multiply each side by 3.}$$
$$2(3x - 5y) = 2(-11) \qquad \text{Multiply each side by 2.}$$

$$\begin{aligned} -6x + 9y &= 18 \\ \underline{6x - 10y} &= \underline{-22} \qquad \text{Add.} \\ -y &= -4 \\ y &= 4 \end{aligned}$$

Note that we could have eliminated y by multiplying by 5 and 3. Now insert $y = 4$ into one of the original equations to find x:

$$-2x + 3(4) = 6 \quad \text{Let } y = 4 \text{ in } -2x + 3y = 6.$$
$$-2x + 12 = 6$$
$$-2x = -6$$
$$x = 3$$

Check that $(3, 4)$ satisfies both equations. The solution set is $\{(3, 4)\}$. ■

We can always use the addition method as long as the equations in a system are in the same form.

EXAMPLE 4

Using the addition method for an inconsistent system

Solve the system:

$$-4y = 5x + 7$$
$$4y = -5x + 12$$

Solution

If these equations are added, both variables are eliminated:

$$\begin{aligned} -4y &= 5x + 7 \\ \underline{4y} &= \underline{-5x + 12} \\ 0 &= 19 \end{aligned}$$

Because this equation is inconsistent, the original equations are inconsistent. The solution set to the system is the empty set, $\varnothing$. ■

Equations Involving Fractions or Decimals

When a system of equations involves fractions or decimals, we can use the multiplication property of equality to eliminate the fractions or decimals.

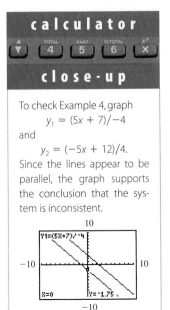

calculator

close-up

To check Example 4, graph
$$y_1 = (5x + 7)/-4$$
and
$$y_2 = (-5x + 12)/4.$$
Since the lines appear to be parallel, the graph supports the conclusion that the system is inconsistent.

EXAMPLE 5

A system with fractions

Solve the system:

$$\frac{1}{2}x - \frac{2}{3}y = 7$$

$$\frac{2}{3}x - \frac{3}{4}y = 11$$

Solution

Multiply the first equation by 6 and the second equation by 12:

$$6\left(\frac{1}{2}x - \frac{2}{3}y\right) = 6(7) \qquad \rightarrow \qquad 3x - 4y = 42$$

$$12\left(\frac{2}{3}x - \frac{3}{4}y\right) = 12(11) \qquad \rightarrow \qquad 8x - 9y = 132$$

To eliminate x, multiply the first equation by -8 and the second by 3:

$$\begin{aligned} -8(3x - 4y) &= -8(42) & \rightarrow & \quad -24x + 32y = -336 \\ 3(8x - 9y) &= 3(132) & \rightarrow & \quad \underline{24x - 27y = 396} \\ & & & \qquad\quad 5y = 60 \\ & & & \qquad\quad\, y = 12 \end{aligned}$$

Substitute $y = 12$ into the first of the original equations:

$$\frac{1}{2}x - \frac{2}{3}(12) = 7$$

$$\frac{1}{2}x - 8 = 7$$

$$\frac{1}{2}x = 15$$

$$x = 30$$

Check (30, 12) in the original system. The solution set is $\{(30, 12)\}$. ■

The strategy for solving a system by addition is summarized as follows.

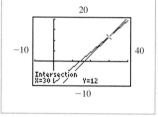

calculator

close-up

To check Example 5, graph

$$y_1 = (7 - (1/2)x)/(-2/3)$$

and

$$y_2 = (11 - (2/3)x)/(-3/4).$$

The lines appear to intersect at (30, 12).

The Addition Method

1. Write both equations in the same form (usually $Ax + By = C$).
2. Multiply one or both of the equations by appropriate numbers (if necessary) so that one of the variables will be eliminated by addition.
3. Add the equations to get an equation in one variable.
4. Solve the equation in one variable.
5. Substitute the value obtained for one variable into one of the original equations to obtain the value of the other variable.
6. Check the two values in both of the original equations.

Applications

Any system of two linear equations in two variables can be solved by either the addition method or substitution. In applications we use whichever method appears to be the simpler for the problem at hand.

E X A M P L E   6

Fajitas and burritos

At the Cactus Cafe the total price for four fajita dinners and three burrito dinners is $48, and the total price for three fajita dinners and two burrito dinners is $34. What is the price of each type of dinner?

Solution

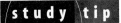

helpful hint

You can see from Example 6 that the standard form $Ax + By = C$ occurs naturally in accounting. This form will occur whenever we have the price of each item and a quantity of two items and want to express the total cost.

Let x represent the price (in dollars) of a fajita dinner, and let y represent the price (in dollars) of a burrito dinner. We can write two equations to describe the given information:

$$4x + 3y = 48$$
$$3x + 2y = 34$$

Because 12 is the least common multiple of 4 and 3 (the coefficients of x), we multiply the first equation by -3 and the second by 4:

$$-3(4x + 3y) = -3(48) \quad \text{Multiply each side by } -3.$$
$$4(3x + 2y) = 4(34) \quad \text{Multiply each side by } 4.$$

$$\begin{array}{r} -12x - 9y = -144 \\ \underline{12x + 8y = 136} \quad \text{Add.} \\ -y = -8 \\ y = 8 \end{array}$$

To find x, use $y = 8$ in the first equation $4x + 3y = 48$:

$$4x + 3(8) = 48$$
$$4x + 24 = 48$$
$$4x = 24$$
$$x = 6$$

So the fajita dinners are $6 each, and the burrito dinners are $8 each. Check this solution in the original problem. ■

E X A M P L E   7

Mixing cooking oil

Canola oil is 7% saturated fat, and corn oil is 14% saturated fat. Crisco sells a blend, Crisco Canola and Corn Oil, which is 11% saturated fat. How many gallons of each type of oil must be mixed to get 280 gallons of this blend?

Solution

study tip

Play offensive math, not defensive math. A student who says, "Give me a question and I'll see if I can answer it," is playing defensive math. The student is taking a passive approach to learning. A student who takes an active approach and knows the usual questions and answers for each topic is playing offensive math.

Let x represent the number of gallons of canola oil, and let y represent the number of gallons of corn oil. Make a table to summarize all facts:

	Amount (gallons)	% fat	Amount of fat (gallons)
Canola oil	x	7	$0.07x$
Corn oil	y	14	$0.14y$
Canola and Corn Oil	280	11	$0.11(280)$ or 30.8

We can write two equations to express the following facts: (1) the total amount of oil is 280 gallons and (2) the total amount of fat is 30.8 gallons. Then we can use multiplication and addition to solve the system.

$$(1) \qquad\qquad x + y = 280 \qquad \text{Multiply by } -0.07. \qquad -0.07x - 0.07y = -19.6$$
$$(2) \quad 0.07x + 0.14y = 30.80 \qquad\qquad\qquad \underline{0.07x + 0.14y = \quad 30.8}$$
$$0.07y = \quad 11.2$$
$$y = \frac{11.2}{0.07} = 160$$

If $y = 160$ and $x + y = 280$, then $x = 120$. Check that $0.07(120) + 0.14(160) = 30.8$. So it takes 120 gallons of canola oil and 160 gallons of corn oil to make 280 gallons of Crisco Canola and Corn Oil. ■

WARM-UPS

True or false? Explain your answer.

Exercises 1–6 refer to the following systems.

a) $3x - y = 9$ **b)** $4x - 2y = 20$ **c)** $x - y = 6$
     $2x + y = 6$ $-2x + y = -10$ $x - y = 7$

1. To solve system (a) by addition, we simply add the equations.

2. To solve system (a) by addition, we can multiply the first equation by 2 and the second by 3 and then add.

3. To solve system (b) by addition, we can multiply the second equation by 2 and then add.

4. Both $(0, -10)$ and $(5, 0)$ are in the solution set to system (b).

5. The solution set to system (b) is the set of all real numbers.

6. System (c) has no solution.

7. Both the addition method and substitution method are used to eliminate a variable from a system of two linear equations in two variables.

8. For the addition method, both equations must be in standard form.

9. To eliminate fractions in an equation, we multiply each side by the least common denominator of all fractions involved.

10. We can eliminate either variable by using the addition method.

8.2   EXERCISES

Reading and Writing   *After reading this section, write out the answers to these questions. Use complete sentences.*

1. What method is presented in this section for solving a system of linear equations?

2. What are we trying to accomplish by adding the equations?

3. What must we sometimes do before we add the equations?

4. How can you recognize an inconsistent system when solving by addition?

5. How can you recognize a dependent system when solving by addition?

6. For which systems is the addition method easier to use than substitution?

Solve each system by the addition method. See Examples 1–3.

7. $x + y = 7$
$x - y = 9$

8. $3x - 4y = 11$
$-3x + 2y = -7$

9. $x - y = 12$
$2x + y = 3$

10. $x - 2y = -1$
$-x + 5y = 4$

11. $2x - y = -5$
$3x + 2y = 3$

12. $3x + 5y = -11$
$x - 2y = 11$

13. $2x - 5y = 13$
$3x + 4y = -15$

14. $3x + 4y = -5$
$5x + 6y = -7$

15. $2x = 3y + 11$
$7x - 4y = 6$

16. $2x = 2 - y$
$3x + y = -1$

17. $x + y = 48$
$12x + 14y = 628$

18. $x + y = 13$
$22x + 36y = 356$

Solve each system by the addition method. Determine whether the equations are independent, dependent, or inconsistent. See Example 4.

19. $3x - 4y = 9$
$-3x + 4y = 12$

20. $x - y = 3$
$-6x + 6y = 17$

21. $5x - y = 1$
$10x - 2y = 2$

22. $4x + 3y = 2$
$-12x - 9y = -6$

23. $2x - y = 5$
$2x + y = 5$

24. $-3x + 2y = 8$
$3x + 2y = 8$

Solve each system by the addition method. See Example 5.

25. $\dfrac{1}{4}x + \dfrac{1}{3}y = 5$
$x - y = 6$

26. $\dfrac{3x}{2} - \dfrac{2y}{3} = 10$
$\dfrac{1}{2}x + \dfrac{1}{2}y = -1$

27. $\dfrac{x}{4} - \dfrac{y}{3} = -4$
$\dfrac{x}{8} + \dfrac{y}{6} = 0$

28. $\dfrac{x}{3} - \dfrac{y}{2} = \dfrac{5}{6}$
$\dfrac{x}{5} - \dfrac{y}{3} = -\dfrac{3}{5}$

29. $\dfrac{1}{8}x + \dfrac{1}{4}y = 5$
$\dfrac{1}{16}x + \dfrac{1}{2}y = 7$

30. $\dfrac{3}{7}x + \dfrac{5}{9}y = 27$
$\dfrac{1}{9}x + \dfrac{2}{7}y = 7$

31. $0.05x + 0.10y = 1.30$
$x + y = 19$

32. $0.1x + 0.06y = 9$
$0.09x + 0.5y = 52.7$

33. $x + y = 1200$
$0.12x + 0.09y = 120$

34. $x - y = 100$
$0.20x + 0.06y = 150$

35. $1.5x - 2y = -0.25$
$3x + 1.5y = 6.375$

36. $3x - 2.5y = 7.125$
$2.5x - 3y = 7.3125$

Write a system of two equations in two unknowns for each problem. Solve each system by the method of your choice. See Examples 6 and 7.

37. *Coffee and doughnuts.* On Monday, Archie paid $2.54 for three doughnuts and two coffees. On Tuesday he paid $2.46 for two doughnuts and three coffees. On Wednesday he was tired of paying the tab and went out for coffee by himself. What was his bill for one doughnut and one coffee?

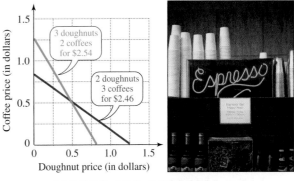

FIGURE FOR EXERCISE 37

38. *Books and magazines.* At Gwen's garage sale, all books were one price, and all magazines were another price. Harriet bought four books and three magazines for $1.45, and June bought two books and five magazines for $1.25. What was the price of a book and what was the price of a magazine?

39. *Boys and girls.* One-half of the boys and one-third of the girls of Freemont High attended the homecoming game, whereas one-third of the boys and one-half of the girls attended the homecoming dance. If there were 570 students at the game and 580 at the dance, then how many students are there at Freemont High?

40. *Girls and boys.* There are 385 surfers in Surf City. Two-thirds of the boys are surfers and one-twelfth of the girls are surfers. If there are two girls for every boy, then how many boys and how many girls are there in Surf City?

41. *Nickels and dimes.* Winborne has 35 coins consisting of dimes and nickels. If the value of his coins is $3.30, then how many of each type does he have?

42. *Pennies and nickels.* Wendy has 52 coins consisting of nickels and pennies. If the value of the coins is $1.20, then how many of each type does she have?

43. *Blending fudge.* The Chocolate Factory in Vancouver blends its double-dark-chocolate fudge, which is 35% fat, with its peanut butter fudge, which is 25% fat, to obtain double-dark-peanut fudge, which is 29% fat.
   a) Use the accompanying graph to estimate the number of pounds of each type that must be mixed to obtain 50 pounds of double-dark-peanut fudge.
   b) Write a system of equations and solve it algebraically to find the exact amount of each type that should be used to obtain 50 pounds of double-dark-peanut fudge.

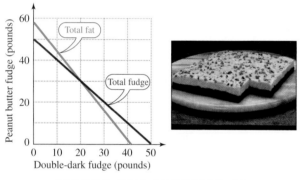

FIGURE FOR EXERCISE 43

44. *Low-fat yogurt.* Ziggy's Famous Yogurt blends regular yogurt that is 3% fat with its no-fat yogurt to obtain low-fat yogurt that is 1% fat. How many pounds of regular yogurt and how many pounds of no-fat yogurt should be mixed to obtain 60 pounds of low-fat yogurt?

45. *Keystone state.* Judy averaged 42 miles per hour (mph) driving from Allentown to Harrisburg and 51 mph

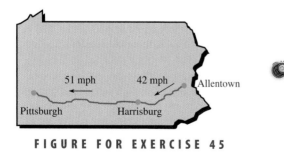

FIGURE FOR EXERCISE 45

driving from Harrisburg to Pittsburgh. If she drove a total of 288 miles in 6 hours, then how long did it take her to drive from Harrisburg to Pittsburgh?

46. *Empire state.* Spike averaged 45 mph driving from Rochester to Syracuse and 49 mph driving from Syracuse to Albany. If he drove a total of 237 miles in 5 hours, then how far is it from Syracuse to Albany?

47. *Probability of rain.* If Valerie Voss states that the probability of rain tomorrow is four times the probability that it doesn't rain, then what is the probability of rain tomorrow? (*Hint:* The probability that it rains plus the probability that it doesn't rain is 1.)

48. *Super Bowl contender.* A Las Vegas odds-maker believes that the probability that San Francisco plays in the next Super Bowl is nine times the probability that they do not play in the next Super Bowl. What is the odds-maker's probability that San Francisco plays in the next Super Bowl?

49. *Rectangular lot.* The width of a rectangular lot is 75% of its length. If the perimeter is 700 meters, then what are the length and width?

50. *Fence painting.* Darren and Douglas must paint the 792-foot fence that encircles their family home. Because Darren is older, he has agreed to paint 20% more than Douglas. How much of the fence will each boy paint?

GETTING MORE INVOLVED

51. *Discussion.* Explain how you decide whether it is easier to solve a system by substitution or addition.

52. *Exploration.* **a)** Write a linear equation in two variables that is satisfied by $(-3, 5)$.
   b) Write another linear equation in two variables that is satisfied by $(-3, 5)$.
   c) Are your equations independent or dependent?
   d) Explain how to select the second equation so that it will be independent of the first.

53. *Exploration.* **a)** Make up a system of two linear equations in two variables such that both $(-1, 2)$ and $(4, 5)$ are in the solution set.
   b) Are your equations independent or dependent?
   c) Is it possible to find an independent system that is satisfied by both ordered pairs? Explain.

8.3 SYSTEMS OF LINEAR EQUATIONS IN THREE VARIABLES

The techniques that you learned in Section 8.2 can be extended to systems of equations in more than two variables. In this section we use elimination of variables to solve systems of equations in three variables.

Definition

The equation $5x - 4y = 7$ is called a linear equation in two variables because its graph is a straight line. The equation $2x + 3y - 4z = 12$ is similar in form, and so it is a linear equation in three variables. An equation in three variables is graphed in a three-dimensional coordinate system. The graph of a linear equation in three variables is a plane, not a line. We will not graph equations in three variables in this text, but we can solve systems without graphing. In general, we make the following definition.

> **Linear Equation in Three Variables**
>
> If A, B, C, and D are real numbers, with A, B, and C not all zero, then
>
> $$Ax + By + Cz = D$$
>
> is called a **linear equation in three variables.**

Solving a System by Elimination

A solution to an equation in three variables is an **ordered triple** such as $(-2, 1, 5)$, where the first coordinate is the value of x, the second coordinate is the value of y, and the third coordinate is the value of z. There are infinitely many solutions to a linear equation in three variables.

The solution to a system of equations in three variables is the set of all ordered triples that satisfy all of the equations of the system. The techniques for solving a system of linear equations in three variables are similar to those used on systems of linear equations in two variables. We eliminate variables by either substitution or addition.

EXAMPLE 1

A linear system with a single solution

Solve the system:

$$
\begin{align}
(1) \qquad & x + y - z = -1 \\
(2) \qquad & 2x - 2y + 3z = 8 \\
(3) \qquad & 2x - y + 2z = 9
\end{align}
$$

Solution

We can eliminate z from Eqs. (1) and (2) by multiplying Eq. (1) by 3 and adding it to Eq. (2):

$$
\begin{array}{ll}
3x + 3y - 3z = -3 & \text{Eq. (1) multiplied by 3} \\
\underline{2x - 2y + 3z = 8} & \text{Eq. (2)} \\
(4) \qquad 5x + y \qquad\quad = 5
\end{array}
$$

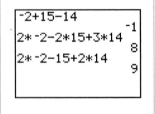

calculator

close-up

You can use a calculator to check that $(-2, 15, 14)$ satisfies all three equations of the original system.

```
-2+15-14
              -1
2*-2-2*15+3*14
              8
2*-2-15+2*14
              9
```

Now we must eliminate the same variable, z, from another pair of equations. Eliminate z from (1) and (3):

$$
\begin{array}{ll}
2x + 2y - 2z = -2 & \text{Eq. (1) multiplied by 2} \\
2x - y + 2z = 9 & \text{Eq. (3)} \\
\hline
(5) \quad 4x + y \quad\quad = 7 &
\end{array}
$$

Equations (4) and (5) give us a system with two variables. We now solve this system. Eliminate y by multiplying Eq. (5) by -1 and adding the equations:

$$
\begin{array}{ll}
5x + y = 5 & \text{Eq. (4)} \\
-4x - y = -7 & \text{Eq. (5) multiplied by } -1 \\
\hline
x \quad\quad = -2 &
\end{array}
$$

Now that we have x, we can replace x by -2 in Eq. (5) to find y:

$$
\begin{aligned}
4x + y &= 7 \\
4(-2) + y &= 7 \\
-8 + y &= 7 \\
y &= 15
\end{aligned}
$$

Now replace x by -2 and y by 15 in Eq. (1) to find z:

$$
\begin{aligned}
x + y - z &= -1 \\
-2 + 15 - z &= -1 \\
13 - z &= -1 \\
-z &= -14 \\
z &= 14
\end{aligned}
$$

Check that $(-2, 15, 14)$ satisfies all three of the original equations. The solution set is $\{(-2, 15, 14)\}$. ∎

The strategy that we follow for solving a system of three linear equations in three variables is stated as follows.

Solving a System in Three Variables

1. Use substitution or addition to eliminate any one of the variables from a pair of equations of the system. Look for the easiest variable to eliminate.
2. Eliminate the same variable from another pair of equations of the system.
3. Solve the resulting system of two equations in two unknowns.
4. After you have found the values of two of the variables, substitute into one of the original equations to find the value of the third variable.
5. Check the three values in all of the original equations.

In the next example we use a combination of addition and substitution.

E X A M P L E 2 **Using addition and substitution**

Solve the system:

$$
\begin{array}{lll}
(1) & x + y & = 4 \\
(2) & 2x - 3z & = 14 \\
(3) & 2y + z & = 2
\end{array}
$$

Solution

From Eq. (1) we get $y = 4 - x$. If we substitute $y = 4 - x$ into Eq. (3), then Eqs. (2) and (3) will be equations involving x and z only.

$$
\begin{aligned}
(3) \qquad 2y + z &= 2 \\
2(4 - x) + z &= 2 \qquad \text{Replace } y \text{ by } 4 - x. \\
8 - 2x + z &= 2 \qquad \text{Simplify.} \\
(4) \qquad -2x + z &= -6
\end{aligned}
$$

Now solve the system consisting of Eqs. (2) and (4) by addition:

$$
\begin{aligned}
2x - 3z &= 14 \qquad \text{Eq. (2)} \\
-2x + z &= -6 \qquad \text{Eq. (4)} \\
\hline
-2z &= 8 \\
z &= -4
\end{aligned}
$$

Use Eq. (3) to find y:

$$
\begin{aligned}
2y + z &= 2 \qquad \text{Eq. (3)} \\
2y + (-4) &= 2 \qquad \text{Let } z = -4. \\
2y &= 6 \\
y &= 3
\end{aligned}
$$

Use Eq. (1) to find x:

$$
\begin{aligned}
x + y &= 4 \qquad \text{Eq. (1)} \\
x + 3 &= 4 \qquad \text{Let } y = 3. \\
x &= 1
\end{aligned}
$$

Check that $(1, 3, -4)$ satisfies all three of the original equations. The solution set is $\{(1, 3, -4)\}$. ■

CAUTION In solving a system in three variables it is essential to keep your work organized and neat. Writing short notes that explain your steps (as was done in the examples) will allow you to go back and check your work.

Graphs of Equations in Three Variables

The graph of any equation in three variables can be drawn on a three-dimensional coordinate system. The graph of a linear equation in three variables is a plane. To solve a system of three linear equations in three variables by graphing, we would have to draw the three planes and then identify the points that lie on all three of them. This method would be difficult even when the points have simple coordinates. So we will not attempt to solve these systems by graphing.

By considering how three planes might intersect, we can better understand the different types of solutions to a system of three equations in three variables. Figure 8.6, on page 414, shows some of the possibilities for the positioning of three planes in three-dimensional space. In most of the problems that we will solve the planes intersect at a single point as in Fig. 8.6(a). The solution set consists of one ordered triple. However, the system may include two equations corresponding to parallel planes that have no intersection. In this case the equations are said to be **inconsistent.** If the system has at least two inconsistent equations, then the solution set is the empty set [see Figs. 8.6(b) and 8.6(c)].

There are two ways in which the intersection of three planes can consist of infinitely many points. The intersection could be a line or a plane. To get a line, we can have either three different planes intersecting along a line, as in Fig. 8.6(d) or two

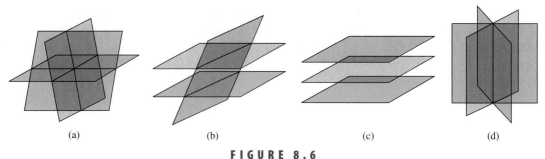

(a) (b) (c) (d)

FIGURE 8.6

equations for the same plane, with the third plane intersecting that plane. If all three equations are equations of the same plane, we get that plane for the intersection. We will not solve systems corresponding to all of the possible configurations described. The following examples illustrate two of these cases.

EXAMPLE 3

An inconsistent system of three linear equations

Solve the system:

$$
\begin{array}{rl}
(1) & x + y - z = 5 \\
(2) & 3x - 2y + z = 8 \\
(3) & 2x + 2y - 2z = 7
\end{array}
$$

Solution

We can eliminate the variable z from Eqs. (1) and (2) by adding them:

$$
\begin{array}{rl}
(1) & x + y - z = 5 \\
(2) & \underline{3x - 2y + z = 8} \\
& 4x - y \quad\;\; = 13
\end{array}
$$

To eliminate z from Eqs. (1) and (3), multiply Eq. (1) by -2 and add the resulting equation to Eq. (3):

$$
\begin{array}{ll}
-2x - 2y + 2z = -10 & \text{Eq. (1) multiplied by } -2 \\
\underline{2x + 2y - 2z = 7} & \text{Eq. (3)} \\
\quad\quad\quad\; 0 = -3 &
\end{array}
$$

Because the last equation is false, there are two inconsistent equations in the system. Therefore the solution set is the empty set. ■

EXAMPLE 4

A dependent system of three equations

Solve the system:

$$
\begin{array}{rl}
(1) & 2x - 3y - z = 4 \\
(2) & -6x + 9y + 3z = -12 \\
(3) & 4x - 6y - 2z = 8
\end{array}
$$

Solution

We will first eliminate x from Eqs. (1) and (2). Multiply Eq. (1) by 3 and add the resulting equation to Eq. (2):

$$
\begin{array}{ll}
6x - 9y - 3z = 12 & \text{Eq. (1) multiplied by 3} \\
\underline{-6x + 9y + 3z = -12} & \text{Eq. (2)} \\
\quad\quad\quad\;\; 0 = 0 &
\end{array}
$$

helpful hint

If you recognize that multiplying Eq. (1) by -3 will produce Eq. (2), and multiplying Eq. (1) by 2 will produce Eq. (3), then you can conclude that all three equations are equivalent and there is no need to add the equations.

The last statement is an identity. The identity occurred because Eq. (2) is a multiple of Eq. (1). In fact, Eq. (3) is also a multiple of Eq. (1). These equations are dependent. They are all equations for the same plane. The solution set is the set of all points on that plane,

$$\{(x, y, z) \mid 2x - 3y - z = 4\}. \qquad \blacksquare$$

Applications

Problems involving three unknown quantities can often be solved by using a system of three equations in three variables.

E X A M P L E   5

Finding three unknown rents

Theresa took in a total of $1240 last week from the rental of three condominiums. She had to pay 10% of the rent from the one-bedroom condo for repairs, 20% of the rent from the two-bedroom condo for repairs, and 30% of the rent from the three-bedroom condo for repairs. If the three-bedroom condo rents for twice as much as the one-bedroom condo and her total repair bill was $276, then what is the rent for each condo?

Solution

Let x, y, and z represent the rent on the one-bedroom, two-bedroom, and three-bedroom condos, respectively. We can write one equation for the total rent, another equation for the total repairs, and a third equation expressing the fact that the rent for the three-bedroom condo is twice that for the one-bedroom condo:

$$x + y + z = 1240$$
$$0.1x + 0.2y + 0.3z = 276$$
$$z = 2x$$

Substitute $z = 2x$ into both of the other equations to eliminate z:

$$x + y + 2x = 1240$$
$$0.1x + 0.2y + 0.3(2x) = 276$$

$$3x + \quad y = 1240$$
$$0.7x + 0.2y = 276$$

$$-2(3x + y) = -2(1240) \qquad \text{Multiply each side by } -2.$$
$$10(0.7x + 0.2y) = 10(276) \qquad \text{Multiply each side by 10.}$$

$$-6x - 2y = -2480$$
$$\underline{7x + 2y = 2760} \qquad \text{Add.}$$
$$x \qquad = 280$$

$$z = 2(280) = 560 \qquad \text{Because } z = 2x$$
$$280 + y + 560 = 1240 \qquad \text{Because } x + y + z = 1240$$
$$y = 400$$

Check that (280, 400, 560) satisfies all three of the original equations. The condos rent for $280, $400, and $560 per week. $\blacksquare$

WARM-UPS

True or false? Explain your answer.

1. The point $(1, -2, 3)$ is in the solution set to the equation $x + y - z = 4$.

2. The point $(4, 1, 1)$ is the only solution to the equation $x + y - z = 4$.

3. The ordered triple $(1, -1, 2)$ satisfies $x + y + z = 2$, $x - y - z = 0$, and $2x + y - z = -1$.

4. Substitution cannot be used on three equations in three variables.

5. Two distinct planes are either parallel or intersect in a single point.

6. The equations $x - y + 2z = 6$ and $x - y + 2z = 4$ are inconsistent.

7. The equations $3x + 2y - 6z = 4$ and $-6x - 4y + 12z = -8$ are dependent.

8. The graph of $y = 2x - 3z + 4$ is a straight line.

9. The value of x nickels, y dimes, and z quarters is $0.05x + 0.10y + 0.25z$ cents.

10. If $x = -2$, $z = 3$, and $x + y + z = 6$, then $y = 7$.

8.3 EXERCISES

Reading and Writing *After reading this section, write out the answers to these questions. Use complete sentences.*

1. What is a linear equation in three variables?

2. What is an ordered triple?

3. What is a solution to a system of linear equations in three variables?

4. How do we solve systems of linear equations in three variables?

5. What does the graph of a linear equation in three variables look like?

6. How are the planes positioned when a system of linear equations in three variables is inconsistent?

Solve each system of equations. See Examples 1 and 2.

7. $x + y + z = 2$
   $x + 2y - z = 6$
   $2x + y - z = 5$

8. $2x - y + 3z = 14$
   $x + y - 2z = -5$
   $3x + y - z = 2$

9. $x - 2y + 4z = 3$
   $x + 3y - 2z = 6$
   $x - 4y + 3z = -5$

10. $2x + 3y + z = 13$
    $-3x + 2y + z = -4$
    $4x - 4y + z = 5$

11. $2x - y + z = 10$
    $3x - 2y - 2z = 7$
    $x - 3y - 2z = 10$

12. $x - 3y + 2z = -11$
    $2x - 4y + 3z = -15$
    $3x - 5y - 4z = 5$

13. $2x - 3y + z = -9$
    $-2x + y - 3z = 7$
    $x - y + 2z = -5$

14. $3x - 4y + z = 19$
    $2x + 4y + z = 0$
    $x - 2y + 5z = 17$

15. $2x - 5y + 2z = 16$
    $3x + 2y - 3z = -19$
    $4x - 3y + 4z = 18$

16. $-2x + 3y - 4z = 3$
    $3x - 5y + 2z = 4$
    $-4x + 2y - 3z = 0$

17. $x + y = 4$
    $y - z = -2$
    $x + y + z = 9$

18. $x + y - z = 0$
    $x - y = -2$
    $y + z = 10$

19. $\begin{aligned} x + y &= 7 \\ y - z &= -1 \\ x + 3z &= 18 \end{aligned}$

20. $\begin{aligned} 2x - y &= -8 \\ y + 3z &= 22 \\ x - z &= -8 \end{aligned}$

Solve each system. See Examples 3 and 4.

21. $\begin{aligned} x - y + 2z &= 3 \\ 2x + y - z &= 5 \\ 3x - 3y + 6z &= 4 \end{aligned}$

22. $\begin{aligned} 2x - 4y + 6z &= 12 \\ 6x - 12y + 18z &= 36 \\ -x + 2y - 3z &= -6 \end{aligned}$

23. $\begin{aligned} 3x - y + z &= 5 \\ 9x - 3y + 3z &= 15 \\ -12x + 4y - 4z &= -20 \end{aligned}$

24. $\begin{aligned} 4x - 2y - 2z &= 5 \\ 2x - y - z &= 7 \\ -4x + 2y + 2z &= 6 \end{aligned}$

25. $\begin{aligned} x - y &= 3 \\ y + z &= 8 \\ 2x + 2z &= 7 \end{aligned}$

26. $\begin{aligned} 2x - y &= 6 \\ 2y + z &= -4 \\ 8x + 2z &= 3 \end{aligned}$

27. $\begin{aligned} 0.10x + 0.08y - 0.04z &= 3 \\ 5x + 4y - 2z &= 150 \\ 0.3x + 0.24y - 0.12z &= 9 \end{aligned}$

28. $\begin{aligned} 0.06x - 0.04y + z &= 6 \\ 3x - 2y + 50z &= 300 \\ 0.03x - 0.02y + 0.5z &= 3 \end{aligned}$

 Use a calculator to solve each system.

29. $\begin{aligned} 3x + 2y - 0.4z &= 0.1 \\ 3.7x - 0.2y + 0.05z &= 0.41 \\ -2x + 3.8y - 2.1z &= -3.26 \end{aligned}$

30. $\begin{aligned} 3x - 0.4y + 9z &= 1.668 \\ 0.3x + 5y - 8z &= -0.972 \\ 5x - 4y - 8z &= 1.8 \end{aligned}$

Solve each problem by using a system of three equations in three unknowns. See Example 5.

31. *Diversification.* Ann invested a total of $12,000 in stocks, bonds, and a mutual fund. She received a 10% return on her stock investment, an 8% return on her bond investment, and a 12% return on her mutual fund. Her total return was $1230. If the total investment in stocks and bonds equaled her mutual fund investment, then how much did she invest in each?

32. *Paranoia.* Fearful of a bank failure, Norman split his life savings of $60,000 among three banks. He received 5%, 6%, and 7% on the three deposits. In the account earning 7% interest, he deposited twice as much as in the account earning 5% interest. If his total earnings were $3760, then how much did he deposit in each account?

33. *Big tipper.* On Monday Headley paid $1.70 for two cups of coffee and one doughnut, including the tip. On Tuesday he paid $1.65 for two doughnuts and a cup of coffee, including the tip. On Wednesday he paid $1.30 for one coffee and one doughnut, including the tip. If he always tips the same amount, then what is the amount of each item?

34. *Weighing in.* Anna, Bob, and Chris will not disclose their weights but agree to be weighed in pairs. Anna and Bob together weigh 226 pounds. Bob and Chris together weigh 210 pounds. Anna and Chris together weigh 200 pounds. How much does each student weigh?

Anna & Bob   Bob & Chris   Anna & Chris

FIGURE FOR EXERCISE 34

35. *Lunch-box special.* Salvador's Fruit Mart sells variety packs. The small pack contains three bananas, two apples, and one orange for $1.80. The medium pack contains four bananas, three apples, and three oranges for $3.05. The family size contains six bananas, five apples, and four oranges for $4.65. What price should Salvador charge for his lunch-box special that consists of one banana, one apple, and one orange?

36. *Three generations.* Edwin, his father, and his grandfather have an average age of 53. One-half of his grandfather's age, plus one-third of his father's age, plus one-fourth of Edwin's age is 65. If 4 years ago, Edwin's grandfather was four times as old as Edwin, then how old are they all now?

37. *Error in the scale.* Alex is using a scale that is known to have a constant error. A can of soup and a can of tuna are placed on this scale, and it reads 24 ounces. Now four identical cans of soup and three identical cans of tuna are placed on an accurate scale, and a weight of 80 ounces is recorded. If two cans of tuna weigh 18 ounces on the bad scale, then what is the amount of error in the scale and what is the correct weight of each type of can?

38. *Three-digit number.* The sum of the digits of a three-digit number is 11. If the digits are reversed, the new number is 46 more than five times the old number. If the hundreds digit plus twice the tens digit is equal to the units digit, then what is the number?

39. *Working overtime.* To make ends meet, Ms. Farnsby works three jobs. Her total income last year was $48,000. Her income from teaching was just $6000 more than her income from house painting. Royalties from her textbook sales were one-seventh of the total money she received from teaching and house painting. How much did she make from each source last year?

40. *Pocket change.* Harry has $2.25 in nickels, dimes, and quarters. If he had twice as many nickels, half as many dimes, and the same number of quarters, he would have $2.50. If he has 27 coins altogether, then how many of each does he have?

GETTING MORE INVOLVED

41. *Exploration.* Draw diagrams showing the possible ways to position three planes in three-dimensional space.

42. *Discussion.* Make up a system of three linear equations in three variables for which the solution set is $\{(0, 0, 0)\}$. A system with this solution set is called a *homogeneous* system. Why do you think it is given that name?

43. *Cooperative learning.* Working in groups, do parts (a)–(d) below. Then write a report on your findings.
  a) Find values of a, b, and c so that the graph of $y = ax^2 + bx + c$ goes through the points $(-1, -2)$, $(1, 0)$, and $(2, 7)$.
  b) Arbitrarily select three ordered pairs and find the equation of the parabola that goes through the three points.
  c) Could more than one parabola pass through three given points? Give reasons for your answer.
  d) Explain how to pick three points for which no parabola passes through all of them.

8.4 SOLVING LINEAR SYSTEMS USING MATRICES

You solved linear systems in two variables by substitution and addition in Sections 8.1 and 8.2. Those methods are done differently on each system. In this section you will learn the Gaussian elimination method, which is related to the addition method. The Gaussian elimination method is performed in the same way on every system. We first need to introduce some new terminology.

In this section

- Matrices
- The Augmented Matrix
- The Gaussian Elimination Method
- Inconsistent and Dependent Equations

Matrices

A **matrix** is a rectangular array of numbers. The **rows** of a matrix run horizontally, and the **columns** of a matrix run vertically. A matrix with m rows and n columns has **order** $m \times n$ (read "m by n"). Each number in a matrix is called an **element** or **entry** of the matrix.

EXAMPLE 1   Order of a matrix

Determine the order of each matrix.

a) $\begin{bmatrix} -1 & 2 \\ 5 & \sqrt{2} \\ 0 & 3 \end{bmatrix}$ **b)** $\begin{bmatrix} 2 & 3 \\ -1 & 5 \end{bmatrix}$ **c)** $\begin{bmatrix} 1 & 2 & 3 \\ 4 & 5 & 6 \\ -1 & 0 & 2 \end{bmatrix}$ **d)** $\begin{bmatrix} 1 & 3 & 6 \end{bmatrix}$

Solution

Because matrix (a) has 3 rows and 2 columns, its order is 3×2. Matrix (b) is a 2×2 matrix, matrix (c) is a 3×3 matrix, and matrix (d) is a 1×3 matrix. ■

The Augmented Matrix

The solution to a system of linear equations such as

$$x - 2y = -5$$
$$3x + y = 6$$

study tip

As soon as possible after class, find a quiet place and work on your homework. The longer you wait, the harder it is to remember what happened in class.

depends on the coefficients of x and y and the constants on the right-hand side of the equation. The matrix of coefficients for this system is the 2×2 matrix

$$\begin{bmatrix} 1 & -2 \\ 3 & 1 \end{bmatrix}.$$

If we insert the constants from the right-hand side of the system into the matrix of coefficients, we get the 2×3 matrix

$$\left[\begin{array}{cc|c} 1 & -2 & -5 \\ 3 & 1 & 6 \end{array}\right].$$

We use a vertical line between the coefficients and the constants to represent the equal signs. This matrix is the **augmented matrix** of the system. Two systems of linear equations are **equivalent** if they have the same solution set. Two augmented matrices are **equivalent** if the systems they represent are equivalent.

E X A M P L E   2   **Writing the augmented matrix**
Write the augmented matrix for each system of equations.

a) $3x - 5y = 7$
$\quad x + \ y = 4$

b) $x + y - \ z = 5$
$\quad 2x \quad + \ z = 3$
$\quad 2x - y + 4z = 0$

c) $x + y \quad = 1$
$\quad y + z = 6$
$\quad\quad\quad z = -5$

Solution

a) $\left[\begin{array}{cc|c} 3 & -5 & 7 \\ 1 & 1 & 4 \end{array}\right]$

b) $\left[\begin{array}{ccc|c} 1 & 1 & -1 & 5 \\ 2 & 0 & 1 & 3 \\ 2 & -1 & 4 & 0 \end{array}\right]$

c) $\left[\begin{array}{ccc|c} 1 & 1 & 0 & 1 \\ 0 & 1 & 1 & 6 \\ 0 & 0 & 1 & -5 \end{array}\right]$ ■

E X A M P L E   3   **Writing the system**
Write the system of equations represented by each augmented matrix.

a) $\left[\begin{array}{cc|c} 1 & 4 & -2 \\ 1 & -1 & 3 \end{array}\right]$

b) $\left[\begin{array}{cc|c} 1 & 0 & 5 \\ 0 & 1 & 1 \end{array}\right]$

c) $\left[\begin{array}{ccc|c} 2 & 3 & 4 & 6 \\ -1 & 0 & 5 & -2 \\ 1 & -2 & 3 & 1 \end{array}\right]$

Solution

a) Use the first two numbers in each row as the coefficients of x and y and the last number as the constant to get the following system:

$$x + 4y = -2$$
$$x - \ y = 3$$

b) Use the first two numbers in each row as the coefficients of x and y and the last number as the constant to get the following system:

$$x = 5$$
$$y = 1$$

c) Use the first three numbers in each row as the coefficients of x, y, and z and the last number as the constant to get the following system:

$$2x + 3y + 4z = 6$$
$$-x \quad\quad + 5z = -2$$
$$x - 2y + 3z = 1$$ ■

study tip

When doing homework or taking notes, use a pencil with an eraser. Everyone makes mistakes. If you get a problem wrong, don't start over. Check your work for errors and use the eraser. It is better to find out where you went wrong than simply to get the right answer.

The Gaussian Elimination Method

When we solve a single equation, we write simpler and simpler equivalent equations to get an equation whose solution is obvious. In the **Gaussian elimination method** we write simpler and simpler equivalent augmented matrices until we get an augmented matrix (like the one in Example 3(b)) in which the solution to the corresponding system is obvious.

Because each row of an augmented matrix represents an equation, we can perform the **row operations** on the augmented matrix. These row operations, which follow, correspond to the usual operations with equations used in the addition method.

Row Operations

The following row operations on an augmented matrix give an equivalent augmented matrix:

1. Interchange two rows of the matrix.
2. Multiply every element in a row by a nonzero real number.
3. Add to a row a multiple of another row.

In the Gaussian elimination method our goal is to use row operations to obtain an augmented matrix that has ones on the **diagonal** in its matrix of coefficients and zeros elsewhere:

$$\begin{bmatrix} 1 & 0 & a \\ 0 & 1 & b \end{bmatrix}$$

The system corresponding to this augmented matrix is $x = a$ and $y = b$. So the solution set to the system is $\{(a, b)\}$.

E X A M P L E   4

Gaussian elimination with two equations in two variables

Use the Gaussian elimination method to solve the system:

$$x - 3y = 11$$
$$2x + y = 1$$

Solution

Start with the augmented matrix:

$$\begin{bmatrix} 1 & -3 & 11 \\ 2 & 1 & 1 \end{bmatrix}$$

Multiply row 1 (R_1) by -2 and add the result to row 2 (R_2). So R_2 is replaced with $-2R_1 + R_2$. In symbols $-2R_1 + R_2 \rightarrow R_2$. Read the arrow as "replaces." Because $-2R_1 = [-2, 6, -22]$ and $R_2 = [2, 1, 1]$, $-2R_1 + R_2 = [0, 7, -21]$. Note that the coefficient of x in the second equation is now 0. We get the following matrix:

$$\begin{bmatrix} 1 & -3 & 11 \\ 0 & 7 & -21 \end{bmatrix} \quad {}^{-2R_1 + R_2 \rightarrow R_2}$$

Multiply each element of row 2 by $\frac{1}{7}$ (in symbols, $\frac{1}{7}R_2 \rightarrow R_2$):

$$\begin{bmatrix} 1 & -3 & | & 11 \\ 0 & 1 & | & -3 \end{bmatrix} \quad \frac{1}{7}R_2 \rightarrow R_2$$

Multiply row 2 by 3 and add the result to row 1. Because $3R_2 = [0, 3, -9]$ and $R_1 = [1, -3, 11]$, $3R_2 + R_1 = [1, 0, 2]$. Note that the coefficient of y in the first equation is now 0. We get the following matrix:

$$\begin{bmatrix} 1 & 0 & | & 2 \\ 0 & 1 & | & -3 \end{bmatrix} \quad 3R_2 + R_1 \rightarrow R_1$$

This augmented matrix represents the system $x = 2$ and $y = -3$. So the solution set to the system is $\{(2, -3)\}$. Check in the original system. ■

The augmented matrix in Example 4 started off with a 1 in the first position on the diagonal. If that position contains a nonzero number other than 1, we can divide the first row by that number. This division might cause fractions to appear in the augmented matrix as shown in the next example.

E X A M P L E   5   **Gaussian elimination involving fractions**

Use the Gaussian elimination method to solve the system

$$2x - 3y = 8$$
$$3x + 2y = -1$$

Solution

Start with the augmented matrix and multiply the first row by $\frac{1}{2}$ (or divide it by 2):

$$\begin{bmatrix} 2 & -3 & | & 8 \\ 3 & 2 & | & -1 \end{bmatrix}$$

$$\begin{bmatrix} 1 & -\frac{3}{2} & | & 4 \\ 3 & 2 & | & -1 \end{bmatrix} \quad \frac{1}{2}R_1 \rightarrow R_1$$

Now multiply the first row by -3 and add the result onto the second row:

$$\begin{bmatrix} 1 & -\frac{3}{2} & | & 4 \\ 0 & \frac{13}{2} & | & -13 \end{bmatrix} \quad -3R_1 + R_2 \rightarrow R_2$$

Multiply the second row by $\frac{2}{13}$:

$$\begin{bmatrix} 1 & -\frac{3}{2} & | & 4 \\ 0 & 1 & | & -2 \end{bmatrix} \quad \frac{2}{13}R_2 \rightarrow R_2$$

Multiply the second row by $\frac{3}{2}$ and add the result onto the first row:

$$\begin{bmatrix} 1 & 0 & | & 1 \\ 0 & 1 & | & -2 \end{bmatrix} \quad \frac{3}{2}R_2 + R_1 \rightarrow R_1$$

The final augmented matrix represents the system $x = 1$ and $y = -2$. So the solution set is $\{(1, -2)\}$. ■

In the next example we use the row operations on the augmented matrix of a system of three linear equations in three variables.

E X A M P L E 6

Gaussian elimination with three equations in three variables

Use the Gaussian elimination method to solve the following system:

$$2x - y + z = -3$$
$$x + y - z = 6$$
$$3x - y - z = 4$$

Solution

Start with the augmented matrix and interchange the first and second rows to get a 1 in the upper left position in the matrix:

$$\begin{bmatrix} 2 & -1 & 1 & | & -3 \\ 1 & 1 & -1 & | & 6 \\ 3 & -1 & -1 & | & 4 \end{bmatrix} \quad \text{The augmented matrix}$$

$$\begin{bmatrix} 1 & 1 & -1 & | & 6 \\ 2 & -1 & 1 & | & -3 \\ 3 & -1 & -1 & | & 4 \end{bmatrix} \quad R_1 \leftrightarrow R_2$$

Now multiply the first row by -2 and add the result onto the second row. Multiply the first row by -3 and add the result onto the third row. These two steps eliminate the variable x from the second and third rows:

$$\begin{bmatrix} 1 & 1 & -1 & | & 6 \\ 0 & -3 & 3 & | & -15 \\ 0 & -4 & 2 & | & -14 \end{bmatrix} \quad \begin{array}{l} -2R_1 + R_2 \rightarrow R_2 \\ -3R_1 + R_3 \rightarrow R_3 \end{array}$$

> **helpful hint**
>
> It is not necessary to perform the row operations in exactly the same order as is shown in Example 6. As long as you use the legitimate row operations and get to the final form, you will get the solution to the system. Of course, you must double check your arithmetic at every step if you want to be successful at Gaussian elimination.

Multiply the second row by $-\frac{1}{3}$ to get 1 in the second position on the diagonal:

$$\begin{bmatrix} 1 & 1 & -1 & | & 6 \\ 0 & 1 & -1 & | & 5 \\ 0 & -4 & 2 & | & -14 \end{bmatrix} \quad -\frac{1}{3}R_2 \rightarrow R_2$$

Use the second row to eliminate the variable y from the first and third rows:

$$\begin{bmatrix} 1 & 0 & 0 & | & 1 \\ 0 & 1 & -1 & | & 5 \\ 0 & 0 & -2 & | & 6 \end{bmatrix} \quad \begin{array}{l} -1R_2 + R_1 \rightarrow R_1 \\ 4R_2 + R_3 \rightarrow R_3 \end{array}$$

Multiply the third row by $-\frac{1}{2}$ to get a 1 in the third position on the diagonal:

$$\begin{bmatrix} 1 & 0 & 0 & | & 1 \\ 0 & 1 & -1 & | & 5 \\ 0 & 0 & 1 & | & -3 \end{bmatrix} \quad -\frac{1}{2}R_3 \rightarrow R_3$$

Use the third row to eliminate the variable z from the second row:

$$\begin{bmatrix} 1 & 0 & 0 & | & 1 \\ 0 & 1 & 0 & | & 2 \\ 0 & 0 & 1 & | & -3 \end{bmatrix} \quad R_3 + R_2 \rightarrow R_2$$

This last augmented matrix represents the system $x = 1$, $y = 2$, and $z = -3$. So the solution set to the system is $\{(1, 2, -3)\}$. ■

    Gaussian elimination may not be as easy to perform on some systems as it was in Example 6. Fractions may appear early in the process, making the computations more difficult. However, computers and even graphing calculators can perform row operations on matrices. So problems with computation can be overcome. The systems that we solve in this section will not be that complicated, but they will increase your understanding of Gaussian elimination.

Inconsistent and Dependent Equations

Inconsistent and dependent equations are easily recognized in using the Gaussian elimination method.

E X A M P L E   7

Gaussian elimination with an inconsistent system

Solve the system:

$$\begin{aligned} x - y &= 1 \\ -3x + 3y &= 4 \end{aligned}$$

helpful hint

The point of Example 7 is to recognize an inconsistent system with Gaussian elimination. We could also observe that -3 times the first equation yields

$$-3x + 3y = -3,$$

which is inconsistent with

$$-3x + 3y = 4.$$

Solution

Start with the augmented matrix:

$$\begin{bmatrix} 1 & -1 & | & 1 \\ -3 & 3 & | & 4 \end{bmatrix}$$

Multiply row 1 by 3 and add the result to row 2. We get the following matrix:

$$\begin{bmatrix} 1 & -1 & | & 1 \\ 0 & 0 & | & 7 \end{bmatrix} \quad 3R_1 + R_2 \rightarrow R_2$$

The second row of the augmented matrix corresponds to the equation $0 = 7$. So the equations are inconsistent, and there is no solution to the system. ■

E X A M P L E   8

Gaussian elimination with a dependent system

Solve the system:

$$\begin{aligned} 3x + y &= 1 \\ 6x + 2y &= 2 \end{aligned}$$

Solution

Start with the augmented matrix:

$$\begin{bmatrix} 3 & 1 & | & 1 \\ 6 & 2 & | & 2 \end{bmatrix}$$

Multiply row 1 by -2 and add the result to row 2. We get the following matrix:

$$\begin{bmatrix} 3 & 1 & | & 1 \\ 0 & 0 & | & 0 \end{bmatrix} \quad -2R_1 + R_2 \rightarrow R_2$$

In the second row of the augmented matrix we have the equation $0 = 0$. So the equations are dependent. Every ordered pair that satisfies the first equation satisfies both equations. The solution set is $\{(x, y) | 3x + y = 1\}$. ∎

WARM-UPS

True or false? Explain your answer.

Statements 1–7 refer to the following matrices:

a) $\begin{bmatrix} 1 & 3 & | & 5 \\ -1 & -3 & | & 2 \end{bmatrix}$

b) $\begin{bmatrix} 1 & 3 & | & 5 \\ 0 & 0 & | & 7 \end{bmatrix}$

c) $\begin{bmatrix} -1 & 2 & | & -3 \\ 2 & -4 & | & 3 \end{bmatrix}$

d) $\begin{bmatrix} 1 & 3 & | & 5 \\ 0 & 0 & | & 0 \end{bmatrix}$

1. The augmented matrix for $x + 3y = 5$ and $-x - 3y = 2$ is matrix (a).

2. The augmented matrix for $2y - x = -3$ and $2x - 4y = 3$ is matrix (c).

3. Matrix (a) is equivalent to matrix (b).

4. Matrix (c) is equivalent to matrix (d).

5. The system corresponding to matrix (b) is inconsistent.

6. The system corresponding to matrix (c) is dependent.

7. The system corresponding to matrix (d) is independent.

8. The augmented matrix for a system of two linear equations in two unknowns is a 2×2 matrix.

9. The notation $2R_1 + R_3 \rightarrow R_3$ means to replace R_3 by $2R_1 + R_3$.

10. The notation $R_1 \leftrightarrow R_2$ means to replace R_2 by R_1.

8.4 EXERCISES

Reading and Writing *After reading this section, write out the answers to these questions. Use complete sentences.*

1. What is a matrix?

2. What is the difference between a row and a column of a matrix?

3. What is the order of a matrix?

4. What is an element of a matrix?

5. What is an augmented matrix?

6. What is the goal of Gaussian elimination?

Determine the order of each matrix. See Example 1.

7. $\begin{bmatrix} 5 & 0 \\ -2 & 3 \end{bmatrix}$

8. $\begin{bmatrix} 1 & 3 & 6 \\ -7 & 0 & 2 \end{bmatrix}$

9. $\begin{bmatrix} a & c \\ 0 & d \\ 3 & w \end{bmatrix}$

10. $\begin{bmatrix} 0 & a & b \\ 5 & 7 & -8 \\ a & b & 2 \end{bmatrix}$ **11.** $\begin{bmatrix} -\sqrt{3} \\ \pi \\ 1 \\ \frac{1}{2} \end{bmatrix}$ **12.** $[3 \quad 0 \quad 4]$

Write the augmented matrix for each system of equations. See Example 2.

13. $2x - 3y = 9$
  $-3x + y = -1$

14. $x - y = 4$
  $2x + y = 3$

15. $x - y + z = 1$
  $x + y - 2z = 3$
  $y - 3z = 4$

16. $x + y = 2$
  $y - 3z = 5$
  $-3x + 2z = 8$

Write the system of equations represented by each augmented matrix. See Example 3.

17. $\begin{bmatrix} 5 & 1 & | & -1 \\ 2 & -3 & | & 0 \end{bmatrix}$ **18.** $\begin{bmatrix} 1 & 0 & | & 4 \\ 0 & 1 & | & -3 \end{bmatrix}$

19. $\begin{bmatrix} 1 & 0 & 0 & | & 6 \\ -1 & 0 & 1 & | & -3 \\ 1 & 1 & 0 & | & 1 \end{bmatrix}$ **20.** $\begin{bmatrix} 1 & 0 & 4 & | & 3 \\ 0 & 2 & 1 & | & -1 \\ 1 & 1 & 1 & | & 1 \end{bmatrix}$

Determine the row operation that was used to convert each given augmented matrix into the equivalent augmented matrix that follows it. See Example 4.

21. $\begin{bmatrix} 3 & 2 & | & 12 \\ 1 & -1 & | & -1 \end{bmatrix}, \begin{bmatrix} 1 & -1 & | & -1 \\ 3 & 2 & | & 12 \end{bmatrix}$

22. $\begin{bmatrix} 1 & -1 & | & -1 \\ 3 & 2 & | & 12 \end{bmatrix}, \begin{bmatrix} 1 & -1 & | & -1 \\ 0 & 5 & | & 15 \end{bmatrix}$

23. $\begin{bmatrix} 1 & -1 & | & -1 \\ 0 & 5 & | & 15 \end{bmatrix}, \begin{bmatrix} 1 & -1 & | & -1 \\ 0 & 1 & | & 3 \end{bmatrix}$

24. $\begin{bmatrix} 1 & -1 & | & -1 \\ 0 & 1 & | & 3 \end{bmatrix}, \begin{bmatrix} 1 & 0 & | & 2 \\ 0 & 1 & | & 3 \end{bmatrix}$

Solve each system using the Gaussian elimination method. See Examples 4–8.

25. $x + y = 3$
  $-3x + y = -1$

26. $x - y = -1$
  $2x - y = 2$

27. $2x - y = 3$
  $x + y = 9$

28. $3x - 4y = -1$
  $x - y = 0$

29. $3x - y = 4$
  $2x + y = 1$

30. $2x - y = -3$
  $3x + y = -2$

31. $6x - 7y = 0$
  $2x + y = 20$

32. $2x + y = 11$
  $2x - y = 1$

33. $2x - 3y = 4$
  $-2x + 3y = 5$

34. $x - 3y = 8$
  $2x - 6y = 1$

35. $x + 2y = 1$
  $3x + 6y = 3$

36. $2x - 3y = 1$
  $-6x + 9y = -3$

37. $x + y + z = 6$
  $x - y + z = 2$
  $2y - z = 1$

38. $x - y - z = 0$
  $-x - y + z = -4$
  $-x + y - z = -2$

39. $2x + y + z = 4$
  $x + y - z = 1$
  $x - y + 2z = 2$

40. $3x - y = 1$
  $x + y + z = 4$
  $x + 2z = 3$

41. $2x - y + z = 0$
  $x + y - 3z = 3$
  $x - y + z = -1$

42. $x - y - z = 0$
  $-x - y + 2z = -1$
  $-x + y - 2z = -3$

43. $-x + 3y + z = 0$
  $x - y - 4z = -3$
  $x + y + 2z = 3$

44. $-x + z = -2$
  $2x - y = 5$
  $y + 3z = 9$

45. $x - y + z = 1$
  $2x - 2y + 2z = 2$
  $-3x + 3y - 3z = -3$

46. $4x - 2y + 2z = 2$
  $2x - y + z = 1$
  $-2x + y - z = -1$

47. $x + y - z = 2$
  $2x - y + z = 1$
  $3x + 3y - 3z = 8$

48. $x + y + z = 5$
  $x - y - z = 8$
  $-x + y + z = 2$

GETTING MORE INVOLVED

49. *Cooperative learning.* Write a step-by-step procedure for solving any system of two linear equations in two variables by the Gaussian elimination method. Have a classmate evaluate your procedure by using it to solve a system.

50. *Cooperative learning.* Repeat Exercise 49 for a system of three linear equations in three variables.

8.5 DETERMINANTS AND CRAMER'S RULE

The Gaussian elimination method of Section 8.4 can be performed the same way on every system. Another method that is applied the same way for every system is Cramer's rule, which we study in this section. Before you learn Cramer's rule, we need to introduce a new number associated with a matrix, called a *determinant*.

Determinants

The determinant of a square matrix is a real number corresponding to the matrix. For a 2×2 matrix the determinant is defined as follows.

Determinant of a 2×2 Matrix

The determinant of the matrix $\begin{bmatrix} a & b \\ c & d \end{bmatrix}$ is defined to be the real number $ad - bc$. We write

$$\begin{vmatrix} a & b \\ c & d \end{vmatrix} = ad - bc.$$

Note that the symbol for the determinant is a pair of vertical lines similar to the absolute value symbol, while a matrix is enclosed in brackets.

EXAMPLE 1

Using the definition of determinant

Find the determinant of each matrix.

a) $\begin{bmatrix} 1 & 3 \\ -2 & 5 \end{bmatrix}$ **b)** $\begin{bmatrix} 2 & 4 \\ 6 & 12 \end{bmatrix}$

Solution

a) $\begin{vmatrix} 1 & 3 \\ -2 & 5 \end{vmatrix} = 1 \cdot 5 - 3(-2)$ **b)** $\begin{vmatrix} 2 & 4 \\ 6 & 12 \end{vmatrix} = 2 \cdot 12 - 4 \cdot 6$

$= 5 + 6$ $= 24 - 24$

$= 11$ $= 0$ ■

calculator

close-up

With a graphing calculator you can define matrix A using MATRX EDIT.

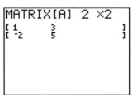

```
MATRIX[A] 2 x2
[ 1      3      ]
[ -2     5      ]
```

Then use the determinant function (det) found in MATRX MATH and the A from MATRX NAMES to find its determinant.

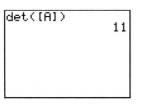

```
det([A])
              11
```

Cramer's Rule (2×2)

To understand Cramer's rule, we first solve a general system of two linear equations in two variables. Consider the system

$$(1) \qquad a_1 x + b_1 y = c_1$$
$$(2) \qquad a_2 x + b_2 y = c_2$$

where a_1, b_1, c_1, a_2, b_2, and c_2 represent real numbers. To eliminate y, we multiply Eq. (1) by b_2 and Eq. (2) by $-b_1$:

$$a_1 b_2 x + b_1 b_2 y = c_1 b_2 \qquad \text{Eq. (1) multiplied by } b_2$$

$$\underline{-a_2 b_1 x - b_1 b_2 y = -c_2 b_1} \qquad \text{Eq. (2) multiplied by } -b_1$$

$$a_1 b_2 x - a_2 b_1 x \qquad = c_1 b_2 - c_2 b_1 \qquad \text{Add.}$$

$$(a_1 b_2 - a_2 b_1)x = c_1 b_2 - c_2 b_1$$

$$x = \frac{c_1 b_2 - c_2 b_1}{a_1 b_2 - a_2 b_1} \qquad \text{Provided that } a_1 b_2 - a_2 b_1 \neq 0$$

Using similar steps to eliminate x from the system, we get

$$y = \frac{a_1 c_2 - a_2 c_1}{a_1 b_2 - a_2 b_1},$$

provided that $a_1 b_2 - a_2 b_1 \neq 0$. These formulas for x and y can be written by using determinants. In the determinant form they are known as **Cramer's rule.**

Cramer's Rule

The solution to the system

$$a_1 x + b_1 y = c_1$$
$$a_2 x + b_2 y = c_2$$

is given by $x = \frac{D_x}{D}$ and $y = \frac{D_y}{D}$, where

$$D = \begin{vmatrix} a_1 & b_1 \\ a_2 & b_2 \end{vmatrix}, \qquad D_x = \begin{vmatrix} c_1 & b_1 \\ c_2 & b_2 \end{vmatrix}, \qquad \text{and} \qquad D_y = \begin{vmatrix} a_1 & c_1 \\ a_2 & c_2 \end{vmatrix},$$

provided that $D \neq 0$.

helpful hint

Notice that Cramer's rule gives us a precise formula for finding the solution to an independent system. The addition and substitution methods are more like guidelines under which we choose the best way to proceed.

Note that D is the determinant made up of the original coefficients of x and y. D is used in the denominator for both x and y. D_x is obtained by replacing the first (or x) column of D by the constants c_1 and c_2. D_y is found by replacing the second (or y) column of D by the constants c_1 and c_2.

E X A M P L E   2   **Solving an independent system with Cramer's rule**

Use Cramer's rule to solve the system:

$$3x - 2y = 4$$
$$2x + \ y = -3$$

Solution

First find the determinants D, D_x, and D_y:

$$D = \begin{vmatrix} 3 & -2 \\ 2 & 1 \end{vmatrix} = 3 - (-4) = 7$$

$$D_x = \begin{vmatrix} 4 & -2 \\ -3 & 1 \end{vmatrix} = 4 - 6 = -2, \qquad D_y = \begin{vmatrix} 3 & 4 \\ 2 & -3 \end{vmatrix} = -9 - 8 = -17$$

By Cramer's rule, we have

$$x = \frac{D_x}{D} = -\frac{2}{7} \qquad \text{and} \qquad y = \frac{D_y}{D} = -\frac{17}{7}.$$

Check in the original equations. The solution set is $\left\{ \left(-\frac{2}{7}, -\frac{17}{7} \right) \right\}$.   ■

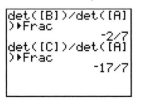

calculator

close-up

Use MATRX EDIT to define D, D_x, and D_y as A, B, and C. Now use Cramer's rule on the home screen to find x and y.

```
det([B])/det([A]
)▶Frac
                -2/7
det([C])/det([A]
)▶Frac
               -17/7
```

C A U T I O N   Cramer's rule works *only* when the determinant D is *not* equal to zero. Cramer's rule solves only those systems that have a single point in their solution set. If $D = 0$, we use elimination to determine whether the solution set is empty or contains all points of a line.

Minors

To each element of a 3 × 3 matrix there corresponds a 2 × 2 matrix that is obtained by deleting the row and column of that element. The determinant of the 2 × 2 matrix is called the **minor** of that element.

E X A M P L E   3   **Finding minors**

Find the minors for the elements 2, 3, and −6 of the 3 × 3 matrix

$$\begin{bmatrix} 2 & -1 & -8 \\ 0 & -2 & 3 \\ 4 & -6 & 7 \end{bmatrix}.$$

Solution

To find the minor for 2, delete the first row and first column of the matrix:

$$\begin{bmatrix} 2 & -1 & -8 \\ 0 & -2 & 3 \\ 4 & -6 & 7 \end{bmatrix}$$

Now find the determinant of $\begin{bmatrix} -2 & 3 \\ -6 & 7 \end{bmatrix}$:

$$\begin{vmatrix} -2 & 3 \\ -6 & 7 \end{vmatrix} = (-2)(7) - (-6)(3) = 4$$

The minor for 2 is 4. To find the minor for 3, delete the second row and third column of the matrix:

$$\begin{bmatrix} 2 & -1 & -8 \\ 0 & -2 & 3 \\ 4 & -6 & 7 \end{bmatrix}$$

Now find the determinant of $\begin{bmatrix} 2 & -1 \\ 4 & -6 \end{bmatrix}$:

$$\begin{vmatrix} 2 & -1 \\ 4 & -6 \end{vmatrix} = (2)(-6) - (4)(-1) = -8$$

The minor for 3 is −8. To find the minor for −6, delete the third row and the second column of the matrix:

$$\begin{bmatrix} 2 & -1 & -8 \\ 0 & -2 & 3 \\ 4 & -6 & 7 \end{bmatrix}$$

Now find the determinant of $\begin{bmatrix} 2 & -8 \\ 0 & 3 \end{bmatrix}$:

$$\begin{vmatrix} 2 & -8 \\ 0 & 3 \end{vmatrix} = (2)(3) - (0)(-8) = 6$$

The minor for −6 is 6.

Evaluating a 3 × 3 Determinant

The determinant of a 3 × 3 matrix is defined in terms of the determinants of minors.

Determinant of a 3 × 3 Matrix

The determinant of a 3 × 3 matrix is defined as follows:

$$\begin{vmatrix} a_1 & b_1 & c_1 \\ a_2 & b_2 & c_2 \\ a_3 & b_3 & c_3 \end{vmatrix} = a_1 \cdot \begin{vmatrix} b_2 & c_2 \\ b_3 & c_3 \end{vmatrix} - a_2 \cdot \begin{vmatrix} b_1 & c_1 \\ b_3 & c_3 \end{vmatrix} + a_3 \cdot \begin{vmatrix} b_1 & c_1 \\ b_2 & c_2 \end{vmatrix}$$

Note that the determinants following a_1, a_2, and a_3 are the minors for a_1, a_2, and a_3, respectively. Writing the determinant of a 3 × 3 matrix in terms of minors is called **expansion by minors.** In the definition we expanded by minors about the first column. Later we will see how to expand by minors using any row or column and get the same value for the determinant.

E X A M P L E   4

Determinant of a 3 × 3 matrix

Find the determinant of the matrix by expansion by minors about the first column.

$$\begin{bmatrix} 1 & 3 & -5 \\ -2 & 4 & 6 \\ 0 & -7 & 9 \end{bmatrix}$$

Solution

$$\begin{vmatrix} 1 & 3 & -5 \\ -2 & 4 & 6 \\ 0 & -7 & 9 \end{vmatrix} = 1 \cdot \begin{vmatrix} 4 & 6 \\ -7 & 9 \end{vmatrix} - (-2) \cdot \begin{vmatrix} 3 & -5 \\ -7 & 9 \end{vmatrix} + 0 \cdot \begin{vmatrix} 3 & -5 \\ 4 & 6 \end{vmatrix}$$

$$= 1 \cdot [36 - (-42)] + 2 \cdot (27 - 35) + 0 \cdot [18 - (-20)]$$

$$= 1 \cdot 78 + 2 \cdot (-8) + 0$$

$$= 78 - 16$$

$$= 62$$

In the next example we evaluate a determinant using expansion by minors about the second row. In expanding about any row or column, the signs of the coefficients of the minors alternate according to the **sign array** that follows:

$$\begin{bmatrix} + & - & + \\ - & + & - \\ + & - & + \end{bmatrix}$$

The sign array is easily remembered by observing that there is a "+" sign in the upper left position and then alternating signs for all of the remaining positions.

EXAMPLE 5

Determinant of a 3 × 3 matrix

Evaluate the determinant of the matrix by expanding by minors about the second row.

$$\begin{bmatrix} 1 & 3 & -5 \\ -2 & 4 & 6 \\ 0 & -7 & 9 \end{bmatrix}$$

calculator
close-up

A calculator is very useful for finding the determinant of a 3 × 3 matrix. Define A using MATRX EDIT.

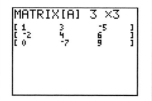

Now use the determinant function from MATRX MATH and the A from MATRX NAMES to find the determinant.

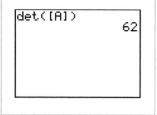

Solution

For expansion using the second row we prefix the signs "$- + -$" from the second row of the sign array to the corresponding numbers in the second row of the matrix, $-2, 4$, and 6. Note that the signs from the sign array are used in addition to any signs that occur on the numbers in the second row.

From the sign array, second row

$$\begin{vmatrix} 1 & 3 & -5 \\ -2 & 4 & 6 \\ 0 & -7 & 9 \end{vmatrix} = -(-2) \cdot \begin{vmatrix} 3 & -5 \\ -7 & 9 \end{vmatrix} + 4 \cdot \begin{vmatrix} 1 & -5 \\ 0 & 9 \end{vmatrix} - 6 \cdot \begin{vmatrix} 1 & 3 \\ 0 & -7 \end{vmatrix}$$

$$= 2(27 - 35) + 4(9 - 0) - 6(-7 - 0)$$

$$= 2(-8) + 4(9) - 6(-7)$$

$$= -16 + 36 + 42$$

$$= 62$$

Note that 62 is the same value that was obtained for this determinant in Example 4. ■

It can be shown that expanding by minors using any row or column prefixed by the corresponding signs from the sign array yields the same value for the determinant. Because we can use any row or column to evaluate a determinant of a 3 × 3 matrix, we can choose a row or column that makes the work easier. We can shorten the work considerably by picking a row or column with zeros in it.

EXAMPLE 6

Choosing the simplest row or column

Find the determinant of the matrix

$$\begin{bmatrix} 3 & -5 & 0 \\ 4 & -6 & 0 \\ 7 & 9 & 2 \end{bmatrix}$$

Solution

We choose to expand by minors about the third column of the matrix because the third column contains two zeros. Prefix the third-column entries $0, 0, 2$ by the signs "$+ - +$" from the third column of the sign array:

$$\begin{vmatrix} 3 & -5 & 0 \\ 4 & -6 & 0 \\ 7 & 9 & 2 \end{vmatrix} = 0 \cdot \begin{vmatrix} 4 & -6 \\ 7 & 9 \end{vmatrix} - 0 \cdot \begin{vmatrix} 3 & -5 \\ 7 & 9 \end{vmatrix} + 2 \cdot \begin{vmatrix} 3 & -5 \\ 4 & -6 \end{vmatrix}$$

$$= 0 - 0 + 2[-18 - (-20)]$$

$$= 4$$ ■

Cramer's Rule (3 × 3)

A system of three linear equations in three variables can be solved by using determinants and Cramer's rule.

Cramer's Rule for Three Equations in Three Unknowns

The solution to the system

$$a_1x + b_1y + c_1z = d_1$$
$$a_2x + b_2y + c_2z = d_2$$
$$a_3x + b_3y + c_3z = d_3$$

is given by $x = \frac{D_x}{D}$, $y = \frac{D_y}{D}$, and $z = \frac{D_z}{D}$, where

$$D = \begin{vmatrix} a_1 & b_1 & c_1 \\ a_2 & b_2 & c_2 \\ a_3 & b_3 & c_3 \end{vmatrix}, \qquad D_x = \begin{vmatrix} d_1 & b_1 & c_1 \\ d_2 & b_2 & c_2 \\ d_3 & b_3 & c_3 \end{vmatrix},$$

$$D_y = \begin{vmatrix} a_1 & d_1 & c_1 \\ a_2 & d_2 & c_2 \\ a_3 & d_3 & c_3 \end{vmatrix}, \qquad D_z = \begin{vmatrix} a_1 & b_1 & d_1 \\ a_2 & b_2 & d_2 \\ a_3 & b_3 & d_3 \end{vmatrix},$$

provided that $D \neq 0$.

Note that D_x, D_y, and D_z are obtained from D by replacing the x-, y-, or z-column with the constants d_1, d_2, and d_3.

E X A M P L E   7   **Solving an independent system with Cramer's rule**

Use Cramer's rule to solve the system:

$$x + y + z = 4$$
$$x - y \, = -3$$
$$x + 2y - z = 0$$

Solution

We first calculate D, D_x, D_y, and D_z. To calculate D, expand by minors about the third column because the third column has a zero in it:

$$D = \begin{vmatrix} 1 & 1 & 1 \\ 1 & -1 & 0 \\ 1 & 2 & -1 \end{vmatrix} = 1 \cdot \begin{vmatrix} 1 & -1 \\ 1 & 2 \end{vmatrix} - 0 \cdot \begin{vmatrix} 1 & 1 \\ 1 & 2 \end{vmatrix} + (-1) \cdot \begin{vmatrix} 1 & 1 \\ 1 & -1 \end{vmatrix}$$

$$= 1 \cdot [2 - (-1)] - 0 + (-1)[-1 - 1]$$
$$= 3 - 0 + 2$$
$$= 5$$

For D_x, expand by minors about the first column:

$$D_x = \begin{vmatrix} 4 & 1 & 1 \\ -3 & -1 & 0 \\ 0 & 2 & -1 \end{vmatrix} = 4 \cdot \begin{vmatrix} -1 & 0 \\ 2 & -1 \end{vmatrix} - (-3) \cdot \begin{vmatrix} 1 & 1 \\ 2 & -1 \end{vmatrix} + 0 \cdot \begin{vmatrix} 1 & 1 \\ -1 & 0 \end{vmatrix}$$

$$= 4 \cdot (1 - 0) + 3 \cdot (-1 - 2) + 0$$
$$= 4 - 9 + 0 = -5$$

For D_y, expand by minors about the third row:

$$D_y = \begin{vmatrix} 1 & 4 & 1 \\ 1 & -3 & 0 \\ 1 & 0 & -1 \end{vmatrix} = 1 \cdot \begin{vmatrix} 4 & 1 \\ -3 & 0 \end{vmatrix} - 0 \cdot \begin{vmatrix} 1 & 1 \\ 1 & 0 \end{vmatrix} + (-1) \cdot \begin{vmatrix} 1 & 4 \\ 1 & -3 \end{vmatrix}$$

$$= 1 \cdot 3 - 0 + (-1)(-7) = 10$$

To get D_z, expand by minors about the third row:

$$D_z = \begin{vmatrix} 1 & 1 & 4 \\ 1 & -1 & -3 \\ 1 & 2 & 0 \end{vmatrix} = 1 \cdot \begin{vmatrix} 1 & 4 \\ -1 & -3 \end{vmatrix} - 2 \cdot \begin{vmatrix} 1 & 4 \\ 1 & -3 \end{vmatrix} + 0 \cdot \begin{vmatrix} 1 & 1 \\ 1 & -1 \end{vmatrix}$$

$$= 1 \cdot 1 - 2(-7) + 0 = 15$$

Now, by Cramer's rule,

$$x = \frac{D_x}{D} = \frac{-5}{5} = -1, \qquad y = \frac{D_y}{D} = \frac{10}{5} = 2, \qquad \text{and} \qquad z = \frac{D_z}{D} = \frac{15}{5} = 3.$$

Check $(-1, 2, 3)$ in the original equations. The solution set is $\{(-1, 2, 3)\}$. ■

If $D = 0$, Cramer's rule does not apply. Cramer's rule provides the solution only to a system of three equations with three variables that has a single point in the solution set. If $D = 0$, then the solution set either is empty or consists of infinitely many points, and we can use the methods discussed in Sections 8.3 or 8.4 to find the solution.

WARM-UPS

True or false? Explain your answer.

1. $\begin{vmatrix} -1 & 2 \\ 3 & -5 \end{vmatrix} = -1$ **2.** $\begin{vmatrix} 2 & 4 \\ -4 & 8 \end{vmatrix} = 0$

3. Cramer's rule solves any system of two linear equations in two variables.

4. The determinant of a 2 × 2 matrix is a real number.

5. If $D = 0$, then there might be no solution to the system.

6. Cramer's rule is used to solve systems of linear equations only.

7. If the graphs of a pair of linear equations intersect at exactly one point, then this point can be found by using Cramer's rule.

8. The determinant of a 3 × 3 matrix is found by using minors.

9. Expansion by minors about any row or any column gives the same value for the determinant of a 3 × 3 matrix.

10. The sign array is used in evaluating the determinant of a 3 × 3 matrix.

8.5   EXERCISES

Reading and Writing   *After reading this section, write out the answers to these questions. Use complete sentences.*

1. What is a determinant?

2. What is Cramer's rule used for?

3. Which systems can be solved using Cramer's rule?

4. What is a minor?

5. How do you find the minor for an element of a 3×3 matrix?

6. What is the purpose of the sign array?

Find the value of each determinant. See Example 1.

7. $\begin{vmatrix} 2 & 5 \\ 3 & 7 \end{vmatrix}$

8. $\begin{vmatrix} -1 & 0 \\ 1 & 1 \end{vmatrix}$

9. $\begin{vmatrix} 0 & 3 \\ 1 & 5 \end{vmatrix}$

10. $\begin{vmatrix} 2 & 4 \\ 6 & 12 \end{vmatrix}$

11. $\begin{vmatrix} -3 & -2 \\ -4 & 2 \end{vmatrix}$

12. $\begin{vmatrix} -2 & 2 \\ -3 & -5 \end{vmatrix}$

13. $\begin{vmatrix} 0.05 & 0.06 \\ 10 & 20 \end{vmatrix}$

14. $\begin{vmatrix} 0.02 & -0.5 \\ 30 & 50 \end{vmatrix}$

Solve each system using Cramer's rule. See Example 2.

15. $2x - y = 5$
$3x + 2y = -3$

16. $3x + y = -1$
$x + 2y = 8$

17. $3x - 5y = -2$
$2x + 3y = 5$

18. $x - y = 1$
$3x - 2y = 0$

19. $4x - 3y = 5$
$2x + 5y = 7$

20. $2x - y = 2$
$3x - 2y = 1$

21. $0.5x + 0.2y = 8$
$0.4x - 0.6y = -5$

22. $0.6x + 0.5y = 18$
$0.5x - 0.25y = 7$

23. $\dfrac{1}{2}x + \dfrac{1}{4}y = 5$
$\dfrac{1}{3}x - \dfrac{1}{2}y = -1$

24. $\dfrac{1}{2}x + \dfrac{2}{3}y = 4$
$\dfrac{3}{4}x + \dfrac{1}{3}y = -2$

Find the indicated minors using the following matrix. See Example 3.

$$\begin{bmatrix} 3 & -2 & 5 \\ 4 & -3 & 7 \\ 0 & 1 & -6 \end{bmatrix}$$

25. Minor for 3

26. Minor for -2

27. Minor for 5

28. Minor for -3

29. Minor for 7

30. Minor for 0

31. Minor for 1

32. Minor for -6

Find the determinant of each 3×3 matrix by using expansion by minors about the first column. See Example 4.

33. $\begin{bmatrix} 1 & 1 & 2 \\ 2 & 3 & 1 \\ 3 & 1 & 5 \end{bmatrix}$

34. $\begin{bmatrix} 2 & 1 & 3 \\ 1 & 1 & 2 \\ 3 & 4 & 6 \end{bmatrix}$

35. $\begin{bmatrix} 2 & 1 & 0 \\ 1 & 0 & 1 \\ 3 & 1 & 2 \end{bmatrix}$

36. $\begin{bmatrix} 1 & 0 & 2 \\ 2 & 1 & 3 \\ 4 & 3 & 0 \end{bmatrix}$

37. $\begin{bmatrix} -2 & 1 & 2 \\ -3 & 3 & 1 \\ -5 & 4 & 0 \end{bmatrix}$

38. $\begin{bmatrix} -2 & 1 & 3 \\ -1 & 4 & 2 \\ 2 & 1 & 1 \end{bmatrix}$

39. $\begin{bmatrix} 1 & 1 & 5 \\ 0 & 3 & 2 \\ 0 & 2 & 3 \end{bmatrix}$

40. $\begin{bmatrix} 1 & 0 & 6 \\ 0 & 1 & 4 \\ 0 & 0 & 9 \end{bmatrix}$

Evaluate the determinant of each 3×3 matrix using expansion by minors about the row or column of your choice. See Examples 5 and 6.

41. $\begin{bmatrix} 3 & 1 & 5 \\ 2 & 0 & 6 \\ 4 & 0 & 1 \end{bmatrix}$

42. $\begin{bmatrix} 2 & 1 & 2 \\ 1 & 2 & 5 \\ 3 & 0 & 0 \end{bmatrix}$

43. $\begin{bmatrix} -2 & 1 & 3 \\ 0 & 1 & -1 \\ 2 & -4 & -3 \end{bmatrix}$

44. $\begin{bmatrix} -2 & 0 & 1 \\ -3 & 2 & -5 \\ 4 & -2 & 6 \end{bmatrix}$

45. $\begin{bmatrix} -2 & -3 & 0 \\ 4 & -1 & 0 \\ 0 & 3 & 5 \end{bmatrix}$

46. $\begin{bmatrix} -2 & 6 & 3 \\ 0 & 4 & 0 \\ -1 & -4 & 5 \end{bmatrix}$

47. $\begin{bmatrix} 2 & 1 & 1 \\ 0 & 0 & 5 \\ 5 & 0 & 4 \end{bmatrix}$

48. $\begin{bmatrix} 2 & 3 & 0 \\ 6 & 4 & 1 \\ 1 & 2 & 0 \end{bmatrix}$

Use Cramer's rule to solve each system. See Example 7.

49. $x + y + z = 6$
$x - y + z = 2$
$2x + y + z = 7$

50. $x + y + z = 2$
$x - y - 2z = -3$
$2x - y + z = 7$

51. $x - 3y + 2z = 0$
$x + y + z = 2$
$x - y + z = 0$

52. $3x + 2y + 2z = 0$
$x - y + z = 1$
$x + y - z = 3$

53. $x + y = -1$
$2y - z = 3$
$x + y + z = 0$

54. $x - y = 8$
$x - 2z = 0$
$x + y - z = 1$

55. $x + y - z = 0$
$2x + 2y + z = 6$
$x - 3y = 0$

56. $x + y + z = 1$
$5x - y = 0$
$3x + y + 2z = 0$

57. $x + y + z = 0$
$2y + 2z = 0$
$3x - y = -1$

58. $x + z = 0$
$x - 3y = 1$
$4y - 3z = 3$

Solve each problem by using two equations in two variables and Cramer's rule.

59. *Peas and beets.* One serving of canned peas contains 3 grams of protein and 11 grams of carbohydrates. One serving of canned beets contains 1 gram of protein and 8 grams of carbohydrates. A dietitian wants to determine the number of servings of each that would provide 38 grams of protein and 187 grams of carbohydrates.
a) Use the accompanying graph to estimate the number of servings of each.
b) Use Cramer's rule to find the number of servings of each.

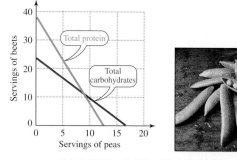

FIGURE FOR EXERCISE 59

60. *Protein and carbohydrates.* One serving of Cornies breakfast cereal contains 2 grams of protein and 25 grams of carbohydrates. One serving of Oaties breakfast cereal contains 4 grams of protein and 20 grams of carbohydrates. How many servings of each would provide exactly 24 grams of protein and 210 grams of carbohydrates?

61. *Milk and a magazine.* Althia bought a gallon of milk and a magazine for a total of $4.65, excluding tax. Including the tax, the bill was $4.95. If there is a 5% sales tax on milk and an 8% sales tax on magazines, then what was the price of each item?

62. *Washing machines and refrigerators.* A truck carrying 3600 cubic feet of cargo consisting of washing machines and refrigerators was hijacked. The washing machines are worth $300 each and are shipped in 36-cubic-foot cartons. The refrigerators are worth $900 each and are shipped in 45-cubic-foot cartons. If the total value of the cargo was $51,000, then how many of each were there on the truck?

63. *Singles and doubles.* Windy's Hamburger Palace sells singles and doubles. Toward the end of the evening, Windy himself noticed that he had on hand only 32 patties and 34 slices of tomatoes. If a single takes 1 patty and 2 slices, and a double takes 2 patties and 1 slice, then how many more singles and doubles must Windy sell to use up all of his patties and tomato slices?

64. *Valuable wrenches.* Carmen has a total of 28 wrenches, all of which are either box wrenches or open-end wrenches. For insurance purposes she values the box wrenches at $3.00 each and the open-end wrenches at $2.50 each. If the value of her wrench collection is $78, then how many of each type does she have?

65. *Gary and Harry.* Gary is 5 years older than Harry. Twenty-nine years ago, Gary was twice as old as Harry. How old are they now?

66. *Acute angles.* One acute angle of a right triangle is 3° more than twice the other acute angle. What are the sizes of the acute angles?

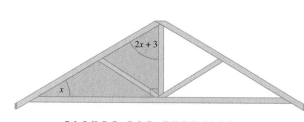

FIGURE FOR EXERCISE 66

67. *Equal perimeters.* A rope of length 80 feet is to be cut into two pieces. One piece will be used to form a square, and the other will be used to form an equilateral triangle.

If the figures are to have equal perimeters, then what should be the length of a side of each?

FIGURE FOR EXERCISE 67

68. Coffee and doughnuts. For a cup of coffee and a doughnut, Thurrel spent $2.25, including a tip. Later he spent $4.00 for two coffees and three doughnuts, including a tip. If he always tips $1.00, then what is the price of a cup of coffee?

69. Chlorine mixture. A 10% chlorine solution is to be mixed with a 25% chlorine solution to obtain 30 gallons of 20% solution. How many gallons of each must be used?

70. Safe drivers. Emily and Camille started from the same city and drove in opposite directions on the freeway. After 3 hours they were 354 miles apart. If they had gone in the same direction, they would have been only 18 miles apart. How fast did each woman drive?

Write a system of three equations in three variables for each word problem. Use Cramer's rule to solve each system.

71. Weighing dogs. Cassandra wants to determine the weights of her two dogs, Mimi and Mitzi. However, neither dog will sit on the scale by herself. Cassandra, Mimi, and Mitzi altogether weigh 175 pounds. Cassandra and Mimi together weigh 143 pounds. Cassandra and Mitzi together weigh 139 pounds. How much does each weigh individually?

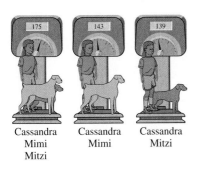

Cassandra   Cassandra   Cassandra
Mimi   Mimi   Mitzi
Mitzi

FIGURE FOR EXERCISE 71

72. Nickels, dimes, and quarters. Bernard has 41 coins consisting of nickels, dimes, and quarters, and they are worth a total of $4.00. If the number of dimes plus the number of quarters is one more than the number of nickels, then how many of each does he have?

73. Finding three angles. If the two acute angles of a right triangle differ by 12°, then what are the measures of the three angles of this triangle?

74. Two acute and one obtuse. The obtuse angle of a triangle is twice as large as the sum of the two acute angles. If the smallest angle is only one-eighth as large as the sum of the other two, then what is the measure of each angle?

GETTING MORE INVOLVED

75. Writing. Explain what to do when you are trying to use Cramer's rule and $D = 0$.

76. Exploration. For what value of a does the system

$$ax - y = 3$$
$$x + 2y = 1$$

have a single solution?

77. Exploration. Can Cramer's rule be used to solve the following system? Explain.

$$2x^2 - y = 3$$
$$3x^2 + 2y = 22$$

78. Writing. For what values of a, b, c, and d is the determinant of the matrix

$$\begin{bmatrix} a & b & 0 \\ c & d & 0 \\ b & a & 0 \end{bmatrix}$$

equal to zero? Explain your answer.

GRAPHING CALCULATOR EXERCISES

79. Use the determinant feature on your graphing calculator to find the determinants in Exercises 7–14 and 33–40 of this section.

80. Solve the systems in Exercises 15–24 and 49–58 of this section by using your graphing calculator to find the necessary determinants.

8.6 GRAPHING LINEAR INEQUALITIES IN TWO VARIABLES

You studied linear inequalities in one variable in Chapter 3. In this section we extend the ideas of linear equations in two variables to study linear inequalities in two variables.

Definition

Linear inequalities in two variables have the same form as linear equations in two variables. An inequality symbol is used in place of the equal sign.

> **Linear Inequality in Two Variables**
>
> If A, B, and C are real numbers with A and B not both zero, then
>
> $$Ax + By < C$$
>
> is called a **linear inequality in two variables.** In place of $<$, we can also use $\leq$, $>$, or $\geq$.

The inequalities

$$3x - 4y \leq 8, \qquad y > 2x - 3, \qquad \text{and} \qquad x - y + 9 < 0$$

are linear inequalities. Not all of these are in the form of the definition, but they could all be rewritten in that form.

An ordered pair is a solution to an inequality in two variables if the ordered pair satisfies the inequality.

EXAMPLE 1

Satisfying a linear inequality

Determine whether each point satisfies the inequality $2x - 3y \geq 6$.

a) $(4, 1)$ **b)** $(3, 0)$ **c)** $(3, -2)$

Solution

a) To determine whether $(4, 1)$ is a solution to the inequality, we replace x by 4 and y by 1 in the inequality $2x - 3y \geq 6$:

$$2(4) - 3(1) \geq 6$$
$$8 - 3 \geq 6$$
$$5 \geq 6 \quad \text{Incorrect}$$

So $(4, 1)$ does not satisfy the inequality $2x - 3y \geq 6$.

b) Replace x by 3 and y by 0:

$$2(3) - 3(0) \geq 6$$
$$6 \geq 6 \quad \text{Correct}$$

So the point $(3, 0)$ satisfies the inequality $2x - 3y \geq 6$.

c) Replace x by 3 and y by -2:

$$2(3) - 3(-2) \geq 6$$
$$6 + 6 \geq 6$$
$$12 \geq 6 \quad \text{Correct}$$

So the point $(3, -2)$ satisfies the inequality $2x - 3y \geq 6$.

Graph of a Linear Inequality

The graph of a linear inequality in two variables consists of all points in the rectangular coordinate system that satisfy the inequality. For example, the graph of the inequality

$$y > x + 2$$

consists of all points where the y-coordinate is larger than the x-coordinate plus 2. Consider the point $(3, 5)$ on the line

$$y = x + 2.$$

The y-coordinate of $(3, 5)$ is equal to the x-coordinate plus 2. If we choose a point with a larger y-coordinate, such as $(3, 6)$, it satisfies the inequality and it is above the line $y = x + 2$. In fact, any point above the line $y = x + 2$ satisfies $y > x + 2$. Likewise, all points below the line $y = x + 2$ satisfy the inequality $y < x + 2$. See Fig. 8.7.

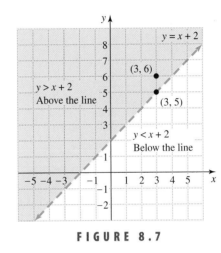

FIGURE 8.7

To graph the inequality, we shade all points above the line $y = x + 2$. To indicate that the line is not included in the graph of $y > x + 2$, we use a dashed line.

The procedure for graphing linear inequalities is summarized as follows.

Strategy for Graphing a Linear Inequality in Two Variables

1. Solve the inequality for y, then graph $y = mx + b$.

   $y > mx + b$ is the region above the line.

   $y = mx + b$ is the line itself.

   $y < mx + b$ is the region below the line.

2. If the inequality involves only x, then graph the vertical line $x = k$.

   $x > k$ is the region to the right of the line.

   $x = k$ is the line itself.

   $x < k$ is the region to the left of the line.

E X A M P L E 2   **Graphing a linear inequality**

Graph each inequality.

a) $y < \dfrac{1}{3}x + 1$ **b)** $y \geq -2x + 3$

c) $2x - 3y < 6$

Solution

a) The set of points satisfying this inequality is the region below the line

$$y = \dfrac{1}{3}x + 1.$$

To show this region, we first graph the boundary line. The slope of the line is $\frac{1}{3}$, and the y-intercept is $(0, 1)$. We draw the line dashed because it is not part of the graph of $y < \frac{1}{3}x + 1$. In Fig. 8.8 the graph is the shaded region.

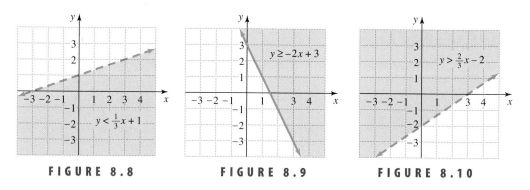

| **FIGURE 8.8** | **FIGURE 8.9** | **FIGURE 8.10** |

b) Because the inequality symbol is $\geq$, every point on or above the line satisfies this inequality. We use the fact that the slope of this line is -2 and the y-intercept is $(0, 3)$ to draw the graph of the line. To show that the line $y = -2x + 3$ is included in the graph, we make it a solid line and shade the region above. See Fig. 8.9.

c) First solve for y:

$$2x - 3y < 6$$
$$-3y < -2x + 6$$
$$y > \dfrac{2}{3}x - 2 \quad \text{Divide by } -3 \text{ and reverse the inequality.}$$

To graph this inequality, we first graph the line with slope $\frac{2}{3}$ and y-intercept $(0, -2)$. We use a dashed line for the boundary because it is not included, and we shade the region above the line. Remember, "less than" means below the line and "greater than" means above the line only when the inequality is solved for y. See Fig. 8.10 for the graph. ■

E X A M P L E 3   **Horizontal and vertical boundary lines**

Graph each inequality.

a) $y \leq 4$ **b)** $x > 3$

Solution

a) The line $y = 4$ is the horizontal line with y-intercept $(0, 4)$. We draw a solid horizontal line and shade below it as in Fig. 8.11.

b) The line $x = 3$ is a vertical line through $(3, 0)$. Any point to the right of this line has an x-coordinate larger than 3. The graph is shown in Fig. 8.12.

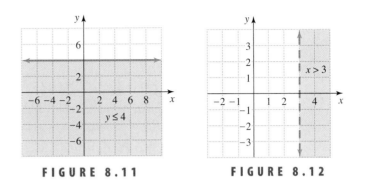

FIGURE 8.11   FIGURE 8.12

Using a Test Point to Graph an Inequality

The graph of a linear equation such as $2x - 3y = 6$ separates the coordinate plane into two regions. One region satisfies the inequality $2x - 3y > 6$, and the other region satisfies the inequality $2x - 3y < 6$. We can tell which region satisfies which inequality by testing a point in one region. With this method it is not necessary to solve the inequality for y.

E X A M P L E   4   **Using a test point**

Graph the inequality $2x - 3y > 6$.

Solution

First graph the equation $2x - 3y = 6$ using the x-intercept $(3, 0)$ and the y-intercept $(0, -2)$ as shown in Fig. 8.13. Select a point on one side of the line, say $(0, 1)$, to test in the inequality. Because

$$2(0) - 3(1) > 6$$

is false, the region on the other side of the line satisfies the inequality. The graph of $2x - 3y > 6$ is shown in Fig. 8.14.

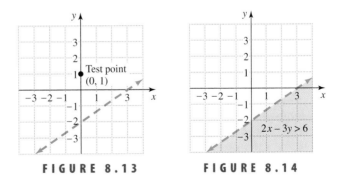

FIGURE 8.13   FIGURE 8.14

Applications

The values of variables used in applications are often restricted to nonnegative numbers. So solutions to inequalities in these applications are graphed in the first quadrant only.

EXAMPLE 5   **Manufacturing tables**

The Ozark Furniture Company can obtain at most 8000 board feet of oak lumber for making two types of tables. It takes 50 board feet to make a round table and 80 board feet to make a rectangular table. Write an inequality that limits the possible number of tables of each type that can be made. Draw a graph showing all possibilities for the number of tables that can be made.

Solution

If x is the number of round tables and y is the number of rectangular tables, then x and y satisfy the inequality

$$50x + 80y \leq 8000.$$

Now find the intercepts for the line $50x + 80y = 8000$:

$$50 \cdot 0 + 80y = 8000 \qquad\qquad 50x + 80 \cdot 0 = 8000$$
$$80y = 8000 \qquad\qquad 50x = 8000$$
$$y = 100 \qquad\qquad x = 160$$

Draw the line through $(0, 100)$ and $(160, 0)$. Because $(0, 0)$ satisfies the inequality, the number of tables must be below the line. Since the number of tables cannot be negative, the number of tables made must be below the line and in the first quadrant as shown in Fig. 8.15. Assuming that Ozark will not make a fraction of a table, only points in Fig. 8.15 with whole-number coordinates are practical.

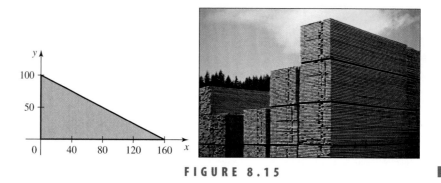

FIGURE 8.15

True or false? Explain your answer.

1. The point $(-1, 4)$ satisfies the inequality $y > 3x + 1$.

2. The point $(2, -3)$ satisfies the inequality $3x - 2y \geq 12$.

3. The graph of the inequality $y > x + 9$ is the region above the line $y = x + 9$.

4. The graph of the inequality $x < y + 2$ is the region below the line $x = y + 2$.

5. The graph of $x = 3$ is a single point on the x-axis.

6. The graph of $y \leq 5$ is the region below the horizontal line $y = 5$.

7. The graph of $x < 3$ is the region to the left of the vertical line $x = 3$.

WARM-UPS

(continued)

8. In graphing the inequality $y \geq x$ we use a dashed boundary line.

9. The point $(0, 0)$ is on the graph of the inequality $y \geq x$.

10. The point $(0, 0)$ lies above the line $y = 2x + 1$.

8.6   EXERCISES

Reading and Writing   *After reading this section, write out the answers to these questions. Use complete sentences.*

1. What is a linear inequality in two variables?

2. How can you tell if an ordered pair satisfies a linear inequality in two variables?

3. How do you determine whether to draw the boundary line of the graph of a linear inequality dashed or solid?

4. How do you decide which side of the boundary line to shade?

5. What is the test point method?

6. What is the advantage of the test point method?

Determine which of the points following each inequality satisfy that inequality. See Example 1.

7. $x - y > 5$ $(2, 3), (-3, -9), (8, 3)$

8. $2x + y < 3$ $(-2, 6), (0, 3), (3, 0)$

9. $y \geq -2x + 5$ $(3, 0), (1, 3), (-2, 5)$

10. $y \leq -x + 6$ $(2, 0), (-3, 9), (-4, 12)$

11. $x > -3y + 4$ $(2, 3), (7, -1), (0, 5)$

12. $x < -y - 3$ $(1, 2), (-3, -4), (0, -3)$

Graph each inequality. See Examples 2 and 3.

13. $y < x + 4$

14. $y < 2x + 2$

15. $y > -x + 3$

16. $y < -2x + 1$

17. $y > \dfrac{2}{3}x - 3$

18. $y < \dfrac{1}{2}x + 1$

19. $y \leq -\dfrac{2}{5}x + 2$

20. $y \geq -\dfrac{1}{2}x + 3$

21. $y - x \geq 0$

22. $x - 2y \leq 0$

23. $x > y - 5$

24. $2x < 3y + 6$

33. $x \leq 100y$

34. $y \geq 600x$

25. $x - 2y + 4 \leq 0$

26. $2x - y + 3 \geq 0$

35. $3x - 4y \leq 8$

36. $2x + 5y \geq 10$

27. $y \geq 2$

28. $y < 7$

Graph each inequality. Use the test point method of Example 4.

37. $2x - 3y < 6$

38. $x - 4y > 4$

29. $x > 9$

30. $x \leq 1$

39. $x - 4y \leq 8$

40. $3y - 5x \geq 15$

31. $x + y \leq 60$

32. $x - y \leq 90$

41. $y - \dfrac{7}{2}x \leq 7$

42. $\dfrac{2}{3}x + 3y \leq 12$

43. $x - y < 5$

44. $y - x > -3$

45. $3x - 4y < -12$

46. $4x + 3y > 24$

FIGURE FOR EXERCISE 50

47. $x < 5y - 100$

48. $-x > 70 - y$

modern rocker requires 12 board feet of maple. Write an inequality that limits the possible number of maple rockers of each type that can be made, and graph the inequality in the first quadrant.

Solve each problem. See Example 5.

49. *Storing the tables.* Ozark Furniture Company must store its oak tables before shipping. A round table is packaged in a carton with a volume of 25 cubic feet (ft^3), and a rectangular table is packaged in a carton with a volume of 35 ft^3. The warehouse has at most 3850 ft^3 of space available for these tables. Write an inequality that limits the possible number of tables of each type that can be stored, and graph the inequality in the first quadrant.

50. *Maple rockers.* Ozark Furniture Company can obtain at most 3000 board feet of maple lumber for making its classic and modern maple rocking chairs. A classic maple rocker requires 15 board feet of maple, and a

51. *Enzyme concentration.* A food chemist tests enzymes for their ability to break down pectin in fruit juices (Dennis Callas, *Snapshots of Applications in Mathematics*). Excess pectin makes juice cloudy. In one test, the chemist measures the concentration of the enzyme, c, in milligrams per milliliter and the fraction of light absorbed by the liquid, a. If $a > 0.07c + 0.02$, then the enzyme is working as it should. Graph the inequality for $0 < c < 5$.

GETTING MORE INVOLVED

52. *Discussion.* When asked to graph the inequality $x + 2y < 12$, a student found that $(0, 5)$ and $(8, 0)$ both satisfied $x + 2y < 12$. The student then drew a dashed line through these two points and shaded the region below the line. What is wrong with this method? Do all of the points graphed by this student satisfy the inequality?

53. *Writing.* Compare and contrast the two methods presented in this section for graphing linear inequalities. What are the advantages and disadvantages of each method? How do you choose which method to use?

8.7

GRAPHING SYSTEMS OF LINEAR INEQUALITIES

In this section

- The Solution to a System of Inequalities
- Graphing a System of Inequalities

In Section 8.6 you learned how to solve a linear inequality. In this section you will solve systems of linear inequalities.

The Solution to a System of Inequalities

A **system of inequalities** consists of two or more inequalities. A point is a solution to a system of inequalities if it satisfies all of the inequalities in the system.

E X A M P L E 1

Satisfying a system of inequalities

Determine whether each point is a solution to the system of inequalities:

$$2x + 3y < 6$$
$$y > 2x - 1$$

a) $(-3, 2)$ **b)** $(4, -3)$ **c)** $(5, 1)$

Solution

a) The point $(-3, 2)$ is a solution to the system if it satisfies both inequalities. Let $x = -3$ and $y = 2$ in each inequality:

$$2x + 3y < 6 \qquad y > 2x - 1$$
$$2(-3) + 3(2) < 6 \qquad 2 > 2(-3) - 1$$
$$0 < 6 \qquad 2 > -7$$

Because both inequalities are satisfied, the point $(-3, 2)$ is a solution to the system.

b) Let $x = 4$ and $y = -3$ in each inequality:

$$2x + 3y < 6 \qquad y > 2x - 1$$
$$2(4) + 3(-3) < 6 \qquad -3 > 2(4) - 1$$
$$-1 < 6 \qquad -3 > 7$$

Because only one inequality is satisfied, the point $(4, -3)$ is not a solution to the system.

c) Let $x = 5$ and $y = 1$ in each inequality:

$$2x + 3y < 6 \qquad y > 2x - 1$$
$$2(5) + 3(1) < 6 \qquad 1 > 2(5) - 1$$
$$13 < 6 \qquad 1 > 9$$

Because neither inequality is satisfied, the point $(5, 1)$ is not a solution to the system. ∎

study tip

Read the text and recite to yourself what you have read. Ask questions and answer them out loud. Listen to your answers to see if they are complete and correct. Would other students understand your answers?

Graphing a System of Inequalities

There are infinitely many points that satisfy a typical system of inequalities. The best way to describe the solution to a system of inequalities is with a graph showing all points that satisfy the system. When we graph the points that satisfy a system, we say that we are graphing the system.

E X A M P L E   2   **Graphing a system of inequalities**

Graph all ordered pairs that satisfy the following system of inequalities:

$$y > x - 2$$
$$y < -2x + 3$$

Solution

We want a graph showing all points that satisfy both inequalities. The lines $y = x - 2$ and $y = -2x + 3$ divide the coordinate plane into four regions as shown in Fig. 8.16. To determine which of the four regions contains points that satisfy the system, we check one point in each region to see whether it satisfies both inequalities. The points are shown in Fig. 8.16.

Check $(0, 0)$:   Check $(0, 5)$:
$0 > 0 - 2$ Correct $5 > 0 - 2$ Correct
$0 < -2(0) + 3$ Correct $5 < -2(0) + 3$ Incorrect

Check $(0, -5)$:   Check $(4, 0)$:
$-5 > 0 - 2$ Incorrect $0 > 4 - 2$ Incorrect
$-5 < -2(0) + 3$ Correct $0 < -2(4) + 3$ Incorrect

The only point that satisfies both inequalities of the system is $(0, 0)$. So every point in the region containing $(0, 0)$ also satisfies both inequalities. The points that satisfy the system are graphed in Fig. 8.17.

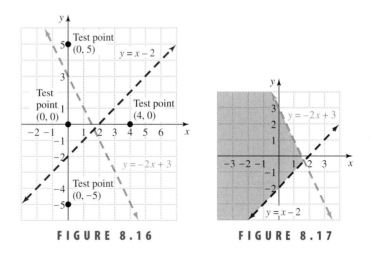

FIGURE 8.16   **FIGURE 8.17**

E X A M P L E   3   **Graphing a system of inequalities**

Graph all ordered pairs that satisfy the following system of inequalities:

$$y > -3x + 4$$
$$2y - x > 2$$

Solution

First graph the equations $y = -3x + 4$ and $2y - x = 2$. Now we select the points $(0, 0)$, $(0, 2)$, $(0, 6)$, and $(5, 0)$. We leave it to you to check each point in the system of inequalities. You will find that only $(0, 6)$ satisfies the system. So only the region containing $(0, 6)$ is shaded in Fig. 8.18.

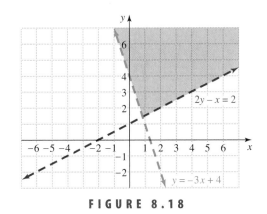

FIGURE 8.18

EXAMPLE 4

Horizontal and vertical boundary lines

Graph the system of inequalities: $x > 4$
$y < 3$

Solution

We first graph the vertical line $x = 4$ and the horizontal line $y = 3$. The points that satisfy both inequalities are those points that lie to the right of the vertical line $x = 4$ and below the horizontal line $y = 3$. See Fig. 8.19 for the graph of the system.

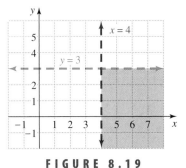

FIGURE 8.19

EXAMPLE 5

Between parallel lines

Graph the system of inequalities: $y < x + 4$
$y > x - 1$

Solution

First graph the parallel lines $y = x + 4$ and $y = x - 1$. These lines divide the plane into three regions. Check $(0, 0)$, $(0, 6)$, and $(0, -4)$ in the system. Only $(0, 0)$ satisfies the system. So the solution to the system consists of all points in between the parallel lines, as shown in Fig. 8.20.

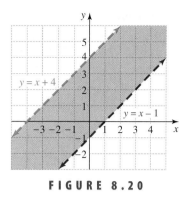

FIGURE 8.20

WARM-UPS

True or false? Explain your answer.

Use the following systems for Exercises 1–7.

a) $y > -3x + 5$
   $y < 2x - 3$

b) $y > 2x - 3$
   $y < 2x + 3$

c) $x + y > 4$
   $x - y < 0$

1. The point $(2, -3)$ is a solution to system (a).

2. The point $(5, 0)$ is a solution to system (a).

3. The point $(0, 0)$ is a solution to system (b).

4. The graph of system (b) is the region between two parallel lines.

5. You can use $(0, 0)$ as a test point for system (c).

6. The point $(2, 2)$ satisfies system (c).

7. The point $(4, 5)$ satisfies system (c).

8. The inequality $x + y > 4$ is equivalent to the inequality $y < -x + 4$.

9. The graph of $y < 2x + 3$ is the region below the line $y = 2x + 3$.

10. There is no ordered pair that satisfies $y < 2x - 3$ and $y > 2x + 3$.

8.7 EXERCISES

Reading and Writing *After reading this section, write out the answers to these questions. Use complete sentences.*

1. What is a system of linear inequalities in two variables?

2. How can you tell if an ordered pair satisfies a system of linear inequalities in two variables?

3. How do we usually describe the solution set to a system of inequalities in two variables?

4. How do you decide whether the boundary lines are solid or dashed?

5. How do you use the test point method for a system of linear inequalities?

6. How do you select test points?

Determine which of the points following each system is a solution to the system. See Example 1.

7. $x - y < 5$ $\quad (4, 3), (8, 2), (-3, 0)$
$\quad 2x + y > 3$

8. $x + y < 4$ $\quad (2, -3), (1, 1), (0, -1)$
$\quad 2x - y < 3$

9. $y > -2x + 1$ $\quad (-3, 2), (-1, 5), (3, 6)$
$\quad y < 3x + 5$

10. $y < -x + 7$ $\quad (-3, 8), (0, 8), (-5, 15)$
$\quad y < -x + 9$

11. $x > 3$ $\quad (-5, 4), (9, -5), (6, 0)$
$\quad y < -2$

12. $y < -5$ $\quad (-2, 4), (0, -7), (6, -9)$
$\quad x < 1$

Graph each system of inequalities. See Examples 2–5.

13. $y > -x - 1$
$\quad y > x + 1$

14. $y < x + 3$
$\quad y < -2x + 4$

15. $y < 2x - 3$
$\quad y > -x + 2$

16. $y > 2x - 1$
$\quad y < -x - 4$

17. $x + y > 5$
$\quad x - y < 3$

18. $2x + y < 3$
$\quad x - 2y > 2$

19. $2x - 3y < 6$
$\quad x - y > 3$

20. $3x - 2y > 6$
$\quad x + y < 4$

21. $x > 5$
$\quad y > 5$

22. $x < 3$
$\quad y > 2$

23. $y < -1$
$\quad x > -3$

24. $y > -2$
$\quad x < 1$

25. $y > 2x - 4$
$\quad y < 2x + 1$

26. $y < -2x + 3$
$\quad y > -2x$

27. $y > x$
    $x > 3$

28. $y < x$
    $y < 1$

35. $x + y > 3$
    $x + y > 1$

36. $x - y < 5$
    $x - y < 3$

29. $y > -x$
    $x < -1$

30. $y < -x$
    $y > -3$

37. $y > 3x + 2$
    $y < 3x + 3$

38. $y > x$
    $y < -x$

31. $x > 1$
    $y - 2x < 3$

32. $y < 2$
    $2x + 3y < 6$

39. $x + y < 5$
    $x - y > -1$

40. $2x - y > 4$
    $x - 5y < 5$

33. $2x - 5y < 5$
    $x + 2y > 4$

34. $3x + 2y < 2$
    $-x - 2y > 4$

41. $2x - 3y < 6$
    $3x + 4y < 12$

42. $x - 3y > 3$
    $x + 2y < 4$

43. $3x - 5y < 15$
$3x + 2y < 12$

44. $x - 4y < 0$
$x + y > 0$

47. ***Allocating resources.*** Wausaukee Enterprises makes yard barns in two sizes. One small barn requires $250 in materials and 20 hours of labor, and one large barn requires $400 in materials and 30 hours of labor. Wausaukee has at most $4000 to spend on materials and at most 300 hours of labor available. Write a system of inequalities that limits the possible number of barns of each type that can be built. Graph the system.

Solve each problem.

45. ***Target heart rate.*** For beneficial exercise, experts recommend that your target heart rate y should be between 65% and 75% of the maximum heart rate for your age x. That is,

$$y > 0.65(220 - x) \quad \text{and} \quad y < 0.75(220 - x).$$

Graph this system of inequalities for $20 < x < 70$.

46. ***Making and storing the tables.*** The Ozark Furniture Company can obtain at most 8000 board feet of oak lumber for making round and rectangular tables. The tables must be stored in a warehouse that has at most 3850 ft³ of space available for the tables. A round table requires 50 board feet of lumber and 25 ft³ of warehouse space. A rectangular table requires 80 board feet of lumber and 35 ft³ of warehouse space. Write a system of inequalities that limits the possible number of tables of each type that can be made and stored. Graph the system.

FIGURE FOR EXERCISE 47

Inequalities can be used to describe limitations on materials used in construction.

8.8 LINEAR PROGRAMMING

In this section we graph the solution set to a system of several linear inequalities. We then use the solution set as the domain of a function for which we are seeking the maximum or minimum value. The method that we use is called **linear programming,** and it can be applied to problems such as finding maximum profit or minimum cost.

Graphing the Constraints

In linear programming we have two variables that must satisfy several linear inequalities. These inequalities are called the **constraints** because they restrict the variables to only certain values. A graph in the coordinate plane is used to indicate the points that satisfy all of the constraints.

E X A M P L E   1   **Graphing the constraints**

Graph the solution set to the system of inequalities and identify each vertex of the region:

$$x \geq 0, \quad y \geq 0$$
$$3x + 2y \leq 12$$
$$x + 2y \leq 8$$

Solution

Graph the line $3x + 2y = 12$ by using its intercepts $(0, 6)$ and $(4, 0)$. Graph $x + 2y = 8$ by using its intercepts $(0, 4)$ and $(8, 0)$ as shown in Fig. 8.21. The points that satisfy $x \geq 0$ are on or to the right of the y-axis. The points on or above the x-axis satisfy $y \geq 0$. The points on or below the line $3x + 2y = 12$ satisfy $3x + 2y \leq 12$. And the points on or below $x + 2y = 8$ satisfy $x + 2y \leq 8$. Points that satisfy all of the inequalities are in the region shaded in the figure. Three of the vertices of the region are easily identified as $(0, 0)$, $(4, 0)$, and $(0, 4)$. To find the fourth vertex, we solve the system $3x + 2y = 12$ and $x + 2y = 8$. Substitute $x = 8 - 2y$ from the second equation into the first equation:

$$3(8 - 2y) + 2y = 12$$
$$24 - 6y + 2y = 12$$
$$-4y = -12$$
$$y = 3$$
$$x = 8 - 2(3) = 2$$

So the fourth vertex is $(2, 3)$.

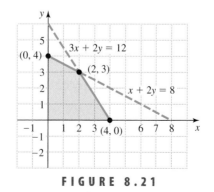

FIGURE 8.21

In linear programming the constraints usually come from physical limitations in some problem. In the next example we write the constraints and then graph the points in the coordinate plane that satisfy all of the constraints.

E X A M P L E   2

Writing the constraints

Jules is in the business of constructing dog houses. A small dog house requires 8 square feet (ft^2) of plywood and 6 ft^2 of insulation. A large dog house requires 16 ft^2 of plywood and 3 ft^2 of insulation. Jules has available only 48 ft^2 of plywood and 18 ft^2 of insulation. Write the constraints on the number of small and large dog houses that he can build with the available supplies and graph the solution set to the system of constraints.

Solution

Let x represent the number of small dog houses and y represent the number of large dog houses. We have two natural constraints, $x \geq 0$ and $y \geq 0$, since he cannot build a negative number of dog houses. A small dog house requires 8 ft^2 of plywood and a large dog house requires 16 ft^2 of plywood. Since only 48 ft^2 of plywood is available, we have $8x + 16y \leq 48$. A small dog house requires 6 ft^2 of insulation and a large dog house requires 3 ft^2 of insulation. Since the total insulation available is 18 ft^2, we have $6x + 3y \leq 18$. Simplify the inequalities to get the following constraints:

$$x \geq 0, \quad y \geq 0$$
$$x + 2y \leq 6$$
$$2x + y \leq 6$$

The line $x + 2y = 6$ has intercepts $(0, 3)$ and $(6, 0)$. The line $2x + y = 6$ has intercepts $(0, 6)$ and $(3, 0)$. These lines intersect at $(2, 2)$. Points that satisfy all of the inequalities are below both lines and in the first quadrant as in Fig. 8.22

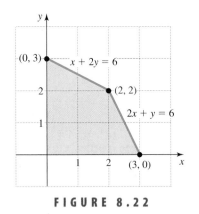

FIGURE 8.22

Maximizing or Minimizing a Linear Function

In Example 2 any ordered pair within the region is a possible solution to the number of dog houses of each type that could be built. If a small dog house sells for $15 and a large dog house sells for $20, then the total revenue in dollars from x small and y large dog houses is $R = 15x + 20y$. Since the revenue is a function of x and y, we write $R(x, y) = 15x + 20y$. The function R is a linear function of x and y. The domain of R is the region graphed in Fig. 8.22.

Linear Function of Two Variables

A function of the form $f(x, y) = Ax + By + C$, where A, B, and C are real numbers, is called a **linear function of two variables.**

Naturally, we are interested in the maximum revenue subject to the constraints on x and y. To investigate some possible revenues, replace R in $R = 15x + 20y$ with, say 35, 50, and 60. The graphs of the parallel lines $15x + 20y = 35$, $15x + 20y = 50$, and $15x + 20y = 60$ are shown in Fig. 8.23. The revenue at any point on the line $15x + 20y = 35$ is \$35. We get a larger revenue on a higher revenue line (and lower revenue on a lower line). The maximum revenue possible will be on the highest revenue line that still intersects the region. Because the sides of the region are straight-line segments, the intersection of the highest (or lowest) revenue line with the region must include a vertex of the region. This is the fundamental principle behind linear programming.

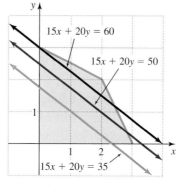

FIGURE 8.23

The Principle of Linear Programming

The maximum or minimum value of a linear function subject to linear constraints occurs at a vertex of the region determined by the constraints.

EXAMPLE 3

Maximizing a linear function with linear constraints

A small dog house requires 8 ft^2 of plywood and 6 ft^2 of insulation. A large dog house requires 16 ft^2 of plywood and 3 ft^2 of insulation. Only 48 ft^2 of plywood and 18 ft^2 of insulation are available. If a small dog house sells for \$15 and a large dog house sells for \$20, then how many dog houses of each type should be built to maximize the revenue and to satisfy the constraints?

Solution

Let x be the number of small dog houses and y be the number of large dog houses. We wrote and graphed the constraints for this problem in Example 2, so we will not repeat that here. The graph in Fig. 8.22 has four vertices: $(0, 0)$, $(0, 3)$, $(3, 0)$, and $(2, 2)$. The revenue function is $R(x, y) = 15x + 20y$. Since the maximum value of this function must occur at a vertex, we evaluate the function at each vertex:

$$R(0, 0) = 15(0) + 20(0) = \$0$$
$$R(0, 3) = 15(0) + 20(3) = \$60$$
$$R(3, 0) = 15(3) + 20(0) = \$45$$
$$R(2, 2) = 15(2) + 20(2) = \$70$$

From this list we can see that the maximum revenue is $70 when two small and two large dog houses are built. We also see that the minimum revenue is $0 when no dog houses of either type are built. ∎

We can summarize the procedure for solving linear programming problems with the following strategy.

Strategy for Linear Programming

Use the following steps to find the maximum or minimum value of a linear function subject to linear constraints.

1. Graph the region that satisfies all of the constraints.
2. Determine the coordinates of each vertex of the region.
3. Evaluate the function at each vertex of the region.
4. Identify which vertex gives the maximum or minimum value of the function.

In the next example we solve another linear programming problem.

EXAMPLE 4   **Minimizing a linear function with linear constraints**

One serving of food A contains 2 grams of protein and 6 grams of carbohydrates. One serving of food B contains 4 grams of protein and 3 grams of carbohydrates. A dietitian wants a meal that contains at least 12 grams of protein and at least 18 grams of carbohydrates. If the cost of food A is 9 cents per serving and the cost of food B is 20 cents per serving, then how many servings of each food would minimize the cost and satisfy the constraints?

Solution

Let x represent the number of servings of food A and y represent the number of servings of food B. Each serving of A contains 2 grams of protein and each serving of B contains 4 grams of protein. If the meal is to contain at least 12 grams of protein, then $2x + 4y \geq 12$. Each serving of A contains 6 grams of carbohydrates and each serving of B contains 3 grams of carbohydrates. If the meal is to contain at least 18 grams of carbohydrates, then $6x + 3y \geq 18$. Simplify each inequality and use the two natural constraints to get the following system:

$$x \geq 0, \quad y \geq 0$$
$$x + 2y \geq 6$$
$$2x + y \geq 6$$

The graph of the constraints is shown in Fig. 8.24. The vertices are $(0, 6)$, $(6, 0)$, and $(2, 2)$. The cost in cents for x servings of A and y servings of B is $C(x, y) = 9x + 20y$. Evaluate the cost at each vertex:

$$C(0, 6) = 9(0) + 20(6) = 120 \text{ cents}$$
$$C(6, 0) = 9(6) + 20(0) = 54 \text{ cents}$$
$$C(2, 2) = 9(2) + 20(2) = 58 \text{ cents}$$

The minimum cost of 54 cents is attained by using six servings of food A and no servings of food B.

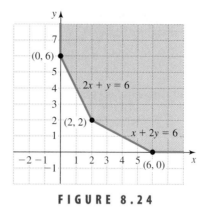

FIGURE 8.24

WARM-UPS

True or false? Explain your answer.

1. The graph of $x \geq 0$ in the coordinate plane consists of the points on or above the x-axis.

2. The graph of $y \geq 0$ in the coordinate plane consists of the points on or to the right of the y-axis.

3. The graph of $x + y \leq 6$ consists of the points below the line $x + y = 6$.

4. The graph of $2x + 3y = 30$ has x-intercept $(0, 10)$ and y-intercept $(15, 0)$.

5. The graph of a system of inequalities is a union of their individual solution sets.

6. In linear programming, constraints are inequalities that restrict the possible values that the variables can assume.

7. The function $F(x, y) = Ax^2 + By^2 + C$ is a linear function of x and y.

8. The value of $R(x, y) = 3x + 5y$ at the point $(2, 4)$ is 26.

9. If $C(x, y) = 12x + 10y$, then $C(0, 5) = 62$.

10. In solving a linear programming problem, we must determine the vertices of the region defined by the constraints.

8.8   EXERCISES

Reading and Writing   *After reading this section, write out the answers to these questions. Use complete sentences.*

1. What is a constraint?

2. What is linear programming?

3. Where do the constraints come from in a linear programming problem?

4. What is a linear function of two variables?

5. Where does the maximum or minimum value of a linear function subject to linear constraints occur?

6. What is the strategy for solving a linear programming problem?

Graph the solution set to each system of inequalities and identify each vertex of the region. See Example 1.

7. $x \geq 0, y \geq 0$
$x + y \leq 5$

8. $x \geq 0, y \geq 0$
$y \leq 5, y \geq x$

9. $x \geq 0, y \geq 0$
$2x + y \leq 4$
$x + y \leq 3$

10. $x \geq 0, y \geq 0$
$x + y \leq 4$
$x + 2y \leq 6$

11. $x \geq 0, y \geq 0$
$2x + y \geq 3$
$x + y \geq 2$

12. $x \geq 0, y \geq 0$
$3x + 2y \geq 12$
$2x + y \geq 7$

13. $x \geq 0, y \geq 0$
$x + 3y \leq 15$
$2x + y \leq 10$

14. $x \geq 0, y \geq 0$
$2x + 3y \leq 15$
$x + y \leq 7$

15. $x \geq 0, y \geq 0$
$x + y \geq 4$
$3x + y \geq 6$

16. $x \geq 0, y \geq 0$
$x + 3y \geq 6$
$2x + y \geq 7$

Solve each problem. See Examples 2–4.

17. *Phase I advertising.* The publicity director for Mercy Hospital is planning to bolster the hospital's image by running a TV ad and a radio ad. Due to budgetary and other constraints, the number of times that she can run the TV ad, x, and the number of times that she can run the radio ad, y, must be in the region shown in the figure on page 457. The function

$$A = 9000x + 4000y$$

gives the total number of people reached by the ads.
a) Find the total number of people reached by the ads at each vertex of the region.
b) What mix of TV and radio ads maximizes the number of people reached?

18. *Phase II advertising.* Suppose the radio station in Exercise 17 starts playing country music and the function for the total number of people changes to

$$A = 9000x + 2000y.$$

a) Find A at each vertex of the region using this function.
b) What mix of TV and radio ads maximizes the number of people reached?

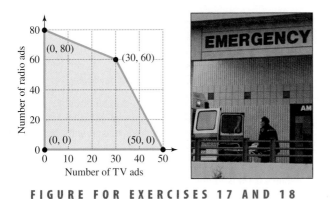

FIGURE FOR EXERCISES 17 AND 18

19. At Burger Heaven a double contains 2 meat patties and 6 pickles, whereas a triple contains 3 meat patties and 3 pickles. Near closing time one day, only 24 meat patties and 48 pickles are available. If a double burger sells for $1.20 and a triple burger sells for $1.50, then how many of each should be made to maximize the total revenue?

20. Sam and Doris manufacture rocking chairs and porch swings in the Ozarks. Each rocker requires 3 hours of work from Sam and 2 hours from Doris. Each swing requires 2 hours of work from Sam and 2 hours from Doris. Sam cannot work more than 48 hours per week, and Doris cannot work more than 40 hours per week. If a rocker sells for $160 and a swing sells for $100, then how many of each should be made per week to maximize the revenue?

21. If a double burger sells for $1.00 and a triple burger sells for $2.00, then how many of each should be made to maximize the total revenue subject to the constraints of Exercise 19?

22. If a rocker sells for $120 and a swing sells for $100, then how many of each should be made to maximize the total revenue subject to the constraints of Exercise 20?

23. One cup of Doggie Dinner contains 20 grams of protein and 40 grams of carbohydrates. One cup of Puppy Power contains 30 grams of protein and 20 grams of carbohydrates. Susan wants her dog to get at least 200 grams of protein and 180 grams of carbohydrates per day. If Doggie Dinner costs 16 cents per cup and Puppy Power costs 20 cents per cup, then how many cups of

each would satisfy the constraints and minimize the total cost?

24. Mammoth Muffler employs supervisors and helpers. According to the union contract, a supervisor does 2 brake jobs and 3 mufflers per day, whereas a helper does 6 brake jobs and 3 mufflers per day. The home office requires enough staff for at least 24 brake jobs and for at least 18 mufflers per day. If a supervisor makes $90 per day and a helper makes $100 per day, then how many of each should be employed to satisfy the constraints and to minimize the daily labor cost?

25. Suppose in Exercise 23 Doggie Dinner costs 4 cents per cup and Puppy Power costs 10 cents per cup. How many cups of each would satisfy the constraints and minimize the total cost?

26. Suppose in Exercise 24 the supervisor makes $110 per day and the helper makes $100 per day. How many of each should be employed to satisfy the constraints and to minimize the daily labor cost?

27. Anita has at most $24,000 to invest in her brother-in-law's laundromat and her nephew's car wash. Her brother-in-law has high blood pressure and heart disease but he will pay 18%, whereas her nephew is healthier but will pay only 12%. So the amount she will invest in the car wash will be at least twice the amount that she will invest in the laundromat but not more than three times as much. How much should she invest in each to maximize her total income from the two investments?

28. Herbert assembles computers in his shop. The parts for each economy model are shipped to him in a carton with a volume of 2 cubic feet (ft^3) and the parts for each deluxe model are shipped to him in a carton with a volume of 3 ft^3. After assembly, each economy model is shipped out in a carton with a volume of 4 ft^3, and each deluxe model is shipped out in a carton with a volume of 4 ft^3. The truck that delivers the parts has a maximum capacity of 180 ft^3, and the truck that takes out the completed computers has a maximum capacity of 280 ft^3. He can receive only one shipment of parts and send out one shipment of computers per week. If his profit on an economy model is $60 and his profit on a deluxe model is $100, then how many of each should he order per week to maximize his profit?

COLLABORATIVE ACTIVITIES

Types of Systems

Assign roles in your groups. Set up the system of equations for each scenario below and attempt to solve it. Analyze your results.

Part I: Talia's Tallitot. Talia plans to make tallitot (prayer shawls) and matching kipot (prayer hats) from a bolt of material. The fabric she plans to use is 250 yards long. She will need 2 yards of material for each tallit (prayer shawl) and $\frac{1}{6}$ of a yard for each kipah (prayer hat). She also plans to put tzitzit (fringes) on the tallit and will need 1 yard of cord for each tallit. The kipot will not have any fringes. She has 100 yards of cord to use for the tzitzit.

1. Write two equations to use to solve for how many kipot and tallitot she can make from the supplies on hand. Let t equal the number of tallitot made and k equal the number of kipot made.

2. Try to solve this system of equations. If you get a solution, state what it is. If not, describe what your result means.

Part II: Karif's Kerchiefs. Karif plans to make red and green kerchiefs in two sizes, medium and large. He has 60 yards of each color of material. He will need $\frac{1}{4}$ of a yard of the appropriate colored fabric for each medium kerchief and $\frac{1}{2}$ of a yard for each large kerchief. He also plans to trim each kerchief with the alternate color of ribbon. He has 240 yards of each color of ribbon. He will need 1 yard of ribbon for each medium kerchief and 2 yards for each large kerchief.

Grouping: 2 to 4 students per group
Topic: Types of solutions to systems of equations

3. Let m equal the number of medium kerchiefs and l equal the number of large kerchiefs. Set up two equations to use to solve for the number of each color of kerchiefs Karif can make from the supplies on hand.

4. Try to solve this system of equations. If you get a solution, state what it is. If not, describe what your result means.

Part III: Maria's Mantillas. Maria plans to make two different kinds of mantillas (veils), some out of lace and some out of a sheer cotton fabric. She also plans to make two sizes of mantillas, one size for women and one size for young girls. She has 150 yards of the lace fabric and 200 yards of the cotton fabric. She will need 2 yards of either fabric for the smaller mantillas and $3\frac{1}{2}$ yards of either fabric for the larger mantillas.

5. Let w equal the number of mantillas for the women and y equal the number of mantillas for the young girls. Set up two equations to use to solve for the number of each mantilla Maria can make from each type of fabric.

6. Try to solve this system of equations. If you get a solution, state what it is. If not, describe what your result means.

Extension: Explain what you would need to do for the systems of equations you could not solve to make them solvable.

WRAP-UP   CHAPTER 8

SUMMARY

Systems of Linear Equations

Methods for solving systems in two variables

Graphing: Sketch the graphs to see the solution.

Substitution: Solve one equation for one variable in terms of the other, then substitute into the other equation.

Addition: Multiply each equation as necessary to eliminate a variable upon addition of the equations.

Examples

The graphs of
$y = x - 1$ and
$x + y = 3$ intersect
at $(2, 1)$.

Substitution:
$x + (x - 1) = 3$

$$\begin{array}{r} -x + y = -1 \\ x + y = 3 \\ \hline 2y = 2 \end{array}$$

Types of linear systems in two variables	Independent: One point in solution set The lines intersect at one point.	$y = x - 5$ $y = 2x + 3$
	Inconsistent: Empty solution set The lines are parallel.	$2x + y = 1$ $2x + y = 5$
	Dependent: Infinite solution set The lines are the same.	$2x + 3y = 4$ $4x + 6y = 8$
Linear equation in three variables	$Ax + By + Cz = D$ In a three-dimensional coordinate system the graph is a plane.	$2x - y + 3z = 5$
Linear systems in three variables	Use substitution or addition to eliminate variables in the system. The solution set may be a single point, the empty set, or an infinite set of points.	$x + y - z = 3$ $2x - 3y + z = 2$ $x - y - 4z = 14$

Matrices and Determinants

Examples

Matrix	A rectangular array of real numbers An $n \times m$ matrix has n rows and m columns.	$\begin{bmatrix} 1 & -3 \\ 2 & 5 \end{bmatrix}, \begin{bmatrix} 1 & 0 & 1 \\ 2 & 1 & 4 \end{bmatrix}$
Augmented matrix	The matrix of coefficients and constants from a system of linear equations	$x - 3y = -7$ $2x + 5y = 19$ Augmented matrix: $\begin{bmatrix} 1 & -3 & -7 \\ 2 & 5 & 19 \end{bmatrix}$
Gaussian elimination method	Use the row operations to get ones on the diagonal and zeros elsewhere for the coefficients in the augmented matrix.	$\begin{bmatrix} 1 & 0 & 2 \\ 0 & 1 & 3 \end{bmatrix}$ $x = 2$ and $y = 3$
Determinant	A real number corresponding to a square matrix	
Determinant of a 2×2 matrix	$\begin{vmatrix} a_1 & b_1 \\ a_2 & b_2 \end{vmatrix} = a_1 b_2 - a_2 b_1$	$\begin{vmatrix} 1 & -3 \\ 2 & 5 \end{vmatrix} = 5 - (-6)$ $= 11$
Determinant of a 3×3 matrix	Expand by minors about any row or column, using signs from the sign array. $\begin{vmatrix} a_1 & b_1 & c_1 \\ a_2 & b_2 & c_2 \\ a_3 & b_3 & c_3 \end{vmatrix} = a_1 \cdot \begin{vmatrix} b_2 & c_2 \\ b_3 & c_3 \end{vmatrix} - a_2 \cdot \begin{vmatrix} b_1 & c_1 \\ b_3 & c_3 \end{vmatrix} + a_3 \cdot \begin{vmatrix} b_1 & c_1 \\ b_2 & c_2 \end{vmatrix}$	Sign array: $\begin{bmatrix} + & - & + \\ - & + & - \\ + & - & + \end{bmatrix}$

Cramer's Rules

| Two linear equations in two variables | The solution to the system

$\begin{aligned} a_1 x + b_1 y &= c_1 \\ a_2 x + b_2 y &= c_2 \end{aligned}$ | |

is given by $x = \dfrac{D_x}{D}$ and $y = \dfrac{D_y}{D}$, where

$$D = \begin{vmatrix} a_1 & b_1 \\ a_2 & b_2 \end{vmatrix}, \qquad D_x = \begin{vmatrix} c_1 & b_1 \\ c_2 & b_2 \end{vmatrix}, \qquad \text{and} \qquad D_y = \begin{vmatrix} a_1 & c_1 \\ a_2 & c_2 \end{vmatrix}$$

provided that $D \neq 0$.

Three linear equations in three variables

The solution to the system

$$a_1 x + b_1 y + c_1 z = d_1$$
$$a_2 x + b_2 y + c_2 z = d_2$$
$$a_3 x + b_3 y + c_3 z = d_3$$

is given by $x = \dfrac{D_x}{D}$, $y = \dfrac{D_y}{D}$, and $z = \dfrac{D_z}{D}$, where

$$D = \begin{vmatrix} a_1 & b_1 & c_1 \\ a_2 & b_2 & c_2 \\ a_3 & b_3 & c_3 \end{vmatrix}, \qquad D_x = \begin{vmatrix} d_1 & b_1 & c_1 \\ d_2 & b_2 & c_2 \\ d_3 & b_3 & c_3 \end{vmatrix},$$

$$D_y = \begin{vmatrix} a_1 & d_1 & c_1 \\ a_2 & d_2 & c_2 \\ a_3 & d_3 & c_3 \end{vmatrix}, \qquad D_z = \begin{vmatrix} a_1 & b_1 & d_1 \\ a_2 & b_2 & d_2 \\ a_3 & b_3 & d_3 \end{vmatrix},$$

provided that $D \neq 0$.

Linear Inequalities in Two Variables

Examples

Graphing the solution to an inequality in two variables

1. Solve the inequality for y, then graph $y = mx + b$.
   $y > mx + b$ is the region above the line.
   $y = mx + b$ is the line itself.
   $y < mx + b$ is the region below the line.

   Remember that "less than" means below the line and "greater than" means above the line only when the inequality is solved for y.

$y > x + 3$
$y = x + 3$
$y < x + 3$

2. If the inequality involves only x, then graph the vertical line $x = k$.
   $x > k$ is the region to the right of the line.
   $x = k$ is the line itself.
   $x < k$ is the region to the left of the line.

$x > 5$
Region to the right of vertical line $x = 5$

Test points

A linear inequality may also be graphed by graphing the equation and then testing a point to determine which region satisfies the inequality.

$x + y > 4$
$(0, 6)$ satisfies the inequality.

Graphing a system of inequalities

Graph the equations and use test points to see which regions satisfy both inequalities.

$x + y > 4$
$x - y < 1$
$(0, 6)$ satisfies the system.

Linear Programming

Use the following steps to find the maximum or minimum value of a linear function subject to linear constraints.

1. Graph the region that satisfies all of the constraints.
2. Determine the coordinates of each vertex of the region.
3. Evaluate the function at each vertex of the region.
4. Identify which vertex gives the maximum or minimum value of the function.

ENRICHING YOUR MATHEMATICAL WORD POWER

For each mathematical term, choose the correct meaning.

1. **system of equations**
   a. a systematic method for classifying equations
   b. a method for solving an equation
   c. two or more equations
   d. the properties of equality

2. **independent linear system**
   a. a system with exactly one solution
   b. an equation that is satisfied by every real number
   c. equations that are identical
   d. a system of lines

3. **inconsistent system**
   a. a system with no solution
   b. a system of inconsistent equations
   c. a system that is incorrect
   d. a system that we are not sure how to solve

4. **dependent system**
   a. a system that is independent
   b. a system that depends on a variable
   c. a system that has no solution
   d. a system for which the graphs coincide

5. **substitution method**
   a. replacing the variables by the correct answer
   b. a method of eliminating a variable by substituting one equation into the other
   c. the replacement method
   d. any method of solving a system

6. **addition method**
   a. adding the same number to each side of an equation
   b. adding fractions
   c. eliminating a variable by adding two equations
   d. the sum of a number and its additive inverse is zero

7. **linear equation in three variables**
   a. $Ax + By + Cz = D$ with A, B, and C not all zero
   b. $Ax + By = C$ with A and B not both zero

c. the equation of a line
   d. $A/x + B/y = C$ with A and B not both zero

8. **linear inequality in two variables**
   a. when two lines are not equal
   b. line segments that are unequal in length
   c. an inequality of the form $Ax + By \geq C$ or with another symbol of inequality
   d. an inequality of the form $Ax^2 + By^2 < C^2$

9. **matrix**
   a. a television screen
   b. a maze
   c. a rectangular array of numbers
   d. coordinates in four dimensions

10. **augmented matrix**
    a. a matrix with a power booster
    b. a matrix with no solution
    c. a square matrix
    d. a matrix containing the coefficients and constants of a system of equations

11. **order**
    a. the length of a matrix
    b. the number of rows and columns in a matrix
    c. the highest power of a matrix
    d. the lowest power of a matrix

12. **determinant**
    a. a number corresponding to a square matrix
    b. a number that is determined by any matrix
    c. the first entry of a matrix
    d. a number that determines whether a matrix has a solution

13. **sign array**
    a. the signs of the entries of a matrix
    b. the sign of the determinant
    c. the signs of the answers
    d. a matrix of $+$ and $-$ signs used in computing a determinant

REVIEW EXERCISES

8.1 *Solve by graphing. Indicate whether each system is independent, inconsistent, or dependent.*

1. $y = 2x - 1$
$x + y = 2$

2. $y = -x$
$y = -x + 3$

3. $y = 3x - 4$
$y = -2x + 1$

4. $x + y = 5$
$x - y = -1$

Solve each system by the substitution method. Indicate whether each system is independent, inconsistent, or dependent.

5. $y = 3x + 11$
$2x + 3y = 0$

6. $x - y = 3$
$3x - 2y = 3$

7. $x = y + 5$
$2x - 2y = 12$

8. $2x - y = 3$
$6x - 9 = 3y$

8.2 *Solve each system by the addition method. Indicate whether each system is independent, inconsistent, or dependent.*

9. $5x - 3y = -20$
$3x + 2y = 7$

10. $-3x + y = 3$
$2x - 3y = 5$

11. $2(y - 5) + 4 = 3(x - 6)$
$3x - 2y = 12$

12. $3x - 4(y - 5) = x + 2$
$2y - x = 7$

8.3 *Solve each system by elimination of variables.*

13. $2x - y - z = 3$
$3x + y + 2z = 4$
$4x + 2y - z = -4$

14. $2x + 3y - 2z = -11$
$3x - 2y + 3z = 7$
$x - 4y + 4z = 14$

15. $x - 3y + z = 5$
$2x - 4y - z = 7$
$2x - 6y + 2z = 6$

16. $x - y + z = 1$
$2x - 2y + 2z = 2$
$-3x + 3y - 3z = -3$

8.4 *Solve each system by using the Gaussian elimination method.*

17. $2x + y = 0$
$x - 3y = 14$

18. $2x - y = 8$
$3x + 2y = -2$

19. $x + y - z = 0$
$x - y + 2z = 4$
$2x + y - z = 1$

20. $2x - y + 2z = 9$
$x + 3y = 5$
$3x + z = 9$

8.5 *Evaluate each determinant.*

21. $\begin{vmatrix} 1 & 3 \\ 0 & 2 \end{vmatrix}$

22. $\begin{vmatrix} -1 & 2 \\ -3 & 5 \end{vmatrix}$

23. $\begin{vmatrix} 0.01 & 0.02 \\ 50 & 80 \end{vmatrix}$

24. $\begin{vmatrix} 1 & 1 \\ \frac{1}{2} & \frac{1}{3} \\ \frac{1}{4} & \frac{1}{5} \end{vmatrix}$

> **study tip**
>
> Note how the review exercises are arranged according to the sections in this chapter. If you are having trouble with a certain type of problem, refer back to the appropriate section for examples and explanations.

Solve each system. Use Cramer's rule.

25. $2x - y = 0$
$3x + y = -5$

26. $3x - 2y = 14$
$2x + 3y = -8$

27. $y = 2x - 3$
$3x - 2y = 4$

28. $2x - y = 7$
$3x + 2y = -7$

29. $x - y = 4$
$x + 2y = 6$

30. $y = 2x - 5$
$y = 3x - 3y$

Evaluate each determinant.

31. $\begin{vmatrix} 2 & 3 & 1 \\ -1 & 2 & 4 \\ 6 & 1 & 1 \end{vmatrix}$

32. $\begin{vmatrix} 1 & -1 & 0 \\ -2 & 0 & 0 \\ 3 & 1 & 5 \end{vmatrix}$

33. $\begin{vmatrix} 2 & 3 & -2 \\ 2 & 0 & 4 \\ -1 & 0 & 3 \end{vmatrix}$

34. $\begin{vmatrix} 3 & -1 & 4 \\ 2 & -1 & 1 \\ -2 & 0 & 1 \end{vmatrix}$

Solve each system. Use Cramer's rule.

35. $x + y = 3$
$x + y + z = 0$
$x - y - z = 2$

36. $x + y = 4$
$y + z = -3$
$x + z = -5$

37. $2x - y + z = 0$
$4x + 6y - 2z = 0$
$x - 2y - z = -9$

38. $x - y + z = 4$
$2x - 3y + z = 2$
$4x - y - z = 18$

8.6 *Graph each inequality.*

39. $y > \dfrac{1}{3}x - 5$

40. $y < \dfrac{1}{2}x + 2$

41. $y \leq -2x + 7$

42. $y \geq x - 6$

43. $y \leq 8$

44. $x \geq -6$

45. $2x + 3y \leq -12$

46. $x - 3y < 9$

8.7 *Graph each system of inequalities.*

47. $x < 5$
$y < 4$

48. $y > -2$
$x < 1$

49. $x + y < 2$
$y > 2x - 3$

50. $x - y > 4$
$2y > x - 4$

51. $y > 5x - 7$
$y < 5x + 1$

52. $y > x - 6$
$y < x - 5$

53. $y < 3x + 5$
$y < 3x$

54. $y > -2x$
$y < -3x$

8.8 *Graph each system of inequalities and identify each vertex of the region.*

55. $x \geq 0, y \geq 0$
$x + 2y \leq 6$
$x + y \leq 5$

56. $x \geq 0, y \geq 0$
$3x + 2y \geq 12$
$x + 2y \geq 8$

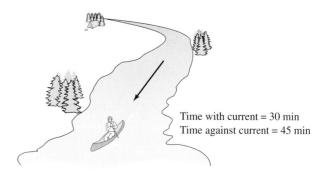

Time with current = 30 min
Time against current = 45 min

FIGURE FOR EXERCISE 61

Solve each problem by linear programming.

57. Find the maximum value of the function $R(x, y) = 6x + 9y$ subject to the following constraints:

$$x \geq 0, y \geq 0$$
$$2x + y \leq 6$$
$$x + 2y \leq 6$$

58. Find the minimum value of the function $C(x, y) = 9x + 10y$ subject to the following constraints:

$$x \geq 0, y \geq 0$$
$$x + y \geq 4$$
$$3x + y \geq 6$$

MISCELLANEOUS

Use a system of equations in two or three variables to solve each word problem. Solve by the method of your choice.

59. *Two-digit number.* The sum of the digits in a two-digit number is 15. When the digits are reversed, the new number is 9 more than the original number. What is the original number?

60. *Two-digit number.* The sum of the digits in a two-digit number is 8. When the digits are reversed, the new number is 18 less than the original number. What is the original number?

61. *Traveling by boat.* Alonzo can travel from his camp downstream to the mouth of the river in 30 minutes. If it takes him 45 minutes to come back, then how long would it take him to go that same distance in the lake with no current?

62. *Driving and dating.* In 4 years Gasper will be old enough to drive. His parents said that he must have a driver's license for 2 years before he can date. Three years ago, Gasper's age was only one-half of the age necessary to

date. How old must Gasper be to drive, and how old is he now?

63. *Three solutions.* A chemist has three solutions of acid that must be mixed to obtain 20 liters of a solution that is 38% acid. Solution A is 30% acid, solution B is 20% acid, and solution C is 60% acid. Because of another chemical in these solutions, the chemist must keep the ratio of solution C to solution A at 2 to 1. How many liters of each should she mix together?

64. *Mixing investments.* Darlene invested a total of $20,000. The part that she invested in Dell Computer stock returned 70% and the part that she invested in U.S. Treasury bonds returned 5%. Her total return on these two investments was $9580.
a) Use the graph to estimate the amount that she put into each investment.
b) Solve a system of equations to find the exact amount that she put into each investment.

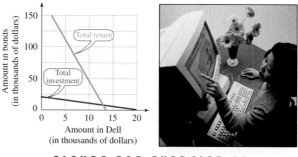

FIGURE FOR EXERCISE 64

65. *Beets and beans.* One serving of canned beets contains 1 gram of protein and 6 grams of carbohydrates. One serving of canned red beans contains 6 grams of protein and 20 grams of carbohydrates. How many servings of each would it take to get exactly 21 grams of protein and 78 grams of carbohydrates?

CHAPTER 8 TEST

Solve the system by graphing.

1. $x + y = 4$
$y = 2x + 1$

Solve each system by substitution.

2. $y = 2x - 8$
$4x + 3y = 1$

3. $y = x - 5$
$3x - 4(y - 2) = 28 - x$

Solve each system by the addition method.

4. $3x + 2y = 3$ **5.** $3x - y = 5$
$4x - 3y = -13$ $-6x + 2y = 1$

Determine whether each system is independent, inconsistent, or dependent.

6. $y = 3x - 5$ **7.** $2x + 2y = 8$
$y = 3x + 2$ $x + y = 4$

8. $y = 2x - 3$
$y = 5x - 14$

Solve the following system by elimination of variables.

9. $x + y - z = 2$
$2x - y + 3z = -5$
$x - 3y + z = 4$

Solve by the Gaussian elimination method.

10. $3x - y = 1$ **11.** $x - y - z = 1$
$x + 2y = 12$ $-x - y + 2z = -2$
                               $-x - 3y + z = -5$

Evaluate each determinant.

12. $\begin{vmatrix} 2 & 3 \\ 4 & -3 \end{vmatrix}$ **13.** $\begin{vmatrix} 1 & -2 & -1 \\ 2 & 3 & 1 \\ 1 & 1 & 0 \end{vmatrix}$

Solve each system by using Cramer's rule.

14. $2x - y = -4$ **15.** $x + y = 0$
$3x + y = -1$ $x - y + 2z = 6$
                              $2x + y - z = 1$

Graph each inequality.

16. $y > 3x - 5$ **17.** $x - y < 3$

18. $x - 2y \geq 4$ **19.** $x < 6$
                                 $y > -1$

20. $2x + 3y > 6$ **21.** $y > 3x - 4$
$3x - y < 3$ $3x - y > 3$

For each problem, write a system of equations in two or three variables. Use the method of your choice to solve each system.

22. One night the manager of the Sea Breeze Motel rented 5 singles and 12 doubles for a total of $390. The next night he rented 9 singles and 10 doubles for a total of $412. What is the rental charge for each type of room?

23. Jill, Karen, and Betsy studied a total of 93 hours last week. Jill's and Karen's study time totaled only one-half as much as Betsy's. If Jill studied 3 hours more than Karen, then how many hours did each one of the girls spend studying?

Solve the following problem by linear programming.

24. Find the maximum value of the function

$$P(x, y) = 8x + 10y$$

subject to the following constraints:

$$x \geq 0, y \geq 0$$
$$2x + 3y \leq 12$$
$$x + y \leq 5$$

Simplify each expression.

1. -3^4

2. $\dfrac{1}{3}(3) + 6$

3. $(-5)^2 - 4(-2)(6)$

4. $6 - (0.2)(0.3)$

5. $5(t - 3) - 6(t - 2)$

6. $0.1(x - 1) - (x - 1)$

7. $\dfrac{-9x^2 - 6x + 3}{-3}$

8. $\dfrac{4y - 6}{2} - \dfrac{3y - 9}{3}$

Solve each equation for y.

9. $3x - 5y = 7$

10. $Cx - Dy = W$

11. $Cy = Wy - K$

12. $A = \dfrac{1}{2}b(w - y)$

Solve each system.

13. $y = x - 5$
    $2x + 3y = 5$

14. $0.05x + 0.06y = 67$
    $x + y = 1200$

15. $3x - 15y = -51$
    $x + 17 = 5y$

16. $0.07a + 0.3b = 6.70$
    $7a + 30b = 67$

Find the equation of each line.

17. The line through $(0, 55)$ and $(-99, 0)$

18. The line through $(2, -3)$ and $(-4, 8)$

19. The line through $(-4, 6)$ that is parallel to $y = 5x$

20. The line through $(4, 7)$ that is perpendicular to
    $y = -2x + 1$

21. The line through $(3, 5)$ that is parallel to the x-axis

22. The line through $(-7, 0)$ that is perpendicular to the x-axis

study tip

Don't wait until the final exam to review material. Do some review on a regular basis. The Making Connections exercises on this page can be used to review, compare, and contrast different concepts that you have studied. A good time to work these exercises is between a test and the start of new material.

Solve.

23. **Comparing copiers.** A self-employed consultant has prepared the accompanying graph to compare the total cost of purchasing and using two different copy machines.

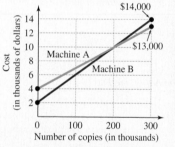

FIGURE FOR EXERCISE 23

a) Which machine has the larger purchase price?

b) What is the per copy cost for operating each machine, not including the purchase price?

c) Find the slope of each line and interpret your findings.

d) Find the equation of each line.

e) Find the number of copies for which the total cost is the same for both machines.

Radicals and Rational Exponents

Wind chill temperature (°F) for 25°F air temperature

Wind velocity (mph)

Just how cold is it in Fargo, North Dakota, in winter? According to local meteorologists, the mercury hit a low of –33°F on January 18, 1994. But air temperature alone is not always a reliable indicator of how cold you feel. On the same date the average wind velocity was 13.8 miles per hour. This dramatically affected how cold people felt when they stepped outside. High winds along with cold temperatures make exposed skin feel colder because the wind significantly speeds up the loss of body heat. Meteorologists use the terms "wind chill factor," "wind chill index," and "wind chill temperature" to take into account both air temperature and wind velocity.

Through experimentation in Antarctica, Paul A. Siple developed a formula in the 1940s that measures the wind chill from the velocity of the wind and the air temperature. His complex formula involving the square root of the velocity of the wind is still used today to calculate wind chill temperatures. Siple's formula is unlike most scientific formulas in that it is not based on theory. Siple experimented with various formulas involving wind velocity and temperature until he found a formula that seemed to predict how cold the air felt. His formula is stated and used in Exercises 101 and 102 of Section 9.1.

9.1 RADICALS

In Section 5.6 you learned the basic facts about powers. In this section you will study roots and see how powers and roots are related.

Roots

We use the idea of roots to reverse powers. Because $3^2 = 9$ and $(-3)^2 = 9$, both 3 and -3 are square roots of 9. Because $2^4 = 16$ and $(-2)^4 = 16$, both 2 and -2 are fourth roots of 16. Because $2^3 = 8$ and $(-2)^3 = -8$, there is only one real cube root of 8 and only one real cube root of -8. The cube root of 8 is 2 and the cube root of -8 is -2.

nth Roots

If $a = b^n$ for a positive integer n, then b is an **nth root of a.** If $a = b^2$, then b is a **square root** of a. If $a = b^3$, then b is the **cube root** of a.

If n is a positive even integer and a is positive, then there are two real nth roots of a. We call these roots **even roots.** The positive even root of a positive number is called the **principal root.** The principal square root of 9 is 3 and the principal fourth root of 16 is 2 and these roots are even roots.

If n is a positive odd integer and a is any real number, there is only one real nth root of a. We call that root an **odd root.** Because $2^5 = 32$, the fifth root of 32 is 2 and 2 is an odd root.

We use the **radical symbol** $\sqrt{}$ to signify roots.

$$\sqrt[n]{a}$$

If n is a positive *even* integer and a is positive, then $\sqrt[n]{a}$ denotes the *principal nth root of a.*

If n is a positive *odd* integer, then $\sqrt[n]{a}$ denotes the nth root of a.

If n is any positive integer, then $\sqrt[n]{0} = 0$.

We read $\sqrt[n]{a}$ as "the nth root of a." In the notation $\sqrt[n]{a}$, n is the **index of the radical** and a is the **radicand.** For square roots the index is omitted, and we simply write $\sqrt{a}$.

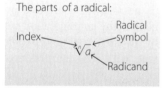
EXAMPLE 1

Evaluating radical expressions

Find the following roots:

a) $\sqrt{25}$ **b)** $\sqrt[3]{-27}$ **c)** $\sqrt[6]{64}$ **d)** $-\sqrt{4}$

Solution

a) Because $5^2 = 25$, $\sqrt{25} = 5$.

b) Because $(-3)^3 = -27$, $\sqrt[3]{-27} = -3$.

c) Because $2^6 = 64$, $\sqrt[6]{64} = 2$.

d) Because $\sqrt{4} = 2$, $-\sqrt{4} = -(\sqrt{4}) = -2$. ∎

CAUTION In radical notation, $\sqrt{4}$ represents the *principal square root of 4*, so $\sqrt{4} = 2$. Note that -2 is also a square root of 4, but $\sqrt{4} \neq -2$.

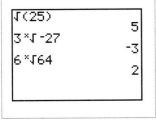

Note that even roots of negative numbers are omitted from the definition of nth roots because even powers of real numbers are never negative. So no real number can be an even root of a negative number. Expressions such as

$$\sqrt{-9}, \quad \sqrt[4]{-81}, \quad \text{and} \quad \sqrt[6]{-64}$$

are not real numbers. Square roots of negative numbers will be discussed in Section 9.6 when we discuss the imaginary numbers.

Roots and Variables

Consider the result of squaring a power of x:

$$(x^1)^2 = x^2, \quad (x^2)^2 = x^4, \quad (x^3)^2 = x^6, \quad \text{and} \quad (x^4)^2 = x^8.$$

When a power of x is squared, the exponent is multiplied by 2. So any even power of x is a perfect square.

Perfect Squares

The following expressions are perfect squares:

$$x^2, \quad x^4, \quad x^6, \quad x^8, \quad x^{10}, \quad x^{12}, \quad \ldots$$

Since taking a square root reverses the operation of squaring, the square root of an even power of x is found by dividing the exponent by 2. Provided x is nonnegative (see Caution below), we have:

$$\sqrt{x^2} = x^1 = x, \quad \sqrt{x^4} = x^2, \quad \sqrt{x^6} = x^3, \quad \text{and} \quad \sqrt{x^8} = x^4.$$

CAUTION   If x is negative, equations like $\sqrt{x^2} = x$ and $\sqrt{x^6} = x^3$ are not correct because the radical represents the nonnegative square root but x and x^3 are negative. That is why we assume x is nonnegative.

If a power of x is cubed, the exponent is multiplied by 3:

$$(x^1)^3 = x^3, \quad (x^2)^3 = x^6, \quad (x^3)^3 = x^9, \quad \text{and} \quad (x^4)^3 = x^{12}.$$

So if the exponent is a multiple of 3, we have a perfect cube.

Perfect Cubes

The following expressions are perfect cubes:

$$x^3, \quad x^6, \quad x^9, \quad x^{12}, \quad x^{15}, \quad \ldots$$

Since the cube root reverses the operation of cubing, the cube root of any of these perfect cubes is found by dividing the exponent by 3:

$$\sqrt[3]{x^3} = x^1 = x, \quad \sqrt[3]{x^6} = x^2, \quad \sqrt[3]{x^9} = x^3, \quad \text{and} \quad \sqrt[3]{x^{12}} = x^4.$$

If the exponent is divisible by 4, we have a perfect fourth power, and so on.

E X A M P L E   2

Roots of exponential expressions

Find each root. Assume that all variables represent nonnegative real numbers.

a) $\sqrt{x^{22}}$ **b)** $\sqrt[3]{t^{18}}$ **c)** $\sqrt[5]{s^{30}}$

Solution

a) $\sqrt{x^{22}} = x^{11}$ because $(x^{11})^2 = x^{22}$.

b) $\sqrt[3]{t^{18}} = t^6$ because $(t^6)^3 = t^{18}$.

c) $\sqrt[5]{s^{30}} = s^6$ because one-fifth of 30 is 6.

Product Rule for Radicals

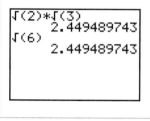
Consider the expression $\sqrt{2} \cdot \sqrt{3}$. If we square this product, we get

$$(\sqrt{2} \cdot \sqrt{3})^2 = (\sqrt{2})^2(\sqrt{3})^2 \quad \text{Power of a product rule}$$
$$= 2 \cdot 3 \quad (\sqrt{2})^2 = 2 \text{ and } (\sqrt{3})^2 = 3$$
$$= 6.$$

The number $\sqrt{6}$ is the unique positive number whose square is 6. Because we squared $\sqrt{2} \cdot \sqrt{3}$ and obtained 6, we must have $\sqrt{6} = \sqrt{2} \cdot \sqrt{3}$. This example illustrates the product rule for radicals.

> **Product Rule for Radicals**
>
> The nth root of a product is equal to the product of the nth roots. In symbols,
> $$\sqrt[n]{ab} = \sqrt[n]{a} \cdot \sqrt[n]{b},$$
> provided all of these roots are real numbers.

E X A M P L E   3

Using the product rule for radicals

Simplify each radical. Assume that all variables represent positive real numbers.

a) $\sqrt{4y}$ **b)** $\sqrt{3y^8}$

Solution

a) $\sqrt{4y} = \sqrt{4} \cdot \sqrt{y}$ Product rule for radicals
      $= 2\sqrt{y}$ Simplify.

b) $\sqrt{3y^8} = \sqrt{3} \cdot \sqrt{y^8}$ Product rule for radicals
      $= \sqrt{3} \cdot y^4$ $\sqrt{y^8} = y^4$
      $= y^4\sqrt{3}$ A radical is usually written last in a product.

Quotient Rule for Radicals

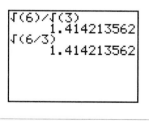
Because $\sqrt{2} \cdot \sqrt{3} = \sqrt{6}$, we have $\sqrt{6} \div \sqrt{3} = \sqrt{2}$, or

$$\sqrt{2} = \sqrt{\frac{6}{3}} = \frac{\sqrt{6}}{\sqrt{3}}.$$

This example illustrates the quotient rule for radicals.

> **Quotient Rule for Radicals**
>
> The nth root of a quotient is equal to the quotient of the nth roots. In symbols,
> $$\sqrt[n]{\frac{a}{b}} = \frac{\sqrt[n]{a}}{\sqrt[n]{b}},$$
> provided that all of these roots are real numbers and $b \neq 0$.

M A T H   A T   W O R K $x^2 + (x+1)^2 = 5^2$

Ernie Godshalk, avid sailor and owner of the sloop *Golden Eye,* has competed in races all over the world. He has learned that success in a race depends on the winds, a good crew, lots of skill, and knowing what makes his boat go fast.

Because of its shape and hull design, the maximum speed (in knots) of a sailboat can be calculated by finding the value 1.3 times the square root of her waterline (in feet). Thus Godshalk knows the maximum possible speed of the *Golden Eye.* But decisions made while she is

YACHTSMAN

under sail make the difference between attaining her maximum speed and only approaching it. For example, Godshalk knows that the pressure on the sails is what makes the boat move at a certain speed. It is especially important to sail where the wind is the strongest. This means that choosing the correct tack, as well as "going where the wind is" can make the difference between coming in first and just finishing the race. In Exercise 105 of this section you will calculate the maximum speed of the *Golden Eye.*

In the next example we use the quotient rule to simplify radical expressions.

E X A M P L E   3   **Using the quotient rule for radicals**
Simplify each radical. Assume that all variables represent positive real numbers.

a) $\sqrt{\dfrac{t}{9}}$

b) $\sqrt[3]{\dfrac{x^{21}}{y^6}}$

Solution

a) $\sqrt{\dfrac{t}{9}} = \dfrac{\sqrt{t}}{\sqrt{9}}$ Quotient rule for radicals

$= \dfrac{\sqrt{t}}{3}$

b) $\sqrt[3]{\dfrac{x^{21}}{y^6}} = \dfrac{\sqrt[3]{x^{21}}}{\sqrt[3]{y^6}}$ Quotient rule for radicals

$= \dfrac{x^7}{y^2}$

■

Rationalizing the Denominator

Square roots such as $\sqrt{2}$, $\sqrt{3}$, and $\sqrt{5}$ are irrational numbers. If roots of this type appear in the denominator of a fraction, it is customary to rewrite the fraction with a rational number in the denominator, or **rationalize** it. We rationalize a denominator by multiplying both the numerator and denominator by another radical that makes the denominator rational.

You can find products of radicals in two ways. By definition, $\sqrt{2}$ is the positive number that you multiply by itself to get 2. So

$$\sqrt{2} \cdot \sqrt{2} = 2.$$

By the product rule, $\sqrt{2} \cdot \sqrt{2} = \sqrt{4} = 2$. Note that $\sqrt[3]{2} \cdot \sqrt[3]{2} = \sqrt[3]{4}$ by the product rule, but $\sqrt[3]{4} \neq 2$. By definition of a cube root,

$$\sqrt[3]{2} \cdot \sqrt[3]{2} \cdot \sqrt[3]{2} = 2.$$

E X A M P L E   5

Rationalizing the denominator

Rewrite each expression with a rational denominator.

a) $\dfrac{\sqrt{3}}{\sqrt{5}}$

b) $\dfrac{3}{\sqrt[3]{2}}$

Solution

a) Because $\sqrt{5} \cdot \sqrt{5} = 5$, multiplying both the numerator and denominator by $\sqrt{5}$ will rationalize the denominator:

$$\frac{\sqrt{3}}{\sqrt{5}} = \frac{\sqrt{3}}{\sqrt{5}} \cdot \frac{\sqrt{5}}{\sqrt{5}} = \frac{\sqrt{15}}{5} \qquad \text{By the product rule, } \sqrt{3} \cdot \sqrt{5} = \sqrt{15}.$$

b) We must build up the denominator to be the cube root of a perfect cube. So we multiply by $\sqrt[3]{4}$ to get $\sqrt[3]{4} \cdot \sqrt[3]{2} = \sqrt[3]{8}$:

$$\frac{3}{\sqrt[3]{2}} = \frac{3}{\sqrt[3]{2}} \cdot \frac{\sqrt[3]{4}}{\sqrt[3]{4}} = \frac{3\sqrt[3]{4}}{\sqrt[3]{8}} = \frac{3\sqrt[3]{4}}{2} \qquad ∎$$

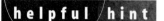

helpful **hint**

If you are going to compute the value of a radical expression with a calculator, it does not matter if the denominator is rational. However, rationalizing the denominator provides another opportunity to practice building up the denominator of a fraction and multiplying radicals.

C A U T I O N   To rationalize a denominator with a single square root, you simply multiply by that square root. If the denominator has a cube root, you build the denominator to a cube root of a perfect cube, as in Example 5(b). For a fourth root you build to a fourth root of a perfect fourth power, and so on.

Simplifying Radicals

When simplifying any expression, we try to make it look "simpler." When simplifying a radical expression, we have three specific conditions to satisfy.

> **Simplified Radical Form for Radicals of Index n**
>
> A radical expression of index n is in **simplified radical form** if it has
>
> 1. *no* perfect nth powers as factors of the radicand,
> 2. *no* fractions inside the radical, and
> 3. *no* radicals in the denominator.

The radical expressions in the next example do not satisfy the three conditions for simplified radical form. To rewrite an expression in simplified form, we use the product rule, the quotient rule, and rationalizing the denominator.

E X A M P L E   6

Writing radical expressions in simplified radical form

Simplify.

a) $\dfrac{\sqrt{10}}{\sqrt{6}}$

b) $\sqrt[3]{\dfrac{5}{9}}$

Solution

a) To rationalize the denominator, multiply the numerator and denominator by $\sqrt{6}$:

$$\frac{\sqrt{10}}{\sqrt{6}} = \frac{\sqrt{10}}{\sqrt{6}} \cdot \frac{\sqrt{6}}{\sqrt{6}} \qquad \text{Rationalize the denominator.}$$

$$= \frac{\sqrt{60}}{6}$$

$$= \frac{\sqrt{4}\sqrt{15}}{6} \qquad \text{Remove the perfect square from } \sqrt{60}.$$

$$= \frac{2\sqrt{15}}{6}$$

$$= \frac{\sqrt{15}}{3} \qquad \text{Reduce } \frac{2}{6} \text{ to } \frac{1}{3}. \text{ Note that } \sqrt{15} \div 3 \neq \sqrt{5}.$$

b) To rationalize the denominator, build up the denominator to a cube root of a perfect cube. Because $\sqrt[3]{9} \cdot \sqrt[3]{3} = \sqrt[3]{27} = 3$, we multiply by $\sqrt[3]{3}$:

$$\sqrt[3]{\frac{5}{9}} = \frac{\sqrt[3]{5}}{\sqrt[3]{9}} \qquad \text{Quotient rule for radicals}$$

$$= \frac{\sqrt[3]{5}}{\sqrt[3]{9}} \cdot \frac{\sqrt[3]{3}}{\sqrt[3]{3}} \qquad \text{Rationalize the denominator.}$$

$$= \frac{\sqrt[3]{15}}{\sqrt[3]{27}}$$

$$= \frac{\sqrt[3]{15}}{3}$$

Simplifying Radicals Involving Variables

In the next example we simplify square roots containing variables. Remember that any even power of a variable is a perfect square.

E X A M P L E   7   **Simplifying square roots with variables**

Simplify each expression. Assume all variables represent positive real numbers.

a) $\sqrt{12x^6}$ **b)** $\sqrt{98x^5y^9}$

Solution

a) Use the product rule to place all perfect squares under the first radical symbol and the remaining factors under the second:

$$\sqrt{12x^6} = \sqrt{4x^6 \cdot 3} \qquad \text{Factor out the perfect squares.}$$

$$= \sqrt{4x^6} \cdot \sqrt{3} \qquad \text{Product rule for radicals}$$

$$= 2x^3\sqrt{3}$$

b) $\sqrt{98x^5y^9} = \sqrt{49x^4y^8} \cdot \sqrt{2xy} \qquad \text{Product rule for radicals}$

$$= 7x^2y^4\sqrt{2xy}$$

In the next example we start with a square root of a quotient.

E X A M P L E   8

Rationalizing the denominator with variables

Simplify each expression. Assume all variables represent positive real numbers.

a) $\sqrt{\dfrac{a}{b}}$

b) $\sqrt{\dfrac{x^3}{y^5}}$

Solution

Write about what you read in the text. Sum things up in your own words. Write out important facts on note cards. When you have a few spare minutes in between classes review your note cards. Try to memorize all the information on the cards.

a) $\sqrt{\dfrac{a}{b}} = \dfrac{\sqrt{a}}{\sqrt{b}}$ Quotient rule for radicals

$\quad\quad = \dfrac{\sqrt{a} \cdot \sqrt{b}}{\sqrt{b} \cdot \sqrt{b}}$ Rationalize the denominator.

$\quad\quad = \dfrac{\sqrt{ab}}{b}$

b) $\sqrt{\dfrac{x^3}{y^5}} = \dfrac{\sqrt{x^3}}{\sqrt{y^5}}$ Quotient rule for radicals

$\quad\quad = \dfrac{\sqrt{x^2} \cdot \sqrt{x}}{\sqrt{y^4} \cdot \sqrt{y}}$ Product rule for radicals

$\quad\quad = \dfrac{x\sqrt{x}}{y^2\sqrt{y}}$ Simplify.

$\quad\quad = \dfrac{x\sqrt{x} \cdot \sqrt{y}}{y^2\sqrt{y} \cdot \sqrt{y}}$ Rationalize the denominator.

$\quad\quad = \dfrac{x\sqrt{xy}}{y^2 \cdot y} = \dfrac{x\sqrt{xy}}{y^3}$

In the next example we simplify cube roots and fourth roots. If the exponent on a variable is a multiple of 3, the expression is a perfect cube. If the exponent is a multiple of 4, then the expression is a perfect fourth power.

E X A M P L E   9

Simplifying higher-index radicals with variables

Simplify. Assume the variables represent positive numbers.

a) $\sqrt[3]{40x^8}$

b) $\sqrt[4]{x^{12}y^5}$

c) $\sqrt[3]{\dfrac{x}{y}}$

Solution

a) Use the product rule to place the largest perfect cube factors under the first radical and the remaining factors under the second:

$$\sqrt[3]{40x^8} = \sqrt[3]{8x^6} \cdot \sqrt[3]{5x^2} = 2x^2\sqrt[3]{5x^2}$$

b) Place the largest perfect fourth power factors under the first radical and the remaining factors under the second:

$$\sqrt[4]{x^{12}y^5} = \sqrt[4]{x^{12}y^4} \cdot \sqrt[4]{y} = x^3y\sqrt[4]{y}$$

c) Multiply by $\sqrt[3]{y^2}$ to rationalize the denominator:

$$\sqrt[3]{\dfrac{x}{y}} = \dfrac{\sqrt[3]{x}}{\sqrt[3]{y}} = \dfrac{\sqrt[3]{x}}{\sqrt[3]{y}} \cdot \dfrac{\sqrt[3]{y^2}}{\sqrt[3]{y^2}} = \dfrac{\sqrt[3]{xy^2}}{\sqrt[3]{y^3}} = \dfrac{\sqrt[3]{xy^2}}{y}$$

WARM-UPS

True or false? Explain your answer.

1. $\sqrt{2} \cdot \sqrt{2} = 2$

2. $\sqrt[3]{2} \cdot \sqrt[3]{2} = 2$

3. $\sqrt[3]{-27} = -3$

4. $\sqrt{-25} = -5$

5. $\sqrt[4]{16} = 2$

6. $\sqrt{9} = 3$

7. $\sqrt{2^9} = 2^3$

8. $\dfrac{\sqrt{10}}{2} = \sqrt{5}$

9. $\dfrac{\sqrt{2}}{2} = \dfrac{1}{\sqrt{2}}$

10. $\dfrac{\sqrt{6}}{\sqrt{3}} = \sqrt{2}$

9.1   EXERCISES

Reading and Writing   *After reading this section, write out the answers to these questions. Use complete sentences.*

1. How do you know if b is an nth root of a?

2. What is a principal root?

3. What is the difference between an even root and an odd root?

4. What symbol is used to indicate an nth root?

5. What is the product rule for radicals?

6. What is the quotient rule for radicals?

For all of the exercises in this section assume that all variables represent positive real numbers.

Find each root. See Example 1.

7. $\sqrt{36}$ 8. $\sqrt{49}$

9. $\sqrt[5]{32}$ 10. $\sqrt[4]{81}$

11. $\sqrt[3]{1000}$ 12. $\sqrt[4]{16}$

13. $\sqrt[4]{-16}$ 14. $\sqrt{-1}$

15. $\sqrt[5]{-32}$ 16. $\sqrt[3]{-125}$

Find each root. See Example 2.

17. $\sqrt{m^2}$ 18. $\sqrt{m^6}$

19. $\sqrt[5]{y^{15}}$ 20. $\sqrt[4]{m^8}$

21. $\sqrt[3]{y^{15}}$ 22. $\sqrt{m^8}$

23. $\sqrt[3]{m^3}$ 24. $\sqrt[4]{x^4}$

25. $\sqrt{3^6}$ 26. $\sqrt{4^2}$

Use the product rule for radicals to simplify each expression. See Example 3.

27. $\sqrt{9y}$ 28. $\sqrt{16n}$

29. $\sqrt{4a^2}$ 30. $\sqrt{36n^2}$

31. $\sqrt{x^4y^2}$ 32. $\sqrt{w^6t^2}$

33. $\sqrt{5m^{12}}$ 34. $\sqrt{7z^{16}}$

35. $\sqrt[3]{8y}$ 36. $\sqrt[3]{27z^2}$

Simplify each radical. See Example 4.

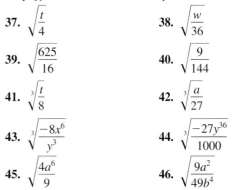

37. $\sqrt{\dfrac{t}{4}}$ 38. $\sqrt{\dfrac{w}{36}}$

39. $\sqrt{\dfrac{625}{16}}$ 40. $\sqrt{\dfrac{9}{144}}$

41. $\sqrt[3]{\dfrac{t}{8}}$ 42. $\sqrt[3]{\dfrac{a}{27}}$

43. $\sqrt[3]{\dfrac{-8x^6}{y^3}}$ 44. $\sqrt[3]{\dfrac{-27y^{36}}{1000}}$

45. $\sqrt{\dfrac{4a^6}{9}}$ 46. $\sqrt{\dfrac{9a^2}{49b^4}}$

Rewrite each expression with a rational denominator. See Example 5.

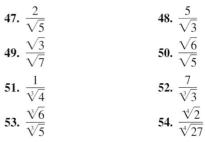

47. $\dfrac{2}{\sqrt{5}}$ 48. $\dfrac{5}{\sqrt{3}}$

49. $\dfrac{\sqrt{3}}{\sqrt{7}}$ 50. $\dfrac{\sqrt{6}}{\sqrt{5}}$

51. $\dfrac{1}{\sqrt[3]{4}}$ 52. $\dfrac{7}{\sqrt[3]{3}}$

53. $\dfrac{\sqrt[3]{6}}{\sqrt[3]{5}}$ 54. $\dfrac{\sqrt[4]{2}}{\sqrt[4]{27}}$

Write each radical expression in simplified radical form. See Example 6.

55. $\dfrac{\sqrt{5}}{\sqrt{12}}$ 56. $\dfrac{\sqrt{7}}{\sqrt{18}}$

57. $\dfrac{\sqrt{3}}{\sqrt{12}}$

58. $\dfrac{\sqrt{2}}{\sqrt{18}}$

59. $\sqrt{\dfrac{1}{2}}$

60. $\sqrt{\dfrac{3}{8}}$

61. $\sqrt[3]{\dfrac{7}{4}}$

62. $\sqrt[4]{\dfrac{1}{5}}$

Simplify. See Examples 7 and 8.

63. $\sqrt{12x^8}$

64. $\sqrt{72x^{10}}$

65. $\sqrt{60a^9b^3}$

66. $\sqrt{63w^{15}z^7}$

67. $\sqrt{\dfrac{x}{y}}$

68. $\sqrt{\dfrac{x^2}{a}}$

69. $\sqrt{\dfrac{a^3}{b^7}}$

70. $\sqrt{\dfrac{w^5}{y^8}}$

Simplify. See Example 9.

71. $\sqrt[3]{16x^{13}}$

72. $\sqrt[3]{24x^{17}}$

73. $\sqrt[4]{x^9y^6}$

74. $\sqrt[4]{w^{14}y^7}$

75. $\sqrt[5]{64x^{22}}$

76. $\sqrt[5]{x^{12}y^5z^3}$

77. $\sqrt[3]{\dfrac{a}{b}}$

78. $\sqrt[3]{\dfrac{a}{w^2}}$

Simplify.

79. $\sqrt[4]{3^{12}}$

80. $\sqrt[3]{2^{-9}}$

81. $\sqrt{10^{-2}}$

82. $\sqrt{-10^{-4}}$

83. $\sqrt{\dfrac{8x}{49}}$

84. $\sqrt{\dfrac{12b}{121}}$

85. $\sqrt[4]{\dfrac{32a}{81}}$

86. $\sqrt[4]{\dfrac{162y}{625}}$

87. $\sqrt[3]{-27x^9y^8}$

88. $\sqrt[4]{32y^8z^{11}}$

89. $\dfrac{\sqrt{ab^3}}{\sqrt{a^3b^2}}$

90. $\dfrac{\sqrt{m^3n^5}}{\sqrt{m^5n}}$

91. $\dfrac{\sqrt[3]{a^2b}}{\sqrt[3]{4ab^2}\sqrt[3]{3ab^5}}$

92. $\dfrac{\sqrt[3]{5xy^2}}{\sqrt[3]{18x^2y}}$

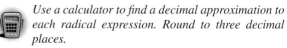 *Use a calculator to find a decimal approximation to each radical expression. Round to three decimal places.*

93. $\dfrac{5}{\sqrt{3}}$

94. $\sqrt{\dfrac{2}{27}}$

95. $\sqrt[3]{\dfrac{1}{3}}$

96. $\sqrt[3]{56}$

97. $\dfrac{\sqrt[3]{9}}{\sqrt[3]{4}}$

98. $\dfrac{\sqrt[4]{25}}{\sqrt{5}}$

99. $\dfrac{\sqrt[6]{16}}{\sqrt[3]{4}}$

100. $\sqrt[5]{2.48832}$

In Exercises 101–108, solve each problem.

101. *Factoring in the wind.* Through experimentation in Antarctica, Paul Siple developed the formula

$$W = 91.4 - \dfrac{(10.5 + 6.7\sqrt{v} - 0.45v)(457 - 5t)}{110}$$

to calculate the wind chill temperature W (in degrees Fahrenheit) from the wind velocity v [in miles per hour (mph)] and the air temperature t (in degrees Fahrenheit). Find the wind chill temperature when the air temperature is 25°F and the wind velocity is 20 mph. Use the accompanying graph to estimate the wind chill temperature when the air temperature is 25°F and the wind velocity is 30 mph.

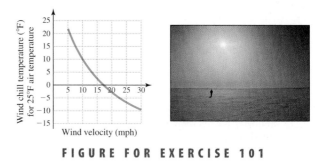

FIGURE FOR EXERCISE 101

102. *Comparing wind chills.* Use the formula from Exercise 101 to determine who will feel colder: a person in

Minneapolis at 10°F with a 15-mph wind or a person in Chicago at 20°F with a 25-mph wind.

103. *Diving time.* The time t (in seconds) that it takes for a cliff diver to reach the water is a function of the height h (in feet) from which he dives:

$$t = \sqrt{\frac{h}{16}}$$

a) Use the properties of radicals to simplify this formula.

b) Find the exact time (according to the formula) that it takes for a diver to hit the water when diving from a height of 40 feet.

c) Use the accompanying graph to estimate the height if a diver takes 2.5 seconds to reach the water?

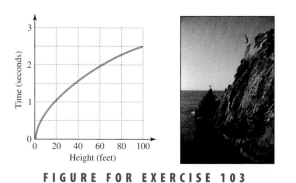

FIGURE FOR EXERCISE 103

104. *Sky diving.* The formula in Exercise 103 accounts for the effect of gravity only on a falling object. According to that formula, how long would it take a sky diver to reach the earth when jumping from 17,000 feet? (A sky diver can actually get about twice as much falling time by spreading out and using the air to slow the fall.)

105. *Maximum sailing speed.* To find the maximum possible speed in knots (nautical miles per hour) for a sailboat, sailors use the formula $M = 1.3\sqrt{w}$, where w is the length of the waterline in feet. If the waterline for the sloop *Golden Eye* is 20 feet, then what is the maximum speed of the *Golden Eye*?

106. *America's Cup.* Since 1988 basic yacht dimensions for the America's Cup competition have satisfied the inequality

$$L + 1.25\sqrt{S} - 9.8\sqrt[3]{D} \le 16.296,$$

where L is the boat's length in meters (m), S is the sail area in square meters, and D is the displacement in cubic meters (*Scientific American*, May 1992). A team

of naval architects is planning to build a boat with a displacement of 21.44 cubic meters (m³), a sail area of 320.13 square meters (m²), and a length of 21.22 m. Does this boat satisfy the inequality? If the length and displacement of this boat cannot be changed, then how many square meters of sail area must be removed so that the boat satisfies the inequality?

 107. *Landing a Piper Cheyenne.* Aircraft design engineers determine the proper landing speed V [in feet per second (ft/sec)] for an airplane from the formula

$$V = \sqrt{\frac{841L}{CS}},$$

where L is the gross weight of the aircraft in pounds (lb), C is the coefficient of lift, and S is the wing surface area in square feet. According to Piper Aircraft of Vero Beach, Florida, the Piper Cheyenne has a gross weight of 8700 lb, a coefficient of lift of 2.81, and a wing surface area of 200 ft². Find the proper landing speed for this plane. What is the landing speed in miles per hour (mph)?

108. *Landing speed and weight.* Because the gross weight of the Piper Cheyenne depends on how much fuel and cargo are on board, the proper landing speed (from Exercise 107) is not always the same. The formula $V = \sqrt{1.496L}$ gives the landing speed as a function of the gross weight only.

a) Find the landing speed if the gross weight is 7000 lb.

b) What gross weight corresponds to a landing speed of 115 ft/sec?

GETTING MORE INVOLVED

109. *Cooperative learning.* Work in a group to determine whether each equation is an identity. Explain your answers.

a) $\sqrt{x^2} = |x|$ **b)** $\sqrt[3]{x^3} = |x|$

c) $\sqrt{x^4} = x^2$ **d)** $\sqrt[4]{x^4} = |x|$

For which values of n is $\sqrt[n]{x^n} = x$ an identity?

110. *Cooperative learning.* Work in a group to determine whether each inequality is correct.

a) $\sqrt{0.9} > 0.9$ **b)** $\sqrt{1.01} > 1.01$

c) $\sqrt[3]{0.99} > 0.99$ **d)** $\sqrt[3]{1.001} > 1.001$

For which values of x and n is $\sqrt[n]{x} > x$?

111. *Discussion.* If your test scores are 80 and 100, then the arithmetic mean of your scores is 90. The geometric mean of the scores is a number h such that

$$\frac{80}{h} = \frac{h}{100}.$$

Are you better off with the arithmetic mean or the geometric mean?

calculator

close-up

You can find the fifth root of 2 using radical notation or exponent notation. Note that the fractional exponent 1/5 must be in parentheses.

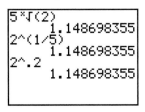

9.2 RATIONAL EXPONENTS

You have learned how to use exponents to express powers of numbers and radicals to express roots. In this section you will see that roots can be expressed with exponents also. The advantage of using exponents to express roots is that the rules of exponents can be applied to the expressions.

Rational Exponents

The *n*th root of a number can be expressed by using radical notation or the exponent $1/n$. For example, $8^{1/3}$ and $\sqrt[3]{8}$ both represent the cube root of 8, and we have

$$8^{1/3} = \sqrt[3]{8} = 2.$$

Definition of $a^{1/n}$

If *n* is any positive integer, then

$$a^{1/n} = \sqrt[n]{a},$$

provided that $\sqrt[n]{a}$ is a real number.

Later in this section we will see that using exponent $1/n$ for *n*th root is compatible with the rules for integral exponents that we already know.

EXAMPLE 1   **Radicals or exponents**

Write each radical expression using exponent notation and each exponential expression using radical notation.

a) $\sqrt[3]{35}$ **b)** $\sqrt[4]{xy}$ **c)** $5^{1/2}$ **d)** $a^{1/5}$

Solution

a) $\sqrt[3]{35} = 35^{1/3}$ **b)** $\sqrt[4]{xy} = (xy)^{1/4}$ **c)** $5^{1/2} = \sqrt{5}$ **d)** $a^{1/5} = \sqrt[5]{a}$ ∎

In the next example we evaluate some exponential expressions.

EXAMPLE 2   **Finding roots**

Evaluate each expression.

a) $4^{1/2}$ **b)** $(-8)^{1/3}$ **c)** $81^{1/4}$ **d)** $(-9)^{1/2}$

Solution

a) $4^{1/2} = \sqrt{4} = 2$

b) $(-8)^{1/3} = \sqrt[3]{-8} = -2$

c) $81^{1/4} = \sqrt[4]{81} = 3$

d) Because $(-9)^{1/2}$ or $\sqrt{-9}$ is an even root of a negative number, it is not a real number. ∎

We now extend the definition of exponent $1/n$ to include any rational number as an exponent. The numerator of the rational number indicates the power, and the denominator indicates the root. For example, the expression

$$8^{2/3} \xleftarrow{\text{Power}} \text{Root}$$

represents the square of the cube root of 8. So we have

$$8^{2/3} = (8^{1/3})^2 = (2)^2 = 4.$$

helpful hint

Note that in $a^{m/n}$ we do not require m/n to be reduced. As long as the nth root of a is real, then the value of $a^{m/n}$ is the same whether or not m/n is in lowest terms.

Definition of $a^{m/n}$

If m and n are positive integers, then

$$a^{m/n} = (a^{1/n})^m,$$

provided that $a^{1/n}$ is a real number.

We define negative rational exponents just like negative integral exponents.

Definition of $a^{-m/n}$

If m and n are positive integers and $a \neq 0$, then

$$a^{-m/n} = \frac{1}{a^{m/n}},$$

provided that $a^{1/n}$ is a real number.

EXAMPLE 3   **Radicals or exponents**

Write each radical expression using exponent notation and each exponential expression using radical notation.

a) $\sqrt[3]{x^2}$

b) $\dfrac{1}{\sqrt[4]{m^3}}$

c) $5^{2/3}$

d) $a^{-2/5}$

Solution

a) $\sqrt[3]{x^2} = x^{2/3}$

b) $\dfrac{1}{\sqrt[4]{m^3}} = \dfrac{1}{m^{3/4}} = m^{-3/4}$

c) $5^{2/3} = \sqrt[3]{5^2}$

d) $a^{-2/5} = \dfrac{1}{\sqrt[5]{a^2}}$ ■

To evaluate an expression with a negative rational exponent, remember that the denominator indicates root, the numerator indicates power, and the negative sign indicates reciprocal:

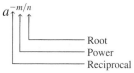

$$a^{-m/n} \quad \begin{array}{l} \text{— Root} \\ \text{— Power} \\ \text{— Reciprocal} \end{array}$$

The root, power, and reciprocal can be evaluated in any order. However, to evaluate $a^{-m/n}$ mentally it is usually simplest to use the following strategy.

Strategy for Evaluating $a^{-m/n}$ Mentally

1. Find the nth root of a.
2. Raise your result to the mth power.
3. Find the reciprocal.

For example, to evaluate $8^{-2/3}$ mentally, we find the cube root of 8 (which is 2), square 2 to get 4, then find the reciprocal of 4 to get $\frac{1}{4}$. In print $8^{-2/3}$ could be written for evaluation as $((8^{1/3})^2)^{-1}$ or $\frac{1}{(8^{1/3})^2}$.

E X A M P L E   4

Rational exponents

Evaluate each expression.

a) $27^{2/3}$ b) $4^{-3/2}$ c) $81^{-3/4}$ d) $(-8)^{-5/3}$

Solution

a) Because the exponent is $2/3$, we find the cube root of 27 and then square it:

$$27^{2/3} = (27^{1/3})^2 = 3^2 = 9$$

b) Because the exponent is $-3/2$, we find the square root of 4, cube it, and find the reciprocal:

$$4^{-3/2} = \frac{1}{(4^{1/2})^3} = \frac{1}{2^3} = \frac{1}{8}$$

c) Because the exponent is $-3/4$, we find the fourth root of 81, cube it, and find the reciprocal:

$$81^{-3/4} = \frac{1}{(81^{1/4})^3} = \frac{1}{3^3} = \frac{1}{27} \qquad \text{Definition of negative exponent}$$

d) $(-8)^{-5/3} = \dfrac{1}{((-8)^{1/3})^5} = \dfrac{1}{(-2)^5} = \dfrac{1}{-32} = -\dfrac{1}{32}$ ■

> **CAUTION** An expression with a negative base and a negative exponent can have a positive or a negative value. For example,
>
> $$(-8)^{-5/3} = -\frac{1}{32} \qquad \text{and} \qquad (-8)^{-2/3} = \frac{1}{4}.$$

Using the Rules of Exponents

All of the rules for exponents hold for rational exponents as well as integral exponents. Of course, we cannot apply the rules of exponents to expressions that are not real numbers.

calculator
close-up

A negative fractional exponent indicates a reciprocal, a root, and a power. To find $4^{-3/2}$ you can find the reciprocal first, the square root first, or the third power first as shown here.

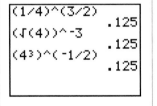

```
(1/4)^(3/2)
               .125
(√(4))^-3
               .125
(4³)^(-1/2)
               .125
```

Rules for Rational Exponents

The following rules hold for any nonzero real numbers a and b and rational numbers r and s for which the expressions represent real numbers.

1. $a^r a^s = a^{r+s}$ Product rule

2. $\dfrac{a^r}{a^s} = a^{r-s}$ Quotient rule

3. $(a^r)^s = a^{rs}$ Power of a power rule

4. $(ab)^r = a^r b^r$ Power of a product rule

5. $\left(\dfrac{a}{b}\right)^r = \dfrac{a^r}{b^r}$ Power of a quotient rule

We can use the product rule to add rational exponents. For example,

$$16^{1/4} \cdot 16^{1/4} = 16^{2/4}.$$

The fourth root of 16 is 2, and 2 squared is 4. So $16^{2/4} = 4$. Because we also have $16^{1/2} = 4$, we see that a rational exponent can be reduced to its lowest terms. If an exponent can be reduced, it is usually simpler to reduce the exponent before we evaluate the expression. We can simplify $16^{1/4} \cdot 16^{1/4}$ as follows:

$$16^{1/4} \cdot 16^{1/4} = 16^{2/4} = 16^{1/2} = 4$$

E X A M P L E   5

Using the product and quotient rules with rational exponents
Simplify each expression.

a) $27^{1/6} \cdot 27^{1/2}$

b) $\dfrac{5^{3/4}}{5^{1/4}}$

Solution

a) $27^{1/6} \cdot 27^{1/2} = 27^{1/6+1/2}$ Product rule for exponents

$\qquad\qquad\qquad = 27^{2/3}$

$\qquad\qquad\qquad = 9$

b) $\dfrac{5^{3/4}}{5^{1/4}} = 5^{3/4-1/4} = 5^{2/4} = 5^{1/2} = \sqrt{5}$ We used the quotient rule to subtract the exponents.

E X A M P L E   6

Using the power rules with rational exponents
Simplify each expression.

a) $3^{1/2} \cdot 12^{1/2}$

b) $(3^{10})^{1/2}$

c) $\left(\dfrac{2^6}{3^9}\right)^{-1/3}$

Solution

a) Because the bases 3 and 12 are different, we cannot use the product rule to add the exponents. Instead, we use the power of a product rule to place the 1/2 power outside the parentheses:

$$3^{1/2} \cdot 12^{1/2} = (3 \cdot 12)^{1/2} = 36^{1/2} = 6$$

b) Use the power of a power rule to multiply the exponents:

$$(3^{10})^{1/2} = 3^5$$

c) $\left(\dfrac{2^6}{3^9}\right)^{-1/3} = \dfrac{(2^6)^{-1/3}}{(3^9)^{-1/3}}$ Power of a quotient rule

$$= \dfrac{2^{-2}}{3^{-3}}$$ Power of a power rule

$$= \dfrac{3^3}{2^2}$$ Definition of negative exponent

$$= \dfrac{27}{4}$$

Simplifying Expressions Involving Variables

When simplifying expressions involving rational exponents and variables, we must be careful to write equivalent expressions. For example, in the equation

$$(x^2)^{1/2} = x$$

it looks as if we are correctly applying the power of a power rule. However, this statement is false if x is negative because the $1/2$ power on the left-hand side indicates the positive square root of x^2. For example, if $x = -3$, we get

$$[(-3)^2]^{1/2} = 9^{1/2} = 3,$$

which is not equal to -3. To write a simpler equivalent expression for $(x^2)^{1/2}$, we use absolute value as follows.

Square Root of x^2

$$(x^2)^{1/2} = |x| \text{ for any real number } x.$$

Note that $(x^2)^{1/2} = |x|$ is also written as $\sqrt{x^2} = |x|$. Both of these equations are identities.

It is also necessary to use absolute value when writing identities for other even roots of expressions involving variables.

EXAMPLE 7

Using absolute value symbols with roots

Simplify each expression. Assume the variables represent any real numbers and use absolute value symbols as necessary.

a) $(x^8 y^4)^{1/4}$ **b)** $\left(\dfrac{x^9}{8}\right)^{1/3}$

Solution

a) Apply the power of a product rule to get the equation $(x^8 y^4)^{1/4} = x^2 y$. The left-hand side is nonnegative for any choices of x and y, but the right-hand side is negative when y is negative. So for any real values of x and y we have

$$(x^8 y^4)^{1/4} = x^2 |y|.$$

b) Using the power of a quotient rule, we get

$$\left(\frac{x^9}{8}\right)^{1/3} = \frac{x^3}{2}.$$

This equation is valid for every real number x, so no absolute value signs are used. ■

Because there are no real even roots of negative numbers, the expressions

$$a^{1/2}, \quad x^{-3/4}, \quad \text{and} \quad y^{1/6}$$

are not real numbers if the variables have negative values. To simplify matters, we sometimes assume the variables represent only positive numbers when we are working with expressions involving variables with rational exponents. That way we do not have to be concerned with undefined expressions and absolute value.

EXAMPLE 8

Expressions involving variables with rational exponents

Use the rules of exponents to simplify the following. Write your answers with positive exponents. Assume all variables represent *positive* real numbers.

a) $x^{2/3}x^{4/3}$

b) $\dfrac{a^{1/2}}{a^{1/4}}$

c) $(x^{1/2}y^{-3})^{1/2}$

d) $\left(\dfrac{x^2}{y^{1/3}}\right)^{-1/2}$

Solution

a) $x^{2/3}x^{4/3} = x^{6/3}$ Use the product rule to add the exponents.

      $= x^2$ Reduce the exponent.

b) $\dfrac{a^{1/2}}{a^{1/4}} = a^{1/2-1/4}$ Use the quotient rule to subtract the exponents.

      $= a^{1/4}$ Simplify.

c) $(x^{1/2}y^{-3})^{1/2} = (x^{1/2})^{1/2}(y^{-3})^{1/2}$ Power of a product rule

      $= x^{1/4}y^{-3/2}$ Power of a power rule

      $= \dfrac{x^{1/4}}{y^{3/2}}$ Definition of negative exponent

d) Because this expression is a negative power of a quotient, we can first find the reciprocal of the quotient, then apply the power of a power rule:

$$\left(\frac{x^2}{y^{1/3}}\right)^{-1/2} = \left(\frac{y^{1/3}}{x^2}\right)^{1/2} = \frac{y^{1/6}}{x} \qquad \frac{1}{3}\cdot\frac{1}{2} = \frac{1}{6}$$

■

WARM-UPS

True or false? Explain your answer.

1. $9^{1/3} = \sqrt[3]{9}$

2. $8^{5/3} = \sqrt[5]{8^3}$

3. $(-16)^{1/2} = -16^{1/2}$

4. $9^{-3/2} = \dfrac{1}{27}$

5. $6^{-1/2} = \dfrac{\sqrt{6}}{6}$

6. $\dfrac{2}{2^{1/2}} = 2^{1/2}$

7. $2^{1/2} \cdot 2^{1/2} = 4^{1/2}$

8. $16^{-1/4} = -2$

9. $6^{1/6} \cdot 6^{1/6} = 6^{1/3}$

10. $(2^8)^{3/4} = 2^6$

9.2 **EXERCISES**

Reading and Writing *After reading this section, write out the answers to these questions. Use complete sentences.*

1. How do we indicate an *n*th root using exponents?

2. How do we indicate the *m*th power of the *n*th root using exponents?

3. What is the meaning of a negative rational exponent?

4. Which rules of exponents hold for rational exponents?

5. In what order must you perform the operations indicated by a negative rational exponent?

6. When is $a^{-m/n}$ a real number?

Write each radical expression using exponent notation and each exponential expression using radical notation. See Example 1.

7. $\sqrt[4]{7}$ **8.** $\sqrt[3]{cbs}$

9. $9^{1/5}$ **10.** $3^{1/2}$

11. $\sqrt{5x}$ **12.** $\sqrt{3y}$

13. $a^{1/2}$ **14.** $(-b)^{1/5}$

Evaluate each expression. See Example 2.

15. $25^{1/2}$ **16.** $16^{1/2}$

17. $(-125)^{1/3}$ **18.** $(-32)^{1/5}$

19. $16^{1/4}$ **20.** $8^{1/3}$

21. $(-4)^{1/2}$

22. $(-16)^{1/4}$

Write each radical expression using exponent notation and each exponential expression using radical notation. See Example 3.

23. $\sqrt[3]{w^7}$ **24.** $\sqrt{a^5}$

25. $\dfrac{1}{\sqrt[3]{2^{10}}}$ **26.** $\sqrt[3]{\dfrac{1}{a^2}}$

27. $w^{-3/4}$ **28.** $6^{-5/3}$

29. $(ab)^{3/2}$ **30.** $(3m)^{-1/5}$

Evaluate each expression. See Example 4.

31. $125^{2/3}$ **32.** $1000^{2/3}$

33. $25^{3/2}$ **34.** $16^{3/2}$

35. $27^{-4/3}$ **36.** $16^{-3/4}$

37. $16^{-3/2}$ **38.** $25^{-3/2}$

39. $(-27)^{-1/3}$ **40.** $(-8)^{-4/3}$

41. $(-16)^{-1/4}$

42. $(-100)^{-3/2}$

Use the rules of exponents to simplify each expression. See Examples 5 and 6.

43. $3^{1/3}3^{1/4}$ **44.** $2^{1/2}2^{1/3}$

45. $3^{1/3}3^{-1/3}$ **46.** $5^{1/4}5^{-1/4}$

47. $\dfrac{8^{1/3}}{8^{2/3}}$ **48.** $\dfrac{27^{-2/3}}{27^{-1/3}}$

49. $4^{3/4} \div 4^{1/4}$ **50.** $9^{1/4} \div 9^{3/4}$

51. $18^{1/2}2^{1/2}$ **52.** $8^{1/2}2^{1/2}$

53. $(2^6)^{1/3}$ **54.** $(3^{10})^{1/5}$

55. $(3^8)^{1/2}$ **56.** $(3^{-6})^{1/3}$

57. $(2^{-4})^{1/2}$ **58.** $(5^4)^{1/2}$

59. $\left(\dfrac{3^4}{2^6}\right)^{1/2}$ **60.** $\left(\dfrac{5^4}{3^6}\right)^{1/2}$

Simplify each expression. Assume the variables represent any real numbers and use absolute value as necessary. See Example 7.

61. $(x^4)^{1/4}$ **62.** $(y^6)^{1/6}$

63. $(a^8)^{1/2}$ **64.** $(b^{10})^{1/2}$

65. $(y^3)^{1/3}$ **66.** $(w^9)^{1/3}$

67. $(9x^6y^2)^{1/2}$ **68.** $(16a^8b^4)^{1/4}$

69. $\left(\dfrac{81x^{12}}{y^{20}}\right)^{1/4}$ **70.** $\left(\dfrac{144a^8}{9y^{18}}\right)^{1/2}$

Simplify. Assume all variables represent positive numbers. Write answers with positive exponents only. See Example 8.

71. $x^{1/2}x^{1/4}$ **72.** $y^{1/3}y^{1/3}$

73. $(x^{1/2}y)(x^{-3/4}y^{1/2})$ **74.** $(a^{1/2}b^{-1/3})(ab)$

75. $\dfrac{w^{1/3}}{w^3}$ **76.** $\dfrac{a^{1/2}}{a^2}$

77. $(144x^{16})^{1/2}$ **78.** $(125a^8)^{1/3}$

79. $\left(\dfrac{a^{-1/2}}{b^{-1/4}}\right)^{-4}$ **80.** $\left(\dfrac{2a^{1/2}}{b^{1/3}}\right)^6$

Simplify each expression. Write your answers with positive exponents. Assume that all variables represent positive real numbers.

81. $(9^2)^{1/2}$ **82.** $(4^{16})^{1/2}$

83. $-16^{-3/4}$

84. $-25^{-3/2}$

85. $125^{-4/3}$

86. $27^{-2/3}$

87. $2^{1/2}2^{-1/4}$

88. $9^{-1}9^{1/2}$

89. $3^{0.26}3^{0.74}$

90. $2^{1.5}2^{0.5}$

91. $3^{1/4}27^{1/4}$

92. $3^{2/3}9^{2/3}$

93. $\left(-\dfrac{8}{27}\right)^{2/3}$

94. $\left(-\dfrac{8}{27}\right)^{-1/3}$

95. $\left(-\dfrac{1}{16}\right)^{-3/4}$

96. $\left(\dfrac{9}{16}\right)^{-1/2}$

97. $(9x^9)^{1/2}$

98. $(-27x^9)^{1/3}$

99. $(3a^{-2/3})^{-3}$

100. $(5x^{-1/2})^{-2}$

101. $(a^{1/2}b)^{1/2}(ab^{1/2})$

102. $(m^{1/4}n^{1/2})^2(m^2n^3)^{1/2}$

103. $(km^{1/2})^3(k^3m^5)^{1/2}$

104. $(tv^{1/3})^2(t^2v^{-3})^{-1/2}$

Use a scientific calculator with a power key (x^y) to find the decimal value of each expression. Round answers to four decimal places.

105. $2^{1/3}$

106. $5^{1/2}$

107. $-2^{1/2}$

108. $(-3)^{1/3}$

109. $1024^{1/10}$

110. $7776^{0.2}$

111. $8^{0.33}$

112. $289^{0.5}$

113. $\left(\dfrac{64}{15,625}\right)^{-1/6}$

114. $\left(\dfrac{32}{243}\right)^{-3/5}$

Simplify each expression. Assume a and b are positive real numbers and m and n are rational numbers.

115. $a^{m/2} \cdot a^{m/4}$

116. $b^{n/2} \cdot b^{-n/3}$

117. $\dfrac{a^{-m/5}}{a^{-m/3}}$

118. $\dfrac{b^{-n/4}}{b^{-n/3}}$

119. $(a^{-1/m}b^{-1/n})^{-mn}$

120. $(a^{-m/2}b^{-n/3})^{-6}$

121. $\left(\dfrac{a^{-3m}b^{-6n}}{a^{9m}}\right)^{-1/3}$

122. $\left(\dfrac{a^{-3/m}b^{6/n}}{a^{-6/m}b^{9/n}}\right)^{-1/3}$

In Exercises 123–130, solve each problem.

123. *Diagonal of a box.* The length of the diagonal of a box can be found from the formula

$$D = (L^2 + W^2 + H^2)^{1/2},$$

where L, W, and H represent the length, width, and height of the box, respectively. If the box is 12 inches long, 4 inches wide, and 3 inches high, then what is the length of the diagonal?

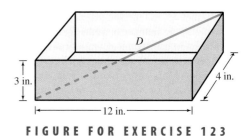

FIGURE FOR EXERCISE 123

124. *Radius of a sphere.* The radius of a sphere is a function of its volume, given by the formula

$$r = \left(\dfrac{0.75V}{\pi}\right)^{1/3}.$$

Find the radius of a spherical tank that has a volume of $\dfrac{32\pi}{3}$ cubic meters.

FIGURE FOR EXERCISE 124

125. *Maximum sail area.* According to the new International America's Cup Class Rules, the maximum sail area in square meters for a yacht in the America's Cup race is given by

$$S = (13.0368 + 7.84D^{1/3} - 0.8L)^2,$$

where D is the displacement in cubic meters (m^3), and L is the length in meters (m). (*Scientific American,* May 1992). Find the maximum sail area for a boat that has a displacement of 18.42 m^3 and a length of 21.45 m.

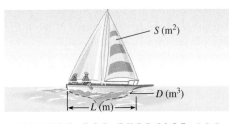

FIGURE FOR EXERCISE 125

126. *Orbits of the planets.* According to Kepler's third law of planetary motion, the average radius R of the orbit of a planet around the sun is determined by $R = T^{2/3}$, where T is the number of years for one orbit and R is measured in astronomical units or AUs (Windows to the Universe, www.windows.umich.edu).

a) It takes Mars 1.881 years to make one orbit of the sun. What is the average radius (in AUs) of the orbit of Mars?

b) The average radius of the orbit of Saturn is 9.05 AU. Use the accompanying graph to estimate the number of years it takes Saturn to make one orbit of the sun.

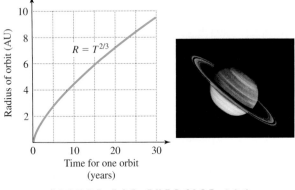

FIGURE FOR EXERCISE 126

127. *Best stock fund.* The average annual return for an investment is given by the formula

$$r = \left(\frac{S}{P}\right)^{1/n} - 1,$$

where P is the initial investment and S is the amount it is worth after n years. The top mutual fund for 1997 in the 3-year category was Fidelity Select-Energy Services (Money Guide to Mutual Funds, 1998), in which an investment of $10,000 grew to $31,895.06 from 1994 to 1997. Find the 3-year average annual return for this fund.

128. *Best bond fund.* The top bond fund for 1997 in the 5-year category was GT Global High Income B. An investment of $10,000 in 1992 grew to $21,830.95 in 1997. Use the formula from the previous exercise to find the 5-year average annual return for this fund.

129. *Overdue loan payment.* In 1777 a wealthy Pennsylvania merchant, Jacob DeHaven, lent $450,000 to the Continental Congress to rescue the troops at Valley Forge. The loan was not repaid. In 1990 DeHaven's descendants filed suit for $141.6 billion (*New York Times,* May 27, 1990). What average annual rate of return were they using to calculate the value of the debt after 213 years? (See Exercise 127.)

130. *California growin'.* The population of California grew from 19.9 million in 1970 to 32.5 million in 2000 (U.S. Census Bureau, www.census.gov). Find the average annual rate of growth for that time period. (Use the formula from Exercise 127 with P being the initial population and S being the population n years later.)

FIGURE FOR EXERCISE 130

GETTING MORE INVOLVED

131. *Discussion.* If we use the product rule to simplify $(-1)^{1/2} \cdot (-1)^{1/2}$, we get

$$(-1)^{1/2} \cdot (-1)^{1/2} = (-1)^1 = -1.$$

If we use the power of a product rule, we get

$$(-1)^{1/2} \cdot (-1)^{1/2} = (-1 \cdot -1)^{1/2} = 1^{1/2} = 1.$$

Which of these computations is incorrect? Explain your answer.

132. *Discussion.* Determine whether each equation is an identity. Explain.

a) $(w^2x^2)^{1/2} = |w| \cdot |x|$

b) $(w^2x^2)^{1/2} = |wx|$

c) $(w^2x^2)^{1/2} = w|x|$

9.3   OPERATIONS WITH RADICALS

In this section

- Adding and Subtracting Radicals
- Multiplying Radicals
- Conjugates

In this section we will use the ideas of Section 9.1 in performing arithmetic operations with radical expressions.

Adding and Subtracting Radicals

To find the sum of $\sqrt{2}$ and $\sqrt{3}$, we can use a calculator to get $\sqrt{2} \approx 1.414$ and $\sqrt{3} \approx 1.732$. (The symbol $\approx$ means "is approximately equal to.") We can then add the decimal numbers and get

$$\sqrt{2} + \sqrt{3} \approx 1.414 + 1.732 = 3.146.$$

We cannot write an exact decimal form for $\sqrt{2} + \sqrt{3}$; the number 3.146 is an approximation of $\sqrt{2} + \sqrt{3}$. To represent the exact value of $\sqrt{2} + \sqrt{3}$, we just use the form $\sqrt{2} + \sqrt{3}$. This form cannot be simplified any further. However, a sum of like radicals can be simplified. **Like radicals** are radicals that have the same index and the same radicand.

To simplify the sum $3\sqrt{2} + 5\sqrt{2}$, we can use the fact that $3x + 5x = 8x$ is true for any value of x. Substituting $\sqrt{2}$ for x gives us $3\sqrt{2} + 5\sqrt{2} = 8\sqrt{2}$. So like radicals can be combined just as like terms are combined.

E X A M P L E   1

Adding and subtracting like radicals

Simplify the following expressions. Assume the variables represent positive numbers.

a) $3\sqrt{5} + 4\sqrt{5}$　　　　　　　　　**b)** $\sqrt[4]{w} - 6\sqrt[4]{w}$

c) $\sqrt{3} + \sqrt{5} - 4\sqrt{3} + 6\sqrt{5}$　　　**d)** $3\sqrt[3]{6x} + 2\sqrt[3]{x} + \sqrt[3]{6x} + \sqrt[3]{x}$

Solution

a) $3\sqrt{5} + 4\sqrt{5} = 7\sqrt{5}$　　　　　**b)** $\sqrt[4]{w} - 6\sqrt[4]{w} = -5\sqrt[4]{w}$

c) $\sqrt{3} + \sqrt{5} - 4\sqrt{3} + 6\sqrt{5} = -3\sqrt{3} + 7\sqrt{5}$

d) $3\sqrt[3]{6x} + 2\sqrt[3]{x} + \sqrt[3]{6x} + \sqrt[3]{x} = 4\sqrt[3]{6x} + 3\sqrt[3]{x}$　　Only like radicals are combined.　■

Remember that *only radicals with the same index and same radicand can be combined by addition or subtraction.* If the radicals are not in simplified form, then they must be simplified before you can determine whether they can be combined.

E X A M P L E   2

Simplifying radicals before combining

Perform the indicated operations. Assume the variables represent positive numbers.

a) $\sqrt{8} + \sqrt{18}$　　　　　　　　**b)** $\sqrt{\dfrac{1}{5}} + \sqrt{20}$

c) $\sqrt{2x^3} - \sqrt{4x^2} + 5\sqrt{18x^3}$　　**d)** $\sqrt[3]{16x^4y^3} - \sqrt[3]{54x^4y^3}$

Solution

a) $\sqrt{8} + \sqrt{18} = \sqrt{4} \cdot \sqrt{2} + \sqrt{9} \cdot \sqrt{2}$

$\qquad\qquad\qquad = 2\sqrt{2} + 3\sqrt{2}$　　Simplify each radical.

$\qquad\qquad\qquad = 5\sqrt{2}$　　　　　　Add like radicals.

Note that $\sqrt{8} + \sqrt{18} \neq \sqrt{26}$.

c a l c u l a t o r

close-up

Check that
$$\sqrt{8} + \sqrt{18} = 5\sqrt{2}.$$

```
√(8)+√(18)
        7.071067812
5√(2)
        7.071067812
```

b) $\sqrt{\dfrac{1}{5}} + \sqrt{20} = \dfrac{\sqrt{5}}{5} + 2\sqrt{5}$ Because $\sqrt{\dfrac{1}{5}} = \dfrac{1}{\sqrt{5}} \cdot \dfrac{\sqrt{5}}{\sqrt{5}} = \dfrac{\sqrt{5}}{5}$ and $\sqrt{20} = 2\sqrt{5}$

$$= \dfrac{\sqrt{5}}{5} + \dfrac{10\sqrt{5}}{5} \quad \text{Use the LCD of 5.}$$

$$= \dfrac{11\sqrt{5}}{5}$$

c) $\sqrt{2x^3} - \sqrt{4x^2} + 5\sqrt{18x^3} = \sqrt{x^2} \cdot \sqrt{2x} - 2x + 5 \cdot \sqrt{9x^2} \cdot \sqrt{2x}$

$$= x\sqrt{2x} - 2x + 15x\sqrt{2x} \quad \text{Simplify each radical.}$$

$$= 16x\sqrt{2x} - 2x \quad\quad\quad \text{Add like radicals only.}$$

d) $\sqrt[3]{16x^4y^3} - \sqrt[3]{54x^4y^3} = \sqrt[3]{8x^3y^3} \cdot \sqrt[3]{2x} - \sqrt[3]{27x^3y^3} \cdot \sqrt[3]{2x}$

$$= 2xy\sqrt[3]{2x} - 3xy\sqrt[3]{2x} \quad \text{Simplify each radical.}$$

$$= -xy\sqrt[3]{2x}$$

Multiplying Radicals

We have already multiplied radicals in Section 9.2, when we rationalized denominators. The product rule for radicals, $\sqrt[n]{a} \cdot \sqrt[n]{b} = \sqrt[n]{ab}$, allows multiplication of radicals with the same index, such as

$$\sqrt{5} \cdot \sqrt{3} = \sqrt{15}, \quad \sqrt[3]{2} \cdot \sqrt[3]{5} = \sqrt[3]{10}, \quad \text{and} \quad \sqrt[5]{x^2} \cdot \sqrt[5]{x} = \sqrt[5]{x^3}.$$

CAUTION The product rule does not allow multiplication of radicals that have different indices. We cannot use the product rule to multiply $\sqrt{2}$ and $\sqrt[3]{5}$.

E X A M P L E 3

Multiplying radicals with the same index

Multiply and simplify the following expressions. Assume the variables represent positive numbers.

a) $5\sqrt{6} \cdot 4\sqrt{3}$ **b)** $\sqrt{3a^2} \cdot \sqrt{6a}$

c) $\sqrt[3]{4} \cdot \sqrt[3]{4}$ **d)** $\sqrt[4]{\dfrac{x^3}{2}} \cdot \sqrt[4]{\dfrac{x^2}{4}}$

Solution

a) $5\sqrt{6} \cdot 4\sqrt{3} = 5 \cdot 4 \cdot \sqrt{6} \cdot \sqrt{3}$

$$= 20\sqrt{18} \quad \text{Product rule for radicals}$$

$$= 20 \cdot 3\sqrt{2} \quad \sqrt{18} = \sqrt{9} \cdot \sqrt{2} = 3\sqrt{2}$$

$$= 60\sqrt{2}$$

b) $\sqrt{3a^2} \cdot \sqrt{6a} = \sqrt{18a^3} \quad \text{Product rule for radicals}$

$$= \sqrt{9a^2} \cdot \sqrt{2a}$$

$$= 3a\sqrt{2a} \quad \text{Simplify.}$$

c) $\sqrt[3]{4} \cdot \sqrt[3]{4} = \sqrt[3]{16}$

$$= \sqrt[3]{8} \cdot \sqrt[3]{2} \quad \text{Simplify.}$$

$$= 2\sqrt[3]{2}$$

helpful hint

Students often write
$$\sqrt{15} \cdot \sqrt{15} = \sqrt{225} = 15.$$

Although this is correct, you should get used to the idea that
$$\sqrt{15} \cdot \sqrt{15} = 15.$$

Because of the definition of a square root, $\sqrt{a} \cdot \sqrt{a} = a$ for any positive number a.

d) $\sqrt[4]{\dfrac{x^3}{2}} \cdot \sqrt[4]{\dfrac{x^2}{4}} = \sqrt[4]{\dfrac{x^5}{8}}$ Product rule for radicals

$\qquad = \dfrac{\sqrt[4]{x^4} \cdot \sqrt[4]{x}}{\sqrt[4]{8}}$ Simplify

$\qquad = \dfrac{x\sqrt[4]{x}}{\sqrt[4]{8}}$

$\qquad = \dfrac{x\sqrt[4]{x} \cdot \sqrt[4]{2}}{\sqrt[4]{8} \cdot \sqrt[4]{2}}$ Rationalize the denominator.

$\qquad = \dfrac{x\sqrt[4]{2x}}{2}$ $\sqrt[4]{8} \cdot \sqrt[4]{2} = \sqrt[4]{16} = 2$

■

We find a product such as $3\sqrt{2}(4\sqrt{2} - \sqrt{3})$ by using the distributive property as we do when multiplying a monomial and a binomial. A product such as $(2\sqrt{3} + \sqrt{5})(3\sqrt{3} - 2\sqrt{5})$ can be found by using FOIL as we do for the product of two binomials.

E X A M P L E   4   **Multiplying radicals**
Multiply and simplify.
a) $3\sqrt{2}\,(4\sqrt{2} - \sqrt{3})$ 　　　　　　**b)** $\sqrt[3]{a}\,(\sqrt[3]{a} - \sqrt[3]{a^2})$
c) $(2\sqrt{3} + \sqrt{5})(3\sqrt{3} - 2\sqrt{5})$ 　　**d)** $(3 + \sqrt{x - 9})^2$

Solution
a) $3\sqrt{2}\,(4\sqrt{2} - \sqrt{3}) = 3\sqrt{2} \cdot 4\sqrt{2} - 3\sqrt{2} \cdot \sqrt{3}$ Distributive property
$\qquad\qquad\qquad\qquad = 12 \cdot 2 - 3\sqrt{6}$ Because $\sqrt{2} \cdot \sqrt{2} = 2$
$\qquad\qquad\qquad\qquad\qquad\qquad\qquad$ and $\sqrt{2} \cdot \sqrt{3} = \sqrt{6}$
$\qquad\qquad\qquad\qquad = 24 - 3\sqrt{6}$

b) $\sqrt[3]{a}\,(\sqrt[3]{a} - \sqrt[3]{a^2}) = \sqrt[3]{a^2} - \sqrt[3]{a^3}$ Distributive property
$\qquad\qquad\qquad\qquad = \sqrt[3]{a^2} - a$

c) $(2\sqrt{3} + \sqrt{5})(3\sqrt{3} - 2\sqrt{5})$

$\qquad\qquad\quad\overbrace{}^{\text{F}}\qquad\overbrace{}^{\text{O}}\qquad\overbrace{}^{\text{I}}\qquad\overbrace{}^{\text{L}}$
$\qquad = 2\sqrt{3} \cdot 3\sqrt{3} - 2\sqrt{3} \cdot 2\sqrt{5} + \sqrt{5} \cdot 3\sqrt{3} - \sqrt{5} \cdot 2\sqrt{5}$
$\qquad = 18 - 4\sqrt{15} + 3\sqrt{15} - 10$
$\qquad = 8 - \sqrt{15}$ Combine like radicals.

d) To square a sum, we use $(a + b)^2 = a^2 + 2ab + b^2$:

$$(3 + \sqrt{x - 9})^2 = 3^2 + 2 \cdot 3\sqrt{x - 9} + (\sqrt{x - 9})^2$$
$$= 9 + 6\sqrt{x - 9} + x - 9$$
$$= x + 6\sqrt{x - 9}$$

■

In the next example we multiply radicals that have different indices.

E X A M P L E   5

Multiplying radicals with different indices

Write each product as a single radical expression.

a) $\sqrt[3]{2} \cdot \sqrt[4]{2}$

b) $\sqrt[3]{2} \cdot \sqrt{3}$

Solution

a) $\sqrt[3]{2} \cdot \sqrt[4]{2} = 2^{1/3} \cdot 2^{1/4}$ Write in exponential notation.

$\qquad\qquad = 2^{7/12}$ Product rule for exponents: $\frac{1}{3} + \frac{1}{4} = \frac{7}{12}$

$\qquad\qquad = \sqrt[12]{2^7}$ Write in radical notation.

$\qquad\qquad = \sqrt[12]{128}$

b) $\sqrt[3]{2} \cdot \sqrt{3} = 2^{1/3} \cdot 3^{1/2}$ Write in exponential notation.

$\qquad\qquad = 2^{2/6} \cdot 3^{3/6}$ Write the exponents with the LCD of 6.

$\qquad\qquad = \sqrt[6]{2^2} \cdot \sqrt[6]{3^3}$ Write in radical notation.

$\qquad\qquad = \sqrt[6]{2^2 \cdot 3^3}$ Product rule for radicals

$\qquad\qquad = \sqrt[6]{108}$ $2^2 \cdot 3^3 = 4 \cdot 27 = 108$

calculator

close-up

Check that
$\sqrt[3]{2} \cdot \sqrt[4]{2} = \sqrt[12]{128}.$

```
2^(1/3)*2^(1/4)
       1.498307077
128^(1/12)
       1.498307077
```

C A U T I O N   Because the bases in $2^{1/3} \cdot 2^{1/4}$ are identical, we can add the exponents [Example 5(a)]. Because the bases in $2^{2/6} \cdot 3^{3/6}$ are not the same, we cannot add the exponents [Example 5(b)]. Instead, we write each factor as a sixth root and use the product rule for radicals.

Conjugates

helpful hint

The word "conjugate" is used in many contexts in mathematics. According to the dictionary, conjugate means joined together, especially as in a pair.

Recall the special product rule $(a + b)(a - b) = a^2 - b^2$. The product of the sum $4 + \sqrt{3}$ and the difference $4 - \sqrt{3}$ can be found by using this rule:

$$(4 + \sqrt{3})(4 - \sqrt{3}) = 4^2 - (\sqrt{3})^2 = 16 - 3 = 13$$

The product of the irrational number $4 + \sqrt{3}$ and the irrational number $4 - \sqrt{3}$ is the rational number 13. For this reason the expressions $4 + \sqrt{3}$ and $4 - \sqrt{3}$ are called **conjugates** of one another. We will use conjugates in Section 9.4 to rationalize some denominators.

E X A M P L E   6

Multiplying conjugates

Find the products. Assume the variables represent positive real numbers.

a) $(2 + 3\sqrt{5})(2 - 3\sqrt{5})$

b) $(\sqrt{3} - \sqrt{2})(\sqrt{3} + \sqrt{2})$

c) $(\sqrt{2x} - \sqrt{y})(\sqrt{2x} + \sqrt{y})$

Solution

a) $(2 + 3\sqrt{5})(2 - 3\sqrt{5}) = 2^2 - (3\sqrt{5})^2$ $(a + b)(a - b) = a^2 - b^2$

$\qquad\qquad\qquad\qquad\quad = 4 - 45$ $(3\sqrt{5})^2 = 9 \cdot 5 = 45$

$\qquad\qquad\qquad\qquad\quad = -41$

b) $(\sqrt{3} - \sqrt{2})(\sqrt{3} + \sqrt{2}) = 3 - 2$

$\qquad\qquad\qquad\qquad\quad = 1$

c) $(\sqrt{2x} - \sqrt{y})(\sqrt{2x} + \sqrt{y}) = 2x - y$

WARM-UPS

True or false? Explain your answer.

1. $\sqrt{3} + \sqrt{3} = \sqrt{6}$
2. $\sqrt{8} + \sqrt{2} = 3\sqrt{2}$
3. $2\sqrt{3} \cdot 3\sqrt{3} = 6\sqrt{3}$
4. $\sqrt[3]{2} \cdot \sqrt[3]{2} = 2$
5. $2\sqrt{5} \cdot 3\sqrt{2} = 6\sqrt{10}$
6. $2\sqrt{5} + 3\sqrt{5} = 5\sqrt{10}$
7. $\sqrt{2}(\sqrt{3} - \sqrt{2}) = \sqrt{6} - 2$
8. $\sqrt{12} = 2\sqrt{6}$
9. $(\sqrt{2} + \sqrt{3})^2 = 2 + 3$
10. $(\sqrt{3} - \sqrt{2})(\sqrt{3} + \sqrt{2}) = 1$

9.3   EXERCISES

Reading and Writing   *After reading this section, write out the answers to these questions. Use complete sentences.*

1. What are like radicals?

2. How do we combine like radicals?

3. Does the product rule allow multiplication of unlike radicals?

4. How do we multiply radicals of different indices?

All variables in the following exercises represent positive numbers.

Simplify the sums and differences. Give exact answers. See Example 1.

5. $\sqrt{3} - 2\sqrt{3}$
6. $\sqrt{5} - 3\sqrt{5}$
7. $5\sqrt{7x} + 4\sqrt{7x}$
8. $3\sqrt{6a} + 7\sqrt{6a}$
9. $2\sqrt[3]{2} + 3\sqrt[3]{2}$
10. $\sqrt[3]{4} + 4\sqrt[3]{4}$
11. $\sqrt{3} - \sqrt{5} + 3\sqrt{3} - \sqrt{5}$
12. $\sqrt{2} - 5\sqrt{3} - 7\sqrt{2} + 9\sqrt{3}$
13. $\sqrt[3]{2} + \sqrt[3]{x} - \sqrt[3]{2} + 4\sqrt[3]{x}$
14. $\sqrt[3]{5y} - 4\sqrt[3]{5y} + \sqrt[3]{x} + \sqrt[3]{x}$

15. $\sqrt[3]{x} - \sqrt{2x} + \sqrt[3]{x}$
16. $\sqrt[3]{ab} + \sqrt{a} + 5\sqrt{a} + \sqrt[3]{ab}$

Simplify each expression. Give exact answers. See Example 2.

17. $\sqrt{8} + \sqrt{28}$
18. $\sqrt{12} + \sqrt{24}$
19. $\sqrt{8} + \sqrt{18}$
20. $\sqrt{12} + \sqrt{27}$
21. $\sqrt{45} - \sqrt{20}$
22. $\sqrt{50} - \sqrt{32}$
23. $\sqrt{2} - \sqrt{8}$
24. $\sqrt{20} - \sqrt{125}$
25. $\dfrac{\sqrt{2}}{2} + \sqrt{2}$
26. $\dfrac{\sqrt{3}}{3} - \sqrt{3}$
27. $\sqrt{80} + \sqrt{\dfrac{1}{5}}$
28. $\sqrt{32} + \sqrt{\dfrac{1}{2}}$
29. $\sqrt{45x^3} - \sqrt{18x^2} + \sqrt{50x^2} - \sqrt{20x^3}$
30. $\sqrt{12x^5} - \sqrt{18x} - \sqrt{300x^5} + \sqrt{98x}$
31. $\sqrt[3]{24} + \sqrt[3]{81}$
32. $\sqrt[3]{24} + \sqrt[3]{375}$
33. $\sqrt[4]{48} - \sqrt[4]{243}$
34. $\sqrt[5]{64} + \sqrt[5]{2}$

35. $\sqrt[3]{54t^4y^3} - \sqrt[3]{16t^4y^3}$

36. $\sqrt[3]{2000w^2z^5} - \sqrt[3]{16w^2z^5}$

Simplify the products. Give exact answers. See Examples 3 and 4.

37. $\sqrt{3} \cdot \sqrt{5}$

38. $\sqrt{5} \cdot \sqrt{7}$

39. $2\sqrt{5} \cdot 3\sqrt{10}$

40. $(3\sqrt{2})(-4\sqrt{10})$

41. $2\sqrt{7a} \cdot 3\sqrt{2a}$

42. $2\sqrt{5c} \cdot 5\sqrt{5}$

43. $\sqrt[4]{9} \cdot \sqrt[4]{27}$

44. $\sqrt[3]{5} \cdot \sqrt[3]{100}$

45. $(2\sqrt{3})^2$

46. $(-4\sqrt{2})^2$

47. $\sqrt[3]{\dfrac{4x^2}{3}} \cdot \sqrt[3]{\dfrac{2x^2}{3}}$

48. $\sqrt[4]{\dfrac{4x^2}{5}} \cdot \sqrt[4]{\dfrac{4x^3}{25}}$

49. $2\sqrt{3}(\sqrt{6} + 3\sqrt{3})$

50. $2\sqrt{5}(\sqrt{3} + 3\sqrt{5})$

51. $\sqrt{5}(\sqrt{10} - 2)$

52. $\sqrt{6}(\sqrt{15} - 1)$

53. $\sqrt[3]{3t}(\sqrt[3]{9t} - \sqrt[3]{t^2})$

54. $\sqrt[3]{2}(\sqrt[3]{12x} - \sqrt[3]{2x})$

55. $(\sqrt{3} + 2)(\sqrt{3} - 5)$

56. $(\sqrt{5} + 2)(\sqrt{5} - 6)$

57. $(\sqrt{11} - 3)(\sqrt{11} + 3)$

58. $(\sqrt{2} + 5)(\sqrt{2} + 5)$

59. $(2\sqrt{5} - 7)(2\sqrt{5} + 4)$

60. $(2\sqrt{6} - 3)(2\sqrt{6} + 4)$

61. $(2\sqrt{3} - \sqrt{6})(\sqrt{3} + 2\sqrt{6})$

62. $(3\sqrt{3} - \sqrt{2})(\sqrt{2} + \sqrt{3})$

Write each product as a single radical expression. See Example 5.

63. $\sqrt[3]{3} \cdot \sqrt{3}$

64. $\sqrt{3} \cdot \sqrt[4]{3}$

65. $\sqrt[3]{5} \cdot \sqrt[4]{5}$

66. $\sqrt[3]{2} \cdot \sqrt[5]{2}$

67. $\sqrt[3]{2} \cdot \sqrt{5}$

68. $\sqrt{6} \cdot \sqrt[3]{2}$

69. $\sqrt[3]{2} \cdot \sqrt[4]{3}$

70. $\sqrt[3]{3} \cdot \sqrt[4]{2}$

Find the product of each pair of conjugates. See Example 6.

71. $(\sqrt{3} - 2)(\sqrt{3} + 2)$

72. $(7 - \sqrt{3})(7 + \sqrt{3})$

73. $(\sqrt{5} + \sqrt{2})(\sqrt{5} - \sqrt{2})$

74. $(\sqrt{6} + \sqrt{5})(\sqrt{6} - \sqrt{5})$

75. $(2\sqrt{5} + 1)(2\sqrt{5} - 1)$

76. $(3\sqrt{2} - 4)(3\sqrt{2} + 4)$

77. $(3\sqrt{2} + \sqrt{5})(3\sqrt{2} - \sqrt{5})$

78. $(2\sqrt{3} - \sqrt{7})(2\sqrt{3} + \sqrt{7})$

79. $(5 - 3\sqrt{x})(5 + 3\sqrt{x})$

80. $(4\sqrt{y} + 3\sqrt{z})(4\sqrt{y} - 3\sqrt{z})$

Simplify each expression.

81. $\sqrt{300} + \sqrt{3}$

82. $\sqrt{50} + \sqrt{2}$

83. $2\sqrt{5} \cdot 5\sqrt{6}$

84. $3\sqrt{6} \cdot 5\sqrt{10}$

85. $(3 + 2\sqrt{7})(\sqrt{7} - 2)$

86. $(2 + \sqrt{7})(\sqrt{7} - 2)$

87. $4\sqrt{w} \cdot 4\sqrt{w}$

88. $3\sqrt{m} \cdot 5\sqrt{m}$

89. $\sqrt{3x^3} \cdot \sqrt{6x^2}$

90. $\sqrt{2t^5} \cdot \sqrt{10t^4}$

91. $\dfrac{1}{\sqrt{2}} - \dfrac{1}{\sqrt{8}} + \dfrac{1}{\sqrt{18}}$

92. $\dfrac{1}{\sqrt{3}} + \sqrt{\dfrac{1}{3}} - \sqrt{3}$

93. $(2\sqrt{5} + \sqrt{2})(3\sqrt{5} - \sqrt{2})$

94. $(3\sqrt{2} - \sqrt{3})(2\sqrt{2} + 3\sqrt{3})$

95. $\dfrac{\sqrt{2}}{3} + \dfrac{\sqrt{2}}{5}$

96. $\dfrac{\sqrt{2}}{4} + \dfrac{\sqrt{3}}{5}$

97. $(5 + 2\sqrt{2})(5 - 2\sqrt{2})$

98. $(3 - 2\sqrt{7})(3 + 2\sqrt{7})$

99. $(3 + \sqrt{x})^2$

100. $(1 - \sqrt{x})^2$

101. $(5\sqrt{x} - 3)^2$

102. $(3\sqrt{a} + 2)^2$

103. $(1 + \sqrt{x + 2})^2$

104. $(\sqrt{x - 1} + 1)^2$

105. $\sqrt{4w} - \sqrt{9w}$

106. $10\sqrt{m} - \sqrt{16m}$

107. $2\sqrt{a^3} + 3\sqrt{a^3} - 2a\sqrt{4a}$

108. $5\sqrt{w^2y} - 7\sqrt{w^2y} + 6\sqrt{w^2y}$

109. $\sqrt{x^5} + 2x\sqrt{x^3}$

110. $\sqrt{8x^3} + \sqrt{50x^3} - x\sqrt{2x}$

111. $\sqrt[3]{-16x^4} + 5x\sqrt[3]{54x}$

112. $\sqrt[3]{3x^5y^7} - \sqrt[3]{24x^5y^7}$

113. $\sqrt[3]{\dfrac{y^7}{4x}}$

114. $\sqrt[4]{\dfrac{16}{9z^3}}$

115. $\sqrt[3]{\dfrac{x}{5}} \cdot \sqrt[3]{\dfrac{x^5}{5}}$

116. $\sqrt[4]{a^3}(\sqrt[4]{a} - \sqrt[4]{a^5})$

117. $\sqrt[3]{2x} \cdot \sqrt[3]{2x}$

118. $\sqrt[3]{2m} \cdot \sqrt[4]{2n}$

In Exercises 119–122, solve each problem.

119. *Area of a rectangle.* Find the exact area of a rectangle that has a length of $\sqrt{6}$ feet and a width of $\sqrt{3}$ feet.

120. *Volume of a cube.* Find the exact volume of a cube with sides of length $\sqrt{3}$ meters.

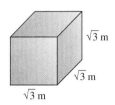

FIGURE FOR EXERCISE 120

121. *Area of a trapezoid.* Find the exact area of a trapezoid with a height of $\sqrt{6}$ feet and bases of $\sqrt{3}$ feet and $\sqrt{12}$ feet.

FIGURE FOR EXERCISE 121

122. *Area of a triangle.* Find the exact area of a triangle with a base of $\sqrt{30}$ meters and a height of $\sqrt{6}$ meters.

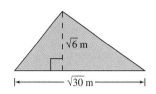

FIGURE FOR EXERCISE 122

GETTING MORE INVOLVED

123. *Discussion.* Is $\sqrt{a} + \sqrt{b} = \sqrt{a + b}$ for all values of a and b?

124. *Discussion.* Which of the following equations are identities? Explain your answers.
   a) $\sqrt{9x} = 3\sqrt{x}$
   b) $\sqrt{9 + x} = 3 + \sqrt{x}$
   c) $\sqrt{x - 4} = \sqrt{x} - 2$
   d) $\sqrt{\dfrac{x}{4}} = \dfrac{\sqrt{x}}{2}$

125. *Exploration.* Because 3 is the square of $\sqrt{3}$, a binomial such as $y^2 - 3$ is a difference of two squares.
   a) Factor $y^2 - 3$ and $2a^2 - 7$ using radicals.
   b) Use factoring with radicals to solve the equations $x^2 - 8 = 0$ and $3y^2 - 11 = 0$.
   c) Assuming a and b are positive real numbers, solve the equations $x^2 - a = 0$ and $ax^2 - b = 0$.

9.4 **MORE OPERATIONS WITH RADICALS**

In this section you will continue studying operations with radicals. We learn to rationalize some denominators that are different from those rationalized in Section 9.2.

Dividing Radicals

In Section 9.3 you learned how to add, subtract, and multiply radical expressions. To divide two radical expressions, simply write the quotient as a ratio and then simplify, as we did in Section 9.2. In general, we have

$$\sqrt[n]{a} \div \sqrt[n]{b} = \frac{\sqrt[n]{a}}{\sqrt[n]{b}} = \sqrt[n]{\frac{a}{b}},$$

provided that all expressions represent real numbers. Note that the quotient rule is applied only to radicals that have the same index.

E X A M P L E   1

Dividing radicals with the same index

Divide and simplify. Assume the variables represent positive numbers.

a) $\sqrt{10} \div \sqrt{5}$ **b)** $(3\sqrt{2}) \div (2\sqrt{3})$ **c)** $\sqrt[3]{10x^2} \div \sqrt[3]{5x}$

Solution

a) $\sqrt{10} \div \sqrt{5} = \dfrac{\sqrt{10}}{\sqrt{5}}$ $\quad a \div b = \dfrac{a}{b}$, provided that $b \neq 0$.

$\qquad\qquad\quad = \sqrt{\dfrac{10}{5}}$ Quotient rule for radicals

$\qquad\qquad\quad = \sqrt{2}$ Reduce.

b) $(3\sqrt{2}) \div (2\sqrt{3}) = \dfrac{3\sqrt{2}}{2\sqrt{3}}$

$\qquad\qquad\qquad\quad = \dfrac{3\sqrt{2}}{2\sqrt{3}} \cdot \dfrac{\sqrt{3}}{\sqrt{3}}$ Rationalize the denominator.

$\qquad\qquad\qquad\quad = \dfrac{3\sqrt{6}}{2 \cdot 3}$

$\qquad\qquad\qquad\quad = \dfrac{\sqrt{6}}{2}$ Note that $\sqrt{6} \div 2 \neq \sqrt{3}$.

c) $\sqrt[3]{10x^2} \div \sqrt[3]{5x} = \dfrac{\sqrt[3]{10x^2}}{\sqrt[3]{5x}}$

$\qquad\qquad\qquad\quad = \sqrt[3]{\dfrac{10x^2}{5x}}$ Quotient rule for radicals

$\qquad\qquad\qquad\quad = \sqrt[3]{2x}$ Reduce. ∎

Note that in Example 1(a) we applied the quotient rule to get $\sqrt{10} \div \sqrt{5} = \sqrt{2}$. In Example 1(b) we did not use the quotient rule because 2 is not evenly divisible by 3. Instead, we rationalized the denominator to get the result in simplified form.

In Chapter 10 it will be necessary to simplify expressions of the type found in the next example.

E X A M P L E   2

Simplifying radical expressions

Simplify.

a) $\dfrac{4 - \sqrt{12}}{4}$ **b)** $\dfrac{-6 + \sqrt{20}}{-2}$

Solution

a) First write $\sqrt{12}$ in simplified form. Then simplify the expression.

$$\dfrac{4 - \sqrt{12}}{4} = \dfrac{4 - 2\sqrt{3}}{4} \quad \text{Simplify } \sqrt{12}.$$

$$= \dfrac{2(2 - \sqrt{3})}{2 \cdot 2} \quad \text{Factor.}$$

$$= \dfrac{2 - \sqrt{3}}{2} \quad \text{Divide out the common factor.}$$

> **helpful hint**
>
> The expressions in Example 2 are the types of expressions that you must simplify when learning the quadratic formula in Chapter 10.

b) $\dfrac{-6 + \sqrt{20}}{-2} = \dfrac{-6 + 2\sqrt{5}}{-2}$

$= \dfrac{-2(3 - \sqrt{5})}{-2}$

$= 3 - \sqrt{5}$ ■

CAUTION To simplify the expressions in Example 2, you must simplify the radical, factor the numerator, and then divide out the common factors. You cannot simply "cancel" the 4's in $\dfrac{4 - \sqrt{12}}{4}$ or the 2's in $\dfrac{2 - \sqrt{3}}{2}$ because they are not common factors.

Rationalizing the Denominator

In Section 9.2 you learned that a simplified expression involving radicals does not have radicals in the denominator. If an expression such as $4 - \sqrt{3}$ appears in a denominator, we can multiply both the numerator and denominator by its conjugate $4 + \sqrt{3}$ to get a rational number in the denominator.

E X A M P L E   3

Rationalizing the denominator using conjugates

Write in simplified form.

a) $\dfrac{2 + \sqrt{3}}{4 - \sqrt{3}}$

b) $\dfrac{\sqrt{5}}{\sqrt{6} + \sqrt{2}}$

Solution

a) $\dfrac{2 + \sqrt{3}}{4 - \sqrt{3}} = \dfrac{(2 + \sqrt{3})(4 + \sqrt{3})}{(4 - \sqrt{3})(4 + \sqrt{3})}$ Multiply the numerator and denominator by $4 + \sqrt{3}$.

$= \dfrac{8 + 6\sqrt{3} + 3}{13}$ $(4 - \sqrt{3})(4 + \sqrt{3}) = 16 - 3 = 13$

$= \dfrac{11 + 6\sqrt{3}}{13}$ Simplify.

b) $\dfrac{\sqrt{5}}{\sqrt{6} + \sqrt{2}} = \dfrac{\sqrt{5}(\sqrt{6} - \sqrt{2})}{(\sqrt{6} + \sqrt{2})(\sqrt{6} - \sqrt{2})}$ Multiply the numerator and denominator by $\sqrt{6} - \sqrt{2}$.

$= \dfrac{\sqrt{30} - \sqrt{10}}{4}$ $(\sqrt{6} + \sqrt{2})(\sqrt{6} - \sqrt{2}) = 6 - 2 = 4$ ■

study tip

Read the text and recite to yourself what you have read. Ask questions and answer them out loud. Listen to your answers to see if they are complete and correct. Would other students understand your answers?

Powers of Radical Expressions

We can use the power of a product rule and the power of a power rule to simplify a radical expression raised to a power. In the next example we also use the fact that a root and a power can be found in either order.

E X A M P L E   4

Finding powers of rational expressions

Simplify. Assume the variables represent positive numbers.

a) $(5\sqrt{2})^3$

b) $(2\sqrt{x^3})^4$

c) $(3w\sqrt[3]{2w})^3$

d) $(2t\sqrt[4]{3t})^3$

Solution

a) $(5\sqrt{2})^3 = 5^3(\sqrt{2})^3$ Power of a product rule

$\qquad = 125\sqrt{8}$ $(\sqrt{2})^3 = \sqrt{2^3} = \sqrt{8}$

$\qquad = 125 \cdot 2\sqrt{2}$ $\sqrt{8} = \sqrt{4}\sqrt{2} = 2\sqrt{2}$

$\qquad = 250\sqrt{2}$

b) $(2\sqrt{x^3})^4 = 2^4(\sqrt{x^3})^4$

$\qquad = 16\sqrt{x^{12}}$

$\qquad = 16x^6$

c) $(3w\sqrt[3]{2w})^3 = 3^3 w^3 (\sqrt[3]{2w})^3$

$\qquad = 27w^3(2w)$

$\qquad = 54w^4$

d) $(2t\sqrt[4]{3t})^3 = 2^3 t^3 (\sqrt[4]{3t})^3 = 8t^3\sqrt[4]{27t^3}$

WARM-UPS

True or false? Explain your answer.

1. $\dfrac{\sqrt{6}}{\sqrt{2}} = \sqrt{3}$

2. $\dfrac{2}{\sqrt{2}} = \sqrt{2}$

3. $\dfrac{4 - \sqrt{10}}{2} = 2 - \sqrt{10}$

4. $\dfrac{1}{\sqrt{3}} = \dfrac{\sqrt{3}}{3}$

5. $\dfrac{8\sqrt{7}}{2\sqrt{7}} = 4\sqrt{7}$

6. $\dfrac{2(2 + \sqrt{3})}{(2 - \sqrt{3})(2 + \sqrt{3})} = 4 + 2\sqrt{3}$

7. $\dfrac{\sqrt{12}}{3} = \sqrt{4}$

8. $\dfrac{\sqrt{20}}{\sqrt{5}} = 2$

9. $(2\sqrt{4})^2 = 16$

10. $(3\sqrt{5})^3 = 27\sqrt{125}$

9.4 EXERCISES

All variables in the following exercises represent positive numbers.

Divide and simplify. See Example 1.

1. $\sqrt{15} \div \sqrt{5}$

2. $\sqrt{14} \div \sqrt{7}$

3. $\sqrt{3} \div \sqrt{5}$

4. $\sqrt{5} \div \sqrt{7}$

5. $(3\sqrt{3}) \div (5\sqrt{6})$

6. $(2\sqrt{2}) \div (4\sqrt{10})$

7. $(2\sqrt{3}) \div (3\sqrt{6})$

8. $(5\sqrt{12}) \div (4\sqrt{6})$

9. $\sqrt[3]{20} \div \sqrt[3]{2}$

10. $\sqrt[4]{48} \div \sqrt[4]{3}$

11. $\sqrt[3]{8x^7} \div \sqrt[3]{2x}$

12. $\sqrt[4]{4a^{10}} \div \sqrt[4]{2a^2}$

Simplify. See Example 2.

13. $\dfrac{6 + \sqrt{45}}{3}$

14. $\dfrac{10 + \sqrt{50}}{5}$

15. $\dfrac{-2 + \sqrt{12}}{-2}$

16. $\dfrac{-6 + \sqrt{72}}{-6}$

Simplify each expression by rationalizing the denominator. See Example 3.

17. $\dfrac{1 + \sqrt{2}}{\sqrt{3} - 1}$

18. $\dfrac{2 - \sqrt{3}}{\sqrt{2} + \sqrt{6}}$

19. $\dfrac{\sqrt{2}}{\sqrt{6} + \sqrt{3}}$

20. $\dfrac{5}{\sqrt{7} - \sqrt{5}}$

21. $\dfrac{2\sqrt{3}}{3\sqrt{2} - \sqrt{5}}$

22. $\dfrac{3\sqrt{5}}{5\sqrt{2} + \sqrt{6}}$

23. $\dfrac{1 + 3\sqrt{2}}{2\sqrt{6} + 3\sqrt{10}}$

24. $\dfrac{3\sqrt{3} + 1}{4 - 5\sqrt{3}}$

Simplify. See Example 4.

25. $(2\sqrt{2})^5$

26. $(3\sqrt{3})^4$

27. $(\sqrt{x})^5$

28. $(2\sqrt{y})^3$

29. $(-3\sqrt{x^3})^3$

30. $(-2\sqrt{x^3})^4$

31. $(2x\sqrt[3]{x^2})^3$

32. $(2y\sqrt[3]{4y})^3$

33. $(-2\sqrt[3]{5})^2$

34. $(-3\sqrt[3]{4})^2$

35. $(\sqrt[3]{x^2})^6$

36. $(2\sqrt[4]{y^3})^3$

In Exercises 37–74, simplify.

37. $\dfrac{\sqrt{3}}{\sqrt{2}} + \dfrac{2}{\sqrt{2}}$

38. $\dfrac{2}{\sqrt{7}} + \dfrac{5}{\sqrt{7}}$

39. $\dfrac{\sqrt{3}}{\sqrt{2}} + \dfrac{3\sqrt{6}}{2}$

40. $\dfrac{\sqrt{3}}{2\sqrt{2}} + \dfrac{\sqrt{5}}{3\sqrt{2}}$

41. $\dfrac{\sqrt{6}}{2} \cdot \dfrac{1}{\sqrt{3}}$

42. $\dfrac{\sqrt{6}}{\sqrt{7}} \cdot \dfrac{\sqrt{14}}{\sqrt{3}}$

43. $(2\sqrt{w}) \div (3\sqrt{w})$

44. $2 \div (3\sqrt{a})$

45. $\dfrac{8 - \sqrt{32}}{20}$

46. $\dfrac{4 - \sqrt{28}}{6}$

47. $\dfrac{5 + \sqrt{75}}{10}$

48. $\dfrac{3 + \sqrt{18}}{6}$

49. $\sqrt{a}(\sqrt{a} - 3)$

50. $3\sqrt{m}(2\sqrt{m} - 6)$

51. $4\sqrt{a}(a + \sqrt{a})$

52. $\sqrt{3ab}(\sqrt{3a} + \sqrt{3})$

53. $(2\sqrt{3m})^2$

54. $(-3\sqrt{4y})^2$

55. $(-2\sqrt{xy^2z})^2$

56. $(5a\sqrt{ab})^2$

57. $\sqrt[3]{m}(\sqrt[3]{m^2} - \sqrt[3]{m^5})$

58. $\sqrt[4]{w}(\sqrt[4]{w^3} - \sqrt[4]{w^7})$

59. $\sqrt[3]{8x^4} + \sqrt[3]{27x^4}$

60. $\sqrt[3]{16a^4} + a\sqrt[3]{2a}$

61. $(2m\sqrt[3]{2m^2})^3$

62. $(-2t\sqrt[6]{2t^2})^5$

63. $\dfrac{4}{2 + \sqrt{8}}$

64. $\dfrac{6}{3 - \sqrt{18}}$

65. $\dfrac{5}{\sqrt{2} - 1} + \dfrac{3}{\sqrt{2} + 1}$

66. $\dfrac{\sqrt{3}}{\sqrt{6} - 1} - \dfrac{\sqrt{3}}{\sqrt{6} + 1}$

67. $\dfrac{1}{\sqrt{2}} + \dfrac{1}{\sqrt{3}}$

68. $\dfrac{4}{2\sqrt{3}} + \dfrac{1}{\sqrt{5}}$

69. $\dfrac{3}{\sqrt{2} - 1} + \dfrac{4}{\sqrt{2} + 1}$

70. $\dfrac{3}{\sqrt{5} - \sqrt{3}} - \dfrac{2}{\sqrt{5} + \sqrt{3}}$

71. $\dfrac{\sqrt{x}}{\sqrt{x} + 2} + \dfrac{3\sqrt{x}}{\sqrt{x} - 2}$

72. $\dfrac{\sqrt{5}}{3 - \sqrt{y}} - \dfrac{\sqrt{5y}}{3 + \sqrt{y}}$

73. $\dfrac{1}{\sqrt{x}} + \dfrac{1}{1 - \sqrt{x}}$

74. $\dfrac{\sqrt{x}}{\sqrt{x} - 3} + \dfrac{5}{\sqrt{x}}$

Replace the question mark by an expression that makes the equation correct. Equations involving variables are to be identities.

75. $\dfrac{\sqrt{2}}{\sqrt{3}} = \dfrac{\sqrt{6}}{?}$

76. $\dfrac{2}{?} = \sqrt{2}$

77. $\dfrac{1}{\sqrt{2} - 1} = \dfrac{\sqrt{2} + 1}{?}$

78. $\dfrac{\sqrt{6}}{\sqrt{6} + 2} = \dfrac{?}{2}$

79. $\dfrac{1}{\sqrt{x} - 1} = \dfrac{?}{x - 1}$

80. $\dfrac{5}{3 - \sqrt{x}} = \dfrac{?}{9 - x}$

81. $\dfrac{3}{\sqrt{2} + x} = \dfrac{?}{2 - x^2}$

82. $\dfrac{4}{2\sqrt{3} + a} = \dfrac{?}{12 - a^2}$

Use a calculator to find a decimal approximation for each radical expression. Round your answers to three decimal places.

83. $\sqrt{3} + \sqrt{5}$

84. $\sqrt{5} + \sqrt{7}$

85. $2\sqrt{3} + 5\sqrt{3}$

86. $7\sqrt{3}$

87. $(2\sqrt{3})(3\sqrt{2})$

88. $6\sqrt{6}$

89. $\sqrt{5}(\sqrt{5} + \sqrt{3})$

90. $5 + \sqrt{15}$

91. $\dfrac{-1 + \sqrt{6}}{2}$

92. $\dfrac{-1 - \sqrt{6}}{2}$

93. $\dfrac{4 - \sqrt{10}}{-2}$

94. $\dfrac{4 + \sqrt{10}}{-2}$

GETTING MORE INVOLVED

95. *Exploration.* A polynomial is prime if it cannot be factored by using integers, but many prime polynomials can be factored if we use radicals.
  a) Find the product $(x - \sqrt[3]{2})(x^2 + \sqrt[3]{2}x + \sqrt[3]{4})$.
  b) Factor $x^3 + 5$ using radicals.
  c) Find the product
  $$(\sqrt[3]{5} - \sqrt[3]{2})(\sqrt[3]{25} + \sqrt[3]{10} + \sqrt[3]{4}).$$
  d) Use radicals to factor $a + b$ as a sum of two cubes and $a - b$ as a difference of two cubes.

96. *Discussion.* Which one of the following expressions is not equivalent to the others?
  a) $(\sqrt[4]{x})^4$ **b)** $\sqrt[4]{x^3}$ **c)** $\sqrt[3]{x^4}$
  d) $x^{4/3}$ **e)** $(x^{1/3})^4$

9.5 SOLVING EQUATIONS WITH RADICALS AND EXPONENTS

One of our goals in algebra is to keep increasing our knowledge of solving equations because the solutions to equations can give us the answers to various applied questions. In this section we will apply our knowledge of radicals and exponents to solving some new types of equations.

The Odd-Root Property

Because $(-2)^3 = -8$ and $2^3 = 8$, the equation $x^3 = 8$ is equivalent to $x = 2$. The equation $x^3 = -8$ is equivalent to $x = -2$. Because there is only one real odd root of each real number, there is a simple rule for writing an equivalent equation in this situation.

Odd-Root Property

If n is an odd positive integer,

$$x^n = k \quad \text{is equivalent to} \quad x = \sqrt[n]{k}$$

for any real number k.

E X A M P L E   1   **Using the odd-root property**

Solve each equation.

a) $x^3 = 27$ **b)** $x^5 + 32 = 0$ **c)** $(x - 2)^3 = 24$

Solution

a) $x^3 = 27$

$\quad x = \sqrt[3]{27}$ Odd-root property

$\quad x = 3$

Check 3 in the original equation. The solution set is $\{3\}$.

b) $x^5 + 32 = 0$

$\quad\quad x^5 = -32$ Isolate the variable.

$\quad\quad x = \sqrt[5]{-32}$ Odd-root property

$\quad\quad x = -2$

Check -2 in the original equation. The solution set is $\{-2\}$.

c) $(x - 2)^3 = 24$

$\quad x - 2 = \sqrt[3]{24}$ Odd-root property

$\quad\quad x = 2 + 2\sqrt[3]{3}$ $\sqrt[3]{24} = \sqrt[3]{8} \cdot \sqrt[3]{3} = 2\sqrt[3]{3}$

Check. The solution set is $\{2 + 2\sqrt[3]{3}\}$.   ■

The Even-Root Property

In solving the equation $x^2 = 4$, you might be tempted to write $x = 2$ as an equivalent equation. But $x = 2$ is not equivalent to $x^2 = 4$ because $2^2 = 4$ and $(-2)^2 = 4$. So the solution set to $x^2 = 4$ is $\{-2, 2\}$. The equation $x^2 = 4$ is equivalent to the compound sentence $x = 2$ or $x = -2$, which we can abbreviate as $x = \pm 2$. The equation $x = \pm 2$ is read "x equals positive or negative 2."

Equations involving other even powers are handled like the squares. Because $2^4 = 16$ and $(-2)^4 = 16$, the equation $x^4 = 16$ is equivalent to $x = \pm 2$. So $x^4 = 16$ has two real solutions. Note that $x^4 = -16$ has no real solutions. The equation $x^6 = 5$ is equivalent to $x = \pm\sqrt[6]{5}$. We can now state a general rule.

Even-Root Property

Suppose n is a positive even integer.

$\quad$ If $k > 0$, then $x^n = k$ is equivalent to $x = \pm\sqrt[n]{k}$.

$\quad$ If $k = 0$, then $x^n = k$ is equivalent to $x = 0$.

$\quad$ If $k < 0$, then $x^n = k$ has no real solution.

E X A M P L E   2

Using the even-root property

Solve each equation.

a) $x^2 = 10$ **b)** $w^8 = 0$ **c)** $x^4 = -4$

Solution

a) $x^2 = 10$

$x = \pm\sqrt{10}$ Even-root property

The solution set is $\{-\sqrt{10}, \sqrt{10}\}$, or $\{\pm\sqrt{10}\}$.

b) $w^8 = 0$

$w = 0$ Even-root property

The solution set is $\{0\}$.

c) By the even-root property, $x^4 = -4$ has no real solution. (The fourth power of any real number is nonnegative.) ■

In the next example the even-root property is used to solve some equations that are a bit more complicated than those of Example 2.

E X A M P L E   3

Using the even-root property

Solve each equation.

a) $(x - 3)^2 = 4$ **b)** $2(x - 5)^2 - 7 = 0$ **c)** $x^4 - 1 = 80$

Solution

a) $(x - 3)^2 = 4$

$x - 3 = 2$ or $x - 3 = -2$ Even-root property

$x = 5$ or $x = 1$ Add 3 to each side.

The solution set is $\{1, 5\}$.

b) $2(x - 5)^2 - 7 = 0$

$2(x - 5)^2 = 7$ Add 7 to each side.

$(x - 5)^2 = \dfrac{7}{2}$ Divide each side by 2.

$x - 5 = \sqrt{\dfrac{7}{2}}$ or $x - 5 = -\sqrt{\dfrac{7}{2}}$ Even-root property

$x = 5 + \dfrac{\sqrt{14}}{2}$ or $x = 5 - \dfrac{\sqrt{14}}{2}$ $\sqrt{\dfrac{7}{2}} = \dfrac{\sqrt{7} \cdot \sqrt{2}}{\sqrt{2} \cdot \sqrt{2}} = \dfrac{\sqrt{14}}{2}$

$x = \dfrac{10 + \sqrt{14}}{2}$ or $x = \dfrac{10 - \sqrt{14}}{2}$

The solution set is $\left\{\dfrac{10 + \sqrt{14}}{2}, \dfrac{10 - \sqrt{14}}{2}\right\}$.

c) $x^4 - 1 = 80$

$x^4 = 81$

$x = \pm\sqrt[4]{81} = \pm 3$

The solution set is $\{-3, 3\}$. ■

In Chapter 6 we solved quadratic equations by factoring. The quadratic equations that we encounter in this chapter can be solved by using the even-root

> **study tip**
>
> Review, review, review! Don't wait until the end of a chapter to review. Do a little review every time you study for this course.

property as in parts (a) and (b) of Example 3. In Chapter 10 you will learn general methods for solving any quadratic equation.

Raising Each Side to a Power

If we start with the equation $x = 3$ and square both sides, we get $x^2 = 9$. The solution set to $x^2 = 9$ is $\{-3, 3\}$; the solution set to the original equation is $\{3\}$. Squaring both sides of an equation might produce a *nonequivalent* equation that has more solutions than the original equation. We call these additional solutions **extraneous solutions.** However, any solution of the original must be among the solutions to the new equation.

> **CAUTION**   When you solve an equation by raising each side to a power, you must check your answers. Raising each side to an odd power will always give an equivalent equation; raising each side to an even power might not.

E X A M P L E   4

Raising each side to a power to eliminate radicals

Solve each equation.

a) $\sqrt{2x - 3} - 5 = 0$ **b)** $\sqrt[3]{3x + 5} = \sqrt[3]{x - 1}$ **c)** $\sqrt{3x + 18} = x$

Solution

a) Eliminate the square root by raising each side to the power 2:

$$\sqrt{2x - 3} - 5 = 0 \qquad \text{Original equation}$$
$$\sqrt{2x - 3} = 5 \qquad \text{Isolate the radical.}$$
$$(\sqrt{2x - 3})^2 = 5^2 \qquad \text{Square both sides.}$$
$$2x - 3 = 25$$
$$2x = 28$$
$$x = 14$$

Check by evaluating $x = 14$ in the original equation:

$$\sqrt{2(14) - 3} - 5 = 0$$
$$\sqrt{28 - 3} - 5 = 0$$
$$\sqrt{25} - 5 = 0$$
$$0 = 0$$

The solution set is $\{14\}$.

b) $\sqrt[3]{3x + 5} = \sqrt[3]{x - 1}$ Original equation
$(\sqrt[3]{3x + 5})^3 = (\sqrt[3]{x - 1})^3$ Cube each side.
$$3x + 5 = x - 1$$
$$2x = -6$$
$$x = -3$$

Check $x = -3$ in the original equation:

$$\sqrt[3]{3(-3) + 5} = \sqrt[3]{-3 - 1}$$
$$\sqrt[3]{-4} = \sqrt[3]{-4}$$

Note that $\sqrt[3]{-4}$ is a real number. The solution set is $\{-3\}$. In this example we checked for arithmetic mistakes. There was no possibility of extraneous solutions here because we raised each side to an odd power.

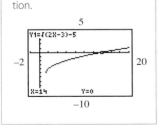

c a l c u l a t o r

c l o s e - u p

If 14 satisfies the equation

$$\sqrt{2x - 3} - 5 = 0,$$

then $(14, 0)$ is an x-intercept for the graph of

$$y = \sqrt{2x - 3} - 5.$$

So the calculator graph shown here provides visual support for the conclusion that 14 is the only solution to the equation.

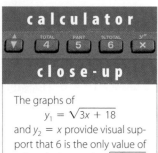

The graphs of
$$y_1 = \sqrt{3x + 18}$$
and $y_2 = x$ provide visual support that 6 is the only value of x for which x and $\sqrt{3x + 18}$ are equal.

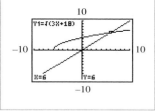

c)

$$\sqrt{3x + 18} = x \qquad \text{Original equation}$$
$$(\sqrt{3x + 18})^2 = x^2 \qquad \text{Square both sides.}$$
$$3x + 18 = x^2 \qquad \text{Simplify.}$$
$$-x^2 + 3x + 18 = 0 \qquad \text{Subtract } x^2 \text{ from each side to get zero on one side.}$$
$$x^2 - 3x - 18 = 0 \qquad \text{Multiply each side by } -1 \text{ for easier factoring.}$$
$$(x - 6)(x + 3) = 0 \qquad \text{Factor.}$$
$$x - 6 = 0 \quad \text{or} \quad x + 3 = 0 \qquad \text{Zero factor property}$$
$$x = 6 \quad \text{or} \quad x = -3$$

Because we squared both sides, we must check for extraneous solutions. If $x = -3$ in the original equation $\sqrt{3x + 18} = x$, we get

$$\sqrt{3(-3) + 18} = -3$$
$$\sqrt{9} = -3$$
$$3 = -3,$$

which is not correct. If $x = 6$ in the original equation, we get

$$\sqrt{3(6) + 18} = 6,$$

which is correct. The solution set is $\{6\}$.

In the next example the radicals are not eliminated after squaring both sides of the equation. In this case we must square both sides a second time. Note that we square the side with two terms the same way we square a binomial.

E X A M P L E   5

Squaring both sides twice
Solve $\sqrt{5x - 1} - \sqrt{x + 2} = 1$.

Solution

It is easier to square both sides if the two radicals are not on the same side, so we first rewrite the equation:

$$\sqrt{5x - 1} - \sqrt{x + 2} = 1 \qquad \text{Original equation}$$
$$\sqrt{5x - 1} = 1 + \sqrt{x + 2} \qquad \text{Add } \sqrt{x + 2} \text{ to each side.}$$
$$(\sqrt{5x - 1})^2 = (1 + \sqrt{x + 2})^2 \qquad \text{Square both sides.}$$
$$5x - 1 = 1 + 2\sqrt{x + 2} + x + 2 \qquad \text{Square the right side like a binomial.}$$
$$5x - 1 = 3 + x + 2\sqrt{x + 2} \qquad \text{Combine like terms on the right side.}$$
$$4x - 4 = 2\sqrt{x + 2} \qquad \text{Isolate the square root.}$$
$$2x - 2 = \sqrt{x + 2} \qquad \text{Divide each side by 2.}$$
$$(2x - 2)^2 = (\sqrt{x + 2})^2 \qquad \text{Square both sides.}$$
$$4x^2 - 8x + 4 = x + 2 \qquad \text{Square the binomial on the left side.}$$
$$4x^2 - 9x + 2 = 0$$
$$(4x - 1)(x - 2) = 0$$
$$4x - 1 = 0 \quad \text{or} \quad x - 2 = 0$$
$$x = \frac{1}{4} \quad \text{or} \quad x = 2$$

Check to see whether $\sqrt{5x - 1} - \sqrt{x + 2} = 1$ for $x = \frac{1}{4}$ and for $x = 2$:

$$\sqrt{5 \cdot \frac{1}{4} - 1} - \sqrt{\frac{1}{4} + 2} = \sqrt{\frac{1}{4}} - \sqrt{\frac{9}{4}} = \frac{1}{2} - \frac{3}{2} = -1$$

$$\sqrt{5 \cdot 2 - 1} - \sqrt{2 + 2} = \sqrt{9} - \sqrt{4} = 3 - 2 = 1$$

Because $\frac{1}{4}$ does not satisfy the original equation, the solution set is $\{2\}$. ■

Equations Involving Rational Exponents

Equations involving rational exponents can be solved by combining the methods that you just learned for eliminating radicals and integral exponents. For equations involving rational exponents, always eliminate the root first and the power second.

EXAMPLE 6

Eliminating the root, then the power

Solve each equation.

a) $x^{2/3} = 4$

b) $(w - 1)^{-2/5} = 4$

Solution

a) Because the exponent $2/3$ indicates a cube root, raise each side to the power 3:

$$x^{2/3} = 4 \qquad \text{Original equation}$$
$$(x^{2/3})^3 = 4^3 \qquad \text{Cube each side.}$$
$$x^2 = 64 \qquad \text{Multiply the exponents: } \frac{2}{3} \cdot 3 = 2.$$
$$x = 8 \quad \text{or} \quad x = -8 \qquad \text{Even-root property}$$

All of the equations are equivalent. Check 8 and -8 in the original equation. The solution set is $\{-8, 8\}$.

b)
$$(w - 1)^{-2/5} = 4 \qquad \text{Original equation}$$
$$[(w - 1)^{-2/5}]^{-5} = 4^{-5} \qquad \text{Raise each side to the power } -5 \text{ to eliminate the negative exponent.}$$
$$(w - 1)^2 = \frac{1}{1024} \qquad \text{Multiply the exponents: } -\frac{2}{5}(-5) = 2.$$
$$w - 1 = \pm\sqrt{\frac{1}{1024}} \qquad \text{Even-root property}$$
$$w - 1 = \frac{1}{32} \quad \text{or} \quad w - 1 = -\frac{1}{32}$$
$$w = \frac{33}{32} \quad \text{or} \quad w = \frac{31}{32}$$

Check the values in the original equation. The solution set is $\left\{\frac{31}{32}, \frac{33}{32}\right\}$. ■

An equation with a rational exponent might not have a real solution because all even powers of real numbers are nonnegative.

helpful hint

Note how we eliminate the root first by raising each side to an integer power, and then apply the even-root property to get two solutions in Example 6(a). A common mistake is to raise each side to the 3/2 power and get $x = 4^{3/2} = 8$. If you do not use the even-root property you can easily miss the solution -8.

calculator

close-up

Check that 31/32 and 33/32 satisfy the original equation.

E X A M P L E   7   **An equation with no solution**

Solve $(2t - 3)^{-2/3} = -1$.

Solution

Raise each side to the power -3 to eliminate the root and the negative sign in the exponent:

$$(2t - 3)^{-2/3} = -1 \qquad \text{Original equation}$$

$$[(2t - 3)^{-2/3}]^{-3} = (-1)^{-3} \qquad \text{Raise each side to the } -3 \text{ power.}$$

$$(2t - 3)^2 = -1 \qquad \text{Multiply the exponents: } -\tfrac{2}{3}(-3) = 2.$$

By the even-root property this equation has no real solution. The square of every real number is nonnegative.

Summary of Methods

The three most important rules for solving equations with exponents and radicals are restated here.

**Strategy for Solving Equations
with Exponents and Radicals**

1. In raising each side of an equation to an even power, we can create an equation that gives extraneous solutions. We must check all possible solutions in the original equation.

2. When applying the even-root property, remember that there is a positive and a negative even root for any positive real number.

3. For equations with rational exponents, raise each side to a positive or negative integral power first, then apply the even- or odd-root property. (Positive fraction—raise to a positive power; negative fraction—raise to a negative power.)

The Distance Formula

Consider the points (x_1, y_1) and (x_2, y_2) as shown in Fig. 9.1. The distance between these points is the length of the hypotenuse of a right triangle as shown in the figure. The length of side a is $y_2 - y_1$ and the length of side b is $x_2 - x_1$. Using the Pythagorean theorem, we can write

$$d^2 = (x_2 - x_1)^2 + (y_2 - y_1)^2.$$

If we apply the even-root property and omit the negative square root (because the distance is positive), we can express this formula as follows.

Distance Formula

The distance d between (x_1, y_1) and (x_2, y_2) is given by the formula

$$d = \sqrt{(x_2 - x_1)^2 + (y_2 - y_1)^2}.$$

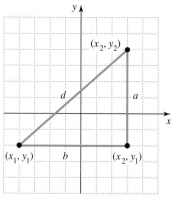

FIGURE 9.1

E X A M P L E   8

Using the distance formula

Find the length of the line segment with endpoints $(-8, -10)$ and $(6, -4)$.

Solution

Let $(x_1, y_1) = (-8, -10)$ and $(x_2, y_2) = (6, -4)$. Now substitute the appropriate values into the distance formula:

$$d = \sqrt{[6 - (-8)]^2 + [-4 - (-10)]^2}$$
$$= \sqrt{(14)^2 + (6)^2}$$
$$= \sqrt{196 + 36}$$
$$= \sqrt{232}$$
$$= \sqrt{4 \cdot 58}$$
$$= 2\sqrt{58} \quad \text{Simplified form}$$

The exact length of the segment is $2\sqrt{58}$. ■

In the next example we find the distance between two points without the distance formula. Although we could solve the problem using a coordinate system and the distance formula, that is not necessary.

E X A M P L E   9

Diagonal of a baseball diamond

A baseball diamond is actually a square, 90 feet on each side. What is the distance from third base to first base?

Solution

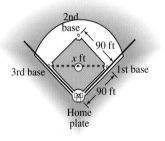

FIGURE 9.2

First make a sketch as in Fig. 9.2. The distance x from third base to first base is the length of the diagonal of the square shown in Fig. 9.2. The Pythagorean theorem can be applied to the right triangle formed from the diagonal and two sides of the square. The sum of the squares of the sides is equal to the diagonal squared:

$$x^2 = 90^2 + 90^2$$
$$x^2 = 8100 + 8100$$
$$x^2 = 16{,}200$$
$$x = \pm\sqrt{16{,}200} = \pm 90\sqrt{2}$$

The length of the diagonal of a square must be positive, so we disregard the negative solution. Checking the answer in the original equation verifies that the *exact* length of the diagonal is $90\sqrt{2}$ feet. ■

WARM-UPS

True or false? Explain your answer.

1. The equations $x^2 = 4$ and $x = 2$ are equivalent.
2. The equation $x^2 = -25$ has no real solution.
3. There is no solution to the equation $x^2 = 0$.
4. The equation $x^3 = 8$ is equivalent to $x = \pm 2$.
5. The equation $-\sqrt{x} = 16$ has no real solution.
6. To solve $\sqrt{x - 3} = \sqrt{2x + 5}$, first apply the even-root property.

(continued)

7. Extraneous solutions are solutions that cannot be found.
8. Squaring both sides of $\sqrt{x} = -7$ yields an equation with an extraneous solution.
9. The equations $x^2 - 6 = 0$ and $x = \pm\sqrt{6}$ are equivalent.
10. Cubing each side of an equation will not produce an extraneous solution.

9.5 EXERCISES

Reading and Writing *After reading this section, write out the answers to these questions. Use complete sentences.*

1. What is the odd-root property?

2. What is the even-root property?

3. What is an extraneous solution?

4. Why can raising each side to a power produce an extraneous solution?

Solve each equation. See Example 1.

5. $x^3 = -1000$
6. $y^3 = 125$

7. $32m^5 - 1 = 0$
8. $243a^5 + 1 = 0$

9. $(y - 3)^3 = -8$
10. $(x - 1)^3 = -1$

11. $\dfrac{1}{2}x^3 + 4 = 0$
12. $3(x - 9)^7 = 0$

Solve each equation. See Examples 2 and 3.

13. $x^2 = 25$
14. $x^2 = 36$

15. $x^2 - 20 = 0$
16. $a^2 - 40 = 0$

17. $x^2 = -9$
18. $w^2 + 49 = 0$

19. $(x - 3)^2 = 16$
20. $(a - 2)^2 = 25$

21. $(x + 1)^2 - 8 = 0$

22. $(w + 3)^2 - 12 = 0$

23. $\dfrac{1}{2}x^2 = 5$
24. $\dfrac{1}{3}x^2 = 6$

25. $(y - 3)^4 = 0$
26. $(2x - 3)^6 = 0$

27. $2x^6 = 128$
28. $3y^4 = 48$

Solve each equation and check for extraneous solutions. See Example 4.

29. $\sqrt{x - 3} - 7 = 0$
30. $\sqrt{a - 1} - 6 = 0$

31. $2\sqrt{w + 4} = 5$
32. $3\sqrt{w + 1} = 6$

33. $\sqrt[3]{2x + 3} = \sqrt[3]{x + 12}$
34. $\sqrt[3]{a + 3} = \sqrt[3]{2a - 7}$

35. $\sqrt{2t + 4} = \sqrt{t - 1}$
36. $\sqrt{w - 3} = \sqrt{4w + 15}$

37. $\sqrt{4x^2 + x - 3} = 2x$
38. $\sqrt{x^2 - 5x + 2} = x$

39. $\sqrt{x^2 + 2x - 6} = 3$
40. $\sqrt{x^2 - x - 4} = 4$

41. $\sqrt{2x^2 - 1} = x$
42. $\sqrt{2x^2 - 3x - 10} = x$

43. $\sqrt{2x^2 + 5x + 6} = x$
44. $\sqrt{5x^2 - 9} = 2x$

Solve each equation and check for extraneous solutions. See Example 5.

45. $\sqrt{x} + \sqrt{x-3} = 3$

46. $\sqrt{x} + \sqrt{x+3} = 3$

47. $\sqrt{x+2} + \sqrt{x-1} = 3$

48. $\sqrt{x} + \sqrt{x-5} = 5$

49. $\sqrt{x+3} - \sqrt{x-2} = 1$

50. $\sqrt{2x+1} - \sqrt{x} = 1$

51. $\sqrt{2x+2} - \sqrt{x-3} = 2$

52. $\sqrt{3x} - \sqrt{x-2} = 4$

53. $\sqrt{4-x} - \sqrt{x+6} = 2$

54. $\sqrt{6-x} - \sqrt{x-2} = 2$

Solve each equation. See Examples 6 and 7.

55. $x^{2/3} = 3$

56. $a^{2/3} = 2$

57. $y^{-2/3} = 9$

58. $w^{-2/3} = 4$

59. $w^{1/3} = 8$

60. $a^{1/3} = 27$

61. $t^{-1/2} = 9$

62. $w^{-1/4} = \dfrac{1}{2}$

63. $(3a-1)^{-2/5} = 1$

64. $(r-1)^{-2/3} = 1$

65. $(t-1)^{-2/3} = 2$

66. $(w+3)^{-1/3} = \dfrac{1}{3}$

67. $(x-3)^{2/3} = -4$

68. $(x+2)^{3/2} = -1$

Find the distance between each given pair of points. See Example 8.

69. $(6, 5), (4, 2)$

70. $(7, 3), (5, 1)$

71. $(3, 5), (1, -3)$

72. $(6, 2), (3, -5)$

73. $(4, -2), (-3, -6)$

74. $(-2, 3), (1, -4)$

Solve each equation.

75. $2x^2 + 3 = 7$

76. $3x^2 - 5 = 16$

77. $\sqrt[3]{2w+3} = \sqrt[3]{w-2}$

78. $\sqrt[3]{2-w} = \sqrt[3]{2w-28}$

79. $(w+1)^{2/3} = -3$

80. $(x-2)^{3/4} = 2$

81. $(a+1)^{1/3} = -2$

82. $(a-1)^{1/3} = -3$

83. $(4y-5)^7 = 0$

84. $(5x)^9 = 0$

85. $\sqrt{x^2 + 5x} = 6$

86. $\sqrt{x^2 - 8x} = -3$

87. $\sqrt{4x^2} = x + 2$

88. $\sqrt{9x^2} = x + 6$

89. $(t+2)^4 = 32$

90. $(w+1)^4 = 48$

91. $\sqrt{x^2 - 3x} = x$

92. $\sqrt[4]{4x^4 - 48} = -x$

93. $x^{-3} = 8$

94. $x^{-2} = 4$

Solve each problem by writing an equation and solving it. Find the exact answer and simplify it using the rules for radicals. See Example 9.

95. *Side of a square.* Find the length of the side of a square whose diagonal is 8 feet.

96. *Diagonal of a patio.* Find the length of the diagonal of a square patio with an area of 40 square meters.

97. *Side of a sign.* Find the length of the side of a square sign whose area is 50 square feet.

98. *Side of a cube.* Find the length of the side of a cubic box whose volume is 80 cubic feet.

99. *Diagonal of a rectangle.* If the sides of a rectangle are 30 feet and 40 feet in length, find the length of the diagonal of the rectangle.

100. *Diagonal of a sign.* What is the length of the diagonal of a rectangular billboard whose sides are 5 meters and 12 meters?

101. *Sailboat stability.* To be considered safe for ocean sailing, the capsize screening value C should be less than 2 (*Sail*, May 1997). For a boat with a beam (or width) b in feet and displacement d in pounds, C is determined by the formula

$$C = 4d^{-1/3}b.$$

a) Find the capsize screening value for the Tartan 4100, which has a displacement of 23,245 pounds and a beam of 13.5 feet.

b) Solve this formula for d.

c) The accompanying graph shows C as a function of d for the Tartan 4100 ($b = 13.5$). For what displacement is the Tartan 4100 safe for ocean sailing?

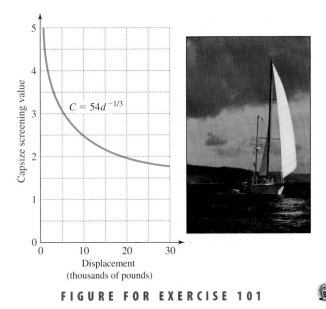

$C = 54d^{-1/3}$

FIGURE FOR EXERCISE 101

102. ***Sailboat speed.*** The sail area-displacement ratio S provides a measure of the sail power available to drive a boat. For a boat with a displacement of d pounds and a sail area of A square feet

$$S = 16Ad^{-2/3}.$$

a) Find S for the Tartan 4100, which has a sail area of 810 square feet and a displacement of 23,245 pounds.

b) Solve the formula for d.

103. ***Diagonal of a side.*** Find the length of the diagonal of a side of a cubic packing crate whose volume is 2 cubic meters.

104. ***Volume of a cube.*** Find the volume of a cube on which the diagonal of a side measures 2 feet.

105. ***Length of a road.*** An architect designs a public park in the shape of a trapezoid. Find the length of the diagonal road marked a in the figure.

106. ***Length of a boundary.*** Find the length of the border of the park marked b in the trapezoid shown in the figure.

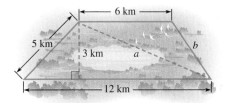

FIGURE FOR EXERCISES 105 AND 106

107. ***Average annual return.*** The formula

$$r = \left(\frac{S}{P}\right)^{1/n} - 1$$

was used to find the average annual return on an investment in Exercise 127 in Section 9.2. Solve the formula for S (the amount). Solve it for P (the original principal).

108. ***Surface area of a cube.*** The formula $A = 6V^{2/3}$ gives the surface area of a cube in terms of its volume V. What is the volume of a cube with surface area 12 square feet?

109. ***Kepler's third law.*** According to Kepler's third law of planetary motion, the ratio $\frac{T^2}{R^3}$ has the same value for every planet in our solar system. R is the average radius of the orbit of the planet measured in astronomical units (AU), and T is the number of years it takes for one complete orbit of the sun. Jupiter orbits the sun in 11.86 years with an average radius of 5.2 AU, whereas Saturn orbits the sun in 29.46 years. Find the average radius of the orbit of Saturn. (One AU is the distance from the earth to the sun.)

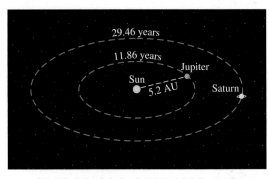

FIGURE FOR EXERCISE 109

110. ***Orbit of Venus.*** If the average radius of the orbit of Venus is 0.723 AU, then how many years does it take for

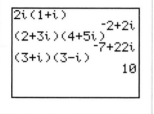

Many graphing calculators can perform operations with complex numbers.

Solution

a) $2i(1 + i) = 2i + 2i^2$ Distributive property

$\qquad\qquad = 2i + 2(-1)$ $i^2 = -1$

$\qquad\qquad = -2 + 2i$

b) Use the FOIL method to find the product:

$$(2 + 3i)(4 + 5i) = 8 + 10i + 12i + 15i^2$$

$$= 8 + 22i + 15(-1) \quad \text{Replace } i^2 \text{ by } -1.$$

$$= 8 + 22i - 15$$

$$= -7 + 22i$$

c) This product is the product of a sum and a difference.

$$(3 + i)(3 - i) = 9 - 3i + 3i - i^2$$

$$= 9 - (-1) \quad i^2 = -1$$

$$= 10$$

We can find powers of i using the fact that $i^2 = -1$. For example,

$$i^3 = i^2 \cdot i = -1 \cdot i = -i.$$

The value of i^4 is found from the value of i^3:

$$i^4 = i^3 \cdot i = -i \cdot i = -i^2 = 1.$$

In the next example we find more powers of imaginary numbers.

E X A M P L E   3

Powers of imaginary numbers

Write each expression in the form $a + bi$.

a) $(2i)^2$ **b)** $(-2i)^2$ **c)** i^6

Solution

a) $(2i)^2 = 2^2 \cdot i^2 = 4(-1) = -4$

b) $(-2i)^2 = (-2)^2 \cdot i^2 = 4i^2 = 4(-1) = -4$

c) $i^6 = i^2 \cdot i^4 = -1 \cdot 1 = -1$

For completeness we give the following symbolic definition of multiplication of complex numbers. However, it is simpler to find products as we did in Examples 2 and 3 than to use this definition.

Multiplication of Complex Numbers

The complex numbers $a + bi$ and $c + di$ are multiplied as follows:

$$(a + bi)(c + di) = (ac - bd) + (ad + bc)i$$

Division of Complex Numbers

To divide a complex number by a real number, divide each term by the real number, just as we would divide a binomial by a number. For example,

$$\frac{4 + 6i}{2} = \frac{2(2 + 3i)}{2}$$

$$= 2 + 3i.$$

To understand division by a complex number, we first look at imaginary numbers that have a real product. The product of the two imaginary numbers in Example 2(c) is a real number:

$$(3 + i)(3 - i) = 10$$

We say that $3 + i$ and $3 - i$ are complex conjugates of each other.

Complex Conjugates

The complex numbers $a + bi$ and $a - bi$ are called **complex conjugates** of one another. Their product is the real number $a^2 + b^2$.

E X A M P L E 4

Products of conjugates

Find the product of the given complex number and its conjugate.

a) $2 + 3i$ **b)** $5 - 4i$

Solution

a) The conjugate of $2 + 3i$ is $2 - 3i$.

$$\begin{aligned}(2 + 3i)(2 - 3i) &= 4 - 9i^2 \\ &= 4 + 9 \\ &= 13\end{aligned}$$

b) The conjugate of $5 - 4i$ is $5 + 4i$.

$$\begin{aligned}(5 - 4i)(5 + 4i) &= 25 + 16 \\ &= 41\end{aligned}$$ ■

We use the idea of complex conjugates to divide complex numbers. The process is similar to rationalizing the denominator. Multiply the numerator and denominator of the quotient by the complex conjugate of the denominator.

E X A M P L E 5

Dividing complex numbers

Find each quotient. Write the answer in the form $a + bi$.

a) $\dfrac{5}{3 - 4i}$ **b)** $\dfrac{3 - i}{2 + i}$ **c)** $\dfrac{3 + 2i}{i}$

Solution

a) Multiply the numerator and denominator by $3 + 4i$, the conjugate of $3 - 4i$:

$$\begin{aligned}\frac{5}{3 - 4i} &= \frac{5(3 + 4i)}{(3 - 4i)(3 + 4i)} \\[2mm] &= \frac{15 + 20i}{9 - 16i^2} \\[2mm] &= \frac{15 + 20i}{25} \qquad 9 - 16i^2 = 9 - 16(-1) = 25 \\[2mm] &= \frac{15}{25} + \frac{20}{25}i \\[2mm] &= \frac{3}{5} + \frac{4}{5}i\end{aligned}$$

b) Multiply the numerator and denominator by $2 - i$, the conjugate of $2 + i$:

$$\frac{3 - i}{2 + i} = \frac{(3 - i)(2 - i)}{(2 + i)(2 - i)}$$

$$= \frac{6 - 5i + i^2}{4 - i^2}$$

$$= \frac{6 - 5i - 1}{4 - (-1)}$$

$$= \frac{5 - 5i}{5}$$

$$= 1 - i$$

c) Multiply the numerator and denominator by $-i$, the conjugate of i:

$$\frac{3 + 2i}{i} = \frac{(3 + 2i)(-i)}{i(-i)}$$

$$= \frac{-3i - 2i^2}{-i^2}$$

$$= \frac{-3i + 2}{1}$$

$$= 2 - 3i$$ ■

The symbolic definition of division of complex numbers follows.

Division of Complex Numbers

We divide the complex number $a + bi$ by the complex number $c + di$ as follows:

$$\frac{a + bi}{c + di} = \frac{(a + bi)(c - di)}{(c + di)(c - di)}$$

Square Roots of Negative Numbers

In Examples 3(a) and 3(b) we saw that both

$$(2i)^2 = -4 \quad \text{and} \quad (-2i)^2 = -4.$$

Because the square of each of these complex numbers is -4, both $2i$ and $-2i$ are square roots of -4. We write $\sqrt{-4} = 2i$. In the complex number system the square root of any negative number is an imaginary number.

Square Root of a Negative Number

For any positive real number b,

$$\sqrt{-b} = i\sqrt{b}.$$

For example, $\sqrt{-9} = i\sqrt{9} = 3i$ and $\sqrt{-7} = i\sqrt{7}$. Note that the expression $\sqrt{7}i$ could easily be mistaken for the expression $\sqrt{7i}$, where i is under the radical. For this reason, when the coefficient of i is a radical, we write i preceding the radical.

E X A M P L E   6

Square roots of negative numbers

Write each expression in the form $a + bi$, where a and b are real numbers.

a) $3 + \sqrt{-9}$ **b)** $\sqrt{-12} + \sqrt{-27}$ **c)** $\dfrac{-1 - \sqrt{-18}}{3}$

Solution

a) $3 + \sqrt{-9} = 3 + i\sqrt{9}$

$= 3 + 3i$

b) $\sqrt{-12} + \sqrt{-27} = i\sqrt{12} + i\sqrt{27}$

$\qquad = 2i\sqrt{3} + 3i\sqrt{3}$ $\sqrt{12} = \sqrt{4}\,\sqrt{3} = 2\sqrt{3}$
$\sqrt{27} = \sqrt{9}\,\sqrt{3} = 3\sqrt{3}$

$\qquad = 5i\sqrt{3}$

c) $\dfrac{-1 - \sqrt{-18}}{3} = \dfrac{-1 - i\sqrt{18}}{3}$

$\qquad = \dfrac{-1 - 3i\sqrt{2}}{3}$

$\qquad = -\dfrac{1}{3} - i\sqrt{2}$ ◼

Imaginary Solutions to Equations

In the complex number system the even-root property can be restated so that $x^2 = k$ is equivalent to $x = \pm\sqrt{k}$ for any $k \neq 0$. So an equation such as $x^2 = -9$ that has no real solutions has two imaginary solutions in the complex numbers.

E X A M P L E   7

Complex solutions to equations

Find the complex solutions to each equation.

a) $x^2 = -9$ **b)** $3x^2 + 2 = 0$

Solution

a) First apply the even-root property:

$$x^2 = -9$$
$$x = \pm\sqrt{-9} \quad \text{Even-root property}$$
$$= \pm i\sqrt{9}$$
$$= \pm 3i$$

Check these solutions in the original equation:

$$(3i)^2 = 9i^2 = 9(-1) = -9$$
$$(-3i)^2 = 9i^2 = -9$$

The solution set is $\{\pm 3i\}$.

b) First solve the equation for x^2:

$$3x^2 + 2 = 0$$
$$x^2 = -\dfrac{2}{3}$$
$$x = \pm\sqrt{-\dfrac{2}{3}} = \pm i\sqrt{\dfrac{2}{3}} = \pm i\dfrac{\sqrt{6}}{3}$$

Check these solutions in the original equation. The solution set is $\left\{\pm i\dfrac{\sqrt{6}}{3}\right\}$. ◼

The basic facts about complex numbers are listed in the following box.

Complex Numbers

1. Definition of i: $i = \sqrt{-1}$, and $i^2 = -1$.
2. A complex number has the form $a + bi$, where a and b are real numbers.
3. The complex number $a + 0i$ is the real number a.
4. If b is a positive real number, then $\sqrt{-b} = i\sqrt{b}$.
5. The numbers $a + bi$ and $a - bi$ are called complex conjugates of each other. Their product is the real number $a^2 + b^2$.
6. Add, subtract, and multiply complex numbers as if they were algebraic expressions with i being the variable, and replace i^2 by -1.
7. Divide complex numbers by multiplying the numerator and denominator by the conjugate of the denominator.
8. In the complex number system $x^2 = k$ for any real number k is equivalent to $x = \pm\sqrt{k}$.

WARM-UPS

True or false? Explain your answer.

1. The set of real numbers is a subset of the set of complex numbers.
2. $2 - \sqrt{-6} = 2 - 6i$
3. $\sqrt{-9} = \pm 3i$
4. The solution set to the equation $x^2 = -9$ is $\{\pm 3i\}$.
5. $2 - 3i - (4 - 2i) = -2 - i$
6. $i^4 = 1$
7. $(2 - i)(2 + i) = 5$
8. $i^3 = i$
9. $i^{48} = 1$
10. The equation $x^2 = k$ has two complex solutions for any real number k.

9.6   EXERCISES

Reading and Writing   *After reading this section, write out the answers to these questions. Use complete sentences.*

1. What are complex numbers?

2. What is an imaginary number?

3. What is the relationship among the real numbers, the imaginary numbers, and the complex numbers?

4. How do we add, subtract, and multiply complex numbers?

5. What is the conjugate of a complex number?

6. How do we divide complex numbers?

Find the indicated sums and differences of complex numbers. See Example 1.

7. $(2 + 3i) + (-4 + 5i)$
8. $(-1 + 6i) + (5 - 4i)$

9. $(2 - 3i) - (6 - 7i)$
10. $(2 - 3i) - (6 - 2i)$

11. $(-1 + i) + (-1 - i)$
12. $(-5 + i) + (-5 - i)$

13. $(-2 - 3i) - (6 - i)$
14. $(-6 + 4i) - (2 - i)$

Find the indicated products of complex numbers. See Example 2.

15. $3(2 + 5i)$

16. $4(1 - 3i)$

17. $2i(i - 5)$

18. $3i(2 - 6i)$

19. $-4i(3 - i)$

20. $-5i(2 + 3i)$

21. $(2 + 3i)(4 + 6i)$

22. $(2 + i)(3 + 4i)$

23. $(-1 + i)(2 - i)$

24. $(3 - 2i)(2 - 5i)$

25. $(-1 - 2i)(2 + i)$

26. $(1 - 3i)(1 + 3i)$

27. $(5 - 2i)(5 + 2i)$

28. $(4 + 3i)(4 + 3i)$

29. $(1 - i)(1 + i)$

30. $(2 + 6i)(2 - 6i)$

31. $(4 + 2i)(4 - 2i)$

32. $(4 - i)(4 + i)$

Find the indicated powers of complex numbers. See Example 3.

33. $(3i)^2$

34. $(5i)^2$

35. $(-5i)^2$

36. $(-9i)^2$

37. $(2i)^4$

38. $(-2i)^3$

39. i^9

40. i^{12}

Find the product of the given complex number and its conjugate. See Example 4.

41. $3 + 5i$

42. $3 + i$

43. $1 - 2i$

44. $4 - 6i$

45. $-2 + i$

46. $-3 - 2i$

47. $2 - i\sqrt{3}$

48. $\sqrt{5} - 4i$

Find each quotient. See Example 5.

49. $\dfrac{3}{4 + i}$

50. $\dfrac{6}{7 - 2i}$

51. $\dfrac{2 + i}{3 - 2i}$

52. $\dfrac{3 + 5i}{2 - i}$

53. $\dfrac{4 + 3i}{i}$

54. $\dfrac{5 - 6i}{3i}$

55. $\dfrac{2 + 6i}{2}$

56. $\dfrac{9 - 3i}{-6}$

Write each expression in the form $a + bi$, where a and b are real numbers. See Example 6.

57. $2 + \sqrt{-4}$

58. $3 + \sqrt{-9}$

59. $2\sqrt{-9} + 5$

60. $3\sqrt{-16} + 2$

61. $7 - \sqrt{-6}$

62. $\sqrt{-5} + 3$

63. $\sqrt{-8} + \sqrt{-18}$

64. $2\sqrt{-20} - \sqrt{-45}$

65. $\dfrac{2 + \sqrt{-12}}{2}$

66. $\dfrac{-6 - \sqrt{-18}}{3}$

67. $\dfrac{-4 - \sqrt{-24}}{4}$

68. $\dfrac{8 + \sqrt{-20}}{-4}$

Find the complex solutions to each equation. See Example 7.

69. $x^2 = -36$

70. $x^2 + 4 = 0$

71. $x^2 = -12$

72. $x^2 = -25$

73. $2x^2 + 5 = 0$

74. $3x^2 + 4 = 0$

75. $3x^2 + 6 = 0$

76. $x^2 + 1 = 0$

Write each expression in the form $a + bi$, where a and b are real numbers.

77. $(2 - 3i)(3 + 4i)$

78. $(2 - 3i)(2 + 3i)$

79. $(2 - 3i) + (3 + 4i)$

80. $(3 - 5i) - (2 - 7i)$

81. $\dfrac{2 - 3i}{3 + 4i}$

82. $\dfrac{-3i}{3 - 6i}$

83. $i(2 - 3i)$

84. $-3i(4i - 1)$

85. $(-3i)^2$

86. $(-2i)^6$

87. $\sqrt{-12} + \sqrt{-3}$

88. $\sqrt{-49} - \sqrt{-25}$

89. $(2 - 3i)^2$

90. $(5 + 3i)^2$

91. $\dfrac{-4 + \sqrt{-32}}{2}$

92. $\dfrac{-2 - \sqrt{-27}}{-6}$

GETTING MORE INVOLVED

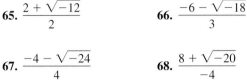

93. *Writing.* Explain why $2 - i$ is a solution to

$$x^2 - 4x + 5 = 0.$$

94. *Cooperative learning.* Work with a group to verify that $-1 + i\sqrt{3}$ and $-1 - i\sqrt{3}$ satisfy the equation

$$x^3 - 8 = 0.$$

In the complex number system there are three cube roots of 8. What are they?

95. *Discussion.* What is wrong with using the product rule for radicals to get

$$\sqrt{-4} \cdot \sqrt{-4} = \sqrt{(-4)(-4)} = \sqrt{16} = 4?$$

What is the correct product?

COLLABORATIVE ACTIVITIES

Laws of Falling Bodies

Jaki Sena is a private investigator working on a case. Her client is accused of killing a woman who had lived in the same apartment building as he did. She pulls over the file of notes from across her desk and looks through it yet again.

"Okay, so my client said he was on the roof of the building just before 4:00 A.M., dropping blocks over into the alleyway. He dropped the last block just as the clock tower began striking the hour. I wonder how high the building is," Jaki says to herself.

During her initial investigation of the crime scene Jaki had asked the building manager whether he knew the height of the five-story building. He had been very condescending and had told her nothing. Another tenant of the building, some sort of math instructor at the city college, had told her about an experiment he had done with shadows and a climbing rope. He had given her a drawing.

Jaki looked through the file until she found the following drawing.

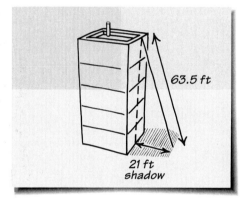

63.5 ft

21 ft
shadow

Grouping: 2 to 4 students per group
Topic: Applications of formulas with square roots

She pulled out a blank piece of paper, and using a calculator, she soon had a number.

"Now I can find out how long it took the block to fall!" Jaki exclaimed. "Okay, now the witness who saw my client on the top of the building also swears he saw the woman walking past the steps next door as the clock tower started to chime 4:00," Jaki said out loud as she got up from her desk and looked out the window. "I have no idea how fast the woman was walking. That may be a hard one to answer. But I do know the steps begin about $9\frac{1}{2}$ feet from where the block fell." Jaki turned back to the file of notes on her desk. She soon found another piece of paper with measurements on it.

"And I do know the woman was 5 feet tall!" Jaki exclaimed, taking out her calculator again.

1. Working in your groups, find the height of the building from the information on Jaki's paper. Round your answer to the nearest tenth of a foot.

2. Find out how many seconds it would take the block to fall from the top of the building to hit the woman on the head. Round your answer to the nearest tenth of a second. (*Hint:* You may need to look in other chapters of the book to find the formulas you need here.)

3. How fast would the woman have to walk to have been hit on the head by the block? Round your answer to the nearest tenth of a foot per second.

4. Could Jaki's client have "done it?" Give a reason for your answer.

WRAP-UP   CHAPTER 9

SUMMARY

Powers and Roots		**Examples**
*n*th roots	If $a = b^n$ for a positive integer n, then b is an nth root of a.	2 and -2 are fourth roots of 16.
Principal root	The positive even root of a positive number.	The principal fourth root of 16 is 2.
Radical notation	If n is a positive even integer and a is positive, then the symbol $\sqrt[n]{a}$ denotes the principal nth root of a.	$\sqrt[4]{16} = 2$ $\sqrt[4]{16} \neq -2$

If n is a positive odd integer, then the symbol $\sqrt[n]{a}$ denotes the nth root of a.

$\sqrt[3]{-8} = -2, \sqrt[3]{8} = 2$

If n is any positive integer, then $\sqrt[n]{0} = 0$.

$\sqrt[5]{0} = 0, \sqrt[6]{0} = 0$

Definition of $a^{1/n}$

If n is any positive integer, then $a^{1/n} = \sqrt[n]{a}$, provided that $\sqrt[n]{a}$ is a real number.

$8^{1/3} = \sqrt[3]{8} = 2$
$(-4)^{1/2}$ is not real.

Definition of $a^{m/n}$

If m and n are positive integers, then $a^{m/n} = (a^{1/n})^m$, provided that $a^{1/n}$ is a real number.

$8^{2/3} = (8^{1/3})^2 = 2^2 = 4$
$(-16)^{3/4}$ is not real.

Definition of $a^{-m/n}$

If m and n are positive integers and $a \neq 0$, then $a^{-m/n} = \frac{1}{a^{m/n}}$, provided that $a^{1/n}$ is a real number.

$8^{-2/3} = \frac{1}{8^{2/3}} = \frac{1}{4}$

Rules for Radicals

Examples

Product rule for radicals

Provided that all roots are real,
$$\sqrt[n]{ab} = \sqrt[n]{a} \cdot \sqrt[n]{b}.$$

$\sqrt{2} \cdot \sqrt{3} = \sqrt{6}$
$\sqrt{4x} = 2\sqrt{x}$

Quotient rule for radicals

Provided that all roots are real and $b \neq 0$,
$$\sqrt[n]{\frac{a}{b}} = \frac{\sqrt[n]{a}}{\sqrt[n]{b}}.$$

$\sqrt{\frac{5}{9}} = \frac{\sqrt{5}}{3}$

$\sqrt{10} \div \sqrt{5} = \sqrt{2}$

Simplified radical form for radicals of index n

A simplified radical of index n has
1. *no* perfect nth powers as factors of the radicand,
2. *no* fractions inside the radical, and

3. *no* radicals in the denominator.

$\sqrt{20} = \sqrt{4 \cdot 5} = 2\sqrt{5}$

$\sqrt{\frac{3}{2}} = \frac{\sqrt{3}}{\sqrt{2}}$

$\frac{\sqrt{3}}{\sqrt{2}} = \frac{\sqrt{3}}{\sqrt{2}} \cdot \frac{\sqrt{2}}{\sqrt{2}} = \frac{\sqrt{6}}{2}$

Rules for Rational Exponents

Examples

If a and b are nonzero real numbers and r and s are rational numbers, then the following rules hold, provided all expressions represent real numbers.

Product rule

$a^r \cdot a^s = a^{r+s}$

$3^{1/4} \cdot 3^{1/2} = 3^{3/4}$

Quotient rule

$\frac{a^r}{a^s} = a^{r-s}$

$\frac{x^{3/4}}{x^{1/4}} = x^{1/2}$

Power of a power rule

$(a^r)^s = a^{rs}$

$(2^{1/2})^{-1/2} = 2^{-1/4}$
$(x^{3/4})^4 = x^3$

Power of a product rule

$(ab)^r = a^r b^r$

$(a^2 b^6)^{1/2} = ab^3$

| Power of a quotient rule | $\left(\dfrac{a}{b}\right)^r = \dfrac{a^r}{b^r}$ | $\left(\dfrac{8}{x^6}\right)^{2/3} = \dfrac{4}{x^4}$ |

Equations

Examples

Equations with radicals and exponents	1. In raising each side of an equation to an even power, we can create an equation that gives extraneous solutions. We must check.	$\sqrt{x} = -3$ $x = 9$
	2. When applying the even-root property, remember that there is a positive and a negative root.	$x^2 = 36$ $x = \pm 6$
	3. For equations with rational exponents, raise each side to a positive or a negative power first, then apply the even- or odd-root property.	$x^{-2/3} = 4$ $(x^{-2/3})^{-3} = 4^{-3}$ $x^2 = \dfrac{1}{64}$ $x = \pm\dfrac{1}{8}$

| Distance formula | The distance between (x_1, y_1) and (x_2, y_2) is $$\sqrt{(x_2 - x_1)^2 + (y_2 - y_1)^2}.$$ | Distance between $(1, -2)$ and $(3, -4)$ is $\sqrt{2^2 + (-2)^2}$ or $2\sqrt{2}$. |

Complex Numbers

Examples

| Complex numbers | Numbers of form $a + bi$, where a and b are real numbers: $i = \sqrt{-1}, i^2 = -1$ | $2 + 3i$
 $-6i$
 $\sqrt{2} + i$ |

| Complex conjugates | Complex numbers of the form $a + bi$ and $a - bi$: Their product is the real number $a^2 + b^2$. | $(2 + 3i)(2 - 3i) = 2^2 + 3^2$
 $= 13$ |

| Complex number operations | Add, subtract, and multiply as algebraic expressions with i being the variable. Simplify using $i^2 = -1$. | $(2 + 5i) + (4 - 2i) = 6 + 3i$
 $(2 + 5i) - (4 - 2i) = -2 + 7i$
 $(2 + 5i)(4 - 2i) = 18 + 16i$ |
| | Divide complex numbers by multiplying numerator and denominator by the conjugate of the denominator. | $(2 + 5i) \div (4 - 2i)$
 $= \dfrac{(2 + 5i)(4 + 2i)}{(4 - 2i)(4 + 2i)}$
 $= \dfrac{-2 + 24i}{20} = -\dfrac{1}{10} + \dfrac{6}{5}i$ |

| Square root of a negative number | For any positive real number b, $\sqrt{-b} = i\sqrt{b}$. | $\sqrt{-9} = i\sqrt{9} = 3i$ |

| Imaginary solutions to equations | In the complex number system, $x^2 = k$ for any real k is equivalent to $x = \pm\sqrt{k}$. | $x^2 = -25$
 $x = \pm\sqrt{-25} = \pm 5i$ |

ENRICHING YOUR MATHEMATICAL WORD POWER

For each mathematical term, choose the correct meaning.

1. **nth root of a**
   a. a square root
   b. the root of a^n
   c. a number b such that $a^n = b$
   d. a number b such that $b^n = a$

2. **square of a**
   a. a number b such that $b^2 = a$
   b. a^2
   c. $|a|$
   d. $\sqrt{a}$

3. **cube root of a**
   a. a^3
   b. a number b such that $b^3 = a$
   c. $a/3$
   d. a number b such that $b = a^3$

4. **principal root**
   a. the main root
   b. the positive even root of a positive number
   c. the positive odd root of a negative number
   d. the negative odd root of a negative number

5. **odd root of a**
   a. the number b such that $b^n = a$, where a is an odd number
   b. the opposite of the even root of a
   c. the nth root of a
   d. the number b such that $b^n = a$, where n is an odd number

6. **index of a radical**
   a. the number n in $n\sqrt{a}$
   b. the number n in $\sqrt[n]{a}$
   c. the number n in a^n
   d. the number n in $\sqrt{a^n}$

7. **like radicals**
   a. radicals with the same index
   b. radicals with the same radicand
   c. radicals with the same radicand and the same index
   d. radicals with even indices

8. **integral exponent**
   a. an exponent that is an integer
   b. a positive exponent
   c. a rational exponent
   d. a fractional exponent

9. **rational exponent**
   a. an exponent that produces a rational number
   b. an integral exponent
   c. an exponent that is a real number
   d. an exponent that is a rational number

10. **radicand**
    a. the expression $\sqrt[n]{a}$
    b. the expression $\sqrt{a}$
    c. the number a in $\sqrt[n]{a}$
    d. the number n in $\sqrt[n]{a}$

11. **complex numbers**
    a. $a + bi$, where a and b are real
    b. irrational numbers
    c. imaginary numbers
    d. $\sqrt{-1}$

12. **imaginary unit**
    a. 1
    b. -1
    c. i
    d. $\sqrt{1}$

13. **imaginary number**
    a. $a + bi$, where a and b are real
    b. i
    c. a complex number
    d. a complex number in which $b \neq 0$

14. **complex conjugates**
    a. i and $\sqrt{-1}$
    b. $a + bi$ and $a - bi$
    c. $(a + b)(a - b)$
    d. i and -1

REVIEW EXERCISES

9.1 *Simplify each radical expression. Assume all variables represent positive real numbers.*

1. $\sqrt[5]{32}$
2. $\sqrt[3]{-27}$
3. $\sqrt[3]{1000}$
4. $\sqrt{100}$
5. $\sqrt{72}$
6. $\sqrt{48}$
7. $\sqrt{x^{12}}$
8. $\sqrt{a^{10}}$
9. $\sqrt[3]{x^6}$
10. $\sqrt[3]{a^9}$
11. $\sqrt{\dfrac{2}{5}}$
12. $\sqrt{\dfrac{1}{6}}$
13. $\sqrt[3]{\dfrac{2}{3}}$
14. $\sqrt[3]{\dfrac{1}{9}}$
15. $\dfrac{2}{\sqrt{3x}}$
16. $\dfrac{3}{\sqrt{2y}}$
17. $\dfrac{\sqrt{10y^3}}{\sqrt{6}}$
18. $\dfrac{\sqrt{5x^5}}{\sqrt{8}}$

19. $\dfrac{3}{\sqrt[3]{2a}}$
20. $\dfrac{a}{\sqrt[3]{a^2}}$
21. $\dfrac{5}{\sqrt[4]{3x^2}}$
22. $\dfrac{b}{\sqrt[4]{a^2b^3}}$

9.2 *Simplify the expressions involving rational exponents. Assume all variables represent positive real numbers. Write your answers with positive exponents.*

23. $(-27)^{-2/3}$
24. $-25^{3/2}$
25. $(2^6)^{1/3}$
26. $(5^2)^{1/2}$
27. $100^{-3/2}$
28. $1000^{-2/3}$

29. $\dfrac{3x^{-1/2}}{3^{-2}x^{-1}}$ **30.** $\dfrac{(x^2y^{-3}z)^{1/2}}{x^{1/2}yz^{-1/2}}$

31. $(a^{1/2}b)^3(ab^{1/4})^2$ **32.** $(t^{-1/2})^{-2}(t^{-2}v^2)$

33. $(x^{1/2}y^{1/4})(x^{1/4}y)$ **34.** $(a^{1/3}b^{1/6})^2(a^{1/3}b^{2/3})$

9.3 *Perform the operations and simplify. Assume the variables represent positive real numbers.*

35. $\sqrt{13} \cdot \sqrt{13}$ **36.** $\sqrt[3]{14} \cdot \sqrt[3]{14} \cdot \sqrt[3]{14}$

37. $\sqrt{27} + \sqrt{45} - \sqrt{75}$

38. $\sqrt{12} - \sqrt{50} + \sqrt{72}$

39. $\sqrt{\dfrac{1}{3}} + \sqrt{27}$ **40.** $\sqrt{\dfrac{1}{2}} - \sqrt{\dfrac{1}{8}}$

41. $3\sqrt{2}\,(5\sqrt{2} - 7\sqrt{3})$

42. $-2\sqrt{a}\,(\sqrt{a} - \sqrt{ab^6})$

43. $(2 - \sqrt{3})(3 + \sqrt{2})$

44. $(2\sqrt{x} - \sqrt{y})(\sqrt{x} + \sqrt{y})$

9.4 *Perform the operations and simplify.*

45. $5 \div \sqrt{2}$ **46.** $(10\sqrt{6}) \div (2\sqrt{2})$

47. $(\sqrt{3})^4$ **48.** $(-2\sqrt{x})^9$

49. $\dfrac{2 - \sqrt{8}}{2}$ **50.** $\dfrac{-3 - \sqrt{18}}{-6}$

51. $\dfrac{\sqrt{6}}{1 - \sqrt{3}}$

52. $\dfrac{\sqrt{15}}{2 + \sqrt{5}}$

53. $\dfrac{2\sqrt{3}}{3\sqrt{6} - \sqrt{12}}$

54. $\dfrac{-\sqrt{xy}}{3\sqrt{x} + \sqrt{xy}}$

55. $(2w\sqrt{2w^2})^6$ **56.** $(m\sqrt[4]{m^3})^8$

9.5 *Find all real solutions to each equation.*

57. $x^2 = 16$ **58.** $w^2 = 100$

59. $(a - 5)^2 = 4$ **60.** $(m - 7)^2 = 25$

61. $(a + 1)^2 = 5$

62. $(x + 5)^2 = 3$

63. $(m + 1)^2 = -8$ **64.** $(w + 4)^2 = 16$

65. $\sqrt{m - 1} = 3$ **66.** $3\sqrt{x + 5} = 12$

67. $\sqrt[3]{2x + 9} = 3$ **68.** $\sqrt[4]{2x - 1} = 2$

69. $w^{2/3} = 4$ **70.** $m^{-4/3} = 16$

71. $(m + 1)^{1/3} = 5$ **72.** $(w - 3)^{-2/3} = 4$

73. $\sqrt{x - 3} = \sqrt{x + 2} - 1$

74. $\sqrt{x^2 + 3x + 6} = 4$

75. $\sqrt{5x - x^2} = \sqrt{6}$

76. $\sqrt{x + 4} - 2\sqrt{x - 1} = -1$

77. $\sqrt{x + 7} - 2\sqrt{x} = -2$

78. $\sqrt{x} - \sqrt{x - 1} = 1$

79. $2\sqrt{x} - \sqrt{x - 3} = 3$

80. $1 + \sqrt{x + 7} = \sqrt{2x + 7}$

9.6 *Perform the indicated operations. Write answers in the form a + bi.*

81. $(2 - 3i)(-5 + 5i)$

82. $(2 + i)(5 - 2i)$

83. $(2 + i) + (5 - 4i)$

84. $(2 + i) + (3 - 6i)$

85. $(1 - i) - (2 - 3i)$

86. $(3 - 2i) - (1 - i)$

87. $\dfrac{6 + 3i}{3}$ **88.** $\dfrac{8 + 12i}{4}$

89. $\dfrac{4 - \sqrt{-12}}{2}$ **90.** $\dfrac{6 + \sqrt{-18}}{3}$

91. $\dfrac{2 - 3i}{4 + i}$ **92.** $\dfrac{3 + i}{2 - 3i}$

Find the imaginary solutions to each equation.

93. $x^2 + 100 = 0$ **94.** $25a^2 + 3 = 0$

95. $2b^2 + 9 = 0$ **96.** $3y^2 + 8 = 0$

MISCELLANEOUS

Determine whether each equation is true or false and explain your answer. An equation involving variables should be marked true only if it is an identity. Do not use a calculator.

97. $2^3 \cdot 3^2 = 6^5$ **98.** $16^{1/4} = 4^{1/2}$

99. $(\sqrt{2})^3 = 2\sqrt{2}$ **100.** $\sqrt[3]{9} = 3$

101. $8^{200} \cdot 8^{200} = 64^{200}$

102. $\sqrt{295} \cdot \sqrt{295} = 295$

103. $4^{1/2} = \sqrt{2}$ **104.** $\sqrt{a^2} = |a|$

105. $5^2 \cdot 5^2 = 25^4$ **106.** $\sqrt{6} \div \sqrt{2} = \sqrt{3}$

107. $\sqrt{w^{10}} = w^5$ **108.** $\sqrt{a^{16}} = a^4$

109. $\sqrt{x^6} = x^3$ **110.** $\sqrt[6]{16} = \sqrt[3]{4}$

111. $\sqrt{x^8} = x^4$ **112.** $\sqrt[9]{2^6} = 2^{2/3}$

113. $\sqrt{16} = 2$ **114.** $2^{1/2} \cdot 2^{1/4} = 2^{3/4}$

115. $2^{600} = 4^{300}$ **116.** $\sqrt{2} \cdot \sqrt[4]{2} = \sqrt[6]{2}$

117. $\dfrac{2 + \sqrt{6}}{2} = 1 + \sqrt{6}$

118. $\dfrac{4 + 2\sqrt{3}}{2} = 2 + \sqrt{3}$

119. $\sqrt{\dfrac{4}{6}} = \dfrac{2}{3}$ **120.** $8^{200} \cdot 8^{200} = 8^{400}$

121. $81^{2/4} = 81^{1/2}$ **122.** $(-64)^{2/6} = (-64)^{1/3}$

123. $(a^4b^2)^{1/2} = |a^2b|$ **124.** $\left(\dfrac{a^2}{b^6}\right)^{1/2} = \dfrac{|a|}{b^3}$

In Exercises 125–138, solve each problem.

125. Find the distance between $(-4, 6)$ and $(2, -8)$.

126. Find the distance between $(-3, -5)$ and $(5, -7)$.

127. *Falling objects.* If we neglect air resistance, the number of feet s that an object falls from rest during t seconds is given by the equation $s = 16t^2$. How long would it take the landing gear of an airplane to reach the earth if it fell off the airplane at 12,000 feet?

128. *Timber.* Anne is pulling on a 60-foot rope attached to the top of a 48-foot tree while Walter is cutting the tree at its base. How far from the base of the tree is Anne standing?

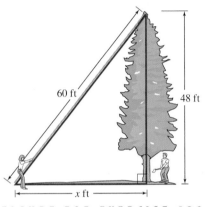

FIGURE FOR EXERCISE 128

129. *Guy wire.* If a guy wire of length 40 feet is attached to an antenna at a height of 30 feet, then how far from the base of the antenna is the wire attached to the ground?

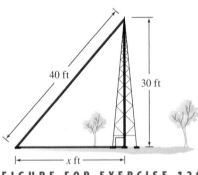

FIGURE FOR EXERCISE 129

130. *Touchdown.* Suppose at the kickoff of a football game, the receiver catches the football at the left side of the goal line and runs for a touchdown diagonally across the field. How many yards would he run? (A football field is 100 yards long and 160 feet wide.)

131. *Long guy wires.* The manufacturer of an antenna recommends that guy wires from the top of the antenna to the ground be attached to the ground at a distance from the base equal to the height of the antenna. How long would the guy wires be for a 200-foot antenna?

132. *Height of a post.* Betty observed that the lamp post in front of her house casts a shadow of length 8 feet when the angle of inclination of the sun is 60 degrees. How tall is the lamp post? (In a 30-60-90 right triangle, the side opposite 30 is one-half the length of the hypotenuse.)

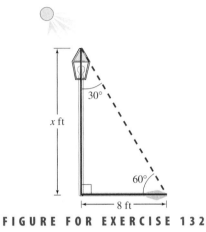

FIGURE FOR EXERCISE 132

133. *Manufacturing a box.* A cubic box has a volume of 40 cubic feet. The amount of recycled cardboard that it takes to make the six-sided box is 10% larger than the surface area of the box. Find the exact amount of recycled cardboard used in manufacturing the box.

134. *Shipping parts.* A cubic box with a volume of 32 cubic feet is to be used to ship some machine parts. All of the parts are small except for a long, straight steel connecting rod. What is the maximum length of a connecting rod that will fit into this box?

135. *Rising costs of health care.* Total annual expenditures on health care in the United States grew from $700 billion in 1990 to $1035 billion in 1996 (Statistical Abstract of the United States, www.census.gov). Find the average annual rate of growth r for that period by solving

$$1035 = 700(1 + r)^6.$$

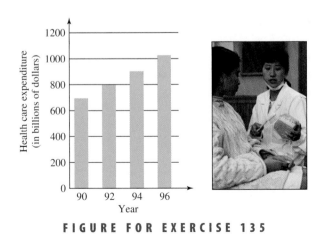

FIGURE FOR EXERCISE 135

136. *Population growth rate.* The formula $P = P_0(1 + r)^n$ gives the population P at the end of an n-year time period, where P_0 is the initial population and r is the average annual growth rate. The U.S. population grew from 248.7 million in 1990 to 270.1 million in 1998 (U.S. Census Bureau). Find the average annual rate of growth for the U.S. population for that period.

137. *Landing speed.* Aircraft engineers determine the proper landing speed V (in feet per second) for an airplane from the formula

$$V = \sqrt{\frac{841L}{CS}},$$

where L is the gross weight of the aircraft in pounds, C is the coefficient of lift, and S is the wing surface area in square feet. Rewrite the formula so that the expression on the right-hand side is in simplified radical form.

138. *Spillway capacity.* Civil engineers use the formula

$$Q = 3.32LH^{3/2}$$

to find the maximum discharge that the dam (a broad-crested weir) shown in the figure can pass before the water breaches its abutments (*Standard Handbook for Civil Engineers,* 1968). In the formula Q is the discharge in cubic feet per second, L is the length of the spillway in feet, and H is the depth of the spillway. Find Q given that $L = 60$ feet and $H = 5$ feet. Find H given that $Q = 3000$ cubic feet per second and $L = 70$ feet.

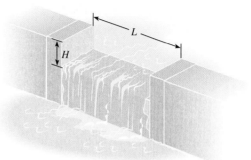

FIGURE FOR EXERCISE 138

CHAPTER 9 TEST

Simplify each expression. Assume all variables represent positive numbers.

1. $8^{2/3}$

2. $4^{-3/2}$

3. $\sqrt{21} \div \sqrt{7}$

4. $2\sqrt{5} \cdot 3\sqrt{5}$

5. $\sqrt{20} + \sqrt{5}$

6. $\sqrt{5} + \dfrac{1}{\sqrt{5}}$

7. $2^{1/2} \cdot 2^{1/2}$

8. $\sqrt{72}$

9. $\sqrt{\dfrac{5}{12}}$

10. $\dfrac{6 + \sqrt{18}}{6}$

11. $(2\sqrt{3} + 1)(\sqrt{3} - 2)$

12. $\sqrt[4]{32a^5y^8}$

13. $\dfrac{1}{\sqrt[3]{2x^2}}$

14. $\sqrt{\dfrac{8a^9}{b^3}}$

15. $\sqrt[3]{-27x^9}$

16. $\sqrt{20m^3}$

17. $x^{1/2} \cdot x^{1/4}$

18. $(9y^4x^{1/2})^{1/2}$

19. $\sqrt[3]{40x^7}$

20. $(4 + \sqrt{3})^2$

Rationalize the denominator and simplify.

21. $\dfrac{2}{5 - \sqrt{3}}$

22. $\dfrac{\sqrt{6}}{4\sqrt{3} + \sqrt{2}}$

Write each expression in the form $a + bi$.

23. $(3 - 2i)(4 + 5i)$

24. $i^4 - i^5$

25. $\dfrac{3 - i}{1 + 2i}$

26. $\dfrac{-6 + \sqrt{-12}}{8}$

Find all real or imaginary solutions to each equation.

27. $(x - 2)^2 = 49$

28. $2\sqrt{x + 4} = 3$

29. $w^{2/3} = 4$

30. $9y^2 + 16 = 0$

31. $\sqrt{2x^2 + x - 12} = x$

32. $\sqrt{x - 1} + \sqrt{x + 4} = 5$

Show a complete solution to each problem.

33. Find the distance between $(-1, 4)$ and $(1, 6)$.

34. Find the exact length of the side of a square whose diagonal is 3 feet.

35. Two positive numbers differ by 11, and their square roots differ by 1. Find the numbers.

36. If the perimeter of a rectangle is 20 feet and the diagonal is $2\sqrt{13}$ feet, then what are the length and width?

37. The average radius R of the orbit of a planet around the sun is determined by $R = T^{2/3}$, where T is the number of years for one orbit and R is measured in astronomical units (AU). If it takes Pluto 248.530 years to make one orbit of the sun, then what is the average radius of the orbit of Pluto? If the average radius of the orbit of Neptune is 30.08 AU, then how many years does it take Neptune to complete one orbit of the sun?

Find all real solutions to each equation or inequality. For the inequalities, also sketch the graph of the solution set.

1. $3(x - 2) + 5 = 7 - 4(x + 3)$

2. $\sqrt{6x + 7} = 4$

3. $|2x + 5| > 1$

4. $8x^3 - 27 = 0$

5. $2x - 3 > 3x - 4$

6. $\sqrt{2x - 3} - \sqrt{3x + 4} = 0$

7. $\dfrac{w}{3} + \dfrac{w - 4}{2} = \dfrac{11}{2}$

8. $2(x + 7) - 4 = x - (10 - x)$

9. $(x + 7)^2 = 25$ **10.** $a^{-1/2} = 4$

11. $x - 3 > 2$ or $x < 2x + 6$

12. $a^{-2/3} = 16$ **13.** $3x^2 - 1 = 0$

14. $5 - 2(x - 2) = 3x - 5(x - 2) - 1$

15. $|3x - 4| < 5$

16. $3x - 1 = 0$ **17.** $\sqrt{y - 1} = 9$

18. $|5(x - 2) + 1| = 3$

19. $0.06x - 0.04(x - 20) = 2.8$

20. $|3x - 1| > -2$ **21.** $\dfrac{3\sqrt{2}}{x} = \dfrac{\sqrt{3}}{4\sqrt{5}}$

22. $\dfrac{\sqrt{x} - 4}{x} = \dfrac{1}{\sqrt{x} + 5}$

23. $\dfrac{3\sqrt{2} + 4}{\sqrt{2}} = \dfrac{x\sqrt{18}}{3\sqrt{2} + 2}$

24. $\dfrac{x}{2\sqrt{5} - \sqrt{2}} = \dfrac{2\sqrt{5} + \sqrt{2}}{x}$

25. $\dfrac{\sqrt{2x} - 5}{x} = \dfrac{-3}{\sqrt{2x} + 5}$

26. $\dfrac{\sqrt{6} + 2}{x} = \dfrac{2}{\sqrt{6} + 4}$

27. $\dfrac{x - 1}{\sqrt{6}} = \dfrac{\sqrt{6}}{x}$ **28.** $\dfrac{x + 3}{\sqrt{10}} = \dfrac{\sqrt{10}}{x}$

29. $\dfrac{1}{x} - \dfrac{1}{x - 1} = -\dfrac{1}{6}$

30. $\dfrac{1}{x^2 - 2x} + \dfrac{1}{x} = \dfrac{2}{3}$

The expression $\dfrac{-b + \sqrt{b^2 - 4ac}}{2a}$ will be used in Chapter 10 to solve quadratic equations. Evaluate it for the given values of a, b, and c.

31. $a = 1, b = 2, c = -15$ **32.** $a = 1, b = 8, c = 12$

33. $a = 2, b = 5, c = -3$ **34.** $a = 6, b = 7, c = -3$

Solve each problem.

35. *Popping corn.* The results of an experiment by D.D. Metzger to determine the relationship between the moisture content of popcorn and the volume of popped corn are shown in the figure (*Cereal Chemistry,* 1989). The formula $v = -94.8 + 21.4x - 0.761x^2$ can be used to model the data shown in the graph. In the formula v is the number of cubic centimeters (cm³) of popped corn that result from popping 1 gram of corn with moisture content $x\%$ in a hot-air popper. Use the formula to find the volume that results when 1 gram of corn with a moisture content of 11% is popped.

FIGURE FOR EXERCISES 35 AND 36

36. *Maximizing the volume of popped corn.* Use the graph to estimate the moisture content that will produce the maximum amount of popped corn. What is the maximum possible volume for popping 1 gram of corn in a hot-air popper?

Quadratic Equations, Functions, and Inequalities

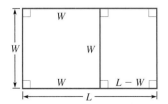

Is it possible to measure beauty? For thousands of years artists and philosophers have been challenged to answer this question. The seventeenth-century philosopher John Locke said, "Beauty consists of a certain composition of color and figure causing delight in the beholder." Over the centuries many architects, sculptors, and painters have searched for beauty in their work by exploring numerical patterns in various art forms.

Today many artists and architects still use the concepts of beauty given to us by the ancient Greeks. One principle, called the Golden Rectangle, concerns the most pleasing proportions of a rectangle. The Golden Rectangle appears in nature as well as in many cultures. Examples of it can be seen in Leonardo da Vinci's *Proportions of the Human Figure* as well as in Indonesian temples and Chinese pagodas. Perhaps one of the best-known examples of the Golden Rectangle is in the façade and floor plan of the Parthenon, built in Athens in the fifth century B.C. In Exercise 63 of Section 10.4 we will see that the principle of the Golden Rectangle is based on a proportion that we can solve using the quadratic formula.

10.1 FACTORING AND COMPLETING THE SQUARE

Factoring and the even-root property were used to solve quadratic equations in Chapters 6, 7, and 9. In this section we first review those methods. Then you will learn the method of completing the square, which can be used to solve any quadratic equation.

Review of Factoring

A quadratic equation is a second-degree polynomial equation of the form

$$ax^2 + bx + c = 0,$$

where a, b, and c are real numbers with $a \neq 0$. If the second-degree polynomial on the left-hand side can be factored, then we can solve the equation by breaking it into two first-degree polynomial equations (linear equations) using the following strategy.

Strategy for Solving Quadratic Equations by Factoring

1. Write the equation with 0 on the right-hand side.
2. Factor the left-hand side.
3. Use the zero factor property to set each factor equal to zero.
4. Solve the simpler equations.
5. Check the answers in the original equation.

EXAMPLE 1

Solving a quadratic equation by factoring

Solve $3x^2 - 4x = 15$ by factoring.

Solution

Subtract 15 from each side to get 0 on the right-hand side:

$$3x^2 - 4x - 15 = 0$$
$$(3x + 5)(x - 3) = 0 \quad \text{Factor the left-hand side.}$$
$$3x + 5 = 0 \quad \text{or} \quad x - 3 = 0 \quad \text{Zero factor property}$$
$$3x = -5 \quad \text{or} \quad x = 3$$
$$x = -\frac{5}{3}$$

The solution set is $\left\{-\frac{5}{3}, 3\right\}$. Check the solutions in the original equation. ∎

helpful hint

After you have factored the quadratic polynomial, use FOIL to check that you have factored correctly before proceeding to the next step.

Review of the Even-Root Property

In Chapter 9 we solved quadratic equations by using the even-root property.

EXAMPLE 2

Solving a quadratic equation by the even-root property

Solve $(a - 1)^2 = 9$.

Solution

By the even-root property $x^2 = k$ is equivalent to $x = \pm\sqrt{k}$.

$$(a - 1)^2 = 9$$

$$a - 1 = \pm\sqrt{9} \quad \text{Even-root property}$$

$a - 1 = 3$	or	$a - 1 = -3$
$a = 4$	or	$a = -2$

Check these solutions in the original equation. The solution set is $\{-2, 4\}$. ■

Completing the Square

We cannot solve every quadratic by factoring because not all quadratic polynomials can be factored. However, we can write any quadratic equation in the form of Example 2 and then apply the even-root property to solve it. This method is called **completing the square.**

The essential part of completing the square is to recognize a perfect square trinomial when given its first two terms. For example, if we are given $x^2 + 6x$, how do we recognize that these are the first two terms of the perfect square trinomial $x^2 + 6x + 9$? To answer this question, recall that $x^2 + 6x + 9$ is a perfect square trinomial because it is the square of the binomial $x + 3$:

$$(x + 3)^2 = x^2 + 2 \cdot 3x + 3^2 = x^2 + 6x + 9$$

Notice that the 6 comes from multiplying 3 by 2 and the 9 comes from squaring the 3. So to find the missing 9 in $x^2 + 6x$, divide 6 by 2 to get 3, then square 3 to get 9. This procedure can be used to find the last term in any perfect square trinomial in which the coefficient of x^2 is 1.

Rule for Finding the Last Term

The last term of a perfect square trinomial is the square of one-half of the coefficient of the middle term. In symbols, the perfect square trinomial whose first two terms are $x^2 + bx$ is $x^2 + bx + \left(\dfrac{b}{2}\right)^2$.

E X A M P L E   3   **Finding the last term**

Find the perfect square trinomial whose first two terms are given.

a) $x^2 + 8x$ b) $x^2 - 5x$

c) $x^2 + \dfrac{4}{7}x$ d) $x^2 - \dfrac{3}{2}x$

Solution

a) One-half of 8 is 4, and 4 squared is 16. So the perfect square trinomial is

$$x^2 + 8x + 16.$$

b) One-half of -5 is $-\dfrac{5}{2}$, and $-\dfrac{5}{2}$ squared is $\dfrac{25}{4}$. So the perfect square trinomial is
$x^2 - 5x + \dfrac{25}{4}$.

c) Since $\frac{1}{2} \cdot \frac{4}{7} = \frac{2}{7}$ and $\frac{2}{7}$ squared is $\frac{4}{49}$, the perfect square trinomial is

$$x^2 + \frac{4}{7}x + \frac{4}{49}.$$

d) Since $\frac{1}{2}\left(-\frac{3}{2}\right) = -\frac{3}{4}$ and $\left(-\frac{3}{4}\right)^2 = \frac{9}{16}$, the perfect square trinomial is

$$x^2 - \frac{3}{2}x + \frac{9}{16}.$$

Another essential step in completing the square is to write the perfect square trinomial as the square of a binomial. Recall that

$$a^2 + 2ab + b^2 = (a + b)^2$$

and

$$a^2 - 2ab + b^2 = (a - b)^2.$$

E X A M P L E   4

Factoring perfect square trinomials

Factor each trinomial.

a) $x^2 + 12x + 36$

b) $y^2 - 7y + \dfrac{49}{4}$

c) $z^2 - \dfrac{4}{3}z + \dfrac{4}{9}$

Solution

a) The trinomial $x^2 + 12x + 36$ is of the form $a^2 + 2ab + b^2$ with $a = x$ and $b = 6$. So

$$x^2 + 12x + 36 = (x + 6)^2.$$

Check by squaring $x + 6$.

b) The trinomial $y^2 - 7y + \frac{49}{4}$ is of the form $a^2 - 2ab + b^2$ with $a = y$ and $b = \frac{7}{2}$. So

$$y^2 - 7y + \frac{49}{4} = \left(y - \frac{7}{2}\right)^2.$$

Check by squaring $y - \frac{7}{2}$.

c) The trinomial $z^2 - \frac{4}{3}z + \frac{4}{9}$ is of the form $a^2 - 2ab + b^2$ with $a = z$ and $b = -\frac{2}{3}$. So

$$z^2 - \frac{4}{3}z + \frac{4}{9} = \left(z - \frac{2}{3}\right)^2.$$

In the next example we use the skills that we practiced in Examples 2, 3, and 4 to solve the quadratic equation $ax^2 + bx + c = 0$ with $a = 1$ by the method of completing the square.

E X A M P L E   5

Completing the square with $a = 1$

Solve $x^2 + 6x + 5 = 0$ by completing the square.

study tip

Most instructors believe that what they do in class is important. If you miss class, then you miss what is important to your instructor and what is most likely to appear on the test.

Solution

The perfect square trinomial whose first two terms are $x^2 + 6x$ is

$$x^2 + 6x + 9.$$

So we move 5 to the right-hand side of the equation, then add 9 to each side:

$$x^2 + 6x \qquad = -5 \qquad \text{Subtract 5 from each side.}$$

$$x^2 + 6x + 9 = -5 + 9 \qquad \begin{array}{l}\text{Add 9 to each side to get} \\ \text{a perfect square trinomial.}\end{array}$$

$$(x + 3)^2 = 4 \qquad \text{Factor the left-hand side.}$$

$$x + 3 = \pm\sqrt{4} \qquad \text{Even-root property}$$

$$x + 3 = 2 \qquad \text{or} \qquad x + 3 = -2$$

$$x = -1 \qquad \text{or} \qquad x = -5$$

Check in the original equation:

$$(-1)^2 + 6(-1) + 5 = 0$$

and

$$(-5)^2 + 6(-5) + 5 = 0$$

The solution set is $\{-1, -5\}$.

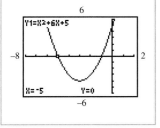

CAUTION All of the perfect square trinomials that we have used so far had a leading coefficient of 1. If $a \neq 1$, then we must divide each side of the equation by a to get an equation with a leading coefficient of 1.

The strategy for solving a quadratic equation by completing the square is stated in the following box.

Strategy for Solving Quadratic Equations by Completing the Square

1. The coefficient of x^2 must be 1.
2. Get only the x^2 and the x terms on the left-hand side.
3. Add to each side the square of $\frac{1}{2}$ the coefficient of x.
4. Factor the left-hand side as the square of a binomial.
5. Apply the even-root property.
6. Solve for x.
7. Simplify.

In our procedure for completing the square the coefficient of x^2 must be 1. We can solve $ax^2 + bx + c = 0$ with $a \neq 1$ by completing the square if we first divide each side of the equation by a.

EXAMPLE 6

Completing the square with $a \neq 1$

Solve $2x^2 + 3x - 2 = 0$ by completing the square.

Solution

For completing the square, the coefficient of x^2 must be 1. So we first divide each side of the equation by 2:

$$\frac{2x^2 + 3x - 2}{2} = \frac{0}{2} \qquad \text{Divide each side by 2.}$$

$$x^2 + \frac{3}{2}x - 1 = 0 \qquad \text{Simplify.}$$

$$x^2 + \frac{3}{2}x = 1 \qquad \begin{array}{l}\text{Get only } x^2 \text{ and } x \text{ terms on the}\\ \text{left-hand side.}\end{array}$$

$$x^2 + \frac{3}{2}x + \frac{9}{16} = 1 + \frac{9}{16} \qquad \text{One-half of } \tfrac{3}{2} \text{ is } \tfrac{3}{4}, \text{ and } \left(\tfrac{3}{4}\right)^2 = \tfrac{9}{16}.$$

$$\left(x + \frac{3}{4}\right)^2 = \frac{25}{16} \qquad \text{Factor the left-hand side.}$$

$$x + \frac{3}{4} = \pm\sqrt{\frac{25}{16}} \qquad \text{Even-root property}$$

$$x + \frac{3}{4} = \frac{5}{4} \qquad \text{or} \qquad x + \frac{3}{4} = -\frac{5}{4}$$

$$x = \frac{2}{4} = \frac{1}{2} \qquad \text{or} \qquad x = -\frac{8}{4} = -2$$

Check these values in the original equation. The solution set is $\left\{-2, \frac{1}{2}\right\}$. ■

In Examples 5 and 6 the solutions were rational numbers, and the equations could have been solved by factoring. In the next example the solutions are irrational numbers, and factoring will not work.

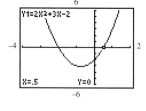

E X A M P L E 7

A quadratic equation with irrational solutions

Solve $x^2 - 3x - 6 = 0$ by completing the square.

Solution

Because $a = 1$, we first get the x^2 and x terms on the left-hand side:

$$x^2 - 3x - 6 = 0$$

$$x^2 - 3x = 6 \qquad \text{Add 6 to each side.}$$

$$x^2 - 3x + \frac{9}{4} = 6 + \frac{9}{4} \qquad \text{One-half of } -3 \text{ is } -\tfrac{3}{2}, \text{ and } \left(-\tfrac{3}{2}\right)^2 = \tfrac{9}{4}.$$

$$\left(x - \frac{3}{2}\right)^2 = \frac{33}{4} \qquad 6 + \tfrac{9}{4} = \tfrac{24}{4} + \tfrac{9}{4} = \tfrac{33}{4}$$

$$x - \frac{3}{2} = \pm\sqrt{\frac{33}{4}} \qquad \text{Even-root property}$$

$$x = \frac{3}{2} \pm \frac{\sqrt{33}}{2} \qquad \text{Add } \tfrac{3}{2} \text{ to each side.}$$

$$x = \frac{3 \pm \sqrt{33}}{2}$$

The solution set is $\left\{\dfrac{3 + \sqrt{33}}{2}, \dfrac{3 - \sqrt{33}}{2}\right\}$. ■

Miscellaneous Equations

The next two examples show equations that are not originally in the form of quadratic equations. However, after simplifying these equations, we get quadratic equations. Even though completing the square can be used on any quadratic equation, factoring and the square root property are usually easier and we can use them when applicable. In the next examples we will use the most appropriate method.

EXAMPLE 8

An equation containing a radical

Solve $x + 3 = \sqrt{153 - x}$.

Solution

Square both sides of the equation to eliminate the radical:

$$x + 3 = \sqrt{153 - x} \quad \text{The original equation}$$
$$(x + 3)^2 = (\sqrt{153 - x})^2 \quad \text{Square each side.}$$
$$x^2 + 6x + 9 = 153 - x \quad \text{Simplify.}$$
$$x^2 + 7x - 144 = 0$$
$$(x - 9)(x + 16) = 0 \quad \text{Factor.}$$
$$x - 9 = 0 \quad \text{or} \quad x + 16 = 0 \quad \text{Zero factor property}$$
$$x = 9 \quad \text{or} \quad x = -16$$

Because we squared each side of the original equation, we must check for extraneous roots. Let $x = 9$ in the original equation:

$$9 + 3 = \sqrt{153 - 9}$$
$$12 = \sqrt{144} \quad \text{Correct}$$

Let $x = -16$ in the original equation:

$$-16 + 3 = \sqrt{153 - (-16)}$$
$$-13 = \sqrt{169} \quad \text{Incorrect because } \sqrt{169} = 13$$

Because -16 is an extraneous root, the solution set is $\{9\}$.

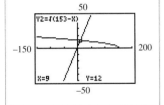

calculator
close-up

You can provide graphical support for the solution to Example 8 by graphing

$$y_1 = x + 3$$

and

$$y_2 = \sqrt{53 - x}.$$

It appears that the only point of intersection occurs when $x = 9$.

EXAMPLE 9

An equation containing rational expressions

Solve $\dfrac{1}{x} + \dfrac{3}{x - 2} = \dfrac{5}{8}$.

Solution

The least common denominator (LCD) for x, $x - 2$, and 8 is $8x(x - 2)$.

$$\frac{1}{x} + \frac{3}{x - 2} = \frac{5}{8}$$
$$8x(x - 2)\frac{1}{x} + 8x(x - 2)\frac{3}{x - 2} = 8x(x - 2)\frac{5}{8} \quad \text{Multiply each side by the LCD.}$$
$$8x - 16 + 24x = 5x^2 - 10x$$
$$32x - 16 = 5x^2 - 10x$$
$$-5x^2 + 42x - 16 = 0$$
$$5x^2 - 42x + 16 = 0 \quad \begin{array}{l}\text{Multiply each side by } -1 \\ \text{for easier factoring.}\end{array}$$
$$(5x - 2)(x - 8) = 0 \quad \text{Factor.}$$
$$5x - 2 = 0 \quad \text{or} \quad x - 8 = 0$$
$$x = \frac{2}{5} \quad \text{or} \quad x = 8$$

Check these values in the original equation. The solution set is $\left\{\frac{2}{5}, 8\right\}$.

Imaginary Solutions

In Chapter 9 we found imaginary solutions to quadratic equations using the even-root property. We can get imaginary solutions also by completing the square.

E X A M P L E 10

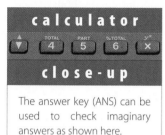

calculator

close-up

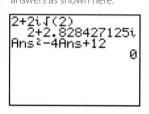

The answer key (ANS) can be used to check imaginary answers as shown here.

An equation with imaginary solutions

Find the complex solutions to $x^2 - 4x + 12 = 0$.

Solution

Because the quadratic polynomial cannot be factored, we solve the equation by completing the square.

$$
\begin{aligned}
x^2 - 4x + 12 &= 0 & &\text{The original equation} \\
x^2 - 4x \quad &= -12 & &\text{Subtract 12 from each side.} \\
x^2 - 4x + 4 &= -12 + 4 & &\text{One-half of } -4 \text{ is } -2, \text{ and } (-2)^2 = 4. \\
(x - 2)^2 &= -8 \\
x - 2 &= \pm\sqrt{-8} & &\text{Even-root property} \\
x &= 2 \pm i\sqrt{8} \\
&= 2 \pm 2i\sqrt{2}
\end{aligned}
$$

Check these values in the original equation. The solution set is $\{2 \pm 2i\sqrt{2}\}$. ■

WARM-UPS

True or false? Explain your answer.

1. Completing the square means drawing the fourth side.
2. The equation $(x - 3)^2 = 12$ is equivalent to $x - 3 = 2\sqrt{3}$.
3. Every quadratic equation can be solved by factoring.
4. The trinomial $x^2 + \frac{4}{3}x + \frac{16}{9}$ is a perfect square trinomial.
5. Every quadratic equation can be solved by completing the square.
6. To complete the square for $2x^2 + 6x = 4$, add 9 to each side.
7. $(2x - 3)(3x + 5) = 0$ is equivalent to $x = \frac{3}{2}$ or $x = \frac{5}{3}$.
8. In completing the square for $x^2 - 3x = 4$, add $\frac{9}{4}$ to each side.
9. The equation $x^2 = -8$ is equivalent to $x = \pm 2\sqrt{2}$.
10. All quadratic equations have two distinct complex solutions.

10.1 EXERCISES

Reading and Writing *After reading this section, write out the answers to these questions. Use complete sentences.*

1. What are the three methods discussed in this section for solving a quadratic equation?

2. Which quadratic equations can be solved by the even-root property?

3. How do you find the last term for a perfect square trinomial when completing the square?

4. How do you complete the square when the leading coefficient is not 1?

Solve by factoring. See Example 1.

5. $x^2 - x - 6 = 0$

6. $x^2 + 6x + 8 = 0$

7. $a^2 + 2a = 15$

8. $w^2 - 2w = 15$

9. $2x^2 - x - 3 = 0$

10. $6x^2 - x - 15 = 0$

11. $y^2 + 14y + 49 = 0$

12. $a^2 - 6a + 9 = 0$

13. $a^2 - 16 = 0$

14. $4w^2 - 25 = 0$

Use the even-root property to solve each equation. See Example 2.

15. $x^2 = 81$

16. $x^2 = \dfrac{9}{4}$

17. $x^2 = \dfrac{16}{9}$

18. $a^2 = 32$

19. $(x - 3)^2 = 16$
20. $(x + 5)^2 = 4$
21. $(z + 1)^2 = 5$
22. $(a - 2)^2 = 8$

23. $\left(w - \dfrac{3}{2}\right)^2 = \dfrac{7}{4}$

24. $\left(w + \dfrac{2}{3}\right)^2 = \dfrac{5}{9}$

Find the perfect square trinomial whose first two terms are given. See Example 3.

25. $x^2 + 2x$

26. $m^2 + 14m$

27. $x^2 - 3x$

28. $w^2 - 5w$

29. $y^2 + \dfrac{1}{4}y$

30. $z^2 + \dfrac{3}{2}z$

31. $x^2 + \dfrac{2}{3}x$

32. $p^2 + \dfrac{6}{5}p$

Factor each perfect square trinomial. See Example 4.
33. $x^2 + 8x + 16$

34. $x^2 - 10x + 25$

35. $y^2 - 5y + \dfrac{25}{4}$

36. $w^2 + w + \dfrac{1}{4}$

37. $z^2 - \dfrac{4}{7}z + \dfrac{4}{49}$

38. $m^2 - \dfrac{6}{5}m + \dfrac{9}{25}$

39. $t^2 + \dfrac{3}{5}t + \dfrac{9}{100}$

40. $h^2 + \dfrac{3}{2}h + \dfrac{9}{16}$

Solve by completing the square. See Examples 5–7. Use your calculator to check.
41. $x^2 - 2x - 15 = 0$
42. $x^2 - 6x - 7 = 0$
43. $x^2 + 8x = 20$
44. $x^2 + 10x = -9$
45. $2x^2 - 4x = 70$
46. $3x^2 - 6x = 24$
47. $w^2 - w - 20 = 0$
48. $y^2 - 3y - 10 = 0$
49. $q^2 + 5q = 14$
50. $z^2 + z = 2$

51. $2h^2 - h - 3 = 0$

52. $2m^2 - m - 15 = 0$

53. $x^2 + 4x = 6$
54. $x^2 + 6x - 8 = 0$
55. $x^2 + 8x - 4 = 0$
56. $x^2 + 10x - 3 = 0$

57. $2x^2 + 3x - 4 = 0$

58. $2x^2 + 5x - 1 = 0$

Solve each equation by an appropriate method. See Examples 8 and 9.
59. $\sqrt{2x + 1} = x - 1$

60. $\sqrt{2x - 4} = x - 14$

61. $w = \dfrac{\sqrt{w + 1}}{2}$

62. $y - 1 = \dfrac{\sqrt{y + 1}}{2}$

63. $\dfrac{t}{t - 2} = \dfrac{2t - 3}{t}$

64. $\dfrac{z}{z + 3} = \dfrac{3z}{5z - 1}$

65. $\dfrac{2}{x^2} + \dfrac{4}{x} + 1 = 0$ **66.** $\dfrac{1}{x^2} + \dfrac{3}{x} + 1 = 0$

Find the complex solutions to each equation. See Example 10.

67. $x^2 + 2x + 5 = 0$ **68.** $x^2 + 4x + 5 = 0$

69. $x^2 + 12 = 0$ **70.** $-3x^2 - 21 = 0$

71. $5z^2 - 4z + 1 = 0$ **72.** $2w^2 - 3w + 2 = 0$

Find all real or imaginary solutions to each equation. Use the method of your choice.

73. $4x^2 + 25 = 0$ **74.** $5w^2 - 3 = 0$

75. $\left(p + \dfrac{1}{2}\right)^2 = \dfrac{9}{4}$ **76.** $\left(y - \dfrac{2}{3}\right)^2 = \dfrac{4}{9}$

77. $5t^2 + 4t - 3 = 0$

78. $3v^2 + 4v - 1 = 0$

79. $m^2 + 2m - 24 = 0$ **80.** $q^2 + 6q - 7 = 0$

81. $\left(a + \dfrac{2}{3}\right)^2 = -\dfrac{32}{9}$

82. $\left(w + \dfrac{1}{2}\right)^2 = -6$

83. $-x^2 + x + 6 = 0$ **84.** $-x^2 + x + 12 = 0$

85. $x^2 - 6x + 10 = 0$ **86.** $x^2 - 8x + 17 = 0$

87. $2x - 5 = \sqrt{7x + 7}$ **88.** $\sqrt{7x + 29} = x + 3$

89. $\dfrac{1}{x} + \dfrac{1}{x - 1} = \dfrac{1}{4}$ **90.** $\dfrac{1}{x} - \dfrac{2}{1 - x} = \dfrac{1}{2}$

If the solution to an equation is imaginary or irrational, it takes a bit more effort to check. Replace x by each given number to verify each statement.

91. Both $2 + \sqrt{3}$ and $2 - \sqrt{3}$ satisfy $x^2 - 4x + 1 = 0$.

92. Both $1 + \sqrt{2}$ and $1 - \sqrt{2}$ satisfy $x^2 - 2x - 1 = 0$.

93. Both $1 + i$ and $1 - i$ satisfy $x^2 - 2x + 2 = 0$.

94. Both $2 + 3i$ and $2 - 3i$ satisfy $x^2 - 4x + 13 = 0$.

Solve each problem.

95. *Approach speed.* The formula $1211.1L = CA^2S$ is used to determine the approach speed for landing an aircraft, where L is the gross weight of the aircraft in pounds, C is the coefficient of lift, S is the surface area of the wings in square feet (ft^2), and A is approach speed in feet per second. Find A for the Piper Cheyenne, which has a gross weight of 8700 lbs, a coefficient of lift of 2.81, and wing surface area of 200 ft^2.

96. *Time to swing.* The period T (time in seconds for one complete cycle) of a simple pendulum is related to the length L (in feet) of the pendulum by the formula $8T^2 = \pi^2 L$. If a child is on a swing with a 10-foot chain, then how long does it take to complete one cycle of the swing?

97. *Time for a swim.* Tropical Pools figures that its monthly revenue in dollars on the sale of x above-ground pools is given by $R = 1500x - 3x^2$, where x is less than 25. What number of pools sold would provide a revenue of $17,568?

98. *Pole vaulting.* In 1981 Vladimir Poliakov (USSR) set a world record of 19 ft $\frac{3}{4}$ in. for the pole vault (Doubleday Almanac). To reach that height, Poliakov obtained a speed of approximately 36 feet per second on the runway. The function $h = -16t^2 + 36t$ gives his height t seconds after leaving the ground.

a) Use the formula to find the exact values of t for which his height was 18 feet.

b) Use the accompanying graph to estimate the value of t for which he was at his maximum height.

c) Approximately how long was he in the air?

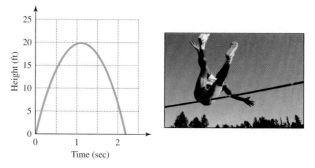

FIGURE FOR EXERCISE 98

GETTING MORE INVOLVED

99. *Discussion.* Which of the following equations is not a quadratic equation?

a) $\pi x^2 - \sqrt{5}x - 1 = 0$ b) $3x^2 - 1 = 0$

c) $4x + 5 = 0$ d) $0.009x^2 = 0$

100. *Exploration.* Solve $x^2 - 4x + k = 0$ for $k = 0, 4, 5,$ and 10.

a) When does the equation have only one solution?

b) For what values of k are the solutions real?

c) For what values of k are the solutions imaginary?

101. *Cooperative learning.* Write a quadratic equation of each of the following types, then trade your equations with those of a classmate. Solve the equations and verify that they are of the required types.

a) a single rational solution

b) two rational solutions

c) two irrational solutions

d) two imaginary solutions

102. *Exploration.* In the next section we will solve $ax^2 + bx + c = 0$ for x by completing the square. Try it now without looking ahead.

GRAPHING CALCULATOR EXERCISES

For each equation, find approximate solutions rounded to two decimal places.

103. $x^2 - 7.3x + 12.5 = 0$

104. $1.2x^2 - \pi x + \sqrt{2} = 0$

105. $2x - 3 = \sqrt{20 - x}$

106. $x^2 - 1.3x = 22.3 - x^2$

10.2　THE QUADRATIC FORMULA

In this section

- Developing the Formula
- Using the Formula
- Number of Solutions
- Applications

Completing the square from Section 10.1 can be used to solve any quadratic equation. Here we apply this method to the general quadratic equation to get a formula for the solutions to any quadratic equation.

Developing the Formula

Start with the general form of the quadratic equation,

$$ax^2 + bx + c = 0.$$

Assume a is positive for now, and divide each side by a:

$$\frac{ax^2 + bx + c}{a} = \frac{0}{a}$$

$$x^2 + \frac{b}{a}x + \frac{c}{a} = 0$$

$$x^2 + \frac{b}{a}x = -\frac{c}{a} \qquad \text{Subtract } \frac{c}{a} \text{ from each side.}$$

One-half of $\frac{b}{a}$ is $\frac{b}{2a}$, and $\frac{b}{2a}$ squared is $\frac{b^2}{4a^2}$:

$$x^2 + \frac{b}{a}x + \frac{b^2}{4a^2} = -\frac{c}{a} + \frac{b^2}{4a^2}$$

Factor the left-hand side and get a common denominator for the right-hand side:

$$\left(x + \frac{b}{2a}\right)^2 = \frac{b^2}{4a^2} - \frac{4ac}{4a^2} \qquad\qquad \frac{c(4a)}{a(4a)} = \frac{4ac}{4a^2}$$

$$\left(x + \frac{b}{2a}\right)^2 = \frac{b^2 - 4ac}{4a^2}$$

$$x + \frac{b}{2a} = \pm\sqrt{\frac{b^2 - 4ac}{4a^2}} \qquad \text{Even-root property}$$

$$x = \frac{-b}{2a} \pm \frac{\sqrt{b^2 - 4ac}}{2a} \qquad \text{Because } a > 0, \sqrt{4a^2} = 2a.$$

$$x = \frac{-b \pm \sqrt{b^2 - 4ac}}{2a}$$

We assumed a was positive so that $\sqrt{4a^2} = 2a$ would be correct. If a is negative, then $\sqrt{4a^2} = -2a$, and we get

$$x = \frac{-b}{2a} \pm \frac{\sqrt{b^2 - 4ac}}{-2a}.$$

However, the negative sign can be omitted in $-2a$ because of the $\pm$ symbol preceding it. For example, the results of $5 \pm (-3)$ and 5 ± 3 are the same. So when a is negative, we get the same formula as when a is positive. It is called the **quadratic formula.**

The Quadratic Formula

The solution to $ax^2 + bx + c = 0$, with $a \neq 0$, is given by the formula

$$x = \frac{-b \pm \sqrt{b^2 - 4ac}}{2a}.$$

Using the Formula

The quadratic formula can be used to solve any quadratic equation.

E X A M P L E   1

Two rational solutions

Solve $x^2 + 2x - 15 = 0$ using the quadratic formula.

Solution

To use the formula, we first identify the values of a, b, and c:

$$1x^2 + 2x - 15 = 0$$
$$\quad\uparrow\qquad\uparrow\qquad\uparrow$$
$$\quad a\qquad b\qquad c$$

The coefficient of x^2 is 1, so $a = 1$. The coefficient of $2x$ is 2, so $b = 2$. The constant term is -15, so $c = -15$. Substitute these values into the quadratic formula:

$$x = \frac{-2 \pm \sqrt{2^2 - 4(1)(-15)}}{2(1)}$$

$$= \frac{-2 \pm \sqrt{4 + 60}}{2}$$

$$= \frac{-2 \pm \sqrt{64}}{2}$$

$$= \frac{-2 \pm 8}{2}$$

$$x = \frac{-2 + 8}{2} = 3 \quad \text{or} \quad x = \frac{-2 - 8}{2} = -5$$

Check 3 and -5 in the original equation. The solution set is $\{-5, 3\}$. ∎

C A U T I O N To identify a, b, and c for the quadratic formula, the equation must be in the standard form $ax^2 + bx + c = 0$. If it is not in that form, then you must first rewrite the equation.

calculator

close-up

Note that the two solutions to

$$x^2 + 2x - 15 = 0$$

correspond to the two x-intercepts for the graph of the function

$$y = x^2 + 2x - 15.$$

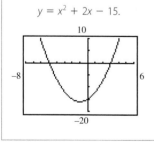

EXAMPLE 2

One rational solution

Solve $4x^2 = 12x - 9$ by using the quadratic formula.

calculator

close-up

Note that the single solution to

$$4x^2 - 12x + 9 = 0$$

corresponds to the single x-intercept for the graph of the function

$$y = 4x^2 - 12x + 9.$$

Solution

Rewrite the equation in the form $ax^2 + bx + c = 0$ before identifying a, b, and c:

$$4x^2 - 12x + 9 = 0$$

In this form we get $a = 4$, $b = -12$, and $c = 9$.

$$x = \frac{12 \pm \sqrt{(-12)^2 - 4(4)(9)}}{2(4)} \qquad \text{Because } b = -12, -b = 12.$$

$$= \frac{12 \pm \sqrt{144 - 144}}{8}$$

$$= \frac{12 \pm 0}{8} = \frac{12}{8} = \frac{3}{2}$$

Check $\frac{3}{2}$ in the original equation. The solution set is $\left\{\frac{3}{2}\right\}$. ■

Because the solutions to the equations in Examples 1 and 2 were rational numbers, these equations could have been solved by factoring. In the next example the solutions are irrational.

EXAMPLE 3

Two irrational solutions

Solve $2x^2 + 6x + 3 = 0$.

Solution

Let $a = 2$, $b = 6$, and $c = 3$ in the quadratic formula:

$$x = \frac{-6 \pm \sqrt{(6)^2 - 4(2)(3)}}{2(2)}$$

$$= \frac{-6 \pm \sqrt{36 - 24}}{4} = \frac{-6 \pm \sqrt{12}}{4}$$

$$= \frac{-6 \pm 2\sqrt{3}}{4} = \frac{2(-3 \pm \sqrt{3})}{2 \cdot 2}$$

$$= \frac{-3 \pm \sqrt{3}}{2}$$

Check these values in the original equation. The solution set is $\left\{\frac{-3 \pm \sqrt{3}}{2}\right\}$. ■

EXAMPLE 4

Two imaginary solutions, no real solutions

Find the complex solutions to $x^2 + x + 5 = 0$.

Solution

Let $a = 1$, $b = 1$, and $c = 5$ in the quadratic formula:

$$x = \frac{-1 \pm \sqrt{(1)^2 - 4(1)(5)}}{2(1)} = \frac{-1 \pm \sqrt{-19}}{2} = \frac{-1 \pm i\sqrt{19}}{2}$$

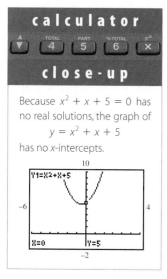
Check these values in the original equation. The solution set is $\left\{\dfrac{-1 \pm i\sqrt{19}}{2}\right\}$. There are no real solutions to the equation. ■

You have learned to solve quadratic equations by four different methods: the even-root property, factoring, completing the square, and the quadratic formula. The even-root property and factoring are limited to certain special equations, but you should use those methods when possible. Any quadratic equation can be solved by completing the square or using the quadratic formula. Because the quadratic formula is usually faster, it is used more often than completing the square. However, completing the square is an important skill to learn. It will be used in the study of conic sections in Chapter 13.

Methods for Solving $ax^2 + bx + c = 0$

Method	Comments	Examples
Even-root property	Use when $b = 0$.	$(x - 2)^2 = 8$ $x - 2 = \pm\sqrt{8}$
Factoring	Use when the polynomial can be factored.	$x^2 + 5x + 6 = 0$ $(x + 2)(x + 3) = 0$
Quadratic formula	Solves any quadratic equation	$x^2 + 5x + 3 = 0$ $x = \dfrac{-5 \pm \sqrt{25 - 4(3)}}{2}$
Completing the square	Solves any quadratic equation, but quadratic formula is faster	$x^2 - 6x + 7 = 0$ $x^2 - 6x + 9 = -7 + 9$ $(x - 3)^2 = 2$

MATH AT WORK

Remodeling a kitchen can be an expensive undertaking. Choosing the correct style of cabinets, doors, and floor covering is just a small part of the process. Joe Prendergast, designer for Lee Kimball Kitchens, is involved in every step of creating a new kitchen for a client.

The process begins with customers visiting the store to see what products are available. Once the client has decided on material and style, he or she

KITCHEN DESIGNER

fills out an extensive questionnaire so that Prendergast can get a sense of the client's lifestyle. Design plans are drawn using a scale of $\frac{1}{2}$ inch to 1 foot. Consideration is given to traffic patterns, doorways, and especially work areas where the cook would want to work unencumbered. Storage areas are a big consideration, as are lighting and color.

A new kitchen can take anywhere from 5 weeks to a few months or more, depending on how complicated the design and construction is. In Exercise 83 of this section we will find the dimensions of a border around a countertop.

Number of Solutions

The quadratic equations in Examples 1 and 3 had two real solutions each. In each of those examples the value of $b^2 - 4ac$ was positive. In Example 2 the quadratic equation had only one solution because the value of $b^2 - 4ac$ was zero. In Example 4 the quadratic equation had no real solutions because $b^2 - 4ac$ was negative. Because $b^2 - 4ac$ determines the kind and number of solutions to a quadratic equation, it is called the **discriminant.**

Number of Solutions to a Quadratic Equation

The quadratic equation $ax^2 + bx + c = 0$ with $a \neq 0$ has
   two real solutions if $b^2 - 4ac > 0$,
   one real solution if $b^2 - 4ac = 0$, and
   no real solutions (two imaginary solutions) if $b^2 - 4ac < 0$.

E X A M P L E   5

Using the discriminant

Use the discriminant to determine the number of real solutions to each quadratic equation.

a) $x^2 - 3x - 5 = 0$ **b)** $x^2 = 3x - 9$ **c)** $4x^2 - 12x + 9 = 0$

Solution

a) For $x^2 - 3x - 5 = 0$, use $a = 1$, $b = -3$, and $c = -5$ in $b^2 - 4ac$:

$$b^2 - 4ac = (-3)^2 - 4(1)(-5) = 9 + 20 = 29$$

Because the discriminant is positive, there are two real solutions to this quadratic equation.

b) For $x^2 - 3x + 9 = 0$, use $a = 1$, $b = -3$, and $c = 9$ in $b^2 - 4ac$:

$$b^2 - 4ac = (-3)^2 - 4(1)(9) = 9 - 36 = -27$$

Because the discriminant is negative, the equation has no real solutions. It has two imaginary solutions.

c) For $4x^2 - 12x + 9 = 0$, use $a = 4$, $b = -12$, and $c = 9$ in $b^2 - 4ac$:

$$b^2 - 4ac = (-12)^2 - 4(4)(9) = 144 - 144 = 0$$

Because the discriminant is zero, there is only one real solution to this quadratic equation.   ■

Applications

With the quadratic formula we can easily solve problems whose solutions are irrational numbers. When the solutions are irrational numbers, we usually use a calculator to find rational approximations and to check.

E X A M P L E   6

Area of a tabletop

The area of a rectangular tabletop is 6 square feet. If the width is 2 feet shorter than the length, then what are the dimensions?

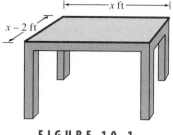

FIGURE 10.1

Solution

Let x be the length and $x - 2$ be the width, as shown in Fig. 10.1. Because the area is 6 square feet and $A = LW$, we can write the equation

$$x(x - 2) = 6$$

or

$$x^2 - 2x - 6 = 0.$$

Because this equation cannot be factored, we use the quadratic formula with $a = 1$, $b = -2$, and $c = -6$:

$$x = \frac{2 \pm \sqrt{(-2)^2 - 4(1)(-6)}}{2(1)}$$

$$= \frac{2 \pm \sqrt{28}}{2} = \frac{2 \pm 2\sqrt{7}}{2} = 1 \pm \sqrt{7}$$

Because $1 - \sqrt{7}$ is a negative number, it cannot be the length of a tabletop. If $x = 1 + \sqrt{7}$, then $x - 2 = 1 + \sqrt{7} - 2 = \sqrt{7} - 1$. Checking the product of $\sqrt{7} + 1$ and $\sqrt{7} - 1$, we get

$$(\sqrt{7} + 1)(\sqrt{7} - 1) = 7 - 1 = 6.$$

The exact length is $\sqrt{7} + 1$ feet, and the width is $\sqrt{7} - 1$ feet. Using a calculator, we find that the approximate length is 3.65 feet and the approximate width is 1.65 feet. ∎

WARM-UPS

True or false? Explain.

1. Completing the square is used to develop the quadratic formula.
2. For the equation $3x^2 = 4x - 7$, we have $a = 3$, $b = 4$, and $c = -7$.
3. If $dx^2 + ex + f = 0$ and $d \neq 0$, then $x = \frac{-e \pm \sqrt{e^2 - 4df}}{2d}$.
4. The quadratic formula will not work on the equation $x^2 - 3 = 0$.
5. If $a = 2$, $b = -3$, and $c = -4$, then $b^2 - 4ac = 41$.
6. If the discriminant is zero, then there are no imaginary solutions.
7. If $b^2 - 4ac > 0$, then $ax^2 + bx + c = 0$ has two real solutions.
8. To solve $2x - x^2 = 0$ by the quadratic formula, let $a = -1$, $b = 2$, and $c = 0$.
9. Two numbers that have a sum of 6 can be represented by x and $x + 6$.
10. Some quadratic equations have one real and one imaginary solution.

10.2   EXERCISES

Reading and Writing   *After reading this section, write out the answers to these questions. Use complete sentences.*

1. What is the quadratic formula used for?

2. When do you use the even-root property to solve a quadratic equation?

3. When do you use factoring to solve a quadratic equation?

4. When do you use the quadratic formula to solve a quadratic equation?

5. What is the discriminant?

6. How many solutions are there to any quadratic equation in the complex number system?

Solve each equation by using the quadratic formula. See Example 1.

7. $x^2 + 5x + 6 = 0$

8. $x^2 - 7x + 12 = 0$

9. $y^2 + y = 6$

10. $m^2 + 2m = 8$

11. $6z^2 - 7z - 3 = 0$

12. $8q^2 + 2q - 1 = 0$

Solve each equation by using the quadratic formula. See Example 2.

13. $4x^2 - 4x + 1 = 0$

14. $4x^2 - 12x + 9 = 0$

15. $9x^2 - 6x + 1 = 0$

16. $9x^2 - 24x + 16 = 0$

17. $9 + 24x + 16x^2 = 0$

18. $4 + 20x = -25x^2$

Solve each equation by using the quadratic formula. See Example 3.

19. $v^2 + 8v + 6 = 0$

20. $p^2 + 6p + 4 = 0$

21. $-x^2 - 5x + 1 = 0$

22. $-x^2 - 3x + 5 = 0$

23. $2t^2 - 6t + 1 = 0$

24. $3z^2 - 8z + 2 = 0$

Solve each equation by using the quadratic formula. See Example 4.

25. $2t^2 - 6t + 5 = 0$

26. $2y^2 + 1 = 2y$

27. $-2x^2 + 3x = 6$

28. $-3x^2 - 2x - 5 = 0$

29. $\frac{1}{2}x^2 + 13 = 5x$

30. $\frac{1}{4}x^2 + \frac{17}{4} = 2x$

Find $b^2 - 4ac$ and the number of real solutions to each equation. See Example 5.

31. $x^2 - 6x + 2 = 0$

32. $x^2 + 6x + 9 = 0$

33. $2x^2 - 5x + 6 = 0$

34. $-x^2 + 3x - 4 = 0$

35. $4m^2 + 25 = 20m$

36. $v^2 = 3v + 5$

37. $y^2 - \frac{1}{2}y + \frac{1}{4} = 0$

38. $\frac{1}{2}w^2 - \frac{1}{3}w + \frac{1}{4} = 0$

39. $-3t^2 + 5t + 6 = 0$

40. $9m^2 + 16 = 24m$

41. $9 - 24z + 16z^2 = 0$

42. $12 - 7x + x^2 = 0$

43. $5x^2 - 7 = 0$

44. $-6x^2 - 5 = 0$

45. $x^2 = x$

46. $-3x^2 + 7x = 0$

Solve each equation by the method of your choice.

47. $\frac{1}{3}x^2 + \frac{1}{2}x = \frac{1}{3}$

48. $\frac{1}{2}x^2 + x = 1$

49. $\frac{w}{w - 2} = \frac{w}{w - 3}$

50. $\frac{y}{3y - 4} = \frac{2}{y + 4}$

51. $\frac{9(3x - 5)^2}{4} = 1$

52. $\frac{25(2x + 1)^2}{9} = 0$

53. $1 + \dfrac{20}{x^2} = \dfrac{8}{x}$

54. $\dfrac{34}{x^2} = \dfrac{6}{x} - 1$

55. $(x - 8)(x + 4) = -42$

56. $(x - 10)(x - 2) = -20$

57. $y = \dfrac{3(2y + 5)}{8(y - 1)}$

58. $z = \dfrac{7z - 4}{12(z - 1)}$

 Use the quadratic formula and a calculator to solve each equation. Round answers to three decimal places and check your answers.

59. $x^2 + 3.2x - 5.7 = 0$

60. $x^2 + 7.15x + 3.24 = 0$

61. $x^2 - 7.4x + 13.69 = 0$

62. $1.44x^2 + 5.52x + 5.29 = 0$

63. $1.85x^2 + 6.72x + 3.6 = 0$

64. $3.67x^2 + 4.35x - 2.13 = 0$

65. $3x^2 + 14{,}379x + 243 = 0$

66. $x^2 + 12{,}347x + 6741 = 0$

67. $x^2 + 0.00075x - 0.0062 = 0$

68. $4.3x^2 - 9.86x - 3.75 = 0$

Solve each problem. See Example 6.

69. Missing numbers. Find two positive real numbers that differ by 1 and have a product of 16.

70. Missing numbers. Find two positive real numbers that differ by 2 and have a product of 10.

71. More missing numbers. Find two real numbers that have a sum of 6 and a product of 4.

72. More missing numbers. Find two real numbers that have a sum of 8 and a product of 2.

73. Bulletin board. The length of a bulletin board is one foot more than the width. The diagonal has a length of $\sqrt{3}$ feet (ft). Find the length and width of the bulletin board.

74. Diagonal brace. The width of a rectangular gate is 2 meters (m) larger than its height. The diagonal brace measures $\sqrt{6}$ m. Find the width and height.

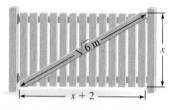

FIGURE FOR EXERCISE 74

75. Area of a rectangle. The length of a rectangle is 4 ft longer than the width, and its area is 10 square feet (ft^2). Find the length and width.

76. Diagonal of a square. The diagonal of a square is 2 m longer than a side. Find the length of a side.

If an object is given an initial velocity of v_0 feet per second from a height of s_0 feet, then its height S after t seconds is given by the formula $S = -16t^2 + v_0 t + s_0$.

77. Projected pine cone. If a pine cone is projected upward at a velocity of 16 ft/sec from the top of a 96-foot pine tree, then how long does it take to reach the earth?

78. Falling pine cone. If a pine cone falls from the top of a 96-foot pine tree, then how long does it take to reach the earth?

79. Penny tossing. If a penny is thrown downward at 30 ft/sec from the bridge at Royal Gorge, Colorado, how long does it take to reach the Arkansas River 1000 ft below?

80. Foul ball. Suppose Charlie O'Brian of the Braves hits a baseball straight upward at 150 ft/sec from a height of 5 ft.

a) Use the formula to determine how long it takes the ball to return to the earth.

b) Use the accompanying graph to estimate the maximum height reached by the ball?

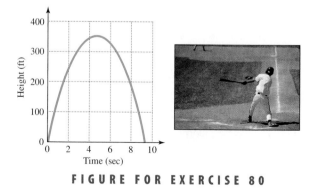

FIGURE FOR EXERCISE 80

In Exercises 81–83, solve each problem.

81. *Recovering an investment.* The manager at Cream of the Crop bought a load of watermelons for $200. She priced the melons so that she would make $1.50 profit on each melon. When all but 30 had been sold, the manager had recovered her initial investment. How many did she buy originally?

82. *Sharing cost.* The members of a flying club plan to share equally the cost of a $200,000 airplane. The members want to find five more people to join the club so that the cost per person will decrease by $2000. How many members are currently in the club?

83. *Kitchen countertop.* A 30 in. by 40 in. countertop for a work island is to be covered with green ceramic tiles, except for a border of uniform width as shown in the figure. If the area covered by the green tiles is 704 square inches (in.²), then how wide is the border?

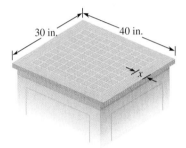

FIGURE FOR EXERCISE 83

GETTING MORE INVOLVED

84. *Discussion.* Find the solutions to $6x^2 + 5x - 4 = 0$. Is the sum of your solutions equal to $-\frac{b}{a}$? Explain why the sum of the solutions to any quadratic equation is $-\frac{b}{a}$. (*Hint:* Use the quadratic formula.)

85. *Discussion.* Use the result of Exercise 84 to check whether $\left\{\frac{2}{3}, \frac{1}{3}\right\}$ is the solution set to $9x^2 - 3x - 2 = 0$. If this solution set is not correct, then what is the correct solution set?

86. *Discussion.* What is the product of the two solutions to $6x^2 + 5x - 4 = 0$? Explain why the product of the solutions to any quadratic equation is $\frac{c}{a}$.

87. *Discussion.* Use the result of the previous exercise to check whether $\left\{\frac{9}{2}, -2\right\}$ is the solution set to $2x^2 - 13x + 18 = 0$. If this solution set is not correct, then what is the correct solution set?

88. *Cooperative learning.* Work in a group to write a quadratic equation that has each given pair of solutions.
  a) -4 and 5 **b)** $2 - \sqrt{3}$ and $2 + \sqrt{3}$
  c) $5 + 2i$ and $5 - 2i$

GRAPHING CALCULATOR EXERCISES

Determine the number of real solutions to each equation by examining the calculator graph of the corresponding function. Use the discriminant to check your conclusions.

89. $x^2 - 6.33x + 3.7 = 0$

90. $1.8x^2 + 2.4x - 895 = 0$

91. $4x^2 - 67.1x + 344 = 0$

92. $-2x^2 - 403 = 0$

93. $-x^2 + 30x - 226 = 0$

94. $16x^2 - 648x + 6562 = 0$

10.3

QUADRATIC FUNCTIONS AND THEIR GRAPHS

We have seen *quadratic functions* on several occasions in this text, but we have not yet defined the term. In this section we study quadratic functions and their graphs.

In this

section

- Definition
- Graphing Quadratic Functions
- The Vertex and Intercepts
- Applications

Definition

If y is determined from x by a formula involving a quadratic polynomial, then we say that y is a *quadratic function of x*.

> **Quadratic Function**
>
> A **quadratic function** is a function of the form
> $$y = ax^2 + bx + c,$$
> where a, b, and c are real numbers and $a \neq 0$.

Without the term ax^2, this function would be a linear function. That is why we specify that $a \neq 0$.

EXAMPLE 1 Finding ordered pairs of a quadratic function

Complete each ordered pair so that it satisfies the given equation.

a) $y = x^2 - x - 6$; $(2, \ )$, $(\ , 0)$

b) $s = -16t^2 + 48t + 84$; $(0, \ )$, $(\ , 20)$

Solution

a) If $x = 2$, then $y = 2^2 - 2 - 6 = -4$. So the ordered pair is $(2, -4)$. To find x when $y = 0$, replace y by 0 and solve the resulting quadratic equation:

$$x^2 - x - 6 = 0$$
$$(x - 3)(x + 2) = 0$$
$$x - 3 = 0 \quad \text{or} \quad x + 2 = 0$$
$$x = 3 \quad \text{or} \quad x = -2$$

The ordered pairs are $(-2, 0)$ and $(3, 0)$.

b) If $t = 0$, then $s = -16 \cdot 0^2 + 48 \cdot 0 + 84 = 84$. The ordered pair is $(0, 84)$. To find t when $s = 20$, replace s by 20 and solve the equation for t:

$$-16t^2 + 48t + 84 = 20$$
$$-16t^2 + 48t + 64 = 0$$
$$t^2 - 3t - 4 = 0$$
$$(t - 4)(t + 1) = 0$$
$$t - 4 = 0 \quad \text{or} \quad t + 1 = 0$$
$$t = 4 \quad \text{or} \quad t = -1$$

The ordered pairs are $(-1, 20)$ and $(4, 20)$. ■

CAUTION When variables other than x and y are used, the independent variable is the first coordinate of an ordered pair, and the dependent variable is the second coordinate. In Example 1(b), t is the independent variable and first coordinate because s depends on t by the formula $s = -16t^2 + 48t + 84$.

Graphing Quadratic Functions

Any real number may be used for x in the formula $y = ax^2 + bx + c$. So the domain (the set of x-coordinates) for any quadratic function is the set of all real numbers, $(-\infty, \infty)$. The range (the set of y-coordinates) can be determined from the graph. All quadratic functions have graphs that are similar in shape. The graph of any quadratic function is called a **parabola.**

EXAMPLE 2 Graphing the simplest quadratic function

Graph the function $y = x^2$, and state the domain and range.

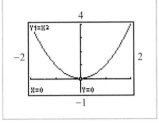

calculator

close-up

This close-up view of $y = x^2$ shows how rounded the curve is at the bottom. When drawing a parabola by hand, be sure to draw it smoothly.

Solution

Make a table of values for x and y:

x	-2	-1	0	1	2
$y = x^2$	4	1	0	1	4

See Fig. 10.2 for the graph. The domain is the set of all real numbers, $(-\infty, \infty)$, because we can use any real number for x. From the graph we see that the smallest y-coordinate of the function is 0. So the range is the set of real numbers that are greater than or equal to 0, $[0, \infty)$.

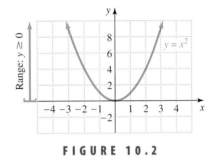

FIGURE 10.2

The parabola in Fig. 10.2 is said to **open upward.** In the next example we see a parabola that **opens downward.** If $a > 0$ in the equation $y = ax^2 + bx + c$, then the parabola opens upward. If $a < 0$, then the parabola opens downward.

E X A M P L E   3

A quadratic function

Graph the function $y = 4 - x^2$, and state the domain and range.

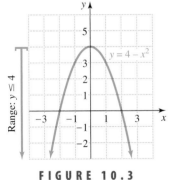

FIGURE 10.3

Solution

We plot enough points to get the correct shape of the graph:

x	-2	-1	0	1	2
$y = 4 - x^2$	0	3	4	3	0

See Fig. 10.3 for the graph. The domain is the set of all real numbers, $(-\infty, \infty)$. From the graph we see that the largest y-coordinate is 4. So the range is $(-\infty, 4]$.

The Vertex and Intercepts

The lowest point on a parabola that opens upward or the highest point on a parabola that opens downward is called the **vertex.** The y-coordinate of the vertex is the **minimum value** of the function if the parabola opens upward, and it is the **maximum value** of the function if the parabola opens downward. For $y = x^2$ the vertex is $(0, 0)$, and 0 is the minimum value of the function. For $g(x) = 4 - x^2$ the vertex is $(0, 4)$, and 4 is the maximum value of the function.

Because the vertex is either the highest or lowest point on a parabola, it is an important point to find before drawing the graph. The vertex can be found by using the following fact.

helpful hint

To draw a parabola or any curve by hand, use your hand like a compass. The two halves of a parabola should be drawn in two steps. Position your paper so that your hand is approximately at the "center" of the arc you are trying to draw.

Vertex of a Parabola

The x-coordinate of the vertex of $y = ax^2 + bx + c$ is $\dfrac{-b}{2a}$, provided that $a \neq 0$.

You can remember $\dfrac{-b}{2a}$ by observing that it is part of the quadratic formula

$$x = \frac{-b \pm \sqrt{b^2 - 4ac}}{2a}.$$

When you graph a parabola, you should always locate the vertex because it is the point at which the graph "turns around." With the vertex and several nearby points you can see the correct shape of the parabola.

E X A M P L E   4

Using the vertex in graphing a quadratic function

Graph $y = -x^2 - x + 2$, and state the domain and range.

Solution

First find the x-coordinate of the vertex:

$$x = \frac{-b}{2a} = \frac{-(-1)}{2(-1)} = \frac{1}{-2} = -\frac{1}{2}$$

Now find y for $x = -\dfrac{1}{2}$:

$$y = -\left(-\frac{1}{2}\right)^2 - \left(-\frac{1}{2}\right) + 2 = -\frac{1}{4} + \frac{1}{2} + 2 = \frac{9}{4}$$

The vertex is $\left(-\dfrac{1}{2}, \dfrac{9}{4}\right)$. Now find a few points on either side of the vertex:

x	-2	-1	$-\dfrac{1}{2}$	0	1
$y = -x^2 - x + 2$	0	2	$\dfrac{9}{4}$	2	0

Sketch a parabola through these points as in Fig. 10.4. The domain is $(-\infty, \infty)$. Because the graph goes no higher than $\dfrac{9}{4}$, the range is $\left(-\infty, \dfrac{9}{4}\right]$.

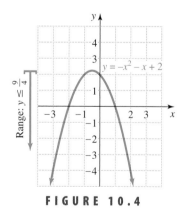

FIGURE 10.4

The y-intercept of a parabola is the point that has 0 as the first coordinate. The x-intercepts are the points that have 0 as their second coordinates.

E X A M P L E   5   **Using the intercepts in graphing a quadratic function**
Find the vertex and intercepts, and sketch the graph of each function.
a) $y = x^2 - 2x - 8$ **b)** $s = -16t^2 + 64t$

Solution

a) Use $x = \dfrac{-b}{2a}$ to get $x = 1$ as the x-coordinate of the vertex. If $x = 1$, then

$$y = 1^2 - 2 \cdot 1 - 8$$
$$= -9.$$

So the vertex is $(1, -9)$. If $x = 0$, then

$$y = 0^2 - 2 \cdot 0 - 8$$
$$= -8.$$

The y-intercept is $(0, -8)$. To find the x-intercepts, replace y by 0:

$$x^2 - 2x - 8 = 0$$
$$(x - 4)(x + 2) = 0$$
$$x - 4 = 0 \quad \text{or} \quad x + 2 = 0$$
$$x = 4 \quad \text{or} \quad x = -2$$

The x-intercepts are $(-2, 0)$ and $(4, 0)$. The graph is shown in Fig. 10.5.

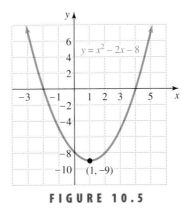

FIGURE 10.5

b) Because s is expressed as a function of t, the first coordinate is t. Use $t = \dfrac{-b}{2a}$ to get

$$t = \frac{-64}{2(-16)} = 2.$$

If $t = 2$, then

$$s = -16 \cdot 2^2 + 64 \cdot 2$$
$$= 64.$$

So the vertex is $(2, 64)$. If $t = 0$, then

$$s = -16 \cdot 0^2 + 64 \cdot 0$$
$$= 0.$$

So the s-intercept is $(0, 0)$. To find the t-intercepts, replace s by 0:

$$-16t^2 + 64t = 0$$
$$-16t(t - 4) = 0$$
$$-16t = 0 \quad \text{or} \quad t - 4 = 0$$
$$t = 0 \quad \text{or} \quad t = 4$$

The t-intercepts are $(0, 0)$ and $(4, 0)$. The graph is shown in Fig. 10.6.

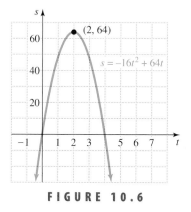

FIGURE 10.6

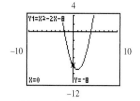

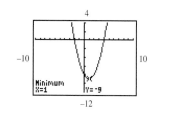
Applications

In applications we are often interested in finding the maximum or minimum value of a variable. If the graph of a quadratic function opens downward, then the maximum value of the second coordinate is the second coordinate of the vertex. If the parabola opens upward, then the minimum value of the second coordinate is the second coordinate of the vertex.

E X A M P L E   6   **Finding the maximum height**

If a projectile is launched with an initial velocity of v_0 feet per second from an initial height of s_0 feet, then its height $s(t)$ in feet is determined by the quadratic function $s(t) = -16t^2 + v_0 t + s_0$, where t is the time in seconds. If a ball is tossed upward with velocity 64 feet per second from a height of 5 feet, then what is the maximum height reached by the ball?

Solution

The height $s(t)$ of the ball for any time t is given by $s(t) = -16t^2 + 64t + 5$. Because the maximum height occurs at the vertex of the parabola, we use $t = \dfrac{-b}{2a}$ to find the vertex:

$$t = \frac{-64}{2(-16)} = 2$$

Now use $t = 2$ to find the second coordinate of the vertex:

$$s(2) = -16(2)^2 + 64(2) + 5 = 69$$

The maximum height reached by the ball is 69 feet. See Fig. 10.7.

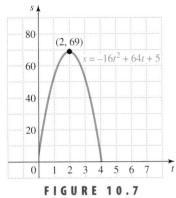

FIGURE 10.7

WARM-UPS

True or false? Explain your answer.

1. The ordered pair $(-2, -1)$ satisfies $y = x^2 - 5$.
2. The y-intercept for $y = x^2 - 3x + 9$ is $(9, 0)$.
3. The x-intercepts for $y = x^2 - 5$ are $(\sqrt{5}, 0)$ and $(-\sqrt{5}, 0)$.
4. The graph of $y = x^2 - 12$ opens upward.
5. The graph of $y = 4 + x^2$ opens downward.
6. The vertex of $y = x^2 + 2x$ is $(-1, -1)$.
7. The parabola $y = x^2 + 1$ has no x-intercepts.
8. The y-intercept for $y = ax^2 + bx + c$ is $(0, c)$.
9. If $w = -2v^2 + 9$, then the maximum value of w is 9.
10. If $y = 3x^2 - 7x + 9$, then the maximum value of y occurs when $x = \dfrac{7}{6}$.

10.3   EXERCISES

Reading and Writing   *After reading this section, write out the answers to these questions. Use complete sentences.*

1. What is a quadratic function?

2. What is a parabola?

3. When does a parabola open upward and when does a parabola open downward?

4. What is the domain of any quadratic function?

5. What is the vertex of a parabola?

6. How can you find the vertex of a parabola?

Complete each ordered pair so that it satisfies the given equation. See Example 1.

7. $y = x^2 - x - 12$ $(3, \quad), (\quad, 0)$

8. $y = -\dfrac{1}{2}x^2 - x + 1$ $(0, \quad), (\quad, -3)$

9. $s = -16t^2 + 32t$ (4,   ), (, 0)

10. $a = b^2 + 4b + 5$ (−2,   ), (, 2)

Graph each quadratic function, and state its domain and range. See Examples 2 and 3.

11. $y = x^2 + 2$

12. $y = x^2 - 4$

13. $y = \frac{1}{2}x^2 - 4$

14. $y = \frac{1}{3}x^2 - 6$

15. $y = -2x^2 + 5$

16. $y = -x^2 - 1$

17. $y = -\frac{1}{3}x^2 + 5$

18. $y = -\frac{1}{2}x^2 + 3$

19. $y = (x - 2)^2$

20. $y = (x + 3)^2$

Find the vertex and intercepts for each quadratic function. Sketch the graph, and state the domain and range. See Examples 4 and 5.

21. $y = x^2 - x - 2$

22. $y = x^2 + 2x - 3$

23. $y = x^2 + 2x - 8$

24. $y = x^2 + x - 6$

25. $y = -x^2 - 4x - 3$

26. $y = -x^2 - 5x - 4$

27. $y = -x^2 + 3x + 4$

28. $y = -x^2 - 2x + 8$

29. $a = b^2 - 6b - 16$

30. $v = -u^2 - 8u + 9$

Find the maximum or minimum value of y for each function.

31. $y = x^2 - 8$ **32.** $y = 33 - x^2$

33. $y = -3x^2 + 14$ **34.** $y = 6 + 5x^2$

35. $y = x^2 + 2x + 3$ **36.** $y = x^2 - 2x + 5$

37. $y = -2x^2 - 4x$ **38.** $y = -3x^2 + 24x$

Solve each problem. See Example 6.

39. *Maximum height.* If a baseball is projected upward from ground level with an initial velocity of 64 feet per second, then its height is a function of time, given by $s(t) = -16t^2 + 64t$. Graph this function for $0 \leq t \leq 4$. What is the maximum height reached by the ball?

40. *Maximum height.* If a soccer ball is kicked straight up with an initial velocity of 32 feet per second, then its height above the earth is a function of time given by $s(t) = -16t^2 + 32t$. Graph this function for $0 \le t \le 2$. What is the maximum height reached by this ball?

41. *Minimum cost.* It costs Acme Manufacturing C dollars per hour to operate its golf ball division. An analyst has determined that C is related to the number of golf balls produced per hour, x, by the equation $C = 0.009x^2 - 1.8x + 100$. What number of balls per hour should Acme produce to minimize the cost per hour of manufacturing these golf balls?

42. *Maximum profit.* A chain store manager has been told by the main office that daily profit, P, is related to the number of clerks working that day, x, according to the equation $P = -25x^2 + 300x$. What number of clerks will maximize the profit, and what is the maximum possible profit?

43. *Maximum area.* Jason plans to fence a rectangular area with 100 meters of fencing. He has written the formula $A = w(50 - w)$ to express the area in terms of the width w. What is the maximum possible area that he can enclose with his fencing?

FIGURE FOR EXERCISE 43

44. *Minimizing cost.* A company uses the function $C(x) = 0.02x^2 - 3.4x + 150$ to model the unit cost in dollars for producing x stabilizer bars. For what number of bars is the unit cost at its minimum? What is the unit cost at that level of production?

45. *Air pollution.* The amount of nitrogen dioxide A in parts per million (ppm) that was present in the air in the city of Homer on a certain day in June is modeled by the function

$$A(t) = -2t^2 + 32t + 12,$$

where t is the number of hours after 6:00 A.M. Use this function to find the time at which the nitrogen dioxide level was at its maximum.

46. *Stabilization ratio.* The stabilization ratio (births/deaths) for South and Central America can be modeled by the function

$$y = -0.0012x^2 + 0.074x + 2.69$$

where y is the number of births divided by the number of deaths in the year $1950 + x$ (World Resources Institute, www.wri.org).
a) Use the graph to estimate the year in which the stabilization ratio was at its maximum.
b) Use the function to find the year in which the stabilization ratio was at its maximum.
c) What was the maximum stabilization ratio from part (b)?
d) What is the significance of a stabilization ratio of 1?

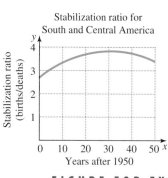

FIGURE FOR EXERCISE 46

47. *Suspension bridge.* The cable of the suspension bridge shown in the accompanying figure hangs in the shape of a parabola with equation $y = 0.0375x^2$, where x and y are in meters. What is the height of each tower above the roadway? What is the length z for the cable bracing the tower?

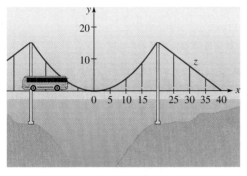

FIGURE FOR EXERCISE 47

GETTING MORE INVOLVED

48. *Exploration.*

    a) Write the function $y = 3(x - 2)^2 + 6$ in the form $y = ax^2 + bx + c$, and find the vertex of the parabola using the formula $x = \frac{-b}{2a}$.

    b) Repeat part (a) with the functions
$y = -4(x - 5)^2 - 9$ and $y = 3(x + 2)^2 - 6$.

    c) What is the vertex for a parabola that is written in the form $y = a(x - h)^2 + k$? Explain your answer.

![calculator icon] **GRAPHING CALCULATOR EXERCISES**

49. Graph $y = x^2$, $y = \frac{1}{2}x^2$, and $y = 2x^2$ on the same coordinate system. What can you say about the graph of $y = kx^2$?

50. Graph $y = x^2$, $y = (x - 3)^2$, and $y = (x + 3)^2$ on the same coordinate system. How does the graph of $y = (x - k)^2$ compare to the graph of $y = x^2$?

51. The equation $x = y^2$ is equivalent to $y = \pm\sqrt{x}$. Graph both $y = \sqrt{x}$ and $y = -\sqrt{x}$ on a graphing calculator. How does the graph of $x = y^2$ compare to the graph of $y = x^2$?

52. Graph each of the following equations by solving for y.

    a) $x = y^2 - 1$

    b) $x = -y^2$

    c) $x^2 + y^2 = 4$

53. Determine the approximate vertex, domain, range, and x-intercepts for each quadratic function.

    a) $y = 3.2x^2 - 5.4x + 1.6$

    b) $y = -1.09x^2 + 13x + 7.5$

10.4 MORE ON QUADRATIC EQUATIONS

In this section we use the ideas and methods of the previous sections to explore additional topics involving quadratic equations.

Using the Discriminant in Factoring

Consider $ax^2 + bx + c$, where a, b, and c are integers with a greatest common factor of 1. If $b^2 - 4ac$ is a perfect square, then $\sqrt{b^2 - 4ac}$ is a whole number, and the solutions to $ax^2 + bx + c = 0$ are rational numbers. If the solutions to a quadratic equation are rational numbers, then they could be found by the factoring method. So if $b^2 - 4ac$ is a perfect square, then $ax^2 + bx + c$ factors. It is also true that if $b^2 - 4ac$ is not a perfect square, then $ax^2 + bx + c$ is prime.

EXAMPLE 1

Using the discriminant

Use the discriminant to determine whether each polynomial can be factored.

a) $6x^2 + x - 15$ **b)** $5x^2 - 3x + 2$

Solution

a) Use $a = 6$, $b = 1$, and $c = -15$ to find $b^2 - 4ac$:

$$b^2 - 4ac = 1^2 - 4(6)(-15) = 361$$

Because $\sqrt{361} = 19$, $6x^2 + x - 15$ can be factored. Using the ac method, we get

$$6x^2 + x - 15 = (2x - 3)(3x + 5).$$

b) Use $a = 5$, $b = -3$, and $c = 2$ to find $b^2 - 4ac$:

$$b^2 - 4ac = (-3)^2 - 4(5)(2) = -31$$

Because the discriminant is not a perfect square, $5x^2 - 3x + 2$ is prime. ■

Writing a Quadratic with Given Solutions

Not every quadratic equation can be solved by factoring, but the factoring method can be used (in reverse) to write a quadratic equation with any given solutions. For example, if the solutions to a quadratic equation are 5 and -3, we can reverse the steps in the factoring method as follows:

$$x = 5 \quad \text{or} \quad x = -3$$
$$x - 5 = 0 \quad \text{or} \quad x + 3 = 0$$
$$(x - 5)(x + 3) = 0 \quad \text{Zero factor property}$$
$$x^2 - 2x - 15 = 0 \quad \text{Multiply the factors.}$$

This method will produce the equation even if the solutions are irrational or imaginary.

EXAMPLE 2

Writing a quadratic given the solutions

Write a quadratic equation that has each given pair of solutions.

a) $4, -6$ **b)** $-\sqrt{2}, \sqrt{2}$ **c)** $-3i, 3i$

Solution

a) Reverse the factoring method using solutions 4 and -6:

$$x = 4 \qquad \text{or} \qquad x = -6$$
$$x - 4 = 0 \qquad \text{or} \qquad x + 6 = 0$$
$$(x - 4)(x + 6) = 0 \quad \text{Zero factor property}$$
$$x^2 + 2x - 24 = 0 \quad \text{Multiply the factors.}$$

b) Reverse the factoring method using solutions $-\sqrt{2}$ and $\sqrt{2}$:

$$x = -\sqrt{2} \qquad \text{or} \qquad x = \sqrt{2}$$
$$x + \sqrt{2} = 0 \qquad \text{or} \quad x - \sqrt{2} = 0$$
$$(x + \sqrt{2})(x - \sqrt{2}) = 0 \quad \text{Zero factor property}$$
$$x^2 - 2 = 0 \quad \text{Multiply the factors.}$$

c) Reverse the factoring method using solutions $-3i$ and $3i$:

$$x = -3i \qquad \text{or} \qquad x = 3i$$
$$x + 3i = 0 \qquad \text{or} \quad x - 3i = 0$$
$$(x + 3i)(x - 3i) = 0 \quad \text{Zero factor property}$$
$$x^2 - 9i^2 = 0 \quad \text{Multiply the factors.}$$
$$x^2 + 9 = 0 \quad \text{Note: } i^2 = -1$$

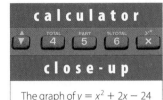

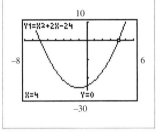

The graph of $y = x^2 + 2x - 24$ supports the conclusion in Example 2(a) because the graph crosses the x-axis at $(4, 0)$ and $(-6, 0)$.

Equations Quadratic in Form

In a quadratic equation we have a variable and its square (x and x^2). An equation that contains an expression and the square of that expression is **quadratic in form** if substituting a single variable for that expression results in a quadratic equation. Equations that are quadratic in form can be solved by using methods for quadratic equations.

E X A M P L E 3

An equation quadratic in form

Solve $(x + 15)^2 - 3(x + 15) - 18 = 0$

Solution

Note that $x + 15$ and $(x + 15)^2$ both appear in the equation. Let $a = x + 15$ and substitute a for $x + 15$ in the equation:

$$(x + 15)^2 - 3(x + 15) - 18 = 0$$
$$a^2 - 3a - 18 = 0$$
$$(a - 6)(a + 3) = 0 \qquad \text{Factor.}$$
$$a - 6 = 0 \qquad \text{or} \qquad a + 3 = 0$$
$$a = 6 \qquad \text{or} \qquad a = -3$$
$$x + 15 = 6 \qquad \text{or} \quad x + 15 = -3 \quad \text{Replace } a \text{ by } x + 15.$$
$$x = -9 \qquad \text{or} \qquad x = -18$$

Check in the original equation. The solution set is $\{-18, -9\}$.

In the next example we have a fourth-degree equation that is quadratic in form. Note that the fourth-degree equation has four solutions.

E X A M P L E   4

A fourth-degree equation

Solve $x^4 - 6x^2 + 8 = 0$.

Solution

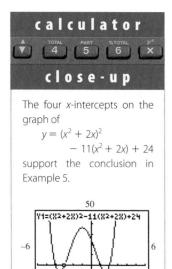

helpful hint

The fundamental theorem of algebra says that the number of solutions to a polynomial equation is less than or equal to the degree of the polynomial. This famous theorem was proved by Carl Friedrich Gauss when he was a young man.

Note that x^4 is the square of x^2. If we let $w = x^2$, then $w^2 = x^4$. Substitute these expressions into the original equation.

$$x^4 - 6x^2 + 8 = 0$$
$$w^2 - 6w + 8 = 0 \qquad \text{Replace } x^4 \text{ by } w^2 \text{ and } x^2 \text{ by } w.$$
$$(w - 2)(w - 4) = 0 \qquad \text{Factor.}$$

$$w - 2 = 0 \qquad \text{or} \qquad w - 4 = 0$$
$$w = 2 \qquad \text{or} \qquad w = 4$$
$$x^2 = 2 \qquad \text{or} \qquad x^2 = 4 \qquad \text{Substitute } x^2 \text{ for } w.$$
$$x = \pm\sqrt{2} \qquad \text{or} \qquad x = \pm 2 \qquad \text{Even-root property}$$

Check. The solution set is $\{-2, -\sqrt{2}, \sqrt{2}, 2\}$. ∎

CAUTION   If you replace x^2 by w, do not quit when you find the values of w. If the variable in the original equation is x, then you must solve for x.

E X A M P L E   5

A quadratic within a quadratic

Solve $(x^2 + 2x)^2 - 11(x^2 + 2x) + 24 = 0$.

Solution

calculator

close-up

The four x-intercepts on the graph of
$y = (x^2 + 2x)^2$
$\qquad - 11(x^2 + 2x) + 24$
support the conclusion in Example 5.

Note that $x^2 + 2x$ and $(x^2 + 2x)^2$ appear in the equation. Let $a = x^2 + 2x$ and substitute.

$$a^2 - 11a + 24 = 0$$
$$(a - 8)(a - 3) = 0 \quad \text{Factor.}$$
$$a - 8 = 0 \quad \text{or} \qquad a - 3 = 0$$
$$a = 8 \quad \text{or} \qquad a = 3$$
$$x^2 + 2x = 8 \quad \text{or} \qquad x^2 + 2x = 3 \quad \text{Replace } a \text{ by } x^2 + 2x.$$
$$x^2 + 2x - 8 = 0 \quad \text{or} \quad x^2 + 2x - 3 = 0$$
$$(x - 2)(x + 4) = 0 \quad \text{or} \quad (x + 3)(x - 1) = 0$$
$$x - 2 = 0 \quad \text{or} \quad x + 4 = 0 \quad \text{or} \quad x + 3 = 0 \quad \text{or} \quad x - 1 = 0$$
$$x = 2 \quad \text{or} \quad x = -4 \quad \text{or} \quad x = -3 \quad \text{or} \quad x = 1$$

Check. The solution set is $\{-4, -3, 1, 2\}$. ∎

The next example involves a fractional exponent. To identify this type of equation as quadratic in form, recall how to square an expression with a fractional exponent. For example, $(x^{1/2})^2 = x$, $(x^{1/4})^2 = x^{1/2}$, and $(x^{1/3})^2 = x^{2/3}$.

E X A M P L E   6

A fractional exponent

Solve $x - 9x^{1/2} + 14 = 0$.

Solution

Note that the square of $x^{1/2}$ is x. Let $w = x^{1/2}$; then $w^2 = (x^{1/2})^2 = x$. Now substitute w and w^2 into the original equation:

$$w^2 - 9w + 14 = 0$$
$$(w - 7)(w - 2) = 0$$
$$w - 7 = 0 \quad \text{or} \quad w - 2 = 0$$
$$w = 7 \quad \text{or} \quad w = 2$$
$$x^{1/2} = 7 \quad \text{or} \quad x^{1/2} = 2 \quad \text{\small Replace } w \text{ by } x^{1/2}.$$
$$x = 49 \quad \text{or} \quad x = 4 \quad \text{\small Square each side.}$$

Because we squared each side, we must check for extraneous roots. First evaluate $x - 9x^{1/2} + 14$ for $x = 49$:

$$49 - 9 \cdot 49^{1/2} + 14 = 49 - 9 \cdot 7 + 14 = 0$$

Now evaluate $x - 9x^{1/2} + 14$ for $x = 4$:

$$4 - 9 \cdot 4^{1/2} + 14 = 4 - 9 \cdot 2 + 14 = 0$$

Because each solution checks, the solution set is $\{4, 49\}$.   ■

CAUTION   An equation of quadratic form must have a term that is the square of another. Equations such as $x^4 - 5x^3 + 6 = 0$ or $x^{1/2} - 3x^{1/3} - 18 = 0$ are not quadratic in form and cannot be solved by substitution.

Applications

Applied problems often result in quadratic equations that cannot be factored. For such equations we use the quadratic formula to find exact solutions and a calculator to find decimal approximations for the exact solutions.

E X A M P L E   7

Changing area

Marvin's flower bed is rectangular in shape with a length of 10 feet and a width of 5 feet (ft). He wants to increase the length and width by the same amount to obtain a flower bed with an area of 75 square feet (ft^2). What should the amount of increase be?

Solution

Let x be the amount of increase. The length and width of the new flower bed are $x + 10$ ft and $x + 5$ ft, as shown in Fig. 10.8. Because the area is to be 75 ft^2, we have

$$(x + 10)(x + 5) = 75.$$

Write this equation in the form $ax^2 + bx + c = 0$:

$$x^2 + 15x + 50 = 75$$
$$x^2 + 15x - 25 = 0 \quad \text{\small Get 0 on the right.}$$
$$x = \frac{-15 \pm \sqrt{225 - 4(1)(-25)}}{2(1)}$$
$$= \frac{-15 \pm \sqrt{325}}{2} = \frac{-15 \pm 5\sqrt{13}}{2}$$

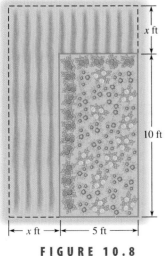

x ft

10 ft

x ft   5 ft

F I G U R E   1 0 . 8

Because the value of x must be positive, the exact increase is

$$\frac{-15 + 5\sqrt{13}}{2} \text{ ft.}$$

Using a calculator, we can find that x is approximately 1.51 ft. If $x = 1.51$ ft, then the new length is 11.51 ft, and the new width is 6.51 ft. The area of a rectangle with these dimensions is 74.93 ft². Of course, the approximate dimensions do not give exactly 75 ft². ■

E X A M P L E   8

Mowing the lawn

It takes Carla 1 hour longer to mow the lawn than it takes Sharon to mow the lawn. If they can mow the lawn in 5 hours working together, then how long would it take each girl by herself?

Solution

If Sharon can mow the lawn by herself in x hours, then she works at the rate of $\frac{1}{x}$ of the lawn per hour. If Carla can mow the lawn by herself in $x + 1$ hours, then she works at the rate of $\frac{1}{x + 1}$ of the lawn per hour. We can use a table to list all of the important quantities.

	Rate	**Time**	**Work**
Sharon	$\frac{1}{x} \frac{\text{lawn}}{\text{hr}}$	5 hr	$\frac{5}{x}$ lawn
Carla	$\frac{1}{x + 1} \frac{\text{lawn}}{\text{hr}}$	5 hr	$\frac{5}{x + 1}$ lawn

Because they complete the lawn in 5 hours, the portion of the lawn done by Sharon and the portion done by Carla have a sum of 1:

$$\frac{5}{x} + \frac{5}{x + 1} = 1$$

$$x(x + 1)\frac{5}{x} + x(x + 1)\frac{5}{x + 1} = x(x + 1)1 \quad \text{Multiply by the LCD.}$$

$$5x + 5 + 5x = x^2 + x$$

$$10x + 5 = x^2 + x$$

$$-x^2 + 9x + 5 = 0$$

$$x^2 - 9x - 5 = 0$$

$$x = \frac{9 \pm \sqrt{(-9)^2 - 4(1)(-5)}}{2(1)}$$

$$= \frac{9 \pm \sqrt{101}}{2}$$

Using a calculator, we find that $\frac{9 - \sqrt{101}}{2}$ is negative. So Sharon's time alone is

$$\frac{9 + \sqrt{101}}{2} \text{ hours.}$$

helpful hint

Note that the equation concerns the portion of the job done by each girl. We could have written an equation about the rates at which the two girls work. Because they can finish the lawn together in 5 hours, they are mowing together at the rate of $\frac{1}{5}$ lawn per hour. So

$$\frac{1}{x} + \frac{1}{x + 1} = \frac{1}{5}.$$

To find Carla's time alone, we add one hour to Sharon's time. So Carla's time alone is

$$\frac{9 + \sqrt{101}}{2} + 1 = \frac{9 + \sqrt{101}}{2} + \frac{2}{2} = \frac{11 + \sqrt{101}}{2} \text{ hours.}$$

Sharon's time alone is approximately 9.525 hours, and Carla's time alone is approximately 10.525 hours. ■

WARM-UPS

True or false? Explain your answer.

1. To solve $x^4 - 5x^2 + 6 = 0$ by substitution, we can let $w = x^2$.
2. We can solve $x^5 - 3x^3 - 10 = 0$ by substitution if we let $w = x^3$.
3. We always use the quadratic formula on equations of quadratic form.
4. If $w = x^{1/6}$, then $w^2 = x^{1/3}$.
5. To solve $x - 7\sqrt{x} + 10 = 0$ by substitution, we let $\sqrt{w} = x$.
6. If $y = 2^{1/2}$, then $y^2 = 2^{1/4}$.
7. If John paints a 100-foot fence in x hours, then his rate is $\frac{100}{x}$ of the fence per hour.
8. If Elvia drives 300 miles in x hours, then her rate is $\frac{300}{x}$ miles per hour (mph).
9. If Ann's boat goes 10 mph in still water, then against a 5-mph current, it will go 2 mph.
10. If squares with sides of length x inches are cut from the corners of an 11-inch by 14-inch rectangular piece of sheet metal and the sides are folded up to form a box, then the dimensions of the bottom will be $11 - x$ by $14 - x$.

10.4   EXERCISES

Reading and Writing   *After reading this section, write out the answers to these questions. Use complete sentences.*

1. How can you use the discriminant to determine if a quadratic polynomial can be factored?

2. What is the relationship between solutions to a quadratic equation and factors of a quadratic polynomial?

3. How do we write a quadratic equation with given solutions?

4. What is an equation quadratic in form?

Use the discriminant to determine whether each quadratic polynomial can be factored, then factor the ones that are not prime. See Example 1.

5. $2x^2 - x + 4$
6. $2x^2 + 3x - 5$
7. $2x^2 + 6x - 5$
8. $3x^2 + 5x - 1$
9. $6x^2 + 19x - 36$
10. $8x^2 + 6x - 27$
11. $4x^2 - 5x - 12$
12. $4x^2 - 27x + 45$

13. $8x^2 - 18x - 45$ **14.** $6x^2 + 9x - 16$

Write a quadratic equation that has each given pair of solutions. See Example 2.

15. $3, -7$ **16.** $-8, 2$

17. $4, 1$ **18.** $3, 2$

19. $\sqrt{5}, -\sqrt{5}$ **20.** $-\sqrt{7}, \sqrt{7}$

21. $4i, -4i$ **22.** $-3i, 3i$

23. $i\sqrt{2}, -i\sqrt{2}$ **24.** $3i\sqrt{2}, -3i\sqrt{2}$

25. $\dfrac{1}{2}, \dfrac{1}{3}$ **26.** $-\dfrac{1}{5}, -\dfrac{1}{2}$

Find all real solutions to each equation. See Example 3.

27. $(2a - 1)^2 + 2(2a - 1) - 8 = 0$

28. $(3a + 2)^2 - 3(3a + 2) = 10$

29. $(w - 1)^2 + 5(w - 1) + 5 = 0$

30. $(2x - 1)^2 - 4(2x - 1) + 2 = 0$

Find all real solutions to each equation. See Example 4.

31. $x^4 - 14x^2 + 45 = 0$ **32.** $x^4 + 2x^2 = 15$

33. $x^6 + 7x^3 = 8$ **34.** $a^6 + 6a^3 = 16$

Find all real solutions to each equation. See Example 5.

35. $(x^2 + 2x)^2 - 7(x^2 + 2x) + 12 = 0$
36. $(x^2 + 3x)^2 + (x^2 + 3x) - 20 = 0$
37. $(y^2 + y)^2 - 8(y^2 + y) + 12 = 0$
38. $(w^2 - 2w)^2 + 24 = 11(w^2 - 2w)$

Find all real solutions to each equation. See Example 6.
39. $x^{1/2} - 5x^{1/4} + 6 = 0$ **40.** $2x - 5\sqrt{x} + 2 = 0$

41. $2x - 5x^{1/2} - 3 = 0$ **42.** $x^{1/4} + 2 = x^{1/2}$

Find all real solutions to each equation.
43. $x^{-2} + x^{-1} - 6 = 0$ **44.** $x^{-2} - 2x^{-1} = 8$

45. $x^{1/6} - x^{1/3} + 2 = 0$ **46.** $x^{2/3} - x^{1/3} - 20 = 0$

47. $\left(\dfrac{1}{y - 1}\right)^2 + \left(\dfrac{1}{y - 1}\right) = 6$

48. $\left(\dfrac{1}{w + 1}\right)^2 - 2\left(\dfrac{1}{w + 1}\right) - 24 = 0$

49. $2x^2 - 3 - 6\sqrt{2x^2 - 3} + 8 = 0$

50. $x^2 + x + \sqrt{x^2 + x} - 2 = 0$

51. $x^{-2} - 2x^{-1} - 1 = 0$ **52.** $x^{-2} - 6x^{-1} + 6 = 0$

Find the exact solution to each problem. If the exact solution is an irrational number, then also find an approximate decimal solution. See Examples 7 and 8.

53. *Country singers.* Harry and Gary are traveling to Nashville to make their fortunes. Harry leaves on the train at 8:00 A.M. and Gary travels by car, starting at 9:00 A.M. To complete the 300-mile trip and arrive at the same time as Harry, Gary travels 10 miles per hour (mph) faster than the train. At what time will they both arrive in Nashville?

54. *Gone fishing.* Debbie traveled by boat 5 miles upstream to fish in her favorite spot. Because of the 4-mph current, it took her 20 minutes longer to get there than to return. How fast will her boat go in still water?

55. *Cross-country cycling.* Erin was traveling across the desert on her bicycle. Before lunch she traveled 60 miles (mi); after lunch she traveled 46 mi. She put in one hour more after lunch than before lunch, but her speed was 4 mph slower than before. What was her speed before lunch and after lunch?

FIGURE FOR EXERCISE 55

56. Extreme hardship. Kim starts to walk 3 mi to school at 7:30 A.M. with a temperature of 0°F. Her brother Bryan starts at 7:45 A.M. on his bicycle, traveling 10 mph faster than Kim. If they get to school at the same time, then how fast is each one traveling?

57. American pie. John takes 3 hours longer than Andrew to peel 500 pounds (lb) of apples. If together they can peel 500 lb of apples in 8 hours, then how long would it take each one working alone?

58. On the half shell. It takes Brent one hour longer than Calvin to shuck a sack of oysters. If together they shuck a sack of oysters in 45 minutes, then how long would it take each one working alone?

59. The growing garden. Eric's garden is 20 ft by 30 ft. He wants to increase the length and width by the same amount to have a 1000-ft² garden. What should be the new dimensions of the garden?

60. Open-top box. Thomas is going to make an open-top box by cutting equal squares from the four corners of an 11 inch by 14 inch sheet of cardboard and folding up the sides. If the area of the base is to be 80 square inches, then what size square should be cut from each corner?

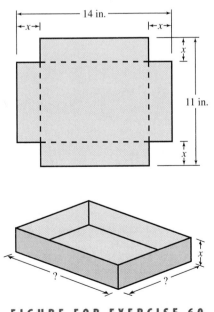

FIGURE FOR EXERCISE 60

61. Pumping the pool. It takes pump A 2 hours less time than pump B to empty a certain swimming pool. Pump A

is started at 8:00 A.M., and pump B is started at 11:00 A.M. If the pool is still half full at 5:00 P.M., then how long would it take pump A working alone?

62. Time off for lunch. It usually takes Eva 3 hours longer to do the monthly payroll than it takes Cicely. They start working on it together at 9:00 A.M. and at 5:00 P.M. they have 90% of it done. If Eva took a 2-hour lunch break while Cicely had none, then how much longer will it take for them to finish the payroll working together?

63. Golden Rectangle. One principle used by the ancient Greeks to get shapes that are pleasing to the eye in art and architecture was the Golden Rectangle. If a square is removed from one end of a Golden Rectangle, as shown in the figure, the sides of the remaining rectangle are proportional to the original rectangle. So the length and width of the original rectangle satisfy

$$\frac{L}{W} = \frac{W}{L - W}.$$

If the length of a Golden Rectangle is 10 meters, then what is its width?

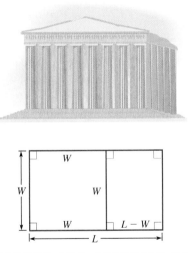

FIGURE FOR EXERCISE 63

GETTING MORE INVOLVED

64. Exploration.
   a) Given that $P(x) = x^4 + 6x^2 - 27$, find $P(3i)$, $P(-3i)$, $P(\sqrt{3})$, and $P(-\sqrt{3})$.
   b) What can you conclude about the values $3i$, $-3i$, $\sqrt{3}$, and $-\sqrt{3}$ and their relationship to each other?

65. Cooperative learning. Work with a group to write a quadratic equation that has each given pair of solutions.
   a) $3 + \sqrt{5}, 3 - \sqrt{5}$ **b)** $4 - 2i, 4 + 2i$
   c) $\dfrac{1 + i\sqrt{3}}{2}, \dfrac{1 - i\sqrt{3}}{2}$

GRAPHING CALCULATOR EXERCISES

Solve each equation by locating the x-intercepts on the graph of a corresponding function. Round approximate answers to two decimal places.

66. $(5x - 7)^2 - (5x - 7) - 6 = 0$

67. $x^4 - 116x^2 + 1600 = 0$

68. $(x^2 + 3x)^2 - 7(x^2 + 3x) + 9 = 0$

69. $x^2 - 3x^{1/2} - 12 = 0$

10.5 QUADRATIC AND RATIONAL INEQUALITIES

In this section we solve inequalities involving quadratic polynomials. We use a new technique based on the rules for multiplying real numbers.

Solving Quadratic Inequalities with a Sign Graph

An inequality involving a quadratic polynomial is called a **quadratic** inequality.

> **Quadratic Inequality**
>
> A quadratic inequality is an inequality of the form
> $$ax^2 + bx + c > 0,$$
> where a, b, and c are real numbers with $a \neq 0$. The inequality symbols $<$, $\leq$, and $\geq$ may also be used.

If we can factor a quadratic inequality, then the inequality can be solved with a **sign graph,** which shows where each factor is positive, negative, or zero.

EXAMPLE 1

Solving a quadratic inequality

Use a sign graph to solve the inequality $x^2 + 3x - 10 > 0$.

Solution

Because the left-hand side can be factored, we can write the inequality as
$$(x + 5)(x - 2) > 0.$$

This inequality says that the product of $x + 5$ and $x - 2$ is positive. If both factors are negative or both are positive, the product is positive. To analyze the signs of each factor, we make a sign graph as follows. First consider the possible values of the factor $x + 5$:

Value	Where	On the number line
$x + 5 = 0$	if $x = -5$	Put a 0 above -5.
$x + 5 > 0$	if $x > -5$	Put $+$ signs to the right of -5.
$x + 5 < 0$	if $x < -5$	Put $-$ signs to the left of -5.

calculator
close-up

Use Y= to set $y_1 = x + 5$ and $y_2 = x - 2$. Now make a table and scroll through the table. The table numerically supports the sign graph in Fig. 10.10.

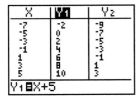

Note that the graph of $y = x^2 + 3x - 10$ is above the x-axis when $x < -5$ or when $x > 2$.

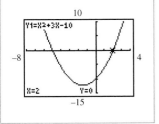

The sign graph shown in Fig. 10.9 for the factor $x + 5$ is made from the information in the preceding table.

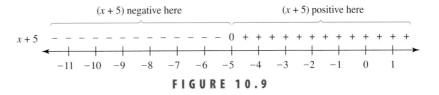

FIGURE 10.9

Now consider the possible values of the factor $x - 2$:

Value	Where	On the number line
$x - 2 = 0$	if $x = 2$	Put a 0 above 2.
$x - 2 > 0$	if $x > 2$	Put + signs to the right of 2.
$x - 2 < 0$	if $x < 2$	Put − signs to the left of 2.

We put the information for the factor $x - 2$ on the sign graph for the factor $x + 5$ as shown in Fig. 10.10. We can see from Fig. 10.10 that the product is positive if $x < -5$ and the product is positive if $x > 2$. The solution set for the quadratic inequality is shown in Fig. 10.11. Note that -5 and 2 are not included in the graph because for those values of x the product is zero. The solution set is $(-\infty, -5) \cup (2, \infty)$.

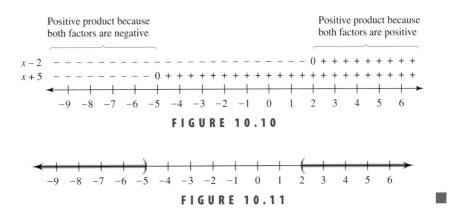

FIGURE 10.10

FIGURE 10.11

In the next example we will make the procedure from Example 1 a bit more efficient.

EXAMPLE 2

Solving a quadratic inequality

Solve $2x^2 + 5x \le 3$ and graph the solution set.

Solution

Rewrite the inequality with 0 on one side:

$$2x^2 + 5x - 3 \le 0$$
$$(2x - 1)(x + 3) \le 0 \quad \text{Factor.}$$

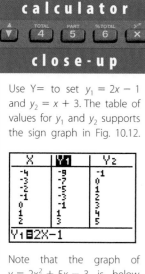

Use Y= to set $y_1 = 2x - 1$ and $y_2 = x + 3$. The table of values for y_1 and y_2 supports the sign graph in Fig. 10.12.

Note that the graph of $y = 2x^2 + 5x - 3$ is below the x-axis when x is between -3 and $\frac{1}{2}$.

Examine the signs of each factor:

$$2x - 1 = 0 \text{ if } x = \frac{1}{2}$$

$$2x - 1 > 0 \text{ if } x > \frac{1}{2}$$

$$2x - 1 < 0 \text{ if } x < \frac{1}{2}$$

$$x + 3 = 0 \text{ if } x = -3$$

$$x + 3 > 0 \text{ if } x > -3$$

$$x + 3 < 0 \text{ if } x < -3$$

Make a sign graph as shown in Fig. 10.12. The product of the factors is negative between -3 and $\frac{1}{2}$, when one factor is negative and the other is positive. The product is 0 at -3 and at $\frac{1}{2}$. So the solution set is the interval $\left[-3, \frac{1}{2}\right]$. The graph of the solution set is shown in Fig. 10.13.

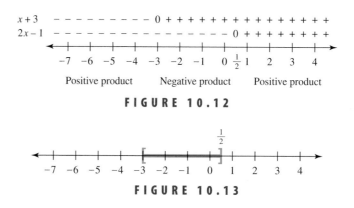

FIGURE 10.12

FIGURE 10.13

We summarize the strategy used for solving a quadratic inequality as follows.

> **Strategy for Solving a Quadratic Inequality with a Sign Graph**
>
> 1. Write the inequality with 0 on the right.
> 2. Factor the quadratic polynomial on the left.
> 3. Make a sign graph showing where each factor is positive, negative, or zero.
> 4. Use the rules for multiplying signed numbers to determine which regions satisfy the original inequality.

Solving Rational Inequalities with a Sign Graph

The inequalities

$$\frac{x + 2}{x - 3} \le 2, \qquad \frac{2x - 3}{x + 5} \le 0 \qquad \text{and} \qquad \frac{2}{x + 4} \ge \frac{1}{x + 1}$$

are called **rational inequalities.** When we solve *equations* that involve rational expressions, we usually multiply each side by the LCD. However, if we multiply each side of any inequality by a negative number, we must reverse the inequality, and when we multiply by a positive number, we do not reverse the inequality. For this reason we generally *do not multiply inequalities by expressions involving variables.* The values of the expressions might be positive or negative. The next two examples show how to use a sign graph to solve rational inequalities that have variables in the denominator.

EXAMPLE 3

Solving a rational inequality

Solve $\frac{x + 2}{x - 3} \le 2$ and graph the solution set.

Solution

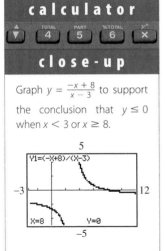

We *do not* multiply each side by $x - 3$. Instead, subtract 2 from each side to get 0 on the right:

$$\frac{x + 2}{x - 3} - 2 \le 0$$

$$\frac{x + 2}{x - 3} - \frac{2(x - 3)}{x - 3} \le 0 \quad \text{Get a common denominator.}$$

$$\frac{x + 2}{x - 3} - \frac{2x - 6}{x - 3} \le 0 \quad \text{Simplify.}$$

$$\frac{x + 2 - 2x + 6}{x - 3} \le 0 \quad \text{Subtract the rational expressions.}$$

$$\frac{-x + 8}{x - 3} \le 0 \quad \begin{array}{l}\text{The quotient of } -x + 8 \text{ and } x - 3 \text{ is less}\\ \text{than or equal to 0.}\end{array}$$

Examine the signs of the numerator and denominator:

$$x - 3 = 0 \text{ if } x = 3 \qquad -x + 8 = 0 \text{ if } x = 8$$
$$x - 3 > 0 \text{ if } x > 3 \qquad -x + 8 > 0 \text{ if } x < 8$$
$$x - 3 < 0 \text{ if } x < 3 \qquad -x + 8 < 0 \text{ if } x > 8$$

Make a sign graph as shown in Fig. 10.14. Using the rule for dividing signed numbers and the sign graph, we can identify where the quotient is negative or zero. The solution set is $(-\infty, 3) \cup [8, \infty)$. Note that 3 is not in the solution set because the quotient is undefined if $x = 3$. The graph of the solution set is shown in Fig. 10.15.

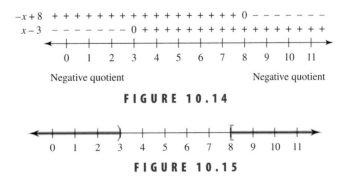

$$
\begin{array}{l}
-x + 8 \quad + + + + + + + + + + + + + + + + \; 0 \; - - - - - - \\
x - 3 \quad \; - - - - - - \; 0 \; + + + + + + + + + + + + + + + + + +
\end{array}
$$

Negative quotient   Negative quotient

FIGURE 10.14

FIGURE 10.15

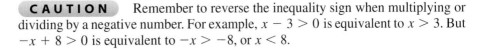

CAUTION   Remember to reverse the inequality sign when multiplying or dividing by a negative number. For example, $x - 3 > 0$ is equivalent to $x > 3$. But $-x + 8 > 0$ is equivalent to $-x > -8$, or $x < 8$.

EXAMPLE 4

Solving a rational inequality

Solve $\dfrac{2}{x + 4} \geq \dfrac{1}{x + 1}$ and graph the solution set.

Solution

We do not multiply by the LCD as we do in solving equations. Instead, subtract $\dfrac{1}{x + 1}$ from each side:

$$\frac{2}{x + 4} - \frac{1}{x + 1} \geq 0$$

$$\frac{2(x + 1)}{(x + 4)(x + 1)} - \frac{1(x + 4)}{(x + 1)(x + 4)} \geq 0 \quad \text{Get a common denominator.}$$

$$\frac{2x + 2 - x - 4}{(x + 1)(x + 4)} \geq 0 \quad \text{Simplify.}$$

$$\frac{x - 2}{(x + 1)(x + 4)} \geq 0$$

Make a sign graph as shown in Fig. 10.16.

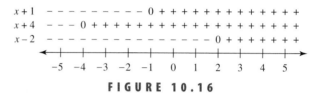

FIGURE 10.16

The computation of

$$\frac{x - 2}{(x + 1)(x + 4)}$$

involves multiplication and division. The result of this computation is positive if all of the three binomials are positive or if only one is positive and the other two are negative. The sign graph shows that this rational expression will have a positive value when x is between -4 and -1 and again when x is larger than 2. The solution set is $(-4, -1) \cup [2, \infty)$. Note that -1 and -4 are not in the solution set because they make the denominator zero. The graph of the solution set is shown in Fig. 10.17.

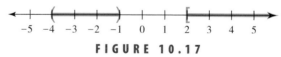

FIGURE 10.17

Solving rational inequalities with a sign graph is summarized below.

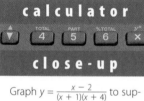

close-up

Graph $y = \dfrac{x - 2}{(x + 1)(x + 4)}$ to support the conclusion that $y \geq 0$ when x is between -4 and -1 or when $x \geq 2$.

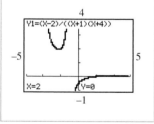

Strategy for Solving a Rational Inequality with a Sign Graph

1. Rewrite the inequality with 0 on the right-hand side.
2. Use only addition and subtraction to get an equivalent inequality.
3. Factor the numerator and denominator if possible.
4. Make a sign graph showing where each factor is positive, negative, or zero.
5. Use the rules for multiplying and dividing signed numbers to determine the regions that satisfy the original inequality.

Another method for solving quadratic and rational inequalities will be shown in Example 5. This method, called the **test point method,** can be used instead of the sign graph to solve the inequalities of Examples 1, 2, 3, and 4.

Quadratic Inequalities That Cannot Be Factored

The following example shows how to solve a quadratic inequality that involves a prime polynomial.

E X A M P L E   5   **Solving a quadratic inequality using the quadratic formula**

Solve $x^2 - 4x - 6 > 0$ and graph the solution set.

Solution

The quadratic polynomial is prime, but we can solve $x^2 - 4x - 6 = 0$ by the quadratic formula:

$$x = \frac{4 \pm \sqrt{16 - 4(1)(-6)}}{2(1)} = \frac{4 \pm \sqrt{40}}{2} = \frac{4 \pm 2\sqrt{10}}{2} = 2 \pm \sqrt{10}$$

As in the previous examples, the solutions to the equation divide the number line into the intervals $(-\infty, 2 - \sqrt{10})$, $(2 - \sqrt{10}, 2 + \sqrt{10})$, and $(2 + \sqrt{10}, \infty)$ on which the quadratic polynomial has either a positive or negative value. To determine which, we select an arbitrary **test point** in each interval. Because $2 + \sqrt{10} \approx 5.2$ and $2 - \sqrt{10} \approx -1.2$, we choose a test point that is less than -1.2, one between -1.2 and 5.2, and one that is greater than 5.2. We have selected -2, 0, and 7 for test points, as shown in Fig. 10.18. Now evaluate $x^2 - 4x - 6$ at each test point.

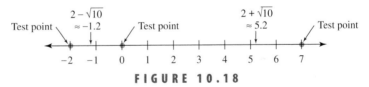

FIGURE 10.18

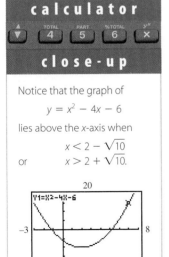
Test point	Value of $x^2 - 4x - 6$ at the test point	Sign of $x^2 - 4x - 6$ in interval of test point
-2	6	Positive
0	-6	Negative
7	15	Positive

Because $x^2 - 4x - 6$ is positive at the test points -2 and 7, it is positive at every point in the intervals containing those test points. So the solution set to the inequality $x^2 - 4x - 6 > 0$ is

$$(-\infty, 2 - \sqrt{10}) \cup (2 + \sqrt{10}, \infty),$$

and its graph is shown in Fig. 10.19.

FIGURE 10.19

The test point method used in Example 5 can be used also on inequalities that do factor. We summarize the strategy for solving inequalities using test points in the following box.

Strategy for Solving Quadratic Inequalities Using Test Points

1. Rewrite the inequality with 0 on the right.
2. Solve the quadratic equation that results from replacing the inequality symbol with the equals symbol.
3. Locate the solutions to the quadratic equation on a number line.
4. Select a test point in each interval determined by the solutions to the quadratic equation.
5. Test each point in the original quadratic inequality to determine which intervals satisfy the inequality.

Applications

The following example shows how a quadratic inequality can be used to solve a problem.

EXAMPLE 6   **Making a profit**

Charlene's daily profit P (in dollars) for selling x magazine subscriptions is determined by the formula

$$P = -x^2 + 80x - 1500.$$

For what values of x is her profit positive?

Solution

We can find the values of x for which $P > 0$ by solving a quadratic inequality:

$$-x^2 + 80x - 1500 > 0$$
$$x^2 - 80x + 1500 < 0 \quad \text{Multiply each side by } -1.$$
$$(x - 30)(x - 50) < 0 \quad \text{Factor.}$$

Make a sign graph as shown in Fig. 10.20. The product of the two factors is negative for x between 30 and 50. Because the last inequality is equivalent to the first, the profit is positive when the number of magazine subscriptions sold is greater than 30 and less than 50.

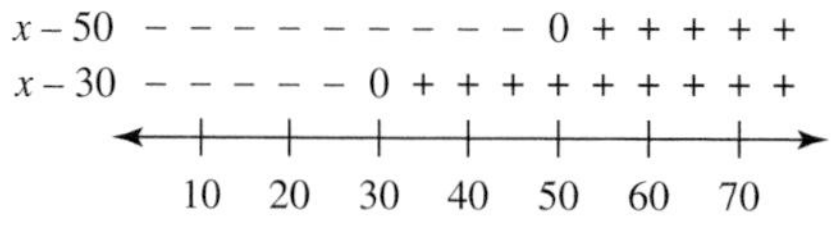

FIGURE 10.20

True or false? Explain.

1. The solution set to $x^2 > 4$ is $(2, \infty)$.
2. The inequality $\frac{x}{x - 3} > 2$ is equivalent to $x > 2x - 6$.
3. The inequality $(x - 1)(x + 2) < 0$ is equivalent to $x - 1 < 0$ or $x + 2 < 0$.

WARM-UPS

(continued)

4. We cannot solve quadratic inequalities that do not factor.

5. One technique for solving quadratic inequalities is based on the rules for multiplying signed numbers.

6. Multiplying each side of an inequality by a variable should be avoided.

7. In solving quadratic or rational inequalities, we always get 0 on one side.

8. The inequality $\frac{x}{2} > 3$ is equivalent to $x > 6$.

9. The inequality $\frac{x-3}{x+2} < 1$ is equivalent to $\frac{x-3}{x+2} - 1 < 0$.

10. The solution set to $\frac{x+2}{x-4} \geq 0$ is $(-\infty, -2] \cup [4, \infty)$.

10.5   EXERCISES

Reading and Writing   After reading this section, write out the answers to these questions. Use complete sentences.

1. What is a quadratic inequality?

2. What is a sign graph?

3. What is a rational inequality?

4. Why don't we usually multiply each side of an inequality by an expression involving a variable?

Solve each inequality. State the solution set using interval notation and graph the solution set. See Examples 1 and 2.

5. $x^2 + x - 6 < 0$

6. $x^2 - 3x - 4 \geq 0$

7. $y^2 - 4 > 0$

8. $z^2 - 16 < 0$

9. $2u^2 + 5u \geq 12$

10. $2v^2 + 7v < 4$

11. $4x^2 - 8x \geq 0$

12. $x^2 + x > 0$

13. $5x - 10x^2 < 0$

14. $3x - x^2 > 0$

15. $x^2 + 6x + 9 \geq 0$

16. $x^2 + 25 < 10x$

Solve each rational inequality. State and graph the solution set. See Examples 3 and 4.

17. $\frac{x}{x-3} > 0$

18. $\frac{a}{a+2} > 0$

19. $\dfrac{x + 2}{x} \le 0$

20. $\dfrac{w - 6}{w} \le 0$

21. $\dfrac{t - 3}{t + 6} > 0$

22. $\dfrac{x - 2}{2x + 5} < 0$

23. $\dfrac{x}{x + 2} > -1$

24. $\dfrac{x + 3}{x} \le -2$

25. $\dfrac{2}{x - 5} > \dfrac{1}{x + 4}$

26. $\dfrac{3}{x + 2} > \dfrac{2}{x - 1}$

27. $\dfrac{m}{m - 5} + \dfrac{3}{m - 1} > 0$

28. $\dfrac{p}{p - 16} + \dfrac{2}{p - 6} \le 0$

29. $\dfrac{x}{x - 3} \le \dfrac{-8}{x - 6}$

30. $\dfrac{x}{x + 20} > \dfrac{2}{x + 8}$

Solve each inequality. State and graph the solution set. See Example 5.

31. $x^2 - 2x - 4 > 0$

32. $x^2 - 2x - 5 \le 0$

33. $2x^2 - 6x + 3 \ge 0$

34. $2x^2 - 8x + 3 < 0$

35. $y^2 - 3y - 9 \le 0$

36. $z^2 - 5z - 7 < 0$

In Exercises 37–60, solve each inequality. State the solution set using interval notation.

37. $x^2 \le 9$

38. $x^2 \ge 36$

39. $16 - x^2 > 0$

40. $9 - x^2 < 0$

41. $x^2 - 4x \ge 0$

42. $4x^2 - 9 > 0$

43. $3(2w^2 - 5) < w$

44. $6(y^2 - 2) + y < 0$

45. $z^2 \ge 4(z + 3)$

46. $t^2 < 3(2t - 3)$

47. $(q + 4)^2 > 10q + 31$

48. $(2p + 4)(p - 1) < (p + 2)^2$

49. $\dfrac{1}{2}x^2 \ge 4 - x$

50. $\dfrac{1}{2}x^2 \le x + 12$

51. $\dfrac{x - 4}{x + 3} \le 0$

52. $\dfrac{2x - 1}{x + 5} \ge 0$

53. $(x - 2)(x + 1)(x - 5) \ge 0$

54. $(x - 1)(x + 2)(2x - 5) < 0$

55. $x^3 + 3x^2 - x - 3 < 0$

56. $x^3 + 5x^2 - 4x - 20 \ge 0$

57. $0.23x^2 + 6.5x + 4.3 < 0$

58. $0.65x^2 + 3.2x + 5.1 > 0$

59. $\dfrac{x}{x - 2} > \dfrac{-1}{x + 3}$

60. $\dfrac{x}{3 - x} > \dfrac{2}{x + 5}$

Solve each problem by using a quadratic inequality. See Example 6.

61. ***Positive profit.*** The monthly profit P (in dollars) that Big Jim makes on the sale of x mobile homes is determined by the formula $P = x^2 + 5x - 50$. For what values of x is his profit positive?

62. ***Profitable fruitcakes.*** Sharon's revenue R (in dollars) on the sale of x fruitcakes is determined by the formula $R = 50x - x^2$. Her cost C (in dollars) for producing x fruitcakes is given by the formula $C = 2x + 40$. For what values of x is Sharon's profit positive? (Profit = revenue − cost.)

If an object is given an initial velocity straight upward of v_0 feet per second from a height of s_0 feet, then its altitude S after t seconds is given by the formula

$$S = -16t^2 + v_0 t + s_0.$$

63. ***Flying high.*** An arrow is shot straight upward with a velocity of 96 feet per second (ft/sec) from an altitude of 6 feet. For how many seconds is this arrow more than 86 feet high?

64. ***Putting the shot.*** In 1978 Udo Beyer (East Germany) set a world record in the shot-put of 72 ft 8 in. If Beyer had projected the shot straight upward with a velocity of 30 ft/sec from a height of 5 ft, then for what values of t would the shot be under 15 ft high?

If a projectile is fired at a 45° angle from a height of s_0 feet with initial velocity v_0 ft/sec, then its altitude S in feet after t seconds is given by

$$S = -16t^2 + \frac{v_0}{\sqrt{2}}t + s_0.$$

65. ***Siege and garrison artillery.*** An 8-inch mortar used in the Civil War fired a 44.5-lb projectile from ground level a distance of 3600 ft when aimed at a 45° angle (Harold R. Peterson, *Notes on Ordinance of the American Civil War*). The accompanying graph shows the altitude of the projectile when it is fired with a velocity of $240\sqrt{2}$ ft/sec.
   a) Use the graph to estimate the maximum altitude reached by the projectile.

   b) Use the graph to estimate approximately how long the altitude of the projectile was greater than 864 ft.

   c) Use the formula to determine the length of time for which the projectile had an altitude of more than 864 ft.

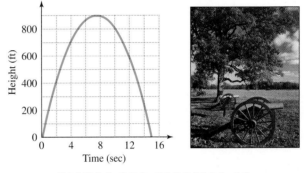

FIGURE FOR EXERCISE 65

66. ***Seacoast artillery.*** The 13-inch mortar used in the Civil War fired a 220-lb projectile a distance of 12,975 ft when aimed at a 45° angle. If the 13-inch mortar was fired from a hill 100 ft above sea level with an initial velocity of 644 ft/sec, then for how long was the projectile more than 800 ft above sea level?

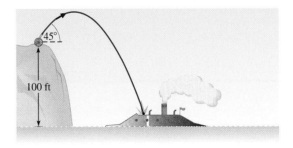

FIGURE FOR EXERCISE 66

GETTING MORE INVOLVED

67. *Cooperative learning.* Work in a small group to solve each inequality for x, given that h and k are real numbers with $h < k$.

a) $(x - h)(x - k) < 0$

b) $(x - h)(x - k) > 0$

c) $(x + h)(x + k) < 0$

d) $(x + h)(x + k) \geq 0$

e) $\dfrac{x - h}{x - k} \geq 0$

f) $\dfrac{x + h}{x + k} \leq 0$

68. *Cooperative learning.* Work in a small group to solve $ax^2 + bx + c > 0$ for x in each case.

a) $b^2 - 4ac = 0$ and $a > 0$

b) $b^2 - 4ac = 0$ and $a < 0$

c) $b^2 - 4ac < 0$ and $a > 0$

d) $b^2 - 4ac < 0$ and $a < 0$

e) $b^2 - 4ac > 0$ and $a > 0$

f) $b^2 - 4ac > 0$ and $a < 0$

GRAPHING CALCULATOR EXERCISES

Match the given inequalities with their solution sets (a through d) by examining a table or a graph.

69. $x^2 - 2x - 8 < 0$ **a.** $(-2, 2) \cup (8, \infty)$

70. $x^2 - 3x > 54$ **b.** $(2, 4)$

71. $\dfrac{x}{x - 2} > 2$ **c.** $(-2, 4)$

72. $\dfrac{3}{x - 2} < \dfrac{5}{x + 2}$ **d.** $(-\infty, -6) \cup (9, \infty)$

COLLABORATIVE ACTIVITIES

Building a Room Addition

Leslie decides to build a rectangular studio on the south side of his house. The studio is attached so that the north side of the studio will be a portion of the current south side of the house. Because Leslie lives in the southwest, he decides to make the room from rammed earth, thus making the walls of the studio 2 feet thick. Because of the cost of building materials and labor, Leslie decides to limit the total area of the studio. He also decides to make the studio's inside south wall twice as long as its inside west wall.

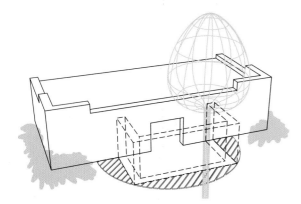

Grouping: 2 to 4 students per group
Topic: Applications of the quadratic formula

Leslie would also like to build a semicircular patio around the studio (sliding glass doors will open from the studio out onto the patio). The studio will be circumscribed by the semicircle. A large tree, which he would like to keep for the summer shade is 20 feet from the south side of the house. See the accompanying sketch Leslie made of the studio and patio.

1. Working in your groups, finish the drawing of the house and studio, adding the given information. Define your variable(s) and find the internal dimensions of the rectangular studio if Leslie wants the *external* area of the studio to be 400 square feet. Round your answers to the nearest foot.

2. Determine how far the semicircular patio extends past the house at the farthest point. Round your answer to the nearest tenth of a foot.

3. Will Leslie be able to build the patio without removing the tree? Give a reason for your answer.

WRAP-UP

C H A P T E R   1 0

SUMMARY

Quadratic Equations		Examples

Quadratic equation

An equation of the form
$ax^2 + bx + c = 0$,
where a, b, and c are real numbers, with $a \neq 0$

$x^2 = 11$
$(x - 5)^2 = 99$
$x^2 + 3x - 20 = 0$

Methods for solving quadratic equations

Factoring:
Factor the quadratic polynomial, then
set each factor equal to 0.

$x^2 + x - 6 = 0$
$(x + 3)(x - 2) = 0$
$x + 3 = 0$ or $x - 2 = 0$

The even-root property:
If $x^2 = k$ ($k > 0$), then $x = \pm\sqrt{k}$.
If $x^2 = 0$, then $x = 0$.
There are no real solutions to $x^2 = k$ for $k < 0$.

$(x - 5)^2 = 10$
$x - 5 = \pm\sqrt{10}$

Completing the square:
Take one-half of middle term, square it,
then add it to each side.

$x^2 + 6x = -4$
$x^2 + 6x + 9 = -4 + 9$
$(x + 3)^2 = 5$

Quadratic formula:
Solves $ax^2 + bx + c = 0$ with $a \neq 0$:

$$x = \frac{-b \pm \sqrt{b^2 - 4ac}}{2a}$$

$2x^2 + 3x - 5 = 0$

$$x = \frac{-3 \pm \sqrt{3^2 - 4(2)(-5)}}{2(2)}$$

Number of solutions

Determined by the discriminant $b^2 - 4ac$:
$b^2 - 4ac > 0$ 2 real solutions

$x^2 + 6x - 12 = 0$
$6^2 - 4(1)(-12) > 0$

$b^2 - 4ac = 0$ 1 real solution

$x^2 + 10x + 25 = 0$
$10^2 - 4(1)(25) = 0$

$b^2 - 4ac < 0$ no real solutions,
2 imaginary solutions

$x^2 + 2x + 20 = 0$
$2^2 - 4(1)(20) < 0$

Factoring

The quadratic polynomial $ax^2 + bx + c$
(with integral coefficients) can be factored
if and only if $b^2 - 4ac$ is a perfect square.

$2x^2 - 11x + 12$
$b^2 - 4ac = 25$
$(2x - 3)(x - 4)$

Writing equations

To write an equation with given solutions,
reverse the steps in solving an equation
by factoring.

$x = 2$ or $x = -3$
$(x - 2)(x + 3) = 0$
$x^2 + x - 6 = 0$

Equations quadratic in form

Use substitution to convert to a quadratic.

$x^4 + 3x^2 - 10 = 0$
Let $a = x^2$
$a^2 + 3a - 10 = 0$

Quadratic Functions		**Examples**
Quadratic function	A function of the form $y = ax^2 + bx + c$, where a, b, and c are real numbers and $a \neq 0$	$y = 3x^2 - 8x + 9$ $p = -3q^2 - 8q + 1$
Graphing a quadratic function	The graph is a parabola opening upward if $a > 0$ and downward if $a < 0$. The first coordinate of the vertex is $\frac{-b}{2a}$. The second coordinate of the vertex is either the minimum value of the function if $a > 0$ or the maximum value of the function if $a < 0$.	$y = x^2 - 2x + 5$ Opens upward Vertex: $(1, 4)$ Minimum value of y is 4.

Quadratic and Rational Inequalities		**Examples**
Quadratic inequality	An inequality involving a quadratic polynomial	$2x^2 - 7x + 6 \geq 0$ $x^2 - 4x - 5 < 0$
Rational inequality	An inequality involving a rational expression	$\dfrac{1}{x - 1} < \dfrac{3}{x - 2}$
Solving quadratic and rational inequalities	Get 0 on one side and express the other side as a product and/or quotient of linear factors. Make a sign graph showing the signs of the factors. Use test points if the quadratic polynomial is prime.	$\overbrace{(x - 5)(x + 1)}^{} < 0$ $x + 1 \quad - - \; 0 + + + + + + + +$ $x - 5 \quad - - - - - - - - \; 0 + +$ $\begin{array}{c} \overset{\longleftarrow}{\vdash\vdash\vdash\vdash\vdash\vdash\vdash\vdash\vdash\vdash\vdash\overset{\longrightarrow}{}} \\ -3\,{-2}\,{-1}\;\;0\;\;1\;\;2\;\;3\;\;4\;\;5\;\;6\;\;7 \end{array}$

ENRICHING YOUR MATHEMATICAL WORD POWER

For each mathematical term, choose the correct meaning.

1. **quadratic equation**
   a. $ax + b = c$ with $a \neq 0$
   b. $ax^2 + bx + c = 0$ with $a \neq 0$
   c. $ax + b = 0$ with $a \neq 0$
   d. $a/x^2 + b/x = c$ with $x \neq 0$

2. **perfect square trinomial**
   a. a trinomial of the form $a^2 + 2ab + b^2$
   b. a trinomial of the form $a^2 + b^2$
   c. a trinomial of the form $a^2 + ab + b^2$
   d. a trinomial of the form $a^2 - 2ab - b^2$

3. **completing the square**
   a. drawing a perfect square
   b. evaluating $(a + b)^2$
   c. drawing the fourth side when given three sides of a square
   d. finding the third term of a perfect square trinomial

4. **quadratic formula**
   a. $x = \dfrac{-b \pm \sqrt{b^2 - 4ac}}{2}$
   b. $x = -b \pm \dfrac{\sqrt{b^2 - 4ac}}{2a}$
   c. $x = \dfrac{-b \pm \sqrt{b^2 - 4ac}}{2a}$
   d. $x = \dfrac{b \pm \sqrt{b^2 - 4ac}}{2a}$

5. **discriminant**
   a. the vertex of a parabola
   b. the radicand in the quadratic formula
   c. the leading coefficient in $ax^2 + bx + c$
   d. to treat unfairly

6. **quadratic function**
   a. $y = ax + b$ with $a \neq 0$
   b. a parabola
   c. $y = ax^2 + bx + c$ with $a \neq 0$
   d. the quadratic formula

7. **quadratic in form**
   a. $ax^2 + bx + c = 0$
   b. a parabola
   c. an equation that is quadratic after a substitution
   d. having four equal sides

8. **quadratic inequality**
   a. $ax^2 + bx + c > 0$ with $a \neq 0$ or with $\geq$, $<$, or $\leq$
   b. $ax + b > 0$ with $a \neq 0$ or with $\geq$, $<$, or $\leq$
   c. completing the square
   d. the Pythagorean theorem

9. **sign graph**
   a. a graph showing the sign of x
   b. a sign on which a graph is drawn
   c. a number line showing the signs of factors
   d. to graph in sign language

10. **rational inequality**
    a. an inequality involving a rational expression(s)
    b. a quadratic inequality

c. an inequality with rational exponents
d. an inequality that compares two fractions

11. **test point**
    a. the end of a chapter
    b. to check if a point is in the right location
    c. a number that is used to check if an inequality is satisfied
    d. a positive integer

REVIEW EXERCISES

10.1 *Solve by factoring.*

1. $x^2 - 2x - 15 = 0$
2. $x^2 - 2x - 24 = 0$
3. $2x^2 + x = 15$
4. $2x^2 + 7x = 4$
5. $w^2 - 25 = 0$
6. $a^2 - 121 = 0$
7. $4x^2 - 12x + 9 = 0$
8. $x^2 - 12x + 36 = 0$

Solve by using the even-root property.

9. $x^2 = 12$
10. $x^2 = 20$
11. $(x - 1)^2 = 9$
12. $(x + 4)^2 = 4$
13. $(x - 2)^2 = \dfrac{3}{4}$
14. $(x - 3)^2 = \dfrac{1}{4}$
15. $4x^2 = 9$
16. $2x^2 = 3$

Solve by completing the square.

17. $x^2 - 6x + 8 = 0$
18. $x^2 + 4x + 3 = 0$
19. $x^2 - 5x + 6 = 0$
20. $x^2 - x - 6 = 0$
21. $2x^2 - 7x + 3 = 0$
22. $2x^2 - x = 6$
23. $x^2 + 4x + 1 = 0$
24. $x^2 + 2x - 2 = 0$

10.2 *Solve by the quadratic formula.*

25. $x^2 - 3x - 10 = 0$
26. $x^2 - 5x - 6 = 0$
27. $6x^2 - 7x = 3$
28. $6x^2 = x + 2$
29. $x^2 + 4x + 2 = 0$
30. $x^2 + 6x = 2$
31. $3x^2 + 1 = 5x$
32. $2x^2 + 3x - 1 = 0$

Find the value of the discriminant and the number of real solutions to each equation.

33. $25x^2 - 20x + 4 = 0$
34. $16x^2 + 1 = 8x$
35. $x^2 - 3x + 7 = 0$
36. $3x^2 - x + 8 = 0$
37. $2x^2 + 1 = 5x$
38. $-3x^2 + 6x - 2 = 0$

Find the complex solutions to the quadratic equations.

39. $2x^2 - 4x + 3 = 0$
40. $2x^2 - 6x + 5 = 0$
41. $2x^2 + 3 = 3x$
42. $x^2 + x + 1 = 0$
43. $3x^2 + 2x + 2 = 0$
44. $x^2 + 2 = 2x$
45. $\dfrac{1}{2}x^2 + 3x + 8 = 0$
46. $\dfrac{1}{2}x^2 - 5x + 13 = 0$

10.3 *Find the vertex and intercepts for each quadratic function, and sketch its graph.*

47. $y = x^2 - 6x$ **48.** $y = x^2 + 4x$

49. $y = x^2 - 4x - 12$ **50.** $y = x^2 + 2x - 24$

51. $y = -2x^2 + 8x$ **52.** $y = -3x^2 + 6x$

53. $y = -x^2 + 2x + 3$ **54.** $y = -x^2 - 3x - 2$

Find the domain and range of each quadratic function.

55. $y = x^2 + 4x + 1$

56. $y = x^2 - 6x + 2$

57. $y = -2x^2 - x + 4$

58. $y = -3x^2 + 2x + 7$

10.4 *Use the discriminant to determine whether each quadratic polynomial can be factored, then factor the ones that are not prime.*

59. $8x^2 - 10x - 3$ **60.** $18x^2 + 9x - 2$

61. $4x^2 - 5x + 2$ **62.** $6x^2 - 7x - 4$

63. $8y^2 + 10y - 25$ **64.** $25z^2 - 15z - 18$

Write a quadratic equation that has each given pair of solutions.

65. $-3, -6$ **66.** $4, -9$

67. $-5\sqrt{2}, 5\sqrt{2}$ **68** $-2i\sqrt{3}, 2i\sqrt{3}$

Find all real solutions to each equation.

69. $x^6 + 7x^3 - 8 = 0$

70. $8x^6 + 63x^3 - 8 = 0$

71. $x^4 - 13x^2 + 36 = 0$

72. $x^4 + 7x^2 + 12 = 0$

73. $(x^2 + 3x)^2 - 28(x^2 + 3x) + 180 = 0$

74. $(x^2 + 1)^2 - 8(x^2 + 1) + 15 = 0$

75. $x^2 - 6x + 6\sqrt{x^2 - 6x} - 40 = 0$

76. $x^2 - 3x - 3\sqrt{x^2 - 3x} + 2 = 0$

77. $t^{-2} + 5t^{-1} - 36 = 0$

78. $a^{-2} + a^{-1} - 6 = 0$

79. $w - 13\sqrt{w} + 36 = 0$

80. $4a - 5\sqrt{a} + 1 = 0$

10.5 *Solve each inequality. State the solution set using interval notation and graph it.*

81. $a^2 + a > 6$

82. $x^2 - 5x + 6 > 0$

83. $x^2 - x - 20 \le 0$

84. $a^2 + 2a \le 15$

85. $w^2 - w < 0$

86. $x - x^2 \le 0$

87. $\dfrac{x - 4}{x + 2} \ge 0$

88. $\dfrac{x - 3}{x + 5} < 0$

89. $\dfrac{x - 2}{x + 3} < 1$

90. $\dfrac{x - 3}{x + 4} > 2$

91. $\dfrac{3}{x + 2} > \dfrac{1}{x + 1}$

92. $\dfrac{1}{x + 1} < \dfrac{1}{x - 1}$

MISCELLANEOUS

In Exercises 93–104, find all real or imaginary solutions to each equation.

93. $144x^2 - 120x + 25 = 0$

94. $49x^2 + 9 = 42x$

95. $(2x + 3)^2 + 7 = 12$

96. $6x = -\dfrac{19x + 25}{x + 1}$

97. $1 + \dfrac{20}{9x^2} = \dfrac{8}{3x}$

98. $\dfrac{x - 1}{x + 2} = \dfrac{2x - 3}{x + 4}$

99. $\sqrt{3x^2 + 7x - 30} = x$

100. $\dfrac{x^4}{3} = x^2 + 6$

101. $2(2x + 1)^2 + 5(2x + 1) = 3$

102. $(w^2 - 1)^2 + 2(w^2 - 1) = 15$
103. $x^{1/2} - 15x^{1/4} + 50 = 0$

104. $x^{-2} - 9x^{-1} + 18 = 0$

Find exact and approximate solutions to each problem.
105. ***Missing numbers.*** Find two positive real numbers that differ by 4 and have a product of 4.

106. ***One on one.*** Find two positive real numbers that differ by 1 and have a product of 1.

107. ***Big screen TV.*** On a 19-inch diagonal measure television picture screen, the height is 4 inches less than the width. Find the height and width.

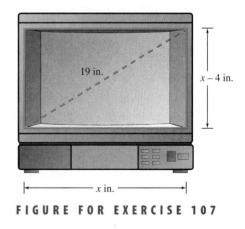

FIGURE FOR EXERCISE 107

108. ***Boxing match.*** A boxing ring is in the shape of a square, 20 ft on each side. How far apart are the fighters when they are in opposite corners of the ring?

109. *Students for a Clean Environment.* A group of environmentalists plans to print a message on an 8 inch by 10 inch paper. If the typed message requires 24 square inches of paper and the group wants an equal border on all sides, then how wide should the border be?

10 in.

8 in.

FIGURE FOR EXERCISE 109

110. *Winston works faster.* Winston can mow his dad's lawn in 1 hour less than it takes his brother Willie. If they take 2 hours to mow it when working together, then how long would it take Winston working alone?

111. *Ping Pong.* The table used for table tennis is 4 ft longer than it is wide and has an area of 45 ft². What are the dimensions of the table?

FIGURE FOR EXERCISE 111

112. *Swimming pool design.* An architect has designed a motel pool within a rectangular area that is fenced on three sides as shown in the figure. If she uses 60 yards of fencing to enclose an area of 352 square yards, then what are the dimensions marked L and W in the figure? Assume L is greater than W.

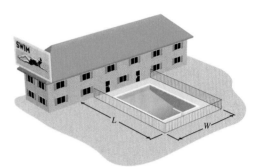

L   W

FIGURE FOR EXERCISE 112

113. *Minimizing cost.* The unit cost in dollars for manufacturing n starters is given by $C = 0.004n^2 - 3.2n + 660$. What is the unit cost when 390 starters are manufactured? For what number of starters is the unit cost at a minimum?

114. *Maximizing profit.* The total profit (in dollars) for sales of x rowing machines is given by $P(x) = -0.2x^2 + 300x - 200$. What is the profit if 500 are sold? For what value of x will the profit be at a maximum?

115. *Decathlon champion.* For 1989 and 1990 Dave Johnson had the highest decathlon score in the world. When Johnson reached a speed of 32 ft/sec on the pole vault runway, his height above the ground t seconds after leaving the ground was given by $h = -16t^2 + 32t$. (The elasticity of the pole converts the horizontal speed into vertical speed.) Find the value of t for which his height was 12 ft.

116. *Time of flight.* Use the information from Exercise 115 to determine how long Johnson was in the air. For how long was he more than 14 ft in the air?

CHAPTER 10 TEST

Calculate the value of $b^2 - 4ac$, and state how many real solutions each equation has.

1. $2x^2 - 3x + 2 = 0$

2. $-3x^2 + 5x - 1 = 0$

3. $4x^2 - 4x + 1 = 0$

Solve by using the quadratic formula.

4. $2x^2 + 5x - 3 = 0$

5. $x^2 + 6x + 6 = 0$

Solve by completing the square.

6. $x^2 + 10x + 25 = 0$

7. $2x^2 + x - 6 = 0$

Solve by any method.

8. $x(x + 1) = 12$

9. $a^4 - 5a^2 + 4 = 0$

10. $x - 2 - 8\sqrt{x - 2} + 15 = 0$

Find the complex solutions to the quadratic equations.

11. $x^2 + 36 = 0$

12. $x^2 + 6x + 10 = 0$

13. $3x^2 - x + 1 = 0$

Graph each quadratic function. State the domain and range.

14. $y = 16 - x^2$ **15.** $y = x^2 - 3x$

Write a quadratic equation that has each given pair of solutions.

16. $-4, 6$

17. $-5i, 5i$

Solve each inequality. State and graph the solution set.

18. $w^2 + 3w < 18$

19. $\dfrac{2}{x - 2} < \dfrac{3}{x + 1}$

Find the exact solution to each problem.

20. The length of a rectangle is 2 ft longer than the width. If the area is 16 ft^2, then what are the length and width?

21. A new computer can process a company's monthly payroll in 1 hour less time than the old computer. To really save time, the manager used both computers and finished the payroll in 3 hours. How long would it take the new computer to do the payroll by itself?

Solve each problem.

22. Find the x-intercepts for the parabola $y = x^2 - 6x + 5$.

23. The height in feet for a ball thrown upward at 48 feet per second is given by $s(t) = -16t^2 + 48t$, where t is the time in seconds after the ball is tossed. What is the maximum height that the ball will reach?

Solve each equation.

1. $2x - 15 = 0$ **2.** $2x^2 - 15 = 0$

3. $2x^2 + x - 15 = 0$

4. $2x^2 + 4x - 15 = 0$

5. $|4x + 11| = 3$

6. $|4x^2 + 11x| = 3$

7. $\sqrt{x} = x - 6$

8. $(2x - 5)^{2/3} = 4$

Solve each inequality.

9. $1 - 2x < 5 - x$

10. $(1 - 2x)(5 - x) \le 0$

11. $\dfrac{1 - 2x}{5 - x} \le 0$

12. $|5 - x| < 3$

13. $3x - 1 < 5$ and $-3 \le x$

14. $x - 3 < 1$ or $2x \ge 8$

Solve each equation for y.

15. $2x - 3y = 9$

16. $\dfrac{y - 3}{x + 2} = -\dfrac{1}{2}$

17. $3y^2 + cy + d = 0$

18. $my^2 - ny = w$

19. $\dfrac{1}{3}x - \dfrac{2}{5}y = \dfrac{5}{6}$

20. $y - 3 = -\dfrac{2}{3}(x - 4)$

Let $m = \dfrac{y_2 - y_1}{x_2 - x_1}$. *Find the value of m for each of the following choices of* $x_1, x_2, y_1,$ *and* y_2.

21. $x_1 = 2, x_2 = 5, y_1 = 3, y_2 = 7$

22. $x_1 = -3, x_2 = 4, y_1 = 5, y_2 = -6$

23. $x_1 = 0.3, x_2 = 0.5, y_1 = 0.8, y_2 = 0.4$

24. $x_1 = \dfrac{1}{2}, x_2 = \dfrac{1}{3}, y_1 = \dfrac{3}{5}, y_2 = -\dfrac{4}{3}$

Solve each problem.

25. *Ticket prices.* In the summer of 1994 the rock group Pearl Jam testified before a congressional committee that Ticketmaster was unfairly raising the prices of the group's concert tickets. One member of the group stated that fans should not have to pay more than $20 to see Pearl Jam. Of course, for any concert, as ticket prices rise, the number of tickets sold decreases, as shown in the figure. If you use the formula $n = 48,000 - 400p$ to predict the number sold depending on the price p, then how many will be sold at $20 per ticket? How many will be sold at $25 per ticket? Use the bar graph to estimate the price if 35,000 tickets were sold.

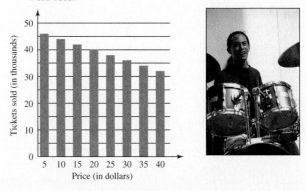

FIGURE FOR EXERCISE 25

26. *Increasing revenue.* Even though the number of tickets sold for a concert decreases with increasing price, the revenue generated does not necessarily decrease. Use the formula $R = p(48,000 - 400p)$ to determine the revenue when the price is $20 and when the price is $25. What price would produce a revenue of $1.28 million? Use the graph to find the price that determines the maximum revenue.

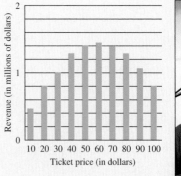

FIGURE FOR EXERCISE 26

Functions

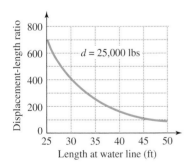

orking in a world of numbers, designers of racing boats blend art with science to design attractive boats that are also fast and safe. If the sail area is increased, the boat will go faster but will be less stable in open seas. If the displacement is increased, the boat will be more stable but slower. Increasing length increases speed but reduces stability. To make yacht racing both competitive and safe, racing boats must satisfy complex systems of rules, many of which involve mathematical formulas.

After the 1988 mismatch between Dennis Conner's catamaran and New Zealander Michael Fay's 133-foot monohull, an international group of yacht designers rewrote the America's Cup rules to ensure the fairness of the race. In addition to hundreds of pages of other rules, every yacht must satisfy the basic inequality

$$\frac{L + 1.25\sqrt{S} - 9.8\sqrt[3]{D}}{0.679} \le 24.000,$$

which balances the length L, the sail area S, and the displacement D.

In the 1979 Fastnet Race 15 sailors lost their lives. After *Exide Challenger*'s carbon-fiber keel snapped off, Tony Bullimore spent 4 days inside the overturned hull before being rescued by the Australian navy. Yacht racing is a dangerous sport. To determine the general performance and safety of a yacht, designers calculate the displacement-length ratio, the sail area-displacement ratio, the ballast-displacement ratio, and the capsize screening value. In Exercises 73 and 74 of Section 11.3 we will see how composition of functions is used to define the displacement-length ratio and the sail area-displacement ratio.

11.1 GRAPHS OF FUNCTIONS AND RELATIONS

Functions were introduced in Section 4.6. In this section we will study the graphs of several types of functions. We graphed linear functions in Chapter 4 and quadratic functions in Chapter 10, but for completeness we will review them here.

Linear and Constant Functions

Linear functions get their name from the fact that their graphs are straight lines.

> **Linear Function**
>
> A **linear function** is a function of the form
> $$f(x) = mx + b,$$
> where m and b are real numbers with $m \neq 0$.

The graph of the linear function $f(x) = mx + b$ is exactly the same as the graph of the linear equation $y = mx + b$. If $m = 0$, then we get $f(x) = b$, which is called a **constant function.** If $m = 1$ and $b = 0$, then we get the function $f(x) = x$, which is called the **identity function.** When we graph a function given in function notation, we usually label the vertical axis as $f(x)$ rather than y.

EXAMPLE 1

Graphing a constant function

Graph $f(x) = 3$ and state the domain and range.

Solution

The graph of $f(x) = 3$ is the same as the graph of $y = 3$, which is the horizontal line in Fig. 11.1. Since any real number can be used for x in $f(x) = 3$ and since the line in Fig. 11.1 extends without bounds to the left and right, the domain is the set of all real numbers, $(-\infty, \infty)$. Since the only y-coordinate for $f(x) = 3$ is 3, the range is $\{3\}$. ■

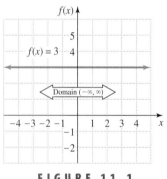

FIGURE 11.1

The domain and range of a function can be determined from the formula or the graph. However, the graph is usually very helpful for understanding domain and range.

EXAMPLE 2

Graphing a linear function

Graph the function $f(x) = 3x - 4$ and state the domain and range.

Solution

The y-intercept is $(0, -4)$ and the slope of the line is 3. We can use the y-intercept and the slope to draw the graph in Fig. 11.2. Since any real number can be used for x in $f(x) = 3x - 4$, and since the line in Fig. 11.2 extends without bounds to the left and right, the domain is the set of all real numbers, $(-\infty, \infty)$. Since the graph extends without bounds upward and downward, the range is the set of all real numbers, $(-\infty, \infty)$.

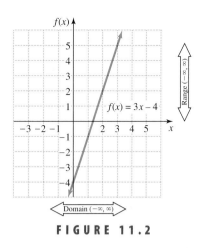

FIGURE 11.2

Absolute Value Functions

The equation $y = |x|$ defines a function because every value of x determines a unique value of y. We call this function the absolute value function.

Absolute Value Function

The **absolute value function** is the function defined by

$$f(x) = |x|.$$

To graph the absolute value function, we simply plot enough ordered pairs of the function to see what the graph looks like.

E X A M P L E   3   **The absolute value function**
Graph $f(x) = |x|$ and state the domain and range.

Solution
To graph this function, we find points that satisfy the equation $f(x) = |x|$.

x	-2	-1	0	1	2
$f(x) = |x|$	2	1	0	1	2

Plotting these points, we see that they lie along the V-shaped graph shown in Fig. 11.3. Since any real number can be used for x in $f(x) = |x|$ and since the graph extends without bounds to the left and right, the domain is $(-\infty, \infty)$. Because the graph does not go below the x-axis and because $|x|$ is never negative, the range is the set of nonnegative real numbers, $[0, \infty)$.

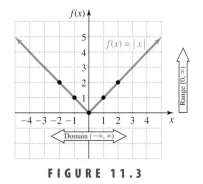

FIGURE 11.3

Many functions involving absolute value have graphs that are V-shaped, as in Fig. 11.3. To graph functions involving absolute value, we must choose points that determine the correct shape and location of the V-shaped graph.

E X A M P L E   4

Other functions involving absolute value

Graph each function and state the domain and range.

a) $f(x) = |x| - 2$ **b)** $g(x) = |2x - 6|$

Solution

a) Choose values for x and find $f(x)$.

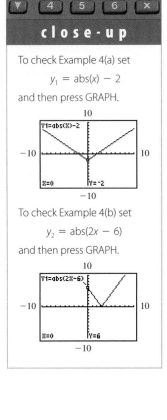
x	-2	-1	0	1	2
$f(x) = |x| - 2$	0	-1	-2	-1	0

Plot these points and draw a V-shaped graph through them as shown in Fig. 11.4. The domain is $(-\infty, \infty)$, and the range is $[-2, \infty)$.

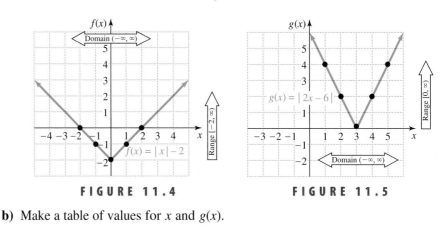

FIGURE 11.4   **FIGURE 11.5**

b) Make a table of values for x and $g(x)$.

x	1	2	3	4	5
$g(x) = |2x - 6|$	4	2	0	2	4

Draw the graph as shown in Fig. 11.5. The domain is $(-\infty, \infty)$, and the range is $[0, \infty)$.

Quadratic Functions

In Chapter 10 we learned that the graph of any quadratic function is a parabola, which opens upward or downward. The vertex of a parabola is the lowest point on a parabola that opens upward or the highest point on a parabola that opens downward. Parabolas will be discussed again when we study conic sections in Chapter 13.

Quadratic Function

A **quadratic function** is a function of the form

$$f(x) = ax^2 + bx + c,$$

where a, b, and c are real numbers, with $a \neq 0$.

E X A M P L E   5   **A quadratic function**

Graph the function $g(x) = 4 - x^2$ and state the domain and range.

Solution

We plot enough points to get the correct shape of the graph.

x	-2	-1	0	1	2
$g(x) = 4 - x^2$	0	3	4	3	0

calculator

close-up

You can find the vertex of a parabola with a calculator. For example, graph

$$y = -x^2 - x + 2.$$

Then use the maximum feature, which is found in the CALC menu. For the left bound pick a point to the left of the vertex; for the right bound pick a point to the right of the vertex; and for the guess pick a point near the vertex.

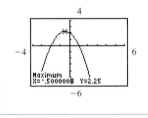

See Fig. 11.6 for the graph. The domain is $(-\infty, \infty)$. From the graph we see that the largest y-coordinate is 4. So the range is $(-\infty, 4]$.

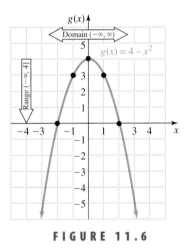

FIGURE 11.6

Square-Root Functions

Functions involving square roots typically have graphs that look like half a parabola.

> **Square-Root Function**
>
> The **square-root function** is the function defined by
> $$f(x) = \sqrt{x}.$$

E X A M P L E   6   **Square-root functions**

Graph each equation and state the domain and range.

a) $y = \sqrt{x}$ **b)** $y = \sqrt{x} + 3$

Solution

a) The graph of the equation $y = \sqrt{x}$ and the graph of the function $f(x) = \sqrt{x}$ are the same. Because $\sqrt{x}$ is a real number only if $x \geq 0$, the domain of this function is the set of nonnegative real numbers. The following ordered pairs are on the graph:

x	0	1	4	9
$y = \sqrt{x}$	0	1	2	3

The graph goes through these ordered pairs as shown in Fig. 11.7. Note that x is chosen from the nonnegative numbers. The domain is $[0, \infty)$ and the range is $[0, \infty)$.

b) Note that $\sqrt{x + 3}$ is a real number only if $x + 3 \geq 0$, or $x \geq -3$. So we make a table of ordered pairs in which $x \geq -3$:

x	-3	-2	1	6
$y = \sqrt{x + 3}$	0	1	2	3

The graph goes through these ordered pairs as shown in Fig. 11.8. The domain is $[-3, \infty)$ and the range is $[0, \infty)$.

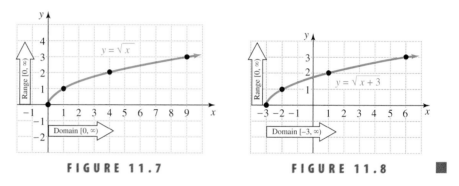

FIGURE 11.7   FIGURE 11.8

Graphs of Relations

A function is a set of ordered pairs in which no two have the same first coordinate and different second coordinates. A **relation** is any set of ordered pairs. Relations and functions can be defined by equations. For example, the set of ordered pairs that satisfies $x = y^2$ is a relation. The equation $x = y^2$ does not define y as a function of x because ordered pairs such as $(4, 2)$ and $(4, -2)$ satisfy $x = y^2$.

The domain of a relation is the set of x-coordinates of the ordered pairs and the range of a relation is the set of y-coordinates. In the next example we graph the relation $x = y^2$ by simply plotting enough points to see the shape of the graph.

E X A M P L E 7

The graph of a relation

Graph $x = y^2$ and state the domain and range.

Solution

Because the equation $x = y^2$ expresses x in terms of y, it is easier to choose the y-coordinate first and then find the x-coordinate.

$x = y^2$	4	1	0	1	4
y	-2	-1	0	1	2

Figure 11.9 shows the graph. The domain is $[0, \infty)$ and the range is $(-\infty, \infty)$.

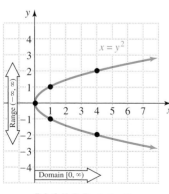

FIGURE 11.9

Vertical-Line Test

Every graph is the graph of a relation, because a relation is any set of ordered pairs. We did not use the term "relation" when we discussed the vertical line test in Section 4.6. So we restate the vertical line test here using that term.

Vertical-Line Test

If it is possible to draw a vertical line that crosses the graph of a relation two or more times, then the graph is not the graph of a function.

   If there is a vertical line that crosses a graph twice (or more), then we have two points (or more) with the same x-coordinate and different y-coordinates, and so the graph is not the graph of a function. If you mentally consider every possible vertical line and none of them crosses the graph more than once, then you can conclude that the graph is the graph of a function.

E X A M P L E   8   **Using the vertical-line test**

Which of the following graphs are graphs of functions?

a)   b)   c)

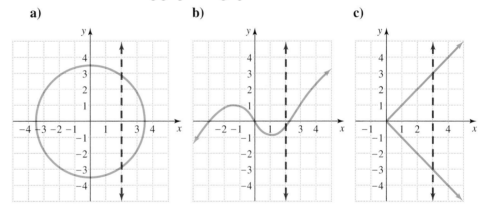

Solution

Neither (a) nor (c) is the graph of a function, since we can draw vertical lines that cross these graphs twice. The graph (b) is the graph of a function, since no vertical line crosses it twice.   ■

WARM-UPS

True or false? Explain your answer.

1. The graph of a function is a picture of all ordered pairs of the function.

2. The graph of every linear function is a straight line.
3. The absolute value function has a V-shaped graph.
4. The domain of $f(x) = \frac{1}{x}$ is $(-\infty, \infty)$.
5. The graph of a quadratic function is a parabola.
6. The range of any quadratic function is $(-\infty, \infty)$.
7. The y-axis and the $f(x)$-axis are the same.
8. The domain of $x = y^2$ is $[0, \infty)$.
9. The domain of $f(x) = \sqrt{x - 1}$ is $(1, \infty)$.
10. The domain of any quadratic function is $(-\infty, \infty)$.

11.1 EXERCISES

Reading and Writing *After reading this section, write out the answers to these questions. Use complete sentences.*

1. What is a linear function?

2. What is a constant function?

3. What is the graph of a constant function?

4. What shape is the graph of an absolute value function?

5. What is the graph of quadratic function called?

6. How can you tell at a glance if a graph is the graph of a function?

Graph each function and state its domain and range. See Examples 1 and 2.

7. $h(x) = -2$

8. $f(x) = 4$

9. $f(x) = 2x - 1$

10. $g(x) = x + 2$

11. $g(x) = \dfrac{1}{2}x + 2$

12. $h(x) = \dfrac{2}{3}x - 4$

13. $y = -\dfrac{2}{3}x + 3$

14. $y = -\dfrac{3}{4}x + 4$

15. $y = -0.3x + 6.5$

16. $y = 0.25x - 0.5$

Graph each absolute value function and state its domain and range. See Examples 3 and 4.

17. $f(x) = |x| + 1$

18. $g(x) = |x| - 3$

19. $h(x) = |x + 1|$

20. $f(x) = |x - 2|$

21. $g(x) = |3x|$

22. $h(x) = |-2x|$

23. $f(x) = |2x - 1|$ **24.** $y = |2x - 3|$

Graph each square-root function and state its domain and range. See Example 6.

33. $g(x) = 2\sqrt{x}$ **34.** $g(x) = \sqrt{x} - 1$

25. $f(x) = |x - 2| + 1$ **26.** $y = |x - 1| + 2$

35. $f(x) = \sqrt{x - 1}$ **36.** $f(x) = \sqrt{x + 1}$

Graph each quadratic function and state its domain and range. See Example 5.

27. $g(x) = x^2 + 2$ **28.** $f(x) = x^2 - 4$

37. $h(x) = -\sqrt{x}$ **38.** $h(x) = -\sqrt{x - 1}$

39. $y = \sqrt{x} + 2$ **40.** $y = 2\sqrt{x} + 1$

29. $f(x) = 2x^2$ **30.** $h(x) = -3x^2$

31. $y = 6 - x^2$ **32.** $y = -2x^2 + 3$

Graph each relation and state its domain and range. See Example 7.

41. $x = |y|$ **42.** $x = -|y|$

43. $x = -y^2$

44. $x = 1 - y^2$

Each of the following graphs is the graph of a relation. Use the vertical-line test on each graph to determine whether y is a function of x. See Example 8.

53.

54.

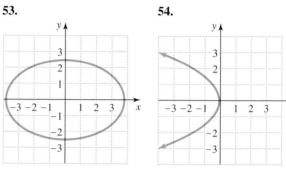

45. $x = 5$

46. $x = -3$

55.

56.

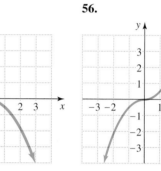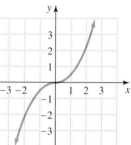

47. $x + 9 = y^2$

48. $x + 3 = |y|$

57.

58.

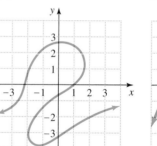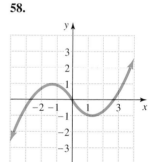

49. $x = \sqrt{y}$

50. $x = -\sqrt{y}$

Graph each function and state the domain and range.

59. $f(x) = 1 - |x|$

60. $h(x) = \sqrt{x - 3}$

51. $x = (y - 1)^2$

52. $x = (y + 2)^2$

61. $y = (x - 3)^2 - 1$ **62.** $y = x^2 - 2x - 3$ **69.** $y = -x^2 + 4x - 4$ **70.** $y = -2|x - 1| + 4$

63. $y = |x + 3| + 1$ **64.** $f(x) = -2x + 4$

GRAPHING CALCULATOR EXERCISES

71. Graph the function $f(x) = \sqrt{x^2}$ and explain what this graph illustrates.

72. Graph the function $f(x) = \frac{1}{x}$ and state the domain and range.

73. Graph $y = x^2$, $y = \frac{1}{2}x^2$, and $y = 2x^2$ on the same coordinate system. What can you say about the graph of $y = kx^2$?

65. $y = \sqrt{x} - 3$ **66.** $y = 2|x|$

74. Graph $y = x^2$, $y = x^2 + 2$, and $y = x^2 - 3$ on the same screen. What can you say about the position of $y = x^2 + k$ relative to $y = x^2$.

75. Graph $y = x^2$, $y = (x + 5)^2$, and $y = (x - 2)^2$ on the same screen. What can you say about the position of $y = (x - k)^2$ relative to $y = x^2$.

67. $y = 3x - 5$ **68.** $g(x) = (x + 2)^2$

76. You can graph the relation $x = y^2$ by graphing the two functions $y = \sqrt{x}$ and $y = -\sqrt{x}$. Try it and explain why this works.

77. Graph $y = (x - 3)^2$, $y = |x - 3|$, and $y = \sqrt{x - 3}$ on the same coordinate system. How does the graph of $y = f(x - k)$ compare to the graph of $y = f(x)$?

11.2 TRANSFORMATIONS OF GRAPHS

We can discover what the graph of almost any function looks like if we plot enough points. However, it is helpful to know something about a graph so that we do not have to plot very many points. In this section we will learn how one graph can be transformed into another by modifying the formula that defines the function.

Reflecting

Consider the graphs of $f(x) = x^2$ and $g(x) = -x^2$ shown in Fig. 11.10. Notice that the graph of g is a mirror image of the graph of f. For any value of x we compute the y-coordinate of an ordered pair of f by squaring x. For an ordered pair of g we square first and then find the opposite because of the order of operations. This gives a correspondence between the ordered pairs of f and the ordered pairs of g. For every ordered pair on the graph of f there is a corresponding ordered pair directly below it on the graph of g, and these ordered pairs are the same distance from the x-axis. We say that the graph of g is obtained by reflecting the graph of f in the x-axis or that g is a reflection of the graph of f.

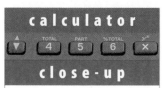

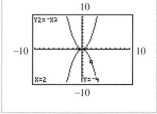

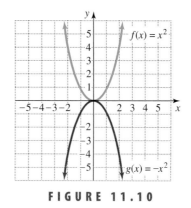

FIGURE 11.10

Reflection

The graph of $y = -f(x)$ is a **reflection** in the x-axis of the graph of $y = f(x)$.

E X A M P L E 1   **Reflection**

Sketch the graphs of each pair of functions on the same coordinate system.

a) $f(x) = \sqrt{x}$, $g(x) = -\sqrt{x}$

b) $f(x) = |x|$, $g(x) = -|x|$

Solution

In each case the graph of g is a reflection of the graph of f. Recall that we graphed the square root function and the absolute value function in the last section. Figures 11.11 and 11.12 show the graphs for these functions.

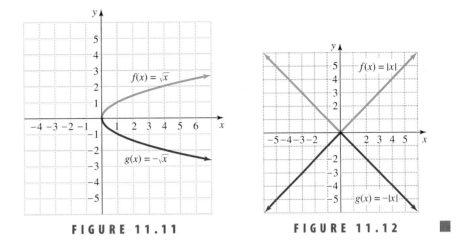

FIGURE 11.11   **FIGURE 11.12**

Translating

Consider the graphs of the functions $f(x) = \sqrt{x}$, $g(x) = \sqrt{x} + 2$, and $h(x) = \sqrt{x} - 6$ shown in Fig. 11.13. In the expression $\sqrt{x} + 2$, adding 2 is the last operation to perform. So every point on the graph of g is exactly two units above a corresponding point on the graph of f, and g has the same shape as the graph of f. Every point on the graph of h is exactly six units below a corresponding point on the graph of f. The graph of g is an upward translation of the graph of f, and the graph of h is a downward translation of the graph of f.

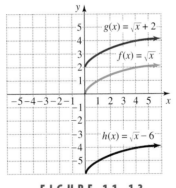

FIGURE 11.13

Translating Upward or Downward

If $k > 0$, then the graph of $y = f(x) + k$ is an **upward translation** of the graph of $y = f(x)$ and the graph of $y = f(x) - k$ is a **downward translation** of the graph of $y = f(x)$.

Consider the graphs of $f(x) = \sqrt{x}$, $g(x) = \sqrt{x - 2}$, and $h(x) = \sqrt{x + 6}$ shown in Fig. 11.14. In the expression $\sqrt{x - 2}$ subtracting 2 is the first operation to perform. So every point on the graph of g is exactly two units to the right of a corresponding point on the graph of f. (We must start with a larger value of x to get the

calculator

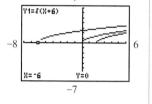

close-up

Note that for a translation of six units to the left, $x + 6$ must be written in parentheses on a graphing calculator.

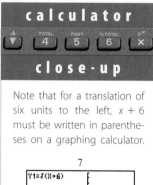

same y-coordinate because we first subtract 2.) Every point on the graph of h is exactly six units to the left of a corresponding point on the graph of f.

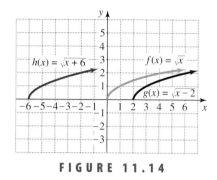

FIGURE 11.14

Translating to the Right or Left

If $h > 0$, then the graph of $y = f(x - h)$ is a **translation to the right** of the graph of $y = f(x)$, and the graph of $y = f(x + h)$ is a **translation to the left** of the graph of $y = f(x)$.

EXAMPLE 2

Translation

Sketch the graph of each function.

a) $f(x) = |x| - 6$ **b)** $f(x) = (x - 2)^2$

c) $f(x) = |x + 3|$

Solution

a) The graph of $f(x) = |x| - 6$ is a translation six units downward of the familiar graph of $f(x) = |x|$. Calculate a few ordered pairs to get an accurate graph. The pairs $(0, -6)$, $(1, -5)$, and $(-1, -5)$ are on the graph shown in Fig. 11.15.

b) The graph of $f(x) = (x - 2)^2$ is a translation two units to the right of the familiar graph of $f(x) = x^2$. Calculate a few ordered pairs to get an accurate graph. The pairs $(2, 0)$, $(0, 4)$, and $(4, 4)$ are on the graph shown in Fig. 11.16.

c) The graph of $f(x) = |x + 3|$ is a translation three units to the left of the familiar graph of $f(x) = |x|$. The pairs $(0, 3)$, $(-3, 0)$, and $(-6, 3)$ are on the graph shown in Fig. 11.17.

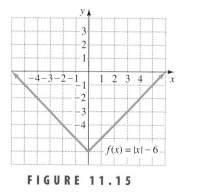

FIGURE 11.15

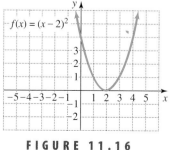

FIGURE 11.16

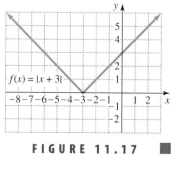

FIGURE 11.17

Stretching and Shrinking

Consider the graphs of $f(x) = x^2$, $g(x) = 2x^2$, and $h(x) = \frac{1}{2}x^2$ shown in Fig. 11.18. Every point on $g(x) = 2x^2$ corresponds to a point directly below on the graph of $f(x) = x^2$. The y-coordinate on g is exactly twice as large as the corresponding y-coordinate on f. This situation occurs because in the expression $2x^2$, multiplying by 2 is the last operation performed. Every point on h corresponds to a point directly above on f, where the y-coordinate on h is half as large as the y-coordinate on f. The factor 2 has stretched the graph of f to form the graph of g, and the factor $\frac{1}{2}$ has shrunk the graph of f to form the graph of h.

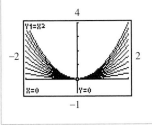
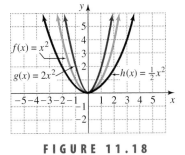

FIGURE 11.18

Stretching and Shrinking

If $a > 1$, then the graph of $y = af(x)$ is obtained by **stretching** the graph of $y = f(x)$. If $0 < a < 1$, then the graph of $y = af(x)$ is obtained by **shrinking** the graph of $y = f(x)$.

Note that the last operation to be performed in stretching or shrinking is multiplication by a. Whereas the function $g(x) = 2\sqrt{x}$ is obtained by stretching $f(x) = \sqrt{x}$ by a factor of 2, $h(x) = \sqrt{2x}$ is not.

E X A M P L E   3

Stretching and shrinking

Graph the functions $f(x) = \sqrt{x}$, $g(x) = 2\sqrt{x}$, and $h(x) = \frac{1}{2}\sqrt{x}$ on the same coordinate system.

Solution

The graph of g is obtained by stretching the graph of f, and the graph of h is obtained by shrinking the graph of f. The graph of f includes the points $(0, 0)$, $(1, 1)$, and $(4, 2)$. The graph of g includes the points $(0, 0)$, $(1, 2)$, and $(4, 4)$. The graph of h includes the points $(0, 0)$, $(1, 0.5)$, and $(4, 1)$. The graphs are shown in Fig. 11.19.

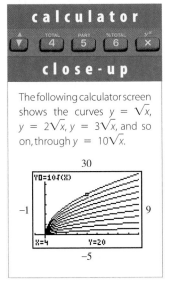
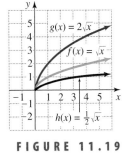

FIGURE 11.19   ■

In the next example we graph a function that results from a combination of translating, stretching, and reflecting.

EXAMPLE 4

Translating, stretching, and reflecting

Graph the function $y = -2\sqrt{x - 3}$.

Solution

The graph of $y = \sqrt{x - 3}$ is a translation three units to the right of the graph of $y = \sqrt{x}$. The graph of $y = 2\sqrt{x - 3}$ is obtained from $y = \sqrt{x - 3}$ by stretching by a factor of 2. The graph of $y = -2\sqrt{x - 3}$ is the mirror image of the graph of $y = 2\sqrt{x - 3}$. All of these graphs are shown in Fig. 11.20.

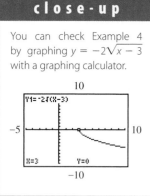

You can check Example 4 by graphing $y = -2\sqrt{x - 3}$ with a graphing calculator.

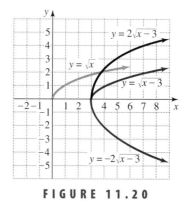

FIGURE 11.20

Graphing Parabolas

The graph of $y = x^2$ is a parabola with vertex $(0, 0)$. The graph of $y = a(x - h)^2 + k$ is a transformation of $y = x^2$ and is also a parabola. It opens upward if $a > 0$ and downward if $a < 0$. Its vertex is (h, k). To graph a parabola in the form $y = a(x - h)^2 + k$, determine its vertex and a couple of points near the vertex.

EXAMPLE 5

Graphing the parabola $y = a(x - h)^2 + k$

Graph $y = -2(x + 3)^2 + 4$.

Solution

Because $x + 3 = x - (-3)$, we have $h = -3$. The vertex is $(-3, 4)$. Since $a = -2$, the parabola opens downward. The points $(-2, 2)$ and $(-4, 2)$ also satisfy the equation. Sketch a parabola through these points as shown in Fig. 11.21.

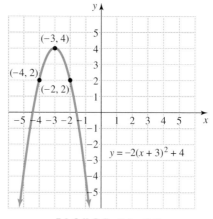

FIGURE 11.21

WARM-UPS

True or false? Explain your answer.

1. The graph of $f(x) = (-x)^2$ is a reflection in the x-axis of the graph of $g(x) = x^2$.

2. The graph of $f(x) = -2$ is a reflection in the x-axis of the graph of $f(x) = 2$.

3. The graph of $f(x) = x + 3$ lies three units to the left of the graph of $f(x) = x$.

4. The graph of $y = |x - 3|$ lies three units to the left of the graph of $y = |x|$.

5. The graph of $y = |x| - 3$ lies three units below the graph of $y = |x|$.

6. The graph of $y = -2x^2$ can be obtained by stretching and reflecting the graph of $y = x^2$.

7. The graph of $f(x) = (x - 2)^2$ is symmetric about the y-axis.

8. For each point on the graph of $y = \sqrt{x}/9$ there is a corresponding point on $y = \sqrt{x}$ that has a y-coordinate three times as large.

9. The graph of $y = \sqrt{x - 3} + 5$ has the same shape as the graph of $y = \sqrt{x}$.

10. The graph of $y = -(x + 2)^2 - 7$ can be obtained by moving $y = x^2$ two units to the left and down seven units and then reflecting in the x-axis.

11.2   EXERCISES

Reading and Writing   After reading this section, write out the answers to these questions. Use complete sentences.

1. What is a reflection in the x-axis of a graph?

2. What is an upward translation of a graph?

3. What is a downward translation of a graph?

4. What is a translation to the right of a graph?

5. What is a translation to the left of a graph?

6. What is stretching and shrinking of a graph?

Sketch the graphs of each pair of functions on the same coordinate system. See Example 1.

7. $f(x) = \sqrt{2x}$,
   $g(x) = -\sqrt{2x}$

8. $y = x, y = -x$

9. $f(x) = x^2 + 1$,
    $g(x) = -(x^2 + 1)$

10. $f(x) = |x| + 1$,
    $g(x) = -|x| - 1$

11. $y = \sqrt{x - 2}$,
    $y = -\sqrt{x - 2}$

12. $y = |x - 1|$,
    $y = -|x - 1|$

13. $f(x) = x - 3$,
    $g(x) = 3 - x$

14. $f(x) = x^2 - 2$,
    $g(x) = 2 - x^2$

Use the ideas of translation to graph each function. See Example 2.

15. $f(x) = x^2 - 4$

16. $f(x) = x^2 + 2$

17. $y = x + 3$

18. $y = x - 1$

19. $f(x) = (x - 3)^2$

20. $f(x) = (x + 1)^2$

21. $y = \sqrt{x} + 1$

22. $y = \sqrt{x} - 3$

23. $f(x) = |x + 2|$

24. $f(x) = |x - 4|$

25. $y = |x| + 2$

26. $y = |x| - 4$

27. $f(x) = \sqrt{x - 1}$

28. $f(x) = \sqrt{x + 6}$

Sketch the graph of each function. See Example 4.

37. $y = \sqrt{x - 2} + 1$

38. $y = -\sqrt{x + 3}$

Use the ideas of stretching and shrinking to graph each function. See Example 3.

29. $f(x) = 3x^2$

30. $f(x) = \dfrac{1}{3}x^2$

39. $y = -|x + 3|$

40. $y = |x - 2| + 1$

31. $y = \dfrac{1}{5}x$

32. $y = 5x$

41. $f(x) = (x + 3)^2 - 5$

42. $f(x) = -2x^2$

43. $y = -\sqrt{x + 1} - 2$

44. $y = -3\sqrt{x + 4} + 6$

33. $f(x) = 3\sqrt{x}$

34. $f(x) = \dfrac{1}{3}\sqrt{x}$

45. $y = -2|x - 3| + 4$

46. $y = 3|x - 1| + 2$

35. $y = \dfrac{1}{4}|x|$

36. $y = 4|x|$

47. $y = -2x + 3$ **48.** $y = 3x - 1$

Find the vertex and graph each parabola. See Example 5.

49. $y = 2(x + 3)^2 + 1$ **50.** $y = 2(x + 1)^2 - 2$

51. $y = -2(x - 4)^2 + 2$ **52.** $y = -2(x - 1)^2 + 3$

53. $y = -3(x - 1)^2 + 6$ **54.** $y = 3(x + 2)^2 - 6$

Match each function with its graph a–h.

55. $y = 2 + \sqrt{x}$ **56.** $y = \sqrt{2 + x}$

57. $y = 2\sqrt{x}$ **58.** $y = \sqrt{\dfrac{x}{2}}$

59. $y = \dfrac{1}{2}\sqrt{x}$ **60.** $y = 2 + \sqrt{x - 2}$

61. $y = -2\sqrt{x}$ **62.** $y = \sqrt{-x}$

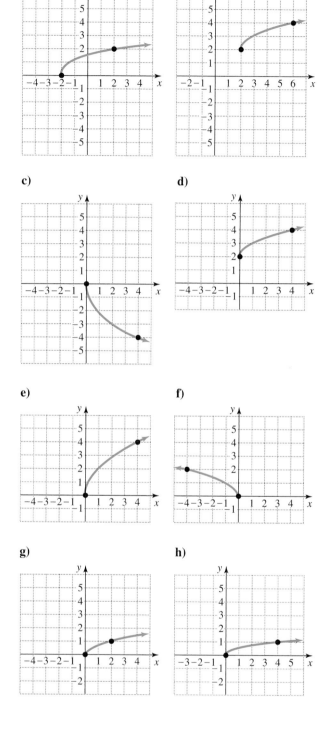

GETTING MORE INVOLVED

63. If the graph of $y = x^2$ is translated eight units upward, then what is the equation of the curve at that location?

64. If the graph of $y = x^2$ is translated six units to the right, then what is the equation of the curve at that location?

65. If the graph of $y = \sqrt{x}$ is translated five units to the left, then what is the equation of the curve at that location?

66. If the graph of $y = \sqrt{x}$ is translated four units downward, then what is the equation of the curve at that location?

67. If the graph of $y = |x|$ is translated three units to the left and then five units upward, then what is the equation of the curve at that location?

68. If the graph of $y = |x|$ is translated four units downward and then nine units to the right, then what is the equation of the curve at that location?

GRAPHING CALCULATOR EXERCISES

69. Graph $f(x) = |x|$ and $g(x) = |x - 20| + 30$ on the same screen of your calculator. What transformations will transform the graph of f into the graph of g?

70. Graph $f(x) = (x + 3)^2$, $g(x) = x^2 + 3^2$, and $h(x) = x^2 + 6x + 9$ on the same screen of your calculator.
   a) Which two of these functions has the same graph? Why are they the same?

   b) Is it true that $(x + 3)^2 = x^2 + 9$ for all real numbers x?
   c) Describe each graph in terms of a transformation of the graph of $y = x^2$.

11.3 COMBINING FUNCTIONS

In this section you will learn how to combine functions to obtain new functions.

In this

section

• Basic Operations with Functions

• Composition

Basic Operations with Functions

An entrepreneur plans to rent a stand at a farmers market for $25 per day to sell strawberries. If she buys x flats of berries for $5 per flat and sells them for $9 per flat, then her daily cost in dollars can be written as a function of x:

$$C(x) = 5x + 25$$

Assuming she sells as many flats as she buys, her revenue in dollars is also a function of x:

$$R(x) = 9x$$

Because profit is revenue minus cost, we can find a function for the profit by subtracting the functions for cost and revenue:

$$P(x) = R(x) - C(x)$$
$$= 9x - (5x + 25)$$
$$= 4x - 25$$

The function $P(x) = 4x - 25$ expresses the daily profit as a function of x. Since $P(6) = -1$ and $P(7) = 3$, the profit is negative if 6 or fewer flats are sold and positive if 7 or more flats are sold.

In the example of the entrepreneur we subtracted two functions to find a new function. In other cases we may use addition, multiplication, or division to combine two functions. For any two given functions we can define the sum, difference, product, and quotient functions as follows.

Sum, Difference, Product, and Quotient Functions

Given two functions f and g, the functions $f + g$, $f - g$, $f \cdot g$, and $\frac{f}{g}$ are defined as follows:

Sum function: $\qquad\qquad (f + g)(x) = f(x) + g(x)$

Difference function: $\qquad (f - g)(x) = f(x) - g(x)$

Product function: $\qquad\quad (f \cdot g)(x) = f(x) \cdot g(x)$

Quotient function: $\qquad\quad \left(\dfrac{f}{g}\right)(x) = \dfrac{f(x)}{g(x)} \qquad$ provided that $g(x) \neq 0$

The domain of the function $f + g$, $f - g$, $f \cdot g$, or $\frac{f}{g}$ is the intersection of the domain of f and the domain of g. For the function $\frac{f}{g}$ we also rule out any values of x for which $g(x) = 0$.

E X A M P L E   1

Operations with functions

Let $f(x) = 4x - 12$ and $g(x) = x - 3$. Find the following.

a) $(f + g)(x)$ 　　　　　　　　　　　　**b)** $(f - g)(x)$

c) $(f \cdot g)(x)$ 　　　　　　　　　　　　**d)** $\left(\dfrac{f}{g}\right)(x)$

Solution

helpful hint

Note that we use $f + g$, $f - g$, $f \cdot g$, and f/g to name these functions only because there is no application in mind here. We generally use a single letter to name functions after they are combined as we did when using P for the profit function rather than $R - C$.

a) $(f + g)(x) = f(x) + g(x)$
$\qquad\qquad = 4x - 12 + x - 3$
$\qquad\qquad = 5x - 15$

b) $(f - g)(x) = f(x) - g(x)$
$\qquad\qquad = 4x - 12 - (x - 3)$
$\qquad\qquad = 3x - 9$

c) $(f \cdot g)(x) = f(x) \cdot g(x)$
$\qquad\qquad = (4x - 12)(x - 3)$
$\qquad\qquad = 4x^2 - 24x + 36$

d) $\left(\dfrac{f}{g}\right)(x) = \dfrac{f(x)}{g(x)} = \dfrac{4x - 12}{x - 3} = \dfrac{4(x - 3)}{x - 3} = 4 \qquad$ for $x \neq 3$.

E X A M P L E   2

Evaluating a sum function

Let $f(x) = 4x - 12$ and $g(x) = x - 3$. Find $(f + g)(2)$.

Solution

In Example 1(a) we found a general formula for the function $f + g$, namely, $(f + g)(x) = 5x - 15$. If we replace x by 2, we get

$$(f + g)(2) = 5(2) - 15$$
$$= -5.$$

We can also find $(f + g)(2)$ by evaluating each function separately and then adding the results. Because $f(2) = -4$ and $g(2) = -1$, we get

$$(f + g)(2) = f(2) + g(2)$$
$$= -4 + (-1)$$
$$= -5.$$ ◼

Composition

A salesperson's monthly salary is a function of the number of cars he sells: $1000 plus $50 for each car sold. If we let S be his salary and n be the number of cars sold, then S in dollars is a function of n:

$$S = 1000 + 50n$$

Each month the dealer contributes $100 plus 5% of his salary to a profit-sharing plan. If P represents the amount put into profit sharing, then P (in dollars) is a function of S:

$$P = 100 + 0.05S$$

Now P is a function of S, and S is a function of n. Is P a function of n? The value of n certainly determines the value of P. In fact, we can write a formula for P in terms of n by substituting one formula into the other:

$$P = 100 + 0.05S$$
$$= 100 + 0.05(1000 + 50n) \quad \text{Substitute } S = 1000 + 50n.$$
$$= 100 + 50 + 2.5n \quad \text{Distributive property}$$
$$= 150 + 2.5n$$

Now P is written as a function of n, bypassing S. We call this idea **composition of functions.**

EXAMPLE 3

The composition of two functions
Given that $y = x^2 - 2x + 3$ and $z = 2y - 5$, write z as a function of x.

Solution
Replace y in $z = 2y - 5$ by $x^2 - 2x + 3$:

$$z = 2y - 5$$
$$= 2(x^2 - 2x + 3) - 5 \quad \text{Replace } y \text{ by } x^2 - 2x + 3.$$
$$= 2x^2 - 4x + 1$$

The equation $z = 2x^2 - 4x + 1$ expresses z as a function of x. ◼

The composition of two functions using f-notation is defined as follows.

Composition of Functions

The **composition** of f and g is denoted $f \circ g$ and is defined by the equation

$$(f \circ g)(x) = f(g(x)),$$

provided that $g(x)$ is in the domain of f.

The notation $f \circ g$ is read as "the composition of f and g" or "f compose g." The diagram in Fig. 11.22 shows a function g pairing numbers in its domain with numbers in its range. If the range of g is contained in or equal to the domain of f, then f pairs the second coordinates of g with numbers in the range of f. The composition function $f \circ g$ is a rule for pairing numbers in the domain of g directly with numbers in the range of f, bypassing the middle set. The domain of the function $f \circ g$ is the domain of g (or a subset of it) and the range of $f \circ g$ is the range of f (or a subset of it).

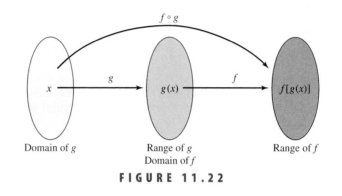

Domain of g Range of g Range of f
Domain of f

FIGURE 11.22

CAUTION The order in which functions are written is important in composition. For the function $f \circ g$ the function f is applied to $g(x)$. For the function $g \circ f$ the function g is applied to $f(x)$. The function closest to the variable x is applied first.

E X A M P L E   4

calculator

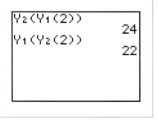

close-up

Set $y_1 = 3x - 2$ and $y_2 = x^2 + 2x$. You can find the composition for Examples 4(a) and 4(b) by evaluating $y_2(y_1(2))$ and $y_1(y_2(2))$. Note that the order in which you evaluate the functions is critical.

Composition of functions

Let $f(x) = 3x - 2$ and $g(x) = x^2 + 2x$. Find the following.

a) $(g \circ f)(2)$

b) $(f \circ g)(2)$

c) $(g \circ f)(x)$

d) $(f \circ g)(x)$

Solution

a) Because $(g \circ f)(2) = g(f(2))$, we first find $f(2)$:

$$f(2) = 3 \cdot 2 - 2 = 4$$

Because $f(2) = 4$, we have

$$(g \circ f)(2) = g(f(2)) = g(4) = 4^2 + 2 \cdot 4 = 24.$$

So $(g \circ f)(2) = 24$.

b) Because $(f \circ g)(2) = f(g(2))$, we first find $g(2)$:

$$g(2) = 2^2 + 2 \cdot 2 = 8$$

Because $g(2) = 8$, we have

$$(f \circ g)(2) = f(g(2)) = f(8) = 3 \cdot 8 - 2 = 22.$$

Thus $(f \circ g)(2) = 22$.

c) $(g \circ f)(x) = g(f(x))$
$$= g(3x - 2)$$
$$= (3x - 2)^2 + 2(3x - 2)$$
$$= 9x^2 - 12x + 4 + 6x - 4 = 9x^2 - 6x$$

So $(g \circ f)(x) = 9x^2 - 6x$.

d) $(f \circ g)(x) = f(g(x))$
$$= f(x^2 + 2x)$$
$$= 3(x^2 + 2x) - 2 = 3x^2 + 6x - 2$$

So $(f \circ g)(x) = 3x^2 + 6x - 2$.   ■

Notice that in Example 4(a) and (b), $(g \circ f)(2) \neq (f \circ g)(2)$. In Example 4(c) and (d) we see that $(g \circ f)(x)$ and $(f \circ g)(x)$ have different formulas defining them. In general, $f \circ g \neq g \circ f$. However, in Section 11.4 we will see some functions for which the composition in either order results in the same function.

It is often useful to view a complicated function as a composition of simpler functions. For example, the function $Q(x) = (x - 3)^2$ consists of two operations, subtracting 3 and squaring. So Q can be described as a composition of the functions $f(x) = x - 3$ and $g(x) = x^2$. To check this, we find $(g \circ f)(x)$:

$$(g \circ f)(x) = g(f(x))$$
$$= g(x - 3)$$
$$= (x - 3)^2$$

We can express the fact that Q is the same as the composition function $g \circ f$ by writing $Q = g \circ f$ or $Q(x) = (g \circ f)(x)$.

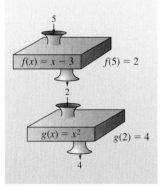
E X A M P L E   5

Expressing a function as a composition of simpler functions

Let $f(x) = x - 2$, $g(x) = 3x$, and $h(x) = \sqrt{x}$. Write each of the following functions as a composition, using f, g, and h.

a) $F(x) = \sqrt{x - 2}$

b) $H(x) = x - 4$

c) $K(x) = 3x - 6$

Solution

a) The function F consists of first subtracting 2 from x and then taking the square root of that result. So $F = h \circ f$. Check this result by finding $(h \circ f)(x)$:

$$(h \circ f)(x) = h(f(x)) = h(x - 2) = \sqrt{x - 2}$$

b) Subtracting 4 from x can be accomplished by subtracting 2 from x and then subtracting 2 from that result. So $H = f \circ f$. Check by finding $(f \circ f)(x)$:

$$(f \circ f)(x) = f(f(x)) = f(x - 2) = x - 2 - 2 = x - 4$$

c) Notice that $K(x) = 3(x - 2)$. The function K consists of subtracting 2 from x and then multiplying the result by 3. So $K = g \circ f$. Check by finding $(g \circ f)(x)$:

$$(g \circ f)(x) = g(f(x)) = g(x - 2) = 3(x - 2) = 3x - 6$$ ■

CAUTION In Example 5(a) we have $F = h \circ f$ because in F we subtract 2 before taking the square root. If we had the function $G(x) = \sqrt{x} - 2$, we would take the square root before subtracting 2. So $G = f \circ h$. Notice how important the order of operations is here.

In the next example we see functions for which the composition is the identity function. Each function undoes what the other function does. We will study functions of this type further in Section 11.4.

E X A M P L E   6

Composition of functions

Show that $(f \circ g)(x) = x$ for each pair of functions.

a) $f(x) = 2x - 1$ and $g(x) = \dfrac{x + 1}{2}$

b) $f(x) = x^3 + 5$ and $g(x) = (x - 5)^{1/3}$

Solution

a) $(f \circ g)(x) = f(g(x)) = f\left(\dfrac{x + 1}{2}\right)$

$$= 2\left(\dfrac{x + 1}{2}\right) - 1$$

$$= x + 1 - 1$$

$$= x$$

b) $(f \circ g)(x) = f(g(x)) = f((x - 5)^{1/3})$

$$= ((x - 5)^{1/3})^3 + 5$$

$$= x - 5 + 5$$

$$= x$$

WARM-UPS

True or false? Explain your answer.

1. If $f(x) = x - 2$ and $g(x) = x + 3$, then $(f - g)(x) = -5$.
2. If $f(x) = x + 4$ and $g(x) = 3x$, then $\left(\dfrac{f}{g}\right)(2) = 1$.
3. The functions $f \circ g$ and $g \circ f$ are always the same.
4. If $f(x) = x^2$ and $g(x) = x + 2$, then $(f \circ g)(x) = x^2 + 2$.
5. The functions $f \circ g$ and $f \cdot g$ are always the same.
6. If $f(x) = \sqrt{x}$ and $g(x) = x - 9$, then $g(f(x)) = f(g(x))$ for every x.
7. If $f(x) = 3x$ and $g(x) = \dfrac{x}{3}$, then $(f \circ g)(x) = x$.
8. If $a = 3b^2 - 7b$, and $c = a^2 + 3a$, then c is a function of b.
9. The function $F(x) = \sqrt{x - 5}$ is a composition of two functions.
10. If $F(x) = (x - 1)^2$, $h(x) = x - 1$, and $g(x) = x^2$, then $F = g \circ h$.

11.3   EXERCISES

Reading and Writing   *After reading this section, write out the answers to these questions. Use complete sentences.*

1. What are the basic operations with functions?

2. How do we perform the basic operations with functions?

3. What is the composition of two functions?

4. How is the order of operations related to composition of functions?

Let $f(x) = 4x - 3$, and $g(x) = x^2 - 2x$. Find the following. See Examples 1 and 2.

5. $(f + g)(x)$

6. $(f - g)(x)$

7. $(f \cdot g)(x)$

8. $\left(\dfrac{f}{g}\right)(x)$

9. $(f + g)(3)$

10. $(f + g)(2)$

11. $(f - g)(-3)$

12. $(f - g)(-2)$

13. $(f \cdot g)(-1)$

14. $(f \cdot g)(-2)$

15. $\left(\dfrac{f}{g}\right)(4)$

16. $\left(\dfrac{f}{g}\right)(-2)$

For Exercises 17–24, use the two functions to write y as a function of x. See Example 3.

17. $y = 3a - 2, a = 2x - 6$

18. $y = 2c + 3, c = -3x + 4$

19. $y = 2d + 1, d = \dfrac{x + 1}{2}$

20. $y = -3d + 2, d = \dfrac{2 - x}{3}$

21. $y = m^2 - 1, m = x + 1$

22. $y = n^2 - 3n + 1, n = x + 2$

23. $y = \dfrac{a - 3}{a + 2}, a = \dfrac{2x + 3}{1 - x}$

24. $y = \dfrac{w + 2}{w - 5}, w = \dfrac{5x + 2}{x - 1}$

Let $f(x) = 2x - 3$, $g(x) = x^2 + 3x$, and $h(x) = \dfrac{x + 3}{2}$. Find the following. See Example 4.

25. $(g \circ f)(1)$

26. $(f \circ g)(-2)$

27. $(f \circ g)(1)$

28. $(g \circ f)(-2)$

29. $(f \circ f)(4)$

30. $(h \circ h)(3)$

31. $(h \circ f)(5)$

32. $(f \circ h)(0)$

33. $(f \circ h)(5)$

34. $(h \circ f)(0)$

35. $(g \circ h)(-1)$

36. $(h \circ g)(-1)$

37. $(f \circ g)(2.36)$

38. $(h \circ f)(23.761)$

39. $(g \circ f)(x)$

40. $(g \circ h)(x)$

41. $(f \circ g)(x)$

42. $(h \circ g)(x)$

43. $(h \circ f)(x)$

44. $(f \circ h)(x)$

45. $(f \circ f)(x)$

46. $(g \circ g)(x)$

47. $(h \circ h)(x)$

48. $(f \circ f \circ f)(x)$

Let $f(x) = \sqrt{x}$, $g(x) = x^2$, and $h(x) = x - 3$. Write each of the following functions as a composition using f, g, or h. See Example 5.

49. $F(x) = \sqrt{x - 3}$

50. $N(x) = \sqrt{x} - 3$

51. $G(x) = x^2 - 6x + 9$

52. $P(x) = x$ for $x \geq 0$

53. $H(x) = x^2 - 3$

54. $M(x) = x^{1/4}$

55. $J(x) = x - 6$

56. $R(x) = \sqrt{x^2 - 3}$

57. $K(x) = x^4$

58. $Q(x) = \sqrt{x^2 - 6x + 9}$

Show that $(f \circ g)(x) = x$ and $(g \circ f)(x) = x$ for each given pair of functions. See Example 6.

59. $f(x) = 3x + 5, g(x) = \dfrac{x - 5}{3}$

60. $f(x) = 3x - 7, g(x) = \dfrac{x + 7}{3}$

61. $f(x) = x^3 - 9, g(x) = \sqrt[3]{x + 9}$

62. $f(x) = x^3 + 1, g(x) = \sqrt[3]{x - 1}$

63. $f(x) = \dfrac{x - 1}{x + 1}, g(x) = \dfrac{x + 1}{1 - x}$

64. $f(x) = \dfrac{x + 1}{x - 3}, g(x) = \dfrac{3x + 1}{x - 1}$

65. $f(x) = \dfrac{1}{x}, g(x) = \dfrac{1}{x}$

66. $f(x) = 2x^3, g(x) = \left(\dfrac{x}{2}\right)^{1/3}$

Solve each problem.

67. *Color monitor.* The CTX CMS-1561 color monitor has a square viewing area that has a diagonal measure of 15 inches (Midwest Micro Catalog). Find the area of the viewing area in square inches (in.²). Write a formula for the area of a square as a function of the length of its diagonal.

68. *Perimeter.* Write a formula for the perimeter of a square as a function of its area.

69. *Profit function.* A plastic bag manufacturer has determined that the company can sell as many bags as it can produce each month. If it produces x thousand bags in a month, the revenue is $R(x) = x^2 - 10x + 30$ dollars, and the cost is $C(x) = 2x^2 - 30x + 200$ dollars. Use the fact that profit is revenue minus cost to write the profit as a function of x.

70. *Area of a sign.* A sign is in the shape of a square with a semicircle of radius x adjoining one side and a semicircle of diameter x removed from the opposite side. If the sides of the square are length $2x$, then write the area of the sign as a function of x.

FIGURE FOR EXERCISE 70

71. *Junk food expenditures.* Suppose the average family spends 25% of its income on food, $F = 0.25I$, and 10%

of each food dollar on junk food, $J = 0.10F$. Write J as a function of I.

72. *Area of an inscribed circle.* A pipe of radius r must pass through a square hole of area M as shown in the figure. Write the cross-sectional area of the pipe A as a function of M.

FIGURE FOR EXERCISE 72

73. *Displacement-length ratio.* To find the displacement-length ratio D for a sailboat, first find x, where $x = (L/100)^3$ and L is the length at the water line in feet (*Sail*, September 1997). Next find D, where $D = (d/2240)/x$ and d is the displacement in pounds.
  a) For the Pacific Seacraft 40, $L = 30$ ft 3 in. and $d = 24{,}665$ pounds. Find D.
  b) For a boat with a displacement of 25,000 pounds, write D as a function of L.
  c) The graph for the function in part (b) is shown in the accompanying figure. For a fixed displacement, does the displacement-length ratio increase or decrease as the length increases?

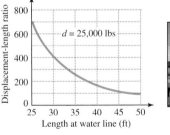

FIGURE FOR EXERCISE 73

74. *Sail area-displacement ratio.* To find the sail area-displacement ratio S, first find y, where $y = (d/64)^{2/3}$ and d is the displacement in pounds. Next find S, where $S = A/y$ and A is the sail area in square feet.
  a) For the Pacific Seacraft 40, $A = 846$ square feet (ft²) and $d = 24{,}665$ pounds. Find S.
  b) For a boat with a sail area of 900 ft², write S as a function of d.

c) For a fixed sail area, does S increase or decrease as the displacement increases?

GETTING MORE INVOLVED

 **75. *Discussion.*** Let $f(x) = \sqrt{x} - 4$ and $g(x) = \sqrt{x}$. Find the domains of f, g, and $g \circ f$.

76. *Discussion.* Let $f(x) = \sqrt{x - 4}$ and $g(x) = \sqrt{x - 8}$. Find the domains of f, g, and $f + g$.

 GRAPHING CALCULATOR EXERCISES

77. Graph $y_1 = x$, $y_2 = \sqrt{x}$, and $y_3 = x + \sqrt{x}$ in the same screen. Find the domain and range of $y_3 = x + \sqrt{x}$ by examining its graph. (On some graphing calculators you can enter y_3 as $y_3 = y_1 + y_2$.)

78. Graph $y_1 = |x|$, $y_2 = |x - 3|$, and $y_3 = |x| + |x - 3|$. Find the domain and range of $y_3 = |x| + |x - 3|$ by examining its graph.

11.4   INVERSE FUNCTIONS

In Section 11.3 we introduced the idea of a pair of functions such that $(f \circ g)(x) = x$ and $(g \circ f)(x) = x$. Each function reverses what the other function does. In this section we explore that idea further.

Inverse of a Function

You can buy a 6-, 7-, or 8-foot conference table in the K-LOG Catalog for \$299, \$329, or \$349, respectively. The set

$$f = \{(6, 299), (7, 329), (8, 349)\}$$

gives the price as a function of the length. We use the letter f as a name for this set or function, just as we use the letter f as a name for a function in the function notation. In the function f, lengths in the domain $\{6, 7, 8\}$ are paired with prices in the range $\{299, 329, 349\}$. The **inverse** of the function f, denoted f^{-1}, is a function whose ordered pairs are obtained from f by interchanging the x- and y-coordinates:

$$f^{-1} = \{(299, 6), (329, 7), (349, 8)\}$$

We read f^{-1} as "f inverse." The domain of f^{-1} is $\{299, 329, 349\}$, and the range of f^{-1} is $\{6, 7, 8\}$. The inverse function reverses what the function does: it pairs prices in the range of f with lengths in the domain of f. For example, to find the cost of a 7-foot table, we use the function f to get $f(7) = 329$. To find the length of a table, that costs \$349, we use the function f^{-1} to get $f^{-1}(349) = 8$. Of course, we could find the length of a \$349 table by looking at the function f, but f^{-1} is a function whose input is price and whose output is length. In general, *the domain of f^{-1} is the range of f, and the range of f^{-1} is the domain of f.* See Fig. 11.23.

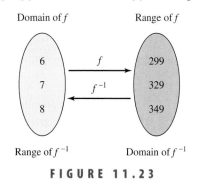

FIGURE 11.23

CAUTION The -1 in f^{-1} is not read as an exponent. It does not mean $\frac{1}{f}$.

The cost per ribbon for Apple Imagewriter ribbons is a function of the number of boxes purchased:

$$g = \{(1, 4.85), (2, 4.60), (3, 4.60), (4, 4.35)\}$$

If we interchange the first and second coordinates in the ordered pairs of this function, we get

$$\{(4.85, 1), (4.60, 2), (4.60, 3), (4.35, 4)\}.$$

This set of ordered pairs is not a function because it contains ordered pairs with the same first coordinates and different second coordinates. So g does not have an inverse function. A function is **invertible** if you obtain a function when the coordinates of all ordered pairs are reversed. So f is invertible and g is not invertible. The function g is not invertible because the definition of function allows more than one number of the domain to be paired with the same number in the range. Of course, when this pairing is reversed, the definition of function is violated.

One-to-One Function

If a function is such that no two ordered pairs have different x-coordinates and the same y-coordinate, then the function is called a **one-to-one** function.

In a one-to-one function each member of the domain corresponds to just one member of the range, and each member of the range corresponds to just one member of the domain. *Functions that are one-to-one are invertible functions.*

Inverse Function

The inverse of a one-to-one function f is the function f^{-1}, which is obtained from f by interchanging the coordinates in each ordered pair of f.

> **helpful hint**
>
> Consider the universal product codes (UPC) and the prices for all of the items in your favorite grocery store. The price of an item is a function of the UPC because every UPC determines a price. This function is not invertible because you cannot determine the UPC from a given price.

E X A M P L E   1

Identifying invertible functions

Determine whether each function is invertible. If it is invertible, then find the inverse function.

a) $f = \{(2, 4), (-2, 4), (3, 9)\}$

b) $g = \left\{ \left(2, \frac{1}{2}\right), \left(5, \frac{1}{5}\right), \left(7, \frac{1}{7}\right) \right\}$

c) $h = \{(3, 5), (7, 9)\}$

Solution

a) Since $(2, 4)$ and $(-2, 4)$ have the same y-coordinate, this function is not one-to-one, and it is not invertible.

b) This function is one-to-one, and so it is invertible.

$$g^{-1} = \left\{ \left(\frac{1}{2}, 2\right), \left(\frac{1}{5}, 5\right), \left(\frac{1}{7}, 7\right) \right\}$$

c) This function is invertible, and $h^{-1} = \{(5, 3), (9, 7)\}$. ■

You learned to use the vertical-line test in Section 4.6 to determine whether a graph is the graph of a function. The **horizontal-line test** is a similar visual test for

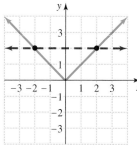

FIGURE 11.24

determining whether a function is invertible. If a horizontal line crosses a graph two (or more) times, as in Fig. 11.24, then there are two points on the graph, say (x_1, y) and (x_2, y), that have different x-coordinates and the same y-coordinate. So the function is not one-to-one, and the function is not invertible.

Horizontal-Line Test

A function is invertible if and only if no horizontal line crosses its graph more than once.

EXAMPLE 2   Using the horizontal-line test

Determine whether each function is invertible by examining its graph.

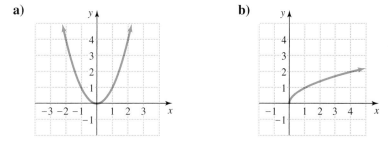

Solution

a) This function is not invertible because a horizontal line can be drawn so that it crosses the graph at $(2, 4)$ and $(-2, 4)$.

b) This function is invertible because every horizontal line that crosses the graph crosses it only once.   ■

Identifying Inverse Functions

Consider the one-to-one function $f(x) = 3x$. The inverse function must reverse the ordered pairs of the function. Because division by 3 undoes multiplication by 3, we could guess that $g(x) = \frac{x}{3}$ is the inverse function. To verify our guess, we can use the following rule for determining whether two given functions are inverses of each other.

Identifying Inverse Functions

Functions f and g are inverses of each other if and only if

$$(g \circ f)(x) = x \text{ for every number } x \text{ in the domain of } f \text{ and}$$
$$(f \circ g)(x) = x \text{ for every number } x \text{ in the domain of } g.$$

In the next example we verify that $f(x) = 3x$ and $g(x) = \frac{x}{3}$ are inverses.

EXAMPLE 3   Identifying inverse functions

Determine whether the functions f and g are inverses of each other.

a) $f(x) = 3x$ and $g(x) = \dfrac{x}{3}$ **b)** $f(x) = 2x - 1$ and $g(x) = \dfrac{1}{2}x + 1$

c) $f(x) = x^2$ and $g(x) = \sqrt{x}$

---

helpful hint

Tests such as the vertical-line test and the horizontal-line test are certainly not accurate in all cases. We discuss these tests to get a visual idea of what graphs of functions and invertible functions look like.

Solution

a) Find $g \circ f$ and $f \circ g$:

$$(g \circ f)(x) = g(f(x)) = g(3x) = \frac{3x}{3} = x$$

$$(f \circ g)(x) = f(g(x)) = f\left(\frac{x}{3}\right) = 3 \cdot \frac{x}{3} = x$$

Because each of these equations is true for any real number x, f and g are inverses of each other. We write $g = f^{-1}$ or $f^{-1}(x) = \frac{x}{3}$.

b) Find the composition of g and f:

$$(g \circ f)(x) = g(f(x))$$
$$= g(2x - 1) = \frac{1}{2}(2x - 1) + 1 = x + \frac{1}{2}$$

So f and g are not inverses of each other.

c) If x is any real number, we can write

$$(g \circ f)(x) = g(f(x))$$
$$= g(x^2) = \sqrt{x^2} = |x|.$$

The domain of f is $(-\infty, \infty)$, and $|x| \neq x$ if x is negative. So g and f are not inverses of each other. Note that $f(x) = x^2$ is not a one-to-one function, since both $(3, 9)$ and $(-3, 9)$ are ordered pairs of this function. Thus $f(x) = x^2$ does not have an inverse. ∎

Switch-and-Solve Strategy

If an invertible function is defined by a list of ordered pairs, as in Example 1, then the inverse function is found by simply interchanging the coordinates in the ordered pairs. If an invertible function is defined by a formula, then the inverse function must reverse or undo what the function does. Because the inverse function interchanges the roles of x and y, we interchange x and y in the formula and then solve the new formula for y to undo what the original function did. This **switch-and-solve** strategy is illustrated in the next two examples.

E X A M P L E   4

The switch-and-solve strategy

Find the inverse of $h(x) = 2x + 1$.

Solution

First write the function as $y = 2x + 1$, then interchange x and y:

$$y = 2x + 1$$
$$x = 2y + 1 \qquad \text{Interchange } x \text{ and } y.$$
$$x - 1 = 2y \qquad \text{Solve for } y.$$
$$\frac{x - 1}{2} = y$$
$$h^{-1}(x) = \frac{x - 1}{2} \qquad \text{Replace } y \text{ by } h^{-1}(x).$$

We can verify that h and h^{-1} are inverses by using composition:

$$(h^{-1} \circ h)(x) = h^{-1}(h(x)) = h^{-1}(2x + 1) = \frac{2x + 1 - 1}{2} = \frac{2x}{2} = x$$

$$(h \circ h^{-1})(x) = h(h^{-1}(x)) = h\left(\frac{x - 1}{2}\right) = 2 \cdot \frac{x - 1}{2} + 1 = x - 1 + 1 = x \quad \blacksquare$$

E X A M P L E   5   **The switch-and-solve strategy**

If $f(x) = \dfrac{x + 1}{x - 3}$, find $f^{-1}(x)$.

Solution

Replace $f(x)$ by y, interchange x and y, then solve for y:

$$y = \frac{x + 1}{x - 3} \qquad \text{Use } y \text{ in place of } f(x).$$

$$x = \frac{y + 1}{y - 3} \qquad \text{Switch } x \text{ and } y.$$

$$x(y - 3) = y + 1 \qquad \text{Multiply each side by } y - 3.$$

$$xy - 3x = y + 1 \qquad \text{Distributive property}$$

$$xy - y = 3x + 1$$

$$y(x - 1) = 3x + 1 \qquad \text{Factor out } y.$$

$$y = \frac{3x + 1}{x - 1} \qquad \text{Divide each side by } x - 1.$$

$$f^{-1}(x) = \frac{3x + 1}{x - 1} \qquad \text{Replace } y \text{ by } f^{-1}(x).$$

You should check that $(f \circ f^{-1})(x) = x$ and $(f^{-1} \circ f)(x) = x$. $\quad \blacksquare$

The strategy for finding the inverse of a function $f(x)$ is summarized as follows.

> **Switch-and-Solve Strategy for Finding f^{-1}**
>
> **1.** Replace $f(x)$ by y.
> **2.** Interchange x and y.
> **3.** Solve the equation for y.
> **4.** Replace y by $f^{-1}(x)$.

helpful hint

You should know from memory the inverses of simple functions that involve one or two operations. For example, the inverse of $f(x) = x + 99$ is $f^{-1}(x) = x - 99$. The inverse of $f(x) = x/33 + 22$ is $f^{-1}(x) = 33(x - 22)$.

If we use the switch-and-solve strategy to find the inverse of $f(x) = x^3$, then we get $f^{-1}(x) = x^{1/3}$. For $h(x) = 6x$ we have $h^{-1}(x) = \frac{x}{6}$. The inverse of $k(x) = x - 9$ is $k^{-1}(x) = x + 9$. For each of these functions there is an appropriate operation of arithmetic that undoes what the function does.

If a function involves two operations, the inverse function undoes those operations in the opposite order from which the function does them. For example, the function $g(x) = 3x - 5$ multiplies x by 3 and then subtracts 5 from that result.

To undo these operations, we add 5 and then divide the result by 3. So

$$g^{-1}(x) = \frac{x+5}{3}.$$

Note that $g^{-1}(x) \neq \frac{x}{3} + 5$.

Even Roots or Even Powers

We need to use special care in finding inverses for functions that involve even roots or even powers. We saw in Example 3(c) that $f(x) = x^2$ is not the inverse of $g(x) = \sqrt{x}$. However, because $g(x) = \sqrt{x}$ is a one-to-one function, it has an inverse. The domain of g is $[0, \infty)$, and the range is $[0, \infty)$. So the inverse of g must have domain $[0, \infty)$ and range $[0, \infty)$. See Fig. 11.25. The only reason that

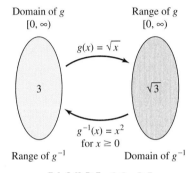

FIGURE 11.25

When a serious automobile accident occurs in Massachusetts, Stephen Benanti, the Commanding Officer of the Accident Reconstruction Section of the State Police, may be called to examine the physical evidence at the scene. Physical evidence can consist of debris, scrapes, and gouges on the road; damage to fixed objects such as utility poles, trees, or guardrails; and the final resting position of and damage to the vehicles.

STATE POLICE OFFICER

One critical type of evidence that is sometimes found are skid marks on the road. Sergeant Benanti can use the lengths of the skid marks to calculate the speeds of accident vehicles. Using a sophisticated laser measuring device, Benanti can store data that later can be downloaded into a computer to reconstruct the accident scene. He must also conduct tests on the road surface to calculate the drag factor, which is the resistance between the tire and the road surface. A smooth, icy surface yields a much lower drag factor than a dry asphalt surface. The minimum speed formula that state troopers use to determine vehicles' speeds is a function of the drag factor and the skid distance: $S = \sqrt{30DF}$, where S = speed, D = distance skidded, and F = drag factor.

In Exercise 73 of this section you will use this minimum speed formula with a given length of skid marks to determine the speed of a vehicle.

$f(x) = x^2$ is not the inverse of g is that it has the wrong domain. So to write the inverse function, we must use the appropriate domain:

$$g^{-1}(x) = x^2 \qquad \text{for} \quad x \geq 0$$

Note that by restricting the domain of g^{-1} to $[0, \infty)$, g^{-1} is one-to-one. With this restriction it is true that $(g \circ g^{-1})(x) = x$ and $(g^{-1} \circ g)(x) = x$ for every nonnegative number x.

E X A M P L E   6   **Inverse of a function with an even exponent**

Find the inverse of the function $f(x) = (x - 3)^2$ for $x \geq 3$.

Solution

Because of the restriction $x \geq 3$, f is a one-to-one function with domain $[3, \infty)$ and range $[0, \infty)$. The domain of the inverse function is $[0, \infty)$, and its range is $[3, \infty)$. Use the switch-and-solve strategy to find the formula for the inverse:

$$y = (x - 3)^2$$
$$x = (y - 3)^2$$
$$y - 3 = \pm\sqrt{x}$$
$$y = 3 \pm \sqrt{x}$$

Because the inverse function must have range $[3, \infty)$, we use the formula $f^{-1}(x) = 3 + \sqrt{x}$. Because the domain of f^{-1} is assumed to be $[0, \infty)$, no restriction is required on x. ∎

Graphs of f and f^{-1}

Consider $f(x) = x^2$ for $x \geq 0$ and $f^{-1}(x) = \sqrt{x}$. Their graphs are shown in Fig. 11.26. Notice the symmetry. If we folded the paper along the line $y = x$, the two graphs would coincide.

If a point (a, b) is on the graph of the function f, then (b, a) must be on the graph of $f^{-1}(x)$. See Fig. 11.27. The points (a, b) and (b, a) lie on opposite sides of the diagonal line $y = x$ and are the same distance from it. For this reason the graphs of f and f^{-1} are symmetric with respect to the line $y = x$.

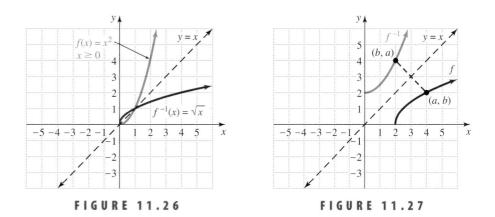

FIGURE 11.26   **FIGURE 11.27**

E X A M P L E 7

Inverses and their graphs

Find the inverse of the function $f(x) = \sqrt{x-1}$ and graph f and f^{-1} on the same pair of axes.

Solution

To find f^{-1}, first switch x and y in the formula $y = \sqrt{x-1}$:

$$x = \sqrt{y-1}$$
$$x^2 = y - 1 \qquad \text{Square both sides.}$$
$$x^2 + 1 = y$$

Because the range of f is the set of nonnegative real numbers $[0, \infty)$, we must restrict the domain of f^{-1} to be $[0, \infty)$. Thus $f^{-1}(x) = x^2 + 1$ for $x \geq 0$. The two graphs are shown in Fig. 11.28.

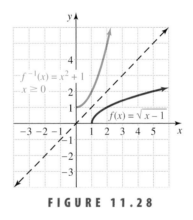

FIGURE 11.28

WARM-UPS

True or false? Explain your answer.

1. The inverse of $\{(1, 3), (2, 5)\}$ is $\{(3, 1), (2, 5)\}$.

2. The function $f(x) = 3$ is a one-to-one function.

3. If $g(x) = 2x$, then $g^{-1}(x) = \frac{1}{2x}$.

4. Only one-to-one functions are invertible.

5. The domain of g is the same as the range of g^{-1}.

6. The function $f(x) = x^4$ is invertible.

7. If $f(x) = -x$, then $f^{-1}(x) = -x$.

8. If h is invertible and $h(7) = -95$, then $h^{-1}(-95) = 7$.

9. If $k(x) = 3x - 6$, then $k^{-1}(x) = \frac{1}{3}x + 2$.

10. If $f(x) = 3x - 4$, then $f^{-1}(x) = x + 4$.

11.4   EXERCISES

Reading and Writing   *After reading this section, write out the answers to these questions. Use complete sentences.*

1. What is the inverse of a function?

2. What is the domain of f^{-1}?

3. What is the range of f^{-1}?

4. What does the -1 in f^{-1} mean?

5. What is a one-to-one function?

6. What is the horizontal-line test?

7. What is the switch-and-solve strategy?

8. How are the graphs of f and f^{-1} related?

Determine whether each function is invertible. If it is invertible, then find the inverse. See Example 1.

9. $\{(-3, 3), (-2, 2), (0, 0), (2, 2)\}$

10. $\{(1, 1), (2, 8), (3, 27)\}$

11. $\{(16, 4), (9, 3), (0, 0)\}$

12. $\{(-1, 1), (-3, 81), (3, 81)\}$

13. $\{(0, 5), (5, 0), (6, 0)\}$

14. $\{(3, -3), (-2, 2), (1, -1)\}$

15. $\{(0, 0), (2, 2), (9, 9)\}$

16. $\{(9, 1), (2, 1), (7, 1), (0, 1)\}$

Determine whether each function is invertible by examining the graph of the function. See Example 2.

17. **18.**

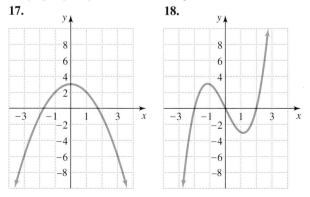

19. **20.**

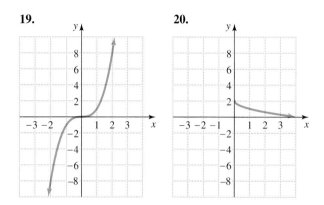

Determine whether each pair of functions f and g are inverses of each other. See Example 3.

21. $f(x) = 2x$ and $g(x) = 0.5x$

22. $f(x) = 3x$ and $g(x) = 0.33x$

23. $f(x) = 2x - 10$ and $g(x) = \dfrac{1}{2}x + 5$

24. $f(x) = 3x + 7$ and $g(x) = \dfrac{x - 7}{3}$

25. $f(x) = -x$ and $g(x) = -x$

26. $f(x) = \dfrac{1}{x}$ and $g(x) = \dfrac{1}{x}$

27. $f(x) = x^4$ and $g(x) = x^{1/4}$

28. $f(x) = |2x|$ and $g(x) = \left|\dfrac{x}{2}\right|$

Determine f^{-1} for each function by using the switch-and-solve strategy. Check that $(f \circ f^{-1})(x) = x$ and $(f^{-1} \circ f)(x) = x$. See Examples 4 and 5.

29. $f(x) = 5x$ **30.** $h(x) = -3x$

31. $g(x) = x - 9$ **32.** $j(x) = x + 7$

33. $k(x) = 5x - 9$ **34.** $r(x) = 2x - 8$

35. $m(x) = \dfrac{2}{x}$ **36.** $s(x) = \dfrac{-1}{x}$

37. $f(x) = \sqrt[3]{x - 4}$ **38.** $f(x) = \sqrt[3]{x + 2}$

39. $f(x) = \dfrac{3}{x - 4}$ **40.** $f(x) = \dfrac{2}{x + 1}$

41. $f(x) = \sqrt[3]{3x + 7}$ **42.** $f(x) = \sqrt[3]{7 - 5x}$

43. $f(x) = \dfrac{x + 1}{x - 2}$

44. $f(x) = \dfrac{1 - x}{x + 3}$

58. $f(x) = x^2 + 3$ for $x \geq 0$

45. $f(x) = \dfrac{x + 1}{3x - 4}$

46. $g(x) = \dfrac{3x + 5}{2x - 3}$

Find the inverse of each function. See Example 6.

47. $p(x) = \sqrt[6]{x}$

48. $v(x) = \sqrt[6]{x}$

59. $f(x) = 5x$

49. $f(x) = (x - 2)^2$ for $x \geq 2$

50. $g(x) = (x + 5)^2$ for $x \geq -5$

51. $f(x) = x^2 + 3$ for $x \geq 0$

52. $f(x) = x^2 - 5$ for $x \geq 0$

53. $f(x) = \sqrt{x + 2}$

54. $f(x) = \sqrt{x - 4}$

In Exercises 55–64, find the inverse of each function and graph f and f^{-1} on the same pair of axes. See Example 7.

55. $f(x) = 2x + 3$

60. $f(x) = \dfrac{x}{4}$

56. $f(x) = -3x + 2$

61. $f(x) = x^3$

57. $f(x) = x^2 - 1$ for $x \geq 0$

62. $f(x) = 2x^3$

63. $f(x) = \sqrt{x - 2}$

64. $f(x) = \sqrt{x + 3}$

For each pair of functions, find $(f^{-1} \circ f)(x)$

65. $f(x) = x^3 - 1$ and $f^{-1}(x) = \sqrt[3]{x + 1}$

66. $f(x) = 2x^3 + 1$ and $f^{-1}(x) = \sqrt[3]{\dfrac{x - 1}{2}}$

67. $f(x) = \dfrac{1}{2}x - 3$ and $f^{-1}(x) = 2x + 6$

68. $f(x) = 3x - 9$ and $f^{-1}(x) = \dfrac{1}{3}x + 3$

69. $f(x) = \dfrac{1}{x} + 2$ and $f^{-1}(x) = \dfrac{1}{x - 2}$

70. $f(x) = 4 - \dfrac{1}{x}$ and $f^{-1}(x) = \dfrac{1}{4 - x}$

71. $f(x) = \dfrac{x + 1}{x - 2}$ and $f^{-1}(x) = \dfrac{2x + 1}{x - 1}$

72. $f(x) = \dfrac{3x - 2}{x + 2}$ and $f^{-1}(x) = \dfrac{2x + 2}{3 - x}$

Solve each problem.

73. *Accident reconstruction.* The distance that it takes a car to stop is a function of the speed and the drag factor. The drag factor is a measure of the resistance between the tire and the road surface. The formula $S = \sqrt{30LD}$ is used to determine the minimum speed S [in miles per hour (mph)] for a car that has left skid marks of length L feet (ft) on a surface with drag factor D.
  a) Find the minimum speed for a car that has left skid marks of length 50 ft where the drag factor is 0.75.

  b) Does the drag factor increase or decrease for a road surface when it gets wet?

  c) Write L as a function S for a road surface with drag factor 1 and graph the function.

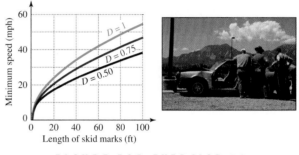

FIGURE FOR EXERCISE 73

74. *Area of a circle.* Let x be the radius of a circle and $h(x)$ be the area of the circle. Write a formula for $h(x)$ in terms of x. What does x represent in the notation $h^{-1}(x)$? Write a formula for $h^{-1}(x)$.

75. *Vehicle cost.* At Bill Hood Ford in Hammond a sales tax of 9% of the selling price x and a \$125 title and license fee are added to the selling price to get the total cost of a vehicle. Find the function $T(x)$ that the dealer uses to get the total cost as a function of the selling price x. Citizens National Bank will not include sales tax or fees in a loan. Find the function $T^{-1}(x)$ that the bank can use to get the selling price as a function of the total cost x.

76. *Carpeting cost.* At the Windrush Trace apartment complex all living rooms are square, but the length of x feet may vary. The cost of carpeting a living room is \$18 per square yard plus a \$50 installation fee. Find the function $C(x)$ that gives the total cost of carpeting a living room of length x. The manager has an invoice for the total cost of a living room carpeting job but does not know in which apartment it was done. Find the function $C^{-1}(x)$ that gives the length of a living room as a function of the total cost of the carpeting job x.

GETTING MORE INVOLVED

77. Discussion. Let $f(x) = x^n$ for n a positive integer. For which values of n is f an invertible function? Explain.

78. Discussion. Suppose f is a function with range $(-\infty, \infty)$ and g is a function with domain $(0, \infty)$. Is it possible that g and f are inverse functions? Explain.

GRAPHING CALCULATOR EXERCISES

79. Most graphing calculators can form compositions of functions. Let $f(x) = x^2$ and $g(x) = \sqrt{x}$. To graph the composition $g \circ f$, let $y_1 = x^2$ and $y_2 = \sqrt{y_1}$. The graph of y_2 is the graph of $g \circ f$. Use the graph of y_2 to determine whether f and g are inverse functions.

80. Let $y_1 = x^3 - 4$, $y_2 = \sqrt[3]{x + 4}$, and $y_3 = \sqrt[3]{y_1 + 4}$. The function y_3 is the composition of the first two functions. Graph all three functions on the same screen. What do the graphs indicate about the relationship between y_1 and y_2?

11.5 VARIATION

If $y = 3x$, then as x varies so does y. Certain functions are customarily expressed in terms of variation. In this section you will learn to write formulas for those functions from verbal descriptions of the functions.

Direct Variation

In a community with an 8% sales tax rate, the amount of tax, t (in dollars), is a function of the amount of the purchase, a (in dollars). This function is expressed by the formula

$$t = 0.08a.$$

If the amount increases, then the tax increases. If a decreases, then t decreases. In this situation we say that t *varies directly with a, or t is directly proportional to a.* The constant tax rate, 0.08, is called the **variation constant** or **proportionality constant.** Notice that t is just a simple linear function of a. We are merely introducing some new terms to express an old idea.

> **Direct Variation**
>
> The statement **y varies directly as x,** or **y is directly proportional to x,** means that
>
> $$y = kx$$
>
> for some constant, k. The constant, k, is a fixed nonzero real number.

Finding the Proportionality Constant

If y varies directly as x and we know corresponding values for x and y, then we can find the proportionality constant.

EXAMPLE 1

calculator

close-up

The graph of $d = 40t$ is a straight line through the origin.

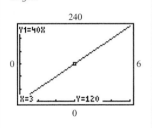

Finding the proportionality constant

Joyce is traveling by car, and the distance she travels, d, varies directly with the amount of time, t, that she drives. In 3 hours she drove 120 miles. Find the proportionality constant and write d as a function of t.

Solution

Because d varies directly as t, we must have a constant k such that

$$d = kt.$$

Because $d = 120$ when $t = 3$, we can write

$$120 = k \cdot 3,$$

or

$$40 = k.$$

So the proportionality constant is 40 mph, and $d = 40t$.

EXAMPLE 2

Direct variation

In a downtown office building the monthly rent for an office is directly proportional to the size of the office. If a 420-square-foot office rents for \$1260 per month, then what is the rent for a 900-square-foot office?

Solution

Because the rent, R, varies directly with the area of the office, A, we have

$$R = kA.$$

Because a 420-square-foot office rents for \$1260, we can substitute to find k:

$$1260 = k \cdot 420$$
$$3 = k$$

Now that we know the value of k, we can write

$$R = 3A.$$

To get the rent for a 900-square-foot office, insert 900 into this formula:

$$R = 3 \cdot 900$$
$$= 2700$$

So a 900-square-foot office rents for \$2700 per month.

calculator

close-up

The graph of $T = 500/R$ shows the time decreasing as the rate increases. For 50 mph the time is 10 hours, whereas for 100 mph the time is 5 hours.

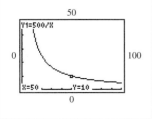

Inverse Variation

In making a 500-mile trip by car, the time it takes is a function of the speed of the car. The greater the speed, the less time it will take. If you decrease the speed, the time increases. We say that the time is *inversely proportional* to the speed. Using the formula $D = RT$ or $T = \frac{D}{R}$, we can write

$$T = \frac{500}{R}.$$

In general, we make the following definition.

> **Inverse Variation**
>
> The statement **y varies inversely as x,** or **y is inversely proportional to x,** means that
>
> $$y = \frac{k}{x}$$
>
> for some nonzero constant, k.

CAUTION Be sure to understand the difference between direct and inverse variation. If y varies directly as x (with $k > 0$), then as x increases, y increases. If y varies inversely as x (with $k > 0$), then as x increases, y decreases.

EXAMPLE 3

Inverse variation

Suppose a is inversely proportional to b, and when $b = 5$, $a = \frac{1}{2}$. Find a when $b = 12$.

Solution

Because a is inversely proportional to b, we have

$$a = \frac{k}{b}$$

for some constant, k. Because $a = \frac{1}{2}$ when $b = 5$, we can find k by substituting these values into the formula:

$$\frac{1}{2} = \frac{k}{5}$$

$$\frac{5}{2} = k \qquad \text{Multiply each side by 5.}$$

Now to find a when $b = 12$, we use the formula with k replaced by $\frac{5}{2}$:

$$a = \frac{\frac{5}{2}}{b}$$

$$a = \frac{\frac{5}{2}}{12} = \frac{5}{2} \cdot \frac{1}{12} = \frac{5}{24}$$

■

Joint Variation

On a deposit of $5000 in a savings account, the interest earned, I, depends on the rate, r, and the time, t. Assuming the interest is simple interest, we can use the formula $I = Prt$ to write

$$I = 5000rt.$$

The variable I is a function of two independent variables, r and t. In this case we say that I *varies jointly* as r and t.

Joint Variation

The statement **y varies jointly as x and z,** or **y is jointly proportional to x and z,** means that

$$y = kxz$$

for some nonzero constant, k.

E X A M P L E   4

Joint variation

Suppose y varies jointly with x and z, and $y = 12$ when $x = 5$ and $z = 2$. Find y when $x = 10$ and $z = -3$.

Solution

Because y varies jointly with x and z, we can write

$$y = kxz$$

for some constant, k. Now substitute $y = 12$, $x = 5$, and $z = 2$, and solve for k:

$$12 = k \cdot 5 \cdot 2$$
$$12 = 10k$$
$$\frac{6}{5} = k$$

Now that we know the value of k, we can rewrite the equation as

$$y = \frac{6}{5}xz.$$

To find y when $x = 10$ and $z = -3$, substitute into the equation:

$$y = \frac{6}{5}(10)(-3)$$
$$y = -36$$

More Variation

We frequently combine the ideas of direct, inverse, and joint variation with powers and roots. A combination of direct and inverse variation is referred to as **combined variation.** Study the examples that follow.

helpful hint

The language of variation is popular in science. Instead of saying $V = kT/P$, a chemist would say that the volume of a gas varies directly with the temperature and inversely with the pressure.

More Variation Examples	
Statement	**Formula**
y varies directly as the square root of x.	$y = k\sqrt{x}$
y is directly proportional to the cube of x.	$y = kx^3$
y is inversely proportional to x^2.	$y = \dfrac{k}{x^2}$
y varies inversely as the square root of x.	$y = \dfrac{k}{\sqrt{x}}$
y varies jointly as x and the square of z.	$y = kxz^2$
y varies directly with x and inversely with the square root of z (combined variation).	$y = \dfrac{kx}{\sqrt{z}}$

> **CAUTION** The variation terms never signify addition or subtraction. We always use multiplication unless we see the word "inversely." In that case we divide.

EXAMPLE 5

Newton's law of gravity

According to Newton's law of gravity, the gravitational attraction F between two objects with masses m_1 and m_2 is directly proportional to the product of their masses and inversely proportional to the square of the distance r between their centers. Write a formula for Newton's law of gravity.

Solution

Letting k be the constant of proportionality, we have

$$F = \frac{km_1 m_2}{r^2}.$$

EXAMPLE 6

House framing

The time t that it takes to frame a house varies directly with the size of the house s in square feet and inversely with the number of framers n working on the job. If three framers can complete a 2500-square-foot house in 6 days, then how long will it take six framers to complete a 4500-square-foot house?

Solution

Because t varies directly with s and inversely with n, we have

$$t = \frac{ks}{n}.$$

Substitute $t = 6$, $s = 2500$, and $n = 3$ into this equation to find k:

$$6 = \frac{k \cdot 2500}{3}$$
$$18 = 2500k$$
$$0.0072 = k$$

Now use $k = 0.0072$, $s = 4500$, and $n = 6$ to find t:

$$t = \frac{0.0072 \cdot 4500}{6}$$
$$t = 5.4$$

So six framers can frame a 4500-square-foot house in 5.4 days.

WARM-UPS

True or false? Explain your answer.

1. If a varies directly as b, then $a = kb$.
2. If a is inversely proportional to b, then $a = bk$.
3. If a is jointly proportional to b and c, then $a = bc$.
4. If a is directly proportional to the square root of c, then $a = k\sqrt{c}$.
5. If b is directly proportional to a, then $b = ka^2$.

(continued)

6. If a varies directly as b and inversely as c, then $a = \dfrac{kb}{c}$.

7. If a is jointly proportional to c and the square of b, then $a = \dfrac{kc}{b^2}$.

8. If a varies directly as c and inversely as the square root of b, then $a = \dfrac{kc}{b}$.

9. If b varies directly as a and inversely as the square of c, then $b = ka\sqrt{c}$.

10. If b varies inversely with the square of c, then $b = \dfrac{k}{c^2}$.

11.5   EXERCISES

Reading and Writing *After reading this section, write out the answers to these questions. Use complete sentences.*

1. What does it mean that y varies directly as x?

2. What is the constant of proportionality in a direct variation?

3. What does it mean that y is inversely proportional to x?

4. What is the difference between direct and inverse variation?

5. What does it mean that y is jointly proportional to x and z?

6. What is the difference between varies directly and directly proportional?

Write a formula that expresses the relationship described by each statement. Use k as a constant of variation. See Examples 1–6.

7. a varies directly as m.

8. w varies directly with P.

9. d varies inversely with e.

10. y varies inversely as x.

11. I varies jointly as r and t.

12. q varies jointly as w and v.

13. m is directly proportional to the square of p.

14. g is directly proportional to the cube of r.

15. B is directly proportional to the cube root of w.

16. F is directly proportional to the square of m.

17. t is inversely proportional to the square of x.

18. y is inversely proportional to the square root of z.

19. v varies directly as m and inversely as n.

20. b varies directly as the square of n and inversely as the square root of v.

Find the proportionality constant and write a formula that expresses the indicated variation. See Example 1.

21. y varies directly as x, and $y = 6$ when $x = 4$.

22. m varies directly as w, and $m = \frac{1}{3}$ when $w = \frac{1}{4}$.

23. A varies inversely as B, and $A = 10$ when $B = 3$.

24. c varies inversely as d, and $c = 0.31$ when $d = 2$.

25. m varies inversely as the square root of p, and $m = 12$ when $p = 9$.

26. s varies inversely as the square root of v, and $s = 6$ when $v = \frac{3}{2}$.

27. A varies jointly as t and u, and $A = 6$ when $t = 5$ and $u = 3$.

28. N varies jointly as the square of p and the cube of q, and $N = 72$ when $p = 3$ and $q = 2$.

29. y varies directly as x and inversely as z, and $y = 2.37$ when $x = \pi$ and $z = \sqrt{2}$.

30. a varies directly as the square root of m and inversely as the square of n, and $a = 5.47$ when $m = 3$ and $n = 1.625$.

Solve each variation problem. See Examples 2–6.

31. If y varies directly as x, and $y = 7$ when $x = 5$, find y when $x = -3$.

32. If n varies directly as p, and $n = 0.6$ when $p = 0.2$, find n when $p = \sqrt{2}$.

33. If w varies inversely as z, and $w = 6$ when $z = 2$, find w when $z = -8$.

34. If p varies inversely as q, and $p = 5$ when $q = \sqrt{3}$, find p when $q = 5$.

35. If A varies jointly as F and T, and $A = 6$ when $F = 3\sqrt{2}$ and $T = 4$, find A when $F = 2\sqrt{2}$ and $T = \frac{1}{2}$.

36. If j varies jointly as the square of r and the cube of v, and $j = -3$ when $r = 2\sqrt{3}$ and $v = \frac{1}{2}$, find j when $r = 3\sqrt{5}$ and $v = 2$.

37. If D varies directly with t and inversely with the square of s, and $D = 12.35$ when $t = 2.8$ and $s = 2.48$, find D when $t = 5.63$ and $s = 6.81$.

38. If M varies jointly with x and the square of v, and $M = 39.5$ when $x = \sqrt{10}$ and $v = 3.87$, find M when $x = \sqrt{30}$ and $v = 7.21$.

Determine whether each equation represents direct, inverse, joint, or combined variation.

39. $y = \dfrac{78}{x}$

40. $y = \dfrac{\pi}{x}$

41. $y = \dfrac{1}{2}x$

42. $y = \dfrac{x}{4}$

43. $y = \dfrac{3x}{w}$

44. $y = \dfrac{4t^2}{\sqrt{x}}$

45. $y = \dfrac{1}{3}xz$

46. $y = 99qv$

In Exercises 47–61, solve each problem.

47. **Lawn maintenance.** At Larry's Lawn Service the cost of lawn maintenance varies directly with the size of the lawn. If the monthly maintenance on a 4000-square-foot lawn is $280, then what is the maintenance fee for a 6000-square-foot lawn?

48. **Weight of the iguana.** The weight of an iguana is directly proportional to its length. If a 4-foot iguana weighs 30 pounds, then how much should a 5-foot iguana weigh?

49. **Gas laws.** The volume of a gas in a cylinder at a fixed temperature is inversely proportional to the weight on the piston. If the gas has a volume of 6 cubic centimeters (cm^3) for a weight of 30 kilograms (kg), then what would the volume be for a weight of 20 kg?

50. **Selling software.** A software vendor sells a software package at a price that is inversely proportional to the number of packages sold per month. When they are selling 900 packages per month, the price is $80 each. If they sell 1000 packages per month, then what should the new price be?

51. **Costly culvert.** The price of an aluminum culvert is jointly proportional to its radius and length. If a 12-foot culvert with a 6-inch radius costs $324, then what is the price of a 10-foot culvert with an 8-inch radius?

52. **Pricing plastic.** The cost of a piece of PVC water pipe varies jointly as its diameter and length. If a 20-foot pipe with a diameter of 1 inch costs $6.80, then what will be the cost of a 10-foot pipe with a $\frac{3}{4}$-inch diameter?

53. **Reinforcing rods.** The price of a steel rod varies jointly as the length and the square of the diameter. If an 18-foot rod with a 2-inch diameter costs $12.60, then what is the cost of a 12-foot rod with a 3-inch diameter?

54. **Pea soup.** The weight of a cylindrical can of pea soup varies jointly with the height and the square of the radius. If a 4-inch-high can with a 1.5-inch radius weighs 16 ounces, then what is the weight of a 5-inch-high can with a radius of 3 inches?

55. *Falling objects.* The distance an object falls in a vacuum varies directly with the square of the time it is falling. In the first 0.1 second after an object is dropped, it falls 0.16 feet.

a) Find the formula that expresses the distance d an object falls as a function of the time it is falling t.

b) How far does an object fall in the first 0.5 second after it is dropped?

c) How long does it take for a watermelon to reach the ground when dropped from a height of 100 feet?

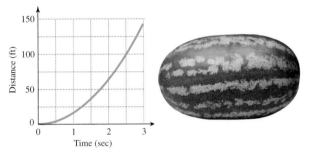

FIGURE FOR EXERCISE 55

56. *Making Frisbees.* The cost of material used in making a Frisbee varies directly with the square of the diameter. If it costs the manufacturer $0.45 for the material in a Frisbee with a 9-inch diameter, then what is the cost for the material in a 12-inch-diameter Frisbee?

57. *Using leverage.* The basic law of leverage is that the force required to lift an object is inversely proportional to the length of the lever. If a force of 2000 pounds applied 2 feet from the pivot point would lift a car, then what force would be required at 10 feet to lift the car?

58. *Resistance.* The resistance of a wire varies directly with the length and inversely as the square of the diameter. If a wire of length 20 feet and diameter 0.1 inch has a resistance of 2 ohms, then what is the resistance of a 30-foot wire with a diameter of 0.2 inch?

59. *Computer programming.* The time t required to complete a programming job varies directly with the complexity of the job and inversely with the number n of programmers working on the job. The complexity c is an arbitrarily assigned number between 1 and 10, with 10 being the most complex. It takes 8 days for a team of three programmers to complete a job with complexity 6. How long will it take five programmers to complete a job with complexity 9?

60. *Shock absorbers.* The volume of gas in a gas shock absorber varies directly with the temperature and inversely

with the pressure. The volume is 10 cubic centimeters (cm^3) when the temperature is 20°C and the pressure is 40 kg. What is the volume when the temperature is 30°C and the pressure is 25 kg?

61. *Bicycle gear ratio.* A bicycle's gear ratio G varies jointly with the number of teeth on the chain ring N (by the pedals) and the diameter of the wheel d, and inversely with the number of teeth on the cog c (on the rear wheel). A bicycle with 27-inch-diameter wheels, 26 teeth on the cog, and 52 teeth on the chain ring has a gear ratio of 54.

a) Find a formula that expresses the gear ratio as a function of N, d, and c.

b) What is the gear ratio for a bicycle with 26-inch-diameter wheels, 42 teeth on the chain ring, and 13 teeth on the cog?

c) A five-speed bicycle with 27-inch-diameter wheels and 44 teeth on the chain ring has gear ratios of 52, 59, 70, 79, and 91. Find the number of teeth on the cog for each gear ratio.

d) For a fixed wheel size and chain ring, does the gear ratio increase or decrease as the number of teeth on the cog increases?

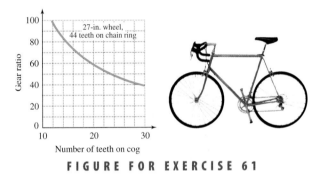

FIGURE FOR EXERCISE 61

 GRAPHING CALCULATOR EXERCISES

62. To see the difference between direct and inverse variation, graph $y_1 = 2x$ and $y_2 = \frac{2}{x}$ using $0 \le x \le 5$ and $0 \le y \le 10$. Which of these functions is increasing and which is decreasing?

63. Graph $y_1 = 2\sqrt{x}$ and $y_2 = \frac{2}{\sqrt{x}}$ by using $0 \le x \le 5$ and $0 \le y \le 10$. At what point in the first quadrant do the curves cross? Which function is increasing and which is decreasing? Which represents direct variation and which represents inverse variation?

11.6 SYNTHETIC DIVISION AND THE FACTOR THEOREM

In this section we study functions defined by polynomials and learn to solve some higher-degree polynomial equations.

Synthetic Division

When dividing a polynomial by a binomial of the form $x - c$, we can use **synthetic division** to speed up the process. For synthetic division we write only the essential parts of ordinary division. For example, to divide $x^3 - 5x^2 + 4x - 3$ by $x - 2$, we write only the coefficients of the dividend $1, -5, 4,$ and -3 in order of descending exponents. From the divisor $x - 2$ we use 2 and start with the following arrangement:

$$2 \,\big|\, 1 \quad -5 \quad 4 \quad -3 \qquad {\scriptstyle (1 \cdot x^3 - 5x^2 + 4x - 3) \div (x - 2)}$$

Next we bring the first coefficient, 1, straight down:

$$
\begin{array}{c|cccc}
2 & 1 & -5 & 4 & -3 \\
 & \downarrow & & \text{Bring down} & \\
\hline
 & 1 & & &
\end{array}
$$

We then multiply the 1 by the 2 from the divisor, place the answer under the -5, and then add that column. Using 2 for $x - 2$ allows us to add the column rather than subtract as in ordinary division:

$$
\begin{array}{c|cccc}
2 & 1 & -5 & 4 & -3 \\
 & & 2 & & \quad\text{Add} \\
\hline
 \text{Multiply} & 1 & -3 & &
\end{array}
$$

We then repeat the multiply-and-add step for each of the remaining columns:

$$
\begin{array}{c|cccc}
2 & 1 & -5 & 4 & -3 \\
 & & 2 & -6 & -4 \\
\hline
 \text{Multiply} & \underbrace{1 \quad -3 \quad -2}_{\text{Quotient}} & & & -7 \quad \leftarrow \text{Remainder}
\end{array}
$$

From the bottom row we can read the quotient and remainder. Since the degree of the quotient is one less than the degree of the dividend, the quotient is $1x^2 - 3x - 2$. The remainder is -7.

The strategy for getting the quotient $Q(x)$ and remainder R by synthetic division can be stated as follows.

Strategy for Using Synthetic Division

1. List the coefficients of the polynomial (the dividend).
2. Be sure to include zeros for any missing terms in the dividend.
3. For dividing by $x - c$, place c to the left.
4. Bring the first coefficient down.
5. Multiply by c and add for each column.
6. Read $Q(x)$ and R from the bottom row.

CAUTION   Synthetic division is used only for dividing a polynomial by the binomial $x - c$, where c is a constant. If the binomial is $x - 7$, then $c = 7$. For the binomial $x + 7$ we have $x + 7 = x - (-7)$ and $c = -7$.

E X A M P L E   1   **Using synthetic division**

Find the quotient and remainder when $2x^4 - 5x^2 + 6x - 9$ is divided by $x + 2$.

Solution

Since $x + 2 = x - (-2)$, we use -2 for the divisor. Because x^3 is missing in the dividend, use a zero for the coefficient of x^3:

$$
\begin{array}{r|rrrrr}
-2 & 2 & 0 & -5 & 6 & -9 \\
   &   & -4 & 8 & -6 & 0 \\
\hline
   & 2 & -4 & 3 & 0 & -9
\end{array}
$$

$\leftarrow 2x^4 + 0 \cdot x^3 - 5x^2 + 6x - 9$

Add

Multiply

$\leftarrow$ Quotient and remainder

Because the degree of the dividend is 4, the degree of the quotient is 3. The quotient is $2x^3 - 4x^2 + 3x$, and the remainder is -9. ∎

The Factor Theorem

Consider the polynomial function

$$P(x) = x^2 + 2x - 15.$$

The values of x for which $P(x) = 0$ are called the **zeros** or **roots** of the function. We can find the zeros of the function by solving the equation $P(x) = 0$:

$$x^2 + 2x - 15 = 0$$
$$(x + 5)(x - 3) = 0$$
$$x + 5 = 0 \quad \text{or} \quad x - 3 = 0$$
$$x = -5 \quad \text{or} \quad x = 3$$

Because $x + 5$ is a factor of $x^2 + 2x - 15$, -5 is a solution to the equation $x^2 + 2x - 15 = 0$ and a zero of the function $P(x) = x^2 + 2x - 15$. We can check that -5 is a zero of $P(x) = x^2 + 2x - 15$ as follows:

$$P(-5) = (-5)^2 + 2(-5) - 15$$
$$= 25 - 10 - 15$$
$$= 0$$

Because $x - 3$ is a factor of the polynomial, 3 is also a solution to the equation $x^2 + 2x - 15 = 0$ and a zero of the polynomial function. Check that $P(3) = 0$:

$$P(3) = 3^2 + 2 \cdot 3 - 15$$
$$= 9 + 6 - 15$$
$$= 0$$

Every linear factor of the polynomial corresponds to a zero of the polynomial function, and every zero of the polynomial function corresponds to a linear factor.

Now suppose $P(x)$ represents an arbitrary polynomial. If $x - c$ is a factor of the polynomial $P(x)$, then c is a solution to the equation $P(x) = 0$, and so $P(c) = 0$. If we divide $P(x)$ by $x - c$ and the remainder is 0, we must have

$$P(x) = (x - c)(\text{quotient}). \quad \text{Dividend equals the divisor times the quotient.}$$

If the remainder is 0, then $x - c$ is a factor of $P(x)$.

The **factor theorem** summarizes these ideas.

The Factor Theorem

The following statements are equivalent for any polynomial $P(x)$.

1. The remainder is zero when $P(x)$ is divided by $x - c$.
2. $x - c$ is a factor of $P(x)$.
3. c is a solution to $P(x) = 0$.
4. c is a zero of the function $P(x)$, or $P(c) = 0$.

To say that statements are equivalent means that the truth of any one of them implies that the others are true.

According to the factor theorem, if we want to determine whether a given number c is a zero of a polynomial function, we can divide the polynomial by $x - c$. The remainder is zero if and only if c is a zero of the polynomial function. The quickest way to divide by $x - c$ is to use synthetic division.

E X A M P L E   2

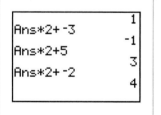

calculator close-up

You can perform the multiply-and-add steps for synthetic division with a graphing calculator as shown here.

```
Ans*2+-3          1
                 -1
Ans*2+5
                  3
Ans*2+-2
                  4
```

Using the factor theorem

Use synthetic division to determine whether 2 is a zero of

$$P(x) = x^3 - 3x^2 + 5x - 2.$$

Solution

By the factor theorem, 2 is a zero of the function if and only if the remainder is zero when $P(x)$ is divided by $x - 2$. We can use synthetic division to determine the remainder. If we divide by $x - 2$, we use 2 on the left in synthetic division along with the coefficients $1, -3, 5, -2$ from the polynomial:

$$
\begin{array}{r|rrrr}
2 & 1 & -3 & 5 & -2 \\
  &   & 2 & -2 & 6 \\
\hline
  & 1 & -1 & 3 & 4
\end{array}
$$

Because the remainder is 4, 2 is not a zero of the function.   ■

E X A M P L E   3

Using the factor theorem

Use synthetic division to determine whether -4 is a solution to the equation $2x^4 - 28x^2 + 14x - 8 = 0$.

Solution

By the factor theorem, -4 is a solution to the equation if and only if the remainder is zero when $P(x)$ is divided by $x + 4$. When dividing by $x + 4$, we use -4 in the synthetic division:

$$
\begin{array}{r|rrrrr}
-4 & 2 & 0 & -28 & 14 & -8 \\
   &   & -8 & 32 & -16 & 8 \\
\hline
   & 2 & -8 & 4 & -2 & 0
\end{array}
$$

Because the remainder is zero, -4 is a solution to $2x^4 - 28x^2 + 14x - 8 = 0$.   ■

In the next example we use the factor theorem to determine whether a given binomial is a factor of a polynomial.

EXAMPLE 4

Using the factor theorem

Use synthetic division to determine whether $x + 4$ is a factor of $x^3 + 3x^2 + 16$.

Solution

According to the factor theorem, $x + 4$ is a factor of $x^3 + 3x^2 + 16$ if and only if the remainder is zero when the polynomial is divided by $x + 4$. Use synthetic division to determine the remainder:

$$-4 \,\big|\, \begin{array}{cccc} 1 & 3 & 0 & 16 \\ & -4 & 4 & -16 \\ \hline 1 & -1 & 4 & 0 \end{array}$$

Because the remainder is zero, $x + 4$ is a factor, and the polynomial can be written as

$$x^3 + 3x^2 + 16 = (x + 4)(x^2 - x + 4).$$

Because $x^2 - x + 4$ is a prime polynomial, the factoring is complete. ∎

Solving Polynomial Equations

The techniques used to solve polynomial equations of degree 3 or higher are not as straightforward as those used to solve linear equations and quadratic equations. The next example shows how the factor theorem can be used to solve a third-degree polynomial equation.

EXAMPLE 5

Solving a third-degree equation

Suppose the equation $x^3 - 4x^2 - 17x + 60 = 0$ is known to have a solution that is an integer between -3 and 3 inclusive. Find the solution set.

Solution

Because one of the numbers $-3, -2, -1, 0, 1, 2,$ and 3 is a solution to the equation, we can use synthetic division with these numbers until we discover which one is a solution. We arbitrarily select 1 to try first:

$$1 \,\big|\, \begin{array}{cccc} 1 & -4 & -17 & 60 \\ & 1 & -3 & -20 \\ \hline 1 & -3 & -20 & 40 \end{array}$$

Because the remainder is 40, 1 is not a solution to the equation. Next try 2:

$$2 \,\big|\, \begin{array}{cccc} 1 & -4 & -17 & 60 \\ & 2 & -4 & -42 \\ \hline 1 & -2 & -21 & 18 \end{array}$$

Because the remainder is not zero, 2 is not a solution to the equation. Next try 3:

$$3 \,\big|\, \begin{array}{cccc} 1 & -4 & -17 & 60 \\ & 3 & -3 & -60 \\ \hline 1 & -1 & -20 & 0 \end{array}$$

The remainder is zero, so 3 is a solution to the equation, and $x - 3$ is a factor of the polynomial. (If 3 had not produced a remainder of zero, then we would have tried $-3, -2, -1,$ and 0.) The other factor is the quotient, $x^2 - x - 20$.

$$x^3 - 4x^2 - 17x + 60 = 0$$
$$(x - 3)(x^2 - x - 20) = 0 \quad \text{Use the results of synthetic division to factor.}$$
$$(x - 3)(x - 5)(x + 4) = 0 \quad \text{Factor completely.}$$
$$x - 3 = 0 \quad \text{or} \quad x - 5 = 0 \quad \text{or} \quad x + 4 = 0$$
$$x = 3 \quad \text{or} \quad x = 5 \quad \text{or} \quad x = -4$$

Check each of these solutions in the original equation. The solution set is $\{3, 5, -4\}$. ■

WARM-UPS

True or false? Explain your answers.

1. To divide $x^3 - 4x^2 - 3$ by $x - 5$, use 5 in the synthetic division.
2. To divide $5x^4 - x^3 + x - 2$ by $x + 7$, use -7 in the synthetic division.

3. The number 2 is a zero of $P(x) = 3x^3 - 5x^2 - 2x + 2$.
4. If $x^3 - 8$ is divided by $x - 2$, then $R = 0$.
5. If $R = 0$ when $x^4 - 1$ is divided by $x - a$, then $x - a$ is a factor of $x^4 - 1$.

6. If -2 satisfies $x^4 + 8x = 0$, then $x + 2$ is a factor of $x^4 + 8x$.
7. The binomial $x - 1$ is a factor of $x^{35} - 3x^{24} + 2x^{18}$.
8. The binomial $x + 1$ is a factor of $x^3 - 3x^2 + x + 5$.
9. If $x^3 - 5x + 4$ is divided by $x - 1$, then $R = 0$.
10. If $R = 0$ when $P(x) = x^3 - 5x - 2$ is divided by $x + 2$, then $P(-2) = 0$.

11.6   EXERCISES

Reading and Writing   *After reading this section, write out the answers to these questions. Use complete sentences.*

1. What is a zero of a function?

2. What is a root of a function?

3. What does it mean that statements are equivalent?

4. What is the quickest way to divide a polynomial by $x - c$?

5. If the remainder is zero when you divide $P(x)$ by $x - c$, then what can you say about $P(c)$?

6. What are two ways to determine whether c is a zero of a polynomial?

Use synthetic division to find the quotient and remainder when the first polynomial is divided by the second. See Example 1.

7. $x^3 - 5x^2 + 6x - 3, \quad x - 2$
8. $x^3 + 6x^2 - 3x - 5, \quad x - 3$
9. $2x^2 - 4x + 5, \quad x + 1$
10. $3x^2 - 7x + 4, \quad x + 2$
11. $3x^4 - 15x^2 + 7x - 9, \quad x - 3$

12. $-2x^4 + 3x^2 - 5, \quad x - 2$

13. $x^5 - 1$, $x - 1$

14. $x^6 - 1$, $x + 1$

15. $x^3 - 5x + 6$, $x + 2$

16. $x^3 - 3x - 7$, $x - 4$

17. $2.3x^2 - 0.14x + 0.6$, $x - 0.32$

18. $1.6x^2 - 3.5x + 4.7$, $x + 1.8$

Determine whether each given value of x is a zero of the given function. See Example 2.

19. $x = 1$, $P(x) = x^3 - x^2 + x - 1$

20. $x = -2$, $P(x) = -2x^3 - 5x^2 + 3x + 10$

21. $x = -3$, $P(x) = -x^4 - 3x^3 - 2x^2 + 18$

22. $x = 4$, $P(x) = x^4 - x^2 - 8x - 16$

23. $x = 2$, $P(x) = 2x^3 - 4x^2 - 5x + 9$

24. $x = -3$, $P(x) = x^3 + 5x^2 + 2x + 1$

Use synthetic division to determine whether each given value of x is a solution to the given equation. See Example 3.

25. $x = -3$, $x^3 + 5x^2 + 2x - 12 = 0$

26. $x = -5$, $x^2 - 3x - 40 = 0$

27. $x = -2$, $x^4 + 3x^3 - 5x^2 - 10x + 5 = 0$

28. $x = -3$, $-x^3 - 4x^2 + x + 12 = 0$

29. $x = 4$, $-2x^4 + 30x^2 + 5x + 12 = 0$

30. $x = 6$, $x^4 + x^3 - 40x^2 - 72 = 0$

31. $x = 3$, $0.8x^2 - 0.3x - 6.3 = 0$

32. $x = 5$, $6.2x^2 - 28.2x - 41.7 = 0$

Use synthetic division to determine whether the first polynomial is a factor of the second. If it is, then factor the polynomial completely. See Example 4.

33. $x - 3$, $x^3 - 6x - 9$

34. $x + 2$, $x^3 - 6x - 4$

35. $x + 5$, $x^3 + 9x^2 + 23x + 15$

36. $x - 3$, $x^4 - 9x^2 + x - 7$

37. $x - 2$, $x^3 - 8x^2 + 4x - 6$

38. $x + 5$, $x^3 + 125$

39. $x + 1$, $x^4 + x^3 - 8x - 8$

40. $x - 2$, $x^3 - 6x^2 + 12x - 8$

41. $x - 0.5$, $2x^3 - 3x^2 - 11x + 6$

42. $x - \dfrac{1}{3}$, $3x^3 - 10x^2 - 27x + 10$

Solve each equation, given that at least one of the solutions to each equation is an integer between -5 and 5. See Example 5.

43. $x^3 - 13x + 12 = 0$

44. $x^3 + 2x^2 - 5x - 6 = 0$

45. $2x^3 - 9x^2 + 7x + 6 = 0$

46. $6x^3 + 13x^2 - 4 = 0$

47. $2x^3 - 3x^2 - 50x - 24 = 0$

48. $x^3 - 7x^2 + 2x + 40 = 0$

49. $x^3 + 5x^2 + 3x - 9 = 0$

50. $x^3 + 6x^2 + 12x + 8 = 0$

51. $x^4 - 4x^3 + 3x^2 + 4x - 4 = 0$

52. $x^4 + x^3 - 7x^2 - x + 6 = 0$

GETTING MORE INVOLVED

53. *Exploration.* We can find the zeros of a polynomial function by solving a polynomial equation. We can also work backward to find a polynomial function that has given zeros.

  a) Write a first-degree polynomial function whose zero is -2.

  b) Write a second-degree polynomial function whose zeros are 5 and -5.

  c) Write a third-degree polynomial function whose zeros are 1, -3, and 4.

  d) Is there a polynomial function with any given number of zeros? What is its degree?

GRAPHING CALCULATOR EXERCISES

54. The x-coordinate of each x-intercept on the graph of a polynomial function is a zero of the polynomial function. Find the zeros of each function from its graph. Use synthetic division to check that the zeros found on your calculator really are zeros of the function.

  a) $P(x) = x^3 - 2x^2 - 5x + 6$

  b) $P(x) = 12x^3 - 20x^2 + x + 3$

55. With a graphing calculator an equation can be solved without the kind of hint that was given for Exercises 43–52. Solve each of the following equations by examining the graph of a corresponding function. Use synthetic division to check.

  a) $x^3 - 4x^2 - 7x + 10 = 0$

  b) $8x^3 - 20x^2 - 18x + 45 = 0$

COLLABORATIVE ACTIVITIES

Betting on Rockets

Gretchen and Rafael are launching model rockets in the park. They have a bet to see whose rocket can go the highest. Both have taken this math course and have some clues on how to use algebra to find out whose went the highest. Below is the formula that they will use to determine whose rocket has gone the highest:

$$h = -16t^2 + v_0 t,$$

where h is the distance (feet) of the rocket above the ground, t is the time passed (seconds) since the rocket was launched, and v_0 is the speed of the rocket when launched (initial velocity).

They have decided to do three trials and to see whose is the highest two out of the three times.

From the following list, have each member of your group choose a different trial.

• **First Trial:** On the first launch Gretchen's rocket has a time in the air of 14 seconds, and Rafael's rocket has a time of 10 seconds.

Grouping: 3 students per group
Topic: Applications of maximum and minimum of a parabola

• **Second Trial:** On the second launch Gretchen's rocket was in the air 11.5 seconds, and Rafael's rocket was in the air 12 seconds.

• **Third Trial:** On the third launch Gretchen's rocket was in the air 11.25 seconds, and Rafael's rocket was in the air 11.3 seconds.

Individually, complete the following for the trial you selected.

1. Find the initial velocity for each rocket using the formula $h = -16t^2 + v_0 t$.

2. Graph the two quadratic functions. What is the h-value for each vertex? Whose rocket went the highest?

Together in your groups, compare your results and decide who wins the bet.

WRAP-UP   CHAPTER 11

SUMMARY

Types of Functions

Type		Examples
Linear function	$y = mx + b$ or $f(x) = mx + b$ for $m \neq 0$. Domain $(-\infty, \infty)$, range $(-\infty, \infty)$. If $m = 0$, $y = b$ is a constant function. Domain $(-\infty, \infty)$, range $\{b\}$	$f(x) = 2x - 3$
Absolute value function	$y = |x|$ or $f(x) = |x|$. Domain $(-\infty, \infty)$, range $[0, \infty)$	$f(x) = |x + 5|$
Quadratic function	$f(x) = ax^2 + bx + c$ for $a \neq 0$	$f(x) = x^2 - 4x + 3$
Square-root function	$f(x) = \sqrt{x}$. Domain $[0, \infty)$, range $[0, \infty)$	$f(x) = \sqrt{x - 4}$
Vertical-line test	If a graph can be crossed more than once by a vertical line, then it is not the graph of a function.	

Transformations of Graphs

Reflecting	The graph of $y = -f(x)$ is a reflection in the x-axis of the graph of $y = f(x)$.	The graph of $y = -x^2$ is a reflection of the graph of $y = x^2$.

Translating	If $k > 0$, the graph of $y = f(x) + k$ is k units above $y = f(x)$ and $y = f(x) - k$ is k units below $y = f(x)$. If $h > 0$, the graph of $y = f(x - h)$ is h units to the right of $y = f(x)$ and $y = f(x + h)$ is h units to the left of $y = f(x)$.	The graph of $y = x^2 + 3$ is three units above $y = x^2$, and $y = x^2 - 3$ is three units below $y = x^2$. The graph of $y = (x - 3)^2$ is three units to the right of $y = x^2$, and $y = (x + 3)^2$ is three units to the left.
Stretching and shrinking	The graph of $y = af(x)$ is obtained by stretching (if $a > 1$) or shrinking (if $0 < a < 1$) the graph of $y = f(x)$.	The graph of $y = 5x^2$ is obtained by stretching $y = x^2$, and $y = 0.1x^2$ is obtained by shrinking $y = x^2$.

Combining Functions

Examples

Sum	$(f + g)(x) = f(x) + g(x)$	For $f(x) = x^2$ and $g(x) = x + 1$ $(f + g)(x) = x^2 + x + 1$
Difference	$(f - g)(x) = f(x) - g(x)$	$(f - g)(x) = x^2 - x - 1$
Product	$(f \cdot g)(x) = f(x) \cdot g(x)$	$(f \cdot g)(x) = x^3 + x^2$
Quotient	$\left(\dfrac{f}{g}\right)(x) = \dfrac{f(x)}{g(x)}$	$\left(\dfrac{f}{g}\right)(x) = \dfrac{x^2}{x + 1}$
Composition of functions	$(g \circ f)(x) = g(f(x))$ $(f \circ g)(x) = f(g(x))$	$(g \circ f)(x) = g(x^2) = x^2 + 1$ $(f \circ g)(x) = f(x + 1)$ $\qquad = x^2 + 2x + 1$

Inverse Functions

Examples

One-to-one function	A function in which no two ordered pairs have different x-coordinates and the same y-coordinate.	$f = \{(2, 20), (3, 30)\}$
Inverse function	The inverse of a one-to-one function f is the function f^{-1}, which is obtained from f by interchanging the coordinates in each ordered pair of f. The domain of f^{-1} is the range of f, and the range of f^{-1} is the domain of f.	$f^{-1} = \{(20, 2), (30, 3)\}$
Horizontal-line test	If there is a horizontal line that crosses the graph of a function more than once, then the function is not invertible.	
Function notation for inverse	Two functions f and g are inverses of each other if and only if both of the following conditions are met. 1. $(g \circ f)(x) = x$ for every number x in the domain of f. 2. $(f \circ g)(x) = x$ for every number x in the domain of g.	$f(x) = x^3 + 1$ $f^{-1}(x) = \sqrt[3]{x - 1}$

Switch-and-solve strategy for finding f^{-1}

1. Replace $f(x)$ by y.
2. Interchange x and y.
3. Solve for y.
4. Replace y by $f^{-1}(x)$.

$$y = x^3 + 1$$
$$x = y^3 + 1$$
$$x - 1 = y^3$$
$$y = \sqrt[3]{x - 1}$$
$$f^{-1}(x) = \sqrt[3]{x - 1}$$

Graphs of f and f^{-1}

Graphs of inverse functions are symmetric with respect to the line $y = x$.

The Language of Variation

Examples

Direct

y varies directly as x, $y = kx$

$z = 5m$

Inverse

y varies inversely as x, $y = \dfrac{k}{x}$

$a = \dfrac{1}{c}$

Joint

y varies jointly as x and z, $y = kxz$

$V = 6LW$

Combined

y varies directly as x and inversely as z, $y = \dfrac{kx}{z}$

$S = \dfrac{3A}{B}$

The Factor Theorem

Examples

Factor theorem

The following are equivalent for $P(x)$, a polynomial in x.
1. The remainder is zero when $P(x)$ is divided by $x - c$.
2. $x - c$ is a factor of $P(x)$.
3. c is a solution to $P(x) = 0$.
4. c is a zero of the function $P(x)$, or $P(c) = 0$.

$P(x) = x^2 - x - 2$

$P(x) = (x - 2)(x + 1)$

$P(-1) = 0$, $P(2) = 0$
-1 and 2 both satisfy
$x^2 - x - 2 = 0$

ENRICHING YOUR MATHEMATICAL WORD POWER

For each mathematical term, choose the correct meaning.

1. **composition of f and g**
   a. the function $f \circ g$ where $(f \circ g)(x) = f(g(x))$
   b. the function $f \circ g$ where $(f \circ g)(x) = g(f(x))$
   c. the function $f \cdot g$ where $(f \cdot g)(x) = f(x) \cdot g(x)$
   d. a diagram showing f and g

2. **sum of f and g**
   a. the function $f \cdot g$ where $(f \cdot g)(x) = f(x) \cdot g(x)$
   b. the function $f + g$ where $(f + g)(x) = f(x) + g(x)$
   c. the function $f \circ g$ where $(f \circ g)(x) = g(f(x))$
   d. the function obtained by adding the domains of f and g

3. **inverse of the function f**
   a. a function with the same ordered pairs as f
   b. the opposite of the function f
   c. the function $1/f$
   d. a function in which the ordered pairs of f are reversed

4. **one-to-one function**
   a. a constant function
   b. a function that pairs 1 with 1
   c. a function in which no two ordered pairs have the same first coordinate and different second coordinates
   d. a function in which no two ordered pairs have the same second coordinate and different first coordinates

5. **vertical-line test**
   a. a visual method for determining whether a graph is a graph of a function
   b. a visual method for determining whether a function is one-to-one

c.   using a vertical line to check a graph

d.   a test on vertical lines

6.   **horizontal-line test**

a.   a test that horizontal lines must pass

b.   a visual method for determining whether a function is one-to-one

c.   a graph that does not cross the x-axis

d.   a visual method for determining whether a graph is a graph of a function

7.   **proportionality constant**

a.   a direct variation

b.   a constant proportion

c.   a ratio that is constant

d.   the constant k in $y = kx$

8.   **y varies directly as x**

a. $y = kx^2$ where k is a constant

b. $y = mx + b$ where m and b are nonzero constants

c. $y = kx$ where k is a nonzero constant

d. $y = k/x$ where k is a nonzero constant

9.   **y varies inversely as x**

a. $y = x/k$ where k is a nonzero constant

b. $y = -x$

c. $y = kx$ where k is a nonzero constant

d. $y = k/x$ where k is a nonzero constant

10.   **zero of a function**

a.   a number a such that $f(0) = a$

b.   a number a such that $f(a) = 0$

c.   the value of $f(0)$

d.   the number zero

11.   **reflection in the x-axis**

a.   the graph of $y = f(-x)$

b.   the graph of $y = -f(x)$

c.   the graph of $y = -f(-x)$

d.   the line of symmetry

12.   **upward translation**

a.   the graph of $y = f(x) + c$ for $c > 0$

b.   the graph of $y = f(x + c)$ for $c < 0$

c.   the graph of $y = f(x - c)$ for $c > 0$

d.   the graph of $y = f(x) + c$ for $c < 0$

13.   **translation to the left**

a.   the graph of $y = f(x) - c$ for $c > 0$

b.   the graph of $y = f(x) + c$ for $c > 0$

c.   the graph of $y = f(x - c)$ for $c > 0$

d.   the graph of $y = f(x + c)$ for $c > 0$

REVIEW EXERCISES

11.1 *Graph each function and state the domain and range.*

1. $f(x) = 3x - 4$

2. $y = 0.3x$

3. $h(x) = |x| - 2$

4. $y = |x - 2|$

5. $y = x^2 - 2x + 1$

6. $g(x) = x^2 - 2x - 15$

7. $k(x) = \sqrt{x} + 2$

8. $y = \sqrt{x - 2}$

9. $y = 30 - x^2$

10. $y = 4 - x^2$

Graph each relation and state its domain and range.

11. $x = 2$

12. $x = y^2 - 1$

13. $x = |y| + 1$

14. $x = \sqrt{y - 1}$

11.2 *Sketch the graph of each function.*

15. $y = \sqrt{x}$

16. $y = -\sqrt{x}$

17. $y = -2\sqrt{x}$

18. $y = 2\sqrt{x}$

19. $y = \sqrt{x - 2}$

20. $y = \sqrt{x + 2}$

21. $y = \frac{1}{2}\sqrt{x}$

22. $y = \sqrt{x - 1} + 2$

23. $y = -\sqrt{x + 1} + 3$

24. $y = 3\sqrt{x + 4} - 5$

11.3 *Let $f(x) = 3x + 5$, $g(x) = x^2 - 2x$, and $h(x) = \frac{x - 5}{3}$.*
Find the following.

25. $f(-3)$

26. $h(-4)$

27. $(h \circ f)(\sqrt{2})$

28. $(f \circ h)(\pi)$

29. $(g \circ f)(2)$

30. $(g \circ f)(x)$

31. $(f + g)(3)$

32. $(f - g)(x)$

33. $(f \cdot g)(x)$

34. $\left(\dfrac{f}{g}\right)(1)$

35. $(f \circ f)(0)$

36. $(f \circ f)(x)$

Let $f(x) = |x|$, $g(x) = x + 2$, and $h(x) = x^2$. Write each of the following functions as a composition of functions, using f, g, or h.

37. $F(x) = |x + 2|$

38. $G(x) = |x| + 2$

39. $H(x) = x^2 + 2$

40. $K(x) = x^2 + 4x + 4$

41. $I(x) = x + 4$

42. $J(x) = x^4 + 2$

11.4 *Determine whether each function is invertible. If it is invertible, find the inverse.*

43. $\{(-2, 4), (2, 4)\}$

44. $\{(1, 1), (3, 3)\}$

45. $f(x) = 8x$

46. $i(x) = -\dfrac{x}{3}$

47. $g(x) = 13x - 6$

48. $h(x) = \sqrt[3]{x - 6}$

49. $j(x) = \dfrac{x + 1}{x - 1}$

50. $k(x) = |x| + 7$

51. $m(x) = (x - 1)^2$

52. $n(x) = \dfrac{3}{x}$

Find the inverse of each function, and graph f and f^{-1} on the same pair of axes.

53. $f(x) = 3x - 1$

54. $f(x) = 2 - x^2$ for $x \geq 0$

55. $f(x) = \dfrac{x^3}{2}$

56. $f(x) = -\dfrac{1}{4}x$

11.5 *Solve each variation problem.*

57. If y varies directly as m and $y = -3$ when $m = \dfrac{1}{4}$, find y when $m = -2$.

58. If a varies inversely as b and $a = 6$ when $b = -3$, find a when $b = 4$.

59. If c varies directly as m and inversely as n, and $c = 20$ when $m = 10$ and $n = 4$, find c when $m = 6$ and $n = -3$.

60. If V varies jointly as h and the square of r, and $V = 32$ when $h = 6$ and $r = 3$, find V when $h = 3$ and $r = 4$.

11.6 *Given that either 2 or -2 is a solution to each of the following equations, solve each equation.*

61. $x^3 - 4x^2 - 11x + 30 = 0$

62. $x^3 - 2x^2 - 6x + 12 = 0$

63. $x^3 - 5x^2 - 2x + 24 = 0$

64. $x^3 + 7x^2 + 4x - 12 = 0$

Determine whether the first polynomial is a factor of the second. If it is, then factor the polynomial completely.

65. $x - 3, \quad x^3 + 4x^2 - 11x - 30$

66. $x + 4, \quad x^3 + x^2 - 10x + 8$

67. $x + 2, \quad x^3 - 5x - 6$

68. $x - 1, \quad 2x^3 - 5x^2 + 2x - 9$

MISCELLANEOUS

69. *Falling object.* If a ball is dropped from a tall building, then the distance traveled by the ball in t seconds varies directly as the square of the time t. If the ball travels 144 feet (ft) in 3 seconds, then how far does it travel in 4 seconds?

70. *Studying or partying.* Evelyn's grade on a math test varies directly with the number of hours spent studying and inversely with the number of hours spent partying during the 24 hours preceding the test. If she scored a 90 on a test after she studied 10 hours and partied 2 hours, then what should she score after studying 4 hours and partying 6 hours?

71. *Inscribed square.* Given that B is the area of a square inscribed in a circle of radius r and area A, write B as a function of A.

72. *Area of a window.* A window is in the shape of a square of side s, with a semicircle of diameter s above it. Write a function that expresses the total area of the window as a function of s.

FIGURE FOR EXERCISE 72

73. *Composition of functions.* Given that $a = 3k + 2$ and $k = 5w - 6$, write a as a function of w.

74. *Volume of a cylinder.* The volume of a cylinder with a fixed height of 10 centimeters (cm) is given by $V = 10\pi r^2$, where r is the radius of the circular base. Write the volume as a function of the area of the base, A.

75. *Square formulas.* Write the area of a square A as a function of the length of a side of the square s. Write the length of a side of a square as a function of the area.

76. *Circle formulas.* Write the area of a circle A as a function of the radius of the circle r. Write the radius of a circle as a function of the area of the circle. Write the area as a function of the diameter d.

CHAPTER 11 TEST

Sketch the graph of each function or relation.

1. $f(x) = -\dfrac{2}{3}x + 1$

2. $y = |x| - 4$

3. $g(x) = x^2 + 2x - 8$

4. $x = y^2$

5. $y = -|x - 2|$

6. $y = \sqrt{x + 5} - 2$

Let $f(x) = -2x + 5$ and $g(x) = x^2 + 4$. Find the following.

7. $f(-3)$ **8.** $(g \circ f)(-3)$

9. $f^{-1}(11)$ **10.** $f^{-1}(x)$

11. $(g + f)(x)$ **12.** $(f \cdot g)(1)$

13. $(f^{-1} \circ f)(1776)$ **14.** $(f/g)(2)$

15. $(f \circ g)(x)$ **16.** $(g \circ f)(x)$

Let $f(x) = x - 7$ and $g(x) = x^2$. Write each of the following functions as a composition of functions using f and g.

17. $H(x) = x^2 - 7$

18. $W(x) = x^2 - 14x + 49$

Determine whether each function is invertible. If it is invertible, find the inverse.

19. $\{(2, 3), (4, 3)\}$

20. $f(x) = \sqrt[3]{x} + 9$

Solve each problem.

21. Find the domain and range of the function $f(x) = |x|$.

22. Find the inverse of $f(x) = \dfrac{2x + 1}{x - 1}$.

23. The volume of a sphere varies directly as the cube of the radius. If a sphere with radius 3 feet (ft) has a volume of 36π cubic feet (ft^3), then what is the volume of a sphere with a radius of 2 ft?

24. Suppose y varies directly as x and inversely as the square root of z. If $y = 12$ when $x = 7$ and $z = 9$, then what is the proportionality constant?

25. The cost of a Persian rug varies jointly as the length and width of the rug. If the cost is $2256 for a 6 foot by 8 foot rug, then what is the cost of a 9 foot by 12 foot rug?

26. Is $x + 1$ a factor of $2x^4 - 5x^3 + 3x^2 + 6x - 4$? Explain.

27. Given that all of the solutions to the equation $x^3 - 12x^2 + 47x - 60 = 0$ are positive integers smaller than 10, find the solution set.

Simplify each expression.

1. $125^{-2/3}$

2. $\left(\dfrac{8}{27}\right)^{-1/3}$

3. $\sqrt{18} - \sqrt{8}$

4. $x^5 \cdot x^3$

5. $16^{1/4}$

6. $\dfrac{x^{12}}{x^3}$

Find the real solution set to each equation.

7. $x^2 = 9$

8. $x^2 = 8$

9. $x^2 = x$

10. $x^2 - 4x - 6 = 0$

11. $x^{1/4} = 3$

12. $x^{1/6} = -2$

13. $|x| = 8$

14. $|5x - 4| = 21$

15. $x^3 = 8$

16. $(3x - 2)^3 = 27$

17. $\sqrt{2x - 3} = 9$

18. $\sqrt{x - 2} = x - 8$

Sketch the graph of each set.

19. $\{(x, y) \mid y = 5\}$

20. $\{(x, y) \mid y = 2x - 5\}$

21. $\{(x, y) \mid x = 5\}$

22. $\{(x, y) \mid 3y = x\}$

23. $\{(x, y) \mid y = 5x^2\}$

24. $\{(x, y) \mid y = -2x^2\}$

Find the missing coordinates in each ordered pair so that the ordered pair satisfies the given equation.

25. $(2, \ ), (3, \ ), (\ , 2), (\ , 16), \quad 2^x = y$

26. $\left(\dfrac{1}{2}, \ \right), (-1, \ ), (\ , 16), (\ , 1), \quad 4^x = y$

Find the domain of each expression.

27. $\sqrt{x}$

28. $\sqrt{6 - 2x}$

29. $\dfrac{5x - 3}{x^2 + 1}$

30. $\dfrac{x - 3}{x^2 - 10x + 9}$

Solve each problem.

31. *Capital cost and operating cost.* To decide when to replace company cars, an accountant looks at two cost components: capital cost and operating cost. The capital cost C (the difference between the original cost and the salvage value) for a certain car is $3000 plus $0.12 for each mile that the car is driven.

a) Write the capital cost C as a linear function of x, the number of miles that the car is driven.

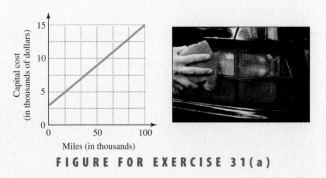

FIGURE FOR EXERCISE 31(a)

b) The operating cost P is \$0.15 per mile initially and increases linearly to \$0.25 per mile when the car reaches 100,000 miles. Write P as a function of x, the number of miles that the car is driven.

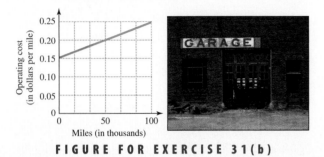

FIGURE FOR EXERCISE 31(b)

32. *Total cost.* The accountant in the previous exercise uses the function $T = \dfrac{C}{x} + P$ to find the total cost per mile.
a) Find T for $x = 20{,}000, 30{,}000$, and $90{,}000$.

b) Sketch a graph of the total cost function.

c) The accountant has decided to replace the car when T reaches \$0.38 for the second time. At what mileage will the car be replaced?

d) For what values of x is T less than or equal to \$0.38?

Exponential and Logarithmic Functions

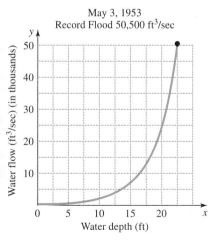

May 3, 1953
Record Flood 50,500 ft³/sec

Water flow (ft³/sec) (in thousands)

Water depth (ft)

Water is one of the essentials of life, yet it is something that most of us take for granted. Among other things, the U.S. Geological Survey (U.S.G.S.) studies freshwater. For over 50 years the Water Resources Division of the U.S.G.S. has been gathering basic data about the flow of both freshwater and saltwater from streams and groundwater surfaces. This division collects, compiles, analyzes, verifies, organizes, and publishes data gathered from groundwater data collection networks in each of the 50 states, Puerto Rico, and the Trust Territories. Records of stream flow, groundwater levels, and water quality provide hydrological information needed by local, state, and federal agencies as well as the private sector.

There are many instances of the importance of the data collected by the U.S.G.S. For example, before 1987 the Tangipahoa River in Louisiana was used extensively for swimming and boating. In 1987 data gathered by the U.S.G.S. showed that fecal coliform levels in the river exceeded safe levels. Consequently, Louisiana banned recreational use of the river. Other studies by the Water Resources Division include the results of pollutants on salt marsh environments and the effect that salting highways in winter has on our drinking water supply.

In Exercises 85 and 86 of Section 12.2 you will see how data from the U.S.G.S. is used in a logarithmic function to measure water quality.

EXPONENTIAL FUNCTIONS AND THEIR APPLICATIONS

12.1

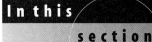

We have studied functions such as

$$f(x) = x^2, \qquad g(x) = x^3, \qquad \text{and} \qquad h(x) = x^{1/2}.$$

For these functions the variable is the base. In this section we discuss functions that have a variable as an exponent. These functions are called *exponential functions*.

Definition

Some examples of exponential functions are

$$f(x) = 2^x, \qquad f(x) = \left(\frac{1}{2}\right)^x, \qquad \text{and} \qquad f(x) = 3^x.$$

> ### Exponential Function
>
> An **exponential function** is a function of the form
>
> $$f(x) = a^x,$$
>
> where $a > 0$ and $a \neq 1$.

We rule out the base 1 in the definition because $f(x) = 1^x$ is the same as the constant function $f(x) = 1$. Zero is not used as a base because $0^x = 0$ for any positive x and nonpositive powers of 0 are undefined. Negative numbers are not used as bases because an expression such as $(-4)^x$ is not a real number if $x = \frac{1}{2}$.

E X A M P L E   1

Evaluating exponential functions

Let $f(x) = 2^x$, $g(x) = \left(\frac{1}{4}\right)^{1-x}$, and $h(x) = -3^x$. Find the following.

a) $f\left(\frac{3}{2}\right)$ **b)** $f(-3)$ **c)** $g(3)$ **d)** $h(2)$

Solution

a) $f\left(\frac{3}{2}\right) = 2^{3/2} = \sqrt{2^3} = \sqrt{8} = 2\sqrt{2}$

b) $f(-3) = 2^{-3} = \dfrac{1}{2^3} = \dfrac{1}{8}$

c) $g(3) = \left(\frac{1}{4}\right)^{1-3} = \left(\frac{1}{4}\right)^{-2} = 4^2 = 16$

d) $h(2) = -3^2 = -9$ Note that $-3^2 \neq (-3)^2$.

For many applications of exponential functions we use base 10 or another base called e. The number e is an irrational number that is approximately 2.718. We will see how e is used in compound interest in Example 10 of this section. Base 10 will be used in the next section. Base 10 is called the **common base,** and base e is called the **natural base.**

E X A M P L E   2

Base 10 and base *e*

Let $f(x) = 10^x$ and $g(x) = e^x$. Find the following and round approximate answers to four decimal places.

a) $f(3)$ b) $f(1.51)$ c) $g(0)$ d) $g(2)$

calculator

close-up

Most graphing calculators have keys for the functions 10^x and e^x.

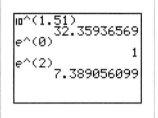

Solution

a) $f(3) = 10^3 = 1000$

b) $f(1.51) = 10^{1.51} \approx 32.3594$ Use the 10^x key on a calculator.

c) $g(0) = e^0 = 1$

d) $g(2) = e^2 \approx 7.3891$ Use the e^x key on a calculator.

Domain

In the definition of an exponential function no restrictions were placed on the exponent x because the domain of an exponential function is the set of all real numbers. So both rational and irrational numbers can be used as the exponent. We have been using rational numbers for exponents since Chapter 9, but we have not yet seen an irrational number as an exponent. Even though we do not formally define irrational exponents in this text, an irrational number such as π can be used as an exponent, and you can evaluate an expression such as 2^π by using a calculator. Try it:

$$2^\pi \approx 8.824977827$$

Graphing Exponential Functions

Even though the domain of an exponential function is the set of all real numbers, we can graph an exponential function by evaluating it for just a few integers.

E X A M P L E   3

Exponential functions with base greater than 1

Sketch the graph of each function.

a) $f(x) = 2^x$ b) $g(x) = 3^x$

Solution

a) We first make a table of ordered pairs that satisfy $f(x) = 2^x$:

x	-2	-1	0	1	2	3
$f(x) = 2^x$	$\frac{1}{4}$	$\frac{1}{2}$	1	2	4	8

As x increases, 2^x increases and 2^x is always positive. Because the domain of the function is $(-\infty, \infty)$, we draw the graph in Fig. 12.1 as a smooth curve through these points. From the graph we can see that the range is $(0, \infty)$.

calculator

close-up

The graph of $f(x) = 2^x$ on a calculator appears to touch the x-axis. When drawing this graph by hand, make sure that it does not touch the x-axis.

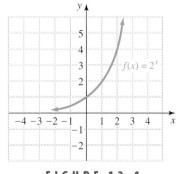

FIGURE 12.1

FIGURE 12.2

b) Make a table of ordered pairs that satisfy $g(x) = 3^x$:

x	-2	-1	0	1	2	3
$g(x) = 3^x$	$\frac{1}{9}$	$\frac{1}{3}$	1	3	9	27

As x increases, 3^x increases and 3^x is always positive. The graph is shown in Fig. 12.2. From the graph we see that the range is $(0, \infty)$. ■

Because $e \approx 2.718$, the graph of $f(x) = e^x$ lies between the graphs of $f(x) = 2^x$ and $g(x) = 3^x$, as shown in Fig. 12.3. Note that all three functions have the same domain and range and the same y-intercept. In general, the function $f(x) = a^x$ for $a > 1$ has the following characteristics:

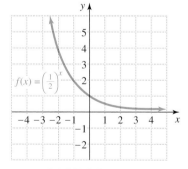

FIGURE 12.3

1. The y-intercept of the curve is $(0, 1)$.
2. The domain is $(-\infty, \infty)$, and the range is $(0, \infty)$.
3. The curve approaches the negative x-axis but does not touch it.
4. The y-values are increasing as we go from left to right along the curve.

E X A M P L E   4

Exponential functions with base between 0 and 1
Graph each function.

a) $f(x) = \left(\dfrac{1}{2}\right)^x$ **b)** $f(x) = 4^{-x}$

Solution

a) First make a table of ordered pairs that satisfy $f(x) = \left(\dfrac{1}{2}\right)^x$:

x	-2	-1	0	1	2	3
$f(x) = \left(\dfrac{1}{2}\right)^x$	4	2	1	$\frac{1}{2}$	$\frac{1}{4}$	$\frac{1}{8}$

As x increases, $\left(\dfrac{1}{2}\right)^x$ decreases, getting closer and closer to 0. Draw a smooth curve through these points as shown in Fig. 12.4.

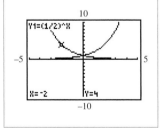

calculator

close-up

The graph of $y = (1/2)^x$ is a mirror image of the graph of $y = 2^x$.

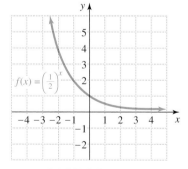

FIGURE 12.4

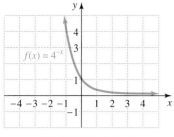

FIGURE 12.5

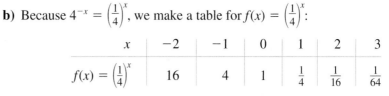

As x increases, $\left(\frac{1}{4}\right)^x$, or 4^{-x}, decreases, getting closer and closer to 0. Draw a smooth curve through these points as shown in Fig. 12.5.

Notice the similarities and differences between the exponential function with $a > 1$ and with $0 < a < 1$. The function $f(x) = a^x$ for $0 < a < 1$ has the following characteristics:

1. The y-intercept of the curve is $(0, 1)$.
2. The domain is $(-\infty, \infty)$, and the range is $(0, \infty)$.
3. The curve approaches the positive x-axis but does not touch it.
4. The y-values are decreasing as we go from left to right along the curve.

CAUTION   An exponential function can be written in more than one form. For example, $f(x) = \left(\frac{1}{2}\right)^x$ is the same as $f(x) = \frac{1}{2^x}$, or $f(x) = 2^{-x}$.

Although exponential functions have the form $f(x) = a^x$, other functions that have similar forms are also called exponential functions. Notice how changing the form $f(x) = a^x$ in the next two examples changes the shape and location of the graph.

EXAMPLE 5

Changing the shape and location
Sketch the graph of $f(x) = 3^{2x-1}$.

Solution
Make a table of ordered pairs:

x	-1	0	$\frac{1}{2}$	1	2
$f(x) = 3^{2x-1}$	$\frac{1}{27}$	$\frac{1}{3}$	1	3	27

The graph through these points is shown in Fig. 12.6.

FIGURE 12.6

EXAMPLE 6

Changing the shape and location
Sketch the graph of $y = -2^{-x}$.

Solution
Because $-2^{-x} = -(2^{-x})$, all y-coordinates are negative. Make a table of ordered pairs:

x	-2	-1	0	1	2
$f(x) = -2^{-x}$	-4	-2	-1	$-\frac{1}{2}$	$-\frac{1}{4}$

The graph through these points is shown in Fig. 12.7.

FIGURE 12.7

Exponential Equations

In Chapter 11 we used the horizontal-line test to determine whether a function is one-to-one. Because no horizontal line can cross the graph of an exponential function more than once, exponential functions are one-to-one functions. For an exponential function one-to-one means that *if two exponential expressions with the same base are equal, then the exponents are equal.*

One-to-One Property of Exponential Functions

For $a > 0$ and $a \neq 1$,

$$\text{if} \quad a^m = a^n, \quad \text{then} \quad m = n.$$

In the next example we use the one-to-one property to solve equations involving exponential functions.

EXAMPLE 7

calculator

close-up

You can see the solution to $2^{2x-1} = 8$ by graphing $y_1 = 2^{2x-1}$ and $y_2 = 8$. The x-coordinate of the point of intersection is the solution to the equation.

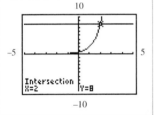

Using the one-to-one property

Solve each equation.

a) $2^{2x-1} = 8$ **b)** $9^{|x|} = 3$ **c)** $\dfrac{1}{8} = 4^x$

Solution

a) Because 8 is 2^3, we can write each side as a power of the same base, 2:

$$2^{2x-1} = 8 \qquad \text{Original equation}$$
$$2^{2x-1} = 2^3 \qquad \text{Write each side as a power of the same base.}$$
$$2x - 1 = 3 \qquad \text{One-to-one property}$$
$$2x = 4$$
$$x = 2$$

Check: $2^{2 \cdot 2 - 1} = 2^3 = 8$. The solution set is $\{2\}$.

b) Because $9 = 3^2$, we can write each side as a power of 3:

$$9^{|x|} = 3 \qquad \text{Original equation}$$
$$(3^2)^{|x|} = 3^1$$
$$3^{2|x|} = 3^1 \qquad \text{Power of a power rule}$$
$$2|x| = 1 \qquad \text{One-to-one property}$$
$$|x| = \frac{1}{2}$$
$$x = \pm\frac{1}{2}$$

calculator

close-up

The equation $9^{|x|} = 3$ has two solutions because the graphs of $y_1 = 9^{|x|}$ and $y_2 = 3$ intersect twice.

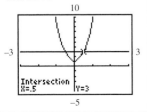

Check $x = \pm\frac{1}{2}$ in the original equation. The solution set is $\left\{-\frac{1}{2}, \frac{1}{2}\right\}$.

c) Because $\dfrac{1}{8} = 2^{-3}$ and $4 = 2^2$, we can write each side as a power of 2:

$$\frac{1}{8} = 4^x \qquad \text{Original equation}$$
$$2^{-3} = (2^2)^x \qquad \text{Write each side as a power of 2.}$$
$$2^{-3} = 2^{2x} \qquad \text{Power of a power rule}$$
$$2x = -3 \qquad \text{One-to-one property}$$
$$x = -\frac{3}{2}$$

Check $x = -\dfrac{3}{2}$ in the original equation. The solution set is $\left\{-\dfrac{3}{2}\right\}$. ∎

The one-to-one property is also used to find the first coordinate when given the second coordinate of an exponential function.

E X A M P L E   8

Finding the x-coordinate in an exponential function

Let $f(x) = 2^x$ and $g(x) = \left(\frac{1}{2}\right)^{1-x}$. Find x if:

a) $f(x) = 32$ **b)** $g(x) = 8$

Solution

a) Because $f(x) = 2^x$ and $f(x) = 32$, we can find x by solving $2^x = 32$:

$$2^x = 32$$
$$2^x = 2^5 \qquad \text{Write both sides as a power of the same base.}$$
$$x = 5 \qquad \text{One-to-one property}$$

b) Because $g(x) = \left(\frac{1}{2}\right)^{1-x}$ and $g(x) = 8$, we can find x by solving $\left(\frac{1}{2}\right)^{1-x} = 8$:

$$\left(\frac{1}{2}\right)^{1-x} = 8$$
$$(2^{-1})^{1-x} = 2^3 \qquad \text{Because } \tfrac{1}{2} = 2^{-1} \text{ and } 8 = 2^3$$
$$2^{x-1} = 2^3 \qquad \text{Power of a power rule}$$
$$x - 1 = 3 \qquad \text{One-to-one property}$$
$$x = 4$$

Applications

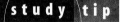

Exponential functions are used to describe phenomena such as population growth, radioactive decay, and compound interest. Here we discuss compound interest. If an investment is earning **compound interest,** then interest is periodically paid into the account and the interest that is paid also earns interest. If a bank pays 6% compounded quarterly on an account, then the interest is computed four times per year (every 3 months) at 1.5% (one-quarter of 6%). Suppose an account has $5000 in it at the beginning of a quarter. We can apply the simple interest formula $A = P + Prt$, with $r = 6\%$ and $t = \frac{1}{4}$, to find how much is in the account at the end of the first quarter.

$$A = P + Prt$$
$$= P(1 + rt) \qquad \text{Factor.}$$
$$= 5000\left(1 + 0.06 \cdot \frac{1}{4}\right) \qquad \text{Substitute.}$$
$$= 5000(1.015)$$
$$= \$5075$$

To repeat this computation for another quarter, we multiply $5075 by 1.015. If A represents the amount in the account at the end of n quarters, we can write A as an exponential function of n:

$$A = \$5000(1.015)^n$$

In general, the amount A is given by the following formula.

Compound Interest Formula

If P represents the principal, i the interest rate per period, n the number of periods, and A the amount at the end of n periods, then

$$A = P(1 + i)^n.$$

EXAMPLE 9

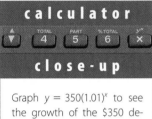

calculator

close-up

Graph $y = 350(1.01)^x$ to see the growth of the $350 deposit in Example 9 over time. After 360 months it is worth $12,582.37.

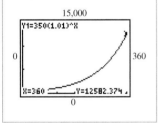

Compound interest formula

If $350 is deposited in an account paying 12% compounded monthly, then how much is in the account at the end of 6 years and 6 months?

Solution

Interest is paid 12 times per year, so the account earns $\frac{1}{12}$ of 12%, or 1% each month, for 78 months. So $i = 0.01$, $n = 78$, and $P = \$350$:

$$A = P(1 + i)^n$$
$$A = \$350(1.01)^{78}$$
$$= \$760.56$$

■

If we shorten the length of the time period (yearly, quarterly, monthly, daily, hourly, etc.), the number of periods n increases while the interest rate for the period decreases. As n increases, the amount A also increases but will not exceed a certain amount. That certain amount is the amount obtained from *continuous compounding* of the interest. It is shown in more advanced courses that the following formula gives the amount when interest is compounded continuously.

helpful hint

Compare Examples 9 and 10 to see the difference between compounded monthly and compounded continuously. Although there is not much difference to an individual investor, there could be a large difference to the bank. Rework Examples 9 and 10 using $50 million as the deposit.

Continuous-Compounding Formula

If P is the principal or beginning balance, r is the annual percentage rate compounded continuously, t is the time in years, and A is the amount or ending balance, then

$$A = Pe^{rt}.$$

CAUTION The value of t in the continuous-compounding formula must be in years. For example, if the time is 1 year and 3 months, then $t = 1.25$ years. If the time is 3 years and 145 days, then

$$t = 3 + \frac{145}{365}$$
$$\approx 3.3973 \text{ years.}$$

EXAMPLE 10

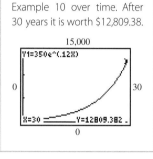

Graph $y = 350e^{0.12x}$ to see the growth of the $350 deposit in Example 10 over time. After 30 years it is worth $12,809.38.

Continuous-compounding formula

If $350 is deposited in an account paying 12% compounded continuously, then how much is in the account after 6 years and 6 months?

Solution

Use $r = 12\%$, $t = 6.5$ years, and $P = \$350$ in the formula for compounding interest continuously:

$$A = Pe^{rt}$$
$$= 350e^{(0.12)(6.5)}$$
$$= 350e^{0.78}$$
$$= \$763.52 \quad \text{Use the } e^x \text{ key on a scientific calculator.}$$

Note that compounding continuously amounts to a few dollars more than compounding monthly did in Example 9. ■

MATH AT WORK $x^2 + (x+1)^2 = 5^2$

Neal Driscoll, a geophysicist at the Lamont-Doherty Earth Observatory of Columbia University, explores both the ocean and the continents to understand the processes that shape the earth. What he finds fascinating is the interaction between the ocean and the land—not just at the shoreline, but underneath the sea as well.

To get a preliminary picture of the ocean floor, Dr. Driscoll has worked with the U.S.G.S., studying the effects of storms on beaches and underwater landscapes. The results of these studies can be used as a baseline to provide help to coastal planners who are building waterfront homes. Other information obtained can be used to direct transporters of dredged material to places where the material is least likely to affect plant, fish, and human life. The most recent study found many different types of ocean floor, ranging from sand and mud to large tracts of algae.

GEOPHYSICIST

Imaging the seafloor is a difficult problem. It can be a costly venture, and there are numerous logistical problems. Recently developed technology, such as towable undersea cameras and satellite position systems, has made the task easier. Dr. Driscoll and his team use this new technology and sound reflection to gather data about how the sediment on the ocean floor changes in response to storm events. This research is funded by the Office of Naval Research (ONR).

In Exercise 28 of the Making Connections exercises you will see how a geophysicist uses sound to measure the depth of the ocean.

WARM-UPS

True or false? Explain your answer.

1. If $f(x) = 4^x$, then $f\left(-\frac{1}{2}\right) = -2$.

2. If $f(x) = \left(\frac{1}{3}\right)^x$, then $f(-1) = 3$.

3. The function $f(x) = x^4$ is an exponential function.

4. The functions $f(x) = \left(\frac{1}{2}\right)^x$ and $g(x) = 2^{-x}$ have the same graph.

5. The function $f(x) = 2^x$ is invertible.

6. The graph of $y = \left(\frac{1}{3}\right)^x$ has an x-intercept.

7. The y-intercept for $f(x) = e^x$ is $(0, 1)$.

8. The expression $2^{\sqrt{2}}$ is undefined.

9. The functions $f(x) = 2^{-x}$ and $g(x) = \frac{1}{2^x}$ have the same graph.

10. If \$500 earns 6% compounded monthly, then at the end of 3 years the investment is worth $500(1.005)^3$ dollars.

12.1 EXERCISES

Reading and Writing After reading this section, write out the answers to these questions. Use complete sentences.

1. What is an exponential function?

2. What is the domain of every exponential function?

3. What are the two most popular bases?

4. What is the one-to-one property of exponential functions?

5. What is the compound interest formula?

6. What does compounded continuously mean?

Let $f(x) = 4^x$, $g(x) = \left(\frac{1}{3}\right)^{x+1}$, and $h(x) = -2^x$. Find the following. See Example 1.

7. $f(2)$

8. $f(-1)$

9. $f\left(\frac{1}{2}\right)$

10. $f\left(-\frac{3}{2}\right)$

11. $g(-2)$

12. $g(1)$

13. $g(0)$

14. $g(-3)$

15. $h(0)$

16. $h(3)$

17. $h(-2)$

18. $h(-4)$

Let $h(x) = 10^x$ and $j(x) = e^x$. Find the following. Use a calculator as necessary and round approximate answers to four decimal places. See Example 2.

19. $h(0)$

20. $h(-1)$

21. $h(2)$

22. $h(3.4)$

23. $j(1)$

24. $j(3.5)$

25. $j(-2)$

26. $j(0)$

Sketch the graph of each function. See Examples 3 and 4.

27. $f(x) = 4^x$

28. $g(x) = 5^x$

29. $h(x) = \left(\frac{1}{3}\right)^x$

30. $i(x) = \left(\frac{1}{5}\right)^x$

31. $y = 10^x$ **32.** $y = (0.1)^x$ **43.** $P = 5000(1.05)^t$ **44.** $d = 800 \cdot 10^{-4t}$

Sketch the graph of each function. See Examples 5 and 6.

33. $y = 10^{x+2}$ **34.** $y = 3^{2x+1}$

Solve each equation. See Example 7.

45. $2^x = 64$ **46.** $3^x = 9$

47. $10^x = 0.001$ **48.** $10^{2x} = 0.1$

49. $2^x = \dfrac{1}{4}$ **50.** $3^x = \dfrac{1}{9}$

51. $\left(\dfrac{2}{3}\right)^{x-1} = \dfrac{9}{4}$ **52.** $\left(\dfrac{1}{4}\right)^{3x} = 16$

53. $5^{-x} = 25$ **54.** $10^{-x} = 0.01$

35. $f(x) = -2^x$ **36.** $k(x) = -2^{x-2}$

55. $-2^{1-x} = -8$ **56.** $-3^{2-x} = -81$

57. $10^{|x|} = 1000$ **58.** $3^{|2x-5|} = 81$

Let $f(x) = 2^x$, $g(x) = \left(\frac{1}{3}\right)^x$, and $h(x) = 4^{2x-1}$. Find x in each case. See Example 8.

59. $f(x) = 4$ **60.** $f(x) = \dfrac{1}{4}$

61. $f(x) = 4^{2/3}$ **62.** $f(x) = 1$

37. $g(x) = 2^{-x}$ **38.** $A(x) = 10^{1-x}$

63. $g(x) = 9$ **64.** $g(x) = \dfrac{1}{9}$

65. $g(x) = 1$ **66.** $g(x) = \sqrt{3}$

67. $h(x) = 16$ **68.** $h(x) = \dfrac{1}{2}$

39. $f(x) = -e^x$ **40.** $g(x) = e^{-x}$

69. $h(x) = 1$ **70.** $h(x) = \sqrt{2}$

Solve each problem. See Example 9.

71. *Compounding quarterly.* If \$6000 is deposited in an account paying 5% compounded quarterly, then what amount will be in the account after 10 years?

72. *Compounding quarterly.* If \$400 is deposited in an account paying 10% compounded quarterly, then what amount will be in the account after 7 years?

41. $H(x) = 10^{|x|}$ **42.** $s(x) = 2^{(x^2)}$

73. *Outstanding performance.* The top stock fund over 10 years was Fidelity Select-Home Finance because it returned an average of 27.6% annually for 10 years (Money's 1998 Guide to Mutual Funds, www.money.com).

a) How much was an investment of $10,000 in this fund in 1988 worth in 1998 if interest is compounded annually?

b) Use the accompanying graph to estimate the year in which the $10,000 investment was worth $75,000.

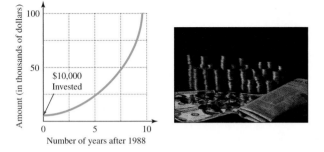

FIGURE FOR EXERCISE 73

74. *Second place.* The Kaufman fund was the second best fund over 10 years with an average annual return of 26.5% (Money's 1998 Guide to Mutual Funds, www.money.com). How much was an investment of $10,000 in this fund in 1988 worth in 1998?

75. *Depreciating knowledge.* The value of a certain textbook seems to decrease according to the formula $V = 45 \cdot 2^{-0.9t}$, where V is the value in dollars and t is the age of the book in years. What is the book worth when it is new? What is it worth when it is 2 years old?

76. *Mosquito abatement.* In a Minnesota swamp in the springtime the number of mosquitoes per acre appears to grow according to the formula $N = 10^{0.1t+2}$, where t is the number of days since the last frost. What is the size of the mosquito population at times $t = 10$, $t = 20$, and $t = 30$?

 In Exercises 77–82, solve each problem. See Example 10.

77. *Compounding continuously.* If $500 is deposited in an account paying 7% compounded continuously, then how much will be in the account after 3 years?

78. *Compounding continuously.* If $7000 is deposited in an account paying 8% compounded continuously, then what will it amount to after 4 years?

79. *One year's interest.* How much interest will be earned the first year on $80,000 on deposit in an account paying 7.5% compounded continuously?

80. *Partial year.* If $7500 is deposited in an account paying 6.75% compounded continuously, then how much will be in the account after 5 years and 215 days?

81. *Radioactive decay.* The number of grams of a certain radioactive substance present at time t is given by the formula $A = 300 \cdot e^{-0.06t}$, where t is the number of years. Find the amount present at time $t = 0$. Find the amount present after 20 years. Use the accompanying graph to estimate the number of years that it takes for one-half of the substance to decay. Will the substance ever decay completely?

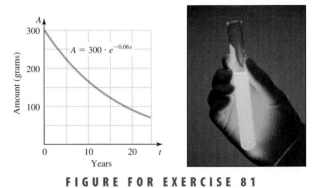

FIGURE FOR EXERCISE 81

82. *Population growth.* The population of a certain country appears to be growing according to the formula $P = 20 \cdot e^{0.1t}$, where P is the population in millions and t is the number of years since 1980. What was the population in 1980? What will the population be in the year 2000?

GETTING MORE INVOLVED

83. *Exploration.* An approximate value for e can be found by adding the terms in the following infinite sum:

$$1 + \frac{1}{1} + \frac{1}{2 \cdot 1} + \frac{1}{3 \cdot 2 \cdot 1} + \frac{1}{4 \cdot 3 \cdot 2 \cdot 1} + \cdots$$

Use a calculator to find the sum of the first four terms. Find the difference between the sum of the first four terms and e. (For e, use all of the digits that your calculator gives for e^1.) What is the difference between e and the sum of the first eight terms?

GRAPHING CALCULATOR EXERCISES

84. Graph $y_1 = 2^x$, $y_2 = e^x$, and $y_3 = 3^x$ on the same coordinate system. Which point do all three graphs have in common?

85. Graph $y_1 = 3^x$, $y_2 = 3^{x-1}$, and $y_3 = 3^{x-2}$ on the same coordinate system. What can you say about the graph of $y = 3^{x-k}$ for any real number k?

LOGARITHMIC FUNCTIONS AND THEIR APPLICATIONS

In Section 12.1 you learned that exponential functions are one-to-one functions. Because they are one-to-one functions, they have inverse functions. In this section we study the inverses of the exponential functions.

In this section

- Definition
- Domain and Range
- Graphing Logarithmic Functions
- Logarithmic Equations
- Applications

Definition

We define $\log_a(x)$ as *the exponent that is used on the base a to obtain x.* Read $\log_a(x)$ as "the base a logarithm of x." The expression $\log_a(x)$ is called a **logarithm.** Because $2^3 = 8$, the exponent is 3 and $\log_2(8) = 3$. Because $5^2 = 25$, the exponent is 2 and $\log_5(25) = 2$. Because $2^{-5} = \frac{1}{32}$, the exponent is -5 and $\log_2\left(\frac{1}{32}\right) = -5$. So the logarithmic equation $y = \log_a(x)$ is equivalent to the exponential equation $a^y = x$.

> **$\log_a(x)$**
>
> For any $a > 0$ and $a \neq 1$,
>
> $$y = \log_a(x) \quad \text{if and only if} \quad a^y = x.$$

EXAMPLE 1 **Using the definition of logarithm**

Write each logarithmic equation as an exponential equation and each exponential equation as a logarithmic equation.

a) $\log_5(125) = 3$ **b)** $6 = \log_{1/4}(x)$

c) $\left(\frac{1}{2}\right)^m = 8$ **d)** $7 = 3^z$

Solution

a) "The base-5 logarithm of 125 equals 3" means that 3 is the exponent on 5 that produces 125. So $5^3 = 125$.

b) The equation $6 = \log_{1/4}(x)$ is equivalent to $\left(\frac{1}{4}\right)^6 = x$ by the definition of logarithm.

c) The equation $\left(\frac{1}{2}\right)^m = 8$ is equivalent to $\log_{1/2}(8) = m$.

d) The equation $7 = 3^z$ is equivalent to $\log_3(7) = z$. ■

The inverse of the base-a exponential function $f(x) = a^x$ is the **base-a logarithmic function** $f^{-1}(x) = \log_a(x)$. For example, $f(x) = 2^x$ and $f^{-1}(x) = \log_2(x)$ are inverse functions as shown in Fig. 12.8. Each function undoes the other.

$$f(5) = 2^5 = 32 \quad \text{and} \quad g(32) = \log_2(32) = 5.$$

To evaluate logarithmic functions remember that a logarithm is an exponent: $\log_a(x)$ is the exponent that is used on the base a to obtain x.

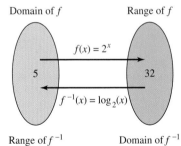

Domain of f Range of f

$f(x) = 2^x$

5   32

$f^{-1}(x) = \log_2(x)$

Range of f^{-1} Domain of f^{-1}

FIGURE 12.8

EXAMPLE 2 **Finding logarithms**

Evaluate each logarithm.

a) $\log_5(25)$ **b)** $\log_2\left(\frac{1}{8}\right)$ **c)** $\log_{1/2}(4)$

d) $\log_{10}(0.001)$ **e)** $\log_9(3)$

Solution

helpful hint

When we write $C(x) = 12x$, we may think of C as a variable and write $C = 12x$, or we may think of C as the name of a function, the cost function. In $y = \log_a(x)$ we are thinking of $\log_a$ only as the name of the function that pairs an x-value with a y-value.

a) The number $\log_5(25)$ is the exponent that is used on the base 5 to obtain 25. Because $25 = 5^2$, we have $\log_5(25) = 2$.

b) The number $\log_2\left(\frac{1}{8}\right)$ is the power of 2 that gives us $\frac{1}{8}$. Because $\frac{1}{8} = 2^{-3}$, we have $\log_2\left(\frac{1}{8}\right) = -3$.

c) The number $\log_{1/2}(4)$ is the power of $\frac{1}{2}$ that produces 4. Because $4 = \left(\frac{1}{2}\right)^{-2}$, we have $\log_{1/2}(4) = -2$.

d) Because $0.001 = 10^{-3}$, we have $\log_{10}(0.001) = -3$.

e) Because $9^{1/2} = 3$, we have $\log_9(3) = \frac{1}{2}$. ■

There are two bases for logarithms that are used more frequently than the others: They are 10 and e. The base-10 logarithm is called the **common logarithm** and is usually written as $\log(x)$. The base-e logarithm is called the **natural logarithm** and is usually written as $\ln(x)$. Most scientific calculators have function keys for $\log(x)$ and $\ln(x)$. The simplest way to obtain a common or natural logarithm is to use a scientific calculator. However, a table of common logarithms can be found in Appendix C of this text.

In the next example we find natural and common logarithms of certain numbers without a calculator or a table.

EXAMPLE 3

Finding common and natural logarithms

Evaluate each logarithm.

a) $\log(1000)$

b) $\ln(e)$

c) $\log\left(\frac{1}{10}\right)$

Solution

a) Because $10^3 = 1000$, we have $\log(1000) = 3$.

b) Because $e^1 = e$, we have $\ln(e) = 1$.

c) Because $10^{-1} = \frac{1}{10}$, we have $\log\left(\frac{1}{10}\right) = -1$. ■

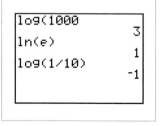

calculator

close-up

A graphing calculator has keys for the common logarithm (LOG) and the natural logarithm (LN).

```
log(1000
                3
ln(e)
                1
log(1/10)
               -1
```

Domain and Range

The domain of the exponential function $y = 2^x$ is $(-\infty, \infty)$, and its range is $(0, \infty)$. Because the logarithmic function $y = \log_2(x)$ is the inverse of $y = 2^x$, the domain of $y = \log_2(x)$ is $(0, \infty)$, and its range is $(-\infty, \infty)$.

CAUTION Because the domain of $y = \log_a(x)$ is $(0, \infty)$ for any $a > 0$ and $a \neq 1$, expressions such as $\log_2(-4)$, $\log_{1/3}(0)$, and $\ln(-1)$ are undefined.

Graphing Logarithmic Functions

In Chapter 11 we saw that the graphs of a function and its inverse function are symmetric about the line $y = x$. Because the logarithm functions are inverses of exponential functions, their graphs are also symmetric about $y = x$.

EXAMPLE 4

A logarithmic function with base greater than 1

Sketch the graph of $g(x) = \log_2(x)$ and compare it to the graph of $y = 2^x$.

calculator
TOTAL PART %TOTAL
4   5   6   ×

close-up

The graphs of $y = \ln(x)$ and $y = e^x$ are symmetric with respect to the line $y = x$. Logarithmic functions with bases other than e and 10 will be graphed on a calculator in Section 12.4.

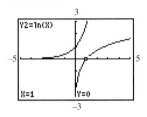

Solution

Make a table of ordered pairs for $g(x) = \log_2(x)$ using positive numbers for x:

x	$\frac{1}{4}$	$\frac{1}{2}$	1	2	4	8
$g(x) = \log_2(x)$	-2	-1	0	1	2	3

Draw a curve through these points as shown in Fig. 12.9. The graph of the inverse function $y = 2^x$ is also shown in Fig. 12.9 for comparison. Note the symmetry of the two curves about the line $y = x$.

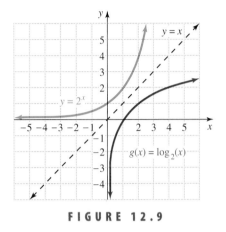

FIGURE 12.9

All logarithmic functions with the base greater than 1 have graphs that are similar to the one in Fig. 12.9. In general, the graph of $f(x) = \log_a(x)$ for $a > 1$ has the following characteristics (see Fig. 12.10):

1. The x-intercept of the curve is $(1, 0)$.
2. The domain is $(0, \infty)$, and the range is $(-\infty, \infty)$.
3. The curve approaches the negative y-axis but does not touch it.
4. The y-values are increasing as we go from left to right along the curve.

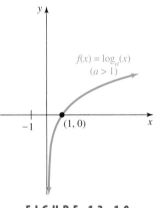

FIGURE 12.10

EXAMPLE 5

A logarithmic function with base less than 1

Sketch the graph of $f(x) = \log_{1/2}(x)$ and compare it to the graph of $y = \left(\frac{1}{2}\right)^x$.

Solution

Make a table of ordered pairs for $f(x) = \log_{1/2}(x)$ using positive numbers for x:

x	$\frac{1}{4}$	$\frac{1}{2}$	1	2	4	8
$f(x) = \log_{1/2}(x)$	2	1	0	-1	-2	-3

The curve through these points is shown in Fig. 12.11. The graph of the inverse function $y = \left(\frac{1}{2}\right)^x$ is also shown in Fig. 12.11 for comparison. Note the symmetry with respect to the line $y = x$.

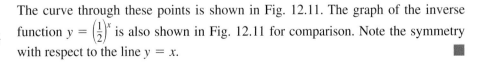

FIGURE 12.11

All logarithmic functions with the base between 0 and 1 have graphs that are similar to the one in Fig. 12.11. In general, the graph of $f(x) = \log_a(x)$ for $0 < a < 1$

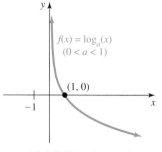

FIGURE 12.12

has the following characteristics (see Fig. 12.12):

1. The x-intercept of the curve is $(1, 0)$.
2. The domain is $(0, \infty)$, and the range is $(-\infty, \infty)$.
3. The curve approaches the positive y-axis but does not touch it.
4. The y-values are decreasing as we go from left to right along the curve.

Figures 12.9 and 12.11 illustrate the fact that $y = \log_a(x)$ and $y = a^x$ are inverse functions for any base a. For any given exponential or logarithmic function the inverse function can be easily obtained from the definition of logarithm.

E X A M P L E   6   **Inverses of logarithmic and exponential functions**

Find the inverse of each function.

a) $f(x) = 10^x$ **b)** $g(x) = \log_3(x)$

Solution

a) To find any inverse function we switch the roles of x and y. So $y = 10^x$ becomes $x = 10^y$. Now $x = 10^y$ is equivalent to $y = \log_{10}(x)$. So the inverse of $f(x) = 10^x$ is $y = \log(x)$ or $f^{-1}(x) = \log(x)$.

b) In $g(x) = \log_3(x)$ or $y = \log_3(x)$ we switch x and y to get $x = \log_3(y)$. Now $x = \log_3(y)$ is equivalent to $y = 3^x$. So the inverse of $g(x) = \log_3(x)$ is $y = 3^x$ or $g^{-1}(x) = 3^x$. ∎

Logarithmic Equations

In Section 12.1 we learned that the exponential functions are one-to-one functions. Because logarithmic functions are inverses of exponential functions, they are one-to-one functions also. For a base-a logarithmic function *one-to-one means that if the base-a logarithms of two numbers are equal, then the numbers are equal.*

One-to-One Property of Logarithms

For $a > 0$ and $a \neq 1$,

$$\text{if} \quad \log_a(m) = \log_a(n), \quad \text{then} \quad m = n.$$

The one-to-one property of logarithms and the definition of logarithms are the two basic tools that we use to solve equations involving logarithms. We use these tools in the next example.

E X A M P L E   7   **Logarithmic equations**

Solve each equation.

a) $\log_3(x) = -2$ **b)** $\log_x(8) = -3$ **c)** $\log(x^2) = \log(4)$

Solution

a) Use the definition of logarithms to rewrite the logarithmic equation as an equivalent exponential equation:

$$\log_3(x) = -2$$
$$3^{-2} = x \quad \text{Definition of logarithm}$$
$$\frac{1}{9} = x$$

Because $3^{-2} = \frac{1}{9}$ or $\log_3\!\left(\frac{1}{9}\right) = -2$, the solution set is $\left\{\frac{1}{9}\right\}$.

b) Use the definition of logarithms to rewrite the logarithmic equation as an equivalent exponential equation:

$$\log_x(8) = -3$$
$$x^{-3} = 8 \qquad \text{Definition of logarithm}$$
$$(x^{-3})^{-1} = 8^{-1} \qquad \text{Raise each side to the } -1 \text{ power.}$$
$$x^3 = \frac{1}{8}$$
$$x = \sqrt[3]{\frac{1}{8}} = \frac{1}{2} \qquad \text{Odd-root property}$$

Because $\left(\frac{1}{2}\right)^{-3} = 2^3 = 8$ or $\log_{1/2}(8) = -3$ the solution set is $\left\{\frac{1}{2}\right\}$.

c) To write an equation equivalent to $\log(x^2) = \log(4)$, we use the one-to-one property of logarithms:

$$\log(x^2) = \log(4)$$
$$x^2 = 4 \qquad \text{One-to-one property of logarithms}$$
$$x = \pm 2 \qquad \text{Even-root property}$$

If $x = \pm 2$, then $x^2 = 4$ and $\log(4) = \log(4)$. The solution set is $\{-2, 2\}$. ■

CAUTION If we have equality of two logarithms with the same base, we use the one-to-one property to eliminate the logarithms. If we have an equation with only one logarithm, such as $\log_a(x) = y$, we use the definition of logarithm to write $a^y = x$ and to eliminate the logarithm.

Applications

When money earns interest compounded continuously, the formula

$$t = \frac{1}{r} \ln\left(\frac{A}{P}\right)$$

expresses the relationship between the time in years t, the annual interest rate r, the principal P, and the amount A. This formula is used to determine how long it takes for a deposit to grow to a specific amount.

EXAMPLE 8

Finding the time for a specified growth

How long does it take $80 to grow to $240 at 12% compounded continuously?

Solution

Use $r = 0.12$, $P = \$80$, and $A = \$240$ in the formula, and use a calculator to evaluate the logarithm:

$$t = \frac{1}{0.12} \ln\left(\frac{240}{80}\right)$$
$$= \frac{\ln(3)}{0.12}$$
$$\approx 9.155$$

It takes approximately 9.155 years, or 9 years and 57 days. ■

WARM-UPS

True or false? Explain your answer.

1. The equation $a^3 = 2$ is equivalent to $\log_a(2) = 3$.
2. If (a, b) satisfies $y = 8^x$, then (a, b) satisfies $y = \log_8(x)$.
3. If $f(x) = a^x$ for $a > 0$ and $a \neq 1$, then $f^{-1}(x) = \log_a(x)$.
4. If $f(x) = \ln(x)$, then $f^{-1}(x) = e^x$.
5. The domain of $f(x) = \log_6(x)$ is $(-\infty, \infty)$.
6. $\log_{25}(5) = 2$
7. $\log(-10) = 1$
8. $\log(0) = 0$
9. $5^{\log_5(125)} = 125$
10. $\log_{1/2}(32) = -5$

12.2 EXERCISES

Reading and Writing *After reading this section, write out the answers to these questions. Use complete sentences.*

1. What is the inverse function for the function $f(x) = 2^x$?

2. What is $\log_a(x)$?

3. What is the difference between the common logarithm and the natural logarithm?

4. What is the domain of $f(x) = \log_a(x)$?

5. What is the one-to-one property of logarithmic functions?

6. What is the relationship between the graphs of $f(x) = a^x$ and $f^{-1}(x) = \log_a(x)$ for $a > 0$ and $a \neq 1$?

Write each exponential equation as a logarithmic equation and each logarithmic equation as an exponential equation. See Example 1.

7. $\log_2(8) = 3$
8. $\log_{10}(10) = 1$
9. $10^2 = 100$
10. $5^3 = 125$
11. $y = \log_5(x)$
12. $m = \log_b(N)$
13. $2^a = b$
14. $a^3 = c$
15. $\log_3(x) = 10$
16. $\log_c(t) = 4$
17. $e^3 = x$
18. $m = e^x$

Evaluate each logarithm. See Examples 2 and 3.

19. $\log_2(4)$
20. $\log_2(1)$
21. $\log_2(16)$
22. $\log_4(16)$
23. $\log_2(64)$
24. $\log_8(64)$
25. $\log_4(64)$
26. $\log_{64}(64)$
27. $\log_2\left(\dfrac{1}{4}\right)$
28. $\log_2\left(\dfrac{1}{8}\right)$
29. $\log(100)$
30. $\log(1)$
31. $\log(0.01)$
32. $\log(10{,}000)$
33. $\log_{1/3}\left(\dfrac{1}{3}\right)$
34. $\log_{1/3}\left(\dfrac{1}{9}\right)$
35. $\log_{1/3}(27)$
36. $\log_{1/3}(1)$
37. $\log_{25}(5)$
38. $\log_{16}(4)$
39. $\ln(e^2)$
40. $\ln\left(\dfrac{1}{e}\right)$

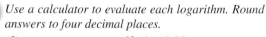 *Use a calculator to evaluate each logarithm. Round answers to four decimal places.*

41. $\log(5)$
42. $\log(0.03)$
43. $\ln(6.238)$
44. $\ln(0.23)$

Sketch the graph of each function. See Examples 4 and 5.

45. $f(x) = \log_3(x)$

46. $g(x) = \log_{10}(x)$

47. $y = \log_4(x)$

48. $y = \log_5(x)$

49. $h(x) = \log_{1/4}(x)$

50. $y = \log_{1/3}(x)$

51. $y = \log_{1/5}(x)$

52. $y = \log_{1/6}(x)$

Find the inverse of each function. See Example 6.

53. $f(x) = 6^x$

54. $f(x) = 4^x$

55. $f(x) = \ln(x)$

56. $f(x) = \log(x)$

57. $f(x) = \log_{1/2}(x)$

58. $f(x) = \log_{1/4}(x)$

Solve each equation. See Example 7.

59. $x = \left(\dfrac{1}{2}\right)^{-2}$

60. $x = 16^{-1/2}$

61. $5 = 25^x$

62. $0.1 = 10^x$

63. $\log(x) = -3$

64. $\log(x) = 5$

65. $\log_x(36) = 2$

66. $\log_x(100) = 2$

67. $\log_x(5) = -1$

68. $\log_x(16) = -2$

69. $\log(x^2) = \log(9)$

70. $\ln(2x - 3) = \ln(x + 1)$

Use a calculator to solve each equation. Round answers to four decimal places.

71. $3 = 10^x$

72. $10^x = 0.03$

73. $10^x = \dfrac{1}{2}$

74. $75 = 10^x$

75. $e^x = 7.2$

76. $e^{3x} = 0.4$

Solve each problem. See Example 8. Use a calculator as necessary.

77. *Double your money.* How long does it take $5000 to grow to $10,000 at 12% compounded continuously?

78. *Half the rate.* How long does it take $5000 to grow to $10,000 at 6% compounded continuously?

79. *Earning interest.* How long does it take to earn $1000 in interest on a deposit of $6000 at 8% compounded continuously?

80. *Lottery winnings.* How long does it take to earn $1000 interest on a deposit of one million dollars at 9% compounded continuously?

The annual growth rate for an investment that is growing continuously is given by

$$r = \frac{1}{t} \ln\left(\frac{A}{P}\right),$$

where P is the principal and A is the amount after t years.

81. *Top stock.* An investment of $10,000 in Dell Computer stock in 1995 grew to $231,800 in 1998.

   a) Assuming the investment grew continuously, what was the annual growth rate?

b) If Dell continues to grow at the same rate, then what will the $10,000 investment be worth in 2002?

82. *Chocolate bars.* An investment of $10,000 in 1980 in Hershey stock was worth $563,000 in 1998. Assuming the investment grew continuously, what was the annual growth rate?

In chemistry the pH *of a solution is defined by*

$$pH = -\log_{10}[H+],$$

where H+ is the hydrogen ion concentration of the solution in moles per liter. Distilled water has a pH *of approximately 7. A solution with a* pH *under 7 is called an acid, and one with a* pH *over 7 is called a base.*

83. *Tomato juice.* Tomato juice has a hydrogen ion concentration of $10^{-4.1}$ mole per liter (mol/L). Find the pH of tomato juice.

84. *Stomach acid.* The gastric juices in your stomach have a hydrogen ion concentration of 10^{-1} mol/L. Find the pH of your gastric juices.

85. *Neuse River* **pH.** The pH of a water sample is one of the many measurements of water quality done by the U.S. Geological Survey. The hydrogen ion concentration of the water in the Neuse River at New Bern, North Carolina, was 1.58×10^{-7} mol/L on July 9, 1998 (Water Resources for North Carolina, wwwnc.usgs.gov). What was the pH of the water at that time?

86. *Roanoke River* **pH.** On July 9, 1998 the hydrogen ion concentration of the water in the Roanoke River at Janesville, North Carolina, was 1.995×10^{-7} mol/L (Water Resources for North Carolina, wwwnc.usgs.gov). What was the pH of the water at that time?

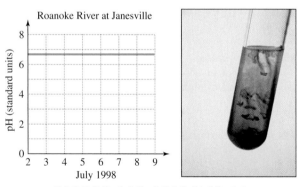

Roanoke River at Janesville

July 1998

FIGURE FOR EXERCISE 86

Solve each problem.

87. *Sound level.* The level of sound in decibels (db) is given by the formula

$$L = 10 \cdot \log(I \times 10^{12}),$$

where I is the intensity of the sound in watts per square meter. If the intensity of the sound at a rock concert is 0.001 watt per square meter at a distance of 75 meters from the stage, then what is the level of the sound at this point in the audience?

88. *Logistic growth.* If a rancher has one cow with a contagious disease in a herd of 1000, then the time in days t for n of the cows to become infected is modeled by

$$t = -5 \cdot \ln\left(\frac{1000 - n}{999n}\right).$$

Find the number of days that it takes for the disease to spread to 100, 200, 998, and 999 cows. This model, called a *logistic growth model,* describes how a disease can spread very rapidly at first and then very slowly as nearly all of the population has become infected.

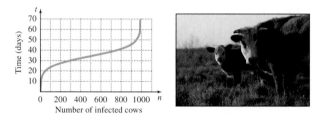

FIGURE FOR EXERCISE 88

GETTING MORE INVOLVED

89. *Discussion.* Use the switch-and-solve method from Chapter 11 to find the inverse of the function $f(x) = 5 + \log_2(x - 3)$. State the domain and range of the inverse function.

90. *Discussion.* Find the inverse of the function $f(x) = 2 + e^{x+4}$. State the domain and range of the inverse function.

GRAPHING CALCULATOR EXERCISES

91. *Composition of inverses.* Graph the functions $y = \ln(e^x)$ and $y = e^{\ln(x)}$. Explain the similarities and differences between the graphs.

92. *The population bomb.* The population of the earth is growing continuously with an annual rate of about 1.6%. If the present population is 6 billion, then the function $y = 6e^{0.016x}$ gives the population in billions x years from now. Graph this function for $0 \le x \le 200$. What will the population be in 100 years and in 200 years?

12.3 PROPERTIES OF LOGARITHMS

The properties of logarithms are very similar to the properties of exponents because *logarithms are exponents.* In this section we use the properties of exponents to write some properties of logarithms. The properties will be used in solving logarithmic equations in Section 12.4.

Product Rule for Logarithms

If $M = a^x$ and $N = a^y$, we can use the product rule for exponents to write

$$MN = a^x \cdot a^y = a^{x+y}.$$

The equation $MN = a^{x+y}$ is equivalent to

$$\log_a(MN) = x + y.$$

Because $M = a^x$ and $N = a^y$ are equivalent to $x = \log_a(M)$ and $y = \log_a(N)$, we can replace x and y in $\log_a(MN) = x + y$ to get

$$\log_a(MN) = \log_a(M) + \log_a(N).$$

So *the logarithm of a product is the sum of the logarithms,* provided that all of the logarithms are defined. This rule is called the **product rule for logarithms.**

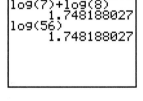

calculator

close-up

You can illustrate the product rule for logarithms with a graphing calculator.

```
log(7)+log(8)
        1.748188027
log(56)
        1.748188027
```

> **Product Rule for Logarithms**
>
> $$\log_a(MN) = \log_a(M) + \log_a(N)$$

E X A M P L E 1

Using the product rule for logarithms

Write each expression as a single logarithm.

a) $\log_2(7) + \log_2(5)$
b) $\ln(\sqrt{2}) + \ln(\sqrt{3})$

Solution

a) $\log_2(7) + \log_2(5) = \log_2(35)$ Product rule for logarithms
b) $\ln(\sqrt{2}) + \ln(\sqrt{3}) = \ln(\sqrt{6})$ Product rule for logarithms

■

Quotient Rule for Logarithms

If $M = a^x$ and $N = a^y$, we can use the quotient rule for exponents to write

$$\frac{M}{N} = \frac{a^x}{a^y} = a^{x-y}.$$

By the definition of logarithm, $\frac{M}{N} = a^{x-y}$ is equivalent to

$$\log_a\left(\frac{M}{N}\right) = x - y.$$

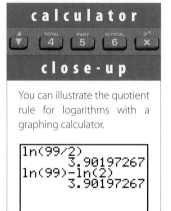

You can illustrate the quotient rule for logarithms with a graphing calculator.

```
ln(99/2)
          3.90197267
ln(99)-ln(2)
          3.90197267
```

Because $x = \log_a(M)$ and $y = \log_a(N)$, we have

$$\log_a\left(\frac{M}{N}\right) = \log_a(M) - \log_a(N).$$

So *the logarithm of a quotient is equal to the difference of the logarithms*, provided that all logarithms are defined. This rule is called the **quotient rule for logarithms.**

Quotient Rule for Logarithms

$$\log_a\left(\frac{M}{N}\right) = \log_a(M) - \log_a(N)$$

E X A M P L E 2

Using the quotient rule for logarithms

Write each expression as a single logarithm.

a) $\log_2(3) - \log_2(7)$ **b)** $\ln(w^8) - \ln(w^2)$

Solution

a) $\log_2(3) - \log_2(7) = \log_2\left(\dfrac{3}{7}\right)$ Quotient rule for logarithms

b) $\ln(w^8) - \ln(w^2) = \ln\left(\dfrac{w^8}{w^2}\right)$ Quotient rule for logarithms

$ = \ln(w^6)$ Quotient rule for exponents

Power Rule for Logarithms

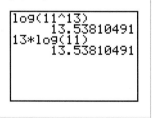

You can illustrate the power rule for logarithms with a graphing calculator.

```
log(11^13)
          13.53810491
13*log(11)
          13.53810491
```

If $M = a^x$, we can use the power rule for exponents to write

$$M^N = (a^x)^N = a^{Nx}.$$

By the definition of logarithms, $M^N = a^{Nx}$ is equivalent to

$$\log_a(M^N) = Nx.$$

Because $x = \log_a(M)$, we have

$$\log_a(M^N) = N \cdot \log_a(M).$$

So *the logarithm of a power of a number is the power times the logarithm of the number,* provided that all logarithms are defined. This rule is called the **power rule for logarithms.**

Power Rule for Logarithms

$$\log_a(M^N) = N \cdot \log_a(M)$$

E X A M P L E 3

Using the power rule for logarithms

Rewrite each logarithm in terms of $\log(2)$.

a) $\log(2^{10})$ **b)** $\log(\sqrt{2})$ **c)** $\log\left(\dfrac{1}{2}\right)$

Solution

a) $\log(2^{10}) = 10 \cdot \log(2)$ Power rule for logarithms

b) $\log(\sqrt{2}) = \log(2^{1/2})$ Write $\sqrt{2}$ as a power of 2.

$\qquad\qquad = \dfrac{1}{2}\log(2)$ Power rule for logarithms

c) $\log\left(\dfrac{1}{2}\right) = \log(2^{-1})$ Write $\dfrac{1}{2}$ as a power of 2.

$\qquad\qquad = -1 \cdot \log(2)$ Power rule for logarithms

$\qquad\qquad = -\log(2)$ ■

Inverse Properties

An exponential function and logarithmic function with the same base are inverses of each other. For example, the logarithm of 32 base 2 is 5 and the fifth power of 2 is 32. In symbols, we have

$$2^{\log_2(32)} = 2^5 = 32.$$

If we raise 3 to the fourth power, we get 81; and if we find the base-3 logarithm of 81, we get 4. In symbols, we have

$$\log_3(3^4) = \log_3(81) = 4.$$

We can state the inverse relationship between exponential and logarithm functions in general with the following inverse properties.

Inverse Properties

1. $\log_a(a^M) = M$ $\qquad\qquad\qquad$ **2.** $a^{\log_a(M)} = M$

E X A M P L E   4

Using the inverse properties

Simplify each expression.

a) $\ln(e^5)$ $\qquad\qquad\qquad\qquad$ **b)** $2^{\log_2(8)}$

Solution

a) Using the first inverse property, we get $\ln(e^5) = 5$.

b) Using the second inverse property, we get $2^{\log_2(8)} = 8$.   ■

Note that there is more than one way to simplify the expressions in Example 4. Using the power rule for logarithms and the fact that $\ln(e) = 1$, we have $\ln(e^5) = 5 \cdot \ln(e) = 5$. Using $\log_2(8) = 3$, we have $2^{\log_2(8)} = 2^3 = 8$.

Using the Properties

We have already seen many properties of logarithms. There are three properties that we have not yet formally stated. Because $a^1 = a$ and $a^0 = 1$, we have $\log_a(a) = 1$ and $\log_a(1) = 0$ for any positive number a. If we apply the quotient rule to $\log_a(1/N)$, we get

$$\log_a\left(\dfrac{1}{N}\right) = \log_a(1) - \log_a(N) = 0 - \log_a(N) = -\log_a(N).$$

So $\log_a\!\left(\frac{1}{N}\right) = -\log_a(N)$. These three new properties along with all of the other properties of logarithms are summarized as follows.

Properties of Logarithms

If M, N, and a are positive numbers, $a \neq 1$, then

1. $\log_a(a) = 1$ **2.** $\log_a(1) = 0$

3. $\log_a(a^M) = M$ Inverse properties   **4.** $a^{\log_a(M)} = M$

5. $\log_a(MN) = \log_a(M) + \log_a(N)$ Product rule

6. $\log_a\!\left(\frac{M}{N}\right) = \log_a(M) - \log_a(N)$ Quotient rule

7. $\log_a\!\left(\frac{1}{N}\right) = -\log_a(N)$ **8.** $\log_a(M^N) = N \cdot \log_a(M)$ Power rule

---

We have already seen several ways in which to use the properties of logarithms. In the next three examples we see more uses of the properties. First we use the rules of logarithms to write the logarithm of a complicated expression in terms of logarithms of simpler expressions.

E X A M P L E   5

Using the properties of logarithms

Rewrite each expression in terms of $\log(2)$ and/or $\log(3)$.

a) $\log(6)$ **b)** $\log(16)$

c) $\log\!\left(\frac{9}{2}\right)$ **d)** $\log\!\left(\frac{1}{3}\right)$

Solution

a) $\log(6) = \log(2 \cdot 3)$

$\qquad\qquad = \log(2) + \log(3)$ Product rule

b) $\log(16) = \log(2^4)$

$\qquad\qquad = 4 \cdot \log(2)$ Power rule

c) $\log\!\left(\frac{9}{2}\right) = \log(9) - \log(2)$ Quotient rule

$\qquad\qquad = \log(3^2) - \log(2)$

$\qquad\qquad = 2 \cdot \log(3) - \log(2)$ Power rule

d) $\log\!\left(\frac{1}{3}\right) = -\log(3)$ Property 7

Examine the values of $\log(9/2)$, $\log(9) - \log(2)$, and $\log(9)/\log(2)$.

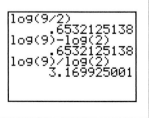

```
log(9/2)
        .6532125138
log(9)-log(2)
        .6532125138
log(9)/log(2)
        3.169925001
```

CAUTION   Do not confuse $\frac{\log(9)}{\log(2)}$ with $\log\!\left(\frac{9}{2}\right)$. We can use the quotient rule to write $\log\!\left(\frac{9}{2}\right) = \log(9) - \log(2)$, but $\frac{\log(9)}{\log(2)} \neq \log(9) - \log(2)$. The expression $\frac{\log(9)}{\log(2)}$ means $\log(9) \div \log(2)$. Use your calculator to verify these two statements.

The properties of logarithms can be used to combine several logarithms into a single logarithm (as in Examples 1 and 2) or to write a logarithm of a complicated expression in terms of logarithms of simpler expressions.

E X A M P L E   6

Using the properties of logarithms

Rewrite each expression as a sum or difference of multiples of logarithms.

a) $\log\left(\dfrac{xz}{y}\right)$

b) $\log_3\left(\dfrac{(x-3)^{2/3}}{\sqrt{x}}\right)$

Solution

a) $\log\left(\dfrac{xz}{y}\right) = \log(xz) - \log(y)$ Quotient rule

$\qquad\qquad\ = \log(x) + \log(z) - \log(y)$ Product rule

b) $\log_3\left(\dfrac{(x-3)^{2/3}}{\sqrt{x}}\right) = \log_3((x-3)^{2/3}) - \log_3(x^{1/2})$ Quotient rule

$\qquad\qquad\qquad\qquad\ = \dfrac{2}{3}\log_3(x-3) - \dfrac{1}{2}\log_3(x)$ Power rule

       In the next example we use the properties of logarithms to convert expressions involving several logarithms into a single logarithm. The skills we are learning here will be used to solve logarithmic equations in Section 12.4.

E X A M P L E   7

Combining logarithms

Rewrite each expression as a single logarithm.

a) $\dfrac{1}{2}\log(x) - 2 \cdot \log(x+1)$

b) $3 \cdot \log(y) + \dfrac{1}{2}\log(z) - \log(x)$

Solution

a) $\dfrac{1}{2}\log(x) - 2 \cdot \log(x+1) = \log(x^{1/2}) - \log((x+1)^2)$ Power rule

$\qquad\qquad\qquad\qquad\qquad = \log\left(\dfrac{\sqrt{x}}{(x+1)^2}\right)$ Quotient rule

b) $3 \cdot \log(y) + \dfrac{1}{2}\log(z) - \log(x) = \log(y^3) + \log(\sqrt{z}) - \log(x)$ Power rule

$\qquad\qquad\qquad\qquad\qquad\quad = \log(y^3 \cdot \sqrt{z}) - \log(x)$ Product rule

$\qquad\qquad\qquad\qquad\qquad\quad = \log\left(\dfrac{y^3 \cdot \sqrt{z}}{x}\right)$ Quotient rule

WARM-UPS

True or false? Explain your answer.

1. $\log_2\left(\dfrac{x^2}{8}\right) = \log_2(x^2) - 3$

2. $\dfrac{\log(100)}{\log(10)} = \log(100) - \log(10)$

(continued)

3. $\ln(\sqrt{2}) = \dfrac{\ln(2)}{2}$

4. $3^{\log_3(17)} = 17$

5. $\log_2\left(\dfrac{1}{8}\right) = \dfrac{1}{\log_2(8)}$

6. $\ln(8) = 3 \cdot \ln(2)$

7. $\ln(1) = e$

8. $\dfrac{\log(100)}{10} = \log(10)$

9. $\dfrac{\log_2(8)}{\log_2(2)} = \log_2(4)$

10. $\ln(2) + \ln(3) - \ln(7) = \ln\left(\dfrac{6}{7}\right)$

12.3 EXERCISES

Reading and Writing *After reading this section, write out the answers to these questions. Use complete sentences.*

1. What is the product rule for logarithms?

2. What is the quotient rule for logarithms?

3. What is the power rule for logarithms?

4. Why is it true that $\log_a(a^M) = M$?

5. Why is it true that $a^{\log_a(M)} = M$?

6. Why is it true that $\log_a(1) = 0$ for $a > 0$ and $a \neq 1$?

Assume all variables involved in logarithms represent numbers for which the logarithms are defined.

Write each expression as a single logarithm and simplify. See Example 1.

7. $\log(3) + \log(7)$

8. $\ln(5) + \ln(4)$

9. $\log_3(\sqrt{5}) + \log_3(\sqrt{x})$

10. $\ln(\sqrt{x}) + \ln(\sqrt{y})$

11. $\log(x^2) + \log(x^3)$

12. $\ln(a^3) + \ln(a^5)$

13. $\ln(2) + \ln(3) + \ln(5)$

14. $\log_2(x) + \log_2(y) + \log_2(z)$

15. $\log(x) + \log(x + 3)$

16. $\ln(x - 1) + \ln(x + 1)$

17. $\log_2(x - 3) + \log_2(x + 2)$

18. $\log_3(x - 5) + \log_3(x - 4)$

Write each expression as a single logarithm. See Example 2.

19. $\log(8) - \log(2)$ **20.** $\ln(3) - \ln(6)$

21. $\log_2(x^6) - \log_2(x^2)$ **22.** $\ln(w^9) - \ln(w^3)$

23. $\log(\sqrt{10}) - \log(\sqrt{2})$

24. $\log_3(\sqrt{6}) - \log_3(\sqrt{3})$

25. $\ln(4h - 8) - \ln(4)$

26. $\log(3x - 6) - \log(3)$

27. $\log_2(w^2 - 4) - \log_2(w + 2)$

28. $\log_3(k^2 - 9) - \log_3(k - 3)$

29. $\ln(x^2 + x - 6) - \ln(x + 3)$

30. $\ln(t^2 - t - 12) - \ln(t - 4)$

Write each expression in terms of $\log(3)$. *See Example 3.*

31. $\log(27)$ **32.** $\log\left(\dfrac{1}{9}\right)$

33. $\log(\sqrt{3})$ **34.** $\log(\sqrt[4]{3})$

35. $\log(3^x)$ **36.** $\log(3^{-99})$

Simplify each expression. See Example 4.

37. $\log_2(2^{10})$ **38.** $\ln(e^9)$

39. $5^{\log_5(19)}$ **40.** $10^{\log(2.3)}$

41. $\log(10^8)$ **42.** $\log_4(4^5)$

43. $e^{\ln(4.3)}$ **44.** $3^{\log_3(5.5)}$

Rewrite each expression in terms of $\log(3)$ *and/or* $\log(5)$. *See Example 5.*

45. $\log(15)$ **46.** $\log(9)$

47. $\log\!\left(\dfrac{5}{3}\right)$ **48.** $\log\!\left(\dfrac{3}{5}\right)$

49. $\log(25)$ **50.** $\log\!\left(\dfrac{1}{27}\right)$

51. $\log(75)$ **52.** $\log(0.6)$

53. $\log\!\left(\dfrac{1}{3}\right)$ **54.** $\log(45)$

55. $\log(0.2)$ **56.** $\log\!\left(\dfrac{9}{25}\right)$

Rewrite each expression as a sum or a difference of multiples of logarithms. See Example 6.

57. $\log(xyz)$

58. $\log(3y)$

59. $\log_2(8x)$

60. $\log_2(16y)$

61. $\ln\!\left(\dfrac{x}{y}\right)$

62. $\ln\!\left(\dfrac{z}{3}\right)$

63. $\log(10x^2)$

64. $\log(100\sqrt{x})$

65. $\log_5\!\left(\dfrac{(x-3)^2}{\sqrt{w}}\right)$

66. $\log_3\!\left(\dfrac{(y+6)^3}{y-5}\right)$

67. $\ln\!\left(\dfrac{yz\sqrt{x}}{w}\right)$

68. $\ln\!\left(\dfrac{(x-1)\sqrt{w}}{x^3}\right)$

Rewrite each expression as a single logarithm. See Example 7.

69. $\log(x) + \log(x-1)$

70. $\log_2(x-2) + \log_2(5)$

71. $\ln(3x-6) - \ln(x-2)$

72. $\log_3(x^2-1) - \log_3(x-1)$

73. $\ln(x) - \ln(w) + \ln(z)$

74. $\ln(x) - \ln(3) - \ln(7)$

75. $3 \cdot \ln(y) + 2 \cdot \ln(x) - \ln(w)$

76. $5 \cdot \ln(r) + 3 \cdot \ln(t) - 4 \cdot \ln(s)$

77. $\dfrac{1}{2}\log(x-3) - \dfrac{2}{3}\log(x+1)$

78. $\dfrac{1}{2}\log(y-4) + \dfrac{1}{2}\log(y+4)$

79. $\dfrac{2}{3}\log_2(x-1) - \dfrac{1}{4}\log_2(x+2)$

80. $\dfrac{1}{2}\log_3(y+3) + 6 \cdot \log_3(y)$

Determine whether each equation is true or false.

81. $\log(56) = \log(7) \cdot \log(8)$ **82.** $\log\!\left(\dfrac{5}{9}\right) = \dfrac{\log(5)}{\log(9)}$

83. $\log_2(4^2) = (\log_2(4))^2$ **84.** $\ln(4^2) = (\ln(4))^2$

85. $\ln(25) = 2 \cdot \ln(5)$ **86.** $\ln(3e) = 1 + \ln(3)$

87. $\dfrac{\log_2(64)}{\log_2(8)} = \log_2(8)$ **88.** $\dfrac{\log_2(16)}{\log_2(4)} = \log_2(4)$

89. $\log\!\left(\dfrac{1}{3}\right) = -\log(3)$ **90.** $\log_2(8 \cdot 2^{59}) = 62$

91. $\log_2(16^5) = 20$ **92.** $\log_2\!\left(\dfrac{5}{2}\right) = \log_2(5) - 1$

93. $\log(10^3) = 3$ **94.** $\log_3(3^7) = 7$

95. $\log(100 + 3) = 2 + \log(3)$ **96.** $\dfrac{\log_7(32)}{\log_7(8)} = \dfrac{5}{3}$

Solve each problem.

97. **Growth rate.** The annual growth rate for continuous growth is given by

$$r = \dfrac{\ln(A) - \ln(P)}{t},$$

where P is the initial investment and A is the amount after t years.

a) Rewrite the formula using a single logarithm.

b) In 1998 a share of Microsoft stock was worth 27 times what it was worth in 1990. What was the annual growth rate for that period?

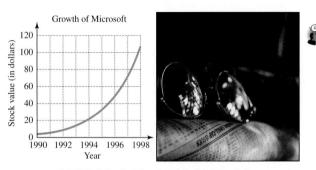

FIGURE FOR EXERCISE 97

98. **Diversity index.** The U.S.G.S. measures the quality of a water sample by using the diversity index d, given by

$$d = -[p_1 \cdot \log_2(p_1) + p_2 \cdot \log_2(p_2) + \cdots + p_n \cdot \log_2(p_n)],$$

where n is the number of different taxons (biological classifications) represented in the sample and p_1 through p_n are the percentages of organisms in each of the n taxons. The value of d ranges from 0 when all organisms in the water sample are the same to some positive number when all organisms in the sample are different. If two-thirds of the organisms in a water sample are in one taxon and one-third of the organisms are in a second taxon, then $n = 2$ and

$$d = -\left[\frac{2}{3} \log_2\left(\frac{2}{3}\right) + \frac{1}{3} \log_2\left(\frac{1}{3}\right)\right].$$

Use the properties of logarithms to write the expression on the right-hand side as $\log_2\left(\frac{3\sqrt[3]{2}}{2}\right)$. (In Section 12.4 you will learn how to evaluate a base-2 logarithm using a calculator.)

GETTING MORE INVOLVED

99. **Discussion.** Which of the following equations is an identity? Explain.

a) $\ln(3x) = \ln(3) \cdot \ln(x)$ b) $\ln(3x) = \ln(3) + \ln(x)$
c) $\ln(3x) = 3 \cdot \ln(x)$ d) $\ln(3x) = \ln(x^3)$

100. **Discussion.** Which of the following expressions is not equal to $\log(5^{2/3})$? Explain.

a) $\frac{2}{3} \log(5)$

b) $\frac{\log(5) + \log(5)}{3}$

c) $(\log(5))^{2/3}$

d) $\frac{1}{3} \log(25)$

GRAPHING CALCULATOR EXERCISES

101. Graph the functions $y_1 = \ln(\sqrt{x})$ and $y_2 = 0.5 \cdot \ln(x)$ on the same screen. Explain your results.

102. Graph the functions $y_1 = \log(x)$, $y_2 = \log(10x)$, $y_3 = \log(100x)$, and $y_4 = \log(1000x)$ using the viewing window $-2 \le x \le 5$ and $-2 \le y \le 5$. Why do these curves appear as they do?

103. Graph the function $y = \log(e^x)$. Explain why the graph is a straight line. What is its slope?

12.4

SOLVING EQUATIONS AND APPLICATIONS

In this

section

- Logarithmic Equations
- Exponential Equations
- Changing the Base
- Strategy for Solving Equations
- Applications

We solved some equations involving exponents and logarithms in Sections 12.1 and 12.2. In this section we use the properties of exponents and logarithms to solve more complex equations.

Logarithmic Equations

The main tool that we have for solving logarithmic equations is the definition of logarithms: $y = \log_a(x)$ if and only if $a^y = x$. We can use the definition to rewrite any equation that has only one logarithm as an equivalent exponential equation.

E X A M P L E   1   **A logarithmic equation with only one logarithm**

Solve $\log(x + 3) = 2$.

Solution

Write the equivalent exponential equation:

$$\log(x + 3) = 2 \qquad \text{Original equation}$$
$$10^2 = x + 3 \qquad \text{Definition of logarithm}$$
$$100 = x + 3$$
$$97 = x$$

Check: $\log(97 + 3) = \log(100) = 2$. The solution set is $\{97\}$. ■

In the next example we use the product rule for logarithms to write a sum of two logarithms as a single logarithm.

E X A M P L E   2   **Using the product rule to solve an equation**

Solve $\log_2(x + 3) + \log_2(x - 3) = 4$.

Solution

Rewrite the sum of the logarithms as the logarithm of a product:

$$\log_2(x + 3) + \log_2(x - 3) = 4 \qquad \text{Original equation}$$
$$\log_2[(x + 3)(x - 3)] = 4 \qquad \text{Product rule}$$
$$\log_2[x^2 - 9] = 4 \qquad \text{Multiply the binomials.}$$
$$x^2 - 9 = 2^4 \qquad \text{Definition of logarithm}$$
$$x^2 - 9 = 16$$
$$x^2 = 25$$
$$x = \pm 5 \qquad \text{Even-root property}$$

To check, first let $x = -5$ in the original equation:

$$\log_2(-5 + 3) + \log_2(-5 - 3) = 4$$
$$\log_2(-2) + \log_2(-8) = 4 \qquad \text{Incorrect}$$

Because the domain of any logarithm function is the set of positive real numbers, these logarithms are undefined. Now check $x = 5$ in the original equation:

$$\log_2(5 + 3) + \log_2(5 - 3) = 4$$
$$\log_2(8) + \log_2(2) = 4$$
$$3 + 1 = 4 \qquad \text{Correct}$$

The solution set is $\{5\}$. ■

CAUTION   Always check that your solutions to a logarithmic equation do not produce undefined logarithms in the original equation.

E X A M P L E   3

Using the one-to-one property of logarithms

Solve $\log(x) + \log(x - 1) = \log(8x - 12) - \log(2)$.

Solution

Apply the product rule to the left-hand side and the quotient rule to the right-hand side to get a single logarithm on each side:

$$\log(x) + \log(x - 1) = \log(8x - 12) - \log(2).$$

$$\log[x(x - 1)] = \log\left(\frac{8x - 12}{2}\right) \quad \text{Product rule; quotient rule}$$

$$\log(x^2 - x) = \log(4x - 6) \quad \text{Simplify.}$$

$$x^2 - x = 4x - 6 \quad \text{One-to-one property of logarithms}$$

$$x^2 - 5x + 6 = 0$$

$$(x - 2)(x - 3) = 0$$

$$x - 2 = 0 \quad \text{or} \quad x - 3 = 0$$

$$x = 2 \quad \text{or} \quad x = 3$$

Neither $x = 2$ nor $x = 3$ produces undefined terms in the original equation. Use a calculator to check that they both satisfy the original equation. The solution set is $\{2, 3\}$. ■

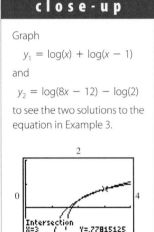

calculator

close-up

Graph

$y_1 = \log(x) + \log(x - 1)$

and

$y_2 = \log(8x - 12) - \log(2)$

to see the two solutions to the equation in Example 3.

CAUTION The product rule, quotient rule, and power rule do not eliminate logarithms from equations. To do so, we use the definition to change $y = \log_a(x)$ into $a^y = x$ or the one-to-one property to change $\log_a(m) = \log_a(n)$ into $m = n$.

Exponential Equations

If an equation has a single exponential expression, we can write the equivalent logarithmic equation.

E X A M P L E   4

A single exponential expression

Find the exact solution to $2^x = 10$.

Solution

The equivalent logarithmic equation is

$$x = \log_2(10).$$

The solution set is $\{\log_2(10)\}$. The number $\log_2(10)$ is the exact solution to the equation. Later in this section you will learn how to use the base-change formula to find an approximate value for an expression of this type. ■

In Section 12.1 we solved some exponential equations by writing each side as a power of the same base and then applying the one-to-one property of exponential functions. We review that method in the next example.

E X A M P L E   5

Powers of the same base

Solve $2^{(x^2)} = 4^{3x-4}$.

Solution

We can write each side as a power of the same base:

$$2^{(x^2)} = (2^2)^{3x-4} \qquad \text{Because } 4 = 2^2$$

$$2^{(x^2)} = 2^{6x-8} \qquad \text{Power of a power rule}$$

$$x^2 = 6x - 8 \qquad \begin{array}{l}\text{One-to-one property of} \\ \text{exponential functions}\end{array}$$

$$x^2 - 6x + 8 = 0$$

$$(x - 4)(x - 2) = 0$$

$$x - 4 = 0 \qquad \text{or} \qquad x - 2 = 0$$

$$x = 4 \qquad \text{or} \qquad x = 2$$

Check $x = 2$ and $x = 4$ in the original equation. The solution set is $\{2, 4\}$. ■

For some exponential equations we cannot write each side as a power of the same base as we did in Example 5. In this case we take a logarithm of each side and simplify, using the rules for logarithms.

E X A M P L E   6

Exponential equation with two different bases

Find the exact and approximate solution to $2^{x-1} = 3^x$.

Solution

We first take the base-10 logarithm of each side:

$$2^{x-1} = 3^x \qquad \text{Original equation}$$

$$\log(2^{x-1}) = \log(3^x) \qquad \text{Take log of each side.}$$

$$(x - 1) \log(2) = x \cdot \log(3) \qquad \text{Power rule}$$

$$x \cdot \log(2) - \log(2) = x \cdot \log(3) \qquad \text{Distributive property}$$

$$x \cdot \log(2) - x \cdot \log(3) = \log(2) \qquad \text{Get all } x\text{-terms on one side.}$$

$$x[\log(2) - \log(3)] = \log(2) \qquad \text{Factor out } x.$$

$$x = \frac{\log(2)}{\log(2) - \log(3)} \qquad \text{Exact solution}$$

$$x \approx -1.7095 \qquad \text{Approximate solution}$$

You can use a calculator to check -1.7095 in the original equation. As the first step of the solution, we could have taken the logarithm of each side using any base. We chose base 10 so that we could use a calculator to find an approximate solution from the exact solution. ■

Changing the Base

Scientific calculators have an x^y key for computing any power of any base, in addition to the function keys for computing 10^x and e^x. For logarithms we have the keys ln and log, but there are no function keys for logarithms using other bases. To solve this problem, we develop a formula for expressing a base-a logarithm in terms of base-b logarithms.

close-up

The base-change formula allows you to graph logarithmic functions with bases other than e and 10. For example, to graph $y = \log_2(x)$, graph $y = \ln(x)/\ln(2)$.

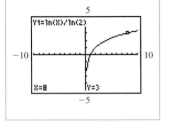

If $y = \log_a(M)$, then $a^y = M$. Now we solve $a^y = M$ for y, using base-b logarithms:

$$a^y = M$$
$$\log_b(a^y) = \log_b(M) \quad \text{Take the base-}b\text{ logarithm of each side.}$$
$$y \cdot \log_b(a) = \log_b(M) \quad \text{Power rule}$$
$$y = \frac{\log_b(M)}{\log_b(a)} \quad \text{Divide each side by } \log_b(a).$$

Because $y = \log_a(M)$, we can write $\log_a(M)$ in terms of base-b logarithms.

Base-Change Formula

If a and b are positive numbers not equal to 1 and M is positive, then

$$\log_a(M) = \frac{\log_b(M)}{\log_b(a)}.$$

In words, we take the logarithm with the new base and divide by the logarithm of the old base. The most important use of the base-change formula is to find base-a logarithms using a calculator. If the new base is 10 or e, then

$$\log_a(M) = \frac{\log(M)}{\log(a)} = \frac{\ln(M)}{\ln(a)}.$$

E X A M P L E   7

Using the base-change formula

Find $\log_7(99)$ to four decimal places.

Solution

Use the base-change formula with $a = 7$ and $b = 10$:

$$\log_7(99) = \frac{\log(99)}{\log(7)} \approx 2.3614$$

Check by finding $7^{2.3614}$ with your calculator. Note that we also have

$$\log_7(99) = \frac{\ln(99)}{\ln(7)} \approx 2.3614.$$

Strategy for Solving Equations

There is no formula that will solve every equation in this section. However, we have a strategy for solving exponential and logarithmic equations. The following list summarizes the ideas that we need for solving these equations.

> #### Solving Exponential and Logarithmic Equations
>
> **1.** If the equation has a single logarithm or a single exponential expression, rewrite the equation using the definition $y = \log_a(x)$ if and only if $a^y = x$.
> **2.** Use the properties of logarithms to combine logarithms as much as possible.
> **3.** Use the one-to-one properties:
>   **a)** If $\log_a(m) = \log_a(n)$, then $m = n$.
>   **b)** If $a^m = a^n$, then $m = n$.
> **4.** To get an approximate solution of an exponential equation, take the common or natural logarithm of each side of the equation.

Applications

In compound interest problems, logarithms are used to find the time it takes for money to grow to a specified amount.

E X A M P L E   8

Finding the time

If $500 is deposited into an account paying 8% compounded quarterly, then in how many quarters will the account have $1000 in it?

Solution

We use the compound interest formula $A = P(1 + i)^n$ with a principal of $500, an amount of $1000, and an interest rate of 2% each quarter:

$$A = P(1 + i)^n$$
$$1000 = 500(1.02)^n \qquad \text{Substitute.}$$
$$2 = (1.02)^n \qquad \text{Divide each side by 500.}$$
$$n = \log_{1.02}(2) \qquad \text{Definition of logarithm}$$
$$= \frac{\ln(2)}{\ln(1.02)} \qquad \text{Base-change formula}$$
$$\approx 35.0028 \qquad \text{Use a calculator.}$$

It takes approximately 35 quarters, or 8 years and 9 months, for the initial investment to be worth $1000. Note that we could also solve $2 = (1.02)^n$ by taking the common or natural logarithm of each side. Try it. ■

helpful hint

When we get $2 = (1.02)^n$, we can use the definition of log as in Example 8 or take the natural log of each side:

$$\ln(2) = \ln(1.02^n)$$
$$\ln(2) = n \cdot \ln(1.02)$$
$$n = \frac{\ln(2)}{\ln(1.02)}$$

In either way we arrive at the same solution.

WARM-UPS

True or false? Explain your answer.

1. If $\log(x - 2) + \log(x + 2) = 7$, then $\log(x^2 - 4) = 7$.
2. If $\log(3x + 7) = \log(5x - 8)$, then $3x + 7 = 5x - 8$.
3. If $e^{x-6} = e^{x^2-5x}$, then $x - 6 = x^2 - 5x$.
4. If $2^{3x-1} = 3^{5x-4}$, then $3x - 1 = 5x - 4$.
5. If $\log_2(x^2 - 3x + 5) = 3$, then $x^2 - 3x + 5 = 8$.
6. If $2^{2x-1} = 3$, then $2x - 1 = \log_2(3)$.
7. If $5^x = 23$, then $x \cdot \ln(5) = \ln(23)$.
8. $\log_3(5) = \dfrac{\ln(3)}{\ln(5)}$
9. $\dfrac{\ln(2)}{\ln(6)} = \dfrac{\log(2)}{\log(6)}$
10. $\log(5) = \ln(5)$

12.4   EXERCISES

Reading and Writing   *After reading this section, write out the answers to these questions. Use complete sentences.*

1. What exponential equation is equivalent to $\log_a(x) = y$?

2. How can you find a logarithm with a base other than 10 or e using a calculator?

Solve each equation. See Examples 1 and 2.

3. $\log_2(x + 1) = 3$
4. $\log_3(x^2) = 4$
5. $3 \log_2(x + 1) - 2 = 13$
6. $4 \log_3(2x) - 1 = 7$
7. $12 + 2 \ln(x) = 14$

8. $23 = 3 \ln(x - 1) + 14$

9. $\log(x) + \log(5) = 1$

10. $\ln(x) + \ln(3) = 0$

11. $\log_2(x - 1) + \log_2(x + 1) = 3$

12. $\log_3(x - 4) + \log_3(x + 4) = 2$

13. $\log_2(x - 1) - \log_2(x + 2) = 2$

14. $\log_4(8x) - \log_4(x - 1) = 2$

15. $\log_2(x - 4) + \log_2(x + 2) = 4$

16. $\log_6(x + 6) + \log_6(x - 3) = 2$

Solve each equation. See Example 3.

17. $\ln(x) + \ln(x + 5) = \ln(x + 1) + \ln(x + 3)$

18. $\log(x) + \log(x + 5) = 2 \cdot \log(x + 2)$

19. $\log(x + 3) + \log(x + 4) = \log(x^3 + 13x^2) - \log(x)$

20. $\log(x^2 - 1) - \log(x - 1) = \log(6)$

21. $2 \cdot \log(x) = \log(20 - x)$

22. $2 \cdot \log(x) + \log(3) = \log(2 - 5x)$

Solve each equation. See Examples 4 and 5.

23. $3^x = 7$ **24.** $2^{x-1} = 5$

25. $e^{2x} = 7$ **26.** $e^{x+3} = 2$

27. $2^{3x+4} = 4^{x-1}$ **28.** $9^{2x-1} = 27^{1/2}$

29. $\left(\dfrac{1}{3}\right)^x = 3^{1+x}$ **30.** $4^{3x} = \left(\dfrac{1}{2}\right)^{1-x}$

 Find the exact solution and approximate solution to each equation. See Example 6.

31. $2^x = 3^{x+5}$

32. $e^x = 10^x$

33. $5^{x+2} = 10^{x-4}$

34. $3^{2x} = 6^{x+1}$

35. $8^x = 9^{x-1}$

36. $5^{x+1} = 8^{x-1}$

Use the base-change formula to find each logarithm to four decimal places. See Example 7.

37. $\log_2(3)$

38. $\log_3(5)$

39. $\log_3\left(\dfrac{1}{2}\right)$

40. $\log_5(2.56)$

41. $\log_{1/2}(4.6)$

42. $\log_{1/3}(3.5)$

43. $\log_{0.1}(0.03)$

44. $\log_{0.2}(1.06)$

For each equation, find the exact solution and an approximate solution when appropriate. Round approximate answers to three decimal places.

45. $x \cdot \ln(2) = \ln(7)$

46. $x \cdot \log(3) = \log(5)$

47. $3x - x \cdot \ln(2) = 1$

48. $2x + x \cdot \log(5) = \log(7)$

49. $3^x = 5$

50. $2^x = \dfrac{1}{3}$

51. $2^{x-1} = 9$

52. $10^{x-2} = 6$

53. $3^x = 20$

54. $2^x = 128$

55. $\log_3(x) + \log_3(5) = 1$

56. $\log(x) - \log(3) = \log(6)$

57. $8^x = 2^{x+1}$

58. $2^x = 5^{x+1}$

In Exercises 59–70, solve each problem. See Example 8.

59. *Finding the time.* How many months does it take for $1000 to grow to $1500 in an account paying 12% compounded monthly?

60. *Finding the time.* How many years does it take for $25 to grow to $100 in an account paying 8% compounded annually?

61. Going with the flow. The flow y [in cubic feet per second (ft^3/sec)] of the Tangipahoa River at Robert, Louisiana, is modeled by the exponential function $y = 114.308e^{0.265x}$, where x is the depth in feet. Find the flow when the depth is 15.8 feet.

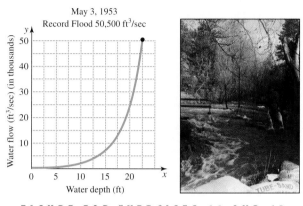

May 3, 1953
Record Flood 50,500 ft^3/sec

Water flow (ft^3/sec) (in thousands)

Water depth (ft)

FIGURE FOR EXERCISES 61 AND 62

62. Record flood. Use the formula of the previous exercise to find the depth of the Tangipahoa River at Robert, Louisiana, on May 3, 1953 when the flow reached an all-time record of 50,500 ft^3/sec (U.S.G.S., waterdata.usgs.gov).

63. Above the poverty level. In a certain country the number of people above the poverty level is currently 28 million and growing 5% annually. Assuming the population is growing continuously, the population P (in millions), t years from now, is determined by the formula $P = 28e^{0.05t}$. In how many years will there be 40 million people above the poverty level?

64. Below the poverty level. In the same country as in Exercise 63, the number of people below the poverty level is currently 20 million and growing 7% annually. This population (in millions), t years from now, is determined by the formula $P = 20e^{0.07t}$. In how many years will there be 40 million people below the poverty level?

65. Fifty-fifty. For this exercise, use the information given in Exercises 63 and 64. In how many years will the number of people above the poverty level equal the number of people below the poverty level?

66. Golden years. In a certain country there are currently 100 million workers and 40 million retired people. The population of workers is decreasing according to the formula $W = 100e^{-0.01t}$, where t is in years and W is in millions. The population of retired people is increasing according to the formula $R = 40e^{0.09t}$, where t is in years

and R is in millions. In how many years will the number of workers equal the number of retired people?

67. Ions for breakfast. Orange juice has a pH of 3.7. What is the hydrogen ion concentration of orange juice? (See Exercises 83–86 of Section 12.2.)

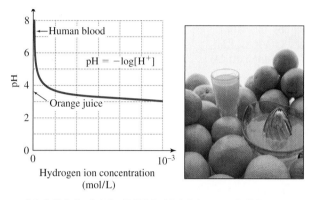

pH

Human blood

$pH = -\log[H^+]$

Orange juice

Hydrogen ion concentration
(mol/L)

FIGURE FOR EXERCISES 67 AND 68

68. Ions in your veins. Normal human blood has a pH of 7.4. What is the hydrogen ion concentration of normal human blood?

69. Diversity index. In Exercise 98 of Section 12.3 we expressed the diversity index d for a certain water sample as

$$d = \log_2\left(\frac{3\sqrt[3]{2}}{2}\right).$$

Use the base-change formula and a calculator to calculate the value of d. Round the answer to four decimal places.

70. Quality water. In a certain water sample, 5% of the organisms are in one taxon, 10% are in a second taxon, 20% are in a third taxon, 15% are in a fourth taxon, 23% are in a fifth taxon, and the rest are in a sixth taxon. Use the formula given in Exercise 98 of Section 12.3 with $n = 6$ to find the diversity index of the water sample.

GETTING MORE INVOLVED

71. Exploration. Logarithms were designed to solve equations that have variables in the exponents, but logarithms can be used to solve certain polynomial equations. Consider the following example:

$$x^5 = 88$$
$$5 \cdot \ln(x) = \ln(88)$$
$$\ln(x) = \frac{\ln(88)}{5} \approx 0.895467$$
$$x = e^{0.895467} \approx 2.4485$$

Solve $x^3 = 12$ by taking the natural logarithm of each side. Round the approximate solution to four decimal places. Solve $x^3 = 12$ without using logarithms and compare with your previous answer.

 72. *Discussion.* Determine whether each logarithm is positive or negative without using a calculator. Explain your answers.
   a) $\log_2(0.45)$
   b) $\ln(1.01)$
   c) $\log_{1/2}(4.3)$
   d) $\log_{1/3}(0.44)$

GRAPHING CALCULATOR EXERCISES

73. Graph $y_1 = 2^x$ and $y_2 = 3^{x-1}$ on the same coordinate system. Use the intersect feature of your calculator to find the point of intersection of the two curves. Round to two decimal places.

74. Bob invested \$1000 at 6% compounded continuously. At the same time Paula invested \$1200 at 5% compounded monthly. Write two functions that give the amounts of Bob's and Paula's investments after x years. Graph these functions on a graphing calculator. Use the intersect feature of your graphing calculator to find the approximate value of x for which the investments are equal in value.

75. Graph the functions $y_1 = \log_2(x)$ and $y_2 = 3^{x-4}$ on the same coordinate system and use the intersect feature to find the points of intersection of the curves. Round to two decimal places. (*Hint:* To graph $y = \log_2(x)$, use the base-change formula to write the function as $y = \ln(x)/\ln(2)$.)

COLLABORATIVE ACTIVITIES

In How Much Space Could We Live?

The formula for population growth is

$$P(t) = P_0 e^{kt},$$

where $P(t)$ is the population after t years, P_0 is the population initially, k is the growth rate per year, and t is the number of years elapsed. We will find out how long it would take to cover the earth completely if the human population were growing exponentially. For all of your calculations, round your answers to one decimal place unless directed otherwise.

1. The population of the world in 1994 was approximately 5.5×10^9 people. If the population was 2.75×10^9 people in 1964, then what is the current growth rate (express as a percent)? Round your answer to the nearest tenth of a percent.

   The earth has a total surface area of approximately 5.1×10^{14} square meters. Seventy percent of this surface area is rock, ice, sand, and open ocean. Another 8% of the total surface area is tundra, lakes and streams, continental shelves, algal beds and reefs, and estuaries. We will consider the remaining area to be suitable for growing food and for living space.

2. Determine the surface area available for growing food and for living space.

Grouping: 2 students per group
Topic: Exponential and logarithmic functions

3. If each person needs 100 square meters of the earth's surface for living space and growing food, then in how many years after 1994 will the livable surface of the earth be used up? (Use the rate and the 1994 population from Question 1 and the surface area from Question 2.)

4. With your partner, think about the following questions and report your conclusions.

   • Does 100 square meters per person for living space and growing food seem reasonable? Take into account that many people live in tall apartment buildings and how that translates into surface area used per person.

   • How much room do you think it takes to grow animals for food (cows, chickens, pigs, etc.)? What about grains, vegetables, nuts, and fruit? Would food grow as well in desert areas, mountainous areas, or jungle areas?

   • Would there be any space left for wild animals or natural plant life? Would there be any space left for shopping malls, movie theaters, concert halls, factories, office buildings, or parking lots?

SUMMARY

Exponential and Logarithmic Functions		**Examples**
Exponential function	A function of the form $f(x) = a^x$ for $a > 0$ and $a \neq 1$	$f(x) = 3^x$
Logarithm function	A function of the form $f(x) = \log_a(x)$ for $a > 0$ and $a \neq 1$	$f(x) = \log_2(x)$
	$y = \log_a(x)$ if and only if $a^y = x$.	$\log_3(8) = x \leftrightarrow 3^x = 8$
Common logarithm	Base-10: $f(x) = \log(x)$	$\log(100) = 2$ because $100 = 10^2$.
Natural logarithm	Base-e: $f(x) = \ln(x)$ $e \approx 2.718$	$\ln(e) = 1$ because $e^1 = e$.
Inverse functions	$f(x) = a^x$ and $g(x) = \log_a(x)$ are inverse functions.	If $f(x) = e^x$, then $f^{-1}(x) = \ln(x)$.

Properties		**Examples**
M, N, and a are positive numbers with $a \neq 1$.		
	$\log_a(a) = 1 \qquad \log_a(1) = 0$	$\log_5(5) = 1, \log_5(1) = 0$
Inverse properties	$\log_a(a^M) = M \qquad a^{\log_a(M)} = M$	$\log(10^7) = 7, e^{\ln(3.4)} = 3.4$
Product rule	$\log_a(MN) = \log_a(M) + \log_a(N)$	$\ln(3x) = \ln(3) + \ln(x)$
Quotient rule	$\log_a\left(\dfrac{M}{N}\right) = \log_a(M) - \log_a(N)$	$\ln\left(\dfrac{2}{3}\right) = \ln(2) - \ln(3)$
	$\log_a\left(\dfrac{1}{N}\right) = -\log_a(N)$	$\ln\left(\dfrac{1}{3}\right) = -\ln(3)$
Power rule	$\log_a(M^N) = N \cdot \log_a(M)$	$\log(x^3) = 3 \cdot \log(x)$
Base-change formula	$\log_a(M) = \dfrac{\log_b(M)}{\log_b(a)}$	$\log_3(5) = \dfrac{\ln(5)}{\ln(3)}$

Equations Involving Logarithms and Exponents		**Examples**
Strategy	1. If there is a single logarithm or a single exponential expression, rewrite the equation using the definition of logarithms: $y = \log_a(x)$ if and only if $a^y = x$.	$2^x = 3$ and $x = \log_2(3)$ are equivalent.

2. Use the properties of logarithms to combine logarithms as much as possible.

$$\log(x) + \log(x - 3) = 1$$
$$\log(x^2 - 3x) = 1$$

3. Use the one-to-one properties:
   a) If $\log_a(m) = \log_a(n)$, then $m = n$.
   b) If $a^m = a^n$, then $m = n$.

$$\ln(x) = \ln(5 - x),$$
$$x = 5 - x$$
$$2^{3x} = 2^{5x-7}, \; 3x = 5x - 7$$

4. To get an approximate solution, take the common or natural logarithm of each side of an exponential equation.

$$2^x = 3, \; \ln(2^x) = \ln(3)$$
$$x \cdot \ln(2) = \ln(3)$$
$$x = \frac{\ln(3)}{\ln(2)}$$

ENRICHING YOUR MATHEMATICAL WORD POWER

For each mathematical term, choose the correct meaning.

1. **exponential function**
   a. $f(x) = a^x$ where $a > 0$ and $a \neq 1$
   b. $f(x) = ax^2$ where $a \neq 0$
   c. $f(x) = ax + b$ where $a \neq 0$
   d. $f(x) = x^n$ where n is an integer

2. **common base**
   a. base 2
   b. base e
   c. base π
   d. base 10

3. **natural base**
   a. base 2
   b. base e
   c. base π
   d. base 10

4. **domain**
   a. the range
   b. the set of second coordinates of a relation
   c. the independent variable
   d. the set of first coordinates of a relation

5. **compound interest**
   a. simple interest
   b. $A = Prt$
   c. an irrational interest rate
   d. interest is periodically paid into the account and the interest earns interest

6. **continuous compounding**
   a. compound interest
   b. using $A = Pe^{rt}$ to compute the amount
   c. frequent compounding
   d. using $A = P(1 + i)^n$ to compute the amount

7. **base-a logarithm of x**
   a. the exponent that is used on the base a to obtain x
   b. the exponent that is used on x to obtain a
   c. the power of 10 that produces x
   d. the power of e that produces a

8. **base-a logarithm function**
   a. $f(x) = a^x$ where $a > 0$ and $a \neq 1$
   b. $f(x) = \log_a(x)$ where $a > 0$ and $a \neq 1$
   c. $f(x) = \log_x(a)$ where $a > 0$ and $a \neq 1$
   d. $f(x) = \log(x)$ where $x > 0$

9. **common logarithm**
   a. $\log_2(x)$
   b. $\log(x)$
   c. $\ln(x)$
   d. $\log_3(x)$

10. **natural logarithm**
    a. $\log_2(x)$
    b. $\log(x)$
    c. $\ln(x)$
    d. $\log_3(x)$

REVIEW EXERCISES

12.1 *Use* $f(x) = 5^x$, $g(x) = 10^{x-1}$, *and* $h(x) = \left(\frac{1}{4}\right)^x$ *for Exercises 1–28. Find the following.*

1. $f(-2)$
2. $f(0)$
3. $f(3)$

4. $f(4)$
5. $g(1)$
6. $g(-1)$

7. $g(0)$
8. $g(3)$
9. $h(-1)$

10. $h(2)$
11. $h\left(\frac{1}{2}\right)$
12. $h\left(-\frac{1}{2}\right)$

Find x in each case.

13. $f(x) = 25$
14. $f(x) = -\frac{1}{125}$
15. $g(x) = 1000$
16. $g(x) = 0.001$
17. $h(x) = 32$
18. $h(x) = 8$
19. $h(x) = \frac{1}{16}$
20. $h(x) = 1$

Find the following.

21. $f(1.34)$

22. $f(-3.6)$

23. $g(3.25)$

24. $g(4.87)$

25. $h(2.82)$

26. $h(\pi)$

27. $h(\sqrt{2})$

28. $h\left(\dfrac{1}{3}\right)$

Sketch the graph of each function.

29. $f(x) = 5^x$

30. $g(x) = e^x$

31. $y = \left(\dfrac{1}{5}\right)^x$

32. $y = e^{-x}$

33. $f(x) = 3^{-x}$

34. $f(x) = -3^{x-1}$

35. $y = 1 + 2^x$

36. $y = 1 - 2^x$

12.2 *Write each exponential equation as a logarithmic equation and each logarithmic equation as an exponential equation.*

37. $10^m = n$

38. $b = a^5$

39. $h = \log_k(t)$

40. $\log_v(5) = u$

Let $f(x) = \log_2(x)$, $g(x) = \log(x)$, and $h(x) = \log_{1/2}(x)$. Find the following.

41. $f\left(\dfrac{1}{8}\right)$ **42.** $f(64)$

43. $g(0.1)$ **44.** $g(1)$

45. $g(100)$ **46.** $h\left(\dfrac{1}{8}\right)$

47. $h(1)$ **48.** $h(4)$

49. x, if $f(x) = 8$ **50.** x, if $g(x) = 3$

51. $f(77)$ **52.** $g(88.4)$

53. $h(33.9)$ **54.** $h(0.05)$

55. x, if $f(x) = 2.475$ **56.** x, if $g(x) = 1.426$

For each function f, find f^{-1} and sketch the graphs of f and f^{-1} on the same set of axes.

57. $f(x) = 10^x$

58. $f(x) = \log_8(x)$

59. $f(x) = e^x$

60. $f(x) = \log_3(x)$

12.3 *Rewrite each expression as a sum or a difference of multiples of logarithms.*

61. $\log(x^2 y)$

62. $\log_3(x^2 + 2x)$

63. $\ln(16)$

64. $\log\left(\dfrac{y}{\sqrt{x}}\right)$

65. $\log_5\left(\dfrac{1}{x}\right)$

66. $\ln\left(\dfrac{xy}{z}\right)$

Rewrite each expression as a single logarithm.

67. $\dfrac{1}{2}\log(x + 2) - 2 \cdot \log(x - 1)$

68. $3 \cdot \ln(x) + 2 \cdot \ln(y) - \dfrac{1}{3}\ln(z)$

12.4 *Find the exact solution to each equation.*

69. $\log_2(x) = 8$ **70.** $\log_3(x) = 0.5$

71. $\log_2(8) = x$ **72.** $3^x = 8$

73. $x^3 = 8$ **74.** $3^2 = x$

75. $\log_x(27) = 3$

76. $\log_x(9) = -\dfrac{1}{3}$

77. $x \cdot \ln(3) - x = \ln(7)$

78. $x \cdot \log(8) = x \cdot \log(4) + \log(9)$

79. $3^x = 5^{x-1}$

80. $5^{(2x^2)} = 5^{3-5x}$

81. $4^{2x} = 2^{x+1}$

82. $\log(12) = \log(x) + \log(7 - x)$
83. $\ln(x + 2) - \ln(x - 10) = \ln(2)$
84. $2 \cdot \ln(x + 3) = 3 \cdot \ln(4)$

85. $\log(x) - \log(x - 2) = 2$

86. $\log_2(x) = \log_2(x + 16) - 1$

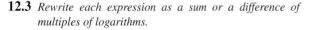
Use a calculator to find an approximate solution to each of the following. Round your answers to four decimal places.

87. $6^x = 12$ **88.** $5^x = 8^{3x+2}$

89. $3^{x+1} = 5$

90. $\log_3(x) = 2.634$

MISCELLANEOUS

Solve each problem.

91. Compounding annually. What does $10,000 invested at 11.5% compounded annually amount to after 15 years?

92. Doubling time. How many years does it take for an investment to double at 6.5% compounded annually?

93. Decaying substance. The amount, A, of a certain radioactive substance remaining after t years, is given by the formula $A = A_0 e^{-0.0003t}$, where A_0 is the initial amount. If we have 218 grams of this substance today, then how much of it will be left 1000 years from now?

94. Wildlife management. The number of white-tailed deer in the Hiawatha National Forest is believed to be growing according to the function

$$P = 517 + 10 \cdot \ln(8t + 1),$$

where t is the time in years from the year 2000.
a) What is the size of the population in 2000?
b) In what year will the population reach 600?
c) Does the population as shown on the accompanying graph appear to be growing faster during the period 2000 to 2005 or during the period 2005 to 2010?
d) What is the average rate of change of the population for each period in part (c)?

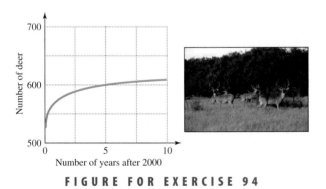

FIGURE FOR EXERCISE 94

95. Comparing investments. Melissa deposited $1000 into an account paying 5% annually; on the same day Frank deposited $900 into an account paying 7% compounded continuously. Find the number of years that it will take for the amounts in the accounts to be equal.

96. Imports and exports. The value of imports for a small Central American country is believed to be growing according to the function

$$I = 15 \cdot \log(16t + 33),$$

and the value of exports appears to be growing according to the function

$$E = 30 \cdot \log(t + 3),$$

where I and E are in millions of dollars and t is the number of years after 2000.
a) What are the values of imports and exports in 2000?
b) Use the accompanying graph to estimate the year in which imports will equal exports.
c) Algebraically find the year in which imports will equal exports.

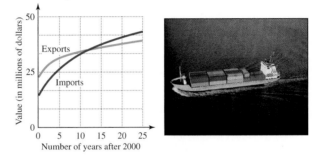

FIGURE FOR EXERCISE 96

97. Finding river flow. The U.S.G.S. measures the water height h (in feet above sea level) for the Tangipahoa River at Robert, Louisiana, and then finds the flow y [in cubic feet per second (ft³/sec)], using the formula

$$y = 114.308 e^{0.265(h - 6.87)}.$$

Find the flow when the river at Robert is 20.6 ft above sea level.

98. Finding the height. Rewrite the formula in Exercise 97 to express h as a function of y. Use the new formula to find the water height above sea level when the flow is 10,000 ft³/sec.

CHAPTER 12 TEST

Let $f(x) = 5^x$ and $g(x) = \log_5(x)$. Find the following.

1. $f(2)$ **2.** $f(-1)$

3. $f(0)$ **4.** $g(125)$

5. $g(1)$ **6.** $g\left(\dfrac{1}{5}\right)$

Sketch the graph of each function.
7. $y = 2^x$

8. $f(x) = \log_2(x)$

Suppose $\log_a(M) = 6$ and $\log_a(N) = 4$. Find the following.

11. $\log_a(MN)$

12. $\log_a\left(\dfrac{M^2}{N}\right)$

13. $\dfrac{\log_a(M)}{\log_a(N)}$

14. $\log_a(a^3M^2)$

15. $\log_a\left(\dfrac{1}{N}\right)$

Find the exact solution to each equation.

16. $3^x = 12$

17. $\log_3(x) = \dfrac{1}{2}$

18. $5^x = 8^{x-1}$

19. $\log(x) + \log(x + 15) = 2$

20. $2 \cdot \ln(x) = \ln(3) + \ln(6 - x)$

 Use a scientific calculator to find an approximate solution to each of the following. Round your answers to four decimal places.

21. Solve $20^x = 5$.

22. Solve $\log_3(x) = 2.75$.

9. $y = \left(\dfrac{1}{3}\right)^x$

23. The number of bacteria present in a culture at time t is given by the formula $N = 10e^{0.4t}$, where t is in hours. How many bacteria are present initially? How many are present after 24 hours?

24. How many hours does it take for the bacteria population of Problem 23 to double?

10. $g(x) = \log_{1/3}(x)$

Find the exact solution to each equation.

1. $(x - 3)^2 = 8$

2. $\log_2(x - 3) = 8$

3. $2^{x-3} = 8$

4. $2x - 3 = 8$

5. $|x - 3| = 8$

6. $\sqrt{x - 3} = 8$

7. $\log_2(x - 3) + \log_2(x) = \log_2(18)$

8. $2 \cdot \log_2(x - 3) = \log_2(5 - x)$

9. $\dfrac{1}{2}x - \dfrac{2}{3} = \dfrac{3}{4}x + \dfrac{1}{5}$

10. $3x^2 - 6x + 2 = 0$

Find the inverse of each function.

11. $f(x) = \dfrac{1}{3}x$

12. $g(x) = \log_3(x)$

13. $f(x) = 2x - 4$

14. $h(x) = \sqrt{x}$

15. $j(x) = \dfrac{1}{x}$

16. $k(x) = 5^x$

17. $m(x) = e^{x-1}$

18. $n(x) = \ln(x)$

Sketch the graph of each equation.

19. $y = 2x$ **20.** $y = 2^x$

21. $y = x^2$ **22.** $y = \log_2(x)$

23. $y = \dfrac{1}{2}x - 4$ **24.** $y = |2 - x|$

25. $y = 2 - x^2$ **26.** $y = e^2$

Solve each problem.

27. *Civilian labor force.* Using data from January 1988 through January 1998, the number of workers in the civilian labor force can be modeled by the linear function

$$n(t) = 1.53t + 113.82$$

or by the exponential function

$$n(t) = 114.0e^{t/80},$$

where t is the number of years since January of 1988 and $n(t)$ is in millions of workers (Bureau of Labor Statistics, stats.bls.gov).

a) Graph both functions on the same coordinate system for $0 \le t \le 15$.

b) What does each model predict for the value of n in January of 1999?

c) Use the Internet or your library to find the actual size of the civilian labor force in January of 1999.

d) Which model's prediction is closer to the actual number for January of 1999?

e) Answer parts (b), (c), and (d) for the present year.

28. *Measuring ocean depths.* In this exercise you will see how a geophysicist uses sound reflection to measure the depth of the ocean. Let v be the speed of sound through the water and d_1 be the depth of the ocean below the ship, as shown in the accompanying figure.

a) The time it takes for sound to travel from the ship at point S straight down to the ocean floor at point B_1 and back to point S is 0.270 second. Write d_1 as a function of v.

b) It takes 0.432 second for sound to travel from point S to point B_2 and then to a receiver at R, which is towed 500 meters behind the ship. Assuming $d_2 = d_3$, write d_2 as a function of v.

c) Use the Pythagorean theorem to find v. Then find the ocean depth d_1.

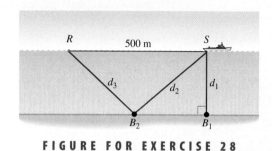

FIGURE FOR EXERCISE 28

Nonlinear Systems and the Conic Sections

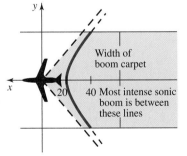

Width of boom carpet

20 40 Most intense sonic boom is between these lines

At a cruising speed of 1540 miles per hour, the Concorde can fly from London to New York in about 3 hours. So why isn't the same aircraft used for a fast flight from New York to Los Angeles? Concerns about cost efficiency and pollution are two reasons. However, most people agree that the biggest problem is noise. Traveling at Mach 2, the supersonic jet is flying faster than the speed of sound. At this speed the Concorde creates a cone-shaped wave in the air, on which there is a momentary change in air pressure. This change in air pressure causes a thunderlike sonic boom. When the jet is traveling parallel to the ground, the cone-shaped wave intersects the ground along one branch of a hyperbola. People on the ground hear the boom as the hyperbola passes them.

Sonic booms not only are noisy, but they have also been known to cause physical destruction such as broken windows and cracked plaster. For this reason supersonic jets are restricted from flying over land areas in the United States and much of the world. Some engineers believe that changing the silhouette of the plane can lessen the sonic boom, but most agree that it is impossible to eliminate the noise altogether.

In this chapter we discuss curves, including the hyperbola, that occur when a geometric plane intersects a cone. In Exercise 54 of Section 13.4 you will see how the altitude of the aircraft is related to the width of the area where the sonic boom is heard.

13.1 NONLINEAR SYSTEMS OF EQUATIONS

In this section

- Solving by Elimination
- Applications

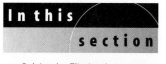

We studied systems of linear equations in Chapter 8. In this section we turn our attention to nonlinear systems of equations.

Solving by Elimination

Equations such as

$$y = x^2, \quad y = \sqrt{x}, \quad y = |x|, \quad y = 2^x, \quad \text{and} \quad y = \log_2(x)$$

are **nonlinear equations** because their graphs are not straight lines. We say that a system of equations is nonlinear if at least one equation in the system is nonlinear. We solve a nonlinear system just like a linear system, by elimination of variables. However, because the graphs of nonlinear equations may intersect at more than one point, there may be more than one ordered pair in the solution set to the system.

EXAMPLE 1 **A parabola and a line**

Solve the system of equations and draw the graph of each equation on the same coordinate system:

$$y = x^2 - 1$$
$$x + y = 1$$

Solution

We can eliminate y by substituting $y = x^2 - 1$ into $x + y = 1$:

$$x + y = 1$$
$$x + (x^2 - 1) = 1 \quad \text{Substitute } x^2 - 1 \text{ for } y.$$
$$x^2 + x - 2 = 0$$
$$(x - 1)(x + 2) = 0$$
$$x - 1 = 0 \quad \text{or} \quad x + 2 = 0$$
$$x = 1 \quad \text{or} \quad x = -2$$

Replace x by 1 and -2 in $y = x^2 - 1$ to find the corresponding values of y:

$$y = (1)^2 - 1 \qquad y = (-2)^2 - 1$$
$$y = 0 \qquad y = 3$$

FIGURE 13.1

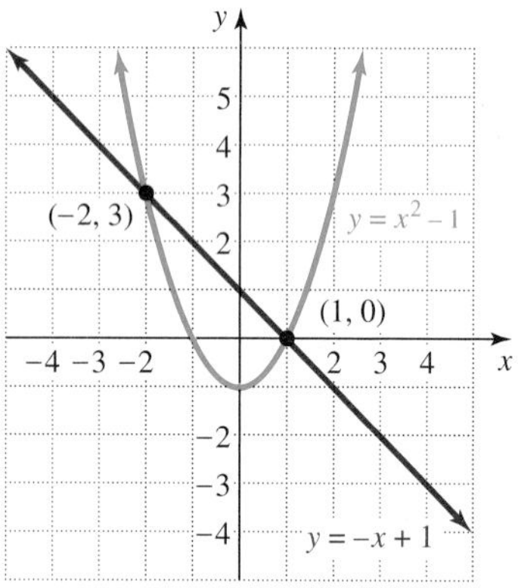

Check that each of the points (1, 0) and $(-2, 3)$ satisfies both of the original equations. The solution set is $\{(1, 0), (-2, 3)\}$. If we solve $x + y = 1$ for y, we get $y = -x + 1$. The line $y = -x + 1$ has y-intercept (0, 1) and slope -1. The graph of $y = x^2 - 1$ is a parabola with vertex $(0, -1)$. Of course, (1, 0) and $(-2, 3)$ are on both graphs. The two graphs are shown in Fig. 13.1. ■

Graphing is not an accurate method for solving any system of equations. However, the graphs of the equations in a nonlinear system help us to understand how many solutions we should have for the system. It is not necessary to graph a system to solve it. Even when the graphs are too difficult to sketch, we can solve the system.

EXAMPLE 2 **Solving a system algebraically with substitution**

Solve the system:

$$x^2 + y^2 + 2y = 3$$
$$x^2 - y = 5$$

Solution

If we substitute $y = x^2 - 5$ into the first equation to eliminate y, we will get a fourth-degree equation to solve. Instead, we can eliminate the variable x by writing $x^2 - y = 5$ as $x^2 = y + 5$. Now replace x^2 by $y + 5$ in the first equation:

$$x^2 + y^2 + 2y = 3$$
$$(y + 5) + y^2 + 2y = 3$$
$$y^2 + 3y + 5 = 3$$
$$y^2 + 3y + 2 = 0$$
$$(y + 2)(y + 1) = 0 \quad \text{Solve by factoring.}$$
$$y + 2 = 0 \quad \text{or} \quad y + 1 = 0$$
$$y = -2 \quad \text{or} \quad y = -1$$

Let $y = -2$ in the equation $x^2 = y + 5$ to find the corresponding x:

$$x^2 = -2 + 5$$
$$x^2 = 3$$
$$x = \pm\sqrt{3}$$

Now let $y = -1$ in the equation $x^2 = y + 5$ to find the corresponding x:

$$x^2 = -1 + 5$$
$$x^2 = 4$$
$$x = \pm 2$$

Check these values in the original equations. The solution set is

$$\{(\sqrt{3}, -2), (-\sqrt{3}, -2), (2, -1), (-2, -1)\}.$$

The graphs of these two equations intersect at four points. ◼

E X A M P L E   3

Solving a system with the addition method

Solve each system:

a) $x^2 - y^2 = 5$
   $x^2 + y^2 = 7$

b) $\dfrac{2}{x} + \dfrac{1}{y} = \dfrac{1}{5}$

   $\dfrac{1}{x} - \dfrac{3}{y} = \dfrac{1}{3}$

Solution

a) We can eliminate y by adding the equations:

$$x^2 - y^2 = 5$$
$$\underline{x^2 + y^2 = 7}$$
$$2x^2 \quad\;\; = 12$$
$$x^2 = 6$$
$$x = \pm\sqrt{6}$$

Since $x^2 = 6$, the second equation yields $6 + y^2 = 7$, $y^2 = 1$, and $y = \pm 1$. If $x^2 = 6$ and $y^2 = 1$, then both of the original equations are satisfied. The solution set is

$$\{(\sqrt{6}, 1)(\sqrt{6}, -1), (-\sqrt{6}, 1), (-\sqrt{6}, -1)\}$$

b) Usually with equations involving rational expressions we first multiply by the least common denominator (LCD), but this would make the given system more complicated. So we will just use the addition method to eliminate y:

$$\frac{6}{x} + \frac{3}{y} = \frac{3}{5} \qquad \text{Eq. (1) multiplied by 3}$$

$$\frac{1}{x} - \frac{3}{y} = \frac{1}{3} \qquad \text{Eq. (2)}$$

$$\frac{7}{x} = \frac{14}{15} \qquad \frac{3}{5} + \frac{1}{3} = \frac{14}{15}$$

$$14x = 7 \cdot 15$$

$$x = \frac{7 \cdot 15}{14} = \frac{15}{2}$$

To find y, substitute $x = \frac{15}{2}$ into Eq. (1):

$$\frac{2}{\frac{15}{2}} + \frac{1}{y} = \frac{1}{5}$$

$$\frac{4}{15} + \frac{1}{y} = \frac{1}{5} \qquad \frac{2}{\frac{15}{2}} = 2 \cdot \frac{2}{15} = \frac{4}{15}$$

$$15y \cdot \frac{4}{15} + 15y \cdot \frac{1}{y} = 15y \cdot \frac{1}{5} \qquad \text{Multiply each side by the LCD, } 15y.$$

$$4y + 15 = 3y$$

$$y = -15$$

Check that $x = \frac{15}{2}$ and $y = -15$ satisfy both original equations. The solution set is $\left\{ \left(\frac{15}{2}, -15 \right) \right\}$. ■

A system of nonlinear equations might involve exponential or logarithmic functions. To solve such systems, you will need to recall some facts about exponents and logarithms.

E X A M P L E   4

A system involving logarithms

Solve the system

$$y = \log_2(x + 28)$$
$$y = 3 + \log_2(x)$$

Solution

Eliminate y by substituting $\log_2(x + 28)$ for y in the second equation:

$$\log_2(x + 28) = 3 + \log_2(x) \qquad \text{Eliminate } y.$$

$$\log_2(x + 28) - \log_2(x) = 3 \qquad \text{Subtract } \log_2(x) \text{ from each side.}$$

$$\log_2\left(\frac{x + 28}{x} \right) = 3 \qquad \text{Quotient rule for logarithms}$$

$$\frac{x + 28}{x} = 8 \qquad \text{Definition of logarithm}$$

$$x + 28 = 8x \qquad \text{Multiply each side by } x.$$

$$28 = 7x \qquad \text{Subtract } x \text{ from each side.}$$

$$4 = x \qquad \text{Divide each side by 7.}$$

If $x = 4$, then $y = \log_2(4 + 28) = \log_2(32) = 5$. Check $(4, 5)$ in both equations. The solution to the system is $\{(4, 5)\}$. ■

Applications

The next example shows a geometric problem that can be solved with a system of nonlinear equations.

E X A M P L E   5   **Nonlinear equations in applications**

A 15-foot ladder is leaning against a wall so that the distance from the bottom of the ladder to the wall is one-half of the distance from the top of the ladder to the ground. Find the distance from the top of the ladder to the ground.

Solution

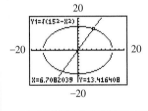

Let x be the number of feet from the bottom of the ladder to the wall and y be the number of feet from the top of the ladder to the ground (see Fig. 13.2). We can write two equations involving x and y:

$$x^2 + y^2 = 15^2 \quad \text{Pythagorean theorem}$$
$$y = 2x$$

Solve by substitution:

$$x^2 + (2x)^2 = 225 \quad \text{Replace } y \text{ by } 2x.$$
$$x^2 + 4x^2 = 225$$
$$5x^2 = 225$$
$$x^2 = 45$$
$$x = \pm\sqrt{45} = \pm 3\sqrt{5}$$

Because x represents distance, x must be positive. So $x = 3\sqrt{5}$. Because $y = 2x$, we get $y = 6\sqrt{5}$. The distance from the top of the ladder to the ground is $6\sqrt{5}$ feet.

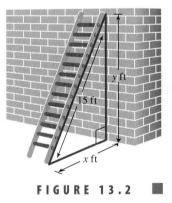

F I G U R E   1 3 . 2   ■

The next example shows how a nonlinear system can be used to solve a problem involving work.

E X A M P L E   6   **Nonlinear equations in applications**

A large fish tank at the Gulf Aquarium can usually be filled in 10 minutes using pumps A and B. However, pump B can pump water in or out at the same rate. If pump B is inadvertently run in reverse, then the tank will be filled in 30 minutes. How long would it take each pump to fill the tank by itself?

Solution

Let a represent the number of minutes that it takes pump A to fill the tank alone and b represent the number of minutes it takes pump B to fill the tank alone. The rate at which pump A fills the tank is $\frac{1}{a}$ of the tank per minute, and the rate at which pump B fills the tank is $\frac{1}{b}$ of the tank per minute. Because the work completed is the product of the rate and time, we can make the following table when the pumps work together to fill the tank:

	Rate	Time	Work
Pump A	$\frac{1}{a} \frac{\text{tank}}{\text{min}}$	10 min	$\frac{10}{a}$ tank
Pump B	$\frac{1}{b} \frac{\text{tank}}{\text{min}}$	10 min	$\frac{10}{b}$ tank

Note that each pump fills a fraction of the tank and those fractions have a sum of 1:

$$(1) \qquad \frac{10}{a} + \frac{10}{b} = 1$$

In the 30 minutes in which pump B is working in reverse, A puts in $\frac{30}{a}$ of the tank whereas B takes out $\frac{30}{b}$ of the tank. Since the tank still gets filled, we can write the following equation:

$$(2) \qquad \frac{30}{a} - \frac{30}{b} = 1$$

Multiply Eq. (1) by 3 and add the result to Eq. (2) to eliminate b:

$$\frac{30}{a} + \frac{30}{b} = 3 \qquad \text{Eq. (1) multiplied by 3}$$

$$\frac{30}{a} - \frac{30}{b} = 1 \qquad \text{Eq. (2)}$$

$$\overline{\qquad \frac{60}{a} \qquad = 4}$$

$$4a = 60$$

$$a = 15$$

Use $a = 15$ in Eq. (1) to find b:

$$\frac{10}{15} + \frac{10}{b} = 1$$

$$\frac{10}{b} = \frac{1}{3} \qquad \text{Subtract } \frac{10}{15} \text{ from each side.}$$

$$b = 30$$

So pump A fills the tank in 15 minutes working alone, and pump B fills the tank in 30 minutes working alone.

True or false? Explain your answer.

1. The graph of $y = x^2$ is a parabola.
2. The graph of $y = |x|$ is a straight line.
3. The point $(3, -4)$ satisfies both $x^2 + y^2 = 25$ and $y = \sqrt{5x + 1}$.
4. The graphs of $y = \sqrt{x}$ and $y = -x - 2$ do not intersect.
5. Substitution is the only method for eliminating a variable when solving a nonlinear system.
6. If Bob paints a fence in x hours, then he paints $\frac{1}{x}$ of the fence per hour.
7. In a triangle whose angles are $30°, 60°,$ and $90°$, the length of the side opposite the $30°$ angle is one-half the length of the hypotenuse.
8. The formula $V = LWH$ gives the volume of a rectangular box in which the sides have lengths L, W, and H.
9. The surface area of a rectangular box is $2LW + 2WH + 2LH$.
10. The area of a right triangle is one-half the product of the lengths of its legs.

13.1   EXERCISES

Reading and Writing   *After reading this section, write out the answers to these questions. Use complete sentences.*

1. Why are some equations called nonlinear?

2. Why do we graph the equations in a nonlinear system?

3. Why don't we solve systems by graphing?

4. What techniques do we use to solve nonlinear systems?

Solve each system and graph both equations on the same set of axes. See Example 1.

5. $y = x^2$
   $x + y = 6$

6. $y = x^2 - 1$
   $x + y = 11$

7. $y = |x|$
   $2y - x = 6$

8. $y = |x|$
   $3y = x + 6$

9. $y = \sqrt{2x}$
   $x - y = 4$

10. $y = \sqrt{x}$
    $x - y = 6$

11. $4x - 9y = 9$
$xy = 1$

12. $2x + 2y = 3$
$xy = -1$

25. $\dfrac{1}{x} - \dfrac{1}{y} = 5$

$\dfrac{2}{x} + \dfrac{1}{y} = -3$

26. $\dfrac{2}{x} - \dfrac{3}{y} = \dfrac{1}{2}$

$\dfrac{3}{x} + \dfrac{1}{y} = \dfrac{1}{2}$

27. $\dfrac{2}{x} - \dfrac{1}{y} = \dfrac{5}{12}$

$\dfrac{1}{x} - \dfrac{3}{y} = -\dfrac{5}{12}$

28. $\dfrac{3}{x} - \dfrac{2}{y} = 5$

$\dfrac{4}{x} + \dfrac{3}{y} = 18$

13. $y = -x^2 + 1$
$y = x^2$

14. $y = x^2$
$y = \sqrt{x}$

29. $x^2y = 20$
$xy + 2 = 6x$

30. $y^2x = 3$
$xy + 1 = 6x$

31. $x^2 + xy - y^2 = -11$
$x + y = 7$

32. $x^2 + xy + y^2 = 3$
$y = 2x - 5$

33. $3y - 2 = x^4$
$y = x^2$

34. $y - 3 = 2x^4$
$y = 7x^2$

Solve each system. See Examples 2 and 3.

15. $x^2 + y^2 = 25$
$y = x^2 - 5$

16. $x^2 + y^2 = 25$
$y = x + 1$

17. $xy - 3x = 8$
$y = x + 1$

18. $xy + 2x = 9$
$x - y = 2$

19. $xy - x = 8$
$xy + 3x = -4$

20. $2xy - 3x = -1$
$xy + 5x = -7$

Solve the following systems involving logarithmic and exponential functions. See Example 4.

35. $y = \log_2(x - 1)$
$y = 3 - \log_2(x + 1)$

36. $y = \log_3(x - 4)$
$y = 2 - \log_3(x + 4)$

37. $y = \log_2(x - 1)$
$y = 2 + \log_2(x + 2)$

38. $y = \log_4(8x)$
$y = 2 + \log_4(x - 1)$

21. $x^2 + y^2 = 8$
$x^2 - y^2 = 2$

22. $y^2 - 2x^2 = 1$
$y^2 + 2x^2 = 5$

39. $y = 2^{3x+4}$
$y = 4^{x-1}$

40. $y = 4^{3x}$
$y = \left(\dfrac{1}{2}\right)^{1-x}$

23. $x^2 + 2y^2 = 8$
$2x^2 - y^2 = 1$

24. $2x^2 + 3y^2 = 8$
$3x^2 + 2y^2 = 7$

Solve each problem by using a system of two equations in two unknowns. See Examples 5 and 6.

41. *Known hypotenuse.* Find the lengths of the legs of a right triangle whose hypotenuse is $\sqrt{15}$ feet and whose area is 3 square feet.

42. *Known diagonal.* A small television is advertised to have a picture with a diagonal measure of 5 inches and a viewing area of 12 square inches (in.²). What are the length and width of the screen?

FIGURE FOR EXERCISE 42

43. *House of seven gables.* Vincent has plans to build a house with seven gables. The plans call for an attic vent in the shape of an isosceles triangle in each gable. Because of the slope of the roof, the ratio of the height to the base of each triangle must be 1 to 4. If the vents are to provide a total ventilating area of 3500 in.², then what should be the height and base of each triangle?

FIGURE FOR EXERCISE 43

44. *Known perimeter.* Find the lengths of the sides of a triangle whose perimeter is 6 feet (ft) and whose angles are 30°, 60°, and 90° (see Appendix A).

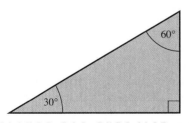

FIGURE FOR EXERCISE 44

45. *Filling a tank.* Pump A can either fill a tank or empty it in the same amount of time. If pump A and pump B are working together, the tank can be filled in 6 hours. When pump A was inadvertently left in the drain position while pump B was trying to fill the tank, it took 12 hours to fill the tank. How long would it take either pump working alone to fill the tank?

46. *Cleaning a house.* Roxanne either cleans the house or messes it up at the same rate. When Roxanne is cleaning with her mother, they can clean up a completely messed up house in 6 hours. If Roxanne is not cooperating, it takes her mother 9 hours to clean the house,

with Roxanne continually messing it up. How long would it take her mother to clean the entire house if Roxanne were sent to her grandmother's house?

47. *Cleaning fish.* Jan and Beth work in a seafood market that processes 200 pounds of catfish every morning. On Monday, Jan started cleaning catfish at 8:00 A.M. and finished cleaning 100 pounds just as Beth arrived. Beth then took over and finished the job at 8:50 A.M. On Tuesday they both started at 8 A.M. and worked together to finish the job at 8:24 A.M. On Wednesday, Beth was sick. If Jan is the faster worker, then how long did it take Jan to complete all of the catfish by herself?

FIGURE FOR EXERCISE 47

48. *Building a patio.* Richard has already formed a rectangular area for a flagstone patio, but his wife Susan is unsure of the size of the patio they want. If the width is increased by 2 ft, then the area is increased by 30 square feet (ft²). If the width is increased by 1 ft and the length by 3 ft, then the area is increased by 54 ft². What are the dimensions of the rectangle that Richard has already formed?

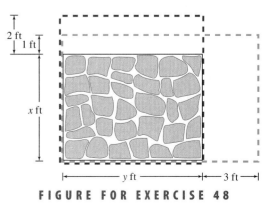

FIGURE FOR EXERCISE 48

49. *Fencing a rectangle.* If 34 ft of fencing are used to enclose a rectangular area of 72 ft^2, then what are the dimensions of the area?

50. *Real numbers.* Find two numbers that have a sum of 8 and a product of 10.

51. *Imaginary numbers.* Find two complex numbers whose sum is 8 and whose product is 20.

52. *Imaginary numbers.* Find two complex numbers whose sum is −6 and whose product is 10.

53. *Making a sign.* Rico's Sign Shop has a contract to make a sign in the shape of a square with an isosceles triangle on top of it, as shown in the figure. The contract calls for a total height of 10 ft with an area of 72 ft^2. How long should Rico make the side of the square and what should be the height of the triangle?

54. *Designing a box.* Angelina is designing a rectangular box of 120 cubic inches that is to contain new Eaties breakfast cereal. The box must be 2 inches thick so that it is easy to hold. It must have 184 square inches of surface area to provide enough space for all of the special offers and coupons. What should be the dimensions of the box?

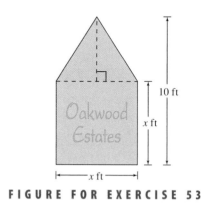

10 ft

x ft

Oakwood Estates

x ft

FIGURE FOR EXERCISE 53

 **GRAPHING CALCULATOR EXERCISES**

55. Solve each system by graphing each pair of equations on a graphing calculator and using the intersect feature to estimate the point of intersection. Find the coordinates of each intersection to the nearest hundredth.

a) $y = e^x - 4$
$y = \ln(x + 3)$

b) $3^{y-1} = x$
$y = x^2$

c) $x^2 + y^2 = 4$
$y = x^3$

13.2 THE PARABOLA

The parabola is one of four different curves that can be obtained by intersecting a cone and a plane as in Fig. 13.3. These curves, called **conic sections,** are the parabola, circle, ellipse, and hyperbola. We graphed parabolas in Sections 10.3 and 11.2. In this section we learn some new facts about parabolas.

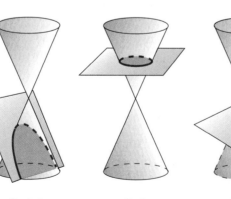
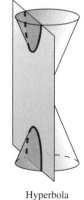

Parabola   Circle   Ellipse   Hyperbola

FIGURE 13.3

The Geometric Definition

In Section 10.3 we called the graph of $y = ax^2 + bx + c$ a parabola. This equation is the **standard equation** of a parabola. In this section you will see that the following geometric definition describes the same curve as the equation.

Parabola

Given a line (the **directrix**) and a point not on the line (the **focus**), the set of all points in the plane that are equidistant from the point and the line is called a **parabola.**

In Section 10.3 we defined the vertex as the highest point on a parabola that opens downward or the lowest point on a parabola that opens upward. We learned that $x = -b/(2a)$ gives the x-coordinate of the vertex. We can also describe the vertex of a parabola as the midpoint of the line segment that joins the focus and directrix, perpendicular to the directrix. See Fig. 13.4.

The focus of a parabola is important in applications. When parallel rays of light travel into a parabolic reflector, they are reflected toward the focus as in Fig. 13.5. This property is used in telescopes to see the light from distant stars. If the light source is at the focus, as in a searchlight, the light is reflected off the parabola and projected outward in a narrow beam. This reflecting property is also used in camera lenses, satellite dishes, and eavesdropping devices.

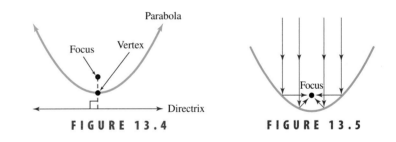

FIGURE 13.4   **FIGURE 13.5**

Developing the Equation

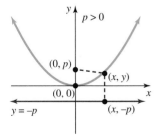

FIGURE 13.6

To develop an equation for a parabola, given the focus and directrix, choose the point $(0, p)$, where $p > 0$ as the focus and the line $y = -p$ as the directrix, as shown in Fig. 13.6. The vertex of this parabola is $(0, 0)$. For an arbitrary point (x, y) on the parabola the distance to the directrix is the distance from (x, y) to $(x, -p)$. The distance to the focus is the distance between (x, y) and $(0, p)$. We use the fact that these distances are equal to write the equation of the parabola:

$$\sqrt{(x - 0)^2 + (y - p)^2} = \sqrt{(x - x)^2 + (y - (-p))^2}$$

To simplify the equation, first remove the parentheses inside the radicals:

$$\sqrt{x^2 + y^2 - 2py + p^2} = \sqrt{y^2 + 2py + p^2}$$

$$x^2 + y^2 - 2py + p^2 = y^2 + 2py + p^2 \quad \text{Square each side.}$$

$$x^2 = 4py \quad \text{Subtract } y^2 \text{ and } p^2 \text{ from each side.}$$

$$y = \frac{1}{4p}x^2$$

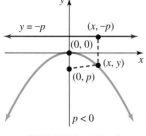

FIGURE 13.7

So the parabola with focus $(0, p)$ and directrix $y = -p$ for $p > 0$ has equation $y = \frac{1}{4p}x^2$. This equation has the form $y = ax^2 + bx + c$, where $a = \frac{1}{4p}$, $b = 0$, and $c = 0$.

If the focus is $(0, p)$ with $p < 0$ and the directrix is $y = -p$, then the parabola opens downward as shown in Fig. 13.7. Deriving the equation using the distance formula again yields $y = \frac{1}{4p}x^2$.

Parabolas in the Form $y = a(x - h)^2 + k$

The simplest parabola, $y = x^2$, has vertex $(0, 0)$. The transformation $y = a(x - h)^2 + k$ is also a parabola and its vertex is (h, k). The focus and directrix of the transformation are found as follows:

> ### Parabolas in the Form $y = a(x - h)^2 + k$
>
> The graph of the equation $y = a(x - h)^2 + k$ $(a \neq 0)$ is a parabola with vertex (h, k), focus $(h, k + p)$, and directrix $y = k - p$, where $a = \frac{1}{4p}$. If $a > 0$, the parabola opens upward; if $a < 0$, the parabola opens downward.

Figure 13.8 shows the location of the focus and directrix for parabolas with vertex (h, k) and opening either upward or downward. Note that the location of the focus and directrix determine the value of a and the shape and opening of the parabola.

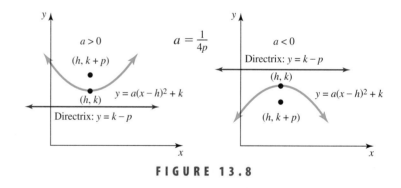

FIGURE 13.8

CAUTION For a parabola that opens upward, $p > 0$, and the focus $(h, k + p)$ is above the vertex (h, k). For a parabola that opens downward, $p < 0$, and the focus $(h, k + p)$ is below the vertex (h, k). In either case the distance from the vertex to the focus and the vertex to the directrix is $|p|$.

Finding the Vertex, Focus, and Directrix

In Example 1 we find the vertex, focus, and directrix from an equation of a parabola. In Example 2 we find the equation given the focus and directrix.

E X A M P L E   1 **Finding the vertex, focus, and directrix, given an equation**
Find the vertex, focus, and directrix for the parabola $y = x^2$.

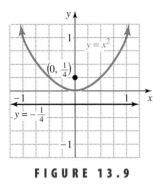

FIGURE 13.9

Solution

Compare $y = x^2$ to the general formula $y = a(x - h)^2 + k$. We see that $h = 0$, $k = 0$, and $a = 1$. So the vertex is (0, 0). Because $a = 1$, we can use $a = \frac{1}{4p}$ to get

$$1 = \frac{1}{4p},$$

or $p = \frac{1}{4}$. Use $(h, k + p)$ to get the focus $\left(0, \frac{1}{4}\right)$. Use the equation $y = k - p$ to get $y = -\frac{1}{4}$ as the equation of the directrix. See Fig. 13.9. ■

E X A M P L E 2

Finding an equation, given a focus and directrix

Find the equation of the parabola with focus $(-1, 4)$ and directrix $y = 3$.

Solution

Because the vertex is halfway between the focus and directrix, the vertex is $\left(-1, \frac{7}{2}\right)$. See Fig. 13.10. The distance from the vertex to the focus is $\frac{1}{2}$. Because the focus is above the vertex, p is positive. So $p = \frac{1}{2}$, and $a = \frac{1}{4p} = \frac{1}{2}$. The equation is

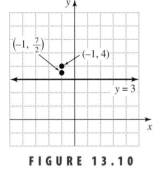

FIGURE 13.10

$$y = \frac{1}{2}(x - (-1))^2 + \frac{7}{2}.$$

Convert to $y = ax^2 + bx + c$ form as follows:

$$y = \frac{1}{2}(x + 1)^2 + \frac{7}{2}$$

$$y = \frac{1}{2}(x^2 + 2x + 1) + \frac{7}{2}$$

$$y = \frac{1}{2}x^2 + x + 4$$

■

Axis of Symmetry

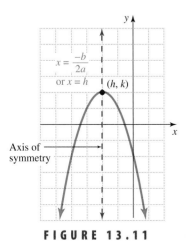

FIGURE 13.11

The graph of $y = x^2$ shown in Fig. 13.9 is **symmetric about the y-axis** because the two halves of the parabola would coincide if the paper were folded on the y-axis. In general, the vertical line through the vertex is the **axis of symmetry** for the parabola. See Fig. 13.11. In the form $y = ax^2 + bx + c$ the x-coordinate of the vertex is $-b/(2a)$ and the equation of the axis of symmetry is $x = -b/(2a)$. In the form $y = a(x - h)^2 + k$ the vertex is (h, k) and the equation for the axis of symmetry is $x = h$.

Changing Forms

Since there are two forms for the equation of a parabola, it is sometimes useful to change from one form to the other. To change from $y = a(x - h)^2 + k$ to the form $y = ax^2 + bx + c$, we square the binomial and combine like terms, as in Example 2. To change from $y = ax^2 + bx + c$ to the form $y = a(x - h)^2 + k$, we complete the square, as in the next example.

E X A M P L E   3

calculator

close-up

The graphs of

$$y_1 = 2x^2 - 4x + 5$$

and

$$y_2 = 2(x - 1)^2 + 3$$

appear to be identical. This supports the conclusion that the equations are equivalent.

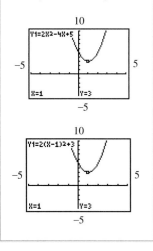

Converting $y = ax^2 + bx + c$ to $y = a(x - h)^2 + k$

Write $y = 2x^2 - 4x + 5$ in the form $y = a(x - h)^2 + k$ and identify the vertex, focus, directrix, and axis of symmetry of the parabola.

Solution

Use completing the square to rewrite the equation:

$$y = 2(x^2 - 2x) + 5$$
$$y = 2(x^2 - 2x + 1 - 1) + 5 \quad \text{Complete the square.}$$
$$y = 2(x^2 - 2x + 1) - 2 + 5 \quad \text{Move } 2(-1) \text{ outside the parentheses.}$$
$$y = 2(x - 1)^2 + 3$$

The vertex is $(1, 3)$. Because $a = \frac{1}{4p}$, we have

$$\frac{1}{4p} = 2,$$

and $p = \frac{1}{8}$. Because the parabola opens upward, the focus is $\frac{1}{8}$ unit above the vertex at $\left(1, 3\frac{1}{8}\right)$, or $\left(1, \frac{25}{8}\right)$, and the directrix is the horizontal line $\frac{1}{8}$ unit below the vertex, $y = 2\frac{7}{8}$ or $y = \frac{23}{8}$. The axis of symmetry is $x = 1$. ■

CAUTION Be careful when you complete a square within parentheses as in Example 3. For another example, consider the equivalent equations

$$y = -3(x^2 + 4x),$$
$$y = -3(x^2 + 4x + 4 - 4),$$

and

$$y = -3(x + 2)^2 + 12.$$

E X A M P L E   4

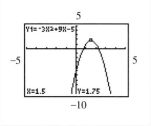

calculator

close-up

A calculator graph can be used to check the vertex and opening of a parabola.

Finding the features of a parabola from standard form

Find the vertex, focus, directrix, and axis of symmetry of the parabola $y = -3x^2 + 9x - 5$, and determine whether the parabola opens upward or downward.

Solution

The x-coordinate of the vertex is

$$x = \frac{-b}{2a} = \frac{-9}{2(-3)} = \frac{-9}{-6} = \frac{3}{2}.$$

To find the y-coordinate of the vertex, let $x = \frac{3}{2}$ in $y = -3x^2 + 9x - 5$:

$$y = -3\left(\frac{3}{2}\right)^2 + 9\left(\frac{3}{2}\right) - 5 = -\frac{27}{4} + \frac{27}{2} - 5 = \frac{7}{4}$$

The vertex is $\left(\frac{3}{2}, \frac{7}{4}\right)$. Because $a = -3$, the parabola opens downward. To find the focus, use $-3 = \frac{1}{4p}$ to get $p = -\frac{1}{12}$. The focus is $\frac{1}{12}$ of a unit below the vertex at $\left(\frac{3}{2}, \frac{7}{4} - \frac{1}{12}\right)$ or $\left(\frac{3}{2}, \frac{5}{3}\right)$. The directrix is the horizontal line $\frac{1}{12}$ of a unit above the vertex, $y = \frac{7}{4} + \frac{1}{12}$ or $y = \frac{11}{6}$. The equation of the axis of symmetry is $x = \frac{3}{2}$. ■

WARM-UPS

True or false? Explain your answer.

1. There is a parabola with focus (2, 3), directrix $y = 1$, and vertex (0, 0).

2. The focus for the parabola $y = \frac{1}{4}x^2 + 1$ is (0, 2).

3. The graph of $y - 3 = 5(x - 4)^2$ is a parabola with vertex (4, 3).

4. The graph of $y = 6x + 3x + 2$ is a parabola.

5. The graph of $y = 2x - x^2 + 9$ is a parabola opening upward.

6. For $y = x^2$ the vertex and y-intercept are the same point.

7. A parabola with vertex (2, 3) and focus (2, 4) has no x-intercepts.

8. The parabola with focus (0, 2) and directrix $y = 1$ opens upward.

9. The axis of symmetry for $y = a(x - 2)^2 + k$ is $x = 2$.

10. If $a = \frac{1}{(4p)}$ and $a = 1$, then $p = \frac{1}{4}$.

13.2   EXERCISES

Reading and Writing   *After reading this section, write out the answers to these questions. Use complete sentences.*

1. What is the definition of a parabola given in this section?

2. What is the location of the vertex?

3. What are the two forms of the equation of a parabola?

4. What is the distance from the focus to the vertex in any parabola of the form $y = ax^2 + bx + c$?

5. How do we convert an equation of the form $y = ax^2 + bx + c$ into the form $y = a(x - h)^2 + k$?

6. How do we convert an equation of the form $y = a(x - h)^2 + k$ into the form $y = ax^2 + bx + c$?

Find the vertex, focus, and directrix for each parabola. See Example 1.

7. $y = 2x^2$

8. $y = \frac{1}{2}x^2$

9. $y = -\frac{1}{4}x^2$

10. $y = -\frac{1}{12}x^2$

11. $y = \frac{1}{2}(x - 3)^2 + 2$

12. $y = \frac{1}{4}(x + 2)^2 - 5$

13. $y = -(x + 1)^2 + 6$

14. $y = -3(x - 4)^2 + 1$

Find the equation of the parabola with the given focus and directrix. See Example 2.

15. Focus (0, 2), directrix $y = -2$

16. Focus (0, −3), directrix $y = 3$

17. Focus $\left(0, -\frac{1}{2}\right)$, directrix $y = \frac{1}{2}$

18. Focus $\left(0, \frac{1}{8}\right)$, directrix $y = -\frac{1}{8}$

19. Focus (3, 2), directrix $y = 1$

20. Focus (−4, 5), directrix $y = 4$

21. Focus (1, −2), directrix $y = 2$

22. Focus (2, −3), directrix $y = 1$

23. Focus (−3, 1.25), directrix $y = 0.75$

24. Focus $\left(5, \dfrac{17}{8}\right)$, directrix $y = \dfrac{15}{8}$

Write each equation in the form $y = a(x - h)^2 + k$. Identify the vertex, focus, directrix, and axis of symmetry of each parabola. See Example 3.

25. $y = x^2 - 6x + 1$

26. $y = x^2 + 4x - 7$

27. $y = 2x^2 + 12x + 5$

28. $y = 3x^2 + 6x - 7$

29. $y = -2x^2 + 16x + 1$

30. $y = -3x^2 - 6x + 7$

31. $y = 5x^2 + 40x$

32. $y = -2x^2 + 10x$

Find the vertex, focus, directrix, and axis of symmetry of each parabola (without completing the square), and determine whether the parabola opens upward or downward. See Example 4.

33. $y = x^2 - 4x + 1$

34. $y = x^2 - 6x - 7$

35. $y = -x^2 + 2x - 3$

36. $y = -x^2 + 4x + 9$

37. $y = 3x^2 - 6x + 1$

38. $y = 2x^2 + 4x - 3$

39. $y = -x^2 - 3x + 2$

40. $y = -x^2 + 3x - 1$

41. $y = 3x^2 + 5$

42. $y = -2x^2 - 6$

Solve each problem.

43. *World's largest telescope.* The largest reflecting telescope in the world is the 6-meter (m) reflector on Mount Pastukhov in Russia. The accompanying figure shows a cross section of a parabolic mirror 6 m in diameter with the vertex at the origin and the focus at (0, 15). Find the equation of the parabola.

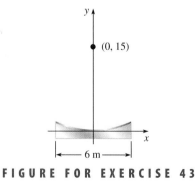

FIGURE FOR EXERCISE 43

44. *Arecibo Observatory.* The largest radio telescope in the world uses a 1000-ft parabolic dish, suspended in a valley in Arecibo, Puerto Rico. The antenna hangs above the vertex of the dish on cables stretching from two towers. The accompanying figure shows a cross section of the parabolic dish and the towers. Assuming the vertex is at (0, 0), find the equation for the parabola. Find the distance from the vertex to the antenna located at the focus.

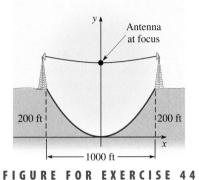

FIGURE FOR EXERCISE 44

49. $y = x^2 + 3x - 4$
  $y = -x^2 - 2x + 8$

50. $y = x^2 + 2x - 8$
  $y = -x^2 - x + 12$

Graph both equations of each system on the same coordinate axes. Use elimination of variables to find all points of intersection.

45. $y = -x^2 + 3$
  $y = x^2 + 1$

46. $y = x^2 - 3$
  $y = -x^2 + 5$

51. $y = x^2 + 3x - 4$
  $y = 2x + 2$

52. $y = x^2 + 5x + 6$
  $y = x + 11$

47. $y = x^2 - 2$
  $y = 2x - 3$

48. $y = x^2 + x - 6$
  $y = 7x - 15$

Solve each problem.

53. Find all points of intersection of the parabola $y = x^2 - 2x - 3$ and the x-axis.

54. Find all points of intersection of the parabola $y = 80x^2 - 33x + 255$ and the y-axis.

55. Find all points of intersection of the parabola $y = 0.01x^2$ and the line $y = 4$.

56. Find all points of intersection of the parabola $y = 0.02x^2$ and the line $y = x$.

57. Find all points of intersection of the parabolas $y = x^2$ and $x = y^2$.

58. Find all points of intersection of the parabolas $y = x^2$ and $y = (x - 3)^2$.

GETTING MORE INVOLVED

59. *Exploration.* Consider the parabola with focus $(p, 0)$ and directrix $x = -p$ for $p > 0$. Let (x, y) be an arbitrary point on the parabola. Write an equation expressing the fact that the distance from (x, y) to the focus is equal to the distance from (x, y) to the directrix. Rewrite the equation in the form $x = ay^2$, where $a = \frac{1}{4p}$.

60. *Exploration.* In general, the graph of $x = a(y - h)^2 + k$ for $a \neq 0$ is a parabola opening left or right with vertex at (k, h).
   a) For which values of a does the parabola open to the right, and for which values of a does it open to the left?
   b) What is the equation of its axis of symmetry?

c) Sketch the graphs $x = 2(y - 3)^2 + 1$ and $x = -(y + 1)^2 + 2$.

GRAPHING CALCULATOR EXERCISES

61. Graph $y = x^2$ using the viewing window with $-1 \leq x \leq 1$ and $0 \leq y \leq 1$. Next graph $y = 2x^2 - 1$ using the viewing window $-2 \leq x \leq 2$ and $-1 \leq y \leq 7$. Explain what you see.

62. Graph $y = x^2$ and $y = 6x - 9$ in the viewing window $-5 \leq x \leq 5$ and $-5 \leq y \leq 20$. Does the line appear to be tangent to the parabola? Solve the system $y = x^2$ and $y = 6x - 9$ to find all points of intersection for the parabola and the line.

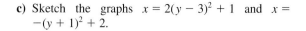

13.3 THE CIRCLE

In this section we continue the study of the conic sections with a discussion of the circle.

Developing the Equation

In this

section

- Developing the Equation
- Equations Not in Standard Form
- Systems of Equations

A circle is obtained by cutting a cone, as was shown in Fig. 13.3. We can also define a circle using points and distance, as we did for the parabola.

> **Circle**
>
> A **circle** is the set of all points in a plane that lie a fixed distance from a given point in the plane. The fixed distance is called the **radius,** and the given point is called the **center.**

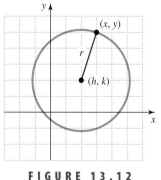

FIGURE 13.12

We can use the distance formula of Section 9.5 to write an equation for the circle with center (h, k) and radius r, shown in Fig. 13.12. If (x, y) is a point on the circle, its distance from the center is r. So

$$\sqrt{(x - h)^2 + (y - k)^2} = r.$$

We square both sides of this equation to get the **standard form** for the equation of a circle.

> **Standard Equation for a Circle**
>
> The graph of the equation
>
> $$(x - h)^2 + (y - k)^2 = r^2$$
>
> with $r > 0$, is a circle with center (h, k) and radius r.

Note that a circle centered at the origin with radius r $(r > 0)$ has the standard equation

$$x^2 + y^2 = r^2.$$

E X A M P L E   1

Finding the equation, given the center and radius

Write the standard equation for the circle with the given center and radius.

a) Center $(0, 0)$, radius 2   **b)** Center $(-1, 2)$, radius 4

Solution

a) The center at $(0, 0)$ means that $h = 0$ and $k = 0$ in the standard equation. So the equation is $(x - 0)^2 + (y - 0)^2 = 2^2$, or $x^2 + y^2 = 4$. The circle with radius 2 centered at the origin is shown in Fig. 13.13.

b) The center at $(-1, 2)$ means that $h = -1$ and $k = 2$. So

$$[x - (-1)]^2 + [y - 2]^2 = 4^2.$$

Simplify this equation to get

$$(x + 1)^2 + (y - 2)^2 = 16.$$

The circle with center $(-1, 2)$ and radius 4 is shown in Fig. 13.14.

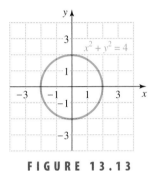

FIGURE 13.13

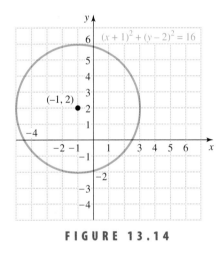

FIGURE 13.14

CAUTION   The equations $(x - 1)^2 + (y + 3)^2 = -9$ and $(x - 1)^2 + (y + 3)^2 = 0$ might look like equations of circles, but they are not. The first equation is not satisfied by any ordered pair of real numbers because the left-hand side is nonnegative for any x and y. The second equation is satisfied only by the point $(1, -3)$.

E X A M P L E   2

Finding the center and radius, given the equation

Determine the center and radius of the circle $x^2 + (y + 5)^2 = 2$.

Solution

We can write this equation as

$$(x - 0)^2 + [y - (-5)]^2 = (\sqrt{2})^2.$$

In this form we see that the center is $(0, -5)$ and the radius is $\sqrt{2}$.

EXAMPLE 3   **Graphing a circle**

Find the center and radius of $(x - 1)^2 + (y + 2)^2 = 9$, and sketch the graph.

Solution

The graph of this equation is a circle with center at $(1, -2)$ and radius 3. See Fig. 13.15 for the graph. ∎

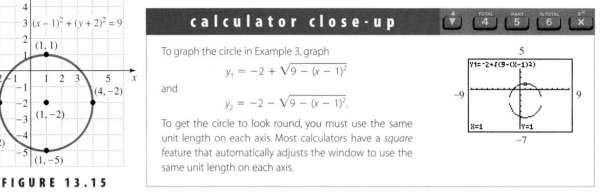

FIGURE 13.15

calculator close-up

To graph the circle in Example 3, graph

$$y_1 = -2 + \sqrt{9 - (x - 1)^2}$$

and

$$y_2 = -2 - \sqrt{9 - (x - 1)^2}.$$

To get the circle to look round, you must use the same unit length on each axis. Most calculators have a *square* feature that automatically adjusts the window to use the same unit length on each axis.

Equations Not in Standard Form

It is not easy to recognize that $x^2 - 6x + y^2 + 10y = -30$ is the equation of a circle, but it is. In the next example we convert this equation into the standard form for a circle by completing the squares for the variables x and y.

EXAMPLE 4   **Converting to standard form**

Find the center and radius of the circle given by the equation

$$x^2 - 6x + y^2 + 10y = -30.$$

Solution

To complete the square for $x^2 - 6x$, we add 9, and for $y^2 + 10y$, we add 25. To get an equivalent equation, we must add on both sides:

$$x^2 - 6x + y^2 + 10y = -30$$
$$x^2 - 6x + 9 + y^2 + 10y + 25 = -30 + 9 + 25 \quad \text{Add 9 and 25 to both sides.}$$
$$(x - 3)^2 + (y + 5)^2 = 4 \quad \text{Factor the trinomials on the left-hand side.}$$

> **helpful hint**
>
> What do circles and lines have in common? They are the two simplest graphs to draw. We have compasses to make our circles look good and rulers to make our lines look good.

From the standard form we see that the center is $(3, -5)$ and the radius is 2. ∎

Systems of Equations

We first solved systems of nonlinear equations in two variables in Section 13.1. We found the points of intersection of two graphs without drawing the graphs. Here we will solve systems involving circles, parabolas, and lines. In the next example we find the points of intersection of a line and a circle.

EXAMPLE 5   **Intersection of a line and a circle**

Graph both equations of the system

$$(x - 3)^2 + (y + 1)^2 = 9$$
$$y = x - 1$$

on the same coordinate axes, and solve the system by elimination of variables.

Solution

The graph of the first equation is a circle with center at $(3, -1)$ and radius 3. The graph of the second equation is a straight line with slope 1 and y-intercept $(0, -1)$. Both graphs are shown in Fig. 13.16. To solve the system by elimination, we substitute $y = x - 1$ into the equation of the circle:

$$(x - 3)^2 + (x - 1 + 1)^2 = 9$$
$$(x - 3)^2 + x^2 = 9$$
$$x^2 - 6x + 9 + x^2 = 9$$
$$2x^2 - 6x = 0$$
$$x^2 - 3x = 0$$
$$x(x - 3) = 0$$

$$x = 0 \qquad \text{or} \qquad x = 3$$
$$y = -1 \qquad\qquad y = 2 \quad \text{Because } y = x - 1$$

Check $(0, -1)$ and $(3, 2)$ in the original system and with the graphs in Fig. 13.16. The solution set is $\{(0, -1), (3, 2)\}$. ■

FIGURE 13.16

The graph shows $y = x - 1$ and $(x - 3)^2 + (y + 1)^2 = 9$.

M A T H   A T   W O R K $x^2 + (x+1)^2 = 5^2$

Friedrich von Huene, a flautist and recorder player, has been crafting woodwind instruments in his family business for over 30 years. Because it is best to play music of earlier centuries on the instruments of their time, von Huene is using many originals as models for his flutes, recorders, and oboes of different sizes.

EARLY MUSICAL INSTRUMENT MAKER

Because museum instruments have many different pitch standards, their dimensions frequently have to be changed to accommodate pitch standards that musicians use today. For a lower pitch the length of the instrument as well as the inside diameter must be increased. For a higher pitch the length has to be shortened and the diameter decreased. However, the factor for changing the length will be different from the factor for changing the diameter. A row of organ pipes demonstrates that the larger and longer pipes are proportionately more slender than the shorter high-pitched pipes. A pipe an octave higher in pitch is about half as long as the pipe an octave lower, but its diameter will be about 0.6 as large as the diameter of the lower pipe.

When making a very large recorder, von Huene carefully chooses the length, the position of the tone holes, and the bore to get the proper volume of air inside the instrument. In Exercises 57 and 58 of this section you will make the kinds of calculations von Huene makes when he crafts a modern reproduction of a Renaissance flute.

True or false? Explain your answer.

1. The radius of a circle can be any nonzero real number.
2. The coordinates of the center must satisfy the equation of the circle.
3. The circle $x^2 + y^2 = 4$ has its center at the origin.
4. The graph of $x^2 + y^2 = 9$ is a circle centered at $(0, 0)$ with radius 9.

5. The graph of $(x - 2)^2 + (y - 3)^2 + 4 = 0$ is a circle of radius 2.
6. The graph of $(x - 3) + (y + 5) = 9$ is a circle of radius 3.
7. There is only one circle centered at $(-3, -1)$ passing through the origin.

8. The center of the circle $(x - 3)^2 + (y - 4)^2 = 10$ is $(-3, -4)$.
9. The center of the circle $x^2 + y^2 + 6y - 4 = 0$ is on the y-axis.
10. The radius of the circle $x^2 - 3x + y^2 = 4$ is 2.

13.3 EXERCISES

Reading and Writing *After reading this section, write out the answers to these questions. Use complete sentences.*

1. What is the definition of a circle?

2. What is the standard equation of a circle?

Write the standard equation for each circle with the given center and radius. See Example 1.

3. Center $(0, 3)$, radius 5
4. Center $(2, 0)$, radius 3
5. Center $(1, -2)$, radius 9
6. Center $(-3, 5)$, radius 4
7. Center $(0, 0)$, radius $\sqrt{3}$
8. Center $(0, 0)$, radius $\sqrt{2}$

9. Center $(-6, -3)$, radius $\dfrac{1}{2}$

10. Center $(-3, -5)$, radius $\dfrac{1}{4}$

11. Center $\left(\frac{1}{2}, \frac{1}{3}\right)$, radius 0.1

12. Center $\left(-\frac{1}{2}, 3\right)$, radius 0.2

Find the center and radius for each circle. See Example 2.

13. $(x - 3)^2 + (y - 5)^2 = 2$
14. $(x + 3)^2 + (y - 7)^2 = 6$

15. $x^2 + \left(y - \dfrac{1}{2}\right)^2 = \dfrac{1}{2}$

16. $5x^2 + 5y^2 = 5$

17. $4x^2 + 4y^2 = 9$

18. $9x^2 + 9y^2 = 49$

19. $3 - y^2 = (x - 2)^2$
20. $9 - x^2 = (y + 1)^2$

Sketch the graph of each equation. See Example 3.

21. $x^2 + y^2 = 9$
22. $x^2 + y^2 = 16$

23. $x^2 + (y - 3)^2 = 9$ **24.** $(x - 4)^2 + y^2 = 16$

35. $x^2 + y^2 = 8y + 10x - 32$

36. $x^2 + y^2 = 8x - 10y$

37. $x^2 - x + y^2 + y = 0$

38. $x^2 - 3x + y^2 = 0$

25. $(x + 1)^2 + (y - 1)^2 = 2$ **26.** $(x - 2)^2 + (y + 2)^2 = 8$

39. $x^2 - 3x + y^2 - y = 1$

40. $x^2 - 5x + y^2 + 3y = 2$

41. $x^2 - \dfrac{2}{3}x + y^2 + \dfrac{3}{2}y = 0$

27. $(x - 4)^2 + (y + 3)^2 = 16$ **28.** $(x - 3)^2 + (y - 7)^2 = 25$

42. $x^2 + \dfrac{1}{3}x + y^2 - \dfrac{2}{3}y = \dfrac{1}{9}$

Graph both equations of each system on the same coordinate axes. Solve the system by elimination of variables to find all points of intersection of the graphs. See Example 5.

43. $x^2 + y^2 = 10$ **44.** $x^2 + y^2 = 4$
       $y = 3x$ $y = x - 2$

29. $\left(x - \dfrac{1}{2}\right)^2 + \left(y + \dfrac{1}{2}\right)^2 = \dfrac{1}{4}$ **30.** $\left(x + \dfrac{1}{3}\right)^2 + y^2 = \dfrac{1}{9}$

45. $x^2 + y^2 = 9$ **46.** $x^2 + y^2 = 4$
       $y = x^2 - 3$ $y = x^2 - 2$

Rewrite each equation in the standard form for the equation of a circle, and identify its center and radius. See Example 4.

31. $x^2 + 4x + y^2 + 6y = 0$

32. $x^2 - 10x + y^2 + 8y = 0$

33. $x^2 - 2x + y^2 - 4y - 3 = 0$

34. $x^2 - 6x + y^2 - 2y + 9 = 0$

47. $(x - 2)^2 + (y + 3)^2 = 4$ **48.** $(x + 1)^2 + (y - 4)^2 = 17$
$\quad y = x - 3$ $\qquad\qquad y = x + 2$

the equation for the bore hole (centered at the origin) and the volume of air in the C# flute.

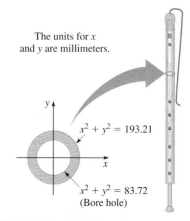

The units for x and y are millimeters.

$x^2 + y^2 = 193.21$

$x^2 + y^2 = 83.72$
(Bore hole)

FIGURE FOR EXERCISES 57 AND 58

In Exercises 49–58, solve each problem.

49. Determine all points of intersection of the circle $(x - 1)^2 + (y - 2)^2 = 4$ with the y-axis.

50. Determine the points of intersection of the circle $x^2 + (y - 3)^2 = 25$ with the x-axis.

51. Find the radius of the circle that has center $(2, -5)$ and passes through the origin.

52. Find the radius of the circle that has center $(-2, 3)$ and passes through $(3, -1)$.

53. Determine the equation of the circle that is centered at $(2, 3)$ and passes through $(-2, -1)$.

54. Determine the equation of the circle that is centered at $(3, 4)$ and passes through the origin.

55. Find all points of intersection of the circles $x^2 + y^2 = 9$ and $(x - 5)^2 + y^2 = 9$.

56. A donkey is tied at the point $(2, -3)$ on a rope of length 12. Turnips are growing at the point $(6, 7)$. Can the donkey reach them?

57. *Volume of a flute.* The volume of air in a flute is a critical factor in determining its pitch. A cross section of a Renaissance flute in C is shown in the accompanying figure. If the length of the flute is 2874 millimeters, then what is the volume of air in the flute (to the nearest cubic millimeter (mm^3))? (*Hint:* Use the formula for the volume of a cylinder.)

58. *Flute reproduction.* To make the smaller C# flute, Friedrich von Huene multiplies the length and cross-sectional area of the flute of Exercise 57 by 0.943. Find

Graph each equation.

59. $x^2 + y^2 = 0$

60. $x^2 - y^2 = 0$ **61.** $y = \sqrt{1 - x^2}$

62. $y = -\sqrt{1 - x^2}$

GETTING MORE INVOLVED

63. *Cooperative learning.* The equation of a circle is a special case of the general equation $Ax^2 + Bx + Cy^2 + Dy = E$, where A, B, C, D, and E are real numbers. Working in small groups, find restrictions that must be placed on *A, B, C, D,* and *E* so that the graph of this equation is a circle. What does the graph of $x^2 + y^2 = -9$ look like?

64. *Discussion.* Suppose lighthouse A is located at the origin and lighthouse B is located at coordinates (0, 6). The captain of a ship has determined that the ship's distance from lighthouse A is 2 and its distance from lighthouse B is 5. What are the possible coordinates for the location of the ship?

GRAPHING CALCULATOR EXERCISES

Graph each relation on a graphing calculator by solving for y and graphing two functions.

65. $x^2 + y^2 = 4$

66. $(x - 1)^2 + (y + 2)^2 = 1$

67. $x = y^2$

68. $x = (y + 2)^2 - 1$

69. $x = y^2 + 2y + 1$

70. $x = 4y^2 + 4y + 1$

13.4 THE ELLIPSE AND HYPERBOLA

In this section

- The Ellipse
- The Hyperbola

In this section we study the remaining two conic sections: the ellipse and the hyperbola.

The Ellipse

An ellipse can be obtained by intersecting a plane and a cone, as was shown in Fig. 13.3. We can also give a definition of an ellipse in terms of points and distance.

> **Ellipse**
>
> An **ellipse** is the set of all points in a plane such that the sum of their distances from two fixed points is a constant. Each fixed point is called a **focus** (plural: foci).

FIGURE 13.17

An easy way to draw an ellipse is illustrated in Fig. 13.17. A string is attached at two fixed points, and a pencil is used to take up the slack. As the pencil is moved around the paper, the sum of the distances of the pencil point from the two fixed points remains constant. Of course, the length of the string is that constant. You may wish to try this.

Like the parabola, the ellipse also has interesting reflecting properties. All light or sound waves emitted from one focus are reflected off the ellipse to concentrate at the other focus (see Fig. 13.18). This property is used in light fixtures where a concentration of light at a point is desired or in a whispering gallery such as Statuary Hall in the U.S. Capitol Building.

The orbits of the planets around the sun and satellites around the earth are elliptical. For the orbit of the earth around the sun, the sun is at one focus. For the elliptical path of an earth satellite, the earth is at one focus and a point in space is the other focus.

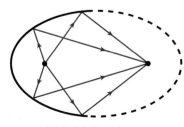

FIGURE 13.18

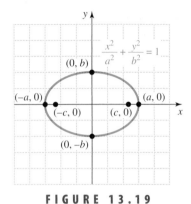

FIGURE 13.19

Figure 13.19 shows an ellipse with foci $(c, 0)$ and $(-c, 0)$. The origin is the center of this ellipse. In general, the **center** of an ellipse is a point midway between the foci. The ellipse in Fig. 13.19 has x-intercepts at $(a, 0)$ and $(-a, 0)$ and y-intercepts at $(0, b)$ and $(0, -b)$. The distance formula can be used to write the following equation for this ellipse. (See Exercise 55.)

Equation of an Ellipse Centered at the Origin

An ellipse centered at $(0, 0)$ with foci at $(\pm c, 0)$ and constant sum $2a$ has equation

$$\frac{x^2}{a^2} + \frac{y^2}{b^2} = 1,$$

where a, b, and c are positive real numbers with $c^2 = a^2 - b^2$.

To draw a "nice-looking" ellipse, we would locate the foci and use string as shown in Fig. 13.17. We can get a rough sketch of an ellipse centered at the origin by using the x- and y-intercepts only.

EXAMPLE 1   Graphing an ellipse

Find the x- and y-intercepts for the ellipse and sketch its graph.

$$\frac{x^2}{9} + \frac{y^2}{4} = 1$$

Solution

To find the y-intercepts, let $x = 0$ in the equation:

$$\frac{0}{9} + \frac{y^2}{4} = 1$$

$$\frac{y^2}{4} = 1$$

$$y^2 = 4$$

$$y = \pm 2$$

To find x-intercepts, let $y = 0$. We get $x = \pm 3$. The four intercepts are $(0, 2)$, $(0, -2)$, $(3, 0)$, and $(-3, 0)$. Plot the intercepts and draw an ellipse through them as in Fig. 13.20.

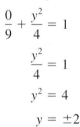

To graph the ellipse in Example 1, graph

$$y_1 = \sqrt{4 - 4x^2/9}$$

and

$$y_2 = -\sqrt{4 - 4x^2/9}.$$

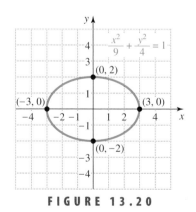

FIGURE 13.20

Ellipses, like circles, may be centered at any point in the plane. To get the equation of an ellipse centered at (h, k), we replace x by $x - h$ and y by $y - k$ in the equation of the ellipse centered at the origin.

Equation of an Ellipse Centered at (h, k)

An ellipse centered at (h, k) has equation

$$\frac{(x - h)^2}{a^2} + \frac{(y - k)^2}{b^2} = 1,$$

where a and b are positive real numbers.

EXAMPLE 2 **An ellipse with center (h, k)**

Sketch the graph of the ellipse:

$$\frac{(x - 1)^2}{9} + \frac{(y + 2)^2}{4} = 1$$

Solution

The graph of this ellipse is exactly the same size and shape as the ellipse

$$\frac{x^2}{9} + \frac{y^2}{4} = 1,$$

which was graphed in Example 1. However, the center for

$$\frac{(x - 1)^2}{9} + \frac{(y + 2)^2}{4} = 1$$

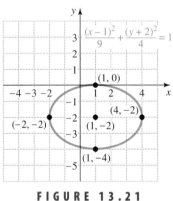

FIGURE 13.21

is $(1, -2)$. The denominator 9 is used to determine that the ellipse passes through points that are three units to the right and three units to the left of the center: $(4, -2)$ and $(-2, -2)$. See Fig. 13.21. The denominator 4 is used to determine that the ellipse passes through points that are two units above and two units below the center: $(1, 0)$ and $(1, -4)$. We draw an ellipse using these four points, just as we did for an ellipse centered at the origin. ∎

The Hyperbola

A hyperbola is the curve that occurs at the intersection of a cone and a plane, as was shown in Fig. 13.3 in Section 13.2. A hyperbola can also be defined in terms of points and distance.

Hyperbola

A **hyperbola** is the set of all points in the plane such that the difference of their distances from two fixed points (foci) is constant.

Like the parabola and the ellipse, the hyperbola also has reflecting properties. If a light ray is aimed at one focus, it is reflected off the hyperbola and goes to the

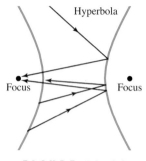

FIGURE 13.22

other focus, as shown in Fig. 13.22. Hyperbolic mirrors are used in conjunction with parabolic mirrors in telescopes.

The definitions of a hyperbola and an ellipse are similar, and so are their equations. However, their graphs are very different. Figure 13.23 shows a hyperbola in which the distance from a point on the hyperbola to the closer focus is N and the distance to the farther focus is M. The value $M - N$ is the same for every point on the hyperbola.

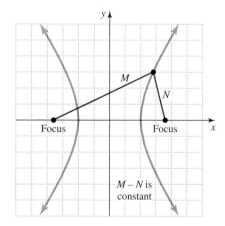

FIGURE 13.23

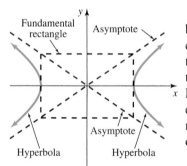

FIGURE 13.24

A hyperbola has two parts called **branches.** These branches look like parabolas, but they are not parabolas. The branches of the hyperbola shown in Fig. 13.24 get closer and closer to the dashed lines, called **asymptotes,** but they never intersect them. The asymptotes are used as guidelines in sketching a hyperbola. The asymptotes are found by extending the diagonals of the **fundamental rectangle,** shown in Fig. 13.24. The key to drawing a hyperbola is getting the fundamental rectangle and extending its diagonals to get the asymptotes. You will learn how to find the fundamental rectangle from the equation of a hyperbola. The hyperbola in Fig. 13.24 opens to the left and right.

If we start with foci at $(\pm c, 0)$ and a positive number a, then we can use the definition of a hyperbola to derive the following equation of a hyperbola in which the constant difference between the distances to the foci is $2a$.

Equation of a Hyperbola Centered at (0, 0)

A hyperbola centered at $(0, 0)$ with foci $(c, 0)$ and $(-c, 0)$ and constant difference $2a$ has equation

$$\frac{x^2}{a^2} - \frac{y^2}{b^2} = 1,$$

where a, b, and c are positive real numbers such that $c^2 = a^2 + b^2$.

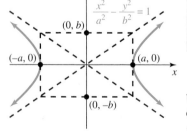

FIGURE 13.25

The graph of a general equation for a hyperbola is shown in Fig. 13.25. Notice that the fundamental rectangle extends to the x-intercepts along the x-axis and extends b units above and below the origin along the y-axis. The facts necessary for graphing a hyperbola centered at the origin and opening to the left and to the right are listed on page 719.

> ### Graphing a Hyperbola Centered at the Origin, Opening Left and Right
>
> To graph the hyperbola $\frac{x^2}{a^2} - \frac{y^2}{b^2} = 1$:
>
> 1. Locate the x-intercepts at $(a, 0)$ and $(-a, 0)$.
> 2. Draw the fundamental rectangle through $(\pm a, 0)$ and $(0, \pm b)$.
> 3. Draw the extended diagonals of the rectangle to use as asymptotes.
> 4. Draw the hyperbola to the left and right approaching the asymptotes.

E X A M P L E   3

A hyperbola opening left and right

Sketch the graph of $\frac{x^2}{36} - \frac{y^2}{9} = 1$, and find the equations of its asymptotes.

Solution

The x-intercepts are $(6, 0)$ and $(-6, 0)$. Draw the fundamental rectangle through these x-intercepts and the points $(0, 3)$ and $(0, -3)$. Extend the diagonals of the fundamental rectangle to get the asymptotes. Now draw a hyperbola passing through the x-intercepts and approaching the asymptotes as shown in Fig. 13.26.

From the graph in Fig. 13.26 we see that the slopes of the asymptotes are $\frac{1}{2}$ and $-\frac{1}{2}$. Because the y-intercept for both asymptotes is the origin, their equations are $y = \frac{1}{2}x$ and $y = -\frac{1}{2}x$.

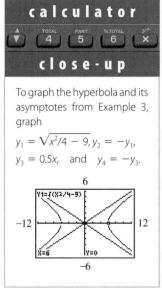
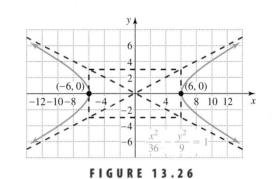

FIGURE 13.26

A hyperbola may open up and down. In this case the graph intersects only the y-axis. The facts necessary for graphing a hyperbola that opens up and down are summarized as follows.

> ### Graphing a Hyperbola Centered at the Origin, Opening Up and Down
>
> To graph the hyperbola $\frac{y^2}{b^2} - \frac{x^2}{a^2} = 1$:
>
> 1. Locate the y-intercepts at $(0, b)$ and $(0, -b)$.
> 2. Draw the fundamental rectangle through $(0, \pm b)$ and $(\pm a, 0)$.
> 3. Draw the extended diagonals of the rectangle to use as asymptotes.
> 4. Draw the hyperbola opening up and down approaching the asymptotes.

E X A M P L E   4

A hyperbola opening up and down

Graph the hyperbola $\frac{y^2}{9} - \frac{x^2}{4} = 1$ and find the equations of its asymptotes.

Solution

If $y = 0$, we get

$$-\frac{x^2}{4} = 1$$

$$x^2 = -4.$$

Because this equation has no real solution, the graph has no x-intercepts. Let $x = 0$ to find the y-intercepts:

$$\frac{y^2}{9} = 1$$

$$y^2 = 9$$

$$y = \pm 3$$

The y-intercepts are $(0, 3)$ and $(0, -3)$, and the hyperbola opens up and down. From $a^2 = 4$ we get $a = 2$. So the fundamental rectangle extends to the intercepts $(0, 3)$ and $(0, -3)$ on the y-axis and to the points $(2, 0)$ and $(-2, 0)$ along the x-axis. We extend the diagonals of the rectangle and draw the graph of the hyperbola as shown in Fig. 13.27. From the graph in Fig. 13.27 we see that the asymptotes have slopes $\frac{3}{2}$ and $-\frac{3}{2}$. Because the y-intercept for both asymptotes is the origin, their equations are $y = \frac{3}{2}x$ and $y = -\frac{3}{2}x$.

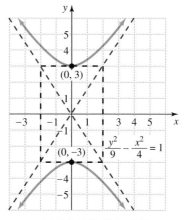

FIGURE 13.27

E X A M P L E   5

A hyperbola not in standard form

Sketch the graph of the hyperbola $4x^2 - y^2 = 4$.

Solution

First write the equation in standard form. Divide each side by 4 to get

$$x^2 - \frac{y^2}{4} = 1.$$

There are no y-intercepts. If $y = 0$, then $x = \pm 1$. The hyperbola opens left and right with x-intercepts at $(1, 0)$ and $(-1, 0)$. The fundamental rectangle extends to the intercepts along the *x*-axis and to the points $(0, 2)$ and $(0, -2)$ along the *y*-axis. We extend the diagonals of the rectangle for the asymptotes and draw the graph as shown in Fig. 13.28.

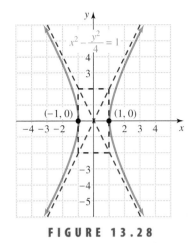

FIGURE 13.28

13.4 EXERCISES

Reading and Writing *After reading this section, write out the answers to these questions. Use complete sentences.*

1. What is the definition of an ellipse?

2. How can you draw an ellipse with a pencil and string?

3. Where is the center of an ellipse?

4. What is the equation of an ellipse centered at the origin?

5. What is the equation of an ellipse centered at (h, k)?

6. What is the definition of a hyperbola?

7. How do you find the asymptotes of a hyperbola?

17. $9x^2 + 16y^2 = 144$

18. $9x^2 + 25y^2 = 225$

8. What is the equation of a hyperbola centered at the origin and opening left and right?

Sketch the graph of each ellipse. See Example 1.

9. $\dfrac{x^2}{9} + \dfrac{y^2}{4} = 1$

10. $\dfrac{x^2}{9} + \dfrac{y^2}{16} = 1$

19. $25x^2 + y^2 = 25$

20. $x^2 + 16y^2 = 16$

21. $4x^2 + 9y^2 = 1$

22. $25x^2 + 16y^2 = 1$

11. $\dfrac{x^2}{9} + y^2 = 1$

12. $x^2 + \dfrac{y^2}{4} = 1$

13. $\dfrac{x^2}{36} + \dfrac{y^2}{25} = 1$

14. $\dfrac{x^2}{25} + \dfrac{y^2}{49} = 1$

Sketch the graph of each ellipse. See Example 2.

23. $\dfrac{(x-3)^2}{4} + \dfrac{(y-1)^2}{9} = 1$ **24.** $\dfrac{(x+5)^2}{49} + \dfrac{(y-2)^2}{25} = 1$

25. $\dfrac{(x+1)^2}{16} + \dfrac{(y-2)^2}{25} = 1$ **26.** $\dfrac{(x-3)^2}{36} + \dfrac{(y+4)^2}{64} = 1$

15. $\dfrac{x^2}{24} + \dfrac{y^2}{5} = 1$

16. $\dfrac{x^2}{6} + \dfrac{y^2}{17} = 1$

27. $(x - 2)^2 + \dfrac{(y + 1)^2}{36} = 1$ **28.** $\dfrac{(x + 3)^2}{9} + (y + 1)^2 = 1$ **35.** $x^2 - \dfrac{y^2}{25} = 1$ **36.** $\dfrac{x^2}{9} - y^2 = 1$

Sketch the graph of each hyperbola and write the equations of its asymptotes. See Examples 3–5.

29. $\dfrac{x^2}{4} - \dfrac{y^2}{9} = 1$ **30.** $\dfrac{x^2}{16} - \dfrac{y^2}{9} = 1$ **37.** $9x^2 - 16y^2 = 144$

31. $\dfrac{y^2}{4} - \dfrac{x^2}{25} = 1$ **32.** $\dfrac{y^2}{9} - \dfrac{x^2}{16} = 1$ **38.** $9x^2 - 25y^2 = 225$

39. $x^2 - y^2 = 1$ **40.** $y^2 - x^2 = 1$

33. $\dfrac{x^2}{25} - y^2 = 1$ **34.** $x^2 - \dfrac{y^2}{9} = 1$

Graph both equations of each system on the same coordinate axes. Use elimination of variables to find all points of intersection.

41. $\dfrac{x^2}{4} + \dfrac{y^2}{9} = 1$

$x^2 - \dfrac{y^2}{9} = 1$

42. $x^2 - \dfrac{y^2}{4} = 1$

$\dfrac{x^2}{9} + \dfrac{y^2}{4} = 1$

43. $\dfrac{x^2}{4} + \dfrac{y^2}{16} = 1$

$x^2 + y^2 = 1$

44. $x^2 + \dfrac{y^2}{9} = 1$

$x^2 + y^2 = 4$

45. $x^2 + y^2 = 4$

$x^2 - y^2 = 1$

46. $x^2 + y^2 = 16$

$x^2 - y^2 = 4$

47. $x^2 + 9y^2 = 9$

$x^2 + y^2 = 4$

48. $x^2 + y^2 = 25$

$x^2 + 25y^2 = 25$

49. $x^2 + 9y^2 = 9$

$y = x^2 - 1$

50. $4x^2 + y^2 = 4$
$y = 2x^2 - 2$

51. $9x^2 - 4y^2 = 36$
$2y = x - 2$

52. $25y^2 - 9x^2 = 225$
$y = 3x + 3$

Solve each problem.

53. *Marine navigation.* The loran (long-range navigation) system is used by boaters to determine their location at sea. The loran unit on a boat measures the difference in time that it takes for radio signals from pairs of fixed points to reach the boat. The unit then finds the equations of two hyperbolas that pass through the location of the boat. Suppose a boat is located in the first quadrant at the intersection of $x^2 - 3y^2 = 1$ and $4y^2 - x^2 = 1$.

a) Use the accompanying graph to approximate the location of the boat.

b) Algebraically find the exact location of the boat.

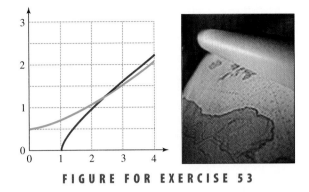

FIGURE FOR EXERCISE 53

54. *Sonic boom.* An aircraft traveling at supersonic speed creates a cone-shaped wave that intersects the ground along a hyperbola, as shown in the accompanying figure. A thunderlike sound is heard at any point on the hyperbola. This sonic boom travels along the ground, following the aircraft. The area where the sonic boom is most noticeable is called the *boom carpet*. The width of the boom carpet is roughly five times the altitude of the aircraft. Suppose the equation of the hyperbola in the figure is

$$\frac{x^2}{400} - \frac{y^2}{100} = 1,$$

where the units are miles and the width of the boom carpet is measured 40 miles behind the aircraft. Find the altitude of the aircraft.

FIGURE FOR EXERCISE 54

GETTING MORE INVOLVED

55. *Cooperative learning.* Let (x, y) be an arbitrary point on an ellipse with foci $(c, 0)$ and $(-c, 0)$ for $c > 0$. The following equation expresses the fact that the distance from (x, y) to $(c, 0)$ plus the distance from (x, y) to $(-c, 0)$ is the constant value $2a$ (for $a > 0$):

$$\sqrt{(x - c)^2 + (y - 0)^2} + \sqrt{(x - (-c))^2 + (y - 0)^2} = 2a$$

Working in groups, simplify this equation. First get the radicals on opposite sides of the equation, then square both sides twice to eliminate the square roots. Finally, let $b^2 = a^2 - c^2$ to get the equation

$$\frac{x^2}{a^2} + \frac{y^2}{b^2} = 1.$$

56. *Cooperative learning.* Let (x, y) be an arbitrary point on a hyperbola with foci $(c, 0)$ and $(-c, 0)$ for $c > 0$. The following equation expresses the fact that the distance from (x, y) to $(c, 0)$ minus the distance from (x, y) to $(-c, 0)$ is the constant value $2a$ (for $a > 0$):

$$\sqrt{(x - c)^2 + (y - 0)^2} - \sqrt{(x - (-c))^2 + (y - 0)^2} = 2a$$

Working in groups, simplify the equation. You will need to square both sides twice to eliminate the square roots.

Finally, let $b^2 = c^2 - a^2$ to get the equation

$$\frac{x^2}{a^2} - \frac{y^2}{b^2} = 1.$$

GRAPHING CALCULATOR EXERCISES

57. Graph $y_1 = \sqrt{x^2 - 1}$, $y_2 = -\sqrt{x^2 - 1}$, $y_3 = x$, and $y_4 = -x$ to get the graph of the hyperbola $x^2 - y^2 = 1$ along with its asymptotes. Use the viewing window $-3 \le x \le 3$ and $-3 \le y \le 3$. Notice how the branches of the hyperbola approach the asymptotes.

58. Graph the same four functions in Exercise 57, but use $-30 \le x \le 30$ and $-30 \le y \le 30$ as the viewing window. What happened to the hyperbola?

13.5 SECOND-DEGREE INEQUALITIES

In this section we graph second-degree inequalities and systems of inequalities involving second-degree inequalities.

In this section

- Graphing a Second-Degree Inequality
- Systems of Inequalities

Graphing a Second-Degree Inequality

A second-degree inequality is an inequality involving squares of at least one of the variables. Changing the equal sign to an inequality symbol for any of the equations of the conic sections gives us a second-degree inequality. Second-degree inequalities are graphed in the same manner as linear inequalities.

E X A M P L E 1

A second-degree inequality

Graph the inequality $y < x^2 + 2x - 3$.

Solution

We first graph $y = x^2 + 2x - 3$. This parabola has x-intercepts at $(1, 0)$ and $(-3, 0)$, y-intercept at $(0, -3)$, and vertex at $(-1, -4)$. The graph of the parabola is drawn with a dashed line, as shown in Fig. 13.29. The graph of the parabola divides the plane into two regions. Every point on one side of the parabola satisfies the inequality $y < x^2 + 2x - 3$, and every point on the other side satisfies the inequality $y > x^2 + 2x - 3$. To determine which side is which, we test a point that is not on the parabola, say $(0, 0)$. Because

$$0 < 0^2 + 2 \cdot 0 - 3$$

is false, the region not containing the origin is shaded, as in Fig. 13.29.

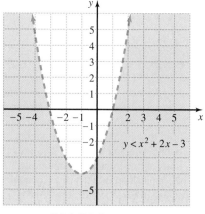

$y < x^2 + 2x - 3$

FIGURE 13.29

E X A M P L E   2   **A second-degree inequality**

Graph the inequality $x^2 + y^2 < 9$.

Solution

The graph of $x^2 + y^2 = 9$ is a circle of radius 3 centered at the origin. The circle divides the plane into two regions. Every point in one region satisfies $x^2 + y^2 < 9$, and every point in the other region satisfies $x^2 + y^2 > 9$. To identify the regions, we pick a point and test it. Select $(0, 0)$. The inequality

$$0^2 + 0^2 < 9$$

is true. Because $(0, 0)$ is inside the circle, all points inside the circle satisfy the inequality $x^2 + y^2 < 9$, as shown in Fig. 13.30. The points outside the circle satisfy the inequality $x^2 + y^2 > 9$.

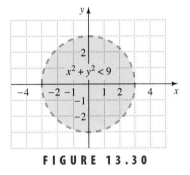

FIGURE   13.30

E X A M P L E   3   **A second-degree inequality**

Graph the inequality $\dfrac{x^2}{4} - \dfrac{y^2}{9} > 1$.

Solution

First graph the hyperbola $\dfrac{x^2}{4} - \dfrac{y^2}{9} = 1$. Because the hyperbola shown in Fig. 13.31 divides the plane into three regions, we select a test point in each region and check to see whether it satisfies the inequality. Testing the points $(-3, 0)$, $(0, 0)$, and $(3, 0)$ gives us the inequalities

$$\frac{(-3)^2}{4} - \frac{0^2}{9} > 1, \qquad \frac{0^2}{4} - \frac{0^2}{9} > 1, \qquad \text{and} \qquad \frac{3^2}{4} - \frac{0^2}{9} > 1.$$

Because only the first and third inequalities are correct, we shade only the regions containing $(3, 0)$ and $(-3, 0)$, as shown in Fig. 13.31.

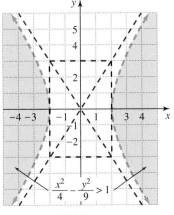

FIGURE   13.31

Systems of Inequalities

A point is in the solution set to a system of inequalities if it satisfies all inequalities of the system. We graph a system of inequalities by first determining the graph of each inequality and then finding the intersection of the graphs.

E X A M P L E   4   **Systems of second-degree inequalities**

Graph the system of inequalities:

$$\frac{y^2}{4} - \frac{x^2}{9} > 1$$
$$\frac{x^2}{9} + \frac{y^2}{16} < 1$$

Solution

Figure 13.32(a) shows the graph of the first inequality. In Fig. 13.32(b) we have the graph of the second inequality. In Fig. 13.32(c) we have shaded only the points that satisfy both inequalities. Figure 13.32(c) is the graph of the system.

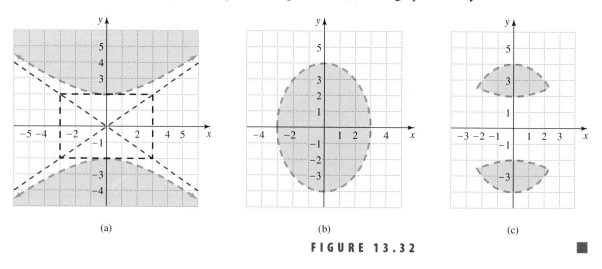

(a)   (b)   (c)

FIGURE 13.32

13.5   EXERCISES

Graph each inequality. See Examples 1–3.

1. $y > x^2$

2. $y \le x^2 + 1$

3. $y < x^2 - x$

4. $y > x^2 + x$

5. $y > x^2 - x - 2$

6. $y < x^2 + x - 6$

7. $x^2 + y^2 \le 9$

8. $x^2 + y^2 > 16$

9. $x^2 + 4y^2 > 4$

10. $4x^2 + y^2 \le 4$

11. $4x^2 - 9y^2 < 36$

12. $25x^2 - 4y^2 > 100$

13. $(x - 2)^2 + (y - 3)^2 < 4$

14. $(x + 1)^2 + (y - 2)^2 > 1$

15. $x^2 + y^2 > 1$

16. $x^2 + y^2 < 25$

17. $4x^2 - y^2 > 4$

18. $x^2 - 9y^2 \le 9$

19. $y^2 - x^2 \leq 1$

20. $x^2 - y^2 > 1$

27. $y > x^2 + x$
$y < 5$

28. $y > x^2 + x - 6$
$y < x + 3$

21. $x > y$

22. $x < 2y - 1$

29. $y \geq x + 2$
$y \leq 2 - x$

30. $y \geq 2x - 3$
$y \leq 3 - 2x$

Graph the solution set to each system of inequalities. See Example 4.

23. $x^2 + y^2 < 9$
$y > x$

24. $x^2 + y^2 > 1$
$x > y$

31. $4x^2 - y^2 < 4$
$x^2 + 4y^2 > 4$

32. $x^2 - 4y^2 < 4$
$x^2 + 4y^2 > 4$

25. $x^2 - y^2 > 1$
$x^2 + y^2 < 4$

26. $y^2 - x^2 < 1$
$x^2 + y^2 > 9$

33. $x - y < 0$
$y + x^2 < 1$

34. $y + 1 > x^2$
$x + y < 2$

35. $y < 5x - x^2$
$x^2 + y^2 < 9$

36. $y < x^2 + 5x$
$x^2 + y^2 < 16$

Solve the problem.

43. ***Buried treasure.***   An old pirate on his deathbed gave the following description of where he had buried some treasure on a deserted island: "Starting at the large palm tree, I walked to the north and then to the east, and there I buried the treasure. I walked at least 50 paces to get to that spot, but I was not more than 50 paces, as the crow flies, from the large palm tree. I am sure that I walked farther in the northerly direction than in the easterly direction." With the large palm tree at the origin and the positive y-axis pointing to the north, graph the possible locations of the treasure.

37. $y \geq 3$
$x \leq 1$

38. $x > -3$
$y < 2$

39. $4y^2 - 9x^2 < 36$
$x^2 + y^2 < 16$

40. $25y^2 - 16x^2 < 400$
$x^2 + y^2 > 4$

FIGURE FOR EXERCISE 43

 GRAPHING CALCULATOR EXERCISES

44. Use graphs to find an ordered pair that is in the solution set to the system of inequalities:

$$y > x^2 - 2x + 1$$
$$y < -1.1(x - 4)^2 + 5$$

Verify that your answer satisfies both inequalities.

45. Use graphs to find the solution set to the system of inequalities:

$$y > 2x^2 - 3x + 1$$
$$y < -2x^2 - 8x - 1$$

41. $y < x^2$
$x^2 + y^2 < 1$

42. $y > x^2$
$4x^2 + y^2 < 4$

COLLABORATIVE ACTIVITIES

Focus on Comets

Conic sections are used to model many different things in the natural world. Astronomers use mirrors in the shape of parabolas in telescopes. They have learned that planets can have elliptical as well as circular orbits around the sun and that comets may have orbits that resemble hyperbolas, parabolas, or ellipses. In this activity we will consider comets with these three types of orbits.

The orbit that a comet will take as it approaches the sun depends on its velocity (as well as other factors). If it has enough velocity to escape from the pull of the sun, it may take either a parabolic orbit or a hyperbolic orbit. If it doesn't have enough velocity, then it will take an elliptical orbit. Of course, a comet that has a parabolic or hyperbolic orbit will not come back again around our sun. Only comets with elliptical orbits do we see again.

For the problems below, round all your answers to two decimal places.

1. Halley's comet is in an elliptical orbit about the sun with the sun at one of the foci. In this problem we will make a scale model of the orbit of Halley's comet about the sun. We will choose our coordinate system so that the ellipse is centered at (0, 0). Halley's comet comes within 0.6 astronomical unit[1] from the sun at its closest point and 35 astronomical units at its farthest point. The position of the comet at these points

[1]An astronomical unit is the distance of the earth from the sun. The earth's orbit is almost circular and so is about the same distance from the sun at any point in its orbit.

Grouping: 2 students per group
Topic: Conic sections

will correspond to the horizontal vertices. Find the equation for the ellipse. Determine where the foci should be. On a piece of cardboard, put thumbtacks at the foci. Determine the length of string you will need to draw the ellipse using the scale 1 centimeter = 1 astronomical unit (AU). Draw the comet's orbit, indicating which focus is the sun.

2. If the velocity of a comet equals the escape velocity (it is going just fast enough to get away), then its orbit will be parabolic. The sun will be at the focus of the parabola. We will place the vertex at the point (0, 0); in this case the vertex will be where the comet is the closest to the sun. Suppose we have a comet that is 0.75 AU from the sun at its closest point. Find the equation for the parabola. Graph the parabola.

3. If the velocity of a comet is greater than the escape velocity (it can easily escape from the sun's gravitational pull), its path will resemble one-half of a hyperbola with the sun as one focus. Assuming we have a left- and a right-opening hyperbola, we can model the path of such a comet along one of its branches. We will center the hyperbola at (0, 0), and place the sun at the left focus and draw the path of the comet as it approaches from the left. Suppose that the comet will be 1.5 AUs from the sun at its closest point. Assume the sun is at the point (−3, 0). Draw a sketch of this scenario. What is the equation for the hyperbola? Graph the hyperbola.

Extension: Graph all three equations on a graphing calculator or computer.

WRAP-UP   CHAPTER 13

SUMMARY

Nonlinear Systems

		Examples
Nonlinear systems in two variables	Use substitution or addition to eliminate variables. Nonlinear systems may have several points in the solution set.	$y = x^2$ $x^2 + y^2 = 4$ Substitution: $y + y^2 = 4$

Parabola

$y = a(x - h)^2 + k$ Opens upward for $a > 0$, downward for $a < 0$
Vertex at (h, k)
To find focus and directrix, use $a = \dfrac{1}{4p}$.
Distance from vertex to focus or directrix is $|p|$.

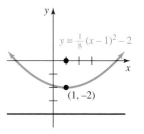
$y = \frac{1}{8}(x - 1)^2 - 2$

$(1, -2)$

$y = ax^2 + bx + c$

Opens upward for $a > 0$, downward for $a < 0$
The x-coordinate of the vertex is $\frac{-b}{2a}$.
Find the y-coordinate of the vertex by evaluating
$y = ax^2 + bx + c$ for $x = \frac{-b}{2a}$.

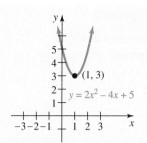

Circle

$(x - h)^2 + (y - k)^2 = r^2$

Center (h, k)
Radius r (for $r > 0$)

Examples

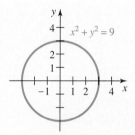

Centered at origin
$x^2 + y^2 = r^2$

Center $(0, 0)$
Radius r (for $r > 0$)

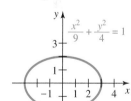

Ellipse

Centered at origin

$\dfrac{x^2}{a^2} + \dfrac{y^2}{b^2} = 1$

Center: $(0, 0)$
x-intercepts: $(a, 0)$ and $(-a, 0)$
y-intercepts: $(0, b)$ and $(0, -b)$
Foci: $(\pm c, 0)$ if $a^2 > b^2$ and $c^2 = a^2 - b^2$
$\quad\quad (0, \pm c)$ if $b^2 > a^2$ and $c^2 = b^2 - a^2$

Examples

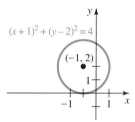

Arbitrary center

$\dfrac{(x - h)^2}{a^2} + \dfrac{(y - k)^2}{b^2} = 1$

Center: (h, k)

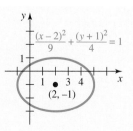

Hyperbola

Centered at origin
Opening left and right

$$\frac{x^2}{a^2} - \frac{y^2}{b^2} = 1$$

Center $(0, 0)$
x-intercepts: $(a, 0)$ and $(-a, 0)$

y-intercepts: none

Examples

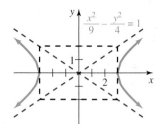

Centered at origin
Opening up and down

$$\frac{y^2}{b^2} - \frac{x^2}{a^2} = 1$$

Centered at $(0, 0)$
x-intercepts: none

y-intercepts: $(0, b)$ and $(0, -b)$

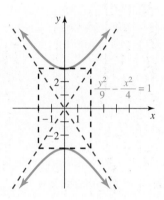

Second-Degree Inequalities

Solution set for
a single inequality

Graph the boundary curve obtained by replacing
the inequality symbol by the equal sign.
Use test points to determine which regions satisfy
the inequality.

Examples

$x^2 + y^2 < 16$

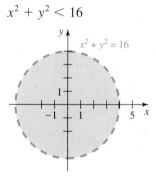

Solution set for a system
of inequalities

Graph the boundary curves. Then select a test
point in each region. Shade only the regions for
which the test point satisfies all inequalities of
the system.

$x^2 + y^2 < 16$
$y > x^2 - 1$

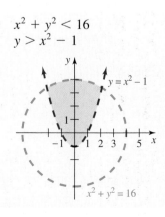

ENRICHING YOUR MATHEMATICAL WORD POWER

For each mathematical term, choose the correct meaning.

1. **nonlinear equation**
   a. an equation that is not lined up
   b. an equation whose graph is a straight line
   c. an equation whose graph is not a straight line
   d. an exponential equation

2. **parabola**
   a. the points in a plane that are equidistant from a point and a line
   b. the points in a plane that are a fixed distance from a fixed point
   c. the points in a plane that are equidistant from two fixed points
   d. the points in a plane the sum of whose distances from two fixed points is a constant

3. **directrix**
   a. the line $y = x$
   b. the line of symmetry of a parabola
   c. the x-axis
   d. the fixed line in the definition of parabola

4. **vertex of a parabola**
   a. the midpoint of the line segment joining the focus and directrix perpendicular to the directrix
   b. the focus
   c. the x-intercept
   d. the endpoint

5. **conic sections**
   a. the two halves of a cone
   b. the vertex and focus
   c. the curves obtained at the intersection of a cone and a plane
   d. the asymptotes

6. **axis of symmetry**
   a. the x-axis
   b. the y-axis
   c. the directrix
   d. the line of symmetry of a parabola

7. **circle**
   a. the points in a plane that are equidistant from a point and a line
   b. the points in a plane that are a fixed distance from a fixed point
   c. the points in a plane that are equidistant from two fixed points
   d. the points in a plane the sum of whose distances from two fixed points is a constant

8. **ellipse**
   a. the points in a plane that are equidistant from a point and a line
   b. the points in a plane that are a fixed distance from a fixed point
   c. the points in a plane that are equidistant from two fixed points
   d. the points in a plane such that the sum of their distances from two fixed points is constant

9. **hyperbola**
   a. the points in a plane that are equidistant from a point and a line
   b. the points in a plane that are a fixed distance from a fixed point
   c. the points in a plane such that the difference of their distances from two fixed points is constant
   d. the points in a plane such that the sum of their distances from two fixed points is a constant

10. **asymptotes**
    a. lines approached by a hyperbola
    b. lines approached by parabolas
    c. tangent lines to a circle
    d. lines that pass through the vertices of an ellipse

REVIEW EXERCISES

13.1 *Graph both equations on the same set of axes, then determine the points of intersection of the graphs by solving the system.*

1. $y = x^2$
   $y = -2x + 15$

2. $y = \sqrt{x}$
   $y = \dfrac{1}{3}x$

3. $y = 3x$

$y = \dfrac{1}{x}$

4. $y = |x|$

$y = -3x + 5$

16. $y = -x^2 - 3x + 4$

17. $y = -\dfrac{1}{2}(x - 2)^2 + 3$

18. $y = \dfrac{1}{4}(x + 1)^2 - 2$

Solve each system.

5. $x^2 + y^2 = 4$

$y = \dfrac{1}{3}x^2$

6. $12y^2 - 4x^2 = 9$

$x = y^2$

7. $x^2 + y^2 = 34$

$y = x + 2$

8. $y = 2x + 1$

$xy - y = 5$

9. $y = \log(x - 3)$

$y = 1 - \log(x)$

10. $y = \left(\dfrac{1}{2}\right)^x$

$y = 2^{x-1}$

11. $x^4 = 2(12 - y)$

$y = x^2$

12. $x^2 + 2y^2 = 7$

$x^2 - 2y^2 = -5$

Write each equation in the form $y = a(x - h)^2 + k$, and identify the vertex of the parabola.

19. $y = 2x^2 - 8x + 1$

20. $y = -2x^2 - 6x - 1$

21. $y = -\dfrac{1}{2}x^2 - x + \dfrac{1}{2}$

22. $y = \dfrac{1}{4}x^2 + x - 9$

13.3 *Determine the center and radius of each circle, and sketch its graph.*

23. $x^2 + y^2 = 100$ **24.** $x^2 + y^2 = 20$

13.2 *Determine the vertex, axis of symmetry, focus, and directrix for each parabola.*

13. $y = x^2 + 3x - 18$

25. $(x - 2)^2 + (y + 3)^2 = 81$ **26.** $x^2 + 2x + y^2 = 8$

14. $y = x - x^2$

15. $y = x^2 + 3x + 2$

27. $9y^2 + 9x^2 = 4$ **28.** $x^2 + 4x + y^2 - 6y - 3 = 0$

38. $\dfrac{y^2}{25} - \dfrac{x^2}{49} = 1$

39. $4x^2 - 25y^2 = 100$

Write the standard equation for each circle with the given center and radius.

29. Center $(0, 3)$, radius 6
30. Center $(0, 0)$, radius $\sqrt{6}$
31. Center $(2, -7)$, radius 5
32. Center $\left(\dfrac{1}{2}, -3\right)$, radius $\dfrac{1}{2}$

13.4 *Sketch the graph of each ellipse.*

33. $\dfrac{x^2}{36} + \dfrac{y^2}{49} = 1$ **34.** $\dfrac{x^2}{25} + y^2 = 1$

40. $6y^2 - 16x^2 = 96$

35. $25x^2 + 4y^2 = 100$ **36.** $6x^2 + 4y^2 = 24$

13.5 *Graph each inequality.*
41. $4x - 2y > 3$ **42.** $y < x^2 - 3x$

Sketch the graph of each hyperbola.

37. $\dfrac{x^2}{49} - \dfrac{y^2}{36} = 1$

43. $y^2 < x^2 - 1$ **44.** $y^2 < 1 - x^2$

45. $4x^2 + 9y^2 > 36$

46. $x^2 + y > 2x - 1$

61. $4y^2 - x^2 = 8$

62. $9x^2 + y = 9$

Sketch the graph of each equation.

63. $x^2 = 4 - y^2$

64. $x^2 = 4y^2 + 4$

Graph the solution set to each system of inequalities.

47. $y < 3x - x^2$
$\quad x^2 + y^2 < 9$

48. $x^2 - y^2 < 1$
$\quad y < 1$

65. $x^2 = 4y + 4$

66. $x = 4y + 4$

49. $4x^2 + 9y^2 > 36$
$\quad x^2 + y^2 < 9$

50. $y^2 - x^2 > 4$
$\quad y^2 + 16x^2 < 16$

67. $x^2 = 4 - 4y^2$

68. $x^2 = 4y - y^2$

69. $x^2 = 4 - (y - 4)^2$

70. $(x - 2)^2 + (y - 4)^2 = 4$

MISCELLANEOUS

Identify each equation as the equation of a straight line, parabola, circle, hyperbola, or ellipse. Try to do these without rewriting the equations.

51. $x^2 = y^2 + 1$

52. $x = y + 1$

53. $x^2 = 1 - y^2$

54. $x^2 = y + 1$

55. $x^2 + x = 1 - y^2$

56. $(x - 3)^2 + (y + 2)^2 = 7$

Write the equation of the circle with the given features.

71. Centered at the origin and passing through $(3, 4)$

57. $x^2 + 4x = 6y - y^2$

58. $4x + 6y = 1$

72. Centered at $(2, -3)$ and passing through $(-1, 4)$

59. $\dfrac{x^2}{3} - \dfrac{y^2}{5} = 1$

60. $x^2 + \dfrac{y^2}{3} = 1$

73. Centered at $(-1, 5)$ with radius 6

74. Centered at $(0, -3)$ and passing through the origin

Write the equation of the parabola with the given features.

75. Focus $(1, 4)$ and directrix $y = 2$

76. Focus $(-2, 1)$ and directrix $y = 5$

77. Vertex $(0, 0)$ and focus $\left(0, \frac{1}{4}\right)$

78. Vertex $(1, 2)$ and focus $\left(1, \frac{3}{2}\right)$

79. Vertex $(0, 0)$, passing through $(3, 2)$, and opening upward

80. Vertex $(1, 3)$, passing through $(0, 0)$, and opening downward

Solve each system of equations.

81. $x^2 + y^2 = 25$
$y = -x + 1$

82. $x^2 - y^2 = 1$
$x^2 + y^2 = 7$

83. $4x^2 + y^2 = 4$
$x^2 - y^2 = 21$

84. $y = x^2 + x$
$y = -x^2 + 3x + 12$

Solve each problem.

85. *Perimeter of a rectangle.* A rectangle has a perimeter of 16 feet and an area of 12 square feet. Find its length and width.

86. *Tale of two circles.* Find the radii of two circles such that the difference in areas of the two is 10π square inches and the difference in radii of the two is 2 inches.

CHAPTER 13 TEST

Sketch the graph of each equation.

1. $x^2 + y^2 = 25$

2. $\dfrac{x^2}{16} - \dfrac{y^2}{25} = 1$

5. $y^2 - 4x^2 = 4$

6. $y = -x^2 - 2x + 3$

3. $y^2 + 4x^2 = 4$

4. $y = x^2 + 4x + 4$

Sketch the graph of each inequality.

7. $x^2 - y^2 < 9$

8. $x^2 + y^2 > 9$

9. $y > x^2 - 9$

Solve each system of equations.

12. $y = x^2 - 2x - 8$
$y = 7 - 4x$

13. $x^2 + y^2 = 12$
$y = x^2$

Solve each problem.

14. Find the center and radius of the circle $x^2 + 2x + y^2 + 10y = 10$.

15. Find the vertex, focus, and directrix of the parabola $y = x^2 + x + 3$. State the axis of symmetry and whether the parabola opens up or down.

Graph the solution set to each system of inequalities.

10. $x^2 + y^2 < 9$
$x^2 - y^2 > 1$

11. $y < -x^2 + x$
$y < x - 4$

16. Write the equation $y = \frac{1}{2}x^2 - 3x - \frac{1}{2}$ in the form $y = a(x - h)^2 + k$.

17. Write the equation of a circle with center $(-1, 3)$ that passes through $(2, 5)$.

18. Find the length and width of a rectangular room that has an area of 108 square feet and a perimeter of 42 ft.

Sketch the graph of each equation.

1. $y = 9x - x^2$

2. $y = 9x$

3. $y = (x - 9)^2$

4. $y^2 = 9 - x^2$

5. $y = 9x^2$

6. $y = |9x|$

7. $4x^2 + 9y^2 = 36$

8. $4x^2 - 9y^2 = 36$

9. $y = 9 - x$

10. $y = 9^x$

Find the following products.

11. $(x + 2y)^2$

12. $(x + y)(x^2 + 2xy + y^2)$

13. $(a + b)^3$

14. $(a - 3b)^2$

15. $(2a + 1)(3a - 5)$

16. $(x - y)(x^2 + xy + y^2)$

Solve each system of equations.

17. $2x - 3y = -4$
$x + 2y = 5$

18. $x^2 + y^2 = 25$
$x + y = 7$

19. $2x - y + z = 7$
$x - 2y - z = 2$
$x + y + z = 2$

20. $y = x^2$
$y - 2x = 3$

Solve each formula for the specified variable.

21. $ax + b = 0$, for x

22. $wx^2 + dx + m = 0$, for x

23. $A = \dfrac{1}{2}h(B + b)$, for B

24. $\dfrac{1}{x} + \dfrac{1}{y} = \dfrac{1}{2}$, for x

25. $L = m + mxt$, for m

26. $y = 3a\sqrt{t}$, for t

Solve each problem.

27. Write the equation of the line in slope-intercept form that goes through the points $(2, -3)$ and $(-4, 1)$.

28. Write the equation of the line in slope-intercept form that contains the origin and is perpendicular to the line $2x - 4y = 5$.

29. Write the equation of the circle that has center $(2, 5)$ and passes through the point $(-1, -1)$.

30. Find the center and radius of the circle $x^2 + 3x + y^2 - 6y = 0$.

Perform the computations with complex numbers.

31. $2i(3 + 5i)$ **32.** i^6

33. $(2i - 3) + (6 - 7i)$ **34.** $(3 + i\sqrt{2})^2$

35. $(2 - 3i)(5 - 6i)$ **36.** $(3 - i) + (-6 + 4i)$

37. $(5 - 2i)(5 + 2i)$ **38.** $(2 - 3i) \div (2i)$

39. $(4 + 5i) \div (1 - i)$ **40.** $\dfrac{4 - \sqrt{-8}}{2}$

Solve.

41. *Going bananas.* Salvadore has observed that when bananas are \$0.30 per pound (lb), he sells 250 lb per day, and when bananas are \$0.40 per lb, he sells only 200 lb per day.

   a) Assume the number of pounds sold, q, is a linear function of the price per pound, x, and find that function.

   b) Salvadore's daily revenue in dollars is the product of the number of pounds sold and the price per pound. Write the revenue as a function of x.

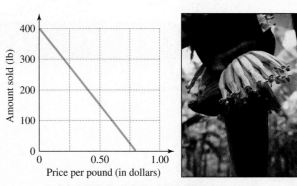

FIGURE FOR EXERCISE 41

   c) Graph the revenue function.

   d) What price per pound maximizes his revenue?

   e) What is his maximum possible revenue?

CHAPTER 14

Sequences and Series

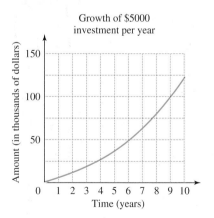

Growth of $5000
investment per year

veryone realizes the importance of investing for the future. Some people go to great pains to study the markets and to make wise investment decisions. Some stay away from investing because they do not want to take chances. However, the most important factor in investing is making regular investments (*Money,* www.money.com). According to *Money,* if you had invested $5000 in the stock market every year at the market high for that year (the worst time to invest) for the last 40 years, your investment would be worth $2.8 million today.

A sequence of periodic investments earning a fixed rate of interest can be thought of as a geometric sequence. In this chapter you will learn how to find the sum of a geometric sequence and to calculate the future value of a sequence of periodic investments. In Exercise 58 of Section 14.4 you will see how *Money* magazine calculated the value of $5000 invested each year for 10 years in Fidelity's Magellan fund.

14.1 SEQUENCES

The word "sequence" is a familiar word. We may speak of a sequence of events or say that something is out of sequence. In this section we give the mathematical definition of a sequence.

Definition

In mathematics we think of a sequence as a list of numbers. Each number in the sequence is called a **term** of the sequence. There is a first term, a second term, a third term, and so on. For example, the daily high temperature readings in Minot, North Dakota, for the first 10 days in January can be thought of as a finite sequence with 10 terms:

$$-9, -2, 8, -11, 0, 6, 14, 1, -5, -11$$

The set of all positive even integers,

$$2, 4, 6, 8, 10, 12, 14, \ldots,$$

can be thought of as an infinite sequence.

To give a precise definition of sequence, we use the terminology of functions. The list of numbers is the range of the function.

> **Sequence**
>
> A **finite sequence** is a function whose domain is the set of positive integers less than or equal to some fixed positive integer. An **infinite sequence** is a function whose domain is the set of all positive integers.

When the domain is apparent, we will refer to either a finite sequence or an infinite sequence simply as a sequence. For the independent variable of the function we will usually use n (for natural number) rather than x. For the dependent variable we write a_n (read "a sub n") rather than y. We call a_n the **nth term,** or the **general term** of the sequence. Rather than use the $f(x)$ notation for functions, we will define sequences with formulas. When n is used as a variable, we will assume it represents natural numbers only.

E X A M P L E   1

Listing terms of a finite sequence

List all of the terms of each finite sequence.

a) $a_n = n^2$ for $1 \leq n \leq 5$

b) $a_n = \dfrac{1}{n + 2}$ for $1 \leq n \leq 4$

Solution

a) Using the natural numbers from 1 through 5 in $a_n = n^2$, we get

$$a_1 = 1^2 = 1,$$
$$a_2 = 2^2 = 4,$$
$$a_3 = 3^2 = 9,$$
$$a_4 = 4^2 = 16,$$

and

$$a_5 = 5^2 = 25.$$

The five terms of this sequence are 1, 4, 9, 16, and 25. We often refer to the listing of the terms of the sequence as the sequence.

b) Using the natural numbers from 1 through 4 in $a_n = \frac{1}{n+2}$, we get the terms

$$a_1 = \frac{1}{1+2} = \frac{1}{3},$$

$$a_2 = \frac{1}{2+2} = \frac{1}{4},$$

$$a_3 = \frac{1}{5},$$

and

$$a_4 = \frac{1}{6}.$$

The four terms of the sequence are $\frac{1}{3}, \frac{1}{4}, \frac{1}{5}$, and $\frac{1}{6}$.

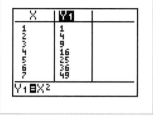

EXAMPLE 2   **Listing terms of an infinite sequence**

List the first three terms of the infinite sequence whose nth term is

$$a_n = \frac{(-1)^n}{2^{n+1}}.$$

Solution

Using the natural numbers 1, 2, and 3 in the formula for the nth term yields

$$a_1 = \frac{(-1)^1}{2^{1+1}} = -\frac{1}{4}, \qquad a_2 = \frac{(-1)^2}{2^{2+1}} = \frac{1}{8}, \qquad \text{and} \qquad a_3 = \frac{(-1)^3}{2^{3+1}} = -\frac{1}{16}.$$

We write the sequence as follows:

$$-\frac{1}{4}, \frac{1}{8}, -\frac{1}{16}, \ldots$$

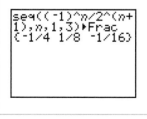

Finding a Formula for the nth Term

We often know the terms of a sequence and want to write a formula that will produce those terms. To write a formula for the nth term of a sequence, examine the terms and look for a pattern. Each term is a function of the term number. The first term corresponds to $n = 1$, the second term corresponds to $n = 2$, and so on.

EXAMPLE 3   **A familiar sequence**

Write the general term for the infinite sequence

$$3, 5, 7, 9, 11, \ldots.$$

Solution

The even numbers are all multiples of 2 and can be represented as $2n$. Because each odd number is 1 more than an even number, a formula for the nth term might be

$$a_n = 2n + 1.$$

To be sure, we write out a few terms using the formula:

$$a_1 = 2(1) + 1 = 3$$
$$a_2 = 2(2) + 1 = 5$$
$$a_3 = 2(3) + 1 = 7$$

So the general term is $a_n = 2n + 1$.   ■

CAUTION   There can be more than one formula that produces the given terms of a sequence. For example, the sequence

$$1, 2, 4, \ldots$$

could have nth term $a_n = 2^{n-1}$ or $a_n = \frac{1}{2}n^2 - \frac{1}{2}n + 1$. The first three terms for both of these sequences are identical, but their fourth terms are different.

EXAMPLE 4   **A sequence with alternating signs**

Write the general term for the infinite sequence

$$1, -\frac{1}{4}, \frac{1}{9}, -\frac{1}{16}, \ldots$$

Solution

To obtain the alternating signs, we use powers of -1. Because any even power of -1 is positive and any odd power of -1 is negative, we use $(-1)^{n+1}$. The denominators are the squares of the positive integers. So the nth term of this infinite sequence is given by the formula

$$a_n = \frac{(-1)^{n+1}}{n^2}.$$

Check this sequence by using this formula to find the first four terms.   ■

In the next example we use a sequence to model a physical situation.

EXAMPLE 5   **The bouncing ball**

Suppose a ball always rebounds $\frac{2}{3}$ of the height from which it falls and the ball is dropped from a height of 6 feet. Write a sequence whose terms are the heights from which the ball falls. What is a formula for the nth term of this sequence?

Solution

On the first fall the ball travels 6 feet (ft), as shown in Fig. 14.1. On the second fall it travels $\frac{2}{3}$ of 6, or 4 ft. On the third fall it travels $\frac{2}{3}$ of 4, or $\frac{8}{3}$ ft, and so on. We write the sequence as follows:

$$6, 4, \frac{8}{3}, \frac{16}{9}, \frac{32}{27}, \ldots$$

The nth term can be written by using powers of $\frac{2}{3}$:

$$a_n = 6\left(\frac{2}{3}\right)^{n-1}$$

■

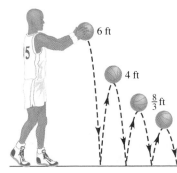

6 ft

4 ft

$\frac{8}{3}$ ft

FIGURE 14.1

$x^2 + (x+1)^2 = 5^2$

Most of us find an upholstered chair, sink into it, and remark on the comfort. Before design consultant Audrey Jordan sits down, she often looks at the fabric to observe the color and texture—and especially to see whether the fabric is one of her original designs. Fabric design is more than just an idea that is printed on a piece of cloth. Consideration must be given to the end product, which could be anything from a handbag to a large sofa. Colors and themes must be chosen with both current trends and styles

FABRIC DESIGNER

in mind. Sometimes a design will be an overall or nondirectional pattern, such as polka dots, which can be cut randomly. More often, it will have a specific theme, such as fruit, which can be cut and sewn in only one direction.

For all products one of the main considerations is the vertical repeat. A good portion of textile machinery is standardized for vertical repeats of 27 inches or fractions thereof. For example, the vertical repeat could be every $13\frac{1}{2}$ inches or every 9 inches. Even though the horizontal repeat can vary, Ms. Jordan must consider both the horizontal and vertical repeats for a particular end product.

In Exercise 45 of this section you will find the standard vertical repeats for a textile machine.

WARM-UPS

True or false? Explain your answer.

1. The nth term of the sequence 2, 4, 6, 8, 10, . . . is $a_n = 2n$.
2. The nth term of the sequence 1, 3, 5, 7, 9, . . . is $a_n = 2n - 1$.
3. A sequence is a function.
4. The domain of a finite sequence is the set of positive integers.
5. The nth term of $-1, 4, -9, 16, -25, . . .$ is $a_n = (-1)^{n+1}n^2$.
6. For the infinite sequence $b_n = \frac{1}{n}$, the independent variable is $\frac{1}{n}$.
7. For the sequence $c_n = n^3$, the dependent variable is c_n.
8. The sixth term of the sequence $a_n = (-1)^{n+1}2^n$ is 64.
9. The symbol a_n is used for the dependent variable of a sequence.
10. The tenth term of the sequence 2, 4, 8, 16, 32, 64, 128, . . . is 1024.

14.1   EXERCISES

Reading and Writing   *After reading this section, write out the answers to these questions. Use complete sentences.*

1. What is a sequence?

2. What is a term of a sequence?

3. What is a finite sequence?

4. What is an infinite sequence?

List all terms of each finite sequence. See Example 1.

5. $a_n = n^2$ for $1 \leq n \leq 8$

6. $a_n = -n^2$ for $1 \leq n \leq 4$

7. $b_n = \dfrac{(-1)^n}{n}$ for $1 \leq n \leq 10$

8. $b_n = \dfrac{(-1)^{n+1}}{n}$ for $1 \leq n \leq 6$

9. $c_n = (-2)^{n-1}$ for $1 \leq n \leq 5$

10. $c_n = (-3)^{n-2}$ for $1 \leq n \leq 5$

11. $a_n = 2^{-n}$ for $1 \leq n \leq 6$

12. $a_n = 2^{-n+2}$ for $1 \leq n \leq 5$

13. $b_n = 2n - 3$ for $1 \leq n \leq 7$

14. $b_n = 2n + 6$ for $1 \leq n \leq 7$

15. $c_n = n^{-1/2}$ for $1 \leq n \leq 5$

16. $c_n = n^{1/2}2^{-n}$ for $1 \leq n \leq 4$

Write the first four terms of the infinite sequence whose nth term is given. See Example 2.

17. $a_n = \dfrac{1}{n^2 + n}$

18. $b_n = \dfrac{1}{(n + 1)(n + 2)}$

19. $b_n = \dfrac{1}{2n - 5}$

20. $a_n = \dfrac{4}{2n + 5}$

21. $c_n = (-1)^n(n - 2)^2$

22. $c_n = (-1)^n(2n - 1)^2$

23. $a_n = \dfrac{(-1)^{2n}}{n^2}$

24. $a_n = (-1)^{2n+1}2^{n-1}$

Write a formula for the general term of each infinite sequence. See Examples 3 and 4.

25. 1, 3, 5, 7, 9, . . .

26. 5, 7, 9, 11, 13, . . .

27. 1, −1, 1, −1, . . .

28. −1, 1, −1, 1, . . .

29. 0, 2, 4, 6, 8, . . .

30. 4, 6, 8, 10, 12, . . .

31. 3, 6, 9, 12, . . .

32. 4, 8, 12, 16, . . .

33. 4, 7, 10, 13, . . .

34. 3, 7, 11, 15, . . .

35. −1, 2, −4, 8, −16, . . .

36. 1, −3, 9, −27, . . .

37. 0, 1, 4, 9, 16, . . .

38. 0, 1, 8, 27, 64, . . .

Solve each problem. See Example 5.

39. *Football penalties.* A football is on the 8-yard line, and five penalties in a row are given that move the ball half the distance to the (closest) goal. Write a sequence of five terms that specify the location of the ball after each penalty.

40. *Infestation.* Leona planted 9 acres of soybeans, but by the end of each week, insects had destroyed one-third of the acreage that was healthy at the beginning of the week. How many acres does she have left after 6 weeks?

41. *Constant rate of increase.* The MSRP for the 1999 Ford F-250 Lariat 4WD Super Duty Super Cab was $32,535 (Edmund's New Car Prices, www.edmunds.com). Suppose the price of this truck increases by 5% each year. Find the prices to the nearest dollar for the 2000 through 2005 models.

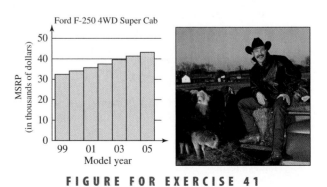

FIGURE FOR EXERCISE 41

42. Constant increase. The MSRP for a new 1999 Mercury Cougar was $21,455 (Edmund's New Car Prices, www.edmunds.com). Suppose the price of this car increases by $1000 each year. Find the prices of the 2000 through 2005 models.

43. Economic impact. To assess the economic impact of a factory on a community, economists consider the annual amount the factory spends in the community, then the portion of the money that is respent in the community, then the portion of the respent money that is respent in the community, and so on. Suppose a garment manufacturer spends $1 million annually in its community and 80% of all money received in the community is respent in the community. Find the first four terms of the economic impact sequence.

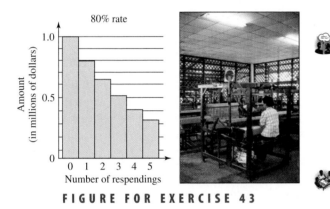

FIGURE FOR EXERCISE 43

44. Less impact. The rate at which money is respent in a community varies from community to community. Find the first four terms of the economic impact sequence for the manufacturer in Exercise 43, assuming only 50% of money received in the community is respent in the community.

45. Fabric design. A fabric designer must take into account the capability of textile machines to produce material with vertical repeats. A textile machine can be set up for a vertical repeat every $\frac{27}{n}$ inches (in.), where n is a natural number. Write the first five terms of the sequence $a_n = \frac{27}{n}$, which gives the possible vertical repeats for a textile machine.

46. Musical tones. The note middle C on a piano is tuned so that the string vibrates at 262 cycles per second, or 262 Hertz (Hz). The C note one octave higher is tuned to 524 Hz. The tuning for the 11 notes in between using the method called *equal temperament* is determined by the sequence $a_n = 262 \cdot 2^{n/12}$. Find the tuning for the 11 notes in between.

GETTING MORE INVOLVED

47. Discussion. Everyone has two (biological) parents, four grandparents, eight great-grandparents, 16 great-great-grandparents, and so on. If we put the word "great" in front of the word "grandparents" 35 times, then how many of this type of relative do you have? Is this more or less than the present population of the earth? Give reasons for your answers.

48. Discussion. If you deposit 1 cent into your piggy bank on September 1 and each day thereafter deposit twice as much as on the previous day, then how much will you be depositing on September 30? The total amount deposited for the month can be found without adding up all 30 deposits. Look at how the amount on deposit is increasing each day and see whether you can find the total for the month. Give reasons for your answers.

49. Cooperative learning. Working in groups, have someone in each group make up a formula for a_n, the nth term of a sequence, but do not show it to the other group members. Write the terms of the sequence on a piece of paper one at a time. After each term is given, ask whether anyone knows the next term. When the group can correctly give the next term, ask for a formula for the nth term.

50. Exploration. Find a real-life sequence in which all of the terms are the same. Find one in which each term after the first is one larger than the previous term. Find out what the sequence of fines is on your campus for your first, second, third, and fourth parking ticket.

51. Exploration. Consider the sequence whose nth term is $a_n = (0.999)^n$.
a) Calculate a_{100}, a_{1000}, and $a_{10,000}$.

b) What happens to a_n as n gets larger and larger?

 **SERIES**

If you make a sequence of bank deposits, then you might be interested in the total value of the terms of the sequence. Of course, if the sequence has only a few terms, you can simply add them. In Sections 14.3 and 14.4 we will develop formulas that give the sum of the terms for certain finite and infinite sequences. In this section you will first learn a notation for expressing the sum of the terms of a sequence.

Summation Notation

To describe the sum of the terms of a sequence, we use **summation notation.** The Greek letter Σ (sigma) is used to indicate sums. For example, the sum of the first five terms of the sequence $a_n = n^2$ is written as

$$\sum_{n=1}^{5} n^2.$$

You can read this notation as "the sum of n^2 for n between 1 and 5, inclusive." To find the sum, we let n take the values 1 through 5 in the expression n^2:

$$\sum_{n=1}^{5} n^2 = 1^2 + 2^2 + 3^2 + 4^2 + 5^2$$
$$= 1 + 4 + 9 + 16 + 25$$
$$= 55$$

In this context the letter n is the **index of summation.** Other letters may also be used. For example, the expressions

$$\sum_{n=1}^{5} n^2, \quad \sum_{j=1}^{5} j^2, \quad \text{and} \quad \sum_{i=1}^{5} i^2$$

all have the same value. Note that i is used as a variable here and not as an imaginary number.

E X A M P L E   1   **Evaluating a sum in summation notation**

Find the value of the expression

$$\sum_{i=1}^{3} (-1)^i (2i + 1).$$

Solution

Replace i by 1, 2, and 3, and then add the results:

$$\sum_{i=1}^{3} (-1)^i (2i + 1) = (-1)^1 [2(1) + 1] + (-1)^2 [2(2) + 1] + (-1)^3 [2(3) + 1]$$
$$= -3 + 5 - 7$$
$$= -5$$

Series

The sum of the terms of the sequence 1, 4, 9, 16, 25 is written as

$$1 + 4 + 9 + 16 + 25.$$

This expression is called a *series*. It indicates that we are to add the terms of the given sequence. The sum, 55, is the sum of the series.

Series

The indicated sum of the terms of a sequence is called a **series.**

Just as a sequence may be finite or infinite, a series may be finite or infinite. In this section we discuss finite series only. In Section 14.4 we will discuss one type of infinite series.

Summation notation is a convenient notation for writing a series.

E X A M P L E   2   **Converting to summation notation**

Write the series in summation notation:

$$2 + 4 + 6 + 8 + 10 + 12 + 14$$

Solution

The general term for the sequence of positive even integers is $2n$. If we let n take the values from 1 through 7, then $2n$ ranges from 2 through 14. So

$$2 + 4 + 6 + 8 + 10 + 12 + 14 = \sum_{n=1}^{7} 2n.$$

■

E X A M P L E   3   **Converting to summation notation**

Write the series

$$\frac{1}{2} - \frac{1}{3} + \frac{1}{4} - \frac{1}{5} + \frac{1}{6} - \frac{1}{7} + \cdots + \frac{1}{50}$$

in summation notation.

helpful **hint**

A series is called an *indicated sum* because the addition is indicated but not actually being performed. The sum of a series is the real number obtained by actually performing the indicated addition.

Solution

For this series we let n be 2 through 50. The expression $(-1)^n$ produces alternating signs. The series is written as

$$\sum_{n=2}^{50} \frac{(-1)^n}{n}.$$

■

Changing the Index

In Example 3 we saw the index go from 2 through 50, but this is arbitrary. A series can be written with the index starting at any given number.

E X A M P L E   4   **Changing the index**

Rewrite the series

$$\sum_{i=1}^{6} \frac{(-1)^i}{i^2}$$

with an index j, where j starts at 0.

Solution

Because i starts at 1 and j starts at 0, we have $i = j + 1$. Because i ranges from 1 through 6 and $i = j + 1$, j must range from 0 through 5. Now replace i by $j + 1$ in the summation notation:

$$\sum_{j=0}^{5} \frac{(-1)^{j+1}}{(j + 1)^2}$$

Check that these two series have exactly the same six terms.

■

True or false? Explain your answer.

1. A series is the indicated sum of the terms of a sequence.

2. The sum of a series can never be negative.

3. There are eight terms in the series $\sum\limits_{i=2}^{10} i^3$.

4. The series $\sum\limits_{i=1}^{9} (-1)^i i^2$ and $\sum\limits_{j=0}^{8} (-1)^j (j + 1)^2$ have the same sum.

5. The ninth term of the series $\sum\limits_{i=1}^{100} \dfrac{(-1)^i}{(i + 1)(i + 2)}$ is $\dfrac{1}{110}$.

6. $\sum\limits_{i=1}^{2} (-1)^i 2^i = 2$

7. $\sum\limits_{i=1}^{5} 3i = 3\left(\sum\limits_{i=1}^{5} i\right)$

8. $\sum\limits_{i=1}^{5} 4 = 20$

9. $\sum\limits_{i=1}^{5} 2i + \sum\limits_{i=1}^{5} 7i = \sum\limits_{i=1}^{5} 9i$

10. $\sum\limits_{i=1}^{3} (2i + 1) = \left(\sum\limits_{i=1}^{3} 2i\right) + 1$

14.2 **EXERCISES**

Reading and Writing *After reading this section, write out the answers to these questions. Use complete sentences.*

1. What is summation notation?

2. What is the index of summation?

3. What is a series?

4. What is a finite series?

Find the sum of each series. See Example 1.

5. $\sum\limits_{i=1}^{4} i^2$

6. $\sum\limits_{j=0}^{3} (j + 1)^2$

7. $\sum\limits_{j=0}^{5} (2j - 1)$

8. $\sum\limits_{i=1}^{6} (2i - 3)$

9. $\sum\limits_{i=1}^{5} 2^{-i}$

10. $\sum\limits_{i=1}^{5} (-2)^{-i}$

11. $\sum\limits_{i=1}^{10} 5i^0$

12. $\sum\limits_{j=1}^{20} 3$

13. $\sum\limits_{i=1}^{3} (i - 3)(i + 1)$

14. $\sum\limits_{i=0}^{5} i(i - 1)(i - 2)(i - 3)$

15. $\sum\limits_{j=1}^{10} (-1)^j$

16. $\sum\limits_{j=1}^{11} (-1)^j$

Write each series in summation notation. Use the index i, and let i begin at 1 in each summation. See Examples 2 and 3.

17. $1 + 2 + 3 + 4 + 5 + 6$

18. $2 + 4 + 6 + 8 + 10$

19. $-1 + 3 - 5 + 7 - 9 + 11$

20. $1 - 3 + 5 - 7 + 9$

21. $1 + 4 + 9 + 16 + 25 + 36$

22. $1 + 8 + 27 + 64 + 125$

23. $\dfrac{1}{3} + \dfrac{1}{4} + \dfrac{1}{5} + \dfrac{1}{6}$

24. $1 - \dfrac{1}{2} + \dfrac{1}{3} - \dfrac{1}{4} + \dfrac{1}{5} - \dfrac{1}{6}$

25. $\ln(2) + \ln(3) + \ln(4)$

26. $e^1 + e^2 + e^3 + e^4$

27. $a_1 + a_2 + a_3 + a_4$

28. $a^2 + a^3 + a^4 + a^5$

29. $x_3 + x_4 + x_5 + \cdots + x_{50}$

30. $y_1 + y_2 + y_3 + \cdots + y_{30}$

31. $w_1 + w_2 + w_3 + \cdots + w_n$

32. $m_1 + m_2 + m_3 + \cdots + m_k$

Complete the rewriting of each series using the new index as indicated. See Example 4.

33. $\displaystyle\sum_{i=1}^{5} i^2 = \sum_{j=0}$

34. $\displaystyle\sum_{i=1}^{6} i^3 = \sum_{j=0}$

35. $\displaystyle\sum_{i=0}^{12} (2i - 1) = \sum_{j=1}$

36. $\displaystyle\sum_{i=1}^{3} (3i + 2) = \sum_{j=0}$

37. $\displaystyle\sum_{i=4}^{8} \frac{1}{i} = \sum_{j=1}$

38. $\displaystyle\sum_{i=5}^{10} 2^{-i} = \sum_{j=1}$

39. $\displaystyle\sum_{i=1}^{4} x^{2i+3} = \sum_{j=0}$

40. $\displaystyle\sum_{i=0}^{2} x^{3-2i} = \sum_{j=1}$

41. $\displaystyle\sum_{i=1}^{n} x^i = \sum_{j=0}$

42. $\displaystyle\sum_{i=0}^{n} x^{-i} = \sum_{j=1}$

Write out the terms of each series.

43. $\displaystyle\sum_{i=1}^{6} x^i$

44. $\displaystyle\sum_{i=1}^{5} (-1)^i x^{i-1}$

45. $\displaystyle\sum_{j=0}^{3} (-1)^j x_j$

46. $\displaystyle\sum_{j=1}^{5} \frac{1}{x_j}$

47. $\displaystyle\sum_{i=1}^{3} i x^i$

48. $\displaystyle\sum_{i=1}^{5} \frac{x}{i}$

A series can be used to model the situation in each of the following problems.

49. *Leap frog.* A frog with a vision problem is 1 yard away from a dead cricket. He spots the cricket and jumps halfway to the cricket. After the frog realizes that he has not reached the cricket, he again jumps halfway to the cricket. Write a series in summation notation to describe how far the frog has moved after nine such jumps.

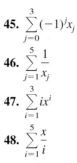

50. *Compound interest.* Cleo deposited $1000 at the beginning of each year for 5 years into an account paying 10% interest compounded annually. Write a series using summation notation to describe how much she has in the account at the end of the fifth year. Note that the first $1000 will receive interest for 5 years, the second $1000 will receive interest for 4 years, and so on.

51. *Total economic impact.* In Exercise 43 of Section 14.1 we described a factory that spends $1 million annually in a community in which 80% of all money received in the community is respent in the community. Use summation notation to write the sum of the first four terms of the economic impact sequence for the factory.

52. *Total spending.* Suppose you earn $1 on January 1, $2 on January 2, $3 on January 3, and so on. Use summation notation to write the sum of your earnings for the entire month of January.

GETTING MORE INVOLVED

53. *Discussion.* What is the difference between a sequence and a series?

54. *Discussion.* For what values of n is $\displaystyle\sum_{i=1}^{n} \frac{1}{i} > 4$?

14.3 ARITHMETIC SEQUENCES AND SERIES

We defined sequences and series in Sections 14.1 and 14.2. In this section you will study a special type of sequence known as an arithmetic sequence. You will also study the series corresponding to this sequence.

Arithmetic Sequences

Consider the following sequence:

$$5, 9, 13, 17, 21, \ldots$$

This sequence is called an arithmetic sequence because of the pattern for the terms. Each term is 4 larger than the previous term.

helpful hint

Arithmetic used as an adjective (ar-ith-met′-ic) is pronounced differently from arithmetic used as a noun (a-rith′-me-tic). Arithmetic (the adjective) is accented similarly to geometric.

> **Arithmetic Sequence**
>
> A sequence in which each term after the first is obtained by adding a fixed amount to the previous term is called an **arithmetic sequence.**

The fixed amount is called the **common difference** and is denoted by the letter d. If a_1 is the first term, then the second term is $a_1 + d$. The third term is $a_1 + 2d$, the fourth term is $a_1 + 3d$, and so on.

> **Formula for the nth Term of an Arithmetic Sequence**
>
> The nth term, a_n, of an arithmetic sequence with first term a_1 and common difference d is
>
> $$a_n = a_1 + (n - 1)d.$$

EXAMPLE 1

The nth term of an arithmetic sequence

Write a formula for the nth term of the arithmetic sequence

$$5, 9, 13, 17, 21, \ldots.$$

Solution

Each term of the sequence after the first is 4 more than the previous term. Because the common difference is 4 and the first term is 5, the nth term is given by

$$a_n = 5 + (n - 1)4.$$

We can simplify this expression to get

$$a_n = 4n + 1.$$

Check a few terms: $a_1 = 4(1) + 1 = 5$, $a_2 = 4(2) + 1 = 9$, and $a_3 = 4(3) + 1 = 13$. ∎

In the next example the common difference is negative.

EXAMPLE 2

An arithmetic sequence of decreasing terms

Write a formula for the nth term of the arithmetic sequence

$$4, 1, -2, -5, -8, \ldots.$$

Solution

Each term is 3 less than the previous term, so $d = -3$. Because $a_1 = 4$, we can write the nth term as

$$a_n = 4 + (n - 1)(-3),$$

or

$$a_n = -3n + 7.$$

Check a few terms: $a_1 = -3(1) + 7 = 4$, $a_2 = -3(2) + 7 = 1$, and $a_3 = -3(3) + 7 = -2$.   ■

In the next example we find some terms of an arithmetic sequence using a given formula for the nth term.

E X A M P L E   3

Writing terms of an arithmetic sequence

Write the first five terms of the sequence in which $a_n = 3 + (n - 1)6$.

Solution

Let n take the values from 1 through 5, and find a_n:

$$a_1 = 3 + (1 - 1)6 = 3$$
$$a_2 = 3 + (2 - 1)6 = 9$$
$$a_3 = 3 + (3 - 1)6 = 15$$
$$a_4 = 3 + (4 - 1)6 = 21$$
$$a_5 = 3 + (5 - 1)6 = 27$$

Notice that $a_n = 3 + (n - 1)6$ gives the general term for an arithmetic sequence with first term 3 and common difference 6. Because each term after the first is 6 more than the previous term, the first five terms that we found are correct.   ■

The formula $a_n = a_1 + (n - 1)d$ involves four variables: a_1, a_n, n, and d. If we know the values of any three of these variables, we can find the fourth.

E X A M P L E   4

Finding a missing term of an arithmetic sequence

Find the twelfth term of the arithmetic sequence whose first term is 2 and whose fifth term is 14.

Solution

Before finding the twelfth term, we use the given information to find the missing common difference. Let $n = 5$, $a_1 = 2$, and $a_5 = 14$ in the formula $a_n = a_1 + (n - 1)d$ to find d:

$$14 = 2 + (5 - 1)d$$
$$14 = 2 + 4d$$
$$12 = 4d$$
$$3 = d$$

Now use $a_1 = 2$, $d = 3$ and $n = 12$ in $a_n = a_1 + (n - 1)d$ to find a_{12}:

$$a_{12} = 2 + (12 - 1)3$$
$$a_{12} = 35$$ ■

Arithmetic Series

The indicated sum of an arithmetic sequence is called an **arithmetic series.** For example, the series

$$2 + 4 + 6 + 8 + 10 + \cdots + 54$$

is an arithmetic series because there is a common difference of 2 between the terms.

We can find the actual sum of this arithmetic series without adding all of the terms. Write the series in increasing order, and below that write the series in decreasing order. We then add the corresponding terms:

$$
\begin{array}{rcccccccc}
S = & 2 + & 4 + & 6 + & 8 + & \cdots & + 52 + & 54 \\
S = & 54 + & 52 + & 50 + & 48 + & \cdots & + & 4 + & 2 \\
\hline
2S = & 56 + & 56 + & 56 + & 56 + & \cdots & + 56 + & 56
\end{array}
$$

Now, how many times does 56 appear in the sum on the right? Because

$$2 + 4 + 6 + \cdots + 54 = 2 \cdot 1 + 2 \cdot 2 + 2 \cdot 3 + \cdots + 2 \cdot 27,$$

there are 27 terms in this sum. Because 56 appears 27 times on the right, we have $2S = 27 \cdot 56$, or

$$S = \frac{27 \cdot 56}{2} = 27 \cdot 28 = 756.$$

If $S_n = a_1 + a_2 + a_3 + \cdots + a_n$ is any arithmetic series, then we can find its sum using the same technique. Rewrite S_n as follows:

$$
\begin{array}{rcccccc}
S_n = & a_1 & + (a_1 + d) & + (a_1 + 2d) & + \cdots + & a_n \\
S_n = & a_n & + (a_n - d) & + (a_n - 2d) & + \cdots + & a_1 \\
\hline
2S_n = & (a_1 + a_n) & + (a_1 + a_n) & + (a_1 + a_n) & + \cdots + & (a_1 + a_n) & \text{Add.}
\end{array}
$$

Because $(a_1 + a_n)$ appears n times on the right, we have $2S_n = n(a_1 + a_n)$. Divide each side by 2 to get the following formula.

Sum of an Arithmetic Series

The sum, S_n, of the first n terms of an arithmetic series with first term a_1 and nth term a_n, is given by

$$S_n = \frac{n}{2}(a_1 + a_n).$$

E X A M P L E 5

The sum of an arithmetic series
Find the sum of the positive integers from 1 to 100 inclusive.

Solution

helpful hint

Legend has it that Carl F. Gauss knew this formula when he was in grade school. Gauss's teacher told him to add up the numbers from 1 through 100 for busy work. He immediately answered 5050.

The described series, $1 + 2 + 3 + \cdots + 100$, has 100 terms. So we can use $n = 100$, $a_1 = 1$, and $a_n = 100$ in the formula for the sum of an arithmetic series:

$$S_n = \frac{n}{2}(a_1 + a_n)$$

$$S_{100} = \frac{100}{2}(1 + 100)$$

$$= 50(101)$$

$$= 5050$$

E X A M P L E 6   **The sum of an arithmetic series**

Find the sum of the series

$$12 + 16 + 20 + \cdots + 84.$$

Solution

This series is an arithmetic series with $a_n = 84$, $a_1 = 12$, and $d = 4$. To get the number of terms, n, we use $a_n = a_1 + (n - 1)d$:

$$84 = 12 + (n - 1)4$$
$$84 = 8 + 4n$$
$$76 = 4n$$
$$19 = n$$

Now find the sum of these 19 terms:

$$S_{19} = \frac{19}{2}(12 + 84) = 912$$

■

WARM-UPS

True or false? Explain your answer.

1. The arithmetic sequence $3, 1, -1, -3, -5, \ldots$ has common difference 2.

2. The sequence $2, 5, 9, 14, 20, 27, \ldots$ is an arithmetic sequence.

3. The sequence $2, 4, 2, 0, 2, 4, 2, 0, \ldots$ is an arithmetic sequence.

4. The nth term of an arithmetic sequence with first term a_1 and common difference d is given by the formula $a_n = a_1 + nd$.

5. If $a_1 = 5$ and $a_3 = 10$ in an arithmetic sequence, then $a_4 = 15$.

6. If $a_1 = 6$ and $a_3 = 2$ in an arithmetic sequence, then $a_2 = 10$.

7. An arithmetic series is the indicated sum of an arithmetic sequence.

8. The series $\sum\limits_{i=1}^{5}(3 + 2i)$ is an arithmetic series.

9. The sum of the first n counting numbers is $\frac{n(n + 1)}{2}$.

10. The sum of the even integers from 8 through 28 inclusive is $5(8 + 28)$.

14.3   EXERCISES

Reading and Writing   After reading this section, write out the answers to these questions. Use complete sentences.

1. What is an arithmetic sequence?

2. What is the nth term of an arithmetic sequence?

3. What is an arithmetic series?

4. What is the formula for the sum of the first n terms of an arithmetic series?

Write a formula for the nth term of each arithmetic sequence. See Examples 1 and 2.

5. $0, 6, 12, 18, 24, \ldots$

6. $0, 5, 10, 15, 20, \ldots$

7. $7, 12, 17, 22, 27, \ldots$

8. $4, 15, 26, 37, 48, \ldots$

9. $-4, -2, 0, 2, 4, \ldots$

10. $-3, 0, 3, 6, 9, \ldots$

11. $5, 1, -3, -7, -11, \ldots$

12. $8, 5, 2, -1, -4, \ldots$

13. $-2, -9, -16, -23, \ldots$

14. $-5, -7, -9, -11, -13, \ldots$

15. $-3, -2.5, -2, -1.5, -1, \ldots$

16. $-2, -1.25, -0.5, 0.25, \ldots$

17. $-6, -6.5, -7, -7.5, -8, \ldots$

18. $1, 0.5, 0, -0.5, -1, \ldots$

In Exercises 19–32, write the first five terms of the arithmetic sequence whose nth term is given. See Example 3.

19. $a_n = 9 + (n - 1)4$

20. $a_n = 13 + (n - 1)6$

21. $a_n = 7 + (n - 1)(-2)$

22. $a_n = 6 + (n - 1)(-3)$

23. $a_n = -4 + (n - 1)3$

24. $a_n = -19 + (n - 1)12$

25. $a_n = -2 + (n - 1)(-3)$

26. $a_n = -1 + (n - 1)(-2)$

27. $a_n = -4n - 3$

28. $a_n = -3n + 1$

29. $a_n = 0.5n + 4$

30. $a_n = 0.3n + 1$

31. $a_n = 20n + 1000$

32. $a_n = -600n + 4000$

Find the indicated part of each arithmetic sequence. See Example 4.

33. Find the eighth term of the sequence that has a first term of 9 and a common difference of 6.

34. Find the twelfth term of the sequence that has a first term of -2 and a common difference of -3.

35. Find the common difference if the first term is 6 and the twentieth term is 82.

36. Find the common difference if the first term is -8 and the ninth term is -64.

37. If the common difference is -2 and the seventh term is 14, then what is the first term?

38. If the common difference is 5 and the twelfth term is -7, then what is the first term?

39. Find the sixth term of the sequence that has a fifth term of 13 and a first term of -3.

40. Find the eighth term of the sequence that has a sixth term of -42 and a first term of 3.

Find the sum of each given series. See Examples 5 and 6.

41. $1 + 2 + 3 + \cdots + 48$

42. $1 + 2 + 3 + \cdots + 12$

43. $8 + 10 + 12 + \cdots + 36$

44. $9 + 12 + 15 + \cdots + 72$

45. $-1 + (-7) + (-13) + \cdots + (-73)$

46. $-7 + (-12) + (-17) + \cdots + (-72)$

47. $-6 + (-1) + 4 + 9 + \cdots + 64$

48. $-9 + (-1) + 7 + \cdots + 103$

49. $20 + 12 + 4 + (-4) + \cdots + (-92)$

50. $19 + 1 + (-17) + \cdots + (-125)$

51. $\displaystyle\sum_{i=1}^{12} (3i - 7)$

52. $\displaystyle\sum_{i=1}^{7} (-4i + 6)$

53. $\displaystyle\sum_{i=1}^{11} (-5i + 2)$

54. $\displaystyle\sum_{i=1}^{19} (3i - 5)$

Solve each problem using the ideas of arithmetic sequences and series.

55. **Increasing salary.** If a lab technician has a salary of \$22,000 her first year and is due to get a \$500 raise each year, then what will her salary be in her seventh year?

56. **Seven years of salary.** What is the total salary for 7 years of work for the lab technician of Exercise 55?

57. **Light reading.** On the first day of October an English teacher suggests to his students that they read five pages

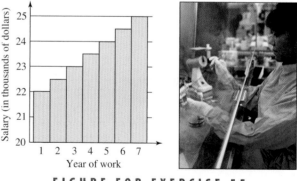

FIGURE FOR EXERCISE 55

of a novel and every day thereafter increase their daily reading by two pages. If his students follow this suggestion, then how many pages will they read during October?

58. *Heavy penalties.* If an air-conditioning system is not completed by the agreed upon date, the contractor pays a penalty of $500 for the first day that it is overdue, $600 for the second day, $700 for the third day, and so on. If the system is completed 10 days late, then what is the total amount of the penalties that the contractor must pay?

GETTING MORE INVOLVED

59. *Discussion.* Which of the following sequences is not an arithmetic sequence? Explain your answer.

a) $\frac{1}{2}, 1, \frac{3}{2}, \ldots$　　　　b) $\frac{1}{2}, \frac{1}{3}, \frac{1}{4}, \ldots$

c) $5, 0, -5, \ldots$　　　　d) $2, 3, 4, \ldots$

60. *Discussion.* What is the smallest value of n for which
$$\sum_{i=1}^{n} \frac{1}{2} > 50?$$

14.4 GEOMETRIC SEQUENCES AND SERIES

In this section

- Geometric Sequences
- Finite Geometric Series
- Infinite Geometric Series
- Annuities

In Section 14.3 you studied the arithmetic sequences and series. In this section you will study sequences in which each term is a *multiple* of the term preceding it. You will also learn how to find the sum of the corresponding series.

Geometric Sequences

In an arithmetic sequence such as 2, 4, 6, 8, 10, ... there is a common difference between consecutive terms. In a geometric sequence there is a common ratio between consecutive terms. The following table contains several geometric sequences and the common ratios between consecutive terms.

Geometric Sequence	Common Ratio
3, 6, 12, 24, 48, ...	2
27, 9, 3, 1, $\frac{1}{3}$, ...	$\frac{1}{3}$
1, −10, 100, −1000, ...	−10

Note that every term after the first term of each geometric sequence can be obtained by multiplying the previous term by the common ratio.

> **Geometric Sequence**
>
> A sequence in which each term after the first is obtained by multiplying the preceding term by a constant is called a **geometric sequence.**

The constant is denoted by the letter r and is called the **common ratio.** If a_1 is the first term, then the second term is $a_1 r$. The third term is $a_1 r^2$, the fourth term is $a_1 r^3$, and so on. We can write a formula for the nth term of a geometric sequence by following this pattern.

> **Formula for the *n*th Term of a Geometric Sequence**
>
> The *n*th term, a_n, of a geometric sequence with first term a_1 and common ratio r is
>
> $$a_n = a_1 r^{n-1}.$$

The first term and the common ratio determine all of the terms of a geometric sequence.

E X A M P L E   1   **Finding the *n*th term**

Write a formula for the *n*th term of the geometric sequence

$$6, 2, \frac{2}{3}, \frac{2}{9}, \dots.$$

Solution

We can obtain the common ratio by dividing any term after the first by the term preceding it. So

$$r = 2 \div 6 = \frac{1}{3}.$$

Because each term after the first is $\frac{1}{3}$ of the term preceding it, the *n*th term is given by

$$a_n = 6\left(\frac{1}{3}\right)^{n-1}.$$

Check a few terms: $a_1 = 6\left(\frac{1}{3}\right)^{1-1} = 6$, $a_2 = 6\left(\frac{1}{3}\right)^{2-1} = 2$, and $a_3 = 6\left(\frac{1}{3}\right)^{3-1} = \frac{2}{3}$.   ■

E X A M P L E   2   **Finding the *n*th term**

Find a formula for the *n*th term of the geometric sequence

$$2, -1, \frac{1}{2}, -\frac{1}{4}, \dots.$$

Solution

We obtain the ratio by dividing a term by the term preceding it:

$$r = -1 \div 2 = -\frac{1}{2}$$

Each term after the first is obtained by multiplying the preceding term by $-\frac{1}{2}$. The formula for the *n*th term is

$$a_n = 2\left(-\frac{1}{2}\right)^{n-1}.$$

Check a few terms: $a_1 = 2\left(-\frac{1}{2}\right)^{1-1} = 2$, $a_2 = 2\left(-\frac{1}{2}\right)^{2-1} = -1$, and $a_3 = 2\left(-\frac{1}{2}\right)^{3-1} = \frac{1}{2}$.   ■

In the next example we use the formula for the *n*th term to write some terms of a geometric sequence.

E X A M P L E   3   **Writing the terms**

Write the first five terms of the geometric sequence whose nth term is

$$a_n = 3(-2)^{n-1}.$$

Solution

Let n take the values 1 through 5 in the formula for the nth term:

$$a_1 = 3(-2)^{1-1} = 3$$
$$a_2 = 3(-2)^{2-1} = -6$$
$$a_3 = 3(-2)^{3-1} = 12$$
$$a_4 = 3(-2)^{4-1} = -24$$
$$a_5 = 3(-2)^{5-1} = 48$$

Notice that $a_n = 3(-2)^{n-1}$ gives the general term for a geometric sequence with first term 3 and common ratio -2. Because every term after the first can be obtained by multiplying the previous term by -2, the terms 3, -6, 12, -24, and 48 are correct. ■

The formula for the nth term involves four variables: a_n, a_1, r, and n. If we know the value of any three of them, we can find the value of the fourth.

E X A M P L E   4   **Finding a missing term**

Find the first term of a geometric sequence whose fourth term is 8 and whose common ratio is $\frac{1}{2}$.

Solution

Let $a_4 = 8$, $r = \frac{1}{2}$, and $n = 4$ in the formula $a_n = a_1 r^{n-1}$:

$$8 = a_1 \left(\frac{1}{2}\right)^{4-1}$$
$$8 = a_1 \cdot \frac{1}{8}$$
$$64 = a_1$$

So the first term is 64. ■

Finite Geometric Series

Consider the following series:

$$1 + 2 + 4 + 8 + 16 + \cdots + 512$$

The terms of this series are the terms of a finite geometric sequence. The indicated sum of a geometric sequence is called a **geometric series.**

We can find the actual sum of this finite geometric series by using a technique similar to the one used for the sum of an arithmetic series. Let

$$S = 1 + 2 + 4 + 8 + \cdots + 256 + 512.$$

Because the common ratio is 2, multiply each side by -2:

$$-2S = -2 - 4 - 8 - \cdots - 512 - 1024$$

Adding the last two equations eliminates all but two of the terms on the right:

$$
\begin{aligned}
S &= 1 + 2 + 4 + 8 + \cdots + 256 + 512 \\
-2S &= \qquad -2 - 4 - 8 - \cdots \qquad\qquad - 512 - 1024 \\
\hline
-S &= 1 \qquad\qquad\qquad\qquad\qquad\qquad - 1024 \quad \text{Add.} \\
-S &= -1023 \\
S &= 1023
\end{aligned}
$$

If $S_n = a_1 + a_1 r + a_1 r^2 + \cdots + a_1 r^{n-1}$ is any geometric series, we can find the sum in the same manner. Multiplying each side of this equation by $-r$ yields

$$-rS_n = -a_1 r - a_1 r^2 - a_1 r^3 - \cdots - a_1 r^n.$$

If we add S_n and $-rS_n$, all but two of the terms on the right are eliminated:

$$
\begin{aligned}
S_n &= a_1 + a_1 r + a_1 r^2 + \cdots \qquad\qquad + a_1 r^{n-1} \\
-rS_n &= \qquad - a_1 r - a_1 r^2 - a_1 r^3 - \cdots \qquad - a_1 r^n \\
\hline
S_n - rS_n &= a_1 \qquad\qquad\qquad\qquad\qquad\qquad - a_1 r^n \quad \text{Add.} \\
(1 - r)S_n &= a_1(1 - r^n)
\end{aligned}
$$
Factor out common factors.

Now divide each side of this equation by $1 - r$ to get the formula for S_n.

Sum of n Terms of a Geometric Series

If S_n represents the sum of the first n terms of a geometric series with first term a_1 and common ratio r $(r \neq 1)$, then

$$S_n = \frac{a_1(1 - r^n)}{1 - r}.$$

E X A M P L E 5

The sum of a finite geometric series

Find the sum of the series

$$\frac{1}{3} + \frac{1}{9} + \frac{1}{27} + \cdots + \frac{1}{729}.$$

Solution

The first term is $\frac{1}{3}$, and the common ratio is $\frac{1}{3}$. So the nth term can be written as

$$a_n = \frac{1}{3}\left(\frac{1}{3}\right)^{n-1}.$$

We can use this formula to find the number of terms in the series:

$$\frac{1}{729} = \frac{1}{3}\left(\frac{1}{3}\right)^{n-1}$$

$$\frac{1}{729} = \left(\frac{1}{3}\right)^{n}$$

Because $3^6 = 729$, we have $n = 6$. (Of course, you could use logarithms to solve for n.) Now use the formula for the sum of six terms of this geometric series:

$$S_6 = \frac{\frac{1}{3}\left[1 - \left(\frac{1}{3}\right)^6\right]}{1 - \frac{1}{3}} = \frac{\frac{1}{3}\left[1 - \frac{1}{729}\right]}{\frac{2}{3}}$$

$$= \frac{1}{3} \cdot \frac{728}{729} \cdot \frac{3}{2}$$

$$= \frac{364}{729}$$

E X A M P L E   6   **The sum of a finite geometric series**

Find the sum of the series

$$\sum_{i=1}^{12} 3(-2)^{i-1}.$$

Solution

This series is geometric with first term 3, ratio -2, and $n = 12$. We use the formula for the sum of the first 12 terms of a geometric series:

$$S_{12} = \frac{3[1 - (-2)^{12}]}{1 - (-2)} = \frac{3[-4095]}{3} = -4095$$

Infinite Geometric Series

Consider how a very large value of n affects the formula for the sum of a finite geometric series,

$$S_n = \frac{a_1(1 - r^n)}{1 - r}.$$

If $|r| < 1$, then the value of r^n gets closer and closer to 0 as n gets larger and larger. For example, if $r = \frac{2}{3}$ and $n = 10, 20,$ and 100, then

$$\left(\frac{2}{3}\right)^{10} \approx 0.0173415, \quad \left(\frac{2}{3}\right)^{20} \approx 0.0003007, \quad \text{and} \quad \left(\frac{2}{3}\right)^{100} \approx 2.460 \times 10^{-18}.$$

Because r^n is approximately 0 for large values of n, $1 - r^n$ is approximately 1. If we replace $1 - r^n$ by 1 in the expression for S_n, we get

$$S_n \approx \frac{a_1}{1 - r}.$$

So as n gets larger and larger, the sum of the first n terms of the infinite geometric series

$$a_1 + a_1 r + a_1 r^2 + \cdots$$

gets closer and closer to $\frac{a_1}{1 - r}$, provided that $|r| < 1$. Therefore we say that $\frac{a_1}{1 - r}$ is the sum of *all* of the terms of the infinite geometric series.

close-up

Experiment with your calculator to see what happens to r^n as n gets larger and larger.

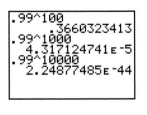

> ### Sum of an Infinite Geometric Series
>
> If $a_1 + a_1 r + a_1 r^2 + \cdots$ is an infinite geometric series, with $|r| < 1$, then the sum S of all of the terms of this series is given by
>
> $$S = \frac{a_1}{1 - r}.$$

EXAMPLE 7

Sum of an infinite geometric series

Find the sum

$$\frac{1}{2} + \frac{1}{4} + \frac{1}{8} + \frac{1}{16} + \cdots.$$

Solution

This series is an infinite geometric series with $a_1 = \frac{1}{2}$ and $r = \frac{1}{2}$. Because $r < 1$, we have

$$S = \frac{\dfrac{1}{2}}{1 - \dfrac{1}{2}} = 1.$$

■

For an infinite series the index of summation i takes the values 1, 2, 3, and so on, without end. To indicate that the values for i keep increasing without bound, we say that i *takes the values from* 1 *through* ∞ (infinity). Note that the symbol "∞" does not represent a number. Using the ∞ symbol, we can write the indicated sum of an infinite geometric series (with $|r| < 1$) by using summation notation as follows:

$$a_1 + a_1 r + a_1 r^2 + \cdots = \sum_{i=1}^{\infty} a_1 r^{i-1}$$

EXAMPLE 8

Sum of an infinite geometric series

Find the value of the sum

$$\sum_{i=1}^{\infty} 8 \left(\frac{3}{4} \right)^{i-1}.$$

Solution

This series is an infinite geometric series with first term 8 and ratio $\frac{3}{4}$. So

$$S = \frac{8}{1 - \dfrac{3}{4}} = 8 \cdot \frac{4}{1} = 32.$$

■

EXAMPLE 9

Follow the bouncing ball

Suppose a ball always rebounds $\frac{2}{3}$ of the height from which it falls and the ball is dropped from a height of 6 feet. Find the total distance that the ball travels.

Solution

The ball falls 6 feet (ft) and rebounds 4 ft, then falls 4 ft and rebounds $\frac{8}{3}$ ft. The following series gives the total distance that the ball falls:

$$F = 6 + 4 + \frac{8}{3} + \frac{16}{9} + \cdots$$

The distance that the ball rebounds is given by the following series:

$$R = 4 + \frac{8}{3} + \frac{16}{9} + \cdots$$

Each of these series is an infinite geometric series with ratio $\frac{2}{3}$. Use the formula for an infinite geometric series to find each sum:

$$F = \frac{6}{1 - \frac{2}{3}} = 6 \cdot \frac{3}{1} = 18 \text{ ft}, \qquad R = \frac{4}{1 - \frac{2}{3}} = 4 \cdot \frac{3}{1} = 12 \text{ ft}$$

The total distance traveled by the ball is the sum of F and R, 30 ft.   ◼

Annuities

One of the most important applications of geometric series is in calculating the value of an annuity. An **annuity** is a sequence of periodic payments. The payments might be loan payments or investments.

EXAMPLE 10   **Value of an annuity**

A deposit of $1000 is made at the beginning of each year for 30 years and earns 6% interest compounded annually. What is the value of this annuity at the end of the thirtieth year?

Solution

The last deposit earns interest for only one year. So at the end of the thirtieth year it amounts to $1000(1.06). The next to last deposit earns interest for 2 years and amounts to $1000(1.06)^2$. The first deposit earns interest for 30 years and amounts to $1000(1.06)^{30}$. So the value of the annuity at the end of the thirtieth year is the sum of the finite geometric series

$$1000(1.06) + 1000(1.06)^2 + 1000(1.06)^3 + \cdots + 1000(1.06)^{30}.$$

Use the formula for the sum of 30 terms of a finite geometric series with $a_1 = 1000(1.06)$ and $r = 1.06$:

$$S_{30} = \frac{1000(1.06)(1 - (1.06)^{30})}{1 - 1.06} = \$83,801.68$$

So 30 annual deposits of $1000 each amount to $83,801.68.   ◼

WARM-UPS

True or false? Explain your answer.

1. The sequence 2, 6, 24, 120, . . . is a geometric sequence.
2. For $a_n = 2^n$ there is a common difference between adjacent terms.
3. The common ratio for the geometric sequence $a_n = 3(0.5)^{n-1}$ is 0.5.
4. If $a_n = 3(2)^{-n+3}$, then $a_1 = 12$.
5. In the geometric sequence $a_n = 3(2)^{-n+3}$ we have $r = \frac{1}{2}$.
6. The terms of a geometric series are the terms of a geometric sequence.

(continued)

7. To evaluate $\sum_{i=1}^{10} 2^i$, we must list all of the terms.

8. $\displaystyle\sum_{i=1}^{5} 6\left(\frac{3}{4}\right)^{i-1} = \frac{9\left[1 - \left(\frac{3}{4}\right)^5\right]}{1 - \frac{3}{4}}$

9. $10 + 5 + \dfrac{5}{2} + \cdots = \dfrac{10}{1 - \dfrac{1}{2}}$

10. $2 + 4 + 8 + 16 + \cdots = \dfrac{2}{1 - 2}$

14.4   EXERCISES

Reading and Writing   *After reading this section, write out the answers to these questions. Use complete sentences.*

1. What is a geometric sequence?

2. What is the nth term of a geometric sequence?

3. What is a geometric series?

4. What is the formula for the sum of the first n terms of a geometric series?

5. What is the approximate value of r^n when n is large and $|r| < 1$?

6. What is the formula for the sum of an infinite geometric series?

Write a formula for the nth term of each geometric sequence. See Examples 1 and 2.

7. $\dfrac{1}{3}, 1, 3, 9, \ldots$

8. $\dfrac{1}{4}, 2, 16, \ldots$

9. $64, 8, 1, \ldots$

10. $100, 10, 1, \ldots$

11. $8, -4, 2, -1, \ldots$

12. $-9, 3, -1, \ldots$

13. $2, -4, 8, -16, \ldots$

14. $-\dfrac{1}{2}, 2, -8, 32, \ldots$

15. $-\dfrac{1}{3}, -\dfrac{1}{4}, -\dfrac{3}{16}, \ldots$

16. $-\dfrac{1}{4}, -\dfrac{1}{5}, -\dfrac{4}{25}, \ldots$

Write the first five terms of the geometric sequence with the given nth term. See Example 3.

17. $a_n = 2\left(\dfrac{1}{3}\right)^{n-1}$

18. $a_n = -5\left(\dfrac{1}{2}\right)^{n-1}$

19. $a_n = (-2)^{n-1}$

20. $a_n = \left(-\dfrac{1}{3}\right)^{n-1}$

21. $a_n = 2^{-n}$

22. $a_n = 3^{-n}$

23. $a_n = (0.78)^n$

24. $a_n = (-0.23)^n$

Find the required part of each geometric sequence. See Example 4.

25. Find the first term of the geometric sequence that has fourth term 40 and common ratio 2.

26. Find the first term of the geometric sequence that has fifth term 4 and common ratio $\frac{1}{2}$.

27. Find r for the geometric sequence that has $a_1 = 6$ and $a_4 = \frac{2}{9}$.

28. Find r for the geometric sequence that has $a_1 = 1$ and $a_4 = -27$.

29. Find a_4 for the geometric sequence that has $a_1 = -3$ and $r = \frac{1}{3}$.

30. Find a_5 for the geometric sequence that has $a_1 = -\frac{2}{3}$ and $r = -\frac{2}{3}$.

Find the sum of each geometric series. See Examples 5 and 6.

31. $\frac{1}{2} + \frac{1}{4} + \frac{1}{8} + \cdots + \frac{1}{512}$

32. $1 + \frac{1}{3} + \frac{1}{9} + \cdots + \frac{1}{81}$

33. $\frac{1}{2} - \frac{1}{4} + \frac{1}{8} - \frac{1}{16} + \frac{1}{32}$

34. $3 - 1 + \frac{1}{3} - \frac{1}{9} + \frac{1}{27} - \frac{1}{81}$

35. $30 + 20 + \frac{40}{3} + \cdots + \frac{1280}{729}$

36. $9 - 6 + 4 - \cdots - \frac{128}{243}$

37. $\sum_{i=1}^{10} 5(2)^{i-1}$

38. $\sum_{i=1}^{7} (10{,}000)(0.1)^{i-1}$

39. $\sum_{i=1}^{6} (0.1)^i$

40. $\sum_{i=1}^{5} (0.2)^i$

41. $\sum_{i=1}^{6} 100(0.3)^i$

42. $\sum_{i=1}^{7} 36(0.5)^i$

Find the sum of each infinite geometric series. See Examples 7 and 8.

43. $\frac{1}{8} + \frac{1}{16} + \frac{1}{32} + \cdots$

44. $\frac{1}{9} + \frac{1}{27} + \frac{1}{81} + \cdots$

45. $3 + 2 + \frac{4}{3} + \cdots$

46. $2 + 1 + \frac{1}{2} + \cdots$

47. $4 - 2 + 1 - \frac{1}{2} + \cdots$

48. $16 - 12 + 9 - \frac{27}{4} + \cdots$

49. $\sum_{i=1}^{\infty} (0.3)^i$

50. $\sum_{i=1}^{\infty} (0.2)^i$

51. $\sum_{i=1}^{\infty} 3(0.5)^{i-1}$

52. $\sum_{i=1}^{\infty} 7(0.4)^{i-1}$

53. $\sum_{i=1}^{\infty} 3(0.1)^i$

54. $\sum_{i=1}^{\infty} 6(0.1)^i$

55. $\sum_{i=1}^{\infty} 12(0.01)^i$

56. $\sum_{i=1}^{\infty} 72(0.01)^i$

Use the ideas of geometric series to solve each problem. See Examples 9 and 10.

57. *Retirement fund.* Suppose a deposit of $2000 is made at the beginning of each year for 45 years into an account paying 12% compounded annually. What is the amount of this annuity at the end of the forty-fifth year?

58. *World's largest mutual fund.* If you had invested $5000 at the beginning of each year for the last 10 years in Fidelity's Magellan fund you would have earned 18.97% compounded annually (Fidelity Investments, www.fidelity.com). Find the amount of this annuity at the end of the tenth year.

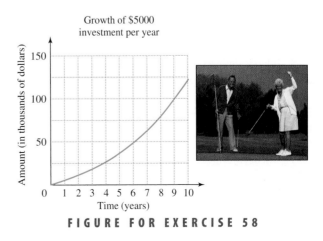

Growth of $5000
investment per year

FIGURE FOR EXERCISE 58

59. Big saver. Suppose you deposit one cent into your piggy bank on the first day of December and, on each day of December after that, you deposit twice as much as on the previous day. How much will you have in the bank after the last deposit?

60. Big family. Consider yourself, your parents, your grandparents, your great-grandparents, your great-great-grandparents, and so on, back to your grandparents with the word "great" used in front 40 times. What is the total number of people you are considering?

61. Total economic impact. In Exercise 43 of Section 14.1 we described a factory that spends $1 million annually in a community in which 80% of the money received is respent in the community. Economists assume the money is respent again and again at the 80% rate. The total economic impact of the factory is the total of all of this spending. Find an approximation for the total by using the formula for the sum of an infinite geometric series with a rate of 80%.

62. Less impact. Repeat Exercise 61, assuming money is respent again and again at the 50% rate.

GETTING MORE INVOLVED

63. Discussion. Which of the following sequences is not a geometric sequence? Explain your answer.
a) $1, 2, 4, \ldots$
b) $0.1, 0.01, 0.001, \ldots$
c) $-1, 2, -4, \ldots$
d) $2, 4, 6, \ldots$

64. Discussion. The repeating decimal number $0.44444\ldots$ can be written as

$$\frac{4}{10} + \frac{4}{100} + \frac{4}{1000} + \cdots,$$

an infinite geometric series. Find the sum of this geometric series.

65. Discussion. Write the repeating decimal number $0.24242424\ldots$ as an infinite geometric series. Find the sum of the geometric series.

14.5 BINOMIAL EXPANSIONS

In Chapter 5 you learned how to square a binomial. In this section you will study higher powers of binomials.

In this section

- Some Examples
- Obtaining the Coefficients
- The Binomial Theorem

Some Examples

We know that $(x + y)^2 = x^2 + 2xy + y^2$. To find $(x + y)^3$, we multiply $(x + y)^2$ by $x + y$:

$$(x + y)^3 = (x^2 + 2xy + y^2)(x + y)$$
$$= (x^2 + 2xy + y^2)x + (x^2 + 2xy + y^2)y$$
$$= x^3 + 2x^2y + xy^2 + x^2y + 2xy^2 + y^3$$
$$= x^3 + 3x^2y + 3xy^2 + y^3$$

The sum $x^3 + 3x^2y + 3xy^2 + y^3$ is called the **binomial expansion** of $(x + y)^3$. If we again multiply by $x + y$, we will get the binomial expansion of $(x + y)^4$. This method is rather tedious. However, if we examine these expansions, we can find a pattern and learn how to find binomial expansions without multiplying.

Consider the following binomial expansions:

$$(x + y)^0 = 1$$
$$(x + y)^1 = x + y$$
$$(x + y)^2 = x^2 + 2xy + y^2$$
$$(x + y)^3 = x^3 + 3x^2y + 3xy^2 + y^3$$
$$(x + y)^4 = x^4 + 4x^3y + 6x^2y^2 + 4xy^3 + y^4$$
$$(x + y)^5 = x^5 + 5x^4y + 10x^3y^2 + 10x^2y^3 + 5xy^4 + y^5$$

Observe that the exponents on the variable x are decreasing, whereas the exponents on the variable y are increasing, as we read from left to right. Also notice that the sum of the exponents in each term is the same for that entire line. For instance, in the fourth expansion the terms x^4, x^3y, x^2y^2, xy^3, and y^4 all have exponents with a sum of 4. If we continue the pattern, the expansion of $(x + y)^6$ will have seven terms containing x^6, x^5y, x^4y^2, x^3y^3, x^2y^4, xy^5, and y^6. Now we must find the pattern for the coefficients of these terms.

Obtaining the Coefficients

If we write out only the coefficients of the expansions that we already have, we can easily see a pattern. This triangular array of coefficients for the binomial expansions is called **Pascal's triangle.**

$$
\begin{array}{ccccccccccc}
 & & & & & 1 & & & & & \\
 & & & & 1 & & 1 & & & & \\
 & & & 1 & & 2 & & 1 & & & \\
 & & 1 & & 3 & & 3 & & 1 & & \\
 & 1 & & 4 & & 6 & & 4 & & 1 & \\
1 & & 5 & & 10 & & 10 & & 5 & & 1
\end{array}
$$

$(x + y)^0 = 1$

$(x + y)^1 = 1x + 1y$

$(x + y)^2 = 1x^2 + 2xy + 1y^2$

$(x + y)^3 = 1x^3 + 3x^2y + 3xy^2 + 1y^3$

$(x + y)^4 = 1x^4 + 4x^3y + 6x^2y^2 + 4xy^3 + 1y^4$

Coefficients in $(x + y)^5$

Notice that each line starts and ends with a 1 and that each entry of a line is the sum of the two entries above it in the previous line. For instance, $4 = 3 + 1$, and $10 = 6 + 4$. Following this pattern, the sixth and seventh lines of coefficients are

$$
\begin{array}{ccccccccccccc}
 & 1 & & 6 & & 15 & & 20 & & 15 & & 6 & & 1 \\
1 & & 7 & & 21 & & 35 & & 35 & & 21 & & 7 & & 1
\end{array}
$$

Pascal's triangle gives us an easy way to get the coefficients for the binomial expansion with small powers, but it is impractical for larger powers. For larger powers we use a formula involving **factorial notation.**

n! (n factorial)

If n is a positive integer, $n!$ (read "n factorial") is defined to be the product of all of the positive integers from 1 through n.

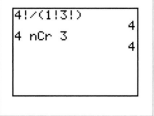

For example, $3! = 3 \cdot 2 \cdot 1 = 6$, and $5! = 5 \cdot 4 \cdot 3 \cdot 2 \cdot 1 = 120$. We also define $0!$ to be 1.

Before we state a general formula, consider how the coefficients for $(x + y)^4$ are found by using factorials:

$$\frac{4!}{4!\,0!} = \frac{4 \cdot 3 \cdot 2 \cdot 1}{4 \cdot 3 \cdot 2 \cdot 1 \cdot 1} = 1 \quad \text{Coefficient of } x^4 \text{ (or } x^4y^0)$$

$$\frac{4!}{3!\,1!} = \frac{4 \cdot 3 \cdot 2 \cdot 1}{3 \cdot 2 \cdot 1 \cdot 1} = 4 \quad \text{Coefficient of } 4x^3y$$

$$\frac{4!}{2!\,2!} = \frac{4 \cdot 3 \cdot 2 \cdot 1}{2 \cdot 1 \cdot 2 \cdot 1} = 6 \quad \text{Coefficient of } 6x^2y^2$$

$$\frac{4!}{1!\,3!} = \frac{4 \cdot 3 \cdot 2 \cdot 1}{1 \cdot 3 \cdot 2 \cdot 1} = 4 \quad \text{Coefficient of } 4xy^3$$

$$\frac{4!}{0!\,4!} = \frac{4 \cdot 3 \cdot 2 \cdot 1}{1 \cdot 4 \cdot 3 \cdot 2 \cdot 1} = 1 \quad \text{Coefficient of } y^4 \text{ (or } x^0y^4)$$

Note that each expression has $4!$ in the numerator, with factorials in the denominator corresponding to the exponents on x and y.

The Binomial Theorem

We now summarize these ideas in the **binomial theorem.**

The Binomial Theorem

In the expansion of $(x + y)^n$ for a positive integer n, there are $n + 1$ terms, given by the following formula:

$$(x + y)^n = \frac{n!}{n!\,0!}x^n + \frac{n!}{(n - 1)!\,1!}x^{n-1}y + \frac{n!}{(n - 2)!\,2!}x^{n-2}y^2 + \cdots + \frac{n!}{0!\,n!}y^n$$

The notation $\binom{n}{r}$ is often used in place of $\frac{n!}{(n - r)!\,r!}$ in the binomial expansion. Using this notation, we write the expansion as

$$(x + y)^n = \binom{n}{0}x^n + \binom{n}{1}x^{n-1}y + \binom{n}{2}x^{n-2}y^2 + \cdots + \binom{n}{n}y^n.$$

Another notation for $\frac{n!}{(n - r)!\,r!}$ is $_nC_r$. Using this notation, we have

$$(x + y)^n = {_nC_0}x^n + {_nC_1}x^{n-1}y + {_nC_2}x^{n-2}y^2 + \cdots + {_nC_n}y^n.$$

E X A M P L E   1　**Calculating the binomial coefficients**

Evaluate each expression.

a) $\dfrac{7!}{4!3!}$　　　　　　　　　　**b)** $\dfrac{10!}{8!2!}$

Solution

a) $\dfrac{7!}{4!3!} = \dfrac{7 \cdot 6 \cdot 5 \cdot \cancel{4} \cdot \cancel{3} \cdot \cancel{2} \cdot \cancel{1}}{\cancel{4} \cdot \cancel{3} \cdot \cancel{2} \cdot \cancel{1} \cdot 3 \cdot 2 \cdot 1} = \dfrac{7 \cdot 6 \cdot 5}{3 \cdot 2 \cdot 1} = 35$

b) $\dfrac{10!}{8!2!} = \dfrac{10 \cdot 9 \cdot \cancel{8} \cdot \cancel{7} \cdot \cancel{6} \cdot \cancel{5} \cdot \cancel{4} \cdot \cancel{3} \cdot \cancel{2} \cdot \cancel{1}}{\cancel{8} \cdot \cancel{7} \cdot \cancel{6} \cdot \cancel{5} \cdot \cancel{4} \cdot \cancel{3} \cdot \cancel{2} \cdot \cancel{1} \cdot 2 \cdot 1} = \dfrac{10 \cdot 9}{2 \cdot 1} = 45$ ▪

E X A M P L E   2

Using the binomial theorem

Write out the first three terms of $(x + y)^9$.

Solution

$$(x + y)^9 = \frac{9!}{9!0!}x^9 + \frac{9!}{8!1!}x^8y + \frac{9!}{7!2!}x^7y^2 + \cdots = x^9 + 9x^8y + 36x^7y^2 + \cdots$$ ▪

E X A M P L E   3

Using the binomial theorem

Write the binomial expansion for $(x^2 - 2a)^5$.

Solution

We expand a difference by writing it as a sum and using the binomial theorem:

$$(x^2 - 2a)^5 = (x^2 + (-2a))^5$$

$$= \frac{5!}{5!0!}(x^2)^5 + \frac{5!}{4!1!}(x^2)^4(-2a)^1 + \frac{5!}{3!2!}(x^2)^3(-2a)^2 + \frac{5!}{2!3!}(x^2)^2(-2a)^3$$

$$+ \frac{5!}{1!4!}(x^2)(-2a)^4 + \frac{5!}{0!5!}(-2a)^5$$

$$= x^{10} - 10x^8a + 40x^6a^2 - 80x^4a^3 + 80x^2a^4 - 32a^5$$ ▪

E X A M P L E   4

Finding a specific term

Find the fourth term of the expansion of $(a + b)^{12}$.

Solution

The variables in the first term are $a^{12}b^0$, those in the second term are $a^{11}b^1$, those in the third term are $a^{10}b^2$, and those in the fourth term are a^9b^3. So

$$\frac{12!}{9!3!}a^9b^3 = 220a^9b^3.$$

The fourth term is $220a^9b^3$.   ▪

Using the ideas of Example 4, we can write a formula for any term of a binomial expansion.

calculator

close-up

Because $_nC_r = \dfrac{n!}{(n - r)!r!}$, we have

$$_{12}C_9 = \frac{12!}{3!9!} \quad \text{and} \quad _{12}C_3 = \frac{12!}{9!3!}.$$

So there is more than one way to compute $12!/(9!3!)$:

```
12!/(9!3!)
              220
12 nCr 9
              220
12 nCr 3
              220
```

Formula for the kth Term of $(x + y)^n$

For k ranging from 1 to $n + 1$, the kth term of the expansion of $(x + y)^n$ is given by the formula

$$\frac{n!}{(n - k + 1)!(k - 1)!}x^{n-k+1}y^{k-1}.$$

E X A M P L E   5 **Finding a specific term**

Find the sixth term of the expansion of $(a^2 - 2b)^7$.

Solution

Use the formula for the kth term with $k = 6$ and $n = 7$:

$$\frac{7!}{(7 - 6 + 1)!(6 - 1)!}(a^2)^2(-2b)^5 = 21a^4(-32b^5) = -672a^4b^5 \qquad \blacksquare$$

We can think of the binomial expansion as a finite series. Using summation notation, we can write the binomial theorem as follows.

The Binomial Theorem (Using Summation Notation)

For any positive integer n,

$$(x + y)^n = \sum_{i=0}^{n} \frac{n!}{(n - i)!i!}x^{n-i}y^i \qquad \text{or} \qquad (x + y)^n = \sum_{i=0}^{n} \binom{n}{i}x^{n-i}y^i.$$

E X A M P L E   6 **Using summation notation**

Write $(a + b)^5$ using summation notation.

Solution

Use $n = 5$ in the binomial theorem:

$$(a + b)^5 = \sum_{i=0}^{5} \frac{5!}{(5 - i)!i!}a^{5-i}b^i \qquad \blacksquare$$

WARM-UPS

True or false? Explain your answer.

1. There are 12 terms in the expansion of $(a + b)^{12}$.
2. The seventh term of $(a + b)^{12}$ is a multiple of a^5b^7.
3. For all values of x, $(x + 2)^5 = x^5 + 32$.
4. In the expansion of $(x - 5)^8$ the signs of the terms alternate.
5. The eighth line of Pascal's triangle is

$$1 \; 8 \; 28 \; 56 \; 70 \; 56 \; 28 \; 8 \; 1.$$

6. The sum of the coefficients in the expansion of $(a + b)^4$ is 2^4.
7. $(a + b)^3 = \sum_{i=0}^{3} \frac{3!}{(3 - i)!i!}a^{3-i}b^i$
8. The sum of the coefficients in the expansion of $(a + b)^n$ is 2^n.
9. $0! = 1!$
10. $\dfrac{7!}{5!2!} = 21$

14.5 EXERCISES

Reading and Writing After reading this section, write out the answers to these questions. Use complete sentences.

1. What is a binomial expansion?

2. What is Pascal's triangle and how do you make it?

3. What does $n!$ mean?

4. What is the binomial theorem?

Evaluate each expression. See Example 1.

5. $\dfrac{5!}{2!\,3!}$ **6.** $\dfrac{6!}{5!\,1!}$

7. $\dfrac{8!}{5!\,3!}$ **8.** $\dfrac{9!}{2!\,7!}$

Use the binomial theorem to expand each binomial. See Examples 2 and 3.

9. $(r + t)^5$

10. $(r + t)^6$

11. $(m - n)^3$

12. $(m - n)^4$

13. $(x + 2a)^3$

14. $(a + 3b)^4$

15. $(x^2 - 2)^4$

16. $(x^2 - a^2)^5$

17. $(x - 1)^7$

18. $(x + 1)^6$

Write out the first four terms in the expansion of each binomial. See Examples 2 and 3.

19. $(a - 3b)^{12}$

20. $(x - 2y)^{10}$

21. $(x^2 + 5)^9$

22. $(x^2 + 1)^{20}$

23. $(x - 1)^{22}$

24. $(2x - 1)^8$

25. $\left(\dfrac{x}{2} + \dfrac{y}{3}\right)^{10}$

26. $\left(\dfrac{a}{2} + \dfrac{b}{5}\right)^8$

Find the indicated term of the binomial expansion. See Examples 4 and 5.

27. $(a + w)^{13}$, 6th term   **28.** $(m + n)^{12}$, 7th term

29. $(m - n)^{16}$, 8th term   **30.** $(a - b)^{14}$, 6th term

31. $(x + 2y)^8$, 4th term   **32.** $(3a + b)^7$, 4th term

33. $(2a^2 - b)^{20}$, 7th term   **34.** $(a^2 - w^2)^{12}$, 5th term

Write each expansion using summation notation. See Example 6.

35. $(a + m)^8$ **36.** $(z + w)^{13}$

37. $(a - 2x)^5$ **38.** $(w - 3m)^7$

GETTING MORE INVOLVED

39. *Discussion.* Find the trinomial expansion for $(a + b + c)^3$ by using $x = a$ and $y = b + c$ in the binomial theorem.

40. *Discussion.* What problem do you encounter when trying to find the fourth term in the binomial expansion for $(x + y)^{120}$? How can you overcome this problem? Find the fifth term in the binomial expansion for $(x - 2y)^{100}$.

COLLABORATIVE ACTIVITIES

Lotteries Are Series(ous)

Roberto and his brother-in-law Horatio each have a child who will be graduating from high school in 5 years. Each would like to buy his child a car for a graduation present. Horatio decides to buy two lottery tickets each week for the next 5 years, hoping to win and buy a new car for his child. The lottery tickets are $2.00 each at the local convenience store. Roberto, who doesn't believe in lotteries, decides to set aside $4.00 each week and to deposit this money in a savings account each quarter for the next 5 years to buy a used car. He finds a bank that will pay 5% yearly interest compounded quarterly.

1. Write a series that represents how much money Horatio will spend on lottery tickets over a 5-year period. Assume there are 52 weeks in a year. Compute the total amount of money spent on lottery tickets.

Grouping: 2 to 4 students per group
Topic: Sequences and series

2. Find the percent paid quarterly on the account. Find the number of compounding periods in 5 years. Find the amount of money Roberto will deposit each quarter.

3. Write a series to show how much money Roberto will have in his savings account at the end of 5 years.

4. Discuss in your groups the chances of Horatio winning the lottery compared to the sure savings Roberto has. Discuss the fact that Horatio would have to pay taxes on his lottery winnings and Roberto pays tax only on the interest earned. Do Horatio's chances increase the longer he buys lottery tickets?

WRAP-UP CHAPTER 14

SUMMARY

Sequences and Series

		Examples
Sequence	Finite—A function whose domain is the set of positive integers less than or equal to a fixed positive integer	$3,\ 5,\ 7,\ 9,\ 11$ $a_n = 2n + 1$ $1 \le n \le 5$
	Infinite—A function whose domain is the set of positive integers	$2,\ 4,\ 6,\ 8,\ \ldots$ $a_n = 2n$
Series	The indicated sum of a sequence	$2 + 4 + 6 + \cdots + 50$
Summation notation	$\displaystyle\sum_{i=1}^{n} a_i = a_1 + a_2 + a_3 + \cdots + a_n$	$\displaystyle\sum_{i=1}^{25} 2i = 2 + 4 + \cdots + 50$

Arithmetic Sequences and Series

		Examples
Arithmetic sequence	Each term after the first is obtained by adding a fixed amount to the previous term.	$6,\ 11,\ 16,\ 21, \ldots$ Fixed amount, d, is 5.
nth term	The nth term of an arithmetic sequence is $a_n = a_1 + (n - 1)d$.	If $a_1 = 6$ and $d = 5$, then $a_n = 6 + (n - 1)5$.
Arithmetic series	The sum of an arithmetic sequence	$6 + 11 + 16 + 21$

| Sum of first n terms | $S_n = \dfrac{n}{2}(a_1 + a_n)$ | $S_4 = \dfrac{4}{2}(6 + 21) = 54$ |

Geometric Sequences and Series

Examples

Geometric sequence Each term after the first is obtained by multiplying the preceding term by a constant.

2, 6, 18, 54, . . .
Constant, r, is 3.

nth term The nth term of a geometric sequence is
$$a_n = a_1 r^{n-1}.$$

$a_1 = 2, r = 3$
$a_n = 2 \cdot 3^{n-1}$

Geometric series (finite) The indicated sum of a finite geometric sequence.
$a_1 + a_1 r + a_1 r^2 + \cdots + a_1 r^{n-1}$

$2 + 6 + 18 + 54 + 162$

Sum of first n terms
$$S_n = \dfrac{a_1(1 - r^n)}{1 - r}$$

$a_1 = 2, r = 3, n = 5$
$S_5 = \dfrac{2(1 - 3^5)}{1 - 3} = 242$

Geometric series (infinite)
$a_1 + a_1 r + a_1 r^2 + a_1 r^3 + \cdots$

$8 + 4 + 2 + 1 + \dfrac{1}{2} + \cdots$

Sum of an infinite geometric series
$S = \dfrac{a_1}{1 - r}$, provided that $|r| < 1$

$a_1 = 8, r = \dfrac{1}{2}$
$S = \dfrac{8}{1 - \dfrac{1}{2}} = 16$

Factorial notation The notation $n!$ represents the product of the positive integers from 1 through n.

$5! = 5 \cdot 4 \cdot 3 \cdot 2 \cdot 1 = 120$

Binomial theorem
$$(x + y)^n = \dfrac{n!}{n!\,0!}x^n + \dfrac{n!}{(n-1)!\,1!}x^{n-1}y$$
$$+ \dfrac{n!}{(n-2)!\,2!}x^{n-2}y^2 + \cdots + \dfrac{n!}{0!\,n!}y^n$$

$(x + y)^3 = x^3 + 3x^2 y + 3xy^2 + y^3$

Using summation notation:
$$(x + y)^n = \sum_{i=0}^{n} \dfrac{n!}{(n-i)!\,i!}x^{n-i}y^i = \sum_{i=0}^{n} \binom{n}{i}x^{n-i}y^i$$

kth term of $(x + y)^n$
$$\dfrac{n!}{(n-k+1)!\,(k-1)!}x^{n-k+1}y^{k-1}$$

Third term of $(a + b)^{10}$ is
$\dfrac{10!}{8!\,2!}a^8 b^2 = 45a^8 b^2.$

ENRICHING YOUR MATHEMATICAL WORD POWER

For each mathematical term, choose the correct meaning.

1. **sequence**
   a. a list of numbers
   b. a procedure for getting the answer
   c. events that happen in order
   d. a linear function

2. **finite sequence**
   a. a short sequence
   b. a sequence of whole numbers
   c. a function whose domain is the set of positive integers
   d. a function whose domain is the set of positive integers less than or equal to a fixed positive integer

3. **infinite sequence**
   a. a short sequence
   b. a sequence of whole numbers
   c. a function whose domain is the set of positive integers
   d. a function whose domain is the set of positive integers less than or equal to a fixed positive integer

4. **series**
   a. a special sequence
   b. the indicated sum of the terms of a sequence
   c. a sequence of positive numbers
   d. a show with many episodes

5. **arithmetic sequence**
   a. a sequence in which each term after the first is obtained by adding a fixed amount to the previous term
   b. a sequence of fractions

   c. a sequence found in arithmetic
   d. a finite sequence

6. **geometric sequence**
   a. a sequence of rectangles
   b. a sequence of geometric formulas
   c. a sequence in which each term after the first is obtained by multiplying the preceding term by a constant
   d. a sequence in which the terms are geometric

7. **geometric series**
   a. a series of geometric shapes
   b. the indicated sum of an arithmetic sequence
   c. a series of ratios
   d. the indicated sum of a geometric sequence

8. **binomial expansion**
   a. the trinomial obtained when a binomial is stretched
   b. the expression obtained from raising a binomial to a whole number power
   c. the coefficients of a binomial
   d. the various powers of a binomial

9. **Pascal's triangle**
   a. an equilateral triangle
   b. a triangle formed by the graphs of three linear equations
   c. the right triangle in the Pythagorean theorem
   d. a triangular array of coefficients for the binomial expansions

10. **$n!$**
    a. the product of the positive integers from 1 through n
    b. the binomial coefficients
    c. the n vertices of Pascal's triangle
    d. 3.141592654

REVIEW EXERCISES

14.1 *List all terms of each finite sequence.*

1. $a_n = n^3$ for $1 \leq n \leq 5$

2. $b_n = (n - 1)^4$ for $1 \leq n \leq 4$

3. $c_n = (-1)^n(2n - 3)$ for $1 \leq n \leq 6$

4. $d_n = (-1)^{n-1}(3 - n)$ for $1 \leq n \leq 7$

Write the first three terms of the infinite sequence whose nth term is given.

5. $a_n = -\dfrac{1}{n}$

6. $b_n = \dfrac{(-1)^n}{n^2}$

7. $b_n = \dfrac{(-1)^{2n}}{2n + 1}$

8. $a_n = \dfrac{-1}{2n - 3}$

9. $c_n = \log_2(2^{n+3})$

10. $c_n = \ln(e^{2n})$

14.2 *Find the sum of each series.*

11. $\displaystyle\sum_{i=1}^{3} i^3$

12. $\displaystyle\sum_{i=0}^{4} 6$

13. $\displaystyle\sum_{n=1}^{5} n(n - 1)$

14. $\displaystyle\sum_{j=0}^{3} (-2)^j$

Write each series in summation notation. Use the index i, and let i begin at 1.

15. $\dfrac{1}{4} + \dfrac{1}{6} + \dfrac{1}{8} + \cdots$

16. $\dfrac{1}{3} + \dfrac{1}{4} + \dfrac{1}{5} + \cdots$

17. $0 + 1 + 4 + 9 + 16 + \cdots$

18. $-1 + 2 - 3 + 4 - 5 + 6 - \cdots$

19. $x_1 - x_2 + x_3 - x_4 + \cdots$

20. $-x^2 + x^3 - x^4 + x^5 - \cdots$

14.3 *Write the first four terms of the arithmetic sequence with the given nth term.*

21. $a_n = 6 + (n - 1)5$

22. $a_n = -7 + (n - 1)4$

23. $a_n = -20 + (n - 1)(-2)$

24. $a_n = 10 + (n - 1)(-2.5)$

25. $a_n = 1000n + 2000$

26. $a_n = -500n + 5000$

Write a formula for the nth term of each arithmetic sequence.

27. $\dfrac{1}{3}, \dfrac{2}{3}, 1, \dfrac{4}{3}, \ldots$

28. $10, 6, 2, -2, \ldots$

29. $2, 4, 6, 8, \ldots$

30. $20, 10, 0, -10, \ldots$

Find the sum of each arithmetic series.

31. $1 + 2 + 3 + \cdots + 24$

32. $-5 + (-2) + 1 + 4 + \cdots + 34$

33. $\dfrac{1}{6} + \dfrac{1}{2} + \dfrac{5}{6} + \dfrac{7}{6} + \cdots + \dfrac{11}{2}$

34. $-3 - 6 - 9 - 12 - \cdots - 36$

35. $\displaystyle\sum_{i=1}^{7} (2i - 3)$

36. $\displaystyle\sum_{i=1}^{6} [12 + (i - 1)5]$

14.4 *Write the first four terms of the geometric sequence with the given nth term.*

37. $a_n = 3\left(\dfrac{1}{2}\right)^{n-1}$

38. $a_n = 6\left(-\dfrac{1}{3}\right)^{n}$

39. $a_n = 2^{1-n}$

40. $a_n = 5(10)^{n-1}$

41. $a_n = 23(10)^{-2n}$

42. $a_n = 4(10)^{-n}$

Write a formula for the nth term of each geometric sequence.

43. $\dfrac{1}{2}, 3, 18, \ldots$

44. $-6, 2, -\dfrac{2}{3}, \dfrac{2}{9}, \ldots$

45. $\dfrac{7}{10}, \dfrac{7}{100}, \dfrac{7}{1000}, \ldots$

46. $2, 2x, 2x^2, 2x^3, \ldots$

Find the sum of each geometric series.

47. $\dfrac{1}{3} + \dfrac{1}{9} + \dfrac{1}{27} + \dfrac{1}{81}$

48. $2 + 4 + 8 + 16 + \cdots + 512$

49. $\displaystyle\sum_{i=1}^{10} 3(10)^{-i}$

50. $\displaystyle\sum_{i=1}^{5} (0.1)^{i}$

51. $\dfrac{1}{4} + \dfrac{1}{12} + \dfrac{1}{36} + \dfrac{1}{108} + \cdots$

52. $12 + (-6) + 3 + \left(-\dfrac{3}{2}\right) + \cdots$

53. $\displaystyle\sum_{i=1}^{\infty} 18\left(\dfrac{2}{3}\right)^{i-1}$

54. $\displaystyle\sum_{i=1}^{\infty} 9(0.1)^{i}$

14.5 *Use the binomial theorem to expand each binomial.*

55. $(m + n)^5$

56. $(2m - y)^4$

57. $(a^2 - 3b)^3$

58. $\left(\dfrac{x}{2} + 2a\right)^5$

Find the indicated term of the binomial expansion.

59. $(x + y)^{12}$, 5th term

60. $(x - 2y)^9$, 5th term

61. $(2a - b)^{14}$, 3rd term

62. $(a + b)^{10}$, 4th term

Write each expression in summation notation.

63. $(a + w)^7$

64. $(m - 3y)^9$

MISCELLANEOUS

Identify each sequence as an arithmetic sequence, a geometric sequence, or neither.

65. $1, 3, 6, 10, 15, \ldots$

66. $9, 12, 16, \dfrac{64}{3}, \ldots$

67. 9, 12, 15, 18, . . .

68. 2, 4, 8, 16, . . .

69. 0, 2, 4, 6, 8, . . .

70. 0, 3, 9, 27, 81, . . .

Solve each problem.

71. Find the common ratio for the geometric sequence with first term 6 and fourth term $\frac{1}{30}$.

72. Find the common difference for an arithmetic sequence with first term 6 and fourth term 36.

73. Write out all of the terms of the series
$$\sum_{i=1}^{5} \frac{(-1)^i}{i!}.$$

74. Write out the first eight rows of Pascal's triangle.

75. Write out all of the terms of the series
$$\sum_{i=0}^{5} \frac{5!}{(5-i)!\,i!}\, a^{5-i} b^i.$$

76. Write out all of the terms of the series
$$\sum_{i=0}^{8} \frac{8!}{(8-i)!\,i!}\, x^{8-i} y^i.$$

77. How many terms are there in the expansion of $(a + b)^{25}$?

78. Calculate $\frac{12!}{8!4!}$.

79. If \$3000 is deposited at the beginning of each year for 16 years into an account paying 10% compounded annually, then what is the value of the annuity at the end of the 16th year?

80. If \$3000 is deposited at the beginning of each year for 8 years into an account paying 10% compounded annually, then what is the value of the annuity at the end of the eighth year? How does the value of the annuity in this exercise compare to that of Exercise 79?

81. If one deposit of \$3000 is made into an account paying 10% compounded annually, then how much will be in the account at the end of 16 years? Note that a single deposit is not an annuity.

CHAPTER 14 TEST

List the first four terms of the sequence whose nth term is given.

1. $a_n = -10 + (n - 1)6$

2. $a_n = 5(0.1)^{n-1}$

3. $a_n = \dfrac{(-1)^n}{n!}$

4. $a_n = \dfrac{2n - 1}{n^2}$

Write a formula for the nth term of each sequence.

5. 7, 4, 1, −2, . . .

6. −25, 5, −1, $\dfrac{1}{5}$, . . .

7. 2, −4, 6, −8, 10, −12, . . .

8. 1, 4, 9, 16, 25, . . .

Write out all of the terms of each series.

9. $\displaystyle\sum_{i=1}^{5} (2i + 3)$

10. $\displaystyle\sum_{i=1}^{6} 5(2)^{i-1}$

11. $\displaystyle\sum_{i=0}^{4} \frac{4!}{(4-i)!\,i!}\, m^{4-i} q^i$

Find the sum of each series.

12. $\displaystyle\sum_{i=1}^{20} (6 + 3i)$

13. $\displaystyle\sum_{i=1}^{5} 10\left(\frac{1}{2}\right)^{i-1}$

14. $\displaystyle\sum_{i=1}^{\infty} 0.35(0.93)^{i-1}$

15. $2 + 4 + 6 + \cdots + 200$

16. $\dfrac{1}{4} + \dfrac{1}{8} + \dfrac{1}{16} + \cdots$

17. $2 + 1 + \dfrac{1}{2} + \dfrac{1}{4} + \cdots + \dfrac{1}{128}$

Solve each problem.

18. Find the common ratio for the geometric sequence that has first term 3 and fifth term 48.

19. Find the common difference for the arithmetic sequence that has first term 1 and twelfth term 122.

20. Find the fifth term in the expansion of $(r - t)^{15}$.

21. Find the fourth term in the expansion of $(a^2 - 2b)^8$.

22. If \$800 is deposited at the beginning of each year for 25 years into an account earning 10% compounded annually, then what is the value of this annuity at the end of the 25th year?

Let $f(x) = x^2 - 3$, $g(x) = 2x - 1$, $h(x) = 2^x$, and $m(x) = \log_2(x)$.
Find the following.

1. $f(3)$ **2.** $f(n)$

3. $f(x + h)$ **4.** $f(x) - g(x)$

5. $g(f(3))$ **6.** $(f \circ g)(2)$

7. $m(16)$ **8.** $(h \circ m)(32)$

9. $h(-1)$ **10.** $h^{-1}(8)$

11. $m^{-1}(0)$ **12.** $(m \circ h)(x)$

Solve each variation problem.

13. If y varies directly as x, and $y = -6$ when $x = 4$, find y when $x = 9$.

14. If a varies inversely as b, and $a = 2$ when $b = -4$, find a when $b = 3$.

15. If y varies directly as w and inversely as t, and $y = 16$ when $w = 3$ and $t = -4$, find y when $w = 2$ and $t = 3$.

16. If y varies jointly as h and the square of r, and $y = 12$ when $h = 2$ and $r = 3$, find y when $h = 6$ and $r = 2$.

Sketch the graph of each inequality or system of inequalities.

17. $x > 3$ and $x + y < 0$ **18.** $|x - y| \geq 2$

19. $y < -2x + 3$ and $y > 2^x$ **20.** $|y + 2x| < 1$

21. $x^2 + y^2 < 4$ **22.** $x^2 - y^2 < 1$

23. $y < \log_2(x)$ **24.** $x^2 + 2y < 4$

25. $\dfrac{x^2}{4} + \dfrac{y^2}{9} < 1$ and $y > x^2$

Perform the indicated operation and simplify. Write answers with positive exponents.

26. $\dfrac{a}{b} + \dfrac{b}{a}$ **27.** $1 - \dfrac{3}{y}$

28. $\dfrac{x - 2}{x^2 - 9} - \dfrac{x - 4}{x^2 - 2x - 3}$ **29.** $\dfrac{x^2 - 16}{2x + 8} \cdot \dfrac{4x^2 + 16x + 64}{x^3 - 16}$

30. $\dfrac{(a^2b)^3}{(ab^2)^4} \cdot \dfrac{ab^3}{a^{-4}b^2}$ **31.** $\dfrac{x^2y}{(xy)^3} \div \dfrac{xy^2}{x^2y^4}$

Simplify.

32. $8^{2/3}$

33. $16^{-5/4}$

34. $-4^{1/2}$

35. $27^{-2/3}$

36. -2^{-3}

37. $2^{-3/5} \cdot 2^{-7/5}$

38. $5^{-2/3} \div 5^{1/3}$

39. $(9^{1/2} + 4^{1/2})^2$

Solve.

40. ***Predicting heights of preschoolers.*** A popular model in pediatrics for predicting the height of preschoolers is the JENNS model. According to this model, if $h(x)$ is the height [in centimeters (cm)] at age x (in years) for $0.25 \leq x \leq 6$, then

$$h(x) = 79.041 + 6.39x - e^{(3.261 - 0.993x)}.$$

a) Find the predicted height in inches for a child of age 4 years, 3 months.

b) If you have a graphing calculator, graph the function as shown in the accompanying figure.

c) Use your graphing calculator to find the age to the nearest tenth of a year for a child who has a height of 80 cm.

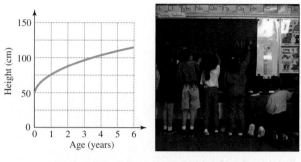

FIGURE FOR EXERCISE 40

Geometric Figures and Formulas

Triangle: A three-sided figure
Area: $A = \frac{1}{2}bh$, Perimeter: $P = a + b + c$
Sum of the measures of the angles is $180°$.

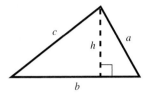

30-60-90 Right Triangle: The side opposite $30°$ is one-half the length of the hypotenuse.

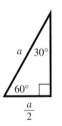

Trapezoid: A four-sided figure with one pair of parallel sides
Area: $A = \frac{1}{2}h(b_1 + b_2)$

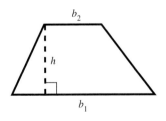

Right Triangle: A triangle with a $90°$ angle
Area $= \frac{1}{2}ab$, Perimeter: $P = a + b + c$
Pythagorean Theorem: $c^2 = a^2 + b^2$

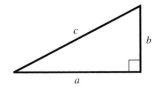

Parallelogram: A four-sided figure with opposite sides parallel
Area: $A = bh$

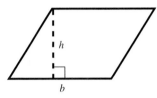

Rhombus: A four-sided figure with four equal sides
Perimeter: $P = 4a$

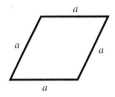

Rectangle: A four-sided figure with four right angles
Area: $A = LW$
Perimeter: $P = 2L + 2W$

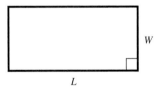

Circle
Area: $A = \pi r^2$
Circumference: $C = 2\pi r$
Diameter: $d = 2r$

Right Circular Cone
Volume: $V = \frac{1}{3}\pi r^2 h$
Lateral Surface Area: $S = \pi r \sqrt{r^2 + h^2}$

Rectangular Solid

Volume: $V = LWH$

Surface Area: $A = 2LW + 2WH + 2LH$

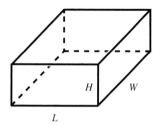

Square: A four-sided figure with four equal sides and four right angles

Area: $A = s^2$

Perimeter: $P = 4s$

Sphere

Volume: $V = \frac{4}{3}\pi r^3$

Surface Area: $S = 4\pi r^2$

Right Circular Cylinder

Volume: $V = \pi r^2 h$

Lateral Surface Area: $S = 2\pi rh$

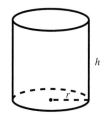

Geometric Terms

An **angle** is a union of two rays with a common endpoint.

A **right angle** is an angle with a measure of 90°.

Two angles are **complementary** if the sum of their measures is 90°.

An **isosceles triangle** is a triangle that has two equal sides.

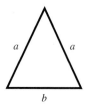

Similar triangles are triangles that have the same shape. Their corresponding angles are equal and corresponding sides are proportional:

$$\frac{a}{d} = \frac{b}{e} = \frac{c}{f}$$

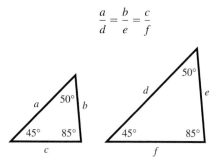

An **acute angle** is an angle with a measure between 0° and 90°.

An **obtuse angle** is an angle with a measure between 90° and 180°.

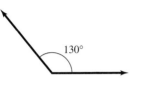

Two angles are **supplementary** if the sum of their measures is 180°.

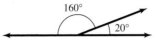

An **equilateral triangle** is a triangle that has three equal sides.

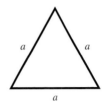

n	n^2	$\sqrt{n}$	n	n^2	$\sqrt{n}$	n	n^2	$\sqrt{n}$
1	1	1.0000	41	1681	6.4031	81	6561	9.0000
2	4	1.4142	42	1764	6.4807	82	6724	9.0554
3	9	1.7321	43	1849	6.5574	83	6889	9.1104
4	16	2.0000	44	1936	6.6332	84	7056	9.1652
5	25	2.2361	45	2025	6.7082	85	7225	9.2195
6	36	2.4495	46	2116	6.7823	86	7396	9.2736
7	49	2.6458	47	2209	6.8557	87	7569	9.3274
8	64	2.8284	48	2304	6.9282	88	7744	9.3808
9	81	3.0000	49	2401	7.0000	89	7921	9.4340
10	100	3.1623	50	2500	7.0711	90	8100	9.4868
11	121	3.3166	51	2601	7.1414	91	8281	9.5394
12	144	3.4641	52	2704	7.2111	92	8464	9.5917
13	169	3.6056	53	2809	7.2801	93	8649	9.6437
14	196	3.7417	54	2916	7.3485	94	8836	9.6954
15	225	3.8730	55	3025	7.4162	95	9025	9.7468
16	256	4.0000	56	3136	7.4833	96	9216	9.7980
17	289	4.1231	57	3249	7.5498	97	9409	9.8489
18	324	4.2426	58	3364	7.6158	98	9604	9.8995
19	361	4.3589	59	3481	7.6811	99	9801	9.9499
20	400	4.4721	60	3600	7.7460	100	10000	10.0000
21	441	4.5826	61	3721	7.8102	101	10201	10.0499
22	484	4.6904	62	3844	7.8740	102	10404	10.0995
23	529	4.7958	63	3969	7.9373	103	10609	10.1489
24	576	4.8990	64	4096	8.0000	104	10816	10.1980
25	625	5.0000	65	4225	8.0623	105	11025	10.2470
26	676	5.0990	66	4356	8.1240	106	11236	10.2956
27	729	5.1962	67	4489	8.1854	107	11449	10.3441
28	784	5.2915	68	4624	8.2462	108	11664	10.3923
29	841	5.3852	69	4761	8.3066	109	11881	10.4403
30	900	5.4772	70	4900	8.3666	110	12100	10.4881
31	961	5.5678	71	5041	8.4261	111	12321	10.5357
32	1024	5.6569	72	5184	8.4853	112	12544	10.5830
33	1089	5.7446	73	5329	8.5440	113	12769	10.6301
34	1156	5.8310	74	5476	8.6023	114	12996	10.6771
35	1225	5.9161	75	5625	8.6603	115	13225	10.7238
36	1296	6.0000	76	5776	8.7178	116	13456	10.7703
37	1369	6.0828	77	5929	8.7750	117	13689	10.8167
38	1444	6.1644	78	6084	8.8318	118	13924	10.8628
39	1521	6.2450	79	6241	8.8882	119	14161	10.9087
40	1600	6.3246	80	6400	8.9443	120	14400	10.9545

This table gives the common logarithms for numbers between 1 and 10. The common logarithms for other numbers can be found by using scientific notation and the properties of logarithms. For example, to find log(1230) we write

$$\log(1230) = \log(1.23 \times 10^3) = \log(1.23) + \log(10^3)$$
$$= 0.0899 + 3 = 3.0899$$

n	0	1	2	3	4	5	6	7	8	9
1.0	.0000	.0043	.0086	.0128	.0170	.0212	.0253	.0294	.0334	.0374
1.1	.0414	.0453	.0492	.0531	.0569	.0607	.0645	.0682	.0719	.0755
1.2	.0792	.0828	.0864	.0899	.0934	.0969	.1004	.1038	.1072	.1106
1.3	.1139	.1173	.1206	.1239	.1271	.1303	.1335	.1367	.1399	.1430
1.4	.1461	.1492	.1523	.1553	.1584	.1614	.1644	.1673	.1703	.1732
1.5	.1761	.1790	.1818	.1847	.1875	.1903	.1931	.1959	.1987	.2014
1.6	.2041	.2068	.2095	.2122	.2148	.2175	.2201	.2227	.2253	.2279
1.7	.2304	.2330	.2355	.2380	.2405	.2430	.2455	.2480	.2504	.2529
1.8	.2553	.2577	.2601	.2625	.2648	.2672	.2695	.2718	.2742	.2765
1.9	.2788	.2810	.2833	.2856	.2878	.2900	.2923	.2945	.2967	.2989
2.0	.3010	.3032	.3054	.3075	.3096	.3118	.3139	.3160	.3181	.3201
2.1	.3222	.3243	.3263	.3284	.3304	.3324	.3345	.3365	.3385	.3404
2.2	.3424	.3444	.3464	.3483	.3502	.3522	.3541	.3560	.3579	.3598
2.3	.3617	.3636	.3655	.3674	.3692	.3711	.3729	.3747	.3766	.3784
2.4	.3802	.3820	.3838	.3856	.3874	.3892	.3909	.3927	.3945	.3962
2.5	.3979	.3997	.4014	.4031	.4048	.4065	.4082	.4099	.4116	.4133
2.6	.4150	.4166	.4183	.4200	.4216	.4232	.4249	.4265	.4281	.4298
2.7	.4314	.4330	.4346	.4362	.4378	.4393	.4409	.4425	.4440	.4456
2.8	.4472	.4487	.4502	.4518	.4533	.4548	.4564	.4579	.4594	.4609
2.9	.4624	.4639	.4654	.4669	.4683	.4698	.4713	.4728	.4742	.4757
3.0	.4771	.4786	.4800	.4814	.4829	.4843	.4857	.4871	.4886	.4900
3.1	.4914	.4928	.4942	.4955	.4969	.4983	.4997	.5011	.5024	.5038
3.2	.5051	.5065	.5079	.5092	.5105	.5119	.5132	.5145	.5159	.5172
3.3	.5185	.5198	.5211	.5224	.5237	.5250	.5263	.5276	.5289	.5302
3.4	.5315	.5328	.5340	.5353	.5366	.5378	.5391	.5403	.5416	.5428
3.5	.5441	.5453	.5465	.5478	.5490	.5502	.5514	.5527	.5539	.5551
3.6	.5563	.5575	.5587	.5599	.5611	.5623	.5635	.5647	.5658	.5670
3.7	.5682	.5694	.5705	.5717	.5729	.5740	.5752	.5763	.5775	.5786
3.8	.5798	.5809	.5821	.5832	.5843	.5855	.5866	.5877	.5888	.5899
3.9	.5911	.5922	.5933	.5944	.5955	.5966	.5977	.5988	.5999	.6010
4.0	.6021	.6031	.6042	.6053	.6064	.6075	.6085	.6096	.6107	.6117
4.1	.6128	.6138	.6149	.6160	.6170	.6180	.6191	.6201	.6212	.6222
4.2	.6232	.6243	.6253	.6263	.6274	.6284	.6294	.6304	.6314	.6325
4.3	.6335	.6345	.6355	.6365	.6375	.6385	.6395	.6405	.6415	.6425
4.4	.6435	.6444	.6454	.6464	.6474	.6484	.6493	.6503	.6513	.6522
4.5	.6532	.6542	.6551	.6561	.6571	.6580	.6590	.6599	.6609	.6618
4.6	.6628	.6637	.6646	.6656	.6665	.6675	.6684	.6693	.6702	.6712
4.7	.6721	.6730	.6739	.6749	.6758	.6767	.6776	.6785	.6794	.6803
4.8	.6812	.6821	.6830	.6839	.6848	.6857	.6866	.6875	.6884	.6893
4.9	.6902	.6911	.6920	.6928	.6937	.6946	.6955	.6964	.6972	.6981
n	0	1	2	3	4	5	6	7	8	9

n	0	1	2	3	4	5	6	7	8	9
5.0	.6990	.6998	.7007	.7016	.7024	.7033	.7042	.7050	.7059	.7067
5.1	.7076	.7084	.7093	.7101	.7110	.7118	.7126	.7135	.7143	.7152
5.2	.7160	.7168	.7177	.7185	.7193	.7202	.7210	.7218	.7226	.7235
5.3	.7243	.7251	.7259	.7267	.7275	.7284	.7292	.7300	.7308	.7316
5.4	.7324	.7332	.7340	.7348	.7356	.7364	.7372	.7380	.7388	.7396
5.5	.7404	.7412	.7419	.7427	.7435	.7443	.7451	.7459	.7466	.7474
5.6	.7482	.7490	.7497	.7505	.7513	.7520	.7528	.7536	.7543	.7551
5.7	.7559	.7566	.7574	.7582	.7589	.7597	.7604	.7612	.7619	.7627
5.8	.7634	.7642	.7649	.7657	.7664	.7672	.7679	.7686	.7694	.7701
5.9	.7709	.7716	.7723	.7731	.7738	.7745	.7752	.7760	.7767	.7774
6.0	.7782	.7789	.7796	.7803	.7810	.7818	.7825	.7832	.7839	.7846
6.1	.7853	.7860	.7868	.7875	.7882	.7889	.7896	.7903	.7910	.7917
6.2	.7924	.7931	.7938	.7945	.7952	.7959	.7966	.7973	.7980	.7987
6.3	.7993	.8000	.8007	.8014	.8021	.8028	.8035	.8041	.8048	.8055
6.4	.8062	.8069	.8075	.8082	.8089	.8096	.8102	.8109	.8116	.8122
6.5	.8129	.8136	.8142	.8149	.8156	.8162	.8169	.8176	.8182	.8189
6.6	.8195	.8202	.8209	.8215	.8222	.8228	.8235	.8241	.8248	.8254
6.7	.8261	.8267	.8274	.8280	.8287	.8293	.8299	.8306	.8312	.8319
6.8	.8325	.8331	.8338	.8344	.8351	.8357	.8363	.8370	.8376	.8382
6.9	.8388	.8395	.8401	.8407	.8414	.8420	.8426	.8432	.8439	.8445
7.0	.8451	.8457	.8463	.8470	.8476	.8482	.8488	.8494	.8500	.8506
7.1	.8513	.8519	.8525	.8531	.8537	.8543	.8549	.8555	.8561	.8567
7.2	.8573	.8579	.8585	.8591	.8597	.8603	.8609	.8615	.8621	.8627
7.3	.8633	.8639	.8645	.8651	.8657	.8663	.8669	.8675	.8681	.8686
7.4	.8692	.8698	.8704	.8710	.8716	.8722	.8727	.8733	.8739	.8745
7.5	.8751	.8756	.8762	.8768	.8774	.8779	.8785	.8791	.8797	.8802
7.6	.8808	.8814	.8820	.8825	.8831	.8837	.8842	.8848	.8854	.8859
7.7	.8865	.8871	.8876	.8882	.8887	.8893	.8899	.8904	.8910	.8915
7.8	.8921	.8927	.8932	.8938	.8943	.8949	.8954	.8960	.8965	.8971
7.9	.8976	.8982	.8987	.8993	.8998	.9004	.9009	.9015	.9020	.9025
8.0	.9031	.9036	.9042	.9047	.9053	.9058	.9063	.9069	.9074	.9079
8.1	.9085	.9090	.9096	.9101	.9106	.9112	.9117	.9122	.9128	.9133
8.2	.9138	.9143	.9149	.9154	.9159	.9165	.9170	.9175	.9180	.9186
8.3	.9191	.9196	.9201	.9206	.9212	.9217	.9222	.9227	.9232	.9238
8.4	.9243	.9248	.9253	.9258	.9263	.9269	.9274	.9279	.9284	.9289
8.5	.9294	.9299	.9304	.9309	.9315	.9320	.9325	.9330	.9335	.9340
8.6	.9345	.9350	.9355	.9360	.9365	.9370	.9375	.9380	.9385	.9390
8.7	.9395	.9400	.9405	.9410	.9415	.9420	.9425	.9430	.9435	.9440
8.8	.9445	.9450	.9455	.9460	.9465	.9469	.9474	.9479	.9484	.9489
8.9	.9494	.9499	.9504	.9509	.9513	.9518	.9523	.9528	.9533	.9538
9.0	.9542	.9547	.9552	.9557	.9562	.9566	.9571	.9576	.9581	.9586
9.1	.9590	.9595	.9600	.9605	.9609	.9614	.9619	.9624	.9628	.9633
9.2	.9638	.9643	.9647	.9652	.9657	.9661	.9666	.9671	.9675	.9680
9.3	.9685	.9689	.9694	.9699	.9703	.9708	.9713	.9717	.9722	.9727
9.4	.9731	.9736	.9741	.9745	.9750	.9754	.9759	.9763	.9768	.9773
9.5	.9777	.9782	.9786	.9791	.9795	.9800	.9805	.9809	.9814	.9818
9.6	.9823	.9827	.9832	.9836	.9841	.9845	.9850	.9854	.9859	.9863
9.7	.9868	.9872	.9877	.9881	.9886	.9890	.9894	.9899	.9903	.9908
9.8	.9912	.9917	.9921	.9926	.9930	.9934	.9939	.9943	.9948	.9952
9.9	.9956	.9961	.9965	.9969	.9974	.9978	.9983	.9987	.9991	.9996
n	0	1	2	3	4	5	6	7	8	9

ANSWERS TO SELECTED EXERCISES

Chapter 1

Section 1.1 Warm-ups T T F F T F T T F F

1. The integers are the numbers in the set $\{\ldots, -3, -2, -1, 0, 1, 2, 3, \ldots\}$.

3. A rational number is a ratio of integers and an irrational number is not.

5. The number a is larger than b if a lies to the right of b on the number line.

7. 6   **9.** 0   **11.** -2 **13.** -12

15. $1, 2, 3, 4, 5$

17. $0, 1, 2, 3, 4$

19. $0, 1, 2, 3, 4$

21. $1, 2, 3, 4, 5, \ldots$

23. $1, 2, 3, 4, 5, \ldots$

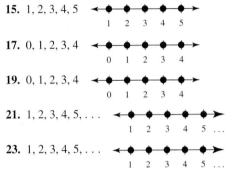

25. True   **27.** False   **29.** True   **31.** True   **33.** True
35. False   **37.** 6   **39.** 0   **41.** 7   **43.** 9   **45.** 45
47. $\frac{3}{4}$ **49.** 5.09   **51.** -16 **53.** $-\frac{5}{2}$ **55.** 2   **57.** 3
59. -9 **61.** 16   **63.** True   **65.** True   **67.** True
69. What is the probability that a tossed coin turns up heads?
71. If a is negative, then $-a$ and $|-a|$ are positive. The rest are negative.

Section 1.2 Warm-ups T T F T T T T T F T

1. If two fractions are identical when reduced to lowest terms, then they are equivalent fractions.

3. To reduce a fraction means to find an equivalent fraction that has no factor common to the numerator and denominator.

5. Convert a fraction to a decimal by dividing the denominator into the numerator.

7. $\frac{6}{8}$ **9.** $\frac{32}{12}$ **11.** $\frac{10}{2}$ **13.** $\frac{75}{100}$ **15.** $\frac{30}{100}$ **17.** $\frac{70}{42}$

19. $\frac{1}{2}$ **21.** $\frac{2}{3}$ **23.** 3   **25.** $\frac{1}{2}$ **27.** 2   **29.** $\frac{3}{8}$

31. $\frac{13}{21}$ **33.** $\frac{12}{13}$ **35.** $\frac{10}{27}$ **37.** 5   **39.** $\frac{7}{10}$ **41.** $\frac{7}{13}$

43. $\frac{3}{5}$ **45.** $\frac{1}{6}$ **47.** 3   **49.** $\frac{1}{15}$ **51.** 4   **53.** $\frac{4}{5}$

55. $\frac{3}{40}$ **57.** $\frac{1}{2}$ **59.** $\frac{1}{3}$ **61.** $\frac{1}{4}$ **63.** $\frac{7}{12}$ **65.** $\frac{1}{12}$

67. $\frac{19}{24}$ **69.** $\frac{11}{72}$ **71.** $\frac{199}{48}$ **73.** $60\%, 0.6$

75. $\frac{9}{100}, 0.09$ **77.** $8\%, \frac{2}{25}$ **79.** $0.75, 75\%$

81. $\frac{1}{50}, 0.02$ **83.** $\frac{1}{100}, 1\%$ **85.** 3   **87.** 1

89. $\frac{71}{96}$ **91.** $\frac{17}{120}$ **93.** $\frac{65}{16}$ **95.** $\frac{69}{4}$

97. $\frac{13}{12}$ **99.** $\frac{1}{8}$ **101.** $54\frac{7}{8}$, up 0.2%

103. a) 1.3 yd^3 **b)** $36\frac{11}{24} \text{ ft}^3$ or $1\frac{227}{648} \text{ yd}^3$

107. Each daughter gets $3 \text{ km}^2 \div 4$ or a $\frac{3}{4} \text{ km}^2$ piece of the farm. Divide the farm into 12 equal squares. Give each daughter an L-shaped piece consisting of 3 of those 12 squares.

Section 1.3 Warm-ups T T T F F F T F T F

1. We studied addition and subtraction of signed numbers.

3. Two numbers are additive inverses of each other if their sum is zero.

5. To find the sum of two numbers with unlike signs, subtract their absolute values. The answer is given the sign of the number with the larger absolute value.

7. 13   **9.** -13 **11.** -1.15 **13.** $-\frac{1}{2}$ **15.** 0   **17.** 0

19. 2   **21.** -6 **23.** 5.6   **25.** -2.9 **27.** $-\frac{1}{4}$

29. $8 + (-2)$ **31.** $4 + (-12)$ **33.** $-3 + 8$
35. $8.3 + (1.5)$ **37.** -4 **39.** -10 **41.** 11   **43.** -11

45. $-\frac{1}{4}$ **47.** $\frac{3}{4}$ **49.** 7   **51.** 0.93   **53.** 9.3

55. -5.03 **57.** 3   **59.** -9 **61.** -120 **63.** 78
65. -27 **67.** -7 **69.** -201 **71.** -322
73. -15.97 **75.** -2.92 **77.** -3.73 **79.** 3.7

81. $\frac{3}{20}$ **83.** $\frac{7}{24}$ **85.** -3.49 **87.** -0.3422

89. -48.84 **91.** -8.85 **93.** $-\$8.85$ **95.** $-7°C$
97. When adding signed numbers, we add or subtract only positive numbers which are the absolute values of the original numbers. We then determine the appropriate sign for the answer.
99. The distance between x and y is given by either $|x - y|$ or $|y - x|$.

Section 1.4 Warm-ups T F T F T T T F T F

1. We learned to multiply and divide signed numbers.

3. To find the product of signed numbers, multiply their absolute values and then affix a negative sign if the two original numbers have opposite signs.

5. To find the quotient of nonzero numbers divide their absolute values and then affix a negative sign if the two original numbers have opposite signs.

7. -27 **9.** 132 **11.** $-\dfrac{1}{3}$ **13.** -0.3 **15.** 144 **17.** 0

19. -1 **21.** 3 **23.** $-\dfrac{2}{3}$ **25.** $\dfrac{5}{6}$ **27.** undefined

29. 0 **31.** -80 **33.** 0.25 **35.** -100 **37.** 27

39. -3 **41.** -4 **43.** -30 **45.** 19 **47.** -0.18

49. 0.3 **51.** -6 **53.** 1.5 **55.** 22 **57.** $-\dfrac{1}{3}$

59. -164.25 **61.** 1529.41 **63.** 16 **65.** -8 **67.** 0

69. 0 **71.** -3.9 **73.** -40 **75.** 0.4 **77.** 0.4

79. -0.2 **81.** -7.5 **83.** $-\dfrac{1}{30}$ **85.** $-\dfrac{1}{10}$ **87.** 7.562

89. 19.35 **91.** 0 **93.** undefined

Section 1.5 Warm-ups F F T F F F F T F T

1. An arithmetic expression is the result of writing numbers in a meaningful combination with the ordinary operations of arithmetic.

3. An exponential expression is an expression of the form a^n.

5. The order of operations tells us the order in which to perform operations when grouping symbols are omitted.

7. -4 **9.** 1 **11.** -8 **13.** -7 **15.** -16 **17.** -4

19. 4^4 **21.** $(-5)^4$ **23.** $(-y)^3$ **25.** $\left(\dfrac{3}{7}\right)^5$ **27.** $5 \cdot 5 \cdot 5$

29. $b \cdot b$ **31.** $\left(-\dfrac{1}{2}\right)\left(-\dfrac{1}{2}\right)\left(-\dfrac{1}{2}\right)\left(-\dfrac{1}{2}\right)\left(-\dfrac{1}{2}\right)$

33. $(0.22)(0.22)(0.22)(0.22)$ **35.** 81 **37.** 0 **39.** 625

41. -216 **43.** $100,000$ **45.** -0.001 **47.** $\dfrac{1}{8}$ **49.** $\dfrac{1}{4}$

51. -64 **53.** -4096 **55.** 27 **57.** -13 **59.** 36

61. 18 **63.** -19 **65.** -17 **67.** -44 **69.** 18

71. -78 **73.** 0 **75.** 27 **77.** 1 **79.** 8 **81.** 7

83. 11 **85.** 111 **87.** 21 **89.** -1 **91.** -11 **93.** 9

95. 16 **97.** 28 **99.** 121 **101.** -73 **103.** 25

105. 0 **107.** -2 **109.** 12 **111.** 82 **113.** -54

115. -79 **117.** -24 **119.** 41.92 **121.** 184.643547

123. 8.0548 **125.** 299.3 million

127. $(-5)^3 = -(5^3) = -5^3 = -1 \cdot 5^3$ and $-(-5)^3 = 5^3$

Section 1.6 Warm-ups T F T F T F F F T F

1. An algebraic expression is the result of combining numbers and variables with the operations of arithmetic in some meaningful way.

3. An algebraic expression is named according to the last operation to be performed.

5. An equation is a sentence that expresses equality between two algebraic expressions.

7. Difference **9.** Cube **11.** Sum **13.** Difference

15. Product **17.** Square **19.** The difference of x^2 and a^2

21. The square of $x - a$ **23.** The quotient of $x - 4$ and 2

25. The difference of $\dfrac{x}{2}$ and 4 **27.** The cube of ab

29. $2x + 3y$ **31.** $8 - 7x$ **33.** $(a + b)^2$

35. $(x + 9)(x + 12)$ **37.** $\dfrac{x - 7}{7 - x}$ **39.** 3 **41.** 3

43. 16 **45.** -9 **47.** -3 **49.** -8 **51.** $-\dfrac{2}{3}$ **53.** 4

55. -1 **57.** 1 **59.** -4 **61.** 0 **63.** Yes **65.** No

67. Yes **69.** Yes **71.** Yes **73.** Yes **75.** No

77. No **79.** $5x + 3x = 8x$ **81.** $3(x + 2) = 12$

83. $\dfrac{x}{3} = 5x$ **85.** $(a + b)^2 = 9$

87. 14.65 **89.** 37.12 **91.** 169.3 cm, 41 cm

93. $12, 19.5, 26.5, 30.5$ **95.** 920 feet

99. For the square of the sum consider $(2 + 3)^2 = 5^2 = 25$. For the sum of the squares consider $2^2 + 3^2 = 4 + 9 = 13$. So $(2 + 3)^2 \neq 2^2 + 3^2$.

Section 1.7 Warm-ups F F T F T T T T T T

1. The commutative property says that $a + b = b + a$ and the associative property says that $(a + b) + c = a + (b + c)$.

3. Factoring is the process of writing an expression or number as a product.

5. The properties help us to understand the operations and how they are related to each other.

7. $r + 9$ **9.** $3(x + 2)$ **11.** $-5x + 4$ **13.** $6x$

15. $-2(x - 4)$ **17.** $4 - 8y$ **19.** $4w^2$ **21.** $3a^2b$

23. $9x^3z$ **25.** -3 **27.** -10 **29.** -21 **31.** 0.6

33. -22.4 **35.** $3x - 15$ **37.** $2a + at$

39. $-3w + 18$ **41.** $-20 + 4y$ **43.** $-a + 7$

45. $-t - 4$ **47.** $2(m + 6)$ **49.** $4(x - 1)$

51. $4(y - 4)$ **53.** $4(a + 2)$ **55.** 2

57. $-\dfrac{1}{5}$ **59.** $\dfrac{1}{7}$ **61.** 1 **63.** -4 **65.** $\dfrac{2}{5}$

67. Commutative property of multiplication

69. Distributive property

71. Associative property of multiplication

73. Inverse properties

75. Commutative property of multiplication

77. Identity property **79.** Distributive property

81. Inverse property **83.** Multiplication property of 0

85. Distributive property **87.** $y + a$

89. $(5a)w$ **91.** $\dfrac{1}{2}(x + 1)$ **93.** $3(2x + 5)$

95. 1 **97.** 0 **99.** $\dfrac{100}{33}$

101. a) 45 bricks/hour **b)** Bricklayer

103. a) 2.63 people/second, $1,591,160$ people/week

105. The perimeter is twice the sum of the length and width.

107. a) Commutative **b)** Not commutative

Section 1.8 Warm-ups T F T T F F F F F T

1. Like terms are terms with the same variables and exponents.

3. We can add or subtract like terms.

5. If a negative sign precedes a set of parentheses, then signs for all terms in the parentheses are changed when the parentheses are removed.

7. 7000 **9.** 1 **11.** 356 **13.** 350 **15.** 36

17. $36,000$ **19.** 0 **21.** 98 **23.** $11w$ **25.** $3x$

27. $5x$ **29.** $-a$ **31.** $-2a$ **33.** $10 - 6t$ **35.** $8x^2$

37. $-4x + 2x^2$ **39.** $-7mw^2$ **41.** $12h$ **43.** $-18b$

45. $-9m^2$ **47.** $12d^2$ **49.** y^2 **51.** $-15ab$

53. $-6a - 3ab$ **55.** $-k + k^2$ **57.** y **59.** $-3y$
61. y **63.** $2y^2$ **65.** $2a - 1$ **67.** $3x - 2$
69. $-2x + 1$ **71.** $8 - y$ **73.** $m - 6$ **75.** $w - 5$
77. $8x + 15$ **79.** $5x - 1$ **81.** $-2a - 1$ **83.** $5a - 2$
85. $3m - 18$ **87.** $-3x - 7$ **89.** $0.95x - 0.5$
91. $-3.2x + 12.71$ **93.** $4x - 4$ **95.** $2y + 4$
97. $2y + m - 1$ **99.** 3 **101.** $0.15x - 0.4$
103. $-14k + 23$ **105.** 45
107. a) $0.28x - 5700.5$ **b)** $\$16,700$
   c) $\$43,000$ **d)** $\$240,000$
109. $4x + 80$, 200 feet
111. If $x = 5$, then $1/2 \cdot 5 = \frac{1}{2} \cdot 5 = 2.5$ because we do division and multiplication from left to right.

Enriching Word Power
1. c **2.** b **3.** a **4.** d **5.** b **6.** d **7.** a **8.** d
9. c **10.** a

Chapter 1 Review
1. $0, 1, 2, 10$ **3.** $-2, 0, 1, 2, 10$ **5.** $-\sqrt{5}, \pi$ **7.** True
9. False **11.** False **13.** True **15.** $\frac{17}{24}$ **17.** 6
19. $\frac{3}{7}$ **21.** $\frac{14}{3}$ **23.** $\frac{13}{12}$ **25.** 2 **27.** -13 **29.** -7
31. -7 **33.** 11.95 **35.** -0.05 **37.** $-\frac{1}{6}$ **39.** $-\frac{11}{15}$
41. -15 **43.** 4 **45.** 5 **47.** $\frac{1}{6}$ **49.** -0.3
51. -0.24 **53.** 1 **55.** 66 **57.** 49 **59.** 41
61. 1 **63.** 50 **65.** -135 **67.** -2 **69.** -16
71. 16 **73.** 5 **75.** 9 **77.** 7 **79.** $-\frac{1}{3}$ **81.** 1
83. -9 **85.** Yes **87.** No **89.** Yes **91.** No
93. Distributive property **95.** Inverse property
97. Identity property **99.** Associative property of addition
101. Commutative property of multiplication
103. Inverse property **105.** Identity property
107. $-a + 12$ **109.** $6a^2 - 6a$
111. $-12t + 39$ **113.** $-0.9a - 0.57$ **115.** $-0.05x - 4$
117. $27x^2 + 6x + 5$ **119.** $-2a$ **121.** $x^2 + 4x - 3$
123. 0 **125.** 8 **127.** -21 **129.** $\frac{1}{2}$ **131.** -0.5
133. -1 **135.** $x + 2$ **137.** $4 + 2x$ **139.** $2x$
141. $-4x + 8$ **143.** $6x$ **145.** x **147.** $8x$
149. 18 memberships per hour

Chapter 1 Test
1. $0, 8$ **2.** $-3, 0, 8$ **3.** $-3, -\frac{1}{4}, 0, 8$
4. $-\sqrt{3}, \sqrt{5}, \pi$ **5.** -21 **6.** -4 **7.** 9 **8.** -7
9. -0.95 **10.** -56 **11.** 978 **12.** 13 **13.** -1
14. 0 **15.** 9740 **16.** $-\frac{7}{24}$ **17.** -20 **18.** $-\frac{1}{6}$
19. -39 **20.** Distributive property
21. Commutative property of multiplication
22. Associative property of addition **23.** Inverse property
24. Identity property **25.** Multiplication property of 0

26. $3(x + 10)$ **27.** $7(w - 1)$ **28.** $6x + 6$
29. $4x - 2$ **30.** $7x - 3$ **31.** $0.9x + 7.5$
32. $14a^2 + 5a$ **33.** $x + 2$ **34.** $4t$ **35.** $54x^2y^2$
36. 41 **37.** 5 **38.** -12 **39.** No **40.** Yes
41. Yes **42.** 9 deliveries per hour
43. $3.66R - 0.06A + 82.205$, 168.905 cm

Chapter 2

Section 2.1 Warm-ups T T F T F T T T T T
1. The addition property of equality says that adding the same number to each side of an equation does not change the solution to the equation.
3. The multiplication property of equality says that multiplying both sides of an equation by the same nonzero number does not change the solution to the equation.
5. You can check a solution to an equation by using a suspected solution to evaluate both sides.
7. $\{1\}$ **9.** $\{9\}$ **11.** $\{1\}$ **13.** $\left\{\frac{2}{3}\right\}$ **15.** $\{-9\}$
17. $\{-19\}$ **19.** $\left\{\frac{1}{4}\right\}$ **21.** $\{0\}$ **23.** $\{-5\}$ **25.** $\{-4\}$
27. $\{3\}$ **29.** $\left\{\frac{1}{4}\right\}$ **31.** $\{-8\}$ **33.** $\{1.8\}$ **35.** $\left\{\frac{2}{3}\right\}$
37. $\left\{\frac{1}{2}\right\}$ **39.** $\{-5\}$ **41.** $\{5\}$ **43.** $\{1.25\}$ **45.** $\left\{\frac{1}{4}\right\}$
47. $\{-2\}$ **49.** $\{120\}$ **51.** $\left\{\frac{5}{9}\right\}$ **53.** $\left\{-\frac{1}{2}\right\}$ **55.** $\{-8\}$
57. $\left\{\frac{1}{3}\right\}$ **59.** $\{-3.4\}$ **61.** $\{99\}$ **63.** $\{-7\}$ **65.** $\{9\}$
67. $\{8\}$ **69.** $\{5\}$ **71.** $\{-5\}$ **73.** $\{-8\}$ **75.** $\{2\}$
77. $\left\{\frac{1}{6}\right\}$ **79.** $\left\{-\frac{1}{3}\right\}$ **81.** $\{44\}$ **83.** $\left\{\frac{3}{4}\right\}$ **85.** $\{7\}$
87. $\{-14\}$ **89.** $\left\{\frac{3}{8}\right\}$
91. 200 packs per capita **93.** 2875 stocks

Section 2.2 Warm-ups T T T F T F T T T T
1. We can solve $ax + b = 0$ with the addition property and the multiplication property of equality.
3. Use the multiplication property of equality to solve $-x = 8$.
5. $\{2\}$ **7.** $\{-2\}$ **9.** $\left\{\frac{2}{3}\right\}$ **11.** $\left\{-\frac{5}{2}\right\}$ **13.** $\{6\}$
15. $\{12\}$ **17.** $\left\{\frac{1}{2}\right\}$ **19.** $\left\{-\frac{1}{6}\right\}$ **21.** $\{4\}$ **23.** $\left\{\frac{5}{6}\right\}$
25. $\{4\}$ **27.** $\{-5\}$ **29.** $\{34\}$ **31.** $\{9\}$ **33.** $\{1.2\}$
35. $\{3\}$ **37.** $\{4\}$ **39.** $\{-3\}$ **41.** $\left\{\frac{1}{2}\right\}$ **43.** $\{30\}$
45. $\{6\}$ **47.** $\{-2\}$ **49.** $\{18\}$ **51.** $\{0\}$ **53.** $\{-2\}$
55. $\left\{\frac{7}{3}\right\}$ **57.** $\{1\}$ **59.** $\{-6\}$ **61.** $\{-12\}$ **63.** $\{-4\}$
65. $\{-13\}$ **67.** $\{1.7\}$ **69.** $\{2\}$ **71.** $\{4.6\}$
73. $\{8\}$ **75.** $\{34\}$ **77.** $\{6\}$ **79.** $\{0\}$ **81.** $\{-10\}$
83. $\{18\}$ **85.** $\{-20\}$ **87.** $\{-3\}$ **89.** $\{-4.3\}$
91. 17 hrs **93.** $20°C$ **95.** 9 ft **97.** $\$14,550$

Section 2.3 Warm-ups T T F F F T T T F T T

1. An identity is an equation that is satisfied by all numbers for which both sides are defined.

3. An inconsistent equation has no solutions.

5. If an equation involves decimals we usually multiply each side by a power of 10 to eliminate all decimals.

7. All real numbers, identity   **9.** $\varnothing$, inconsistent

11. $\{0\}$, conditional   **13.** $\varnothing$, inconsistent

15. $\varnothing$, inconsistent   **17.** $\{1\}$, conditional

19. $\varnothing$, inconsistent   **21.** All real numbers, identity

23. All nonzero real numbers, identity

25. All real numbers, identity

27. $\{7\}$ **29.** $\{24\}$ **31.** $\{16\}$ **33.** $\{-12\}$ **35.** $\{60\}$

37. $\{24\}$ **39.** $\{90\}$ **41.** $\{6\}$ **43.** $\{-2\}$

45. $\{80\}$ **47.** $\{60\}$ **49.** $\{200\}$ **51.** $\{800\}$

53. $\left\{\dfrac{9}{2}\right\}$ **55.** $\{3\}$ **57.** $\{25\}$ **59.** $\{-2\}$ **61.** $\{-3\}$

63. $\{5\}$ **65.** $\{-10\}$ **67.** $\{2\}$ **69.** $\{-4\}$

71. All real numbers   **73.** All real numbers   **75.** $\{100\}$

77. $\left\{-\dfrac{3}{2}\right\}$ **79.** $\{30\}$ **81.** $\{6\}$ **83.** $\{0.5\}$

85. $\{19,608\}$ **87.** \$128,000

89. a) \$140,000 **b)** \$139,210

Section 2.4 Warm-ups F F F F F T T T F T

1. A formula is an equation with two or more variables.

3. To solve for a variable means to find an equivalent equation in which the variable is isolated.

5. To find the value of a variable in a formula, we can solve for the variable then insert values for the other variables, or insert values for the other variables and then solve for the variable.

7. $R = \dfrac{D}{T}$ **9.** $D = \dfrac{C}{\pi}$ **11.** $P = \dfrac{I}{rt}$ **13.** $C = \dfrac{5}{9}(F - 32)$

15. $h = \dfrac{2A}{b}$ **17.** $L = \dfrac{P - 2W}{2}$ **19.** $a = 2A - b$

21. $r = \dfrac{S - P}{Pt}$ **23.** $a = \dfrac{2A - hb}{h}$ **25.** $x = \dfrac{b - a}{2}$

27. $x = -7a$ **29.** $x = 12 - a$ **31.** $x = 7ab$

33. $y = -x - 9$ **35.** $y = -x + 6$ **37.** $y = 2x - 2$

39. $y = 3x + 4$ **41.** $y = -\dfrac{1}{2}x + 2$ **43.** $y = x - \dfrac{1}{2}$

45. $y = 3x - 14$ **47.** $y = \dfrac{1}{2}x$ **49.** $y = \dfrac{3}{2}x + 6$

51. 2   **53.** 7   **55.** $-\dfrac{9}{5}$ **57.** 1   **59.** 1.33

61. 4%   **63.** 4 years   **65.** 7 yards   **67.** 225 feet

69. \$300   **71.** 20%   **73.** 160 feet   **75.** 24 cubic feet

77. 4 inches   **79.** 8 feet   **81.** 12 inches

83. 640 milligrams, age 13   **85.** 3.75 milliliters

87. $L = F\sqrt{S} - 2D + 5.688$

Section 2.5 Warm-ups T T T F T F F F T F

1. To express addition we use words such as plus, sum, increased by, and more than.

3. Complementary angles have degree measures with a sum of 90°.

5. Distance is the product of rate and time.

7. $x + 3$ **9.** $x - 3$ **11.** $5x$ **13.** $0.1x$ **15.** $\dfrac{x}{3}$

17. $\dfrac{1}{3}x$ **19.** x and $x + 15$ **21.** x and $6 - x$

23. x and $-4 - x$ **25.** x and $x + 3$ **27.** x and $0.05x$

29. x and $1.30x$ **31.** x and $90 - x$ **33.** x and $120 - x$

35. n and $n + 2$ **37.** x and $x + 1$

39. $x, x + 2$, and $x + 4$ **41.** $x, x + 2, x + 4$, and $x + 6$

43. $3x$ miles   **45.** $0.25q$ dollars   **47.** $\dfrac{x}{20}$ hour

49. $\dfrac{x - 100}{12}$ meters per second   **51.** $5x$ square meters

53. $2w + 2(w + 3)$ inches   **55.** $150 - x$ feet

57. $2x + 1$ feet   **59.** $x(x + 5)$ square meters

61. $0.18(x + 1000)$ **63.** $\dfrac{16.50}{x}$ dollars per pound

65. $90 - x$ degrees   **67.** $x(x + 5) = 8$

69. $x - 0.07x = 84{,}532$ **71.** $500x = 100$

73. $0.05x + 0.10(x + 2) = 3.80$ **75.** $x + 5 = 13$

77. $x + (x + 1) + (x + 2) = 42$ **79.** $x(x + 1) = 182$

81. $0.12x = 3000$ **83.** $0.05x = 13$

85. $x(x + 5) = 126$ **87.** $5n + 10(n - 1) = 95$

89. $x + x - 38 = 180$

91. a) $r + 0.6(220 - (30 + r)) = 144$

b) Target heart rate increases as resting heart rate increases.

93. $x(x + 3) = 24$ **95.** $w(w - 4) = 24$

Section 2.6 Warm-ups F T T F F T T T F T

1. In this section we studied number, geometric, and uniform motion problems.

3. Uniform motion is motion at a constant rate of speed.

5. Complementary angles are angles whose degree measures have a sum of 90°.

7. 46, 47, 48   **9.** 75, 77   **11.** 47, 48, 49, 50

13. Length 50 meters, width 25 meters

15. Width 42 inches, length 46 inches

17. 13 inches   **19.** 35°   **21.** 65 miles per hour

23. 55 miles per hour   **25.** 4 hours, 2048 miles

27. Raiders 32, Vikings 14   **29.** 3 hours, 106 miles

31. Crawford 1906, Wayne 1907, Stewart 1908

33. 7 ft, 7 ft, 16 ft

Section 2.7 Warm-ups T T F T F T

1. We studied discount, investment, and mixture problems in this section.

3. The product of the rate and the original price gives the amount of discount. The original price minus the discount is the sale price.

5. A table helps us to organize the information given in a problem.

7. \$320   **9.** \$400

11. \$80,000   **13.** \$30.24

15. 100 Fund \$10,000, 101 Fund \$13,000

17. Fidelity \$14,000, Price \$11,000

19. 30 gallons

21. 20 liters of 5% alcohol, 10 liters of 20% alcohol

23. 55,700　　**25.** $15,000　　**27.** 75%　　**29.** 600

31. 42 private rooms, 30 semiprivate rooms　　**33.** 12 pounds

35. 4 nickels, 6 dimes　　**37.** 800 gallons　　**39.** $\frac{2}{3}$ gal

Enriching Word Power

1. b　　**2.** d　　**3.** c　　**4.** c　　**5.** d　　**6.** d　　**7.** a　　**8.** b
9. c　　**10.** d

Chapter 2 Review

1. {35}　　**3.** {−6}　　**5.** {−7}　　**7.** {13}　　**9.** {7}
11. {2}　　**13.** {7}　　**15.** {0}　　**17.** {−8}

19. ∅, inconsistent　　**21.** All real numbers, identity
23. All nonzero real numbers, identity　　**25.** 24, conditional
27. 80, conditional　　**29.** 1000, conditional

31. $\left\{\frac{1}{4}\right\}$　　**33.** $\left\{\frac{21}{8}\right\}$　　**35.** $\left\{-\frac{4}{5}\right\}$　　**37.** {4}　　**39.** {24}

41. {−100}　　**43.** $x = -\frac{b}{a}$　　**45.** $x = \frac{b+2}{a}$

47. $x = \frac{V}{LW}$　　**49.** $x = -\frac{b}{3}$　　**51.** $y = -\frac{5}{2}x + 3$

53. $y = -\frac{1}{2}x + 4$　　**55.** $y = -2x + 16$　　**57.** −13

59. $-\frac{2}{5}$　　**61.** 17　　**63.** $x + 9$　　**65.** x and $x + 8$

67. $0.65x$　　**69.** $x(x+5) = 98$　　**71.** $2x = 3(x − 10)$
73. $x + x + 2 + x + 4 = 88$
75. $t + 2t + t − 10 = 180$　　**77.** 77, 79, 81
79. Betty 45 mph, Lawanda 60 mph
81. Wanda $36,000, husband $30,000
83. a) $543　**b)** $25　　**85.** 400　　**87.** 31°

Chapter 2 Test

1. $\left\{\frac{7}{6}\right\}$　　**2.** {−7}　　**3.** {2}　　**4.** {−9}　　**5.** ∅

6. All real numbers　　**7.** {700}　　**8.** {2}
9. Inconsistent　　**10.** Identity

11. Conditional　　**12.** $y = \frac{2}{3}x − 3$　　**13.** $a = \frac{m+w}{P}$

14. $x = \frac{3}{2-a}$　　**15.** 14 meters　　**16.** 9 in.

17. 150 liters　　**18.** 30°, 60°, 90°

Making Connections Chapters 1–2

1. $8x$　　**2.** $15x^2$　　**3.** $2x + 1$　　**4.** $4x − 7$
5. $-2x + 13$　　**6.** 60　　**7.** 72　　**8.** −10
9. $-2x^3$　　**10.** −1　　**11.** 1　　**12.** All real numbers
13. 0　　**14.** 1　　**15.** 2　　**16.** 2　　**17.** $\frac{13}{2}$　　**18.** 200

19. a) $13,600　**b)** $10,000　**c)** $12,000

Chapter 3

Section 3.1 Warm-ups F F T F F F T T T T
1. An inequality is a sentence that expresses inequality between two algebraic expressions.
3. If a is less than b, then a lies to the left of b on the number line.

5. When you multiply or divide by a negative number, the inequality symbol is reversed.
7. F　　**9.** T　　**11.** T　　**13.** T　　**15.** Yes
17. No　　**19.** No
21. $(-\infty, -1]$　　**23.** $(20, \infty)$

25. $[3, \infty)$　　**27.** $(-\infty, 2.3)$

29. $(1, \infty)$　　**31.** $(-\infty, -3]$　　**33.** $(-\infty, 5)$
35. $[-4, \infty)$　　**37.** >　　**39.** >　　**41.** ≤　　**43.** >
45. $(-2, \infty)$　　**47.** $[-4, \infty)$

49. $(5, \infty)$　　**51.** $[-3, \infty)$

53. $(13, \infty)$　　**55.** $[-1, \infty)$

57. $(-\infty, 4]$　　**59.** $\left(\frac{2}{3}, \infty\right)$

61. $\left(\frac{13}{3}, \infty\right)$　　**63.** $(-\infty, \infty)$

65. ∅　　**67.** $(1, \infty)$

69. $(-\infty, 2.397]$　　**71.** $(-\infty, -17)$

73. $x =$ Tony's height, $x > 6$ feet
75. $s =$ Wilma's salary, $s < $80,000
77. $v =$ speed of the Concorde, $v \le 1450$ mph
79. $a =$ amount Julie can afford, $a \le $400
81. $b =$ Burt's height, $b \le 5$ feet
83. $t =$ Tina's hourly wage, $t \le $8.20
85. $x =$ price of car, $x < $9100
87. $x =$ price of truck, $x \ge $9100.92　　**89.** $[0, 40)$

91. a) Increasing   **b)** 2005
93. x = final exam score, $x \geq 77$
95. x = the price of A-Mart jeans, $x < \$16.67$

Section 3.2   Warm-ups T T F T T T F T F T

1. A compound inequality consists of two inequalities joined with the words "and" or "or."
3. A compound inequality using or is true when either one or the other or both inequalities is true.
5. The inequality $a < b < c$ means that $a < b$ and $b < c$.
7. No   **9.** Yes   **11.** No
13. No   **15.** Yes   **17.** Yes
19.

21.

23.   **25.** $\varnothing$

27.   **29.** $\varnothing$

31. $(-\infty, 1) \cup (10, \infty)$ **33.** $(9, \infty)$

35. $(-6, \infty)$ **37.** $(1, 4]$

39. $(-\infty, \infty)$ **41.** $\varnothing$

43. $(4, 7)$ **45.** $[-3, 2)$

47. $\left(-\dfrac{7}{3}, 3\right]$ **49.** $(-1, 5)$

51. $[2, 3]$

53. $(2, \infty)$ **55.** $(-\infty, 5)$ **57.** $[2, 4]$ **59.** $(-\infty, \infty)$
61. $\varnothing$ **63.** $[4, 5)$ **65.** $[1, 6]$ **67.** $x > 2$
69. $x < 3$ **71.** $x > 2$ or $x \leq -1$
73. $-2 \leq x < 3$ **75.** $x \geq -3$ **77.** $(5, 7)$
79. $(-1, 1] \cup (10, \infty)$ **81.** $(-3, 3)$
83. x = final exam score, $73 \leq x \leq 86.5$
85. x = price of truck, $\$11,033 \leq x \leq \$13,811$
87. x = number of cigarettes on the run, $4 \leq x \leq 18$
89. a) 1,144,700   **b)** 2002   **c)** 2013   **d)** 2005
91. $-b < x < -a$ provided $a < b$
93. a) $(12, 32)$ **b)** $(-20, 10]$ **c)** $(0, 9)$ **d)** $[-3, -1]$

Section 3.3   Warm-ups T F F T F T F F T F

1. Absolute value of a number is the number's distance from 0 on the number line.
3. Since both 4 and -4 are four units from 0, $|x| = 4$ has two solutions.
5. Since the distance from 0 for every number on the number line is greater than or equal to 0, $|x| \geq 0$.
7. $\{-5, 5\}$ **9.** $\{2, 4\}$ **11.** $\{-3, 9\}$ **13.** $\left\{-\dfrac{8}{3}, \dfrac{16}{3}\right\}$
15. $\{12\}$ **17.** $\{-20, 80\}$ **19.** $\varnothing$ **21.** $\{0, 5\}$
23. $\{0.143, 1.298\}$ **25.** $\{-2, 2\}$ **27.** $\{-11, 5\}$
29. $\{0, 3\}$ **31.** $\left\{-6, \dfrac{4}{3}\right\}$ **33.** $\{1, 3\}$ **35.** $(-\infty, \infty)$
37. $|x| < 2$ **39.** $|x| > 3$ **41.** $|x| \leq 1$ **43.** $|x| \geq 2$
45. No   **47.** Yes   **49.** No   **51.** Yes
53. $(-\infty, -6) \cup (6, \infty)$ **55.** $(-3, 3)$

57. $(-\infty, -1] \cup [5, \infty)$ **59.** $\left(-\dfrac{1}{2}, \dfrac{9}{2}\right)$

61. $[-2, 12]$ **63.** $\left(-\infty, -\dfrac{9}{2}\right] \cup \left[\dfrac{15}{2}, \infty\right)$

65. $(-\infty, 2) \cup (2, \infty)$

67. $(-\infty, \infty)$ **69.** $\varnothing$ **71.** $(-\infty, \infty)$
73. $(-\infty, -3) \cup (-1, \infty)$ **75.** $(-4, 4)$ **77.** $(-1, 1)$
79. $(0.255, 0.847)$ **81.** $\varnothing$ **83.** $(4, 5)$
85. 1401 or 1429   **87.** Between 121 and 133 pounds
89. a) 1 second   **b)** 1 second   **c)** $0.5 < t < 1.5$
91. a) $(-\infty, \infty)$ **b)** $(-\infty, \infty)$ **c)** all reals except $n = 0$

Enriching Word Power

1. c   **2.** d   **3.** d

Chapter 3 Review

1. $(-3, \infty)$ **3.** $(0, \infty)$

5. $(-\infty, -8]$ **7.** $\left(-\infty, \dfrac{11}{2}\right)$

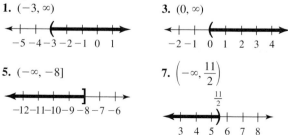

9. $[48, \infty)$

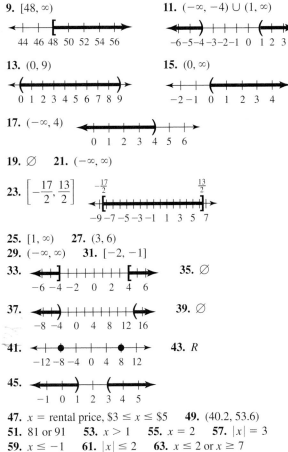

11. $(-\infty, -4) \cup (1, \infty)$

13. $(0, 9)$

15. $(0, \infty)$

17. $(-\infty, 4)$

19. $\varnothing$ **21.** $(-\infty, \infty)$

23. $\left[-\dfrac{17}{2}, \dfrac{13}{2}\right]$

25. $[1, \infty)$ **27.** $(3, 6)$
29. $(-\infty, \infty)$ **31.** $[-2, -1]$

33.   **35.** $\varnothing$

37.   **39.** $\varnothing$

41.   **43.** R

45.

47. x = rental price, $\$3 \le x \le \$5$ **49.** $(40.2, 53.6)$
51. 81 or 91   **53.** $x > 1$ **55.** $x = 2$ **57.** $|x| = 3$
59. $x \le -1$ **61.** $|x| \le 2$ **63.** $x \le 2$ or $x \ge 7$
65. $|x| > 3$ **67.** $5 < x < 7$ **69.** $|x| > 0$

Chapter 3 Test
1. $-3 < x \le 2$ **2.** $x > 1$ **3.** $[3, \infty)$ **4.** $(1, 6)$
5. $(-\infty, 5) \cup (9, \infty)$ **6.** $(-3, 3)$ **7.** $(-\infty, -2) \cup (2, \infty)$
8. $(-1, \infty)$ **9.** $[4, 8]$

10. $(-\infty, -7) \cup (13, \infty)$

11. $(5, \infty)$ **12.** $\left(-8, -\dfrac{1}{2}\right)$

13. $[-5, 3)$ **14.** $(-\infty, 15)$

15. $\{2, 5\}$ **16.** $(-\infty, \infty)$ **17.** $\varnothing$ **18.** $\{2.5\}$ **19.** $\varnothing$
20. R **21.** At most $\$2000$ **22.** Less than 6 feet
23. $|x - 28{,}000| > 3000$, Brenda makes more than $\$31{,}000$ or less than $\$25{,}000$.

Making Connections Chapters 1–3
1. $11x$ **2.** $30x^2$ **3.** $3x + 1$ **4.** $4x - 3$ **5.** 899
6. 961 **7.** 841 **8.** 25 **9.** 13 **10.** -25 **11.** 5
12. -4 **13.** $-2x + 13$ **14.** 60 **15.** 72 **16.** -9
17. $-3x^3$ **18.** 1 **19.** $\{0\}$ **20.** R **21.** $\{0\}$ **22.** $\{1\}$
23. $\left\{-\dfrac{1}{3}\right\}$ **24.** $\{1\}$ **25.** R **26.** $\{1000\}$ **27.** $\left\{-\dfrac{17}{5}, 1\right\}$
28. a) $87{,}500$ **b)** $C_r = 4500 + 0.06x$, $C_b = 8000 + 0.02x$
c) $87{,}500$ **d)** Buying is $\$1300$ cheaper.
e) $(75{,}000, 100{,}000)$

Chapter 4

Section 4.1 Warm-ups F F F F T T T F F T
1. An ordered pair is a pair of numbers in which there is a first number and a second number, usually written as (a, b).
3. The origin is the point of intersection of the x-axis and y-axis.
5. A linear equation in two variables is an equation of the form $Ax + By = C$, where A and B are not both zero.
7. $(0, 9), (5, 24), (2, 15)$ **9.** $(0, -7), (-4, 5), (-2, -1)$
11. $(0, 5), (10, -115)$, and $(-1, 17)$
13. $(3, 0), (0, -2), (12, 6)$ **15.** $(5, -3), (5, 5), (5, 0)$
17–31 odd   **33.**

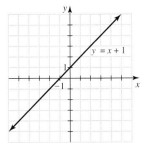

35.   **37.**

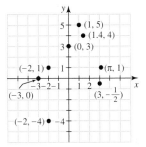

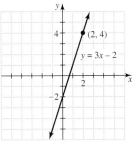

39.   **41.**

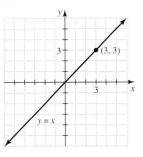

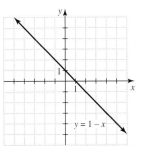

43.

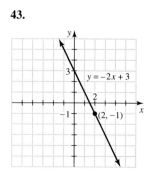

45.

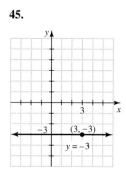

73.

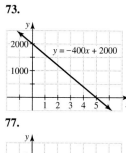

75.

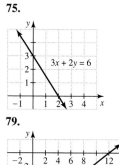

47.

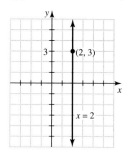

49.

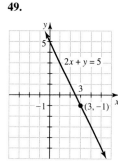

77.

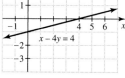

79.

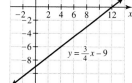

81.

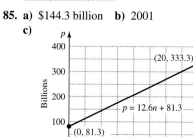

83. 75%, 67, 68, and up

85. a) $144.3 billion   **b)** 2001

c)

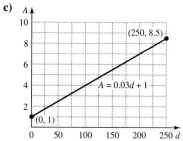

51.

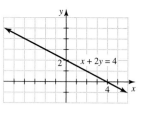

53.

55.

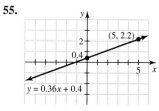

57. Quadrant II
59. x-axis
61. Quadrant III
63. Quadrant I
65. Quadrant II
67. y-axis

87. a) 4 atm   **b)** 130 ft

c)

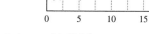

69.

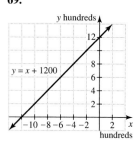

71.

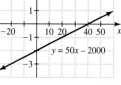

89. x = the number of radio ads, y = the number of TV ads, 21 solutions

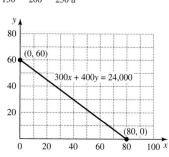

91.

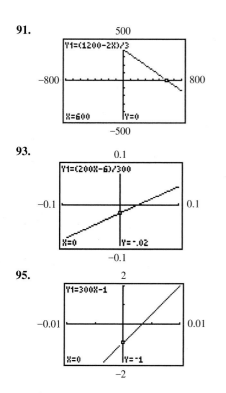

93.

95.

Section 4.2 Warm-ups T T F T F F F F T T
1. The slope of a line is the ratio of its rise and run.
3. Slope is undefined for vertical lines.
5. Lines with positive slope are rising as you go from left to right, while lines with negative slope are falling as you go from left to right.

7. $-\dfrac{2}{3}$ **9.** $\dfrac{2}{3}$ **11.** $\dfrac{3}{2}$ **13.** 0 **15.** $\dfrac{2}{5}$ **17.** Undefined

19. 2 **21.** $-\dfrac{5}{3}$ **23.** $\dfrac{5}{7}$ **25.** $-\dfrac{4}{3}$ **27.** -1

29. 1 **31.** Undefined **33.** 0 **35.** 3
37. **39.**

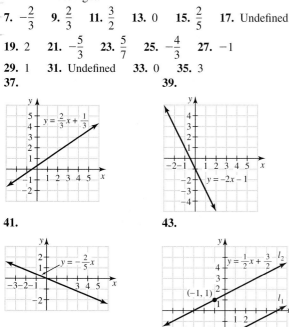

41. **43.**

45.

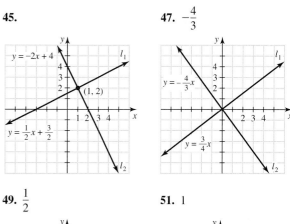

47. $-\dfrac{4}{3}$

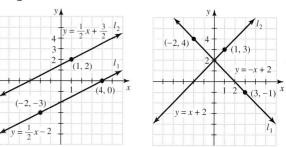

49. $\dfrac{1}{2}$ **51.** 1

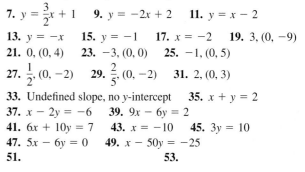

53. a) 100,000 slope, average yearly increase is $100,000
   b) $1.2 million
   c) $2 million

55. $\dfrac{11}{14}$, the percentage increases 0.79% per year

Section 4.3 Warm-ups T F T T T F F T T F
1. Slope-intercept form is $y = mx + b$.
3. The standard form is $Ax + By = C$.
5. The slope-intercept form allows us to write the equation from the y-intercept and the slope.

7. $y = \dfrac{3}{2}x + 1$ **9.** $y = -2x + 2$ **11.** $y = x - 2$

13. $y = -x$ **15.** $y = -1$ **17.** $x = -2$ **19.** 3, $(0, -9)$

21. 0, $(0, 4)$ **23.** -3, $(0, 0)$ **25.** -1, $(0, 5)$

27. $\dfrac{1}{2}$, $(0, -2)$ **29.** $\dfrac{2}{5}$, $(0, -2)$ **31.** 2, $(0, 3)$

33. Undefined slope, no y-intercept **35.** $x + y = 2$
37. $x - 2y = -6$ **39.** $9x - 6y = 2$
41. $6x + 10y = 7$ **43.** $x = -10$ **45.** $3y = 10$
47. $5x - 6y = 0$ **49.** $x - 50y = -25$
51. **53.**

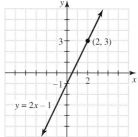

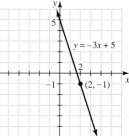

55.

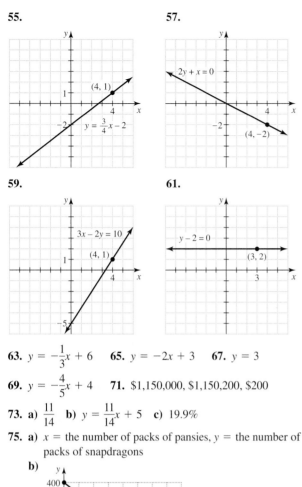

57.

59.

61.

63. $y = -\dfrac{1}{3}x + 6$ **65.** $y = -2x + 3$ **67.** $y = 3$

69. $y = -\dfrac{4}{5}x + 4$ **71.** $1,150,000, $1,150,200, $200

73. a) $\dfrac{11}{14}$ **b)** $y = \dfrac{11}{14}x + 5$ **c)** 19.9%

75. a) $x =$ the number of packs of pansies, $y =$ the number of packs of snapdragons

b)

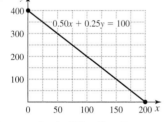

c) $y = -2x + 400$ **d)** -2
e) If the number of packs of pansies goes up by 1, then the number of packs of snapdragons goes down by 2.

77.

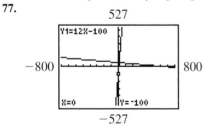

Section 4.4 Warm-ups F F T T F T T T T T
1. Point-slope form is $y - y_1 = m(x - x_1)$.
3. If you know two points on a line, find the slope. Then use it along with a point in point-slope form to write the equation of the line.
5. Nonvertical parallel lines have equal slopes.

7. $y = 5x + 11$ **9.** $y = \dfrac{3}{4}x - 20$ **11.** $y = \dfrac{2}{3}x + \dfrac{1}{3}$

13. $y = \dfrac{1}{3}x + \dfrac{7}{3}$ **15.** $y = -\dfrac{1}{2}x + 4$ **17.** $y = -6x - 13$

19. $2x - y = 7$ **21.** $x - 2y = 6$ **23.** $2x - 3y = 2$
25. $3x - 2y = -1$ **27.** $3x + 5y = -11$

29. $x - y = -2$ **31.** $y = -x + 4$ **33.** $y = \dfrac{5}{3}x - 1$

35. $y = -\dfrac{1}{3}x + 5$ **37.** $y = x + 3$ **39.** $y = -\dfrac{2}{3}x + \dfrac{5}{3}$

41. $y = -2x - 5$ **43.** $y = \dfrac{1}{3}x + \dfrac{7}{3}$ **45.** $y = 2x - 1$

47. a) $y = 0.9x - 75.8$ **b)** 18.7 billion
49. a) $y = 181.818x - 344,181.46$
    b) $x =$ year, $y =$ per capita gross domestic product in dollars
    c)

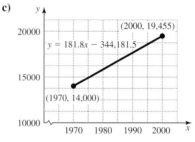

    d) $181.82   **e)** $19,454.54

51. $-\dfrac{2}{3}, \dfrac{4}{5}, -\dfrac{A}{B}$ **53.** They look parallel, but they are not.

55. They will look perpendicular in the right window.

Section 4.5 Warm-ups T T F T T F T F T T
1. Values of the dependent variable are determined from values of the independent variable.
3. We use letters other than x and y so that it is easier to remember what the letters represent.
5. To write a linear function from two points, find the slope and then use point-slope form.
7. $F = 3Y$ **9.** $D = 65T$ **11.** $C = \pi D$ **13.** $P = 7.8H$
15.

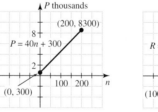

17.

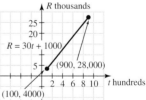

19.

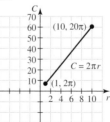

21.

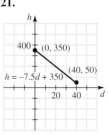

23. $7.05

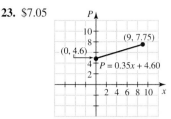

25. $C = 20n + 30$, $170   **27. a)** $S = 3L - \dfrac{41}{4}$ **b)** 8.5

29. $v = 32t + 10$, 122 ft/sec   **31.** $w = -\dfrac{1}{120}t + \dfrac{3}{2}$, $\dfrac{5}{6}$ inch

33. $A = 0.6w$, 3.6 inches

35. $a = 0.08c$, 0.24, 6.25 mg/ml

37. a) Female   **b)** Male   **c)** 65   **d)** 235

Section 4.6 Warm-ups F T F T T F F T F T
1. A function is a set of ordered pairs in which no two have the same first coordinate and different second coordinates.
3. All descriptions of functions involve ordered pairs that satisfy the definition.
5. The domain is the set of all first coordinates of the ordered pairs.
7. $C = 0.50t + 5$ **9.** $T = 1.09S$ **11.** $C = 2\pi r$
13. $P = 4s$ **15.** $A = 5h$ **17.** Yes   **19.** Yes
21. No   **23.** Yes   **25.** Yes   **27.** Yes   **29.** No
31. Yes   **33.** Yes   **35.** No   **37.** Yes   **39.** No
41. Yes   **43.** Yes   **45.** Yes   **47.** No   **49.** Yes
51. No   **53.** No   **55.** $\{1, 2, 3\}$, $\{3, 5, 7\}$
57. R (all reals), $[0, \infty)$ **59.** R, R   **61.** R, $[0, \infty)$
63. $(0, \infty)$, $(0, \infty)$ **65.** -1 **67.** 0   **69.** 13
71. -2.75 **73.** 2   **75.** 1   **77.** 150.988   **79.** 0.31
81. a) 100 ft/sec, 68 ft/sec, 36 ft/sec   **b)** Decreasing
83. 183 pounds, 216 pounds   **85.** $C = \dfrac{r}{0.96}$ **87.** Both
89. Neither   **91.** Neither   **93.** b is a function of a

Enriching Word Power
1. d   **2.** a   **3.** b   **4.** c   **5.** b   **6.** a   **7.** c
8. c   **9.** a   **10.** d

Chapter 4 Review
1. Quadrant II   **3.** x-axis   **5.** y-axis
7. Quadrant IV   **9.** $(0, -5)$, $(-3, -14)$, $(4, 7)$

11. $\left(0, -\dfrac{8}{3}\right), \left(3, -\dfrac{2}{3}\right), \left(-6, -\dfrac{20}{3}\right)$

13.   **15.**

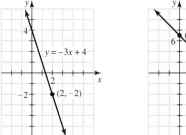

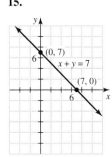

17. 1   **19.** $\dfrac{3}{2}$ **21.** $\dfrac{3}{7}$ **23.** 3, $(0, -18)$
25. 2, $(0, -3)$ **27.** 2, $(0, -4)$
29.   **31.**

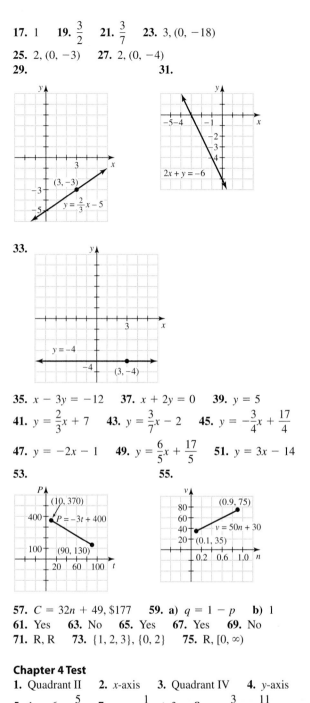

33.

35. $x - 3y = -12$ **37.** $x + 2y = 0$ **39.** $y = 5$
41. $y = \dfrac{2}{3}x + 7$ **43.** $y = \dfrac{3}{7}x - 2$ **45.** $y = -\dfrac{3}{4}x + \dfrac{17}{4}$
47. $y = -2x - 1$ **49.** $y = \dfrac{6}{5}x + \dfrac{17}{5}$ **51.** $y = 3x - 14$
53.   **55.**

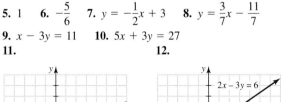

57. $C = 32n + 49$, $177   **59. a)** $q = 1 - p$ **b)** 1
61. Yes   **63.** No   **65.** Yes   **67.** Yes   **69.** No
71. R, R   **73.** $\{1, 2, 3\}$, $\{0, 2\}$ **75.** R, $[0, \infty)$

Chapter 4 Test
1. Quadrant II   **2.** x-axis   **3.** Quadrant IV   **4.** y-axis
5. 1   **6.** $-\dfrac{5}{6}$ **7.** $y = -\dfrac{1}{2}x + 3$ **8.** $y = \dfrac{3}{7}x - \dfrac{11}{7}$
9. $x - 3y = 11$ **10.** $5x + 3y = 27$
11.   **12.**

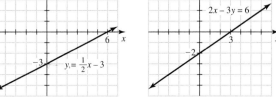

13.

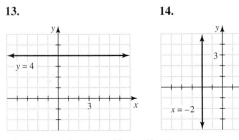

$y = 4$

14.

$x = -2$

15. Yes   **16.** No   **17.** R, $[0, \infty)$
18. R, $[0, \infty)$ **19.** 1   **20.** 5
21. $S = 0.75n + 2.50$
22. $P = 3v + 20$, 80 cents

Making Connections Chapters 1–4

1. -1 **2.** -34 **3.** 1   **4.** 72   **5.** -4
6. -28 **7.** $-\dfrac{7}{2}$ **8.** 0.4   **9.** $\dfrac{1}{10}$ **10.** 15
11. $13x$ **12.** $3x - 36$ **13.** $-x^2 + 13x$
14. $-4x + 32$ **15.** $-13x + 72$
16. $x - 3$ **17.** $2x - 4$ **18.** $x + 2$
19. x **20.** $\dfrac{3}{5}x + 1$

21.

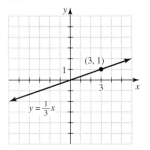

$(3, 1)$
$y = \dfrac{1}{3}x$

22.

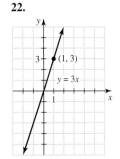

$(1, 3)$
$y = 3x$

23.

$y = -3x$
$(1, -3)$

24.

$y = -\dfrac{1}{3}x$
$(3, -1)$

25.

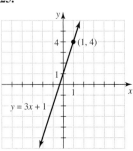

$(1, 4)$
$y = 3x + 1$

26.

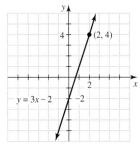

$(2, 4)$
$y = 3x - 2$

27.

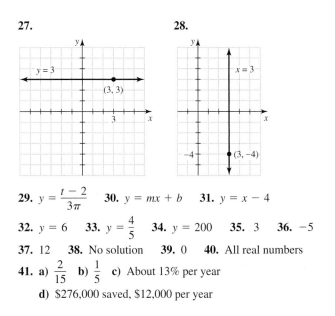

$y = 3$
$(3, 3)$

28.

$x = 3$
$(3, -4)$

29. $y = \dfrac{t - 2}{3\pi}$ **30.** $y = mx + b$ **31.** $y = x - 4$
32. $y = 6$ **33.** $y = \dfrac{4}{5}$ **34.** $y = 200$ **35.** 3   **36.** -5
37. 12   **38.** No solution   **39.** 0   **40.** All real numbers
41. a) $\dfrac{2}{15}$ **b)** $\dfrac{1}{5}$ **c)** About 13% per year
d) \$276,000 saved, \$12,000 per year

Chapter 5

Section 5.1 Warm-ups F F T T F T F T T F
1. A term is a single number or the product of a number and one or more variables raised to powers.
3. The degree of a polynomial in one variable is the highest power of the variable in the polynomial.
5. Polynomials are added by adding the like terms.
7. $-3, 7$ **9.** $0, 6$ **11.** $\dfrac{1}{3}, \dfrac{7}{2}$ **13.** Monomial, 0
15. Monomial, 3   **17.** Binomial, 1   **19.** Trinomial, 10
21. Binomial, 6   **23.** Trinomial, 3   **25.** 6   **27.** -85
29. 71   **31.** -4.97665 **33.** $4x - 8$ **35.** $2q$
37. $x^2 + 3x - 2$ **39.** $x^3 + 9x - 7$ **41.** $3a^2 - 7a - 4$
43. $-3w^2 - 8w + 5$ **45.** $9.66x^2 - 1.93x - 1.49$
47. $-4x + 6$ **49.** -5 **51.** $-z^2 + 2z$
53. $w^5 + w^4 - w^3 - w^2$ **55.** $2t + 13$ **57.** $-8y + 7$
59. $-22.85x - 423.2$ **61.** $4a + 2$ **63.** $-2x + 4$
65. $2a$ **67.** $-5m + 7$ **69.** $4x^2 + 1$
71. $a^3 - 9a^2 + 2a + 7$ **73.** $-3x + 9$
75. $2y^3 + 7y^2 - 4y - 14$ **77.** $5m - 5$ **79.** $2y + 1$
81. $2x^2 + 7x - 5$ **83.** $-3m + 3$ **85.** $-11y - 3$
87. $2x^2 - 6x + 12$ **89.** $-5z^4 - 8z^3 + 3z^2 + 7$
91. $100x + 500$ dollars, \$5500
93. $6x + 3$ meters, 27 meters
95. $5x + 40$ miles, 140 miles   **97.** 800 feet, 800 feet
99. $0.17x + 74.47$ dollars, \$244.47   **101.** 1321.39 calories
103. Yes, yes, yes

Section 5.2 Warm-ups F F T F T T T T T F
1. The product rule for exponents says that $a^m \cdot a^n = a^{m+n}$.
3. To multiply a monomial and a polynomial we use the distributive property.
5. To multiply any two polynomials we multiply each term of the first polynomial by every term of the second polynomial.
7. $27x^5$ **9.** $14a^{11}$ **11.** $-30x^4$ **13.** $27x^{17}$

15. $-54s^2t^2$ **17.** $24t^7w^8$ **19.** $25y^2$ **21.** $4x^6$
23. $4y^7 - 8y^3$ **25.** $-18y^2 + 12y$
27. $-3y^3 + 15y^2 - 18y$ **29.** $-xy^2 + x^3$
31. $15a^4b^3 - 5a^5b^2 - 10a^6b$ **33.** $-2t^5v^3 + 3t^3v^2 + 2t^2v^2$
35. $x^2 + 3x + 2$ **37.** $x^2 + 2x - 15$
39. $t^2 - 13t + 36$ **41.** $x^3 + 3x^2 + 4x + 2$
43. $6y^3 + y^2 + 7y + 6$ **45.** $2y^8 - 3y^6z - 5y^4z^2 + 3y^2z^3$
47. $2a^2 + 7a - 15$ **49.** $14x^2 + 95x + 150$
51. $20x^2 - 7x - 6$ **53.** $2am - 6an + mb - 3nb$
55. $x^3 + 9x^2 + 16x - 12$ **57.** $-4a^4 + 9a^2 - 8a - 12$
59. $x^2 - y^2$ **61.** $x^3 + y^3$ **63.** $u - 3t$ **65.** $-3x - y$
67. $3a^2 + a - 6$ **69.** $-3v^2 - v + 6$ **71.** $-6x^2 + 27x$
73. $-6x^2 + 27x + 2$ **75.** $-x - 7$ **77.** $36x^{12}$
79. $-6a^3b^{10}$ **81.** $25x^2 + 60x + 36$ **83.** $25x^2 - 36$
85. $6x^7 - 8x^4$ **87.** $m^3 - 1$ **89.** $3x^3 - 5x^2 - 25x + 18$
91. $x^2 + 4x$ square feet, 140 square feet
93. $x^2 + \frac{1}{2}x$ square feet, 27.5 square feet
95. $x^2 + 5x$ **97.** $8.05x^2 + 15.93x + 6.12$ square meters
99. 30,000, $300,000, $40,000p - 1000p^2$
101. $10x^5 + 10x^4 + 10x^3 + 10x^2 + 10x$, $67.16

Section 5.3 Warm-ups F T T T T F T F F F

1. We use the distributive property to find the product of two binomials.
3. The purpose of FOIL is to provide a faster method for finding the product of two binomials.
5. $x^2 + 6x + 8$ **7.** $a^2 - a - 6$ **9.** $2x^2 - 5x + 2$
11. $2a^2 - a - 3$ **13.** $w^2 - 60w + 500$
15. $y^2 - ay + 5y - 5a$ **17.** $5w - w^2 + 5m - mw$
19. $10m^2 - 9mt - 9t^2$ **21.** $45a^2 + 53ab + 14b^2$
23. $x^4 - 3x^2 - 10$ **25.** $h^6 + 10h^3 + 25$
27. $3b^6 + 14b^3 + 8$ **29.** $y^3 - 2y^2 - 3y + 6$
31. $6m^6 + 7m^3n^2 - 3n^4$ **33.** $12u^4v^2 + 10u^2v - 12$
35. $b^2 + 9b + 20$ **37.** $x^2 + 6x - 27$
39. $a^2 + 10a + 25$ **41.** $4x^2 - 4x + 1$ **43.** $z^2 - 100$
45. $a^2 + 2ab + b^2$ **47.** $a^2 - 3ab + 2b^2$
49. $2x^2 + 5xy - 3y^2$ **51.** $5t^2 - 7t + 2$
53. $h^2 - 16h + 63$ **55.** $h^2 + 14hw + 49w^2$
57. $4h^4 - 4h^2 + 1$ **59.** $8a^2 + a - \frac{1}{4}$
61. $\frac{1}{8}x^2 + \frac{1}{6}x - \frac{1}{6}$ **63.** $-12x^6 - 26x^5 + 10x^4$
65. $x^3 + 3x^2 - x - 3$ **67.** $9x^3 + 45x^2 - 4x - 20$
69. $2x + 10$ **71.** $2x^2 + 5x - 3$ square feet
73. $5.2555x^2 + 0.41095x - 1.995$ square meters
75. 12 ft^2, $3h$ ft^2, $4h$ ft^2, h^2 ft^2, $h^2 + 7h + 12$ ft^2, $(h + 3)(h + 4) = h^2 + 7h + 12$

Section 5.4 Warm-ups F T T T F T T T F F

1. The special products are $(a + b)^2$, $(a - b)^2$, and $(a + b)(a - b)$.
3. It is faster to do by the new rule than with FOIL.
5. $(a + b)(a - b) = a^2 - b^2$
7. $x^2 + 2x + 1$ **9.** $y^2 + 8y + 16$ **11.** $9x^2 + 48x + 64$
13. $s^2 + 2st + t^2$ **15.** $4x^2 + 4xy + y^2$
17. $4t^2 + 12ht + 9h^2$ **19.** $a^2 - 6a + 9$

21. $t^2 - 2t + 1$ **23.** $9t^2 - 12t + 4$ **25.** $s^2 - 2st + t^2$
27. $9a^2 - 6ab + b^2$ **29.** $9z^2 - 30yz + 25y^2$
31. $a^2 - 25$ **33.** $y^2 - 1$ **35.** $9x^2 - 64$ **37.** $r^2 - s^2$
39. $64y^2 - 9a^2$ **41.** $25x^4 - 4$ **43.** $x^3 + 3x^2 + 3x + 1$
45. $8a^3 - 36a^2 + 54a - 27$
47. $a^4 - 12a^3 + 54a^2 - 108a + 81$
49. $a^4 + 4a^3b + 6a^2b^2 + 4ab^3 + b^4$
51. $a^2 - 400$ **53.** $x^2 + 15x + 56$ **55.** $16x^2 - 1$
57. $81y^2 - 18y + 1$ **59.** $6t^2 - 7t - 20$
61. $4t^2 - 20t + 25$ **63.** $4t^2 - 25$ **65.** $x^4 - 1$
67. $4y^6 - 36y^3 + 81$ **69.** $4x^6 + 12x^3y^2 + 9y^4$
71. $\frac{1}{4}x^2 + \frac{1}{3}x + \frac{1}{9}$ **73.** $0.04x^2 - 0.04x + 0.01$
75. $a^3 + 3a^2b + 3ab^2 + b^3$ **77.** $2.25x^2 + 11.4x + 14.44$
79. $12.25t^2 - 6.25$
81. $x^2 - 25$ square feet, 25 square feet smaller
83. $3.14b^2 + 6.28b + 3.14$ square meters
85. $v = k(R^2 - r^2)$ **87.** $P + 2Pr + Pr^2$, $242
89. $20,230.06
91. The first is an identity and the second is a conditional equation.

Section 5.5 Warm-ups F F T F T F T T T T

1. The quotient rule is used for dividing monomials.
3. When dividing a polynomial by a monomial the quotient should have the same number of terms as the polynomial.
5. The long division process stops when the degree of the remainder is less than the degree of the divisor.
7. 1 **9.** 1 **11.** 1 **13.** 1 **15.** x^6 **17.** $\frac{3}{a^5}$
19. $\frac{-4}{x^4}$ **21.** $-y$ **23.** $-3x$ **25.** $\frac{-3}{x^3}$ **27.** $x - 2$
29. $x^3 + 3x^2 - x$ **31.** $4xy - 2x + y$ **33.** $y^2 - 3xy$
35. $x + 2, 7$ **37.** $2, -10$ **39.** $a^2 + 2a + 8, 13$
41. $x - 4, 4$ **43.** $h^2 + 3h + 9, 0$ **45.** $2x - 3, 1$
47. $x^2 + 1, -1$ **49.** $3 + \frac{15}{x - 5}$ **51.** $-1 + \frac{3}{x + 3}$
53. $1 - \frac{1}{x}$ **55.** $3 + \frac{1}{x}$ **57.** $x - 1 + \frac{1}{x + 1}$
59. $x - 2 + \frac{8}{x + 2}$ **61.** $x^2 + 2x + 4 + \frac{8}{x - 2}$
63. $x^2 + \frac{3}{x}$ **65.** $-3a$ **67.** $\frac{4t^4}{w^5}$ **69.** $-a + 4$
71. $x - 3$ **73.** $-6x^2 + 2x - 3$ **75.** $t + 4$
77. $2w + 1$ **79.** $4x^2 - 6x + 9$ **81.** $t^2 - t + 3$
83. $v^2 - 2v + 1$ **85.** $x - 5$ meters
87. $x^8 + x^7 + x^6 + x^5 + x^4 + x^3 + x^2 + x + 1$
89. $10x \div 5x$ is not equivalent to the other two.

Section 5.6 Warm-ups F F F F F F T F F T

1. The product rule says that $a^m a^n = a^{m+n}$.
3. These rules do not make sense without identical bases.
5. The power of a product rule says that $(ab)^n = a^n b^n$.
7. 128 **9.** $6u^{10}$ **11.** a^4b^{10} **13.** $\frac{-1}{2a^4}$ **15.** $\frac{2a^6}{5b^4}$
17. 200 **19.** x^6 **21.** $2x^{12}$ **23.** $\frac{1}{t^2}$ **25.** $\frac{1}{2}$ **27.** x^3y^6

29. $-8t^{15}$ **31.** $-8x^6y^{15}$ **33.** $a^9b^2c^{14}$ **35.** $\dfrac{x^{12}}{64}$

37. $\dfrac{16a^8}{b^{12}}$ **39.** $-\dfrac{x^6}{8y^3}$ **41.** $\dfrac{4z^{12}}{x^8}$ **43.** 45   **45.** 81

47. -19 **49.** -1 **51.** $\dfrac{8}{125}$ **53.** 200   **55.** 128

57. $\dfrac{1}{16}$ **59.** $15x^{11}$ **61.** $-125x^{12}$ **63.** $-27y^6z^{19}$

65. $\dfrac{3v}{u}$ **67.** $-16x^9t^6$ **69.** $\dfrac{8}{x^6}$ **71.** $\dfrac{-32a^{15}b^{20}}{c^{25}}$

73. $\dfrac{y^5}{32x^5}$ **75.** $P(1+r)^{15}$

Section 5.7 Warm-ups T F F T F T T T T F
1. A negative exponent means "reciprocal," as in $a^{-n} = \dfrac{1}{a^n}$.
3. The new quotient rule is $a^m/a^n = a^{m-n}$ for any integers m and n.
5. Convert from standard notation by counting the number of places the decimal must move so that there is one nonzero digit to the left of the decimal point.

7. $\dfrac{1}{3}$ **9.** $\dfrac{1}{16}$ **11.** $-\dfrac{1}{16}$ **13.** 4   **15.** $\dfrac{8}{125}$ **17.** $\dfrac{1}{3}$

19. 1250   **21.** 82   **23.** x **25.** $-\dfrac{16}{x^4}$ **27.** $\dfrac{6}{a^5}$ **29.** $\dfrac{1}{u^8}$

31. $-4t^2$ **33.** $2x^{11}$ **35.** $\dfrac{1}{x^{10}}$ **37.** a^9 **39.** $\dfrac{x^{12}}{16}$

41. $\dfrac{y^6}{16x^4}$ **43.** $\dfrac{x^2}{4y^6}$ **45.** $\dfrac{a^{16}}{16c^8}$ **47.** $\dfrac{1}{6}$ **49.** $\dfrac{3}{2}$

51. $-14x^6$ **53.** $\dfrac{2a^4}{b^2}$ **55.** 9,860,000,000

57. 0.00137   **59.** 0.000001   **61.** 600,000   **63.** 9×10^3
65. 7.8×10^{-4} **67.** 8.5×10^{-6} **69.** 5.25×10^{11}
71. 6×10^{-10} **73.** 2×10^{-38} **75.** 5×10^{27}
77. 9×10^{24} **79.** 1.25×10^{14} **81.** 2.5×10^{-33}
83. 8.6×10^9 **85.** 2.1×10^2 **87.** 2.7×10^{-23}
89. 3×10^{15} **91.** 9.135×10^2 **93.** 5.715×10^{-4}
95. 4.426×10^7 **97.** 1.577×10^{182}
99. 4.910×10^{11} feet   **101.** 4.65×10^{-28} hours
103. 9.040×10^8 feet
105. a) $1 per pound and 1%
     b) $1,000,000 per pound
     c) No
107. $10,727.41
109. a) $w < 0$ **b)** m is odd   **c)** $w < 0$ and m odd

Enriching Word Power
1. a   **2.** d   **3.** b   **4.** c   **5.** d   **6.** b   **7.** a   **8.** b
9. c   **10.** a   **11.** a   **12.** c

Chapter 5 Review
1. $5w - 2$ **3.** $-6x + 4$
5. $2w^2 - 7w - 4$ **7.** $-2m^2 + 3m - 1$
9. $-50x^{11}$ **11.** $121a^{14}$ **13.** $-4x + 15$
15. $3x^2 - 10x + 12$ **17.** $15m^5 - 3m^3 + 6m^2$
19. $x^3 - 7x^2 + 20x - 50$ **21.** $3x^3 - 8x^2 + 16x - 8$
23. $q^2 + 2q - 48$ **25.** $2t^2 - 21t + 27$

27. $20y^2 - 7y - 6$ **29.** $6x^4 + 13x^2 + 5$ **31.** $z^2 - 49$
33. $y^2 + 14y + 49$ **35.** $w^2 - 6w + 9$ **37.** $x^4 - 9$
39. $9a^2 + 6a + 1$ **41.** $16 - 8y + y^2$ **43.** $-5x^2$
45. $\dfrac{-2a^2}{b^2}$ **47.** $-x + 3$ **49.** $-3x^2 + 2x - 1$ **51.** -1
53. $m^3 + 2m^2 + 4m + 8$ **55.** $m^2 - 3m + 6, 0$
57. $b - 5, 15$ **59.** $2x - 1, -8$ **61.** $x^2 + 2x - 9, 1$
63. $2 + \dfrac{6}{x - 3}$ **65.** $-2 + \dfrac{2}{1 - x}$ **67.** $x - 1 - \dfrac{2}{x + 1}$
69. $x - 1 + \dfrac{1}{x + 1}$ **71.** $6y^{30}$ **73.** $\dfrac{-5}{c^6}$ **75.** b^{30}
77. $-8x^9y^6$ **79.** $\dfrac{8a^3}{b^3}$ **81.** $\dfrac{8x^6y^{15}}{z^{18}}$ **83.** $\dfrac{1}{32}$ **85.** $\dfrac{1}{1000}$
87. $\dfrac{1}{x^3}$ **89.** a^4 **91.** a^{10} **93.** $\dfrac{1}{x^{12}}$ **95.** $\dfrac{x^9}{8}$
97. $\dfrac{9}{a^2b^6}$ **99.** 5×10^3 **101.** 340,000
103. 4.61×10^{-5} **105.** 0.00000569   **107.** 7×10^{-4}
109. 1.6×10^{-15} **111.** 8×10^1 **113.** 3.2×10^{-34}
115. $x^2 + 10x + 21$ **117.** $t^2 - 7ty + 12y^2$ **119.** 2
121. $-27h^3t^{18}$ **123.** $2w^2 - 9w - 18$ **125.** $9u^2 - 25v^2$
127. $9h^2 + 30h + 25$ **129.** $x^3 + 9x^2 + 27x + 27$
131. $14s^5t^6$ **133.** $\dfrac{k^8}{16}$ **135.** $x^2 - 9x - 5$
137. $5x^2 - x - 12$ **139.** $x^3 - x^2 - 19x + 4$
141. $x + 6$
143. $P = 4w + 88, A = w^2 + 44w, P = 288$ ft, $A = 4700$ ft^2
145. $R = -15p^2 + 600p$, $5040, $20

Chapter 5 Test
1. $7x^3 + 4x^2 + 2x - 11$ **2.** $-x^2 - 9x + 2$
3. $-2y^2 + 3y$ **4.** -1 **5.** $x^2 + x - 1$
6. $15x^5 - 21x^4 + 12x^3 - 3x^2$ **7.** $x^2 + 3x - 10$
8. $6a^2 + a - 35$ **9.** $a^2 - 14a + 49$
10. $16x^2 + 24xy + 9y^2$ **11.** $b^2 - 9$ **12.** $9t^4 - 49$
13. $4x^4 + 5x^2 - 6$ **14.** $x^3 - 3x^2 - 10x + 24$
15. $2 + \dfrac{6}{x - 3}$ **16.** $x - 5 + \dfrac{15}{x + 2}$ **17.** $-35x^8$
18. $12x^5y^9$ **19.** $-2ab^4$ **20.** $15x^5$ **21.** $\dfrac{-32a^5}{b^{10}}$
22. $\dfrac{3a^4}{b^2}$ **23.** $\dfrac{3}{t^{16}}$ **24.** $\dfrac{1}{w^2}$ **25.** $\dfrac{s^6}{9t^4}$ **26.** $\dfrac{-8y^3}{x^{18}}$
27. 5.433×10^6 **28.** 6.5×10^{-6} **29.** 4.8×10^{-1}
30. 8.1×10^{-27} **31.** $x - 2, 3$ **32.** $-2x^2 + x + 15$
33. $x^2 + 4x$ ft^2, $4x + 8$ ft, 32 ft^2, 24 ft
34. $R = -150q^2 + 3000q$, $14,400

Making Connections Chapters 1–5
1. 8   **2.** -9 **3.** 41   **4.** 2^{25} **5.** 32   **6.** 992
7. 144   **8.** -1 **9.** 64   **10.** 34   **11.** 899   **12.** 961
13. $x^2 + 8x + 15$ **14.** $x + 3$ **15.** $4x + 15$
16. $x^3 + 13x^2 + 55x + 75$ **17.** $-15t^5v^7$ **18.** $5tv$
19. $3y - 4$ **20.** $4y^2 - 5y - 3$ **21.** $-\dfrac{1}{2}$ **22.** 7
23. $\dfrac{3}{2}$ **24.** 4   **25.** -3 **26.** $-\dfrac{2}{3}$

27. $\dfrac{2.25n + 100,000}{n}$, \$102.25, \$3.25, \$2.35, It averages out to 10 cents per disk.

Chapter 6

Section 6.1 Warm-ups F F F T T T T F F T
1. To factor means to write as a product.
3. You can find the prime factorization by dividing by prime factors until the result is prime.
5. The GCF for two monomials consists of the GCF of their coefficients and every variable that they have in common raised to the lowest power that appears on the variable.
7. $2 \cdot 3^2$ **9.** $2^2 \cdot 13$ **11.** $2 \cdot 7^2$ **13.** $2^2 \cdot 5 \cdot 23$
15. $2^2 \cdot 3 \cdot 7 \cdot 11$ **17.** 4   **19.** 12   **21.** 8   **23.** 4
25. 1   **27.** $2x$ **29.** $2x$ **31.** xy **33.** $12ab$ **35.** 1
37. $6ab$ **39.** $9(3x)$ **41.** $8t(3t)$ **43.** $4y^2(9y^3)$
45. $uv(u^3v^2)$ **47.** $2m^4(-7n^3)$ **49.** $-3x^3yz(11xy^2z)$
51. $x(x^2 - 6)$ **53.** $5a(x + y)$ **55.** $h^3(h^2 - 1)$
57. $2k^3m^4(-k^4 + 2m^2)$ **59.** $2x(x^2 - 3x + 4)$
61. $6x^2t(2x^2 + 5x - 4t)$ **63.** $(x - 3)(a + b)$
65. $(y + 1)^2(a + b)$ **67.** $9a^2b^4(4ab - 3 + 2b^5)$
69. $8(x - y), -8(-x + y)$
71. $4x(-1 + 2x), -4x(1 - 2x)$
73. $1(x - 5), -1(-x + 5)$ **75.** $1(4 - 7a), -1(-4 + 7a)$
77. $8a^2(-3a + 2), -8a^2(3a - 2)$
79. $6x(-2x - 3), -6x(2x + 3)$
81. $2x(-x^2 - 3x + 7), -2x(x^2 + 3x - 7)$
83. $2ab(2a^2 - 3ab - 2b^2), -2ab(-2a^2 + 3ab + 2b^2)$
85. $x + 2$ hours
87. a) $S = 2\pi r(r + h)$ **b)** $S = 2\pi r^2 + 10\pi r$ **c)** 3 in.
89. The GCF is an algebraic expression.

Section 6.2 Warm-ups F T F F T T F F T T
1. A perfect square is a square of an integer or an algebraic expression.
3. A perfect square trinomial is of the form $a^2 + 2ab + b^2$ or $a^2 - 2ab + b^2$.
5. A polynomial is factored completely when it is a product of prime polynomials.
7. $(a - 2)(a + 2)$ **9.** $(x - 7)(x + 7)$
11. $(y + 3x)(y - 3x)$ **13.** $(5a + 7b)(5a - 7b)$
15. $(11m + 1)(11m - 1)$ **17.** $(3w - 5c)(3w + 5c)$
19. Perfect square trinomial   **21.** Neither
23. Perfect square trinomial   **25.** Neither
27. Difference of two squares
29. Perfect square trinomial
31. $(x + 6)^2$ **33.** $(a - 2)^2$ **35.** $(2w + 1)^2$
37. $(4x - 1)^2$ **39.** $(2t + 5)^2$ **41.** $(3w + 7)^2$
43. $(n + t)^2$ **45.** $5(x - 5)(x + 5)$
47. $-2(x - 3)(x + 3)$ **49.** $a(a - b)(a + b)$
51. $3(x + 1)^2$ **53.** $-5(y - 5)^2$ **55.** $x(x - y)^2$
57. $-3(x - y)(x + y)$ **59.** $2a(x - 7)(x + 7)$
61. $3a(b - 3)^2$ **63.** $-4m(m - 3n)^2$ **65.** $(b + c)(x + y)$
67. $(x - 2)(x + 2)(x + 1)$ **69.** $(3 - x)(a - b)$
71. $(a^2 + 1)(a + 3)$ **73.** $(a + 3)(x + y)$
75. $(c - 3)(ab + 1)$ **77.** $(a + b)(x - 1)(x + 1)$

79. $(y + b)(y + 1)$ **81.** $6ay(a + 2y)^2$
83. $6ay(2a - y)(2a + y)$ **85.** $2a^2y(ay - 3)$
87. $(b - 4w)(a + 2w)$
89. $h = -16(t - 20)(t + 20)$, 6336 feet   **91.** $y - 3$ inches

Section 6.3 Warm-ups T T F F T F T F F F
1. We factored $ax^2 + bx + c$ with $a = 1$.
3. If there are no two integers that have a product of c and a sum of b, then $x^2 + bx + c$ is prime.
5. A polynomial is factored completely when all of the factors are prime polynomials.
7. $(x + 3)(x + 1)$ **9.** $(x + 3)(x + 6)$
11. $(a - 3)(a - 4)$ **13.** $(b - 6)(b + 1)$
15. $(y + 2)(y + 5)$ **17.** $(a - 2)(a - 4)$
19. $(m - 8)(m - 2)$ **21.** $(w + 10)(w - 1)$
23. $(w - 4)(w + 2)$ **25.** Prime
27. $(m + 16)(m - 1)$ **29.** Prime
31. $(z - 5)(z + 5)$ **33.** Prime
35. $(m + 2)(m + 10)$ **37.** Prime   **39.** $(m - 18)(m + 1)$
41. Prime   **43.** $(t + 8)(t - 3)$ **45.** $(t - 6)(t + 4)$
47. $(t - 20)(t + 10)$ **49.** $(x - 15)(x + 10)$
51. $(y + 3)(y + 10)$ **53.** $(x + 3a)(x + 2a)$
55. $(x - 6y)(x + 2y)$ **57.** $(x - 12y)(x - y)$
59. Prime   **61.** $w(w - 8)$ **63.** $2(w - 9)(w + 9)$
65. $x^2(w^2 + 9)$ **67.** $(w - 9)^2$ **69.** $6(w - 3)(w + 1)$
71. $2x^2(4 - x)(4 + x)$ **73.** $3(w + 3)(w + 6)$
75. $w(w^2 + 18w + 36)$ **77.** $8v(w + 2)^2$
79. $6xy(x + 3y)(x + 2y)$ **81.** $x + 4$ feet
83. 3 feet and 5 feet   **85.** d

Section 6.4 Warm-ups T F T F T F F F F T
1. We factored $ax^2 + bx + c$ with $a \neq 1$.
3. If there are no two integers whose product is ac and whose sum is b, then $ax^2 + bx + c$ is prime.
5. 2 and 10   **7.** -6 and 2   **9.** 3 and 4
11. -2 and -9 **13.** -3 and 4   **15.** $(2x + 1)(x + 1)$
17. $(2x + 1)(x + 4)$ **19.** $(3t + 1)(t + 2)$
21. $(2x - 1)(x + 3)$ **23.** $(3x - 1)(2x + 3)$
25. $(2x - 3)(x - 2)$ **27.** $(5b - 3)(b - 2)$
29. $(4y + 1)(y - 3)$ **31.** Prime
33. $(4x + 1)(2x - 1)$ **35.** $(3t - 1)(3t - 2)$
37. $(5x + 1)(3x + 2)$ **39.** $(5x - 1)(3x - 2)$
41. $(x + 2)(3x + 1)$ **43.** $(5x + 1)(x + 2)$
45. $(3a - 1)(2a - 5)$ **47.** $(5a + 1)(a + 2)$
49. $(2w + 3)(2w + 1)$ **51.** $(5x - 2)(3x + 1)$
53. $(4x - 1)(2x - 1)$ **55.** $(15x - 1)(x - 2)$
57. $2(x^2 + 9x - 45)$ **59.** $(3x - 5)(x + 2)$
61. $(5x + y)(2x - y)$ **63.** $(6a - b)(7a - b)$
65. $w(9w - 1)(9w + 1)$ **67.** $2(2w - 5)(w + 3)$
69. $3(2x + 3)^2$ **71.** $(3w + 5)(2w - 7)$
73. $3z(x - 3)(x + 2)$ **75.** $y^2(10x - 9)(x + 1)$
77. $(a + 5b)(a - 3b)$ **79.** $-t(3t + 2)(2t - 1)$
81. $2t^2(3t - 2)(2t + 1)$ **83.** $y(2x - y)(2x - 3y)$
85. $-1(w - 1)(4w - 3)$
87. $-2a(2a - 3b)(3a - b)$
89. $h = -8(2t + 1)(t - 3)$, 0 feet
91. a) $-4, 4$ **b)** $\pm 8, \pm 16$ **c)** $\pm 1, \pm 7, \pm 13, \pm 29$

Section 6.5 Warm-ups F F T T T F T F T T F

1. If there is no remainder, then the dividend factors as the divisor times the quotient.

3. If you divide $a^3 + b^3$ by $a + b$ there will be no remainder.

5. $a^3 + b^3 = (a + b)(a^2 - ab + b^2)$

7. $(x + 4)(x - 3)(x + 2)$ **9.** $(x - 1)(x + 3)(x + 2)$

11. $(x - 2)(x^2 + 2x + 4)$ **13.** $(x + 5)(x^2 - x + 2)$

15. $(x + 1)(x^2 + x + 1)$ **17.** $(m - 1)(m^2 + m + 1)$

19. $(x + 2)(x^2 - 2x + 4)$ **21.** $(2w + 1)(4w^2 - 2w + 1)$

23. $(2t - 3)(4t^2 + 6t + 9)$ **25.** $(x - y)(x^2 + xy + y^2)$

27. $(2t + y)(4t^2 - 2ty + y^2)$ **29.** $2(x - 3)(x + 3)$

31. $4(x + 5)(x - 3)$ **33.** $x(x + 2)^2$ **35.** $5am(x^2 + 4)$

37. $(3x + 1)^2$ **39.** $y(3x + 2)(2x - 1)$ **41.** Prime

43. $2(4m + 1)(2m - 1)$ **45.** $(3a + 4)^2$

47. $2(3x - 1)(4x - 3)$ **49.** $3a(a - 9)$

51. $2(2 - x)(2 + x)$ **53.** $x(6x^2 - 5x + 12)$

55. $ab(a - 2)(a + 2)$ **57.** $(x - 2)(x + 2)^2$

59. $2w(w - 2)(w^2 + 2w + 4)$ **61.** $3w(a - 3)^2$

63. $5(x - 10)(x + 10)$ **65.** $(2 - w)(m + n)$

67. $3x(x + 1)(x^2 - x + 1)$ **69.** $4(w^2 + w - 1)$

71. $a^2(a + 10)(a - 3)$ **73.** $aw(2w - 3)^2$

75. $(t + 3)^2$ **77.** Length $x + 5$ cm, width $x + 3$ cm

79. $(-1 + 1)^3 = (-1)^3 + 1^3, (1 + 2)^3 \neq 1^3 + 2^3$

Section 6.6 Warm-ups F F T T T F T T T F

1. A quadratic equation has the form $ax^2 + bx + c = 0$ with $a \neq 0$.

3. The zero factor property says that if $ab = 0$ then $a = 0$ or $b = 0$.

5. Dividing each side by a variable is not usually done because the variable might have a value of zero.

7. $-4, -5$ **9.** $-\dfrac{5}{2}, \dfrac{4}{3}$ **11.** $2, 7$ **13.** $-4, 6$

15. $-1, \dfrac{1}{2}$ **17.** $0, -7$ **19.** $-5, 4$ **21.** $\dfrac{1}{2}, -3$

23. $0, -8$ **25.** $-\dfrac{9}{2}, 2$ **27.** $\dfrac{2}{3}, -4$ **29.** 5 **31.** $\dfrac{3}{2}$

33. $0, -3, 3$ **35.** $-4, -2, 2$ **37.** $-1, 1, 3$

39. $0, 4, 5$ **41.** $-4, 4$ **43.** $-3, 3$ **45.** $0, -1, 1$

47. $-3, -2$ **49.** $-\dfrac{3}{2}, -4$ **51.** $-6, 4$ **53.** $-1, 3$

55. $-4, 2$ **57.** $-5, -3, 5$ **59.** Length 12 ft, width 5 ft

61. Width 5 ft, length 12 ft **63.** 2 and 3, or -3 and -2

65. 5 and 6 **67. a)** 25 sec **b)** last 5 sec **c)** increasing

69. 6 sec **71.** Base 6 in., height 13 in.

73. 20 ft by 20 ft **75.** 80 ft

77. 3 yd by 3 yd, 6 yd by 6 yd **79.** 12 mi **81.** 25%

Enriching Word Power

1. a **2.** d **3.** c **4.** a **5.** c **6.** b **7.** c **8.** a
9. d **10.** c

Chapter 6 Review

1. $2^4 \cdot 3^2$ **3.** $2 \cdot 29$ **5.** $2 \cdot 3 \cdot 5^2$ **7.** 18 **9.** $4x$

11. $3(x + 2)$ **13.** $-2(-a + 10)$ **15.** $a(2 - a)$

17. $3x^2y(2y - 3x^3)$ **19.** $3y(x^2 - 4x - 3y)$

21. $(y - 20)(y + 20)$ **23.** $(w - 4)^2$ **25.** $(2y + 5)^2$

27. $(r - 2)^2$ **29.** $2t(2t - 3)^2$ **31.** $(x + 6y)^2$

33. $(x - y)(x + 5)$ **35.** $(b + 8)(b - 3)$

37. $(r - 10)(r + 6)$ **39.** $(y - 11)(y + 5)$

41. $(u + 20)(u + 6)$ **43.** $3t^2(t + 4)$

45. $5w(w^2 + 5w + 5)$ **47.** $ab(2a + b)(a + b)$

49. $x(3x - y)(3x + y)$ **51.** $(7t - 3)(2t + 1)$

53. $(3x + 1)(2x - 7)$ **55.** $(3p + 4)(2p - 1)$

57. $-2p(5p + 2)(3p - 2)$ **59.** $(6x + y)(x - 5y)$

61. $2(4x + y)^2$ **63.** $5x(x^2 + 8)$ **65.** $(3x - 1)(3x + 2)$

67. $(x + 2)(x - 1)(x + 1)$ **69.** $xy(x - 16y)$

71. $(a + 1)^2$ **73.** $(x^2 + 1)(x - 1)$

75. $(a + 2)(a + b)$ **77.** $-2(x - 6)(x - 2)$

79. $(m - 10)(m^2 + 10m + 100)$

81. $(x + 2)(x^2 - 2x + 5)$ **83.** $(x + 4)(x + 5)(x - 3)$

85. $0, 5$ **87.** $0, 5$ **89.** $-\dfrac{1}{2}, 5$ **91.** $-4, -3, 3$

93. $-2, -1$ **95.** $-\dfrac{1}{2}, \dfrac{1}{4}$ **97.** $5, 11$

99. 6 in. by 8 in. **101.** $v = k(R - r)(R + r)$ **103.** 6 ft

Chapter 6 Test

1. $2 \cdot 3 \cdot 11$ **2.** $2^4 \cdot 3 \cdot 7$ **3.** 16 **4.** 6 **5.** $3y^2$

6. $6ab$ **7.** $5x(x - 2)$ **8.** $6y^2(x^2 + 2x + 2)$

9. $3ab(a - b)(a + b)$ **10.** $(a + 6)(a - 4)$

11. $(2b - 7)^2$ **12.** $3m(m^2 + 9)$ **13.** $(a + b)(x - y)$

14. $(a - 5)(x - 2)$ **15.** $(3b - 5)(2b + 1)$

16. $(m + 2n)^2$ **17.** $(2a - 3)(a - 5)$

18. $z(z + 3)(z + 6)$ **19.** $(x - 1)(x - 2)(x - 3)$

20. $\dfrac{3}{2}, -4$ **21.** $0, -2, 2$ **22.** $-2, \dfrac{5}{6}$

23. Length 12 ft, width 9 ft **24.** -4 and 8

Making Connections Chapters 1–6

1. -1 **2.** 2 **3.** -3 **4.** 57 **5.** 16 **6.** 7 **7.** $2x^2$

8. $3x$ **9.** $3 + x$ **10.** $6x$ **11.** $24yz$ **12.** $6y + 8z$

13. $4z - 1$ **14.** t^6 **15.** t^{10} **16.** $4t^6$

17. $x < -9$ **18.** $x \geq 3$

19. $x > 12$ **20.** $x < 600$

21. $\dfrac{3}{2}$ **22.** $-\dfrac{1}{2}$ **23.** $3, -5$ **24.** $\dfrac{3}{2}, -\dfrac{1}{2}$ **25.** $0, 3$

26. $0, 1$ **27.** R **28.** No solution **29.** 10 **30.** 40

31. $-3, 3$ **32.** $-5, \dfrac{3}{2}$ **33.** Length 21 ft, width 13.5 ft

Chapter 7

Section 7.1 Warm-ups F T T F F T T F F T

1. A rational number is a ratio of two integers with the denominator not 0.

3. A rational number is reduced to lowest terms by dividing the numerator and denominator by the GCF.

5. The quotient rule is used in reducing ratios of monomials.

7. -3 **9.** 5 **11.** $-0.6, 9, 401, -199$ **13.** -1 **15.** $\frac{5}{3}$

17. $4, -4$ **19.** Any number can be used. **21.** $\frac{2}{9}$

23. $\frac{7}{15}$ **25.** $\frac{2a}{5}$ **27.** $\frac{13}{5w}$ **29.** $\frac{3x+1}{3}$ **31.** $\frac{2}{3}$

33. $w-7$ **35.** $\frac{a-1}{a+1}$ **37.** $\frac{x+1}{2x-2}$ **39.** $\frac{x+3}{7}$

41. x^3 **43.** $\frac{1}{z^5}$ **45.** $-2x^2$ **47.** $\frac{-3m^3n^2}{2}$ **49.** $\frac{-3}{4c^3}$

51. $\frac{5c}{3a^4b^{16}}$ **53.** $\frac{35}{44}$ **55.** $\frac{11}{8}$ **57.** $\frac{21}{10x^4}$ **59.** $\frac{33a^4}{16}$

61. -1 **63.** $-h-t$ **65.** $\frac{-2}{3h+g}$ **67.** $\frac{-x-2}{x+3}$

69. -1 **71.** $\frac{-2y}{3}$ **73.** $\frac{x+2}{2-x}$ **75.** $\frac{-6}{a+3}$ **77.** $\frac{x^4}{2}$

79. $\frac{x+2}{2x}$ **81.** -1 **83.** $\frac{-2}{c+2}$ **85.** $\frac{x+2}{x-2}$

87. $\frac{-2}{x+3}$ **89.** q^2 **91.** $\frac{u+2}{u-8}$ **93.** $\frac{a^2+2a+4}{2}$

95. $y+2$ **97.** $\frac{300}{x+10}$ hr **99.** $\frac{4.50}{x+4}$ dollars/lb

101. $\frac{1}{x}$ pool/hr

103. a) \$0.75 **b)** \$0.75, \$0.63, \$0.615
   c) Approaches \$0.60

Section 7.2 Warm-ups T T T F T F F F T T T

1. Rational numbers are multiplied by multiplying their numerators and their denominators.

3. Reducing can be done before multiplying rational numbers or expressions.

5. $\frac{7}{9}$ **7.** $\frac{18}{5}$ **9.** $\frac{42}{5}$ **11.** $\frac{a}{44}$ **13.** $\frac{-x^5}{a^3}$ **15.** $\frac{18t^8y^7}{w^4}$

17. $\frac{2a}{a-b}$ **19.** $3x-9$ **21.** $\frac{8a+8}{5a^2+5}$ **23.** 30 **25.** $\frac{2}{3}$

27. $\frac{10}{9}$ **29.** $\frac{7x}{2}$ **31.** $\frac{2m^2}{3n^6}$ **33.** -3 **35.** $\frac{2}{x+2}$

37. $\frac{1}{4t-20}$ **39.** x^2-1 **41.** $2x-4y$ **43.** $\frac{x+2}{2}$

45. $\frac{x^2+9}{15}$ **47.** $9x+9y$ **49.** -3 **51.** $\frac{a+b}{a}$

53. $\frac{2b}{a}$ **55.** $\frac{y}{x}$ **57.** $\frac{-a^6b^8}{2}$ **59.** $\frac{1}{9m^3n}$ **61.** $\frac{x^2+5x}{3x-1}$

63. $\frac{a^3+8}{2a-4}$ **65.** 1 **67.** $\frac{(m+3)^2}{(m-3)(m+k)}$ **69.** $\frac{13.1}{x}$ mi

71. 5 m^2 **73. a)** $\frac{1}{8}$ **b)** $\frac{4}{3}$ **c)** $\frac{2x}{3}$ **d)** $\frac{3x}{4}$

Section 7.3 Warm-ups F F T T F F F F T T

1. We can build up a denominator by multiplying the numerator and denominator of a fraction by the same nonzero number.

3. For fractions, the LCD is the smallest number that is a multiple of all of the denominators.

5. $\frac{9}{27}$ **7.** $\frac{14x}{2x}$ **9.** $\frac{15t}{3bt}$ **11.** $\frac{-36z^2}{8awz}$ **13.** $\frac{10a^2}{15a^3}$

15. $\frac{8xy^3}{10x^2y^5}$ **17.** $\frac{-20}{-8x-8}$ **19.** $\frac{-32ab}{20b^2-20b^3}$

21. $\frac{3x-6}{x^2-4}$ **23.** $\frac{3x^2+3x}{x^2+2x+1}$ **25.** $\frac{y^2-y-30}{y^2+y-20}$

27. 48 **29.** 180 **31.** $30a^2$ **33.** $12a^4b^6$

35. $(x-4)(x+4)^2$ **37.** $x(x+2)(x-2)$

39. $2x(x-4)(x+4)$ **41.** $\frac{4}{24}, \frac{9}{24}$ **43.** $\frac{9b}{252ab}, \frac{20a}{252ab}$

45. $\frac{2x^3}{6x^5}, \frac{9}{6x^5}$ **47.** $\frac{4x^4}{36x^3y^5z}, \frac{3y^6z}{36x^3y^5z}, \frac{6xy^4z}{36x^3y^5z}$

49. $\frac{2x^2+4x}{(x-3)(x+2)}, \frac{5x^2-15x}{(x-3)(x+2)}$ **51.** $\frac{4}{a-6}, \frac{-5}{a-6}$

53. $\frac{x^2-3x}{(x-3)^2(x+3)}, \frac{5x^2+15x}{(x-3)^2(x+3)}$

55. $\frac{w^2+3w+2}{(w-5)(w+3)(w+1)}, \frac{-2w^2-6w}{(w-5)(w+3)(w+1)}$

57. $\frac{-5x-10}{6(x-2)(x+2)}, \frac{6x}{6(x-2)(x+2)},$
$\frac{9x-18}{6(x-2)(x+2)}$

59. $\frac{2q+8}{(2q+1)(q-3)(q+4)}, \frac{3q-9}{(2q+1)(q-3)(q+4)},$
$\frac{8q+4}{(2q+1)(q-3)(q+4)}$

61. Identical denominators are needed for addition and subtraction.

Section 7.4 Warm-ups F T T T T F T F T F

1. We can add rational numbers with identical denominators as follows: $\frac{a}{c} + \frac{b}{c} = \frac{a+b}{c}$.

3. The LCD is the smallest number that is a multiple of all denominators.

5. $\frac{1}{5}$ **7.** $\frac{3}{4}$ **9.** $-\frac{2}{3}$ **11.** $-\frac{3}{4}$ **13.** $\frac{5}{9}$ **15.** $\frac{103}{144}$

17. $-\frac{31}{40}$ **19.** $\frac{5}{24}$ **21.** $\frac{5}{w}$ **23.** 3 **25.** -2 **27.** $\frac{3}{h}$

29. $\frac{x-4}{x+2}$ **31.** $\frac{17}{10a}$ **33.** $\frac{w}{36}$ **35.** $\frac{b^2-4ac}{4a}$

37. $\frac{2w+3z}{w^2z^2}$ **39.** $\frac{-x-3}{x(x+1)}$ **41.** $\frac{3a+b}{(a-b)(a+b)}$

43. $\frac{15-4x}{5x(x+1)}$ **45.** $\frac{a^2+5a}{(a-3)(a+3)}$ **47.** 0 **49.** $\frac{7}{2a-2}$

51. $\frac{-2x+1}{(x-5)(x+2)(x-2)}$ **53.** $\frac{7x+17}{(x+2)(x-1)(x+3)}$

55. $\frac{2x^2-x-4}{x(x-1)(x+2)}$ **57.** $\frac{a+51}{6a(a-3)}$ **59.** $\frac{11}{x}$ feet

61. $\frac{315x+600}{x(x+5)}$ hours, 5 hours **63.** $\frac{4x+6}{x(x+3)}, \frac{5}{9}$ job

Section 7.5 Warm-ups F T F F F F F T T T

1. A complex fraction is a fraction that has fractions in its numerator, denominator, or both.

3. $-\frac{10}{3}$ **5.** $\frac{22}{7}$ **7.** $\frac{14}{17}$ **9.** $\frac{45}{23}$ **11.** $\frac{3a+b}{a-3b}$

13. $\frac{5a-3}{3a+1}$ **15.** $\frac{x^2-4x}{6x^2-2}$ **17.** $\frac{10b}{3b^2-4}$ **19.** $\frac{y-2}{3y+4}$

21. $\dfrac{x^2 - 2x + 4}{x^2 - 3x - 1}$ **23.** $\dfrac{5x - 14}{2x - 7}$ **25.** $\dfrac{a - 6}{3a - 1}$

27. $\dfrac{-3m + 12}{4m - 3}$ **29.** $\dfrac{-w + 5}{9w + 1}$ **31.** -1 **33.** $\dfrac{6x - 27}{4x - 6}$

35. $\dfrac{2x^2}{3y}$ **37.** $\dfrac{a^2 + 7a + 6}{a + 3}$ **39.** $1 - x$ **41.** $\dfrac{32}{95}, \dfrac{11}{35}$

Section 7.6 Warm-ups F F F F F T T T T T
1. The first step is usually to multiply each side by the LCD.
3. An extraneous solution is a number that appears when we solve an equation, but it does not check in the original equation.
5. 12 **7.** 30 **9.** 5 **11.** 4 **13.** 4 **15.** 4 **17.** 3
19. 2 **21.** $-5, 2$ **23.** 2, 3 **25.** $-3, 3$ **27.** 2
29. No solution **31.** No solution **33.** 3 **35.** 10
37. 0 **39.** $-5, 5$ **41.** 3, 5 **43.** 1 **45.** 3 **47.** 0
49. 4 **51.** -20 **53.** 3 **55.** 3 **57.** $54\dfrac{6}{11}$ mm

Section 7.7 Warm-ups T F F T T T F F F T
1. A ratio is a comparison of two numbers.
3. Equivalent ratios are ratios that are equivalent as fractions.
5. In the proportion $\dfrac{a}{b} = \dfrac{c}{d}$ the means are b and c and the extremes are a and d.

7. $\dfrac{5}{7}$ **9.** $\dfrac{8}{15}$ **11.** $\dfrac{7}{2}$ **13.** $\dfrac{9}{14}$ **15.** $\dfrac{5}{2}$ **17.** $\dfrac{15}{1}$

19. 3 to 2 **21.** 9 to 16 **23.** 31 to 1 **25.** 2 to 3 **27.** 6

29. $-\dfrac{2}{5}$ **31.** $-\dfrac{27}{5}$ **33.** 10 **35.** 3 **37.** 5 **39.** $-\dfrac{3}{4}$

41. $\dfrac{5}{4}$ **43.** 108 **45.** 176,000 **47.** Lions 85, Tigers 51

49. 40 luxury cars, 60 sports cars **51.** 84 in. **53.** 15 min
55. 536.7 mi **57.** 3920 lbs, 2000 lbs **59.** 6000
61. a) 3 to 17 **b)** 14.4 lbs **63.** 4074

Section 7.8 Warm-ups T T F T T F F T F T

1. $y = 2x - 5$ **3.** $y = -\dfrac{1}{2}x - 2$

5. $y = mx - mb - a$ **7.** $y = -\dfrac{1}{3}x - \dfrac{1}{3}$ **9.** $C = \dfrac{B}{A}$

11. $p = \dfrac{a}{1 + am}$ **13.** $m_1 = \dfrac{r^2 F}{km_2}$ **15.** $a = \dfrac{bf}{b - f}$

17. $r = \dfrac{S - a}{S}$ **19.** $P_2 = \dfrac{P_1 V_1 T_2}{T_1 V_2}$ **21.** $h = \dfrac{3V}{4\pi r^2}$

23. $\dfrac{5}{12}$ **25.** $-\dfrac{6}{23}$ **27.** $\dfrac{128}{3}$ **29.** -6 **31.** $\dfrac{6}{5}$

33. Marcie 4 mph, Frank 3 mph
35. Bob 25 mph, Pat 20 mph **37.** 5 mph **39.** 6 hours
41. 40 minutes **43.** 1 hour 36 minutes
45. Bananas 8 pounds, apples 10 pounds **47.** 80 gallons

Enriching Word Power
1. b **2.** a **3.** a **4.** d **5.** a **6.** b **7.** d **8.** a
9. c **10.** d

Chapter 7 Review

1. $\dfrac{6}{7}$ **3.** $\dfrac{c^2}{4a^2}$ **5.** $\dfrac{2w - 3}{3w - 4}$ **7.** $-\dfrac{x + 1}{3}$ **9.** $\dfrac{1}{2}k$ **11.** $\dfrac{2x}{3y}$

13. $a^2 - a - 6$ **15.** $\dfrac{1}{2}$ **17.** 108 **19.** $24a^7 b^3$

21. $12x(x - 1)$ **23.** $(x + 1)(x - 2)(x + 2)$ **25.** $\dfrac{15}{36}$

27. $\dfrac{10x}{15x^2 y}$ **29.** $\dfrac{-10}{12 - 2y}$ **31.** $\dfrac{x^2 + x}{x^2 - 1}$ **33.** $\dfrac{29}{63}$

35. $\dfrac{3x - 4}{x}$ **37.** $\dfrac{2a - b}{a^2 b^2}$ **39.** $\dfrac{27a^2 - 8a - 15}{(2a - 3)(3a - 2)}$

41. $\dfrac{3}{a - 8}$ **43.** $\dfrac{3x + 8}{2(x + 2)(x - 2)}$ **45.** $-\dfrac{3}{14}$

47. $\dfrac{6b + 4a}{3a - 18b}$ **49.** $\dfrac{-2x + 9}{3x - 1}$ **51.** $\dfrac{x^2 + x - 2}{-4x + 13}$

53. $-\dfrac{15}{2}$ **55.** 9 **57.** -3 **59.** $\dfrac{21}{2}$ **61.** 5 **63.** 8

65. 56 cups water, 28 cups rice **67.** $y = mx + b$

69. $m = \dfrac{1}{F - v}$ **71.** $y = 4x - 13$ **73.** 200 hours

75. Bert 60 cars, Ernie 50 cars **77.** 29.7 million metric tons

79. $\dfrac{10}{2x}$ **81.** $\dfrac{-2}{5 - a}$ **83.** $\dfrac{3x}{x}$ **85.** $2m$ **87.** $\dfrac{1}{6}$

89. $\dfrac{1}{a + 1}$ **91.** $\dfrac{5 - a}{5a}$ **93.** $\dfrac{a - 2}{2}$ **95.** $b - a$

97. $\dfrac{1}{10a}$ **99.** $\dfrac{3}{2x}$ **101.** $\dfrac{4 + y}{6xy}$ **103.** $\dfrac{8}{a - 5}$

105. $-1, 2$ **107.** $-\dfrac{5}{3}$ **109.** 6 **111.** $\dfrac{1}{2}$

113. $\dfrac{3x + 7}{(x - 5)(x + 5)(x + 1)}$ **115.** $\dfrac{-5a}{(a - 3)(a + 3)(a + 2)}$

117. $\dfrac{2}{5}$

Chapter 7 Test

1. $-1, 1$ **2.** $\dfrac{2}{3}$ **3.** 0 **4.** $-\dfrac{14}{45}$ **5.** $\dfrac{1 + 3y}{y}$

6. $\dfrac{4}{a - 2}$ **7.** $\dfrac{-x + 4}{(x + 2)(x - 2)(x - 1)}$ **8.** $\dfrac{2}{3}$ **9.** $\dfrac{-2}{a + b}$

10. $\dfrac{a^3}{18b^4}$ **11.** $-\dfrac{4}{3}$ **12.** $\dfrac{3x - 4}{-2x + 6}$ **13.** $\dfrac{15}{7}$ **14.** 2, 3

15. 12 **16.** $y = -\dfrac{1}{5}x + \dfrac{13}{5}$ **17.** $c = \dfrac{3M - bd}{b}$

18. 7.2 minutes
19. Brenda 15 mph and Randy 20 mph, or Brenda 10 mph and Randy 15 mph **20.** $72 billion

Making Connections Chapters 1–7

1. $\dfrac{7}{3}$ **2.** $-\dfrac{10}{3}$ **3.** -2 **4.** No solution **5.** 0

6. $-4, -2$ **7.** $-1, 0, 1$ **8.** $-\dfrac{15}{2}$ **9.** $-6, 6$ **10.** $-2, 4$

11. 5 **12.** 3 **13.** $y = \dfrac{c - 2x}{3}$ **14.** $y = \dfrac{1}{2}x + \dfrac{1}{2}$

15. $y = \dfrac{c}{2 - a}$ **16.** $y = \dfrac{AB}{C}$ **17.** $y = 3B - 3A$

18. $y = \dfrac{6A}{5}$ **19.** $y = \dfrac{8}{3 - 5a}$ **20.** $y = 0$ or $y = B$

21. $y = \dfrac{2A - hb}{h}$ **22.** $y = -\dfrac{b}{2}$ **23.** 64 **24.** 16

25. 49   **26.** 121   **27.** $-2x - 2$ **28.** $2a^2 - 11a + 15$

29. x^4 **30.** $\dfrac{2x + 1}{5}$ **31.** $\dfrac{1}{2x}$ **32.** $\dfrac{x + 2}{2x}$ **33.** $\dfrac{x}{2}$

34. $\dfrac{x - 2}{2x}$ **35.** $-\dfrac{7}{5}$ **36.** $\dfrac{3a}{4}$ **37.** $x^2 - 64$ **38.** $3x^3 - 21x$

39. $10a^{14}$ **40.** x^{10} **41.** $k^2 - 12k + 36$ **42.** $j^2 + 10j + 25$

43. -1 **44.** $3x^2 - 4x$ **45.** $P = \dfrac{r + 2}{(1 + r)^2}$, \$1.81, \$7.72

Chapter 8

Section 8.1 Warm-ups T F F T T T T T T T

1. The intersection point of the graphs is the solution to an independent system.

3. The graphing method can be very inaccurate.

5. If the equation you get after substituting turns out to be incorrect, like $0 = 9$, then the system has no solution.

7. $\{(1, 2)\}$ **9.** $\{(0, -1)\}$ **11.** $\{(2, -1)\}$ **13.** $\varnothing$

15. $\{(x, y) \mid x + 2y = 8\}$ **17.** c   **19.** b

21. $\{(8, 3)\}$, independent   **23.** $\{(-3, 2)\}$, independent

25. $\varnothing$, inconsistent   **27.** $\{(x, y) \mid y = 2x - 5\}$, dependent

29. $\{(5, -1)\}$, independent   **31.** $\{(7, 7)\}$, independent

33. $\{(15, 25)\}$, independent   **35.** $\{(20, 10)\}$, independent

37. $\left\{\left(\dfrac{9}{2}, -\dfrac{1}{2}\right)\right\}$, independent   **39.** $\varnothing$, inconsistent

41. $\{(x, y) \mid 3y - 2x = -3\}$, dependent

43. $\{(0.75, 1.125)\}$, independent

45. Width 13 feet, length 28 feet

47. \$14,000 at 5%, \$16,000 at 10%   **49.** -12 and 14

51. 94 toasters, 6 vacation coupons

53. State tax \$3553, federal tax \$28,934

55. \$20,000

57. a) \$500,000   **b)** \$300,000   **c)** 20,000   **d)** \$400,000

59. a   **61. a)** $(2.8, 2.6)$ **b)** $(1.0, -0.2)$

Section 8.2 Warm-ups T F T T F T T F T T

1. In this section we learned the addition method.

3. In some cases we multiply one or both of the equations on each side to change the coefficients of the variable that we are trying to eliminate.

5. If an identity, such as $0 = 0$, results from addition of the equations, then the equations are dependent.

7. $\{(8, -1)\}$ **9.** $\{(5, -7)\}$ **11.** $\{(-1, 3)\}$

13. $\{(-1, -3)\}$ **15.** $\{(-2, -5)\}$ **17.** $\{(22, 26)\}$

19. $\varnothing$, inconsistent   **21.** $\{(x, y) \mid 5x - y = 1\}$, dependent

23. $\left\{\left(\dfrac{5}{2}, 0\right)\right\}$, independent   **25.** $\{(12, 6)\}$ **27.** $\{(-8, 6)\}$

29. $\{(16, 12)\}$ **31.** $\{(12, 7)\}$ **33.** $\{(400, 800)\}$

35. $\{(1.5, 1.25)\}$ **37.** \$1.00   **39.** 1380 students

41. 31 dimes, 4 nickels

43. a) $(20, 30)$ **b)** 20 pounds chocolate, 30 pounds peanut butter   **45.** 4 hours   **47.** 80%

49. Width 150 meters, length 200 meters

Section 8.3 Warm-ups F F T F F T T F F F

1. A linear equation in three variables is an equation of the form $Ax + By + Cz = D$ where A, B, and C cannot all be zero.

3. A solution to a system of linear equations in three variables is an ordered triple that satisfies all of the equations in the system.

5. The graph of a linear equation in three variables is a plane in a three-dimensional coordinate system.

7. $\{(1, 2, -1)\}$ **9.** $\{(1, 3, 2)\}$ **11.** $\{(1, -5, 3)\}$

13. $\{(-1, 2, -1)\}$ **15.** $\{(-1, -2, 4)\}$ **17.** $\{(1, 3, 5)\}$

19. $\{(3, 4, 5)\}$ **21.** $\varnothing$ **23.** $\{(x, y, z) \mid 3x - y + z = 5\}$

25. $\varnothing$ **27.** $\{(x, y, z) \mid 5x + 4y - 2z = 150\}$

29. $\{(0.1, 0.3, 2)\}$

31. \$1500 stocks, \$4500 bonds, \$6000 mutual fund

33. Coffee \$0.40, doughnut \$0.35, tip \$0.55   **35.** \$0.95

37. Soup 14 ounces, tuna 8 ounces, error 2 ounces

39. \$24,000 teaching, \$18,000 painting, \$6000 royalties

Section 8.4 Warm-ups T T T F T F F F T F

1. A matrix is a rectangular array of numbers.

3. The order of a matrix is the number of rows and columns.

5. An augmented matrix is a matrix where the entries in the first column are the coefficients of x, the entries in the second column are the coefficients of y, and the entries in the third column are the constants from a system of two linear equations in two unknowns.

7. 2×2 **9.** 3×2 **11.** 3×1 **13.** $\begin{bmatrix} 2 & -3 & 9 \\ -3 & 1 & -1 \end{bmatrix}$

15. $\begin{bmatrix} 1 & -1 & 1 & 1 \\ 1 & 1 & -2 & 3 \\ 0 & 1 & -3 & 4 \end{bmatrix}$

17. $\begin{aligned} 5x + y &= -1 \\ 2x - 3y &= 0 \end{aligned}$ **19.** $\begin{aligned} x &= 6 \\ -x + z &= -3 \\ x + y &= 1 \end{aligned}$

21. $R_1 \leftrightarrow R_2$ **23.** $\dfrac{1}{5} R_2 \to R_2$ **25.** $\{(1, 2)\}$ **27.** $\{(4, 5)\}$

29. $\{(1, -1)\}$ **31.** $\{(7, 6)\}$ **33.** $\varnothing$

35. $\{(x, y) \mid x + 2y = 1\}$ **37.** $\{(1, 2, 3)\}$ **39.** $\{(1, 1, 1)\}$

41. $\{(1, 2, 0)\}$ **43.** $\{(1, 0, 1)\}$

45. $\{(x, y, z) \mid x - y + z = 1\}$ **47.** $\varnothing$

Section 8.5 Warm-ups T F F T T T T T T T

1. A determinant is a real number associated with a square matrix.

3. Cramer's rule works on systems that have exactly one solution.

5. A minor for an element is obtained by deleting the row and column of the element and finding the determinant of the 2×2 matrix that remains.

7. -1 **9.** -3 **11.** -14 **13.** 0.4 **15.** $\{(1, -3)\}$

17. $\{(1, 1)\}$ **19.** $\left\{\left(\dfrac{23}{13}, \dfrac{9}{13}\right)\right\}$ **21.** $\{(10, 15)\}$

23. $\left\{\left(\dfrac{27}{4}, \dfrac{13}{2}\right)\right\}$ **25.** 11   **27.** 4   **29.** 3   **31.** 1

33. -7 **35.** -1 **37.** 9   **39.** 5   **41.** 22   **43.** 6

45. 70   **47.** 25   **49.** $\{(1, 2, 3)\}$ **51.** $\{(-1, 1, 2)\}$

53. $\{(-3, 2, 1)\}$ **55.** $\left\{\left(\dfrac{3}{2}, \dfrac{1}{2}, 2\right)\right\}$ **57.** $\{(0, 1, -1)\}$

59. a) $(9, 11)$ **b)** 9 servings peas, 11 servings beets

61. Milk \$2.40, magazine \$2.25   **63.** 12 singles, 10 doubles

65. Gary 39, Harry 34   **67.** Square 10 feet, triangle $\frac{40}{3}$ feet
69. 10 gallons of 10% solution, 20 gallons of 25% solution
71. Mimi 36 pounds, Mitzi 32 pounds, Cassandra 107 pounds
73. 39°, 51°, 90°

Section 8.6 Warm-ups T T T F F F T F T F
1. A linear inequality has the same form as a linear equation except that an inequality symbol is used.
3. If the inequality symbol includes equality, then the boundary line is solid; otherwise it is dashed.
5. In the test point method we test a point to see which side of the boundary line satisfies the inequality.
7. $(-3, -9)$ **9.** $(3, 0), (1, 3)$
11. $(2, 3), (0, 5)$
13.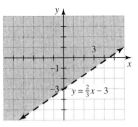
15.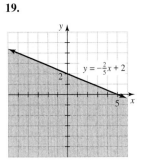
17.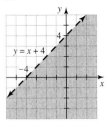
19.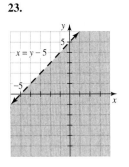
21.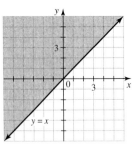
23.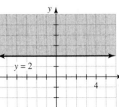
25.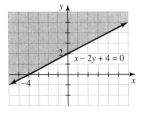
27.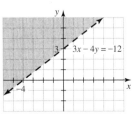

29.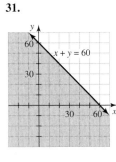
31.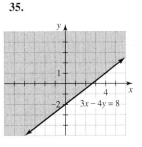
33.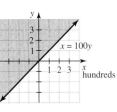
35.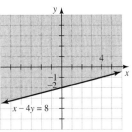
37.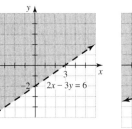
39.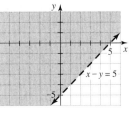
41.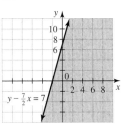
43.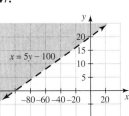
45.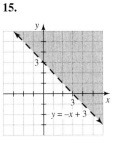
47.

49. $5x + 7y \le 770$

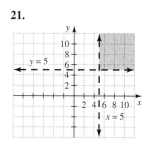

51.

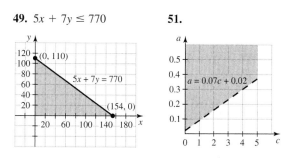

25.

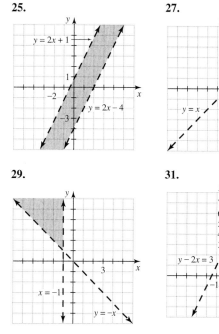

27.

Section 8.7 Warm-ups F T T T F F T F T T

1. A system of linear inequalities in two variables is a pair of linear inequalities in two variables.
3. The solution set to a system of inequalities is usually described with a graph.
5. To use the test point method, select a point in each region determined by the graphs of the boundary lines.
7. $(4, 3)$ **9.** $(3, 6)$
11. $(9, -5)$

29.

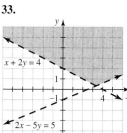

31.

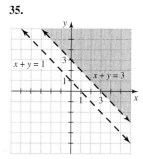

13.

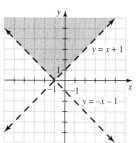

15.

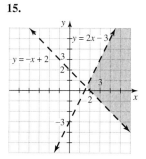

33.

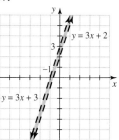

35.

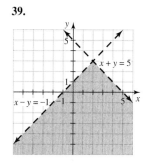

17.

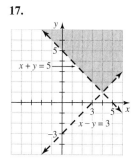

19.

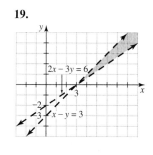

37.

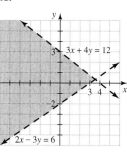

39.

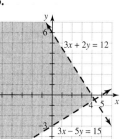

21.

23.

41.

43.

45.

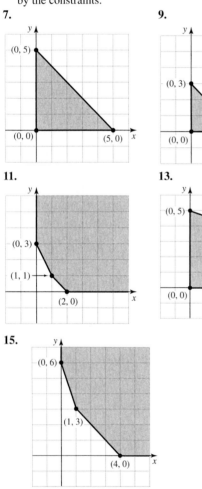

47. $5x + 8y \leq 80$
$2x + 3y \leq 30$

Section 8.8 Warm-ups F F F F F T F T F T
1. A constraint is an inequality that restricts the values of the variables.
3. Constraints may be limitations on the amount of available supplies, money, or other resources.
5. The maximum or minimum of a linear function subject to linear constraints occurs at a vertex of the region determined by the constraints.

7.

9.

11.

13.

15.

17. a) 0, 320,000, 510,000, 450,000
   b) 30 TV ads and 60 radio ads
19. 6 doubles, 4 triples   **21.** 0 doubles, 8 triples
23. 1.75 cups Doggie Dinner, 5.5 cups Puppie Power

25. 10 cups Doggie Dinner, 0 cups Puppie Power
27. Laundromat $8000, carwash $16,000

Enriching Word Power
1. c   **2.** a   **3.** a   **4.** d   **5.** b   **6.** c   **7.** a   **8.** c
9. c   **10.** d   **11.** b   **12.** a   **13.** d

Chapter 8 Review
1. $\{(1, 1)\}$, independent   **3.** $\{(1, -1)\}$, independent
5. $\{(-3, 2)\}$, independent   **7.** $\varnothing$, inconsistent
9. $\{(-1, 5)\}$, independent
11. $\{(x, y) \mid 3x - 2y = 12\}$, dependent   **13.** $\{(1, -3, 2)\}$
15. $\varnothing$ **17.** $\{(2, -4)\}$ **19.** $\{(1, 1, 2)\}$ **21.** 2
23. -0.2 **25.** $\{(-1, -2)\}$ **27.** $\{(2, 1)\}$
29. $\left\{\left(\dfrac{14}{3}, \dfrac{2}{3}\right)\right\}$ **31.** 58   **33.** -30 **35.** $\{(1, 2, -3)\}$
37. $\{(-1, 2, 4)\}$
39.

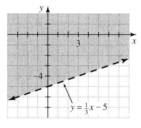

41.

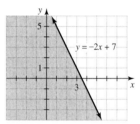

43.

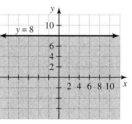

45.

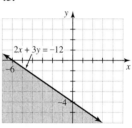

47.

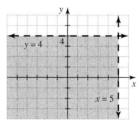

49.

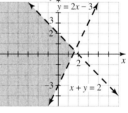

51.

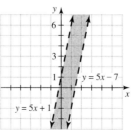

53.

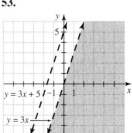

55. $(0, 0), (0, 3), (4, 1), (5, 0)$

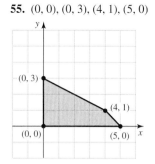

57. 30   **59.** 78   **61.** 36 minutes
63. 4 liters of 30% solution A, 8 liters of 20% solution B, 8 liters of 60% solution C
65. Three servings of each

Chapter 8 Test

1. $\{(1, 3)\}$ **2.** $\left\{\left(\frac{5}{2}, -3\right)\right\}$ **3.** $\{(x, y) \mid y = x - 5\}$
4. $\{(-1, 3)\}$ **5.** $\varnothing$ **6.** Inconsistent   **7.** Dependent
8. Independent   **9.** $\{(1, -2, -3)\}$ **10.** $\{(2, 5)\}$
11. $\{(3, 1, 1)\}$ **12.** -18 **13.** -2 **14.** $\{(-1, 2)\}$
15. $\{(2, -2, 1)\}$

16.   **17.**

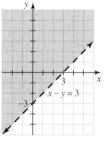

18.   **19.**

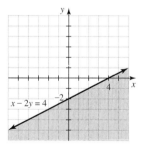

20.   **21.**

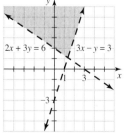

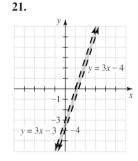

22. Singles \$18, doubles \$25
23. Jill 17 hours, Karen 14 hours, Betsy 62 hours   **24.** 44

Making Connections Chapters 1–8
1. -81 **2.** 7   **3.** 73   **4.** 5.94   **5.** $-t - 3$
6. $-0.9x + 0.9$ **7.** $3x^2 + 2x - 1$ **8.** y **9.** $y = \frac{3}{5}x - \frac{7}{5}$
10. $y = \frac{C}{D}x - \frac{W}{D}$ **11.** $y = \frac{K}{W - C}$ **12.** $y = \frac{bw - 2A}{b}$
13. $\{(4, -1)\}$ **14.** $\{(500, 700)\}$ **15.** $\{(x, y) \mid x + 17 = 5y\}$
16. $\varnothing$ **17.** $y = \frac{5}{9}x + 55$ **18.** $y = -\frac{11}{6}x + \frac{2}{3}$
19. $y = 5x + 26$ **20.** $y = \frac{1}{2}x + 5$ **21.** $y = 5$
22. $x = -7$
23. a) Machine A
b) Machine B \$0.04 per copy, machine A \$0.03 per copy
c) The slopes 0.04 and 0.03 are the per copy cost for each machine
d) B: $y = 0.04x + 2000$, A: $y = 0.03x + 4000$
e) 200,000

Chapter 9

Section 9.1 Warm-ups T F T F T T F F T T
1. If $b^n = a$, then b is an nth root of a
3. If $b^n = a$, then b is an even root provided n is even or an odd root provided n is odd.
5. The product rule for radicals says that $\sqrt[n]{a} \cdot \sqrt[n]{b} = \sqrt[n]{ab}$ provided all of these roots are real.
7. 6   **9.** 2   **11.** 10   **13.** Not a real number   **15.** -2
17. m **19.** y^3 **21.** y^5 **23.** m **25.** 27   **27.** $3\sqrt{y}$
29. $2a$ **31.** x^2y **33.** $m^6\sqrt{5}$ **35.** $2\sqrt[3]{y}$ **37.** $\frac{\sqrt{t}}{2}$
39. $\frac{25}{4}$ **41.** $\frac{\sqrt[3]{t}}{2}$ **43.** $\frac{-2x^2}{y}$ **45.** $\frac{2a^3}{3}$ **47.** $\frac{2\sqrt{5}}{5}$
49. $\frac{\sqrt{21}}{7}$ **51.** $\frac{\sqrt[3]{2}}{2}$ **53.** $\frac{\sqrt[3]{150}}{5}$ **55.** $\frac{\sqrt{15}}{6}$ **57.** $\frac{1}{2}$
59. $\frac{\sqrt{2}}{2}$ **61.** $\frac{\sqrt[3]{14}}{2}$ **63.** $2x^4\sqrt{3}$ **65.** $2a^4b\sqrt{15ab}$
67. $\frac{\sqrt{xy}}{y}$ **69.** $\frac{a\sqrt{ab}}{b^4}$ **71.** $2x^4\sqrt[3]{2x}$ **73.** $x^2y\sqrt[4]{xy^2}$
75. $2x^4\sqrt[5]{2x^2}$ **77.** $\frac{\sqrt[3]{ab^2}}{b}$ **79.** 27   **81.** $\frac{1}{10}$
83. $\frac{2\sqrt{2x}}{7}$ **85.** $\frac{2\sqrt[4]{2a}}{3}$ **87.** $-3x^3y^2\sqrt[3]{y^2}$ **89.** $\frac{\sqrt{b}}{a}$
91. $\frac{\sqrt[3]{18}}{6b^2}$ **93.** 2.887   **95.** 0.693   **97.** 1.310   **99.** 1

101. $-4°F, -10°F$ **103. a)** $t = \frac{\sqrt{h}}{4}$ **b)** $\frac{\sqrt{10}}{2}$ sec
c) 100 ft   **105.** 5.8 knots   **107.** 114.1 ft/sec, 77.8 mph

Section 9.2 Warm-ups T F F T T T T F T T
1. The nth root of a is $a^{1/n}$.
3. The expression $a^{-m/n}$ means $\frac{1}{a^{m/n}}$.
5. The operations can be performed in any order, but the easiest is usually root, power, and then reciprocal.

7. $7^{1/4}$ **9.** $\sqrt[5]{9}$ **11.** $(5x)^{1/2}$ **13.** $\sqrt{a}$ **15.** 5

17. -5 **19.** 2 **21.** Not a real number **23.** $w^{7/3}$

25. $2^{-10/3}$ **27.** $\sqrt[4]{\dfrac{1}{w^3}}$ **29.** $\sqrt{(ab)^3}$ **31.** 25 **33.** 125

35. $\dfrac{1}{81}$ **37.** $\dfrac{1}{64}$ **39.** $-\dfrac{1}{3}$ **41.** Not a real number

43. $3^{7/12}$ **45.** 1 **47.** $\dfrac{1}{2}$ **49.** 2 **51.** 6 **53.** 4

55. 81 **57.** $\dfrac{1}{4}$ **59.** $\dfrac{9}{8}$ **61.** $|x|$ **63.** a^4 **65.** y

67. $|3x^3y|$ **69.** $\left|\dfrac{3x^3}{y^5}\right|$ **71.** $x^{3/4}$ **73.** $\dfrac{y^{3/2}}{x^{1/4}}$ **75.** $\dfrac{1}{w^{8/3}}$

77. $12x^8$ **79.** $\dfrac{a^2}{b}$ **81.** 9 **83.** $-\dfrac{1}{8}$ **85.** $\dfrac{1}{625}$

87. $2^{1/4}$ **89.** 3 **91.** 3 **93.** $\dfrac{4}{9}$ **95.** Not a real number

97. $3x^{9/2}$ **99.** $\dfrac{a^2}{27}$ **101.** $a^{5/4}b$ **103.** $k^{9/2}m^4$

105. 1.2599 **107.** -1.4142 **109.** 2 **111.** 1.9862

113. 2.5 **115.** $a^{3m/4}$ **117.** $a^{2m/15}$ **119.** a^nb^m

121. $a^{4m}b^{2n}$ **123.** 13 inches **125.** 274.96 m²

127. 47.2% **129.** 6.1%

Section 9.3 Warm-ups F T F F T F T F F T

1. Like radicals are radicals with the same index and the same radicand.

3. In the product rule the radicals must have the same index but do not have to have the same radicand.

5. $-\sqrt{3}$ **7.** $9\sqrt{7x}$ **9.** $5\sqrt[3]{2}$ **11.** $4\sqrt{3}-2\sqrt{5}$

13. $5\sqrt[3]{x}$ **15.** $2\sqrt[3]{x}-\sqrt{2x}$ **17.** $2\sqrt{2}+2\sqrt{7}$

19. $5\sqrt{2}$ **21.** $\sqrt{5}$ **23.** $-\sqrt{2}$ **25.** $\dfrac{3\sqrt{2}}{2}$

27. $\dfrac{21\sqrt{5}}{5}$ **29.** $x\sqrt{5x}+2x\sqrt{2}$ **31.** $5\sqrt[3]{3}$ **33.** $-\sqrt[3]{3}$

35. $ty\sqrt[3]{2t}$ **37.** $\sqrt[3]{15}$ **39.** $30\sqrt{2}$ **41.** $6a\sqrt{14}$

43. $3\sqrt[4]{3}$ **45.** 12 **47.** $\dfrac{2x\sqrt[3]{3x}}{3}$ **49.** $6\sqrt{2}+18$

51. $5\sqrt{2}-2\sqrt{5}$ **53.** $3\sqrt[3]{t^2}-t\sqrt[3]{3}$ **55.** $-7-3\sqrt{3}$

57. 2 **59.** $-8-6\sqrt{5}$ **61.** $-6+9\sqrt{2}$ **63.** $\sqrt[6]{3^5}$

65. $\sqrt[6]{5^7}$ **67.** $\sqrt[6]{500}$ **69.** $\sqrt[12]{432}$ **71.** -1 **73.** 3

75. 19 **77.** 13 **79.** $25-9x$ **81.** $11\sqrt{3}$

83. $10\sqrt{30}$ **85.** $8-\sqrt{7}$ **87.** $16w$ **89.** $3x^2\sqrt{2x}$

91. $\dfrac{5\sqrt{2}}{12}$ **93.** $28+\sqrt{10}$ **95.** $\dfrac{8\sqrt{2}}{15}$ **97.** 17

99. $9+6\sqrt{x}+x$ **101.** $25x-30\sqrt{x}+9$

103. $x+3+2\sqrt{x+2}$ **105.** $-\sqrt{w}$ **107.** $a\sqrt{a}$

109. $3x^2\sqrt{x}$ **111.** $13x\sqrt[3]{2x}$ **113.** $\dfrac{y^2\sqrt[3]{2x^2y}}{2x}$

115. $\dfrac{x^2\sqrt[3]{5}}{5}$ **117.** $\sqrt[3]{32x^5}$ **119.** $3\sqrt{2}$ square feet (ft²)

121. $\dfrac{9\sqrt{2}}{2}$ ft² **123.** No

Section 9.4 Warm-ups T T F T F T F T T T

1. $\sqrt{3}$ **3.** $\dfrac{\sqrt{15}}{5}$ **5.** $\dfrac{3\sqrt{2}}{10}$ **7.** $\dfrac{\sqrt{2}}{3}$ **9.** $\sqrt[3]{10}$

11. $x^2\sqrt[3]{4}$ **13.** $2+\sqrt{5}$ **15.** $1-\sqrt{3}$

17. $\dfrac{1+\sqrt{6}+\sqrt{2}+\sqrt{3}}{2}$ **19.** $\dfrac{2\sqrt{3}-\sqrt{6}}{3}$

21. $\dfrac{6\sqrt{6}+2\sqrt{15}}{13}$ **23.** $\dfrac{18\sqrt{5}+3\sqrt{10}-2\sqrt{6}-12\sqrt{3}}{66}$

25. $128\sqrt{2}$ **27.** $x^2\sqrt{x}$ **29.** $-27x^4\sqrt{x}$ **31.** $8x^5$

33. $4\sqrt[3]{25}$ **35.** x^4 **37.** $\dfrac{\sqrt{6}+2\sqrt{2}}{2}$ **39.** $2\sqrt{6}$

41. $\dfrac{\sqrt{2}}{2}$ **43.** $\dfrac{2}{3}$ **45.** $\dfrac{2-\sqrt{2}}{5}$ **47.** $\dfrac{1+\sqrt{3}}{2}$

49. $a-3\sqrt{a}$ **51.** $4a\sqrt{a}+4a$ **53.** $12m$ **55.** $4xy^2z$

57. $m-m^2$ **59.** $5x\sqrt[3]{x}$ **61.** $8m^4\sqrt[4]{8m^2}$

63. $2\sqrt{2}-2$ **65.** $2+8\sqrt{2}$ **67.** $\dfrac{3\sqrt{2}+2\sqrt{3}}{6}$

69. $7\sqrt{2}-1$ **71.** $\dfrac{4x+4\sqrt{x}}{x-4}$ **73.** $\dfrac{x+\sqrt{x}}{x-x^2}$

75. $\dfrac{\sqrt{6}}{3}$ **77.** $\dfrac{\sqrt{2}+1}{1}$ **79.** $\dfrac{\sqrt{x}+1}{x-1}$ **81.** $\dfrac{3\sqrt{2}-3x}{2-x^2}$

83. 3.968 **85.** 12.124 **87.** 14.697 **89.** 8.873

91. 0.725 **93.** -0.419

95. a) x^3-2 **b)** $(x+\sqrt[3]{5})(x^2-\sqrt[3]{5}x+\sqrt[3]{25})$ **c)** 3

d) $(\sqrt[3]{a}+\sqrt[3]{b})(\sqrt[3]{a^2}-\sqrt[3]{ab}+\sqrt[3]{b^2})$
$(\sqrt[3]{a}-\sqrt[3]{b})(\sqrt[3]{a^2}+\sqrt[3]{ab}+\sqrt[3]{b^2})$

Section 9.5 Warm-ups F T F F T F F T T T

1. The odd-root property says that if n is an odd positive integer, then $x^n=k$ is equivalent to $x=\sqrt[n]{k}$ for any real number k.

3. An extraneous solution is a solution that appears when solving an equation but does not satisfy the original equation.

5. $\{-10\}$ **7.** $\left\{\dfrac{1}{2}\right\}$ **9.** $\{1\}$ **11.** $\{-2\}$ **13.** $\{-5,5\}$

15. $\{-2\sqrt{5},2\sqrt{5}\}$ **17.** $\varnothing$ **19.** $\{-1,7\}$

21. $\{-1-2\sqrt{2},-1+2\sqrt{2}\}$ **23.** $\{-\sqrt{10},\sqrt{10}\}$

25. $\{3\}$ **27.** $\{-2,2\}$ **29.** $\{52\}$ **31.** $\left\{\dfrac{9}{4}\right\}$ **33.** $\{9\}$

35. $\varnothing$ **37.** $\{3\}$ **39.** $\{-5,3\}$ **41.** $\{1\}$ **43.** $\varnothing$

45. $\{4\}$ **47.** $\{2\}$ **49.** $\{6\}$ **51.** $\{7\}$ **53.** $\{-5\}$

55. $\{-3\sqrt{3},3\sqrt{3}\}$ **57.** $\left\{-\dfrac{1}{27},\dfrac{1}{27}\right\}$ **59.** $\{512\}$

61. $\left\{\dfrac{1}{81}\right\}$ **63.** $\left\{0,\dfrac{2}{3}\right\}$ **65.** $\left\{\dfrac{4-\sqrt{2}}{4},\dfrac{4+\sqrt{2}}{4}\right\}$ **67.** $\varnothing$

69. $\sqrt{13}$ **71.** $2\sqrt{17}$ **73.** $\sqrt{65}$ **75.** $\{-\sqrt{2},\sqrt{2}\}$

77. $\{-5\}$ **79.** $\varnothing$ **81.** $\{-9\}$ **83.** $\left\{\dfrac{5}{4}\right\}$ **85.** $\{-9,4\}$

87. $\left\{-\dfrac{2}{3}, 2\right\}$ **89.** $\left\{-2 - 2\sqrt[4]{2}, -2 + 2\sqrt[4]{2}\right\}$ **91.** $\{0\}$

93. $\left\{\dfrac{1}{2}\right\}$ **95.** $4\sqrt{2}$ feet   **97.** $5\sqrt{2}$ feet   **99.** 50 feet

101. a) 1.89   **b)** $d = \dfrac{64b^3}{C^3}$ **c)** $d > 19{,}683$ pounds

103. $\sqrt[5]{32}$ meters   **105.** $\sqrt{73}$ kilometers (km)

107. $S = P(1 + r)^n, P = S(1 + r)^{-n}$ **109.** 9.5 AU

111. $\{-1.8, 1.8\}$ **113.** $\{4.993\}$ **115.** $\{-26.372, 26.372\}$

Section 9.6 Warm-ups T F F T T T T F T F

1. A complex number is a number of the form $a + bi$, where a and b are real numbers.

3. The union of the real numbers and the imaginary numbers is the set of complex numbers.

5. The conjugate of $a + bi$ is $a - bi$.

7. $-2 + 8i$ **9.** $-4 + 4i$ **11.** -2 **13.** $-8 - 2i$

15. $6 + 15i$ **17.** $-2 - 10i$ **19.** $-4 - 12i$

21. $-10 + 24i$ **23.** $-1 + 3i$ **25.** $-5i$ **27.** 29

29. 2   **31.** 20   **33.** -9 **35.** -25 **37.** 16   **39.** i

41. 34   **43.** 5   **45.** 5   **47.** 7   **49.** $\dfrac{12}{17} - \dfrac{3}{17}i$

51. $\dfrac{4}{13} + \dfrac{7}{13}i$ **53.** $3 - 4i$ **55.** $1 + 3i$

57. $2 + 2i$ **59.** $5 + 6i$ **61.** $7 - i\sqrt{6}$ **63.** $5i\sqrt{2}$

65. $1 + i\sqrt{3}$ **67.** $-1 - \dfrac{1}{2}i\sqrt{6}$ **69.** $\{\pm 6i\}$

71. $\{\pm 2i\sqrt{3}\}$ **73.** $\left\{\pm\dfrac{i\sqrt{10}}{2}\right\}$ **75.** $\{\pm i\sqrt{2}\}$

77. $18 - i$ **79.** $5 + i$ **81.** $-\dfrac{6}{25} - \dfrac{17}{25}i$ **83.** $3 + 2i$

85. -9 **87.** $3i\sqrt{3}$ **89.** $-5 - 12i$ **91.** $-2 + 2i\sqrt{2}$

Enriching Word Power

1. d   **2.** b   **3.** b   **4.** b   **5.** d   **6.** b   **7.** c   **8.** a
9. d   **10.** c   **11.** a   **12.** c   **13.** d   **14.** b

Chapter 9 Review

1. 2   **3.** 10   **5.** $6\sqrt{2}$ **7.** x^6 **9.** x^2 **11.** $\dfrac{\sqrt{10}}{5}$

13. $\dfrac{\sqrt[3]{18}}{3}$ **15.** $\dfrac{2\sqrt{3x}}{3x}$ **17.** $\dfrac{y\sqrt{15y}}{3}$ **19.** $\dfrac{3\sqrt[3]{4a^2}}{2a}$

21. $\dfrac{5\sqrt[3]{27x^2}}{3x}$ **23.** $\dfrac{1}{9}$ **25.** 4   **27.** $\dfrac{1}{1000}$ **29.** $27x^{1/2}$

31. $a^{7/2}b^{7/2}$ **33.** $x^{3/4}y^{5/4}$ **35.** 13   **37.** $3\sqrt{5} - 2\sqrt{3}$

39. $\dfrac{10\sqrt{3}}{3}$ **41.** $30 - 21\sqrt{6}$

43. $6 - 3\sqrt{3} + 2\sqrt{2} - \sqrt{6}$ **45.** $\dfrac{5\sqrt{2}}{2}$ **47.** 9

49. $1 - \sqrt{2}$ **51.** $\dfrac{-\sqrt{6} - 3\sqrt{2}}{2}$ **53.** $\dfrac{3\sqrt{2} + 2}{7}$

55. $256w^{10}$ **57.** $\{-4, 4\}$ **59.** $\{3, 7\}$

61. $\{-1 - \sqrt{5}, -1 + \sqrt{5}\}$ **63.** $\varnothing$ **65.** $\{10\}$ **67.** $\{9\}$

69. $\{-8, 8\}$ **71.** $\{124\}$ **73.** $\{7\}$ **75.** $\{2, 3\}$ **77.** $\{9\}$

79. $\{4\}$ **81.** $5 + 25i$ **83.** $7 - 3i$ **85.** $-1 + 2i$

87. $2 + i$ **89.** $2 - i\sqrt{3}$ **91.** $\dfrac{5}{17} - \dfrac{14}{17}i$ **93.** $\{\pm 10i\}$

95. $\left\{\pm\dfrac{3i\sqrt{2}}{2}\right\}$ **97.** False   **99.** True   **101.** True

103. False   **105.** False   **107.** False   **109.** False
111. True   **113.** False   **115.** True   **117.** False
119. False   **121.** True   **123.** True   **125.** $2\sqrt{58}$

127. $5\sqrt{30}$ seconds   **129.** $10\sqrt{7}$ feet   **131.** $200\sqrt{2}$ feet

133. $26.4\sqrt[3]{25}$ ft^2 **135.** 6.7%   **137.** $V = \dfrac{29\sqrt{LCS}}{CS}$

Chapter 9 Test

1. 4   **2.** $\dfrac{1}{8}$ **3.** $\sqrt{3}$ **4.** 30   **5.** $3\sqrt{5}$ **6.** $\dfrac{6\sqrt{5}}{5}$

7. 2   **8.** $6\sqrt{2}$ **9.** $\dfrac{\sqrt{15}}{6}$ **10.** $\dfrac{2 + \sqrt{2}}{2}$ **11.** $4 - 3\sqrt{3}$

12. $2ay^2\sqrt[4]{2a}$ **13.** $\dfrac{\sqrt[3]{4x}}{2x}$ **14.** $\dfrac{2a^4\sqrt{2ab}}{b^2}$ **15.** $-3x^3$

16. $2m\sqrt{5m}$ **17.** $x^{3/4}$ **18.** $3y^2x^{1/4}$ **19.** $2x^2\sqrt[3]{5x}$

20. $19 + 8\sqrt{3}$ **21.** $\dfrac{5 + \sqrt{3}}{11}$ **22.** $\dfrac{6\sqrt{2} - \sqrt{3}}{23}$

23. $22 + 7i$ **24.** $1 - i$ **25.** $\dfrac{1}{5} - \dfrac{7}{5}i$ **26.** $-\dfrac{3}{4} + \dfrac{1}{4}i\sqrt{3}$

27. $\{-5, 9\}$ **28.** $\left\{-\dfrac{7}{4}\right\}$ **29.** $\{-8, 8\}$ **30.** $\left\{\pm\dfrac{4}{3}i\right\}$ **31.** $\{3\}$

32. $\{5\}$ **33.** $2\sqrt{2}$ **34.** $\dfrac{3\sqrt{2}}{2}$ feet   **35.** 25 and 36

36. Length 6 ft, width 4 ft   **37.** 39.53 AU, 164.97 years

Making Connections Chapters 1–9

1. $\left\{-\dfrac{4}{7}\right\}$ **2.** $\left\{\dfrac{3}{2}\right\}$

3. $(-\infty, -3) \cup (-2, \infty)$

```
     ←———)———(——→
   -5 -4 -3 -2 -1  0
```

4. $\left\{\dfrac{3}{2}\right\}$ **5.** $(-\infty, 1)$

```
   ←————————)—+—+—→
   -3 -2 -1  0  1  2  3
```

6. $\varnothing$ **7.** $\{9\}$ **8.** $\varnothing$ **9.** $\{-12, -2\}$ **10.** $\left\{\dfrac{1}{16}\right\}$

11. $(-6, \infty)$

```
   ←+—+—(———————→
  -8 -7 -6 -5 -4 -3 -2
```

12. $\left\{-\dfrac{1}{64}, \dfrac{1}{64}\right\}$ **13.** $\left\{-\dfrac{\sqrt{3}}{3}, \dfrac{\sqrt{3}}{3}\right\}$ **14.** R

15. $\left(-\dfrac{1}{3}, 3\right)$

```
           -⅓
   ←+—+—(———————)—+—→
  -2 -1  0  1  2  3  4
```

16. $\left\{\dfrac{1}{3}\right\}$ **17.** $\{82\}$ **18.** $\left\{\dfrac{6}{5}, \dfrac{12}{5}\right\}$ **19.** $\{100\}$ **20.** R

21. $\{4\sqrt{30}\}$ **22.** $\{400\}$ **23.** $\left\{\dfrac{13 + 9\sqrt{2}}{3}\right\}$

24. $\{-3\sqrt{2}, 3\sqrt{2}\}$ **25.** $\{5\}$ **26.** $\{7 + 3\sqrt{6}\}$

27. $\{-2, 3\}$ **28.** $\{-5, 2\}$ **29.** $\{-2, 3\}$ **30.** $\left\{\dfrac{1}{2}, 3\right\}$

31. 3 **32.** -2 **33.** $\dfrac{1}{2}$ **34.** $\dfrac{1}{3}$ **35.** 48.5 cm^3

36. 14%, 56 cm^3

Chapter 10

Section 10.1 Warm-ups F F F F T F F T F F

1. In this section, quadratic equations are solved by factoring, the even root property, and completing the square.
3. The last term is the square of one-half the coefficient of the middle term.

5. $\{-2, 3\}$ **7.** $\{-5, 3\}$ **9.** $\left\{-1, \dfrac{3}{2}\right\}$ **11.** $\{-7\}$

13. $\{-4, 4\}$ **15.** $\{-9, 9\}$ **17.** $\left\{-\dfrac{4}{3}, \dfrac{4}{3}\right\}$ **19.** $\{-1, 7\}$

21. $\{-1 - \sqrt{5}, -1 + \sqrt{5}\}$ **23.** $\left\{\dfrac{3 - \sqrt{7}}{2}, \dfrac{3 + \sqrt{7}}{2}\right\}$

25. $x^2 + 2x + 1$ **27.** $x^2 - 3x + \dfrac{9}{4}$ **29.** $y^2 + \dfrac{1}{4}y + \dfrac{1}{64}$

31. $x^2 + \dfrac{2}{3}x + \dfrac{1}{9}$ **33.** $(x + 4)^2$ **35.** $\left(y - \dfrac{5}{2}\right)^2$

37. $\left(z - \dfrac{2}{7}\right)^2$ **39.** $\left(t + \dfrac{3}{10}\right)^2$ **41.** $\{-3, 5\}$

43. $\{-10, 2\}$ **45.** $\{-5, 7\}$ **47.** $\{-4, 5\}$ **49.** $\{-7, 2\}$

51. $\left\{-1, \dfrac{3}{2}\right\}$ **53.** $\{-2 - \sqrt{10}, -2 + \sqrt{10}\}$

55. $\{-4 - 2\sqrt{5}, -4 + 2\sqrt{5}\}$

57. $\left\{\dfrac{-3 - \sqrt{41}}{4}, \dfrac{-3 + \sqrt{41}}{4}\right\}$ **59.** $\{4\}$

61. $\left\{\dfrac{1 + \sqrt{17}}{8}\right\}$ **63.** $\{1, 6\}$ **65.** $\{-2 - \sqrt{2}, -2 + \sqrt{2}\}$

67. $\{-1 - 2i, -1 + 2i\}$ **69.** $\{-2i\sqrt{3}, 2i\sqrt{3}\}$

71. $\left\{\dfrac{2 \pm i}{5}\right\}$ **73.** $\left\{-\dfrac{5}{2}i, \dfrac{5}{2}i\right\}$ **75.** $\{-2, 1\}$

77. $\left\{\dfrac{-2 - \sqrt{19}}{5}, \dfrac{-2 + \sqrt{19}}{5}\right\}$ **79.** $\{-6, 4\}$

81. $\left\{\dfrac{-2 - 4i\sqrt{2}}{3}, \dfrac{-2 + 4i\sqrt{2}}{3}\right\}$ **83.** $\{-2, 3\}$

85. $\{3 - i, 3 + i\}$ **87.** $\{6\}$ **89.** $\left\{\dfrac{9 - \sqrt{65}}{2}, \dfrac{9 + \sqrt{65}}{2}\right\}$

95. 136.9 ft/sec **97.** 12 **99.** c **103.** $\{4.56, 2.74\}$
105. $\{3.53\}$

Section 10.2 Warm-ups T F T F T T T T F F

1. The quadratic formula can be used to solve any quadratic equation.
3. Factoring is used when the quadratic polynomial is simple enough to factor.

5. The discriminant is $b^2 - 4ac$.

7. $\{-3, -2\}$ **9.** $\{-3, 2\}$ **11.** $\left\{-\dfrac{1}{3}, \dfrac{3}{2}\right\}$ **13.** $\left\{\dfrac{1}{2}\right\}$

15. $\left\{\dfrac{1}{3}\right\}$ **17.** $\left\{-\dfrac{3}{4}\right\}$ **19.** $\{-4 \pm \sqrt{10}\}$

21. $\left\{\dfrac{-5 \pm \sqrt{29}}{2}\right\}$ **23.** $\left\{\dfrac{3 \pm \sqrt{7}}{2}\right\}$ **25.** $\left\{\dfrac{3 \pm i}{2}\right\}$

27. $\left\{\dfrac{3 \pm i\sqrt{39}}{4}\right\}$ **29.** $\{5 \pm i\}$ **31.** 28, 2 **33.** $-23, 0$

35. 0, 1 **37.** $-\dfrac{3}{4}, 0$ **39.** 97, 2 **41.** 0, 1 **43.** 140, 2

45. 1, 2 **47.** $\left\{-2, \dfrac{1}{2}\right\}$ **49.** $\{0\}$ **51.** $\left\{\dfrac{13}{9}, \dfrac{17}{9}\right\}$

53. $\{4 \pm 2i\}$ **55.** $\{2 \pm i\sqrt{6}\}$ **57.** $\left\{-\dfrac{3}{4}, \dfrac{5}{2}\right\}$

59. $\{-4.474, 1.274\}$ **61.** $\{3.7\}$ **63.** $\{-2.979, -0.653\}$
65. $\{-4792.983, -0.017\}$ **67.** $\{-0.079, 0.078\}$

69. $\dfrac{1 + \sqrt{65}}{2}$ and $\dfrac{-1 + \sqrt{65}}{2}$ **71.** $3 + \sqrt{5}$ and $3 - \sqrt{5}$

73. Width $\dfrac{-1 + \sqrt{5}}{2}$ ft, length $\dfrac{1 + \sqrt{5}}{2}$ ft

75. Width $-2 + \sqrt{14}$ ft, length $2 + \sqrt{14}$ ft

77. 3 sec **79.** 7.02 sec **81.** 80 **83.** 4 in.
89. $\{0.652, 5.678\}$ **91.** $\varnothing$ **93.** $\varnothing$

Section 10.3 Warm-ups T F T T F T T T T F

1. A quadratic function is a function of the form $y = ax^2 + bx + c$ with $a \neq 0$.
3. If $a > 0$ then the parabola opens upward. If $a < 0$ then the parabola opens downward.
5. The vertex is the highest point on a parabola that opens downward or the lowest point on a parabola that opens upward.
7. $(3, -6), (4, 0), (-3, 0)$ **9.** $(4, -128), (0, 0), (2, 0)$
11. Domain $(-\infty, \infty)$, range $[2, \infty)$

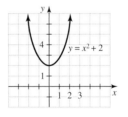

13. Domain $(-\infty, \infty)$, range $[-4, \infty)$

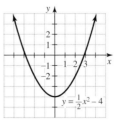

15. Domain $(-\infty, \infty)$, range $(-\infty, 5]$

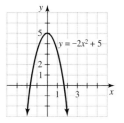

17. Domain $(-\infty, \infty)$, range $(-\infty, 5]$

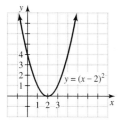

19. Domain $(-\infty, \infty)$, range $[0, \infty)$

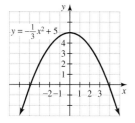

21. Vertex $\left(\dfrac{1}{2}, -\dfrac{9}{4}\right)$, intercepts $(0, -2)$, $(-1, 0)$, $(2, 0)$, domain $(-\infty, \infty)$, range $\left[-\dfrac{9}{4}, \infty\right)$

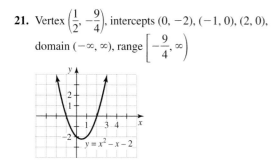

23. Vertex $(-1, -9)$, intercepts $(0, -8)$, $(-4, 0)$, $(2, 0)$, domain $(-\infty, \infty)$, range $[-9, \infty)$

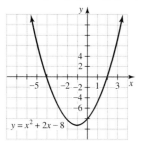

25. Vertex $(-2, 1)$, intercepts $(0, -3)$, $(-1, 0)$, $(-3, 0)$, domain $(-\infty, \infty)$, range $(-\infty, 1]$

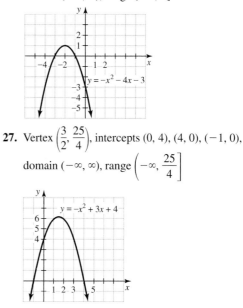

27. Vertex $\left(\dfrac{3}{2}, \dfrac{25}{4}\right)$, intercepts $(0, 4)$, $(4, 0)$, $(-1, 0)$, domain $(-\infty, \infty)$, range $\left(-\infty, \dfrac{25}{4}\right]$

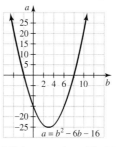

29. Vertex $(3, -25)$, intercepts $(0, -16)$, $(8, 0)$, $(-2, 0)$, domain $(-\infty, \infty)$, range $[-25, \infty)$

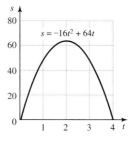

31. Minimum -8 **33.** Maximum 14 **35.** Minimum 2
37. Maximum 2
39. Maximum 64 feet

41. 100   **43.** 625 square meters
45. 2 P.M.   **47.** 15 meters, 25 meters
49. The graph of $y = kx^2$ gets narrower as k gets larger.
51. The graph of $y = x^2$ has the same shape as $x = y^2$.
53. a) Vertex $(0.84, -0.68)$, domain $(-\infty, \infty)$, range $[-0.68, \infty)$, $(1.30, 0)$, $(0.38, 0)$
b) Vertex $(5.96, 46.26)$, domain $(-\infty, \infty)$, range $(-\infty, 46.26]$, $(12.48, 0)$, $(-0.55, 0)$

Section 10.4   Warm-ups T F F T F F F T F F

1. If the coefficients are integers and the discriminant is a perfect square, then the quadratic polynomial can be factored.
3. If the solutions are a and b, then the quadratic equation $(x - a)(x - b) = 0$ has those solutions.
5. Prime   **7.** Prime   **9.** $(3x - 4)(2x + 9)$
11. Prime   **13.** $(4x - 15)(2x + 3)$
15. $x^2 + 4x - 21 = 0$ **17.** $x^2 - 5x + 4 = 0$
19. $x^2 - 5 = 0$ **21.** $x^2 + 16 = 0$ **23.** $x^2 + 2 = 0$
25. $6x^2 - 5x + 1 = 0$ **27.** $\left\{-\dfrac{3}{2}, \dfrac{3}{2}\right\}$ **29.** $\left\{\dfrac{-3 \pm \sqrt{5}}{2}\right\}$
31. $\{\pm\sqrt{5}, \pm 3\}$ **33.** $\{-2, 1\}$ **35.** $\{-1 \pm \sqrt{5}, -3, 1\}$
37. $\{-3, -2, 1, 2\}$ **39.** $\{16, 81\}$ **41.** $\{9\}$ **43.** $\left\{-\dfrac{1}{3}, \dfrac{1}{2}\right\}$
45. $\{64\}$ **47.** $\left\{\dfrac{2}{3}, \dfrac{3}{2}\right\}$ **49.** $\left\{\pm\dfrac{\sqrt{14}}{2}, \pm\dfrac{\sqrt{38}}{2}\right\}$
51. $\{-1 + \sqrt{2}, -1 - \sqrt{2}\}$ **53.** 2 P.M.
55. Before $-5 + \sqrt{265}$ or 11.3 mph, after $-9 + \sqrt{265}$ or 7.3 mph
57. Andrew $\dfrac{13 + \sqrt{265}}{2}$ or 14.6 hours, John $\dfrac{19 + \sqrt{265}}{2}$ or 17.6 hours
59. Length $5 + 5\sqrt{41}$ or 37.02 ft, width $-5 + 5\sqrt{41}$ or 27.02 ft
61. $14 + 2\sqrt{58}$ or 29.2 hours
63. $-5 + 5\sqrt{5}$ or 6.2 meters
67. $\{-10, -4, 4, 10\}$ **69.** $\{4.27\}$

Section 10.5   Warm-ups F F F F T T T T T F

1. A quadratic inequality has the form $ax^2 + bx + c > 0$. In place of $>$ we can also use $<$, $\leq$, or $\geq$.
3. A rational inequality is an inequality involving a rational expression.
5. $(-3, 2)$

7. $(-\infty, -2) \cup (2, \infty)$

9. $(-\infty, -4] \cup \left[\dfrac{3}{2}, \infty\right)$

11. $(-\infty, 0] \cup [2, \infty)$

13. $(-\infty, 0) \cup \left(\dfrac{1}{2}, \infty\right)$

15. $(-\infty, \infty)$

17. $(-\infty, 0) \cup (3, \infty)$

19. $[-2, 0)$

21. $(-\infty, -6) \cup (3, \infty)$

23. $(-\infty, -2) \cup (-1, \infty)$

25. $(-13, -4) \cup (5, \infty)$

27. $(-\infty, -5) \cup (1, 3) \cup (5, \infty)$

29. $[-6, 3) \cup [4, 6)$

31. $(-\infty, 1 - \sqrt{5}) \cup (1 + \sqrt{5}, \infty)$

33. $\left(-\infty, \dfrac{3 - \sqrt{3}}{2}\right] \cup \left[\dfrac{3 + \sqrt{3}}{2}, \infty\right)$

35. $\left[\dfrac{3 - 3\sqrt{5}}{2}, \dfrac{3 + 3\sqrt{5}}{2}\right]$

37. $[-3, 3]$ **39.** $(-4, 4)$ **41.** $(-\infty, 0] \cup [4, \infty)$
43. $\left(-\dfrac{3}{2}, \dfrac{5}{3}\right)$ **45.** $(-\infty, -2] \cup [6, \infty)$
47. $(-\infty, -3) \cup (5, \infty)$ **49.** $(-\infty, -4] \cup [2, \infty)$
51. $(-3, 4]$ **53.** $[-1, 2] \cup [5, \infty)$
55. $(-\infty, -3) \cup (-1, 1)$ **57.** $(-27.58, -0.68)$
59. $(-\infty, -2 - \sqrt{6}) \cup (-3, -2 + \sqrt{6}) \cup (2, \infty)$
61. $6, 7, 8, \ldots$ **63.** 4 seconds
65. a) 900 ft   **b)** 3 sec   **c)** 3 sec
67. a) (h, k) **b)** $(-\infty, h) \cup (k, \infty)$
    c) $(-k, -h)$ **d)** $(-\infty, -k] \cup [-h, \infty)$
    e) $(-\infty, h] \cup (k, \infty)$ **f)** $(-k, -h]$
69. c   **71.** b

Enriching Word Power

1. b   **2.** a   **3.** d   **4.** c   **5.** b   **6.** c
7. c   **8.** a   **9.** c   **10.** a   **11.** c

Chapter 10 Review

1. $\{-3, 5\}$ **3.** $\left\{-3, \dfrac{5}{2}\right\}$ **5.** $\{-5, 5\}$ **7.** $\left\{\dfrac{3}{2}\right\}$
9. $\{\pm 2\sqrt{3}\}$ **11.** $\{-2, 4\}$ **13.** $\left\{\dfrac{4 \pm \sqrt{3}}{2}\right\}$ **15.** $\left\{\pm\dfrac{3}{2}\right\}$
17. $\{2, 4\}$ **19.** $\{2, 3\}$ **21.** $\left\{\dfrac{1}{2}, 3\right\}$ **23.** $\{-2 \pm \sqrt{3}\}$
25. $\{-2, 5\}$ **27.** $\left\{-\dfrac{1}{3}, \dfrac{3}{2}\right\}$ **29.** $\{-2 \pm \sqrt{2}\}$
31. $\left\{\dfrac{5 \pm \sqrt{13}}{6}\right\}$ **33.** $0, 1$ **35.** $-19, 0$ **37.** $17, 2$
39. $\left\{\dfrac{2 \pm i\sqrt{2}}{2}\right\}$ **41.** $\left\{\dfrac{3 \pm i\sqrt{15}}{4}\right\}$ **43.** $\left\{\dfrac{-1 \pm i\sqrt{5}}{3}\right\}$
45. $\{-3 \pm i\sqrt{7}\}$

47. Vertex $(3, -9)$, intercepts $(0, 0)$, $(6, 0)$

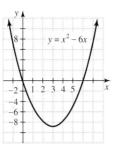

49. Vertex $(2, -16)$, intercepts $(0, -12)$, $(-2, 0)$, and $(6, 0)$

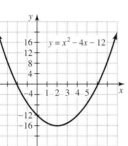

51. Vertex $(2, 8)$, intercepts $(0, 0)$, $(4, 0)$

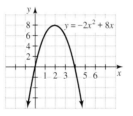

53. Vertex $(1, 4)$, intercepts $(0, 3)$, $(-1, 0)$, $(3, 0)$

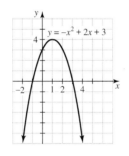

55. Domain $(-\infty, \infty)$, range $[-3, \infty)$
57. Domain $(-\infty, \infty)$, range $(-\infty, 4.125]$
59. $(4x + 1)(2x - 3)$ **61.** Prime
63. $(4y - 5)(2y + 5)$ **65.** $x^2 + 9x + 18 = 0$
67. $x^2 - 50 = 0$ **69.** $\{-2, 1\}$
71. $\{\pm 2, \pm 3\}$ **73.** $\{-6, -5, 2, 3\}$ **75.** $\{-2, 8\}$

77. $\left\{-\dfrac{1}{9}, \dfrac{1}{4}\right\}$ **79.** $\{16, 81\}$

81. $(-\infty, -3) \cup (2, \infty)$ **83.** $[-4, 5]$

$\xleftarrow{\hspace{0.5em}} \quad \overset{(\hspace{3em})}{\underset{-4\,-3\,-2\,-1\ 0\ 1\ \ 2\ 3}{}} \quad \xrightarrow{\hspace{0.5em}}$ $\xleftarrow{\hspace{0.5em}} \quad \overset{[\hspace{3.5em}]}{\underset{-4\,-3\,-2\,-1\ 0\ 1\ 2\ 3\ 4\ 5}{}} \quad \xrightarrow{\hspace{0.5em}}$

85. $(0, 1)$ **87.** $(-\infty, -2) \cup [4, \infty)$

$\xleftarrow{\hspace{0.5em}} \quad \overset{(\hspace{1.5em})}{\underset{-2\,-1\ \ 0\ 1\ \ 2\ 3}{}} \quad \xrightarrow{\hspace{0.5em}}$ $\xleftarrow{\hspace{0.5em}} \quad \underset{-4\,-3\,-2\,-1\ 0\ 1\ 2\ 3\ 4\ 5\ 6}{} \quad \xrightarrow{\hspace{0.5em}}$

89. $(-3, \infty)$ **91.** $(-2, -1) \cup \left(-\dfrac{1}{2}, \infty\right)$

$\xleftarrow{\hspace{0.5em}} \quad \overset{(\hspace{4em}}{\underset{-5\,-4\ -3\,-2\,-1\ \ 0\ \ 1}{}} \quad \xrightarrow{\hspace{0.5em}}$ $\xleftarrow{\hspace{0.5em}} \quad \overset{-\frac{1}{2}}{\overset{(\ )(\hspace{3em}}{\underset{-3\ -2\,-1\ \ 0\ \ 1\ \ 2}{}}} \quad \xrightarrow{\hspace{0.5em}}$

93. $\left\{\dfrac{5}{12}\right\}$ **95.** $\left\{\dfrac{-3 \pm \sqrt{5}}{2}\right\}$ **97.** $\left\{\dfrac{4 \pm 2i}{3}\right\}$ **99.** $\left\{\dfrac{5}{2}\right\}$

101. $\left\{-2, -\dfrac{1}{4}\right\}$ **103.** $\{625, 10{,}000\}$

105. $-2 + 2\sqrt{2}$ and $2 + 2\sqrt{2}$, or 0.83 and 4.83

107. Width $\dfrac{4 + \sqrt{706}}{2}$ or 15.3 inches, height $\dfrac{-4 + \sqrt{706}}{2}$ or 11.3 inches

109. 2 inches   **111.** Width 5 ft, length 9 ft
113. \$20.40, 400   **115.** 0.5 second and 1.5 seconds

Chapter 10 Test

1. $-7, 0$ **2.** $13, 2$ **3.** $0, 1$ **4.** $\left\{-3, \dfrac{1}{2}\right\}$

5. $\left\{-3 \pm \sqrt{3}\right\}$ **6.** $\{-5\}$ **7.** $\left\{-2, \dfrac{3}{2}\right\}$ **8.** $\{-4, 3\}$

9. $\{\pm 1, \pm 2\}$ **10.** $\{11, 27\}$ **11.** $\{\pm 6i\}$ **12.** $\{-3 \pm i\}$

13. $\left\{\dfrac{1 \pm i\sqrt{11}}{6}\right\}$

14. Domain $(-\infty, \infty)$, range $(-\infty, 16]$

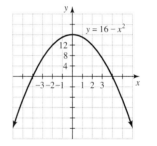

15. Domain $(-\infty, \infty)$, range $\left[-\dfrac{9}{4}, \infty\right)$

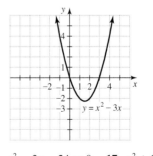

16. $x^2 - 2x - 24 = 0$ **17.** $x^2 + 25 = 0$
18. $(-6, 3)$ **19.** $(-1, 2) \cup (8, \infty)$

$\xleftarrow{\hspace{0.5em}} \quad \overset{(\hspace{4em})}{\underset{-6\,-5\,-4\,-3\,-2\,-1\ 0\ 1\ \ 2\ 3}{}} \quad \xrightarrow{\hspace{0.5em}}$ $\xleftarrow{\hspace{0.5em}} \quad \overset{(\hspace{1.5em})\hspace{3em}(}{\underset{-1\,0\ 1\ 2\ 3\ 4\ 5\ 6\ 7\ 8\,9\,10}{}} \quad \xrightarrow{\hspace{0.5em}}$

20. Width $-1 + \sqrt{17}$ ft, length $1 + \sqrt{17}$ ft

21. $\dfrac{5 + \sqrt{37}}{2}$ or 5.5 hours   **22.** $(1, 0), (5, 0)$ **23.** 36 feet

Making Connections Chapters 1–10

1. $\left\{\dfrac{15}{2}\right\}$ **2.** $\left\{\pm\dfrac{\sqrt{30}}{2}\right\}$ **3.** $\left\{-3, \dfrac{5}{2}\right\}$

4. $\left\{\dfrac{-2 \pm \sqrt{34}}{2}\right\}$ **5.** $\left\{-\dfrac{7}{2}, -2\right\}$

6. $\left\{-3, \dfrac{1}{4}, \dfrac{-11 \pm \sqrt{73}}{8}\right\}$ **7.** $\{9\}$

8. $\left\{-\dfrac{3}{2}, \dfrac{13}{2}\right\}$ **9.** $(-4, \infty)$ **10.** $\left[\dfrac{1}{2}, 5\right]$

11. $\left[\dfrac{1}{2}, 5\right)$ **12.** $(2, 8)$ **13.** $[-3, 2)$ **14.** $(-\infty, \infty)$

15. $y = \dfrac{2}{3}x - 3$ **16.** $y = -\dfrac{1}{2}x + 2$

17. $y = \dfrac{-c \pm \sqrt{c^2 - 12d}}{6}$

18. $y = \dfrac{n \pm \sqrt{n^2 + 4mw}}{2m}$ **19.** $y = \dfrac{5}{6}x - \dfrac{25}{12}$

20. $y = -\dfrac{2}{3}x + \dfrac{17}{3}$ **21.** $\dfrac{4}{3}$ **22.** $-\dfrac{11}{7}$ **23.** -2

24. $\dfrac{58}{5}$ **25.** 40,000, 38,000, \$32.50

26. \$800,000, \$950,000, \$40 or \$80, \$60

Chapter 11

Section 11.1 Warm-ups T T T F T F T T F T

1. A linear function is a function of the form $f(x) = mx + b$, where m and b are real numbers with $m \neq 0$.

3. The graph of a constant function is a horizontal line.

5. The graph of a quadratic function is a parabola.

7. $(-\infty, \infty), \{-2\}$ **9.** $(-\infty, \infty), (-\infty, \infty)$

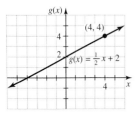 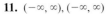   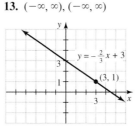

11. $(-\infty, \infty), (-\infty, \infty)$ **13.** $(-\infty, \infty), (-\infty, \infty)$

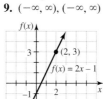   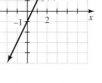

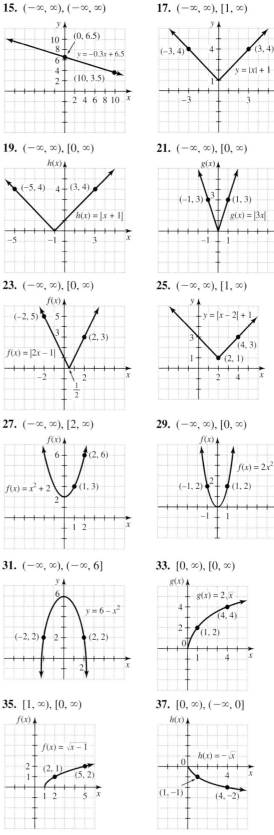

15. $(-\infty, \infty), (-\infty, \infty)$ **17.** $(-\infty, \infty), [1, \infty)$

19. $(-\infty, \infty), [0, \infty)$ **21.** $(-\infty, \infty), [0, \infty)$

23. $(-\infty, \infty), [0, \infty)$ **25.** $(-\infty, \infty), [1, \infty)$

27. $(-\infty, \infty), [2, \infty)$ **29.** $(-\infty, \infty), [0, \infty)$

31. $(-\infty, \infty), (-\infty, 6]$ **33.** $[0, \infty), [0, \infty)$

35. $[1, \infty), [0, \infty)$ **37.** $[0, \infty), (-\infty, 0]$

39. $[0, \infty)$, $[2, \infty)$

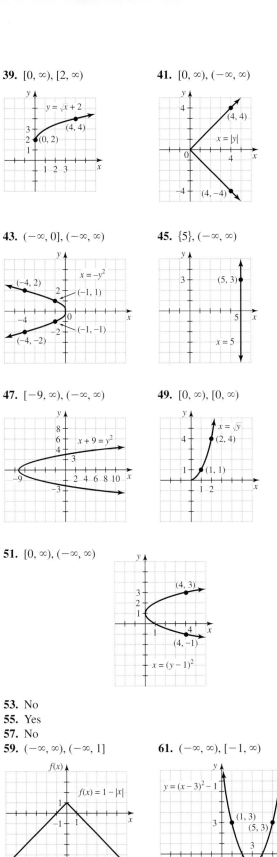

41. $[0, \infty)$, $(-\infty, \infty)$

43. $(-\infty, 0]$, $(-\infty, \infty)$

45. $\{5\}$, $(-\infty, \infty)$

47. $[-9, \infty)$, $(-\infty, \infty)$

49. $[0, \infty)$, $[0, \infty)$

51. $[0, \infty)$, $(-\infty, \infty)$

53. No

55. Yes

57. No

59. $(-\infty, \infty)$, $(-\infty, 1]$

61. $(-\infty, \infty)$, $[-1, \infty)$

63. $(-\infty, \infty)$, $[1, \infty)$

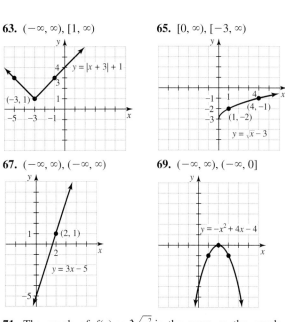

65. $[0, \infty)$, $[-3, \infty)$

67. $(-\infty, \infty)$, $(-\infty, \infty)$

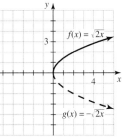

69. $(-\infty, \infty)$, $(-\infty, 0]$

71. The graph of $f(x) = \sqrt{x^2}$ is the same as the graph of $f(x) = |x|$.

73. For large values of k the graph gets narrower and for smaller values of k the graph gets broader.

75. The graph of $y = (x - k)^2$ moves to the right for $k > 0$ and to the left for $k < 0$.

77. The graph of $y = f(x - k)$ lies to the right of the graph of $y = f(x)$ when $k > 0$.

Section 11.2 Warm-ups F T T F T T F T T F

1. The graph of $y = -f(x)$ is a reflection in the x-axis of the graph of $y = f(x)$.

3. The graph of $y = f(x) + k$ for $k < 0$ is a downward translation of $y = f(x)$.

5. The graph of $y = f(x + k)$ for $k > 0$ is a translation to the left of $y = f(x)$.

7.

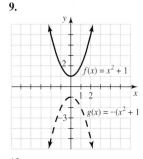

9.

11.

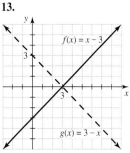

13.

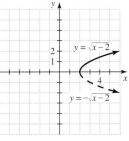

15.

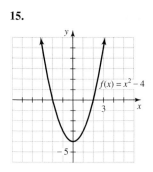

17.

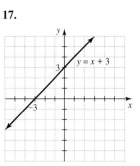

19.

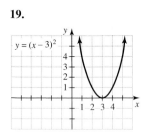

21.

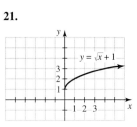

23.

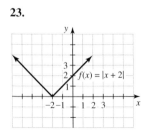

25.

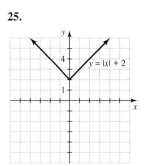

27.

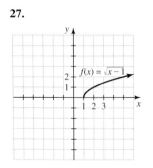

29.

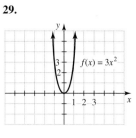

31.

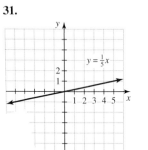

33.

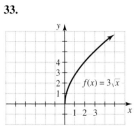

35.

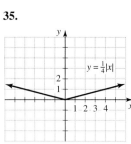

37.

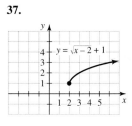

39.

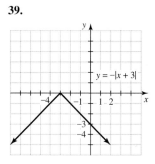

41.

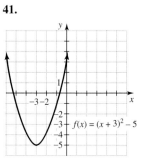

43.

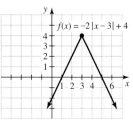

45.

47.

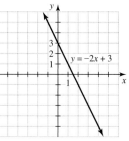

49.

51.

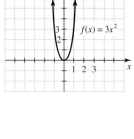

53.

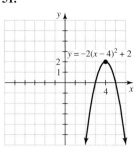

55. d   **57.** e   **59.** h   **61.** c   **63.** $y = x^2 + 8$

65. $y = \sqrt{x + 5}$ **67.** $y = |x + 3| + 5$
69. a) $y = 0, x = 0$ **b)** $y = 30, x = 20$
   c) move f to the right 20 units and upward 30 units

Section 11.3 Warm-ups T T F F F F T T T T
1. The basic operations of functions are addition, subtraction, multiplication, and division.
3. In the composition of functions the second function is evaluated on the result of the first function.
5. $x^2 + 2x - 3$ **7.** $4x^3 - 11x^2 + 6x$ **9.** 12   **11.** -30
13. -21 **15.** $\dfrac{13}{8}$ **17.** $y = 6x - 20$ **19.** $y = x + 2$
21. $y = x^2 + 2x$ **23.** $y = x$ **25.** -2 **27.** 5   **29.** 7
31. 5   **33.** 5   **35.** 4   **37.** 22.2992   **39.** $4x^2 - 6x$
41. $2x^2 + 6x - 3$ **43.** x **45.** $4x - 9$ **47.** $\dfrac{x + 9}{4}$
49. $F = f \circ h$ **51.** $G = g \circ h$ **53.** $H = h \circ g$
55. $J = h \circ h$ **57.** $K = g \circ g$ **67.** $112.5 \text{ in.}^2, A = \dfrac{d^2}{2}$
69. $P(x) = -x^2 + 20x - 170$ **71.** $J = 0.025I$
73. a) 397.8   **b)** $D = \dfrac{1.116 \times 10^7}{L^3}$ **c)** decreases
75. $[0, \infty), [0, \infty), [16, \infty)$ **77.** $[0, \infty), [0, \infty)$

Section 11.4 Warm-ups F F F T T T F T T T F
1. The inverse of a function is a function with the same ordered pairs except that the coordinates are reversed.
3. The range of f^{-1} is the domain of f.
5. A function is one-to-one if no two ordered pairs have the same second coordinate with different first coordinates.
7. The switch-and-solve strategy is used to find a formula for an inverse function.
9. No   **11.** Yes, $\{(4, 16), (3, 9), (0, 0)\}$ **13.** No
15. Yes, $\{(0, 0), (2, 2), (9, 9)\}$ **17.** No   **19.** Yes   **21.** Yes
23. Yes   **25.** Yes   **27.** No   **29.** $f^{-1}(x) = \dfrac{x}{5}$
31. $g^{-1}(x) = x + 9$ **33.** $k^{-1}(x) = \dfrac{x + 9}{5}$ **35.** $m^{-1}(x) = \dfrac{2}{x}$
37. $f^{-1}(x) = x^3 + 4$ **39.** $f^{-1}(x) = \dfrac{3}{x} + 4$
41. $f^{-1}(x) = \dfrac{x^3 - 7}{3}$ **43.** $f^{-1}(x) = \dfrac{2x + 1}{x - 1}$
45. $f^{-1}(x) = \dfrac{1 + 4x}{3x - 1}$ **47.** $p^{-1}(x) = x^4$ for $x \geq 0$
49. $f^{-1}(x) = 2 + \sqrt{x}$ **51.** $f^{-1}(x) = \sqrt{x - 3}$
53. $f^{-1}(x) = x^2 - 2$ for $x \geq 0$
55. $f^{-1}(x) = \dfrac{1}{2}x - \dfrac{3}{2}$ **57.** $f^{-1}(x) = \sqrt{x + 1}$

59. $f^{-1}(x) = \dfrac{x}{5}$ **61.** $f^{-1}(x) = \sqrt[3]{x}$

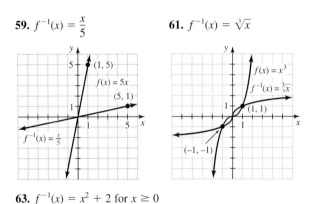

63. $f^{-1}(x) = x^2 + 2$ for $x \geq 0$

65. $(f^{-1} \circ f)(x) = x$ **67.** $(f^{-1} \circ f)(x) = x$
69. $(f^{-1} \circ f)(x) = x$ **71.** $(f^{-1} \circ f)(x) = x$
73. a) 33.5 mph   **b)** decreases   **c)** $L = \dfrac{S^2}{30}$
75. $T(x) = 1.09x + 125, T^{-1}(x) = \dfrac{x - 125}{1.09}$
77. An odd positive integer   **79.** Not inverses

Section 11.5 Warm-ups T F F T F T F F F T
1. If y varies directly as x, then $y = kx$ for some constant k.
3. If y is inversely proportional to x, then $y = k/x$.
5. If y is jointly proportional to x and z, then $y = kxz$ for some constant k.
7. $a = km$ **9.** $d = k/e$ **11.** $I = krt$ **13.** $m = kp^2$
15. $B = k\sqrt[3]{w}$ **17.** $t = \dfrac{k}{x^2}$ **19.** $v = \dfrac{km}{n}$ **21.** $y = \dfrac{3}{2}x$
23. $A = \dfrac{30}{B}$ **25.** $m = \dfrac{36}{\sqrt{p}}$ **27.** $A = \dfrac{2}{5}tu$
29. $y = \dfrac{1.067x}{z}$ **31.** $-\dfrac{21}{5}$ **33.** $-\dfrac{3}{2}$ **35.** $\dfrac{1}{2}$
37. 3.293   **39.** Inverse   **41.** Direct   **43.** Combined
45. Joint   **47.** \$420   **49.** 9 cm^3 **51.** \$360
53. \$18.90   **55. a)** $d = 16t^2$ **b)** 4 feet   **c)** 2.5 seconds
57. 400 pounds   **59.** 7.2 days
61. a) $G = \dfrac{Nd}{c}$ **b)** 84   **c)** 23, 20, 17, 15, 13   **d)** decreases
63. $(1, 1)$, y_1 increasing, y_2 decreasing, y_1 direct variation, y_2 inverse variation

Section 11.6 Warm-ups T T F T T T T T T T
1. A zero of the function f is a number a such that $f(a) = 0$.
3. Two statements are equivalent means that they are either both true or both false.

5. If the remainder is zero when $P(x)$ is divided by $x - c$, then $P(c) = 0$.

7. $x^2 - 3x, -3$ **9.** $2x - 6, 11$

11. $3x^3 + 9x^2 + 12x + 43, 120$

13. $x^4 + x^3 + x^2 + x + 1, 0$ **15.** $x^2 - 2x - 1, 8$

17. $2.3x + 0.596, 0.79072$ **19.** Yes **21.** Yes **23.** No

25. Yes **27.** No **29.** Yes **31.** Yes

33. $(x - 3)(x^2 + 3x + 3)$ **35.** $(x + 5)(x + 3)(x + 1)$

37. No **39.** $(x + 1)(x - 2)(x^2 + 2x + 4)$

41. $(2x - 1)(x - 3)(x + 2)$ **43.** $\{-4, 1, 3\}$ **45.** $\left\{-\frac{1}{2}, 2, 3\right\}$

47. $\left\{-4, -\frac{1}{2}, 6\right\}$ **49.** $\{-3, 1\}$ **51.** $\{-1, 1, 2\}$

53. a) $f(x) = x + 2$ **b)** $f(x) = x^2 - 25$
   c) $f(x) = (x - 1)(x + 3)(x - 4)$
   d) Yes, the degree is the same as the number of zeros

55. a) $\{-2, 1, 5\}$ **b)** $\left\{-\frac{3}{2}, \frac{3}{2}, \frac{5}{2}\right\}$

Enriching Word Power

1. a **2.** b **3.** d **4.** d **5.** a **6.** b **7.** d **8.** c
9. d **10.** b **11.** b **12.** a **13.** d

Chapter 11 Review

1. $(-\infty, \infty), (-\infty, \infty)$

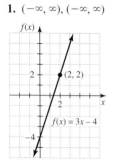

3. $(-\infty, \infty), [-2, \infty)$

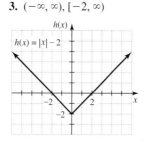

5. $(-\infty, \infty), [0, \infty)$

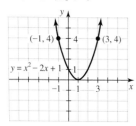

7. $[0, \infty), [2, \infty)$

9. $(-\infty, \infty), (-\infty, 30]$

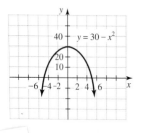

11. $\{2\}, (-\infty, \infty)$

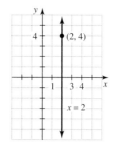

13. $[1, \infty), (-\infty, \infty)$

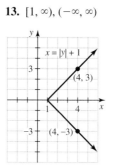

15.

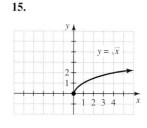

17.

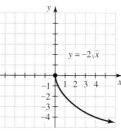

19.

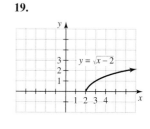

21.

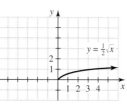

23.

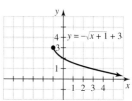

25. -4 **27.** $\sqrt{2}$ **29.** 99 **31.** 17
33. $3x^3 - x^2 - 10x$ **35.** 20 **37.** $F = f \circ g$
39. $H = g \circ h$ **41.** $I = g \circ g$ **43.** No
45. Yes, $f^{-1}(x) = x/8$ **47.** Yes, $g^{-1}(x) = \dfrac{x + 6}{13}$
49. Yes, $j^{-1}(x) = \dfrac{x + 1}{x - 1}$ **51.** No
53. $f^{-1}(x) = \dfrac{1}{3}x + \dfrac{1}{3}$ **55.** $f^{-1}(x) = \sqrt[3]{2x}$

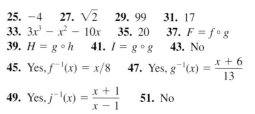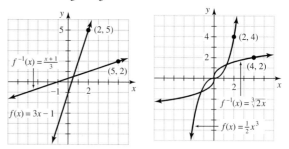

57. 24 **59.** -16 **61.** $\{-3, 2, 5\}$
63. $\{-2, 3, 4\}$ **65.** $(x - 3)(x + 2)(x + 5)$
67. No **69.** 256 ft **71.** $B = \dfrac{2A}{\pi}$
73. $a = 15w - 16$ **75.** $A = s^2, s = \sqrt{A}$

Chapter 11 Test

1.

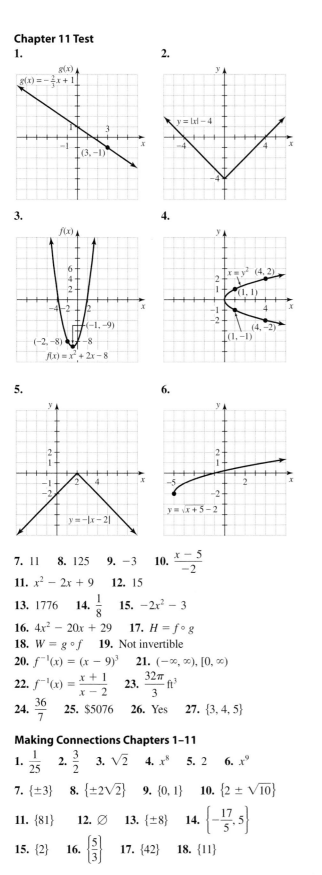

2.

3.

4.

5.

6.

7. 11   **8.** 125   **9.** -3 **10.** $\dfrac{x-5}{-2}$

11. $x^2 - 2x + 9$ **12.** 15

13. 1776   **14.** $\dfrac{1}{8}$ **15.** $-2x^2 - 3$

16. $4x^2 - 20x + 29$ **17.** $H = f \circ g$

18. $W = g \circ f$ **19.** Not invertible

20. $f^{-1}(x) = (x-9)^3$ **21.** $(-\infty, \infty), [0, \infty)$

22. $f^{-1}(x) = \dfrac{x+1}{x-2}$ **23.** $\dfrac{32\pi}{3}$ ft^3

24. $\dfrac{36}{7}$ **25.** \$5076   **26.** Yes   **27.** $\{3, 4, 5\}$

Making Connections Chapters 1–11

1. $\dfrac{1}{25}$ **2.** $\dfrac{3}{2}$ **3.** $\sqrt{2}$ **4.** x^8 **5.** 2   **6.** x^9

7. $\{\pm 3\}$ **8.** $\{\pm 2\sqrt{2}\}$ **9.** $\{0, 1\}$ **10.** $\{2 \pm \sqrt{10}\}$

11. $\{81\}$ **12.** $\varnothing$ **13.** $\{\pm 8\}$ **14.** $\left\{-\dfrac{17}{5}, 5\right\}$

15. $\{2\}$ **16.** $\left\{\dfrac{5}{3}\right\}$ **17.** $\{42\}$ **18.** $\{11\}$

19.

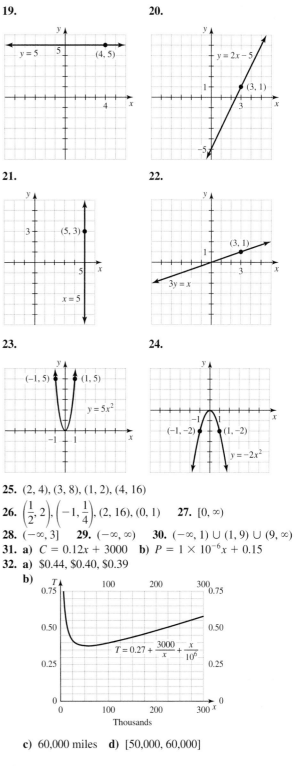

20.

21.

22.

23.

24.

25. $(2, 4), (3, 8), (1, 2), (4, 16)$

26. $\left(\dfrac{1}{2}, 2\right), \left(-1, \dfrac{1}{4}\right), (2, 16), (0, 1)$ **27.** $[0, \infty)$

28. $(-\infty, 3]$ **29.** $(-\infty, \infty)$ **30.** $(-\infty, 1) \cup (1, 9) \cup (9, \infty)$

31. a) $C = 0.12x + 3000$ **b)** $P = 1 \times 10^{-6}x + 0.15$

32. a) \$0.44, \$0.40, \$0.39

b)

c) 60,000 miles   **d)** [50,000, 60,000]

Chapter 12

Section 12.1 Warm-ups F T F T T F T F T F

1. An exponential function has the form $f(x) = a^x$ where $a > 0$ and $a \neq 1$.

3. The two most popular bases are e and 10.

5. The compound interest formula is $A = P(1 + i)^n$.

7. 16   **9.** 2   **11.** 3   **13.** $\dfrac{1}{3}$ **15.** -1 **17.** $-\dfrac{1}{4}$

19. 1   **21.** 100   **23.** 2.718   **25.** 0.135

27.

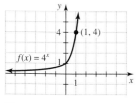

29.

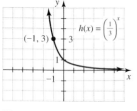

31.

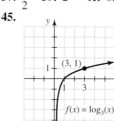

33.

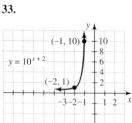

35

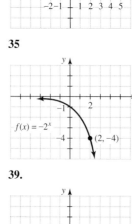

37.

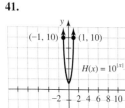

39.

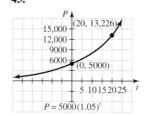

41.

43.

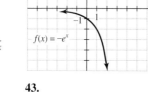

45. $\{6\}$ **47.** $\{-3\}$ **49.** $\{-2\}$ **51.** $\{-1\}$ **53.** $\{-2\}$

55. $\{-2\}$ **57.** $\{-3, 3\}$ **59.** 2   **61.** $\dfrac{4}{3}$ **63.** -2 **65.** 0

67. $\dfrac{3}{2}$ **69.** $\dfrac{1}{2}$ **71.** \$9861.72

73. a) \$114,421.26   **b)** 1996   **75.** \$45, \$12.92

77. \$616.84   **79.** \$6230.73

8̶1̶. 300 grams, 90.4 grams, 12 years, no

83. 2.66666667, 0.0516, 2.8×10^{-5}

85. The graph of $y = 3^{x-k}$ lies k units to the right of $y = 3^x$ when $k > 0$ and $|k|$ units to the left of $y = 3^x$ when $k < 0$.

Section 12.2   Warm-ups T F T T F F F F T T

1. If $f(x) = 2^x$, then $f^{-1}(x) = \log_2(x)$.

3. The common logarithm uses the base 10 and the natural logarithm uses base e.

5. The one-to-one property for logarithmic functions states that if $\log_a(m) = \log_a(n)$, then $m = n$.

7. $2^3 = 8$ **9.** $\log(100) = 2$ **11.** $5^y = x$ **13.** $\log_2(b) = a$

15. $3^{10} = x$ **17.** $\ln(x) = 3$ **19.** 2   **21.** 4   **23.** 6

25. 3   **27.** -2 **29.** 2   **31.** -2 **33.** 1   **35.** -3

37. $\dfrac{1}{2}$ **39.** 2   **41.** 0.6990   **43.** 1.8307

45.

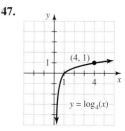

47.

49.

51.

53. $f^{-1}(x) = \log_6(x)$ **55.** $f^{-1}(x) = e^x$ **57.** $f^{-1}(x) = \left(\dfrac{1}{2}\right)^x$

59. $\{4\}$ **61.** $\left\{\dfrac{1}{2}\right\}$ **63.** $\{0.001\}$ **65.** $\{6\}$ **67.** $\left\{\dfrac{1}{5}\right\}$

69. $\{\pm 3\}$ **71.** $\{0.4771\}$ **73.** $\{-0.3010\}$ **75.** $\{1.9741\}$

77. 5.776 years   **79.** 1.9269 years

81. a) 104.8%   **b)** \$15,320,208

83. 4.1   **85.** 6.8   **87.** 90 db

89. $f^{-1}(x) = 2^{x-5} + 3$, $(-\infty, \infty)$, $(3, \infty)$

91. $y = \ln(e^x) = x$ for $-\infty < x < \infty$, $y = e^{\ln(x)} = x$ for $0 < x < \infty$

Section 12.3   Warm-ups T F T T F T F F F T

1. The product rule for logarithms states that $\log_a(MN) = \log_a(M) + \log_a(N)$.

3. The power rule for logarithms states that $\log_a(M^N) = N \cdot \log_a(M)$.

5. Since $\log_a(M)$ is the exponent you would use on a to obtain M, using $\log_a(M)$ as the exponent produces M: $a^{\log_a(M)} = M$.

7. $\log(21)$ **9.** $\log_3(\sqrt{5x})$ **11.** $\log(x^5)$ **13.** $\ln(30)$

15. $\log(x^2 + 3x)$ **17.** $\log_2(x^2 - x - 6)$ **19.** $\log(4)$

21. $\log_2(x^4)$ **23.** $\log(\sqrt{5})$ **25.** $\ln(h - 2)$

27. $\log_2(w - 2)$ **29.** $\ln(x - 2)$ **31.** $3\log(3)$

33. $\dfrac{1}{2}\log(3)$ **35.** $x\log(3)$ **37.** 10   **39.** 19   **41.** 8

43. 4.3   **45.** $\log(3) + \log(5)$ **47.** $\log(5) - \log(3)$

49. $2\log(5)$ **51.** $2\log(5) + \log(3)$ **53.** $-\log(3)$

55. $-\log(5)$ **57.** $\log(x) + \log(y) + \log(z)$

59. $3 + \log_2(x)$ **61.** $\ln(x) - \ln(y)$ **63.** $1 + 2\log(x)$

65. $2\log_5(x - 3) - \dfrac{1}{2}\log_5(w)$

67. $\ln(y) + \ln(z) + \dfrac{1}{2}\ln(x) - \ln(w)$ **69.** $\log(x^2 - x)$

71. $\ln(3)$ **73.** $\ln\!\left(\dfrac{xz}{w}\right)$ **75.** $\ln\!\left(\dfrac{x^2y^3}{w}\right)$

77. $\log\!\left(\dfrac{(x-3)^{1/2}}{(x+1)^{2/3}}\right)$ **79.** $\log_2\!\left(\dfrac{(x-1)^{2/3}}{(x+2)^{1/4}}\right)$ **81.** False

83. True   **85.** True   **87.** False   **89.** True   **91.** True

93. True   **95.** False   **97.** $r = \ln((A/P)^{1/t})$, 41.2%   **99.** b

101. The graphs are the same because
$$\ln(\sqrt{x}) = \ln(x^{1/2}) = \dfrac{1}{2}\ln(x).$$

103. The graph is a straight line because $\log(e^x) = x\log(e) \approx$
$0.434x$. The slope is $\log(e)$ or approximately 0.434.

Section 12.4 Warm-ups T T T F T T T F T F

1. The exponential equation $a^y = x$ is equivalent to
$\log_a(x) = y$.

3. $\{7\}$ **5.** $\{31\}$ **7.** $\{e\}$ **9.** $\{2\}$ **11.** $\{3\}$ **13.** $\varnothing$

15. $\{6\}$ **17.** $\{3\}$ **19.** $\{2\}$ **21.** $\{4\}$ **23.** $\{\log_3(7)\}$

25. $\left\{\dfrac{\ln(7)}{2}\right\}$ **27.** $\{-6\}$ **29.** $\left\{-\dfrac{1}{2}\right\}$

31. $\dfrac{5\ln(3)}{\ln(2) - \ln(3)}$, -13.548 **33.** $\dfrac{4 + 2\log(5)}{1 - \log(5)}$, 17.932

35. $\dfrac{\ln(9)}{\ln(9) - \ln(8)}$, 18.655   **37.** 1.5850   **39.** -0.6309

41. -2.2016 **43.** 1.5229   **45.** $\dfrac{\ln(7)}{\ln(2)}$, 2.807

47. $\dfrac{1}{3 - \ln(2)}$, 0.433   **49.** $\dfrac{\ln(5)}{\ln(3)}$, 1.465

51. $1 + \dfrac{\ln(9)}{\ln(2)}$, 4.170   **53.** $\log_3(20)$, 2.727   **55.** $\dfrac{3}{5}$ **57.** $\dfrac{1}{2}$

59. 41 months   **61.** 7524 ft^3/sec   **63.** 7.1 years

65. 16.8 years   **67.** 2.0×10^{-4} **69.** 0.9183   **71.** 2.2894

73. (2.71, 6.54)   **75.** (1.03, 0.04), (4.73, 2.24)

Enriching Word Power

1. a   **2.** d   **3.** b   **4.** d   **5.** d   **6.** b   **7.** a   **8.** b

9. b   **10.** c

Chapter 12 Review

1. $\dfrac{1}{25}$ **3.** 125   **5.** 1   **7.** $\dfrac{1}{10}$ **9.** 4   **11.** $\dfrac{1}{2}$ **13.** 2

15. 4   **17.** $-\dfrac{5}{2}$ **19.** 2   **21.** 8.6421   **23.** 177.828

25. 0.02005   **27.** 0.1408

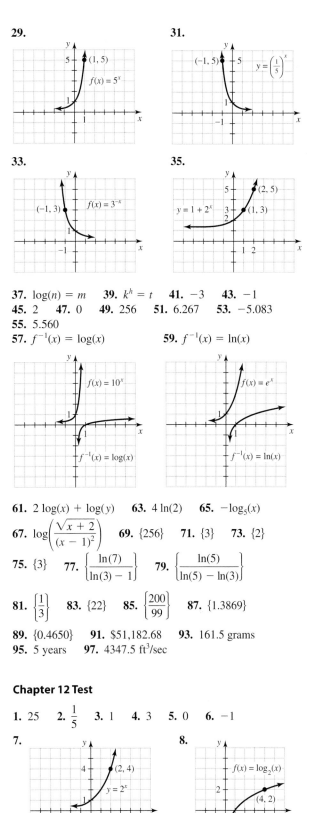

29.

31.

33.

35.

37. $\log(n) = m$ **39.** $k^h = t$ **41.** -3 **43.** -1

45. 2   **47.** 0   **49.** 256   **51.** 6.267   **53.** -5.083

55. 5.560

57. $f^{-1}(x) = \log(x)$ **59.** $f^{-1}(x) = \ln(x)$

61. $2\log(x) + \log(y)$ **63.** $4\ln(2)$ **65.** $-\log_5(x)$

67. $\log\!\left(\dfrac{\sqrt{x+2}}{(x-1)^2}\right)$ **69.** $\{256\}$ **71.** $\{3\}$ **73.** $\{2\}$

75. $\{3\}$ **77.** $\left\{\dfrac{\ln(7)}{\ln(3) - 1}\right\}$ **79.** $\left\{\dfrac{\ln(5)}{\ln(5) - \ln(3)}\right\}$

81. $\left\{\dfrac{1}{3}\right\}$ **83.** $\{22\}$ **85.** $\left\{\dfrac{200}{99}\right\}$ **87.** $\{1.3869\}$

89. $\{0.4650\}$ **91.** \$51,182.68   **93.** 161.5 grams

95. 5 years   **97.** 4347.5 ft^3/sec

Chapter 12 Test

1. 25   **2.** $\dfrac{1}{5}$ **3.** 1   **4.** 3   **5.** 0   **6.** -1

7.

8.

9.
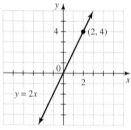

10.
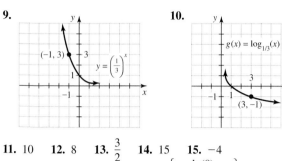

11. 10   **12.** 8   **13.** $\dfrac{3}{2}$ **14.** 15   **15.** -4

16. $\{\log_3(12)\}$ **17.** $\{\sqrt{3}\}$ **18.** $\left\{\dfrac{\ln(8)}{\ln(8)-\ln(5)}\right\}$

19. $\{5\}$ **20.** $\{3\}$ **21.** $\{0.5372\}$ **22.** $\{20.5156\}$

23. 10, 147,648   **24.** 1.733 hours

Making Connections Chapters 1–12

1. $\left\{3 \pm 2\sqrt{2}\right\}$ **2.** $\{259\}$ **3.** $\{6\}$ **4.** $\left\{\dfrac{11}{2}\right\}$ **5.** $\{-5, 11\}$

6. $\{67\}$ **7.** $\{6\}$ **8.** $\{4\}$ **9.** $\left\{-\dfrac{52}{15}\right\}$ **10.** $\left\{\dfrac{3 \pm \sqrt{3}}{3}\right\}$

11. $f^{-1}(x) = 3x$ **12.** $g^{-1}(x) = 3^x$ **13.** $f^{-1}(x) = \dfrac{x+4}{2}$

14. $h^{-1}(x) = x^2$ for $x \geq 0$ **15.** $j^{-1}(x) = \dfrac{1}{x}$

16. $k^{-1}(x) = \log_5(x)$ **17.** $m^{-1}(x) = 1 + \ln(x)$

18. $n^{-1}(x) = e^x$

19.

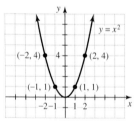

20.
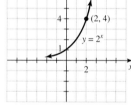

21.

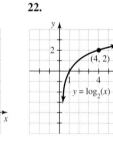

22.

23.

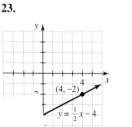

24.
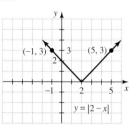

25.
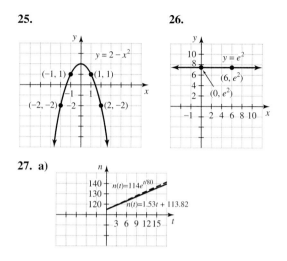

26.

27. a)

b) linear 130.7 million, exponential 130.8 million

28. a) $d_1 = 0.135v$ **b)** $d_2 = 0.216v$ **c)** $d_1 = 200.2$ meters

Chapter 13

Section 13.1 Warm-ups T F F T F T T T T T

1. If the graph of an equation is not a straight line, then it is called nonlinear.

3. Graphing is not an accurate method for solving a system and the graphs might be difficult to draw.

5. $\{(2, 4), (-3, 9)\}$ **7.** $\{(-2, 2), (6, 6)\}$

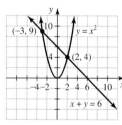
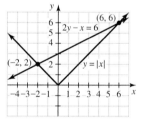

9. $\{(8, 4)\}$ **11.** $\left\{\left(-\dfrac{3}{4}, -\dfrac{4}{3}\right), \left(3, \dfrac{1}{3}\right)\right\}$

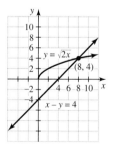
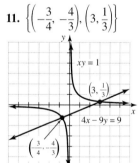

13. $\left\{\left(\dfrac{\sqrt{2}}{2}, \dfrac{1}{2}\right), \left(-\dfrac{\sqrt{2}}{2}, \dfrac{1}{2}\right)\right\}$

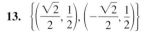
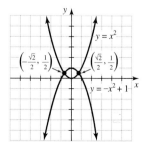

15. $\{(0, -5), (3, 4), (-3, 4)\}$ **17.** $\{(4, 5), (-2, -1)\}$

19. $\left\{\left(-3, -\dfrac{5}{3}\right)\right\}$

21. $\{(\sqrt{5}, \sqrt{3}), (\sqrt{5}, -\sqrt{3}), (-\sqrt{5}, \sqrt{3}), (-\sqrt{5}, -\sqrt{3})\}$

23. $\{(\sqrt{2}, \sqrt{3}), (\sqrt{2}, -\sqrt{3}), (-\sqrt{2}, \sqrt{3}), (-\sqrt{2}, -\sqrt{3})\}$

25. $\left\{\left(\dfrac{3}{2}, -\dfrac{3}{13}\right)\right\}$ **27.** $\{(3, 4)\}$ **29.** $\left\{\left(-\dfrac{5}{3}, \dfrac{36}{5}\right), (2, 5)\right\}$

31. $\{(2, 5), (19, -12)\}$

33. $\{(\sqrt{2}, 2), (-\sqrt{2}, 2), (1, 1), (-1, 1)\}$

35. $\{(3, 1)\}$ **37.** $\varnothing$ **39.** $\{(-6, 4^{-7})\}$

41. $\sqrt{3}$ ft and $2\sqrt{3}$ ft

43. Height $5\sqrt{10}$ inches, base $20\sqrt{10}$ inches

45. Pump A 24 hours, pump B 8 hours

47. 40 minutes **49.** 8 ft by 9 ft **51.** $4 - 2i$ and $4 + 2i$

53. Side of square 8 ft, height of triangle 2 ft

55. a) $(1.71, 1.55), (-2.98, -3.95)$
　　b) $(1, 1), (0.40, 0.16)$
　　c) $(1.17, 1.62), (-1.17, -1.62)$

Section 13.2 Warm-ups F T T F F T T T T T

1. A parabola is the set of all points in a plane that are equidistant from a given line and a fixed point not on the line.

3. A parabola can be written in the forms $y = ax^2 + bx + c$ or $y = a(x - h)^2 + k$.

5. We use completing the square to convert $y = ax^2 + bx + c$ into $y = a(x - h)^2 + k$.

7. Vertex $(0, 0)$, focus $\left(0, \dfrac{1}{8}\right)$, directrix $y = -\dfrac{1}{8}$.

9. Vertex $(0, 0)$, focus $(0, -1)$, directrix $y = 1$

11. Vertex $(3, 2)$, focus $(3, 2.5)$, directrix $y = 1.5$

13. Vertex $(-1, 6)$, focus $(-1, 5.75)$, directrix $y = 6.25$

15. $y = \dfrac{1}{8}x^2$ **17.** $y = -\dfrac{1}{2}x^2$ **19.** $y = \dfrac{1}{2}x^2 - 3x + 6$

21. $y = -\dfrac{1}{8}x^2 + \dfrac{1}{4}x - \dfrac{1}{8}$

23. $y = x^2 + 6x + 10$

25. $y = (x - 3)^2 - 8$, vertex $(3, -8)$, focus $(3, -7.75)$, directrix $y = -8.25$, $x = 3$

27. $y = 2(x + 3)^2 - 13$, vertex $(-3, -13)$, focus $(-3, -12.875)$, directrix $y = -13.125$, $x = -3$

29. $y = -2(x - 4)^2 + 33$, vertex $(4, 33)$, focus $\left(4, 32\dfrac{7}{8}\right)$, directrix $y = 33\dfrac{1}{8}$, $x = 4$

31. $y = 5(x + 4)^2 - 80$, vertex $(-4, -80)$, focus $\left(-4, -79\dfrac{19}{20}\right)$, directrix $y = -80\dfrac{1}{20}$, $x = -4$

33. Vertex $(2, -3)$, focus $\left(2, -2\dfrac{3}{4}\right)$, directrix $y = -3\dfrac{1}{4}$, $x = 2$, upward

35. Vertex $(1, -2)$, focus $\left(1, -2\dfrac{1}{4}\right)$, directrix $y = -1\dfrac{3}{4}$, $x = 1$, downward

37. Vertex $(1, -2)$, focus $\left(1, -1\dfrac{11}{12}\right)$, directrix $y = -2\dfrac{1}{12}$, $x = 1$, upward

39. Vertex $\left(-\dfrac{3}{2}, \dfrac{17}{4}\right)$, focus $\left(-\dfrac{3}{2}, 4\right)$, directrix $y = \dfrac{9}{2}$, $x = -\dfrac{3}{2}$, downward

41. Vertex $(0, 5)$, focus $\left(0, 5\dfrac{1}{12}\right)$, directrix $y = 4\dfrac{11}{12}$, $x = 0$, upward

43. $y = \dfrac{1}{60}x^2$

45. $\{(-1, 2), (1, 2)\}$ **47.** $\{(1, -1)\}$

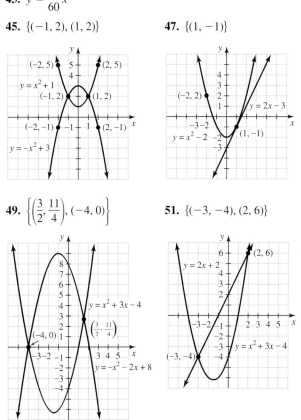

49. $\left\{\left(\dfrac{3}{2}, \dfrac{11}{4}\right), (-4, 0)\right\}$ **51.** $\{(-3, -4), (2, 6)\}$

53. $(3, 0), (-1, 0)$ **55.** $(20, 4), (-20, 4)$ **57.** $(0, 0), (1, 1)$

61. The graphs have identical shapes.

Section 13.3 Warm-ups F F T F F F T F T F

1. A circle is the set of all points in a plane that lie at a fixed distance from a fixed point.

3. $x^2 + (y - 3)^2 = 25$ **5.** $(x - 1)^2 + (y + 2)^2 = 81$

7. $x^2 + y^2 = 3$ **9.** $(x + 6)^2 + (y + 3)^2 = \dfrac{1}{4}$

11. $\left(x - \dfrac{1}{2}\right)^2 + \left(y - \dfrac{1}{3}\right)^2 = 0.01$ **13.** $(3, 5), \sqrt{2}$

15. $\left(0, \dfrac{1}{2}\right), \dfrac{\sqrt{2}}{2}$ **17.** $(0, 0), \dfrac{3}{2}$ **19.** $(2, 0), \sqrt{3}$

21. 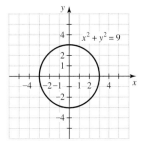 **23.**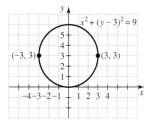

25.

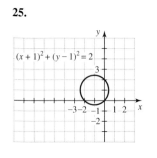

27.

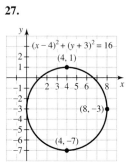

29.

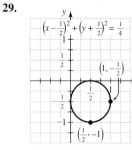

31. $(x + 2)^2 + (y + 3)^2 = 13, (-2, -3), \sqrt{13}$

33. $(x - 1)^2 + (y - 2)^2 = 8, (1, 2), 2\sqrt{2}$

35. $(x - 5)^2 + (y - 4)^2 = 9, (5, 4), 3$

37. $\left(x - \dfrac{1}{2}\right)^2 + \left(y + \dfrac{1}{2}\right)^2 = \dfrac{1}{2}, \left(\dfrac{1}{2}, -\dfrac{1}{2}\right), \dfrac{\sqrt{2}}{2}$

39. $\left(x - \dfrac{3}{2}\right)^2 + \left(y - \dfrac{1}{2}\right)^2 = \dfrac{7}{2}, \left(\dfrac{3}{2}, \dfrac{1}{2}\right), \dfrac{\sqrt{14}}{2}$

41. $\left(x - \dfrac{1}{3}\right)^2 + \left(y + \dfrac{3}{4}\right)^2 = \dfrac{97}{144}, \left(\dfrac{1}{3}, -\dfrac{3}{4}\right), \dfrac{\sqrt{97}}{12}$

43. $\{(1, 3), (-1, -3)\}$

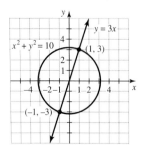

45. $\{(0, -3), (\sqrt{5}, 2), (-\sqrt{5}, 2)\}$

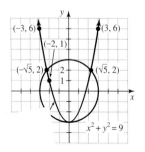

47. $\{(0, -3), (2, -1)\}$

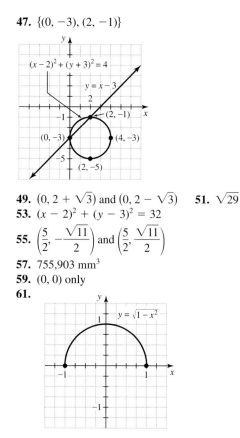

49. $(0, 2 + \sqrt{3})$ and $(0, 2 - \sqrt{3})$ **51.** $\sqrt{29}$

53. $(x - 2)^2 + (y - 3)^2 = 32$

55. $\left(\dfrac{5}{2}, -\dfrac{\sqrt{11}}{2}\right)$ and $\left(\dfrac{5}{2}, \dfrac{\sqrt{11}}{2}\right)$

57. 755,903 mm³

59. $(0, 0)$ only

61.

63. B and D can be any real numbers, but A must equal C, and $4AE + B^2 + D^2 > 0$. No ordered pairs satisfy $x^2 + y^2 = -9$.

65. $y = \pm\sqrt{4 - x^2}$

67. $y = \pm\sqrt{x}$

69. $y = -1 \pm \sqrt{x}$

Section 13.4 Warm-ups F F T T T F F T T T

1. An ellipse is the set of all points in a plane such that the sum of their distances from two fixed points is constant.

3. The center of an ellipse is the point that is midway between the foci.

5. The equation of an ellipse centered at (h, k) is
$$\dfrac{(x - h)^2}{a^2} + \dfrac{(y - k)^2}{b^2} = 1.$$

7. The asymptotes of a hyperbola are the extended diagonals of the fundamental rectangle.

9.   **11.**

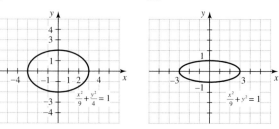

13.

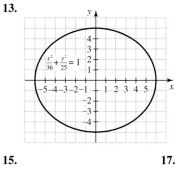

15.

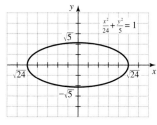

17.

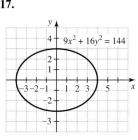

19.

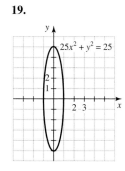

21.

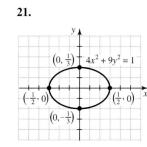

23.

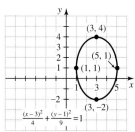

25.

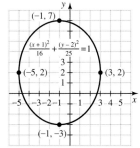

27.

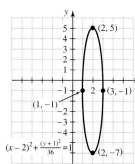

29. $y = \pm\frac{3}{2}x$

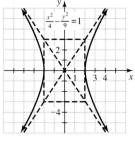

31. $y = \pm\frac{2}{5}x$

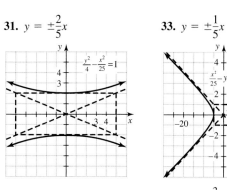

33. $y = \pm\frac{1}{5}x$

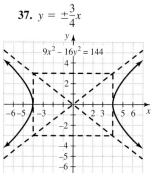

35. $y = \pm 5x$

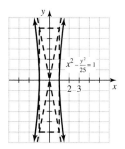

37. $y = \pm\frac{3}{4}x$

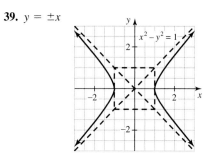

39. $y = \pm x$

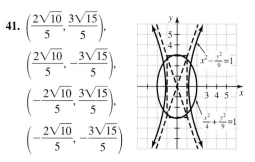

41. $\left(\dfrac{2\sqrt{10}}{5}, \dfrac{3\sqrt{15}}{5}\right),$ $\left(\dfrac{2\sqrt{10}}{5}, -\dfrac{3\sqrt{15}}{5}\right),$ $\left(-\dfrac{2\sqrt{10}}{5}, \dfrac{3\sqrt{15}}{5}\right),$ $\left(-\dfrac{2\sqrt{10}}{5}, -\dfrac{3\sqrt{15}}{5}\right)$

43. No points of intersection

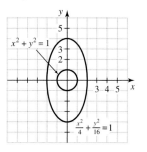

45. $\left(\dfrac{\sqrt{10}}{2}, \dfrac{\sqrt{6}}{2}\right),$

$\left(\dfrac{\sqrt{10}}{2}, -\dfrac{\sqrt{6}}{2}\right),$

$\left(-\dfrac{\sqrt{10}}{2}, \dfrac{\sqrt{6}}{2}\right),$

$\left(-\dfrac{\sqrt{10}}{2}, -\dfrac{\sqrt{6}}{2}\right)$

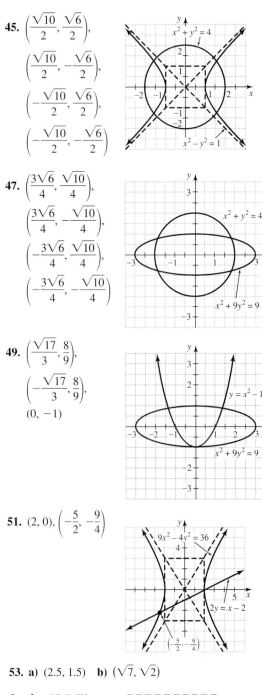

47. $\left(\dfrac{3\sqrt{6}}{4}, \dfrac{\sqrt{10}}{4}\right),$

$\left(\dfrac{3\sqrt{6}}{4}, -\dfrac{\sqrt{10}}{4}\right),$

$\left(-\dfrac{3\sqrt{6}}{4}, \dfrac{\sqrt{10}}{4}\right),$

$\left(-\dfrac{3\sqrt{6}}{4}, -\dfrac{\sqrt{10}}{4}\right)$

49. $\left(\dfrac{\sqrt{17}}{3}, \dfrac{8}{9}\right),$

$\left(-\dfrac{\sqrt{17}}{3}, \dfrac{8}{9}\right),$

$(0, -1)$

51. $(2, 0), \left(-\dfrac{5}{2}, -\dfrac{9}{4}\right)$

53. a) $(2.5, 1.5)$ **b)** $(\sqrt{7}, \sqrt{2})$

Section 13.5 Warm-ups F T T T F F F T T T
1.
3.

5.
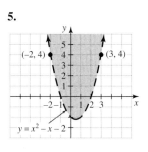

7.
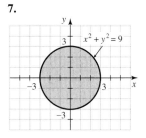

9.
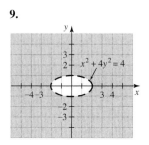

11.
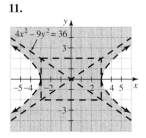

13.
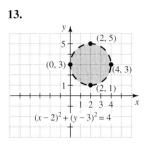

15.
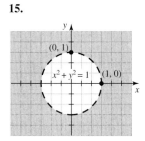

17.
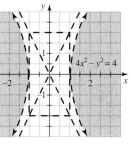

19.
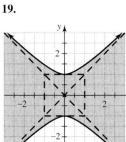

21.
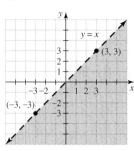

23.
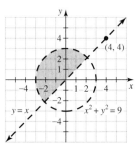

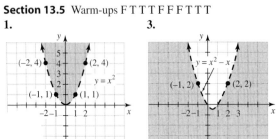

25.

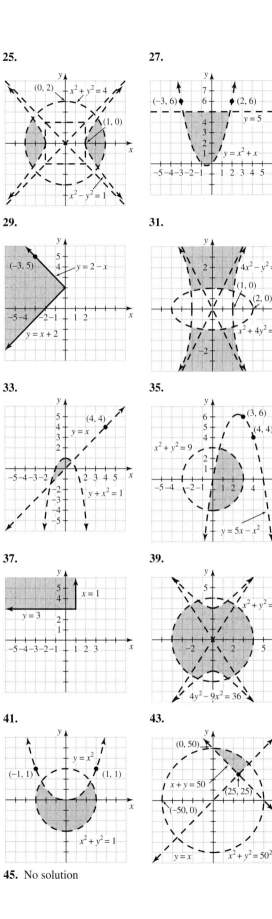

27.

29.

31.

33.

35.

37.

39.

41.

43.

45. No solution

Enriching Word Power
1. c   **2.** a   **3.** d   **4.** a   **5.** c   **6.** d   **7.** b   **8.** d
9. c   **10.** a

Chapter 13 Review
1. $\{(3, 9), (-5, 25)\}$

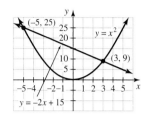

3. $\left\{\left(\dfrac{\sqrt{3}}{3}, \sqrt{3}\right),\right.$
$\left.\left(-\dfrac{\sqrt{3}}{3}, -\sqrt{3}\right)\right\}$

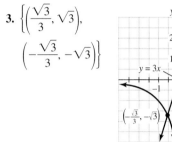

5. $\{(\sqrt{3}, 1), (-\sqrt{3}, 1)\}$ **7.** $\{(-5, -3), (3, 5)\}$
9. $\{(5, \log(2))\}$ **11.** $\{(2, 4), (-2, 4)\}$
13. Vertex $\left(-\dfrac{3}{2}, -\dfrac{81}{4}\right)$, axis of symmetry $x = -\dfrac{3}{2}$, focus $\left(-\dfrac{3}{2}, -20\right)$, directrix $y = -\dfrac{41}{2}$
15. Vertex $\left(-\dfrac{3}{2}, -\dfrac{1}{4}\right)$, axis of symmetry $x = -\dfrac{3}{2}$, focus $\left(-\dfrac{3}{2}, 0\right)$, directrix $y = -\dfrac{1}{2}$
17. Vertex $(2, 3)$, axis of symmetry $x = 2$, focus $\left(2, \dfrac{5}{2}\right)$, directrix $y = \dfrac{7}{2}$
19. $y = 2(x - 2)^2 - 7$, $(2, -7)$
21. $y = -\dfrac{1}{2}(x + 1)^2 + 1$, $(-1, 1)$
23. $(0, 0)$, 10   **25.** $(2, -3)$, 9

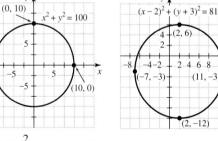

27. $(0, 0)$, $\dfrac{2}{3}$

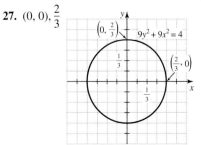

29. $x^2 + (y - 3)^2 = 36$ **31.** $(x - 2)^2 + (y + 7)^2 = 25$

33.

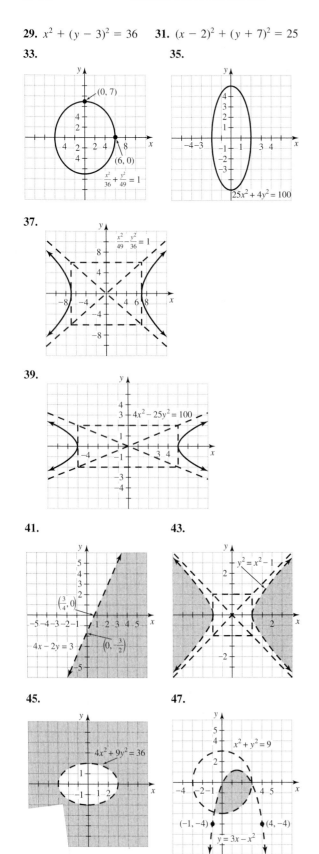

35.

37.

39.

41.

43.

45.

47.

49.

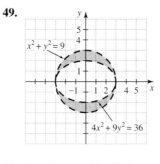

51. Hyperbola   **53.** Circle
55. Circle   **57.** Circle
59. Hyperbola   **61.** Hyperbola

63.

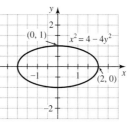

65.

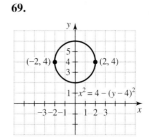

67.

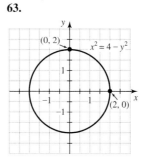

69.

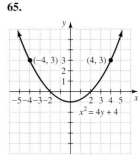

71. $x^2 + y^2 = 25$
73. $(x + 1)^2 + (y - 5)^2 = 36$
75. $y = \dfrac{1}{4}(x - 1)^2 + 3$
77. $y = x^2$ **79.** $y = \dfrac{2}{9}x^2$
81. $\{(4, -3), (-3, 4)\}$ **83.** $\varnothing$
85. 6 ft, 2 ft

Chapter 13 Test

1.

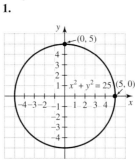

2.

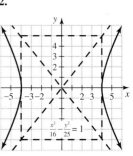

3.

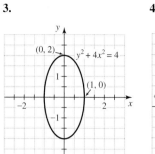

4.

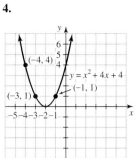

12. $\{(-5, 27), (3, -5)\}$
13. $\{(\sqrt{3}, 3), (-\sqrt{3}, 3)\}$
14. $(-1, -5), 6$
15. Vertex $\left(-\frac{1}{2}, \frac{11}{4}\right)$, focus $\left(-\frac{1}{2}, 3\right)$, directrix $y = \frac{5}{2}$, axis of symmetry $x = -\frac{1}{2}$, upward
16. $y = \frac{1}{2}(x - 3)^2 - 5$
17. $(x + 1)^2 + (y - 3)^2 = 13$
18. 12 ft, 9 ft

Making Connections Chapters 1–13

1.

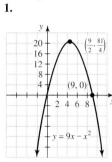

2.

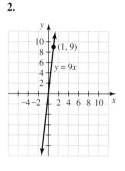

5.

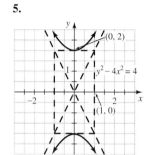

6.

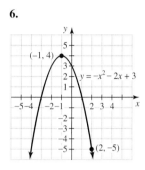

3.

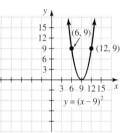

4.

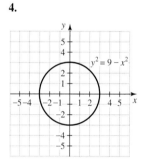

7.

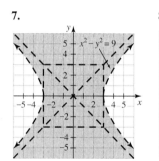

8.

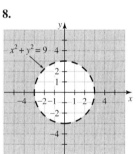

5.

6.

9.

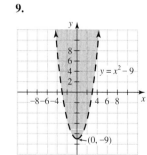

10.

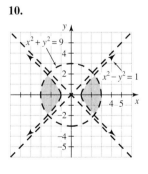

7.

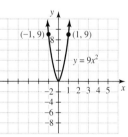

8.

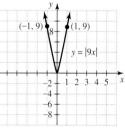

11.

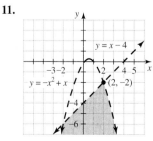

12.

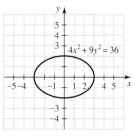

17.

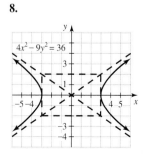

9.   **10.**

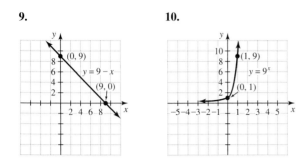

11. $x^2 + 4xy + 4y^2$ **12.** $x^3 + 3x^2y + 3xy^2 + y^3$
13. $a^3 + 3a^2b + 3ab^2 + b^3$ **14.** $a^2 - 6ab + 9b^2$
15. $6a^2 - 7a - 5$ **16.** $x^3 - y^3$ **17.** $\{(1, 2)\}$
18. $\{(3, 4), (4, 3)\}$ **19.** $\{(1, -2, 3)\}$ **20.** $\{(-1, 1), (3, 9)\}$

21. $x = -\dfrac{b}{a}$ **22.** $x = \dfrac{-d \pm \sqrt{d^2 - 4wm}}{2w}$

23. $B = \dfrac{2A - bh}{h}$ **24.** $x = \dfrac{2y}{y - 2}$ **25.** $m = \dfrac{L}{1 + xt}$

26. $t = \dfrac{y^2}{9a^2}$ **27.** $y = -\dfrac{2}{3}x - \dfrac{5}{3}$ **28.** $y = -2x$

29. $(x - 2)^2 + (y - 5)^2 = 45$ **30.** $\left(-\dfrac{3}{2}, 3\right), \dfrac{3\sqrt{5}}{2}$

31. $-10 + 6i$ **32.** -1 **33.** $3 - 5i$ **34.** $7 + 6i\sqrt{2}$

35. $-8 - 27i$ **36.** $-3 + 3i$ **37.** 29 **38.** $-\dfrac{3}{2} - i$

39. $-\dfrac{1}{2} + \dfrac{9}{2}i$ **40.** $2 - i\sqrt{2}$

41. a) $q = -500x + 400$ **b)** $R = -500x^2 + 400x$
    c)   **d)** \$0.40 per pound   **e)** \$80

Chapter 14

Section 14.1 Warm-ups T T T F F F T F T T
1. A sequence is a list of numbers.
3. A finite sequence is a function whose domain is the set of positive integers less than or equal to some fixed positive integer.
5. 1, 4, 9, 16, 25, 36, 49, 64

7. $-1, \dfrac{1}{2}, -\dfrac{1}{3}, \dfrac{1}{4}, -\dfrac{1}{5}, \dfrac{1}{6}, -\dfrac{1}{7}, \dfrac{1}{8}, -\dfrac{1}{9}, \dfrac{1}{10}$ **9.** 1, -2, 4, -8, 16

11. $\dfrac{1}{2}, \dfrac{1}{4}, \dfrac{1}{8}, \dfrac{1}{16}, \dfrac{1}{32}, \dfrac{1}{64}$ **13.** -1, 1, 3, 5, 7, 9, 11

15. $1, \dfrac{\sqrt{2}}{2}, \dfrac{\sqrt{3}}{3}, \dfrac{1}{2}, \dfrac{\sqrt{5}}{5}$ **17.** $\dfrac{1}{2}, \dfrac{1}{6}, \dfrac{1}{12}, \dfrac{1}{20}$

       $-1, 1, \dfrac{1}{3}$ **21.** $-1, 0, -1, 4$ **23.** $1, \dfrac{1}{4}, \dfrac{1}{9}, \dfrac{1}{16}$

       $2n - 1$ **27.** $a_n = (-1)^{n+1}$ **29.** $a_n = 2n - 2$
       $3n$ **33.** $a_n = 3n + 1$ **35.** $a_n = (-1)^n 2^{n-1}$

37. $a_n = (n - 1)^2$ **39.** 4, 2, 1, $\dfrac{1}{2}, \dfrac{1}{4}$
41. \$34,162, \$35,870, \$37,663, \$39,546, \$41,524, \$43,600
43. \$1,000,000, \$800,000, \$640,000, \$512,000
45. 27 in., 13.5 in., 9 in., 6.75 in., 5.4 in.
47. 137,438,953,500, larger
51. a) 0.9048, 0.3677, 0.00004517   **b)** a_n goes to zero

Section 14.2 Warm-ups T F F F F T T T T F
1. Summation notation provides a way to write a sum without writing out all of the terms.
3. A series is the indicated sum of the terms of a sequence.
5. 30   **7.** 24   **9.** $\dfrac{31}{32}$ **11.** 50   **13.** -7 **15.** 0

17. $\displaystyle\sum_{i=1}^{6} i$ **19.** $\displaystyle\sum_{i=1}^{6} (-1)^i(2i - 1)$ **21.** $\displaystyle\sum_{i=1}^{6} i^2$

23. $\displaystyle\sum_{i=1}^{4} \dfrac{1}{2 + i}$ **25.** $\displaystyle\sum_{i=1}^{3} \ln(i + 1)$ **27.** $\displaystyle\sum_{i=1}^{4} a_i$ **29.** $\displaystyle\sum_{i=1}^{48} x_{i+2}$

31. $\displaystyle\sum_{i=1}^{n} w_i$ **33.** $\displaystyle\sum_{j=0}^{4} (j + 1)^2$ **35.** $\displaystyle\sum_{j=1}^{13} (2j - 3)$

37. $\displaystyle\sum_{j=1}^{5} \dfrac{1}{j + 3}$ **39.** $\displaystyle\sum_{j=0}^{3} x^{2j+5}$ **41.** $\displaystyle\sum_{j=0}^{n-1} x^{j+1}$

43. $x + x^2 + x^3 + x^4 + x^5 + x^6$ **45.** $x_0 - x_1 + x_2 - x_3$

47. $x + 2x^2 + 3x^3$ **49.** $\displaystyle\sum_{i=1}^{9} 2^{-i}$ **51.** $\displaystyle\sum_{i=1}^{4} 1{,}000{,}000(0.8)^{i-1}$

53. A sequence is basically a list of numbers. A series is the indicated sum of the terms of a sequence.

Section 14.3 Warm-ups F F F F F F T T T F
1. An arithmetic sequence is one in which each term after the first is obtained by adding a fixed amount to the previous term.
3. An arithmetic series is an indicated sum of an arithmetic sequence.
5. $a_n = 6n - 6$ **7.** $a_n = 5n + 2$ **9.** $a_n = 2n - 6$
11. $a_n = -4n + 9$ **13.** $a_n = -7n + 5$
15. $a_n = 0.5n - 3.5$ **17.** $a_n = -0.5n - 5.5$
19. 9, 13, 17, 21, 25   **21.** 7, 5, 3, 1, -1
23. -4, -1, 2, 5, 8   **25.** -2, -5, -8, -11, -14
27. -7, -11, -15, -19, -23 **29.** 4.5, 5, 5.5, 6, 6.5
31. 1020, 1040, 1060, 1080, 1100   **33.** 51   **35.** 4
37. 26   **39.** 17   **41.** 1176   **43.** 330   **45.** -481
47. 435   **49.** -540 **51.** 150   **53.** -308
55. \$25,000   **57.** 1085   **59. b**

Section 14.4 Warm-ups F F T T T T F F T F
1. A geometric sequence is one in which each term after the first is obtained by multiplying the preceding term by a constant.
3. A geometric series is an indicated sum of a geometric sequence.
5. The approximate value of r^n when n is large and $|r| < 1$ is 0.

7. $a_n = \dfrac{1}{3}(3)^{n-1}$ **9.** $a_n = 64\left(\dfrac{1}{8}\right)^{n-1}$ **11.** $a_n = 8\left(-\dfrac{1}{2}\right)^{n-1}$

13. $a_n = 2(-2)^{n-1}$ **15.** $a_n = -\dfrac{1}{3}\left(\dfrac{3}{4}\right)^{n-1}$

17. $2, \dfrac{2}{3}, \dfrac{2}{9}, \dfrac{2}{27}, \dfrac{2}{81}$ **19.** $1, -2, 4, -8, 16$

21. $\dfrac{1}{2}, \dfrac{1}{4}, \dfrac{1}{8}, \dfrac{1}{16}, \dfrac{1}{32}$

23. 0.78, 0.6084, 0.4746, 0.3702, 0.2887 **25.** 5 **27.** $\dfrac{1}{3}$

29. $-\dfrac{1}{9}$ **31.** $\dfrac{511}{512}$ **33.** $\dfrac{11}{32}$ **35.** $\dfrac{63050}{729}$ **37.** 5115

39. 0.111111 **41.** 42.8259 **43.** $\dfrac{1}{4}$ **45.** 9 **47.** $\dfrac{8}{3}$

49. $\dfrac{3}{7}$ **51.** 6 **53.** $\dfrac{1}{3}$ **55.** $\dfrac{4}{33}$ **57.** \$3,042,435.27

59. \$21,474,836.47 **61.** \$5,000,000 **63.** d **65.** $\dfrac{8}{33}$

Section 14.5 Warm-ups F F F T T T T T T T
1. The sum obtained for a power of a binomial is called a binomial expansion.
3. The expression $n!$ is the product of the positive integers from 1 through n.
5. 10 **7.** 56
9. $r^5 + 5r^4t + 10r^3t^2 + 10r^2t^3 + 5rt^4 + t^5$
11. $m^3 - 3m^2n + 3mn^2 - n^3$
13. $x^3 + 6ax^2 + 12a^2x + 8a^3$
15. $x^8 - 8x^6 + 24x^4 - 32x^2 + 16$
17. $x^7 - 7x^6 + 21x^5 - 35x^4 + 35x^3 - 21x^2 + 7x - 1$
19. $a^{12} - 36a^{11}b + 594a^{10}b^2 - 5940a^9b^3$
21. $x^{18} + 45x^{16} + 900x^{14} + 10500x^{12}$
23. $x^{22} - 22x^{21} + 231x^{20} - 1540x^{19}$

25. $\dfrac{x^{10}}{1024} + \dfrac{5x^9y}{768} + \dfrac{5x^8y^2}{256} + \dfrac{5x^7y^3}{144}$ **27.** $1287a^8w^5$

29. $-11440m^9n^7$ **31.** $448x^5y^3$ **33.** $635,043,840a^{28}b^6$

35. $\displaystyle\sum_{i=0}^{8} \dfrac{8!}{(8-i)!\,i!} a^{8-i}m^i$ **37.** $\displaystyle\sum_{i=0}^{5} \dfrac{5!(-2)^i}{(5-i)!\,i!} a^{5-i}x^i$

39. $a^3 + b^3 + c^3 + 3a^2b + 3a^2c + 3ab^2 + 3ac^2 + 3b^2c + 3bc^2 + 6abc$

Enriching Word Power
1. a **2.** d **3.** c **4.** b **5.** a **6.** c **7.** d **8.** b
9. d **10.** a

Chapter 14 Review
1. 1, 8, 27, 64, 125 **3.** 1, 1, −3, 5, −7, 9

5. $-1, -\dfrac{1}{2}, -\dfrac{1}{3}$ **7.** $\dfrac{1}{3}, \dfrac{1}{5}, \dfrac{1}{7}$ **9.** 4, 5, 6 **11.** 36 **13.** 40

15. $\displaystyle\sum_{i=1}^{\infty} \dfrac{1}{2(i+1)}$ **17.** $\displaystyle\sum_{i=1}^{\infty} (i-1)^2$ **19.** $\displaystyle\sum_{i=1}^{\infty} (-1)^{i+1}x_i$

21. 6, 11, 16, 21 **23.** −20, −22, −24, −26

25. 3000, 4000, 5000, 6000 **27.** $a_n = \dfrac{n}{3}$ **29.** $a_n = 2n$

31. 300 **33.** $\dfrac{289}{6}$ **35.** 35 **37.** $3, \dfrac{3}{2}, \dfrac{3}{4}, \dfrac{3}{8}$ **39.** $1, \dfrac{1}{2}, \dfrac{1}{4}, \dfrac{1}{8}$

41. 0.23, 0.0023, 0.000023, 0.00000023

43. $a_n = \dfrac{1}{2}(6)^{n-1}$ **45.** $a_n = 0.7(0.1)^{n-1}$ **47.** $\dfrac{40}{81}$

49. 0.3333333333 **51.** $\dfrac{3}{8}$ **53.** 54

55. $m^5 + 5m^4n + 10m^3n^2 + 10m^2n^3 + 5mn^4 + n^5$
57. $a^6 - 9a^4b + 27a^2b^2 - 27b^3$ **59.** $495x^8y^4$

61. $372,736a^{12}b^2$ **63.** $\displaystyle\sum_{i=0}^{7} \dfrac{7!}{(7-i)!\,i!} a^{7-i}w^i$ **65.** Neither

67. Arithmetic **69.** Arithmetic **71.** $\dfrac{1}{\sqrt[3]{180}}$

73. $-1 + \dfrac{1}{2} - \dfrac{1}{6} + \dfrac{1}{24} - \dfrac{1}{120}$

75. $a^5 + 5a^4b + 10a^3b^2 + 10a^2b^3 + 5ab^4 + b^5$
77. 26 **79.** \$118,634.11 **81.** \$13,784.92

Chapter 14 Test
1. −10, −4, 2, 8 **2.** 5, 0.5, 0.05, 0.005 **3.** $-1, \dfrac{1}{2}, -\dfrac{1}{6}, \dfrac{1}{24}$

4. $1, \dfrac{3}{4}, \dfrac{5}{9}, \dfrac{7}{16}$ **5.** $a_n = 10 - 3n$ **6.** $a_n = -25\left(-\dfrac{1}{5}\right)^{n-1}$

7. $a_n = (-1)^{n-1}2n$ **8.** $a_n = n^2$ **9.** $5 + 7 + 9 + 11 + 13$
10. $5 + 10 + 20 + 40 + 80 + 160$
11. $m^4 + 4m^3q + 6m^2q^2 + 4mq^3 + q^4$ **12.** 750 **13.** $\dfrac{155}{8}$

14. 5 **15.** 10,100 **16.** $\dfrac{1}{2}$ **17.** $\dfrac{511}{128}$ **18.** ±2

19. 11 **20.** $1365r^{11}t^4$ **21.** $-448a^{10}b^3$ **22.** \$86,545.41

Making Connections Chapters 1–14
1. 6 **2.** $n^2 - 3$ **3.** $x^2 + 2xh + h^2 - 3$ **4.** $x^2 - 2x - 2$

5. 11 **6.** 6 **7.** 4 **8.** 32 **9.** $\dfrac{1}{2}$ **10.** 3 **11.** 1

12. x **13.** $-\dfrac{27}{2}$ **14.** $-\dfrac{8}{3}$ **15.** $-\dfrac{128}{9}$ **16.** 16

17.

18.

19.

20.

21.

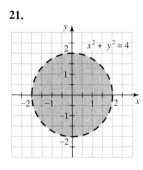

22.

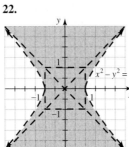

25.

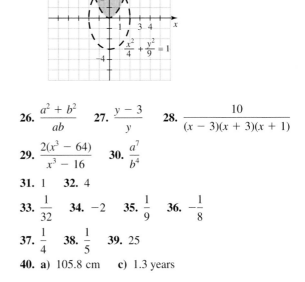

23.

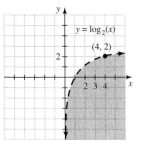

24.

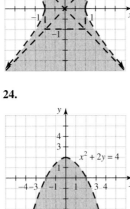

26. $\dfrac{a^2 + b^2}{ab}$ **27.** $\dfrac{y - 3}{y}$ **28.** $\dfrac{10}{(x - 3)(x + 3)(x + 1)}$

29. $\dfrac{2(x^3 - 64)}{x^3 - 16}$ **30.** $\dfrac{a^7}{b^4}$

31. 1   **32.** 4

33. $\dfrac{1}{32}$ **34.** -2 **35.** $\dfrac{1}{9}$ **36.** $-\dfrac{1}{8}$

37. $\dfrac{1}{4}$ **38.** $\dfrac{1}{5}$ **39.** 25

40. a) 105.8 cm   **c)** 1.3 years

INDEX

DEFINITIONS, RULES, AND FORMULAS

Subsets of the Real Numbers

Natural Numbers $= \{1, 2, 3, \ldots\}$

Whole Numbers $= \{0, 1, 2, 3, \ldots\}$

Integers $= \{\ldots -3, -2, -1, 0, 1, 2, 3, \ldots\}$

Rational $= \left\{\dfrac{a}{b}\,\middle|\, a \text{ and } b \text{ are integers with } b \neq 0\right\}$

Irrational $= \{x \mid x \text{ is not rational}\}$

Properties of the Real Numbers

For all real numbers a, b, and c

$a + b = b + a;\ a \cdot b = b \cdot a$ Commutative

$(a + b) + c = a + (b + c);\ (ab)c = a(bc)$ Associative

$a(b + c) = ab + ac;\ a(b - c) = ab - ac$ Distributive

$a + 0 = a;\ 1 \cdot a = a$ Identity

$a + (-a) = 0;\ a \cdot \dfrac{1}{a} = 1\ (a \neq 0)$ Inverse

$a \cdot 0 = 0$ Multiplication property of 0

Absolute Value

$|a| = \begin{cases} a & \text{for } a \geq 0 \\ -a & \text{for } a < 0 \end{cases}$

$\sqrt{x^2} = |x|$ for any real number x.

$|x| = k \leftrightarrow x = k \text{ or } x = -k$ $(k > 0)$

$|x| < k \leftrightarrow -k < x < k$ $(k > 0)$

$|x| > k \leftrightarrow x < -k \text{ or } x > k$ $(k > 0)$

(The symbol $\leftrightarrow$ means "if and only if.")

Interval Notation

$(a, b) = \{x \mid a < x < b\}$

$[a, b] = \{x \mid a \leq x \leq b\}$

$(a, b] = \{x \mid a < x \leq b\}$

$[a, b) = \{x \mid a \leq x < b\}$

$(-\infty, a) = \{x \mid x < a\}$

$(a, \infty) = \{x \mid x > a\}$

$(-\infty, a] = \{x \mid x \leq a\}$

$[a, \infty) = \{x \mid x \geq a\}$

Exponents

$a^0 = 1$

$a^{-1} = \dfrac{1}{a}$

$a^{-r} = \dfrac{1}{a^r} = \left(\dfrac{1}{a}\right)^r$

$\dfrac{1}{a^{-r}} = a^r$

$a^r a^s = a^{r+s}$

$\dfrac{a^r}{a^s} = a^{r-s}$

$(a^r)^s = a^{rs}$

$(ab)^r = a^r b^r$

$\left(\dfrac{a}{b}\right)^r = \dfrac{a^r}{b^r}$

$\left(\dfrac{a}{b}\right)^{-r} = \left(\dfrac{b}{a}\right)^r$

Roots and Radicals

$a^{1/n} = \sqrt[n]{a}$

$a^{m/n} = \left(\sqrt[n]{a}\right)^m = \sqrt[n]{a^m}$

$\sqrt[n]{ab} = \sqrt[n]{a} \cdot \sqrt[n]{b}$

$\sqrt[n]{\dfrac{a}{b}} = \dfrac{\sqrt[n]{a}}{\sqrt[n]{b}}$

Factoring

$a^2 + 2ab + b^2 = (a + b)^2$

$a^2 - 2ab + b^2 = (a - b)^2$

$a^2 - b^2 = (a + b)(a - b)$

$a^3 - b^3 = (a - b)(a^2 + ab + b^2)$

$a^3 + b^3 = (a + b)(a^2 - ab + b^2)$

Rational Expressions

$\dfrac{a}{b} + \dfrac{c}{b} = \dfrac{a + c}{b}$

$\dfrac{a}{b} - \dfrac{c}{b} = \dfrac{a - c}{b}$

$\dfrac{ac}{bc} = \dfrac{a}{b}$

$\dfrac{a}{b} + \dfrac{c}{d} = \dfrac{ad + bc}{bd}$

$\dfrac{a}{b} \cdot \dfrac{c}{d} = \dfrac{ac}{bd}$

$\dfrac{a}{b} \div \dfrac{c}{d} = \dfrac{a}{b} \cdot \dfrac{d}{c}$

If $\dfrac{a}{b} = \dfrac{c}{d}$, then $ad = bc$.

Quadratic Formula

The solutions to $ax^2 + bx + c = 0$ with $a \neq 0$ are

$$x = \frac{-b \pm \sqrt{b^2 - 4ac}}{2a}.$$

Distance Formula

The distance from (x_1, y_1) to (x_2, y_2), is

$$\sqrt{(x_2 - x_1)^2 + (y_2 - y_1)^2}.$$

Midpoint Formula

The midpoint of the line segment with endpoints (x_1, y_1) and (x_2, y_2) is

$$\left(\frac{x_1 + x_2}{2}, \frac{y_1 + y_2}{2}\right).$$

Slope Formula

The slope of the line through (x_1, y_1) and (x_2, y_2) is

$$\frac{y_2 - y_1}{x_2 - x_1} \quad \text{(for } x_1 \neq x_2\text{)}.$$

Linear Function

$f(x) = mx + b$ with $m \neq 0$

Graph is a line with slope m.

A constant function
$f(x) = 2$

The identity function
$f(x) = x$

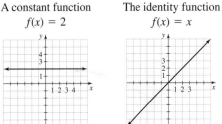

Absolute Value Function

$f(x) = |x|$

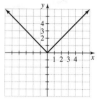

Quadratic Function

$f(x) = ax^2 + bx + c$ with $a \neq 0$

Graph is a parabola.

$f(x) = x^2$ (the squaring function)

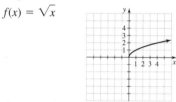

Square-Root Function

$f(x) = \sqrt{x}$

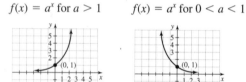

Exponential Function

$f(x) = a^x$ for $a > 0$ and $a \neq 1$

One-to-one property: $a^m = a^n \leftrightarrow m = n$

$f(x) = a^x$ for $a > 1$ $f(x) = a^x$ for $0 < a < 1$

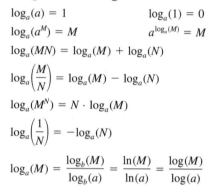

Logarithmic Function

$f(x) = \log_a(x)$ for $a > 0$ and $a \neq 1$

One-to-one property: $\log_a(m) = \log_a(n) \leftrightarrow m = n$

Base-a logarithm: $y = \log_a(x) \leftrightarrow a^y = x$

Natural logarithm: $y = \ln(x) \leftrightarrow e^y = x$

Common logarithm: $y = \log(x) \leftrightarrow 10^y = x$

$f(x) = \log_a(x)$
for $a > 1$

$f(x) = \log_a(x)$
for $0 < a < 1$

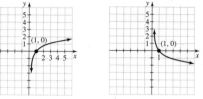

Properties of Logarithms

$\log_a(a) = 1$ $\qquad\qquad$ $\log_a(1) = 0$

$\log_a(a^M) = M$ $\qquad\qquad$ $a^{\log_a(M)} = M$

$\log_a(MN) = \log_a(M) + \log_a(N)$

$\log_a\left(\dfrac{M}{N}\right) = \log_a(M) - \log_a(N)$

$\log_a(M^N) = N \cdot \log_a(M)$

$\log_a\left(\dfrac{1}{N}\right) = -\log_a(N)$

$\log_a(M) = \dfrac{\log_b(M)}{\log_b(a)} = \dfrac{\ln(M)}{\ln(a)} = \dfrac{\log(M)}{\log(a)}$

Interest Formulas

A = amount and P = principal

Compound interest: $A = P(1 + i)^n$, where n = number of periods and i = interest rate per period

Continuous compounding: $A = Pe^{rt}$, where r = annual interest rate and t = time in years

Variation

Direct: $y = kx$ $(k \neq 0)$

Inverse: $y = \dfrac{k}{x}$ $(k \neq 0)$

Joint: $y = kxz$ $(k \neq 0)$

Straight Line

Slope-intercept form: $y = mx + b$

Slope: m y-intercept: $(0, b)$

Point-slope form: $y - y_1 = m(x - x_1)$

Slope: m Point: (x_1, y_1)

Standard form: $Ax + By = C$

Horizontal: $y = k$ Vertical: $x = k$